FINANCE FOR THE MINERALS INDUSTRY

Co-Editors

C. Richard Tinsley
European Banking Company
London, England

Mark E. Emerson
Resource Exchange Corp.
New York, New York

W. Durand Eppler
Crocker National Bank
San Francisco, California

Sponsored by the Minerals Resource Management Committee
of the Society of Mining Engineers of AIME

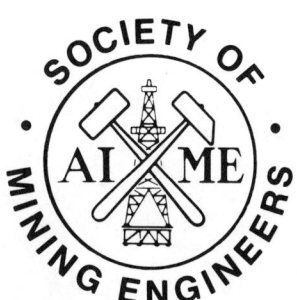

Published by the
Society of Mining Engineers
of the
American Institute of Mining, Metallurgical, and Petroleum Engineers, Inc.
New York, New York • 1985

Copyright © 1985 by the
American Institute of Mining, Metallurgical, and Petroleum Engineers, Inc.

Printed in the United States of America
by Port City Press, Baltimore, Maryland

**All rights reserved. This book, or parts thereof, may not be
reproduced in any form without permission of the publisher.**

**Library of Congress Catalog Card Number 84-52556
ISBN 0-89520-435-5**

FINANCE VOLUME ORGANIZATION

Chapter Editors

William R. Bush
Duval Corporation
Tucson, Arizona

Michael N. Cramer
Manufacturers Hanover Trust
New York, New York

Mark E. Emerson
Resource Exchange Corporation
New York, New York

W. Durand Eppler
Crocker National Bank
San Francisco, California

Wolfgang O. Gluschke
United Nations
New York, New York

Simon D. Handelsman
Canadian Geoscience Corporation
Vancouver BC, Canada

Leons Kovisars
Chase Econometrics
Bala Cynwyd, Pennsylvania

John C. Robison III
First National Bank of Chicago
Chicago, Illinois

C. Richard Tinsley
European Banking Company
London, England

Production Coordinators

Ta M. Li
Thyssen Mining Construction Inc.
Lakewood, Colorado

William J. Potter
Prudential-Bache Securities, Inc.
New York, New York

Marianne Snedeker
Society of Mining Engineers of AIME
Littleton, Colorado

Finance for the Minerals Industry Symposium

Sponsored by the Minerals Resource Management Committee of SME-AIME
During the SME-AIME Annual Meeting, New York, New York, February 24-28, 1985

Chairmen: **C. Richard Tinsley, Mark E. Emerson, W. Durand Eppler**

Sunday, February 24, PM

FINANCIAL EVALUATION FOR THE MINING INDUSTRY

Chairmen

L. Kovisars
Chase Econometrics
Bala Cynwyd, Pennsylvania

M.N. Cramer
Manufacturers Hanover Trust Co.
New York, New York

Monday, February 25, AM

STRATEGIC AND FINANCIAL PLANNING

Chairmen

W.R. Bush
Duval Corporation
Tucson, Arizona

H.J. Sandri, Jr.
Burlington Northern, Incorporated
Seattle, Washington

Monday, February 25, PM

TAX AND ACCOUNTING ISSUES

Chairmen

I. Murray
Ernst and Whinney
Toronto, Ontario, Canada

W.D. Eppler
Crocker National Bank
San Francisco, California

Tuesday, February 26, AM

MINERAL DEVELOPMENT AGREEMENTS AND ECONOMIC RENT

Chairmen

W.O. Gluschke
United Nations
New York, New York

W.D. Eppler
Crocker National Bank
San Francisco, California

Tuesday, February 26, PM

ANALYSIS OF RISK I

Chairmen

J.C. Robison, III
First National Bank of Chicago
Chicago, Illinois

R.W. Phelps
Continental Illinois National Bank
Chicago, Illinois

Wednesday, February 27, AM

ANALYSIS OF RISK II

Chairmen

J.C. Robison, III
First National Bank of Chicago
Chicago, Illinois

G.H. Espley
Bank of Montreal
Toronto, Ontario, Canada

Wednesday, February 27, PM

SOURCES OF FUNDING

Chairmen

C.R. Tinsley
European Banking Company, Limited
London, England

J.H. Boettcher
Vickers, Boettcher,
Massy-Greene Partners
San Francisco, California

Thursday, February 28, AM

CASE STUDIES

Chairmen

S.D. Handelsman
Canadian Geoscience Corporation
Vancouver, British Columbia, Canada

J.K. Hammes
Citibank N.A.
New York, New York

FOREWORD
DE RE HISTORICA

The idea for a comprehensive book on the subject of finance for the minerals industry arose out of a very successful luncheon meeting on Project Finance for the Mining Industry that served as the first mini-symposium sponsored by the (then) newly formed Economic and Finance Subsection of the New York Section of AIME. Following the positive response to this meeting, Mark Emerson and W.D. (Randy) Eppler had the idea of organizing and offering a short course on project finance to be given at the AIME Annual Meeting.

Randy drew up a short course outline, and he and Mark drafted a proposal which was approved by the SME-AIME Continuing Education Committee. They began to recruit participants and teachers for the course. Later, Mark Emerson and Tomek Ulatowski were sitting in the shade of a palm tree in Paradise Island (1978 SME-AIME Fall Meeting in the Bahamas). Tomek, approached with the idea of participating, responded positively, adding that he felt that a definitive book on the subject was very desirable to fill a gap in the literature. Gary Castle and Hans Schreiber were also recruited and the first short course was given in conjunction with the 1979 SME-AIME Annual Meeting in New Orleans.

The course was well received and repeated several times with revisions and additions at various SME meetings including Las Vegas (1980), Chicago (1981), and Dallas (1982). Randy recruited new lecturers including Richard Tinsley and Jim Boettcher.

In addition to the short courses on Project Finance, the Minerals Resource Management Committee (MRM) of SME has organized and sponsored other short courses, and MRM's regular session programming has included papers on various aspects of mineral finance at the SME Annual and Fall Meetings.

Building on the material developed for the short course Jim Boettcher and Randy prepared an initial outline for a comprehensive book on mineral finance that was discussed at the SME Hawaii meeting in 1982 and that was used to obtain SME-AIME Board of Directors approval for the symposium and this book.

A working subcommittee was formed in Honolulu, and a considerable amount of thought and work was put into the structuring of the book and Symposium at various meetings. Richard Tinsley developed the current book structure and its contents, and as the project proceeded "polishes" were made and useful ideas contributed by the MRM committee including the various chapter editors and session chairmen. Production support from the unnamed word processing operators merits the readers thanks.

Numerous review meetings, involving at times the entire committee, were held in the elegant dining room of the Emersons' Brownstone in New York, and these long sessions were sustained by Mary Ellen's delicious preparations in the kitchen—good ideas come from a good repast.

The purpose of the Symposium is to round out the short courses and programming in the area of finance for the mining industry and to produce a book which will be useful to all SME and AIME members and others in the industry. This book (available at the meeting) includes, in addition to the other sources mentioned, especially prepared papers, and all of the papers to be presented at the Symposium.

Space limitations preclude mentioning all those who made the book and symposium possible. The chief credit goes to the writers who made it all possible.

Everyone involved with this project hopes that this book will be useful to students and practitioners alike. The breadth and depth of the coverage of various aspects of mine finance is a reflection of the care and attention, arm twisting, and teamwork and scrutiny given by the committee.

Simon D. Handelsman, *Chairman*
Minerals Resource Management Committee

TABLE OF CONTENTS

Finance Volume Organization	iii
Finance for the Minerals Industry Symposium	iv
Foreword	vii
Introduction	xiii

Chapter

1 FINANCIAL REQUIREMENTS OF THE MINERALS INDUSTRY — 1

	Preface. C.R. Tinsley	3
1.1	What Does Financing Mean for the Mining Industry? J.K. Hammes	5
1.2	Changes in the Financial Structure of the Mining Industry Over Time. E.L. Robinson	10
1.3	Capital Requirements of the Mineral Industry. S.D. Strauss	17
1.4	The History of Financing a Multinational Mining Company. A. Tuke	23
	Bibliography	29

2 CAPITAL STRUCTURE FOR A MINERALS COMPANY — 31

	Preface. L. Kovisars	33
2.1	An Introduction to Capital Structure. W.L. Langdon	35
2.2	Financial Objectives of a Mining Company. E.K. Cork	42
2.3	Financial Aspects of Joint Venture and Multiparty Structures. D.C.N. Cockburn	47
2.4	Financial Evaluation of a Coal Mine Acquisition: A Case Study. J.H. Boettcher and W.D. Eppler	53
2.5	Financing of the Small, Independent Mining Enterprise. G. Caraghiaur	65
	Bibliography	71

3 FINANCIAL EVALUATION FOR THE MINERALS INDUSTRY — 73

	Preface. M. Cramer	75
3.1	Methods of Investment Analysis for the Minerals Industries. G.H.K. Schenck	77
3.2	Tailoring the Financing Decision to Project Economics. F.B. Sotelino and M.A. Gustafson	94
3.3	Measuring the Economic Viability of Resource Projects (A Western Surface Coal Project). G.R. Castle, D.M. Higgins, and B.J. Klimley	103
3.4	Optimum Production Rate for High-Grade/Low Tonnage Mines. R. Glanville	114
3.5	Sensitivity Analysis for Mining Projects. J.C. Robison III	126
3.6	Sensitivity of Mining Projects to Capital, Operating and Debt Cost Variations. R.A. Boulay	129

3.7 Considerations in Leveraged Studies for Mineral Ventures.
    W.P. Lohden 134
3.8 The Impact of Inflation on Hurdle Rates for Project Selection.
    N.H. Cole 140
3.9 Evaluation of the Lease or Buy Decision.
    J.R. Caldon 150
3.10 Theory and Practice of Post-Installation Appraisals of Projects.
    R.D. Mills 160
    Bibliography 165

4 ACCOUNTING FOR MINERALS COMPANIES — 167

Preface. *M.E. Emerson* 169
4.1 Accounting for Minerals Companies — A Guide to Understanding Financial Statements of Mining Companies in an International Environment.
    Ernst and Whinney 171
    Bibliography 225

5 MINERALS INDUSTRY TAXATION — 227

Preface. *M.E. Emerson* 229
5.1 A Comparison of Local, State and Federal Taxes in Eight U.S. States.
    R.L. Davidoff 231
5.2 Current Trends in Canadian Mining Taxation.
    R.B. Parsons 241
5.3 Overview of U.S. Taxation of Mining Companies.
    D.J. McCarthy 254
5.4 The Tax Structure of Foreign Mining Investments in the United States.
    Nicasio del Castillo 263
5.5 Comparison of International Mining Tax Regimes.
    Coopers & Lybrand 274
5.6 Tax Planning Through the Use of Multiple Corporations.
    J.J. McCabe 278
5.7 The Crush on Australian Mining Profits.
    V.R. Forbes 283
    Bibliography 290

6 PROFIT AND ECONOMIC RENT — 291

Preface. *S. Handelsman* 293
6.1 Economic Rent and Its Relationship to Finance.
    H.D. Drechsler 295
6.2 Resource Rent Tax Proposals in Australia.
    P.H. Fletcher 300
6.3 Economic Rent Considerations in International Mineral Development Finance.
    J.K. Hammes 305
    Bibliography 309

7 IMPACT OF MINERAL DEVELOPMENT AND OPERATING AGREEMENTS ON FINANCING — 311

Preface. *W.O. Gluschke* 313
7.1 Mineral Exploration and Development Agreements: An Overview.
    W.O. Gluschke 315

7.2	Sharing Risks and Rewards in International Contract Negotiations. *P.L. Lizaur*	326
7.3	Alternatives for Developing Country Mining Finance. *S. Zorn*	331
7.4	Three Recent Mineral Development Agreements in South America. *T.W. Walde*	342
7.5	The Effect of Host Government Attitude Upon Foreign Investment in Mining. *G.E. Pralle*	353
	Bibliography	359

8 STRATEGIC AND FINANCIAL PLANNING — 361

	Preface. *W.R. Bush*	363
8.1	The Evolution of the Planning Process in the U.S. *R.A. Arnold*	365
8.2	Competitive Analysis. *L. Kovisars*	369
8.3	Assessing Strategies for Natural Resource Companies. *B.B. Castleman*	377
8.4	Mineral Industry Acquisition Analysis. *H.J. Sandri, Jr.*	383
8.5	Financial Management of Diversified Companies. *P.J. Maxworthy*	390
8.6	Long-Range Planning in a Copper Mining Company. *F.A. Buttazzoni and J.C. Munita*	401
	Bibliography	413

9 RISK ANALYSIS FOR MINE FINANCE — 415

	Preface. *J.C. Robison III*	417
9.1	Analysis of Risk Sharing. *C.R. Tinsley*	419
9.2	Understanding the Risks in Coal Reserve Estimates. *J.E. McNulty*	427
9.3	Operating Risk (Improving the Odds in a High Risk Industry). *C.A. Born*	433
9.4	Operational Risk Assessment of Mining Enterprises. *P.J. Szabo*	436
9.5	Dealing with Interest Rate and Exchange Rate Risks. *J.L. Poole*	440
9.6	A Systematic Approach to Political Risk Analysis. *W.D. Coplin and M.K. O'Leary*	445
9.7	Country Risk Analysis. *J.R. Stuermer and P. Allen*	454
9.8	OPIC Insurance Programs for the Mining Sector. *B.T. Mansbach*	457
9.9	Feasibility Studies and Other Pre-Project Estimates: How Reliable Are They? *G.R. Castle*	461
9.10	Cost Overrun Risk and How It Was Minimized During Construction of the Mt. Gunnison Mine. *L.P. Haldane*	466
9.11	Conventional and Nonconventional Risks Insurance for Mining Projects. *C. Berry*	472
	Bibliography	477

10 INFORMATION REQUIREMENTS — 479

Preface. *C.R. Tinsley* — 481
- 10.1 Independent Engineering Information for Project Financing. *M.E. Emerson* — 483
- 10.2 The Role of the Independent Consulting Firm in Project Financing. *H.W. Schreiber* — 488
- 10.3 What Bankers Look For In Project Loan Applications. *N.J. Gibbs and J. Sroka* — 493
- 10.4 Technical and Financial Elements of a Mining Project Loan Request: Preparing a Complete Information Memorandum. *T.P. Bispham* — 498
- 10.5 Information Requirements for Equity Issues. *E.L. Affleck and W.G. Stevenson* — 501
- 10.6 Understanding the Loan Approval Process. *G.P. Thomason* — 507
- 10.7 Banking Needs for Project Development Financing. *C.R. Tinsley* — 512

11 SOURCES OF FINANCE — 517

Preface. *C.R. Tinsley* — 519
- 11.1 Sources of Funding for Mineral Projects. *T. Ulatowski* — 521
- 11.2 Basic Sources of Mine Financing. *K.G. Arne* — 527
- 11.3 Mining Finance. *H.D. Drechsler* — 530
- 11.4 Project Financing Supports and Structuring. *C.R. Tinsley* — 549
- 11.5 Project Financing—Guidelines for the Commercial Banker. *G.R. Castle* — 559
- 11.6 Project Finance—Does It Exist in the Mining Industry: CRA's Experience. *K. Wightman* — 568
- 11.7 How to Finance Mineral Prospects. *E.F. Cruft* — 574
- 11.8 Financing the Juniors. *M. Chender* — 579
- 11.9 Funding a Promotional Exploration Company. *J.S. Brock* — 584
- 11.10 Financing the Development of Small Mining Projects—An Operator's Viewpoint. *F.H. Brooks* — 590
- 11.11 Venture Capital for the Mining Industry. *J.H. Boettcher* — 592
- 11.12 Public Financing as a Source of Funding for the Canadian Mineral Industry. *B.J. Gorval and R.L. Kemeny* — 600
- 11.13 Commercial Paper: An Innovative Source of Financing for Mining Projects. *V. Yablonsky, R. Gillham, and G.R. Castle* — 611
- 11.14 Financing the Industrial Minerals Industry. *C.F. Baiz III* — 614
- 11.15 Special Considerations in Project Finance for the Industrial Minerals Industry. *C.R. Tinsley* — 618

11.16	Trade Financing Supplement.	622
11.17	Forfaiting Export Finance in the Free Market.	
	M. Bradbury.	624
11.18	European Export Credit Programmes.	
	E.A. Rides.	630
11.19	North American Export Credit Programs Supplement.	636
11.20	World Bank Group Financing.	
	L.H. Cash.	639
11.21	International Finance Corporation Supplement.	641
11.22	Leveraged Leasing for the Minerals Industry.	
	R.G. Ravenscroft.	645
11.23	Using a Gold Loan as a Financing Mechanism: Case Studies.	
	R.M.M. Rowe and S.D. Handelsman.	657
	Bibliography.	661

12 CASE STUDIES 663

	Preface. S.D. Handelsman.	665
12.1	Financing Three Golds.	
	P.A. Allen.	667
12.2	A Case Study of the Strathcona Sound Project (A Non Recourse Project Financing).	
	H.W. Schreiber and G.R. Castle.	671
12.3	International Mining Company Case Study: Industrial Minera Mexico, S.A.	
	T. Ulatowski.	683
12.4	Financing of Teck's Investment in the Bullmoose Coal Project.	
	N.R. Macmillan.	718
12.5	Financing the Ok Tedi Mine—Case Study of the Process from a Government Perspective.	
	S. McGill.	724
12.6	Inco Limited's Soroako Nickel Project: A Case Study in Financing Large Overseas Mining Projects.	
	R.T. De Gavre.	736
12.7	Development of the Ranger Uranium Financing from Banks, Customers, Shareholders, and the Stock Market.	
	S.J. Hodge, N. Miskelly, and C.R. Tinsley.	755
12.8	Financing a Government-Owned Industrial Mineral Company.	
	D.A. Karvonen.	769
12.9	Minera Real de Angeles—A Case Study.	
	D.J. Worth.	781
12.10	Mergers and Acquisitions in the Mining Industry (Brascan/Noranda) with Particular Emphasis on the Hedging of Financial Risk.	
	J.T. Eyton.	786
12.11	Evaluation of the 16-to-1 Mine as a Candidate for Project Financing—A Case Study.	
	H.W. Schreiber and D.W. Neuhaus.	790
12.12	The Impact of a Recessionary Environment on Private Company Financing.	
	W.J. Potter and R.N. Pyle.	810
12.13	Risk Capital: Financing Junior Mining Companies.	
	M. Pezim.	816
	Author Index.	835
	Subject Index.	837

INTRODUCTION

The key to a successful mining operation is an exceptional ore body, a well designed and capably executed mining plan, and careful maintenance of the equipment. Good fortune is a major factor in the first of these elements: even prospecting requires planning and money. The development and operation of the smallest mine is surrounded by financial considerations—investment, flows of revenue and expenditure, and the repayment of debt.

The goal of the enterprise is to reward the owners by increasing their equity while limiting risks to a level commensurate with the desired yields. And that is not easy, particularly in the uncertain economic times that have characterized the mining industry in the 1970s and the 1980s.

The minerals industry consists of approximately 1,250 metalliferous mining operations around the world producing in excess of 136 kt/a (150,000 stpy) of ore. These are about equally divided between underground operations and surface mines, the latter including some 200 alluvial operations. The coal industry has a similar total number of mining operations at a production in excess of 454 kt/a (500,000 stpy).

The purpose of this book is to draw together the key aspects for financing the minerals industry. Some 100 papers are presented representing the collaboration of as many individuals. Contributors include thirty-five practitioners with financial institutions, twenty-three representatives of minerals companies, and twenty-eight mineral consultants, accountants, and other specialists providing techno-economic services to the minerals industry. Thirteen authors are with governments, the United Nations, universities, and other institutions involved with mineral policy.

One of the objectives of this book was to provide the reader with an international perspective of mineral finance. The editors believe that they were successful in this respect. Thirty-eight contributors are residents of mineral-producing countries outside the United States. These include Canada (nineteen), the United Kingdom (seven), Australia (ten), and some representatives of mineral-rich lesser developed countries (LDCs).

In addition to the authors, the editors wish to recognize the professional efforts of the SME staff in preparing these papers for publication.

The papers are organized into twelve chapters. These include the financial requirements and the structure of mineral companies, financial evaluation and accounting, taxation, profit and economic rent, overseas operating agreements, financial planning and management, risk analysis, sources of information and finance, and case studies.

In preparing these papers, the contributors had the objective that this book was intended to be a general reference for SME-AIME practicing mining engineers, geologists, and others, especially those engaged in economic work or management. Judging from the response to SME short courses and sessions on mine finance over the past few years, from which certain of these papers and case studies are drawn, *Finance for the Minerals Industry* is expected to be as widely used as a minerals economics textbook. The insights into the analytical and information requirements required to determine the financial structure of a minerals company as well as the selection of the appropriate financing vehicle can be of value to practitioners and students alike.

Chapter 1 begins with a review of the financial requirements of the minerals industry and describes some disturbing trends in the future availability of finance. More carefully tuned financing strategies are indicated for the minerals industry.

The capital structure for a minerals company is discussed in Chapter 2, contributing the viewpoint of large and small companies as well as multiparty ventures.

Financial evaluation for the minerals industry requires a combination of methods. As covered in Chapter 3, sensitivity analysis and hurdle rate consideration warrant close attention in the present business environment.

The techniques of financial evaluation are founded on an understanding of financial accounting. Chapter 4 presents a detailed and thorough review of this subject for major minerals producing countries.

Mineral industry taxation, the subject of Chapter 5, presents an increasing burden on industry profitability. Optimization of tax considerations is becoming a key element in the choice of one financing versus another.

Industry profitability is largely sustained by its ability to extract *economic rent* from the wasting asset that is a mine. Chapter 6, "Profit and Economic Rent," lays out this linkage, one which has been adopted by governments around the world seeking to improve their tax take from the industry.

The impact of mineral development and operating agreements on financing the minerals industry is described in Chapter 7. Taxation, either direct, indirect, or economic-rent based, is the main issue besides local ownership, training, or control.

The considerations for financial planning and management follow in Chapter 8. Financial management of diversified companies and the integration of financial and operating planning are stressed. This important aspect is too often relegated to the back office by an industry dominated by engineers or geoscientists.

Risk analysis applied to the technical, economic, and financial components of project development is a precursor to understanding the types of supports required to raise finance for any mineral project. Financial risks, including the analysis of interest rate and foreign exchange risks and the methods of dealing with such risks, are covered in Chapter 9. Political risks are also addressed, noting that political risk insurance can be mandatory for financings in many LDCs.

The information requirements for different types of financing are given in Chapter 10, concentrating on bank and stock exchange financing. Chapter 11 provides a description of some 35-40 different sources of finance for the minerals industry, from raising funds for exploration drilling through bank and supranational agency financing. The trend is toward a package of different sources of finance and an understanding of the approaches and information requirements of each is important to the financiers in the minerals industry today.

A significant part of the text is dedicated to case studies in Chapter 12. The cases presented cover a broad range of minerals, countries, and financing situations, and scale of projects.

The Minerals Resource Management Committee of SME-AIME is pleased to be able to bring this collection of papers into the first comprehensive book on finance for the minerals industry. The editors believe that this volume will be one which SME-AIME members, and others associated with the industry, will wish to retain as a useful reference and a stimulating source of ideas with which to develop even more creative financing methods in the challenging years which lie ahead for the minerals industry.

CHIEF EDITORS

C. Richard Tinsley
European Banking Company Limited
London, England

Mark E. Emerson
Resource Exchange Corp.
New York, New York
United States of America

W. Durand Eppler
Crocker National Bank
San Francisco, California
United States of America

Chapter 1

Financial Requirements of the Minerals Industry

CHAPTER 1: FINANCIAL REQUIREMENTS OF THE MINERALS INDUSTRY

PREFACE

It is widely recognized that the internal cashflow-generating capability of the minerals industry has declined steeply since the late 1960's. The prime reasons cited are environmental costs, poor market conditions caused by indiscipline by government-controlled producers, higher taxation of "resource" profits (particularly during cyclical periods of high prices), higher debt service costs (record "real" interest rates), higher operating costs, new requirements to contribute to "social" rather than productive assets, and overinvestment in some industries. However, a structural change is occurring in the minerals industry which is making it less able to attract financing.

Lesser developed countries (LDCs) with major minerals potential are finding it increasingly difficult to raise money for new mineral developments since almost all new capital is required to pay interest on their national debt. To cope with an international environment, characterized by a record high cost of capital brought about by the strong U.S. dollar, most LDCs have responded by devaluing their currencies and have maintained competitive operating costs from their existing production base. Most industrialized nations, the U.S. in particular, have become correspondingly uncompetitive and many production facilities have either been cut back or closed. This trend is even more insidious than the wave of nationalizations in the early 1970's. Simon Strauss, in his paper "Capital Requirements of the Mineral Industry," notes that a "curious treadmill has developed" where "the LDC's borrow money from the IMF and then, to meet their debt requirements, continue to maximize production, keeping pressure on price."

The situation has been compounded by a rapid improvement in mineral exploration technology such as the Canadian-driven geophysical dynamo and satellite imagery. New geological theories are providing important leads to major minerals discoveries. No longer can anyone point to a shortage of mineral reserves to exploit. Exploration expenditures, traditionally taking 25% of annual mineral companies' net cash flow, are being sharply trimmed.

The industry's financing requirements include funding of exploration, new and replacement capital expenditures, maintenance costs, working capital, and debt service. Various authors have attempted to come to grips with the industry's capital requirements (see Table 1) with the actual figure being around $10 - $15 billion per annum if all categories are included. This is quite a small percentage of the worldwide capital requirements of industry and governments, and the issue is not whether capital will be available but whether the industry can afford to get what it needs.

Accurate statistics on the financial requirements of the minerals industry are difficult to obtain because there are few solely minerals companies left. Many have diversified and others have become the object of the diversification of other companies, primarily petroleum companies. However, the industry's capital requirements are generally broken down as follows: internally generated funds 15% - 25%, new equity 15% - 25%, and debt 50% - 60%. Consequently, most minerals companies have today crossed the 50:50 debt:equity ratio threshold and can only look forward to selling assets to reduce debt or to growing debt: equity ratios in the face of the difficult equity market now prevalent for minerals producers. The importance of financing for the mining industry is addressed in John Hammes' paper "What does Financing mean for the Mining Industry" while Beth Robinson's paper "Changes in the Financial Structure of the Mining Industry over Time" presents a review of how the financial profile of U.S. companies has changed over time, reflecting the trend to higher debt and illustrating the poor financial returns which have characterized the industry.

The providers of debt to the minerals industry are primarily banks and supranational agencies. Both are suffering from a collection of problem credits entered into in the go-go period of the mid to late 1970's, a period when "project financing" and "country" credits reached their pinnacles of popularity. In some cases, the banks and supranational agencies were irresponsible in failing to grasp the risks associated with the mining industry, in particular market risks. In today's environment of high government-owned production of many mineral products the situation is complicated by some bank financed projects which are forced to operate to produce cash flow at any cost because producing represents the only means of (partly) meeting the debt-service requirements of these project financings, and the high percentage of debt far exceeds the value realizable from disposal of the assets or sale of the venture to another company. This is not always the case, however, and Sir Anthony Tuke's paper, "History of Financing a Mining Company over Time," shows how judicious and careful co-ordination between the consumer and miner was important for raising finance for Rio Tinto Zinc projects around the world. Strategic investments made by this company have made it a major success in the capital intensive minerals industry.

Some companies are reverting to the strategy adopted in the first third of this century when small operations were established with equity and any expansions thereof financed out of cash flow. This trend to "thinking small" is partly a result of the dearth of easy bank and equity finance but also a realization of the downward spiral faced when operating successively higher tonnages at ever declining grades.

Consuming companies rarely feel the need to support the financing of a minerals company or its projects because today a buyers' market prevails. If banks, supranational agencies and the mining industry itself decrease their ability or willingness to finance the industry and if little help can be expected from governments, then one must conclude that the capital requirements of the industry are being trimmed due to a shortfall in adequate returns on investment. The only relief will come from the consuming public via higher prices or the consuming companies efforts to secure a source of supply - a timetable that seems to be a long way away from 1985. Currency and capital markets will not likely help the industry's already weakened ability to attract the $6 - $10 billion in new financing needed worldwide each year.

Table 1

CAPITAL REQUIREMENTS PER ANNUM

(US$ billion)

Source	Date	Scope	Industry	Amount p.a.	Units
Haworth, Aimone	1977	Free world	All minerals	$14.1	$1977
Macgregor, Vickers	1974	Free world	All minerals	3.3 - 6.1	$1974
West	1971	Free world	All minerals	6.7	$1970
Dodson, Johnson	1969	U.S.	Non-ferrous metals	2.2	$current
Beckerdite	1968	Free world	Major metals + U	4.5	$current

C. Richard Tinsley
Chapter Editor
European Banking Company Limited

1.1

WHAT DOES FINANCE MEAN FOR THE MINING INDUSTRY?

John K. Hammes

Vice President
Citibank N.A.
New York, New York

INTRODUCTION

This introductory paper presents a description and definition of what the finance function is and what it specifically means for the mining industry. In its simplest terms, finance is, "management of money matters". As an area within the discipline of business administration it draws heavily on the related fields of economics and accounting, and to a lesser extent on marketing and production.

MINING FINANCE DEFINED

In preparing to write about what finance means to the mining industry, I asked a number of mining people the question, "What does mining finance mean to you?" Most responded that mining finance was "How a company borrowed money for mine development", and perhaps half those questioned included "raising equity" in the definition.

It is not surprising that the mining industry concept of finance is the providing of funds needed for the objectives of the business, for that was once a view shared by authorities in the field of business administration. Further, it is certainly the area where much of the emphasis has been placed in the past ten to fifteen years as the industry struggled with the development of more remote and more refractory or lower-grade orebodies while profitability and internal cashflow were below expectations. No wonder project finance became such a popular topic. Twenty years ago it is probable that the emphasis would have been on mine valuation, for that subject was the area of finance then most often examined in industry publications. Papers concerning the pros and cons of payback, internal rate of return, and discounted cash flow appeared regularly.

Let's examine the following situations, each the result of a financial decision:

- Funding the exploration program in a developing country was accomplished by buying the next six-months budget needs in local currency. One week later the currency devalued 30 percent. If the purchase had been delayed or the local currency borrowed short term, a substantial saving would have been made in funding the program.

- A major mining company borrows heavily to finance both expansion of its primary mining business and its entry into several new industry segments. When markets for all of the commodities it produces turn down for a prolonged period, there is concern over the adequacy of cashflow to cover debt service in addition to other company needs. The company's financial officers, however, had arranged very large committed credit facilities from numerous financial institutions which assured the availability of funds for several years. This enabled the company to implement a strategy to conserve cash, sell assets and survive until conditions improved. Further, short term suppliers of credit also were comfortable with the assurance that longer term credit was there to support them.

- A medium size mining company financed several mines independently with limited obligations on the part of the parent company. The company decided to refinance the operations under one agreement. More flexible terms, lower interest cost and a longer tenor financing were achieved in return for assuming the risk as a direct obligor.

- The exploration and development group of a large mining company proposes to develop a new mine in a country in which the company does not presently operate. When the investment is rejected they are advised that one reason was that the reinvestment assumption inherent in the discounted cashflow - rate of return (DCF ROR) calculation was held invalid for the project because of the country's regulatory policy regarding remittence of dividends and returns of capital.

- A precious metals producer borrowed gold metal under an agreement to repay the borrowed metal and interest in gold. The borrowed gold was sold and existing floating rate bank debt was repaid. The interest rate saving between floating rate debt and the gold borrowing cost saved the company more than 10% per annum. In addition, there was not a significant opportunity cost involved because the company had previously hedged the gold price and realized the gain by closing out its position when the gold borrowing transaction was completed.

- A large foreign-base metal producer entered into an electronic banking arrangement with a multinational bank which resulted in a five day decrease in the time of its receivables collections. This resulted in a direct savings through the reduction in working capital financing requirements and indirectly through an improvement in the administration of receivables.

- A domestic coal company structured an acquisition financing untilizing separate production payment and a land holding company financings. The resultant structure reduces future income taxes and provides favorable accounting treatment in that the borrowings did not appear on the company's balance sheet as long term debt.

These are just a few examples of the issues and problems dealt with under the subject mining finance. Providing the company with adequate funds to accomplish its objectives, and seeing that these funds are provided on the best terms possible is a principle part of the finance function. However, as the examples demonstrate, the finance job is much broader than just that of supplying funds. The finance function is concerned with all the decisions that go into maximizing a company's values by effectively using the funds, not just supplying them. For the mining industry, this includes such areas as mine valuation, acquisition and divestiture strategy, managing working capital, hedging financial risk, tax planning, and managing capital structure.

AREAS OF FINANCIAL MANAGEMENT

Management of Liquidity

Figure 1 depicts the areas of financial management in the context of a simple balance sheet. The first area, management of liquidity, is the management of working capital i.e. the company's money position. It encompasses managing the cash, the investments that can readily be converted into cash, and the available sources of credit that can be drawn when needed such as bank lines. It includes forecasting the working capital needs, reducing surplus funds and investing those necessary for the business, and reducing the cost of carrying required receivables and inventory. This is an area where electronic banking, and the increased array of financial products such as selling receivables, commercial paper borrowing and the wider range of short term investment opportunities have improved the financial managers options.

FIGURE 1 AREAS OF FINANCIAL MANAGEMENT

```
                    I LIQUIDITY

        x x x x x x x x x x | x x x x x x x x x x
        x x x x x x x x x x | x x x x x x x x x x
        x Current Assets  x | Current Liabilities
        x x x x x x x x x x | x x x x x x x x x x
  III   / / / / / / / / / / | o o o o o o o o o o    II
        / / / / / / / / / / | o o o o o o o o o o
  L     / / / / / / / / / / | o o o o o o o o o o    C
  O     / / / / / / / / / / | o o o o o o o o o o    A
  N     / / / / / / / / / / | o o o o o o o o o o    P
  G     / / / / / / / / / / | o o o o o o o o o o    I
        / / / / / / / / / / | o Long Term Debt  o    T
  T     / / / / / / / / / / | o o o o o o o o o o    A
  E     / / / / / / / / / / | o o o o o o o o o o    L
  R     / / / / / / / / / / | o o o o o o o o o o
  M     / / / / / / / / / / | o o o o o o o o o o    S
        / / Fixed Assets / / | o o o o o o o o o o    T
  I     / / / / / / / / / / | o o o o o o o o o o    R
  N     / / / / / / / / / / | o o o o o o o o o o    U
  V     / / / / / / / / / / | o o o o o o o o o o    C
  E     / / / / / / / / / / | o o o o o o o o o o    T
  S     / / / / / / / / / / | o o o o o o o o o o    U
  T     / / / / / / / / / / | o o o Equity o o o     R
  M     / / / / / / / / / / | o o o o o o o o o o    E
  E     / / / / / / / / / / | o o o o o o o o o o
  N     / / / / / / / / / / | o o o o o o o o o o
  T     / / / / / / / / / / | o o o o o o o o o o
  S     / / / / / / / / / / | o o o o o o o o o o
```

Management of Capital Structure

The second area, management of capital structure, deals with the remaining liabilities on the balance sheet. This area includes the determination of debt capacity, the sources and cost of long term debt and the securities instruments used in raising long term debt and equity. The concepts of cost of capital, dividend policy and the assessment of financial risk are concerns managed in this area. I would include sovereign risk, comprised of expropriation and nationalization, foreign exchange convertibility and transferability, and war, riot and insurrection risk in this area. Clearly, transferability risk is a financial one while certain aspects of sovereign risk might better be under the category of investment management.

Management of Long Term Investments

The third area, management of long term investments, is concerned with the capital budgeting decision. No firm has unlimited resources, and as capital is employed its cost varies in reaction to the company's success with its prior investments. This area includes the "yardstick problem", i.e. the measurement of economic returns, the concept of hurdle rates or minimum cutoff points for investment profitability, and the analysis of operating risk. Operating risk includes product price and volume risk, inflation in operating costs or supplies and technology risk.

FINANCIAL OBJECTIVES

The financial manager operates within these three areas of financial management to achieve the company's objectives. These objectives include the maximization of profits, the maximization of the shareholders wealth, and the recognition of social responsibility.

Profit

Profit maximization seems a clear objective, yet several problems occur in applying this concept. Just as mine operators may "high grade" or neglect mine development to maximize short term profit, so may the financial manager make decisions which produce favorable short term results at the risk of longer term profitability. For example, borrowing a currency which has low interest rates may result in low current borrowing costs but may lead to a foreign exchange loss if the currency revalues. Then there is the question of how profits are measured. Should the maximization be that of accounting profits or of profit including some measure of opportunity costs? Should the measurement be profit as a percent of sales, as a percent of assets, or as a percent of equity? What weight has the rate of increase in profits in this determination? Finally, what is the degree of risk associated with the decision. If our example of a low-cost foreign exchange borrowing can be hedged to reduce the risk does this change our perspective?

Shareholder Wealth

The maximization of shareholder wealth is the concept of maximizing the present value of future results. For a public company this is reflected in the market value of shareholders stock. Financial decisions regarding capital structure and investment decisions in addition to short term profitability measurement affect market value. Thus the manager may consider timing and risk in making decisions as the market will reflect a concensus judgement on the decision. Identifying and tracking the impact of financing and other decisions on market value is more difficult to quantify than is the effect on profitability, and many intangibles enter into the perception of a company's value. Perhaps the industry analyst's opinion of the financial management's skills is as an important factor as some of the bottom-line results.

Social Responsibility

Finally, there is the consideration of social responsibility. The company has a social responsibility in dealing with the needs of its professional managers, employees, consumers, suppliers, community and nation. These responsibilities may not be seen as clearly compatible with profit maximization but, in fact, are consistent with wealth maximization. Although much of this responsibility is covered through law and contractual obligations, the financial manager must still see that these obligations are funded in the most efficient manner.

AREAS RELATED TO FINANCE

Earlier I stated that finance draws on other areas of business administration. Let's examine the dependence on these:

Economics

An understanding of monetary and fiscal policy, and a general knowledge of economics is crucial to the success of a financial manager. The performance of the economy is reflected in the level of business activity. Demand for mineral commodities, inflation rates and foreign exchange movements are all impacted by economic conditions. The demand for money, and therefore the availability and cost of credit and equity are a function of the level of the economy. Capital budgeting, debt capacity, and tenor of borrowing are impacted by the economic outlook. The concept of economic rent is applicable in examining taxation policy and dealing with owners of mineral reserves.

Accounting

In a sense, accounting is the language of business and finance. Accounting data are analyzed to understand the past and are used to analyze the effects of current decisions. Accounting measures performance, and is used to prepare financial statements and cash flow projections. On the one hand accounting is used by the finance manager to help assess long-term investments, to determine the optimum tenor of debt instruments used to finance, to measure the increased working capital needed to support a new project, to disclose the timing and extent of foreign exchange exposure, and in many additional ways to assist in defining and measuring financing requirements and risk. On the other hand, the financial manager knows that the affect of financing decisions on the company's future tax accounting and on its public financial statements will directly affect the company's tax liabilities and investors valuation of the company's market value. Then too, rating agency decisions on ranking of debt instruments will also be affected. Thus accounting is indispensible both as a tool in financial analysis and as an area critically affected by those decisions. The financial manager uses both accounting and is subjected to it.

Marketing and Production

Marketing and production not only impact the financial decisions because they are components of operating risk assessment but also because they may have a direct impact on financing. For example, much of the limited recourse project financing that has been done in recent years has been tied to the existence of long-term sales contracts. In certain segments of the minerals industry long term contracts with price escalators are common. In others, contracts with price reopeners are prevalent. But in many cases, long term contracts are not usually available. Obtaining contracts in order to achieve project financing may put the mining company in an unfavorable position opposite the market, and result in reducing the potential improvement in profitability as markets strengthen. Similarly, borrowers required to execute long term smelting or refining contracts, either as a supplier or buyer of concentrates, may suffer depending on which side of the transaction the company is on and what the outlook if for the availability of concentrates. Another example is the forecast needs for working capital. The financial manager must be aware of the usual payment terms for sales of concentrates and commodities in order to plan for financing the receivables associated with these sales.

In summary the finance function requires that management has a knowledge of finance together with a good grasp of economics and accounting, and the skill to anticipate and identify the impact of marketing and production considerations on financing. One is not likely to encounter a well managed mining company whose financial management doesn't have a good understanding of mining technology and commodity markets.

MINING FINANCE TRENDS

Increased Use of Debt

The most significant trend in mine finance during the past 10-15 years is the large increase in the use of debt financing. Various explanations are given for this. Profitability within the mining industry has not been adequate to attract the equity needed to fund the industry capital needs which far exceeded internal cash generation. Earlier forecasts of consumption growth rates for many minerals led to large increases in mine capacity. Often, new mines developed low grade orebodies in remote locations. Already high capital costs per unit of production were neverless underestimated because project scale and location did not make it practical to do definitive engineering and establish good estimating criteria prior to committing to development. Meanwhile, it seemed higher and higher inflation rates were adding to the costs of the inputs to production at a faster rate than those of the prices of the mineral commodities produced.

As the burden of capital expenditures become excessive, mining companies turned to joint ventures to share the risk, looked for limited recourse financing of projects, and generally increased the level of debt in the industry.

Joint venture participants often had different objectives or different tax positions and this led to the development of ownership structures which often complicated financing. Limited recourse finance took a variety of forms but was generally a term loan to the project company or to a financing vehicle where sponsors limited their obligations to providing performance guarantees and/or total or partial financial guarantees until a set of conditions was achieved which terminated the sponsor obligations. Key issues included balance sheet treatment and risk sharing. Production payment financing was commonly used because that structure was not reported as debt and because the principal obligation was an operating covenant rather than a financial guarantee. Ore reserve risk was usually assessed and taken by lenders provided there were adequate reserves to support operations for 1.5-2 times the life of the loan. Marketing risk was usually reduced through execution of sales contracts with end consumers, and other operating risks were often taken by lenders after completion tests had demonstrated project viability. In foreign projects, sovereign risk sharing became an issue, and clarifying whether sponsors, lenders or third parties underwrote risks of expropriation, currency convertibility or insurrection became a key negotiation.

Deteriorating Financial Condition

Without taking into account project financing, mining certainly became a more leveraged industry during the past two decades. For this exercise I looked at a composite balance sheet of nine major mining companies in the early 1960's. Long term debt was less than 5% of total capitalization, the working capital position was good, with a ratio of current assets to current liabilities in excess of 4:1; and reported income averaged about 8% on assets and equity. By the early 1980's, four of the companies I surveyed had been acquired so it was difficult to prepare or compare a composite financial statement for that group. Looking at the most recent available financial statements, however, indicates that debt as a percent of capitalization for the group rose to over 30%. In addition, a number of contingent liabilities related to project financings appear in those company's financial statements. Working capital ratios have declined and return on assets varies considerably between them, but probably averages less than 5% on assets. One fact is evident; the two or three companies in the composit that have reasonably good records of profitability are those which maintained a low debt ratio. As a generalization, it seems that an industry with high operating leverage should opt for low financial leverage. One hears many comments and concerns on Banker's irresponsible lending to developing countries; perhaps the mining industry would have benefited had more lenders turned down that industry's borrowing requests rather than sucumb to the pressures of competition.

Financial Product Developments

Another trend is the proliferation of financial products available and the development of electronic banking. These are not unique to the mining industry, since they provide information and a range of products which can improve any financial managers performance. Electronic banking provides on-line, real-time access to account balances, and simplifies the transfer of funds. A company's money position can be monitored and surplus funds invested. The range of financial products includes commercial paper and floating rate note borrowing, foreign exchange swaps and options, zero coupon bonds, non recourse receivables financing, commodity linked borrowing, interest rate swaps, etc. The list continues to grow with new products or variation on old ones introduced to the market each month. These products make it possible for the financial manager to access alternative sources of money to achieve lower interest costs and also insure against interest rate and foreign exchange movements to reduce financial risk.

Government Intervention

Yet another external influence is the increased financial impact of government intervention. This includes both the direct cost of payments to government and the indirect costs associated with environmental and other government action. State and federal royalty and severance tax payments, and the economic rent sought by foreign governments under the concept of "permanent sovereignty over natural resources" are rising. Government is becoming both a de facto and a real, direct partner in the mining industry. This has implications for lenders as to who has preference in the event funds generation is inadequate to service both debt and meet other franchise payments. Financial obligations of government versus private shareholders to contribute equity and fund development are also issues negotiated in international mining development agreements. The financing of pollution control equipment, additional mining equipment to meet new standards of reclaimation, and liabilities under pension and health and safety legislation add to an already burdensome financing requirement.

CONCLUSION

Mining finance is a multifaceted discipline. Far from answering the question "where do we borrow funds for mine development", it covers investment decisions, and the management of financial assets. It interfaces closely with the marketing and accounting functions. In essence finance is maximizing a company's values and improving its performance by effectively managing the use and supply of funds. As a contributing editor I will be satisfied if this book has made this definition of finance more meaningful to the reader.

1.2

CHANGES IN THE FINANCIAL STRUCTURE OF THE
MINING INDUSTRY OVER TIME

by

Elizabeth L. Robinson
Morgan Guaranty Trust Company
Assistant Vice President
New York, New York

INTRODUCTION

The financial structure of the mining industry has changed strikingly over the last thirty years. The changes that have taken place in the last ten years include both the acceleration of adverse trends in familiar balance sheet numbers and ratios, and the subtle but extensive increase in the complexity and diversity of funding sources, both on and off the balance sheet.

The objectives of this paper are:

1. to briefly trace the evolution of trends in the mining industry's financial structure;

2. to focus in more detail on changes in the last decade; and

3. to analyze those trends both within the framework of capital structure theory, and in the context of changes in business and economic conditions.

This survey focuses primarily on the mining companies listed on the major U.S. stock exchanges, since financial data for these firms are readily available. The producers of minerals are a diverse lot, however, and the quantitative data on "the mineral industry" as characterized in this survey fails to encompass the financial arrangements for the bulk of production worldwide, whether owned by governments or private-sector enterprises. Even large portions of the U.S. minerals industry - subsidiaries of still larger firms, medium-sized privately-held companies, and the small miners - were not included in the quantitative part of the analysis.

THE EVOLUTION OF THE INDUSTRY'S FINANCIAL STRUCTURE

In the aftermath of the business failures of the Great Depression, most borrowers and lenders considered minimal debt on the balance sheet the soundest policy. By the 1950's, however, a body of literature was developing which demonstrated that the use of leverage could improve the value of the firm, given a corporate tax structure which allows interest, but not dividends, to be deducted from income.

Optimal Capital Structure

The objective of undertaking a new project is to improve the wealth of the current owners of the firm. If internally-generated funds are insufficient to allow growth at the desired rate, the company can raise funds either by issuing new equity, or by incurring debt. The most attractive source is the one which will provide the funds for the lowest cost. Both stockholders and lenders expect to receive compensation for use of their funds at a rate which reflects two things: (1) the return they could earn in alternative investments, and (2) the riskiness of the projected return.

Factors which increase business risk include the variability of demand, sales prices and input prices, and "operating leverage", or the proportion of fixed to variable costs. In addition, from the investors' point of view, the use of financial leverage concentrates risk onto the shareholders, because lenders enjoy a prior claim in the event of bankruptcy. So although financial leverage can increase a project's expected rate of return, and therefore, presumably, earnings per share, too much debt will be perceived as uncomfortably risky by equity investors, and have a negative effect on stock prices. In theory, then, there should be an optimal level of debt which the firm's managers should seek to obtain in its financial structure. The desirable proportion of debt should in general be lower for a firm with a higher inherent business risk.

In addition to their own judgement, financial managers are sensitive to the attitudes of lenders, rating agencies, and investment bankers in setting financial targets (Scott and Johnson). In practice, financial structure will also depend upon the company's growth rate and level of profitability, and conditions in the capital markets at a given time.

Pre-1973 Financial Structure

The 1950's and 1960's were a prosperous time and as industrial output grew the demand for metals followed. However, during this period, much of expansion of the mining industry was self-financed. With robust demand for mineral commodities reflected in prices, retained earnings could provide funds for expansion and new projects. Moreover, mining is perceived as having a high level of business risk - highly variable product demand and prices and large fixed costs which create high operating leverage. Investors and lenders therefore require a higher rate of return to compensate for the additional risk, so external funds are relatively costly. Mining company managers, sensitive to the volatility of mineral prices, generally adopted conservative attitudes towards financial leverage.

A notable exception was the aluminum industry. Characterized by enormous capital requirements, and a small number of market participants who could presumably exercise price power, the aluminum companies relied heavily on debt financing. Coal companies, offered volume and price protection in the form of long-term contracts with their customers, also used more leverage than non-fuel mineral firms.

Changes in the Business Environment

Several important trends in the mining industry's business environment which began in the 1960's increased both the cost and uncertainty of new mining projects. First, the ownership of mineral resources and producton began passing to larger, more diversified entities. Large non-mining corporations entered the U.S. domestic industry through the acquisition of coal operations and reserves. Oil companies such as Conoco, Occidental, Sohio and Gulf all acquired major coal companies while others began acquiring reserves. For example, many of the federal coal leases in Wyoming's Powder River Basin went to oil companies. A trend toward governmental involvement was exemplified by the nationalization of the copper industries of Zaire, Zambia and Chile. Mexicanization of natural resource industries had begun earlier. Countries which did not seek ownership began imposing tax and royalty regimes designed to claim a larger portion of economic rents.

At the same time, mining companies were using new mining and metallurgical technologies to obtain economies of scale and meet growing demand through the development of massive, low-grade ore deposits. These projects had enormous capital requirements and were often located in remote areas which required investments in supporting infrastructure.

Project Finance

In the late 1960's these trends converged in the evolution of a new attitude and technique which took advantage of the existence of an asset in the ground suitable for security. The financial community's focus had shifted from an emphasis on the ratios of debt and equity in a firm's capital structure, toward the concept that the ability to generate future cash flow determines debt capacity. Project lending - looking to the ability of the cash flows of an individual project to repay a loan - was extended from oil field financing to large-scale mining projects. Project financing became fashionable not only because it enabled an ore body to be developed, but because it satisfied other needs of the participants. For newer entrants into mineral development, such as foreign governments or consumers, their asset strength or lack thereof need not constitute a barrier. For mining companies, expansion could be funded without adding on paper at least, to the liabilities side of the balance sheet. For both borrowers there was often the prospect of a limited or non-recourse loan to reduce their risk. For big international banks which were aggressively seeking new lending opportunities, this specialized form of loan offered up-front fees and higher margins than straight corporate loans.

THE PAST DECADE: 1973-1983

A 1979 Arthur D. Little report, commenting upon the copper industry, described a decade of "modest growth in sales, sharply increasing production costs, low return on invested capital, eroding profit margins, and higher debt." It explained, "these trends have reflected the combined pressures of inflation, higher cost of capital, declining ore grades, stagnant productivity, steep rises in energy and other factor costs, increased capital requirements for environmental control, and the worst recession during the postwar period." The depressed mineral prices which began with the 1982 recession have yet to recover as this paper is being written in 1984. The low 1982 prices brought unprecedented drops in earnings to the mining industry, including the first losses in history for some well-established corporations. Thus the picture painted in the ADL report remains essentially intact despite relatively healthy years in 1979 and 1980. While some of these factors have had similar effects on the financial performance and structure of other U.S. industries, some factors are peculiar and detrimental to the mining sector. The financial indicators summarized in the following section show, the patterns of change are more pronounced for the mining industry.

Despite some basic similarities that exist in the requirements of firms mining different commodities, the differences in norms of financial structure among district segments of the industry are not slight. So while there are truisms which pertain to the industry as a whole, whether looked at over time or relative to industrial averages, patterns within some segments may be very different. In the discussion which follows, standard financial indicators are compared for the Dow Jones Industrial average firms, (the "DJI") and a group of 17 publicly-traded independent mining companies which, for lack of data on privately held companies of mining subsidiaries of large corporations, are a proxy for the mining industry (Table 1). The primary minerals produced by members of the group are aluminum, copper, iron ore, molybdenum, nickel, and coal.

TABLE 1. MINING COMPANY SAMPLE

	Asarco, Inc.
	Cleveland - Cliffs Iron Co.
	Cominco Ltd.
	Falconbridge Ltd.
	Hanna Mining Co.
	Inco Ltd.
	Newmont Mining Corp.
	Phelps Dodge Corp.
Aluminum:	Aluminum Co. of America
	Alcan Aluminum Ltd.
	Kaiser Aluminum & Chemical
	Reynolds Metals Co.
Coal:	Eastern Gas & Fuel Assoc.
	North American Coal
	Pittston Co.
	Westmoreland Coal Co.

Industrial Organization

In looking at a list of companies which could have been included in the 1973 figures, but were no longer independent firms in 1983, it is clear that the major transformation took place in the base metal industry (Table 2).

TABLE 2. MINING COMPANY ACQUISITIONS

Company	Acquired By	Year
Anaconda	Atlantic Richfield	1977
Copper Range	Louisiana Land & Explor.	1977
Inspiration	Hudson Bay	1978
Kennecott	Sohio	1981
St. Joe	Fluor	1981
Molycorp	Union Oil	1976
Utah Intl.	General Electric	1976
	B.H.P.	1983
Cyprus	Amoco	1979

In copper, Anaconda, Copper Range, Inspiration and Kennecott would certainly have been included. St. Joe Minerals, Cerro, Molycorp, and Texasgulf would also have been on the list. All of these companies were acquired by oil companies or conglomerates during the decade. In fact, "merger mania" is one factor accounting for the much higher asset growth rate seen for the DJI group in comparison with the industry during this period. Average assets grew 162% for the DJI and only 105% for the mining companies (Figure 1). Only the coal subgroup matched the pace, benefitting from the oil price shocks. In 1980, Fortune magazine, reviewing changes in their list of 500 largest U.S. Industrials from its first appearance in 1955, notes that the basic industry contingent has been losing position over the years, but also explains that most of the 238 companies which have disappeared were actually swallowed up by mergers and acquisitions (Hayes, 1980).

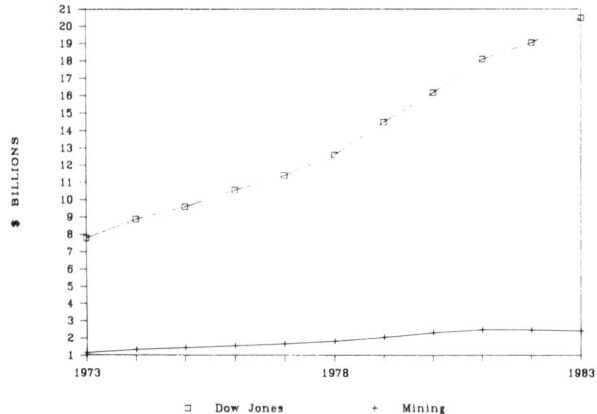

FIGURE 1. ASSET SIZE

The group of four independent coal companies, however, does not reflect the changes that have taken place in the composition of the industry. All four ranked among the 15 largest producers in 1973; one has since slipped. All of the new names among the 1983 top 15 are subsidiaries of oil or utility companies, primarily having achieved their high production by opening large Western surface mines. Eastern miners, particularly metallugical coal-producing subsidiaries of steel firms, have lost production.

Thus the petroleum companies expanded their coal company acquisitions of the 1960's and followed with further diversification into nonfuel minerals in the late 1970's and early 1980's, via grass roots projects and mergers.

Profitability

The mining industry is known to be highly cyclical. It is clear from the measures of profitability in Figure 2 that at its peaks over the last decade the mining industry's results were at best only slightly better than the returns for the DJI and the extent to which earnings fell in between was much greater. Over the entire period, the independent mining companies have been less profitable than the U.S. average. Return on invested capital averaged 8% versus 11% for the DJI, and earnings before interest and taxes averaged 13% versus 20%. The coal group was the only one to outperform the average, and then only for the three year period 1974-1976.

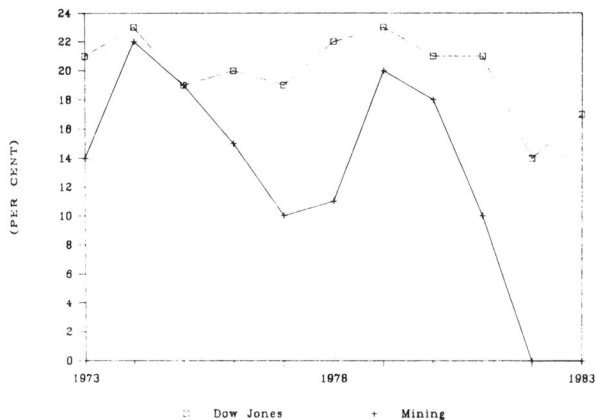

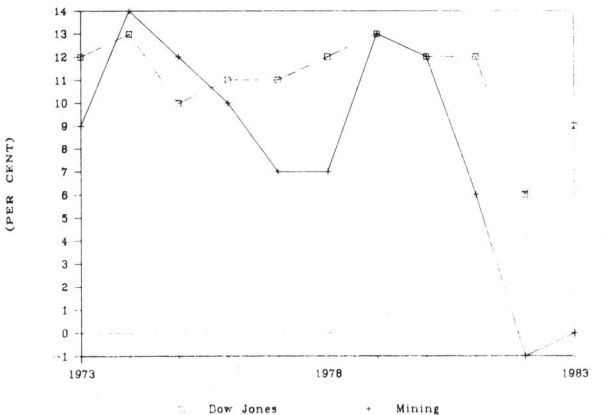

FIGURE 2. RETURNS ON INVESTED CAPITAL

Leverage and Coverage

Leverage for the DJI, as measured by the ratio of Total Borrowed Funds to Net Worth, showed a dramatic growth between 1973 and 1983 from 43% to 72%. For the mining industry as a composite, leverage increased only slightly from 84% to 97% (Figure 3). However, the timing of the increase in debt, and the reasons for additional borrowings are significantly different.

Within the mining industry, aluminum and coal companies have traditionally financed with a high level of debt, while base metal companies, leary of market price fluctuations, kept target debt levels well below that of the DJI average. Average figures during the period are 80% for aluminum and 165% for coal; 44% for copper and 13% for iron ore; and 70% for the Dow Jones Industrials. While leverage has actually fallen within the aluminum sector, and increased only slightly for the coal group, the financial leverage of iron, copper, molybdenum and nickel producers has increased steadily since 1973. The reasons include:

- capital outlays for government-mandated pollution control projects, usually financed in whole or part with tax-exempt securities;

- cyclical borrowings to fund normal capital expenditures, interest on existing debt, dividends, and operating losses during troughs, often from nominally short-term committed facilities;

- the failure of prices to rebound following the 1983 economic recovery, which ordinarily would have enabled mining companies to reduce debt closer to target levels.

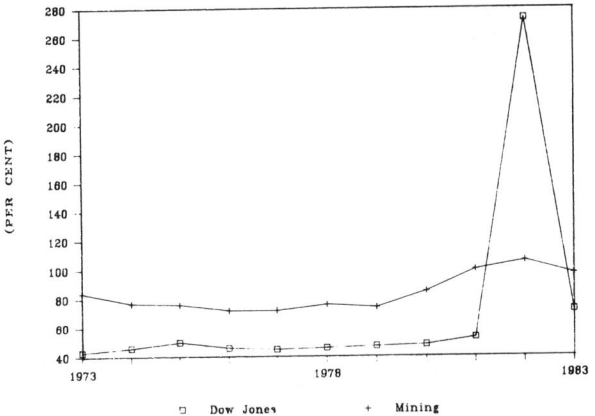

FIGURE 3. DEBT/NET WORTH

The big increase in debt for the DJI companies came in 1982, a year characterized by a "borrowing binge" on the part of the largest U.S. companies. Much of the borrowing was for takeovers, and by 1983 a large portion of this debt had already been repaid. Some of the increase was also due to runaway borrowings by a few large troubled companies such as International Harvester which, like many mining companies, required cash infusions to cover losses.

Coverage ratios, to which bankers look in determining a firm's debt capacity and rating agencies look in evaluating credit risk, were not strong for the capital-intensive mining sector even ten years ago, but have deteriorated to pathetic levels (Figure 4,5). While in 1982-1983 earnings before interest and taxes (EBIT) were 5-6 times interest charges for the DJI, EBIT for the mining companies was essentially zero. Many firms were forced into both the equity and debt markets to cover interest payments. The averages for the 1973-1983 period were a Times Interest

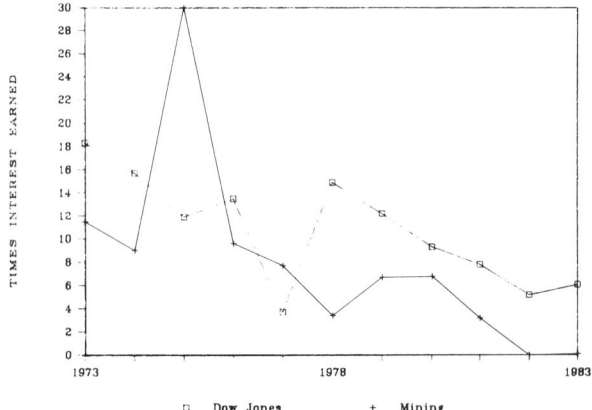

FIGURE 4. INTEREST COVERAGE

Earned ratio of 11 for the DJI, and only 8 for the mining industry. Similarly, annual cash flow approximately equaled total long term debt for the DJI, but would cover only 59% of principal for the mining industry.

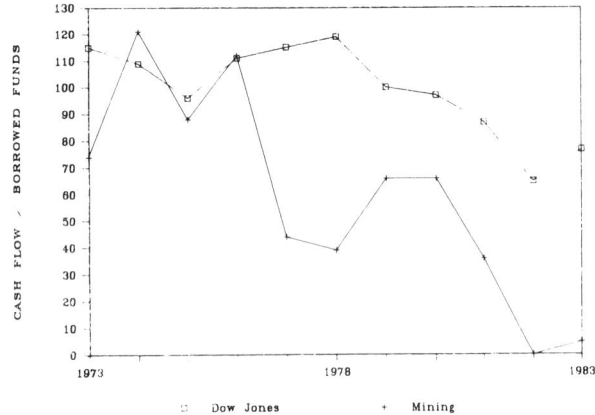

FIGURE 5. DEBT SERVICE COVERAGE

Stock Performance

The results of diminishing profitability, higher debt service requirements, and dilution of shares, are reflected in the common dividends paid by the two groups of companies. Whereas the dividend payout percentage has been fairly constant over the period for the DJI at about 50%, mining companies on the whole cut or omitted dividends beginning in 1982, and had not restored them two years later.

While the average P/E ratio for the DJI stood in 1983 almost exactly where it was in 1973, the shrunken earnings of the mining companies inflated their average P/E ratio, despite stock prices approaching their five-year lows. The time-weighted average for the market value of equity just equaled the book value of equity for the mining companies and in 1983 stood at .94; the DJI has fared somewhat better with a ratio of 1.37: 1 in 1983.

The Hidden Industry

The profitability picture for mining subsidiaries of oil companies and conglomerates undoubtedly parallels that of listed companies. With perhaps the exception of some Western coal producers, few of the investments in minerals have been consistently profitable. The international oil sector has nevertheless been amongst the most profitable industries, and almost certainly did not require capital infusions to offset mining losses. Their heavy borrowings went to finance cheap reserve acquisitions through giant takeovers of other established companies.

In light of the drag on earnings presented by mining operations, many of the earlier acquisitions are now being divested, or at least offered for sale, and exploration offices closed or greatly reduced. Occidental, Arco, General Electric and Getty are all examples of recent or expected divestors.

Precious Metals - The Small Bright Spot

Gold and silver prices, which in 1984 are depressed from peaks reached in the middle of the period, have nonetheless risen considerably from their 1973 levels. However, two major precious metals producers, Homestake and Hecla, which could have been included in the mining industry composite were excluded to avoid distorting the averages. Homestake's statistics would have improved the picture a bit; however, Hecla incurred a huge debt attempting to finance Lakeshore, a large copper project which was ultimately abandoned, resulting in a negative net worth on Hecla's 1978 balance sheet.

High precious metals prices during the 1970's and early 1980's revived interest in exploration for gold and silver, and in reworking or leaching of previously mined areas. In addition to projects undertaken by major mining companies, many small companies were formed, and a number have been brought public on the over-the-counter markets in the U.S. or on the Canadian exchanges. Some of these producers are now being acquired by other, larger mining companies. Others recently have sought debt financing to bring prospects into production.

New Financial Vehicles

The volatile economic environment of the 1973-1983 period brought about rapid changes in the financial services industry itself. Perhaps even more remarkable than the trends in debt vs. equity within the typical firm's capital structure, is the diversity and complexity of financial products and sources to meet very specific needs. With high inflation and interest rates, commercial banks found it increasingly difficult to compete with less regulated institutions. In 1973, banks made more than 40%

of all loans to non-financial borrowers; by 1983, they provided only 25% (Roos, 1983). As different financial intermediaries began to compete and as the financial marketplace became more global, a wider variety of products and services were offered at lower costs to borrowers.

The innovations have been responses to demands the volatility of the environment placed on both lenders and borrowers, and to the need of prospective lenders to tailor products to customers in the more competitive market. For example, lenders became increasingly unwilling to offer fixed interest rates in an era of rapidly rising rates, so companies found themselves with variable rate debt on their balance sheet. However, the uncertainty of interest costs creates a problem for the financial manager. As a result, instruments have been developed to swap floating rate debt for fixed rate debt with no balance sheet impact.

Another set of financial vehicles has developed to tap unconventional sources of capital or types of investors for whom a particular instrument may be attractive even though the issuing company or industry may hold no special appeal. Floating and adjustable rate issues and zero-coupon notes target pools of capital looking for interest rate protection and tax-exempt income, respectively. The investor base can also be broadened through commercial paper and private placement programs.

Mining companies are taking advantage of all of these options. In addition, natural resources industries are using the value of their assets in the ground in new ways. A greater range of commodity and foreign exchange futures and options can be used in various combinations to manage inventories and to hedge price and exchange risks. Shares paying precious metal dividends have been offered, and projects financed by bank loans repayable in gold after the start of production.

In addition, revisions in the U.S. tax code designed to encourage investment provided opportunities, not only for financing of assets and sheltering income, but for generating cash. Equipment leasing had been an important source of financing for mining companies. "Safe-harbor" rules allowed a company to "sell" tax incentives which due to its own unprofitability, it was unable to use. Many mining firms found this an important source of cash flow as operating income fell in the early 1980's.

Another example of the influence of tax laws on financial structure is the debt-for-equity swap. Debt which can be retired at a discount from face value by an exchange of stock is not subject to tax on the gain, and reduces leverage on the balance sheet.

Finally, tax laws also encourage merger and acquisition activity of some types. For example, acquired assets can be revalued and depreciated at newer, faster depreciation rates in some cases.

OUTLOOK: 1985 AND BEYOND

At the mid-point of the 1980's, U.S. inflation and nominal interest rates have fallen from the high levels of the previous decade, energy costs are stable, and industrial production recovering. However, with a strong U.S. dollar, and excess capacity overhanging most mineral markets, prices are not expected to show a sustained improvement for several years. Structural changes in the economies of the major consuming nations, such as a deemphasis of heavy industry, conservation of materials, and substitution of non-metals, will slow the rate of demand growth from the historical trend. On the supply side, producers seem to be tending to restrain production insufficiently at low price levels, and to overreact to small improvements in the market. Foreign producers have foreign exchange and employment objectives; North American miners may find losses minimized by selling at low prices rather than taking temporary or permanent closures.

A decade ago a theoretical argument was set forth that to the extent that natural resource markets look at flow transactions rather than asset values, the markets could easily be unstable (Solow, 1984). The argument is that if net prices are expected to rise at less than the prevailing interest, resource deposits are a poor way to hold wealth. Thus production should be increased to convert ore into dollars which can then earn at the higher alternative rate. As production rises, prices are pushed still further down the demand curve in a vicious cycle.

In Solow's agrument, asset markets provide the counteracting force. Existing assets will be written down in value until the rise in net price will provide an appropriate return on the lower value. This is in fact what appears to be taking place in the industry. Operations are being closed indefinitely, reengineered to a lower capacity, sold at or below book value, or written down to more accurately reflect market prices. Common stocks are trading at close to historical lows in many segments of the industry.

Debt levels are unlikely to increase significantly, as cash flow management and maintaining agency ratings have become major concerns. Many recent asset sales have provided cash for the purpose of reducing debt to levels which are more comfortable within tight operating margins. Restructuring of existing debt is likely to continue where opportunities to reduce debt service costs and volatility are available.

Project financings will probably be less in vogue. Fewer companies are looking at new megaprojects which could add to already glutted markets, and commercial banks are not eager to take on additional exposure to many debtor nations, nor to assume price risk on projects.

Sponsoring mining companies are perceived to be weaker borrowers than in the past. Funds do still seem to be available, however, for low cost projects with reasonably certain markets for their products.

Most exploration in recent years has been for small precious metals properties, many of which are being brought to the development stage by enterpreneurs, then acquired by or merged with larger, more established companies. Mining companies seeking diversification may also increasingly opt for the merger route rather than grass roots developments in new areas. Bankers may see fewer lending opportunities, and investment bankers feel the competition for advisory services.

REFERENCES

Cincotta, George A, et al, 1982, "The Changing World of Corporate Finance", Dun's Business Month, June, pp. 92-110.

Financial World, 1982, "The Borrowing Binge", September 1, pp. 41-43.

Hayes, Linda Snyder, 1980, "Twenty-Five Years of Change in the Fortune 500," Fortune, May 5, pp. 88-96.

Houseman, T., Rosborough, B.W., and Denner, R.L., 1970, "Commercial Banks' Role in Mine Financing," Engineering/Mining Journal, December, pp. 82-83.

Nadkarni, Ravindra, Bozdogan, Kirkor and Korn, Donald H., 1979, "Economic Impact of Environmental Regulations on the U.S. Copper Industry", ADL Impact Services Company, March, 13 pp.

Roos, Lawrence K., 1983, "The Blitzkrieg of Financial Innovations", Bankers Monthly Magazine, March, pp. 15-22.

Scott, Jr., David F., and Johnson, Dara J., 1982, "Financing Policies and Practices in Large Corporations," Financial Management, Summer, pp. 51-59.

Solow, Robert M., 1974, "The Economics of Resources of Economics", The American Economic Review, 64 (2), May, pp. 1-14.

1.3

CAPITAL REQUIREMENTS OF THE MINERAL INDUSTRY

Simon D. Strauss

Consultant, New York, New York

INTRODUCTION

The changes in ownership, management, and operation of the non-fuel minerals industry since the Second World War have been dramatic. World-leading enterprises have lost their identity; host countries have taken over operations from multi-national firms; marketing practices have been drastically altered - and yet the world industry is supplying minerals on a scale far greater than had been anticipated in projections made at mid-century. The world's known mineral resources have been enormously expanded, despite the doom sayers who for decades have been predicting that depletion of non-renewable resources would impose limits to growth.

PALEY REPORT

One can set the stage by reviewing how the minerals outlook was regarded in a thoughtful and comprehensive analysis made in the years 1950 and 1951. During the War years, President Harry Truman of the United States had come to appreciate the decisive role played by minerals resources through his chairmanship as a U.S Senator of a committee that investigated the procurement and allocation of natural resources. In 1950, in the Cold War era that resulted in actual military conflict in Korea, he appointed a prestigious five-man commission to study the problems of Materials Policy over the following quarter century.

Their findings, published in five impressive volumes in early 1952, became known as the Paley report. The text was prepared by expert scientists and economists who spent 18 months assembling statistics, making projections of future demand, and interviewing leaders in industry, academia, and government.

Table 1

PALEY REPORT

(million short tons)

Western World Consumption	1950	Projection mid-1970's	Actual 1973	Projection Exceeded by
Iron ore	233	365	679	86%
Aluminum metal	1.39	6.00	13.20	120%
Refined copper	2.55	3.85	7.11	85%

The Commission had correctly foreseen that aluminum would replace copper as the largest volume non-ferrous metal, but it had fallen short of reality by a wide margin indeed for all three materials. Even though the Commission's expectations for growth proved to be extremely understated, it had concluded its report with grave foreboding about the effect of its estimates of demand in relation to supply, as follows:

"Consumption of almost all materials is expanding at compound rates and is thus pressing harder and harder against resources which, whatever else they may be doing, are not similarly expanding. This materials problem is thus not the kind of 'shortage' problem, local and transient, which in the past has found its solution in price changes which have brought supply and demand back into balance. The terms of the materials problem we face today are larger and more pervasive."

With respect to some minerals, the Paley Report expressed scepticism that, in fact, supplies would be forthcoming regardless of price. Because the Paley demand forecasts understated the actual growth in demand for virtually every mineral, one could logically have expected frequent and protracted shortages in the years since 1950.

Presented at the International Conference on Recent Advances in Mineral Science and Technology, Johannesburg, S. Africa, March 1984.

PALEY POSTSCRIPT

Although occasional shortage of supplies of some non-fuel minerals did occur, they were of limited duration, as the Commission would label "local and transient." If one uses a deflator to offset the effects of inflation, prices of most minerals today are at or below levels prevailing at the time of the Commission's report. Thus the pervasive price problems that the report anticipated did not occur during the time period covered by its study and have not yet occurred, almost a decade later.

The expected shortages have not developed because of the remarkable progress made by the minerals industries in identifying, developing and exploiting new mineral resources through the new world.

A striking example from the Paley Report is the absence of Australia from considerations of potential sources of iron ore, bauxite, nickel and manganese. In 1950 the Australian government was so concerned about what it considered to be its limited domestic iron-ore resources that an embargo had been placed on iron-ore exports. Currently, Australia is the Western World's largest producer of iron ore and bauxite, the third-largest miner of nickel, and the fifth-largest miner of manganese.

Other instances could be cited, but the point should be clear. Contrary to the scare headlines about mineral depletion in some of the popular press, even though the world today is using far larger quantities of these essential materials than was the case forty years ago in wartime, the mining industry has been able to find, develop and produce all the minerals for which there is current demand. Major unexploited deposits have been identified that will be available to provide additional supplies when needed — providing, of course, that the finance can be found for the massive capital required.

CAPITAL REQUIREMENTS

Projects that would have cost tens of millions thirty years ago today require hundreds of millions in terms of the accepted international yardstick for currencies — that is the United States dollar. But the dollar itself is a changing yardstick.

Some tabulations are shown below to compare the capital requirements of roughly equivalent enterprises launched at various dates since World War II in US dollar terms. The dollar is selected as the common denominator for several reasons. Under the Bretton Woods agreement of 1944 the US dollar was designated as one of two accepted reserve assets for central banks — the other being gold. The dollar is the currency of the country with the greatest gross national product. A large part of the finance for most mining projects is raised in US dollar terms — both debt and equity (there are, of course, exceptions — notably in the case of Australia and the South African gold mining industry). World market prices of most mineral products — gold, copper, nickel, aluminum and others — are usually denominated in US dollars.

Table 2

CAPITAL COSTS

COPPER MINE AND MILL

Mine	Location	Date of Start-up	Initial Capacity 000 stpy Cu content	Cost of Facilities US$ million	US$ per stpy Capacity
Silver Bell	Arizona	1953	20	$18	$900
Tyrone	New Mexico	1969	55	118	2,150
Andina	Chile	1970	65	139	2,100
Lornex	B.C, Canada	1972	60	138	2,300
Caridad	Mexico	1980	150	673	4,500
Copper Flat	New Mexico	1982	20	103	5,150
Tintaya	Peru	1985E	58	320	5,500

E = Estimated

Perhaps the best way to look at capital requirements is to compare the capital investment required to produce a unit of metals or minerals with the going price for that unit. In this way, some of the distorting influence of currency depreciation is eliminated.

Copper

Table 2 on copper mines and mills shows that a modest US project — the Silver Bell mine — represented an investment in 1953 of $900 per short ton of copper metal capacity. The price of copper in that year averaged $580 a short ton, so the investment was 1.55 times the price of copper that year.

A comparable US project, launched in 1982, was the Copper Flat mine which required an investment of $5,150 per ton of capacity. The average price of copper in 1982 was $1,460 a ton, so the investment was 3.53 times the price of copper that year.

Table 3

CAPITAL COSTS - COPPER MINE AND MILL

Mine	Location	Date of Start-up	Initial Capacity 000 stpy Cu content	Cost of Facilities US$ million	US$ per stpy Capacity
Toquepala	Peru	1959	145	$237	$1,635
Cuajone	Peru	1976	130	726	4,000
Sar Chesmeh	Iran	1982	160	1,400	8,750

At Peublo Viejo, the largest gold mine in the Western Hemisphere today, the investment of US$50 million was equivalent to $125 per ounce of annual capacity (see Table 4). In 1975, the year it started producing, the average price of gold was $161 an ounce, so the investment was only 0.78 times the value of the product.

Table 4

CAPITAL COSTS - GOLD MINE AND MILL

Mine	Location	Date of Start-up	Annual Capacity 000 ozpy Gold	Capital Cost US$ million	Cost per oz. of Capacity
Pueblo Viejo	Dom. Rep.	1975	400	$50	$125
El Indio	Chile	1981	350	200	570
Detour L.	Ont., Canada	1983	200	131	655
McLaughlin	Cal., USA	1985E	200	187-237	935-1200

E = Estimated

MINERALS INDUSTRY FINANCIAL REQUIREMENTS

A more recent Latin American operation is the El Indio mine in Chile, which started in 1981 and which represented an investment of $570 an ounce of annual capacity. The average price of gold in 1981 was $464. Thus the investment was 1.23 times the value of the product. Both Pueblo Viejo and El Indio have substantial byproduct silver outputs, and El Indio is also recovering a significant amount of copper. Taking this into consideration, the investments are roughly comparable.

Not so the projected McLaughlin gold mine in California, scheduled to be in production with a capacity of 200,000 ounces annually in 1985. Present estimates of investment are between $187 million and $237 million or from $935 - $1,200 per ounce to be produced. Unless the price of gold rises significatnly, it is obvious the operators will have invested more than double the value of the annual product.

Co-products: Copper/Gold/Silver

The Bougainville project (see Table 5) is representative of a group of recent major undertakings that involve ores in which more than one metal is of sufficient economic significance so that the output must be described as involving co-products rather than by-products of a major metal. In analyzing the investment at Bougainville, because the value of gold and copper produced tends to be roughly comparable, it seems appropriate to calculate the capital required as being roughly half for gold output and half for copper output.

Table 5
CAPITAL COSTS
COPPER AND PRECIOUS METAL MINE AND MILL

Mine	Location	Date of Start-up	Annual Capacity 000	Cost of Facilities US$ million	Cost per ton of Capacity
Bougainville	Papua New Guinea	1972	200 st Cu 600 oz Au	$430	$1,075 Cu $ 358 Au
Troy	Montana, US	1981	18 st Cu 4,250 oz Ag	90	$2,500 Cu $ 10.60 Ag
Ok Tedi	Papua New Guinea	1984 E	125 st Cu 450 oz Au	1,400	$5,600 Cu $1,500 Au

E = Estimated

Note: On these three projects capital costs are divided equally between gold or silver and copper to arrive at capital costs in terms of annual unit of metal output.

Bougainville required an investment of US$430 million. Taking half of this as pertaining to the gold product of 600,000 ounces annually, the per-ounce investment was $358 an ounce of gold; while to obtain the 200,000 short tons of copper annually the investment was $1,075 per ton of copper capacity. Metal prices in 1972 were quite different from those currently prevailing, with the big change having occurred in the gold price. The average gold price in 1972 was $58 an ounce and the average London copper price was $970 a short ton. Based on those prices, investment was equal to 6.17 times the annual value of gold produced and 1.11 times the annual value of copper produced. Annual output, based on today's prices, has a gross value somewhat in excess of the original investment, which has turned out very well for the investors and the government of Papua New Guinea.

Now under way in that country is a second major gold-copper project, Ok Tedi (see S.C. McGill's paper in Chapter 12 - Ed.). Its more isolated location is requiring much heavier investment than at Bougainville. And, of course, the effects of inflation in the twelve years that have elapsed are also driving up the amount of capital required. Based on current estimates, Ok Tedi will start off with well over three times as much sunk capital as Bougainville. Its production of both gold and copper will average out at less than Bougainville's although, in the first few years of operation, gold output may exceed that of Bougainville.

Molybdenum

Molybdenum and nickel, two metals widely used in the steel industry, have seen enormous changes in the post-war era. During the fifties the markets for each were dominated by a single seller - American Metal Climax (now AMAX) in molybdenum and International Nickel (now Inco) in nickel. Both companies achieved outstanding success in developing markets. This encouraged other companies to explore for, develop and equip deposits. Today both markets are highly competitive and equipped capacity is far in excess of current demand.

Because molybdenum deposits are low grade and for the most part in remote locations, capital requirements are enormous. The current value of molybdenum is in the order of $4 a pound, so the capital required is from two and a half to almost five times the value of annual output.

Table 6
CAPITAL COSTS
MOLYBDENUM MINE AND MILL

Mine	Location	Date of Start-up	Annual Capacity million lb. Mo	Cost of Facilities US$ million	Cost per lb of Capacity
Henderson	Colo., US	1976	50	$500	$10
Tonopah	Nev., US	1982	12	$230	$19.25
Questa	N.M., US	1983	20	$250	$12.50
Thompson Ck.	Ida., US	1984 E	20	$375	$18.75

E = Estimated

Nickel

Among the laterite deposits, the Nicaro mine in Cuba was equipped for production during World War II at a cost of about $1,400 per ton of annual capacity when the price of nickel was about $700 a short ton. The most recent major laterite project to start up - the Cerro Matoso mine in Colombia - represents an investment of over $18,000 a ton of annual capacity. Nickel prices are nominally over $6,000 a ton but the real market is currently at $5,000 a ton or less.

Table 7

CAPITAL COSTS

LATERITIC NICKEL PROJECTS – MINE AND PLANT

Mine	Location	Date of Start-up	Annual Capacity 000 stpy Ni	Cost of Facilities US$ million	Cost per ton of Capacity
Nicaro	Cuba	1944	22	$400	$1,400
Dominicana	Dom. Rep.	1971	31	180	5,800
Marinduque	Phil. Isl.	1975	37	315	8,400
Greenvale	Australia	1976	22	260	11,800
Exmibal	Guatemala	1977	14	225	26,500
Soroako	Indonesia	1977	40	813	20,300
Cerro Matoso	Colombia	1982	22	400	18,200

CAPITAL INTENSITY

The relationship between the cost per unit of annual capacity and price per unit is the capital intensity. In manufacturing industries, a usual yardstick for an investment is that annual gross revenues from product sales should exceed the amount of capital invested.

If an investment is as much as four times the expected gross value of annual output, then annual sales will be only 25% of the funds originally committed. After deducting operating costs and taxes, the chances of obtaining an acceptable return on investment are slim indeed.

The position of individual projects may differ significantly from the overall trend, but the tabulations indicate that capital requirements for new mineral projects have been escalating at a faster rate than the value of mineral production, as reflected in commodity prices. Several factors have contributed to the capital cost pressure:

(i) Lower grades of ore are being exploited. Capital requirements are related to tonnage of material extracted to a far greater extent than to tonnage of final product.

(ii) A higher proportion of the new ventures are located in remote areas lacking infrastructure. This adds significantly to capital costs for transport, power, housing and other facilities.

(iii) More sophisticated equipment is being developed. While this tends to reduce operating costs, it adds to capital investment.

Government regulations related to the environment and to safety have added significantly to investment without expanding productive capacity. These costs apply in larger measure to mining ventures in the developed countries, but some of the developing countries are now beginning to follow suit.

CAPITAL COSTS

LEAD-ZINC MINES AND MILLS

Mine	Location	Date of Start-up	Annual Capacity st metal in Concs	Cost of Facilities US$ million	Cost per ton of metal Capacity (a)
Anvil	Yukon, Can.	1969	130,000 Zn 90,000 Pb	$61	$280
Mattabi	Ont., Can.	1972	60,000 Zn 5,000 Pb	41	630
Black Mtn.	South Africa	1979	100,000 Pb 19,000 Zn	190	1,600 (b)
Polaris	N.W.T., Can.	1982	130,000 Zn 30,000 Pb	144	900
Elura	N.S.W., Aus.	1983	70,000 Zn 45,000 Pb	170	1,480 (b)

(a) Cost per ton is calculated on basis of combined lead-zinc capacity
(b) Silver production will be about 4 million ozs annually

Table 9

CAPITAL COSTS

SILVER MINE AND MILL

Mine	Location	Date of Start-up	Annual Capacity 000 Oz Ag	Cost of Facilities US$ million	Cost per oz of Capacity
Galena	Idaho, U.S	1955	4,000	$3	$0.90
Coeur	Idaho, U.S	1976	2,500	23	9.20
Equity	B.C., Canada	1981	7,000	144	20.00 (a)

(a) Substantial by-product values in gold and copper

CAPITAL PRESSURE ON MINING COMPANIES

The impact of the greater capital requirements on individual private enterprises has been severe. In the United States, Canada and Australia the leading mining companies thirty years ago were largely debt free and many of them were, in fact, cash rich. Retained earnings, depreciation, and depletion allowances generated adequate flow to undertake most new projects without recourse to lending institutions.

Inflation has changed that. Not only has it escalated the costs of new undertakings to levels that mandate access to the banks and other lenders, but it has meant that depreciation and depletion allowances (based on investments made in the era of limited inflation) no longer generate sufficient cash to replace obsolescent equipment in existing operations. Thus a worn-out shovel cannot be replaced with a new shovel if the only funds available are the depreciation allowance on the original machine.

The financial problems of some companies have been compounded by the nationalization of their foreign holdings. The loss of once-profitable foreign subsidiaries helped to undermine the stability of many parent companies and contributed to their loss of independence.

The rising level of capital requirements for mining enterprises has also had its influence on the financial stability of some developing countries that took over ownership of deposits and processing plants. This has been particularly true in Latin America. The spirit of nationalism and the desire for industrialization has motivated many of these governments to undertake enterprises of dubious economic viability. This has added to debt burdens which, compounded by rising interest rates, they now find difficult to carry. Part of the collapse of the Mexican peso may be attributed, for instance, to huge unprofitable investments in steel production. Bolivia's ownership of tin mines has been a losing venture financially, and this has been further exacerbated by construction of tin and lead smelters that yield a negative return on the investment. Peru has recently built a zinc smelter at high cost that is operating at a tremendous loss.

OPERATING COSTS

Operating costs are affected by the manner in which inflation has eroded the values of many currencies. From the end of World War II until 1971, devaluations or revaluations of currencies were monitored by the International Monetary Fund under the Bretton Woods agreement. The standard of measurement was the United States dollar and that currency was linked to gold at $35 an ounce. During this period relatively infrequent changes in monetary values were made by the principal developed nations. For example, the rate of exchange between the pound sterling and the dollar changed only twice in 27 years. The pound had been worth US$4 at the end of the War. In late 1949, the pound was devalued to $2.80. In late 1967 it fell further to $2.40. After the United States cut the tie between the dollar and gold at $35 in 1971, however, an era of floating-exchange rates ensued. Despite occasional periods of strength, for the most part the quotation for the pound has sagged in US dollar terms.

In the area of mineral production, the three developed countries of greatest consequence are Canada, Australia and South Africa. Their currencies were relatively stable during the 1944-71 era, but under the current phase of floating rates they have weakened substantially in relation to the US dollar. However, they have not undergone the drastic changes which have affected the currencies of the Third World countries. In fact, devaluation of currencies of some Third World countries has been endemic throughout the entire post-World War II period, especially in Latin America and Africa.

A simple illustration will make the point clear. Under the ruling military junta, the value of the Chilean peso was held at a rate of 39 to the US dollar from 1971 until 1982. Costs of production by the four large mines operated by Codelco, the state-owned copper company, were estimated in 1981 to have been above US$0.60 a pound. Since 1982 the peso has been steadily devalued. Currently, Codelco costs of producing copper are calculated at under US$0.50 a pound, even though some costs for equipment, interest and imported supplies must be paid in US dollars.

In the past, gains from devaluation were gradually eroded as internal costs rose. Thus, in the fifties and sixties the Mexican mining industry enjoyed temporary gains from devaluation but in time these were offset as wages, internal supply costs, and export taxes were adjusted upward. As a consequence, despite devaluation, Mexican mining remained in the doldrums until the government changed its tax structure.

And even with the temporary help of devalued currency in some countries, production costs have nevertheless risen, in US dollar terms, due to inefficiencies that have developed. Some state-owned enterprises in developing countries have failed to maintain adequate stocks of spare parts; equipment that is obsolescent has not been replaced; advance development work has not matched the tonnage of ore extracted; transport facilities have deteriorated; and insistence on processing locally in expensive new smelters and refineries has involved higher conversion costs than were incurred when ores and concentrates were exported.

ENVIRONMENTAL PRESSURE ON CAPITAL

Environmental concerns have had a greater impact on the mining in the developed countries than in the developing nations, but recently they have begun to surface even in such remote areas as, for instance, Papua New Guinea. The Ok Tedi project and the government have been debating the best means of tailings disposal to avoid contamination of surface waters in a mountain area with scant population.

Perhaps the most striking example of public environmental concerns affecting mineral operations has been in the United States. Air and water pollution moved to the centre stage of national consciousness in the late sixties. Legislation was enacted, administrative agencies were established, and comprehensive regulations were promulgated by both federal and state governments. Political campaigns were largely influenced by debates on this issue. The public was bombarded by media reports - both by television and by newspapers - of the health problems created by pollution, including contamination of food products, particularly those treated with chemical additives.

As a consequence, the mining industry was confronted with a maze of laws and rules. These imposed strict limitations on emissions that affected the quality of air and water; required new measures to enhance worker safety; controlled the way in which waste products were disposed of; and spawned endless litigation arising from worker and public claims of injury or death resulting from past operations.

A specific example of how the environmental issue affected mining is afforded by copper. Nine major US mining companies, all engaged in copper production and some engaged in other mining activities, at the end of 1964 had an aggregate long-term debt of US$252 million - an average of less than $30 million a company. In the 19 years following the end of World War II, they had been able to fund their capital expenses from retained profits plus the allowances for depreciation and depletion. Of the nine companies, five either were free of any debt or had debts of less than $6 million each.

Some 12 years later at the end of 1976, the same nine companies had aggregate debt of $3,204 million - an average of more than $350 million per company. The two smallest companies had debts of about $30 million each. All of the others owed over $200 million and four owed more than $500 million each. Some of the debt was due to attempts to diversify; some of it was the consequence of inflation - that is the depreciation allowances on equipment during 1964-76 were not adequate to generate funds to purchase replacement units. But the largest single factor was the high cost of capital expenditures on pollution controls. These added nothing to capacity and materially increased operating costs. A survey by the US Department of Commerce indicated that non-ferrous metals had the highest percentage of capital expenditures for pollution controls of any United States industry.

The tabulation of debt for these companies can no longer be updated because five no longer exist as independent firms. Since the end of 1976, four have been taken over by larger and financially stronger oil companies. A fifth was merged into a Canadian company, which in turn has been restructured by its majority stockholder - Anglo-American Corporation of South Africa. Thus separate balance sheets are no longer published. The loss of independent status was undoubtedly largely attributable to the weakened financial posture of these companies, brought about in part by capital costs mandated under the environmental laws.

NATIONALISM

The Chilean example relfects the pressure of nationalism on the operations of the minerals industry. This has been most pronounced in the developing countries, but it is also quite clear in such countries as Canada, Australia and Ireland - where foreign participation in mineral ventures is now carefully circumscribed.

In the late 1960's two of the nine companies referred to above, Anaconda and Kennecott, had negotiated to relinquish control of their large copper mines in Chile to the government - but they expected to retain a minority interest and to be compensated for the shares they sold. Instead, following the election of a strongly leftist government in 1970, all their properties in Chile were expropriated and they were denied compensation. Subsequently, when the military junta took over, some compensation was negotiated - but the sums paid were small in relation to the previous financial returns enjoyed by the two companies.

With a large share of the world's production now in the hand of public-sector companies, mineral supply has become less responsive to demand. When the domestic or world economy falters, output is nevertheless continued at or close to capacity. The imbalance creates price weakness. Thus, despite high production, foreign exchange revenues are reduced. For Third World Countries there is now available on offsetting source of hard currencies. They have recourse to the compensatory financing mechanism of the International Monetary Fund (IMF). During the period of copper-price weakness that developed in 1982 and 1983, six copper-producing countries among the less developed countries obtained well over $1,000 million from the IMF. These are loans, which must eventually be paid off.

Thus the curious treadmill has developed that because the copper price is weak, the LDCs borrow money from the IMF and then, to meet their debt requirements, they continue to maximise production, keeping pressure on the price. Yet is is a fair guess that if the LDCs had curtailed production, prices would have been higher, their foreign-exchange earnings would have been greater, and they would not have had to borrow from the IMF.

FUTURE CAPITAL AVAILABILITY

In the past, if host governments were prepared to guarantee loans to specific projects, finance was readily available from lending institutions both private and public. Today those guarantees no longer seem as secure as they once did. During the last two years, many governments have been unable to meet existing interest and debt-amortization schedules. Country after country has had to apply to its lenders or the IMF to renegotiate existing loans.

In the light of this experience, new projects will be reviewed with a more critical eye - both by private lenders and by public-sector lending institutions as well. Among the capital-exporting countries that provide the great bulk of the funds for such projects, the tax-paying public has grown increasingly restive. Questions are raised over the size of contributions to organizations such as the World Bank, the IMF, and the various regional international banks. As for the private banks, faced with the possibility of sizable write-offs on existing loans, they are definitely turning more cautious.

CONCLUSION

In consequence, funds for new projects may not readily be forthcoming unless they can pass stricter standards of viability than have hitherto been applied. Lenders will consider both the level of demand for the commodity to be produced, the prices that can reasonably be anticipated, the country where the project is located, and the capital intensity and other pressures encountered for the specific project.

1.4

THE HISTORY OF FINANCING A MULTINATIONAL MINING COMPANY

SIR ANTHONY TUKE

CHAIRMAN, THE RIO TINTO-ZINC CORPORATION PLC
EX-CHAIRMAN, BARCLAYS BANK PLC

Members of the Society of Mining Engineers may well regard it as rather unusual that a paper on this subject is being presented by someone whose first taste of mining came at the age of 60 or so - someone who last heard of cathodes and anodes in 1938. On the other hand, the title does refer to the financing of a mine and at least I can claim to have spent all my business life since the war until 1981 in the world of banking, though this inevitably means that I have been used to looking at these and other problems through what you might regard as the wrong end of the telescope.

The lending of money, and more particularly the provision of finance for commerce and industry, is the central job of a banker and I have been involved in this in various ways over many years, but regrettably only very marginally in the world of mining. As you will see from what follows, the rather specialised form of finance which RTZ required was, to my regret, not provided by the British banking community since the great majority of the finance came from American banks and, to a lesser extent, from German banks.

I am most indebted to Mr. Roy Wright for a great deal of the nuts and bolts in this paper. He was one of the three central figures in the growth and expansion of RTZ during the 1950's and 1960's and was right at the centre of the negotiations with bankers, with governments and with many others. The going, as you will see, was far from easy.

The essential basis for the financing of each of the RTZ group's mining projects was a firm long-term sales contract for sufficient of the output of the particular mine to produce the cash flow necessary to service the loans. Wherever possible, loans were raised in the same currency as the sales contracts were made. Each sales contract was designed to give the customer the product he wanted with a long-term assurance of supply and at the same time to give the bankers the protection they required. Thus the marketing concept and the financing concept for a particular mine formed one overall plan, and talks with the customers and the bankers were conducted in parallel until a satisfactory marketing/financial package was agreed in principle.

Preliminary talks were begun soon after the discovery of a potentially viable orebody. As confidence about a discovery increased, customers were persuaded to enter into contingent sales contracts before the very detailed and costly business of the economic and technical feasibility study of the mine was launched; such a study might cost 10% or more of the estimated total capital cost. At the same time provisional understandings were reached with the bankers. The contingent sales contracts usually gave a period of grace of eighteen months to two years, during which time the decision whether or not to continue with the mine had to be made. This decision rested mainly on the result of the feasibility study and also on coming to a firm agreement with the bankers. If all went well and the mine went ahead, then the contingent sales contracts became firm, but they fell away if the decision was negative.

The Company was fortunate that during the whole period when its major mines were being developed, world trade was increasing rapidly and Japan was becoming a dominant industrial power. The Japanese supported many of the mines, including Hamersley Iron Ore, Lornex Copper, Bougainville Copper to name a few, with the basic long-term sales contracts that enabled their development. The Germans supported Palabora and to a lesser extent Bougainville.

The mines were financed as individual projects generally with a loan/equity ratio of about 65/35, without legal recourse to RTZ for the loans. RTZ was, however, responsible for providing any capital overruns required to bring the mines into commercial operations. Nevertheless, the RTZ management regarded itself as morally responsible to the lenders because, Palabora aside, the sales contracts had either a fixed selling price (like Hamersley), an indexed price (like the Elliot Lake Uranium Mines and Mary Kathleen), or a floor price (as in Lornex, Bougainville and Rossing). Thus the Company believed that any financial failure would have been due to poor management or technical shortcomings, and not to events outside RTZ's control, hence the moral responsibility to the lenders. In retrospect that was somewhat naive, but in those days inflation was very low and currency exchange variations few and far between.

Perhaps in the very early days lenders were not fully aware that the RTZ Group had little in the way of management strengths or competent technical resources. Also perhaps they did not appreciate that RTZ's financial resources could not have matched the moral obligations it assumed or that the failure of just one of its early major endeavours would have strained the Company to the limit.

Moral responsibility was never precisely defined, but in general it meant that RTZ would do its utmost to meet the loan obligations irrespective of the legal situation. At the very outset, in Canada in 1957, the Company was put to the test and rapidly learnt some important lessons. The Northspan uranium mines in Elliot Lake had a serious capital overrun which was concerned not only with plant or equipment, but was caused by a host of problems that arose partly from the extraordinary crash programme imposed by the terms of the uranium supply contracts, and partly from a lack of management and technical experience. The Company never doubted that it had to stand behind the Northspan commitments and earned the respect of all the lenders by taking this attitude. Funds from a rights issue in London, plus funds so bravely provided by the Canadian bankers, gave the breathing space needed to organise the mines. Even so, some re-scheduling of loan repayments was necessary but with the overall period of repayment unchanged.

During its build-up phase RTZ in London financed its share of equity required for its operations by loans to 100%-owned Local Holding Companies in the various overseas countries. Dividends from the Operating Companies flowed to the Local Parent Companies, whose dividends in turn flowed to the Local Holding Company, either to be retained there for further local investment or remitted to London in repayment of the loans. The Local Holding Company in a particular country held its shareholding in the so-called Local Parent Company, which in turn held its shareholding in the various Operating Companies.

Thus it was the Local Holding Company that subscribed to equity issues by the Local Parent Company. The Local Parent Company in turn subscribed to equity issues by the Mining Companies.

It should be remembered that during the whole of the growth period, namely 1955 to 1970, RTZ as a UK based Company was constrained by Exchange regulations. These made it extremely difficult and complex for the Company to continue to grow as it would have wished, particularly in the United States. Until 1966 it could invest freely in what was then known as the sterling area, but it was under strict control so far as the rest of the world was concerned. The controls were tightened in 1965-6 and in the 1966 Budget so-called 'voluntary' controls were introduced on investment in developed sterling area countries, including Australia and South Africa.

Turning now to the specific examples, the Elliot Lake uranium mines in 1956 were Rio Tinto's first major post-war endeavour under its new management team, and its first experience of project financing. Uranium was the boom material of the mid-1950's and there was no lack of lenders for those mines that had obtained Canadian Government contracts. A contract inferred that the mine had the necessary ore reserves of suitable grade and could operate profitably, but this was not necessarily true. Essentially the contracts were on a cost plus basis with the costs being partially indexed. The selling price was negotiated for each individual mine after a scrutiny by the Canadian Government agency of the capital and operating cost estimates.

These mines were something of a risk to the lenders because at that stage Rio Tinto's management had so little experience. Perhaps only ignorance led them to attempt to develop eight mines in a two-year crash programme under appalling physical and climatic conditions with such slender resources. The reason for the crash programme was that the contracts had a fixed terminal date, which meant that the mines had to be in production within a certain time if they were to fulfil their deliveries. No experienced company would have attempted it then and certainly none would today.

Money-raising for the uranium mines was quite straightforward project financing. Northspan, underwritten by a US/Canadian banking group headed by Morgan Stanley was the most difficult because it was the largest and also because it was at the time of the Suez crisis when anti-British feeling was rife. One special feature of the financing of all the Elliot Lake mines was that the bonds had warrants attached, which entitled the subscriber to buy a certain number of ordinary shares at set prices at pre-determined dates. There was no scientific basis for deciding the number of warrants and so the true value of the bonds varied considerably from one mine to another. The Toronto market was, however, relatively unsophisticated in those days, and surprisingly no detailed analysis of what was being offered was made by investment analysts.

Another feature of the Elliot Lake mines was that the mines were not 'proved' in any detail in regard to tonnage, grade and conditions underground. Again this was because of the crash programme and also because it was considered that a relatively few widely spaced drill holes gave a good enough indication. This was generally true, but led to problems at two of Northspan's five mines. It was much the same practice as is adopted in South Africa with the gold-bearing reef, but the significant difference was that the South African gold mines had a long history and there was a tremendous background of geological knowledge. The other difference was that the gold mines were generally financed by equity, whereas Elliot Lake mines were financed largely by loans.

The next major mine to be financed was Palabora. It was difficult for two reasons. First, it was the first major open-cast, low-grade copper mine in Southern Africa. To the South Africans accustomed to the average 2-3% or higher grade of the Zambian Copper Belt, the low copper grade of Palabora (about 0.6% initially) made it seem most improbable that it would ever be profitable. Secondly, investment in South Africa was unpopular at the time as a hangover from the Sharpesville riots and South Africa's withdrawal from the Commonwealth. All this meant more generous 'sweeteners' to the lenders than the Company would have wished.

The first move was to obtain a contract from Norddeutsche Affinerie of Hamburg for 36,000 tons of blister copper a year for the first five years, then dropping to 30,000 tons for the balance of the twenty-year contract. The production plan was then for 80,000 tons per annum (tpa) of blister for five or six years, falling away to 60,000 tpa as the grade reduced. In fact, the Company knew that it could maintain, or even increase, the 80,000 tpa by installing more plant and increasing the throughput. Clearly a decision on that would have to wait until the mine's track record was established. To encourage Norddeutsche Affinerie to take up the contract, they were offered copper anodes to their design which could go directly to their refinery, instead of copper bars which would have to be re-melted and cast.

In return for security of supply in the form of a long-term contract for a product precisely tailored to its needs, Norddeutsche Affinerie was asked for some financial contribution to the project. Initially, it was asked for a loan of US$1m for every 1,000 tons of copper that Palabora proposed delivering in the first year. In fact, the loan eventually received was US$27m rather than the requested US$36m. Norddeutsche Affinerie were a little surprised at this formula but they were used to providing financial support to their suppliers, although on a much smaller scale. They themselves did not have the financial resources, but they claimed they could arrange a loan from the German banks.

At that time the German 'economic miracle' was under way, and the German Government was anxious to ensure national security of supply of raw materials. On this score, therefore, Palabora's timing was just right. Ignorance of the German banks and of copper markets proved expensive, because Norddeutsche Affinerie claimed an option to buy a small percentage of Palabora shares in exchange for organising the loan. Furthermore, it struck a very hard bargain on blister treatment terms. Norddeutsche Affinerie brought in the Kreditanstalt fur Wiederaufbau (K.f.W) who provided a loan of DM108m, or at the then rate of exchange, R19.3m (US$27m) to be repaid over seven years, commencing two years after the start of drawdown. The German Federal Government gave guarantees to K.f.W for proportions of the loan for both economic and political risks - 80% and 90% of the loan respectively for an additional interest rate of 0.75%. Before the sales and K.f.W loan agreements could be signed, Rio Tinto had to secure the balance of the loan finance required, a further R20-25m (US$28-30m) which it proposed raising in South Africa. Also, the Company had to obtain undertakings from the South African Government that the contract for the export of copper to Norddeutsche Affinerie would be fulfilled "without restriction" and that Palabora would be able to obtain the deutschmarks to repay K.f.W. These are unusual conditions for a Government to insist upon when dealing with a mining company in a country with such an excellent record in international banking circles, but the South African Government, nevertheless, gave those undertakings.

Rio Tinto had started talks in South Africa with the Government-sponsored agency, the Industrial Development Corporation (IDC), about the required balance of the loan finance. The IDC was concerned with the development of South African industry and with the raw materials that fed those industries. At that time refined copper was imported from the Zambian Copper Belt and the South African Government was concerned about the potential adverse impact of an export embargo by the new Zambian Government. There was thus a favourable atmosphere for the loan negotiations; the IDC made it very clear that there was to be some South African participation in Palabora in return for its financial support.

This suited Rio Tinto because it was the Company's policy to invite local participation in its ventures. Indeed, it was more than a policy: it was a fundamental philosophy that local investors should have the opportunity to participate. Their participation was not wanted in exploration, or even during the proving stage, but once the loan finance and sales contracts had been negotiated. The thinking behind this was that it would prevent any feeling that the Company was taking a nation's wealth from the ground and exporting it without allowing the local people to share in that wealth. At that stage, in the early 1960's, this was very advanced thinking. Few countries then - and certainly no developed country - insisted on local participation, but they were only too pleased to see the development of their natural resources by foreign corporations.

Whilst the IDC's ideas about local participation were fine in principle, there was a great gulf over the size of that participation. Their opening shot was that 25% should be made available, but this was eventually whittled down to about 12½%.

The success of the earlier Canadian uranium financing, where warrants were issued with the bonds, prompted Rio Tinto to support a similar package for Palabora. The IDC felt that warrants would be unacceptable, but agreed to offer a package of debentures and ordinary shares that had not been attempted before in South Africa. The end result was that the IDC and certain Insurance Companies made a private placing of a package consisting of R8.625m of R100 debentures at par, together with 2.5 million R1 shares at R1.25 (R3.125m), making a total of R11.750m (US$16m). In addition, the IDC and the same Insurance Companies underwrote an issue to the public of R100 debentures at par and R1 ordinary shares at R1.25 per share. The subscribers had to buy four debentures and 100 shares as a package in any multiples. The issue was for 40,000 debentures (R4m), together with 1 million shares (R1.25m), making a total of R5.25m (US$7.3m).

This public offering was fully subscribed which was, in retrospect, a little surprising as it was neither fish nor fowl to the South African public, but an unaccustomed mixture of fixed interest debentures with what to many seemed high risk ordinary shares. As a reaction to the Northspan capital overruns, Rio Tinto also wanted capital available to meet any overrun. In response, the IDC worked out a scheme whereby the Insurance Companies agreed to take Preference Shares to the value of R3m should Palabora decide to issue those shares - in other words, if there was an overrun. Fortunately, there was no overrun and these shares were never issued.

Thus the total amount committed by the IDC and its friends amounted to R20m (US$27.8m). This was, however, to be increased. The overall financing plan was based on the expectation of a loan from the United States' Exim Bank of about R7m (US$9.8m) to cover the purchase of American heavy mining equipment. In the event, the Exim Bank refused to make the loan because of the South African political scene, but the IDC advanced the necessary funds, and the financing of Palabora was completed.

The next major project was Hamersley Iron Ore mines in Western Australia, in which RTZ was in partnership with Kaiser Steel. The prime market was obviously Japan, as Australian iron ore mining appeared to dovetail perfectly with Japan's plans for steel. The marketing discussions with the Japanese were nonetheless lengthy and arduous. It was then unusual for the world's steelmakers to enter into long-term contracts. The Japanese understood why they were needed for financing purposes, but European steelmakers were, in contrast, unwilling to contemplate more than one year renewable contracts.

The US financial institutions were initially discouraging because the mine was to be based on Japanese contracts, and there was a general feeling that Japan was over borrowed. That it was to be in a remote undeveloped area of Western Australia, unknown to the American financial markets, was also unhelpful. A highly tentative package was put together from UK sources, but it was cumbersome and very expensive.

Until the Eurodollar market developed, the UK banks played no part in any RTZ projects. This was mainly because the UK banks had no understanding or experience of long-term loans on an off-balance sheet, project-financing basis.

In the end, Kaiser Steel arranged the loan finance through First Boston and Bank of America, without undue difficulty. It was nearly caught up in proposed US regulations to restrict foreign loans by US banks, because sales contracts had not been firmed up. The original view was that a sales contract for lump and fines would be followed by a further contract for pellets, which necessitated additional finance for a pellet plant. The banks were therefore asked to provide loans in two sections, for the iron ore complex with infrastructure, and for the pellet plant, at the same time as an offer was made to the Japanese. This enabled the Company to establish its position in the US money market ahead of the proposed regulations, although the interest rate was increased from 6% to 7% under the interest equalisation tax. The loan of US$120m was for seven years, with no repayment for the first three. Later, because of some concern over the projected capital costs, the banks were persuaded to treat their loans as applying to the entire project. This gave the necessary flexibility to build the entire complex without detailing where each dollar was spent.

The next projects were two large open pit copper mines in British Columbia and Papua New Guinea. It seemed a little unfortunate that both Lornex and Bougainville Copper were discovered at about the same time in the early 1960's, and both reached a stage of investigation for the testing of possible markets and financial sources by 1967. Lornex was a very low grade (0.42%) copper ore body. Bougainville was a much more promising ore body and, moreover, had a by-product of gold.

Once again, fortuitously, the timing was right inasmuch as the Japanese were expanding their copper smelting industry. RTZ did not wish to give the impression that it had two competing mines, and mounted a joint mission in January 1968 to introduce the projects and to show that they both came under the RTZ umbrella.

The proposed output of Lornex was 50-60,000 tpa of copper in concentrate form. Bougainville's output at that time was undecided as the investigation of the orebody was far from complete, but it seemed likely to exceed 100,000 tpa of copper in concentrate form. The initial marketing plan was that the entire output of Lornex would go to Japan but that only half Bougainville's production would go to Japan, with the balance mainly to Europe, so as to avoid too great a dependence on Japan. From the beginning, Bougainville looked likely to be one of the finest, if not the finest, mines in the world, but it was not clear that Lornex was a viable project. The copper price at that time was low and most economists were forecasting an increasing oversupply of the metal with a further softening of the price. Rio Algom, RTZ's Canadian subsidiary, attached great importance to the project, however, as a means of diversifying away from uranium.

It seemed that the only assurance that Lornex would be able to service the loan finance was to secure a floor price from the customers. Unfortunately, the required floor price was around 40 cents per lb (cents/lb), which was not far below the market price. There had been one or perhaps two small contracts for copper with a floor price and a ceiling price, but none with such a high floor price as Lornex needed, and a ceiling was unacceptable. The objective seemed unrealistic to experts in the copper industry, but the idea was floated with the Japanese. RTZ argued that Lornex was going to be virtually a tied mine for the Japanese copper smelting industry, which warranted a special form of contract.

In exchange for a 40 cents floor price, RTZ agreed to regard any difference between a lower market price and the floor price as a loan to Lornex to be repaid when the market price exceeded 40 cents, always provided it never received less than 40 cents while repaying the loan. Further, on the same basis that customers should assist with the financing, the Company asked for a contribution towards the loans required. The Japanese smelters eventually accepted, without too much difficulty, the concept of a floor price but insisted it should be 38 cents/lb and this was the figure ultimately agreed.

Further, in return for a twelve-year sales contract, they agreed to provide loans of US$25m towards the total capital requirement which was estimated at US$130m.

While the Lornex financing was being agreed, negotiations were progressing with customers and bankers in regard to Bougainville. By mid-1968 drilling had indicated much larger reserves of ore than were initially appreciated and this made possible a considerable increase in the proposed contracts with the Japanese smelters. Bougainville was likely to be an outstanding mine and did not really need a floor price. Having secured one for Lornex, it seemed wise, for many reasons, to seek the same for Bougainville. The smelters eventually agreed on 34 cents/lb, later reduced to 30 cents.

Talks were also begun with Norddeutsche Affinerie, and with RTZ's Spanish associates, Rio Tinto Patino, and, in the usual fashion, parallel talks started with the bankers.

A production rate for the first few years' operation of 150,000 tpa was by now envisaged, which would gradually drop during the life of the mine, because of falling grades unless more plant was installed. On the basis of starting up at 150,000 tpa, however, the projected cash flow showed that a very quick redemption of the total loan requirement, then estimated at about US$200m, was unlikely. The only solution appeared to be a mixture of shorts holding a first mortgage and the longs with a second mortgage. The long category of loans would also have the right to take up 2% or 3% of the equity.

The initial financing plan was for loans from three or four sources with about US$60m from the Japanese, split between the two categories of loans, and US$40m from the German K.f.W, similarly split. The Bank of America felt it might join in for a modest amount. Early contacts with the UK banks were discouraging. They pointed to all the likely problems such as Bank of England restrictions, the possibility of the Australian mandate ending before loan repayment had been completed, and enquired about a parent company guarantee. Thus it increasingly appeared that financing from several separate sources was becoming far too complex. As the responses from the Japanese and German banks showed there was a resistance to taking an equal share of two different types of loans, the Company moved away from the original idea of having two different mortgages and proposed offering the same security to each. The proposed security was the shareholdings of Conzinc Riotinto of Australia (CRA) and New Broken Hill Consolidated in Bougainville. A similar arrangement had been made for Hamersley. The thinking was that, as RTZ did not and would not own outright the land and the mineral rights but only leases over them, it made more sense for the security to be the shareholding in the Company that not only held the leases, but the sales contracts, etc.

CRA recommended the Australian Administration should be offered an interest of 20% in the mine (at par) to be held in trust until independence of the Territory. The principal reason was to demonstrate the Company's enlightened good faith to the Administration and the United Nations. A by-product was, however, to give the bankers and the customers a greater sense of security, and the 20% might be regarded as part of an overall deal when tax arrangements were negotiated.

All this conceptual financial planning occurred prior to October 1968, a period of ten months from the start of the sales and financing negotiations. It was still unclear what the ultimate output of the mine would be or the capital cost, but the hope was that the interim feasibility studies would show that all indebtedness could be cleared in eight years of operation, with the possibility too of some dividends in that period. At this stage, the Bank of America suggested they might act as lead Bank.

It was now clear that the seven Japanese smelters would enter into contracts for an annual delivery of 95,000 tpa of copper in concentrate for the first five years of operation and 80,000 tpa for the next five years, but with an RTZ 'put' option of a further 16,500 tpa. The German contract was for 52,500 tpa for fifteen years, and the Spanish contract was for 15,000 tpa. In fact, actual output proved somewhat higher, and additional contracts were made for smaller tonnages elsewhere in the world.

By January 1969, the talks with the Bank of America were progressing, but a serious snag arose over the question of the long-term loans. The bank said these loans would be raised through the Societe Financière Europeénne, a club of eight major international banks, or a similar syndicate in Eurodollars, and a period of seven years was envisaged. Anything over five years was, however, regarded as too risky politically. The Bank of America could not, therefore, support the long-term section of the loan package without guarantees against appropriation and legislation that might restrict debt servicing. This came as a complete surprise to the RTZ team who could not understand the reasoning but, in spite of its strong protest, the Bank remained adamant.

So far as the short-term loans (five years) were concerned, the Bank was prepared to accept the full commercial and political risks, but beyond that "as the role of the entrepreneur".

Neither RTZ nor CRA would consider guarantees, and there was no likelihood of the Australian Government doing so. Then CRA came up with the idea that it might be possible to arrange for an Australian bank to lend the money to Bougainville Copper, if the Bank of America would organise long-term finance and deposit it in Australia. After meetings with the Governor of the Australian Reserve Bank, it was agreed that the Commonwealth Trading Bank of Australia would be the lender to Bougainville of the long-term money organised by the Bank of America.

There was considerable haggling over terms with the Bank of America, who wanted a total of 5% of the equity – 1% for them as managers, and 4% to the long-term lenders. This was not well received as the whole concept of the long-term loans had now changed and all risks removed and, in any event, the Company had never envisaged more than 2% or 3%. Eventually, the Bank of America received 1% and the long-term lenders had an option of 2%.

By December 1969, final agreement was reached, and 22 banks subscribed to the short-term (they were now called intermediate-term) loans of US$154m and 16 banks to the long-term loans of US$92.4m. The intermediate loans carried an interest rate of $1\frac{1}{2}$% over the six-month interbank Eurodollar rate, and the longs $1\frac{3}{8}$% above a Eurobond rate.

The agreement allowed the Company very considerable flexibility, including permission to substitute cheaper finance if it could be obtained. In other words, the Company had secured its financial umbrella, but at somewhat higher interest rates than it had expected, and could now look around for cheaper sources.

In the next few months, a major effort was made to secure Exim Bank and other loans by agreement with the Bank of America, and the final picture that emerged showed a dramatic change in the make-up of the loans. The intermediate loans dropped to US$55m and the long-term to US$67.4m. Equipment loans from US, Japanese, and Australian sources totalled US$86.2m. There was an Australian housing loan of US$14m, and a Japanese cash loan of US$30m.

There were, of course, the usual conditions regarding dividends and mandatory repayments, but the principal feature of this overall exercise was the initial securing of the financial umbrella provided by the Bank of America and the Commonwealth Trading Bank which made possible the subsequent loan substitution.

Bougainville came into production in about April 1972 and, with the good fortune that favoured so many of RTZ's projects, did so at a time of rising copper prices.

RTZ has been involved in other major financings such as the Churchill Falls hydroelectric scheme, and Rossing Uranium. There have also been many smaller projects throughout the past thirty years. Today's economic and market conditions appear far more difficult than those that prevailed during the Company's period of rapid growth, and the financing of new projects much harder. Nonetheless, attractive mines can always be financed even in unpromising circumstances when a company and its financial advisors are prepared to be imaginitive, and not be hamstrung by apparently insuperable problems.

CHAPTER 1 : FINANCIAL REQUIREMENTS OF THE MINERALS INDUSTRY

BIBLIOGRAPHY

Avery, D., Not on Queen Victoria's Birthday, The Story of the Rio Tinto Zinc Group, Collins, London 1974.

Dodson, C.R. and Johnson, D.K., "Future Capital Requirements for the Expansion of the Non-Ferrous Metals Industry", presented at the American Mining Congress, San Francisco, California, 21 October 1969.

Haworth, G.R. and Aimone, J.T., "How Major New Mines will be Financed in the Future", Mining Engineering, September 1977.

Mikesell, R.F., New Patterns for World Mineral Development, British North America Committee, 1979.

Mikesell, R.F. "Financing for Expanding Free World Mine Producing Capacity through 1990", Mining Congress Journal, July 1978.

Radetzki, M. and Zorn, S., Financing Mining Projects in Developing Countries, Mining Journal Books, London, 1979.

Rendell, R.S. ed, Internal Financial Law - Lending, Capital Transfers, and Institutions, 1980

West, M. "Capital Requirements of the Worldwide Mining Industry", presented at American Mining Congress, Las Vegas, Nevada, October 1971.

Wocol, Coal : Bridge to the Future, Chapter 8, Ballinger Publishing Company, Cambridge, Mass., 1980.

Chapter 2

Capital Structure For a Minerals Company

CHAPTER 2: CAPITAL STRUCTURE FOR A MINERALS COMPANY

PREFACE

Capital structure within the minerals industry can be segregated into two primary categories: the capital structure associated with mineral corporations, reflecting the mix of their assets, liabilities and owners equity and the capital structure of individual mineral projects, ranging from unlevered, 100% equity investments to 100% financed projects. Although the two categories are related, the microeconomic decision concerning the capital structure of an individual project may or may not be reflected in the capital structure of the project's sponsors.

As discussed in Chapter 1, the capital structure of the minerals industry, in particular mining, has undergone significant changes in the past decade. Poor commodity prices, brought about by the substantial increases to low-cost productive capacity made by many developing nations and the strong U.S. dollar have put severe pressure on U.S. producers and contributed to a structural change in the industry's capital structure as evidenced by declining capitalization ratios. This growing debt burden has prompted companies to develop "non-recourse" or "project" financing structures, spreading the risks of project development among other parties and to seek "off-balance sheet" treatment of projects to reduce the perceived amount of leverage carried by mineral producers.

The principal objectives of financial structuring are to allocate the rewards of project development in relation to the risks assumed by the sponsors and to maximize the return on equity of the venture while maintaining an acceptable level of debt coverage. So long as the after tax cost of borrowing funds is lower than the cost of capital of a firm, the return on equity of an investment will increase with additional leverage and the optimal financial structure for a stand-alone venture would be the maximum amount of leverage that the project can support and still maintain a level of debt coverage acceptable to lenders.

In today's economic environment of inflationary uncertainty, characterized by record high "real" interest rates, the cost of capital obtained from the financial markets is extremely difficult to predict and this represents a substantial risk to companies developing new projects. As a result, the use of internally generated funds may have a lower "risk-adjusted" cost of capital than borrowed funds and a few companies with cash resources are today financing new projects entirely out of internal cash. Although such a strategy might imply a greater risk to the developer, one of the principal benefits to the developer investing his own funds is the greater autonomy and control over the project development. With an equity financed capital structure there are no requirements to provide voluminous information to lenders and to comply with a seemingly unending list of questions and restrictive covenants imposed on an undeveloped project. The developer, being extremely close to the project, may have a thorough understanding of the inherent risks while it may be impossible or uneconomical to fully explain and allocate project risks to the satisfaction of prospective lenders.

After completion of the project and demonstration of an operating cost profile for the enterprise, the sponsor may decide to raise financing supported by the established project cash flows. At this stage, the borrower is in a much stronger negotiating position with prospective lenders and the cost and terms of the financial package will reflect the lower risk associated with a completed project. Additional sources of financing (and higher degrees of leverage) become available to completed projects and in this way the timing of when to seek fianancing for a project can have a significant impact on the resulting capital structure. Many other factors also contribute to the determination of the appropriate capital structure for a mineral project including the risk-adjusted estimates of project profitability, the degree of sponsor commitment, third party support in the form of guarantees or marketing agreements, and legal, tax and accounting issues affecting the venture.

Financial structures can also be created to recruit equity and debt from outside sources for companies seeking to develop mineral ventures with little or no capital of their own. Joint ventures are the most common mechanism for attracting outside capital, through "earn-in" formulas or via direct investment, although the ownership interest demanded by the company supplying the capital may be excessively high. Limited partnerships and leasing transactions are alternative mechanisms to raise outside capital which enable the sponsoring company to maintain greater operating control over the venture. The trade-off in utilizing these financing vehicles is that the high cost of recruiting both equity and debt may postpone any significant returns to the sponsor until the debt has been repaid and the providers of equity compensated to achieve a specified return on equity. Depending on many factors, including the cost of the project, the resource life and the expectation of future mineral prices, this may or may not be an acceptable wait.

The papers in this chapter have been selected to present a basic understanding of capital structuring applied to the minerals industry. In the first paper, Professor Langdon discusses the recent history and development of capital structure for mining projects pointing out two salient characteristics of our industry: high leverage and high dividend payout. Are these desirable? Are there alternatives? What are the factors impacting the determination of the optimal structure for a mining venture? Langdon suggests that the industry re-examine traditional financing methods and suggests several alternatives.

Ken Cork, offering the viewpoint of a major mining company, echoes some of Professor Langdon's comments and enlarges their perspective. Mr. Cork delineates the operational constraints imposed on financial managers in minerals companies who must balance the desires of their investors with the realities of the financial marketplace. Compounding this difficult task is price volatility; the double-edged sword which constantly hangs over the head of the financial planner. The volatility of metal prices is infamous: year-to-year changes of 50 percent are not uncommon, and even a cursory glance at the charts accompanying Mr. Cork's paper will convince us that long-term financial planning in mining is a daunting task. The presence of long-term external influences compounds this difficulty and sometimes causes us to wonder if it is even possible.

With full recognition of the uncertainties involved in projecting future price trends and endless analysis striving to determine the optimal capital structure, the actual work of structuring and implementing mining projects and company financing must proceed. The last three papers in this chapter discuss areas that may be considered sub-topics within capital structure. David Cockburn's paper focuses on the "Financial Aspects of Joint Venture and Multiparty Structures" while Jim Boettcher and Randy Eppler's paper "Financing a Coal Mine Acquisition" describes the analytical process that a commercial banker employs in structuring the leveraged buy-out of an easter coal mine. George Caraghiaur's paper, "Financing of the Small, Independent Mining Enterprise", concentrates on the various capital structures and financing sources available to independent mineral producers.

These papers illustrate that in developing a project capital structure one may have to consider alternatives which lead to solutions that may not be optimal, but which provide workable alternatives that are acceptable to all parties involved in the project development. These articles also point out several key elements of sensitivity arising in multiparty agreements, the range of differences and grounds for agreement between parties with differing goals and the very real difference between financing large mining companies and smaller enterprises. Financing for the minerals industry is rapidly evolving and alternative capital structures and financial instruments not available a few years ago are in common use today. New solutions will arise during the 1980's to cope with the unique problems of this decade and the papers included in this chapter and book are evidence of this trend.

Leons Kovisars
Chapter Editor
Chase Econometrics

AN INTRODUCTION TO CAPITAL STRUCTURE

by

Dr. William L. Langdon
State University of New York at Utica/Rome
Professor of Finance/Management Science
Utica, New York

PRELIMINARY COMMENTS

Recent years have been difficult for the mining industry, as reported in the most recently published <u>Minerals Yearbook (Volume III)</u>, the estimated value of world crude mineral production has been generally declining since an historical peak in 1979 (16).* The costs of mineral extraction have, however, been rising rapidly. Individual project costs are, in general, exceeding the growth rate of mining companies, while operating and technical risks continue to increase (16). Risks resulting from currency flotation and political actions in the host country are also on the increase. Labor difficulties (such as those experienced by Phelps Dodge) and the Reagan Administration's recent rejection of a proposal by the International Trade Commission for higher tariffs and/or quotas on copper imports have also contributed to the mining industry's difficulties.

The net result of these factors has been financial difficulties for mining companies with respect to paying for their current financial obligations as well as trouble finding new financing. It may be that one of the main factors in determining the future of the mining industry will be centered upon capital structure, i.e. how its assets are financed.

Historical Perspective on Capital Structure In The Mining Industry

As discussed by Isreal Borenstein in (4) and Thomas Navin in (7), mining firms typically have used a financial model based upon high financial leverage, i.e. large amounts of debt relative to equity. This has been especially true for American firms developing orebodies discovered overseas.

Problems sometimes occurred for those who were forced to commit themselves to completion guarantees in order to obtain bank loans. Mines under these circumstances must be brought into operation regardless of market conditions. Politicians in overseas countries also may use the high profits required by these high risk ventures to argue for nationalization of mines.

Coupled with the frequent use of high leverage, mining companies traditionally have paid out a higher percentage of earnings in the form of dividends than manufacturing firms. This is significant, in that such funds paid out in dividends are not available for other uses. Hence, a greater reliance upon debt and/or additional stock issues than would ordinarily be the case for manufacturing firms. Thomas Navin has stated in (7), that the theory behind the high dividend payout was that, "most mining investors thought of themselves as investing in a wasting asset, one that would gradually deplete over time until it ultimately became valueless." Investors, therefore, required a receipt, in the form of dividends, of a percentage of current earnings and a return on original capital. This view of reality, apparently, has not changed in some quarters, even though modern mine management ordinarily replaces depleted mines with new ones. It may be that this practice of high dividend payout coupled with relatively high financial leverage should be reexamined. Current capital structure theory could prove interesting and useful for those concerned with current conditions. Capital structure patterns and related practices in this industry are not, in many cases, congruent with those thought by some to be theoretically optimal.

Prelude to Capital Structure Discussion

With a few notable exceptions (such as the Guggenheims and Anaconda's chief operating officer in 1971) mining companies' chief executives and boards of directors have not been persons with financial backgrounds. This, one could conjecture, might have accounted for the relatively non-sophisticated nature of capital structures of mining companies. In order to appreciate the possibilities with respect to alternative influence on a firm's performance, one must be acquainted with the basics of financial theory. This paper is an attempt to familiarize the reader with (1) capital structure in general, (2) the usual components of this structure, (3) the general theory behind the use of financial leverage, and

---
*References are listed in the Bibliography, p. 71.

(4) selected theoretical positions with respect to determining an optimal capital structure. It is assumed that the reader has little or no background in accounting and/or finance.[1] Subsequent sections are therefore designed to convey the general nature of the elements discussed, without an in-depth exploration of the many nuances involved. One is encouraged to pursue a more in-depth analysis via the references contained herein.

Sections on (1) Capital Structure Definition, (2) the theory of financial leverage as well as that on (3) whether an optimal capital structure exists, are presented in the abstract, i.e. no specific industry linkage is made. The intent is to encourage an open mind on the part of the reader with respect to "how things ought to be done."

DEFINITION OF CAPITAL STRUCTURE

General Nature of a Balance Sheet

A corporate entity may be visualized as an equality between assets on one side of a balance and liabilities and equity on the other. This relationship is depicted as:

$$\text{Assets} = \text{Liabilities} + \text{Equity}$$

Capital Structure refers to the relationship between liabilities and equity. Liabilities may be viewed as creditors' rights to assets and equity as owners' rights to these assets. Figure 1 depicts the principal components of a corporate balance sheet.

Figure 1

Assets	Liabilities	Equity

(1) Current Assets
  (a) cash
  (b) marketable securities
  (c) accounts receivable
  (c) inventory
	(1) Current Liabilities
	  (a) accounts payable
	  (b) notes payable
		(1) Preferred Stock
		(2) Common Stock
		(3) Paid in Capital In Excess of Par
		(4) Retained Earnings

(2) Long Term Assets
  (a) Building
  (b) Plant
  (c) Equipment
  (d) Land, Minerals, etc.
	(2) Long Term Debt
	  (a) Bonds
	  (b) Subordinated bonds
	  (c) Debentures
	  (d) Subordinated Debentures
	  (e) Long Term Notes	

Current assets, in theory, are those assets expected to be in the form of cash, or converted into cash, within one fiscal period. Long term Assets (usually referred to as fixed assets) are not expected to be converted into cash within one fiscal period. A brief definition for the major capital structure components is presented in Figure 2.

Figure 2

Capital Structure Component	Definition
1. Bonds	A long term promissory note usually secured by a pledge of any specific property.
2. Debenture	A long-term promissory note that is not secured by a pledge of any specific property.
3. Subordinated Bonds and Debentures	Bonds and debentures that have claims on assets after unsubordinated bonds and debentures in liquidation proceedings.
4. Preferred Stock	A hybrid form of financing normally with no voting rights, a fixed dividend and claims on assets in liquidation after creditors, but before common stockholders. The usual limitation on payment to preferred stock in liquidation is the par value of the stock.
5. Convertible securities	A share of preferred stock or bond that is convertible, at the option of the holder, into common stock.
6. Common stock	Evidence of ownerships of a fraction of a corporation's assets. These evidences of ownership normally are accompanied by the right to vote, the right to maintain the same fractional ownership in the corporation and the right to share in assets upon liquidation.
7. Callable security	A security that may be redeemed, at the option of the corporation, in exchange for cash.

For a more complete discussion of the various alternative financial instruments, the interested reader is referred to (7, 15). Current liabilities are generally expected to be paid within one fiscal period while long-term debt may have maturities from more than one fiscal period to greater than twenty years. Returns to preferred and common stockholders are in the form of dividends and, in the case of common stock, capital gains resulting from market price increases in stock prices.

---

[1] Readers with backgrounds in these areas are likely to derive minimal benefit from time invested reading this paper.

The Matching Principle

A theoretically correct relationship between asset structure and overall capital structure has been succinctly stated in the "Matching Principle". This principle states that one should finance short-term needs with short-term sources and finance long-term needs with long-term sources (7,15). The underlying rationale is that the timing of cash inflows should roughly match the timing of cash outflows required by alternative means of financing. An example of an extreme mismatch in this sense was that of the financial structure of Chrysler Motors a few years ago. This corporation had approximately 140 million dollars of short-term debt (technically to be paid within one fiscal period), matched against near zero dollars of expected net cash inflow for the year. The theoretically correct relationship inferred by the "Matching Principle" is presented in Figure 3.

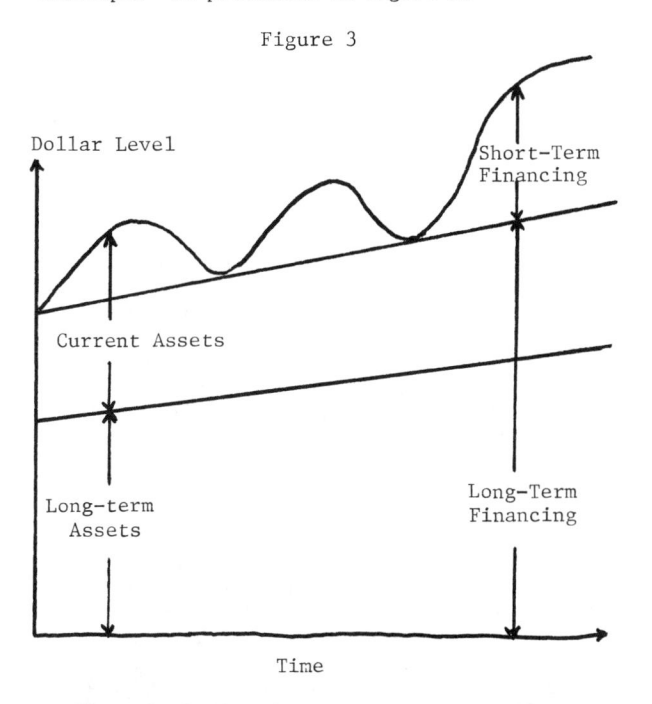

Figure 3

[Graph showing Dollar Level vs Time with Current Assets (oscillating curve), Long-term Assets, Short-Term Financing, and Long-Term Financing]

A number of features supplementary to the matching principle are inferred in Figure 3. First of these is the assumption that growth in both current and long-term assets is assumed. This assumed growth (usually termed "the growing concern principle") has a directly related assumption of an increase in overall levels of required financing overtime (7). A second inference is that, although current assets are technically convertible into cash within one fiscal period, the actual dollar level of a portion of these assets increases overtime and hence requires permanent financing. One may infer from this that short-term financing, theoretically, should be used only for those current assets actually varying in value over a fiscal period; these are termed fluctuating current assets. Current assets not varying in total balance sheet value over time are called permanent current assets.

As might be expected, exceptions to the matching principle are numerous. One of these occurs when interest rates are expected to drop within the fiscal period. During such periods, financial managers may choose to use short-term means to finance permanent needs with an anticipated substitution of long-term financing subsequent to the decrease in interest rates. For a more complete discussion of this and related topics, the interested reader is referred to (9, 12, 13, 15).

THE THEORY OF FINANCIAL LEVERAGE

The Intent of Using Financial Leverage

Financial leverage refers to the use of debt with the intended purpose of increasing returns to owners of common stock. Such leverage may, however, have the opposite effect during periods of declining revenues. This concept is demonstrated in Figure 4.

Section I of Figure 4 presents a very simple balance sheet with the assumption that no debt is contained within the capital structure. Section II is a depiction of the impact on (1) percent return on owner's equity, and (2) earnings per share, resulting from various assumed levels of return on total assets. Section III illustrates a balance sheet representing a corporation with the same amount of assets as those represented in Section I, but with use made of financial leverage, i.e. 50% of the assets are financed with debt. Section IV depicts the results of changes in income levels on (1) percent return on owner's equity and (2) earnings per share for the levered firm.

Figure 4

I. <u>Balance Sheet with All Equity Financing</u>

Assets $100 = Liabilities $0 + Equity* $100
  *2 shares at 50% assumed

II. <u>Income Statements with Alternative Pre-Tax Returns on Assets and No Debt in the Capital Structure.</u>

Return on Assets	5%	10%	15%	20%	25%
Equivalent Pre-Tax Dollar Returns	$5	$10	$15	$20	$25
Less Taxes (at an Assumed 50%)	2.50	5	7.50	10	12.50
Net Income after Taxes	$2.50	$5.00	$7.50	$10.00	$12.50
% Return on Equity	2.5%	5%	7.5%	10%	12.5%
Earnings per share	$1.25	$2.50	$3.75	$5.00	$6.25

III. <u>Balance Sheet with 50% Debt Financing.</u>

Assets $100 = Liabilities* $50 + Equity** $50
  *(interest cost assumed to be at 10% or $5.
  **(1 share outstanding at $50 assumed).

(Figure 4 continued on next page)

IV. Income Statements for 50% Debt Financing With Alternative Pre-Tax Return on Assets.

Return on Assets	5%	10%	15%	20%	25%
Equivalent Pre-Tax Dollar Returns	$5	$10	$15	$20	$25
Less Interest Expense	$5	$ 5	$ 5	$ 5	$ 5
Taxable Income	$0	$ 5	$10	$15	$20
Less Taxes	0	2.50	5	7.50	10
Net Income After Taxes	$0	$ 2.50	$ 5	$ 7.50	$10
Percent Return on Equity	0%	5%	10%	15%	20%
Earnings Per Share	$0	$ 2.50	$ 5	$ 7.50	$10

It is intended that the reader, after an examination of this example, clearly see how financial leverage may magnify returns to owners in times of prosperity while having the opposite effect when revenues are low. An overall view of this concept, in the form of a graph, is presented in Figure 5. At revenue levels beyond a given amount, e.t. Point A, Figure 5, the use of financial leverage is clearly advantageous to stockholders. Lower levels of revenue, however, would result in stockholders being better of if less debt was included in the capital structure. Accurate forecasts of revenue are implicitly required when determining the optimal level of debt. It is also implicit, that the range (in terms of time into the future) of relatively accurate forecasts is monotonically related to the maturity structure of debt. We shall now turn to a brief discussion of general considerations with respect to the use of various financial instruments with financial leverage as a background.

Figure 5

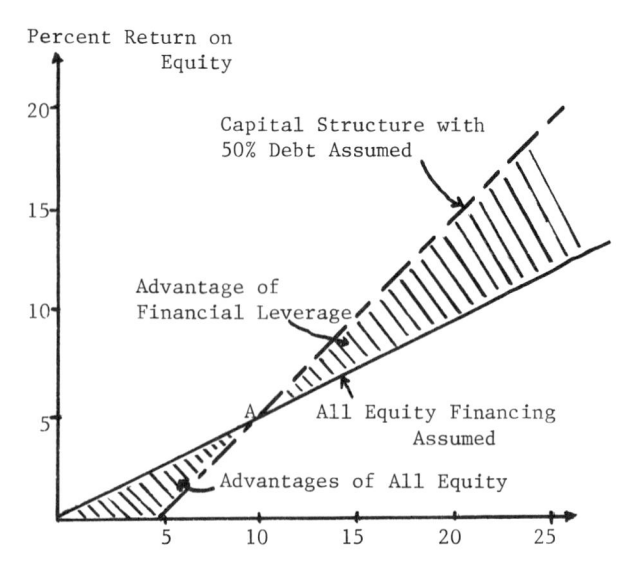

General Considerations

Long-term debt is usually advisable when earnings are relatively stable and/or future earnings are expected, with relative certainty, to be large. This advisability accrues from the leverage factor as discussed hereinbefore. Anticipated high rates of inflation also favor the use of debt as repayment of such obligations are made with dollars having less purchasing power than when borrowed. Temporary depression of a firm's common stock's market price also points toward the use of debt for new financing. Other considerations with respect to the relative level of debt in a corporation's capital structure shall be addressed in a subsequent section on optimal capital structure.

Preferred stock usage is currently undergoing some changes across industrial categories. Many corporations are retiring their preferred stock issues primarily as a result of the nondeductibility of preferred stock dividends for determining taxable income (10). The usual practice has been a substitution of bonds, or debentures, since the interest on these instruments is a tax deductible expense. These exchanges are accomplished by calling in callable securities or offering attractive exchange propositions for non-callable preferred stock.

Some corporations using preferred stock as a means of financing argue that forecasted income levels are sufficient to cover dividend payments yet lack the earnings stability over time to justify the risk of increased debt levels. These corporations may gain the advantage of leverage without the level of risk associated with debt issues; preferred stock dividends may be unpaid without a formal default as would be the case with many forms of debt contracts.

Financing with common stock, while resulting in more risk for the owners of these securities, has a distinct advantage for corporations with uncertain earnings prospects. As illustrated in Figures 4 and 5, losses are not magnified as would be the case if debt financing were involved. There are no fixed charges with the use of common stock which might cause a firm with declining earnings into immediate bankruptcy or reorganization.

The intent to' this point has been to convey the (1) definition, (2) importance, from a leverage viewpoint, (3) principal components and (4) general considerations with respect to the use of the principal components, of capital structure. These elements are quite straightforward and, from a theoretical perspective, relatively noncontroversial. It is to the controversial elements of the topic that we shall now turn.

Some theoreticians argue that capital structure is an irrelevant consideration. The arguments purport that to view the mixture of debt and equity instruments as having an influence on the fortunes of a firm is, in some respects, irrational. Others disagree with this approach, while providing theoretical and empirical support for the viewpoint that the management of a firm's capital structure may "make or break" it.

DOES AN OPTIMAL CAPITAL STRUCTURE EXIST?

In order to appreciate fully the importance of the arguments to follow, one should have an understanding of how decisions with respect to the purchase of fixed assets are made. Categorically such decision making is termed capital budgeting. Since an adequate discussion of this topic usually requires one or more volumes, such a discussion is not possible herein. The reader, in the short term at least, is asked to accept the work of theoreticians such as Weston (15) and Van Horne (9). These scholars have indicated that a capital budgeting method termed the Net Present Value (NPV) method is theoretically correct. Although more sophisticated methods involving stocastic simulations and/or linear programming are superior in many ways, surveys of firms across industrial categories[2] have indicated (Circa 1982) that the NPV method is the most predominant.

The NPV method suggests that alternative long-term investment alternatives should be ranked according to the Net Present Value of the forecasted receipts from the alternative investment proposals. This can be stated as

$$NPV = \sum_{i=1}^{n} R_i / (1 + C_w)^n - I$$

Where R_i = the forecasted net dollar return in period i

C_w = the weighted average or marginal cost of capital

I = the original investment

n = the expected life of the project

This overly simplistic version of the NPV is presented only as a device to illustrate the importance of various implications resulting from the arguments to follow. As can be noted from an examination of the model, NPV decreases (for given levels of anticipated receipts and Investment) as Capital Cost increases and vice-versa. The following discussion of capital structure is thought be more meaningful when viewed from this perspective, i.e. higher capital costs limit investment alternatives: Corporations cannot profitably invest in projects whose returns are less than their own costs of financing.

Net Income Approach

Arguments with respect to an optimal capital structure may be conveniently summarized by three theoretical positions, (1) the Net Income Approach, (3) the Net Operating Income Approach, and (3) the Traditional approach. Each approach views the outcomes of increases in debt relative to equity differently. Figure 6 presents a summary of the terms and assumed relationships employed in the discussion hereinafter.

Figure 6

(1) $C_i = \dfrac{I}{D} = \dfrac{\text{Annual Interest Charges}}{\text{Market Value of Outstanding Debt}}$

Where: C_i is the yield on the Company's debt.

(2) $C_s = \dfrac{E}{S} = \dfrac{\text{Available Earnings to Common Stock-Holders}}{\text{Market Value of Outstanding Stock}}$

Where: C_s is the required rate of return for investors, assuming 100 percent dividend payout and zero growth in earnings.

(3) $C_w = \dfrac{N}{M} = \dfrac{\text{Net Operating Earnings}}{\text{Total Market Value of the Firm}}$

Where: (1) $M = D + S$
(2) C_w is the weighted-average cost of capital and sometimes referred to as a firm's overall capitalization rate.
(3) C_w may also be expressed as

$$C_w = C_i \frac{D}{D+S} + C_s \frac{E}{D+S}$$

The Net Income approach may be viewed in Figure 7. Initial debt is assumed to be $1,000 at an annual cost of 18 percent. Section A of Figure 7 assumes that the market value of common stock is determined by discounting in perpetuity the earning available after interest charges, i.e. market value of common stock is equal to the earnings available divided by an assumed capitalization rate of twenty-five percent. An assumed increase in the level of debt to $2,000 is assumed in Part B. The usual assumption, when using this approach as an illustration, is that the proceeds of the debt issue are used to repurchase common stock. Interest charges are assumed to remain constant at 18 percent per year.

Figure 7
(Net Income Approach)

A. Original Debt Position
N Net Operating Earnings	$1,000
I Interest Charges (18% of $1,000)	180
E Earnings Available to Stockholders	$ 820
C_s Required Rate of Return for Investors	.25
S Market Value of Common Stock	$3,280
D Market Value of Debt	1,000
Total Market Value of Company	$4,280

The implied weighted average cost of capital is

$$C_w = \frac{N}{M} = \frac{\$1,000}{\$4,280} = 23.36\%$$

Figure 7 continued on next page

---
[2] See empirical work compiled by Serraino, Singhui and Soldofsky in (13) for a more complete discussion of these empirical findings.

B. Incremental Debt Position
    N Net Operating Earnings $1,000
    I Interest Charges (18% of $2,000)   360
    E Earnings Available to Stockholders $ 640
    C_s Required Rate of Return for Investors .25
    S Market Value of Common Stock* $2,560
    D Market Value of Debt $2,000
    M Total Market Value of Company $4,560

The implied weighted average cost of capital is now:

$$C_w = \frac{N}{M} = \frac{\$1,000}{\$4,560} = 21.93\%$$

*$S = E/C_s$

It is important to note that this approach assumes increases in debt levels result in (1) decreases in overall capital cost and (2) increases in corporate market values. Pivotal to these conclusions are the assumptions that the rate of interest on debt (C_i) and the required rate of return for common stockholders (C_s) are unaffected by changes in financial leverage. A summary of the inferences using this approach are presented in graphical form in Figure 8.

Figure 8

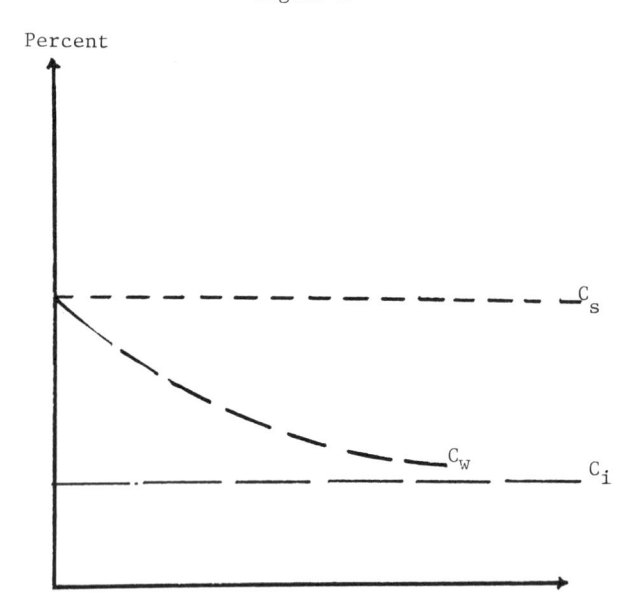

[Graph: Percent vs Financial Leverage, showing C_s as flat dashed line, C_w as decreasing curve, C_i as flat line]

Net Operating Income Approach

This approach assumes that a firm's total value is obtained by capitalizing net operating income at a fixed rate. The elements from the previous example are to remain unchanged with the exception of a constant 25 percent weighted average cost of capital (overall capitalization rate) which is assumed to be unaffected by the level of financial leverage. Figure 9 summarizes the net results of increased financial leverage under the Net Operating Income Approach.

Figure 9
(Net Operating Income Approach)

A. Original Debt Position
    N Net Operating Earnings $1,000
    C_w Overall Capitalization Rate   .25
    M Total Market Value of Company $4,000
    D Market Value of Debt   1,000
    S Market Value of Stock $3,000

The implied cost of common stock financing is

$$C_s = \frac{\$\ 640}{\$3,000} = 21.33\%$$

B. Incremental Debt Position
    N Net Operating Earnings $1,000
    C_w Overall Capitalization Rate   .25
    M Total Market Value of Company $4,000
    D Market Value of Debt   2,000
    S Market Value of Stock $2,000

The implied cost of common stock financing is now

$$C_s = \frac{\$\ 840}{\$2,000} = 42\%$$

A graph depicting the Net Operating Income approach is presented in Figure 10. The overall implication of this approach is that an optimal capital structure does not exist and is not of major importance with respect to overall capital cost (C_w) or a firm's value. This position was first put forth in 1958 by Modigliani and Miller and is discussed at some length in (9,15).

Figure 10

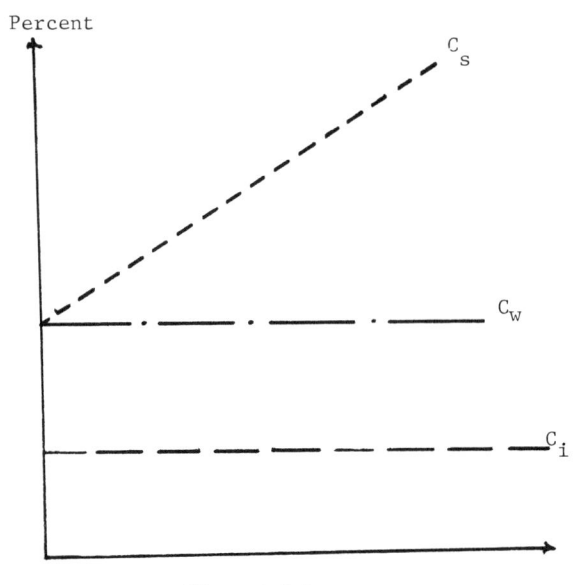

[Graph: Percent vs Financial Leverage, showing C_s as increasing dashed line, C_w as flat line, C_i as flat dashed line]

Traditional Approach

A position between the extremes of the NI and NoI approaches is that put forth in what has been termed the "Traditional Approach." This viewpoint is more congruent with the bulk of financial literature and practice to date. This view notes that after a critical value of financial leverage has been reached (usually at or near the industry average debt to equity ratio) the markets for the corporation's stocks and debt instruments perceives an increase in risk. The net result of this preception on bonds is a decrease in market value for a given promised coupon. This results in an increase in the cost of debt (C_i).

Stock prices (C_s) are argued to reflect the increased perceived risk by dropping for a given earnings level, therefore also increasing the cost of this type of financing. The overall cost of financing (C_w) decreases as debt is increased into the capital structure up to a critical level. This, a direct result of the increased weighting of the lower cost debt as financial leverage is increased. Beyond the critical level, however, the increasing costs of debt and equity financing result in an increase of the overall cost of capital. The aforementioned "critical level" of financial leverage is believed to be the optimal level of debt relative to equity, i.e. the traditional approach argues that an optimal Capital Structure does exist. Figure 11 is a summary of the traditional approach in graphic format.

Figure 11

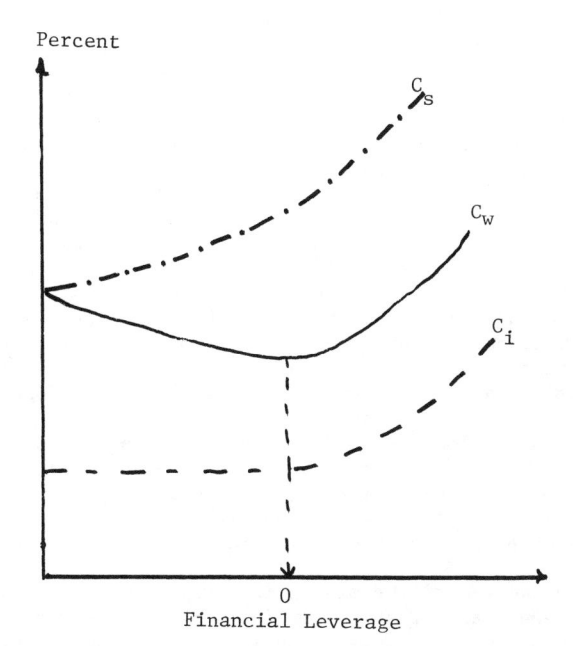

Point "0" is the minimum weighted average cost and therefore the optimal capital structure.

Although the Traditional Approach may have a strong appeal from a common sense standpoint, the reader is advised to read the literature suggested hereinbefore, at some length, before drawing conclusions with respect to the relative merits of the three positions discussed. The topic is complex and has been introduced in its most elementary format, thereby omitting the rigorous theoretical and empirical findings supporting each approach.

SUMMARY

Selected Normative Considerations

Having proceeded to this point, one may have given some thought to the wisdom behind using traditional instrument to achieve high financial leverage coupled with high dividend payout in an industry typified by large levels of risk and capital requirements. It is perhaps the appropriate time for the industry as a whole to re-examine traditional financial instruments and structures. New, more flexible financial arrangements, congruent with the characteristics of mining appear to be necessary. This, of course, is particularly true with respect to international financing arrangements. As noted by Ferguson and Haclin, one of the main factors in determining the pattern of future mining developments will be the flexibility which the international banking community can show in devising and funding capital structures for mines which allow operating freedom to mining companies, while maintaining adequate security to lenders (14). Recent increases in lending to the mining sector by the World Bank Group have been a step in this direction.

Recent Developments

Manufacturing firms have recently turned to medium-term notes (unsecured promissory notes) to more closely match the maturity of its debt with the firm's longevity of assets. These and other new financial alternatives allow various industries as well as specific firms within industrial categories to operate more effectively.

It may well be that theoretically optimal capital structures should vary within the large industrial category of mining. New financing in the area of zinc production (which seems to be regaining popularity in the automobile industry causing a rise in zinc prices) may be accomplished using different financial arrangements than those for copper and lead, where weak demand and over-capacity are evident. The point is, financial management within the mining industry might benefit from a re-examination of their current financial practices and beliefs. This conceivably could extend to the acquisition and financing of assets complementary, in a counter-cyclical sense, to the operation of mines.[3] It is the opinion of this author that the only limits to the financial possibilities within the mining sector are the imagination and determination of those directly involved.

[3]Considerable literature is available with respect to the portfolio theory of assets and related empirical evidence. For example, see (13).

2.2

FINANCIAL OBJECTIVES OF A MINING COMPANY

Mr. E. Kendall Cork

Noranda Inc.
Toronto, Ontario

The traditional financial objective for a single mine company has been to operate as frugally as possible and to pay out most of the earnings as dividends. If the business is cyclical (as it is for most metals) the dividends might fluctuate quite widely. When the mine is exhausted the company disappears. This is still quite a viable strategy for a single mine company.

It is not however a viable strategy for the world as a whole. The mining industry is built by mine development companies who can mobilize the people and capital to bring new mines into production. Their skills must include marketing, engineering, finance and other politics. It is very rare for a property to be brought in without the support of a major company that can provide all these services. The exceptions will usually have some other form of big brother support, for example the U.S. government uranium contracts at guaranteed generous prices.

The mine development company will seek as a minimum to perpetuate itself by developing new mines in order to replace those which are running out. The more common and more ambitious objective is to grow -- that is to add to its ore reserves and current production by developing more new mines. The financial objectives for that company are very different. Obviously if all the earnings were paid out in dividends there would be nothing left to work with. The first financial policy then is to spend an appropriate amount on exploration for new properties. The next is to retain enough of the earnings to provide the capital for new projects at least sufficient for the equity.

There is no magic formula as to what proportion of earnings should properly be distributed as dividends by a growth-oriented mine development company. As a rough rule of thumb distributing half or more will probably leave too little to work on and 30% or so is probably a good balance. However the circumstances differ widely from company to company.

It may be useful to set an objective for the rate of growth of a company's earnings. Some have picked rates such as 15% per annum compounded. Others have set a target in real terms which might appear as 10 or 11% plus inflation. Obviously the arithmetic of compound interest is very attractive; however in practice there is much variation. Indeed current returns from existing operations swing widely with the business cycle and there is no assurance that economic new properties will be found according to someone's arbitrary time schedule. For example, Western Mining Corporation Limited in Australia explored for 30 years with little to show for it, but then found the great Australian nickel deposits and more recently the huge Roxby Downs copper. That long dry spell could not have fitted anyone's arbitrary calendar of growth and yet they would not have found such orebodies without that long period of effort. Should they have abandoned the search?

Once a new property has been found or acquired there has to be a threshold rate of return on the new capital to be invested against which to evaluate the property's economics. Conventionally this seems to be 15% after tax, a number common in other heavy industries as well. In some cases it is expressed as a lower number plus allowance for inflation. Discounted cash flow analysis is a very useful tool but it does not make the decision. In the end a "go" decision depends on judgment of many factors some of which are numbers used in the DCF calculation whose credibility must be examined. It is curious how frequently investment proposals come in with the rates of return very close to 15%. The project advocates know that a number much less than 15% will not fly and that a number much more is not necessary. With much higher nominal and real interest rates of recent years, even though before tax, logic suggests that the hurdle rate should also rise. The power of compound interest is so great that 20% is very hard to achieve in any cash flow projection but 18% may be a sensible yard-stick. Once again it is remarkable how many project proposals come in with an 18% return.

On the record the mining industry as a whole has not been overly restrictive in choosing its hurdle rates of return. This is shown by the abundance of metals in recent years and the failure of metal prices to keep up with inflation.

All of the foregoing is standard text book stuff.

It is not at all funny to observe that the major financial objective for a good many mining companies today is simply survival, and that means survival in meeting this week's payroll let alone satisfying the bank. In an industry which is still depressed the text book subjects of rate of return on new investment seem rather academic.

The problem is partly cyclical. The United States has been experiencing its best ever economic recovery but from the depths of its worst post war recession. The rest of the industrialized world is filling in nicely behind the United States but the benefit has not yet trickled down very plentifully to the mining industries. In the normal course the prices of most mine products lag the upturn as excess stocks are gradually consumed and excess capacity put to work. If the current upturn continues long enough one can expect history will be repeated once again and there will be some period of prosperity.

However, in many ways the current upturn is different. Most observers perceive a secular change against the abundant use of metals and minerals.

IS THERE A LONG WAVE IN ECONOMIC GROWTH?

In recent years a non-Keynesian view of economic development has gained some belated recognition. The idea of a long wave of technological innovation lasting 55 to 60 years on which is superposed a capital goods cycle of 9 or 10 years and a consumer cycle of perhaps 3 years is now widespread but it carries still quite a bit of mysticism. The great analyst and advocate of the Kondratieff wave was Joseph Schumpeter a professor at Harvard from 1932 to 1950. In his book "Business Cycles"* he observed that there is no reason why every long wave had to be of the same length but the average period of 57 years fitted neatly his explanation of the onset of the North American depression in 1932. Extrapolating the same 57 year period moves the shadow of 1929 ahead to 1986 and the shadow of 1932 to 1989. What a gloomy thought! It is quite arguable however that the hard times since 1980 have already accomplished enough of the structural changes necessary. If so the worst is now behind us providing the banking system can hang together.

Inventions are created all the time but they may not at first have any major impact. The basis of the long wave is that broad innovations in the economy come in clusters when labour and financial markets are supportive and resources have been released from obsolete industries. At first the innovations put extra pressure on construction and capital goods industries to build the plants to produce the new products. For example to make and use automobiles society had to build auto plants, rubber tire plants, build and pave highways, find oil and build refineries and gasoline service stations. In turn these required new steelmills, metal mines, and smelters, and they facilitated business and recreational travel and urban sprawl.

After a while there is adequate capacity to produce the new products and the market becomes over-supplied. The emphasis then shifts to product improvement, lower unit costs and share-of-market economics. In time the market is saturated, the goods freely available to consumers, and the producers compete each other out of business. Inefficient or overextended companies close. In this phase, if the consumer/inventory cycle and the replacement capital cycle also move down together, there is greatest risk of credit collapse and a downward spiral as happened following the crises of 1929 and 1873. Thereafter comes gradual recovery as resources are deployed to newer industries that are in demand. Society is ready to put accumulated inventions to work in a new wave of innovation. This is a much abbreviated sketch of Schumpeter's view of business cycles!

The pattern is familiar enough industry by industry. Schumpeter's point is that several major industries follow the pattern in unison, stamping their long wave on the whole economy.

The strength of Schumpeter's analysis can be summarized by examining the prediction made in the last paragraph of his book. Writing in 1938 he said "recovery and prosperity phases (of the business cycle) should be more, and recession and depression phases less strongly marked during the next three decades than they have been in the last two". How right those words proved to be! Who else in 1938, after all the vicissitudes of the inter-war years, was taking such an optimistic view?

In similar vein, who in 1968 would have said that for most of the rest of the century the recovery and prosperity phases would be rather weak and the recession and depression phases more severe than they were in the 1950's and 1960's? In fact the last 15 years have obviously been tough. Any mining exeuctive who foresaw such an adverse trend in 1968 and turned his company's policy towards the defence should be a hero today. Being now half way through that difficult span one must accept that there are more than random factors at work and that happy times for mining will be rather brief as the business cycles peak from time to time.

Schumpeter was also vividly aware of the march of state intervention in the economy and increasing public hostility to business. One can only imagine how caustic he would be against the massive state interventions of the past 20 years diverting production away from the goods and services people really want enough to pay for, and supporting uneconomic old industries. Mining has been jerked around by such policies.

---

* The writer used "Business Cycles", by Joseph A. Schumpeter, abridged by Rendigs Fels, McGraw-Hill, Inc. 1964. Alternatively see the original "Business Cycle: A Theoretical, Historical and Statistical Analysis of the Capitalist Process", New York, 1939, 2 volumes. There is also a summary in "The Schumpeterian System" by Richard V. Clemence and Francis S. Doody, Cambridge 1950, Part II.

Whether or not the long wave theory is accepted, mining has obviously been having a hard time!

OTHER FUNDAMENTALS

Five other trends also have affected the mining industry. Perhaps they derive from the long wave.

1) The cartel price for oil has stopped the growth in world energy consumption. This in turn has impeded the production and use of all sorts of things including things made from metals. A collapse in energy prices would be marvelous for the mining industry in the medium term.

2) The leading industrialized countries have evolved towards consuming services more than things, a natural shift which has been accelerated by the high price of energy. Even industries which are growing are likely to manufacture light weight products. Most of the hardware being used in the information revolution is light in weight.

3) Substitution of other materials, notably plastics, glass and ceramics, has shrunk the market for metals. Metal makers have not developed new markets to replace the lost ones.

4) Mining became heavily socialized between 1960 and 1980 as Third World governments took over ownership of the mines and First World governments (the industrialized Western democracies) took so much of the earnings of mines through aggressive taxes and environmental costs.

5) Disinflation has followed the nasty experiment with inflationary policies from 1965 to 1980. Whether the huge fiscal deficits of 1980-1984 are inflationary remains to be seen, but severly high real interest rates are clearly depressing metal markets and eroding what equity remains in many mining companies. The late 1970's, and particularly 1979 and 1980, saw a fundamental change in monetary management. Accelerating inflation became an untenable policy because financial markets rebelled and voters turned against it as immoral and harmful. Just as real interest rates were ridiculously low for a decade leading up to that crisis, so we have found real interest rates can rise very high until savers and spenders are convinced that the policy will not relapse back into inflation. There have been huge government borrowings in many countries and sharp differences in the degree of reliance on free market forces to eliminate obsolete facilities and encourage new ones. The resulting contrast of forces between countries has caused wild movement in exchange rates with drastic effects on the mines' operating costs and revenues and on the Treasurer's debt portfolio. Miners can no longer expect inflation to bring bad projects on side because the interest cost is too heavy. Growth in consumption of metals in some of the third world countries was startling, but the banking crisis has hit them on the head particularly those countries in Latin America.

Some of these pressures could ease sooner than others, but none seems merely temporary. On the other hand one need not lose heart totally. The long wave does not fall forever. We are probably half way through the downward phase. After a time the long wave will rise again, carried in part by old industries having another turn as well as by new industries. In the meantime the third world needs goods of every kind, if only they can make their social systems more productive. Energy prices are not rising and could fall.

If all this describes the world we are going to continue to work in then metals will continue to be very cyclical but with severe downs and unsatisfying ups for some years yet. Prices will not often achieve levels required for new greenfields capacity. The financial objective of a mining company in such a world must be to make the most of every circumstance.

ASSET MANAGEMENT

The market for some products is better than others. To some degree this reflects the adverse participation of governments in each product as well as the circumstances of innovation in its use. The point for a large mining company is to be diversified across the spectrum of base and precious metals and to sponsor research and development of new uses for each one. In Noranda's case we have diversified into other resources too, forest products and natural gas. The test for each operation is its break-even point compared to the competition, taking into account all its competitive advantages and disadvantages including grade, by-products, cost and transportation. The average cost of production of some metals has declined sharply since 1981 partly through better efficiency and partly through high-grading. The latter reduces the life of the orebody and someday those ore reserves so sacrificed will have to be replaced.

The role of public terminal and futures markets has been expanding. Their effect is to allow the industry very little control over the pricing of its product. The producers' choice is either to accept the market's price or withhold deliveries. The marginal producer will probably operate until the price is below his cash break-even cost of production. The lower-cost producer can choose whether to cut back or to produce at full speed in the hope of forcing out permanently some of the competition. This predatory option will depress everyone's prices and profits in the short run. If the competing capacity so attacked can shut down now but start up again later, or be easily replaced, then the predator will never recoup those lost profits.

As markets fluctuate the financial objective of commercial management to achieve the best prices over time must include consideration of using the forward markets. This is a thankless job because the manager cannot possibly be completely right.

The challenge is to sell forward when prices are reasonably high, or find put options at acceptable cost. Another benefit of terminal markets is that when prices are sufficiently low the operator can sell his production to speculators who will carry it at no interest cost to the producer.

Getting the best prices over time may also include using the forward foreign exchange market. Fluctuations in currency prices can have a big effect on a mine's revenues directly. The indirect effect may be even greater as currency fluctuation changes the economics of the competition. For example consider the aggressive devaluations in Chile and Sweden.

New projects are still going forward in some products, notably gold. Project financing is available where the economics of the property are very strong. Whether to use it depends on the pricing of the loan package and the nature of the commitments that the bank requires from the developing company.

The new project has to have a very low operating cost, since it must amortize its capital whereas the competitive existing project has its capital already sunk.

LIABILITY MANAGEMENT

The title of Treasurer suggests an officer who looks after the riches in the Treasury. Back in 1962 I added up the holdings of cash and equivalent for the major pools of capital in Canadian mining and they totalled $400 million. That doesn't sound like much but those were heavy dollars back then. Total bank loans to the mining industry were only $100 million and there were few other debts.

Today an equivalent list of major Canadian mining companies would show almost no cash and perhaps $6 billion of debt. Today the Treasurer looks after somebody else's money that he has borrowed.

Obviously the Treasurer's first duty is to provide enough funds to support the operations that are viable. In many cases it is argued that the cost of money is less than it seems since interest is deductible from taxable income. But there can be few mining companies anywhere that are paying taxes so that the full burden of the interest must be borne currently. Most of the debt is at floating rates because Tresurers cannot bear the thought of locking in such levels of interest for a long term.

The most critical choice is between equity and debt. Common shares and their equivalent are forever. The conundrum is to know whether such equity capital is expensive in the current market or cheap, relative to the high real interest rates on debt.

CONCLUSION

In summary, those of us middle-aged treasurers who grew up in the mining business are not likely to repeat the experience of our youth before we retire. We had the joys of the 1950's, 60's and early 70's and our children will have the next turn. Younger people in the business can look forward to an improving trend eventually. Exploration people who despair of finding new base metal deposits that are economic in today's environment can take heart. We are probably half way through the downswing and lead times on new projects are long. Consistent exploration is needed to have deposits for production when the world needs them.

Above all metal prices will continue to fluctuate widely and we will have some good moments. Financial objectives should be towards diversification, product enhancement, low breakeven points, hedging of prices and exchange rates and strong balance sheets.

APPENDIX

In the following charts the thin lines show the nominal prices for oil and various metals in this century. The thick lines show the prices in constant 1984 U.S. dollars, computed by dividing the nominal prices by the GNE deflator, and prior to 1947, by the U.S. Producer Price Index.

These illustrate the historical context in which a treasurer must judge whether prices for his products are high or low.

The prices shown reflect markets fairly closely except for aluminum where the producer price was often discounted, sometimes severely. The aluminum price used since 1972 is the Metals Week U.S. Market Price.

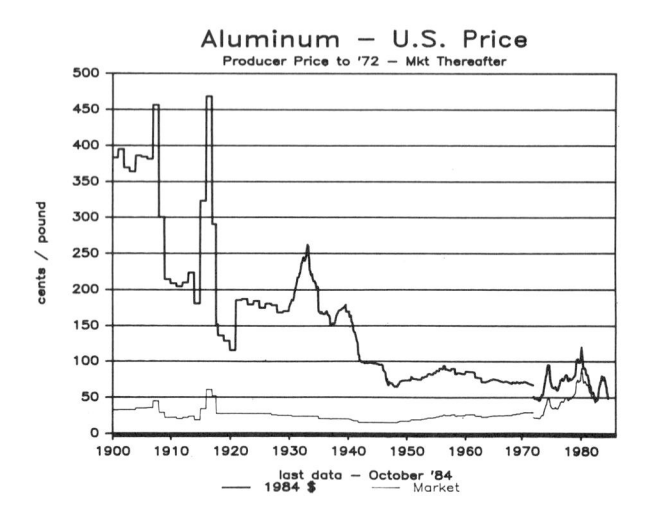

FINANCE FOR THE MINERALS INDUSTRY

Aluminum — U.S. Price
Producer Price to '72 — Mkt Thereafter
last data — October '84
— 1984 $ —— Market

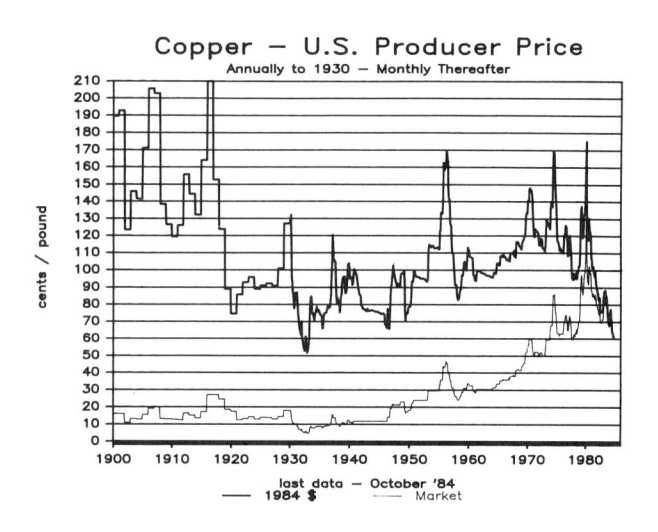

Copper — U.S. Producer Price
Annually to 1930 — Monthly Thereafter
last data — October '84
— 1984 $ —— Market

Gold — London Final Fix
Annually to 1930 — Monthly Thereafter
last data — October '84
— 1984 $ —— Market

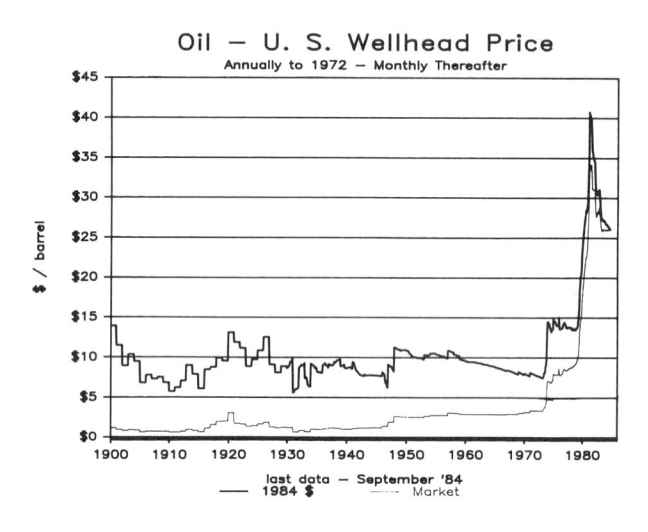

Oil — U. S. Wellhead Price
Annually to 1972 — Monthly Thereafter
last data — September '84
— 1984 $ —— Market

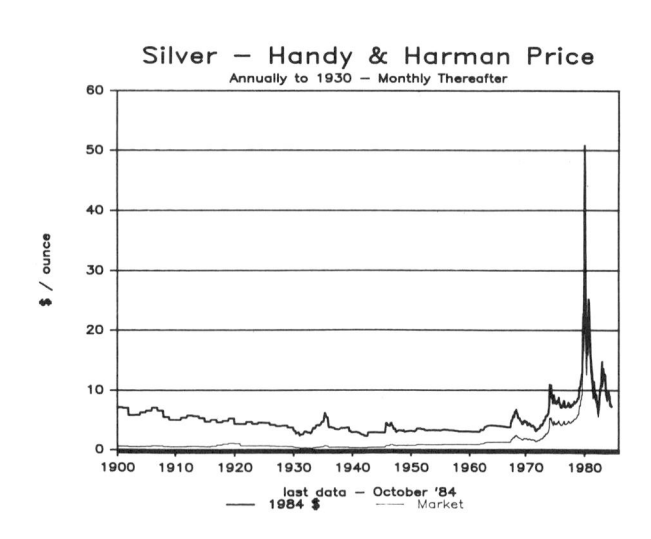

Silver — Handy & Harman Price
Annually to 1930 — Monthly Thereafter
last data — October '84
— 1984 $ —— Market

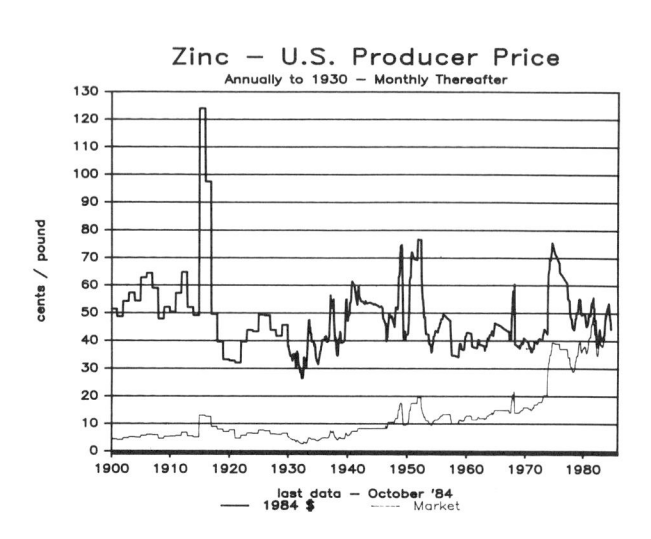

Zinc — U.S. Producer Price
Annually to 1930 — Monthly Thereafter
last data — October '84
— 1984 $ —— Market

2.3

FINANCIAL ASPECTS OF JOINT VENTURE
AND
MULTIPARTY STRUCTURES

Mr. David C. N. Cockburn

Senior Manager, Project Finance Group
European Banking Company Limited, London

INTRODUCTION

This paper analyses certain of the financial considerations which will commonly apply in relation to the following questions:-

- what equity structure or project vehicle should be adopted for the exploration and/or exploitation of a project by more than one unrelated participant?

- what vehicle should be adopted by a participant for its participation in a multiparty project?

Due to local, politically motivated laws in certain jurisdictions, or due to the overriding requirements of a dominant participant, there may be little choice of the underlying multiparty structure which may be adopted for the exploration, appraisal and/or development phases of a project. Nevertheless, even in such circumstances, attention should be given, at least in the preparation of the related documentation, or in considering methods of participating or funding the project, to certain of the points discussed below.

This discussion of financial considerations in the context of the more commonly adopted multiparty structures concentrates primarily on fiscal, financial accounting, financing and funding aspects.

TYPES OF PROJECT VEHICLE FOR MULTIPARTY INTERESTS

The well known types of vehicle adopted for the participation in a project by more than one participant (assuming that the participants are effectively unrelated) are as follows.

Joint Venture (unincorporated joint venture)

By far the most commonly adopted structure for mining ventures, a joint venture is usually defined and documented in terms to distinguish it from a partnership for important tax and legal reasons. In essence, this is a <u>contractual</u> association of more than one participant to carry out a specific commercial project or purpose. The relationship between the separate participants and their respective rights, interests, obligations and liabilities are agreed between the participants in one or more contracts (viz Joint Venture Agreements or Joint Operating Agreements together with ancillary documentation). Unlike other project vehicles of participation, in most jurisdictions there is no specific body of law governing joint ventures. The participants thus enjoy contractual freedom to negotiate whatever deal they may wish between themselves.

Jointly-Owned Project Company (corporate joint venture)

A very familiar vehicle, this is a separate and distinct legal and tax entity incorporated under the company laws of the place of incorporation to undertake specific objects, and granted specific powers, in relation to the Project. Although certain rights, interests and obligations may be circumscribed in the governing corporate instruments by the promoter participants, the corporation laws of the relevant place of incorporation, and the selection of such governing laws (assuming freedom of choice) and future political changes in such corporate laws, should be carefully considered.

The use of a company in a multiparty structure may be peripheral to the project itself as a specific purpose vehicle (for example, a joint sales agent or marketing company or a joint borrowing vehicle) or it may form a central function such as in the case of a tolling company which will own the project for legal and taxation purposes and raise financing with the backing of the shareholder participants commonly by the adoption of take-off, throughput and/or user agreements.

Partnership

This is a legal relationship, subject to the often wide-sweeping restrictions of the relevant local partnership laws, under which the participants have agreed to carry on a business venture together with a view to profit. The effect of the relevant partnership laws, as well as the various types of partnerships and the contractual arrangements mutually entered into, will have an important bearing on the participants' relationship between themselves and with regard to third parties.

Trusts

A trust is a concept available in many jurisdictions whereby a trustee may be vested with property in such a manner that the trustee is required to hold the property for the benefit of others or for purposes other than those of the trustee. The important distinction lies in the fact that the technical legal ownership of the relevant property is with the trustee whilst the beneficial economic enjoyment of such property lies with the beneficiaries of the trust.

A trust is more commonly adopted as a device to create a desired accounting, tax or legal result rather than a vehicle for the actual exploitation of a project. For example, it may be adopted to minimise the full impact of taxation; to raise inexpensive funding from investors; to avoid the effects of the consolidation of liabilities or debt incurred in relation to the project onto a participant's balance sheet; or to separate legal title and ownership from the participant beneficiaries. It is primarily a tool used by individual participants within a multiparty structure and has, in general, a limited application for all of the participants in a multiparty project venture.

Examples of the use of trusts, which have been particularly prevalent in Australia recently, are The Western Australia Diamond Trust, The Queensland Coal Trust and the Delhi Australia Fund (see below).

TAX CONSIDERATIONS

The selection of a particular multiparty structure and, almost inevitably, the participant's choice of vehicle in entering into such a multiparty structure will be primarily dictated by tax considerations. Specific jurisdictional tax rules and their direct impact and effect on various forms of equity participation structures are ever-changing factors and are broadly covered in Chapter 5. Expert and early advice at the formative stage of the multiparty structure or when considering participation in an existing multiparty structure is an obvious necessity. Such considerations are equally all important at any subsequent stage where consideration is being given by a participant to raising external funds or entering into other ventures in the same tax jurisdiction and/or creating a more tax effective long-term position. In this context, it is important to bear in mind not only the vehicle to be adopted and local tax rules but also (where relevant) tax effects in the jurisdiction of the participant's (or its parent company's) tax domicile or residence if elsewhere.

There are certain tax considerations worthy of note which will commonly be found in many jurisdictions and which may be relevant in considering the best type of structure.

Project Companies

Group tax relief provisions or the adoption of a group consolidated tax return by a participant may be available only at comparatively high percentages of share ownership (75-80% being common). At first sight it is, thus, not possible for more than one shareholder participant (if any) to enjoy group tax relief provisions in relation to the project company. If the creation of different classes of shares (possibly combined with tax equalisation arrangements), indirect ownership devices, the adoption of a trust or lease or the use of a specific type of "company" (for example, a "transparent" partnership company) cannot solve this problem, a jointly-owned project company may result in the loss, or costly carry forward, of tax losses, depreciation, capital allowances, investment grants and interest expense, with materially adverse present value and cash flow consequences for the project company as well as its shareholders.

Nevertheless, the loss of tax deductions or the delay in their utilisation may only materially affect one participant. The other participant(s) may not initially have other profits against which immediate tax deductions could be applied. In such circumstances it may be possible for the tax-paying participant to be given the requisite degree of share ownership for group tax relief purposes and for such participant to share the cash flow advantages so gained with the other less fortunate participants by tax equalisation or other similar arrangements. Furthermore, the inability to utilise tax deductions immediately may be circumvented to a great extent by third party or participant leasing - at least in relation to capital assets.

Partnerships

A partnership is not a separate taxable entity and tax effects and results flow directly through to the individual partnership participants. Thus, individual partners may apply their respective share of losses and income attributable to the partnership against their general tax position as and when the same arise. As a result, partnerships and, in particular, limited partnerships (which avoid the joint and several liability problems created by general partnerships by setting a cap on the limited partners' liabilities), have flourished in many jurisdictions for mainly tax-related reasons. However, although limited partnerships have proved to be popular methods of raising tax-sheltered funds from investors (often during the exploration phase where tax deductions are available and project financing is not) the liabilities of the general partner can be particularly onerous (although a nominally capitalised limited liability company may be adopted as a "notional" general partner) and there are frequently legal restrictions on the ability of a partner to transfer its interest.

Although a partnership is not a separate taxable entity in most jurisdictions it is often required to make a single partnership return as if it were such a tax-paying body. As such, whilst the tax effects flow through separately to the participants, the ability of each participant to make its own separate tax elections will be lost in the absence of unanimity. Nevertheless, the significance or possibility of such a loss must be projected in the light of the relevant possible elections, means of circumventing such differences and the participants' own forecasts.

Joint Ventures

The participants in an unincorporated joint venture will be entitled to treat their proportionate interest in depreciation, tax deductions, investment grants etc. as well as revenues as their own for all tax purposes. Similarly, each participant may make its own separate elections for tax purposes. Thus, most of the potential disadvantages of a project company and of a partnership referred to above may be avoided. However, the avoidance of such potential disadvantages may to some extent be illusory if, for example, due to local legal reasons, each participant in a multiparty structure must participate through a locally incorporated participant vehicle in circumstances where group tax relief is not possible or where favourable double taxation treaties do not exist.

FUNDING CONSIDERATIONS

As the project proceeds through its various phases and the future funding commitments of the participants crystalise, each participant will evaluate and re-evaluate its general financial strategy. Some typical questions which will arise, when the scale and timing of the relevant development begins to crystalise and subsequently when technical studies and cash flow projections have been carried out, are as follows:-

- Should the participant farm-out part of its interest in a project and realise cash for injection in other projects by way of strategic diversification and/or obtain some form of carried interest from the farmee? When would such a step be most beneficial? What restrictions, pre-emption rights or legal constraints apply to the ability to transfer part or all of the participant's interest?

- To what extent are medium or long-term sales contracts achievable and/or desirable? What is the risk that sales terms will be the subject of "unilateral" renegotiation?

- Would the introduction of customers, suppliers or contractors as minority participants be beneficial?

- What are the objectives on equity/debt funding? If high gearing is sought to allow more efficient, alternative use of equity what is the maximum ratio of debt to equity that can be supported by the participant or from its interest in the project? Should debt or other funds be raised on a corporate or limited or non-recourse or equity-kicker basis?

- To what extent can non-debt funding be achieved - for example, leasing, export credit subsidies and incentives, equity or quasi-equity funding (convertible debentures, hedging arrangements, preferred capital etc)?

In appraising such considerations and the restrictions faced, or advantages afforded, by the relevant multiparty structure, a final key question will relate to the desirability of appointing a financial adviser for the participants and/or the relevant participant. Such an adviser may vary from a corporate finance adviser for a public flotation to the manpower resources of a large project finance team.

FINANCING

In considering participation in a project involving more than one unrelated participant, it is obviously important to ensure so far as possible that both the relevant participant as well as each of the other participants is capable of funding its respective obligations in relation to the project. If one participant is capable of funding all or part of its project obligations from external cash resources or from general corporate borrowings, it may be unwilling to accept a multiparty structure with comparatively weaker co-participants which would require the joint project vehicle to raise project financing at a greater cost than that which such participant would otherwise have borne.

Conversely, if a structure is adopted which contemplates certain participants raising project financing for their separate underlying interests and obligations, it is equally important to ensure that the documentation and the underlying structure clearly allow such financing options to be effected and that the financial strength and funding ability (or lack of such strength or ability) of such participants may be enhanced and made more certain through the backing of project financiers or cost effective leasing or corporate finance techniques.

These are multitudinous financing techniques and funding options which have been, and no doubt will be, adopted in this area. These may vary from a commercial paper programme (such as CRA's for its Argyle Diamond Mine participation) to a joint project financing (such as that for the Bamboo Creek Joint Venture in Western Australia). However, at the outset it is important to stress the need for a flexible approach, in recognising future financing requirements, in drafting the documentation underlying the relationship between the participants and in appreciating the mutual benefit in assuring greater ease in accessing differing sources of committed funding provided directly by project financiers or under a letter of credit or guarantee provided by project financiers.

Financing Considerations for Different Multiparty Structures

<u>Joint Ventures</u> Under an unincorporated joint venture each joint venturer is in general free to select and follow its own financing options. As each joint venturer will usually have an undivided legal interest in the relevant project assets, it may offer security over its individual assets (or provide for leasing over such assets) and its interest in the project product and cash flows as the basis for project financing.

A considerable amount has been written about project lenders' requirements in relation to financing the individual interest of a joint venturer and the particular advantages and flexibility of a joint venture structure and it is not intended to reiterate the same here. In general, however, such works emphasise the flexibility of the joint venture structure for financing options and the importance of ensuring that the joint venture documentation at the outset recognises the needs and requirements of project financers. Nevertheless, whilst a joint venture may in general provide the greatest flexibility for the individual participants, the success or not of the relevant project is, by its nature, that of the joint venture itself. The considerable savings in management time, financing costs and administration achieved by a joint financing involving several or all of the joint venturers should not be forgotten in the rush to select individual funding options.

<u>Project Companies</u> A joint venture corporation may borrow in its own name. The project assets belong to the company and not the several participant shareholders. Although a shareholder participant may pledge its shares, it cannot create security over a specific part of the project assets or give undertakings in relation to future actions to be taken by the company or its Board of Directors (including the payment of dividends). It is thus not usually possible to create a true project financing over a shareholder's interest in a jointly-owned project company. However, many examples exist of project financing being made available to the jointly-owned project company itself. For example, El Indio and Sociedad Chilena de Litio projects in Chile.

The relative credit strengths or differing funding objectives of the participants may be a crucial factor against the adoption of a jointly-owned project company. If any participant wishes to raise financing from non-project financing sources to fund its commitments or a participant has insufficient independent credit standing in its own right to support project financing debt, in circumstances where non-recourse debt is not available from the outset, a project financing by the project company may not be possible.

Credit support for a project financing raised by the project company may be required beyond the project and its assets from the participants themselves, particularly prior to completion. This may take the form of a completion guarantee or top-up support and will usually be granted by the shareholders on a several basis. As such, the cost of the financing, which will reflect the several nature of the contingent support of the participants, will increase disproportionately, particularly for the more creditworthy participant(s).

Nevertheless, where satisfactory mutual financing objectives are agreed and are obtainable, the advantages and savings of joint financing through a single project vehicle will often be significant.

<u>Partnerships</u> Partnerships may also borrow in their own name but, unlike project companies, it is also possible for the individual partner participants to raise project-related financing in their individual names. The joint and several liability of partners beyond the partnership assets (which would improve the credit-rating of the partnership on a joint partnership financing) may be avoided to a satisfactory extent by the use of interposed nominally capitalised subsidiaries, acknowledgements from the lenders that they will limit their right of recourse solely to the relevant partner's share of the partnership assets, insurance policies, cross-indemnities from credit-worthy partners and/or limited partnership structures. This ability to raise limited or non-recourse debt by individual participants has been frequently unrecognised and underutilised.

General Financing Considerations

There are a vast number of differing financing techniques and sources of funds which may be utilised for financing a project (see Chapter 11). Although there appear to be distinct fashions for a specific financing technique or structure from time to time (where complexity may arise from attempting to fit a particularly ingenious structure created for one financing to a wholly different project or inapplicable participant circumstances), participants should avoid being confused by complex structures which stress certain advantages at the cost of other disadvantages and should bear the following overlapping points carefully to the fore:-

(a) What is the cost of the structure as compared to the advantages sought and obtained? For example:-

- if banks are willing to take project risks, would it be better to access cheaper or tax effective sources of funding and request the project financiers to provide letters of credit or guarantees to the providers of such alternative more favourable sources?

- would private or quasi-public insurance be cheaper in relation to certain project risks (for example, political risk insurance)?

- are export or subsidised trade or vendor credits etc. available?

(b) What risks does the participant want the providers of finance to take? For example:-

- does the participant feel comfortable with certain risks which need not therefore be taken by the providers of finance?

(c) What is the financial accounting/balance sheet impact of the proposed financing structure?

(d) How complex is the structure both in terms of management and professional time and cost at the time of negotiation and in terms of administration throughout the life of the transaction?

(e) Can the financing structure be made more tax effective?

(f) Does the financing structure provide flexibility to access other financing techniques (on a specific or general basis) at a subsequent time - for example, leasing, fixed rate funding, preference share capital, Eurobonds, etc.?

Taxation and Financing Considerations

Tax considerations should remain closely in mind when evaluating participation in other potential projects in the same or even other jurisdictions or when considering different methods of raising funds for the development of the project. Advance tax planning can have significant effects on cash flows.

Cash flows and cash projections may be significantly enhanced by a realisation of the different value of tax paying or tax loss properties to different participants with varying tax positions. For example, a high tax-paying property may be worth more to a company with unutilised tax losses which it is presently carrying forward or a forward sale of project product may allow tax deductions to be applied immediately and for effective short-term or medium-term deferred credit financing to be obtained from a forward purchaser. The ability to farm-in and out of multiparty structures in this context will obviously be an important consideration.

Tax considerations in relation to financing may be illustrated by reference to the project financings where the use of tax assumptions by the project lenders in calculating future cash flow, based on actual or forecast capital and exploration expenditure, has allowed considerably larger amounts of limited recourse funds to be made available. Such assumptions are often based on the freedom of the participant to make its own elections for tax purposes and/or upon group tax relief provisions being applicable. Similarly, the recent spate of unit trust structures in Australia (for example, the Queensland Coal Trust for the acquisition by BHP of Utah's properties and the Western Australia Diamond Trust for the minority stake of Northern Mining Corporation in the Argyle Diamond Mine Joint Venture), has resulted in considerable tax savings and access to comparatively inexpensive funds.

FINANCIAL ACCOUNTING AND BALANCE
SHEET CONSIDERATIONS

It is often at best ironic that the considerable efforts and cost expended in assuring that a project financing is without, or with only limited, recourse to the relevant participant usually will result in the project debt being directly consolidated on to the participant's group balance sheet with consequent adverse effects on credit rating, financial ratios etc.

This will be so despite the fact that the project lenders cannot look to the participant's other group assets and may be further aggravated if the cost of the "non-recourse" financing for the development is significant in the context of the participant's or its group's overall assets and size. It is, thus, important to consider the effect of funding the development of the relevant project from a financial accounting point of view. It may, therefore, be important to consider methods whereby consolidation rules may be avoided or minimised in relation to financing. The aim, of course, should be to emphasise and divorce the non-recourse nature of the debt, not to hide debt completely from general investors or creditors.

Project Companies

In most jurisdictions, statutory or accounting rules will only require consolidation of liabilities and assets on a line-by-line basis where a participant/shareholder owns more than 50% or otherwise "controls" the relevant company. The relevant tests of control and definitions of the requisite share ownership will differ between jurisdictions and may even vary as between consolidation rules and taxation rules for group tax relief purposes. Thus, in some jurisdictions it has been possible to create a wholly off-balance sheet project vehicle where the depreciation and tax losses may nevertheless be claimed by the underlying participant's group under group tax relief rules - an interesting result!

Nevertheless, if such techniques are not available and a participant/shareholder owns more than 50% of a project company, all of the assets and, often more significantly, all of the liabilities and debts of the project company will be consolidated upwards often with significantly adverse effects upon the group balance sheet, financial ratios and the perspective of credit-rating agencies, investors and creditors.

Partnerships

Typically, the participant's percentage interest in a partnership is reflected through to such participant's balance sheet under equity accounting principles. However, a control test may apply requiring line-by-line consolidation where more than 50% control lies in the hand of the participant. Nevertheless, even in these circumstances it may be possible to avoid the full effects arising from consolidation where, for example, project finance lenders agree to limit their recourse in relation to a borrowing raised by the partnership itself solely to each of the relevant partner's interests in the partnership assets.

Joint Ventures

In general, the financial accounting position for an unincorporated joint venture is the same as for an interest in a partnership. However, the significant advantage of a joint venture in this context lies in the fact that each participant under such a structure is, or should be, under properly drafted joint venture documentation, capable of selecting its own financing options. A joint venturer is thus free to create a financing structure with favourable balance sheet treatment - for example, a forward purchase or non-monetary production payment which would be treated as a deferred liability or an off-balance sheet trust financing such as achieved by the Delhi Australia Fund. Indeed one financing structure recently created actually improved the overall debt to equity ratio of the underlying "borrower."

GROUP DEBT DOCUMENTATION

In the context generally of raising financing for the group coffers on a corporate basis, participants which may wish at a later stage to raise project financing should be particularly careful to avoid wide sweeping cross-defaults, negative pledges and financial covenants which impinge upon the group where the "Group" is defined in the underlying documentation to include all subsidiaries. Specific exceptions should be sought, and should usually be granted, for project financings where the lenders have recourse for debt service solely to the cash flow generated from the project and/or to security over the relevant project assets. However, if such provisions already exist, participation in a project under certain limited partnership structures, off-balance sheet corporate or trust vehicles or simply by a less than 50% shareholding in a project company or borrowing vehicle may circumvent such problems.

CONCLUSION

Unincorporated joint ventures for commercial, legal and financial reasons have proved to be the most flexible structure for multiparty participation in a project. However, corporate joint ventures, partnerships and trusts will continue to afford greater advantages in certain circumstances and should not be rejected without due consideration of the mutual and individual aims and objectives of the participants. In any event, there will continue to be an important need for close and careful examination of the commercial and management requirements and objectives sought from time to time.

Above all, expert early advice should be sought from tax, legal, accounting and/or financial advisers to highlight the comparative advantages and disadvantages which are afforded in selecting the relevant structure and, subsequently, in considering the funding requirements of the participants jointly as well as individually.

FINANCIAL EVALUATION OF A COAL MINE ACQUISITION
A CASE STUDY

by

James H. Boettcher
Vice President

W. Durand Eppler
Vice President

Specialized Finance Group
Crocker National Bank

INTRODUCTION

In the second quarter of 1979, Rapid Mining Co., Inc. ("Rapid") inquired whether Crocker National Bank ("CNB") would provide $4 million in term loan financing to help ("Dry Creek") acquire the outstanding shares of the Dry Creek Coal Company ("Dry Creek") in West Virginia. Prior to approaching CNB, Rapid had sought financing with the assistance of a financial advisor from a number of commercial banks without success. A summary of the initial terms requested by Rapid is shown in **Table 1** compared with the actual terms on which the transaction closed.

It is not difficult to note that the two sets of terms differ significantly. This is often the case when financing requests based on the expectations and vision of a mine developer are juxtaposed with more pragmatic requirements of a commercial bank. The "art" and fund of structuring financing plans for mining ventures (highly leveraged acquisitions or grass roots) comes in working with mine developers to arrive at a final financial structure that meets their objectives and constraints as well as those of the bank or other financial institutions providing much of the financing. Beginning with the terms requested by Rapid and ending with the terms that closed the transaction, this article briefly describes the process we went through with Rapid Mining.

The Seller and Buyer

The Dry Creek Coal Company was established in 1961, after its owner acquired the coal leases from Republic Steel. Republic has mined the primary coal seam from 1948-1961. Dry Creek was owned as a subchapter S corporation (i.e., for federal income tax purposes, income is taxed at the shareholder level) and had been run on a low key basis. Since the company had virtually no debt and limited cash needs, the owner produced primarily when rail cars were available and income was required. Output was therefore erratic as is shown in the top of **Table II**, but the result was a profitable little company with an attractive free cash position. Because the owner wished to retire and enjoy his wealth in a more readily spendable form, he had put the mine, with stated recoverable reserves of approximately 13 million tons, up for sale.

In contrast, Rapid Coal's investors consisted of a number of investors of disparate financial situations. They had agreed upon a primary objective of income maximization and planned an annual production rate of 120,000 TPA which they and the current owner believed realizable.

The Formulation of a Financing Structure

Figure 1 presents a simplified overview of the sequence of steps taken in the financing process which lasted approximately three months from start to finish. Since banks see many requests for funding and considerble time is required to follow up on each, an important step was determining, after a brief review of the materials provided to us, whether the basic "eonomic entity" was viable. In this case, the size and quality of the reserves, the inherent operating cost characteristics, the sales arrangements, the management/labor situation and Rapid's principles all appeared to be sound enough to warrant investing the time necessary to do a thorough analysis of the project. Once this decision was made, the evaluation and analytic process shown in **Figure 2** was initiated.

The first and most fundamental concern of any banker is whether the loan will be paid back as scheduled from the cash flow generated by the project. The amount of debt relative to the owner's equity is of course crucial in such a determination. Before the debt/equity issue was addressed, however, a number of more qualitative factors had to be studied and if

Presented at the Coal Age Coal Management Symposium
October 1981, Louisville, KY

Table I

Rapid Mining Company Acquisition of the

Dry Creek Coal Company

Comparison of Financing Terms

Item	As Initially Proposed	As in Final Documents
Loan Amount:	$4,000,000	$3,700,000
Equity:	800,000	1,150,000
Total "Equity at Risk"	800,000	1,600,000
Final Maturity:	7 years	8 years
Cash Flow Interruption Protection:	No	Yes (up to $450,000)
Sales Arrangements:		
- Account Administation:	No	Yes (all revenues)
- Contract Life	3 years	5 years
- Advanced Payments in event of rail car shortage	No	Yes
- 1st year & Base Price	Yes	Yes
- 80% payment upon loading	No	Yes
Initial Budget for Capital Expenditures	$50,000	$200,000
Initial Working Capital Budget	$50,000	$300,000
Cash Flow Hierarchy Tied to Debt Service (Including Mandatory Cash recapture)	No	Yes
Continuing Shareholder Commitment	No	Yes (strong comfort letter)

Table II

Rapid Mining Co. Acquisition
of the
Dry Creek Coal Company

Financial Summary of Operations, 1974 - 1979

(Dollars in Thousands)

	1974	1975	1976	1977	1978	1979
Tons Production	83.9M	72.1M	77.0M	64.4M	42.4M	59.1M (1)
Average Price @Ton	20.5	52.8	48.0	44.1	43.9	44.0
INCOME STATEMENT						
Coal Sales	1,721	3,811	3,696	2,841	1,862	2,597
Operating Expenses	1,143	1,418	1,708	1,723	1,460	1,911
Cost @ Ton	13.6	19.7	22.2	26.8	34.4	32.3 (5)
Net Profit before Taxes	589	2,403	2,004	1,127	466	827
Depreciation	115	140	207	194	148	119
Cash Flow before Taxes	704	2,543	2,211	1,321	614	946
BALANCE SHEET						
Cash	922	1,750	1,551	1,379	1,657	2,503
A/C Rec., Inventory	235	718	300	293	643	464
Current Assets	1,157	2,168	1,851	1,672	2,300	2,967
Mining Equipment Gross	1,017	1,214	1,641	1,643	1,651	1,699
Net of Accum. Depr.	360	447	694	517	390	327
Other Fixed Assets Gross	147	170	179	180	173	223
Net of Accum. Depr.	107	114	111	96	18	62
Net Fixed Assets	467	561	805	613	409	399
Other Assets	29	35	39	43	70	94

CAPITAL STRUCTURE FOR MINERALS COMPANY

FIGURE 1
Overview of Steps Taken In The Financial Structuring Process

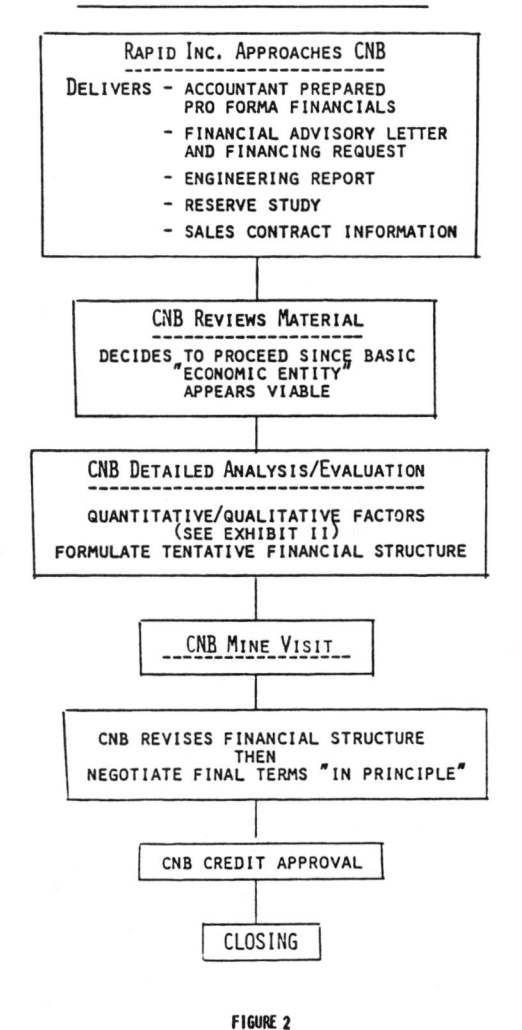

```
┌─────────────────────────────────────────┐
│       RAPID INC. APPROACHES CNB         │
│       ───────────────────────           │
│  DELIVERS - ACCOUNTANT PREPARED         │
│              PRO FORMA FINANCIALS       │
│           - FINANCIAL ADVISORY LETTER   │
│              AND FINANCING REQUEST      │
│           - ENGINEERING REPORT          │
│           - RESERVE STUDY               │
│           - SALES CONTRACT INFORMATION  │
└─────────────────────────────────────────┘
                    │
                    ▼
┌─────────────────────────────────────────┐
│         CNB REVIEWS MATERIAL            │
│         ───────────────────             │
│   DECIDES TO PROCEED SINCE BASIC        │
│       "ECONOMIC ENTITY"                 │
│          APPEARS VIABLE                 │
└─────────────────────────────────────────┘
                    │
                    ▼
┌─────────────────────────────────────────┐
│      CNB DETAILED ANALYSIS/EVALUATION   │
│      ─────────────────────────────      │
│    QUANTITATIVE/QUALITATIVE FACTORS     │
│           (SEE EXHIBIT II)              │
│   FORMULATE TENTATIVE FINANCIAL STRUCTURE│
└─────────────────────────────────────────┘
                    │
                    ▼
┌─────────────────────────────────────────┐
│             CNB MINE VISIT              │
└─────────────────────────────────────────┘
                    │
                    ▼
┌─────────────────────────────────────────┐
│      CNB REVISES FINANCIAL STRUCTURE    │
│                  THEN                   │
│   NEGOTIATE FINAL TERMS "IN PRINCIPLE"  │
└─────────────────────────────────────────┘
                    │
                    ▼
┌─────────────────────────────────────────┐
│           CNB CREDIT APPROVAL           │
└─────────────────────────────────────────┘
                    │
                    ▼
              ┌──────────┐
              │ CLOSING  │
              └──────────┘
```

FIGURE 2
Overview of Bank Evaluation/Analysis

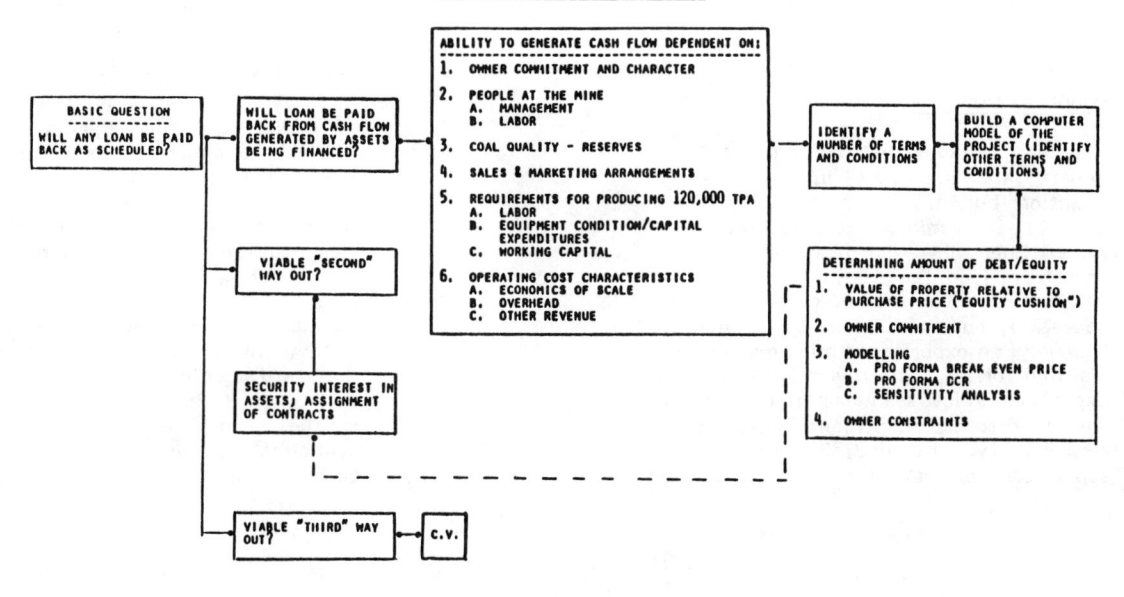

BASIC QUESTION
WILL ANY LOAN BE PAID BACK AS SCHEDULED?

WILL LOAN BE PAID BACK FROM CASH FLOW GENERATED BY ASSETS BEING FINANCED?

ABILITY TO GENERATE CASH FLOW DEPENDENT ON:
1. OWNER COMMITMENT AND CHARACTER
2. PEOPLE AT THE MINE
   A. MANAGEMENT
   B. LABOR
3. COAL QUALITY - RESERVES
4. SALES & MARKETING ARRANGEMENTS
5. REQUIREMENTS FOR PRODUCING 120,000 TPA
   A. LABOR
   B. EQUIPMENT CONDITION/CAPITAL EXPENDITURES
   C. WORKING CAPITAL
6. OPERATING COST CHARACTERISTICS
   A. ECONOMICS OF SCALE
   B. OVERHEAD
   C. OTHER REVENUE

IDENTIFY A NUMBER OF TERMS AND CONDITIONS

BUILD A COMPUTER MODEL OF THE PROJECT (IDENTIFY OTHER TERMS AND CONDITIONS)

VIABLE "SECOND" WAY OUT?

SECURITY INTEREST IN ASSETS; ASSIGNMENT OF CONTRACTS

DETERMINING AMOUNT OF DEBT/EQUITY
1. VALUE OF PROPERTY RELATIVE TO PURCHASE PRICE ("EQUITY CUSHION")
2. OWNER COMMITMENT
3. MODELLING
   A. PRO FORMA BREAK EVEN PRICE
   B. PRO FORMA DCR
   C. SENSITIVITY ANALYSIS
4. OWNER CONSTRAINTS

VIABLE "THIRD" WAY OUT? → C.V.

55

concerns arose, they had to be resolved. These qualitative issues were described briefly as follows:

1. Owner Commitment and Character

    After several meetings with Rapid, a thorough inspections of the mine and extensive background checks on Rapid's owners, it was decided to go along with Rapid's request of not requiring any shareholder guarantees for any loans. This decision was based primarily on the quantity and grade of proven reserves and the already well developed mining infrastructure which, if the loan to equity relationship was determined "prudently", should provide adequate collateral coverage of the debt should the bank ever have to initiate foreclosure.

    Still, however, we probably would not have proceeded if we did not believe that the new owners would stand firmly behind the mine and back it with their resources as necessary in times of need, regardless of whether they issued written gurantees. This was especially true since, although several of Rapid's executives would be traveling regularly to the mine, none would be living there.

    A strong comfort letter was used to formalize their support, but our real decision in this regard was based on numerous meetings, credit checks and references which gave us considerable confidence in the integrity and character of the people involved. This confidence in management proved to be a very important aspect of the transaction -- both prior to and subsequent to the loan closing, and the importance in developing a strong rapport between the mine owner, the mine manager and their bankers cannot be overestimated.

2. People at the Mine

    Capable, reliable managers and skilled miners are important in any mining situation but crucial in a small, West Virginia coal mine. Rapid's managers shared this view and first obtained an agreement from the existing owner to stay on for a year after the takeover to provide operating and management continuity. Secondly, an experienced mine manager in his mid-forties, with over 20 years of experience in low-seam eastern coal mines was hired to assume primary management responsiblity. Finally, all current shift supervisors and other miners agreed to stay on with the new owners.

    Even though the mine was non-union and UMW contract negotiations were then several years off, the possibility of a strike had to be considered. Aside from encouraging good management, we required Rapid to put up a $450,000 standby letter of credit ("L/C") that could be drawn upon for debt service in the event operating cash flow was interrupted by a strike (or protracted rail car absence). Calculated to cover at least 6 months of maximum debt service, the amount was also used to increase the amount of the owner's "equity at risk" as is discussed later.

3. Coal Quantity - Reserves

    The Dry Creek's tract consisted of seven local seams on property which includes two separate lease areas with different mining rights applying to each. **Table III** (A) indicates how the initial 8,905,831 estimate of recoverable reserves referenced by the owners were reduced first in (B) by eliminating small and/or poor quality seams; second in (C) by eliminating small remanants with high recovery costs; and third in (D) by using a 28" recovery line to devote the minimum mining height to arrive at a total quantity of recoverable 2,065,000 tons of clean coal. At an estimated production rate of 120,000 TPA, these reserves were sufficient for approximately 17 years of operation and were sufficient to support the eight-year term loan finally agreed upon.

4. Sales and Marketing Arrangement/Transportation

    The willingness of one of Europe's largest steel companies to enter into a strong coal purchase contract was a major factor in our deciding to work with Rapid on a financing plan. After a series of discussions with Rapid and the steel company's U.S. purchasing agents, it was agreed that the contract would contain the following provisions:

    - An extention in term from three years to five years with a renewal option thereafter.

    - Payment terms of 80 percent upon rail car loading and 20 percent upon shipboard sampling.

    - In the event rail cars were not available, it would call for 70 percent cash payment for up to

CAPITAL STRUCTURE FOR MINERALS COMPANY

Table III

Rapid Mining Inc. Acquisition of the Dry Creek Coal Company

Coal Reserves (A)

Area #1

Coal Bed	In-Place Tons	Recoverable Tons
Benchly	778,869	512,266
Storm Creek	476,765	212,373
Pocahontas No. 9	129,412	81,529
Pocahontas No. 7	5,505,833	1,491,731
Pocahontas No. 6		
Lower Bench	1,668,420	1,051,105
Upper Bench	1,835,356	963,562
Sub Total	10,293,644	4,312,566

Area #2

Coal Bed	In-Place Tons	Recoverable Tons
Benchly	405,718	284,002
Storm Creek	801,750	376,054
Pocahontas No. 9	482,632	230,049
Pocahontas No. 7	5,942,475	1,640,123
Lower Bench	485,296	300,000
Sub Total	10,773,486	4,593,265
Total	21,067,130	8,905,831

Coal Reserves (B)

Coal Bed	Area #1	Area #2	Total
Benchly	430,000	—	430,000
Storm Creek	212,000	—	212,000
Pocahontas No.6	1,051,105	1,673,037	2,724,142
TOTAL	1,693,105	1,673,037	3,366,142

Coal Reserves (C)

Area	Acres	Thickness	Percent Recovery	Tonnage
A	13.15	30"	85	48,690
B	20.53	30"	85	76,015
C	2.28	30"	85	8,442
D	9.73	30"	85	36,026
E	27.91	30"	85	103,340
Main Bl.	587.99	30"	70	1,792,900
TOTAL				2,065,413

15,000 tons (i.e., 1-1/2 months estimated production equivalent to $451,000 @ $43.00/T)

- A first-year fixed price of $43.00/T renegotiated thereafter with a 50/50 split of market prices exceeding $43.00/T.

- A floor price of $37.00/T.

- A specific acknowledgement that the steel company was aware that CNB was entering into the financing largely on the basis of its sales contract.

Transportation facilities were good as the mine was located on a siding with a loading facility. Although both Rapid and the buyer's purchasing agents believed that rail cars could be obtained when needed, the various provisions referenced above were made in order that the buyer would assume the mine's cash flow if cars were temporarily unavailable.

5. **Production of 120,000 TPA**

Since the mine had never produced at the expanded rate of 10,000 tones/month, reaching this point was seen as one of the greatest "risks" in the transation. A well-known coal mine consulting/engineering firm recommended operating four unit shifts per production day and various other manpower changes as shown in **Table IV**. The firm also recommended that a third operating section be opened up as well as a surface storage area to provide maximum production and marketing flexibility.

In addition, it recommended $688,000 of first-year capital expenditure as shown in **Table V**. Rapid and several other mining

TABLE IV

Rapid Mining Co. Acquisition of the Dry Creek Coal Company

MANPOWER NEEDS

Description	Present	Required
o Officers	1	
o Salaried - General Manager	0	1
- Superintendent	1	1
- Section Foreman	3	4
- Outside Foreman	0	3
- Engineer	1	1
- Maintenance	1	1
- Training	0	1
- Clerks	1	2
	8	14

	Present	Required
o Wage Earners		
- Outside	8	10
- Inside		
1) Face Production - Per Unit Shift		
a) Miner Operator	2	2
b) Mobile Bridge Conveyor	2	2
c) Repairman	0	1
d) Timber	1	1
e) Utility	1	1
	6	7
	Three operating unit shifts 18	Four unit Shifts 28
2) General Inside		
a) Belt	3	4
b) Surge Bin	2	2
c) Track	6	8
	11	14
Total Wage Earners	37	52

By Shifts: Days 25
Afternoons 25
Midnight 2

Table V

Rapid Mining Company Acquisition of the Dry Creek Coal Company

CAPITAL EXPENDITURES
($000)

	Description	Quantity	Proj. Cost Per Unit ($000)
1.	Surface Storage	one	150
2.	Battery Operated Personnel Carriers	three	22
3.	Battery Operated Mantrip Carriers	three	35
4.	Six (6) ton locomotives	two	66
5.	Face Equipment		
	a) 101MC miner	one	248*
	b) 506-C5 Double Bridge Conveyor	one	75
	c) 94L Bridge Conveyor	one	22
	d) 30" Belt Conveyor with 1000' Structure and related hardware	one	70
	TOTAL		688

*Onsite shortly after the engineering report was completed.

engineers regarded this amount as high, and by repairing existing equipment and purchasing reconditioned equipment, the amount of first-year capital expenditures was reduced to $200,000. In a difficult, highly leveraged situation such as this, where a company's ability to service debt depends on realizing targeted production levels, we can say with some retrospet that corners should not be cut when planning the equipment needed to meet production targets. (In this case, sometime after the close of financing, the owners ended up putting in virtually all of the $888,000 that had been "pared" from the consulting engineering firm's expectation of capital requirements at the time of acquisition.

6. Operating Cost Characteristics

Since it had been run on an ad hoc, informal way, as indicated by the production record shown in line one of Table II, it was difficult to estimate the appropriate operating cost/ton to use for preparing pro forma cash flow. Line (5) of Table II shows a current dollar cost/ton ranging from $13.60 in 1974 to $32.30 in 1979. Based on detailed analysis of past costs plus several consulting engineering reports, it was decided to use the estimates shown in **Table VI**, working out to $24.24/ton based on the "Base Case" production assumptions of 92,000 tons in 1980 and 115,200 tons annually thereafter.

TABLE VI

RAPID MINING INC. ACQUISITION OF THE DRY CREEK COAL COMPANY

ESTIMATED MINING COSTS - AS OF JUNE, 1980

1.	Officers Salaries		$ 125,000
2.	Mining - Salaried		245,000
	- Wage Earners (Includes Vacation Pay)		$ 701,484
3.	Mine Supplies		135,720
4.	Electricity		40,700
5.	Maintenance		73,290
6.	Engineering		12,000
7.	Legal & Accounting		15,000
8.	Fringe Benefits		
	a)	Supplies - Soap	372
	b)	Clothing Allowance	5,625
	c)	UMWA H&R	181,220
	d)	Insurance	116,600
	e)	Emp. Exams - X-ray	5,400
	f)	Instructions & Meet.	5,120
9.	Taxes - Fees - Insurance		
	a)	NICOA Dues	450
	b)	Reclamation Fees	13,570
	c)	Workmen's Comp.	192,500
	d)	State Black Lung	21,010
	e)	B&O	158,340
	f)	FICA	77,207
	g)	Employment	12,870
	h)	Personal Property	12,000
	i)	Black Lung - Excise	22,620
	j)	Excise & License	130
	k)	Corporate License	50
	l)	Insurance - Equipment	11,762
10.	Other		
	a)	Employers Ser. Corp.	1,200
	b)	Office Rent	1,800
	c)	Telephone	1,810
	d)	Misc. Expense	905
	e)	Fines & Penalties	2,400
	TOTALS		$2,193,155
	Per Ton		$ 24.24

When sensitivity analysis was later performed on production levels, it was assumed that the $24.24/T was a "fixed" cost within +1 - 20 percent for the 115,200 TPA. This assumption is not too unrealistic based on a review of **Table VII** which indicates that, short of laying off one or two shifts, most of the costs would not change significantly with minor fluctuations in production.

Another important consideration in any mine financing is the relationship of a particular mine's unit operating costs to the costs of competing operatings. Ideally, we would like to see a mine's costs in the lower one-third for the industry and in this case, Dry Creek's operation seemed to meet this test. More importantly, however, is the all-in breakeven price (BEP) including debt service. The BEP itself is a function of the amount of debt, its repayment schedule and its interest rate. The determination of these parameters is therefore discussed next.

Table VII

Rapid Mining Company Acquisition of the
Dry Creek Coal Company

A) Base Case Sources and Uses of Funds

Uses of Funds

"Net" Purchase Price	4,350,000
Net Working Capital	300,000
Initial Capital Expenditures	200,000
Standby Letters-of-Credit	450,000
TOTAL	5,300,000

Sources of Funds

Senior Bank Loan	3,700,000
Equity	800,000
Subordinated Shareholder Loan	350,000
Standby Letter-of-Credit	450,000
TOTAL	5,300,000

B) Base Case - High - Low   Debt/Equity Structures

Base Case	High D/E	Low D/E
Before L/C		
Debt = $3.7MM	Debt = $4.35MM	Debt = 3.06
Equity = 1.15MM	Equity = $0.50MM	Equity = 1.79
D/E = 3.2	D/E = 8.7	D/E = 1.7
Debt as % of Total = 76%	Debt as % of Total = 90%	Debt as % of Total = 63%
After L/C		
Debt = $3.7MM	Debt = $4.33MM	Debt = 3.06
Equity = $1.60MM	Equity = 0.95MM	Equity = 2.24
D/E = 2.31	D/E = 4.6	D/E = 1.4
Debt as % of Total = 70%	Debt as % of Total = 82%	Debt as % of Total = 58%

Determining the "Right" Debt/Equity Ratio: Structuring of Financing

The "right" debt/equity for a mine acquisition can be an elusive and much contested number. A banker has simultaneous, but opposite, forces pulling toward "safety" on one hand and "doing the transaction" on the other; e.g., 99 percent equity funding rule may assure "safe" loans but will probably result in little business, but a short career! The "right" ratio is obviously somewhere in between and depends primarily on the four factors that follow.

1. The "Value" of the Property Relative to the Purchase Price

    Every purchaser or developer of a mining property is almost by definition convinced that he is getting "a good deal" relative to what the property is "worth". This indeed was the rationale for Rapid requesting their initial approximately 92 percent financing. Based on a $4,350,000 purchase price, the approximate cost per ton of reserves was as follows:

Tons of Reserves Considered	Price per Ton of Reserves
8,905,831	$0.48
3,366,142	1.29
2,264,142	1.67
2,065,413	2.11

    These numbers are, of course, difficult to compare with other situations because the properties vary significantly. However, based on looking at other sales at that time and discussions with several property appraisers, we concluded that 70 to 75 percent financing would probably provide an adequate "equity cushion" in the event the property ever had to be foreclosed upon in order to repay the loan.

2. Owner Commitment

    Even if Rapid was getting such a great bargain that 100 percent secured financing could be justified, it is unlikely that any bank would consider it. The reason is that with no "cash at risk" it is far too easy for the buyers to justify walking away from the property in times of difficulty. A certain amount of equity will therefore always be required. In this case, because of the considerable amount of owner commitment that is needed to ensure the success in operating small eastern determined that 30 percent of the total funding requirement should be "at risk". As shown in Table VII and as discussed below, this amount was broken down for a variety of reasons.

3. Modelling and Sensitivity Analysis

    A computer model of the mine is highly valuable in determining the impact of a particular financing structure on the mine economics. In this case, a comprehensive model was built based on the various assumptions defined up to now. Approximately four simulations of the model were run for each of the three different debt/equity ratios defined in Table VII, as well as four different final loan maturities ranging from 6 to 9 years. In order to allow time for bringing production levels up to expectations, in each case no principal payments were assumed for the first 18 months following start-up.

For each of the simulations, two different numbers were recorded. These were:

- The Cash Flow Breakeven Price ("BEP") which is equal to all cash outflows -- including operating expenses, interest, principal payments and capital expenditures -- divided by the number of tons sold, represents the average price/ton that must be obtained to break-even on a cash flow basis. To allow for uncertainty and error in our assumptions, it was decided to target a highest acceptable "Base Case" BEP of $35.00/Ton. This provided a minimum of $8.00/Ton between the BEP and the first year's contract price, and the sensitivity analysis that followed showed this to be a reasonable margin.

- The Debt Coverage Ratio ("DCR") which is defined as:

$$DCR = \frac{(\text{Profit After Tax}) + (\text{Non-Cash Charges}) + (\text{Interest}) - (\text{Capital Expenditures})}{(\text{Interest} + \text{Principal})}$$

If this number equals 1.0, it means that just enough cash is generated to meet debt service requirements. The minimum targeted DCR for a particular project is a subjective determination, depending on a variety of factors including the expected variability of cash flow in that industry, and the confidence one has in the assumptions to the Base Case Model. For this project, a minimum pro forma DCR of 1.5 was selected as a screening criteria.

The highest BEP and lowest DCR for each combination of debt/equity and final maturity were then plotted as shown in **Figure 3**. The constraints were then stated and superimposed on the graph (**Figures 4, 5, and 6**), thus outlining what combinations of debt/equity and final maturity would meet the criteria established (i.e., the target debt/equity and final loan maturity).

In the next stage of the analysis, we applied the financial structure suggested by the preceeding analysis to the Base Case and performed a number of "sensitivity" runs. The purpose of this exercise was to get an appreciation for

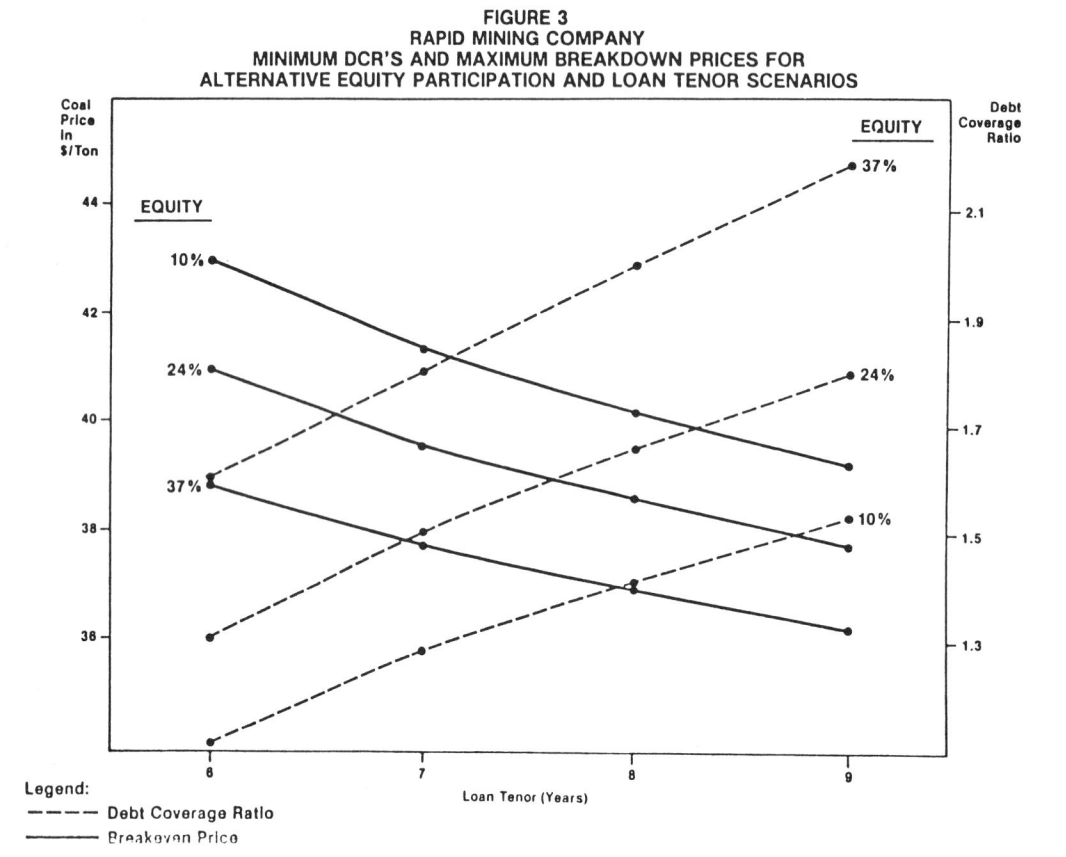

FIGURE 3
RAPID MINING COMPANY
MINIMUM DCR'S AND MAXIMUM BREAKDOWN PRICES FOR
ALTERNATIVE EQUITY PARTICIPATION AND LOAN TENOR SCENARIOS

Legend:
----- Debt Coverage Ratio
─── Breakeven Price

CAPITAL STRUCTURE FOR MINERALS COMPANY

FIGURE 4
RAPID MINING COMPANY
MINIMUM DCR'S AND MAXIMUM BREAKDOWN PRICES FOR
ALTERNATIVE EQUITY PARTICIPATION AND LOAN TENOR SCENARIOS

Legend:
- - - - Debt Coverage Ratio
——— Breakeven Price

▨ Above Maximum Breakeven Price ($39/Ton)

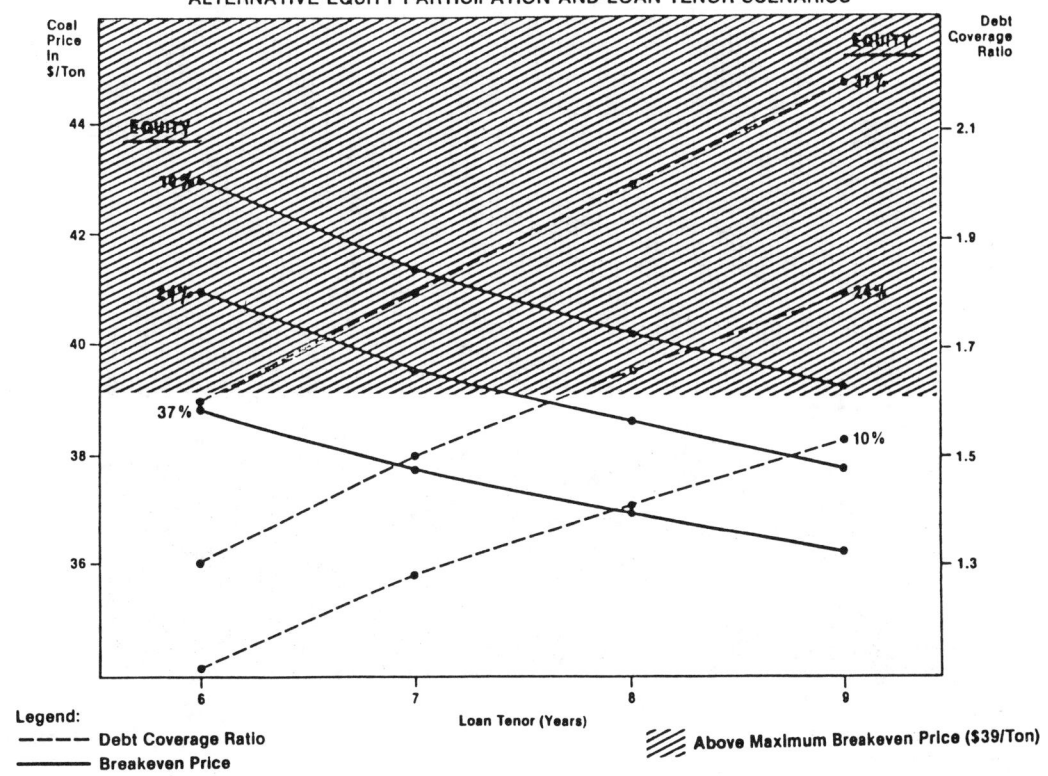

FIGURE 5
RAPID MINING COMPANY
MINIMUM DCR'S AND MAXIMUM BREAKDOWN PRICES FOR
ALTERNATIVE EQUITY PARTICIPATION AND LOAN TENOR SCENARIOS

Legend:
- - - - Debt Coverage Ratio
——— Breakeven Price

▨ Below Minimum Debt Coverage Ratio (1.5)

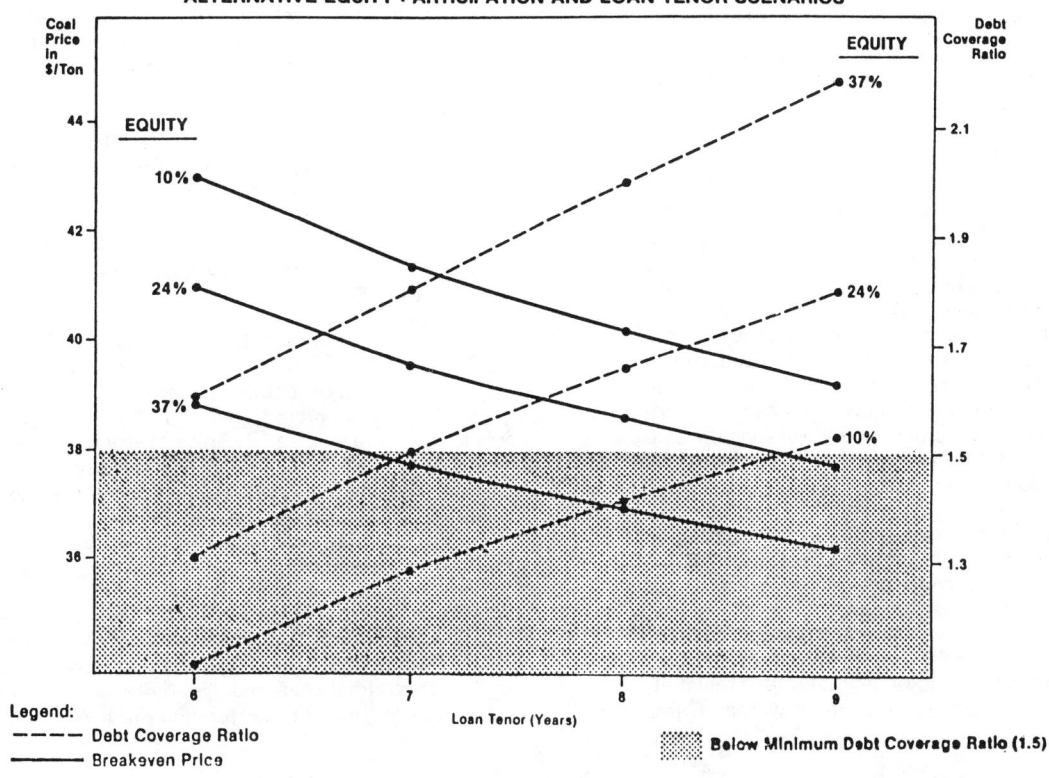

**FIGURE 6
RAPID MINING COMPANY
MINIMUM DCR'S AND MAXIMUM BREAKDOWN PRICES FOR
ALTERNATIVE EQUITY PARTICIPATION AND LOAN TENOR SCENARIOS**

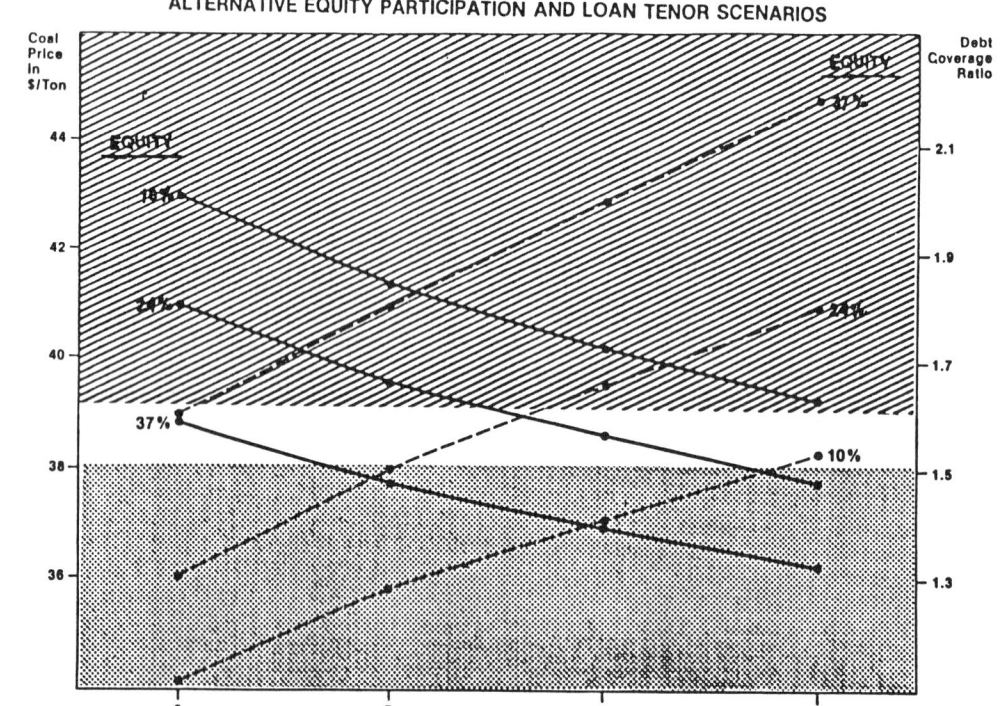

Legend:
- - - - Debt Coverage Ratio
——— Breakeven Price

▨ Above Maximum Breakeven Price ($39/Ton)
▦ Below Minimum Debt Coverage Ratio (1.5)

the amount of variation in key variables that could be tolerated before the project got into trouble. The results of this analysis is shown in **Figure 7**.

4. The Effect of Owner Constraints and Resources

As is typical of most mine financings, the bankers and the owners often initially disagree on the amount of equity required. In this case, the owners believed $800,000 of equity to be more than sufficient equity based on their own forecasts and strong belief in the "bargain" purchase price they had negotiated. We, on the other hand, were adamant (partly for reasons of "owner commitment", "equity cushion" and primarily to realize more reasonable pro forma BEPs and DCRs) that another $350,000 was needed.

After much discussion, a compromise acceptable to both sides was reached to use a subordinated shareholder loan ("SSL") for the required $350,000. The payment of interest and principal on this SSL, however, was strictly tied to the availability of an acceptable cash residual service payments were made. The "Cash Flow Hierarchy" mechanism that dictated how available cash was to be disbursed is shown in **Table VIII**. The beneficial feature of using a SSL was that it satisfied the needs of both Rapid and its bankers as follows:

> From a banker's perspective: It was a source of capital that was "at risk" as a form of equity that would only be paid if the company was doing as well as Rapid anticipated (and we certainly hoped but could not count on).

> From Rapid's perspective: It was additional funding committed but, unlike equity, it earned a definite committed but, unlike equity, it earned a definite return (even if deferred, it was compounded) and the principal would be returned long before any true equity, and even fairly soon if anticipated cash flows were realized.

In short, the SSL was a convenient mechanism for us to say to the owners "since there are definite risks present at this time that require more equity, why don't you put up funding in a way (i.e.,

**FIGURE 7
RAPID MINING COMPANY
IMPACT OF ALTERNATE PRODUCTION, OPERATING COST AND INTEREST COST
SCENARIOS ON MAXIMUM BREAKEVEN PRICE (1981) FOR DCR = 1.3**

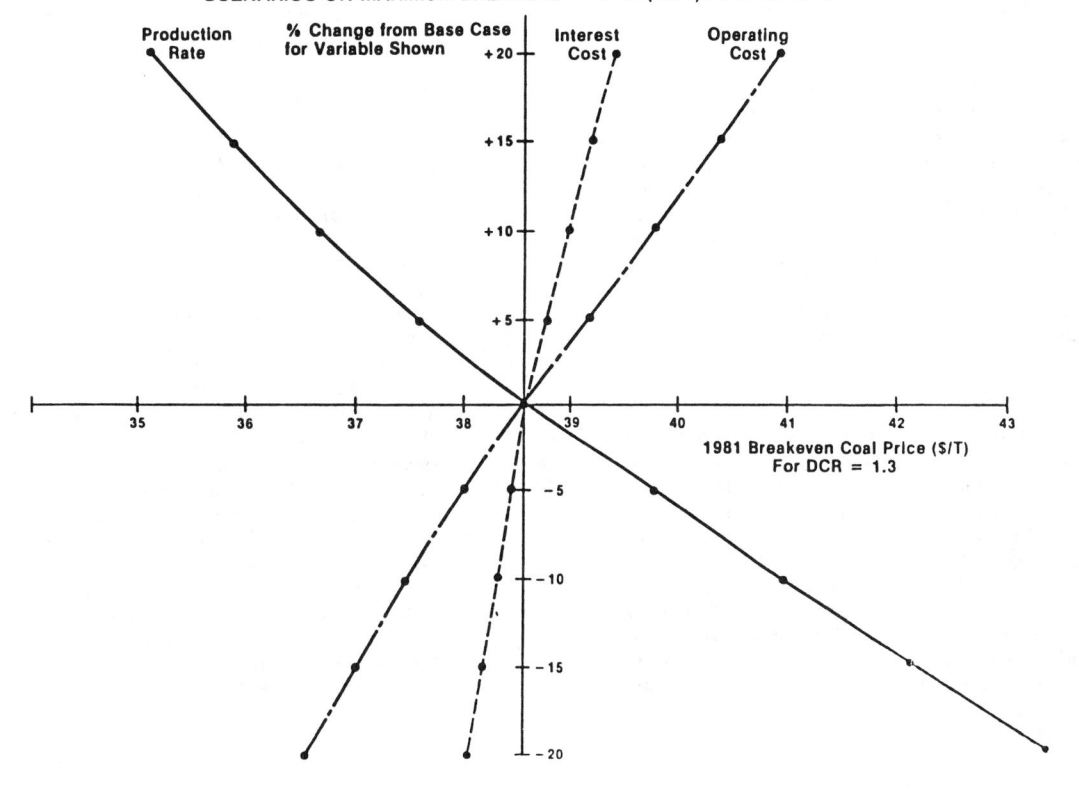

Table VIII

Rapid Coal Company Acquisition of the Dry Creek Coal Company

Cash Debt Coverage Ratio & Hierarchy of Payments

$$CDCRB = \frac{PAT_T + NCC_T + INT_T - CAPEX_T + CFWC_T - COUT_T}{INT_T + PRIN_T}$$

Where,   T = Fiscal year just ended
PAT = Profit after tax
NCC = Total non-cash charges
(depreciation & depletion)
INT = Total senior interest payments
PRIN = Repayment of principal on other than
subordinated debt
CAPEX = Capital expenditures on plant & equipment
CFWC = Change in free working capital
COUT = Cash outflow, from accumulated cash balances,
under "Hierarchy of Payments."

No cash distributions permitted unless CDCRB 1.30.

Cash distributions when permitted, are made according to the following Hierarchy of Payments:

1. To pay deferred interest on subordinated debt
2. To pay current interest on subordinated debt
3. –80% of remaining excess cash split pro rata between the principal payments on the subordinated debt and bank notes in the inverse order of maturity.
–20% of the remaining excess cash may be paid out as cash dividends.

SSL) that will, if the risks we perceive never materialized, be returned to you quickly and with an attractive return."

Once the level of equity had been established which gave the project some greater flexibility, a reexamination of the cash flows (in light of the various debt structures considered) revealed that improved viability (a lower breakeven price and improved debt coverage ratio) could be obtained by extending the final maturity of the debt by one year from seven years (which Rapid had requested) to eight years. This extension was acceptable to the bank only because of the increased equity in the project and further worked to lessen the impact of adverse operating scenarios on project cash flow.

Having thus negotiated with Rapid an advantageous mix of debt and equity, the remaining structuring task of the acquisition financing was to develop meaningful and realistic loan convenants and cash flow tests to monitor the performance of the project. The goal in deriving these covenants was twofold. First, it was important to allow the new operation to function with some freedom. Second, it was essential that the bank develop tests which were tight enough that the bank would learn early-on if any problems were developing at the mine, and then to be in a position to act to protect their interests if the situation continued to deteriorate.

CONCLUSION

This article has introduced the analysis, procedures, and evaluation performed by a bank in structuring the highly leveraged acquisition of a small eastern coal mine. Considerably more work was of course done than could be summarized here. The structuring stages outlined here (as well as our subsequent experiences with the company) reiterated the importance of good, constructive communications between the importance of good, constructive communications between small mine owners (as well as operators, if not the same person) and the bankers to the mine. Running a small mine with high debt levels can be a risky business, particularly at today's record-setting interest rtes, but the risk can be significantly reduced through the kind of analysis and structuring presented here. This methodology seeks to first understand the constraints and risks of a given project and to use this knowledge to develop a structure which blends the objectives of the owners and the banks with the realities of the mining operation.

2.5

FINANCING OF THE SMALL, INDEPENDENT MINING ENTERPRISE

GEORGES CARAGHIAUR

Engineering Geologist
Transalta Utilities Corporation
Calgary, Alberta

INTRODUCTION

A small, independent mining enterprise (SIME) is an independently owned and operated business concern engaged primarily in mineral development, from the exploration to the production stage. By definition, a SIME is not dominant in its field and has no long-term financial backing. It is not a major producer of minerals, nor is it a petroleum company, a government organization or a subsidiary of such organizations. A SIME may or may not be listed on stock exchanges.

The SIME is also called small miner, junior mining company, speculative mining company, and penny-stock mining company.

The survival and growth of a SIME is most often controlled by an independently minded entrepreneur, who not only conceives and carries out new projects, but is materially affected by the success or failure of these projects. Usually, a SIME does not have a large or constant cash flow. Such a lack of long-term financial backing forces the SIME to meet the challenge of finding capital for its activities on a step-by-step basis. Its survival strategy is thus based on finding funds at the right time, of the right type and in the right quantities.

The purpose of this paper is to show that the type of venture in which a SIME is involved affects the difficulty of raising new capital and the choice of possible sources of funds.

The Small, Independent Mining Enterprise (SIME)

The SIME is not a figment of the imagination of financial speculators. In Canada, there are more than 1,800 small mineral enterprises which spend on the average less than $1,000,000 per year on activity which is directed at bringing forth new mineral supply. In the United States, 64% of the non-fuel mining operations are conducted by nine or fewer men, and the coal industry is characterized by several thousand smaller independent producers, still predominantly family-owned or controlled. In Chile, 80% of the mines can be classified as small mines. In the Philippines, approximately 60% of metal mining operations are classified as small mines. The SIME is actively engaged in mining all over the world, and very much so in the Pacific Rim.

The Mineral Supply Process

The whole process of bringing forth new mineral supply involves a dynamic sequence of activities, ranging from the initial exploration for a mineral occurrence to the production of a marketable product. A typical sequence is outlined in Figure 1. At the grass-roots exploration stage, a series of geological, geophysical and geochemical tests is aimed at locating a mineral occurrence. Especially in the early phases of such activities, the uncertainty (risk) surrounding a venture is high, since only general geological information exists on which to base a judgement of the economic viability of the venture. Such risk is attractive to few investors.

As the information-gathering process continues, uncertainty diminishes. Once exploration efforts result in the discovery of a mineral occurrence, i.e. when indications of mineralization and mineable widths are obtained by drilling, there is still uncertainty concerning the extent of mineralization and the probability of having located an economic mineral deposit. The expected return on investment at such a stage may well be negative but the slight possibility of a very high positive return will make a number of financial sources available to an exploration company. At the development stage, much of the uncertainty

---

Presented at the Minerals Resource Management Committee Symposium, Mineral Resources of the Pacific Rim, at the First International SME-AIME Fall Meeting, September 4-9, 1982.

associated with exploration is absent, but another type of risk associated with the variability in return persists. As the available information increases and investor uncertainty decreases, more conservative investors become interested, hence increasing the reservoir from which funds can be sought.

```
        Regional appraisal
     Selection of a favourable area
                │
                ▼
         Reconnaissance
     Definition of anomalous areas
                │                      ┐
                ▼                      │
       Ground investigation            │
          of target areas              │  Grass-roots
     Definition of drilling targets    │  Exploration
                │                      │
                ▼                      │
     First-stage exploratory drilling  │
     Discovery of a mineral occurrence │
                │                      ┘
                ▼                      ┐
      Second-stage exploratory and     │
          delineation drilling         │
       Delineation of an ore body      │
                │                      │  On-site
                ▼                      │  Exploration
      Underground exploration and      │
             bulk sampling             │
       Definition of an economic       │
             mineral reserve           │
                │                      ┘
                ▼
       Mine development and
          construction of
        processing facilities
        Establishment of
        productive capacity
                │
                ▼
            Production
```

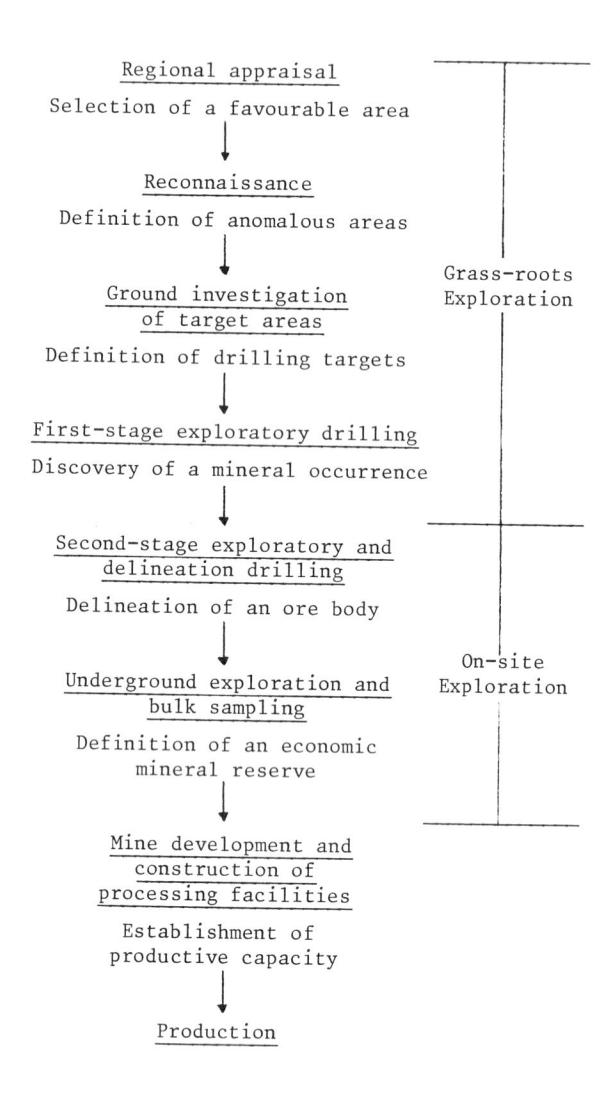

FIGURE I

Typical SIME Sequence in Establishing
New Mineral Supply

As one progresses from one stage of development to another in the mineral supply sequence, the amount of funds necessary increases. For example, the average cost of discovering a mineral occurrence in the Canadian Shield is $450,000 but the development of a small, base metal mine, with a capacity of 500 tons per day, requires an initial investment of at least $100,000,000.

The mineral supply process is thus characterized both by decreasing uncertainty and increasing costs as one progresses from one stage to another. Since the costs, risks and expected returns substantially differ at each stage, it is appropriate for the SIME to finance each stage differently.

FINANCING ALTERNATIVES

Many SIMEs, like other businesses, are started with the investment in an enterprise of capital from an individual or group of individuals. There usually comes a time, however, when a SIME cannot continue to survive on such limited sources of funds. They become insufficient for the SIME to reach a critical minimum scale of activity.

The sources of capital examined in this paper include only the major existing and potential private sources available to the SIME. They include: stock issues in the public market, investment partnerships and tax shelters, venture capital companies, major mining companies, banks and leasing companies. Examples are taken from the United States and Canada but can be generalized to other countries where similar regulatory, fiscal and tax environments exist. Government loan and assistance programs, which admittedly are of some importance in some countries, have been excluded from this paper.

Figure 2 summarises the situation. The columns show the stages in the mineral supply process at which a SIME is active with regard to a particular mineral property. The rows display possible sources of funds.

SIME's Activity in the Mineral Supply Process	Grass-roots exploration		On-site exploration			
Appropriate Source of Funds ↓	Ground investigation of target areas	First-stage exploratory drilling	Second-stage exploratory and delineation drilling	Underground exploration and bulk sampling	Mine development and construction of processing facilities	Production
Investment Partnership	x	x	x			
Venture Capital Companies			x	x		
Major Mining Companies		x	x	x	x	x
Public Markets			x	x	x	
Banks					x	x
Leasing Companies					x	x

FIGURE 2

Financing Alternatives for the SIME

The choice of a financing alternative by a SIME, however, does not depend solely on the nature of its activity or the needs and goals of the different investors. As suggested in Figure 2, two or more financing alternatives may be appropriate at any selected stage. In such cases, the final choice will depend on a SIME's goals and the relative effectiveness of a given source, as measured by the costs and difficulty involved in the process of raising capital from the different sources mentioned.

Grass Roots Exploration

A grass-roots exploration program usually begins with the selection of favourable areas of land. The next step embodies a series of geological, geophysical and geochemical reconnaissance surveys. Regional appraisal and the decision to pursue reconnaissance activities are based strictly on geological concepts. The capital required for such seed projects usually comes from the initial investment by the entrepreneurs in a SIME. Indeed, entrepreneurs who have faith in a project must be willing to risk their own capital until enough information is gathered to convince other potential investors of the merits of the project.

Detailed Ground Investigation. Reconnaissance work leads to the selection of anomalies which are favourable target areas for more detailed ground investigation. This stage of activity includes mineral rights acquisition and detailed geological, ground-geophysical and geochemical mapping. A SIME usually finances this stage on its own, but if enough information has been obtained through previous work, a SIME can possibly convince other investors of the project's merits.

In these early stages of a project, however, the geological uncertainty is very high. There is little information on which to judge the economic potential of a project. Because of this, external capital can be obtained only through negotiations with a small group of specialized and informed investors. This enables a SIME to make flexible financing arrangements tailored to the investor's financial and tax needs, therefore facilitating the task of raising funds.

INVESTMENT PARTNERSHIPS AND TAX SHELTERS: For a SIME, private placements can be effective in raising funds for high-risk projects. In the U.S. and Canada, for example, private placements are a relatively simple and rapid way to raise capital because they provide an exemption from the lengthy and costly registration process under federal, blue-sky and provincial securities laws. Investment partnerships between a SIME and a small group of high-income tax bracket investors can transfer the tax advantages of mineral exploration to the investors, thus providing the incentive - in the form of tax shelters - to compensate for the high risks.

A typical partnership arrangement for a SIME is the limited partnership which functions as follows: a SIME enters into an agreement under which it contributes properties and the limited partners contribute cash. The agreement provides that all intangible exploration and development costs should be borne by the participants contributing cash. Such expenses are attributed to the individual partners according to their pro-rata interest in the partnership. The individual partners can thus consolidate their pro-rata share of partnership expenses to reduce their own income and thus their taxes, rather than having to wait until the partnership's properties start providing income.

Under such an arrangement, the limited partners bear most of the risk of a venture. These risks, however, are offset by the tax shelter provided, as well as the potential for substantial returns if the program is successful. Investment partnerships can hence provide the SIME with a source of funds not otherwise available, and also leave the SIME with complete control over an exploration program. Indeed, limited partners, for example, have no voice in the management of a partnership - unlike shareholders in a corporation.

Exploration Drilling Stage. At the exploratory drilling stage, until a mineral occurrence has been discovered, a SIME can raise capital through investment partnerships for the reasons noted above.

MAJOR MINING CORPORATIONS: Alternatively, a SIME can raise capital from a large mining corporation at such a stage. Indeed, it is sometimes possible for a SIME to do a limited amount of exploration on a property and then submit a well-researched proposal to a major mining corporation (MMC) for financing. In this way, the large companies benefit from the efficient exploration techniques of the SIMEs and complement their in-house exploration programs. A joint-venture agreement with a major mining company has substantial advantages for the SIME. First, it can provide a SIME with a cash flow (following the negotiation of an option agreement and contracted work, for example). Second, MMCs have the financial means to bring a project to completion, thus eliminating for the SIME a major part of the problems involved in finding more capital for the subsequent exploration and development stages. Furthermore, a SIME can benefit from the technical know-how and operating experience of major mining corporations, which can prove to be important if the exploration leads to the discovery and exploitation of a mineral deposit.

A typical arrangement between a SIME and a MMC is as follows:

- A MMC options a property from a SIME for a given amount of time (usually one year with the right of first refusal in subsequent years). In exchange, the SIME receives a downpayment, usually expenses to date, plus a rate of return.

- The MMC guarantees money, usually anywhere from $50,000 to what the MMC thinks it is worth spending, to continue exploration of the prospect for the period of the option.

- At the end of the option period, depending on results, the MMC will renew its option and guarantee more money for exploration.

- The option can be renewed two to four times. At that time, the equity interest of the SIME is figured up (assuming that a decision to develop the property is made). The carried-free interest retained by the SIME is usually between 10-25%.

Because it is not the large mining companies' business to provide equity capital to small ones, the SIME must expect that a large company will usually demand management control of the project in which it invests. Therefore, a SIME whose owners want to maintain management control over a project should try to finance its activities through alternative methods.

VENTURE CAPITAL COMPANIES: These organizations do not usually finance a project before initial capital has been expended to develop an ongoing organization and some evidence of commercial feasibility is indicated. Because of this, they are not an appropriate source of capital for a SIME at the grass-roots exploration stages.

PUBLIC STOCK ISSUES: Stock issues, as a way of raising capital, are not always appropriate for a SIME before it reaches the late exploration or development stages of a project. First, regulations by government agencies seriously affect the feasibility, ease and costs of making public offerings of speculative mining stocks. Indeed, the prime concern of most securities regulations is to assure complete and accurate disclosure of all material facts relating to a proposed stock issuance, in order to allow investors a realistic basis for appraisal and an informed judgement in a purchase decision. This translates into the production of a registration statement which is relatively costly to produce, given the small amount of money usually raised by SIMEs (typically between $75,000 and $250,000). Often the review of the registration statement for early-stage exploration projects is long, fraught with delays, and can prevent a SIME from completing its financing efforts during favourable phases in the market.

Second, stringent access conditions to most stock exchanges have restrained their use by SIMEs. As securities markets have matured in North America, for example, their role has shifted more towards secondary trading of securities and the SIME has been effectively denied access to many stock exchanges. Indeed, stock-exchange listing requirements are now too stringent for many SIMEs to meet, especially those whose projects are not near the development stage. For example, listing requirements on the Pacific Stock Exchange, which includes the former San Francisco Stock Exchange (once an important trading center in mining stocks) include a requirement for outstanding shares with a minimum market value of $1,000,000. On the Toronto Stock Exchange (the most important market for mining stocks in Canada) the working capital requirement of $500,000 is generally hard to meet until a SIME has established the existence of an ore body.

The consequences of not being listed on a stock exchange are important for a SIME. It means that shares have to be sold in the Over-The-Counter (OTC) market, where a substantial selling effort is required, thereby raising the issuance costs for the SIME. In Ontario, for example, on average only 36% of the money paid by the public for shares issued in the OTC accrues to the benefit of the exploration company's treasury.

Finally, the high risk associated with mineral exploration makes it difficult for the SIME to find investment bankers willing to assist in the complicated process of issuing stocks, especially the small, speculative issues. A consequence is that "best-effort" offerings, instead of firm underwritings, will characterize speculative mining issues, i.e. a SIME will receive only the net proceeds of the securities which can be sold to the public.

On-site Exploration

Favourable discoveries provide justification for subsequent stages of on-site exploration. The purpose of these stages is to provide information for estimating the size, grade and other relevant characteristics of a discovery. Detailed exploratory drilling, followed by delineation drilling, can in some cases be sufficient to delineate an ore body. In more difficult cases, however, underground exploration (e.g. shaft sinking, drifts, cross-cuts, raises) is necessary to confirm the estimated grade and quantity of reserves and to obtain bulk samples, thus defining an economic mineral deposit with confidence.

INVESTMENT PARTNERSHIPS: Even when a mineral occurrence has been discovered, the probability that such an occurrence is part of an economic deposit is still low. Investment to be made in further exploration is still of a speculative nature. Investment partnerships are a good way to raise the risk capital needed for this second-phase exploratory and delineation drilling stage. This is especially true when the general partner (SIME) puts high priority on management control, because limited partners have no voice in the day-to-day management of a partnership. The usefulness of a limited partnership in the form of a private placement is limited, however, as a project progresses toward its final stages. In Canada and the U.S. for example, in order to be granted the registration exemption and other advantages provided by private placements, a SIME must comply with another set of rigid securities regulations. Most important is the fact that only a small number of investors can participate in a private placement. Therefore, when large amounts of capital - in excess of about $1,000,000 - are necessary, as in the underground exploration and development stages, investment partnerships are usually not a feasible source of funds for the SIME.

Recent experience, however, shows that public offerings of tax shelter programs may provide viable mineral financing plans in the future, as they have in the oil and gas industry for many years.

MAJOR MINING CORPORATIONS: MMCs usually prefer to enter in joint-venture agreements with SIMEs who have properties ready for the more expensive stages of the mineral supply sequence, such as on-site exploration. This not only lets them benefit from the efficient exploration techniques of the SIMEs but also puts them in a stronger bargaining situation to obtain controlling interest in the property, if the SIME is unable to tap other sources of development capital. The amount of equity a SIME will be able to retain will depend on:

- The amount of money already spent on the property.

- The attractiveness of the property.

- Degree of competition for the property by the large companies.

- Negotiating ability of the SIME.

VENTURE CAPITAL COMPANIES: At the on-site exploration stages, a SIME may be able to find a venture capital firm ready to inject capital in a mineral project. Financial arrangements with sophisticated investors, such as venture capital companies, can be costly for a SIME, given the return that such companies seek in exchange for their participation in a project. And even though venture capital companies do not usually get involved in the day-to-day operation of a project, they usually place strong restrictions with regard to the decisions which can affect the financial status of a project. These disadvantages must be viewed in the context of the availability of other sources of capital and should be weighed against the advantages of on-going managerial and financial assistance provided by venture capital companies.

PUBLIC STOCK ISSUES: When financial markets are in a buoyant phase and speculative capital is readily available, a public stock issue should be considered by a SIME. In the past, SIMEs in the U.S. and Canada have been able to raise funds successfully in the public markets for on-site exploration projects. Admittedly, raising funds from the public involves a longer and more complicated process than negotiations with small groups of investors. The legal, engineering, accounting and other issuance costs, as well as commission fees, can amount to a relatively high percentage of the money raised in public markets. Timing of stock issues is also important, since it may be impossible to raise speculative capital when the financial markets are in a bearish state. Nevertheless, public stock issues can provide the means to raise large amounts of capital without giving up controlling interest.

When the situation warrants its use, a public stock issue is a natural complement to alternative financing schemes used by a SIME in previous stages of activity. Indeed, partners in a limited partnership can easily convert their pro-rata interest in the partnership into the common stock of a SIME at a negotiated price per share. This arrangment can have significant tax advantages for the financially successful partnership, since rates of taxation for corporations are usually much lower than for individuals. For a venture capital company, helping a company make a public share issue is a natural way to increase the liquidity of its investment in the company and initiate the process of recouping the investment.

Development and Production

The preparation of a detailed feasibility study for the development of a mine follows the collection of data in the grass-roots and on-site exploration stages. The feasibility study is the basis on which the decision to go ahead with the project is made.

Raising equity capital from one or a small group of investors is not appropriate at the development and production stages. First, the capital required for these capital-intensive stages is likely to be in excess of what such sources would wish to expend. Most importantly, it is preferable to raise large sums of equity from a very large group of investors, e.g. through public stock issues. Indeed, such a procedure eliminates the reliance on one or a few large holders who could exercise an undesirable degree of control. At this stage, much of the geological uncertainty is removed overcoming much of the speculative aspect of a venture. Access to securities exchanges is easier, information is more readily available to investors, and the costs of public financing are relatively lower.

The exception to this rule is that a SIME may find it preferable to take advantage of the technical know-how, operating experience, and financial strength of a major mining company and thus proceed with a joint venture. Even in such cases, however, a SIME may want to make a public offering to pay for its share of development expenses.

Strict reliance on equity to finance the development and construction phases, however, could be a mistake from a financial management point of view. Indeed, the use of some kind of debt would provide financial leverage and hence increase the potential return to equity.

Although it may be impossible for a SIME to borrow capital on the strength of its balance sheet, non-recourse financing may be available at the development stage. Indeed, once ore reserves have been proven and the economic viability of a project shown in a detailed feasibility study, a SIME can, under certain conditions, get a non-recourse project financing loan. Of the financing conditions required for project financing the most critical is the necessity to offer a completion guarantee, i.e. a bridging guarantee which expires after a period of time sufficient to ensure the lender that the project operates as presented in the feasibility study.

Understandably, to be of any value to the lender, the completion guarantee must be provided by a company which is financially strong. Very often, however, it is the period in which an ore body has been thoroughly delineated and the project is brought to the point where the banks will consider getting involved that can be the most difficult for the SIME. Indeed, the SIME, generally speaking, will be out of funds at that time, having spent most of the previously raised capital on the earlier steps of resource development.

A lender may accept, under cetain conditions, a guarantee provided by a SIME, as long as the equity interest in the project is very high. When equity participation is low, however, a SIME must find a third-party guarantor who is financially strong. Such guarantors are motivated by economic needs such as assuring a source of supply. Guarantees are considered to be contingent debt and are shown as footnotes on the guarantor's balance sheet. On the other hand, guarantees may be outside existing loan covenants restricting debt and the guarantor's capital is preserved for other uses. Guarantees will thus be a popular support of a project financing, as long as the guarantor's balance sheet impact is less than for a loan.

The challenge facing the SIME, however, is to secure these guarantees without giving up too much equity in the project or at worst, control over the project. Without doubt, the requirements for a meaningful completion undertaking is the most formidable task for the SIME. Indeed, the novice SIME, with no proven financial and managerial track record, with no outstanding assets (besides its project) or balance sheet, and which has a firm determination of keeping control of the project, may very well find the task of securing financial backing impossible.

In some cases, another kind of project financing, in the form of leasing, may be preferable to bank loans. It can provide a lower cost of capital to the SIME and increase the amount of debt financing not otherwise available.

One characteristic of mining is the long lead time to a fully taxed position for any project. Leasing can overcome this limitation by transferring to the lessor - who is in a better position to use them - tax benefits related to the project. A major portion of the tax benefits to the lessor are then passed through to the lessee (SIME) in the form of a low lease rate. Sale and lease-back transactions of coal preparation plants in the U.S., where the tax benefits of depreciation and investment tax credits are claimed by leasing companies, are an example of the benefits of lease financing for the SIME.

Depending on the language and intent of covenants in existing loan and note agreements, a lease may provide a SIME with a method of financing when other types of financing are not permitted under such restrictions. In addition, only a small proportion of lending officers and analysts for lending institutions treat leases as the equivalent of debt for both balance sheet and coverage ratios. Therefore, there are advantages for a SIME to use leasing as an off-balance sheet method of financing, since it may permit the firm to obtain greater amounts of debt financing.

CONCLUSION

Because of different conditions prevailing at different stages in the mineral supply process, only certain sources of financing are appropriate for each stage.

In general, a SIME should finance its high-risk activities through negotiations with a small number of specialized and informed investors. This way easier and more flexible arrangements can be made, thus facilitating the task of raising funds. When the uncertainty surrounding a project is lower and large sums of capital are needed, the use of public stock issues is preferable, as it can make it possible for a SIME to maintain control over a project. When most risks have been removed, the use of debt favorably leverages a project and prevents further dilution of equity.

The above generalization holds true when only the risk factors of a project are considered. As stated above, however, the choice of a financing alternative depends also heavily on the goals and needs of a SIME, as well as the feasibility of raising capital from given sources at any time.

CHAPTER 2: CAPITAL STRUCTURE FOR A MINERALS COMPANY

BIBLIOGRAPHY

1. Business Week, August 13, 1984, "What's Behind the Boom in Medium-Term Notes", pps. 108-109.

2. The Bonanza Kings, Richard H. Peterson, University of Nebraska Press, Lincoln, Nebraska, 1973.

3. Capital and Operating Cost Estimating System Handbook. Straam Engineers, Incorporated. Arcadia, California. U.S. Department of Commerce, National Technical Information Service. July 1978.

4. Capital and Output Trends in Mining Industries, 1870 - 1948, Studies in Capital Formation and Financing, Occasional Paper 45; Israel Borenstein. National Bureau of Economic Research, Inc. 1954.

5. Census of Mineral Industries, Department of Commerce, 1982.

6. "Concern Over the Financial Structure." V. 19, No. 34. The Manufacturers Hanover Financial Digest. Manufacturers Hanover Trust Company. August 23, 1982.

7. Copper Mining and Management, Thomas R. Navin, University of Arizona Press, Tucson, Arizona. 1978.

8. "The Crisis In American Mining." Vital Speeches, Vol. 5, No. 8, Pgs. 242-245, Feb. 1, 1984.

9. Financial Management and Policy, Sixth Edition, James C. Van Horne. Prentice-Hall, Inc., Englewood Cliffs, New Jersey 1983.

10. Financial Times International Year Book, 1984, Edited by D. Russell. United Kingdom, London, Longman, 1984.

11. Financial World, January 10, 1984, pgs. 70-71.

12. "Financing the Mining Sector - The Bank's Role," David M. Sassoon, Finance & Development, Vol. 12, No. 3, Pgs. 21-23, 41. Sept. 1975.

13. Frontiers of Financial Management, 3rd ed; Serraino, Singhvi, Soldofsky. Southwestern Publishing Co., 1981.

14. "Is There Enough Money in Mining," Ferguson, Nicholas and Graham Haclin; Banker (UK), Vol. 126, No. 607, Pgs. 1011-1015, Sept. 1976.

15. Managerial Finance, Seventh Edition, J. Fred Weston, Eugene F. Brigham. The Dryden Press, Hinsdale, Illinois, 1981.

16. Minerals Yearbook, 1982; Volumes I, II and III. United States Department of the Interior, 1982.

17. Politics, Minerals and Survival, edited by Ralph W. Marsden. The University of Wisconsin Press. 1975.

18. "Up From the Pits: But Stockholders Are Still Teed Off at AMAX." Thomas Wrona, Barron's, Vol. 63, No. 22, Pgs. 16-23, May 30, 1983.

ADDITIONAL REFERENCES

Caraghiaur, G., 1981, "Alternatives for Financing the Small, Independent Mining Enterprise in the United States and Canada," unpublished M.S. Thesis, The Pennsylvania State University, University Park, PA, 90 pp.

Mayer, R.F., ed., 1980, The Future of Small Scale Mining, United Nations Institute for Training and Research (UNITAR), McGraw-Hill, New York, 501 pp.

Chapter 3

Financial Evaluation for the Minerals Industry

CHAPTER 3: FINANCIAL EVALUATION FOR THE MINERALS INDUSTRY

PREFACE

Financial evaluation is the third leg of the triumvirate of evaluation applied to the minerals or any other industry. The other pillars are technical and economic evaluation and any successful mineral project will be the result of a combination of favorable technical, economic and financial conditions. Stated most simply, financial evaluation focuses on the analysis of the cash flow generating capacity of the venture under alternative scenarios of leverage, timing of debt drawdowns and repayments, cost of financing and sources of funding. Whereas the parameters of technical evaluation are objectively determined, subject to the interpretation of the technical analyst, economic and especially financial parameters of project evaluation are more subjectively determined and there is considerable discussion about the "art" of financing and of financial evaluation.

Although mineral companies have always applied some analytical techniques to evaluate new projects, the techniques of financial evaluation utilized by lenders were traditionally limited to a review of the firm's financial statements and loans were advanced as general corporate obligations with little accompanying analysis conducted on the underlying strengths or weaknesses of the venture. With the advent of "non-recourse" or "project" financing in the 1970's (in which the project cash flows represent the only means of loan repayment) this situation changed drastically and financial institutions were forced to develop the techniques of financial evaluation in order to keep pace with the changing financial requirements of the minerals industry.

Project finance groups were established by lenders to the minerals industry in an attempt to focus the appropriate technical, economic and financial resources of the financial community and to establish a closer link between financial and mineral companies. The principal contribution made by these analysts has been the improved ability of the financial community to identify discreet risks associated with project implementation and to develop project financial structures which allocate the proceeds of the venture among the project participants in relation to the risks allocated among these parties.

The emphasis of financial analysis is also changing in response to the current economic climate of scarce capital resources and inflationary uncertainty. Whereas in the not too distant past the financial analyst's objective was to seek allocation of the project risks while striving to maximize the return to project sponsors, in today's environment the analyst often seeks less to maximize the return to the sponsors and focuses principally on reducing financial risk. As an indication of this, projects are evaluated today on the basis of their competitive cost structure within the industry and on their ability to withstand the lows of business cycles rather than on the expectation of future profits. Although profits are important, the general inability of experts within and outside of the minerals industry to predict future commodity rates has demonstrated that expectations of future profits are at best a guess, whereas capital and operating costs can be estimated with much higher degrees of confidence.

The papers in this chapter have been selected to present a sample of the current approaches to financial evaluation, focusing on the determination of the appropriate degree of leverage for a mineral project, sensitivity analysis and specific related areas of financial analysis. George Schenk's paper "Methods of Investment Analysis for the Minerals Industry" leads off the chapter with a discussion of the basic principles of financial analysis and the quantitative tools employed by the analyst, including internal rate of return and net present value analysis. Fernando Sotelino and Michael Gustafson, in "Tailoring the Financial Decision to Project Economics," apply many of these analytical tools in presenting the financial evaluation of a mine expansion. Their analysis focuses first on understanding the underlying strengths of the expansion on an unlevered basis and then moves to a discussion of the search for the most appropriate funding structure. Gary Castle's paper, "Measuring the Economic Viability of Resource Projects," describes the analytical process that the commercial banker employs in conducting a financial evaluation of a western U.S. surface coal project. Next, Ross Glanville utilizes techniques of financial analysis to develop a methodology for determining the "Optimal Production Rate for High Grade/Low-Tonnage Mines."

Sensitivity analysis, employed to determine the impact which changes in key technical, economic and financial parameters have on the financial and economic performance of a venture, is one of the principal techniques of financial evaluation. John Robison's paper "Sensitivity Analysis for Mining Projects" and Richard Boulay's paper on "Sensitivity of Mining Projects to Capital, Operating and Debt Cost Variations" present a basic introduction to this important area of analysis.

Whether to borrow funds for mineral projects and when to borrow are both important decisions to be faced by companies developing projects. William Lohden's paper "Considerations in Leveraged Studies for Mineral Ventures" addresses these issues from the standpoint of the mineral company.

Two specific areas of financial analysis are next discussed. Neil Cole's paper "The Impact of Inflation on Hurdle Rates for Project Selection" has particular relevance in today's environment of high inflation uncertainty in which the financial analyst must seek to understand and deal with the various impacts which domestic and global inflation and inflationary expectations will have on the development of mineral projects. John Caldon's paper, "Evaluation of the Lease or Buy Decision," presents an examination of the basic financial considerations involved in the choice between purchased and leased equipment, distinguishing between the two basic types of leases and presenting the tax and accounting issues pertaining to leases within various industrialized nations.

The final paper, "Post Installation Appraisal of Mineral Projects," by Richard Mills, presents an important area of financial analysis which is the benefit hindsight adds in reviewing projects which have already been placed into service. By carefully evaluating the analytical process which led to a project's development many valuable lessons can be learned and applied to the analysis of future projects.

The subject of financial evaluation for the minerals industry is extremely broad and it is not intended that the papers presented in this chapter comprehensively cover this rapidly changing and expanding field. The papers are intended to present a general understanding of the methodologies of financial analysis and their application to project evaluation. Many other examples of financial evaluation applied to the minerals industry are to be found elsewhere in this book, particularly among the Case Studies presented in Chapter 12.

Michael N. Cramer
Chapter Editor
Manufacturers Hanover Trust

METHODS OF INVESTMENT ANALYSIS FOR THE MINERALS INDUSTRIES

George H. K. Schenck

The Pennsylvania State University
Associate Professor
Department of Mineral Economics

Abstract. The investment analysis methods that are most generally accepted in the mineral industries in the mid-80's are presented along with their advantages and disadvantages. Also covered are current widely-used practices for introducing the uncertain nature of most of the important variables in an investment analysis, which is contrasted to the former dominance of deterministic analysis. Finally, the discussion turns to current practice for handling the two critical areas: cost of capital and inflation. The advice contained is reinforced with data from a survey of current practices in the petroleum industry.

INTRODUCTION

The goal of investment is to increase the present value of the investing organization over what that value would have been had the particular investment not been made. This goal provides the framework of what investment analysis should accomplish.

The decade of the 80's has seen broad acceptance in the mineral industries of discounting methods for investment analysis (net present value, internal rate of return, and present value ratio) and a decline in use of undiscounted methods (payback period, and return on investment or capitalization rate). Also the emerging concern regarding the unwarranted assumption that variables can be represented by their most likely value in a deterministic analysis along with easy access to computers has led to much wider use of analytical methods that reflect the uncertainty associated with most variables important in an investment analysis. Finally, arguments regarding the proper measure of cost of capital (for discounting) and the correct treatment of inflation point to acceptable methods that now are receiving wide utilization. The following disucssion of these matters is reinforced with data from a survey of the investment practices of the petroleum industry.

BACKGROUND[1]

Investment analysis of a mineral venture is fundamentally the same as analysis of any other commercial investment. It involves the application of several simple concepts: (1) estimation of the periodic cash flows expected to be generated by the mineral project, and (2) adjustment of those cash flows to a common basis by discounting them at some appropriate interest rate. These simple concepts must then be variously modified to reflect more closely the complex interrelationships that impact on value in the real world. Data and information drawn from other types of studies such as financial analysis is useful in investment-decision analysis, but if used incorrectly, such information can be misleading.

Investment analysis balances cash flow data against the investment required at a mineral property in order to establish a relative measure of the investment return for the particular project. In the process, quantitative information on financing and tax aspects become important because such data are required to forecast accurately the project's cash flow. It is at this

Table No. 1

TYPES OF ANALYSES USEFUL FOR VALUATION OF MINERAL VENTURES

Type of Analysis	Type of Project Decision Intended
1. Investment Analysis	Should the firm invest in a project?
2. Financing Analysis	What is the best way to finance a project?
3. Cash Budget Analysis	What is the timing of money requirements and cash throw-off?
4. Accounting and Tax Analysis	Is an accounting profit being made and how should the after-tax flow be maximized?
5. Financial Securities Analysis	How might the investment affect the value of the company's securities?
6. Managerial Economics	What is the optimal plant size; at what volume will the mine: (a) be most profitable, (b) shutdown or (c) breakeven?

point, when information and data are drawn from other types of studies, that the analyst must be clear as to exactly what data is needed for the investment analysis, and in what form. Analysts should be selective so as not to unintentionally use similar but inconsistent data developed in other types of business analyses.

There are at least five other types of studies done in business that resemble investment analysis and use much the same information. The type of information provided by each of the types of project studies is summarized in Table 1.

The investment analyst should be careful not directly to use data from one type of study in another type of study without careful consideration of the compatability of the data. The analyst will often find it necessary to adjust data when using it for a different type of study.

Although a firm making an investment decision must carefully prepare various analyses that account for the major factors affecting project revenues, there are ten essential considerations at the core of any mineral venture analysis. These are:

(1) GEOLOGIC EVALUATION--Both the mineral reserves of the project and the geologic environment are key considerations in developing a property. They impact significantly on costs and provide the major constraint on the length and rate of production.

(2) REVENUE ESTIMATION--Production rate, commodity price, and plant location are the principal determinants of this aggregate measure of the monetary benefits of a mineral project.

(3) CASH FLOW--Money required and generated by the venture should be evaluated in terms of the periodic after-tax cash flows, not income or profits.

(4) TAXES--Cash flow from a mineral project can be significantly increased by optimizing the interrelation between allowable deductions from taxable income such as (a) investment tax credit, (b) depreciation, (c) depletion, (d) exploration expense, and (e) interest expense.

(5) COST OF CAPITAL--This refers to the selection of the appropriate after-tax, weighted-average interest (discount) rate for new funds to the company. Such a discount rate should reflect the particular company's blended cost for the money it will employ in a venture.

(6) TIMING--Usually estimates for the time necessary to produce revenue from a project are over-optimistic, an error that can result in the overvaluation of projects.

(7) DISCOUNTING--Because cash flows from a project are received periodically and are deferred to occur through time (perhaps 20 or more years), they should be discounted at the firm's cost of capital to one common time basis--usually the present. Note: this is different than adjusting for inflation (consideration 8), though the two may be combined for ease of computation.

(8) INFLATION--It is necessary to explicitly provide for the differential effects of inflation on the various revenue and cost categories. Inflation is different from and reduces the discount rate from the current (nominal) dollar rate if the analysis is done in constant (real) dollars.

(9) INTERDEPENDENCE--A financial study should consider the effects of interdependence among revenues, costs, income taxes, financing arrangements, and between the planned project and other activities of the mineral producer.

(10) UNCERTAINTY--The analysis should specifically take into account that cash flows are predictions based on current estimates of such things as future commodity prices and future tax regulations. There is considerable risk that actual results will differ significantly from the forecasts.

A mineral firm makes a capital investment hoping to realize benefits (profits) over a reasonably long period of time. Mineral investment analysis is the multifacted activity of searching for and deciding upon the value of various investment proposals, and includes the making of financing, marketing, and production cost analyses to determine the profit potential of each investment proposal.

An example of the winnowing of mineral projects that occurs between initial discussion to final development of a mineral project is provided by recent data from a major mineral producer which shows that for 100 mineral development proposals submitted, they chose to look at ten. Of those ten only two were optioned and drilled and after two years development had begun on only one of these.

Interdependence and Mutual Exclusivity

In project analysis it is important to recognize if investment choices are economically independent or if they are not. Management will not always have free choice to select one investment project or another because projects can be dependent on one another or mutually exclusive.

For example, a coal company may be able to choose between a mine in Kentucky or one in West Virginia--independent investments A and B. However, if because mine A's coal is high sulfur, and it therefore requires an additional investment in a coal processing plant, then investment in mine A is dependent on investment in the coal processing plant. It also is possible for two investments to be mutually exclusive. For example, a particular 250 acre tract can't be used both for a quarry and a shopping mall, because these two uses are mutually exclusive.

Extending the example of Case B (the free-standing mine in West Virginia) a little further, it is possible to build there a unit-train loading terminal to increase the net revenue realized from the sale of the coal, though the project's coal is readily shipped by truck without the terminal. Despite the fact that the Case B mine doesn't require a terminal, this project's revenues will be increased if the terminal is built. Therefore, in the case of the West Virginia property, we are really considering two projects. Project "B," a stand-alone mine and project "C," a mine and rail terminal. In project C, investment in the terminal is economically dependent on investment in Project B, the mine.

If investment in the terminal increases the benefits from the overall project, then the second investment (the terminal) is considered to complement the first. Also in this case, the mine investment is a prerequisite of the second investment because without the mine the terminal would be virtually valueless.

When evaluating investments that are dependent (or complementary), the most effective way to compare the investments is to combine them into sets that are either independent or mutually exclusive. For example, in the case of the Kentucky and West Virginia coal mines, the proposals could be grouped as follows: Project A - Kentucky mine with processing plant, Project B - West Virginia stand-alone mine and Project C - West Virginia mine with rail terminal. Now, the management decision is to choose which among the independent projects A, B, and C it wishes to develop. Making choices among projects is discussed a few pages later.

Feasibility and Valuation--Both Analyses Based on Same Concepts

The concepts covered here have proved useful for accomplishing two similar yet different types of analyses--(1) mineral project feasibility studies and (2) mineral property valuation. Both types of these analyses are similar because the purpose is to establish the economic value to a business of a particular mineral project. The analyses differ in the problem each seeks to resolve. The first type deals with investment decisions and the second type with estimating a price for a mineral property.

Feasibility Studies. These specifically deal with presenting sufficient investment data regarding projects so that management can make a DECISION as to whether or not the firm should invest in that particular mineral project. The fundamental problem is to determine whether the financial return expected from the project justifies the investment required. Feasibility studies also are used to determine the best (optimal) combination of equipment for a project or to evaluate proposed new equipment acquisitions. The purpose of these studies is to SELECT PROJECTS THAT WILL ADD TO THE PRESENT VALUE OF THE FIRM after all project costs are covered.

Feasibility studies range from the very preliminary (roughed out on a few sheets of paper over several hours) to a final, definitive estimate that can cost several million dollars and usually is based on exhaustive drilling and testing. Such a hierarchy of studies is used so that unsatisfactory projects can be eliminated as early and cheaply as possible in the screening process. Pre-screening based on simple criteria should eliminate projects that don't fit the company philosophy regarding such key elements as growth rate, diversification, skill, and risk.

The value of a mineral-producing operation derives from three separate sources: (1) the qualities of the mineral deposit itself, (2) payment for use of the capital invested such as for exploration and for mining equipment, and (3) the quality of the technical and the entrepreneurial ability utilized, such as new extractive technology or especially skilled management.

The discussion to this point should have created an awareness that investment analysis of mineral resource projects is a multi-faceted activity. For example, the ten essential considerations listed as being at the core of any analysis indicates that specialized information may be required from geologists (geologic evaluation), market analysts (revenue estimation), mining engineers (mine planning and cost estimation), financial advisers (cost of capital, taxes and inflation), lawyers (land title, taxes and sales contracts), computer system modelers and programmers (handling data input and output), and land men (land purchase and property market value). A mineral engineer or a mineral economist is usually experienced in several of the above areas and will often manage the study and preparation of the valuation report or investment study.

FIVE COMMON DETERMINISTIC METHODS OF ANALYSIS

A business investment represents a commitment of financial resources usually for a medium to long term. Typically, investment analysis relates to accepting, rejecting or comparing alternative business ventures.

Mineral investment evaluation relies on estimating either the net present value of a project or the rate of return the project will yield on the total investment required. Both measures require a forecast of net cash flows and of capital outlays as the basis for accepting or rejecting an investment opportunity. The five common deterministic, no-risk models of investment worth are discussed here and risk adjusted methods of investment analysis are provided later.

Evaluation of the relative desirability or value of a mineral project or property requires calculation of some type of quantitative measure of the economic merits of the project. The measures used fall into two general categories: discounted and undiscounted. The distinction between the two is that discounting specifically recognizes the "time-value" cost of money invested in a project and undiscounted measures do not.

At one extreme, the investment decision may require only the choice between one business alternative and leaving the money deposited in a bank. The more common type of decision though is when a company considers that it has more alternative ventures in which to invest than funds to invest (i.e., it is capital constrained).[2] In this case, a particular project will be funded on the basis that it: (1) ranks among the company's highest profit projects, and (2) meets some minimum profitability criteria (hurdle rate). The first consideration requires the rank ordering of investment projects with the most profitable projects funded first, then going down the ranking with succeeding projects funded until the money runs out. The second, a cutoff or minimum hurdle-rate type criterion, requires that no project will be funded which doesn't improve the company's net return. The minimum required return r (r = percent return) is usually measured by the firm's aveage cost of obtaining new (additional) capital funds.

Table No. 2

FIVE COMMON MEASURES USED FOR INVESTMENT ANALYSIS

Measure	Rule for Accepting an Investment Venture
DISCOUNTED MEASURES	
1. Net present value* (NPV)	A positive NPV
2. Internal Rate of Return (IRR) (also known as DCF-ROI)	An IRR greater than company's cost of funds
3. Present Value Ratio (PVR)	Ratio of revenues to expenditures greater than 1
NOT DISCOUNTED	
4. Payback Period (PbP)	PbP no greater than company's maximum
5. Return on Investment (ROI or CR)**	ROI (CR) above company's minimum

Notes: *Known to appraisers as the Inwood Method. This general method includes two specialized formulations used for mining: Hoskold and Morkill.

**Capitalization rate (CR) is a synonymous term used in real estate appraisal.

This section presents the five common deterministic (no-risk) measures of project profitability that are used to evaluate investment projects. The basic operating rule for each of the five is provided in Table 2.

The rule for accepting a project in each case is based on a conclusion that the company will be better off financially by accepting the project than it would be if the project were rejected. Because the purpose for using each investment decision measure is the same, you would expect that, when comparing a set of various projects using each of the measures, the same decision will result. That is not the case. The five measures will, in fact, result in different decisions and rank orderings. The discounted measures provide better decisions than do the others.

It is important that any decision rule be used in a consistent manner. For example, when a company uses the NPV method, it should be consistent in the discount rate (r) that it uses in evaluations. For another company which uses payback period (PbP), it is important that it be consistent in its definition of the costs that are deducted from revenue (example: analyses by the Eastern Division should not include depreciation if analyses by the Western Division when calculating PbP excludes depreciation).

The most well-known method of analyzing a project's investment merit is based on a deterministic model, $P = \sum_{o}^{n} (R - E)$ which assigns a single value or point estimate to each of the critical parameters.

This no-risk model disregards the uncertainty associated with cash flows, and because of this weakness it should not be used alone for investment analysis, although it often is.

The value P of the above formula is calculated by adjusting the stream of periodic cash flows (R - E) to a uniform base by discounting at a suitable interest rate (r). The investor's return from an investment is the cumulative net cash flow left after subtracting expenses (outlays) from revenues (receipts). Revenue usually is a roughly continuous—though not necessarily a uniform—stream of cash derived from the sale of products during the life of a project. Expenditures are of two basically different types: (1) capital costs, and (2) operating costs.

Future Monetary Benefits are Measured by Discounted Cash Flows (CF)

A pervasive theme in investment analysis is that the value (or relative merit) of an investment is based on the future monetary benefits the investment can provide. The FIRST PRINCIPLE of evaluation is: monetary benefits are measured as net cash flows (CF) not as profit or income. Cash flows are identified as to their timing so that the individual cash flows can be adjusted by discounting to one specified time-datum base, usually the present (n = 0). For investment analysis of a project, net cash flow

(CF) in any period is calculated as: CF = Revenue - (Operating Expenses* + Taxes + Capital Investment). *Financing expenses of dividends, interest and repayment are EXCLUDED.

Cash Flow. Cash flows differ from net after-tax income. Remember that after-tax cash flow is calculated either: (a) by subtracting only CASH expenditures including taxes from net revenues on an Income Statement or (b) by restoring (adding back) to after-tax income the non-cash write-offs of: depreciation, depletion and amortization. In investment analysis any EXPENSES RELATED TO FINANCING that may have been deducted in an accounting or a cash budget analysis such as interest, dividends and loan repayments ARE ALSO ADDED BACK to the Income Statement.

Interest is a fundamental concept of economics often referred to as the time value of money. Future amounts discounted back to the present are called the Net Present Value, or NPV. Discounted Cash Flows are abbreviated as DCF.

The time value of money leads to the SECOND PRINCIPLE of evaluation: adjust all monetary amounts to a common point in time before combining them. The purpose of this adjustment is to assure that all monetary amounts in an economic evaluation are shown in equivalent terms. Discounting is the mathematical technique that reduces an amount due in the future by the interest which would otherwise accumulate on the initial investment during the period between the present (or a base year) and the future due date "n". Discounting is not an adjustment for inflation or for risk; theoretically both of these should be taken care of separately, but often are combined into one joint rate.

Discounting Equation:

$$P = \frac{F}{(1+r)^n} \text{ ; or } P = F(1+r)^{-n}$$

Discounting is much more frequently used in economic evaluation than is compound interest because with discounting a future revenue (F) is adjusted back to the time-datum base used for an investment evaluation, usually the present (P). Compound interest is less frequently used in evaluation because with it a present amount (P), such as the investment in a project, is adjusted forward to a future (F) time-datum base.

Timing. In actual practice, cash flows occur nearly continuously as: products are sold by a mineral producer, customers pay for the products, and the producing company pays its operating expenses. Timing of capital expenditures is much more discontinuous or "lumpy", occurring primarily at the beginning of a project. However, there may be later infusions of capital to replace worn-out equipment on a cycle such as every seven years for pit trucks at an open pit mine, every 15 years for a dragline, or every four years for reworking an oil well. Timing is usually specified by a subscript as in CF_4 that stands for cash flow in year 4.

There are several choices that can be made when specifying the timing of cash flows: (1) beginning (or end) or period, (2) middle of, and (3) continuous. Generally, for initial calculations the first is adopted, beginning (or end) of period, and the period is usually one year of project life.

Consistency is an important virtue in project evaluation and this is adequate reason to adopt a timing convention. Capital investments are considered to occur at the beginning of a period. Net cash flows (CF) are considered to occur at the end of a period, a convention that is viewed as financially conservative.

Deferment Period. Most mineral projects have a pre-production (deferment) period from the base-date used for investment analysis. Correctly estimating the length of this deferment period is important. For mineral projects, it would be unusual for the deferment period to be less than one year, and for large projects it can easily reach seven or more years.

Cash Flow Precautions

After-Tax Income. Cash flows are always after-tax cash flows unless otherwise noted. It is unusual for before-tax cash flows to be used.

Exclusions. Do not deduct from revenue: (a) non-cash write-offs such as depreciation and depletion or (b) financing outlays such as dividends, interest, or repayment of loan principal. If you have deducted such values (for example, to calculate the income tax), be sure to add them back before computing an investment analysis.

Conventional and Unconventional Cash Flows. The sequence of negative and positive cash flows can affect the usefulness of the various measures used in investment analysis. No problems are created with conventional investments in which a series of outlays is followed by a series of positive cash flows. An unconventional investment is an investment for which the regular pattern "outlay then cash flow" is not met. For example, a pattern first of receipts followed by an investment outlay, followed by more receipts, would be labeled unconventional.

Unconventional investments can create problems for the analyst trying to compare one project to another. For example, while there is only one value for the internal rate of return (IRR) of a conventional project there can be two or more IRR values for an unconventional investment.

DISCOUNT MEASURES FOR INVESTMENT ANALYSIS

There are three methods for using discounting for decision making in mineral industries' investment analysis but only one, net present value (NPV), that is useful for calculating a property's value by the Income Method. The other two discounting methods used for investment analysis are internal rate of return (IRR)[3] and present value ratio (PVR). Each of these uses net cash flow as a measure of the return in evaluating the profitability of an investment and each is discussed separately in the following pages.

A. Net Present Value (NPV)

Net present value analysis requires that a project's value be calculated using the INCOME method. In the NPV technique, all cash flows are brought to the common time-datum basis of the present, using an appropriate discount rate (r). The formula for discounting each individual cash flow (CF) is $P = CF(1+r)^{-n}$, and project NPV is the discounted sum of all project cash flows: $P_{proj} = \Sigma_0^n\ CF_1 + CF_2 \ldots CF_n$

Decision Rules with NPV:

1. Accept - Fund all projects with a positive NPV, reject projects with a negative NPV.

2. Rank Order - NPV is not useful for rank ordering the relative desirability of projects. It may give a rough ranking if projects require about the same investment.

3. Project Valuation - The value of the project is its NPV. If the cost of acquiring the mineral property was not included in the original outlay, NPV indicates the maximum amount that can be paid for the property and still provide the required yield, r, on the project.

Regarding the advantages of the NPV method and the disadvantages, these are grouped--for all five investment analysis criterions--at the end of this section, after the methodology of each method has been presented. The next step is to briefly discuss how the appropriate discount rate(r) is determined for NPV calculations.

Determination of the Investment Discount Rate (r). Theoretically, when an investment is considered riskless, the discount rate for investment analysis should be the highest, risk-free rate available to the company on the money markets at the time the analysis is made.[4]

However, practically speaking, mineral investments are not risk-free. Therefore, because companies wish consistency in their evaluation of investments, their planning personnel tend to establish a standard discount rate that they change only periodically. This investment discount rate should be based on the long-run weighted average cost of money to the particular firm. At the beginning of the 1980s the investment discount rate was set at 15 percent or above at many major U.S.-based petroleum and mining firms.

Example of an NPV Calculation: Find the NPV of a proposed investment in the new strip mine of Sommers Coal, given a discount rate (r) of 15% and the cash flows (CF) as shown:

	GIVEN		WORK SHEET: 15% discount factor	
n	description	CF	discount factor	discounted cash flow (DCF)
0	investment	(200,000)	1.0000	($200,000)
1	net cash flow	100,000	.8696	87,000
2	net cash flow	100,000	.7561	76,000
3	net cash flow	100,000	.6575	66,000
4	net cash flow	100,000	.5718	57,000
4	net shut-down cost	(50,000)	.5718	(29,000)
			NPV	$ 57,000

Specialized Formulations Used for Mining: Many years ago several special NPV-type formulations were developed for use in mining valuations. Most notable of those was the Hoskold method, which is infrequently used now except for a few special instances where its use is favored by legal requirements or tradition.

B. Internal Rate of Return (IRR)

The intent of the IRR procedure is to find the discounted yield (r) of a particular investment. It is this yield that the investor can use to measure one particular investment opportunity against the yield from other competing opportunities. The procedure to calculate IRR is to find the discount rate (r) that will make the present value of the revenues from the venture equal to the present value of the expenditures for it, which of course is the yield on the expenditure. This is the point where NPV = 0. Symbolically: Given: P, a series of cash flows: $CF_0 \to n$, and n; find r using the equation

$$P = \Sigma_0^n\ [CF_0 + CF_1(1+r)^{-1} + CF_2(1+r)^{-2} \ldots CF_n(1+r)^{-n}]$$

If there is only one investment expenditure and a uniform cash flow (CF), this equation can be solved. If there are additional expenditures or varying cash flows, then there are too many variables for a mathematical solution and the equation must be solved by trial and error. Solution is quite simple because all that needs to be done is to substitute different values of (r) until that r is found for which NPV of the venture equals zero. To solve for IRR, one set of discounted cash flows (DCF) is calculated and summed and the process repeated until the sum of one set approximately equals zero. The procedure is easily presented in an example, which is a continuation of the Sommers Coal case just used in the example NPV calculation.

Example of an IRR Calculation: Find the IRR of an investment in the new Sommers Coal strip mine given the following cash flows.

GIVEN			WORK SHEET	
n	description	CF @ 0%	DCF @ 15%	DCF @ 25%
0	investment	($200,000)	($200,000)	($200,000)
1	net cash flow	100,000	87,000	80,000
2	net cash flow	100,000	76,000	64,000
3	net cash flow	100,000	66,000	51,000
4	net cash flow	100,000	57,000	41,000
4	net shut-down cost	(50,000)	(29,000)	(21,000)
	total DCF	$150,000	$ 57,000	$ 15,000

To this point in the example, it should be clear that the IRR (yield on the expenditure) is greater than 25%. (Note that as a larger r is used, NPV is getting closer to 0). The question now arises should the next trial be 26%, 30%, or some other value. There are two ways to roughly estimate an IRR. One is a graphical solution, the second is a rule of thumb; both ways are illustrated below. Normally, a rough estimate would be calculated before starting the trial solutions for IRR.

Rule of Thumb Approximation of IRR. To approximate the IRR, divide 75% by the payback period (PbP).[5] For short-life investments use 65% as the numerator and for long-life investments use 85% as the numerator. Sommers Coal is clearly a short-life investment and the payback period is two years, in this case therefore approximate IRR = $\frac{65\%}{2}$ = 32.5%.

Graphical Approximation of IRR. Plotting the NPV amounts of a project for several values of r permits interpolation of that value of r for which NPV = 0 (the value of NPV which corresponds to the IRR). In the Sommers Coal example, using the two NPV values already calculated, the graph is drawn (Figure 1).

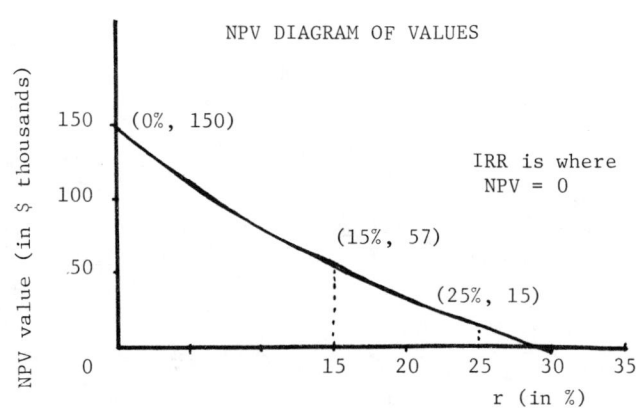

NPV DIAGRAM OF VALUES

FIGURE 1

Note in plotting known points that NPV is $150,000 at the discount rate r = 0,[6] and the NPV for 15% and 25% were calculated just above. These points are plotted and the IRR is estimated by the intersection with the ordinate. The plotted IRR is about 30%.

Returning to the example problem and trying r = 30%, NPV is computed to be just slightly negative at minus $1,000. This means that the point where NPV of the project equals zero is at discount rate (r) just slightly less than 30%. If you now tried 29% you would find that NPV was a positive $3,000. Therefore, by interpolation, the internal rate of return (IRR) for the coal property is about 29.75%, calculated from 29% + $\frac{3000}{3000 + 1000}$; you may round the answer to 30%.

Decision Rules with IRR:

1. Accept - Fund projects with an IRR at least equal to the firm's minimum, "target," discount rate (r), reject projects with IRR less than r.

2. Rank Order - Rank projects from the highest IRR to lowest.

3. Project Valuation: IRR is not helpful in direct valuation of a project.

C. Present Value Ratio (PVR)

PVR is sometimes called the Profitability Index (PI). PVR is a dimensionless number calculated by taking the discounted net cash flow revenues ($DCF_{rev.}$) for a venture and dividing them by the discounted net cash flow expenditures ($DCF_{exp.}$). PVR is utilized to eliminate some of the disadvantages of NPV and IRR but has disadvantages of its own as discussed in a few pages. Continuing with the example of Sommers Coal and using the calculations already presented:

Formula: PVR = $\frac{\text{NPV revenues}}{\text{NPV expenditures}}$

PVR at 15% r = $\frac{\$286,000}{229,000}$ = 1.25

PVR at 25% r = $\frac{\$236,000}{221,000}$ = 1.07

PVR at 0% r = $\frac{\$400,000}{250,000}$ = 1.6

Decision Rules with PVR:

1. Accept - Fund projects with PVR at least equal to 1, reject projects with PVR below 1.

2. Rank Order - Rank order projects from largest PVR to smallest.

3. Project Valuation - PVR is not helpful in direct valuation of projects.

SIMPLE UNDISCOUNTED MEASURES FOR
INVESTMENT ANALYSIS

Two widely accepted measures for investment analysis and decision-making are payback period (PbP), and return on investment (ROI) (or capitalization rate, CR). When investments are short-lived, occur at the beginning of the venture, and returns are uniform--or the returns can be considered a perpetuity--undiscounted measures may be adequate for choosing among alternative investments. Investments in the mineral industries usually do not meet such simple conditions.

The merit of the undiscounted measures is that they are simple to apply and to explain to others. However, the analyst must be careful because of the many different definitions in use for the variables used to calculate these measures. PbP and ROI are still widely used and PbP often will be reported as part of an investment analysis if only as a secondary indicator. CR is important in appraisal and real estate work.

A. Payback Period (or payout period)

Payback period is simple to calculate because it is defined as the number of years it takes for the net cash flow to return the original investment. Knowing the investment, the analyst merely checks the cumulative total of net cash flow to see at what point the cumulative total equals the investment. If the net cash flows are almost uniform, then it is only necessary to divide the investment by the average cash flow. Payback period begins when sales begin. The period before which revenues begin is generally ignored. Usually depreciation and depletion are also ignored (or if subtracted, are added back) because net cash flow (CF) is considered a more appropriate denominator than is net income (which, however, is used by a few analysts).

Example: Repeating the case of the Sommers Coal mine already evaluated by the discounted methods, the analyst would see:

	GIVEN		Cumulative
n	description	CF	payback
0	investment	($200,000)	(200,000)
1	net cash flow	100,000	(100,000)
2	net cash flow	100,000	0

Payback period (PbP) for the new coal mine is CORRECTLY FOUND BY INSPECTION OF THE CUMULATIVE DATA; it is two years, or PbP can be approximated by the formula:

$$PbP = \frac{investment}{average\ net\ CF} = \frac{\$200,000}{\$100,000/yr.} = 2\ yr.$$

The payback period will remain unchanged no matter how long the venture continues after payback and no matter what the cost of money. Some analysts who use PbP do discount the cash flows. For Sommers Coal the discounted PbP would be about 2.5 yr. at r = 15% and about 3 yr. at r = 25%, see second and third year "Worksheet" values used in the earlier "Example of an IRR calculation." The period it takes for the "Cumulative Payback" to turn positive is the correct PbP.

Decision Rules with PbP:

1. Accept - Fund projects with a PbP up to the firm's target PbP, reject projects with a higher PbP.

2. Rank Order - Rank projects from the shortest PbP to the longest.

3. Project Valuation - PbP is not useful in valuation of projects.

B. Return on Investment (ROI) (also known as return on assets and accounting rate of return)

This investment measure is a simple arithmetic ratio (given in percent) calculated by dividing a venture's annual, after-tax, net income (return) by the cost of the original investment employed to achieve that income. This measure is frequently called accounting rate of return (ARR) and average rate of return (also ARR). The definitions of ROI and ARR are so many and so varied that they no longer have a single widely accepted meaning.[7] The term capitalization rate (CR) commonly used in real estate appraisals is equivalent to ROI.

A variation of ROI is to use average investment (original investment ÷ 2) rather than the total original investment. The net income used is after-tax and therefore includes the deductions for non-cash write-offs such as depreciation and depletion. Be careful, as some ROI calculations are before-tax, which, if ROI is so defined, should be specially noted. Time is disregarded in calculating ROI.

Formula: $$ROI = \frac{average\ net\ income}{investment} \times 100\%$$

$$average\ ROI = \frac{average\ net\ income}{average\ investment} \times 100\%$$

Example: continuing with the Sommers Coal mine example and using straight line depreciation of assets ($200,000 ÷ 4 yr.), return is calculated as follows:

			CALCULATION FOR ROI	
	GIVEN		WORK SHEET	
n	description	CF	depreciation	net income*
0	investment	(200,000)		
1	net cash flow	100,000	(50,000)	50,000
2	net cash flow	100,000	(50,000)	50,000
3	net cash flow	100,000	(50,000)	50,000
4	net cash flow	100,000	(50,000) 000	
4	reclaim + salv.	(50,000)	. . . . . .	

*This last column is after-tax net income. To calculate net income (return) after tax when given CF, only non-cash write-offs, in this case depreciation, need be subtracted from CF.

Calculation of ROI

$$\text{Average net income} = \frac{\$50,000 \times 4 \text{ yr.} - \$50,000}{4 \text{ yr. life}}$$
$$= \$37,500$$

Then:
$$\text{ROI} = \frac{\$37,500}{\$200,000} \times 100\% = 19\%$$

Alternate: disregarding net shutdown costs of reclamation and salvage
average net income = $50,000
$$\text{ROI} = \frac{\$50,000}{\$200,000} \times 100\% = 25\%$$

CAUTION: Because of the numerous ways to calculate ROI, be sure to be consistent among projects in the formula used for ROI.

Decision Rules with ROI:

1. Accept - Fund projects with ROI at least equal to firm's target ROI, reject projects with ROI below target.

2. Rank Order - Rank projects from highest ROI to lowest.

3. Project Valuation - ROI is not helpful in direct valuation of projects.

EVALUATING PROJECTS THAT AREN'T EASILY MATCHED

Mutually exclusive projects are the result of limited resources--usually limited availability of capital funds to invest but perhaps limited mineral reserves (for example, too small a deposit to support both a cement plant and a construction aggregates plant or too small an oil refinery to produce both adequate ethylene and kerosene feedstocks for a petrochemical plant).

Investments with Different Lives

With mutually exclusive investments, there is an added problem when selecting one alternative (A) over another (B) because the investor can only select one. With independent or complementary investment projects, the life of the investment doesn't create a problem because we can accept one or more projects, whereas mutual exclusivity means we can accept only one.

In some cases of mutual exclusivity, the NPV method gives a clear choice. For example, as between selling a mineral prospect for $150,000 today or developing it ourselves and reaping a NPV of $135,000 at 12% over its 15-year life, it is clear that sale of the prospect is the preferred alternative.[8]

Where there are mutually exclusive investments with different lives, then the method is to assume that at the expiration of the life of the shorter-lived investment "B," the funds will be invested at the company's cost of capital, r, for "X," the length of time necessary to equal the life of the longer-lived investment "A," so that B + X = A. For example, consider a quarry (investment A) that can be operated as a road stone producer for nine years at an NPV of $350,000 or as a joint-product road stone and chemical limestone property for five years (investment B) at an NPV of $300,000. To compare A and B we need only add to the $300,000 NPV of investment B an additional period from the end of the fifth year to nine years at the firm's cost of capital. This now makes the life of B also nine years (5 years plus 4 years) and equal to the nine-year life of A. But the nine year NPV of B remains $300,000 as compared to the nine-year NPV of A of $350,000. The choice goes to alternative A, the road stone quarry.[9]

Incremental Analysis

Another method for comparing mutually exclusive projects is to evaluate the incremental benefit of selecting one project over another. Take for example the case shown next in which Project A is the Hidden Valley strip mine and Project B is the Hidden Valley tennis club. These are mutually exclusive uses for the land. The question is which project is it preferable to fund.

Hidden Valley Property

Project	Investment Required, $	Project Life, Yr.	Annual Cash Flow, $	NPV @ 9%	IRR	PVR
A	1,000,000	5	280,000	89,100	12.5%	1.09
B	1,500,000	5	414,000	113,000	11.8%	1.07
B-A	500,000	5	134,000	21,000	10.8%	1.04

Using the NPV criterion, Project B seems more profitable, whereas using the IRR or the PVR criteria Project A seems more profitable. By comparing the incremental return, B-A, on the added $500,000 investment it is possible for management to decide which of the two mutually exclusive projects they prefer. The decision is now highlighted as to whether the return on the additional $500,000 investment required for Project B (the tennis club) is enough to compensate for the added risks in establishing a tennis club versus operating the strip mine. To some the 10% clearly wouldn't be enough to cover the risk associated with a tennis club.

INVESTMENT MEASURES COMPARED

In general, there are two major types of financial measurements; those that recognize the time value of money and those that do not. Other differences include use of cash flow versus net income. The measurements that recognize the time value of money use discounted cash flows (DCF) rather than net income.

In discounting (or compounding) there is an explicit recognition of the divergent time patterns of the investment expenditures (cash drain) and the positive net cash flows in such a way as to reflect the time value differences of near-term versus distant cash flows. For investment purposes, it is cash flow (CF) rather than net income that is more important because it is lack of cash that can severely limit a firm's ability to meet its financial obligations or to invest in additional projects.

For property valuations, NPV (the Income Method) is clearly the appropriate technique to use. For making investment decisions, multiple methods are usually used in practice, the intent being to offset the disadvantages of one method with the advantages of a second method. The RELATIVE SIZE CRITERION and the INDUSTRY COST CRITERION, both of which are discussed one page later, should not be forgotten as useful additional screening devices.

Payback

PbP uses cash flows, but ignores cash flows subsequent to the payback period. A project that pays back in five years and stops is rated just as highly as one that pays back in the same five years and then continues for 20 more years. PbP also ignores the timing of all cash flows within the payback period. It is used mainly as a screening device, the shorter the payback period, the more desirable the project. PbP can be useful for appraising high risk projects where the economic life is difficult to predict. Because it is measured in time units, payback cannot be correlated with the discount rate (r).

Return on Investment and Capitalization Rate

ROI and CR use net income rather than cash flow and do not recognize the timing of either the investment or the earnings. Because these measures ignore the length of time earnings will be received, ROI and CR do not distinguish an investment that will yield, for example, 22% per year for only five years from another that will yield 22% per year for 20 years.

There is no definite correlation between projects accepted using ROI (or CR) and those selected by NPV with the investment rate (r) or by the internal rate of return (IRR). The predicted annual earnings used for ROI (CR) are important in situations where a venture is relatively large compared to the overall corporation because the market price of a corporation's stock may be affected by changes in the firm's total annual earnings or earnings per share due from the project. In such a case ROI (or CR) can be a useful measure for financial securities analysis.

Net Present Value

NPV uses cash flows and explicitly recognizes the time value of money with the discount rate (r). It is particularly useful in appraising the market value of a property or other asset. If the NPV of a project is positive, the rate of return will be greater than the discount rate. With the same set of projects, relative value may be changed by changing the rate (r) used to evaluate that set of projects.

NPV is a poor tool for ranking projects, but ranking is not necessary if capital rationing is not considered to be a problem. For choosing among projects requiring different absolute investment outlays (a $1 million project versus a $50 million project) NPV is not useful.

Hoskold. This variant of NPV was once very widely used in the mining industry. Its popularity has declined because a basic premise, the sinking fund concept, is no longer consistent with modern corporate financing practices. The Hoskold approach is giving way to net present value (NPV) except where Hoskold method is institutionalized as in the tax statutes of some states. In addition, it has the disadvantages of NPV.

Internal Rate of Return

The IRR method recognizes time value of money and uses cash flows rather than earnings. The mathematical formulation of the method means that cash flows generated by the project are treated as though they were reinvested at the IRR rate rather than at the firm's cost of capital (r). Because of this undesirable mathematical distortion, the high yield of relatively good projects with IRRs significantly better than the firm's average investment return (r) is unintentionally increased, and because of the same mathematical problem, the yield of poor projects is reduced further.

Reinvestment of project cash flows (CF) at the internal rate, which is inherent in the IRR formula, is a major flaw in the method and inconsistent with the assumption of investment analysis that cash flows from a project are reinvested at the firm's cost of capital (r). Because IRR does not use the cost of capital (r), ranking of ventures by IRR will be independent of the firm's discount rate. With unconventional investments IRR has the additional disadvantage that there can be multiple solutions.

IRR is related to both NPV and PVR. When the firm's discount rate equals the internal rate of return, NPV will equal zero and PVR will equal one.

Present Value Ratio

PVR uses the same concept underlying NPV in comparing projects requiring different investments. Because it is a ratio, PVR does allow rank ordering of investments, with the highest PVR considered the best. However, it has the same drawback as NPV when comparing alternatives having different lives. Also, because PVR is a ratio, it cannot be compared with the cost of capital (r), except that a PVR greater than one yields a return better than the cost of capital while a PVR less than one yields a return worse than the cost of capital. The NPV and PVR measurements are related; when NPV is positive, PVR is greater than one, when NPV is negative, PVR is less than one.

TWO ADDITIONAL INVESTMENT TESTS

Recently a few larger mineral producers have adopted one of the two following criterions as an additional test that proposed investments should be able to pass:

1. A project's average total cost should be in the lower-cost half of the industry cost curve.

2. A project should increase the firm's overall average net income by about five percent or more.

Low Cost Criterion

Some producers are at the low cost end of their industry's average cost curve because of favorable ore grade, geological, locational, or processing cost advantages. The rising industry cost curve means that some proportion (perhaps 30 percent) of mines and plants in a mineral industry will be at the low cost end and perhaps an equal proportion will be marginal--having costs close to the product's selling price--with the balance somewhere between.

During economic downturns when commodity prices tend to fall as demand shifts to lower levels, the high cost producers will find that their average costs exceed price. They are losing money and if the bad times last too long, they will fail. Low cost producers probably will continue to return a relatively acceptable cash flow in bad times and face little risk of financial failure because it is unlikely that demand will drop to a level where their revenue doesn't cover costs.

The test is that new ventures in order to be of interest to several well-known major American mineral producers must have average costs that are expected to be in the low-cost half of the industry cost curve. These major firms have decided not to sink millions of dollars in a relatively high cost producer even if the venture does have a positive present value (NPV) or its rate of return is above the firm's target rate of return (r). For some large companies there still appear to be more mineral projects that pass the "low-cost producer" test than these companies can finance. Example: A major oil company is planning to invest in the copper industry and has carried out extensive studies to estimate the copper industry's worldwide cost curve. It has found that the lowest-cost 50% of the copper industry has a cost of 90¢ per pound (or less) of blister copper. This firm will not invest in a copper project with a cost above 90¢ per pound.

Relative Size Criterion

By examining past decisions, the planning staff at one major mineral producer estimated that management was implicitly applying the following test on proposed new ventures: reject ventures that are not expected to increase overall average earnings by about 5% or more. A $20,000 new venture can be too small to be of interest to a major firm even if it meets all the company's criteria other than size. The reason is that a venture must have enough of an impact on corporate earnings to justify the management time required to launch the venture and get it running. It is a matter of opportunity cost. Management has only so much time and time spent on a relatively small project cannot be spent on a large project, which would have a greater impact on earnings.

If the venture is one to be managed by the corporate level, the earnings test of five percent relates to corporate earnings. If the project is to be managed at a lower divisional level, the five percent earnings test relates to the earnings of that divisional group. Example: The Urex Corporation has an opportunity to develop the Blue Gulch property. The investment analysis shows that the property will yield $2 million per year representing a 30% return on investment which is 10% greater than the firm's IRR "hurdle" rate. The company's total annual earnings of the division are $75 million. The property is rejected as too small under the relative size criterion ($2,000,000 < $75,000,000 x 5%).

DISCOUNT RATE MEASURED AS FIRM'S COST OF CAPITAL

The average cost of money used by a business is referred to as its cost of capital. It is this cost that is the appropriate measure of the discount rate that should be used in investment analysis. The cost of capital is of crucial importance for calculating the acceptability of investments and the value of mineral properties. It is given as a percent (example: r = 16%) rather than as a dollar amount, the usual measure for other costs. There are five general categories of financial instruments specifically used by mineral producers to acquire capital: debt, equity, internal sources, leases, and participation (sharing by others) in a specific project.

Future cost of capital, not historical cost, provides the discount rate that is relevant for new projects. The market price expected to be realized when the new financial instruments are sold should be used when calculating the return on capital expected by investors, not book value of existing securities.

There are two techniques used for calculating the cost of capital. Both involve a weighted average but one is based on straightforward calculation. The second, the capital asset pricing model, takes into consideration the manner in which the market value of the financial securities of the company vary over time and against a measure of a much broader financial market.

A firm should use a cost of capital that is normalized (averaged) over time rather than a spot average. The normalized cost will dampen short-term variations due to cyclical elements in the general economy and financial markets. The underwriting expense for raising new capital should be included specifically in the cash flow calculations rather than by reducing the investment by the underwriting expense.

Weighted Average Cost of Capital in Current Dollars

PROCEDURE: The cost of capital (r) in current dollars is correctly calculated as the <u>normalized, average cost of new capital funds</u> to be acquired for use in the business <u>weighted for the proportion that each component is expected to provide</u>. The separate components are valued and the nominal cost is calculated for their <u>expected future values</u> when the funds are to be acquired. The computation of the weighted cost of capital will appear as shown in Table 4. Mathematically

$$r = a(r_1) + b(r_2) + \ldots + i(r_n)$$

where average r is the weighted average cost of capital, r_1 is the cost of the first type of capital; r_2 the cost of the second type and so on and a, b et cetera represent the weighting appropriate for each of the types of capital, see example of Sommers Coal in Table 4.

Table No. 4

Weighted Cost of New Capital for Sommers Coal in Current Dollars

Source	Amount of New Capital	(%)	Nominal* Cost (r_n)	After-tax Cost**	Contribution to Weight
Bank loans, unsecured	$ 300,000	15%	16%(r_1)	9.6%	1.4
Mortgage bond	400,000	20%	14%(r_2)	8.4%	1.7
Equipment notes and leases	300,000	15%	16%(r_3)	9.6%	1.4
Preferred stock, new	80,000	4%	17%(r_4)	17%	2.9
Common stock, new	100,000	5%	22%(r_5)	22%	1.1
Internal funds	820,000	41%	19%(r_6)	19%	7.7
	$2,000,000	100%	Weighted average cost, r =		16.2% after tax

*The nominal rate is the <u>current</u> market rate.

**Effective income tax rate assumed to be 40%.

The weighted average cost of capital for Sommers Coal is much less than the cost of new stock. Why? Because of the deductibility of the interest (r_1) on debt and on leases when the firm's income from the income tax is calculated.

<u>Cost of Common Stock Funds</u>. Calculating the cost of capital raised from selling common stock is most controversial and it still is an unsettled issue among financial theorists. The key to calculating the cost of capital from selling new common stock is that buyers of common stock expect that their return from the stock will increase through time from both dividends <u>and from an increase in the market value of the stock</u>. For example, a buyer of stock in the Sommers Coal Company at a market price of $10.00 with a $1.00 per share dividend might be expecting that after a year the market price may increase to $11 per share. In such a case his annual return, r_5, would be:

$$r_5 = \frac{\overbrace{\$1}^{\text{dividend}} + \overbrace{(\$11 - \$10)}^{\text{increase value}}}{\$10} \times 100\%$$

$$= 20\%$$

This example is intended to show that the cost of capital from selling new common stock is made up of a dividend <u>and an expectation of a capital gain of $1 for the stockholder</u> in Sommers Coal Company.

Many small mineral producers do not pay a dividend but can still sell stock. In this case the investor expects a gain due to an increase in value of the company perhaps such as a sellout to a larger firm in a few years as earnings increase. For example, if Sommers Coal didn't pay a dividend but the investor expected earnings to go up enough to support a 20% increase in the value of the stock in one year from a $10 to a $12 market value then:

$$r_5 = \frac{\$12 - \$10}{\$10} \times 100\%$$

$$= 20\%$$

<u>Cost of Internal Funds</u>. Internally generated funds such as retained earnings, depreciation and depletion are not "free" of cost to the company. They have an opportunity cost. If they were not available for use in new ventures the firm would have to provide equivalent funds using one of the other sources of funds such as debt. Financial theory suggests that the market return on a company's own stock be used as the measure of the cost of internally generated funds. The logic is quite simple; the company can always buy its own stock with its internally generated funds. Because no underwriting costs are involved for internal funds this cost is less than the cost of funds to be derived from the sale of new common stock. A ten to fifteen percent discount from the cost of capital for common stock provides a rough "rule of thumb" for calculating the cost of internal funds.

Example: The cost of capital from common stock has been estimated to be 22% and entails a 15% discount to the underwriter (stockbroker), find the cost (r_6) of internal funds:

$$r_6 = (22\%)(1 - .15) = 19\%$$

The Capital Asset Pricing Model Alternative

The capital asset pricing model (CAP model) is a sophisticated technique preferred by financial analysts for calculating the relative value of financial securities. The CAP model can be used to estimate the required return (r) on each class of securities used to finance a new project if the project's relative riskiness is first estimated. As before, "return" includes both: (1) the cash payment (dividend, interest) and (2) the expected gain (increase) in the security's price.

To calculate cost of capital using the CAP model, first the required return (r) for each source of capital is calculated using the CAP model and then a weighted average is calculated for the mix of capital sources in the same weighted-cost manner as was just discussed.

The CAP model is preferred by financial analysts because it explicitly accounts for the riskiness of earning a return from a particular security by separately assessing:

1. the risk of the general securities' market
2. the risk of the particular security as distinct from the general market but in comparison to the general market (this measure is called the beta coefficient, (β)

The operational rule of the CAP model is that the higher the beta coefficient, the higher will be the return required for that particular security. A beta of unity ($\beta = 1$) means the security tracks the general market risk. A beta higher than one means the security is more risky. The beta coefficients for many firms are available from several sources including Merrill Lynch, Pierce, Fenner and Smith, stockbrokers, and Value Line, a financial periodical. The beta coefficients for some mineral companies in Fall 1981 were:

AMAX - .95; North American Coal - 1.2; Lone Star Cement - 1.10
ALCOA - 1.15; Exxon - 0.90; and Newmont - .95.

The formula for using the CAP model to calculate the cost of capital, r, for a financial security is:

$r = r_s + (r_m - r_s)\beta$, where r_s =

safe rate (treasury bills), r_m = general market rate (i.e. $\beta = 1.0$) and the Beta coefficient β for the selected security.

Example: Find the cost of common stock capital for ALCOA given:

$\beta = 1.15$, $r_s = 12\%$ and $r_m = 16\%$.

Calculation: $r = 12\% + (16\% - 12\%) 1.15$
$r = 17\%$

The cost of other capital for ALCOA would be calculated in the same manner, and finally the weighted average cost of all capital would be calculated as shown in Table 4.

INFLATION

Inflation affects the cost of capital, r. Inflation should not be ignored in investment analysis because inflation can seriously distort the returns acutally received from a project as compared to the returns anticipated in project studies.

The policies suggested here recognize two distinctly different levels of investment analysis: (I) Quick, order-of-magnitude (or preliminary) analyses that are intended as an initial screening, and (II) Comprehensive, definitive (or final) and budget (or intermediate) analyses which are done prior to the commitment of significant expenditures. Hyper or extreme inflation where the value of local currency may change from 50% to 100% or more in a year is another problem that is handled separately.

Procedures to Follow

I. Quick estimates: (a) Dollars: use constant,[10] present-year dollars for estimating all money amounts whether for present year outlays or cash flows 20 or 30 years in the future. Do not adjust any money amounts such as revenues or capital costs for inflation. (b) Cost of capital: use the constant-$ cost of capital (about 7%, see below) which is lower than the current-$ cost of capital.

II. Comprehensive estimates: (a) Dollars: use current,[11] inflation-bloated dollars at their future value for estimating all money amounts expected actually to be received (or disbursed) in each period. (b) Cost of capital: use current-$, weighted-average, cost of capital to discount cash flows. The current-$ cost of capital is higher than the constant-$ cost of capital used for "quick" estimates. It is the current market rate.

REMEMBER: Quick estimate--use constant ("real") dollars
Comprehensive estimate--use current ("nominal") dollars

Justification. The comprehensive recognition of the impact of inflation on an investment project will be reflected better by using Procedure II (current, inflated dollars and discounting at the current-$ cost of capital), but a significant effort in time and estimating is required. Such effort is not warranted at the early, initial-effort stages of an analysis when only a "quick" evaluation is needed. Note that other data used at this early stage are also only rough estimates and, in addition, because it is common for most projects to be rejected at the initial stage, it is not reasonable to increase the amount of work to the level required for a comprehensive inflation-adjusted analysis.

IMPORTANT: No project rejected by the quick Procedure I test would be accepted by the comprehensive Procedure II current-$ test provided the relative cost of capital and the inflation rate for each test are roughly correct.

It is useful to understand the reasons for using current dollars for inflation-adjusted analysis. It is incorrect to conduct an investment analysis on the basis that the impact of inflation is neutral. The reason is: inflation does not bear with the same force on each of the numerous cost and revenue items that are considered in a comprehensive economic evaluation of a mineral investment, especially taxes. Tax regulations do not provide for an inflated value to be used for depreciation of capital items. The effect is that when products are sold an income tax is paid both on actual income and on the "mirage" income produced by inflation. Internal revenue service rules regarding inventory accounting practices add to this problem.

Constant-$ estimates are QUICK because the analyst uses all receipts and outlays in terms of today's prices of these items. Today's dollars are the constant-$'s used for this type of analysis. The resulting constant-$ cash flows must be discounted with a constant-$ discount rate. The constant-$ discount rate is about 7% which is below the current-market interest rate by the inflation rate. It is calculated for a firm as its weighted- average, current cost of capital minus the expected inflation rate. For the mineral production sector of our economy, the current-$ (1984), weighted-average, after-tax cost of capital is about 14%, and the expected inflation rate is about 7%. In strongly inflationary times, the constant-$ rate will be well below the current-$ rate. While most analysts are ucomfortable with an evaluation that uses a discount rate below the current-$ rate, the only correct alternative is the cumbersome, comprehensive (Procedure II) method.

INCLUDING UNCERTAINTY IN MINERAL INVESTMENT ANALYSIS

Any investment is subject to uncertainty. Mineral investments are open to special hazards that often loom larger for them than for other types of investment. The major hazards for mineral investment are: geological, production, marketing, and political. Each of these can effect costs, revenues and timing and therefore will have an impact on each of the measures of investment worth. Probability and statistics provide useful concepts and techniques for dealing with uncertainty. While some authors differentiate between risk and uncertainty, we will not stress the distinction.[12]

There are two basically different concepts for recognizing uncertainty in investment analysis. The first and traditional way is INDIRECT. There are three devices used: (1) risk-adjusted discount rate, (2) shorter payback period and (3) sensitivity analysis. The percentile-change version of sensitivity analysis is a hybrid that incorporates uncertainty but one variable at a time. There are two methods that DIRECTLY incorporate uncertainty into the investment analysis: (1) expected value (as decision trees) and (2) simulation. Here information on direct methods focuses on pointing out differences with the indirect methods. Indirect methods are discussed because of their similarities to the material already presented. The direct methods are the subject of a later chapter.

Relative Impact of Variances

The numerous variables involved in an investment analysis or property valuation are going to differ from their most likely (expected) value used in deterministic methods. A central question of measuring uncertainty is what (variance) error does each variable exhibit and what is the relative impact of the variance. Variance is a probability measure for departure of a variable from its expected value and impact is measured in monetary (dollar) units (adjusted by discounting). Because of discounting, the timing of variances is important.

Variance of CAPITAL expenditures can have a significant impact on the worth of a project because such expenditures are relatively large and occur early in a project. Variances in SHUT DOWN costs are much less significant because they occur late in the project and are usually relatively small compared to the initial investment. Variance in product PRICE has a major impact because it is both continuous throughout the project and is multiplied by output quantity. Geological factors effecting the OUTPUT RATE such as grade can also be significant because their impact can be of long duration (many years) and they can at the same time both reduce revenue and increase cost (a double negative on cash flow). Errors in quantity of RESERVES are not nearly as important because their impact is felt at the end of a project. Experience has shown that the most important variables in a mineral project are: (a) sales level, (b) grade to and from the processing plant, (c) product price, (d) investment required, and (e) start-up date. This is because actual values differ so often from values for these variables used in the project analysis.

Expressing Probability

Probability is the long-term regularity (frequency) that underlies random events. Probability density functions (PDF) and cumulative probability density functions (CDF) are useful for expressing and communicating probability. A PDF shows for a portion of a value range of a variable the probability that the variable will occur in that range. The CDF for a specified value of the variable provides the probability that the value will be less than or equal to a value specified by the analyst.

Expected value is another device used. With this technique the analyst multiplies a specific value of a variable by the probability that value will occur. The probabilities must add to 1.0. (Example: The probabilities for sales are: 50% they will be $50,000,000, 10% - $100,000,000 and 40% - $30,000,000. The expected value of sales then is $47,000,000). The decision-tree technique is an extension in which expected values are used to examine a series of sequential investment decisions.

Recognizing uncertainty in investment analyses uses subjective probability. The quantities used in subjective probability estimates usually are not mathematically derived from empirical data but instead express an analyst's or expert's quantified judgment of the likelihood of the occurrence of some event of interest to the analyst.

Indirect Methods of Recognizing Uncertainty

We now turn from the general discussion of

uncertainty in mineral projects to examine the methods of recognizing that uncertainty.

Risk-adjusted Discount Rate. The most common method used for recognizing uncertainty in the outcome of an investment is to raise the discount rate for more risky investments (lower it for less risky investments). Inherent in this concept is that the returns from successful risky projects must be sufficiently above the normal return so as to cover the losses of risky projects that fail. Example: if the cost of capital is 15%, the discount rate for projects that have only a 50% chance of success should be approximately double the rate for projects whose success is almost certain.

The risk-adjusted rate is also consistent with the concept utilized in the capital asset pricing model (CAP model). The capital market line of the CAP model shows that as riskiness (β) increases so does the return required by investors.

Where the β coefficient (riskiness) is low, the required rate of return is low--the safe rate of return for a government security is perhaps 10%. A rate of return consistent with the present average market portfolio would return perhaps 17% with a β coefficient of 1.0. A more risky portfolio (higher β) would have to return perhaps 20% or more.

The difficulty with the risk-adjusted discount rate is that seldom can an accurate measure of the aggregate investment risk in a particular project be provided. The risk adjusted return is not usually selected based on an explicit evaluation of project risk and it does not separately evaluate the uncertainty associated with each of the many variables. The risk-adjusted discount rate is instead a general number that is subject to much debate.

Shorter Payback Period

By shortening the payback period required before funding more risky investments versus the payback for less risky investments, the analyst is in effect raising the required return. The result is much the same as with the risk-adjusted rate. Risk is not explicitly evaluated and therefore the reasonableness of the adjustment cannot be tested.

Sensitivity Analysis

Sensitivity analysis can also be thought of as "what if" analysis. It includes two principal formats, the key element in each is that one or more of the major variables of the common, deterministic (certainty) method is shifted from its fixed value used in the deterministic analysis. The two major formats for handling each variable in a sensitivity analysis are: (1) most likely, best case and worst case, and (2) sensitivity. A more detailed examination is provided later in this book.

Most Likely, Best Case and Worst Case. In this format all the major variables are shifted simultaneously by the analyst in a manner that usually reflects past observed extremes. Thus in an analysis each of the variables would first be fixed at what the analyst believes is the most likely level--this is probably identical with the original analysis--and the investment evaluation is calculated. In the next step, the analyst shifts the variables to reflect what he believes is their best probable juxtaposition (note that this is not the same as the best probable value of each variable). The next step is to carry out a similar calculation for the worst case that the analyst believes is probable. These values are then presented together to show a three point range of values from best, to most likely, worst that the analyst believes could be realized.

Equal Changes in Variables. Following the initial deterministic study in which variables usually are given their most likely values each major variable is then shifted an equal percentage plus and minus (example 5%)--the variables are shifted one at a time. The percentage impact such a shift has on the investment criteria is then noted. In an example case, a five percent shift in quantity sold can change profits ten percent whereas a five percent shift in unit price can change profits by 32 percent.

WHAT THE PETROLEUM INDUSTRY IS DOING

The appropriateness of the above concepts was tested in a recent survey we conducted of investment analysis practices in the oil and gas industry. Responses to our survey were received from 19 of the 20 largest firms in the industry as measured by sales and the total of 103 responses represented 92 percent of annual sales by American oil and gas producers.

Ninety-eight percent of the companies used at least one form of discounted cash flow (DCF) analysis. Internal Rate of Return (IRR) was the most popular primary method which 69 percent used, easily outdistancing its closest undiscounted challenger, Payback Period which 49 percent used as a primary measure. Third choice went to Net Present Value (NPV), which one-third used as a primary measure but was also the most popular secondary measure. The most popular combination of measures (chosen by 45 percent) formed a triad: internal rate of return and payback for primary analysis, with net present value employed as a secondary measure.

Firms were asked how they computed the cash flow stream of a potential project. Forty reported using after-tax cash flow while 34 firms used pre-tax cash flow. Nine firms checked both categories. Only one firm investing more than $125 million annually reported using pre-tax cash flow only.

Discount Rate. Respondents were asked what they used as their discount rate (or cost of capital) for investment analysis. There was so much confusion and ambiguity in the response that a meaningful answer could only be derived when we restricted the sample size to those firms having capital budgets in excess of $500 million--the largest companies in the industry. For these 19, the average current-dollar, after-tax cost of capital was reported as 16 percent in late 1983. The most popular method for calculating the rate was weighted average cost of capital.

Despite theoretical arguments that a firm can always raise capital for good projects, ninety-two percent of the companies experienced capital rationing. When asked how often capital rationing proved to be a restriction with which they had to contend, 44 percent of the firms replied that it was always a restriction. Forty-three percent said it was a restriction only in certain years, while five percent said that it was a restriction with which they had to contend more than once a year. Only eight percent of the respondents never experienced capital rationing.

Survey participants were asked how they accounted for risk in capital expenditure analysis. The response indicated significant sophistication in this area. (Sixty-three firms used a quantitative approach: sensitivity analysis, probability factors, or computer simulation and one hundred percent of the respondents explicitly incorporate risk in investment analysis.

Sensitivity analysis was the most common method of accounting for risk. Raising the discount rate followed as a close second, with 38 percent of the responding firms employing it. Subjective adjustments and probability factors ranked third and fourth. A majority of firms, 59 percent, used only one technique to analyze risk. The most popular single method used was expected value in which cash flows are multiplied by probability factors. This technique was followed closely by subjective nonquantitative adjustments and raising of the required rate of return. The largest firms in the oil and gas industry selected sensitivity analysis as their primary technique by a wide margin. Monte Carlo computer simulation was commonly used by the larger oil and gas corporations. In fact, for firms with capital budgets in excess of $500 million, it ranked third behind sensitivity analysis and raising the discount rate. Computer simulation was least popular with the smallest firms.

Questioned as to how they analyze cash flow streams in investment analysis, 57 percent replied that they use a current-dollar method for inflation accounting, while 34 percent reported using constant-dollars. Eight percent, which are among the industry's largest, indicated that they used both current-dollar and constant-dollar methods. Two firms did not answer this question.

Survey participants were asked if they had an internal, formal method of dealing with inflation. The responses showed a wide diversity of prevalent techniques. Sixty firms account for inflation, with general price level indexing being the most popular single method. One hundred percent of those firms with annual capital budgets in excess of $500 million--the industry's largest--explicitly account for inflation in their investment analysis. Thirty-two percent of the responding firms do not consider inflation. They use the same method as for external reporting--historical cost.

The results of our survey when matched against similar but more broadly based investment surveys show that the oil and gas industry ranks among the most sophisticated in its practical use of investment theory. However, our survey has also revealed several areas where improvements are possible at a number of firms. The smaller companies still using before-tax cash flows and payback should take steps to change; the mathematical problem with high IRRs should be recognized; there should be greater use of quantitative risk analyses; inflation's impact on the discount rate should be adjusted for; and a formal procedure requiring post-completion audits at each company should be considered.

Probably the most startling difference between investment analysis practices at large and small oil and gas firms was the use of pre-tax cash flow by the smaller firms. Tax considerations are so important in oil and gas investment that it was surprising to find that the vast majority of firms investing less than $125 million annually reported using only <u>before-tax</u> cash flows in their analyses. This practice misses the significant tax advantages offered by certain mineral projects. It should be avoided because before-tax analysis can lead to rejecting projects that have a better after-tax return than some alternative projects.

FOOTNOTES

[1] Much of the material in this section was first written for and used in the author's courses on property appraisal and venture analysis.

[2] In financial theory, this concept is referred to as capital rationing. Theory shows that it is an illusory problem and that a firm will always be able to acquire funds for a project with a return above its cost of capital.

[3] Also known as the discounted cash flow return on investment (DCF-ROI).

[4] The risk free rate is usually taken to be the rate on long-term U.S. government bonds.

[5] Payback period equals the number of years for the net undiscounted cash flow to equal the investment, see CF column in the example where you can calculate that the payback period is two years. This calculation is described in detail in the text.

6/ Calculated from Sommers Coal as:
4 x $100,000 - (200,000 + 50,000) = $150,000.

7/ Return on Equity (ROE) is another similar measure. The investment divisor is reduced by the amount of any borrowings and that is divided into the net income.

8/ The choice is clear because: (a) $150,000 is > $135,000 and (b) 12% is a rather low discount rate. If the discount rate were to drop further to 8% in one year, the decision might not look so clear in that year.

9/ The reason that NPV of the shorter-lived project B remains at $300,000 is because the funds reinvested for the additional four years compound at "r" but then must be discounted at "r," which neutralizes the effect of the additional period added to the life of B.

10/ Constant dollars are dollars adjusted (deflated) to the constant value of a dollar at the date of the study. Many authors refer to constant dollars as "real" dollars in contrast to current (nominal) dollars.

11/ "Current dollars" represent money valued at the time it is used. Therefore, the purchasing power of a current dollar of one year will not be equivalent to the purchasing power of a current dollar of another year. Current-dollar comparisons cannot therefore be freely made from one year to another in an investment analysis. For example, a 1980 current dollar may well buy twice the mine supplies that an inflation-bloated 1990 current dollar will buy. Some authors refer to current dollars as "nominal" dollars.

12/ The difference between risk and uncertainty is based on risk being used to refer to those variable outcomes for which the probability of occurrence is known but any single event is unpredictable. Uncertainty deals with situations where not all the variable outcomes are known nor are their probabilities known either.

3.2

TAILORING THE FINANCING DECISION TO PROJECT ECONOMICS

FERNANDO B. SOTELINO AND MICHAEL A. GUSTAFSON

CROCKER NATIONAL BANK

INTRODUCTION

The degree of success of any new project will ultimately depend on two factors: (i) the underlying economic strength of the project; and (ii) how successfully the parties involved can coordinate their interests to establish a financing structure which indeed capitalizes on the project's inherent economic attractiveness.

A project's underlying economic attractiveness is determined by its inherent comparative advantage in accessing the factors of production, combining these factors of production to create a product and marketing this product. Initially, the decision regarding whether to go ahead with the project - the go/no go decision - has to be based on its total ability to generate cash flow returns to project participants to compensate them for resources committed and risks undertaken.

The extent to which project sponsors will actually benefit from the project's inherent economic attractiveness is fundamentally dependent on their ability to establish the long-term financing arrangement which best meets the objectives and constraints of each project participant and sources the most advantageous funding. The decision regarding the financing of the project, the "how-to-go" decision, is then an attempt to find the capital structure, consistent with market terms and conditions and sound financial management policies which will best: (i) match planned outflows with cash inflows; and (ii) maximize the return for project sponsors.

This article makes use of a case study to illustrate the analytical process involved in dealing with the go/no go and the how-to-go decisions on a project. Though most commonly applied to project finance situations, the guideline described herein is equally applicable to other financing situations, such as leveraged buyouts and debt restructurings, where timely repayment of debt depends largely on the future cash flows to be generated by the entity being financed.

Our case study is of a hypothetical expansion of an existing copper mining operation in Latin America. The company, Latin American Copper (LAC) currently has 147 million metric tons of proven sulphide ore reserves. The existing mine complex consists of an open pit mining operation and a conventional flotation plant with a processing capacity of 20,000 metric tons of ore per day. LAC sells copper concentrates primarily to Japanese custom smelters. The mine has been in operation for 5 years, fine copper production last year was 108,240 metric tons, and LAC currently has assets of $175 million and net worth of $135 million.

Due to a recent increase in proven reserves, LAC is considering doubling the treatment capacity of its flotation plant to 40,000 tons per day, and expanding its mining capacity to meet the additional concentrator capacity (the "Project"). Project capital expenditures and incremental working capital have been estimated at approximately $210 million. The construction period is expected to cover one year, including a three-month start-up phase.

The Project is to be funded by a combination of equity, senior debt, including both supplier credits and/or commercial bank term loans, and subordinated debt. LAC's shareholders are willing to commit a maximum of $50 million to the Project. A government development agency, which is also a minority shareholder, might be willing to lend approximately $40 million to LAC in the form of subordinated notes, depending on the ability of the Project to generate foreign exchange reserves. Commercial banks cannot be expected to offer loans with maturities longer than ten years.

The Project is expected to increase the value of the firm because:

(i) It will solidify the Company's position as a low-cost copper producer by taking advantage of economies of scale. As a result of the expansion, cash production costs are expected to decrease 7.0¢/lb. and all-in costs 11.0¢/lb;

(ii) The potential earnings stream from the sulphide ore reserves will be realized sooner, even though a substantial capital investment will be required.

MAKING THE GO/NO GO DECISION

From an analytical standpoint, making the go/no go decision on an expansion situation such as the one faced by LAC involves:

(i) Generation of unlevered cash flow forecasts for each, the expansion and the no expansion cases;

(ii) Calculation of the incremental unlevered cash flow, representing the difference between the unlevered cash flows resulting from the expansion and the projected cash flows without the expansion project;

(iii) Assessment of the impact of changes on key parameters on the net present value of the incremental unlevered cash flow calculated at a discount rate which incorporates the risk free opportunity cost of money, LAC's business risk, LAC's financial risk and any other perceived additional risks of the expansion.

If the decision is to proceed with the expansion, then careful assessment is required of the impact of alternative funding schemes. In order to save time and facilitate the analysis, one general cash flow model can be built to represent all cases: the "Unlevered No Expansion", the "Unlevered Expansion" and, in case of a go decision, the "Levered Expansion". The Unlevered Expansion can be viewed as a special case of the Levered Expansion where equity capital is the only source of funds, and the Unlevered No-Expansion can be viewed as a special case of the Unlevered Expansion where a different mining plan is followed and certain capital expenditures and expenses are avoided or delayed. Cash flows should be defined in the most general terms, i.e. in accordance with the Levered Expansion, so that they remain valid when analyzing all cases.

The cash flow model used to analyze the LAC expansion has the following general characteristics:

(i) Detailed capital expenditure schedules with appropriate allowances for real and inflationary price increases and cost overrun contingencies. Ongoing capital expenditure requirements are separated from those expenditures tied to the expansion.

(ii) Copper concentrate and by-product silver production schedules, which are based on the Company's mining and stockpile usage plans and data on ore grades, recovery factors, and product grades. Exhibit I illustrates the production schedules generated for the LAC analysis.

(iii) Revenue flows based on a concentrate product value, which takes into account precious metals prices, smelting and refining charges and losses, transportation charges and losses, and real and inflationary price adjustments.

(iv) Direct costs are broken out for the mining and ore treatment operations and split into fixed and variable components where applicable Indirect costs are estimated for both current operations and expected increases due to the expansion. All costs are broken down into the currency of occurrence, escalated at the appropriate inflation rate, and then converted to U.S. dollars.

(v) Working capital such as cash, receivables, inventories, and payables should be related to production, sales, or cost figures using historical financial statements as a starting point when possible. The following working capital accounts were included in the LAC analysis: cash on hand, accounts receivable, raw materials and supplies, work in process, stockpile inventory, finished goods, prepaid expenses, accounts payable, and accrued expenses. The resulting working capital amount is referred to as required working capital, as opposed to working capital which includes short-term investments and short-term debt.

On the basis of the above, the total equity cash flow can be calculated for the Unlevered Expansion case, as illustrated in Exhibit II. As can be observed in Exhibit II under the sub-title External Funding, drawdowns of bank debt and subordinated debt are made equal to zero, since we are, at this point, interested in the calculation of the unlevered equity cash flow. The interest charges and principal repayments shown in Exhibit II refer to the "old" debt, already existing in the books of LAC. The same exhibit, if generated for the Unlevered No-Expansion case, would also show such interest charges and principal repayments, but would show equity investment equal to zero, as no expansion would be taking place.

The go/no go decision is to be made on the basis of the incremental unlevered total equity cash flow, representing the difference between the Unlevered Expansion total equity cash flow (last line of Exhibit II) and the Unlevered No-Expansion total equity cash flow, as shown in Table 1 for the first 10 years.

Table 1
INCREMENTAL UNLEVERED EQUITY CASH FLOW
('000 $)

Year	Expansion	No Expansion	Increment
1	-192.6	1.4	-191.2
2	22.5	5.2	17.3
3	63.3	12.3	51.0
4	60.3	19.2	41.1
5	70.3	25.5	44.8
6	62.2	10.4	51.8
7	66.3	12.2	54.1
8	68.7	12.5	56.2
9	79.0	-	-
10	80.8	12.5	68.3
11	81.8	13.5	68.3
12	142.6*	27.8	114.8
13	-	29.5	- 29.5
14	-	30.7	- 30.7
15	-	31.9	- 31.9
16	-	33.1	- 33.1
17	-	34.4	- 34.4
18	-	35.8	- 35.8
19	-	60.0	- 60.0
20	-	59.0	- 59.0
21	-	56.2	- 56.2
22	-	117.0	-117.0

Assessment of the project's inherent economic attractiveness should be conducted through calculation of the net present values of the incremental cash flows. Many complex methods exist for estimating discount rates, but in general the rates utilized should take into account the risk-free opportunity cost of money, the general business risk of the company, financial risk, and any perceived additional new venture risks. In the case of the LAC expansion, the discount rate utilized for assessing the incremental unlevered equity cash flow was 15%, resulting from a risk-free rate of 10% and an additional 5% for business risk. Although the operation of the Project was to be no different from current operations, a small increment of 2% was added to the discount rate in years 1 - 2 to account for any start-up risks. The resulting net present value (NPV) of the incremental unlevered total equity cash flow is $15 million, showing that the Project should indeed add to the value of the firm.

---
* The Expansion alternative would lead to full depletion of ore reserves by the end of year 12. The significantly larger amount for the equity cash flow in this year results from the liquidation of then existing current assets, mainly receivables and inventories.

Calculation of internal rates of return (IRR) can also be useful, but certain pitfalls exist, e.g.:

(i) IRR's give no indication of the amount of the investment; and

(ii) IRR's assume that the required return for resources committed and risks undertaken remains constant from year to year.

Sensitivity analysis should be performed to determine the degree to which project returns are affected by changes in key technical and economic parameters. For example, Table 2 below shows the impact of variation in copper prices on the net present value of the incremental unlevered equity cash flow.

Table 2
IMPACT OF COPPER PRICE ON GO/NO GO DECISION

Copper Price (¢/lb)	Net Present Value ('000 000 $)	Decision
Base Case	15.3	Go
-10%	-15.9	No Go
- 5%	2.0	Almost indifferent
+15%	30.6	Go

As can be observed, copper prices 5% below the base case would make the expansion alternative a marginal project when compared to the No-Expansion option. It should be noted, however, that the small net present value obtained for the 5% below base case copper price scenario does not necessarily mean that LAC would become marginally profitable if it were to undertake the expansion and this copper price scenario were to occur. This small net present value was calculated from the _incremental_ unlevered equity cash flow and simply means that, under this copper price scenario, one should be virtually indifferent between the expansion and no expansion alternatives.

SEARCHING FOR THE MOST APPROPRIATE
FUNDING STRUCTURE

Several different cash flows are relevant in determining the appropriate capital structure of the project over time and the risk/return relationships among the various project participants. They can be categorized as follows:

(i) Those cash flows which allow for assessment of the Project's overall ability to service debt, namely:

(a) The cash flows available prior to servicing each type of debt;

(b) The cash flows pertaining to debt service.

(ii) The incremental cash flows accruing to different project participants, such as:

(a) The total equity cash flow, which is calculated as in the unlevered analysis adding in the effects of debt financing;

(b) Individual equity sponsor cash flows, which are easily derived from the total equity cash flow;

(c) Individual lender cash flows, which

include drawdowns, debt service, and any profit-sharing or equity interest due the lender; and

(d) Government cash flows, for example the incremental change in the country's foreign exchange reserves due to either avoided or actual foreign exchange flows.

The organization of Exhibit III, the Cash Flow Analysis for the Levered Expansion case deserves some explanation here. The first subtotal in Exhibit III, Unfunded Operating Cash, represents the cash generated by normal operations before any external funding, including any required increases in working capital or fixed assets, and the second subtotal, Funded Operating Cash, includes external long-term funding obtained for the Project. The third subtotal, Cash for Debt Service, is obtained by adding back interest charges and represents the cash flow which is available to meet all debt service payments. The line item Beginning Excess Funds and Reserves may or may not be included in the calculation of Cash for Debt Service. In the LAC example it represents a cash reserve required by lenders, and is therefore considered cash available to service debt; however, it should not be included in the determination of the cash generated <u>in a given year</u> which is available to service debt. Cash available after debt service is then distributed among short-term debt, short-term assets, and dividends. Note also that most of Exhibit IV, the Project Participant Cash Flow Summary for the Levered Expansion Scenario breaks out the individual cash flow of each project participant and is easily derived from line items contained in Exhibit III.

Searching for the most appropriate funding structure requires:

(i) determining the amount of debt service the Project can support in each year; then

(ii) fitting the available financing alternatives into this optimal debt service schedule, in such a way as to minimize financial cost.

A good indication of a project's ability to service debt is the debt coverage ratio (DCR), which may be different for the Project as a whole and for each lender. The DCR is defined as the ratio between cash available to service debt and debt service cash flow.

A DCR of less than 1.0 indicates that cash flow is insufficient to service debt. The larger the DCR, the greater the amount of comfort afforded to lenders and other project participants.

Given the risks inherent in the Project and LAC's past financial performance, it was expected that lenders would require a DCR, taking into account all senior and subordinated debt service, of 1.5 or higher. With an equity investment of $50 million and subordinated debt of $40 million,

the balance to be financed with senior debt amounts to $120 million. Table 3 illustrates the calculation of the maximum amount of new senior debt which LAC could support under the base case scenario with the constraint of maintaining a minimum DCR of 1.5.

Table 3
MAXIMUM NEW SENIOR DEBT SERVICE PROFILE
('000 000 $)

Year	3	4	5	6	7	8	9	10
Cash Available for Debt Service (1)	81.5	92.2	76.0	66.6	69.4	70.6	79.3	79.1
Required DCR (2)	1.5	1.5	1.5	1.5	1.5	1.5	1.5	1.5
Maximum Debt Service Service (3)=(1)/(2)	54.3	61.5	50.7	44.4	46.3	47.1	52.9	52.7
Existing Debt Service (4)	14.5	13.3	12.1	10.9	–	–	–	–
Maximum New Debt Service (5)=(3)-(4)	39.9	48.2	38.6	33.5	46.3	47.1	52.9	52.7
Subordinated Debt Service (6)	4.0	4.0	4.0	4.0	4.0	4.0	23.5	21.5
Maximum New Senior Debt Service (7)=(5)-(6)	35.8	44.2	34.6	29.5	42.3	43.1	29.6	31.2

On the basis of the maximum new senior debt service the project can support (line 7 of Table 3) and keeping in mind a maximum term for the senior debt of 10 years, a repayment schedule was preliminarily designed calling for no principal repayments in years 1 to 3 and principal repayments of 16.67% of the amount borrowed in years 4 to 9 (see Exhibit III). This debt service profile characterizes a standard type of financing which, at a first glance, seems to be consistent with the project's cash generating capacity.

Breakeven analysis, however, shows that this is not the case. Breakeven values and the percentage change from base case which would result in the minimum cash flow generation necessary to meet debt service (DCR=1.0) are shown in Table 4 for the preliminary funding structure proposed.

Table 4
LEVERED EXPANSION CASE –
BREAKEVEN VALUES FOR COPPER PRICES

Year	Projected Base Case Copper Price (¢/lb)	Breakeven Copper Price for DCR-1.0	% Change
3	93.2	80.4	-13.7
4	100.0	89.2	-10.3
5	106.1	90.3	-14.9
6	112.6	96.9	-13.9
7	118.4	100.5	-15.1
8	124.4	105.1	-15.5
9	130.7	112.8	-13.7
10	137.4	112.0	-18.5

As can be observed, the breakeven price of copper in year 4 is only 10% below the base case copper price indicating that, increasing the grace period by one year to start principal repayments in year 5 would be advisable. This financing structure (4 years of grace and 6 of repayment) would allow for approximately a 14% change from projected base case copper prices in all years without shortfall on debt service payments.

It is a relatively common characteristic of capital-intensive projects that the project's cash generating capacity is lowest in the early years of operation, growing over time as, for example, operating performance targets are reached. This problem can be dealt with by delaying debt financing by funding initial-and relatively small-capital expenditures with equity, which allows for pushing principal payments further out in the life of the project, as well as by tailoring repayment schedules to the project's expected cash generating capacity. While strong contractual supports do mitigate construction, operating, input availability and marketing risks and are key in any project financing situation, careful design of a funding structure and associated debt servicing schedules leased on detailed analysis of project economics minimizes the need for actual reliance on such contracts, reducing the risk and of contractual renegotiations and debt reschedulings.

CONCLUSION

This article has focused on the analytical process to be undertaken by project sponsors or, on their behalf, by their financial advisors, in the search for the most appropriate funding scheme based on careful assessment of the impact of financing on the project's profitability and ability to service debt. An approach has been recommended and illustrated by means of a case study, consisting of essentially three stages:

(i) Determining the project's total equity cash flow, defined as the total amount of funds available for distribution among project participants to compensate them for resources committed and risks undertaken;

(ii) Searching for the most appropriate capital structure for the project through tailoring of a debt service profile which takes into consideration the objectives and constraints of sponsors and lenders and which is consistant with the project's total equity cash flow;

(iii) Fitting available funding sources into the optimal debt service schedule designed, in such a way as to minimize financial cost.

The problem of heavy debt service burdens in the early years of a project's life can be dealt with by approaches such as delaying debt financing by funding initial expenditures with equity and utilizing techniques which allow for gradually increasing repayment schedules. While the risk of default can be mitigated by strong construction, operation, input purchase and output sales agreements, careful tailoring of debt service to the project's cash generating capacity minimizes reliance on such contracts and reduces the risk of contract renegotiations or debt reschedulings.

Exhibit 1
Unlevered Expansion Case
ECONOMIC AND PRODUCTION ASSUMPTIONS

	0	1	2	3	4	5	6	7	8	9	10
Economic Assumptions											
US Inflation (%)	4.2	4.2	4.9	5.1	5.1	5.1	5.1	5.1	5.1	5.1	5.1
Domestic Inflation (%)	15.0	15.0	15.0	15.0	15.0	15.0	15.0	15.0	15.0	15.0	15.0
Copper Real Escalation (%)	−2.4	3.0	3.0	2.0	2.0	1.0	1.0	—	—	—	—
Silver Real Escalation (%)	−2.4	0.5	2.7	2.0	—	—	—	—	—	—	—
Copper Price (¢/lb)	75.0	80.5	87.0	93.2	100.0	106.1	112.6	118.4	124.4	130.7	137.4
Silver Price ($/troz)	9.20	9.63	10.37	11.12	11.69	12.28	12.91	13.57	14.26	14.99	15.75
Local Currency Devaluation (%)	9.4	9.4	8.8	8.6	8.6	8.6	8.6	8.6	8.6	8.6	8.6
Local Currency/US $ FX Rate	50.0	54.7	59.5	64.6	70.2	76.2	82.8	89.9	97.7	106.1	115.2
Production Assumptions											
Ore Mined:											
Sulphide Mined OP (000 MT)	6,000	6,000	13,000	13,000	13,000	13,000	14,000	14,000	14,000	14,000	14,000
Waste Ore Mined (000 MT)	12,000	12,000	26,000	26,000	26,000	26,000	28,000	28,000	28,000	28,000	28,000
Ore Treated:											
Sulphide Mined (000 MT)	6,000	6,000	13,000	13,000	13,000	13,000	14,000	14,000	14,000	14,000	14,000
Sulph. from Stockpile (000 MT)	1,000	1,000	1,000	1,000	1,000	1,000	—	—	—	—	—
Sulph. to Concentrator (000 MT)	7,000	7,000	14,000	14,000	14,000	14,000	14,000	14,000	14,000	14,000	14,000
Ore Grades:											
OP Sulphide Grade (% Cu)	1.80	1.80	1.80	1.80	1.80	1.80	1.80	1.80	1.80	1.80	1.80
Sulph. from Stock Grade (% Cu)	1.50	1.50	1.50	1.50	1.50	1.50	—	—	—	—	—
Sulphide Grade (% Cu)	1.76	1.76	1.78	1.78	1.78	1.78	1.80	1.80	1.80	1.80	1.80
Sulphide Ag Content (gpt)	20.0	20.0	20.0	20.0	20.0	20.0	20.0	20.0	20.0	20.0	20.0
Recovery:											
Sulphide Cu Recovery (%)	88.0	88.0	88.0	88.0	88.0	88.0	88.0	88.0	88.0	88.0	88.0
Sulphide Ag Recovery (%)	80.0	80.0	80.0	80.0	80.0	80.0	80.0	80.0	80.0	80.0	80.0
Product Grades:											
Concentrate Grade (% Cu)	32.0	32.0	32.0	32.0	32.0	32.0	32.0	32.0	32.0	32.0	32.0
Concentrate Ag Content (gpt)	331.1	331.1	327.1	327.1	327.1	327.1	323.2	323.2	323.2	323.2	323.2
Production											
Concentrate Production (MT)	338,250	338,250	684,750	684,750	684,750	684,750	693,000	693,000	693,000	693,000	693,000
Silver in Concentrate (kg)	112,000	112,000	224,000	224,000	224,000	224,000	224,000	224,000	224,000	224,000	224,000
Fine Silver Production (MT)	108,240	108,240	219,120	219,120	219,120	219,120	221,760	221,760	221,760	221,760	221,760

Exhibit II
Unlevered Expansion Case
CASH FLOW ANALYSIS ('000 $)

	1	2	3	4	5	6	7	8	9	10
Operating Cash										
Net Earnings	−18,527	33,437	34,947	31,839	40,942	51,217	53,880	56,653	59,539	63,885
Noncash Charges	25,000	43,468	44,289	44,204	34,480	24,773	25,081	25,405	25,745	23,522
Change in Req'd Work. Cap.	−3,384	38,927	195	−265	−1,245	7,129	5,661	5,948	6,250	6,568
Capital Expenditures	192,520	5,465	5,744	6,037	6,345	6,668	7,009	7,366	—	—
Unfunded Operating Cash	−182,662	32,512	73,298	70,272	70,322	62,193	66,292	68,744	79,034	80,840
External Funding										
New Bank Debt Drawdown	—	—	—	—	—	—	—	—	—	—
Sub. Debt Drawdown	—	—	—	—	—	—	—	—	—	—
Equity Investment	195,000	—	—	—	—	—	—	—	—	—
Funded Operating Cash	12,338	32,512	73,298	70,272	70,322	62,193	66,292	68,744	79,034	80,840
Cash Available to Service Debt										
Interest Charges & Fees	4,500	3,300	2,100	900	—	—	—	—	—	—
Begin. Excess Funds & Reserves	—	—	—	—	—	—	—	—	—	—
Cash for Dept Service	16,838	35,812	75,398	71,172	70,322	62,193	66,292	68,744	79,034	80,840
Debt Service										
Senior Debt Interest & Fees	4,500	3,300	2,100	900	—	—	—	—	—	—
Senior Debt Repayments	10,000	10,000	10,000	10,000	—	—	—	—	—	—
Cash for Sub. Debt Service	2,338	22,512	63,298	60,272	70,322	62,193	66,292	68,744	79,034	80,840
Sub. Debt Interest & Fees	—	—	—	—	—	—	—	—	—	—
Sub. Debt Repayment	—	—	—	—	—	—	—	—	—	—
Cash After Debt Service	2,338	22,512	63,298	60,272	70,322	62,193	66,292	68,744	79,034	80,840
Distribution of Remaining Cash										
Change in Short Term Debt	—	—	—	—	—	—	—	—	—	—
Ending Cash Reserve	—	—	—	—	—	—	—	—	—	—
Dividends	2,338	22,512	63,298	60,272	70,322	62,193	66,292	68,744	79,034	80,840
Ending Excess Funds	—	—	—	—	—	—	—	—	—	—
Equity Cash Flow										
Equity Investment	195,000	—	—	—	—	—	—	—	—	—
Dividends	2,338	22,512	63,298	60,272	70,322	62,193	66,292	68,744	79,034	80,840
Total Equity Cash Flow	−192,662	22,512	63,298	60,272	70,322	62,193	66,292	68,744	79,034	80,840

Exhibit III
Levered Expansion Case
CASH FLOW ANALYSIS ('000 $)

	1	2	3	4	5	6	7	8	9	10
Operating Cash										
Net Earnings	−23,783	9,016	22,415	35,246	31,470	42,872	46,655	50,540	54,790	60,024
Noncash Charges	25,000	44,125	44,947	44,862	35,137	25,430	25,739	26,062	26,403	24,180
Change in Req'd Work. Cap.	−3,498	38,921	188	−219	−1,250	7,123	5,655	5,943	6,244	6,562
Capital Expenditures	199,095	5,465	5,744	6,037	6,345	6,668	7,009	7,366	—	—
Unfunded Operating Cash	−194,380	8,755	61,430	74,290	61,512	54,511	59,730	63,294	74,949	77,642
External Funding										
New Bank Debt Drawdown	115,000	5,000	—	—	—	—	—	—	—	—
Sub. Debt Drawdown	40,000	—	—	—	—	—	—	—	—	—
Equity Investment	50,000	—	—	—	—	—	—	—	—	—
Funded Operating Cash	10,620	13,755	61,430	74,290	61,512	54,511	59,730	63,294	74,949	77,642
Cash Available to Service Debt										
Interest Charges & Fees	4,500	21,563	20,500	18,700	15,400	13,000	10,600	8,200	5,300	1,500
Begin. Excess Funds & Reserves	—	620	4,376	20,000	20,000	20,000	20,000	20,000	20,000	—
Cash for Dept Service	15,120	35,938	86,306	112,990	96,912	87,511	90,330	91,494	100,249	79,142
Debt Service										
Senior Debt Interest & Fees	4,500	17,563	16,500	14,700	11,400	9,000	6,600	4,200	1,800	—
Senior Debt Repayments	10,000	10,000	10,000	30,000	20,000	20,000	20,000	20,000	20,000	—
Cash for Sub. Debt Service	620	8,376	59,806	68,290	65,512	58,511	63,730	67,294	78,449	79,142
Sub. Debt Interest & Fees	—	4,000	4,000	4,000	4,000	4,000	4,000	4,000	3,500	1,500
Sub. Debt Repayment	—	—	—	—	—	—	—	—	20,000	20,000
Cash After Debt Service	620	4,376	55,806	64,290	61,512	54,511	59,730	63,294	54,949	57,642
Distribution of Remaining Cash										
Change in Short Term Debt	—	—	—	—	—	—	—	—	—	—
Ending Cash Reserve	—	4,376	20,000	20,000	20,000	20,000	20,000	20,000	—	—
Dividends	—	—	35,806	44,290	41,512	34,511	39,730	43,294	54,949	57,642
Ending Excess Funds	620	—	—	—	—	—	—	—	—	—
Equity Cash Flow										
Equity Investment	50,000	—	—	—	—	—	—	—	—	—
Dividends	—	—	35,806	44,290	41,512	34,511	39,730	43,294	54,949	57,642
Total Equity Cash Flow	−50,000	—	35,806	44,290	41,512	34,511	39,730	43,294	54,949	57,642

Exhibit IV
Levered Expansion Case
PROJECT PARTICIPANT CASH FLOW SUMMARY ('000 $)

	1	2	3	4	5	6	7	8	9	10
Shareholders										
Equity Investment	50,000	—	—	—	—	—	—	—	—	—
Dividends	—	—	35,806	44,290	41,512	34,511	39,730	43,294	54,949	57,642
Total Equity Cash Flow	−50,000	—	35,806	44,290	41,512	34,511	39,730	43,294	54,949	57,642
Majority Share. Equity Invest.	37,500	—	—	—	—	—	—	—	—	—
Majority Share. Dividends	—	—	28,645	35,432	33,210	27,609	31,784	34,635	43,959	46,114
Majority Share. Cash Flow	−37,500	—	28,645	35,432	33,210	27,709	31,784	34,635	43,959	46,114
Lenders										
New Bank Debt Drawdown	115,000	5,000	—	—	—	—	—	—	—	—
New Bank Debt Repayment	—	—	—	20,000	20,000	20,000	20,000	20,000	20,000	—
Interest on New Bank Debt	2,400	14,100	14,400	13,800	11,400	9,000	6,600	4,200	1,800	—
New Bank Debt Fees	1,775	163	—	—	—	—	—	—	—	—
New Bank Debt Cash Flow	−110,825	9,263	14,400	33,800	31,400	29,000	26,600	24,200	21,800	—
Sub. Debt Drawdown	40,000	—	—	—	—	—	—	—	—	—
Sub. Debt Repayment	—	—	—	—	—	—	—	—	20,000	20,000
Interest on Sub. Debt	2,000	4,000	4,000	4,000	4,000	4,000	4,000	4,000	3,500	1,500
Sub. Debt Fees	400	—	—	—	—	—	—	—	—	—
Sub. Lender Cash Flow	−37,600	4,000	4,000	4,000	4,000	4,000	4,000	4,000	23,500	21,500
Existing Bank Debt Cash Flow	14,500	13,300	12,100	10,900	—	—	—	—	—	—
Lender Cash Flow	−133,925	26,563	30,500	48,700	35,400	33,000	30,600	28,200	45,300	21,500
Gov't Development Agency										
Minority Share. Cash Flow	−12,500	—	7,161	8,858	8,302	6,902	7,946	8,659	10,990	11,528
Sub. Lender Cash Flow	−37,600	4,000	4,000	4,000	4,000	4,000	4,000	4,000	23,500	21,500
Gov't Agency Cash Flow	−50,100	4,000	11,161	12,858	12,302	10,902	11,946	12,659	34,490	33,028
Foreign Exchange										
Trade Balance	−180,416	135,161	177,486	195,700	209,686	221,326	233,814	245,740	266,016	279,584
Services Balance	−26,575	−28,207	−53,812	−60,290	−55,112	−45,711	−48,530	−49,694	−58,449	−59,142
FX Effect Before Capital Flows	−206,991	106,954	123,674	135,410	154,574	175,615	185,284	196,046	207,567	220,442
Capital Flows Balance	194,380	−8,755	−25,624	−30,000	−20,000	−20,000	−20,000	−20,000	−20,000	−20,000
FX Effect After Capital Flows	−12,612	98,199	98,049	105,410	134,574	155,615	165,284	176,046	187,567	200,442

3.3

MEASURING THE ECONOMIC VIABILITY
OF RESOURCE PROJECTS
(A WESTERN SURFACE COAL PROJECT)

Grover R. Castle, Daniel M. Higgins, Brooks J. Klimley

Chemical Bank
New York, New York

Measuring The Economic Viability Of Resource Projects
(A Western Surface Coal Project)

The bankers approach to the analysis of a new "green fields" mining project is similar but different from the approach that the mining company might take. The purpose of this paper shall be to discuss the techniques that are used by mining companies and bankers as they try to measure the economic viability of mining projects.

The analysis of these new projects (by both the mining companies and the banker) is based on a projected revenue stream and not on historical earnings or past perfomance because these projects are start up situations with no history to look to. The starting point in the analysis of these projections is the choice of the assumptions to be used in preparing these projections. After the assumptions have been determined and the projections are prepared then the projections are submitted to various tests in order to determine whether the project measures up to previously agreed upon minimum standards.

In order to illustrate the analytical procedures used in evaluating a mineral project the writer has constructed a hypothetical case involving the development of a western surface coal mine (Property). The basic data in connection with this hypothetical new project is described in the next section of this paper. After the Description of the Project is a section which shows the type of analysis that a mining company might prepare in order to approach a bank for a loan and the final section illustrates the type of analysis that a bank might prepare in evaluating this loan request.

Description Of The Project or Basic Data In Connection With Undeveloped Western Surface Coal Property

An independent mining/geological Consultant's report has been completed on the Property which confirms that the Property contains 175 million tons of economically recoverable coal subdivided into the following reserve categories in accordance with U.S. Bureau of Mines and U.S. Geological Survey definitions:

Measured or Proven	90,000,000
Indicated or Probable	50,000,000
Inferred or Possible	35,000,000

The owners of the reserves have chosen to drill up those reserves that are nearest to the subcrop and have not spent as much time on those portions of the Property which are on the fringes of the deposit where the overburden is thicker because the stripping ratio and hence the operating costs and capital requirements are higher down dip. The proven reserves are based on an average thickness of 80 feet, of which a mining recovery of 90% is projected. The overburden thickness varies from 0' to 280' with an average thickness of 127'. The overburden consists mainly of scoria, sandstone, gravel, shale and silty clay. 134 holes were cored (and analysis done on five foot increments) in the measured reserve on the following spacing:

100	foot centers to define the burn line;
1,320	foot centers to define the early mining areas;
2,640	foot centers to identify the later mining areas.

An analysis of the samples reveals the following average analysis:

Moisture	26%
Volatile Matter	31%
Fixed Carbon	35%
Ash	8.%
Sulphur (all forms)	.6%
BTU	8500

The Development Plan has taken full advantage of the coal's uniform quality and minimal overburden to minimize operating costs. Conventional truck and shovel mining techniques will prepare a uniform bench of optimal height for the dragline to operate in the simple side cast mode. After the first five years, the pit configuration allows future mining to proceed at or near the average strip ratio.

In accordance with the definitive mine and development plan, a two year period is scheduled for the construction of facilities, equipment delivery and erection, and mine development at a projected cost (not including interest during construction) of $125,700,000 based on mid-1984 quotations or $138,040,000 in actual dollars. The breakdown of this cost is summarized below:

Development Cost

Major Categories of Development Cost	Projected Cost*
Dragline - one machine w/60 cubic yard nominal bucket capacity	$ 25,900
Other major mining equipment (including shovels, front end loaders, dozers, drills, powder truck, haulers and scrapers)	13,750
Maintenance equipment (cranes, forklifts, welding, electrical and mechanics trucks)	750
Support equipment and systems (graders, water truck, power supply, etc.)	4,300
Coal handling facilities (rail loop, silos, truck dump and crusher, conveyor and sampling system, batch weighing system and scales)	32,250
Office, warehouse and shops	9,350
Preproduction stripping, development drilling and permitting	19,400
Working Capital	20,000
	$125,700

*All costs are in thousands of 1984 dollars including an 8% provision for contingency.

The design capacity of the Project calls for 6 million tons of coal per year with a two year buildup to full production.

Mining Company Analysis

Tables I and II illustrate the type of analysis that a mining company might prepare in connection with a project similar to the western surface coal mine that has been described in this Paper. Based on this analysis the total project development cost has a rate of return of 23%, and the rate of return for the proposed equity (10% of projected cost) to be invested by the mining company is 42%. However, any such analysis is only as good as the assumptions that have been made in the preparation of the analysis.

The assumptions used by the mining company in the preparation of their analysis and loan request are listed below:

1. Project development shall commence in January 1, 1985 with onsite work beginning in the Spring. Development of the Mine shall take two years before the commencement of commercial production.

2. Commercial production from the mine shall commence January 1, 1987 and the mine shall produce 2,000,000 tons of shippable coal in 1987, 3,000,000 tons in 1988 and 6,000,000 tons in 1989.

FINANCIAL EVALUATION FOR MINERALS INDUSTRY

TABLE I

Company Analysis - Western Surface Coal Property
Projection of Comparative Balance Sheets
(000's Omitted)

YEARS ENDED DECEMBER 31,

ASSETS	1987	1988	1989	1990	1991	1992	1993	1994	1995	1996
Cash	$ 3,750	$ 4,220	$ 11,900	$18,080	$20,000	$20,000	$20,000	$20,000	$20,000	$20,000
Accounts Receivable	2,770	4,160	8,320	8,320	8,320	13,260	14,060	14,900	15,790	16,740
Inventories	1,660	2,500	4,990	4,990	4,990	7,960	8,430	8,940	9,480	10,040
Current Assets	8,180	10,880	25,210	31,390	33,310	41,220	42,490	43,840	45,270	46,780
Plant, property & equipment	132,350	132,350	132,350	132,350	132,350	132,350	132,350	132,350	132,350	132,350
Less depreciation & depletion	(5,440)	(10,880)	(16,320)	(21,760)	(27,150)	(32,550)	(37,940)	(43,340)	(48,730)	(54,130)
Total Assets	$135,090	$132,350	$141,240	$141,980	$138,510	$141,020	$136,900	$132,850	$128,890	$125,000

LIABILITIES AND NET WORTH	1987	1988	1989	1990	1991	1992	1993	1994	1995	1996
Current Portion Bank Term Loan	$ -	$ 8,570	$ 8,570	$ 8,570	$22,280	$22,280	$22,280	$22,280	$22,270	$ -
Accounts Payable	4,160	6,250	12,480	12,480	12,480	19,890	21,080	22,350	23,690	25,110
Accrued Liabilities	670	1,010	2,020	2,020	2,020	3,180	3,370	3,580	3,790	4,020
Total Current Liabilities	4,830	15,830	23,070	23,070	36,780	45,350	46,730	48,210	49,750	29,130
Bank Term Loan	137,100	128,530	119,960	111,360	89,110	66,830	44,550	22,270	-	-
Total Liabilities	$141,930	$144,360	$143,030	$134,430	$125,890	$112,180	$91,280	$70,480	$49,750	$29,130
Deferred Taxes	-	-	5,110	9,780	13,950	37,430	51,830	57,320	63,260	69,580
Common Stock	15,230	15,230	15,230	15,230	15,230	15,230	15,230	15,230	15,230	15,230
Retained Earnings: Beginning	(11,620)	(22,070)	(27,240)	(22,130)	(17,460)	(16,560)	(23,820)	(21,440)	(10,180)	650
Add: Net Income (Loss)	(10,450)	(5,170)	5,110	4,670	4,180	23,480	26,790	30,230	33,810	37,510
Less: Divid.	-	-	-	-	(3,280)	(30,740)	(24,410)	(18,970)	(22,980)	(27,100)
Ending	(22,070)	(27,240)	(22,130)	(17,460)	(16,560)	(23,820)	(21,440)	(10,180)	650	11,060
Total Liabilities And Net Worth	$135,090	$132,350	$141,240	$141,980	$138,510	$141,020	$136,900	$132,850	$128,890	$125,000
Net Current Assets	$3,350	$(4,950)	$2,140	$8,320	$(3,470)	$(4,130)	$(4,240)	$(4,370)	$(4,480)	$17,650
Current Ratio	1.69	.69	1.09	1.36	.91	.91	.91	.91	.91	1.61
Net Worth		$(6,840)	$(12,010)	$(6,900)	$(2,230)	$(1,330)	$(8,590)	$(6,210)	$5,050	$15,880	$26,290

TABLE II

Company Analysis - Western Surface Coal Property
Projection of Comparative Income Accounts
(000's Omitted)

	YEARS ENDED DECEMBER 31,									
	1987	1988	1989	1990	1991	1992	1993	1994	1995	1996
Production - Tons	2,000	3,000	6,000	6,000	6,000	6,000	6,000	6,000	6,000	6,000
Gross Revenue from sale of coal	$26,000	$39,000	$78,000	$78,000	$78,000	$124,320	$131,760	$139,680	$148,080	$156,960
Pass through of severance tax	7,280	10,920	21,840	21,840	21,840	34,800	36,900	39,120	41,460	43,920
Total Gross Revenue	$33,280	$49,920	$99,840	$99,840	$99,840	159,120	$168,660	$178,800	$189,540	$200,880
Operating Costs (including corporate overhead)	$9,550	$15,190	$32,190	$34,120	$36,170	$38,340	$40,640	$43,080	$45,670	$48,410
Maintenance and Repairs of Property, and Equipment	850	850	1,730	1,730	1,730	1,730	1,730	1,730	1,730	1,730
Government Royalty	4,160	6,240	12,480	12,480	12,480	19,890	21,080	22,350	23,690	25,110
Severance tax	7,280	10,920	21,840	21,840	21,840	34,800	36,900	39,120	41,460	43,920
Depreciation and depletion	5,440	5,440	5,440	5,440	5,390	5,390	5,390	5,390	5,390	5,390
Total	$27,280	$38,640	$73,680	$75,610	$77,610	$100,150	$105,740	$111,670	$117,940	$124,560
Net Operating Income	$6,000	$11,280	$26,160	$24,230	$22,230	$58,970	$62,920	$67,130	$71,600	$76,320
Interest Expense	16,450	16,450	15,940	14,900	13,870	12,020	9,340	6,670	3,990	1,310
Net Income (Loss) before Taxes on Income	$(10,450)	$(5,170)	$10,220	$9,330	$8,360	$46,950	$53,580	$60,460	$67,610	$75,010
Taxes on Income - Current	$ -	$ -	$ -	$ -	$ -	$ -	$12,390	$24,750	$27,860	$31,170
Deferred	-	-	5,110	4,660	4,180	23,470	14,400	5,480	5,940	6,330
	$ -	$ -	$5,110	$4,660	$4,180	$23,470	$26,790	$30,230	$33,800	$37,500
Net Income	$(10,450)	$(5,170)	$5,110	$4,670	$4,180	$23,480	$26,790	$30,230	$33,810	$37,510
Cash Flow Available for Principal	$(5,010)*	$270	$15,660	$14,770	$13,750	$52,340	$46,580	$41,100	$45,140	$49,230
Required Principal Payments	-	-	8,570	8,570	8,570	22,280	22,280	22,280	22,280	22,270
Annual Coverage	-	-	1.83	1.72	1.60	2.35	2.09	1.84	2.03	2.21
Cash Flow Available for Debt Service	$11,440	$16,720	$31,600	$29,670	$27,620	$64,360	$55,920	$47,770	$49,130	$50,540
Debt Service (Principal plus Interest)	16,450	16,450	24,510	23,470	22,440	34,300	31,620	28,950	26,270	23,580
Annual Coverage	.70*	1.02	1.29	1.26	1.23	1.88	1.77	1.65	1.87	2.14
Excess Cash Flow Paid Out in Dividends	$ -	$ -	$ -	$ -	$3,280	$30,740	$24,410	$18,970	$22,980	$27,100

*The $5,010,000 deficit of cash flow available to cover interest in 1987 is covered by borrowings under the Bank Loan.

The Mine has a design capacity of 6,000,000 tons of shippable coal per year and this level of production is achieved in 1989. Even though the independent geology report indicates a total of 175 million tons of reserves, the projection assumes only 136,250,000 tons. The company has given full value to the 90 million tons of proven reserves but the probable and possible reserves have been cut in order to reflect the greater degree of uncertainty involved with these two classifications. The 50 million tons of probable has been cut to 37-1/2 million tons (75%) and the 35 million tons of possible has been carried at 8-3/4 million tons (25%).

3. One hundred percent of the production from the Mine is dedicated to a long term contract (Sales Contract) with a major utility which provides for a fixed price of $13 per ton (F.O.B. the Mine loaded in rail cars) through December 31, 1991. Thereafter, the Sales Contract provides for the sale price to be redetermined on a year to year basis based on the market price for similar coal.

The Company projection assumes that the price received remains at $13 per ton during the first five years of the Sales Contract and thereafter the market price for coal escalates with inflation at 6% per year (starting mid-1984) from the $13 per ton provided in the Sales Contract. Therefore starting in 1992 (the first year after the $13. fixed price in the Sales Contract converts to a market price) the sale price is $20.72 per ton and this price then escalates at 6% per year.

4. The Company projection also assumes that pursuant to the Sales Contract the purchasing utility pays the severance tax on the production from the Mine.

5. The average cash operating costs for the Mine have been assumed to be $3.86 per ton excluding royalties and severance tax (the severance tax has been assumed to be a pass through to the purchasing utility). This $3.86 cost is mid-1984 dollars and has been escalated at 6% per year.

6. Corporate Overhead has been charged at a rate of 16¢ per ton and is included with operating costs on Table II.

7. The Accelerated Cost Recovery System (ACRS) has been adopted for tax depreciation purposes. Book lives have been selected to reflect actual experience with the type of asset being depreciated. The tax lives and book lives are listed below according to the dollar amount of assets acquired:

(000's Omitted)

Dollar Amount Of Depreciation*	Tax Life Under ACRS	Book Life
$ 150	4	4
11,150	5	12
62,000	5	Life Of Mine
16,700	15	Life Of Mine
10,000	Expensed	Life Of Mine
25,000	Cost depletion basis	

*All costs are in mid-1984 dollars.

The depreciation charge shown on Table II has been computed on a straight line basis.

8. Percentage depletion has been charged at statutory rates.

9. Taxes on income have been charged at an overall rate of 50% in order to allow for some state and local taxes.

10. It is assumed that $850,000 per year is spent for replacement of equipment and other facilities in 1987 and 1988 and that $1,730,000 per year is spent for this purpose thereafter. On Table II these amounts have been expensed.

11. $137,100,000 (90%) of the total Development Costs has been financed under a term loan that matures in 32 quarterly installments in years 1989-1996. In years 1989-1991 the four quarterly payments total $8,570,000 per year and in years 1992-1996 the four quarterly payments total $22,280,000 in each year. The $137,100,000 represents 90% of the Total Projected Development Cost as broken down below:

	(000's Omitted)
Development Cost (including Working Capital)	$138,040
Interest Capitalized during Construction	14,293
	$152,333

12. After Cash on hand has been built up to a total of $20 million all excess cash flow not needed for capital expenditures, principal payments on long term debt or to maintain cash at the $20 million level, shall be paid out in dividends.

13. A receivable turnover rate of 30 days has been assumed and inventory has been computed as 5% of sales. Accrued liabilities are 2% of sales and accounts payable have been determined based on a 45 day turnover.

Bank Analysis

A bank would use the company analysis as a starting point, and might or might not base its decision on the same projection that the mining company has used. However, quite often it is necessary for the bank to revise the mining company's forecast because there is a fundamental or basic difference in the position of the bank and the company. The company is in an equity position which means that the company can and should give effect to and take into consideration any pluses or "romance" that might be involved in the project. The reason for this is that if the project is a success the company stands to gain all of the benefits, and, therefore, it is only right for the owner to take an optimistic approach to the Project.

On the other hand, the bank is a lender and as such is not in an equity position. No matter how successful the venture is, the bank's return on its investment is limited to only interest on its loan. The bank does not share in the profits of the venture as does the mining company through its equity position. Therefore, the bank's approach is often more conservative than the company. Usually the bank will prepare one or more Revised projections each of which will show a more conservative approach to the Project.

Table III shows a Revised Projection that the bank might prepare as a "Base Case" or a "most reasonable" case. The assumptions used in the preparation of Table III are the same as the company's assumptions except as listed below:

1. The projection assumes only 90 million tons of coal reserves because the bank only gives value to the reserves classified as proven or measured reserves.

2. For the purposes of this Revised Projection, the market price for coal is assumed to escalate at 5% per year from a mid-1984 price of $9. per ton. Therefore, starting in 1992 (the first year after the $13. fixed price in the Sales Contract converts to a market price) the sale price is $13.30 per ton and this price then escalates at 5% per year thereafter.

3. $99,200,000 (about 65%) of the Total Development Cost has been financed under a bank term loan that matures in 32 equal quarterly installments in years 1989-1996. In years 1989-1991 the four quarterly payments in each year total $6,200,000 per year and in years 1992-1996 the four quarterly payments amount to $16,120,000 in each year. The $99,200,000 represents about 65% of the total Projected Development Costs rather than 90% as proposed by the mining company because many banks prefer to see the sponsor of the project have a significant investment in the Project. They like to have this investment large enough so that it indicates a commitment to the Project. Furthermore, the Revised Projection does not provide an adequate coverage for a loan covering 90% of the Development Cost.

Based on this Revised Projection the rate of return on the total amount spent on the development of the Project ($152,333,000) is 14% and the rate of return on the Mining Company's equity investment (from Excess Cash Flow available for dividends) in the Project ($53,133,000) is 17%. In addition to these computations there are a number of other ratios and tests that a banker considers in the determination of the amount of loan that the project will support. Some of these tests are listed below:

FINANCIAL EVALUATION FOR MINERALS INDUSTRY 109

TABLE III

Revised Projection - Base Case - Western Surface Coal Property
Projection of Comparative Income Accounts
(000's Omitted)

	1987	1988	1989	1990	1991	1992	1993	1994	1995	1996	Thereafter 1997 thru 2,003	Cumulative Total
Production - Tons	2,000	3,000	6,000	6,000	6,000	6,000	6,000	6,000	6,000	6,000	37,000	90,000
Gross Revenue from sale of coal	$26,000	$39,000	$78,000	$78,000	$78,000	$79,800	$83,760	$87,960	$92,340	$96,960	$715,380	$1,455,200
Pass through of severance tax	7,280	10,920	21,840	21,840	21,840	22,320	23,450	24,600	25,860	27,180	198,970	406,110
Total Gross Revenue	$33,280	$49,920	$99,840	$99,840	$99,840	$102,120	$107,220	$112,560	$118,200	$124,140	$914,350	$1,861,310
Operating Costs (including Corporate Overhead)	$9,550	$15,190	$32,190	$34,120	$36,170	$38,340	$40,640	$43,030	$45,670	$48,410	$370,040	$713,400
Maintenance and Repairs of Property, and Equipment	850	850	1,730	1,730	1,730	1,730	1,730	1,730	1,730	1,730	12,110	27,650
Government Royalty	4,160	5,240	12,480	12,480	12,480	12,770	13,400	14,070	14,780	15,520	114,450	232,830
Severance tax	7,280	10,920	21,840	21,840	21,840	22,340	23,460	24,630	25,860	27,150	200,310	407,470
Depreciation and depletion	7,060	7,060	7,060	7,060	7,020	7,020	7,020	7,020	7,020	7,020	43,390	114,250
Total	$28,900	$40,260	$75,300	$77,230	$79,240	$82,200	$86,250	$90,530	$95,060	$99,830	$740,300	$1,495,600
Net Operating Income	$4,380	$9,660	$24,540	$22,610	$20,600	$19,920	$20,970	$22,030	$23,140	$24,310	$173,550	$365,710
Interest Expense (12%)	11,900	11,900	11,530	10,790	10,040	8,710	6,770	4,840	2,910	970	-	80,350
Net Income (Loss) before Taxes on Income	$(7,520)	$(2,240)	$13,010	$11,820	$10,560	$11,210	$14,200	$17,190	$20,230	$23,340	$173,550	$285,350

(Cont'd. Table III)

	1987	1988	1989	1990	1991	1992	1993	1994	1995	1996	Thereafter 1997 thru 2,003	Cumulative Total
Taxes on Income - Current	$ -	$ -	$ -	$ -	$ -	$ -	$ -	$ 660	$ 8,050	$ 9,430	$68,130	$86,270
Deferred	-	-	6,500	5,910	5,280	5,610	7,100	7,950	2,080	2,240	21,080	63,750
	$ -	$ -	$ 6,500	$ 5,910	$ 5,280	$ 5,610	$ 7,100	$ 8,610	$10,130	$11,670	$89,210	$150,020
Net Income (Loss)	$(7,520)	$(2,240)	$5,510	$5,910	$5,280	$5,500	$7,100	$8,580	$10,100	$11,670	$84,340	$135,330
Cash Flow Available for Principal	$ (460)*	$ 4,820	$20,070	$18,800	$17,580	$18,230	$21,220	$23,550	$19,200	$20,930	$149,310	$313,330
Required Principal Payments	-	-	6,200	6,200	6,200	16,120	16,120	16,120	16,120	16,120	-	99,200
Annual Coverage	.96*	1.41	3.24	3.05	2.84	1.13	1.32	1.46	1.19	1.30	-	3.16
Cash Flow Available for Debt Service	$11,440	$16,720	$31,600	$22,670	$27,520	$26,940	$27,990	$28,390	$22,110	$21,900	$149,310	$393,690
Debt Service (Principal plus Interest)	11,900	11,900	17,730	16,990	16,240	24,830	22,890	20,960	19,030	17,090	-	179,560
Annual Coverage	.96*	1.41	1.78	1.75	1.70	1.09	1.22	1.36	1.16	1.28	-	2.19
Excess Cash Flow (Available For Prepayments or dividends)	$(460)	$4,820	$13,870	$12,680	$11,380	$ 2,110	$ 5,100	$ 7,430	$ 3,080	$ 4,810	$149,310	$214,130
Cumulative Excess Cash Flow	(460)	4,360	18,230	30,910	42,290	44,400	49,500	56,930	60,010	64,820	214,130	
Cash Flow Available for Debt Service	$11,440	$16,720	$31,600	$29,670	$27,620	$26,940	$27,990	$28,390	$22,110	$21,900	$149,310	$393,690
Cumulative Cash Flow available for Debt Service	11,440	28,160	59,760	89,430	117,050	143,990	171,980	200,370	222,480	244,380	393,690	
Present Worth Cash Flow available Debt Service	$10,212	$13,331	$22,497	$18,360	$15,670	$13,550	$12,653	$11,464	$7,974	$7,048	32,450	165,814
Cumulative Present Worth Cash Flow Available Debt Service	10,212	23,543	46,040	64,900	80,570	94,220	106,878	118,342	126,316	133,364	165,814	

*The $460,000 deficit of cash flow available to cover interest in 1987 is covered by a borrowing under the Bank Loan.

1. **Amount Being Loaned Per Ton of Coal in the Ground** - This is one of the most important tests for many banks. Many banks have a rule of thumb for different types of mines as to how much they will be willing to loan for each ton of coal in the ground. In the case of the western surface coal mine, the total amount of the loan is $99,200,000 and the total amount of recoverable proven (measured) reserves is 90 million tons which means that the banks are loaning $1.10 per ton as computed below:

$$\frac{\text{Amount Of Loan}}{\text{Total Recoverable Proven (Measured) Reserves of Coal}} = \frac{\$99,200,000}{90,000,000 \text{ tons}} = \$1.10 \text{ per ton}$$

$1.10 per ton is a rather full loan for a western surface coal mine. However, possibly this is mitigated by the fact that:

a) the mine has a five year sales contract at a favorable $13. a ton price, and

b) the mine has substantial additional probable and possible reserves.

2. **Initial Development Cost Per Annual Ton of Production** - The total cost of developing this project is $152,333,000 and the design capacity of the mine is 6 million tons per year and, therefore, the initial development cost for each annual ton of production is $25.39 per ton computed as shown below:

$$\frac{\text{Total Initial Development Cost}}{\text{Annual Production Capacity}} = \frac{\$152,333,000}{6,000,000 \text{ Tons}} = \$25.39 \text{ per ton}$$

This is another of the common ways of comparing one project with other projects and it gives an indication of how competitive the project is with other projects in a particular area.

Many bankers develop "rules of thumb" as to what a competitive mine should cost for each ton of production. For example, some bankers believe that for a mine in the United States a reasonable initial development cost per annual ton should be under $45. per ton for a surface mine and under $65. per ton for an underground mine. Therefore, $25.39 per ton indicates that the hypothetical mine in this illustration would compare favorably with other mines.

This measure can become distorted when mines in different locations (or parts of the world) are compared on this basis. For example, it is not fair to compare a U.S. property on this basis against a mine that is being developed in a remote location where schools, railroads and port facilities need to be built in order to support the mine. Obviously the U.S. property will look better because the costs will not include these kinds of infrastructure.

3. **Initial Development Cost Per Ton of Coal in Place** - Again the cost of developing the Western Surface Coal Property is $152,333,000 and the property has a total of 90 million tons of proven measured reserves. Therefore, the cost of developing this project shall be $1.69 for each ton of coal in the ground as indicated below:

$$\frac{\text{Total Initial Development Cost}}{\text{Total Recoverable Proven (Measured) Reserves Of Coal}} = \frac{\$152,333,000}{90 \text{ Million Tons}} = \$1.69 \text{ for each Ton Of Coal In Ground}$$

4. **Annual Coverage of Principal** - After the Bank has prepared a Base Case projection (Table III) which the Bank believes shows a reasonable estimate of the cash flow expected to be generated at the mine this projected cash flow in each year is often compared with the expected principal payments that must be made in that year in order to determine an annual coverage for these annual principal payments.

These annual coverage computations for the Western Surface Coal Property are shown below the Table III Revised Projection. A coverage of 1.25 times would be sufficient to satisfy most banks and even a lower coverage in an occasional year (such as 1.13 in 1992 and 1.19 in 1995) would probably be acceptable in a situation where the output is dedicated to a favorable sales contract as in the Western Surface Coal Property.

5. Annual Coverage of Debt Service - The annual coverage can also be computed by comparing cash flow available for debt service (i.e. cash flow before deduction for interest) to annual debt service (principal plus interest). This annual coverage computation for the Western Surface Coal Property is also shown below the Table III Revised Projection.

6. Annual Direct Cash Operating Costs Per Annual Ton of Production - Another method of comparing one project against another is the operating cost per ton of production. This is one of the key measures that are used to determine how competitive a property is.

7. Reserve Coverage Ratio - All Projections are based on estimates of the amount of reserves in the ground and for this reason most banks do not want their loan to pay off at the same time that the last ton of coal is being mined. As a result of this concern, many banks focus on the amount of mineral reserves remaining in the ground at the point in time when the loan is scheduled to pay off. There is a rule of thumb that developed from project financings based on oil properties which suggest that half of the oil (or coal in this case) or half the future cash flow should still be in the ground at the point in time when the loan pays out. The Bank Revised Projection assumes a loan of $99,200,000 which pays out December 31, 1996 and based on the Table III Revised Projection, there is $149,310,000 of cash flow available for debt service remaining at December 31, 1996. This $149,310,000 is 38% of the total projected $393,690,000 of cash flow available for debt service.

38% is probably a reasonable coverage in this hypothetical case because this analysis gives no credit for probable and possible reserves and also because this type of a coal reserve can be predicted with a greater degree of certainty than oil reserves.

8. Present Worth Coverage Ratio - The present value of the cash flow available for debt service has been computed and is shown on the bottom of Table III. The cumulative present value (discounted at 12%) for the cash flow available for debt service covering the life of the property is $165,814,000. This means that the cash flow stream indicated on Table III could retire a $165,814,000 loan plus 12% interest over the life of the mine. Another coverage test that is sometimes considered is to relate this $165,814,000 present value to the amount of the loan. Based on the $99,200,000 loan suggested by Table III the coverage would be 1.67 computed as described below:

$$\frac{\text{Cumulative Present Value Of Cash Flow Available For Debt Service}}{\text{Amount Of Loan}} = \frac{\$165,814,000}{\$99,200,000} = 1.67$$

9. Sensitivity Studies - Table III has been described as a base case or most realistic case. Banks will also prepare one or more other revised projections in order to determine how sensitive the project is to various contingencies. For example, Table IV illustrates how sensitive the base case economics are to changes in interest rates. The projection in Table IV is identical to Table III except that a 16% interest rate has been assumed for the life of the Loan and the annual coverages are positive in every year except 1992 where there is a very slight deficit. However, over the life of the Loan there is $49,690,000 of cumulative excess cash flow available for debt service and if the projection is carried out through 2,003 then $149,630,000 of Excess Cash Flow is available.

FINANCIAL EVALUATION FOR MINERALS INDUSTRY

TABLE IV

Revised Projection - Sensitivity (Assumes Interest Rate of 16%) - Western Surface Coal Property
Projection of Comparative Income Accounts
(000's Omitted)

	YEARS ENDED DECEMBER 31,									
	1987	1988	1989	1990	1991	1992	1993	1994	1995	1996
Production - Tons	2,000	3,000	6,000	6,000	6,000	6,000	6,000	6,000	6,000	6,000
Gross Revenue from sale of coal	$26,000	$39,000	$78,000	$78,000	$78,000	$79,800	$83,760	$87,960	$92,340	$96,960
Pass through of severance tax	7,280	10,920	21,840	21,840	21,840	22,320	23,460	24,600	25,860	27,180
Total Gross Revenue	$33,280	$49,920	$99,840	$99,840	$99,840	102,120	$107,220	$112,560	$118,200	$124,140
Operating Costs (including Corporate Overhead)	$9,550	$15,190	$32,190	$34,120	$36,170	$38,340	$40,640	$43,080	$45,670	$48,410
Maintenance and Repairs of Property, and Equipment	850	850	1,730	1,730	1,730	1,730	1,730	1,730	1,730	1,730
Government Royalty	4,160	6,240	12,480	12,480	12,480	12,770	13,400	14,070	14,780	15,520
Severance tax	7,280	10,920	21,840	21,840	21,840	23,340	23,460	24,630	25,860	27,150
Depreciation and depletion	7,230	7,230	7,230	7,230	7,190	7,190	7,190	7,190	7,190	7,190
Total	$29,070	$40,430	$75,470	$77,400	$79,410	$82,370	$86,420	$90,700	$95,230	$100,000
Net Operating Income	$4,210	$9,490	$24,370	$22,440	$20,430	$19,750	$20,800	$21,860	$22,970	$24,140
Interest Expense (16%)	16,290	16,290	15,780	14,750	13,730	11,900	9,260	6,620	3,980	1,330
Net Income (Loss) before Taxes on Income	$(12,080)	$(6,800)	$8,590	$7,690	$6,700	$7,850	$11,540	$15,240	$18,990	$22,810
Taxes on Income										
Current	$ -	$ -	$ -	$ -	$ -	$ -	$ -	$ -	$ -	$3,700
Deferred	-	-	4,300	3,840	3,350	3,930	5,780	7,640	9,510	7,700
Total Tax	$	$	$4,300	$3,840	$3,350	$3,930	$5,780	$7,640	$9,510	$11,400
Net Income (Loss)	$(12,080)	($6,800)	$4,290	$3,850	$3,350	$3,920	$5,760	$7,600	$9,480	$11,410
Cash Flow available for Principal	$(4,850)	$430	$15,820	$14,920	$13,890	$15,040	$18,730	$22,430	$26,180	$26,300
Required Principal Payments	-	-	6,200	6,200	6,200	16,120	16,120	16,120	16,120	16,120
Annual Coverage	-	-	2.55	2.41	2.24	.93	1.16	1.39	1.62	1.63
Cash Flow Available for Debt Service	$11,440	$16,720	$31,600	$29,670	$27,620	$26,940	$27,990	$29,050	$30,160	$27,630
Debt Service (Principal plus Interest)	16,290	16,290	21,980	20,950	19,930	28,020	25,380	22,740	20,100	17,450
Annual Coverage	.70	1.03	1.44	1.42	1.39	.96	1.10	1.28	1.50	1.58
Excess Cash Flow (Available for Prepayments Or Dividends)	($4,850)	$430	$9,620	$8,720	$7,690	$(1,080)	$2,610	$6,310	$10,060	$10,180
Cumulative Excess Cash Flow	($4,850)	($4,420)	$5,200	$13,920	$21,610	$20,530	$23,140	$29,450	$39,510	$49,690

3.4

OPTIMUM PRODUCTION RATE FOR HIGH-GRADE/LOW TONNAGE MINES

ROSS GLANVILLE

WRIGHT ENGINEERS LTD., VANCOUVER, B.C. CANADA

INTRODUCTION

The Optimum Production Rate (OPR) is one of the most important parameters in the evaluation of a mineral deposit. The OPR can also be expressed as the Optimum Mine Life (OML) in years since the expected mine life is determined by dividing the OPR per year into the estimated ore reserves.

Unfortunately, very little time and effort has been directed towards the determination of the OPR. Instead, "rules of thumb" are often applied to select a mine life without due consideration of the economic implications of such a selection. A "justification" for a particular production rate is often based on a pre-conceived arbitrary requirement for a mine life of 5, 10, or 15 years, for example. In this paper it is demonstrated that such arbitrary selections for high-grade/low-tonnage mines (such as many underground gold/silver deposits) often lead to sub-optimal results. In fact, an apparently uneconomic deposit at an arbitrary production rate may be economic at the OPR. Consequently, investment opportunities may be overlooked if one analyzes mine properties based on arbitrary production rates.

This paper analyzes the interrelationships of variables such as production rates, capital and operating costs, cut-off grades, discount rates, metal prices, etc. The analysis shows that the OML for high-grade/low-tonnage deposits is often in the range of two to four years. Although such mine lives intuitively appear too short to many individuals, it should be noted that, in the past, much of the mining industry became familiar with the economics of large scale mining operations. The rules of thumb developed for such large scale operations should not be applied to high-grade/low-tonnage mines such as many of the underground gold deposits.

DEFINITION OF OPTIMUM PRODUCTION RATE

The OPR is selected on the basis of maximizing the net present value of the after-tax cash flows. However, as discussed later in this paper, the production rate selected for a particular mine may be somewhat higher or lower than the theoretical OPR as a result of other variables such as:
1) the probability or expectation of additional reserves being discovered

2) the likelihood of being able to custom-mill nearby deposits owned by others

3) the configuration or attitude of the orebody, which may place upper limits on the production rate

4) the residual, or salvage values, of the mine/mill facilities

COMPARISON BETWEEN DIFFERENT PRODUCTION RATES

For an orebody with definite physical cutoffs*, or boundaries (such as faults or unconformities), the total ore reserves at different production rates may be identical. Consequently, for such an orebody, varying production rates result in differing mine lives. Although the metallurgical recoveries, the mine dilution, and other similar physical parameters may vary slightly at different production rates, the major differences are the capital costs and operating costs per ton. The ramifications of these differences are especially important at low production rates (up to approximately 1000 tons per day), but can also be significant at higher production rates. In

---
*Later in this paper, the aspect of an economic cut-off, as opposed to a physical cut-off, is introduced.

addition to the cost differences between various production rates is the fact that the cash inflows (return to invested capital) are realized much sooner at higher production rates.

The combination of the above factors cause significantly different cash flow profiles, which, in turn, result in substantial differences in net present values between different production rates. Before presenting and discussing the graphs of net present values versus production rates, the capital and operating costs at several different production rates for a specific deposit will be analyzed.

CHARACTERISTICS OF THE ORE DEPOSIT

For purposes of illustration, we have assumed an underground gold mine in an accessible area of central British Columbia with diluted mining reserves of 500,000 tons grading 0.45 ounces per ton. We also assumed good mining conditions, a 500-ft shaft, long hole open stopes, primary crusher underground, average ore hardness, rod and ball grinding, and 90% gold recovery. The remaining assumptions are provided in the detailed cash flow outputs. However, a summary of the 400 ton per day case is presented in Table 13.

Although this particular example is used for purposes of illustration, we have anylyzed many other high-grade/low-tonnage deposits in different parts of the world, and found that the principles presented in this paper also apply to these other properties.

CAPITAL COSTS

Based on Wright Engineer's extensive experience and computerized data bases, capital costs were developed for operations with production rates of 100, 200, 400, 600, 800, and 1000 tons per day. With mining reserves of 500,000 tons, these daily tonnages translate into mine live of approximately 15, 7, 3 1/2, 2 1/2, and 1 1/2 years, respectively. The actual "breakdown" of the various capital cost components are presented in Tables 1 through 6, with the total capital costs versus production rate illustrated in Figure 1. As the total capital costs do not increase very rapidly as the production rate increases (or mine life decreases). This occurs because there is a large component of fixed costs (such as access roads, power supply, administrative buildings, water supply and other infrastructures) in the total capital costs to low production capacities.

Another way of expressing the relationship of capital cost to production rate is to show the percentage increases in the total capital costs for a doubling of capacity. This relationship is presented in graphical form in Figure 2. The key aspect of this graph is that the percentages are very low at low production capacities, but increase relatively rapid. For example, the graph shows the percentage capital cost increase for a doubling of capacity from 100 (to 200) tons per day to only 15%. However, a doubling of capacity

Table 1
Capital Cost Estimate
Hypothetical Gold Mine

100 stpd

1984 Cdn.$
(millions)

Mine development	6.60
Site development	0.56
Crushing	1.01
Processing	4.93
Water supply	0.39
Tailings disposal and water reclamation	0.71
Power supply and distribution (hydro)	1.34
Ancillary buildings	2.75
Access road, surface vehicles and fuel storage	1.66
Engineering and field supervision	2.01
Administration costs	1.83
TOTAL CAPITAL COST	24.38

Table 2
Capital Cost Estimate
Hypothetical Gold Mine

200 stpd

1984 Cdn.$
(millions)

Mine development	8.96
Site development	0.59
Crushing	1.14
Processing	5.47
Water supply	0.41
Tailings disposal and water reclamation	0.75
Power supply and distribution (hydro)	1.84
Ancillary buildings	3.13
Access road, surface vehicles and fuel storage	1.66
Engineering and field supervision	2.25
Administration costs	1.88
TOTAL CAPITAL COST	28.08

Table 3
Capital Cost Estimate
Hypothetical Gold Mine

400 stpd

1984 Cdn.$
(millions)

Mine development	12.38
Site development	0.64
Crushing	1.22
Processing	6.48
Water supply	0.44
Tailings disposal and water reclamation	0.83
Power supply and distribution (hydro)	1.84
Ancillary buildings	3.28
Access road, surface vehicles and fuel storage	1.67
Engineering and field supervision	2.74
Administration costs	1.99
TOTAL CAPITAL COST	33.50

Table 4
Capital Cost Estimate
Hypothetical Gold Mine

600 stpd

1984 Cdn.$
(millions)

Mine development	15.35
Site development	0.69
Crushing	1.30
Processing	7.44
Water supply	0.47
Tailings disposal and water reclamation	0.91
Power supply and distribution (hydro)	1.84
Ancillary buildings	3.42
Access road, surface vehicles and fuel storage	1.68
Engineering and field supervision	3.11
Administration costs	2.10
TOTAL CAPITAL COST	38.30

Table 5
Capital Cost Estimate
Hypothetical Gold Mine

800 stpd

1984 Cdn.$
(millions)

Mine development	18.20
Site development	0.73
Crushing	1.38
Processing	8.34
Water supply	0.50
Tailings disposal and water reclamation	0.99
Power supply and distribution (hydro)	1.84
Ancillary buildings	3.56
Access road, surface vehicles and fuel storage	1.69
Engineering and field supervision	3.58
Administration costs	2.20
TOTAL CAPITAL COST	43.00

Table 6
Capital Cost Estimate
Hypothetical Gold Mine

1000 stpd

1984 Cdn.$
(millions)

Mine development	21.01
Site development	0.78
Crushing	1.45
Processing	9.20
Water supply	0.54
Tailings disposal and water reclamation	1.06
Power supply and distribution (hydro)	1.84
Ancillary buildings	3.70
Access road, surface vehicles and fuel storage	1.70
Engineering and field supervision	3.88
Administration costs	2.31
TOTAL CAPITAL COST	47.47

Figure 1

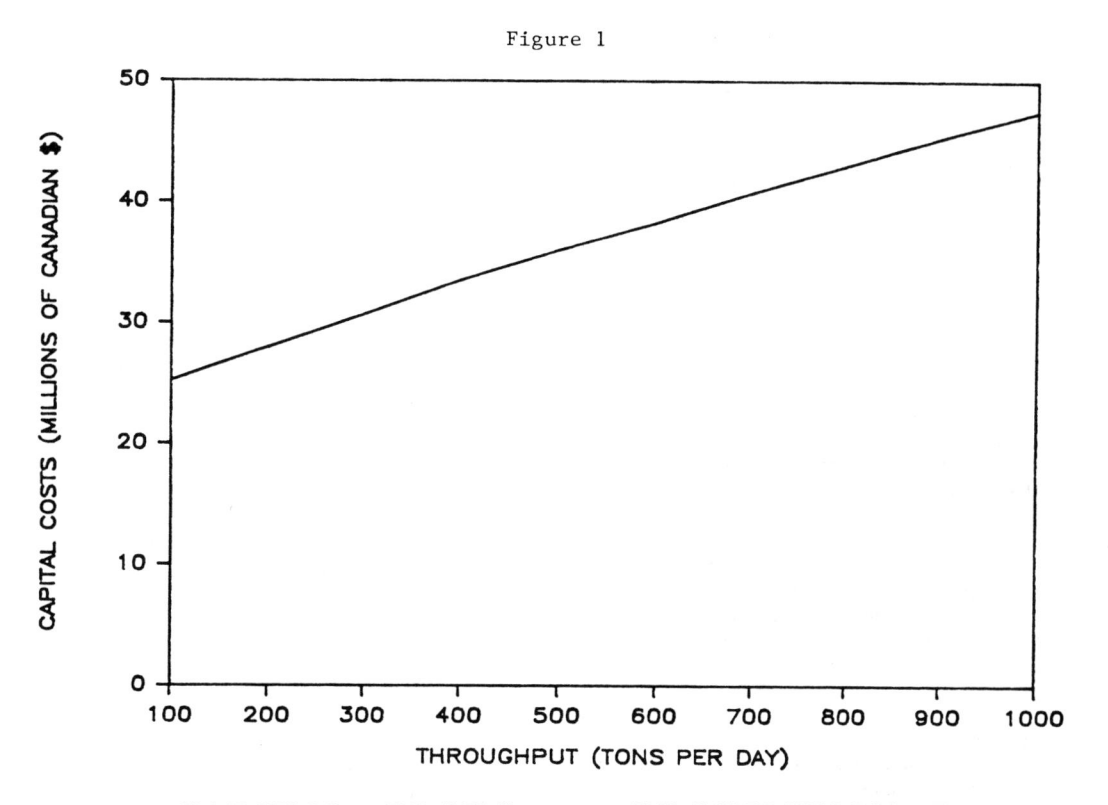

CAPITAL COSTS vs PRODUCTION RATE

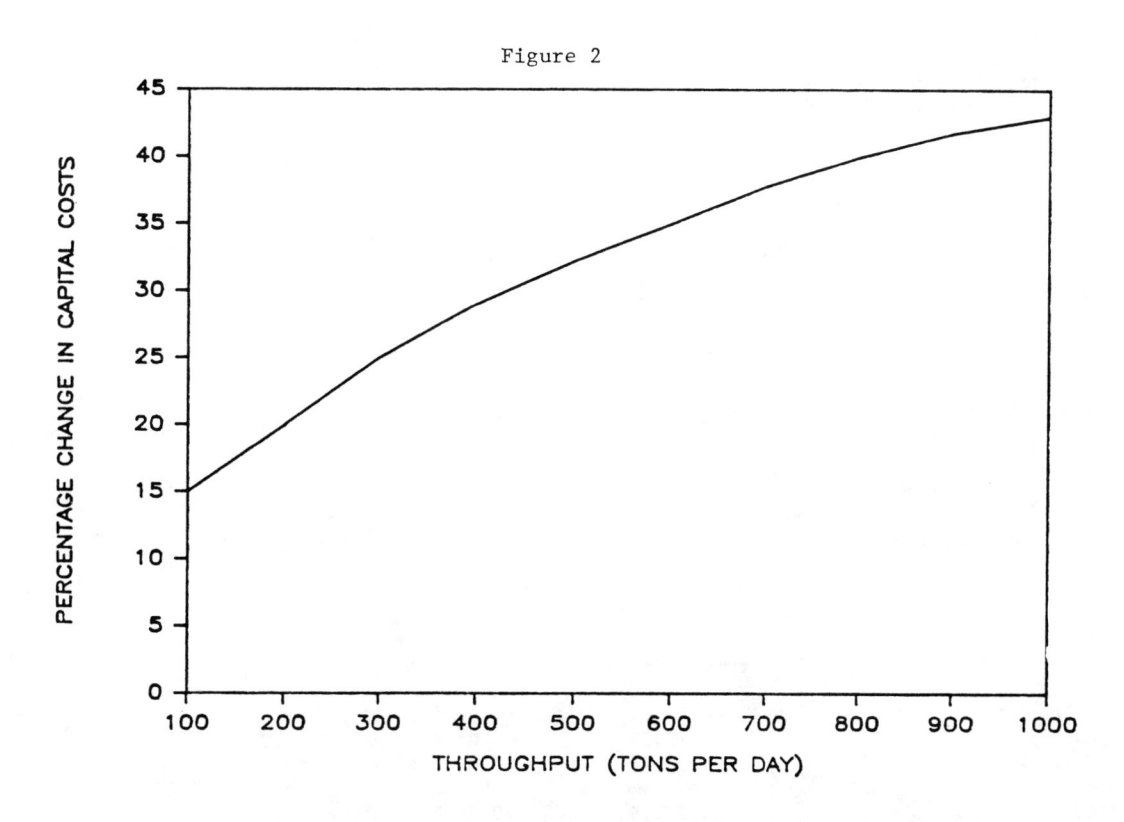

Figure 2

FOR A DOUBLING OF CAPACITY

from 400 (to 800) tons per day requires an increase in capital of almost 30%. At significantly higher production levels, a doubling of capacity requires well over 50% higher capital costs. In fact, as production rates became very high, the economies of scale diminish quite rapidly and a doubling of capacity requires capital costs almost twice as high.

OPERATING COSTS

As in the case with the capital costs, operating costs were developed for operations with production rates of 100, 200, 400, 600, 800, and 1000 tons per day. These operating costs, along with the major cost classifications, are presented in Tables 7 to 12, and shown in graphical form in Figure 3. This latter graph shows the dramatic decrease in operating costs as the production rate increases from very low levels. This dramatic decrease again results from the fact that there is a larger component of fixed costs in the total operating costs at low production levels. As the tonnage throughput increase, these fixed costs are "spread" over a greater number of tons, thus lowering the cost per ton dramatically.

For example, the operating cost decreases by almost $20 per ton when production is increased from 200 to 400 tons per day. However, a similar 200 tpd increases in production from 800 to 1000 tons per day results in the operating cost per ton decreasing by less than $4 per ton.

Table 7
Operating Cost Estimate
Hypothetical Gold Mine

100 stpd

1984 Cdn.$
per ton Milled

A. Mining Cost	49.63
B. Milling Cost	15.96
C. General Overheads	37.46
TOTAL COST	$103.05

Table 8
Operating Cost Estimate
Hypothetical Gold Mine

200 stpd

1984 Cdn.$
per ton Milled

A. Mining Cost	37.61
B. Milling Cost	12.09
C. General Overheads	26.49
TOTAL COST	$ 76.19

Table 9
Operating Cost Estimate
Hypothetical Gold Mine

400 stpd

1984 Cdn.$
per ton Milled

A. Mining Cost	28.50
B. Milling Cost	9.16
C. General Overheads	18.73
TOTAL COST	$ 56.39

Table 10
Operating Cost Estimate
Hypothetical Gold Mine

600 stpd

1984 Cdn.$
per ton Milled

A. Mining Cost	24.24
B. Milling Cost	7.79
C. General Overheads	15.30
TOTAL COST	$ 47.33

Table 11
Operating Cost Estimate
Hypothetical Gold Mine

800 stpd

1984 Cdn.$
per ton Milled

A. Mining Cost	21.60
B. Milling Cost	6.95
C. General Overheads	13.24
TOTAL COST	$ 41.79

Table 12
Operating Cost Estimate
Hypothetical Gold Mine

1000 stpd

1984 Cdn.$
per ton Milled

A. Mining Cost	19.76
B. Milling Cost	6.35
C. General Overheads	11.84
TOTAL COST	$ 37.95

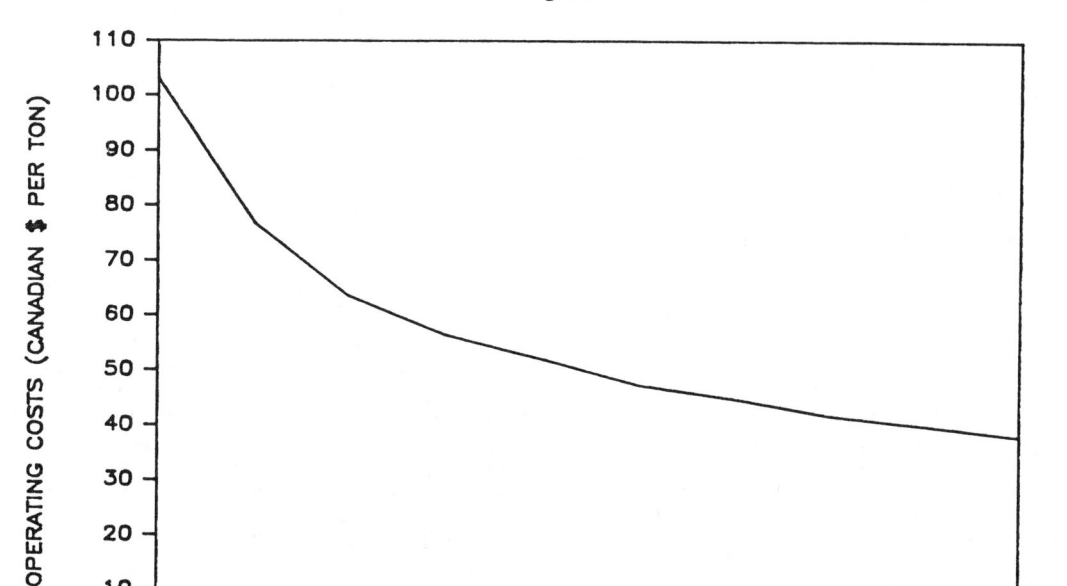

Figure 3

OPERATING COSTS vs PRODUCTION RATE

TRADE-OFF BETWEEN CAPITAL AND OPERATING COSTS

As noted above, the operating costs per ton decrease dramatically as production capacities increase from low levels. Consequently, the operating margin (revenue minus operating costs) increases significantly. Nevertheless, we must recognize that the capital costs also increase at higher production rates. The most important point, however, is that up to a particular production rate (the OPR) the impact of the decrease in operating costs per ton is much greater than the increase in capital costs. This occurs because operating capital costs per ton decline so rapidly, while capital costs increase only slightly, when increasing from low production levels, an increase in the production rate results in the operating costs per ton decreasing only very slightly and the capital costs increasing more significantly.

Ignoring the time value of money and risk (both of which must be incorporated in the discount rate) for the moment*, some simple calculations will illustrate the trade-off between capital and operating costs for two production levels - one at 100 tons per day, and one at 400 tons per day. For reserves of 500,000 these throughput capacities translate into mine lives of approximately 15 and 3 1/2 years, respectively (assuming a 350 day per year operation). The undiscounted costs are present below:

	100 tpd	400 tpd
Op costs/ton	$103	$56
Op costs in tl.	$ 52.0 min.	$28.9 min.
Cap. costs in tl.	$ 24.4 min.	$33.5 min.
Total Costs	$ 76.4 min.	$62.4 min.

Another way of viewing the impact of producing at 400 tons per day instead of 100 tons per day is to consider the incremental, or marginal, costs. In this case the incremental capital costs are $9.1 million ($33.5 - $24.4) while the incremental operating costs saving is $23.1 million ($52.0 - $28.9). Although the incremental capital cost is expended earlier than the "receipt" of the operating cost savings, at 400 tons per day the operating cost savings are realized over the first 3 1/2 years of production (since the mine life is only 3 1/2 years at 400 tpd). Consequently, it should be immediately obvious that the incremental investment is very profitable on a present value

---
*The impact of discount rate is discussed later in this paper.

basis (even ignoring the fact that the revenue is received much sooner at 400 tpd than at 100 tpd). This is confirmed by the detailed cash flow analysis, as discussed in the next section. It should be noted here that we have ignored any salvage value or disposal value of the mine/mill facilities at the end of the mine life.

CONSTRUCTION PERIOD/START-UP

Although the capital and operating costs have been emphasized to this point, other factors which are presented in the next section, must be incorporated into the cash flow financial analyses.

One of the important factors, is that the construction time for larger projects will be greater than that for smaller projects. To incorporate this consideration, we have assumed that the 100 tpd project would take 1 year to construct while the 1000 tpd project would take 2 years to construct. The times for the other production rates between 100 and 1000 tpd have been interpolated between the 1 and 2 years, respectively.

Another factor that is often significant is the fact that there may be start-up and shut-down inefficiencies due to a variety of factors, including personnel training, mill tune-up, environmental clean-up, labor turnover, etc. No matter how high the presumed production rate, it cannot be maintained uniformly, with instantaneous start-up and die-down, and without sequencing problems or the constraints of continuous critical-path elements in the production process. As a result, for purposes of this analysis we have assumed that the production rates for the first six (6) months and last six (6) months of operation would be 30% lower, and that the operating costs per ton over these same periods would be 20% higher. As a result, the higher production rates (shorter mine lives) are impacted more severely than the lower production rates.

RESULTS OF FINANCIAL ANALYSIS

A graph summarizing the after tax net present values at production levels of 100, 200, 400, 600, 800, and 1000 tons per day is presented in Figure 4. A summary computer printout of one cash flow financial analysis at a production level of 400 tons per day is presented in Table 13. The detailed cash flow analyses, including all of the income and mining tax information, have been omitted to save space.

Although the assumptions behind the graph in Figure 4 included a $350 U. S. gold price and a 12% real after-tax discount rate, large variations in the gold price and discount rate do not significantly affect the optimum production level.*

---
*See Figure 5 and Figure 6 for confirmation of this statement.

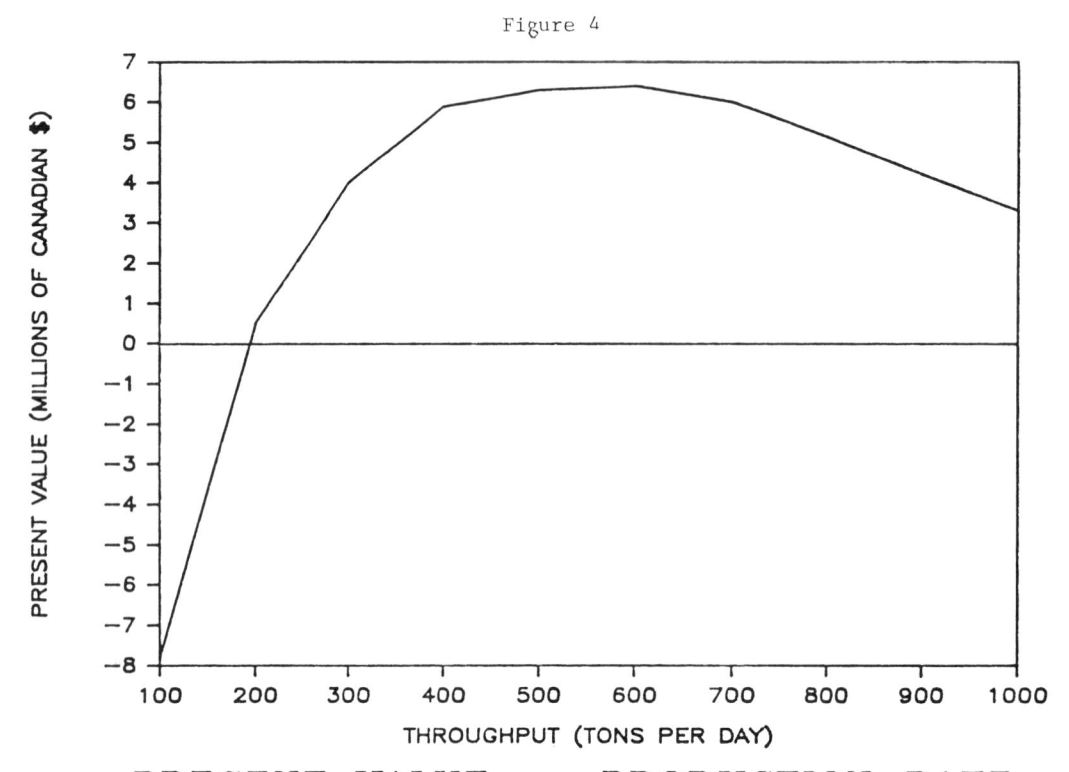

Figure 4

PRESENT VALUE vs PRODUCTION RATE
AT $350 US GOLD

FINANCIAL EVALUATION FOR MINERALS INDUSTRY

Table 13

CASHFLOW SUMMARY

	1985	1986	1987	1988	1989	1990	ACCUM
ORE MILLED (000 S TONS)	0.00	71.00	140.00	140.00	149.00	0.00	500.00
NET SMELTER RETURN	0.000	14.953	29.484	29.484	31.379	0.000	105.300
-OPERATING COSTS	0.000	4.335	7.895	7.895	8.797	0.000	28.921
MINE SITE INCOME	0.000	10.617	21.589	21.589	22.582	0.000	76.378
-FEDERAL INCOME TAX PAID	-0.261	-0.110	-0.000	0.000	6.795	0.000	6.424
-B.C. INCOME TAX PAID	0.000	0.000	0.000	0.000	5.371	0.000	5.371
-B.C. MIN RES TAX PAID	0.000	0.000	0.000	0.000	4.784	0.000	4.784
TOTAL TAXES PAID	-0.261	-0.110	0.000	0.000	16.949	0.000	16.578
CASH FLOW BEFORE CAPITAL COSTS	0.261	10.727	21.589	21.589	5.633	0.000	59.800
-CAPITAL COSTS - CLASS 28	10.820	4.390	0.000	0.000	0.000	0.000	15.210
-CAPITAL COSTS - PROCESSING	6.240	2.710	0.000	0.000	0.000	0.000	8.950
-CAPITAL COSTS - DEVELOPMENT	6.670	2.670	0.000	0.000	0.000	0.000	9.340
-WORKING CAPITAL REQUIRED	0.000	2.495	1.453	0.000	0.190	0.000	4.138
-CAPITALIZED INTEREST	0.000	0.000	0.000	0.000	0.000	0.000	0.000
INITIAL CAPITAL COSTS	23.730	12.265	1.453	0.000	0.190	0.000	37.638
+WORKING CAPITAL RECOVERY	0.000	0.000	0.000	0.000	0.000	4.138	4.138
+SALVAGE	0.000	0.000	0.000	0.000	0.000	0.000	0.000
TOTAL CAPITAL COSTS	23.730	12.265	1.453	0.000	0.190	-4.138	33.500
CASH FLOW BEFORE FINANCING	-23.469	-1.538	20.136	21.589	5.444	4.138	26.300
+PRIMARY BANK LOAN DRAWDOWN	0.000	0.000	0.000	0.000	0.000	0.000	0.000
+OPTIONAL LOAN DRAWDOWN	0.000	0.000	0.000	0.000	0.000	0.000	0.000
-SCHEDULED LOAN REPAYMENT	0.000	0.000	0.000	0.000	0.000	0.000	0.000
-OPTIONAL LOAN REPAYMENT	0.000	0.000	0.000	0.000	0.000	0.000	0.000
-INTEREST EXPENSE	0.000	0.000	0.000	0.000	0.000	0.000	0.000
NET EQUITY CASH AVAILABLE (REQUIRED	-23.469	-1.538	20.136	21.589	5.444	4.138	26.300
ACCUMULATIVE TOTAL	-23.469	-25.007	-4.871	16.719	22.162	26.300	0.000
DISCOUNTED NCF (8.0 PCT)	-22.583	-1.370	16.612	16.491	3.850	2.710	15.710
DISCOUNTED NCF (10.0 PCT)	-22.377	-1.333	15.867	15.466	3.545	2.450	13.618
DISCOUNTED NCF (12.0 PCT)	-22.176	-1.298	15.168	14.520	3.269	2.219	11.702
DISCOUNTED NCF (14.0 PCT)	-21.981	-1.264	14.512	13.648	3.019	2.013	9.947
DISCOUNTED NCF (16.0 PCT)	-21.790	-1.231	13.894	12.842	2.791	1.829	8.336
DISCOUNTED NCF (18.0 PCT)	-21.605	-1.200	13.313	12.096	2.585	1.665	6.854
DISCOUNTED NCF (20.0 PCT)	-21.424	-1.170	12.765	11.405	2.396	1.518	5.491
BEFORE TAX PAYBACK PERIOD (YEARS	2.2	0.0	0.0	0.0	0.0	0.0	0.0
AFTER TAX PAYBACK PERIOD (YEARS)	2.2	0.0	0.0	0.0	0.0	0.0	0.0
PRE-TAX RATE OF RETURN (PCT)	39.67	0.00	0.00	0.00	0.00	0.00	39.67
AFTER TAX RATE OF RETURN (PCT)	30.22	0.00	0.00	0.00	0.00	0.00	30.22

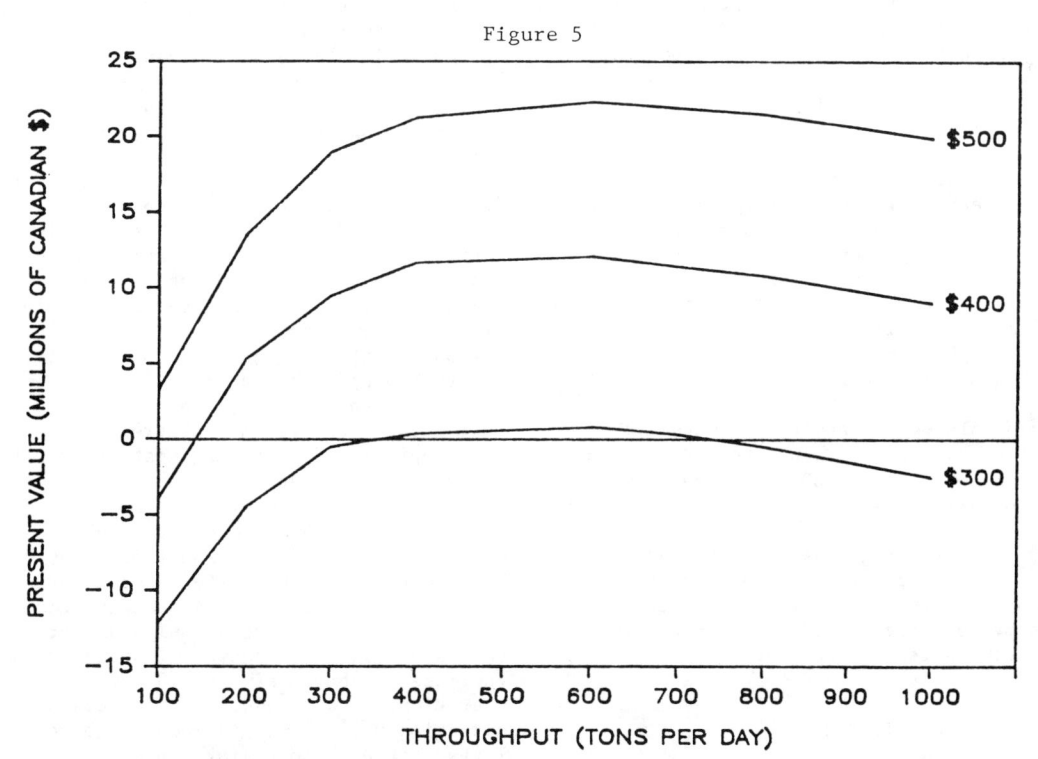

Figure 5

PRESENT VALUE vs PRODUCTION RATE
AT $300, $400 & $500 US GOLD

Figure 6

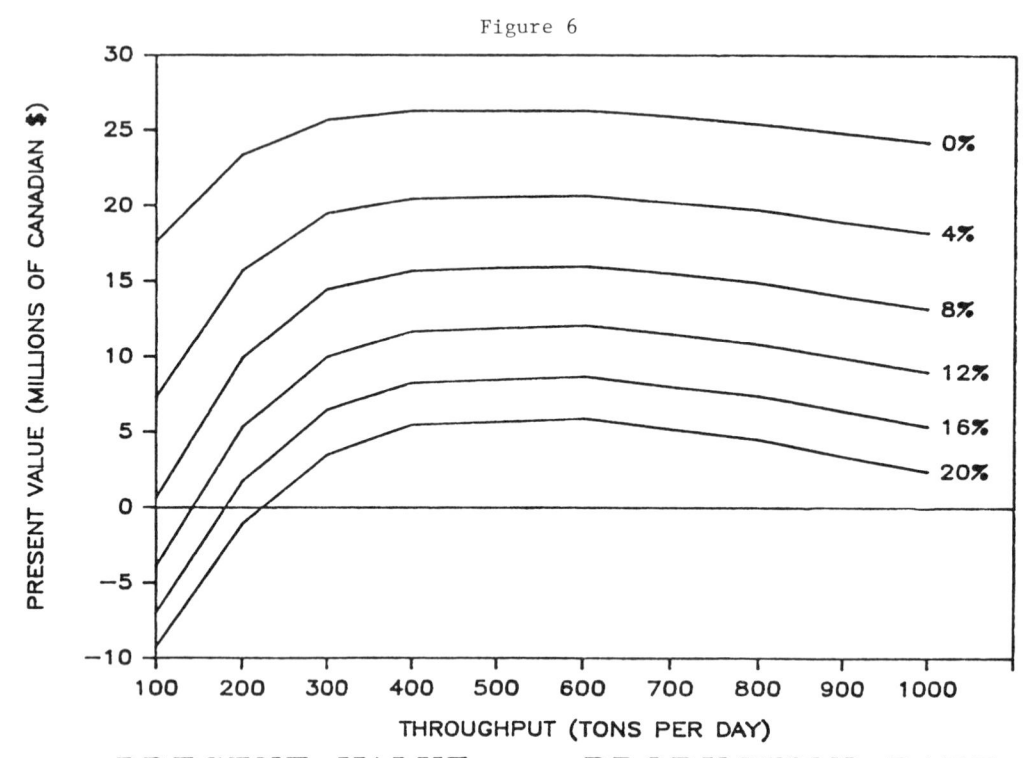

PRESENT VALUE vs PRODUCTION RATE
AT $400 US GOLD

In our example, the optimum production rate is near 600 tons per day, (a mine life of approximately 2 1/2 years) since the present value is the highest at that rate. In fact, the net present value is $6.4 million at 600 tons per day, whereas it is negative (meaning it is uneconomic) at 100 tpd. Consequently, if one arbitrarily picked a mine production rate of 100 tpd (15 years), the project would appear uneconomic. If one selects a production rate of 200 tpd (7 years), the project would just barely be economic with a net present value of only $0.5 million. At 400 tpd the net present value increases dramatically to $6.0 million. It is slightly more attractive at 600 tpd (2 1/2 years) with a net present value of $6.4 million. The present value then declines to $5.1 million at 800 tons per day and to only $3.3 million at 1000 tpd.

Although many people would intuitively expect a 7 year mine live (200 tpd) to be more profitable than a 2 1/2 year mine live (600 tpd), the results show just the opposite.

Consequently, a detailed analysis of the OPR must be carried out. More money can be "saved" or "made" by spending very small dollars to determine the OPR than by spending considerably more time, effort, and money in less productive areas.

SENSITIVITIES

In order to determine the impact on the OML of changing some of the key variables, a variety of "sensitivities" were carried out. Sensitivities (tabulated in Table 14) of net present value to changes in the price of gold are presented in Figure 5, while sensitivities to the discount rate are shown in Figure 6. As can be seen from Figure 5, the OML does not change significantly as a result of metal prices ranging from $300 to $500 per ounce of gold. However, at higher priced gold, the optimum shifts slightly to an even higher production rate. As can be seen from Figure 6, the discount rate does not significantly affect the OPR. However, at very low discount rates, the "range of reasonable production rates" (where the net present values are not significantly lower than the comparable range at higher discount rates. For example, at a 0% discount rate the range is approximately from 250 tpd to 850 tpd, whereas at a 20% discount rate the range is approximately from 350 tpd to 750 tpd. This occurs because the interrelationship between the capital and operating costs dwarfs the impact of the discount rate.

It should be noted here that the discount rate chosen can be an important factor in determining the optimum mine life for larger mining operations. This occurs because the operating costs per ton decrease only slightly and the capital costs increase significantly as one increases from, say a 40,000 tpd open pit operation, to a 60,000 tpd operation. This situation will result in much longer optimum mine lives than those for high-grade/low-tonnage properties. Consequently, the impact of the discount rate becomes an important determinant of

FINANCIAL EVALUATION FOR MINERALS INDUSTRY

Table 14
Net Present Values

tpd		0	4	Discount Rate 8	12	16	20
1000	$300	8.7	4.2	0.6	-2.5	-5.0	-7.2
	$400	24.2	18.2	13.2	9.0	5.4	2.4
	$500	38.8	31.4	25.2	19.9	15.4	11.6
800	$300	10.4	6.1	2.5	-0.5	-3.0	-5.1
	$400	25.4	19.7	14.9	10.8	7.4	4.5
	$500	39.4	32.4	26.5	21.5	17.2	13.5
600	$300	11.3	7.2	3.7	0.8	-1.7	-3.7
	$400	26.3	20.7	16.0	12.1	8.7	5.9
	$500	39.7	32.9	27.2	22.3	18.1	14.5
400	$300	11.5	7.0	3.4	0.4	-1.0	-4.1
	$400	26.3	20.5	15.7	11.7	8.3	5.5
	$500	38.7	31.8	26.1	21.3	17.2	13.7
200	$300	8.8	3.0	-1.3	-4.5	-7.1	-9.0
	$400	23.4	15.7	9.9	5.3	1.7	-1.1
	$500	35.3	26.1	19.1	13.5	9.1	5.5
100	$300	2.2	-4.9	-9.4	-12.2	-14.1	-15.4
	$400	17.6	7.3	0.6	-3.9	-7.0	-9.3
	$500	30.4	17.7	9.2	3.3	-0.9	-4.0

mine life in these situations.

Although the OPR has been selected on the basis of maximizing the net present values of the after-tax cash flows, an OPR selected on the basis of the after-tax discounted cash flow rate of return (or internal rate of return) would be almost identical in most situations. This can be seen in Table 14 and Figure 7, where the throughput rate is plotted against the rate of return at gold prices of U. S. $300, $400, and $500.

IS THE OPTIMAL MINE LIFE ALWAYS BEST?

Although the financial theory tells us that the theoretical optimum production rate (or optimum mine life) should be chosen, practical experience suggests that this may not always be the case. However, one should quantify this "practical experience" and thus be able to justify a different production rate.

Figure 5 shows that at the optimum production level of 600 tpd, the net present value is $6.4 million. However, the net present values of $6.2 million and $5.1 million at 400 tpd and 800 tpd, respectively, are not significantly lower. In other words, over a fairly wide range (400 to 800 tpd) of production rates, the net present values change very little. In fact, the differences are

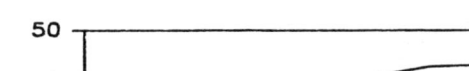

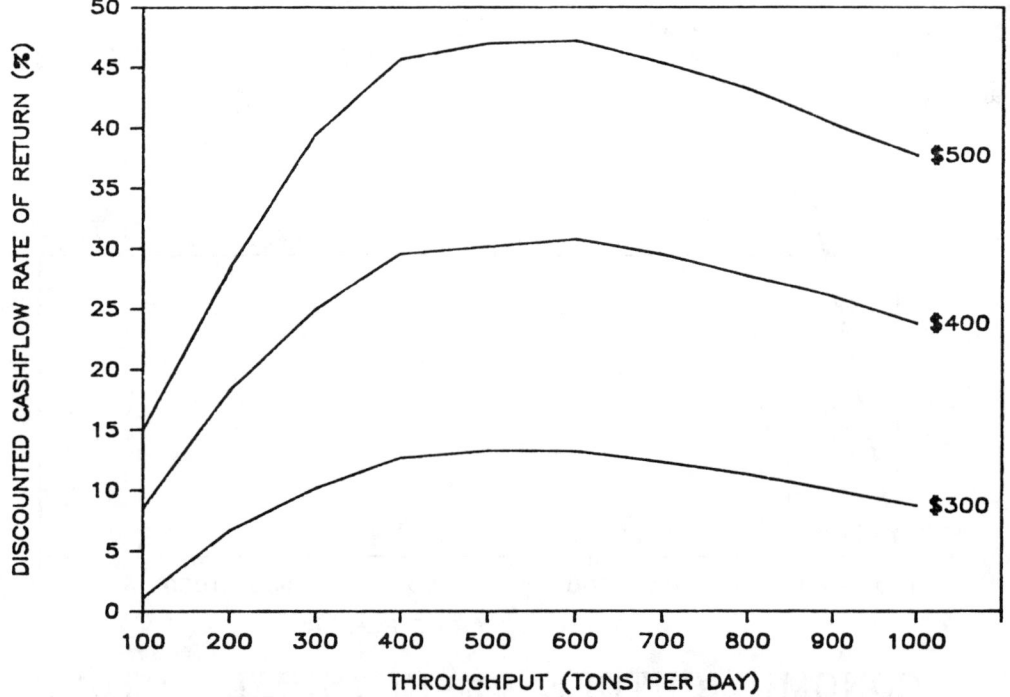

Figure 7

RATE OF RETURN vs THROUGHPUT
AT THREE GOLD PRICES

relatively insignificant when one recognizes the potential of error of estimating in a variety of input parameters. However, at production levels of less than 400 tpd, the net present value falls off rapidly.

If one felt that there was practically no possibility of finding additional ore, or felt that the 500,000 ton estimate was perhaps too high, then it appears that a "practical optimum" level might be 400 tpd, while the net present value is almost as high as it is at 600 tpd. Conversely, if it was felt that there was a high probability of substantially increasing the reserves, then the higher production level of 800 tpd would be recommended. If there were other deposits in the area, they might be custom milled after your own ore is all milled in less than 2 years (at 800 tpd) after start-up. In addition, with a lower cut-off grade possible at a higher production rate, there are improved chances that future peripheral discoveries of low grade or marginal ore might be profitably mined. However, if one could not develop enough working faces in the small orebody to enable production at the "optimum level," then the "optimum" level should be reduced because of this physical constraint.

IMPACT OF CUT-OFF GRADE DETERMINATION

In many cases where reserves are approximately 500,000 tons, for example, the production level is arbitrarily set at, say, 200 tons per day (7 year mine life). If this were done in this case, the cutoff grade would have been based on the operating costs for a 200 tpd operation. Since the operating costs at 200 tpd ($76 per ton) are much higher that those of 600 tpd ($48) the cut-off grade at 200 tpd would be much to high for the optimum production level of 600 tpd.

If we assumed an economic cutoff (that is, ore gradually decreasing in grade) as opposed to a physical cutoff (such as a fault or unconformity) which we have assumed up to now, the cutoffs at 200 tpd and 600 tpd would be about 0.20 ounces and 0.13 ounces respectively at $350 U. S. gold. Since there might be a considerable amount of tonnage between grades of 0.13 and 0.20, the optimum production rate would permit this to be mined at a profit. Thus, the net present value at the OPR would be higher than what we have assumed in the example to date. Figure 8 shows that the effect of optimizing the cutoff will be to shift the optimum point to a higher production level.

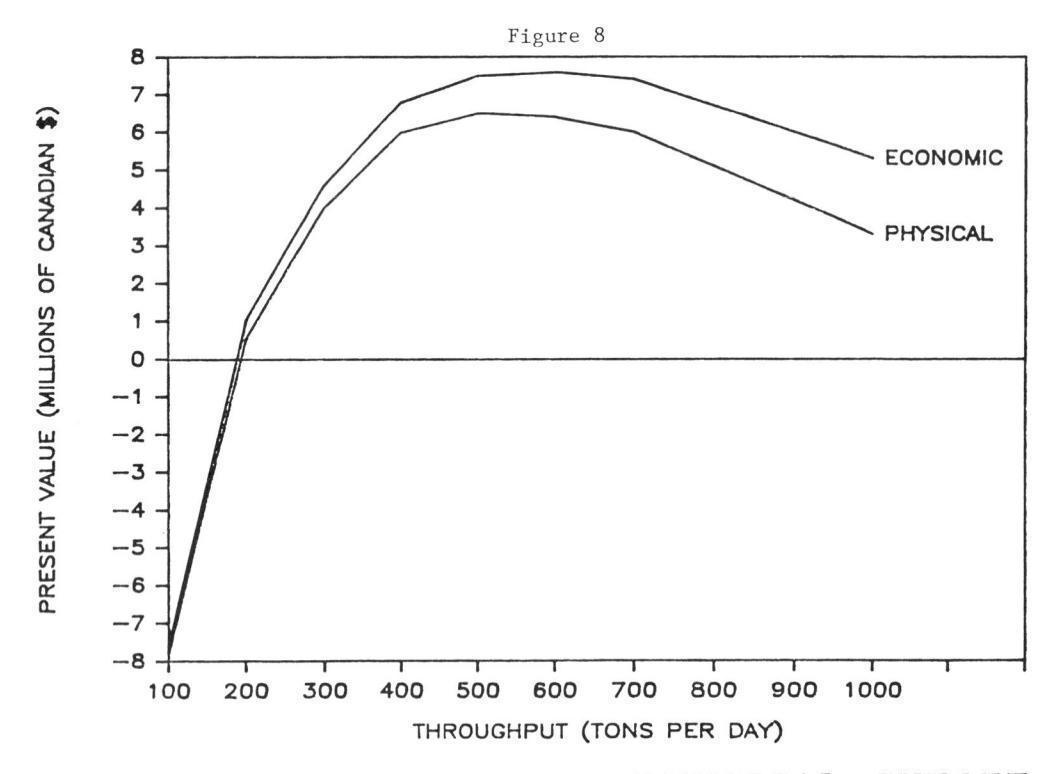

Figure 8

ECONOMIC CUTOFF vs PHYSICAL CUTOFF
AT $350 US GOLD

The amount of "shift" will obviously depend on the amount of tonnage (and grade) between the economic cutoffs at different production levels.

CONCLUSION

As can be seen from the graphs and the analysis presented in this paper, considerable attention should be directed towards determining the optimum production rate, since dramatic increases in value can result from relatively insignificant expenditures. Once a decision to design and build at a particular scale is made, it is often too difficult and costly to change. Consequently, a detailed analysis should be carried out to determine some of the key variables, at several different production rates (which should "bracket" the optimum rate), such as:
1) operating costs
2) capital costs
3) tonnage and grade at different cutoff grades
4) salvage value of the mine/mill facilities

Although the optimum production level will vary, depending on the relationship between the key variables, in different situations, the principles outlined in this paper still apply. In many cases, the optimum mine life for high-grade/low-tonnage deposits is between 2 and 4 years. As a result one cannot arbitrarily assume a production level; since, in many cases an otherwise economic deposit (at the OPR) will often be seen to be uneconomic. Consequently significant investment opportunities may be overlooked.

Table 15
Discounted Cashflow Return On Investment

Production Rate (tons per day)	Gold Price				
	300	350	400	450	500
100	1.1	4.9	8.5	12.1	15.0
200	6.7	12.6	18.3	23.6	28.4
400	12.7	21.3	30.2	38.4	46.1
600	13.2	19.5	30.8	39.0	47.0
800	11.3	21.9	27.7	35.8	43.2
1000	8.7	16.4	23.8	30.8	37.7

3.5

SENSITIVITY ANALYSIS FOR MINING PROJECTS

JOHN C. ROBISON III

FIRST NATIONAL BANK OF CHICAGO

INTRODUCTION

Sensitivity analysis is a means of gauging the impact of individual risks on a financing. Key risks can occur in three time periods:

- Feasibility, engineering and construction phase;
- Start up phase (usually through completion);
- Operating phase (post completion).

If a company has agreed to cover these risks prior to completion, then less attention will be paid to the sensitivity of the project in the first two time periods. Since the operating phase of the project generates cash for both operations and debt service and generally does not begin until two to six years after the loan is signed, it is wise to look at scenarios other than the projected "base-case" scenario in order to determine the project's capability to operate successfully under a broad range of probable future events.

Lenders are compensated at a fixed spread over their borrowing base and generally are not compensated if the project is successful beyond expectation. They are also subject to the loss of principal plus funding costs if the project performs below expectation. Because of this, lenders establish sensitivities around the parameters of conservative projections developed in the project feasibility studies and the cash flows resulting from these projections. The initial cash flow projections should demonstrate the project's ability to meet cash costs, debt service requirements (principal and interest), and still have sufficient cushion to cover contingencies such as strikes, price fluctuations, force majeur situations, etc. (See Schreiber, Chapter 10). The selection of the level of sensitivity analysis to be performed is the result of both objective criteria, including probabilistic analysis applied to the technical parameters, and, of equal importance, subjective criteria applied by the analysts, especially with regard to the economic and financial parameters impacting the project cash flows.

RISK SENSITIVITY

Assumptions used in the cash flow projections should be realistic. Risks identified by the lenders should be identified and cash flows prepared to assess whether these project risks are acceptable. Typical project risks include the following (see also Chapter 9):

- Sufficient reserves - quantity and quality
- Product price
- Production level
- Operating costs
- Availability of energy and supplies
- Transportation and infrastructure requirements and costs
- Interest rate
- Completion risks
- Tax level
- Political risks
- Cost over-run risks
- Operator experience
- Management
- Technological risks
- Environmental permits and risks
- Foreign exchange risks
- Insurance coverage
- Equity of sponsors
- Inflation rate
- Capital expenditure during project life

In order to see the effect on the cash flow projections, each risk should be analyzed for its impact on the cash flow generating capacity of the project. Sensitivities can be examined to see if the resultant cash flow projection will meet the lenders minimum credit standards.

1. Reserve Sensitivity

    In general, lenders will not assume the reserve risk. However, it should also be noted that lenders receive loan proposals in which reserves range anywhere from proven to possible. Sensitivities are not generally run for various reserve levels and reserve estimates are either considered acceptable or not acceptable by the lenders.

2. Price Sensitivity

If sales contracts are applicable, the lender would expect the term of the contract to either extend throughout the life of the loan or have a reasonable expectation, based on past operating history, that the contract will be renewed prior to the expiration date. Sensitivities on the price should reflect downward movement at any renegotiation. For example, the Japanese coking and steam coal contracts recently have been significantly adjusted downward both in terms of price and tonnage. Products that are considered commodities and are traded on an exchange have a market for whatever tonnage can be produced, although not at a specific price. Because these commodities are freely traded, the price tends to be cyclical and lenders generally will not accept cash flow projections from borrowers based on price forecasts steadily inflating with time, particularly if prices are escalated at a faster rate than costs. Sensitivities can be made assuming:

1. Constant growth of prices, but at less than cost inflation;
2. Constant prices and costs;
3. Constant prices with inflated costs;
4. Increasing historical price trends projected in the future;
5. Cyclical prices based on historical trends.

Any increasing spread of profit margin will increase cash availability to service debt, but may not be realistic. Lenders can, however, face the loss of a portion of the outstanding loan or a forced extension of the term (length) of the loan by constantly squeezing margins due to lesser than expected price inflation.

3. Production Sensitivity

Volume or tonnage is estimated for new projects and should include sufficient downtime for scheduled maintenace and normal delays. With on-going projects, historical volumes can be used if the historical factors are applicable to the future estimates. Sensitivities may need to be run at less than stated capacities (80% to 95%) due to industry experience or the impact of cyclical demand.

4. Operating Sensitivity

The operator should be experienced in bringing this type of project on stream. Operating costs can increase substantially over estimates used in the base-case scenario. Operating cost sensitivities are conventionally plus or minus 10% and 20% although attention to specific cost inputs may be more relevant in particular situations, e.g. power cost for an aluminium smelter.

5. Interest Sensitivity

Interest rates are generally projected from relatively short historical experience. If interest rates have been rising quickly at the time of analysis, then future interest-rate projections generally will be high. A base interest rate for the next 10 years should be prepared with sensitivities run on incremental levels (we use 3% steps) to analyze the impact which increased interest rates will have on the project economics. Interest rates of 3-5% above the assumed underlying inflation rates are a common guideline for the base case, with the higher end of this range more closely approximating the actual spread during the recent environment of high rates and low inflation rates in the U.S.

6. Tax Sensitivity

For domestic U.S. projects the income tax levels of 48% may not be the appropriate level for the payer of taxes as a function of ownership structure. An analysis of the project is necessary to determine the effect of annual depletion, depreciation and tax-losses carried forward. Little sensitivity anslysis is conducted in this area for U.S. projects as lenders have traditionally borne the project's tax risks even in overseas ventures.

7. Cost Over-Run Sensitivity

Most projects include a 10% over-run contingency and care is needed to determine if this is a to-be-spent contingency or simply a contingency cushion above the to-be-spent budget. Experience suggests that this contingency is usually consumed during construction due to unforeseen circumstances by the estimator and does not inlcude delay in construction or higher than expected costs of supplies. Sensitivities should be made showing the impact of further capital cost over-runs or additional interest costs to be capitalized due to a delay in start-up. Conventional over-run sensitivities range to plus 20-30% of the to-be-spent budget.

8. Environmental Sensitivity

Environmental risks may increase costs and reduce production. Acid rain issues, waste dumps, and the political impact of the project in the immediate area may have a dramatic effect on the project economics. Sensitivities can be run reducing production levels or increasing operating costs associated with maintaining compliance with environmental standards.

9. Foreign Exchange Sensitivity

   Devaluation of the local currency is the equivalent of reducing operating costs, if the product is priced in US$. Conversely, if the local currency appreciates in value, the net effect is a reduced profit margin. This sensitivity can be simulated by changing operating costs that are incurred in the local currency. It is very difficult to assess the impact of currency fluctuations, especially over the usually long life, say 10 years, of most mine financings.

10. Inflation Sensitivity

    Inflation should be projected for the base case. Sensitivities should be run showing no inflation and different inflation levels for price and operating cost. Differential inflation rates between different countries or between the costs, prices, and loan currency will filter back through the foreign exchange sensitivity over the life of the loan, again complicating the analysis.

11. Capital Expenditure Sensitivity

    Replacement capital expenditures must be paid in order to maintain the specified production level. Sensitivities can be run showing increases due to changes in operations or additional replacements.

CONCLUSION

Individual risks require different measures of sensitivity to determine a project's robustness in the face of the many pressures and risks associated with attaining the expected future cash flow. Lenders predictably will take a cautious view, which caution can to some extent be ameliorated by sound project studies and independent consultants' reviews.

3.6

SENSITIVITY OF MINING PROJECTS TO CAPITAL, OPERATING AND DEBT COST VARIATIONS

RICHARD A. BOULAY

ABSTRACT

The effects of capital, operating and debt cost variations on project performance criteria are examined using a theoretical model and an actual case history. On an after-tax basis, financial measurements are found to be most sensitive to operating cost changes, somewhat sensitive to reasonable changes in capital costs and relatively insensitive to interest rate fluctuations.

PROJECT EVALUATION MODEL

This paper discusses relationships between variations in important forecasting areas -- capital costs, operating costs, debt interest rates -- and the financial performance of mining projects. Two computerized project evaluation models are discussed. The first, used to investigate the sensitivity of rate of return to interest rate and debt/equity variations, is a generalized capital investment model which reflects Canadian tax law. The second, used to evaluate the inter-relationships between operating costs, capital costs and interest rate fluctuations and their effects upon rate of return and other project measurement criteria, is a model which was actually used to evaluate a coal reserve acquisition and operating loan in the Eastern United States.

A simple project simulation model is outlined in Table 1. It represents a project in isolation from a corporation's overall financial position. The arbitrary assumptions incorporated into the Base Case outlined in Table 1 are:

- A total capital investment of $100 million, including accrued pre-production interest and any cost overruns to the beginning of Year 1.

- Project life is in excess of 10 years, although these calculations consider only the first 10 years.

- Debt interest rate of 10.0%. This is an arbitrarily chosen rate.

- Debt to equity ratio is 75/25, that is, $75 million debt and $25 million equity.

- Depreciation is calculated on a straight line basis over 10 years.

- Capital Cost Allowance (write-off of capital expenditures for tax purposes) is assumed to be 30% on a declining balance basis.

- Deferred taxes are calculated using the depreciation and capital cost allowance rates.

- The project's total tax rate is assumed to be 48%.

- 85% of the project cash flow is dedicated to debt repayment. The remainder is assumed to be allocated for other corporate uses.

The annual operating profit has been arbitrarily set at $22.5 million per year in order to force a rate of return on equity of approximately 15% (actually 14.6%). Payback of the $75 million debt component of the total capital expenditure occurs in Year 6. The rate of return on total funds employed (a true measure of the project's efficiency) is 6.8% in the Base Case.

SENSITIVITY RESULTS

Table 2

% DEVIATION

	INTEREST	TOTAL INTEREST PAID
I = 7.5%	-25%	-30%
I = 10.0%	-	-
I = 12.5%	+25%	+35%

Table 2 summarizes one result of varying the interest rate. The Base Case assumes the cost of funds at 10% per year. Table 2 indicates that the total dollar amount of interest paid over the

---

Based on a paper in AIME Council of Economics, Proceedings, 1979.

TABLE 1
CAPITAL PROJECT MODEL
INTEREST RATE VARIATION SERIES   CASE 0 : BASECASE NO. 1

	1	2	3	4	5	6	7	8	9	10
<<INPUT PARAMETERS>>										
INTEREST RATE	0.100	0.100	0.100	0.100	0.100	0.100	0.100	0.100	0.100	0.100
TOTAL FUNDS REQUIRED	10,000	0	0	0	0	0	0	0	0	0
DEBT/EQUITY RATIO	0.75	0.75	0.75	0.75	0.75	0.75	0.75	0.75	0.75	0.75
OPERATING PROFIT	2,250	2,250	2,250	2,250	2,250	2,250	2,250	2,250	2,250	2,250
*/• C.P. TO DEBT	0.85	0.85	0.85	0.85	0.85	0.85	0.85	0.85	0.85	0.85
<<OUTPUT RESULTS>>										
DEPRECIATION	1,000	1,000	1,000	1,000	1,000	1,000	1,000	1,000	1,000	1,000
DEBT INTEREST(•/•)	750	623	484	334	171	18	0	0	0	0
PRE-TAX INCOME	500	628	766	916	1,079	1,232	1,250	1,250	1,250	1,250
C C A	1,500	1,628	1,766	1,916	1,510	504	353	247	173	121
TAXABLE INCOME	0	0	0	0	569	1,728	1,897	2,003	2,077	2,129
DEFERRED TAX	240	301	368	440	245	0	0	0	0	0
TAX PAID (CURRENT)	0	0	0	0	273	829	911	961	997	1,022
TAX PAID (FROM DEFERRED)	0	0	0	0	0	238	311	361	397	286
TOTAL TAX PAID	0	0	0	0	273	1,067	1,221	1,323	1,394	1,308
CUM TOTAL TAX PAID	0	0	0	0	273	1,340	2,562	3,884	5,278	6,587
NET PROFIT	500	628	766	916	806	165	29	-73	-144	-58
DEBT O/S BEGIN YEAR	7,500	6,225	4,842	3,341	1,712	177	0	0	0	0
C.P. GENERATED	1,500	1,628	1,766	1,916	1,806	1,165	1,029	927	856	942
DEBT O/S END YEAR	6,225	4,842	3,341	1,712	177	0	0	0	0	0
NET C.P. TO PROJECT	225	244	265	287	271	988	1,029	927	856	942
CUM NET C.P. TO PROJECT	225	469	734	1,021	1,292	2,280	3,309	4,236	5,092	6,034
R.O.R. ON EQUITY	14.6	0.0	0.0	0.0	0.0	0.0	0.0	0.0	0.0	0.0
R.O.R. ON TOTAL FUNDS	6.8	0.0	0.0	0.0	0.0	0.0	0.0	0.0	0.0	0.0
P.V. OF N.C.P. AT 12•/•	2,873	0	0	0	0	0	0	0	0	0
GOVT TAX TAKE(PV 8•/•)	4,542	0	0	0	0	0	0	0	0	0
TOTAL INT PAID	2,380	0	0	0	0	0	0	0	0	0

life of the project is sensitive to changes in the debt interest rate. Specifically, a decrease of 25% in the interest rate causes a decrease of 30% in the total amount of interest paid. Conversely, an increase of 25% results in an increase of 35% in the dollar amount of interest paid. The relationship is not quite linear due to the amortization and tax effects of varying interest rates.

Table 3

% DEVIATION

	INTEREST	TOTAL INTEREST AND TAXES PAID
I = 7.5%	-25%	-4%
I = 10.0%	-	-
I = 12.5%	+25%	+4%

Table 3 illustrates the sensitivity of the total amount of funds used for both interest and tax payments. As indicated in the Table, we can conclude that this project measurement is relatively insensitive to wide interest-rate fluctuations.

Table 4

INTEREST	RATE OF RETURN ON TOTAL FUNDS EMPLOYED
7.5%	7.6%
10.0%	6.8%
12.5%	6.0%

Table 4 continues to use, as an evaluation criterion, the changing interest rate (-25%, +25%) and this illustration indicates the sensitivity of the rate of return on total funds employed. One can conclude that the rate of return on total funds is somewhat sensitive to interest rate fluctuations but not so sensitive as to constitute an over-riding consideration in the evaluation of this project.

Table 5 shows the sensitivity of the equity rate of return to interest rate changes. Again, one can conclude that the equity rate of return is only somewhat sensitive to wide fluctuations in debt interest rate.

If the limitations inherent in this generalized

Table 5

INTEREST RATE	RATE OF RETURN ON EQUITY
7.5%	15.9%
10.0%	14.6%
12.5%	13.3%

capital project estimating model are accepted, one is forced to conclude that measures of project performance are relatively insensitive to large variations in the cost of funds. Project sponsors should still try to negotiate the most favourable interest rates, however, the significance of this cost item should be kept in perspective when considering an overall financing package.

Table 6

INTEREST RATE - 10.0%

DEBT/EQUITY	RATE OF RETURN ON EQUITY
90/10	26.0%
80/20	16.3%
75/25	14.6%
70/30	13.5%
50/50	10.9%

Table 6 summarizes the results of five debt to equity simulations including the Base Case simulation which assumes a debt to equity ratio of 75/25 and yields a rate of return on equity of 14.6%. At a debt to equity ratio of 50/50, the rate of return drops to 10.9%. The 90/10 debt to equity ratio yields an equity rate of return of 26.0% and at debt to equity ratios above 90%, the rate of return on equity will begin to climb very steeply. It is obvious that a project sponsor's return on a capital project is extremely sensitive to changes in the debt to equity structure.

These observations logically dictate that a project sponsor negotiating for the financing of capital funds will quickly recognize the wisdom of yielding on interest rate matters in order to negotiate increases in the debt to equity ratio. Unfortunately, this conclusion must be qualified by a practical caveat, namely that higher debt/equity ratios usually require increased loan security and that this increased loan security will utlimately translate as a cost to the project sponsor.

CASE STUDY

The unformatted, operating, cash flow and debt repayment pro-forma in Table 7 represents an Eastern U.S. coal project with the results of the simulations presented as Figure 1. A full description of input parameters and output results is given in Appendix 1. The key input parameters as they effect the sensitivity evaluations discussed here are Tons Sold (Line 3), Coal Price (Line 4), Operating Costs in $/ton (Line 5) and the loan amounts detailed in Lines 54 and 112, the latter constituting debt financing for plant and equipment emplacement as outlined in Line 51.

Figure 1 shows the rate of return on total funds employed (Line 63) on the vertical axis as a function of percentage deviations for each input parameter scenario as measured on the horizontal axis. The intensity of the negative slope of each plotted line is a direct measure of the project's sensitivity to variations in the input parameter represented by each line.

The steepness of the operating cost line indicates an extreme project sensitivity to variations in this input quantity. The flatter capital cost and interest lines show a much reduced sensitivity effect. Specifically, the slope of the operating cost line is 0.99:1, that is, a 1% (approximately) change in operating costs will result in a 1% change in the project rate of return. This compares with a capital cost line slope of 0.34:1 and an interest line slope of 0.08:1.

These results correspond to the results derived by similar analytical methods applied to other projects. The plots usually display a linear relationship between input parameter variances and output results except when large negative input values do not permit amortization of debt. In these unusual cases, the compounding effect of interest causes severe dislocations in the plotted results. The results of simulations which test higher and higher capital cost values are particularly susceptible to this phenomenon. Normally, however, these conditions do not exist when testing reasonable variations from Base Case values, that is, values which reflect current cost levels being experienced by the mining industry.

CONCLUSION

It can be concluded that bottom line project results are insensitive to interest rate variations even at high debt to equity ratios. This assumes, of course, that the project under consideration is a viable one and attractive to the extent that management will elect to proceed with the construction, management and financing arrangements necessary to activate the operation. Consequently, project sponsors would do well to capitalize on this insensitivity in negotiating for more favourable terms on matters which have higher impact on financing results.

Table 7

```
CASE 0 : U.S. COAL PROJECT
YEARS 1 THRU 10
DIVISION D1
```

	1 PERIOD	1	2	3	4	5	6	7	8	9	10
3	TONS SOLD (*1000)	200	400	600	600	600	600	600	600	600	600
4	COAL PRICE ($/TON)	26.50	30.00	32.00	32.00	32.00	32.00	32.00	32.00	32.00	32.00
5	OPERATING COST ($/TON)	17.50	19.00	21.00	21.00	21.00	21.00	21.00	21.00	21.00	21.00
6	*GROSS SALES (1000*S)	5300	12000	19200	19200	19200	19200	19200	19200	19200	19200
7	OPERATING COST ($000*S)	3500	7600	12600	12600	12600	12600	12600	12600	12600	12600
8	S.G.AND A ($000*S)	260	520	780	780	780	780	780	780	780	780
9	ROYALTIES ($000*S)	150	300	450	450	450	450	450	450	450	450
11	DEPRECIATION ($000*S)	500	800	1140	1192	1034	907	806	724	660	520
12	STATE TAX	265	600	960	960	960	960	960	960	960	960
13	*TOTAL COST OF SALES	4675	9820	15930	15982	15824	15697	15596	15514	15450	15310
14	*OPERATING INCOME($000*S)	625	2180	3270	3218	3376	3503	3604	3686	3750	3882
15	*INTEREST EXPENSE 1	658	678	558	438	318	198	78	78	78	78
16	*INTEREST EXPENSE 2	200	400	500	500	400	300	200	100	0	0
17	*INCOME AFTER INT($000*S)	-232	1103	2213	2281	2659	3006	3327	3508	3673	3805
18	*DEPLETION ALLOWANCE	-116	551	1106	1140	1329	1503	1663	1754	1836	1875
19	*TAXABLE INCOME	0	551	1106	1140	1329	1503	1663	1754	1836	1930
20	TAX AT 48*/*($000*S)	0	265	531	547	638	721	798	842	882	926
21	*TAX AFTER I.T.C.	0	132	266	315	598	681	758	802	882	926
22	*NET INCOME	-232	970	1947	1965	2061	2324	2569	2706	2791	2879
23	CASH FLOW	268	1770	3087	3157	3094	3231	3374	3431	3451	3406
25	WORKING CAPITAL REQU*MENT	575	200	0	0	0	0	0	0	0	0
51	PLANT AND EQUIPMENT	2000	2000	1000	0	0	0	0	0	0	0
52	EQUIPMENT REPLACEMENT	0	500	400	400	400	400	400	400	0	0
53	NON*ITC CAP. EXPEND.	0	0	0	0	0	0	0	0	0	0
54	PURCHASE LOAN (1)	6000	0	0	0	0	0	0	0	0	0
112	CAP. EQUIP. LOAN (2)	2000	2000	1000	0	0	0	0	0	0	0
27	WORKING CAPITAL INJECTION	575	200	0	0	0	0	0	0	0	-775
113	TOTAL LOANS	8575	2200	1000	0	0	0	0	0	0	-775
28	PAYMENTS LOAN 1 ($000*S)	0	1200	1200	1200	1200	1200	0	0	0	0
29	PAYMENTS LOAN 2 ($000*S)	0	0	0	1000	1000	1000	1000	1000	0	0
30	TOTAL LOAN PAYMENTS	0	1200	1200	2200	2200	2200	1000	1000	0	0
31	CURRENT AVAIL FOR PMTS	268	1270	2687	2757	2694	2831	2974	3031	3451	2631
32	NET C.F. TO PROJECT	268	70	1487	557	494	631	1974	2031	3451	2631
33	CUM NET C.F. TO PROJECT	268	338	1825	2382	2876	3508	5482	7512	10963	13594
62	P.V. OF NET C.F. (12*/*)	6112	0	0	0	0	0	0	0	0	0
63	R.O.R. ON FUNDS EMPLOYED	18.4	.0	.0	.0	.0	.0	.0	.0	.0	.0

Fig. 1 - Percentage deviation versus rate of return.

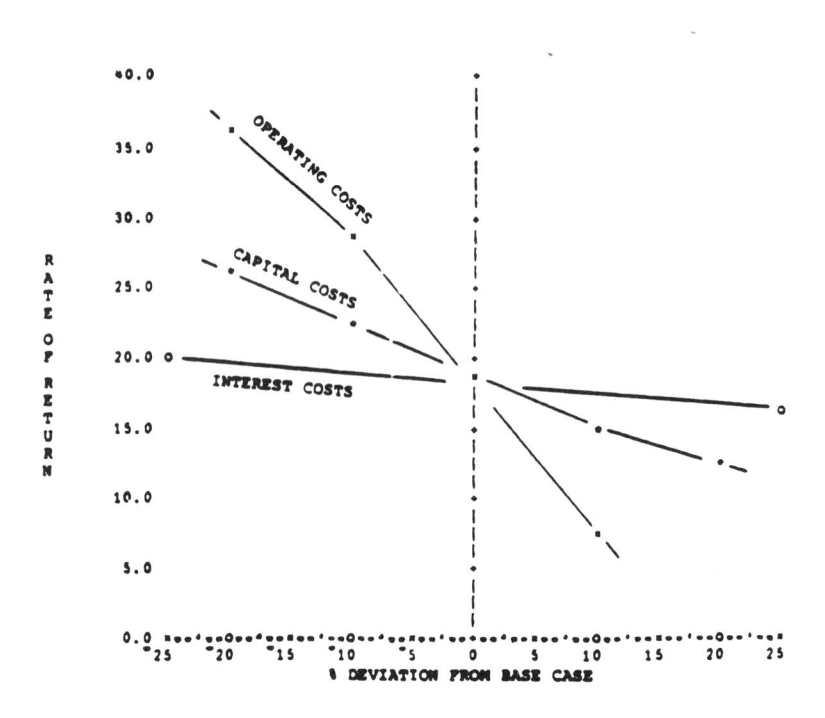

APPENDIX 1

DESCRIPTION OF INPUT PARAMETERS
AND OUTPUT PARAMETERS

Input Parameters
Line #

Line #	Parameter
107	TONS SOLD
108	MINING COST
4	COAL PRICE
25	WORKING CAPITAL REQUIREMENT
51	PLANT AND EQUIPMENT
52	EQUIPMENT REPLACEMENT
53	NON-ITC CAPITAL EXPENDITURE
54	PURCHASE LOAN (1)
112	CAPITAL EQUIPMENT LOAN (2)
27	WORKING CAPITAL INJECTION
28	REPAYMENTS LOAN 1
29	REPAYMENTS LOAN 2

Derivation of Output Results:
Line #

6 GROSS SALES — tons sold x coal price.

7 OPERATING COST — tons sold x mining cost. Sensitivities run at +5, +10, -5, -10, -20%.

8 SG & A — Sales general and administrative expenses calculated on the basis of $1.30 per ton sold.

9 ROYALTIES — Calculated at a rate of $0.75 per ton sold.

11 DEPRECIATION — Calculated using the double declining balance method and assuming initial depreciable asset base of $2.50 million.

12 STATE TAX — Calculated at a rate of 5.0% of gross sales.

13 TOTAL COST OF SALES — The sum of all costs outlined in lines 7 through 12 inclusive.

14 OPERATING INCOME — Derived by subtracting total cost of sales from gross sales.

15 INTEREST EXP.1 — Calculated at a rate of 10% of the outstanding balance of loans indicated in lines 54 and 27 as reduced by payments indicated in line 28. Moreover, the interest is calculated on the basis of drawdowns being made on the first day of the year and payments being made on the last day.

16 INTEREST EXP.2 — Calculated at a rate of 10% of the outstanding balance of the loan indicated in line 112 as reduced by payments indicated in line 29. Sensitivities at 7.5% and 12.5% p.a.

17 INCOME AFTER INTEREST — Derived by subtracting interest expenses from operating income.

18 DEPLETION ALLOWANCE — Calculated in accordance with I.R.S. regulations.

19 TAXABLE INCOME — Derived by subtracting depletion allowance from income after interest.

20 TAX AT 48% — Calculated on basis of 48% of taxable income.

21 TAX AFTER I.T.C. — Derived by subtracting the investment tax credit from the tax liability indicated in line 20. The income tax credit is derived according to I.R.S. regulations.

22 NET INCOME — Derived by subtracting tax after I.T.C. from taxable income.

23 CASH FLOW — Derived by adding net income and depreciation as indicated in lines 22 and 11, respectively.

113 TOTAL LOANS — Derived by adding the loans indicated in lines 54, 112 & 27. Sensitivities at +10%, +20%, -10%, -20%.

30 TOTAL LOAN PAYMENTS — Derived by adding the loan payments as detailed in lines 28 and 29.

31 CURRENT AVAILABLE FOR PAYMENTS — Derived according to the following formula: (L23 + L27 + L112) - (L25 + L51 + L52 + L53).

32 NET CASH FLOW TO PROJECT — Derived by subtracting total loan payments from current available for payments as indicated in lines 30 and 31, respectively.

33 CUMULATIVE NET CASH FLOW TO PROJECT — The cumulative amount of net cash flow is indicated in line 32.

62 PRESENT VALUE NET CASH FLOW (12%) — The present value of the cash flow stream as detailed in line 32 discounted at a rate of 12%.

63 R.O.R. ON FUNDS EMPLOYED — The true rate of return of the cash flow stream defined by the formula L31 - (L54 + L112 + L27). The rate indicated in column 1 of line 63 is the discount rate at which the cash flow stream reduces to 0 on a present value basis.

3.7

CONSIDERATIONS IN LEVERAGED STUDIES FOR MINERAL VENTURES

WILLIAM P. LOHDEN

Treasurer, Callahan Mining Corporation
Darien, Connecticut

INTRODUCTION

It is recognized that, for a variety of reasons, some companies in the mineral industry today are by no means cash-rich and, in fact, are reporting a growing proportion of long-term debt in their capital structures. A review of recent balance sheets of small, medium-size, and some larger companies, indicates that a fair number still have substantial positions in cash equivalents such as U.S. government issues and other short-term money market instruments.

In part, this paper concerns itself with companies which have sufficient cash and equivalents and/or cash generation to internally finance mining projects or expansions but which, for some reason, choose to borrow funds. Misconceptions which appear to frequently underlie so-called "leveraging" decisions are explored in this paper.

Also considered are those circumstances under which any company, including one that is cash rich, should consider borrowing for a project in an amount beyond what would otherwise be necessary based upon its internal financing ability. I do not attempt to cover the raising of equity capital in public markets in this paper.

OVERALL RETURN ON INVESTABLE CASH

If you approach an individual with an investment which can be financed from savings, or by borrowing at a rate higher than one is earning on savings, that individual will quickly tell you that to pay interest at a rate higher than is being earned on savings would reduce overall income. Assuming that additional investment opportunities are not currently being considered, the individual concludes that a draw-down on savings, leaving a cushion for contingencies suited to individual circumstances, is the correct financial decision. For reasons discussed later, the simple mathematical logic leading to this conclusion tends to become obscured at the corporate level.

When Leveraged Studies Are Inappropriate or Premature

Surprisingly, I have heard of a number of companies that could afford to finance an investment solely with internal funds, which have made DCFROR (i.e. discounted-cash-flow rate-of-return) studies not only on a 100% equity basis but also on a leveraged basis. Certainly, a leveraged study would be appropriate if additional projects were competing for more capital than was available internally. Often, however, other projects are not far enough along to warrant a leveraged study and there is some danger that a project may be undertaken because it has an estimated acceptable rate of return, on a leveraged basis, that is not ultimately earned because some of the other projects are postponed or abandoned and borrowing is less than expected.

For the above reasons, until it has been determined that (1) borrowing is a reasonable certainty for a particular project and (2) the timing and amount to be borrowed has been determined within a reasonable range of accuracy, I believe it is inappropriate and misleading to calculate DCFROR for a project other than on a 100% equity basis.

Even when there are several projects currently competing for funds and borrowing will be required to finance one or more of them, using the common yardstick of calculating DCFROR on a 100% equity basis is a safer method to initially evaluate and compare the attractiveness of the projects. Many companies, whether they would have to borrow or not, will reject projects that on a 100% equity basis do not meet their minimum rate of return standards. Some companies which have to borrow will undertake a project provided the project's DCFROR on a leveraged basis meets its minimum return standards. The danger of this approach is that even moderate shrinkages of estimated cash flows could plunge return below

---

Presented at the Joint MMIJ-AIME Meeting, Denver, Colorado, September 1-3, 1976.

the company's minimum standard. In other words, while leveraging can escalate DCFROR, if a project is less profitable than estimated, leveraging will reduce profits just as quickly as it will enhance them. Thus, borrowing adds financial risk above and beyond the operating risk of how the project itself will perform.

Intercompany Versus External Loans

A second example follows of where borrowing may be either unnecessary or the need for it overestimated, resulting in a reduction in overall return on shareholders' equity. This example concerns the parent company philosophy occasionally encountered which holds that, despite ample cash in the parent's treasury, an operating subsidiary should demonstrate its ability to "stand on its own two feet" and borrow from outside the corporate framework when it has a need for financing. It should be noted that where an entity is controlled by two or more corporations, the same principle could apply. Let me make it clear that I am referring to a situation where the foregoing philosophy exists for reasons other than the parent desiring to isolate itself from potential claims of the subsidiary's creditors. Rather, I am addressing those situations where the philosophy exists because it is believed to put greater pressure on the subsidiary to more tightly manage its funds and/or to repay its borrowings more quickly.

Assuming the parent's surplus cash is invested in relatively low-yielding short-term money market instruments, the result of the subsidiary's external borrowing at a rate higher than being earned on the short-term investments results in a net outflow of dollars from the consolidated corporate entity. In my view, a less expensive and more results-oriented alternative would be to strengthen operating controls at the subsidiary and improve the communication of financial information to the parent. In addition, of course, the subsidiary should not get a "free-ride", and the parent should charge an appropriate rate of interest on funds advanced. Since many corporations manage their short-term investment portfolio to provide for maturities of a portion of the portfolio at least monthly, loans can be made at maturity dates. If loans must be made prior to maturities by liquidating a portion of the portfolio, losses on sale, if any, would, of course, have to be weighed against the savings that would otherwise result from the loan being intercompany rather than external.

Why Overall Return on Investable Cash is the Key Consideration

The principle involved in the foregoing two examples is that the return on investment of the total pool of investable cash of a consolidated entity must be considered, not merely the results of the placement of a portion of such cash in a particular investment.

As illustrated in Tables 1 and 2, borrowing funds for a portion of the cost of a project can increase the DCFROR on the equity investment in the project. However, if the balance of surplus cash is invested in the money/stock/bond markets at an after-tax rate less than the cost of borrowed money, the overall DCFROR on the total pool of cash invested in the project, plus the market, will be lower than if the project had been financed on a 100% equity basis. Not only will the DCFROR be less but it follows that the overall growth in dollars from the project and the money market (as measured at the end of the project's life) will be less than if a 100% equity investment had been made. It also follows that return on shareholders' equity will be lowered as a result. Four case examples in Appendix 1 illustrate the overall growth-rate/return-on-investment concept.

Reasons for Misconceptions Pertaining to Leveraging

Earlier I referred to factors which appear to most frequently underlie misconceptions pertaining to leveraging. Those factors which I have been able to discern in my experience include:

1. Considering separate investment opportunities, with respect to the equity-debt financing mix, by:

   (a) viewing them as if they were financed from separate pools of capital rather than from total available funds; and

   (b) placing undue emphasis on the return on equity from the separate investments when, in fact, such emphasis should be subordinated to the primary objective of maximizing overall return on the corporation's total investable funds.

2. The notion that leveraging for the corporation as a whole, which is what really matters in the final analysis, can occur regardless of the corporation's cash position and the yield earned on its surplus cash. Hopefully, Appendix 1 will be helpful in dispelling this notion.

WHEN TO BORROW

When a company lacks sufficient funds to fully finance an attractive project, debt financing is obviously appropriate and will enhance DCFROR. Apart from this fundamental need, there are circumstances under which a company should consider borrowing for a project in an amount beyond what would otherwise be necessary based upon its internal financing ability.

Reserving Cash for Contingencies and Cost Overruns

An appropriate cash reserve, tailored to individual corporate needs, should be maintained. As long as borrowing is utilized above this amount, leveraging can occur once the amount of borrowing exceeds the reserve.

Non-recourse Loans

Non-recourse loans from institutional lenders and banks should be investigated. Provided loan covenants and obligations are complied with, such loans look solely to the cash flow from a project for payment of debt service thereby reducing risk to the mining company and possibly increasing borrowing capacity. An additional advantage could include flexible loan amortization based on actual production rather than on a fixed schedule.

In order to grant a loan on a non-recourse basis, the lenders will require independently prepared technical and economic feasibility studies. They must be convinced that the project is capable of yielding cash flows that generously cover debt service. A guarantee of completion will also be required. Typically, this guarantee will involve completion of the project by a specified date (usually somewhat later than estimated completion); coverage by the guarantor of cost overruns with equity; and a period of production during which the project will have performed reasonably in accord with the estimates of a feasibility study. Sales contracts for the output of the project are usually required as a condition of the lending. Further, once the project is completed, the company must agree to maintain the facility and equipment in good condition and continue to operate unless the project clearly becomes uneconomic.

When Financing More Than One Project

When work on an additional project in the development stage may be getting underway in the near future, which together with a current project might require more funds than would be available internally, debt financing obviously must be considered. If the current project had a significantly higher ratio of cash flow to debt service than indicated for the second project, it might ultimately be more profitable to borrow for the current project and reserve internal funds for the second project on the basis that more favourable borrowing terms would be available for the first project.

Lenders, of course, look to a corporation's overall ability to service debt; therefore, the foregoing observation becomes more valid as the size of one or both of the contemplated projects increases in relation to the company's existing size. Apart from this aspect, it should be noted that assuming (a) constant money market conditions, (b) that the company had not approached its maximum borrowing capacity, and (c) that the second project held promise of meeting a company's minimum DCFROR requirement, lending terms for the second project might not be too different from those of the first.

Surplus Funds Invested at a Higher Return

Overall profitability is increased when surplus internal funds can be invested at a higher return than the cost to borrow money. One of the means of achieving this objective is for a corporation to invest in preferred stocks. Although I know of several mining companies which have preferred stock portfolios, let me make it clear that I am not advocating investments in preferred stocks, but merely citing one of the few examples of how surplus funds can be invested at a return higher than the cost of borrowed money. The yields on preferred stocks are augmented by the 85% tax exemption on dividends applicable to a corporate recipient on the majority of such issues. Most corporate borrowers would pay an after-tax interest cost less than the after-tax yield from preferred stocks. Thus, rather than liquidate these investments, leveraging would result from borrowing funds to aid in financing a project in this example.

Industrial Revenue Bonds

Industrial revenue bond (IRB) financing is offered by agencies established by polical subdivisions of states to induce companies to undertake projects which will increase employment and otherwise benefit local and state economics. IRB's can be issued in certain states to finance mining projects. For example, proceeds from the sales of IRB's may be dedicated to the purchase of plant and equipment for a new mining project, or expansion of an existing project, which is leased by the agency to the mining company. With this type of financing, the mining company incurs long-term debt in the form of a capitalized lease obligation. The company will have the option to purchase the facilities at nominal cost following retirement of the bonds. The agency's debt is guaranteed by the company to the purchasers of the bonds and rentals under the lease with the agency are equal to an amount which repays the principal of and interest on the bonds. In practice, lease payments (debt service) are commonly paid directly to the holder of the bonds rather than to the agency of the state subdivision.

The type of IRB financing which I am describing limits capital expenditures to $5,000,000 within a given political jurisdiction during a six-year period beginning three years prior to the bond issue and ending three years after the bond issue. The $5,000,000 limitation includes any expenditure which could be capitalized under Internal Revenue Service regulations, whether the mining company does so or not, and also includes all capital expenditures whether from IRB proceeds or otherwise. Because many new mining projects require capital expenditures of over $5,000,000, the more likely use for IRB financing is in the expansion of existing facilities.

There is another form of IRB financing under which a maximum of $1,000,000 of bonds can be issued without any restrictions on total capital expenditures.

There is another form of IRB financing under which a maximum of $1,000,000 of bonds can be issued without any restrictions on total capital expenditures.

Notwithstanding the in-house ability to finance a project, leveraging can result from IRB financing because of the following:

1. During the period of time that the agency of the political subdivision owns the facility, no property taxes are payable since the agency is a non-taxpaying entity. As part of the inducement to proceed with the project, an agreement may be reached with the political subdivision under which the mining company pays an annual amount in lieu of property taxes for the lease period which is less than the amount of property taxes which would be incurred if the company owned the facilities.

2. Since the issuer of the IRB's is an agency of a state political subdivision, the interest on the bonds is tax-exempt to the purchasers.

When private placements of the bonds are made, an insurance company or bank is most frequently involved. Because the interest on the IRB's is tax-exampt, these institutional lenders will make IRB-related loans to companies at an attractive interest rate. For example, some banks are currently charging interest at a rate equal to 70-75% of their prime lending rate to better credit risks. This rate is an all-inclusive rate and does not require any compensating balances.

Leveraging can occur with this type of financing because property tax savings coupled with the current low interest rate on IRB-related loans may well result in an effective cost below that which surplus cash can earn in money market investments.

Leasing

Leasing is a form of borrowing with interest costs incorporated within the leasing terms. By "leasing", I am referring to a true lease rather than to a conditional purchase contract.

Straight across-the-board leasing of all facilities and equipment would typically be an expensive form of financing. Normally, when a company has a choice, a less expensive alternative would be to borrow funds to purchase the needed equipment. However, there could be circumstances under which selective leasing would increase DCFROR and, in these instances, the effect of using this form of financing should be evaluated as against purchasing. In making this evaluation, care should be taken to include all costs, for example, maintenance, that would be common to both alternatives.

"Selective leasing" might involve circumstances where certain equipment is needed for a relatively short period of time; for example, somewhere between when a firm commitment is made to put a project into production and when operations begin. Leasing, of course, eliminates the uncertainty inherent in purchasing of when, and at what price, equipment can be resold. An important input, therefore, into a study of leasing versus purchasing, would involve an appraisal of resale timing and value, including consideration of possible changes in technology. (See John Caldon's paper "Evaluation of the Lease or Buy Decision" in Chapter 3).

Instances of where leasing would leverage DCFROR could vary depending upon particular circumstances attendant to individual projects. Therefore, I am not attempting in this paper to cite numerous instances of where leasing might be considered as an alternative to purchasing, but merely suggesting that leasing, on a selective basis, should not be overlooked as a means of enhancing return on the equity investment in certain projects.

SUMMARY

In summary, one of the principal purposes of this paper has been an attempt to shed some light on distinguishing fact from the fiction that too often enshrouds the concept of leveraging and results in erroneous financing decisions. In addition, I hope that some of the circumstances I have listed (there are undoubtedly more), where even a cash-rich company can leverage or obtain some other benefit by borrowing for a project, prove useful in planning for the financing of mineral projects.

APPENDIX 1

Descriptions, Assumptions and Comments Pertaining to Table 1.

1. The four cases in Table 1 cover variations of how an $8,000,000 project with a ten-year operating life is financed when a total of $8,000,000 of investable cash is available. The $8,000,000 amount is arbitrary but the reasons for the project

cost and the investable cash amounts being the same is: (a) to eliminate the situation where with a lesser amount of cash there would be a definite need to raise additional funds and, (b) to compare overall results in four cases where the total amount available for investment is constant.

2. The project generates annual cash flows of $1,700,000 (recovery of investment plus return on investment) before debt service. The borrowing cases (II through IV) assume that debt is repaid in ten equal annual installments. A different pattern of debt repayment could have been assumed but in no event would the results of Cases II and III exceed Case I. In Cases II and III, however, the faster the debt is repaid, the higher the DCFROR will be because the interest rate on the debt is higher than the money-market interest rate.

3. A tax rate of 50% is assumed. Cash flows are after-tax dollars except it is assumed that the $8,000,000 cost to finance the project (whether from 100% equity funds or from a combination of equity and borrowed finds) results in no tax benefits at time zero. If part of the $8,000,000 total cost was immediately tax deductible, cash savings from tax benefits would be the same in all four cases and the order of ranking of the DCFROR for the cases would remain the same.

4. For simplicity in calculating, it is assumed that all investments and cash flows are end-of-period amounts and that interest is compounded annually. The cases lend themselves to the use of standard compound interest formulae and factors.

5. Interest on debt determined to have been incurred for the project would be a deduction in determining "taxable income from the property" under IRS regulations. If U.S. statutory percentage depletion was subject to the limitation of 50% of taxable income from the property, the amount of allowable tax depletion would be reduced, thereby reducing tax savings and cash flow. This possibility has not been taken into account in comparing the results of Case I versus the results of borrowing in Cases II through IV, but if it were a factor the DCFROR's and cash accumulations for the borrowing cases would be reduced to some extent.

6. A description of the four cases, including a summary of results of investing $8,000,000 under the conditions of Cases I, II, III and IV is as follows:

OVERALL RESULTS

CASE	CASH ACCUMULATION AT END OF PROJECT	DCF-ROR
I. $8,000,000 of equity, that is 100% of investable cash, is used to finance the project.	$19,489,000	9.3%
II. $1,000,000 of equity is invested in the project; $7,000,000 of surplus cash is kept in short-term investments yielding 3% after-tax; $7,000,000 is borrowed at 5% after-tax for use in the project.	$18,568,000	8.8%
III. $7,000,000 of equity is invested in the project; $1,000,000 of surplus cash is kept in short-term investments yielding 3% after-tax; $1,000,000 is borrowed at 5% after-tax for use in the project.	$19,357,000	9.2%
IV. Same as Case II except that $7,000,000 of surplus cash earns 7% after-tax in a preferred stock investment program.	$22,931,000	11.1%

Table 1 indicates that other than Case IV, Case I results in the highest overall DCFROR and cash accumulation. That is, $8,000,000 grows to $19,489,000 in ten years. Case II illustrates that by borrowing $7,000,000 for use in the project, the DCFROR on the project itself increases dramatically, but after taking into consideration the low 3% after-tax yield on the $7,000,000 which is kept in short-term investments, overall DCFROR falls below Case I and after-tax cash accumulation is $921,000 less than in Case I. Case III illustrates that by getting closer to a 100% equity investment, results improve over Case II but still lag behind those of Case I. Case IV results are the best and prove the point made earlier that if it is possible to earn more in a surplus cash investment program (e.g. preferred stocks) than the after-tax cost of borrowed money, overall profitability is enhanced.

(Editor's Note: The impact of interest costs and investment income will alter the tax profile for these Cases. The tax position of the company is also an important consideration when considering leasing. Statements about US tax deductions and IRB procedures have changed since this paper was written in 1976.)

FINANCIAL EVALUATION FOR MINERALS INDUSTRY

Table 1 Cash Flow Diagrams
After-Tax Cash Flows (000's omitted)

	Time Zero	1	2	3	4	5	6	7	8	9	10	Cash Accumulation	DCF-ROR	Explanation
Case I														
Project (Investment) cash flow	$(8,000)	$1,700	$1,700	$1,700	$1,700	$1,700	$1,700	$1,700	$1,700	$1,700	$1,700	$ -0-	16.7%	(1) There is no "cash accumulation" shown on the project line because each year the cash flow is transferred to an investment in the money market rather than remaining idle. DCFROR is as calculated on an "H-P 27".
(Reinvestment) in money market		(1,700)												
			→at end of each year									19,489	3.0%	(2) Uniform series annual compound interest factor applicable to 3.0% is 11.464 x 1,700 = 19,489.
Overall result	(8,000)	-0-		→at end of each year								19,489	9.3%	(3) Present value of a single future sum, 8,000 ÷ 19,489 = .4105 which factor in the compound interest tables interpolates to 9.3% interest.
Case II														
Project (Investment) cash flow before debt interest and repayment	(1,000)	1,700 {350} {700}	1,700 {315} {700}	1,700 {280} {700}	1,700 {245} {700}	1,700 {210} {700}	1,700 {175} {700}	1,700 {140} {700}	1,700 {105} {700}	1,700 {70} {700}	1,700 {35} {700}			
Debt Interest														
Debt repayment														
Project (Investment) cash flow	(1,000)	650	685	720	755	790	825	860	895	930	965			(1) See explanation No. (1) in Case I.
(Reinvested) in money market		(650)	(685)	etc., increasing at a constant gradient of $35 each year								9,161	3.0%	(2) To convert a gradient series to an equivalent series of uniform annual amounts, formula is (650 + 35(4.26)) = 799.10 x 11.464 = 9,161.
				→at end of each year										
(Invest) surplus cash in money market	(1,000)	-0-												
	(7,000)											9,407	3.0%	(3) 7,000 x single payment annual compound amount factor of 1.3439 for 3.0% = 9,407.
Overall result	(8,000)	-0-		→at end of each year								18,568	8.8%	(4) See explanation No. (3) in Case I. 8,000 ÷ 18,568 = .4308 which factor interpolates to 8.8% interest.
Case III														
Project (Investment) cash flow before debt interest and repayment	(7,000)	1,700 {50} {100}	1,700 {45} {100}	1,700 {40} {100}	1,700 {35} {100}	1,700 {30} {100}	1,700 {25} {100}	1,700 {20} {100}	1,700 {15} {100}	1,700 {10} {100}	1,700 {5} {100}			
Debt Interest														
Debt repayment														
Project (Investment) cash flow	(7,000)	1,550	1,555	1,560	1,565	1,570	1,575	1,580	1,585	1,590	1,595		18.2%	(1) See explanation No. (1) in Case I.
(Reinvested) in money market		(1,550)	(1,555)	etc., increasing at a constant gradient of $5 each year								18,013	3.0%	(2) See explanation No. (2) in Case II. Formula in Case III is (1,550 + 5(4.26)) = 1,571.30 x 11.464 = 18,013.
				→at end of each year										
(Invest) surplus cash in money market	(7,000)	-0-												
	(1,000)											1,344	3.0%	(3) 1,000 x factor of 1.3439 as in explanation No. (3) in Case II.
Overall result	(8,000)	-0-		→at end of each year								19,357	9.2%	(4) See explanation No. (3) in Case I. 8,000 ÷ 19,357 = .4133 which factor interpolates to 9.2% interest.
Case IV														
Project (Investment) cash flow after debt service as in Case II	(1,000)	650	685	720	755	790	825	860	895	930	965			(1) See explanation No. (1) in Case I.
(Reinvested) in money market		(650)	(685)	etc., increasing at a constant gradient of $35 each year								9,161	3.0%	(2) See explanation No. (2) in Case II.
				→at end of each year										
(Invest) surplus cash in preferred stocks	(1,000)	-0-												
	(7,000)											13,770	7.0%	(3) 7,000 x single payment annual compound amount factor of 1.9672 for 7.0% = 13,770.
Overall result	(8,000)	-0-		→at end of each year								22,931	11.1%	(4) See explanation No. (3) in Case I. 8,000 ÷ 22,931 = .3489 which factor interpolates to 11.1% interest.

3.8

THE IMPACT OF INFLATION ON HURDLE RATES FOR PROJECT SELECTION

by

Neil H Cole

N H Cole and Associates Pty Ltd,
Managing Director, Sydney

Abstract Cost and price inflation and financial gearing through loans are characteristic of modern resources projects. Conventional discounted cash flow and IRR analyses, in real terms, and without financing, have inadequacies which can yield misleading results. The degree of improvement in value for project equity owners is limited to the value of the additional tax benefit accorded through the financing instrument used. Debt financing may be low in price and value, but its priority means that the equity owners assume a higher risk, thereby increasing the cost, or price, of their capital. The cost of equity capital, or investment hurdle rate, is related to the Capital Asset Pricing Model. Distortions do occur, with inflation, due to depreciation effects, which can cause the adoption of misleading project rate of return or hurdle rate criteria. For investment decisions, the greatest accuracy possible in forecasting is essential, especially for inflation rates, foreign exchange rates, and commodity prices.

INTRODUCTION

Project selection methodology is a cornerstone amongst the building blocks of mining industry financing techniques. While the usual view is that it is of primary interest to mine owners and equity participants, it is just as relevant, for their own particular interests for others involved with the project, such as bankers, government onlookers, export credit agencies, corporate analysts and share investors and others.

HISTORICAL TRENDS

Discounted cash flow analysis for mining projects came of age in the 1950's and 1960's, before the advent of routine and simplified computer use for project investment analysis. As most commonly interpreted, the hurdle rate has come to be accepted as that threshold rate of return, above which, equity investment in a particular project can be seriously considered.

Over the past 15 years, there are a number of trends which, directly and indirectly, impact on the hurdle rate and project selection process.

Inflation Considerable volatility in inflation rates, both shorter and longer term, has been experienced. Internationally, inflation rates have occurred differentially, and have peaked at different times.

Price Volatility Commodity prices have continued to respond to economic cycles. Price cycles have not consistently mirrored inflation cycles. Some commodities have entered long term patterns of real price growth or decline. Near term metal price expectations have been moderated.

Exchange Rates Foreign currency exchange rate volatility has been marked in the past 15 years, and this has been interwoven with a greater worldwide mobility of capital. Interrelated with differential inflation rates, exchange rate forecasting is now as critical in project economic analysis as is any other single revenue determinant. However, mining companies in general are not allocating as much executive manpower to this subject as to other revenue or cost forecasting aspects.

Lead Times Project lead times are typically substantially longer, partly due to regulatory, environmental, and other such pressures, and partly due to the technological challenge of larger resources projects such as offshore oil recovery, oil-shale, and very large coal developments e.g. the Selby pit in England.

Project Finance 15 years ago, it was exceptional to see loans to mining projects or mining corporations. Pioneering limited recourse loans had been arranged around 1966 for the Hamersley iron ore project, in the then atmosphere of total belief as to consumer acceptance of contracted tonnages according to plan. This attitude was jolted in 1971, with cutbacks forced on concentrate deliveries from the Bougainville copper operations, bringing mineral economists into a new era of risk analysis.

Project finance, often on a limited recourse basis, has now become a standard tool for development. Mining company gearing ratios have substantially increased, and there is now the necessity for projects to be examined with and without financial leverage, and to understand how project selection techniques and hurdle rates need to be reconsidered in the light of financial gearing.

Investor Expectations

In the contracting environment of the past 15 years, with resources companies having less room to move, the emerging presence of oil revenues has seen all of the oil majors move more actively into mining projects and mining companies. Investor expectations have declined. The rule of the 1960's, requiring a real discounted cash flow rate of return of 15 per cent per annum, has now been largely rewritten by mining companies, especially by the larger ones.

While there are undoubtedly those that through a combination of superior management, hard work, and luck have consistently achieved better rates of return, the view of the chief financial executive of one large, successful international enterprise engaged in a wide array of new mining projects, in many countries, is that only two projects have arisen in the past few years that give indications of a real equity rate of return of over 10 per cent per annum. One US oil group is on record as indicating that it seeks a minimum 7 per cent real return for investment undertaking by its affiliated base metal mining company.

PROJECT RANKING AND SELECTION

For any type of investor, the ranking of investment alternatives is a major prerequisite in the capital allocation process. In terms of equity investment, the two main viewpoints are (a) that of the corporation, and (b) that of a stockholder.

The holder of common stock is most interested in total return over a period, from a combination of capital gain and dividend income, moderated by the particular taxation circumstances of the shareholder.

The corporation's motive in classical terms is the generation of wealth for its stockholders, which tends to be reflected very largely through its performance in terms of five financial ratios:

1. Return on shareholders' funds ("ROSF")
2. Earnings per share growth
3. Price/earnings ratios (and hence share prices) related to earnings per share growth
4. Dividend policy and payout ratio
5. Net asset backing

Of these, the most critical can be argued to relate to the return on shareholders' funds, and it is towards this objective to which discounted cash flow analysis can be tied. To the extent that the real life reporting, in a corporation's annual accounts, is in escalated terms, a strong argument for methods of project investment analysis inclusive of inflation can be made.

Examination of international corporate performance tends to indicate that companies with superior performance are sometimes able to achieve ROSF performance, after tax, of 20 per cent. Sound performance is frequently in the 12-15 per cent ROSF range. This means that the typical and recurrent return on shareholders's equity retained within the corporation, inclusive of the inflation effect, is 12-15 per cent. At the level of the single project corporation, ROSF can be compared with a project return, even though one is an accounting return and the other is a cash flow return.

Ranking Methods

It is well known that project ranking by internal rate of return will not always be consistent with project ranking according to net present values discounted at a specific hurdle rate.

Notwithstanding the difficulties later discussed in this chapter, an optimal choice of methodology will have certain features:

1. At or near the threshold of acceptable rates of return, the method or methods used will tend to show a consistent rating for different projects. Within a choice of say 12 projects, the method(s) should show consistently that a certain number tend to be either accepts, rejects, or marginal. If analysis by one method shows 6 projects marginal, and another method shows only 1 project marginal, then at least one method is unsatisfactory and should be changed or rejected.

2. Apart from ranking, the hurdle rate selected must be such as to be able to determine a realistic valuation, without undue conservatism or exaggeration. This is especially important when project assets are

being acquired or sold, rather than in the case of a project development.

Typical Hurdle Rates

From the experience of this author in recent years, real after tax discount rates as indicated in Table 1 have been applied by different classes of investors.

TABLE 1
Typical Real After-tax Hurdle Rates

Investor Type	Typical Range	Comments
Oil company	20+%p.a.	compensation for high-cost exploration risks
Small producing gold company	15-20%p.a.	opportunistic, flexible
Mining house	8-12%p.a.	restricted, large opportunities
Investor joint venturer	7-10%p.a.	buying into existing projects

Risk Attributes

The trend in hurdle rates in Table 1 reflects the risk exposure to which different classes of investors expose themselves, and also the competitive pressure of the marketplace. In particular, it also indicates the prior risks undertaken and the sunk costs of abortive exploration and development efforts. Whereas a large resources group might expend $50 million prior to finding a project in which it can spend $100 million on development, the more passive institutional investor may incur less than 5 man years of work and an aggregate of $1 million on investigations before finding an opportunity to invest $30 million in a project.

The Principal Valuation Alternatives

Cash flow analysis methods, for the purposes of the present discussion, fall into four principal categories:

1. Equity basis alone, no gearing, in real values

2. Project with financial gearing, in real values

3. Equity basis alone, no gearing, with escalated or inflated revenues and costs

4. Project with financial gearing, with escalated or inflated revenues and costs

Of these 4 categories, the first and last are most realistic. The second is sometimes applied, with real interest rates taken at low levels e.g 2-3 per cent per annum. The third method can give distorted, and in some cases, an unrealistic view of the projected picture, but it can be correct also in particular circumstances.

EMERGING TRENDS

Selection of a method depends on the use to which the particular analysis or valuation is being put.

Types and Uses of Valuations

A wide range of applications is known. Only a few of the major uses are set out below.

Preliminary Project Appraisal For typical mining projects, a category 1 approach is considered satisfactory for preliminary appraisal, farm-in consideration or pre-feasibility study. It may not be satisfactory, however, for processing or smelter evaluations.

Major Investment Decisions An ungeared, real prices approach will be satisfactory in those cases where the decision is relatively obvious, i.e. an internal rate of return to equity above 15 per cent or below 5 per cent per annum. However, the majority of opportunities seem to tend to fall in the 7-12 per cent range, and a more sophisticated approach will be needed.

Banking Analysis The ability to service debt is based on best forward estimated cash flows, with inflation included. The objective of a banking analysis is more towards debt coverage ratios than net present value analysis. Project finance analysis almost universally involves escalated cases, with the financial leverage included, although present value analyses may also be used to determine, especially for petroleum ventures, if there is a sufficent additional net present worth in the reserves to cover contingencies.

The feature emerging is the relative need for precision and confidence levels, depending on the perspective and risk/reward position of different classes of investors towards the project in question.

Capital Assets Pricing Model ("CAPM")

A growing trend in recent years has been towards the more widespread adoption of CAPM for capital budgeting decisions relating to resources projects. This method utilises the cost of capital concept, with computerised systems making available such cost estimates for different industry sectors.

The derivation of such cost of capital estimates is through interpretation of beta values, which are historically derived measures of the volatility of total rates of return on investment.

CAPM can be expressed in the following form, to determine the cost of equity for a particular investment under consideration:

Cost of equity
　capital　　= cost or price attributed to riskfree assets (e.g. government securities)

　　　　　　+ premium for average market risk, weighted by the relative risk attributed to the project compared to the market average

The price or rate of return on risk free assets is available from the yield on government securities. A large part of this yield is attributable to long run inflation expectations. For example, this value might be 9.0 per cent per annum.

Proponents of CAPM determine the average market risk premium by assessing the total rate of return available longrun, typically on a 4 year basis, through investment in common stocks. After deduction of 9.0 per cent, the risk premium might be estimated at 6.0 per cent, of which a part will also be an inflation rate of return expectation.

The relative risk weighting or beta factor is interpreted as the relative volatility of total return, in the market, of the sector most representative of the project, compared to the market as a whole. The market has a beta value of unity. For example, utilities and railroad stocks may be less volatile than the average, and would be assessed as having a beta factor below 1. For a group of gold mining stocks recently sampled, the total group beta was 1.4, and within the group, the range was 1.1 to 2.1. It should be noted that beta values will drift over time in relation to shifts in the fundamental and market circumstances.

CAPM and Inflation   As it is market derived, CAPM automatically includes inflation, and thereby offers analysts the apparent convenience of not requiring to be unduly concerned about the derivation of specific inflation estimates. This in itself is an aspect in favour of using CAPM, and it also does not suffer from the distortion of depreciation or similar effects discussed in a later section of this chapter. However, analysts still have to project escalated revenues and costs, and it is arguable if the analyst's views would normally coincide with the unknown market view of inflation rates.

Application of the Model   Using the example figures cited above, the form would be:

Cost of equity capital = 9.0% + (6.0% X 1.4)
　　　　　　　　　　　 = 9.0% + 8.4%
　　　　　　　　　　　 = 17.4%

In the case of a gold project investment decision, that reflects the average market spread in terms of its risk and development profile, an analyst using the CAPM approach would use a cost of capital discount rate of 17.4 per cent per annum as his hurdle rate, whereas for an "average" investment (beta = 1) the cost of capital would be 15.0 per cent.

NET PRESENT VALUE METHODOLOGY

As a preliminary to isolating the impacts of inflation on project selection and hurdle rates, it is convenient to restate the form in which net cash flow is usually derived. The availability of depreciation provides a tax deduction, which in the hands of a tax-paying owner, has a benefit equal to the amount of such a deduction multiplied by the tax rate, taken as 40 per cent in the example shown. Similar treatment is applied to interest.

In the example shown in Table 2, the usual net cash flow derivation is as shown in Row 17, cross tallied. This example has been worked with a constant inflation assumption of 5 per cent for both revenues and costs.

Alternative Presentation

The example can be reworked in the following form:

	CROSS TOTAL	12% NPV
Operating profit (Row 8)	19003	12007
assume fully taxed at 40% (Row 23)	8123	5171
	10880	6836
+ depreciation tax benefit (Row 21)	4000	2575
+ interest tax benefit (Row 22)	522	369
	15402	9782
- capital expenditure (Row 15)	10000	8929
+/- loans received/repaid (Row 16)	0	1450
Reworked NCF (equals Row 17)	5402	2303

TABLE 2
NPV Methodology Example : Escalation 5%pa

Row No.	Project year	1985	1986	1987	1988	1989	1990	TOTALS	12% NPV
1	Net mine revenues		6000	6000	6000	6000	6000	30000	19311
2	Opcosts & admin.		2500	2500	2500	2500	2500	12500	8046
3	Escalation fctr	5%	1.05	1.10	1.16	1.22	1.28		
4	Esc'd revenues		6300	6615	6946	7293	7658	34811	22163
5	- esc'd costs		2625	2756	2894	3039	3191	14505	9234
6	Surplus		3675	3859	4052	4254	4467	20307	12928
7	- interest paid		458	389	297	160	0	1304	921
8	Op. profit		3218	3470	3754	4094	4467	19003	12007
9	- depreciation		2000	2000	2000	2000	2000	10000	6437
10	Taxable		1218	1470	1754	2094	2467	9003	5570
11	- tax at 40%		487	588	702	838	987	3601	2228
12	Net profit		731	882	1053	1256	1480	5402	3342
13	Op. profit		3218	3470	3754	4094	4467	19003	12007
14	- tax paid		487	588	702	838	987	3601	2228
15	- cap expend	10000						10000	8929
16	+/-loans in/out	6000	-900	-1200	-1800	-2100		0	1450
17	Net cash flow	-4000	1831	1682	1253	1156	3480	5402	2300
18	12% NPV factor	0.893	0.797	0.712	0.636	0.567	0.507		
19	12% NPV amount	-3571	1459	1197	796	656	1763	2300	
20	Loan o/s yr end	6000	5100	3900	2100				
21	Dep'n tax benefit		800	800	800	800	800	4000	2575
22	Int. tax benefit		183	156	119	64		522	369
23	Notional tax		1470	1544	1621	1702	1787	8123	5171
	UNGEARED PROJECT								
24	Surplus (row 6)	0	3675	3859	4052	4254	4467	20307	12928
25	- depreciation		2000	2000	2000	2000	2000	10000	6437
26	Taxable		1675	1859	2052	2254	2467	10307	6491
27	- tax at 40%		670	744	821	902	987	4123	2596
28	Profit		1005	1115	1231	1353	1480	6184	3895
29	Surplus	0	3675	3859	4052	4254	4467	20307	12928
30	- tax paid		670	744	821	902	987	4123	2596
31	- cap expend	10000						10000	8929
32	Net cash flow	-10000	3005	3115	3231	3353	3480	6184	1403
33	12% NPV factor	0.893	0.797	0.712	0.636	0.567	0.507		
34	12% NPV amount	-8929	2396	2217	2053	1902	1763	1403	

Note that the totals both before and after discounting are equal with both methods of analysis, indicating the validity of the reworked approach. For convenience in showing the reconciliation, both the reworked example and Table 2 have the cross tallies discounted at a 12 per cent hurdle rate. However, as later discussed, this is not the rate most applicable to all cash flows, as some should be discounted at a rate equal to the market cost of debt. At a level escalated for inflation, this is 7.63 per cent per annum for the worked example in Table 2.

The reworked example and the full model can be used to show the specific inflation impacts more conveniently than in the case of the traditional net cash flow presentation.

Early Cash Flow Derivatives The revised presentation shows that these can be restricted to those items, largely capital related or derived, which are incurred early in the project, during the period for which inflation estimates can be provided with greatest confidence.

These are the depreciation tax benefit (Row 21), the interest tax benefit (Row 22), the capital expenditure (Row 15) and loan moneys (Row 16).

Later Cash Flows The other cash flows arise entirely from escalated revenues and costs, Rows 4 and 5, from which the operating surplus, Row 6, is calculated. It is these items to which greatest uncertainty must be attributed, due to the lack of confidence in projecting inflation estimates long term.

As later discussed, it is the distortion of capital related items, especially depreciation, which causes an inconsistency between hurdle rate selection in valuations undertaken separately with real incomes and costs, and with inflation.

The extent of this distortion is indicated in the impact analysis hereunder.

FINANCIAL EVALUATION FOR MINERALS INDUSTRY

DIFFERENTIAL IMPACT ANALYSIS

The simple cash flow analysis presented in Table 2 has a number of standard assumptions. It uses a 5 per cent revenue and cost inflation projection, and a hurdle rate of 12 per cent per annum for analysis purposes. The interest rate is at 7.6 per cent, derived from a real interest rate reference point of 2.5 per cent per annum.

Table 3 has been prepared to show whether the front-end related items, and the depreciation tax benefit are significant in a relative sense.

TABLE 3
Relative Impact Analysis

	Before Discounting	After 12%pa Discounting	As % of NPV
Total net cash flow	6,706	3,222	100%
A. EARLY CASH FLOWS			
Depreciation tax benefit	4,000	2,575	80%
Interest tax benefit	522	369	11%
Capital expend.	10,000	8,929	277%
Loans, net	0	1,450	45%
B. LATER CASH FLOWS			
Escalated revenue	34,811	22,163	688%
Escalated costs	14,505	9,234	287%

Analysis Table 3 shows that the NPV impact of capital related items is quite secondary to the impact of the revenue and operating cost items incurred throughout the project. It is not possible to be more specific, as the difference in any case depends on all of those ratios and factors whhich go to make up the cash flow profile for any case under review.

However, what this type of impact analysis does highlight is the scope for analytical benefits and accuracy that will arise from getting revenue estimates, inclusive of inflation, to the highest degree of accuracy possible.

Price and Cost Inflation Particularly in the present regime of fluctuating international exchange rates, it is increasingly difficult to extrapolate or otherwise estimate price or cost escalation. There is an inadequate body of evidence to justify a parallelism of inflation rates, but this is the solution most often adopted. For each commodity and each project in each country, as much work as possible must be done to forecast these variables.

As mentioned in an earlier section of this chapter, exchange rate movements can have a dramatic effect on the achievement or otherwise of results anticipated when any initial project analysis was undertaken. Unfortunately, exchange rate forecasting is not amongst the usual skills available in-house to mining companies, but many have found it profitable to take external advice on foreign currency exposure and management.

INFLATION ADJUSTED DISCOUNT RATES

Table No. 4 hereunder presents a range of inflation adjusted discount rates, derived from the formula

$$\text{Adjusted rate} = (1 + \text{real discount rate}) \times (1 + \text{inflation rate}) - 1$$

TABLE 4
Inflation Adjusted Discount Rates

Rate*	Inflation Rates % p.a.					
	3%	5%	8%	10%	12%	15%
6%	9.2%	11.3%	14.5%	16.6%	18.7%	21.9%
8%	11.2%	13.4%	16.6%	18.8%	21.0%	24.2%
10%	13.3%	15.5%	18.8%	21.0%	23.2%	26.5%
12%	15.4%	17.6%	21.0%	23.2%	25.4%	28.8%
15%	18.5%	20.8%	24.2%	26.5%	28.8%	32.3%

* Real discount rates before inflation

The same data as in Table No 4 are shown graphically in Figure No. 1.

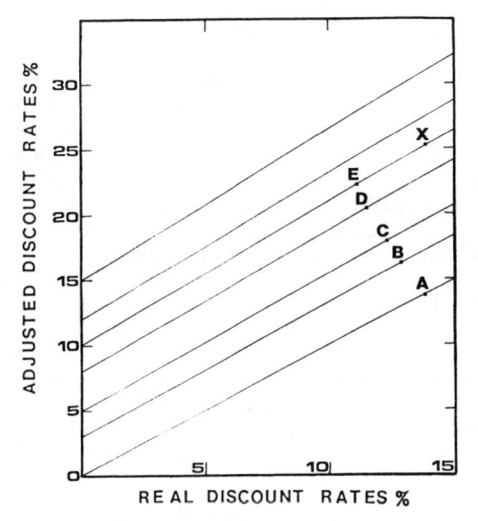

Figure 1. Inflation Adjusted Discount Rates. Curves shown are for 0% (A), 3% (B), 5% (C), 8%(D), 10% (EX), 12% and 15% inflation rates.

Applying Adjusted Discount Rates

Without financial gearing, the development case already presented in the major cash flow analysis (Table 2) has been separately analysed with a range of projected inflation rates. The same inflation rate was used for revenues and costs. Results of these discounted cash flow analyses are shown in Table 5.

TABLE 5
Inflation Adjusted DCF Analyses
Showing net present values

Discount Rates	..........Inflation Rates..........				
	0%	3%	5%	8%	10%
6%	2090				
8%	1462	2151			
10%	903	1537	1986		
12%		989	1403	2063	
15%			644	1229	2531
20%				110	451
25%					-461
IRR	13.8%	16.3%	18.0%	20.6%	22.3%

Obviously, under nil inflation, the internal rates of return to equity ("IRR") rates will be equal. If there were no distortions, and the relationships applied accurately as shown on Figure 1, then at a 10% inflation rate, the return that started out as 13.8 per cent (point A on Figure 1) should move vertically up to the 10% inflation rate line, under inflation, to a value, at point X, of 25.2 per cent.

However, as shown in the far right column of Table 5, due to the depreciation distortion, the actual IRR value computed is 22.3 per cent (point E), not 25.2 per cent. Moving stepwise from nil to 10 per cent inflation, the calculated IRR follows the path ABCDE in Figure 1.

In an effort to get to a real world solution, is either answer more correct, and what causes the discrepancy? The more correct answer will be that which most closely models the true situation, and if 10 per cent is the best-guess inflation figure, then the better answer - not because it is lower and therefore "safer" or "more conservative" (it is not) - is the 22.3 per cent.

Discrepancy Adjustment The discrepancy can be proven by a simple inflation adjustment of the depreciation figure, where upon the 25.5 per cent result again emerges, but this is a false adjustment.

To overcome the apparent limitation so arising, an inflation adjusted hurdle rate that has been derived by formula application, will have to be reduced by a weighting factor. This factor will be specific to the cash flow characterisitics of the particular project under examination, influenced by factors such as project life, lead time, capital intensity, and operating margins.

INFLATION AND GEARING

Particular caution must be taken with the interpretation of any financial analysis that is undertaken on a net present value basis with a moderate or higher level of gearing. The risk spectrum is distorted and a casual reading tends to ignore the associated liability brought onto the corporate balance sheet when any financial leverage is present.

Table 6 sets out the result of reworking the major base case earlier presented, with a range of escalation rates, discount rates and gearing ratios. It is important to note that these are project value figures, which do not necessarily coincide with the value of an investor's equity position in the project.

TABLE 6
Geared Net Present Values for Projects

Escalation Rate	Real Hurdle Rates	Adjusted Hurdle Rates*	Geared Project NPV's*					
			nil debt	Debt to equity ratio				
				1:3	1:2	1:1	2:1	3:1
0%	8%	8%	1462	1827	1949	2193	2436	2558
	10%	10%	903	1355	1505	1807	2108	2259
	12%	12%	405	934	1110	1463	1815	1991
	IRR	= 13.8%						
3%	8%	11.2%	1190	1596	1731	2001	2272	2407
	10%	13.3%	665	1150	1311	1634	1957	2118
	12%	15.4%	199	752	937	1306	1675	1860
	IRR	= 16.3%						
5%	8%	13.4	1033	1461	1604	1889	2174	2317
	10%	15.5	529	1031	1198	1532	1867	2034
	12%	17.6	80	647	836	1214	1592	1781
	IRR	= 18.0%						
8%	8%	16.6	828	1284	1435	1739	2043	2195
	10%	18.8	352	875	1049	1398	1746	1921
	12%	21.0	-72	510	705	1093	1481	1676
	IRR	= 20.6%						
10%	8%	18.8	709	1180	1337	1651	1965	2122
	10%	21.0	249	783	961	1318	1674	1852
	12%	23.2	-160	431	628	1021	1415	1612
	IRR	= 22.3%						

* based on formula adjustment of discount rates

A difficult concept which must be understood relates to the difference between the value of the project itself as in Table 6, and the value to its equity investors, when financial gearing is involved.

Project Participants and Risk Allocation

The value of a project is no more or less than the aggregate value to all of those parties that fund its development. The number of project participants can be simplified to be three only, for the generalised case.

The three categories of participants are the project equity owners, the lenders of debt funds, and the Inland Revenue Service, through the provision of a tax shelter or tax shield, which is the benefit amount attributed to interest payments which was described previously.

The value of project participation is different to the equity owners and lenders, largely reflecting the level of risk to which they want their allocated capital exposed. In an analogous concept to that of leveraged leasing, which is comprehensively covered in another chapter of this book, the project risk and project return is balanced, as on a fulcrum or lever between

1. DEBT CAPITAL - priority and secured return, at a low rate, in return for security; low risk

and

2. EQUITY CAPITAL - subordinated position after debt (thus higher risk), but expecting to use gearing to get better return than pure project, i.e. leverage.

Specifically, the equity investor deliberately moves to a position of greater relative risk than the project as a whole. In terms of CAPM, the beta factor increases, and his cost of capital thereby is increased.

The contribution available from tax authorities is limited, as shown in various of the tabulated values herein. The tax benefit also is a low risk component.

The project value consists of the value to the owner, calculated at the owner's hurdle rate or cost of capital considered for such risk exposure, the value to the lender at his requisite rate of return on funds made available, and the value of the interest tax benefit.

Value to Equity

In a source-and-allocation-of-funds manner, the sources of cash are from operation of the pure project and from the interest tax shelter. The uses are to provide returns to the lenders and the owners. This can be expressed in the following form, related to the example in this chapter, from Table 2 reworked at a nil inflation rate.

```
Net cash flow to project, on
  ungeared basis                    4500
plus tax benefit from
  interest (60% gearing)             171
                                    ----
                                    4671
less cash flows to debt
  loans received       6000
  less loans repaid    6000
  net principal           0
  plus interest paid    428
                                     428
                                    ----
Net cash flow to equity             4243
                                    ====
```

These are before discounting to determine net present values. The discounting factors used for the debt cash flows and the tax benefit, which arises solely due to the debt and is thus linked, are the market price for debt, in this case related to a real interest rate of 2.5 per cent. For cases with inflation, interest rates are adjusted by a formula identical to that used for discount rate adjustment, namely

Adjusted rate = (1 + real interest rate)
               X (1 + inflation rate) - 1

A range of cases has been examined using different gearing ratios and excalation rates.

Gearing and Equity Values

By application of the above formula showing the net cash flow to equity calculation, the equity values for the standard example project have been determined, as shown in Tables 7, 8, and 9, for different escalation rates, and using basic discount rates of 8, 10, and 12 per cent prior to formula adjustment.

TABLE 7
Gearing and Equity NPV's : Escalation 0%pa

GEARING AMOUNT	PROFIT AMOUNT	NET CASH FLOW	TAX BENEFIT	8.0%	10.0%	12.0%	EQUITY IRR pa
0	4500	4500	0	1462	903	405	13.8%
1000	4457	4457	29	1488	929	432	15.0%
2000	4415	4415	57	1515	956	458	16.6%
3000	4372	4372	86	1541	982	484	18.5%
4000	4329	4329	114	1567	1009	511	21.0%
5000	4286	4286	143	1594	1035	537	24.4%
6000	4244	4244	171	1620	1061	564	29.4%
7000	4201	4201	200	1647	1088	590	37.5%
8000	4158	4158	228	1673	1114	616	53.7%
9000	4115	4115	257	1699	1141	643	105.1%
10000	4073	4073	285	1726	1167	669	

TABLE 8
Gearing and Equity NPV's : Escalation 5%pa

GEARING AMOUNT	PROFIT AMOUNT	NET CASH FLOW	TAX BENEFIT	<---DISCOUNTED NPV'S--->			EQUITY IRR pa
				13.4%	15.5%	17.6%	
0	6184	6184	0	1033	529	80	18.0%
1000	6054	6054	87	1249	745	296	19.3%
2000	5923	5923	174	1464	960	512	20.9%
3000	5793	5793	261	1680	1176	728	22.9%
4000	5662	5662	348	1896	1392	944	25.4%
5000	5532	5532	435	2112	1608	1159	28.9%
6000	5402	5402	522	2327	1824	1375	33.8%
7000	5271	5271	608	2543	2039	1591	41.7%
8000	5141	5141	695	2759	2255	1807	56.8%
9000	5011	5011	782	2975	2471	2022	100.9%
10000	4880	4880	869	3190	2687	2238	

NOTE: The 3 discount rates are adjusted by the inflation formula

TABLE 9
Gearing and Equity NPV's : Escalation 10%pa

GEARING AMOUNT	PROFIT AMOUNT	NET CASH FLOW	TAX BENEFIT	<---DISCOUNTED NPV'S--->			EQUITY IRR pa
				18.8%	21.0%	23.2%	
0	8103	8103	0	709	249	-160	22.3%
1000	7885	7885	145	1114	654	246	23.7%
2000	7667	7667	291	1519	1059	651	25.3%
3000	7449	7449	436	1924	1465	1056	27.4%
4000	7231	7231	581	2329	1870	1461	30.0%
5000	7013	7013	727	2734	2275	1866	33.5%
6000	6795	6795	872	3139	2680	2271	38.5%
7000	6577	6577	1017	3545	3085	2676	46.2%
8000	6359	6359	1163	3950	3490	3082	60.4%
9000	6141	6141	1308	4355	3895	3487	
10000	5923	5923	1454	4760	4301	3892	

NOTE: The 3 discount rates are adjusted by the inflation formula

Although the far right column shows the calculated internal rates of return to equity, this is not advanced as proper practice, but rather to illustrate how dangerous and misleading the mis-quotation of IRR values can be.

Discussion

Tables 7, 8 and 9 illustrate a number of points:

1. Financial gearing has a favourable impact on equity values.

2. The extent of the improvement depends on the hurdle rate adopted for discounting purposes.

3. The NPV improvement through gearing is quite secondary in magnitude compared to the importance of best determining the cost of capital or hurdle rate.

4. Quotation of a leveraged IRR to equity is likely to lead to misleading conclusions, except in comparing projects with very similar risk spectra or those analysed without gearing.

5. At moderate and high inflation rates, the effect of a greater level of gearing on equity valuation is more pronounced.

Implementation in Practice

Corporate managers of mining projects must use their best endeavours to make recommendations and take decisions based on financial modelling techniques which most closely parallel the anticipated funding and fiscal framework within which any project is to be structured.

Sensitivity analysis has not been discussed herein, but is a standard discipline for analysts and corporate modellers. Based on a retrospective 10 year view, the areas of greatest weakness may be in exchange rate forecasting and in the determination of a firm's cost of capital.

Inflation, especially under moderate to high levels of financial leverage can have a dramatic effect on equity value. It can be misleading to take the "safe" or "conservative" course by tending to overstate inflation. At real discount rates of 8.0 per cent, in the examples shown in Table 8 and Table 9, an overvaluation of 35 per cent (3139 ÷ 2327) would result for a 60 per cent geared project, if, with good intention, a "safe" inflation rate was estimated at 10 per cent rather than 5 per cent.

As with all investment analysis, the critical input factors are those relating to revenue determinants. Almost as much as exchange rates, but not in the forgotten basket, is the consideration of commodity price estimation. Managers must become disciplined in price forecasting and should be encouraged to avoid historical statistics as the primary basis : to project prices based around a market quotation as at the end of February 1985 is of itself an action which adopts a price forecast.

Limited Recourse Loans

In the corporate management area, a growing feature of mining projects has been the emergence in recent years of financing by way of limited recourse loans to specific projects. Such loans differ from corporate borrowing facilities insofar as limited only credit support is necessary from the project owners once the project has, to the satisfaction of lenders, proven its ability to produce to feasibility study levels of performance, both physical and financial.

The valuation of projects having limited recourse borrowing arrangements, and the associated hurdle rates, should be addressed in the same way as with projects having standard corporate loans. On a probability distribution basis, it may be argued in theory that the owner can be detached from the very minor risk, and financial consequences, of project failure, but there are no banking analysts who would admit to making limited recourse privileges available to any project to which any tangible, even if quite minor, risk could be attributed.

CONCLUSION

The equity valuation of mining projects is not unduly influenced by the distortion of depreciation and related impacts. It is however strongly influenced by the need to adopt the most accurate forecasting basis for inflation and gearing levels. Project sponsors can eliminate the uncertainty of one of these variables by entering discussions with bankers about the extent of loan gearing and indicative terms, at the early project planning stages.

Standardised techniques are available for the valuation of equity interests under inflation and with financial gearing. These have to be rigorously applied to ensure that better projects are consistently selected, and also, of equal importance, to ensure that satisfactory projects are not inadvertently eliminated or otherwise rejected as a result of the chosen decision making methodology.

3.9

EVALUATION OF THE LEASE OR BUY DECISION

Mr. John R. Caldon

Tax Partner
Price Waterhouse, Australia

INTRODUCTION

Leasing has become a major industry in many countries because of the advantages it affords. Whilst the accounting, legal and tax treatment of leasing varies from country to country, there is far more similarity than difference. The general approach to the subject is largely the same. This paper looks to the basic <u>financial</u> considerations in the decision to lease or buy.

TYPES OF LEASES

Before embarking upon the subject, it would seem appropriate to define a "lease". A lease may be defined as:

> "an agreement granting one party the right to use, for a specified period of time, property owned by another party in return for a series of payments of rental by the user to the owner".

Ownership, of course, generally entitles the owner to unrestricted use of the asset over its useful life in contrast to a lease which gives the lessee use over a limited period. The distinction between the status of owner and a lessee often is fairly small. In this case we are dealing with what is generally termed a "finance lease". A finance lease is one which substantially transfers the risks and benefits of ownership from the legal owner to the lessee. For example, a finance lease would exist where the lessee enters into a lease of an asset for most of its useful life and has an option to buy the asset at the end of the lease for a price which takes cognisance of the rentals already paid.

Where a lease does not substantially transfer the benefits and risks of ownership to the lessee, i.e. those benefits and risks remain with the owner and lessor, the lease is termed an "operating lease" and is an agreement for the renting and usage by the lessee of an asset owned by the lessor for a limited period.

An example of an operating lease might be where a tenant leases a building for a defined period say 6 years with the rent and occupancy subject to renegotiation at the end of the lease term.

The economic consequences of the distinction between a finance and an operating lease is very important. With a finance lease, the lessee has a de facto economic interest in the asset and the alternative to a finance lease is some form of purchase; with an operating lease, the lessee does not have a basic economic interest in the asset and consequently a comparison with a purchase option involves far more than simply a funding comparison.

Some of the principal differences between finance and operating leases can be summarised as follows:

(a) finance leases:
 - the risks and benefits of ownership are substantially transferred to the lessee;
 - the lessor normally recovers his investment during the term of the lease;
 - the lessor in entering into the transaction is relying more on the creditworthiness of the lessee than the underlying value of the asset;
 - the lease is either non-cancellable or a large penalty will apply for unilateral cancellation by the lessee;

(b) operating leases
 - the risks and benefits of ownership remain essentially with the lessor;
 - the lease rental income would not normally cover the recovery of the lessor's investment;
 - the lessor evaluates his risk in terms of the underlying value of the assets;

- the lease term is usually only for an insignificant part of the estimated useful economic life of the leased asset.

A crucial concept in the comparison of operating and of finance leases is that of residual value. It is an agreed value of the asset to be found by its sale, or otherwise by the lessee at the end of a finance lease term. The rent paid to the lessor by the lessee is therefore calculated through determination of the interest factor which he includes in calculating the lease rentals plus the recovery of capital (being the difference between original cost and the residual value). The higher the residual value the lower the rent and vice versa. Lessors, of course, generally favor short rental periods and low residuals to minimise their risk.

The concept of a residual value is not normally present with operating leases.

UNDERLYING TAX CONSIDERATIONS

There seems little difference in treatment of operating leases throughout the world. The lessor (i.e. the owner) of the plant is entitled to tax depreciation (where available); the lessee obtains a tax deduction for the lease rentals; the lease rentals are reported as income by the lessor.

There are differences in the treatment of finance leases in various countries. For example, in Australia and U.K. a finance lease is generally treated in exactly the same way as an operating lease, that is to say, the lessor is entitled to claim tax depreciation and the lessee obtains a tax deduction for the lease rentals.

This contrasts with the position in Canada where many lease transactions are treated as purchases, the lessee rather than the lessor being entitled to tax depreciation and the lessor simply reporting the interest content of the rentals received. This distinction in treatment explains the enormous popularity of finance leasing in Australia and the U.K. and its lack of popularity in Canada.

In the U.S.A. we have seen a change in the treatment of finance leases. The Economic Recovery Tax Act of 1981 saw major changes in depreciation rates and the treatment of leases. For Federal income tax purposes depreciation had, in the past, been based on the useful life of the property. To stimulate new capital formation, The Economic Recovery Tax Act replaced existing depreciation methods and procedures in many categories with ACRS. Under ACRS, depreciation deductions are specified for predetermined "recovery" periods which ordinarily are shorter than the existing useful life of most properties.

ACRS applied to "recovery property" placed in service after December 31, 1980.

Four "general" cost recovery classes were created for tangible personal property, based on the nature of the assets. These general classes are referred to as the "3-year", "5-year", "10-year" and "15-year public-utility" classes. Most plant falls into the five-year category and in broad terms is depreciated fairly evenly for tax over the five years.

The Act also created a "safe harbor" election that guarantees that an arrangement which, in form, is a lease will be characterised as a lease for purposes of allowing ACRS deductions and investment tax credit (ITC) even to a nominal lessor. It is a precondition that all parties to the lease must characterise the arrangement as a lease and agree to treat the lessor as the owner.

The effect of these measures was that equipment leasing became an attractive investment for many corporations. The combined impact of available ITC and the liberal depreciation under ACRS was particularly favourable. Additionally, taxpayers unable to utilise investment tax credits, (because of net operating losses, foreign tax credits, and heavy investment depreciable property) found it advantageous to enter into an agreement with a tax-paying corporate lessor, whereby some of the tax savings realised by the lessor were reflected in lower lease rentals.

Arguably, the safe-harbor leasing provisions were too successful because the succeeding 1982 Tax Act made numerous changes to leasing transactions, reducing the benefits of, and finally repealing after 1983, safe-harbor leasing and expanding the non-safe-harbor leasing by a new category of "finance" leases. This new category, called finance leases, is subject to many of the restrictions and requirements applicable to safe-harbor leases.

There are many detailed requirements which must be met to qualify as a "finance lease". For example, there is a requirement prohibiting a lessor from reducing its tax liability by more than 50 percent. This limitation expires for property placed in service after September 30, 1985 for a lessor's year beginning after such date.

Similarly, the recovery periods for ACRS deductions are extended to 5 years for 3 year property, 8 years for 5 year property and 15 years for 10 year property.

Finance leases must be characterised by the parties as leases and must have economic substance.

If these and other conditions are met, the lease will be categorised as a finance lease.

If an agreement qualifies as a finance lease, the following factors will not be taken into account in determining whether the agreement is a lease:

- the lessee has the right to purchase the property at a fixed price (so long as the price is 10 percent or more of the property's original cost to the lessor);

- the property is not readily usable by any person other than the lessee.

As stated above, the difference in tax treatment between say the U.K. and Canada is largely responsible for the popularity of leasing in the U.K. and its lack of popularity in Canada. In the U.K. the lessor is entitled to the available depreciation deductions. Until recently, these amounted to a 100 percent deduction for the cost of the plant item in the first year. In other words a U.K. lessor could deduct all the cost of leased plant against his other income in the year he enters into the lease. Recognition of these substantial tax benefits in the hands of the lessor is reflected in the incorporation of a low interest rate in the computation of the lease rental. Many companies in the U.K. are in tax loss and are therefore unable to currently benefit from the accelerated write-off for tax of expenditure on plant. For them a low interest factor incorporated in the lease rental is preferable to the tax benefit of owning the plant itself.

It should be mentioned in passing that the U.K. is now in the process of phasing out the current accelerated depreciation rates. Over the next couple of years, the first year allowances will be phased down from the present full first year write-off to leave an annual allowance of 25 percent (diminishing balance) as the normal depreciation method.

The effect of these changes will be to place the U.K. in the more "normal" position where depreciation rates are in line with those of other OECD countries. When fully implemented, the general U.K. rate will be 25 percent on a diminishing value basis.

As stated above, under the U.K. old scale of depreciation, a tax loss company was virtually certain to lease rather than buy its plant. A tax paying company might take a different view. The decision for the tax paying companies is not straightforward - it basically depends on a comparison of the deal offered by the leasing company with the cost of another form of funding. A methodology for evaluating the decision is set out below. But before going on to that stage, I would emphasise that for companies in tax loss, leasing is likely to be the cheaper financing mechanism in a country with favourable depreciation rates.

BASIC DISCOUNTED CASH FLOW EXAMPLE

Historically, there have been three sources of funds from which a business can fund a new asset:

- internal working capital

- further borrowing

- leasing

Whilst the latter two have an obvious implicit interest rate, there is some difficulty in determining the cost of the former. Initially we will concentrate on a comparison of the latter two. To do so, we need to apply discounted cash flow techniques to compare the cash flow costs of the two options.

This raises the question of the appropriate discount rate. This is perhaps the hardest thing to get to grips with in respect of discounted cash flow modelling. Some options available are:

- the cost of the marginal rate of borrowing of the business

- the average rate of interest paid by the business

- the average rate of return derived by the business on its investments

- the rate of return set by the business for new projects

- the rate of return derived by the business on the investment of surplus monies.

It will be appreciated that the discount rate will vary enormously within the above parameters from presumably a low being the rate of return on surplus funds to a high being the rate of return required on new investments. Which is the correct discount rate? The answer ultimately lies in the company's corporate philosophy and individual situation. Today's conventional wisdom, I believe, is to take the discount rate used to evaluate new investments. In other words, if new projects are required to yield an after tax return of say 15%, this rate is used to evaluate the respective cash flows of the lease/buy proposal. If the lease proposal shows a lesser outgoing in the early years than the purchase proposal, then on this theory the money initially "saved" would be invested in other businesses (or aspects of the business) to achieve the desired yield.

Looking at a single, small capital purchase such as that set out on the following pages, this theory appears manifestly utopian and unreal. There would be no new business competing for the few thousand dollars of cash flow available. But this rebuttal too suffers in its simplicity. If the "small capital purchase" were one of many thousand such investments made by the business, in overall terms the "cash flow" savings might be substantial and merit the "business opportunity foregone discounting" approach. Furthermore, if a business adopted this policy regularly, the policy would produce regular "savings" each year which would in consequence arguably be available to sustain a new business.

Imagine on the other hand a fairly dormant company with stores of cash available on deposit. Perhaps the deposit interest reinvestment rate is the appropriate discount rate here.

In summary then the discount rate used would appear to depend upon the company's profile and philosophy.

Another consideration is whether the discount rate should be calculated on a pre tax or after tax basis. I believe the answer is straight-forward in most cases.

For companies in a current tax paying position an after-tax discount rate should be used; for companies which are in substantial tax loss, a pre-tax rate should be used. As an example, if a deposit rate discount rate were used, a taxpaying company might use a discount rate of around 5%, a non taxpayer say 10% (10% being assumed to be the deposit rate).

The answer for some businesses is more complex - they pay tax in one year and are in tax loss in another year. Perhaps a discount rate weighted to take cognisance of their position should be used. On the above example, if the business paid tax in one year out of every two perhaps a discount rate of 7.5% would be used.

The preferred discount rate may vary from evaluation to evaluation depending upon any market movements in available rates between evaluations. It will need to be reviewed for each comparison.

Having briefly discussed the all important question of the discount rate we will proceed to a model of a lease evaluation. To illustrate the methodology I have produced this fairly simple model. In terms of the current example, a discount rate of 6% is used. This is calculated by taking a return of 12% and adjusting it by a tax rate of 50%. For ease of computation, the discounting has been applied at the end of each year and not as a monthly basis.

Assume:

Cost of equipment	$30,000
Date of acquisition	1 July 1983
Tax rate	50% and firm is in a tax paying position
Company year end	30 June
Tax payment date	Quarterly after year end
Depreciation rate allowable for tax purposes	10% straight line
Market value of equipment at the end of firm's use	$3,000

Lease proposal:

Lease term	5 years
Lease payment period	Monthly in advance
Lease payment amount per annum	$8,328
Residual payment in lease contract	$3,000

Borrowing proposal:

Loan term	5 years
Loan payment period	Monthly in arrears
Loan amount per annum	$8,008
Balance at end of term	Nil

Cost of the lease

To ascertain the cost of the lease the following points need to be considered:

- The timing and amount of each cash outflow; for example on 1 July 1983 the first lease payment is made, the second is on 1 August and so on.

- The tax factors associated with the lease and when these benefits are due from the company's point of view; for example the lease payments are fully deductible and the tax shelter associated with these will first appear in the quarterly tax payment due at the end of September 1984.

- Any further outflows required under the lease and when they occur; for example payment of the residual value of the lease contract.

Given this information, the after-tax outflow associated with the lease for each period can be derived and converted to a present value cost by discounting at the appropriate rate, which as discussed is an after-tax rate. The steps involved in determining the cost of the lease are depicted below:

1 End of year June	2 Lease Payment $	3 Tax benefit Col 2 x 50% $	4 Net cash outflow Col 2 - Col 3 $
1984	8,328	–	8,328
1985	8,328	4,164	4,164
1986	8,328	4,164	4,164
1987	8,328	4,164	4,164
1988	8,328	4,164	4,164
1989		4,164	(4,164)

End of year June	5 Present value factor @ 6%	6 Present value cost of lease $
1984	0.9434	7,857
1985	0.8900	3,706
1986	0.8396	3,496
1987	0.7921	3,298
1988	0.7473	3,112
1989	0.7050	(2,936)
		$18,533

Cost of borrowing proposal

To evaluate the borrowing proposal we must work out:

- the interest content of the loan payments (which are tax deductible)

- the tax deductions and benefits of the proposal

- and then the after tax cash flows of the proposal.

1 End of year	2 Loan Repayments	3 Loan Interest	4 Depreciation 10% pa	5 Allowable deductions Col 3 + 4
June	$	$	$	$
1984	8,008	3,140	3,000	6,140
1985	8,008	2,630	3,000	5,630
1986	8,008	2,067	3,000	5,067
1987	8,008	1,445	3,000	4,445
1988	8,008	759	3,000	15,759
1989	(3,000)*		12,000**	

1 End of year	6 Tax benefit on Col 5 at 50%	7 Net cash outflow Col 2 - Col 6	8 Present value factor at 6%	9 Present value cost of loan
June	$	$		$
1984	–	8,008	0.9434	7,555
1985	3,070	4,938	0.8900	4,395
1986	2,815	5,193	0.8396	4,360
1987	2,534	5,474	0.7921	4,336
1988	2,223	2,785	0.7473	2,081
1989	7,879	(7,879)	0.7050	(5,555)
				$17,172

* proceeds on sale of plant
** loss on sale of plant for tax purposes

Loan interest determined as follows:
(NB Repayment schedule implies a rate of 10.464%)

End of year	Loan repayment	Outstanding principal	Interest @ 10.464%	Principal portion of repayment
June	$	$	$	$
1984	8,008	30,000	3,140	4,869
1985	8,008	25,131	2,630	5,378
1986	8,008	19,753	2,067	5,941
1987	8,008	13,812	1,445	6,563
1988	8,008	7,249	759	7,249

On the model shown above, purchase is preferable to leasing. This answer arises out of the interaction of the following factors:

- most importantly, the discount factor – a higher discount factor favours the approach involving lower payments in the earlier years

- the annual quantum of the rental payments – the longer the period and the higher the residual, the lower the rentals which normally assists the D.C.F. comparison

- the annual quantum of the payments on the purchase option

- the tax depreciation rate available – higher rates will generally favor the purchase option.

To demonstrate the effects of these factors let us suppose we use a 15 percent discount rate instead of a discount rate of 6 percent. Furthermore, let us assume that the lease involves a 50 percent residual (i.e. $15,000) which turns out to equal the proceeds on the sale of the equipment after 5 years. The lease rentals have been calculated using the same implicit rate of interest as in the previous model.

Cost of the lease

1 End of -year	2 Lease Payment	3 Tax Benefit at 50%	4 Net cash outflow Col 2 - Col 3	5 Present value factor @ 15%	6 Present value cost of lease
June	$	$	$		$
1984	5,928	–	5,928	0.8696	5,155
1985	5,928	2,964	2,964	0.7561	2,241
1986	5,928	2,964	2,964	0.6575	1,949
1987	5,928	2,964	2,964	0.5718	1,695
1988	5,928	2,964	2,964	0.4972	1,474
1989		2,964	(2,964)	0.4323	(1,281)
					$11,233

As can be seen the effect of changing the above parameters is to reduce the net present value of the cost of the lease option by over $7,000. The effect of the changed discount rate and changed proceeds of sale on the borrowing proposal is as follows:

Cost of borrowing proposal

1 End of year	2 Loan Repayments	3 Loan Interest	4 Depreciation 10% straight line	5 Allowable deduction Col 3 + 4
June	$	$	$	$
1984	8,008	3,140	3,000	6,140
1985	8,008	2,630	3,000	5,630
1986	8,008	2,067	3,000	5,067
1987	8,008	1,445	3,000	4,445
1988	8,008	759	3,000	3,759
1989	(15,000)[a]			

a – proceeds of sale

End of Year	6 Tax Benefit on Col 5 at 50%	7 Net Cash outflows Col 2 - Col 6	8 Present value factor @ 15%	9 Present value cost of loan
June	$	$		$
1984		8,008	0.8696	6,964
1985	3,070	4,938	0.7561	3,734
1986	2,815	5,193	0.6575	3,414
1987	2,533	5,475	0.5718	3,131
1988	2,223	(9,215)	0.4972	(4,582)
1989	1,879	(1,879)	0.4323	(812)
				$11,849

Whilst the effect of new parameters was to reduce the lease model cost by $7,000 the changed parameters impact the purchase model by only around $5,000. Whilst purchasing was clearly preferred under the earlier model, leasing is demonstrably superior under the latter.

What the above models clearly demonstrate is that the answer to the lease versus buy decision does not relate to the question of the underlying interest rate implicit in the lease or purchase option. Rather it is a complex result dependent upon discount rate, term of lease, residual value, expected proceeds on final disposal of the plant et al.

We might pause for a moment to examine the situation of a company in tax loss. Looking at the first example, the lease cost is as follows:

1 End of year	2 Lease payment	3 Present value factor at 6%	4 Present value cost of lease
June	$		$
1984	8,328	0.9434	7,857
1985	8,328	0.8900	7,412
1986	8,328	0.8396	6,992
1987	8,328	0.7921	6,597
1988	8,328	0.7473	6,224
			$35,082

The cost of the loan is:

End of year	2 Loan repayments	3 Present value factor at 6%	4 Present value cost of loan
June	$		$
1984	8,008	0.9434	7,555
1985	8,008	0.8900	7,127
1986	8,008	0.8396	6,724
1987	8,008	0.7921	6,343
1988	8,008	0.7473	3,742
	(3,000)		$31,491

The absence of tax benefits clearly makes the lease approach even less desirable under the first model. Of course, it is arguable that the 6% discount rate should not be used since this was formulated as an after tax rate. The effect of the discount rate is not crucial as shown below. Under the second model we do not find leasing cheaper than purchase as it was in a post tax model. The cost of the borrowing approach is as follows:

1 End of year	2 Loan repayments	3 Present value factor at 15%	4 Present value of loan
June	$		$
1984	8,008	0.8696	6,964
1985	8,008	0.7561	6,055
1986	8,008	0.6575	5,265
1987	8,008	0.5718	4,579
1988	8,008	0.4972	(3,476)
	(15,000)		
			$19,387

In contrast the cost of the lease is as follows:

1 End of year	2 Lease payments	3 Present value factor at 15%	4 Present value of lease
June	$		$
1984	5,928	0.8696	5,155
1985	5,928	0.7561	4,482
1986	5,928	0.6575	3,898
1987	5,928	0.5718	3,390
1988	5,928	0.4972	2,947
			$19,872

The above models serve to demonstrate that the tax position of the borrower is an important aspect of evaluating the position. A lease was favoured in a tax payable position but not when in tax loss.

Of course, it will be realised that these models are very simplistic. The structure of financing arrangements both lease and purchase especially with big ticket items is likely to be far more complex than the models set out above. For example, it may provide for rentals which escalate from a low base in the early years; the lease might be interest only; the lease might provide for an initial grace period.

There are many fine tuning points which would have a marked impact on the discounted cash flow exercise.

At the beginning of this section I suggested that there were three funding options. Two are compared above. The third relates to internal funding.

The evaluation of this option, I believe, basically relates in general terms to the available cash of the business and the business philosophy on its cash balances. If the business is awash with cash, it may be hard to justify any course of option other than internal funding. Such funding is likely to have an exceedingly low after tax cost.

If the business is less liquid and sees the preservation of some liquidity as a goal, the evaluation basically gets back to the sort of technique described above in determining the pros and cons of leasing and borrowing options. As discussed above, the more liquid a business is the more likely it is to use a low discount rate in the evaluation (say its marginal borrowing cost); equally a business without plans of expansion might adopt a similar posture.

OTHER FACTORS RELEVANT IN LEASE v. BUY DECISION

What the above models show is the basic financial comparison between leasing and buying. There are many other factors which bear on this decision. Briefly summarised, the generally perceived advantages of leasing are:

- Leasing is generally easier to obtain than other forms of funding

- Leasing, by not requiring an initial outlay of part of the purchase price of an asset, conserves a firm's cash resources

- Leasing is a form of "off balance sheet" financing and is not counted as an external liability when determining borrowing limitations under debenture trust deeds e.g. in Australia

The generally perceived disadvantages are:

- The interest cost of lease financing is usually higher than the interest cost of debt financing

- As a leased asset is owned by the lessor, the lessee has no right or interest in the residual value of the leased asset at the expiry of the lease. (Whilst this may be technically correct at law, in Australia at least, the practice exists for the lessee to acquire the plant at the residual value at the end of lease term. There is an unwritten gentleman's agreement to this effect. I am not sure how widespread is this practice in other countries).

AVAILABILITY OF CREDIT

This area is a most difficult one to evaluate.

In a perfect capital market it does not matter how an asset is financed. The same amounts of money would be available for the same periods at the same interest rate. But in the imperfect capital markets of the real world we can readily observe that this is not the case.

A very important aspect of this consideration of this subject is the lender's approach to the financing. Whilst in a perfect market, a lease should be perceived no differently from a loan secured by a chattel mortgage over the equipment, very often the lender does have a different perception.

In Australia, where the writer has the most experience, we now have the interesting position that a major trading bank can be approached with say several different financing propositions (which are economically equivalent) and the bank will quote divergent rates. For example, it would not be uncommon to receive widely differing quotes for a leveraged lease, an equity lease and a loan secured by chattel mortgage even if the security, the term, the tax effects on the proposed lender/lessor are very similar. Perhaps this is caused by the bank's having different departments dealing with different areas of financing.

What this emphasises, however, is that the market is far from perfect and that a source of certain funding may well be available whilst a comparable source is not.

I believe it fair to say that lease funding is generally more readily available than other forms of debt. This may well be a reflection of pricing because leasing in straight interest terms is generally more expensive than a conventional debt financing.

A very important advantage of leasing which must not be overlooked is that generally the lessor funds the entire cost of the equipment.

Against the advantages of availability and of quantum must be measured the disadvantage of period. Leasing generally is short term (but see separate section on leveraged leasing) whilst conventional debt is often longer. More often than not, leasing is for a term of say 3 to 5 years with 8 years being a fairly long lease. Conventional debt funding, of course, quite often is for longer periods.

It seems that lease funding is more readily available than money borrowed by conventional means. This brings forward another perceived advantage of leasing. The point is that leasing monies are more generally available and accordingly the use of leasing results in the "release" of other funds for more profitable use elsewhere.

Does leasing affect the availability of other sources of funds? Logically it should not because leasing is simply another debt form.

Accordingly, leasing should not result in an increase of the lessee's credit pool.

However, to the extent that a lessee's borrowing is restricted by external factors, the claim that leasing increases the credit pool may have some substance. Lenders frequently impose covenants on a company, which have the effect of restricting the amount of debt it can incur. These restrictive covenants are based upon balance-sheet relationships. In general, a company's lease obligations are not included on its balance sheet, and as a result will not be included in the calculation of its borrowing limit. In contrast, if the asset was purchased and financed by a debenture issue, both the asset and liability would be shown on the balance sheet and included in the calculation of the company's borrowing limit.

On many occasions leasing would appear to increase the pool of credit available. Of course a prudent business manager must determine whether the business can function properly with the added effective debt created by these additional leasing obligations.

ACCOUNTING IMPLICATIONS

As set out above there is a major difference between an operating and a financing lease. This distinction is becoming of increasing importance for accounting purposes. This is not the place to discuss the accounting tests in depth which determine whether a lease is of an operating or a financing nature (these criteria are very briefly summarised at the end of this section).

What should be readily apparent is that from the stance of lessee there are considerable advantages in the lease not being treated as a financing device and, being "off balance sheet". Many companies' existing borrowings are limited by trust deeds. These deeds specify that the ratio of debt to equity funds will not exceed certain limits. The classification off balance sheet may therefore be crucial importance in avoiding the breach of a trust deed. Along the same lines, companies may suffer an interest "penalty" if their debt/equity ratio exceeds a specified figure - again, off balance sheet treatment may help. The interest penalty referred to above can take many forms. A straightforward example would be the case of a company which declines from an "A" to "B" rating with the rating agencies because of changes in its debt/equity ratio, thereby increasing the cost of future borrowings.

To summarise briefly those companies with trust deed or other restrictions may find leasing to be a very worthwhile "borrowing" avenue. This is particularly the case for Australian companies at present although moves to change the accounting rules to count capitalised lease obligations as debt are in train.

SUMMARY OF ACCOUNTING STANDARDS FOR LEASES

1. Australia

AAS 17 "Accounting for Leases" - applicable for periods ending on or after 31 March 1985.

Definition: Finance lease means a lease which effectively transfers from the lessor to the lessee substantially all the risks and benefits incident to ownership of the leased property.

Criteria: The effective passing of the benefits of ownership could normally be assumed where the following criteria are satisfied:

(a) the lease is non cancellable; and

(b) either of the following tests is met:

(i) the lease term is for 75% or more of the useful life of the leased property (unless the beginning of the lease term falls within the last 25% of the total useful life); or

(2) the present value at the beginning of the lease term of the minimum lease payments equals or exceeds 90% of the fair value of the leased property to the lessor at the inception of the lease.

2. Canada

Accounting Recommendation 3065 "Leases" applicable for financial periods commencing on or after 1 January 1979.

Definition: Capital lease is a lease that from the point of view of the lessee transfers substantially all the benefits and risks incidental to ownership of the property to the lessee.

Criteria: A lease would normally transfer substantially all of the benefit and risks of ownership to the lessee when at the inception of the lease one or more of the following conditions are present:

(a) there is reasonable assurance that the lessee will obtain ownership of the leased property at the end of the lease term (either through the terms of the lease itself or through the provision of a bargain purchase option)

(b) the lease term is of such a duration that the lessee will receive substantially all of the economic benefits expected to be derived from the use of the leased property over the life span - i.e. where the lease term is equal to 75% or more of the economic life

(c) the present value at the beginning of the

lease term of the minimum lease payments is equal to 90% or more of the fair value of the leased property at the inception of the lease.

3. U.S.A.

FAS 13 "Accounting for Leases"

Definition: A capital lease is one which meets one or more of the following criteria.

Criteria: If, at its inception, a lease meets one or more of the following criteria the lease shall be classified a capital lease by the lessee:

(a) the lease transfers ownership of the property to the lessee by the end of the lease term

(b) the lease contains a bargain purchase option

(c) the lease term is equal to 75% or more of the estimated economic life of the lease property

(d) the present value at the beginning of the lease term of the minimum lease payments equals or exceeds 90% of the excess of the fair value of the leased property.

4. U.K.

Proposed statement of standard accounting practice ED 29 - Accounting for leases and hire purchase contracts.

Definition: Finance lease transfers substantially all the risks and rewards of ownership of an asset to the lessee.

Criteria: A transfer of risks and rewards occurs if at the start of the lease the present value of the minimum lease payments amounts to substantially all (more than 90%) of the fair value of the leased asset to the lessor at the start of the lease.

The PV should be calculated by using the rate of interest implicit in the lease or a commercial rate.

5. International

IAS17 Accounting for Leases

Definition: A lease that transfers substantially all the risks and rewards incident to ownership of an asset. Title may or may not pass.

Criteria:

(a) The lease is non-cancellable

(b) Secures for the lessor the recovery of his capital outlay plus a return for the funds invested: i.e. the PV at the inception of the lease of the minimum lease payments is greater than or equal to substantially all the fair value of the leased asset

(c) The lease term is for the major part of the useful life of the asset. Title may or may not pass.

LEVERAGE LEASING/CROSS BORDER LEASING

As stated earlier under the section headed "underlying tax considerations", a major advantage of leasing arises out of the tax depreciation rates available on the relevant plant. Particularly in the case of new mining developments, there are considerable advantages in leasing which has the effect of moving the benefit of tax deductions from the lessee to the lessor and the lessor reflecting that benefit in a lower interest rate.

In the case of "big ticket" leasing, one frequently sees leveraged leases which bring together a lender(s) with the available finance, a lessor(s) with the available tax base, and, of course, the lessee who wishes to use the plant.

From the stance of the lessee the criteria for evaluating such a lease are no different from the criteria for evaluating a conventional lease - what are the relevant cash flows and how do they compare in discounted cash flow forms with alternative financing mechanisms. It should be noted that with leveraged leases it is quite common to see lease terms of similar length to conventional borrowing.

These same criteria apply equally to another form of leasing which is extremely complex but if put into operation can result in extremely cheap money. It is referred to as "double-dip" or "cross-border" leasing.

"Double-dip" leasing is an arrangement that, because of a mismatch of leasing rules in the lessor and lessee countries, allows both the lessor and lessee to be treated as the owner of the property for tax purposes. Thus the lessee (as well as the lessor) is entitled to benefits such as tax depreciation and tax credits. Otherwise a lessee's deductions would be limited to the rent paid. In a cross-border setting, the double-dip lease generally will be considered a true (or "operating") lease in not only the lessor's country but also in the lessee's country. Typically the lessor's country will characterise leases solely on the basis of legal ownership, while the lessee's country will look to the economic reality of the arrangement (e.g. an option to purchase for a nominal price at the end of the lease-term is strong evidence that the lease is in fact an instalment sale).

It is the lack of symmetry in the way countries treat lease arrangements that makes the double dip-possible. Double-dip arrangements have been very common between the U.S. and the U.K. but they are now less advantageous since the U.K. scaled down its first-year depreciation write-off on equipment to ten percent (from 100 percent).

Another example would involve the use of a Swiss leasing company to lease assets to a U.K. company under a hire purchase arrangement that is treated as an operating lease for Swiss tax purposes and a purchase for U.K. tax purposes. Thus both the U.K. corporation and the captive Swiss leasing company get depreciation deductions. The hire purchase arrangement has lease payments for a term of years with an option to buy for a nominal amount.

The U.K. company also gets a deduction for the interest factor in the hire purchase agreement. The U.K. company gets a front-end deduction while the Swiss company must take its depreciation on a straight-line method. But the real advantage is that the U.K. depreciation generates a 52 percent tax benefit while the payments received in Switzerland are taxed at only 14 percent and are kept within the corporate group.

Double-dip leasing is plainly a very involved area. Because of the expense of establishing this form of lease, its use is restricted to "big ticket" items.

SOME SPECIAL PROBLEMS OF THE MINING INDUSTRY

Like all industries, the mining industry has its own unique problems. Its expenditure can be characterised as follows in broad headings:

- Exploration

- Intangible development and civil works (such as site clearance, shafts, roads)

- Plant

- Buildings

- Provision of employee infrastructure (schools, housing, police etc.)

- Contributions to State where mining company operates.

In many countries, expenditures by mining companies attract special tax concessions. One point which the mining company must address, is whether the relevant asset will attract the same favourable tax concessions if owned by a non-mining finance company. If not, the leasing option is at a grave disadvantage. Similarly, several of the items listed above by their very nature do not readily lend themselves to leasing (e.g. exploration expenditure).

SOME OTHER CONSIDERATIONS

Stamp Duties

Very often stamp duties of some kind are imposed on the lease agreement or upon the lease rentals. Frequently these are fairly immaterial but this is not always the case. For example, in Australia lease rentals attract stamp duty at a rate of $1\frac{1}{2}\%$ - which must be factored into the cash analysis of the leasing option.

Exchange gains/losses

In the case of leases which are treated as operating leases for tax purposes, their effect is to change the nature of the transaction from a capital purchase to a revenue outgoing. This may have a significant tax consequence in relation to the tax treatment of exchange gains/losses for the borrower/lessee. In many countries exchange gains/losses on plant purchases are treated as capital and treated differently for tax from a normal trading expense. Accordingly if a major plant item is financed by an offshore borrowing, an exchange movement on the repayment of the debt may be ignored for tax purposes. If the plant instead is funded through a lease denominated in foreign currency, the effect would generally be to make the exchange movements deductible/assessable (as the case may be) to the lessee for tax purposes.

Withholding tax

In the case of overseas funding, the withholding tax aspects need careful consideration. Many countries impose withholding tax on interest payments. If a leasing approach is adopted instead, does the lessor suffer withholding tax or income tax on his rental receipts?

CONCLUSION

Leasing often has advantages in terms of easier access to funds and creating the ability for 100% gearing. It may confer accounting advantages. Through the use of conventional tax deductions (or on a more sophisticated level through leveraged leasing or through cross-border leasing) it may involve substantial cash flow savings compared with traditional debt form of funding.

At the same time it normally involves higher implicit interest rates and therefore the "borrower" must resort to a full analysis on each major purchase of the respective merits of leasing versus borrowing.

The tax and evaluation issues are complex and advice should be sought from leasing and tax specialists. The paper plainly contains many generalisations which I hope are not misleading in specific circumstances.

* * * * * *

3.10

Theory and Practice of Post-Installation Appraisals of Projects

Mr. Richard D. Mills

Cyprus Coal Company

Throughout the mining industry, many companies perform sophisticated business and financial analyses of proposed capital investments. However, once the project is approved, it is never reevaluated. At best, there is an attempt to control capital costs and minimize overruns. The post-installation appraisal, a formalized review of investment performance, may prove helpful.

Typical Project Review

A typical capital investment is approved after business and financial benefits have been reviewed with management.* The financial data shown to management often include a net income statement, cash flow analysis and summary economic data similar to those shown in Table 1 overleaf.

These financials report the expected impact of the project on the corporation financial statements. This project is a rudimentary example of a capital investment decision. An actual project would undergo a more sophisticated analysis incorporating depletion benefits, deferred taxes and a detailed cost breakdown. The example project has several financial benefits that could result in management approval. For example, it provides an immediate net income benefit to the company, generates positive cash flows every year of the project life and has a discounted cash flow rate of return of 25 percent. However, it also has the disadvantage of a long payback period. A sensitivity analysis would show which areas of the analysis have the greatest impacts on profitability.

*The level of management approval will vary depending on each company's delegation of authority levels.

Post-Installation Program Objectives

The post-installation appraisal goes beyond the analysis shown in Table 1 by comparing actual results to those projected in the original economic analysis. The post-installation program has three primary objectives:
1. To provide a periodic performance review of major spending programs.
2. To review decision-making and evaluation procedures.
3. To provide a means of controlling individual project investments.

The first major function of the post-installation program is the <u>periodic review of actual performance</u> of projects and investment programs, identifying those factors which proved critical to the health of investments. This review will serve to keep management abreast of progress, provide a comparison of actual performance against expectations, and provide valuable information that can be used for improving future economics.

The second major function of the program is to perform a <u>continuous review of the decision-making process</u>. This includes testing of evaluation procedures, forecasting techniques, and logic used in making decisions. An information system should be set up whereby original evaluations and historical performance data gathered under the appraisal program are documented for use by evaluators and management when assessing past and future investment proposals. Such data should be particularly useful in evaluating new and developing types of operations and in activities where companies repeatedly reinvest. The primary purpose of this feedback system is to sharpen estimates to assure that they are realistic and attainable. The appraisal program also provides assurance that consistent evaluation procedures are being used throughout a company, leading to improved procedures

Table No. 1
EXAMPLE ACQUISITION ECONOMICS, $

	Year 0	1	2	3	4	5	6	7	8	9	10	TOTAL
Production, Units	–	100	150	200	200	200	200	200	200	200	200	1850
Price, $/Unit	–	10	10	10	11	11	12	13	14	15	16	
Profit and Loss												
Sales Revenue	–	1000	1500	2000	2200	2200	2400	2600	2800	3000	3200	22900
Cash Operating Cost	–	(700)	(1050)	(1400)	(1400)	(1400)	(1400)	(1400)	(1400)	(1400)	(1400)	(12950)
Depr., Depl. & Amortization	–	(220)	(240)	(260)	(280)	(300)	(100)	(100)	(90)	(80)	(150)	(1820)
General, Sell & Administrative	–	(100)	(150)	(200)	(200)	(200)	(200)	(200)	(200)	(200)	(200)	(1850)
Income Before Tax	–	(20)	60	140	320	300	700	900	1110	1320	1450	6280
Taxes: Current @ 50%	–	10	(30)	(70)	(160)	(150)	(350)	(450)	(555)	(660)	(725)	(3140)
Inv. Tax Credit	–	110	10	10	10	10	10	10	5	5	–	180
Net Income After Taxes	–	100	40	80	170	160	360	460	560	665	725	3320
Cash Flow												
Net Income After Taxes	–	100	40	80	170	160	360	460	560	665	725	3320
Non-cash Charges	–	220	240	260	280	300	100	100	90	80	150	1820
Capital Expenditures	(1000)	(100)	(100)	(100)	(100)	(100)	(100)	(100)	(50)	(50)	(20)	(1820)
Working Capital Changes	–	(100)	(50)	(50)	(20)	–	(20)	(20)	(20)	(20)	300	–
Net Cash Flow	(1000)	120	130	190	330	360	340	440	580	675	1155	3320

Net Cash Flow 3320
Present Value
 -Discounted at 15% 695
 -Discounted at 20% 291
DCF Rate of Return 25
Payback, Year 5TH

where needed. Information developed under the appraisal program will provide improved measures of investment performance.

Third, the post-installation program provides a control on investments by individual project or program, similar to the control reporting system used by many companies. Such a control on investment profitability is a part of the capital budgeting process. It is helpful if management can be supplied such reviews of operational, financial and economic consequences of investment decisions after the act, especially when the planning and investment evaluation functions of the company are largely decentralized. To make this control function most effective, all major investments should be covered in the appraisal program.

Prerequisites to an Effective Appraisal Program

The success of the post-installation program hinges upon close working relationships and cooperation among the personnel working on project evaluations. An environment conducive to free information flow is essential in making investment appraisals objective and realistic, and in communicating the results most effectively. Appraisals should provide data which will permit an assessment of the degree of risks being considered, and lead to the development of a realistic risk-taking attitude. Investment productivity can best be enhanced by taking risks and then carefully assessing and capitalizing on the results. The reason for looking back as is done in post-appraisals is to sharpen estimates and shape future plans, not to "point the finger" at someone's past mistakes. To make an appraisal most effective, considerable emphasis should be placed on analysis of results and development of significant conclusions. In order to make the appraisals more penetrating, the technical as well as the economic features of projects should be studied. Personnel in departments responsible for conception, implementation and operation of the project should participate actively in developing data, interpreting results, and developing recommendations for improving future performance.

Responsibilities

Primary responsibility for coordinating and putting together post-appraisals could be assigned to financial or planning departments within a company. The operating departments should share responsibility for the preparation of appraisals. They should be as closely involved as they are when investment proposals are being prepared for management. In this way, all concerned will benefit from experience and be able to apply that experience.

Recommendations which result from appraisal studies should cover both operational and financial aspects. Since operating management is responsible for implementing changes, they should shape specific operational recommendations. Recommendations contained in appraisals should be reviewed by management and appropriate action taken.

Selection and Timing

The appraisal program should cover all major investments. Investment projects should be post-appraised individually except where there are other similar investments and a logical combination can be made. At a minimum, appraisal reports should cover investments reviewed by parent company management prior to investment, whether these investments were presented as individual items or as part of the capital expenditure budget. Continuing investment programs could be periodically appraised, as performance trends develop.

There is no single best time for a post-appraisal. The results of appraisals can be most useful in improving future performance if they are timely and recommendations are followed up. In general, these studies should be made after enough operating experience has been gained to permit a meaningful comparison with original expectations. Where an investment is extraordinarily successful, this fact should be brought to management's attention as early as possible to capitalize on the information. Where performance is significantly below expectations, plans for enhancing profitability should be developed. Results should be checked after such plans have been implemented.

Study Depth

Preset objectives would be a preliminary guide to determining depth of an appraisal. Depth may range anywhere from a short report on investment performance covering one or two key factors, to a more comprehensive report including historical performance to date, a comparison with original expectations, and a revised forecast of future expectations. The depth of study will depend on the potential value of the knowledge to be gained from further analysis.

Table No. 2
EXAMPLE POST-INSTALLATION APPRAISAL, $

	Actual				Revised Estimate							TOTAL
	0	1	2	3	4	5	6	7	8	9	10	
Production, Units	–	50	100	150	175	200	200	200	200	200	200	1675
Price, $/Unit	–	10	10	10	10	11	11	12	13	14	15	
Profit and Loss												
Sales Revenue	–	500	1000	1500	1750	2200	2200	2400	2600	2800	3000	19950
Cash Operating Cost	–	(370)	(750)	(1130)	(1310)	(1500)	(1500)	(1500)	(1500)	(1500)	(1500)	(12560)
Depr., Depl. $ Amortization	–	(220)	(240)	(260)	(280)	(300)	(100)	(100)	(90)	(80)	(150)	(1820)
General, Selling & Admin.	–	(100)	(150)	(200)	(200)	(200)	(200)	(200)	(200)	(200)	(200)	(1850)
Income Before Taxes	–	(190)	(140)	(90)	(40)	200	400	600	810	1020	1150	3720
Taxes: Current @ 50%	–	95	70	45	20	(100)	(200)	(300)	(405)	(510)	(575)	(1860)
Inv. Tax Credit	–	120	10	10	10	10	10	10	5	5	–	190
Net Income After Taxes	–	25	(60)	(35)	(10)	110	210	310	410	515	575	2050
Cash Flow												
Net Income After Taxes	–	25	(60)	(35)	(10)	110	210	310	410	515	575	2050
Non-cash Charges	–	220	240	260	280	300	100	100	90	80	150	1820
Capital Expenditures	(1000)	(100)	(100)	(100)	(100)	(100)	(100)	(100)	(50)	(50)	(20)	(1820)
Working Capital Changes	–	(150)	(75)	(75)	(50)	(50)	(20)	(20)	(20)	(20)	480	–
Net Cash Flow	(1000)	(5)	5	50	120	260	190	290	430	525	1185	2050

Net Cash Flow   2050
Present Value
 -Discounted at 15%   2
 -Discounted at 20%   (293)
DCF Rate of Return   15
Payback, Years   8th

A stepwise approach should be applied. The first step in an appraisal would be to establish historical performance to date in P&L format, including key performance items such as volume, price, etc., as appropriate. Where data are available, original estimates should be tabulated and a comparison made with actual performance to date.

The next step is an analysis of both key technical and economic factors, relating one to the other, and reconciling significant differences from original estimates to determine what can be learned from actual performance and where improvements can be made. The analysis would also include evaluation of the logic of the original decision.

The appraisal should be limited to an analysis of these key factors unless it appears to be worthwhile to evaluate future plans in light of past experience. A forecast can be made in approximate fashion and a range of expectations stated where sharper estimates are not practical. A more definitive reforecast of future expectations and reevaluation of the project should be made only when it will serve a useful purpose. Such complete appraisals can be used as a sound basis for analyzing various alternatives to determine future plans. The role of the project or program in the company's overall investment strategy can also be analyzed. Interactions with other investments and the competitive environment should be stated to the extent known.

Example Post-Installation Appraisal

Let us assume that three years have gone by since the earlier example project was approved and it is time for a post-installation appraisal of the project. The first three years of actual results as well as new estimates for the remainder of the project have been incorporated in Table 2.

Four major differences from the original evaluation have occurred during the first three years: (1) slower production buildup, (2) lower prices, (3) higher cash operating costs on a $ per unit basis, and (4) increased working capital requirements. The economic impact lowered the discounted cash flow rate of return from 25 percent in the original evaluation to 15 percent in the post appraisal. The payback was also delayed from the 5th to the 8th year. At this stage of analysis, we would determine the discounted cash flow rate of return impact resulting from each of the four differences. There would also be an analysis of the operating reasons for the slower production buildup, high cash operating costs and increased work capital. For example, the lower economics could result from markets growing at slower rates than originally projected.

Of course, not all of a company's capital spending decisions will have such problems but this example is meant to show the benefits of a post appraisal program. In addition to providing a review of <u>actual versus expected performance</u>, the post installation financial appraisal also provides data that can help management <u>control investments</u>. In this case, the project economics have significantly deteriorated. If similar capital projects also have lower economics, the company's future capital spending may need to be reallocated with a lower percentage of spending for this area of the business. This analysis could also reveal shortcomings in the <u>decision making and evaluation process</u>. It is possible that inadequate attention was given to working capital requirements and the time required to reach optimal sales levels. Operating personnel may better understand future pricing trends within the industry and learn to project more realistic operating costs.

Conclusion

A post-installation appraisal of projects can be an effective management tool. Its goal is to review past capital investment decisions in order to improve future performance and decisions.

CHAPTER 3: FINANCIAL EVALUATION FOR THE MINERALS INDUSTRY

BIBLIOGRAPHY

Blecke, C.J. and Gotthilf, D.L., 1980, *Financial Analysis for Decision Making*, Prentice-Hall.

Burke, F.M., Jr. and Bowhay, R.W., 1984, *Income Taxation of Natural Resources*, Prentice-Hall.

Burke, F.M., Jr., 1981, *Valuation and Valuation Planning for Closely Held Businesses*, Prentice-Hall.

Daems, H.E.A., 1980, "SIM Simulated Investment Management. A Simple Do It Yourself Kit for Financial Analysis of Mining Projects and Other Production Studies," *Engineering and Mining Journal*, September.

Gilbertson, B., 1980, "Beta Coefficients and the Discount Rate in Project Evaluation," *Journal of the South African Institute of Mining and Metallurgy*, May.

Hoskins, J.R. and Green, W.R., 1977, "Mineral Industry Costs," Northwest Mining Association, Spokane, 1977.

Levy, H. and Sarnat, M., 1978, *Capital Investment and Financial Decisions*, Prentice-Hall.

Mackenzie, B.W., 1970, 1971, "Evalulating the Economics of Mine Development, Part I and II," *Canadian Mining Journal*, December and March.

Manssen, L.B., 1983, "Financial Evaluation of Mining Projects: Is 'Common Practice' Enough?", *Mining Engineering*, June.

Mathews, A.A., 1977, "Mining and Beneficiation of Metallic and Nonmetallic Minerals, Except Fossil Fuels, in the United States and Canada," *Capital and Operating Cost Estimating System Handbook*, U.S. Department of Commerce, National Technical Information Service, Publication PB-27734C, October.

Sani, E., 1977, "The Role of Weighted Average Cost of Capital in Evaluating a Mining Venture," *Mining Engineering*, May.

Schwab, B. and Drechsler, H.D., 1978, "Evaluation of New Mining Ventures: Average Cost Versus Net Present Value," *CIM Bulletin*, January.

Chapter 4

Accounting for Minerals Companies

CHAPTER 4: ACCOUNTING FOR MINING COMPANIES:
A GUIDE TO UNDERSTANDING FINANCIAL STATEMENTS
OF MINING COMPANIES IN AN INTERNATIONAL ENVIRONMENT

PREFACE

This chapter on accounting was prepared by Ernst and Whinney especially for this book on finance for the minerals industry. The objective was to provide a comprehensive but practical review of accounting concepts for the general membership of SME-AIME and the business community associated with the minerals industry.

The accounting principles and methods presented are applicable to all entities operating in the hard-rock extractive industry, whether precious metals, coal, uranium, or industrial minerals. In keeping with the international scope of the book, the accounting concepts provide comparisons of accounting standards for four major mining countries: Canada, the United States, the United Kingdom, and Australia. In addition, numerous examples from the financial reports of international mining companies are cited as illustrations for the concepts presented.

The chapter is divided into eleven subchapters, as follows:

1) Financial Statements; Elements and Concepts

2) Financial Statement Analysis

3) Long-term Investments and Business Combinations

4) Foreign Currency Translation

5) Cost Capitalization

6) Depreciation, Depletion and Amortization

7) Measurement and Valuation of Inventories

8) Financing Transactions

9) Recognition of Revenue

10) Income Taxes

11) Other Important Areas of Presentation and Disclosure

In addition, twenty-two controversial accounting issues in the four countries are illustrated for comparison in easy-to-follow, tabular format, in Appendix 1.

Appendix 2 discusses methods and approaches for accounting for transactions denominated in foreign currencies.

Mark E. Emerson
Chapter Editor
Resource Exchange Corporation

4.1

ACCOUNTING FOR MINERALS COMPANIES--

A GUIDE TO UNDERSTANDING FINANCIAL STATEMENTS OF MINING COMPANIES IN AN INTERNATIONAL ENVIRONMENT

ERNST & WHINNEY, CANADA

INTRODUCTION

The principal objective of this chapter is to provide guidance in understanding and making effective use of published financial statements of mining companies. Although financial statements only represent a gathering together of information relevant to an enterprise's financial position and results of operations, they increasingly reflect the complexities of the modern day business environment. In the mining industry, a wide variety of accounting practices have evolved to deal with the unique aspects of the business in the absence of authoritative accounting pronouncements from standard-setting bodies. Given such complexities and divergent practices, financial statements of mining companies are not easily understood and cannot be used effectively as a tool in decision-making without a basic knowledge of fundamental accounting concepts, generally accepted accounting principles and their application in the mining industry.

As such, this chapter is not restricted to a discussion of the accounting for the unique aspects of the mining industry. These aspects are dealt with at length but the text also includes a discussion of general accounting concepts and practices which have wide application in the mining industry.

Frequent references are made in the text to international practices and to specific rules and requirements of individual countries. In this respect, the scope is restricted to four major mining environments, Canada, the UK, the US and Australia, all of which have made significant contributions to the development of generally accepted accounting principles.

Acknowledgements

This chapter was prepared under the direction of Ian Murray of Ernst & Whinney in Toronto, Canada. Contributions were received from Graham Butt, Mark O'Sullivan and Dominic Robinson, all from Ernst & Whinney in Toronto and Bill Duvall and Tom Schoenbaechler from Ernst & Whinney in Louisville, Kentucky, US. The authors would also like to acknowledge the assistance received from David Ferguson, Don MacLean, David Taylor and Susan Zoutman from Ernst & Whinney, Toronto, from the UK and Australian practices of Ernst & Whinney and from Ernst & Whinney International.

FINANCIAL STATEMENTS - ELEMENTS AND CONCEPTS

Financial statements were originally a stewardship report to owners to satisfy them that management had given a proper account of the resources with which it had been entrusted. Today, the number of users of financial statements has expanded to include the following groups:

- shareholders, present and prospective;
- lenders, present and prospective;
- financial analysts;
- regulatory authorities;
- taxing authorities;
- labour unions and employees; and
- creditors and customers.

The expectations of these user groups differ. For example, investors and lenders require information that will indicate the adequacy of their security and the profitability of the enterprise; taxing authorities will be concerned with the concept of income; labour unions and employees will focus on the profitability of the enterprise.

Ground Rules for Preparing Financial Statements

<u>Responsibility for Financial Statements</u> Financial statements are the representations of <u>management</u>. This may strike the reader as a rather obvious statement. It is important to remember, however, the role of judgment in accounting. Notwithstanding the promulgation of accounting standards in the developed countries, considerable flexibility remains in the selection of accounting practices and methods and in their presentation. Arguably, there is more flexibility in the mining industry as few countries have developed standards to deal with its unique aspects. Accordingly, financial statements are as much a product of management's perspective and philosophy as they are a compliance document.

Underlying Preparation Concepts   Information included in financial statements is prepared on the basis of certain fundamental accounting concepts. The fundamental accounting concepts which underlie the preparation of financial statements are:

- going concern;
- accrual;
- matching; and
- consistency.

These concepts are not usually referred to in financial statements because their acceptance and applicability are assumed. If they are not followed, disclosure of the departure is necessary. When the going concern concept is applied, the enterprise is viewed as continuing in operation for the foreseeable future. Financial statements are presented on the basis that the enterprise does not intend or need to liquidate its assets or materially curtail the scale of its operations. Under the accrual concept, revenues and expenses are recognized in the accounts as they are earned or incurred, rather than as the cash is received or paid. The matching concept also comes into play here, whereby expenses are included in the income statement of the same period as the revenues to which they relate. Under the consistency concept, the accounting principles selected and their method of application are the same from period to period.

Another general principle worth mentioning because of its wide applicability to the mining industry is that relating to cost deferral. In the mining industry, more so than in other industries, costs are capitalized on the balance sheet and carried forward to offset anticipated revenues in future periods. Almost every cost has some value extending beyond the period in which it is incurred. No hard and fast rule can be made between costs that are expensed as soon as they are incurred and those that may be deferred on the balance sheet. The dividing line is established by practical judgment involving examination of the nature of the cost, the reliability of the estimate of its potential contribution to future revenues, the materiality of the cost, and so on. Costs to be deferred will usually be readily identifiable in the cost system, frequently being costs of a specific project collected on individual work orders. In contrast, the case for deferment of arbitrary percentages of an ordinary administrative cost is rarely convincing.

Since they are expressed in monetary terms, financial statements can give an impression of accuracy. In fact, they are a combination of factual information and the subjective judgments of management about the adequacy of such items as a provision to write down exploration and development costs to realizable value. Judgment is also required in selecting from among acceptable alternatives, for example, whether the straight-line method or unit-of-production method should be used for depreciation of capitalized costs. The level of detail of financial statement disclosure is another matter for the judgment of management.

Lastly, financial statements are generally prepared using the historical cost concept or convention and do not attempt to measure current value. In recent years, the accounting profession has taken steps to introduce the concept of current value reporting through requirements to disclose information relating to the effects of changing prices. For many reasons, the introduction of these requirements has not been successful and historical cost continues to be used as the principal basis for preparing financial statements.

Of course, the difficulty with the historical cost concept is that assets may be worth considerably more than cost, and management may wish to see these value increments reflected in the financial statements. In response to these concerns, the accounting profession has reached a compromise and some countries, such as the UK and Australia, now permit the use of revaluation accounting. This concept is discussed in more detail in the Financing Transactions section of this chapter.

Generally Accepted Accounting Principles   Most financial statements are prepared in accordance with generally accepted accounting principles (GAAP). These are broad rules concerning the measurement, allocation and disclosure of financial events and transactions. The most important role that GAAP plays is to provide guidance on the selection of appropriate accounting principles and of the methods used to apply these principles. There are alternative principles and methods available and management must use its judgment to select those most appropriate. This selection is governed by considerations of industry practice, prudence, substance over form and materiality. The selection of one accounting method over another will be governed by those practices which are commonly followed throughout the particular industry. For example, exploration costs can be accounted for by any one of several methods ranging from expensing all exploration costs as incurred to capitalizing all such costs and carrying them forward on the balance sheet. Prudence dictates that where uncertainty is involved, reasonable conservatism should be exercised. Substance over form requires that transactions should be accounted for giving consideration to their financial reality and not merely according to their legal form. Materiality requires management to be the judge of whether the matter concerned is significant enough to affect the decision of the user of financial statements.

GAAP DEFINED: No precise definition of GAAP exists. GAAP has as its basis the pronouncements of each country's standard-setting bodies, for example, the Financial Accounting Standards Board (FASB) in the US, the Australian Accounting Board, the Canadian Institute of Chartered Accountants (CICA) and, in the UK, the Institutes of Chartered Accountants of Scotland, Ireland and

England and Wales. In addition, GAAP includes practice which has not been codified, but which has been accepted by the business community; industry accounting practices which make sense given certain industry conditions; and International Accounting Standards (IAS), which pronouncements are the results of the deliberations of the International Accounting Standards Committee (IASC), a committee composed of representatives from the principal standard-setting bodies, including the US, Australia, Canada and the UK. Legislation is also another part of the framework for GAAP. Statutes such as the UK Companies Acts and the Directives of the European Economic Community contain a great number of specific provisions relating to financial presentation and disclosure. All of these sources combine to form GAAP.

GAAP'S MANDATE: The framework for GAAP is provided by the preparation concepts discussed above. In addition, with respect to the presentation of accounting information, there are three main facets. These are measurement, allocation and disclosure. Measurement is concerned with determining the amount to be ascribed to a transaction or event. In most cases, the measurement issue is clear-cut, such as in the case of determining the cost of a purchased property. More difficult questions of measurement can arise, such as the rate to be used to translate amounts denominated in a foreign currency. Allocation is concerned with apportioning revenues and costs among accounting periods. This process can be reasonably precise, as when allocating deferred financing costs over the period of the debt to maturity, or may involve estimates, such as the economic life of certain fixed assets, over which their cost is to be depreciated. Disclosure is concerned with the amount of detail presented in the financial statements. In particular, GAAP is concerned with disclosure which is relevant, informative and useful to a reader of financial statements.

GAAP IN DIFFERENT COUNTRIES: The local regulations of each country govern, to a great degree, financial reporting. These regulations include the pronouncements of the standard-setting bodies described above as well as statutory requirements. The extent and degree to which the standard-setting bodies participate in the development of GAAP differs between countries. For example, the US has a more regulated approach to the setting of accounting standards than Canada, Australia or the UK. The guidance contained in US standards is more extensive than that available in any other country and when US standards do overlap with those of other countries, the US pronouncements are generally more specific and detailed in the amount of direction provided. The UK, Canada and Australia take a less rule-oriented approach to the codification of GAAP with more discretion left to the preparer of financial statements as to the appropriate principles to be applied. Often, the experience or formal standards of countries such as the US, will be drawn on to determine the propriety of a particular accounting treatment. In terms of accounting principles relating to the mining industry, few specifically developed standards exist. Statement of Accounting Standards of Australia (AAS) No. 7 deals with accounting for the extractive industries, including accounting for pre-production costs, inventories and revenues. No other country has developed standards specific to the mining industry. More extensive guidance exists relating to the oil and gas industry, largely because the FASB in the US has published three separate standards relating to accounting and financial statement disclosure in the oil and gas industry.

GAAP IN THE INTERNATIONAL ENVIRONMENT: The form, scope and content of accounting standards and principles do differ between countries. The objective of International Accounting Standards is to produce standards which meet with worldwide acceptance, in the interests of harmonizing as far as possible the diversity of accounting standards and principles that exist. The IASC also exists to provide guidance to those countries which do not have any formal standards in a particular area of accounting.

International Accounting Standards are designed to concentrate on essentials. They are not so complex that they cannot be applied effectively worldwide and any entity which operates or reports outside the jurisdiction of its incorporation is advised to conform to such standards. In addition, they are in the nature of general recommendations and do not override local regulations.

At the end of this chapter, a comparison of International Accounting Standards with accounting rules and principles applicable in the US, Canada, Australia and the UK is provided to facilitate an understanding of situations in which financial statements prepared in accordance with local rules and principles conflict with certain provisions of International Accounting Standards. In addition, throughout this chapter reference has been made to differences in GAAP between the respective countries.

There is an increasing trend towards the international harmonization of accounting standards and principles. The new standards for foreign currency translation (see separate section in this chapter) are a good example of the importance given to the harmonization of accounting standards and principles by the respective standard-setting bodies.

The Components of Financial Statements

The main elements of financial statements are:
- the balance sheet or statement of financial position;
- the statement of income or earnings;
- the statement of changes in financial position or sources and applications of funds; and
- notes to the financial statements.

In addition, statements of retained earnings and equity are presented frequently, to provide a continuity of amounts for a period. A brief description of the composition of each statement is provided below.

Balance Sheet:
- It presents the assets, liabilities and equity of an enterprise as at a point in time and, as such, is a representation of the financial position of the enterprise.
- Assets are a representation of the economic resources and the estimated future economic benefits obtained or controlled by an enterprise.
- Assets are classified as current - those generally realizable within twelve months - or long-term.
- Liabilities are a representation of the obligations and the estimated future sacrifices of economic benefits.
- As for assets, liabilities are classified as current or long-term.
- Equity represents the owner's interest in the enterprise.

Income Statement:
- It presents the revenues, expenses and profit recorded by an enterprise since the last reported financial statements.
- It should show separately:
  . the results of ongoing operations,
  . the results of discontinued operations,
  . large non-recurring operating items (unusual items),
  . non-operating gains and losses,
  . income taxes,
  . large, non-recurring, non operating items (extraordinary items),
  . net income, and
  . earnings per share.

Statement of Changes in Financial Statements
- It explains how the activities of the enterprise have been financed and how the enterprise's economic resources have been used during the period.
- It can be prepared from either a working capital or a cash perspective.
- It will generally consist of four elements:
  . funds from operations,
  . other sources of funds (financing activities),
  . applications of funds (investing activities), and
  . the change in working capital or cash.

The main components of the balance sheet, particularly as they affect mining enterprises, are discussed throughout this chapter. The components of the statements of income and changes in financial position are not dealt with elsewhere and it is appropriate to expand briefly at this time on some of the presentation concepts discussed above.

Statement of Income An example best illustrates the concepts of income statement presentation. For this purpose, we have selected the 1981 income statement of Inco, a large North American mining company, as reproduced in Figure 1. Methods of presentation in other environments do differ from that illustrated here but the underlying principles are generally the same.

INCO LIMITED AND SUBSIDIARIES

CONSOLIDATED STATEMENT OF EARNINGS

Year Ended December 31	1981 (in thousands)
REVENUES	
Net sales	$1,885,923
Other income	25,677
	1,911,600
COSTS AND EXPENSES	
Cost of sales and operating expenses	1,454,430
Selling, general and administrative expenses	144,378
Research and development	36,996
Exploration	32,653
Interest expense	147,130
Currency translation adjustments	(5,616)
	1,809,971
Earnings from continuing operations before income and mining taxes	101,629
Income and mining taxes	81,162
EARNINGS FROM CONTINUING OPERATIONS BEFORE EXTRAORDINARY CHARGES	20,467
Loss from operations of discontinued business segments, net of applicable income taxes	(25,347)
Loss before extraordinary charges	(4,880)
Extraordinary charges	
Provision for loss on disposition of discontinued business segments, net of applicable income taxes	(245,000)
Loss on revaluation of investment in Exmibal	(219,638)
NET LOSS	$ (469,518)
Net loss per common share	
Continuing operations	$ (.10)
Discontinued business segments	(.33)
Extraordinary charges	(6.08)
NET LOSS PER COMMON SHARE	$ (6.51)

Note:

Comparative amounts for preceding periods have not been presented in this illustration.

FIGURE 1

The magnitude of the numbers in Inco's income statement conveniently illustrates the concepts of continued versus discontinued operations and extraordinary items. A reader of financial statements wants to evaluate the future potential of an enterprise. This evaluation can only be performed if the results of discontinued operations are shown separately from the results of ongoing activities. In Inco's case, the operating loss of $25 million from discontinued business segments has a significant bearing on that evaluation as it transformed earnings from ongoing operations of $20 million to an overall loss from operations of approximately $5 million.

The reader should also be able to distinguish between the results of operations and gains and losses on non-operating items which will often be non-recurring and which therefore should be evaluated separately. The provision for the loss on disposition of discontinued business segments of $245 million is non-operating in the sense that it did not result from operational activities but is the consequence of a decision to withdraw from a line of business without recovery of the company's monetary investment in that line of business. This loss is classified as extraordinary because it is not part of normal operations to dispose of a large business segment, it is non-recurring and should not be considered as a recurring factor in an evaluation of the ordinary operations of the company.

Inco's income statement provides a good example of an informative and concise summary of the company's operating activities. One item not shown is what was referred to above as an unusual item. As defined, unusual items typically relate to operations but are disclosed separately because of their size. Examples might be a large write-off of exploration expenditures or a provision, in addition to the regular charge, for depreciation in respect of idle mining equipment. Another income statement item not illustrated, and which is discussed in the long-term investments section, is the share of earnings from associated companies. Generally, this item is shown separately from results of operations, unless a significant portion of an enterprise's activities is conducted through associates.

Statement of Changes in Financial Position The statement of changes in financial position (the funds statement) has historically been prepared on a working capital basis, the main emphasis being on presenting changes in working capital rather than changes in cash and the cash flow generated from operations. Growing disillusionment with the working capital basis of presentation, mainly because of its perceived failure to pinpoint liquidity problems, has seen a shift in practice to the cash basis of presentation. As the cash basis of presentation is currently only in an evolutionary stage of implementation in practice, it may be helpful to indicate some of the perceived difficulties with the working capital basis of presentation of funds statements and what simple adjustments can be made to that basis to determine if an enterprise is generating positive cash flow from operations.

ABSENCE OF CASH FLOW INFORMATION: A business, to survive, needs to be profitable and liquid. Profitability must be accompanied by cash flow which meets the liquidity needs of the enterprise. It is possible to earn profits without necessarily generating equivalent cash flow. What ultimately enables an enterprise to prosper is the tangible resource of cash and not profit which is the result of applying certain accounting methods and conventions. The reporting of profit is important to financial statement users but so too is the disclosure of the cash flow of the enterprise. The conventional basis of presentation of the funds statement does not identify the generation of cash flow from operations.

DIVERSITY IN PREPARATION: In accounting standards little has been prescribed concerning the structure and content of a funds statement. The result is that there is a considerable diversity in funds statement reporting practices. In addition, a mixture of cash and non-cash data is reported, often in no apparent order and not reconcilable with the related balance sheet and income statement amounts. Note, however, that these deficiencies will not all necessarily be cured by using a cash basis presentation.

ADJUSTING TO A CASH BASIS: The main adjustment to convert to a cash basis is to remove the impact of the change in non-cash working capital items, normally receivables, inventories and accounts payable from the conventional funds from operations amount. In the cash basis funds statement of Canada's Noranda Mines, in Figure 2, the first two captions are funds from operations and funds from operating working capital. The amount relating to funds from operations reflects the conventional presentation. If the funds from operating working capital amount is added to the funds from operations amount the resultant amount is an approximation of cash flow from operations.

NORANDA MINES LIMITED

CONSOLIDATED STATEMENT OF CHANGES IN FINANCIAL POSITION

Year Ended December 31	1983	1982
	(in thousands)	
FUNDS FROM (TO) OPERATIONS		
Loss	$(34,599)	$(82,944)
Mining properties write-down	94,546	--
Depreciation and amortization	191,583	169,105
Taxes provided not currently payable	(98,726)	(47,541)
Minority interests in earnings of subsidiaries	9,673	6,457
Losses of associated companies net of dividends received	8,001	68,611
	170,478	113,688
FUNDS FROM (TO) OPERATING WORKING CAPITAL		
Change in accounts, advances and tolls receivable	(240,089)	(58,447)
Change in inventories	18,322	(89,554)
Change in accounts and taxes payable	278,247	(35,338)
	56,480	(183,339)
USES OF FUNDS		
Fixed asset additions	358,903	662,747
Deferred expenditures	39,923	42,921
Investments and advances	98,960	29,061
Dividends - shareholders	91,022	117,673
- minority shareholders of sub.	6,526	3,252
Payment of debt	79,815	99,826
	675,149	955,480
FINANCING REQUIRED	448,191	1,025,131
SOURCES OF FINANCING		
Brenda Mines Limited	--	62,863
Issue of common shares	3,850	17,076
Long-term financing	383,888	899,957
Fixed asset disposals	56,440	31,945
Other	(1,821)	42,621
	$442,357	$1,054,462
(Increase) decrease in bank advances less cash, short-term notes and marketable investments	$(5,834)	$29,331

FIGURE 2

On this basis, it can be seen that Noranda Mines had positive cash flow from operations in 1983 but not in 1982. Such information would not have been readily identifiable using the conventional basis of presentation as the impact on funds from operations of changes in non-cash working capital items is not presented in this fashion.

Noranda's funds statement is also presented so as to highlight the amount of financing required and how that financing was accomplished. Using 1983 amounts, the aggregate uses of funds of $675 million are deducted from the $170 million provided by operations and $57 million by working capital to leave a net financing requirement of $448 million. Other forms of presentation may be equally informative and it is difficult to say if more rigidity in the presentation of funds statements would represent an improvement in financial reporting.

Notes to Financial Statements The last main element of financial statements are the notes to financial statements. They are an integral part of the financial statements and are also management's representations. Notes are concerned with disclosure of significant accounting policies, especially when a selection has been made from alternative accounting principles or methods. Notes also provide detail concerning the composition of specific financial statement items, for example, an analysis of exploration and development expenditures. They would include additional information concerning financial statement items such as details of security for liabilities and additional information pertaining to financial position such as information concerning contingent liabilities. As well, notes would disclose details of significant events that occurred after the date of the financial statements; information concerning a change in accounting policy; and disclosures required by statute.

Development Stage Enterprises Certain entities in the mining business that are described as being development stage enterprises, may not prepare an income statement as they have no operations. An entity is generally described as being in the development stage if it is devoting substantially all its efforts to establishing a new business and there has been no significant revenue from planned, principal operations (FAS No. 7). The same accounting principles that apply to established operating enterprises govern the recognition of revenue and capitalization of costs by a development stage enterprise. If in applying these principles it is determined that it is more appropriate to defer or capitalize all costs since inception rather than write off all, or a portion, to expense, then a statement of deferred costs will be presented in place of an income statement. A statement of deferred costs will present the major cost components capitalized such as exploration, development and administration costs. A development stage enterprise in the mining industry will typically be devoting its efforts to raising capital and exploring for and developing mineral resources.

Summary

In this section, we have discussed the ground rules that are fundamental to any set of financial statements, the major elements of the financial statements and some of the important concepts underlying the typical presentation of each of these elements. The discussion has been general in nature rather than specific to mining companies. It is appropriate to emphasize that, due to the absence of authoritative pronouncements relating to the mining industry, management plays a large role in the selection of accounting principles and methods. This fact underlies the need for users of financial statements to understand the philosophy and motivations of management. Without this understanding, decisions regarding the information presented in financial statements may be uninformed.

FINANCIAL STATEMENT ANALYSIS

Readers can use certain tools to perform an analysis of financial statements, the results of which can indicate the performance of the enterprise. These tools are accounting ratios and they can be used as indications of the profitability, activity, liquidity and equity position of an enterprise.

The ratios described below are those commonly used in the mining industry.

Current Ratio

The current ratio is calculated as the ratio of current assets to current liabilities. Current assets typically include cash, short-term investments, accounts receivable, inventories and prepaid expenses. Current liabilities include short-term bank borrowings, accounts payable and accrued expenses, income taxes payable and the current portion of long-term debt. Current assets less current liabilities are referred to as working capital.

The current ratio is one of two commonly used measures of liquidity and shows the relationship of current assets to current liabilities. The implications to an enterprise of the trend of this ratio over a period of time will depend on the individual circumstances of the enterprise and can only be evaluated in that context. Clearly, a declining trend indicates that the enterprise is short of working capital and may have difficulty in meeting its current obligations.

The current ratio is used extensively by lenders as a means of monitoring an enterprise's liquidity position. A debt agreement will often include a covenant requiring a minimum level of working capital. Should such a covenant not be met by the enterprise, there should be disclosure of the breach in its financial statements unless a waiver has been received from the lender.

Classification as current generally envisages realization of an asset, or the payment or discharging of a liability, within one year.

Acid-Test Ratio

The acid-test ratio is the ratio of current assets, excluding inventory and prepaid expenses, to current liabilities. Also known as the liquidity or quick ratio, it supplements the current ratio. It is a more severe test of financial safety because it deals only with assets that are quickly available for paying short-term creditors.

Return on Capital Employed

The return on capital employed is the ratio of net income to average equity and long-term debt. Both the terms equity and long-term debt are defined in the Financing Transactions section of this chapter.

The ratio of net income to capital employed is the basic ratio that many investment analysts use in measuring an enterprise's performance from period to period and in comparing one enterprise with another. It is a yardstick of how well management has used the funds invested in the business and in that respect it can be an important tool for management. Net income for the purposes of this ratio usually is measured as net income before interest expense, income taxes and extraordinary items. Capital employed usually is taken as the sum of equity and long-term debt on an average basis. In making internal comparisons between segments of a particular business, this ratio serves as a useful starting point. It is also a useful general indicator of progress over a period of years.

There are many different ways that can be used to show how profitably funds are employed. Depending on the purpose of the exercise, or on the particular circumstances of a business or industry, the composition of the profit figure and of the assets employed may vary. For example, it might be desirable to include in capital employed all or a part of bank indebtedness and other short-term obligations if these were of a semi-permanent, renewable nature. It might be appropriate to go even further and relate profits to total operating assets only.

Debt to Equity Ratio

The debt to equity ratio is the ratio of long-term debt to equity.

One way to test the fundamental financial stability of a company is to consider how the operating assets have been financed. The two main sources of funds are equity and debt. The maintenance of a proper balance between these two sources is of prime importance and so the ratio of debt to equity is very significant. A company financed by a high proportion of borrowed money, often described as highly leveraged, has the potential for producing a high rate of return on its equity funds. On the other hand, it has the potential of incurring the burden of significant interest charges and a higher risk of insolvency should the business falter as interest rates rise.

In computing the debt to equity ratio, or any other ratio involving a measure of debt or equity, it is important to understand the components of each. For example, in the US, Securities and Exchange Commission (SEC) registrants are required to report preferred stock with mandatory redemption requirements outside the equity section of the balance sheet on the basis that such stock has characteristics akin to debt. In addition, companies in the UK and Australia may have asset revaluation reserves within the equity section. These reserves are rarely seen in Canada and are not permitted in the US.

Gross Profit Margin Percentage

The ratio of gross profit as a percentage of net revenues provides a measure of profitability. Gross profit is the excess of net sales over the cost of goods sold. Its effectiveness as a measurement of performance depends greatly on the user's knowledge of the business and of the impact of external economic conditions on operating results. For example, a gold mining enterprise may decide to mine low-grade ore when prices are low so that revenues can be maximized when prices recover. In this situation, operating results will not only be impacted by low prices but also by the low yield from mined ore.

Inventory Turnover Ratio

The inventory turnover ratio is the ratio of the cost of sales to inventory and indicates the rate at which inventory is being converted into sales. It is calculated as the cost of goods sold divided by average inventories. The ratio also indicates if the level of inventories held is justified in relation to sales revenues. An unduly slow rate of inventory turnover means that working capital is being tied up and that liquidity is being reduced.

In using this ratio the components of inventory should first be determined. The turnover ratio can be distorted through the inclusion of supplies inventories which will generally have a significantly longer turnover period.

Cash Flow Per Share

Cash flow per share is calculated as funds from operations, adjusted for the change in the non-cash components of working capital divided by the weighted average number of shares outstanding.

The importance of cash flow from operations as a measure of performance was discussed in regard to funds statements in the preceding section of this chapter. Translated to a per share amount, it provides a useful comparison to earnings per share before extraordinary items.

Summary

The preceding discussion on financial statement analysis is not intended to be comprehensive but rather it has covered those ratios which can be regarded as the most important and which will be useful in the context of the mining industry. Other key ratios such as operating costs per ton produced may also be relevant.

In using accounting ratios, their limitations should be recognized. Ratios serve only to highlight certain operating characteristics of a business. They are not an end in themselves and further detailed analysis is normally required before important decisions about a business are made. It also should be noted that ratios can sometimes hide operating characteristics of a business rather than highlight them. In addition, a ratio says little about the operations of a company when considered in isolation for a particular period. Rather, the trend of a ratio over a number of periods should be looked at in order to draw the most useful conclusions about the information presented. Otherwise it is not known if the ratio is normal, expected or unusual unless it is compared to that of preceding periods or those of other companies within the same industry.

In making inter-company and country by country ratio comparisons, several considerations are necessary. Firstly, it is necessary to consider the economic environment in which a company operates. For example, Canadian companies are generally more highly leveraged than those in the US with a resultant difference in debt to equity ratios. Secondly, throughout this chapter we have referred to differences in accounting practices and in methods of presentation between companies, and from one country to another. These differences also impact ratio comparisons.

These considerations should be kept in mind in applying ratio analysis to financial statement amounts and in their use as a basis for financial decision-making.

LONG-TERM INVESTMENTS AND BUSINESS COMBINATIONS

An enterprise may conduct some or all of its activities through the ownership of interests in other entities. These investments, usually considered long-term investments, will include:

- wholly-owned subsidiaries;
- majority-owned subsidiaries;
- investments where the investor has the ability to exert significant influence over the activities of the investee, otherwise known as associated companies; and
- joint ventures.

When an enterprise conducts a significant amount of its activities through such interests, there is a need to prepare financial statements which present the results of operations and financial position of the enterprise including its subsidiaries and other investees, rather than those of the enterprise alone.

This section discusses the manner in which the results of operations and financial position of these various investments are reflected in the financial statements of the investor and is organized according to the nature of the investment interest. That interest may be a parent-subsidiary relationship, an associate company relationship, or a joint venture. In addition, a section on business combinations is included which describes the methods used to account for the acquisition of a controlling interest in an entity.

Accounting for Subsidiaries

In the North American environment, a subsidiary is defined for accounting purposes as a company in which another company, the parent, holds the majority of the shares which have the right to elect a majority of the directors. Thus, the emphasis is on legal control. The ownership by one company of more than 50% of the voting shares of another would require the inclusion of the results of operations and the financial position of the subsidiary in the financial statements of the parent. International Accounting Standard No. 3, Consolidated Financial Statements, takes a more liberal view and also allows consolidation of a company in which a group:

a) owns more than half the equity capital, but less than half the voting power; or

b) has the power to control, by statute or by agreement, the financial and operating policies.

The broader principles of IAS No. 3 are supported in varying degrees in the UK and Australia. Accordingly, certain investees in those areas may be consolidated which would not be consolidated in the US or Canada.

Consolidated financial statements present financial information concerning the group as if it were a single economic entity without regard for the legal form of the separate entities that comprise the group. They normally include the results of operations and the financial position of the parent company and all its subsidiaries on a line-by-line basis. That is, all of the assets, liabilities, revenues and expenses of subsidiary companies are included in the financial statements of the parent company. This method of presentation contrasts with the accounting for associated companies where the investment and income therefrom are shown as single-line items in the financial statements of the parent company.

The Computational Process The process of preparing consolidated financial statements is straightforward. All of the assets, liabilities, revenues and expenses of the subsidiaries are added to like items in the financial statements of the parent company. Even if the subsidiary is not wholly-owned, the computation of assets and liabilities is nevertheless the same. The only difference is that ownership interests in subsidiaries held by stockholders outside the group are described as minority interests and are shown as a single-line item outside the equity section of the balance sheet. In addition, a deduction is made on the income statement for the interest of the minority stockholders in the income of the subsidiaries.

As part of the consolidation process certain adjustments have to be made to the financial statements of the parent and subsidiaries prior to combination. Intercompany balances and intercompany transactions between the companies in the group are eliminated including sales, charges and dividends. Unrealized profits resulting from transactions between companies within the group, for example, involving inventory or fixed assets, are also eliminated. Profits on such items are considered only to be realized when they are sold outside the group. Depreciation and amortization expense in respect of a subsidiary's fixed assets will need to be adjusted on consolidation so that the consolidated expense is computed on the basis of the amounts paid for such assets on acquisition by the parent company. It may also be necessary to make adjustments to subsidiary company amounts to ensure that such amounts are prepared using accounting policies which conform with those of the parent company.

<u>Differing Reporting Periods</u> It is preferable that the parent company and its subsidiaries have a common financial reporting period. However, the fact that a subsidiary has a different reporting period does not preclude its consolidation in the financial statements of the parent company. When the financial statements of subsidiaries with reporting dates different from that of the parent are consolidated, significant transactions or events which have occurred in the intervening period are recognized through adjustments or disclosure. In addition, the dates to which the financial statements of the subsidiaries have been prepared should be disclosed.

<u>Exclusion of Subsidiaries from Consolidation</u> In certain circumstances, subsidiaries can be excluded from consolidation. Commonly, subsidiaries will be excluded when control by the parent is likely to be temporary or the ability of the parent company to control the assets of the subsidiary is impaired. For example, when a foreign-based subsidiary is subject to nationalization by the foreign government, that subsidiary will normally be excluded from consolidation. Falconbridge Nickel Mines, a Canadian company, has a Zimbabwe investment called Blanket Mines which is reported on the cost basis in Falconbridge's consolidated financial statements despite 100% ownership of Blanket. Income from the investment is reported only to the extent of dividends received and none of the assets and liabilities of the Zimbabwe operation are combined with those of Falconbridge. No explanation for this treatment is given in the financial statements but it is probable that control and/or access to Blanket's equity is considered to be restricted.

It is generally considered appropriate to exclude from consolidation a subsidiary whose business activities are dissimilar from those of other companies within the group. In such cases, the presentation of separate subsidiary financial information in addition to the consolidated financial statements would provide more useful information. IAS No. 3 specifies further grounds for exclusion as impracticability of consolidation, the likelihood of disproportionate expense or delay, or the opinion of the directors that the effect of consolidation would be misleading or harmful.

A subsidiary excluded from consolidation would normally be accounted for using the equity method of accounting, except when control of the subsidiary is impaired, in which case the cost method should be used. These methods of accounting are discussed below. The practice of excluding a subsidiary from consolidation due to dissimilar activities is commonly seen with large conglomerates whose activities are extremely diverse but is rarely seen in the financial statements of mining companies.

Significantly Influenced Investees

When a company holds investments in entities which are not subsidiaries there are two principal methods of accounting for such investments, the equity method and the cost method. The equity method of accounting generally is required if a company holds an investment over which it exercises significant influence but does not control. A presumption of significant influence normally exists when a company holds an interest of 20 percent or more in the voting equity of another company.

Under the equity method, the carrying amount of an investment in the shares of an investee is increased or decreased to recognize the investor's share of the profits or losses of the investee from the date of acquisition. Dividends received reduce the carrying amount of the investment.

The application of this method of accounting results in more informative reporting of the net assets and income of the investor than would be the case if only dividends received or receivable from the associate were reported because the investor's proportionate share of the associate's earnings and equity are reported in the financial statements of the investor.

It should be noted that conceptually the difference between accounting for a subsidiary and the accounting for a significantly influenced investee or associate is one of presentation only. The computation of an investor's share of income from an investee is the same whether the investment is consolidated as a subsidiary or accounted for by the equity method. The different method of presentation is justified by the concept of control. When an investment is consolidated, the investor legally controls the assets and liabilities of the investee. When the equity method is used, the investor does not have control, but there is a presumption that it is in a position to exercise significant influence over the investee.

When an investment in an associated company is significant to the overall results of operations and financial position of the consolidated group, it is useful to provide additional financial information of the investee. This can be done by providing condensed financial information by way of note to the investor's financial statements or by including the separate financial statements of the investee as part of the investor's annual report.

It may be evident that the rules governing accounting for long-term investments can be somewhat arbitrary and leave little scope for discretion on the part of the investor. In many cases, the investor may have effective control over the assets and liabilities of an investee but is not permitted to consolidate. In such situations, the disclosure of financial information relating to the investee is necessary for a proper understanding of the overall results and financial standing of the investor. For example, if a company conducts a large portion of its overall activities through associated companies over whom the company has effective but not strictly legal control, then ratio analysis – such as debt to equity or return on capital employed – by reference to the consolidated financial statements of the investor alone may not reveal the true situation. Through effective control, the investor will have control over the resources of the investees and will no doubt allocate these resources within the overall group to ensure their optimum use. In such circumstances, it may be more useful to analyze the results of operations and financial condition of the overall group taking into consideration the individual assets, liabilities, revenues and expenses of equity accounted investees.

Cost Method of Accounting

Under the cost method, an investor records an investment in the shares of an investee at cost and income is recognized only to the extent of dividends received or receivable. The cost method is generally appropriate when an investor holds less than 20 percent of the voting equity of a company, when the investor has only temporary control (for example, if disposal of the investment is planned) or when the ability of the investor to control the assets is impaired.

Valuation of Long-Term Investments   When a long-term investment is accounted for using the equity or cost methods, there is a presumption that the carrying value will ultimately be realized. Many jurisdictions require disclosure in the financial statements of quoted market values in addition to the carrying value, and to the extent that the quoted value of an investment is lower than the carrying value for an extended period of time, the underlying investment should be written down. In other words, a loss in value of an investment that is other than a temporary decline sometimes will occur. The actual value of the investment to the investor may become lower than the carrying value and the impairment is expected to remain for a prolonged period. A non-temporary decline is obvious in some cases, such as bankruptcy or an agreement to sell an investment at an amount which will result in a loss. Less obviously, permanent impairment of the value of an investment may be indicated by conditions such as a prolonged period during which the quoted market value of the investment is less than its carrying value, severe losses by the investee in the current year or current and prior years, liquidity or going concern problems of the investee or the appraised value of the investment is less than its carrying value. In Canada, when one of the above conditions persists for a period of three or four years, there is a general presumption that there has been a loss in value of the investment which is other than a temporary decline. This presumption can be rebutted by persuasive evidence to the contrary.

When accounting for a permanent impairment in the value of an investment when the investment is consolidated with the financial statements of the parent, the carrying values of specific tangible and intangible assets would be reduced to the extent of the decline. When either the equity or cost method is used, the permanent decline in value would be recognized by writing down the investment in the financial statements of the investor.

Many companies argue, with some justification, that quoted market values are irrelevant if there are no plans to sell the investment and even if an investment was to be sold, quoted market value may not necessarily be indicative of realizable value for a large block of shares. Accordingly, the use of such values as an indicator of realizable value should be approached with caution.

Aside from the consideration of control discussed above, international practice with regard to the consolidation, equity and cost methods of accounting is generally uniform. Disclosures in the financial statements should include:

- a description of the accounting policy for long-term investments;
- a listing of subsidiaries and other long-term investments;
- reasons for not consolidating subsidiaries, if any;
- summarized financial information or separate financial statements of non-consolidated subsidiaries and significant equity accounted investments;
- market values of both equity and cost accounted investments; and
- assets of subsidiaries and equity accounted investees restricted as to distribution to the parent company.

Joint Ventures

Recent years have seen an increasing number of companies engaging in joint activities to accomplish specific objectives. This is particularly evident in the capital intensive extractive industries where advances in technology have accelerated capital replacement costs. While companies may prefer complete control, a number of factors have resulted in the formation of jointly-owned projects. These factors can generally be summarized as follows:

- financially a project may be beyond one company's resources;
- sharing of risks - particularly when developing new processes or techniques such as coal liquefaction;
- improving operating efficiencies with a substantially larger project; and
- combining technical skills and/or expertise with property interests as is commonly done in developing countries with natural resources.

Definition An investment made in the form of a joint venture may be a corporation, partnership or undivided interest and may be incorporated or unincorporated. Regardless of its form, the characteristic of a joint venture which distinguishes it from other forms of business arrangements is that each venturer has an equal vote (regardless of ownership) in all major decisions affecting the venture. A definition commonly used in the US and Canada of a joint venture is that in Section 3055 of the CICA Handbook:

"A joint venture is an arrangement whereby two or more parties (the venturers) jointly control a specific business undertaking and contribute resources towards its accomplishments. The life of the joint venture is limited to that of the undertaking which may be of short or long-term duration depending on the circumstances.

A distinctive feature of a joint venture is that the relationship between the venturers is governed by an agreement (usually in writing) which establishes joint control. Decisions in all areas essential to the accomplishment of a joint venture require the consent of the venturers, as provided by the agreement; none of the individual venturers is in a position to unilaterally control the venture. This feature of joint control distinguishes investments in joint ventures from investments in other enterprises where control of decisions is related to the proportion of voting interest held."

Accounting Issues Although there are a number of methods currently used in accounting for an investment in a joint venture, four are prevalent: consolidation, equity, cost and proportionate consolidation. The first three methods were discussed previously.

Under the proportionate consolidation method each venturer records its proportionate share of each asset, liability, revenue and expense of the joint venture and includes such amounts in the respective captions in its financial statements.

For example, if a venturer owns a 50% interest in an exploration joint venture, then the venturer would record 50% of the carrying values of the joint venture's assets, liabilities, revenues and expenses in its financial statements.

Although any of the above methods are acceptable in accounting for an investment in a joint venture, they are not interchangeable. The method selected should be the one which reflects the nature of the investment. Although practice varies, the equity method is the most prevalent method of accounting for investments in joint ventures in the US because voting control is generally not limited to one participant but is shared equally. In Australia, and Canada proportionate consolidation for investments in joint ventures is used extensively. Arguably, proportionate consolidation is a more informative method of presentation as it sets out the investor's proportionate interest. This presentation may be particularly relevant when the joint venture is financed externally as opposed to by the venture participants. If equity accounting is used in such circumstances, the investor's obligations with respect to the external financing arrangement do not appear on the investor's balance sheet. Proportionate consolidation may also be appropriate when the activities of the venture are a significant part and merely an extension of the investor's principal activities.

An example of the use of the proportionate consolidation method is the treatment by Canada's Denison Mines of its investment in the Quintette coal project in British Columbia. Denison has a 50% interest in this large project which has an estimated total cost of $950 million for which separate financing was arranged. Denison is required to make an equity contribution of up to $175 million in connection with the project financing and, if this amount was accounted for by the equity method rather than recognizing its proportionate share of assets and liabilities, the impact on Denison's balance sheet would be dramatic. Based on reported 1983 amounts, long-term debt would decrease by $309 million, assets would increase by an equivalent amount and the debt to equity ratio would be 0.8:1 rather than 1.4:1, a significant improvement.

Regardless of the method followed in accounting for an investment in a joint venture, the financial statements of the venture may have to be adjusted for any inconsistencies with the venturer's financial statements in significant accounting policies. For example, the accounting policies for inventory valuation, depreciation, or development costs may have to be made to conform and profits or losses from transactions between the entities eliminated as in the case of intercompany profits in inventory, transfers of property, or management fees.

As with any long-term investment, continued evaluation of a venturer's investment in a joint venture is an ongoing process and appropriate allowances should be made when a permanent diminution in value occurs.

Disclosure Issues   To enhance a reader's understanding of all important aspects of a company's operations, it is important that all significant activities carried out through joint ventures be disclosed. Both the nature and method of accounting for an investment in a joint venture should normally be disclosed in the accounting policies note of the investor's financial statements. Additionally, disclosure should be made of any direct or contingent liabilities of the venturer with respect to their participation in the venture. Such disclosures could include the potential liability of the venturer with respect to their share of the venture's liabilities or obligation to fund their share of any cash deficiencies. In this respect, adequate disclosure can remedy the perceived deficiency of using equity accounting rather than proportionate consolidation.

To the extent an investment is accounted for using either the equity or proportionate consolidation methods, disclosure of summarized financial information of the venture in the notes to financial statements of the investor may be appropriate for the same reasons as outlined for other significant investments.

The following example is a comprehensive disclosure note relating to long-term investments.

Example of Disclosure of Accounting for Long-Term Investments.
Peko-Walsend Ltd. - 1983 Financial Statements.

"Principles of consolidation

The consolidated accounts encompass the accounts of the parent company and of the subsidiary companies of the group. All intercompany accounts and transactions have been eliminated including realized profit relating to inventories transferred within the group. Subsidiary companies in the group are listed in Note 21.

The consolidated accounts reflect the group's interest in associated companies where the investment is 20% to 50% inclusive in the associated company, according to the equity method of accounting. The group's share of the undistributed profits of the associated companies, as disclosed by the latest audited or management accounts, is not available for distribution except to the extent that those profits are realized as dividends received and after deducting any taxes payable thereon. The investment in the associated companies is valued at cost adjusted for the group's share of profits or losses.

The group's interest in other listed and unlisted companies which are not subsidiary or associated companies are shown as investments. The major investments are shown in Note 7. In these cases dividend income only is included in profit.

The group's interest in joint ventures is set out in Note 20, and is taken into the consolidated balance sheet by including the group's share of the relevant assets and liabilities. The group's share of profits or losses is incorporated in the profit and loss account."

Legal Entity Financial Statements

Throughout this section we have discussed the concept of consolidated financial statements which attempt to meet the needs of those readers interested in the fortunes of the group as a whole. The needs of those interested in the financial position of the parent company or of individual subsidiaries, such as creditors, regulatory authorities and minority interests, are served by separate financial statements of the parent company or of those subsidiaries. These financial statements may not be publicly available depending on the jurisdiction in which the entity is reporting. For example, in Australia, parent company financial statements are required by law and will often be presented alongside the consolidated financial statements. In North America, parent company financial statements are generally not required in public filings except by certain Securities and Exchange Commission registrants when a significant portion of a registrant's total assets are those of subsidiaries whose assets are restricted as to distribution to the parent company.

Business Combinations

The discussion in this section refers to acquisitions of subsidiary companies through business combinations. Similar concepts apply to the acquisition of associated companies over which the investor exerts significant influence.

Accounting Issues   The method of accounting for an acquisition can have a significant bearing on the future reported results and earnings of the consolidated entity. The two generally accepted methods of accounting for a business combination are the pooling of interests method and the purchase method. These methods produce different results in the consolidated financial statements of the parent company, so consideration must be given as to which method is appropriate inasmuch as discretion is available within the framework of GAAP for the reporting enterprise.

The pooling method, or merger method as it is known in the UK, is suitable for specifically defined circumstances. It is rarely used in Canada but is more common in the United States. The general rule is that pooling should be followed only when there is an exchange of voting shares and an acquirer cannot be identified. In other words, two entities continue business in parallel and neither is taking over the other operationally. In these circumstances, when one entity cannot be considered the acquirer of another, the assets and liabilities are combined and are accounted for in the combined company's financial statements at their carrying values in the combining companies. Asset values are not adjusted to fair value and revenues and expenses of the combining entities are reported for the entire fiscal period of the combination. In effect, the fact of the combination does not disturb the continuity of the reported earnings and the financial position of the combining entities.

If one of the parties to a business combination can be identified as the acquirer of the other, the purchase method of accounting for the combination is appropriate. Under the purchase method, income and expenses of the acquired subsidiary are consolidated from the date of acquisition only. At that date, an allocation of the purchase price consideration to the identifiable assets and liabilities of the subsidiary takes place so that reported net assets in the consolidated financial statements will include those of the acquired company at their cost to the acquiring company rather than at their carrying value in the accounts of the subsidiary. Any discrepancy between the purchase price and the fair value of the net assets of the acquired entity will be treated as goodwill arising on consolidation. The practice for dealing with goodwill varies from country to country. In the US and Canada, goodwill is considered to be an intangible asset and is usually required to be amortized over a period not exceeding 40 years. In this sense it is considered to represent the economic benefit of expected future profits from the operations of the subsidiary. If the value of the goodwill is regarded as having been permanently impaired in future periods, for example because the acquired subsidiary incurs substantial operating losses, then it is required to be written off against earnings in the consolidated financial statements.

In Australia, practice has varied in the past between charging the amount of goodwill to reserves in equity to recording it as an intangible asset as in North America. A new standard, AAS No. 18, Accounting for Goodwill, which is effective for accounting periods ending on or after March 31, 1985, requires that purchased goodwill be accounted for as an intangible asset with a maximum amortization period of 20 years.

In the UK, practice has also varied between charging goodwill to reserves and recording it as an intangible asset. Under the terms of Exposure Draft 30, also entitled Accounting for Goodwill, it appears that this practice will be permitted to continue. Companies are to be given the choice of immediate write-off of goodwill to reserves or recording it as an intangible asset with amortization to income on a systematic basis over the estimated useful economic life.

Within the purchase method there are accounting alternatives available. These alternatives are the <u>parent company</u> and the <u>entity</u> approaches. Both approaches view the combination as an investment by one company in the net assets of another so that consolidation of the acquired net assets at fair value takes place. It is the extent to which the net assets are consolidated when less than 100% of the equity of a company is acquired which marks the difference between the alternatives.

Under the parent company method, which is prescribed by Canadian standards, if a company acquires, say 65% of the equity of another, the combination is accounted for as if 65% of the net assets stated at fair value had been acquired. The remaining 35% of the net assets, which represent the interests of minority stockholders, are included in the combined entity at their carrying values in the accounts of the subsidiary and not at their fair value. On the other hand, the entity method, used in the US and in the UK, permits the consolidation of 100% of the fair value of the net assets on the basis that 100% of the assets are controlled not just 65%. Under the entity method, there is none of the split valuation of the fair value of the acquired assets which makes critics of the parent company method feel uncomfortable.

In contrasting the purchase and pooling methods, the simplicity of the pooling method is evident. In view of this simplicity, why then has the use of the pooling method become relatively restricted? Aside from the theoretical justification for the use of the purchase method, there is some interesting historical background to the current limited use of the pooling method.

In the 1960's, some creative United States enterprises used the pooling method for business combinations as a means of artificially improving their results of operations and financial position. In simple terms, this was achieved dramatically as described below.

A company with a poor earnings performance acquired a profitable company with a good record of current and accumulated earnings. The pooling method was used to account for the business combination. The application of this method had the following impact:

a) The results of operations of the acquired company were included in the financial statements of the acquiring company not only from the beginning of the reporting period in which the combination took place, but also for the comparative period. Accordingly, the earnings picture looked rosy. If the purchase method had been used, the results of operations of the acquired company would only have been included from the date of acquisition.

b) The accumulated retained earnings of the acquired company were added to the retained earnings of the acquiring company. Thus, the consolidated balance sheet gave the appearance of a solid earnings performance over a period of years. If the purchase method had been followed, the accumulated retained earnings of the acquired company at the date of acquisition would have been capitalized and included in capital stock.

c) In the period following the acquisition, the acquiring company often sold off some of the profitable assets of the acquired company. These assets had not been revalued on acquisition, as would have been the case under the purchase method, and, consequently, substantial gains were realized boosting the earnings performance.

Following these abuses, the US rules for using the pooling method, as set out in APB Opinion No. 16, were strengthened considerably and the method is now used less frequently to account for business combinations between autonomous entities. The method does continue to be used extensively in reorganizations or amalgamations between, for example, a parent company and its subsidiaries. In these circumstances, the transaction is viewed as one of form rather than substance between entities under common control and the pooling method is considered to be the most appropriate accounting.

Disclosure Issues   Acquisitions are commonplace in the mining industry, particularly in the smaller growth-oriented enterprises where acquisition is often seen as the fastest way to grow. To gain a full understanding of the accounting for a business combination the following disclosures should be present:

- the method used of accounting for the combination;

- a summary of the net assets acquired;

- if the purchase method has been used, the treatment of goodwill, if any;

- if the purchase method has been used, the date from which the results of operations of the acquired company have been included in the consolidated financial statements;

- whichever method is used, a discussion of the impact of the business combination on the results of operations and the financial position of the combined entity; and

- if the consideration given is contingent upon some future event, disclosure of that fact and how such contingent consideration would be accounted for.

Summary

This completes our discussion of accounting for long-term investments. The concepts of consolidation and business combination accounting are theoretical in nature. It is necessary to have a basic grasp of this theory to understand how business combinations are accounted for and why the purchase and pooling methods are used. In addition, it is necessary to gain some insight into some of the practical difficulties with these theoretical concepts. Fortunately, with some minor exceptions, international practice in this area is uniform.

FOREIGN CURRENCY TRANSLATION

For those grappling with theoretical accounting concepts and perhaps approaching this section with some trepidation due to the theoretical nature of accounting for foreign currency translation, we bring some positive news. After many years of controversy and discussion, the international environment has recently achieved a unified front with the issue of standards in the US, the UK and Canada and by the IASC, all of which have adopted a similar approach to the major issues. While some critics have greeted the new standards with dismay – because they are seen as imposing yet another set of complex rules on the business community, most have greeted them with relief – because they represent the result of a sincere attempt to deal, in a businesslike way, with a baffling problem that seemed insoluble. For companies in the mining industry, these new standards undoubtedly have an impact on the reported financial statements as many operate in an international environment – through investment activity, participation in capital markets and earning of revenues which are often based on world commodity prices denominated in US dollars.

A discussion of the implications of the different concepts of and approaches to accounting for foreign currency translation is provided in Appendix II to this chapter. In addition, the IAS comparison in Appendix I summarizes the remaining conflicting issues between IAS No. 21, Accounting for the Effects of Changes in Foreign Exchange Rates, and the respective standards of the US, the UK and Canada. In this section, we will confine our discussion to the implications of the new standards on the financial statements of mining companies and to the use of foreign exchange hedging activities, which have been given greater recognition in the new standards.

The New Way

The major developments in accounting for foreign currency translation relate to the translation of the financial statements of foreign-based entities and the treatment of the resultant unrealized exchange gains and losses. Under the new standards now in effect, the financial statements of foreign-based entities are permitted to be translated to the reporting currency using the current rate method. When this method is used, the resultant unrealized exchange gain or loss is not included in income but is taken directly to equity. The use of the current rate method is based on the philosophy of the foreign operation as a self-sustaining unit, separate from the home operation and undoubtedly will apply to many foreign-based entities.

The impact of these developments for mining companies has varied according to the rules or practice that previously existed. In the US, for example, the implications were significant. The previous rules, set out in FAS No. 8, required certain non-monetary assets of foreign-based entities, such as mining properties, to be translated using the historic rate. In contrast, long-term liabilities were translated at current rates together with current assets and liabilities. The application of this method of translation, known as the temporal method, generally resulted in significant unrealized gains or losses which were required to be recognized in income. To aggravate the situation, these rules were applied at a time when the US dollar was declining in value relative to currencies of foreign-based entities, with the result that significant unrealized exchange losses were required to be recognized in income. Thus in the US, the new approach, embodied in FAS No. 52, has had a major impact, both on balance sheet amounts, where non-monetary assets of self-sustaining operations are now translated using the current rate and on the income statement in that unrealized gains and losses relating to the translation of the financial statements of self-sustaining operations are now taken directly to equity.

Conversely, in Canada, the most common practice prior to the introduction of the revised S.1650 of the CICA Handbook, was to use the current/non-current method. Using this method, both non-current assets and liabilities of foreign entities were translated using the historic rate and to the extent that non-current assets were financed by non-current liabilities, the resultant unrealized exchange gains and losses offset each other. The net gain or loss was often recognized in income but generally was not as great as that recorded under FAS No. 8. Thus, in Canada the impact of the new developments on the financial statements of mining companies may be more significant on balance sheet amounts than income statement amounts but will depend on the previously adopted practice.

When the foreign operation is deemed to be merely an extension of the parent company's operation such as when the foreign operation sells all its product to the parent company and is wholly financed by the parent company, then the financial statements of the foreign-based operation (hereafter referred to as integrated operations) are required to be translated using the temporal method. The resultant unrealized exchange gain or loss is recognized in income. In some countries such as Canada, the portion of the unrealized gain or loss relating to long-term monetary items, such as long-term debt, is deferred and amortized over the life of the related item.

Foreign Currency Transactions

The new standards relating to accounting for foreign currency transactions, as distinct from those relating to the translation of financial statements of foreign-based entities, did not result in major changes to previous practice or rules. As indicated in Appendix II, transactions denominated in a foreign currency are generally translated at the rate of exchange in effect at the transaction date rather than the settlement date. Exchange gains and losses which result from movements in exchange rates between the transaction and settlement dates are generally recognized through income. One exception, in some countries, to this general rule is the treatment of unrealized gains and losses relating to long-term monetary items. As discussed above, in relation to integrated operations, Canada requires such gains and losses to be deferred and amortized over the life of the related item. Canada may have adopted the deferral and amortization approach because of the significance of foreign debt in the balance sheets of Canadian companies. Compared to US companies, Canadian companies look more often to foreign sources for borrowings to finance domestic operations. This is particularly true for mining companies. The Canadian approach is predicated on the assumption that any exchange gain or loss is, in essence, a financing gain or loss which should be spread over the term of the debt.

Hedging Activities   One aspect of the new standards, which can be viewed as more of a development than a change, relates to foreign exchange hedging activities. The new standards give greater recognition to the use of hedging activities by permitting certain unrealized gains and losses to be offset against those arising on the hedged item.

Many companies in the mining industry seek to protect themselves against exposure to foreign exchange rates through hedging activities. Common examples are:

1. Forward Contracts
   A forward exchange contract is an agreement to exchange currency of different countries at an agreed rate (the forward rate) on a future date. Assume that on December 1, 1984, a company agrees to exchange $100,000 for 500,000 units of a foreign currency at a forward rate of $0.20 on March 1, 1985. The contract is to operate as a hedge of a foreign debt repayment of 500,000 foreign currency units also due on March 1, 1985. By entering into the forward contract, the company has effectively fixed the exchange rate for the repayment of the debt at $0.20 and therefore "hedged" the foreign exchange exposure related to the debt.

2. Using a Future Revenue Stream to Hedge Debt Repayments
   Assume that a Canadian gold mining company has US dollar debt and its main revenue stream is gold revenue in US dollars. In this situation, the future revenue stream can be viewed as hedging the interest and principal payments on the debt.

3. Hedge of a Net Investment in a Foreign Operation
   Assume a UK company invests in a Canadian gold mining company and decides to finance the purchase with Canadian dollar debt. In this situation, if the Canadian dollar drops in value, then the investment will depreciate in value in terms of sterling but the debt servicing payments will also decrease in terms of ster-

ling. In this respect, the Canadian dollar financing can be viewed as a hedge of the investment in the Canadian company.

In any of these situations, unrealized gains and losses will arise relating to both the hedge and the hedged item. The new standards permit, in varying degrees, the netting of these unrealized gains and losses. For example, assume that a US company has an investment in an Australian company which is considered to be a self-sustaining operation. The gain or loss arising on the translation of the financial statements of the Australian company to US dollars will be taken directly to equity. If the US company has Australian dollar debt in its own accounts, then that debt can be considered as a hedge, or a partial hedge, of the net investment in the Australian company and the gain or loss arising on translation of the debt can be offset against the amount taken to equity. Had the debt not been considered as a hedge, the gain or loss would have been recorded in income under US principles. Clearly the amount of the hedge cannot exceed the hedged item. If the US dollar investment in the Australian company is $10 million, then the amount of the debt that can be treated as a hedge cannot exceed $10 million.

The extent to which each of the new standards recognizes hedging activities varies. In addition, in any situation where a hedge is identified, including all of the situations described above, the hedge has to be identified as such and there should be reasonable assurance as to its effectiveness. For example, a future revenue stream can only be considered as a hedge of long-term debt if there is reasonable assurance that sufficient amounts will be received, at the times expected, to service the debt. In general terms, the recognition of hedging activities can be viewed as a positive step by the accounting profession.

Financial Statement Disclosure

The degree of exposure to foreign exchange risk can be a significant factor in an enterprise's future well-being. Accordingly, there generally should be disclosure of significant activities in other countries and of significant transactions in currencies in other than the reporting currency. In addition, with respect to the accounting for foreign currency translation, the following disclosures are either required or would be useful:

- a description of the accounting policy, particularly the method used for the translation of financial statements of foreign-based entities;

- the adjustment arising from the translation of the financial statements of self-sustaining operations, as a separate component of equity;

- significant elements giving rise to exchange gains and losses accumulated in the separate component of equity;

- the method used to amortize exchange gains and losses on long-term monetary items (where this practice is permitted or required);

- the deferred amount of exchange gains and losses relating to the translation of long-term monetary items should be disclosed as a deferred charge or deferred credit (where this practice is permitted or required); and

- the net exchange gain or loss recorded in the income statement.

Summary

Foreign currency translation is a concept which has application to business enterprises generally. There are no aspects which are unique to mining enterprises, except, perhaps, in the extent to which they are involved in activities involving foreign exchange. With regard to the new accounting standards, it is too early to tell how successful these will be. They have been developed in response to criticism from the business community and, the fact that a common approach between countries has been adopted, should ensure their acceptance in the international business environment. These changes, together with the greater recognition afforded to hedging activities, should result in the presentation of foreign currency transactions and the translation of financial statements of foreign-based entities on a basis which reflects, to a greater degree, economic reality.

COST CAPITALIZATION

One of the most significant issues in financial reporting in the mining industry concerns the identification of costs incurred by mining companies that should be capitalized (or deferred) and those that should be expensed to income as incurred.

The capitalization issue is most significant in relation to accounting for costs incurred in prospecting for, acquiring, exploring and developing mineral reserves.

Australia is the only country which has issued guidance in this area in the form of AAS No. 7, Accounting for the Extractive Industries. The lack of authoritative literature defining GAAP makes the issue both significant and controversial as there are wide variations in accounting practices, not only between mining companies operating in different countries, but also between mining companies operating in the same country. Consequently, two companies with identical operations, revenues and costs could theoretically report significantly different results. It is therefore critical for any user of financial statements of mining companies to understand the various alternatives available, and also to understand the effect that a particular alternative has on reported results.

The objectives of this section are as follows:

- to define terms;

- to outline the accounting concepts underlying the capitalization decision;

- to describe the various accounting practices in use in the mining industry and the factors influencing a company's choice of a particular policy;

- to demonstrate the effect of different accounting practices on reported results;

- to discuss practices common to countries; and

- to note other issues relating to cost capitalization.

Definition of Terms

Phases of operation There are five phases of any mining operation which are defined in AAS No. 7 as follows:

Exploration: The search for a mineral deposit, including topographical, geological, geochemical and geophysical studies and exploratory drilling.

Evaluation: The determination of the technical feasibility and commercial viability of a particular project.

It includes the determination of the volume and grade of the deposit, examination and testing of extraction methods and metallurgical or treatment processes, surveys of transportation and infrastructure requirements, and market and finance studies.

For the purposes of this section of the chapter exploration costs include evaluation costs.

The exploration and evaluation phases normally end when the availability of financing and the existence of markets are established and the decision to proceed to both development and production is made.

Development: The establishment of access to the deposit and other preparation for commercial production.

It includes shafts, underground drives and permanent excavations, roads and tunnels and advance re-removal of overburden and waste rock.

Construction: The establishment and commissioning of facilities for the extraction, treatment and transportation of product from the deposit.

Such facilities include infrastructure, buildings, machinery and equipment.

The development and construction phases of a project normally end when commercial levels of production are achieved.

Production: The day-to-day activities directed to obtaining a saleable product from the deposit on a commercial scale.

It includes extraction and any processing prior to sale.

Capitalize To defer or carryforward costs in the balance sheet, normally to be amortized against income of future accounting periods.

The Capitalization Decision

It is generally agreed that all costs incurred during the development and construction phases of a project should be capitalized and subsequently amortized against revenue derived from production in future accounting periods. Such costs will normally include not only direct development and construction costs, but also related general and administrative expenses and any losses incurred from the sale of product before commercial levels of production are achieved and the mining operation enters into the production phase.

Costs relating to development during the production phase are normally expensed to income as incurred, unless they clearly relate to future levels of production activity in which case they may be capitalized.

The capitalization decision therefore is more involved in regard to costs incurred during the exploration phase of a mining operation. A discussion follows of some of the concepts involved in making the capitalization decision and of the reasons why such a wide divergence in the accounting for exploration costs exists in practice.

To Capitalize or Not to Capitalize Exploration Costs?

In the section of this chapter on financial statement elements and concepts, we discussed some of the ground rules for preparing financial statements. Most relevant to the capitalization decision are the concepts of matching, prudence and conservatism and the principles underlying cost deferral.

Under the matching concept, costs incurred are matched against the revenues generated by these costs in order to obtain a measure of profitability. At the time costs are incurred, the revenues may not be determinable to any degree of accuracy in which case an estimate has to be made of the anticipated benefits likely to be derived from incurring the costs. At any reporting date, a company will attempt to determine which costs incurred to date can be related in whole or in part to anticipated future benefits. Those costs which can be so related can be deferred, or capitalized, and matched with the related revenues when they arise.

For example, a manufacturing company purchases a machine which is expected to have an economic life of five years at which time it will have no value. In order to match properly the cost of the machine with the expected benefits, a company will capitalize the cost of the machine and write this cost off to income over the economic life of five years.

As illustrated above, the capitalization decision is normally straightforward for a manufacturing company. Capital costs will generally be identified with tangible assets whose future benefits can be estimated. Thus, the costs to be capitalized can be identified with relative ease.

The capitalization decision for companies in the extractive industries is more complex particularly with respect to exploration costs. Companies incur such costs with the expectation of deriving future benefits in the form of economically recoverable reserves. However, a significant delay normally exists between the time that such costs are incurred and the time that the related benefits, if any, become apparent. On account of the delay, a company may not be able to evaluate the benefits that will be derived from such costs at a reporting date. Should such costs be capitalized or should they be charged to income in the current period?

Strict application of the prudency and conservatism concepts arguably would dictate that such costs should be written off as incurred until a future benefit can be estimated. As will be seen later in this section, these concepts are the basis for the accounting policies of some mining companies. The application of these concepts is subjective and arguably not realistic for companies such as those in the development stage (see discussion in the section on Financial Statements - Elements and Concepts) or in the initial stage of operations. As a result of these and other concerns, the concept has evolved of permitting the capitalization of costs at least until a decision can be made as to their future benefit.

A further factor in the capitalization decision is the degree to which costs can be associated with a particular project or project area.

In this regard, some companies only capitalize costs that are directly related to a project. Often, all prospecting and general geological and geophysical costs are expensed as incurred because of the difficulty in associating them to specific projects. In addition, any exploration department administration costs are expensed as incurred because of the lack of direct association with specific projects. Again, this is an area of considerable subjectivity and other companies argue that such costs are necessarily a part of the exploration effort and charge each project with its proportionate share. A rule of thumb is that if the cost is identifiable with a project within the cost system, then capitalization is usually acceptable.

In addition, it is necessary to consider the extent to which costs are common to a group of projects rather than specific projects in the geographic area. To the extent that a company is conducting its exploration effort in an area comprising many different projects, many of the costs can be attributed to the overall area only and any allocation beyond that to specific projects is purely arbitrary. In this situation, the concept has evolved of capitalizing all costs relating to an area or "area of interest."

Selecting an Appropriate Accounting Policy for Exploration Costs

Up to this point, we have discussed some of the concepts surrounding the capitalization decision, the nature of costs which may be capitalized and the methods by which they can be allocated. From these approaches, a number of accounting policies have evolved for accounting for exploration costs. The discussion below relates to those which are used most frequently in practice. AAS No. 7 describes five accounting policies commonly used by companies for exploration costs:

- costs written off;
- costs written off and reinstated;
- successful efforts;
- full cost;
- area of interest.

A description follows of each policy together with an example of how companies disclose such policies in practice.

Costs Written Off   Under this method, all exploration costs are written off as incurred. No reinstatement is made if economically recoverable reserves are discovered. This method is said to be justified by the low probability of success and reflects a strict application of the prudency and conservatism concepts. Additionally, it is argued that a sunk cost of this nature has ordinarily no market value and cannot therefore be considered to be an asset. The opponents of this method claim it is excessively conservative, and precludes a proper matching of costs with revenues.

Example: Canada's Inco - "Expenditures for mineral exploration are expensed as incurred."

Costs Written Off and Reinstated   Under this method, all exploration and evaluation costs are written off as incurred. Should economically recoverable reserves be confirmed, then all costs pertaining to their discovery and evaluation are reinstated. As with the previous method, this method is advocated in recognition of the low probability of success in the exploration phase. Its opponents claim it is too conservative and that it is subject to distortions in the reporting of periodic income.

Example: The Hudson Bay Mining & Smelting Co. - "...mineral exploration costs are expensed as incurred. Expenditures for projects deemed commercially productive are capitalized with a corresponding credit to earnings at the time this determination is made."

Successful Efforts Under this method, exploration costs are written off as incurred until it is deemed that a project is successful. Subsequent exploration costs incurred in a successful area of interest are capitalized. Proponents of the successful efforts method claim that only costs of successful operations ought to be carried forward. Its opponents argue that it fails to recognize the total costs of establishing the existence of economically recoverable reserves and thus fails to match costs with revenues.

Example: Canada's Denison Mines - "The Company accounts separately for each group of permits, licenses or leases in a designated exploration or development area as a separate area of interest. All exploration costs relating to each area of interest are written off in the year incurred. If it is determined that an area of interest contains economically recoverable reserves, all costs relating to that area for the current and subsequent years are deferred."

Full Cost This method relates all exploration costs, wherever incurred and without distinguishing between particular areas of interest, to the total economically recoverable reserves discovered by the entity. In effect, it recognizes only one comprehensive cost centre or area of interest. The proponents of this method argue that all exploration costs are essential, whether or not related to successful efforts, in order to add to the entity's pool of economically recoverable reserves. Accordingly, they argue that these costs should be matched with the revenues produced from this pool of reserves, subject to the constraint that the aggregate of costs carried forward should not exceed the net realizable value of all economically recoverable reserves so as to avoid an overstatement of assets. Opponents of the full cost method point out that it is not feasible to apply this constraint to an entity searching for, but not yet possessing economically recoverable reserves, nor is it always practicable to apply it even when such reserves exist because of problems in determining their value. The main argument against the full cost method is that it fails to properly match revenues and costs, because current exploration and evaluation costs are charged against revenues produced from previously discovered reserves. Accordingly a cause and effect relationship is said to be lacking. In addition, by permitting costs of abandoned areas to be carried forward as assets, the method ignores the accepted criteria for cost deferral.

Normally, the full cost method is employed only by petroleum and natural gas concerns. In the US, SEC reporting requirements for oil and gas enterprises restrict the size of area of interest to a country-wide basis.

Area of Interest This method represents an approach part way between the full cost method and the costs written off method. It differs from the full cost method in that it is based on a much more restricted area of interest. It differs from the costs written off method in that it permits, subject to certain constraints imposed by consideration of prudence, the carrying forward of exploration and evaluation costs, so as to achieve as far as possible a proper matching of revenues and related costs. In most cases, the area of interest will comprise a single mine or deposit. For any one area of interest, the exploration costs may be carried forward so long as a reasonable probability of success in that area exists. If the search is unsuccessful or evaluation produces a negative result, the costs associated with the area are written off.

Example: Australia's Seltrust Holdings - "Exploration and evaluation costs related to an area of interest are carried in the balance sheet where:

a) rights to tenure of the area of interest are current; and

b) one of the following conditions is met:

  i) such costs are expected to be recouped through successful development and exploitation of the area of interest or alternatively, by its sale; or
  ii) exploration and/or evaluation activities in the area of interest have not yet reached a stage which permits a reasonable assessment of the existence or otherwise of economically recoverable reserves, and active and significant operations in, or in relation to, the area are continuing.

Accumulated expenditure on areas which have been abandoned, or are considered to be of no value, is written off in the year in which such a decision is made."

An Illustration of Methods Used In order to demonstrate the effect that the use of different accounting policies has on reported results, a numerical example follows. Assume four different companies with identical operations. Each company uses a different accounting policy for exploration expenditures as follows:

Company	Accounting Policy
A	Costs written off
B	Costs written off with reinstatement
C	Successful efforts
D	Area of interest

The companies have three exploration areas with the following costs:

Area	1	2	3	Total
Costs incurred in previous periods	$100	$350	$	$450
Costs incurred in current period	50	100	50	200
	$150	$450	$ 50	$650

Area 1: During the current period it was determined that the area would not produce any economically recoverable reserves. Area abandoned at end of period.

Area 2: Exploration determined to be successful at beginning of period, development to follow.

Area 3: Exploration is ongoing - no determination of eventual success is possible at this stage.

The determination of the exploration cost charged to the current year's income for each of the companies is as follows:

COMPANY Accounting Policy	A Costs Written Off	B Rein- Statement	C Successful Efforts	D Area of Interest
Company A Areas 1, 2 and 3	$200			
Company B Areas 1 and 3 Area 2 reinstated		$100 (350)		
Company C Area 1 and 3			$100	
Company D Area 1				$150
Charge (credit) to income for period	$200	$(250)	$100	$150

<u>Factors Influencing the Selection of an Accounting Policy</u> Ultimately, the selection of one method of accounting for exploration costs over another will be a reflection of management's philosophy. Other factors influencing the choice of policy may be a company's stage of evolution and the materiality of exploration expenditures in relation to total operations.

Companies in the development stage with little or no revenue, insignificant proved reserves and substantial exploration costs are likely to capitalize as much as possible. The alternative of writing these costs off to income may not be acceptable to management who justifiably want to present the best picture of the company's performance and prospects. Conversely, companies with established reserves and revenues, and who have exploration costs which are less significant in relation to total operations, are more likely to write off all exploration costs as incurred. In this situation, the costs can be absorbed against the results from other operations without a dramatic impact on net income.

Practice in Different Countries

<u>North America</u> An informal survey of 32 companies in the mining industry revealed the use of each method to be as follows:

Costs written off	16
Successful efforts	7
Area of interest	5
Reinstatement	4
	32

Within North America, therefore, it seems that the most conservative accounting policy, writing off all exploration expenses as incurred, is also predominant in practice.

<u>UK</u> An informal survey of UK companies in the mining industry did not produce any identifiable pattern.

<u>Australia</u> Following the introduction of AAS No. 7 in 1977, companies in Australia no longer have a choice in the manner in which they account for exploration costs. The standard, which covers not only mineral but also petroleum and natural gas extraction entities, requires that the area of interest method be used.

Although AAS No. 7 requires the use of the area of interest method, there is still flexibility in the application of the method. AAS No. 7 permits the carrying forward of costs in the stated circumstances but this does not preclude a policy of writing off exploration costs as incurred until the commercial potential of an area of interest can be assessed. For example, the Peko-Walsend policy for exploration costs is as follows:

"Exploration expenditure is written off up to the point of demonstration of commercial resource. Subsequent exploration expenditure on the resource is capitalized."

Financial Statement Disclosure

As in the choice of exploration accounting policy, there are wide variations between companies in the extent of disclosure of capitalized costs.

Detailed disclosure ideally should take the form of presenting capitalized costs in the following categories (assuming each to be significant):

- land;
- buildings;
- machinery and equipment;
- mining properties;
- exploration costs;
- development costs; and
- other capitalized costs.

The amount of such costs relating to properties both in the non-productive and the pre-production stages also should be disclosed. The level of disclosure is dependent on the country in which a company reports. Disclosure of gross cost and accumulated depreciation for each major class of depreciable asset is generally required. In Canada, there is no requirement to do so but it is considered desirable.

AAS No. 7 includes specific requirements in relation to the disclosure of exploration and development costs and the related amortization as follows:

- the amount of exploration or development costs written off in the period;
- the amount charged in the period for amortization of exploration or development costs carried forward, irrespective of how such costs are classified or described in the balance sheet;
- costs carried forward in respect of areas of interest still in the exploration phase, with an explanation that ultimate recoupment of such costs is dependent on successful development and commercial exploration or, alternatively, sale of the respective areas;
- costs carried forward in respect of areas of interest in the development phase in which production has not yet commenced, with an explanation that amortization is not being charged pending the commencement of production; and
- costs carried forward in respect of areas of interest in which production has commenced with accumulated amortization charges being shown separately as a deduction.

Other Capitalization Issues

Incremental Revenues Incremental revenues are derived from sales of products obtained from exploration or development activity. Companies account for such revenues in one of two ways.

Firstly, some companies credit all such revenues, net of processing costs, to exploration or development costs as applicable. Secondly, some companies include such revenues in sales, and charge to income the estimated extraction cost of the minerals related to those revenues. AAS No. 7 requires that the latter method be used if such revenues are material.

Capitalized Interest Companies often capitalize interest costs relating to exploration (if the related costs are deferred) and development projects. Capitalization of interest on borrowings is widely accepted in practice under these general conditions:

- borrowings need not be specifically identified with the project;
- capitalization should cease when the project becomes operational;
- capitalization should cease during extended delays; and
- only net interest should be capitalized; interest income from the temporary investment of borrowed funds should be offset against interest expense.

The capitalization of interest is covered by IAS No. 23 and FAS No. 34. There are no pronouncements relating to the capitalization of interest in the UK, Canada or Australia. One difference between IAS No. 23 and FAS No. 34 is that FAS No. 34 requires capitalization of interest when its effect is material whereas IAS No. 23 only requires a policy of capitalizing or not capitalizing interest to be established.

When interest is capitalized, it is generally allocated to the individual asset to which it relates and amortized or depreciated as part of that asset.

Major Repairs and Maintenance Costs Companies account for the cost of major repairs and maintenance costs in one of three ways:

- provision is made in advance of incurring the expenditure;
- costs are written off as incurred; or
- costs are capitalized and written off over some future period.

These variations in treatment are not unique to the mining industry. The policy of providing for such repairs is obviously more conservative than the capitalization of such costs for subsequent amortization.

Coal Mining Industry The principles discussed in this section apply equally to companies in the coal mining industry. In general, exploration costs incurred in the coal mining industry are not as significant as those incurred in other mining environments and consequently, such costs are often written off as incurred.

Summary

This concludes our discussion of cost capitalization in the mining industry. The main theme arising from the discussion is that there are significant variations in practice between mining companies in the international environment. These divergencies in practice highlight the importance of being aware of, and understanding such variations when comparing the financial position and results of operations of mining companies.

DEPRECIATION, DEPLETION AND AMORTIZATION

The preceding section discussed the capitalization of costs by mining companies. This section involves the depreciation, depletion and amortization of these costs.

To the extent that revenues from an area of interest will be earned over several reporting periods, it is necessary to devise a method of allocating capitalized costs to these periods in a manner which best reflects the diminution in value of the related asset. The term amortization is used in this section to describe the method of allocating costs to the income statement and encompasses both the depreciation of tangible assets and the depletion and amortization of other capitalized costs. Depletion is an expense to income for that portion of the value of wasting assets, such as ore reserves, consumed or removed in the period.

There are two main bases for amortization, namely:

- production output, and

- time.

For a mining company on a production output basis the cost of an asset is amortized during an accounting period in the proportion of the ore produced during the period to total ore reserves. On a time basis, the cost of an asset is amortized by a fixed percentage each period in such a way that, at the end of the asset's useful life, the whole cost of that asset has been written off.

Amortization Over Time

The straight-line and the declining balance methods are the two principal methods of amortizing capitalized costs on a time basis.

Under the straight-line method, the cost of an asset, less any residual value, is amortized in equal instalments over its estimated useful life.

Example: A company purchases a machine for $11,000. The machine has a life of five years after which the residual value is estimated to be $1,000.

The depreciable cost of the asset is $10,000 being the total cost of $11,000 less the estimated residual value of $1,000. The equal annual amortization charge for five years is therefore $2,000 ($10,000 / 5).

The use of the straight-line method assumes that the economic benefit derived from an asset is constant throughout its useful life. The main advantage of the method is its simplicity in application.

Under the declining balance method, cost less any accumulated amortization brought forward is multiplied by a pre-determined percentage each period so as to reduce the book value of the asset to residual value at the end of its useful life.

Example: A company purchases an excavator for $100,000 with an expected life of four years after which time the estimated residual value is $13,000. The amortization rate is determined to be 40% and is calculated as follows:

Year	Book Value at Beginning of Year	Amortization For Year	Book Value at End of Year
1	$100,000	$40,000	$60,000
2	60,000	24,000	36,000
3	36,000	14,400	21,600
4	21,600	8,640	12,960

Notes:
(i) Amortization for year equals 40% of opening book value
(ii) Book value at end of year four approximates estimated residual value of $13,000.

The declining balance method incorporates the concept that an asset's repair and maintenance costs increase steadily during its useful life. Consequently, the aggregate of depreciation expense and repair and maintenance costs to income should be relatively equal over the life of the asset. Amortization charged to income reduces or declines over the life of the asset.

When using either of these methods to amortize a mine facility, it is necessary to estimate the life of the mine as it may effectively limit the life of the mine facility. If a mine facility cannot be transported to a new site when a mine is closed, then its economic life for accounting purposes cannot exceed the estimated life of the mine.

Amortization Based on Production

Under the unit of production basis, amortization of capitalized costs is calculated by apportioning such costs in the proportion of the production output for the period to economically recoverable reserves at the beginning of the period.

Unamortized development costs at beginning of year	$300,000
Development costs capitalized during year	$100,000
Economically recoverable reserves at end of year	150,000 tons
Production during year	50,000 tons

Amortization for the year is calculated to be:

$$\text{Unamortized costs at end of year} \times \frac{\text{Production for Year}}{\text{Reserves at Beginning of Year}}$$

$$= (300{,}000 + 100{,}000) \times \frac{50{,}000}{150{,}000 + 50{,}000}$$

$$= 400{,}000 \times \frac{50{,}000}{200{,}000}$$

$$= \underline{\$100{,}000}$$

Estimating Economically Recoverable Reserves The estimation of economically recoverable ore reserves is a matter of great technical difficulty and uncertainty. AAS No. 7 provides guidance in this area and requires the assessment of such factors as:

- security of tenure of the property (including special conditions attaching to leases or permits);

- the possibility that technological developments or discoveries may make the product obsolete or uneconomical at some future point in time;
- changes in technology, market or economic conditions affecting either sales prices or production costs, with a consequent impact on production grades; and
- likely future changes in factors such as recovery rate, dilution rate, and production efficiencies during extraction, processing and transportation of products.

Because of the uncertainties involved and the possibility that some of the factors set out above may act as future constraints, companies will often put an arbitrary limit on the economically recoverable reserves used in amortization calculations.

Economically recoverable ore reserves are normally determined at least once every reporting period. Since the four factors noted above may change from period to period, the estimates of economic reserves may fluctuate significantly during the life of the mine. Such fluctuations could in turn cause wide variations in the periodic amortization charge. The impact of these changes is generally accounted for prospectively since they result from new information becoming available which was not available at the time the previous estimate was made. This means that prior period amounts are not adjusted and the change is accounted for over the remaining life of the mine.

Two main variations exist between companies in their determination of amortization using the unit of production method, namely the categories of reserves included, and the costs taken into the amortization calculation.

Economically recoverable reserves may include one or all of the following categories:

<u>Proved Reserves</u> Ore that has been blocked out in three dimensions. Such ore may be readily accessible (developed) or may require further development before becoming readily accessible (undeveloped).

<u>Probable Reserves</u> Ore that includes extensions near at hand to proved ore when conditions are such that ore will probably be found but when the extent and limiting conditions cannot be defined as precisely as for proven ore.

<u>Possible Reserves</u> When some presumption as to the existence of ore is warranted from the relation of the land to adjacent ore bodies and geological structures, but when exploration and evaluation data is not sufficient for it to be classified as probable.

Mining companies differ as to which of these reserves are taken into the amortization calculation. A company will employ one of the following bases:

- Proved developed;
- Proved (developed and undeveloped);
- Proved and probable; or
- Proved, probable and possible.

The greater the basis of reserves taken into the amortization, the less conservative a particular accounting policy is. Most companies in the US, the UK, Canada and Australia use proved and probable reserves.

The second variation in practice between mining companies is the costs included in the amortization calculation. Some companies, as in the example noted above, include only costs incurred to date while others take into account all estimated future costs that are likely to be achieved in developing ore reserves included in the amortization base. Clearly, it is more conservative to include an estimate of such additional costs in the amortization calculation.

Amortization by Time or Production?

It is generally accepted that the production basis of amortization most appropriately reflects a proper matching of revenue with the costs incurred to derive that revenue because the using up of assets in the mining industry can be associated more directly with production rather than time.

In practice, most companies use the production basis to amortize property acquisition costs and capitalized exploration and development costs and the time basis to amortize construction costs such as tangible equipment and plant. The time basis used for construction costs is usually the straight-line method. The time basis is used rather than the production basis for reasons of simplicity and also because some items within the category of construction costs may have a shorter life span than the life of the mine.

It should be emphasized that the estimation of economically recoverable reserves is an extremely complex process, especially at times of high inflation and widely fluctuating mineral prices. In recognition of this uncertainty, the practice by many companies of limiting mine lives arbitrarily to a certain number of years may be a sound one.

Financial Statement Disclosure

The extent of disclosure of accounting policies relating to amortization varies widely between companies. Some companies will give a comprehensive description of the basis of amortization, asset lives and methods of application while others disclose minimal information.

An example of comprehensive disclosure of an amortization policy is given by Canada's Echo Bay:

> "<u>Property, plant and equipment</u> ... are recorded at cost. Depreciation is provided using the straight-line method based on estimated economic life to a

maximum of 15 years for tangible fixed assets. Upon sale or retirement, the property, plant and equipment and related depreciation are removed from the accounts and any gains or losses are taken into income as they occur...

Deferred mine expenditures ... include acqui- sition, development, interest and certain construction costs ... These expenditures are charged to operations over the estimated life of the mine by the unit of production method, based on proven and probable ore reserves ...

Ore reserves ... ore reserves as at December 31, 1983 have been estimated under the super- vision of independent geologists to be:

	Tons	Grade	Gold Content (ozs.)
Proven	2,201,100	.399	878,400
Probable	1,208,200	.387	470,500
	3,409,300	.396	1,348,900

In arriving at rates for amortization under the unit of production method the proven and probable ore reserves as indicated above are used. Such ore reserve estimates are revised as data becomes available to warrant revision."

An example of less detailed disclosure is given by Canada's Noranda Mines:

"Depreciation and development charges
Depreciation of property, plant and equipment and amortization of development expenditures are based on the estimated service lives of the assets calculated using the method appropriate to the circumstances, for the most part straight-line for fixed assets and unit of production for development."

Disclosure of accumulated amortization and amortization expense also varies significantly from country to country and from company to company.

IAS No. 4, Depreciation Accounting, states that disclosures for each major class of depreciable asset should include:

- the method of amortization;
- useful lives or amortization rates;
- amortization expense for the period; and
- gross depreciable assets and accumulated amortization thereof.

IAS No. 4 does not apply to expenditures for exploration and excavation of non-regenerative resources, but the general principles can be considered as useful guidance. The requirements of AAS No. 7 with regard to disclosure of the amortization of exploration and development costs are discussed in the cost capitalization section.

Other Factors Relating To The Amortization Of Mining Assets

Commencement of Amortization During the development and pre-production phases of a new mining venture, many assets are used before the mine reaches commercial levels of production. Generally, companies do not charge amortization on those assets until the date when production in commercial quantities is achieved. In the early years of operations, however, some companies charge amortization on a pro rata basis with the level of production achieved.

Diminution in Value It may happen that estimates made by a mining company show that net revenue from a particular cost centre or cost centres is no longer expected to be sufficient to cover total unamortized capitalized costs. This may be the result of falling metal prices, revised estimates of the extent or grade of the ore body or high production costs. In this situation, it is generally accepted practice in most countries to write down the carrying value of unamortized costs to their estimated realizable value. If the write down is significant, it may require disclosure in the income statement as an unusual item.

Fully Amortized Assets In the later stages of the life of a mine, a significant proportion of its assets may have become fully depreciated. Normally, mining companies do not make any adjustment in this situation. Such fully depreciated assets are retained in the cost and accumulated amortization accounts until the closing of the mine, at which time they are removed. If the cost of such fully amortized assets is material in relation to the company as a whole, then it is desirable to disclose both the cost and the normal annual amortization charges appropriate to these assets.

Care and Maintenance As a result of changed economic circumstances, a mine may be put on what is called a care and maintenance basis for accounting purposes. Such circumstances may be depressed metal prices, fluctuating exchange rates or increased production costs. It is only appropriate to put a mine on a care and maintenance basis if circumstances are expected to change in such a way that the mine will be able to operate profitably in the future. Companies normally account for such a situation as follows:

- cost and accumulated amortization remain on the balance sheet. Companies may write off that portion of the total unamortized cost that is projected to be irrecoverable as a result of a permanent change in circumstances;

- costs incurred during care and maintenance are normally written off as incurred;

- amortization normally continues on assets amortized on the basis of time. Amortization is not normally charged for assets amortized on a production basis.

<u>Non-Depreciable Assets</u> The cost of an asset should only be amortized if the asset's useful life is in some way limited. An asset's useful life may be:

- pre-determined, for example leaseholds;
- directly governed by extraction or consumption;
- dependent on the extent of use; or
- reduced by obsolescence or physical deterioration.

If the life of an asset is not restricted by any of these factors, then that asset is non-depreciable. The most common type of non-depreciable asset is freehold land.

Summary

There is reasonable conformity in practice in amortization policies in the mining industry. The amount of amortization recorded by a company will depend on the cost capitalization policy. Companies which write off all exploration costs as incurred will have high write-offs recorded in the income statement but lower amortization charges as the amortization base is lower. Accordingly, it is important to consider both the cost capitalization and amortization policies in order to evaluate the impact on the results of operations of the application of different policies.

MEASUREMENT AND VALUATION OF INVENTORIES

The principal purpose of this section is to consider the concepts and principles underlying the measurement and valuation of a mining company's product inventories of ores, concentrates, refined metals and minerals.

In earlier sections of this chapter, we discussed the matching concept where the intention is to match costs with revenues. This concept is the central issue in inventory accounting. The objective is to defer those costs that relate to product inventory until it is sold, when the costs can be charged to income to offset the associated revenues. A further consideration relates to the conservatism concept. If the costs relating to product inventory are greater than estimated realizable value, it is necessary to write down the related inventory to its realizable value. The difficulties inherent in applying these basic principles can be classified into three broad categories:

- determination of the costs that are assignable to product inventory,
- separating those accumulated inventory costs between the inventory that has been sold and the inventory remaining on hand, and
- identifying realizable value.

More specifically, further complications are introduced when considering measurement and valuation questions within the mining industry. These complications are caused by the nature and location of the product.

This section examines the basic theoretical background of inventory accounting. The discussion will then focus on the specific problems associated with its application in the mining industry.

Assigning Costs to Product Inventory

Cost, in practice, is generally determined by two different methods - namely direct costing and absorption costing. There are both practical and theoretical issues associated with each method which are described in detail below.

Costs incurred in production can be broadly split into two categories - costs associated with creating the capacity to produce, such as production facilities and capital equipment; and production costs that are incurred in the use of those facilities. In the long term, all these costs are ultimately variable in nature as capacity and production can be expanded and contracted. In the short term, capacity costs are substantially fixed in nature and are commonly referred to as fixed costs while production costs vary with throughput and are commonly referred to as variable costs. Capacity costs include those relating to the operation of the facility on a day-to-day basis such as insurance, heat and light. Production costs include those directly relating to extraction such as labour and consumable materials and all processing or conversion costs.

There is no definitive classification for each type of cost. In many instances the classification will be clear-cut, in other cases the distinction will not be so clear. When the distinction is not particularly clear, these costs are regarded as being semi-variable.

Under direct costing, only the variable production costs are included in the valuation of inventory. All other costs are regarded as period costs and are charged to income when incurred. Under absorption costing, an additional amount in respect of fixed costs is added to inventory. The essential difference between the two methods is that under direct costing, the cost of sales amount reported in financial statements will fluctuate in line with sales volume only, whereas under absorption costing both the volume of sales and the level of production affect the cost of sales amount. If an amount in respect of fixed costs is added to inventory, the more inventory produced, the greater is the amount of fixed costs included in inventory with the consequence that less remains to be charged to income. Under absorption costing it is, therefore, possible to raise the level of profits merely by increasing production.

The advocates of absorption costing claim that the method is more accurate than direct costing because it considers the company as a complete economic entity in that profit is impacted not only by the volume of sales but also by the volume of production. They argue that under direct costing, only the level of sales activity is considered and that consequently, only a part of the economic activities of the corporate enterprise are reflected in the income statement. Further, absorption costing proponents claim that a better matching of revenues and costs is achieved if fixed costs are retained in inventory and only charged to income when the product is sold rather than when the costs are incurred.

Supporters of direct costing point out that no income can be made in a period until all fixed costs have been recovered. Once those costs have been recovered, each additional sale will contribute an amount to income equal to the difference between its selling price and variable cost.

Over a period of time, direct and absorption costing will produce substantially the same results on an aggregate basis, as both sales and production volumes tend to equal. In the short term the effect of the two methods of costing can produce widely differing results under various combinations of sales and production. This is illustrated in the following example:

EFFECT OF ABSORPTION AND DIRECT
COSTING IN COMPUTING PERIODIC INCOME

	Period 1	Period 2	Period 3	Period 4
Tons sold	3,000	5,000	10,000	9,000
Tons produced	9,000	3,000	8,000	9,000

Selling price - $400 per ton
Variable production cost - $200 per ton
Fixed production costs - $1,000,000 per period ($100 per ton)
Normal capacity - 10,000 tons per period
Administrative and distribution costs - $150,000 per period

($'000)

ABSORPTION COSTING				
Sales	$1,200	$2,000	$4,000	$3,600
Cost of Sales:				
Inventory at beginning	$ nil	$1,800	$1,200	$ 600
Cost of production	2,700	900	2,400	2,700
	$2,700	$2,700	$3,600	$3,300
Inventory at end	1,800	1,200	600	600
	$ 900	$1,500	$3,000	$2,700
Gross Profit	$ 300	$ 500	$1,000	$ 900
Expenses:				
Idle capacity	$ 100	$ 700	$ 200	$ 100
Administrative and distribution	150	150	150	150
	$ 250	$ 850	$ 350	$ 250
Income or (loss)	$ 50	$ (350)	$ 650	$ 650
DIRECT COSTING				
Sales	$1,200	$2,000	$4,000	$3,600
Cost of Sales:				
Inventory at beginning	$ nil	$1,200	$ 800	$ 400
Cost of production	1,800	600	1,600	1,800
	$1,800	$1,800	$2,400	$2,200
Inventory at end	1,200	800	400	400
	$ 600	$1,000	$2,000	$1,800
Contribution margin	$ 600	$1,000	$2,000	$1,800
Expenses:				
Fixed production	$1,000	$1,000	$1,000	$1,000
Administrative and distribution	150	150	150	150
	$1,150	$1,150	$1,150	$1,150
Income or (loss)	$ (550)	$ (150)	$ 850	$ 650

Notes:
1. The cost of idle capacity is determined by taking the difference between normal capacity and actual production in tons times the fixed production cost per ton.
2. No amortization of capitalized costs is included in this example. Under absorption costing, a portion of such amortization may be charged to inventory with the remainder to cost of sales. Under direct costing, the full amount is charged to income. This aspect is discussed later in the section.

In practice, both methods of costing are used. In the US and Canada it is not acceptable to exclude all fixed costs from inventory. Accordingly, it can be concluded that the use of direct costing is not permitted in these countries for external reporting purposes. In addition, IAS No. 2, Valuation and Presentation of Inventories in the Context of the Historical Cost System, requires that fixed costs of production that relate to putting inventory into its present location and condition, should be allocated to inventory. Thus, IAS No. 2 also appears to prohibit the use of direct costing although there may be subjectivity in determining those fixed costs of production that can be directly related to the location and condition of inventory.

Separation of Costs

Each time product inventory is sold, it is necessary to determine the cost so that it can be matched with the associated revenue. As we have seen, the elements of cost assigned will depend on which method of costing is used - direct or absorption costing. In addition, it is also necessary to determine whether the cost removed from inventory should be based on current, past or average costs. That is, production costs may be $300 per ton currently, $275 per ton at the time the inventory now sold was produced and somewhere in between on an average basis. The three most common approaches used to remove costs from inventory are referred to as last-in, first-out (LIFO), first-in, first-out (FIFO) and average costs and all are used in practice.

Under both the LIFO and FIFO methods an attempt is made to match the cost removed from inventory with the physical flow of inventory. With FIFO, an assumption is made that when inventory is sold it is the oldest inventory and costs will be removed based on production costs at the time it was produced. With LIFO, an assumption is made that when inventory is sold it is the most recent inventory and the cost removed from inventory will be based on the most recent production cost. The average cost method values all like inventory at the same amount and thus represents something in between LIFO and FIFO cost.

There are differing views on the merits of the FIFO and LIFO methods. The central issue is whether costs removed from inventory should be based on current production costs (LIFO) or past production costs (FIFO). Further, whichever method is chosen should correspond with the normal physical flow of inventory. A practical consideration is that in some countries, such as the US, the use of LIFO is permitted for taxation purposes. As the use of LIFO generally will result in lower income for taxation purposes than if FIFO were used, LIFO is used frequently for this reason alone.

It is easy to see that the use of LIFO and FIFO methods could result in significantly different reported earnings, especially in periods of rapidly changing price levels. The existence of these various methods is an inconsistency in accounting theory which varies by country depending upon the uniformity of adoption. In Australia, the LIFO method is not permitted and under IAS No. 2, LIFO is permitted only if there is disclosure of the difference between the use of LIFO versus the use of other acceptable methods.

Identification of Realizable Value

Under GAAP, inventory is valued at the lower of cost and realizable value. The broad objective of this principle is to ensure that any losses that may be incurred in the future, but which are the result of current events, are provided for in the results of the current period. Put another way, if some part of the cost at which inventory is carried is not going to be recovered, then it should be written off.

The most common definition of realizable value is estimated sales proceeds less any significant costs of selling. The determination of realizable value may not be easy as it will involve a subjective estimation of future price levels. The term net realizable value is often used to emphasize that inventory is stated net of selling costs.

Unique Aspects of Inventory Measurement and Valuation in the Mining Industry

As discussed briefly in the introduction to this section, a number of complications arise when applying general inventory accounting techniques to the mining industry. These problems can be categorized into two basic components - measurement and valuation. Measurement problems are primarily caused by the fact that the extractive industry is the only industry in which the raw materials have to be found, rather than purchased. When found, the question then arises as to what is the appropriate point in time to measure the quantity of material on hand as inventory. The material on hand may consist of the ore deposit, broken ore, crushed ore, concentrate or refined metal. The point of measurement of the product for accounting purposes depends on the type of mining activity together with the extent of processing activity carried out by the producer.

For example, many coal companies inventory both processed coal and uncleaned coal at the mine site. In addition, a number of surface coal mining companies inventory uncovered coal if significant costs have been incurred in removing the overburden. Thus, within the coal industry, measurement of inventory can take place as soon as the seam has been uncovered as relatively little processing is required in order to make it saleable.

Similarly, iron ore is usually measured as inventory when the primary crushing has taken place as it usually requires little processing before reaching the final product stage.

Conversely, ore with low metal content such as copper, gold and silver is measured at the concentrate stage as it is normally impractical to measure such quantities with any reasonable degree of accuracy prior to that.

The measurement techniques employed can vary. Options are physically weighing the product, estimating the quantities or employing survey techniques.

To summarize, the appropriate point in time to measure the quantity of material on hand should be determined by considering the extent of processing that is carried out by the producer, the relative value of the product and, probably most important, the point at which it can be measured with a reasonable degree of accuracy. The only standard dealing with this issue is AAS No. 7 which states that inventories should be brought to account at the earliest stage at which materials representing saleable product, or which can be converted to saleable product, can be recognized and the quantities of such materials can be determined by physical measurement or reasonable estimate.

As discussed, inventory is valued at the lower of cost and realizable value, cost usually being determined by either the direct or the absorption method of costing. The major problem in arriving at the cost of inventory in the mining industry is that the assessment of which costs to include often involves considerable subjectivity. This is particularly true when there is a wide range of costs such as mine costs, and costs relating to transportation, port facilities and administrative facilities.

In situations where absorption costing is used, there appears to be a tendency to exclude the amortization of capitalized costs. This treatment is supported as being not only prudent, but also as an attempt at eliminating further problems relating to the lower of cost and realizable value test, given the difficulty and uncertainty involved in determining realizable value. An example of the disclosure of this treatment for amortization is included later in this section.

With respect to the many distinct cost elements within an operation, most entities using the absorption method include all administrative costs associated with the mine location, together with any transportation costs, in their determination of the fixed costs that are included in inventory. In common with general practice, head office administrative costs are excluded from this determination as they do not relate directly to production.

Miscellaneous Related Topics

<u>Joint Products and By-Products</u> Joint products are two or more products that are produced simultaneously from a common raw material, each having a significant relative sales value. One of the products cannot be produced without the other and are not distinct until the split-off point has been reached. By-products are similar in all

respects to joint products except that the by-product, through relatively low unit value or quantity produced, does not have a significant relative sales value.

In general, single product valuation concepts apply equally to joint products, except that, at the split-off point, the cost has to be apportioned between each product. In most cases, this is a straightforward exercise and is carried out by reference to other products relative sales volume or volume of production. After separation, costs would be allocated in the normal manner to each product.

Varying practices are used to value by-products. These practices include reducing main product costs by the estimated realizable value of the by-products and giving no recognition of value until realization.

Spares and Stores Spares and stores comprise those items of equipment and parts that are necessary to ensure the continued operation of production facilities. These items are especially critical in remote mining operations. The general basis of accounting for these items is the purchase cost to the enterprise. Some items of spares are regarded as insurance spares in that they are major items that must be on hand in the event of a breakdown. Frequently, such items are included as a part of fixed assets, rather than as inventory, and are consequently depreciated over the expected life of the related equipment.

Financial Statement Presentation and Disclosure
Generally, inventory amounts are classified as a current asset on the balance sheet. Spares and stores inventories may be classified separately from product inventories within current assets or, in certain circumstances as discussed above, may be classified as fixed assets. Aside from basic classification, the most important aspects of the presentation and disclosure of inventory are the accounting policy followed and an analysis of inventory components.

At a minimum, accounting policy disclosure should include the basis of valuation (for example, lower of cost and realizable value), and the method in which costs are removed from inventory (for example, the FIFO basis). A useful, but often not required, disclosure is a description of the elements of cost included in inventory, particularly of those elements for which an allocation has been made to inventory such as administration cost.

Most countries require, or encourage, disclosure of the major components of inventory, that is, finished goods, work-in-process and raw materials. In Australia, there is a statutory requirement to distinguish between finished goods and work-in-process.

Illustrative examples of recent financial statement disclosures of inventory accounting policies are as follows:

Canada's Echo Bay:

"Inventories of gold, which includes bullion and concentrates in process, are valued at the lower of cost, using the 'first-in, first-out-method, and net realizable value. Materials and supplies are valued at the lower of average cost and net realizable value."

Australia's Seltrust Holdings:

"Stocks of ore, concentrate and matte, are valued at the lower of cost and net realizable value.
Stores are valued at cost, with due consideration taken for obsolescence.
"Cost" of ore, concentrate and matte, includes all relevant expenditure incurred in bringing such inventories to their current status. Fixed and variable overheads based on the level of activity achieved have been, where appropriate, included in the cost of production and year-end inventory. Cost is determined using the average method on a "first-in, first-out" basis.
"Net realizable value" comprises estimated sales proceeds at balance sheet date converted, where applicable, at the year-end exchange rate, or the applicable forward exchange contract rate less applicable selling expenses."

It is not normal practice to state the use of direct or absorption costing. Rather, the description of the costs included should be sufficient to indicate which method is being used.

Summary

To summarize, GAAP for inventory consists of many alternatives. Inventory can be valued at direct cost or absorption cost, either of which can be computed under the FIFO, LIFO, average cost or other conventions. Within this array of options, the overriding objective is to fairly match costs with their associated revenues and to provide a valuation of inventory which most closely relates to its "cost". This is not always a simple task and disclosure plays an important role in communicating to the reader of financial statements the way in which an enterprise attempts to achieve this objective.

FINANCING TRANSACTIONS

In order to develop an appreciation of the way in which an enterprise is financed, it is necessary to have a good understanding of the components of debt and equity and their significance. This section briefly explains the distinctive features of both equity and debt financing arrangements and how these arrangements are presented in financial statements. It also addresses the accounting issues relating to certain financing activities unique to the mining industry and the concepts of lease financing and off balance sheet financing.

Equity

In simple terms, the equity section of a balance sheet presents all contributions received or deemed to be received from stockholders, whether the related shares are common or preferred, voting or non-voting, and all accumulated earnings reinvested in the business. As such, it is a statement of the stockholders' or owners' interests in the enterprise and should exclude any transaction or amount not received from or accruing to the stockholders. This definition holds true for most North American companies. In other countries such as the UK and Australia, asset revaluation is permitted, which results in a third component of equity commonly described as asset revaluation reserves.

The interest of owners ranks after those of creditors, therefore their interest in the assets of an enterprise is what is left after all others have made their claims and as such, represents a residual interest. In monetary terms, equity represents the source of distributions by an enterprise to its owners. Contrasted with liabilities, distributions to owners are discretionary, whereas liabilities, once incurred, represent obligations which must be satisfied on demand at a specified or determinable date, or on the occurrence of a specified event.

Components of Equity   Equity consists of a minimum of two components: amounts received or deemed to be received in respect of issues of common and preferred stock ("capital stock") and accumulated earnings reinvested in the business ("retained earnings").

In most jurisdictions, the legal concept of par value is still used, or has been used until recently, with the result that there is often another caption representing the difference between amounts received for shares issued and the underlying par value. The description given to this component varies from country to country. The terms most commonly used are contributed or paid-in surplus in North America and share premium accounts in the UK and Australia.

Another major component seen frequently in the UK and Australia are reserves relating to revalued assets. By definition, these amounts are unrealized and represent the other side of an accounting entry to increase the value of certain assets to reflect the difference between estimated fair value and book value. In the US the revaluation of assets is not permitted and in Canada it is only rarely seen.

It is important to be aware of this in comparing debt pressure ratios of companies in different countries. For example, the Australian company, Western Mining Corporation, has shareholders' equity of $650 million of which $159 million relates to an asset revaluation reserve. In the US, such a reserve would not exist.

Revaluation accounting, as it is commonly known, is useful in providing the reader with a more realistic picture of the value of an entity, particularly when current value, however determined, is so significantly different from book value as to make the latter largely meaningless. Caution, however, is necessary. Considerable subjectivity is used in arriving at a current value and, as previously indicated, any resultant reserve is not realized and generally should not be viewed as distributable.

It is also worth noting that revaluation accounting is usually applied to assets such as land and buildings and not to mining properties, where the high degree of subjectivity in valuing an ore body renders the resultant valuation unreliable.

Finally, a new component of equity has recently appeared on the balance sheets of many companies, as a result of the issue of new accounting standards relating to foreign currency translation. As discussed in the foreign currency translation section of this chapter, adjustments relating to the translation of financial statements of foreign-based entities are taken directly to equity and disclosed as a separate component.

In summary, the components of equity will generally comprise:

- capital stock which may be split between par value and a surplus category;
- retained earnings;
- asset valuation reserves, the use of which will vary from country to country; and
- translation adjustments relating to the translation of financial statements of certain foreign-based entities.

Presentation of Equity   With the exception of asset revaluation reserves, there are no significant differences from country to country in the components of equity. There are, however, differences in presentation. Typical presentations of equity in each of North America, the UK and Australia are as follows:

North America	UK	Australia
Capital stock	Share capital	Issued capital
Contributed surplus	Share premium account	Reserves
Retained earnings	Reserves	Retained earnings or profits

The major presentation differences relate to the reserves category. In the UK, retained earnings are included in reserves whereas in Australia these are often disclosed separately. In addition, share premium accounts are typically presented separately by UK companies whereas in Australia they are included in the reserves caption.

In either case, the components of reserves are generally disclosed by way of a note to the financial statements.

DISTRIBUTABLE VERSUS NON-DISTRIBUTABLE AMOUNTS: In analyzing an enterprise's equity position, care should be taken to distinguish between distributable versus non-distributable components. In Canada, for example, capital stock and retained earnings are considered to be distributable - the former through a return of capital and the latter by payment of dividends. In most jurisdictions, asset revaluation reserves are considered to be non-distributable for the simple reason that they represent unrealized amounts. In addition, in consolidated financial statements there may be restrictions on the ability of the parent company to influence the distribution of retained earnings of subsidiaries and associates. Accordingly, disclosure of such amounts is useful.

This breakdown is often supplemented by a note regarding limits on the distribution of reserves by subsidiaries and associates and the tax implications in the event of a distribution by an overseas subsidiary or associate.

What Price Equity? In most stock issues, the amount allocated to capital stock is easily determined by reference to cash received, before or after deducting the costs of selling the issue. Some stock issues are less straightforward and involve, for example:

- receipt of non-cash assets as consideration;

- the selling of tax benefits to investors as in the case of flow-through shares; and

- commodity-backed issues.

NON-CASH CONSIDERATION: When non-cash assets are received as consideration for stock issued, it is necessary to value the stock either by reference to what was given up - the stock, or by reference to what was received - assets. In either case, there is an element of subjectivity involved and ultimately management or the directors will be responsible for determining a reasonable value. When the issuer is a publicly-traded entity, the value usually can be determined by reference to trading value of stock at or around the transaction date. In volatile market conditions such value may not be readily apparent and, in the event of a major issue of stock, trading value may not be a good indicator. If the entity's shares are not publicly traded, the valuation process becomes even more subjective unless the value of the assets received is readily determinable.

In these non-cash transactions, the financial statements should disclose the basis of valuation to facilitate the reader's understanding of the transaction. An example of the implications of such transactions is Canada's Campbell Resources who completed a major acquisition in 1983, issuing 3.6 million shares as consideration for interests acquired. An amount of $10 per share was allocated to the shares issued for a total deemed acquisition cost of $36 million. Based on trading values at that time, the shares could have been valued as high as $15 and as low as $5 for an aggregate consideration of $54 million and $18 million respectively - a substantial range in values.

The view has been taken that when there is discretion involved, management will always elect to value non-cash consideration as high as possible to boost asset values and improve the debt to equity ratio. Such an approach does have a cost - the higher the values allocated to assets the greater the impact on future years' net income when those assets are depreciated or amortized. Thus, it is not always in management's best interests to maximize assigned values.

SELLING TAX BENEFITS TO INVESTORS - FLOW-THROUGH SHARES: This section primarily relates to the Canadian environment where legislation has been passed enabling enterprises to sell unused tax benefits to investors.

In the normal course of business, a mining company has to fund its exploration and development activities. When exploration and development costs are incurred, tax deductions are available to offset taxable income. Many mining enterprises are not imminently taxable with the result that they are unable to use available tax deductions. To assist such enterprises in funding their exploration and development activities, legislation now permits the associated tax deductions to be sold to investors through the issuance of flow-through shares. These shares are identical to any other share issued by an enterprise except that the proceeds must be applied by the issuer to eligible exploration and development activities. In addition, because the investor is acquiring the right to a tax deduction as well as an investment in the issuing enterprise, the shares are issued at a premium over underlying market value. This premium may be in excess of 50% of market value.

The principal accounting issue relating to flow-through shares is whether any recognition should be given to the tax deductions forgone by the issuing enterprise. Should any part of the proceeds received be attributed to the giving up of the tax deduction or should the full amount received be allocated to equity? In the former case, the argument is that the enterprise has forgone a tax benefit which would otherwise have been reflected in income as the related expenditures are written off for accounting purposes. In the latter case, if no part of the proceeds is attributed to the giving up of this right, then, when the expenditures are written off to income, there will be no associated tax recovery. The argument raised in this situation is that an enterprise is unlikely to use the flow-through share financing concept if it is currently or imminently taxable, thus no part of the proceeds should be allocated to tax deductions given up.

Example

Assumptions:

(1) Enterprise issues 10,000 flow-through shares
    at $25 each. $250,000
(2) Effective tax rate.   35%
(3) Option 1 - tax relating to the tax deductions
    forgone provided at 35%.   87,500
(4) Option 2 - full proceeds allocated to equity.

	OPTION 1		OPTION 2	
	DR	CR	DR	CR
Issue of Shares:				
Cash	$250,000		$250,000	
Share Capital		$162,500		$250,000
Deferred Credit		87,500		
Expenditure Incurred and Capitalized:				
Exploration and development expenditures	250,000		250,000	
Cash		250,000		250,000
Write-Off of Exploration and Development Expenditures:				
Net Income	162,500		250,000	
Deferred Credit	87,500			
Exploration and development expenditures		250,000		250,000

This example sets out two alternatives for accounting for flow-through shares. Other options may be equally acceptable such as recording the tax effect as a reduction in exploration and development expenditures or valuing the shares by reference to what the investor received rather than what the issuing enterprise gave up.

The approach in Option 2 has more practical merit in that the accounting is easily understood and involves no subjective determinations as in the case of Option 1. Option 2 is most likely to be used in practice until an authoritative pronouncement is forthcoming.

In any event, as any of the above approaches can produce a significantly different result, there should be disclosure by way of an accounting policy note and if Option 2 is used, it is helpful to state by way of note that the related exploration and development expenditures carry no tax deductibility.

COMMODITY-BACKED ISSUES: Commodity-backed issues provide the investor with an opportunity to benefit from increases in commodity prices by linking the return on investment to future commodity prices or by granting options to acquire a specified commodity at a fixed price. In the US in 1980, Sunshine Mining Company issued bonds indexed to silver, that is, the face value of each bond was payable at maturity at the greater of US $1,000 or the market price of 50 ounces of silver. The issue was attractive to the investor as it provided a hedge against inflation and to Sunshine Mining Company as it didn't have to market the issue at a large discount. Though not an equity issue, the accounting implications in dealing with such instruments are similar.

In the Canadian environment several equity issues have been marketed in recent years in the form of a combination of common or preferred stock plus warrants to purchase gold at a future date.

These financing instruments are somewhat unique to the mining industry and there is little authoritative literature which discusses the appropriate accounting. Without providing a theoretical discussion of the accounting implications of commodity-backed issues, the major questions can be summarized as follows:

1) When an issue provides the opportunity to participate in an entity's commodity production or to benefit from future price increases in commodities, should any part of the resultant proceeds be allocated to this feature?

2) If an allocation of the proceeds is made, how should it be accounted for?

The principles of US APB Opinion No. 14, Convertible Debt and Debt with Stock Purchase Warrants are generally used for guidance in this area. APB No. 14 is based on the principle that when debt is issued with the option to convert such debt to equity at a future date, no value should be allocated to the conversion feature as it is not detachable from the debt instrument. Likewise, with debt carrying a stock warrant purchase feature, no value should be allocated to the warrant unless it is detachable, that is, unless the warrant can be sold as a separate legal instrument.

These principles appear to have been followed in accounting for commodity-backed issues in the North American environment. Sunshine Mining Company allocated the full proceeds to the debt presumably because, inter alia, the silver index feature was inseparable from the debt instrument. Similarly, in the Canadian environment, it appears to be common practice to allocate part of the proceeds from an issue to detachable warrants only.

Generally, if the issue involves an allocation of proceeds to detachable warrants, it appears to be practice to account for this portion of the proceeds as deferred revenue. In that scenario, the transaction is viewed as a sale of future production and the deferred revenue is taken into income as production takes place. In this respect, the drawing-down of the deferred credit over the production period to the income statement will offset the cost to the issuer of selling gold to warrant-holders at prices less than market.

Reproduced below is an example of the deferred revenue treatment used by Canada's Echo Bay Mines to account for an issue of gold purchase warrants.

Echo Bay Mines

> "Deferred Revenue
>
> There were 6,400,000 gold purchase warrants issued on February 5, 1981. One million, six hundred thousand are exercisable on each of January 31, 1986, 1987, 1988 and 1989. Each warrant will entitle the holder to purchase, from the

company, 0.01765 of a troy ounce of gold for U.S. $10.50 (of which U.S. $5.25 was prepaid at the time of issue of the warrant) plus a 1% delivery fee.

If a warrant is not exercised on its exercise date, the holder may tender the warrant for cancellation at anytime during the two years after the exercise date in return for (i) the market value at the exercise date of the gold purchasable under the warrant less the unpaid purchase price, the 1% delivery fee and a U.S. $0.25 service fee, if the market value of gold was equal to or greater than U.S. $625.00 per troy ounce on the exercise date, or (ii) U.S. $5.25, if the market value of gold was less than U.S. $625.00 per troy ounce on the exercise date.

In the event that production of gold is suspended, the exercise dates of the warrants will be extended. Notwithstanding such extension, a warrant may be tendered for cancellation in return for U.S. $5.25 on or after its original exercise date. No warrant will be exercisable unless production has been resumed on or before December 31, 1992.

Under certain circumstances related to the company's gold reserves, the holders of warrants have the right to accelerate the exercise of warrants."

The following illustrates the application of the accounting using the amounts disclosed above:

- The warrant exercise price approximates US $600 per troy ounce.

- If a warrant holder exercises 100 warrants on January 31, 1986 when the market price of gold is US $700, the following accounting ensues:

Receipt of Cash from Warrant Holder

	DR	CR
Cash (100 x $5.25)	$525.	
Deferred revenue (100 x $5.25)		525.
Revenue		$1,100

- The $1,100 amount recorded as revenue represents the total received from the warrant holder in connection with the purchase of gold, being a combination of cash received on the exercise of the warrants plus the amount allocated to the warrants on issuance. This compares to an amount of $1,236 (100 x 0.01765 x $700) that would have been received had the company sold the gold on the open market. The difference of $136 represents the notional cost to the company of the arrangement.

- If the price of gold does not exceed US $625 during the exercise period, the warrant holders are entitled to a return of the $5.25 per warrant. In that event, the amount allocated

as deferred revenue becomes a cash obligation of the company.

- Conversely, if the price of gold rises significantly in excess of the total exercise price of US $600 per troy ounce, the cost to the company, in terms of "lost" revenue by dedicating future production to warrant holders, increases.

Clearly these arrangements are not straightforward from an accounting perspective and the accounting outlined cannot be viewed as definitive. Apart from disclosure of the accounting for the proceeds on issuance, there should be disclosure of the terms under which investors can participate in these arrangements and of the amount of dedicated future production. With this information, it should be possible to assess the likely impact of these arrangements at any given level of prices.

Classification of Redeemable Preferred Stock

Many preferred stock issues have all the characteristics of debt-instruments - dividends are cumulative, the shares are non-participating and redemption is mandatory. In the past, all such issues have been classified as part of equity. SEC registrants are now required to classify such issues outside the equity caption. An example is the redeemable preference shares of Echo Bay (issued in connection with the sale of the gold purchase warrants discussed above) which must be redeemed at specified rates and amounts commencing on December 31, 1986. The amount allocated on issuance of $40 million is described as redeemable preferred shares and is classified in the balance sheet immediately above the equity caption. Note that the related dividends continue to be afforded the usual treatment as a charge to retained earnings.

Long-Term Debt

Long-term debt generally consists of:
- long-term bank loans
- debentures
- bonds
- mortgages
- capitalized leases
- other obligations of a long-term nature

With the exception of capitalized leases, which are discussed later in this section, there are few problems associated with the accounting for long-term debt. Amounts due within one year are classified in current liabilities and any premium or discount on issuance of bonds and debentures is amortized to income over the term of the debt as a reduction of, or increase in, interest expense respectively.

When debt is convertible at a future date to equity, the principles of APB Opinion No. 14, Convertible Debt and Debt with Stock Purchase Warrants, are applied and no amount is attached to the conversion feature as it is not separable from the debt instrument. In addition, when debt is acquired and the rate attached to the debt instrument is significantly lower than the market rate at the date of acquisition, then consideration should be given to discounting the face amount of the debt. When this is done, the amount of discount is recorded in income over the life of the related debt as additional interest expense.

Generally, the most important aspect of long-term debt from an accounting perspective is the note disclosure which should include:

- interest rates;
- terms of repayment including scheduled repayments generally for a minimum of five years;
- terms of conversion, if any;
- conditions attached to the debt such as debt covenants and restrictions on the payment of dividends;
- extent to which each issue is secured;
- contingencies affecting payment of interest or principal; and
- unused lines of credit (both short and long term).

Lease Financing The concept of lease financing is explored in a separate paper in this text. In that paper, reference is made to the accounting implications of lease financing and a summary of accounting standards for leases is provided. From an accounting perspective, the main issue brought out in the paper is that there may be advantages from the perspective of the lessee in having a lease classified as operating rather than as a financing device.

As a supplement to that discussion, the following observations may be useful.

Accounting for leases is a relatively new area, FASB being the first accounting body to issue a standard in the form of FAS No. 13, Accounting for Leases, in 1976. This statement has been followed by the issuance of standards in Canada and Australia and a standard is imminent in the UK. A comparison of these standards and proposed approaches to IAS No. 17, Accounting for Leases, is provided in Appendix I to this chapter.

Perhaps the most pertinent question is the impact of these new standards on the reported financial statements of mining companies. In Canada, it would appear, based on a review of a cross-section of mining company financial statements, that mining enterprises currently do not finance a significant portion of their mining activities through leasing arrangements. On that basis, the impact of the introduction of CICA Handbook Section 3065 on the financial statements of Canadian mining companies can be said to be minimal. It remains to be seen if a similar conclusion will be reached in Australia and the UK. Due to the limited use of leveraged leasing by UK mining companies, it appears that the impact of the new standard in that country will be minimal. There is potential for greater impact in Australia, where this financing alternative is more frequently used. In the US, leveraged leasing is used frequently in the mining industry, particularly in the coal industry to finance equipment purchases for coal mining operations. At first glance then, one would surmise that the US standard has had an impact on mining companies in the US. Practice does indicate, however, that many enterprises are taking steps to structure lease arrangements to avoid their classification as finance leases on the balance sheet. In this respect, it is unfortunate that the introduction of this standard has resulted in some enterprises restructuring their leasing arrangements, with a potential economic cost, merely to avoid an accounting classification.

In regard to leasing arrangements in the coal industry, coal companies commonly lease land or mining rights rather than buying the land outright. In these circumstances, most lease agreements provide for royalty payments to the lessor when the coal is sold and may also provide for cash payments to the lessor at the inception of the lease - prior to any mining activity taking place. In addition, annual minimum royalty payments typically are required.

For accounting purposes, these payments are considered advance or prepaid royalties and are charged against income as the coal reserves are mined and sold. When such payments are deferred, they are generally classified as non-current assets unless the payments are expected to be charged against income in the succeeding year.

Off Balance Sheet Financing

Off balance sheet financing is a term used to describe those financing arrangements by an enterprise which do not appear on the balance sheet and which may not enter into debt pressure or net worth calculations.

In the course of this section, we have alluded directly and indirectly to the concept of off balance sheet financing on several occasions. In our discussion of equity accounted investees, including joint ventures, we noted that one of the potential problems with equity accounting is that the true debt obligations of a company may not be presented due to the single-line presentation of certain investments. Good disclosure of financial information relating to significant investees including joint ventures can compensate for this perceived deficiency. Secondly, in regard to lease financing, the introduction of lease accounting standards should minimize the ability of enterprises to finance significant activities through financing type leases without recording the related obligation on the balance sheet. As discussed above success in this area will be limited if enterprises feel it is necessary to restructure leasing arrangements in order to avoid the impact of these standards.

Thus, it can be concluded that while potential for off balance sheet financing still exists the introduction of new accounting standards and improved standards of disclosure should reduce considerably the number of examples in practice of "hidden" debt.

Summary

An understanding of the financing arrangements of an enterprise is key to an overall analysis of its financial position and profit potential. It is difficult to generalize about the financing arrangements of mining enterprises; the financing of some mining companies is straightforward but in many cases is very complex. These complexities, together with varied methods of presentation, the impact of government assistance, the potential for off balance sheet financing and the flexibility available in dealing with innovative financing instruments adds up to a formidable package for the average user of financial statements to absorb and understand. Accounting standards have played a role in ensuring that the presentation of financing arrangements reflects the substance of the arrangement and in ensuring that minimum disclosures are made. Thereafter, it is up to the reporting enterprise to strike a balance in presentation and disclosure that provides the necessary information in an understandable form.

RECOGNITION OF REVENUE

The accounting principles relating to the recognition of revenue within the mining industry encompass a number of unique concepts. These arise for a variety of reasons which include the nature of the product, its quantification for sales value purposes and the numerous methods by which sales contracts are established. Other factors which can influence the revenue recognition question are the often significant time delays between the production of the mineral and its arrival at the end user, frequently due to the remoteness of the extractive operation and the availability of transport, together with prices that are generally set by market forces and are not controllable by the producer.

While many areas of accounting and reporting have been standardized, there are few standards that relate to the question of revenue recognition. General principles of revenue recognition are set out in IAS No. 17, Revenue Recognition. The broad principles are that revenue should be recognized from the sale of goods or services when the terms have been effectively completed and it is reasonable to expect ultimate collection. A comparison of these principles with local rules and requirements is provided in Appendix I. These rules and requirements provide little guidance for dealing with the unique aspects of revenue recognition in the mining industry.

Although Australia has not issued a general standard relating to revenue recognition, AAS No. 7, Accounting for the Extractive Industries, does include a section on revenue recognition specifically designed for companies in the extractive industries. This part of AAS No. 7 emphasizes the concepts of saleable form and passage of physical control which are discussed in more detail below. The UK, the US and Canada have not issued standards relating to revenue recognition in the mining industry.

Despite the absence of definitive guidance for mining companies in this area, a number of concepts have evolved over time which give appropriate recognition to the economic substance of a sale. These concepts are discussed below.

When Does a Sale Take Place?

Generally, the recognition of a transaction as a sale would take place when the seller has transferred to the buyer the significant rights and obligations of ownership of the product sold. Within the mining industry, it is generally accepted that a sale takes place when all of the following conditions have been met:

- the product is in a form such that it can be delivered to the purchaser with no further work or processing necessary by the seller;

- the actual physical quantity and quality of the product can be determined with reasonable accuracy;

- the selling price of the product can be determined with reasonable accuracy; and

- physical control over the product has passed from the producer.

These specific conditions, which result from the unique characteristics of the industry, tend to expand the narrow definition of determining the point at which a sale takes place to encompass the broader question of when does the substance of the transaction indicate that a sale has taken place. To answer this question, a closer examination of the above conditions is necessary.

Saleable Form Terms and conditions of sales contracts can vary greatly and this, in turn, results in numerous methods of accounting for these sales. For example, contracts can specify that a product is in saleable form at a number of different points in the processing cycle. These points could be when the ore has been mined, when the product is in pellet or concentrate form, or in the form of refined metal. Alternatively, a contract may specify that the point of sale is when the product is loaded for transportation or when the product has reached a designated point toward its destination.

As a general rule, a product would meet the condition of saleable form if no additional work or processing needs to be carried out by the producer in order to meet the requirements of the sales contract. This principle is brought out in AAS No. 7.

Determination of Quantities and Qualities Frequently, the quantity and quality of shipments of product cannot be precisely determined until the results of the final assay or other tests are known and the delivery weights have been agreed between the buyer and seller. In addition, particular specifications may be required by the sales contract in terms of, for example, the size and composition of the product.

In these circumstances it is generally acceptable to recognize a sale provided that an estimate of the out-turn can be made with a reasonable degree of accuracy. Whether this can be achieved primarily depends on individual circumstances and usually is determined based on past experience in such estimates. These estimates normally should include allowances for any penalties that are likely to be suffered.

When the final weights and assays have been determined, the previously estimated revenues would be adjusted in that period to reflect the final out-turn.

Determination of Price   Within the mining industry, it is generally accepted that the point in time at which a contract price is triggered does not usually influence the timing of the recognition of the revenue relating to that contract, provided that, prior to that time, the price can be estimated with a reasonable degree of accuracy. To the extent that the initial estimate is not accurate, an adjustment would be made to revenue at the time of the final settlement.

The question as to whether the contract price can be reasonably estimated depends on a number of factors. The factors that would most likely affect this estimate could be as follows:

- the volatility of the markets;

- the presence of contract escalation clauses;

- the time delay between production or delivery and the point of price determination;

- the degree to which the contract is based on fixed or spot prices; and

- the effect of charges, such as smelting, not directly under the control of the producer.

In individual circumstances there may well be other significant factors that would be taken into consideration.

Complex Pricing Arrangements   There are many examples in the industry where complex pricing structures exist. For example, in the coal industry the basis for recognizing revenue is generally the contractual base price per ton plus actual escalations indicated by indices or formulae. These escalations, in certain instances, are a function of the costs incurred by the producer, or the result of complicated multi-variable calculations. In many cases amounts relating to escalations are only recognized as revenue when they have been agreed and the purchaser has been billed the additional amounts. Other contracts exist where minerals are sold to a purchaser at a base price with provision for an excess payment should the purchaser realize a higher price on resale. General practice, in these circumstances, would be to recognize the base price as revenue at the point of sale and, should any excess be payable, recognize this excess as revenue at the point of resale.

Another common method of price determination is for the producer to enter into a forward contract whereby the producer commits to selling a portion of its production at some determinable point in the future for a fixed price. In this way the producer obtains a hedge against future price changes. Here, revenue is recognized when the usual conditions for a sale have been met. The price to be used in recognizing that revenue would be the contract price with the usual subsequent adjustment for weight and assay. A discussion of sales of future production when related to obtaining financing is included in the financing transactions section of this chapter.

Physical Control Over the Product   Physical control over the product is generally regarded as being relinquished when the product has passed into the control of an independent carrier for delivery to a purchaser or when the product has been shipped to a custom smelter for processing. In the latter situation, practice varies with regard to revenue recognition.

Custom smelters operate on either a purchase or a toll basis. On a purchase basis, the smelter is responsible for the final sale of the metal produced; on a toll basis, the smelter is entitled only to a treatment (toll) charge. This is usually fixed by contract or based on a formula related to the selling price of the metal. If the smelter is the purchaser, then revenue is generally recognized at time of shipment. If the smelter operates on a toll basis, the timing of the revenue recognition is not clear. Some would argue that no revenue should be recognized until delivery is made to the ultimate purchaser as, until the smelting process is complete, the product is not in saleable form. Others would argue that as long as all the other conditions are met, it is appropriate to recognize revenue at the time of shipment or delivery to the smelter.

It also should be noted that under the concept of relinquishing physical control there is no necessity for physical delivery to have taken place for revenue to be recognized.

Inasmuch as the concept described above has a reasonably straightforward application, a number of grey areas do exist. For example, say a significant advance payment had been made for a specific product which has been earmarked for a purchaser. Would revenue be recognized in these circumstances? In general, current practice would be not to recognize the revenue because the fact that a significant advance payment has been received in itself would not affect the determination. Other grey areas include circumstances when delivery cannot be made due to, for example, infrequent transportation, requests for delays in shipments, and transportation problems outside the control of the producer. Again, in these circumstances it is not generally acceptable to recognize revenue.

Summary   In reviewing the above conditions, one can clearly see that the substance of the transaction takes precedence over its legal form. If these conditions are satisfied, then practice in the mining industry would be to recognize revenue notwithstanding that legal title may not have passed. Further, it would appear that such practice is consistent with the principles of IAS No. 18.

Long-Term Contracts

One area that has particular significance when considering revenue recognition is that of long-term contracts. These contracts are very common in the mining industry, particularly in the coal industry. Long-term contracts can run from anywhere between five and thirty-five years. However, recent trends indicate that the maximum length of such contracts is declining to around twenty years with the average length at roughly ten years. Long-term contracts offer significant advantages to both the producer and purchaser. Generally, for the purchaser, the advantages are security of quality and quantity together with stability and predictability of price. To the producers, advantages are guaranteed future sales and the consequent protection of profits. In many cases, a long-term contract is the foundation upon which many mining operations are taken through the development stage and into the production stage.

Sales under long-term contracts are accounted for under the same general principles already described but also take into consideration the provisions of the particular sales agreement. Special consideration may be required where losses - temporary or otherwise - are being incurred on a particular contract. Although in neither situation is the recognition of the revenue affected, it is generally accepted that where losses are expected to be incurred throughout the remainder of a contract, a provision against all anticipated future losses should be recognized. Many companies would consider their long-term contracts in the aggregate for the purposes of this test, and consequently losses on an individual contract may not be recognized if, in the aggregate, contracts are expected to be profitable. These loss situations on contracts are not particularly common due to the presence of escalation clauses or other such provisions for the producer to recover all costs. Further, numerous practical difficulties exist in precisely determining the extent of losses that may arise in the future.

Pre-Production Payments   Certain long-term sales contracts provide for the purchaser to make capital contributions toward the equipment or initial startup costs. The accounting for amounts received under such an arrangement is generally by one of the following methods, all of which are acceptable:

- the amount is treated as deferred revenue and taken into income over the life of the contract, usually in proportion to the quantity sold;

- the amount is taken to income over the life of the asset to which it relates; or

- the amount is applied against the capital expenditure to which it relates, which consequently reduces the future charge to income for amortization.

Foreign Currency Transactions

Frequently, contracts are fixed in a currency other than that of the producer's country. In these situations, the revenue is generally recorded at the rate of exchange in effect at the date the revenue is recognized. In many of these circumstances settlement is at a later date. This generally results in an exchange gain or loss that arises from exchange rate movements between the date that the transaction is initially recorded (date of revenue recognition) and the date of settlement. Treatment of this difference varies among countries. In North America, the difference is identified as an exchange gain or loss whereas in the UK and Australia the difference can be similarly treated but also can be treated as an adjustment to revenue. Reported earnings are identical under either treatment.

Financial Statement Disclosure

Because there is a variety in types of sales contracts in the mining industry and the variety of practices for recognizing revenue under these arrangements, disclosure of an entity's policy for revenue recognition is useful information to the reader. Disclosure should include the conditions under which revenue is recognized, particularly in those circumstances where recognition takes place at a point prior to the passage of legal title.

Following are a number of illustrative examples which display the divergency in practice among mining companies. On reviewing these examples it is worth noting that it is not always clear if the conditions of sale discussed in this section have been met.

Canada's Falconbridge

"Revenues from the sale of refined metals and industrial minerals are recorded in the accounts when the rights and obligations of ownership pass to the buyer."

Canada's Campbell Resources

"The sale of metals processed through the flotation circuit at the mine are recorded in income three months after arrival of the concentrate at the custom smelter which is the normal time required to smelt, refine and sell the metals contained in the concentrate. The sales of precious metals produced through the cyanide circuit are recorded in income when the metals are poured at the mine site."

Australia's Seltrust Holdings

"Revenue is recognized when the substance of a transaction indicates that a sale has taken place. A sale will occur when title to the goods passes, or alternatively, where the product is in the form in which it is to be sold and it is within the purchaser control pursuant to an enforceable sales contract."

Summary

To summarize, we have examined some of the concepts which have evolved for the recognition of revenue in mining companies and some of the unique business arrangements in the mining industry which give rise to revenue recognition problems. The examples of financial statement disclosure illustrate a great divergency of treatment in practice. In time, this divergency will probably diminish as greater effort is directed towards standardizing practice.

INCOME TAXES

Financial reporting, and more specifically income determination, is designed to provide information that is useful to those who make economic decisions about business enterprises, and is generally dictated by GAAP in the reporting environment. Income determination for taxation purposes is designed to aid taxing authorities in raising revenue consistent with the achievement of social, economic, and political goals, and is dictated by rules and regulations established by the government assessing the tax. Due to these conceptual differences, financial reporting income (reported income) and taxable income are usually different.

This section discusses the nature of these differences and the approach taken in each country to deal with them. References to specific deductions and credits available for tax purposes relate to the US environment and are referred to for purposes of illustration only. The nature and number of differences between reported income and taxable income in an enterprise vary greatly according to the environment in which it operates.

Discussion of Differences Between Reported Income and Taxable Income

Within the mining industry, differences often arise between reported income and taxable income. Reasons for such differences include legislated tax incentives to promote activities in this high risk industry, for example, special depletion allowances; tax rules allowing companies to deduct certain expenses only when paid, such as reclamation costs; and accounting rules which require expensing certain expenditures over the period of economic benefit that are deductible for tax purposes when incurred, such as mine development costs.

Regardless of why differences occur between reported income and taxable income, they generally can be classified as either timing differences or permanent differences. Transactions and events that enter into the determination of reported income in periods different from when they enter into the determination of taxable income create timing differences. Timing differences originate in one period and reverse in another. For example, if they increase taxable income in one period, they reduce taxable income in a later period or periods. An example is mine development costs which may be expensed as incurred under rules established by US taxing authorities, but are typically capitalized and expensed over mine production for financial accounting purposes. Permanent differences, on the other hand, result from transactions and events that enter into the determination of either reported income or taxable income but not both. Permanent differences do not reverse. For example, the amortization of goodwill recorded in connection with the purchase of another company, which is generally not deductible for income tax purposes, is a permanent difference.

Permanent differences do not present a problem in accounting for income taxes. If the only differences between reported income and taxable income for a period are permanent, that is, there are no timing differences, reported tax expense should equal taxes actually payable for the period. To illustrate, assume that reported income of $1 million before taxes excludes $200,000 of excess depletion deductible as a result of statutory tax laws, and as a result taxable income is $800,000. At a tax rate of 50%, taxes payable for income taxation purposes and tax expense for financial reporting purposes are $400,000 ($800,000 x 50%).

If timing differences exist, the appropriate accounting is not so clear and different practices are followed from one country to another.

Interperiod Tax Allocation

The varying practices regarding timing differences centre around to what extent the tax effect of timing differences should follow the period in which those transactions are recognized for financial reporting purposes. The process by which taxes payable, currently or in the future, are allocated to the period in which the transaction is recognized for financial reporting purposes is referred to as interperiod tax allocation.

Figures 3 and 4 provide an example of interperiod tax allocation:

	Year 1	Year 2	Total
Reported income before tax	$1,000,000	$1,000,000	$2,000,000
Less deduction for excess statutory depletion (permanent difference)	200,000	200,000	400,000
	800,000	800,000	1,600,000
Mine development costs (timing difference)	(300,000)	300,000	-0-
Taxable income	$ 500,000	$1,100,000	$1,600,000
Taxes payable (assume 50%)	$ 250,000	$ 550,000	$ 800,000

FIGURE 3

The mine development costs relate to expenditures made in preparation for mining that would occur in year two (assume mining is started and completed in year two). Thus, it is a timing difference that originates and reverses within a two-year period. Using interperiod tax allocation, the tax effect of the mine development cost deduction ($150,000 or $300,000 x 50%) is recorded in year two, the period that the deduction is recognized for financial accounting purposes, resulting in reported tax expense for each period as illustrated in Figure 4 below:

	Year 1	Year 2	Total
Reported income before tax	$1,000,000	$1,000,000	$2,000,000
Less deduction for excess statutory depletion (permanent difference)	200,000	200,000	400,000
	$ 800,000	$ 800,000	$1,600,000
Taxes payable (as per Figure 1)	$ 250,000	$ 550,000	$ 800,000
Tax effect of timing difference	150,000	(150,000)	-0-
Reported tax expense	$ 400,000	$ 400,000	$ 800,000

FIGURE 4

The example demonstrates that interperiod tax allocation uses taxes currently payable as the starting point for computing reported tax expense. Total reported tax expense for year one is determined by increasing the taxes actually payable by the tax effect of the mine development costs recognized as a deduction for tax purposes but deferred for financial accounting purposes.

The offset to the income statement charge is a balance sheet amount called a deferred tax credit that remains in the balance sheet until year two when the timing difference reverses and the mine development costs enter into the determination of reported income. Then, the tax effect of the reversing timing differences is subtracted from taxes payable to arrive at total reported tax expense.

The differences in opinion regarding interperiod tax allocation, and differences in practice in certain parts of the world, focus on to what extent interperiod tax allocation should be applied. Differing views are:

- no allocation (the taxes payable method) - Income tax expense for financial accounting purposes should equal income taxes actually payable;

- partial allocation - Interperiod tax allocation should be applied only to those timing differences between reported income and taxable income which are expected to reverse in the foreseeable future;

- comprehensive allocation - Interperiod tax allocation should be applied to all timing differences between reported income and taxable income irrespective of the projected reversal period.

Proponents of the taxes payable method believe that income tax expense for financial reporting purposes results only from taxable income. They believe that interperiod tax allocation artificially smoothes income and ignores contingencies and uncertainties regarding future payment of income taxes which are deferred. They believe the taxes payable method recognizes income tax when the related benefits provided by the government are received and provides a more exact measurement of the entity's current expense.

Allocation advocates believe that the tax effect of a given transaction or event can be determined and that it should be recognized when the transaction or event enters into the determination of reported income. They contend that the flow-through method erroneously destroys the relationship of pre-tax income and income tax expense shown in financial statements, and that interperiod tax allocation achieves a better matching of revenues and expenses.

Proponents of partial allocation emphasize that interperiod tax allocation should be applied only to timing differences that can be reasonably expected to reverse in the future. For example, partial allocation advocates believe that differences arising from accelerated depreciation for tax purposes should not be subject to interperiod tax allocation, that is, no liability should be established for the difference in accelerated and normal depreciation deductions. They believe that if based on budgets, forecasts, and other estimates and judgments, the cumulative amount of the difference is expected to increase, interperiod tax allocation creates a liability which will never become payable.

Proponents of comprehensive allocation believe that by substantially eliminating the need for judgment, that is, by including all timing differences, integrity and consistency of results is furthered, and the method provides a more thorough and consistent matching of revenues and expenses.

In practice, there is little support for the taxes payable method described above. The IASC has stated in IAS No. 12 that comprehensive interperiod tax allocation should normally be applied, but does permit use of the partial allocation approach if timing differences will not reverse in at least the next three years and there is no indication that they will reverse after that time. In the US, Australia and Canada, comprehensive interperiod tax allocation is currently required. In the UK, the partial allocation approach is used and is similar in application to that permitted by IAS No. 12.

Tax Credits

Many governments allow various credits for certain items in determining taxes payable for a period. These credits result chiefly from the government's aim to achieve its policy objectives and differ from tax deductions in the sense that they reduce taxes payable rather than taxable income. Common credits in the US include the investment tax credit designed to promote business expansion and capital spending, foreign tax credits designed to promote foreign business expansion and avoid excess taxation on foreign earnings, and various specialty credits related to research and development, energy conservation and increased employment spending.

In the US, such credits are generally accounted for as reductions of tax expense for the period in which they arise - the flow-through method. In accounting for the investment tax credit, however, there is some support that the credit relates to the qualified property and that it should be deferred in the period it arises and subsequently amortized to income over the depreciable life of the property - the deferral method. Although both methods have support and are used in practice, the flow-through method is most widely utilized in the US. In Canada, a recent CICA pronouncement requires the use of the deferral method for such credits.

Income Tax Disclosures

IAS No. 12 requires the following disclosures:

- tax expense related to income from the ordinary activities of the enterprise;

- tax expense relating to unusual items, to prior period items, and to changes in accounting policy;

- tax effects, if any, related to assets that have been revalued to amounts in excess of historical cost or previous revaluation; and

- the relationship between tax expense and accounting income if not explained by the tax rates effective in the country of the reporting enterprise.

Income tax disclosure requirements in local jurisdictions are generally more comprehensive than that set forth above and normally would include:

- estimated taxes actually payable for the period, in total and the amount payable as of the end of the period;

- details regarding timing differences, including the effect on taxes payable in the current year and amounts deferred and included as assets and liabilities in the balance sheet;

- details regarding the effect of tax credits, including the method used to account for such credits; and

- unused deductions, tax credits, and operating loss carryforwards not currently utilized and available to reduce taxable income in future periods, along with expiration dates.

Disclosure requirements for contingencies also generally require disclosure of significant tax examinations in progress and the probable effect on an entity's financial position as a result of such examinations.

Current Developments

Currently, the main controversy in accounting for income taxes is the partial versus comprehensive tax allocation argument. There is a body of evidence that, in those countries where comprehensive tax allocation is used, the deferred tax amount on the balance sheet has continued to increase, with no amount ever becoming payable.

As a result of this build-up of deferred taxes on the balance sheet, the prescribed use of comprehensive tax allocation is under review in both Canada and the US. This issue is pertinent for mining companies and, in analyzing a mining company's financial statements, it is worthwhile to determine to what extent the total income tax expense relates to deferred tax and whether or not the deferred tax account on the balance sheet has continued to increase historically. This type of analysis might provide an indication of the impact of using the comprehensive allocation approach. As indicated above, disclosure in the financial statements generally includes details of significant timing differences. These differences should also be reviewed to determine if they are short or long term, recurring or non-recurring. Such review can assist in assessing the future income tax position of an enterprise as well as assisting in the understanding of how income tax expense has been calculated.

Summary

This section has been restricted to a brief discussion of approaches to income tax accounting and current practice in that regard. The taxation of companies in the extractive industries generally tends to be complex in most countries and this in turn complicates the related accounting. Accordingly, it is not unusual for the reported income tax expense of mining companies to bear little relationship to pre-tax income. As indicated, a reconciliation of income tax expense based on normal tax rates to actual income tax expense is often provided by way of disclosure. Such are the complexities of fiscal systems, however, that an understanding of these disclosures may defeat all but those intimately familiar with the subject.

OTHER IMPORTANT AREAS OF PRESENTATION AND DISCLOSURE

This section discusses important areas of presentation and disclosure not dealt with elsewhere in this chapter. These areas of disclosure are necessary for the reader to have a proper understanding of the basis upon which the financial statements of an enterprise are prepared and presented and an understanding of the impact of specific events and transactions on the results of operations and financial position. It is organized by topic as follows:

- Accounting Policy Disclosures
- Changes in Accounting Policies
- Contingencies and Commitments
- Subsequent Events
- Related Party Transactions
- Financial Information in Public Offering Documents

Accounting Policy Disclosures

The accounting policies of a reporting entity are the specific accounting principles and the methods of applying those principles that are judged by management to be most appropriate to present fairly the entity's financial position and operations. Information about the accounting policies adopted by a company is essential for financial statement users to gain the fullest understanding of the principles underlying the preparation of and presentation of financial statements. Depending on management philosophy, policies may be conservative, aggressive or somewhere in between. If, for example, there are two otherwise comparable mining companies, one of which expenses and one of which capitalizes all exploration costs, then clearly an understanding of the impact of the policies of each is necessary to compare their relative performance and potential. A form of sensitivity analysis is necessary to determine what the impact would be if different policies were to be applied. This exercise is particularly important in mining companies where there are many diverse practices in the absence of authoritative pronouncements.

Given the importance of understanding an entity's accounting policies, there is generally a mandatory reporting requirement. CICA Handbook Section 1505 requires the following:

"As a minimum, disclosure of information on accounting policies should be provided in the following situations:

a) when a selection has been made from alternative acceptable accounting principles and methods;

b) when there are accounting principles and methods used which are peculiar to an industry in which an enterprise operates, even if such accounting principles and methods are predominantly followed in that industry."

The Canadian requirements are similar to those of IAS No. 1, Disclosure of Accounting Policies, and those of the US, the UK and Australia.

While the disclosure of a company's accounting policies is generally prescribed, there is some flexibility in the method of presentation in the financial statements. The most common presentation is as a separate section immediately before or after the financial statements or as the first note to the financial statements. As a general rule, the presentation of accounting policies should be consistent with their being an integral part of the financial statements.

Canada's Inco provides a good example of accounting policy disclosures common to the mining industry. The notes to Inco's financial statements are headed "Explanatory Financial Section", and the summary of significant accounting policies is included as the first note. The first paragraph of the summary is interesting in that it contains a statement of the objective of the summary:

"Summary of Significant Accounting Policies.

This summary of the major accounting policies of Inco is presented to assist the reader in evaluating the financial statements ... These policies apply to the continuing operations of the company and have been followed consistently in all material respects for the periods covered in the financial statements, except as described ..."

Note that the going concern and consistency concepts are referred to explicitly. As discussed in the section on financial statement elements and concepts, these concepts are fundamental to the preparation of financial statements and as they are assumed, explicit reference is only necessary when there is a departure therefrom. In Inco's case, there was an accounting change during the year so there has been a departure from consistent application of the company's accounting policies. Changes in accounting policies are discussed further below.

Following the introductory paragraph there are a number of policies described:

- Principles of consolidation
- Translation of financial statements into US dollars
- Inventories
- Property, plant and equipment
- Depreciation and depletion
- Exploration
- Income and mining taxes
- Pension plans
- Net loss per common share

Two aspects of the foreign currency translation policy merit comment. Firstly, notwithstanding that Inco is a Canadian company, it reports in US amounts. This practice is unusual but quite acceptable. Secondly, the company provides a detailed explanation of the method of translation of financial statements into US dollars.

Although the company is merely following the Canadian requirements with regard to the translation of financial statements of foreign-based entities, the disclosure provides the reader with an explanation of a complex area of accounting.

Versions of the Inco policies relating to the valuation of inventories, property, plant and equipment, depreciation and depletion, exploration and income and mining taxes generally will be encountered in any mining company financial statements because of the policy that there are alternatives available. A policy dealing with the treatment of pension costs will apply only to those companies which sponsor pension plans for the benefit of their employees.

One aspect of policy disclosure, typically seen in UK and Australian companies, is the preparation of financial statements under the historical cost convention, except for the revaluation of certain assets. In these countries, as discussed in the financing transactions section of this chapter, asset revaluations are permitted and this fact should be stated in the policy notes.

Changes in Accounting Policies

Perhaps surprisingly, one of the more frequently occurring items seen in mining company financial statements is a change in accounting policy. There is a presumption that accounting policies do not change from period to period and that when a change is made, it is only made to adopt a preferable practice. The number of changes noted is perhaps a function of the continuing evolution of acceptable mining accounting practices in the absence of authoritative pronouncements. Alternatively, some would argue that the number of changes is a function of greater pressures on management to achieve good results in poor economic conditions. Whatever the reasons, there is no doubt that frequent changes in accounting policies by an enterprise, other than to comply with a new accounting standard, ultimately can be viewed negatively by users of financial statements.

A change in accounting policy is the result of a choice between two or more accounting bases, for example, the writing off of exploration expenditures as incurred rather than carrying forward such expenditures on the balance sheet. An accounting change does not arise from the adoption or modification of a basis of accounting necessitated by transactions or events that are clearly different in substance from those previously occurring.

A change in an accounting policy can be introduced into the financial statements in different ways. The new policy may be applied:

- To the current and future financial statements. In other words, the change is applied prospectively.

- By restating the financial statements for all periods presented. This is known as retroactive application of the change as if the new policy had always been in use and results in changes to previously reported amounts.

- By presenting as a single item in the income statement for the current period the amount of the cumulative effect on retained earnings at the beginning of the period in which the change is made. This is sometimes known as the cumulative effect presentation.

Local standards vary in approach to this issue. In Canada and the UK, retroactive application is generally required. In the US, the cumulative effect presentation is used except for two specific situations, one of which is a change to or from the full cost method of accounting for oil and gas enterprises. Australia also requires the use of the cumulative effect method unless otherwise stipulated.

Given these divergent practices and the apparent frequency in occurrence of accounting changes, comparison of financial statements is made that much more difficult. Required disclosure, however, will usually provide information on the impact of the change on reported income for the period in which the change is made. In addition, cumulative effect changes are usually presented as separate line items in the income statement so that income is reported before and after the effect of the change.

The Hudson Bay Mining & Smelting Co. provides an example of the disclosure of a retroactive application of a change in accounting policy:

"Change in Accounting Policy

Effective January 1, 1983, HBMS adopted the policy of capitalizing interest costs incurred during the construction of certain assets as part of the historical cost of such assets in accordance with FAS No. 34. Previously, such interest costs were charged to earnings as incurred. This change was made retroactively and reduced previously reported losses in the years ended December 31, 1982, and 1981 by $6,222,000 and $2,576,000, respectively. This change resulted in cumulative increases in retained earnings of $9,627,000 and $3,405,000 at December 31, 1982 and 1981 respectively."

Echo Bay's 1983 financial statements also contains note disclosure of an accounting policy change. In this case, Echo Bay changed its depreciation policy of fixed assets from the unit of production method to the straight-line method on the basis that more appropriate matching of costs with revenues results and to conform with industry practice. The change was reflected only in the current period because of the immaterial effect of the change on the 1982 results.

A distinction should be made between a change in an accounting policy, the accounting for which is described above and a change in an accounting estimate, which is accounted for in the period of change. The preparation of financial statements involves making estimates which are based on the circumstances existing at the time when the financial statements are prepared. For example, estimates are required on the useful lives of depreciable assets. To the extent that the actual amount, or revised estimate is different from the original estimate, the difference is not considered to be an error as it usually results from new information becoming available and would be accounted for in the period of the change. It is often difficult to distinguish between a change in an accounting policy and a change in an accounting estimate. For example, an enterprise may change from deferring and amortizing a cost to reporting it as an expense when incurred because the estimated future benefits have become uncertain. In cases where it is difficult to draw a clear distinction, such changes are generally treated as changes in accounting estimates, with appropriate disclosure.

Contingencies and Commitments

Contingencies   "A contingency is an existing condition, situation or set of circumstances involving uncertainty as to possible gain or loss to an enterprise that will ultimately be resolved when one or more future events occur or fail to occur" (FAS No. 5). This definition closely resembles that contained in IAS No. 10, Contingencies and Events Occurring After the Balance Sheet Date, and is also similar to definitions used in Canadian and UK standards.

The general rule established by these standards is that an estimated loss for a contingency should be recorded against income if it is probable that an asset has been impaired or a liability has been incurred at the date of the financial statements and the amount of the loss can be reasonably estimated. Contingent losses which do not meet both these conditions should be disclosed in the notes to financial statements if there is a reasonable possibility, or it is likely that a loss of an indeterminable amount may have arisen. Contingent gains can be disclosed but are never recorded as income as, to do so, would be contrary to the conservatism concept that income should not be anticipated but recorded only when realized. Disclosure of a contingency should include:

- the nature of the contingency;
- the factors that are uncertain; and
- an estimate of the financial effect of the contingency or a statement that such an estimate cannot be made.

An example of a contingency disclosure is provided below relating to Energy Resources of Australia. Note that this disclosure indicates that amounts have already been recorded in respect of probable losses as well as providing an estimate of the maximum additional loss. In addition, the disposition of the additional loss, if it occurs, is noted.

"Contingency

Claims have been lodged against the company and others by contractors in connection with the construction of the Mine Plant, Equipment and Facilities. To the extent that it is expected that the company may eventually be required to meet these claims, they have been included in these accounts. The maximum additional contingent liability at 30 June 1983 was $2,800,000 (1982 $4,200,000). Any difference between the amount in the accounts and the amount eventually paid will be reflected in fixed assets."

Commitments   Any contractual obligations that are significant in relation to the current financial position or future operations of an entity should be disclosed. These include significant obligations of the following types:

- commitments that involve a high degree of speculative risk, where the taking of such risks is not inherent in the nature of the business;

- commitments to make expenditures that are abnormal in relation to the financial position or usual business operations, for example, commitments for substantial exploration expenditures;

- commitments to issue shares; and

- commitments that will govern the level of a certain type of expenditure for a considerable period into the future.

Typical commitments arising in the mining industry which should be disclosed are:

- required participation in joint ventures;
- forward contracts entered into;
- minimum lease payments in respect of non-cancellable operating leases;
- minimum royalty payments under non-cancellable royalty lease agreements; and
- gold and silver production committed to be sold at fixed prices.

Subsequent Events

Events which arise subsequent to the balance sheet date but prior to issuance of financial statements should be either recorded in the financial statements or disclosed in the notes, depending on the nature of the event. Standards are comparatively uniform in this area and all require the subsequent event to be recorded in the financial statements when it provides additional evidence of a condition or conditions existing at the balance sheet date. Otherwise, the occurrence of the event requires disclosure in the notes only. If, for example, subsequent to year end an entity decides to shut down a mine because the mine is no longer commercially viable, should the resultant loss be recorded at the balance sheet date? If, at the balance sheet date, there was uncertainty as to the future operation of the mine because of marginal operations and that management's decision to shut the mine down confirmed or clarified that uncertainty, then the resultant loss should be recorded. On the other hand, if the decision to shut down the mine can be clearly attributed to a significant fall in metal prices since the balance sheet date, transforming the mine from a profitable to a loss operation, then the resultant loss can be attributed to economic events arising after the balance sheet date and therefore can be recorded in that period. There is a fine distinction between events that should be recorded as an adjustment and those that should be disclosed by way of a note. Ultimately, a degree of judgment is involved when deciding upon the appropriate treatment. Events not required to be recorded but which would require disclosure in the notes include an issue of debt or business combination subsequent to the balance sheet date. The disclosure should describe the impact of the event on the financial position and future operations of the entity.

Related Party Transactions

Related party transactions are those transactions which take place between parties who, because of their relationship, are not independent of one another. These transactions may be entered into on terms differing from those that might have prevailed if the parties had been independent of each other. Transactions between related parties may have a significant effect on the financial position and results of operations of an enterprise.

Standards in this area are found only in Canada and the US. The Canadian rules state that parties are considered to be related when one party has the ability to exercise, directly or indirectly, control or significant influence over the operating and financial decisions of the other. This definition is similar to that in the US. Examples of transactions between related parties include transactions between parent company and subsidiaries, subsidiaries of a common parent, an enterprise and its principal owners, management or members of their immediate families. Many related party transactions occur in the normal course of business, including sales, purchases and transfers of real or personal property; services received or provided, for example, management services; borrowing, lending and guarantees; inter-company billings based on allocations of common costs; and the use of property and equipment by lease or otherwise. The fact that a transaction may have the appearance of being in the normal course of business does not preclude its disclosure. The Canadian and US accounting standards do not attempt to deal with the problems of measurement and accounting treatment of related party transactions. They require only that certain information about such transactions is disclosed in the notes to the financial statements. This information should include the nature of the relationship(s) involved; a description of the nature and the extent of transactions; amounts due from or to related parties at the balance sheet date; and, if not otherwise apparent, the terms of settlement.

Disclosure of related party transactions should be useful to an overall understanding of how a company conducts its activities. For example, Canada's Campbell Resources provides information on its management of certain mines owned by associated companies. This is perhaps a useful indication of the extent to which the company is in a position to significantly influence the activities of these associates. Information is also provided on the amounts recognized as income, the method by which the amounts are determined and the amounts outstanding at the end of the period.

Financial Statements in Public Offering Documents

The discussion so far has dealt primarily with information in annual financial statements. It is also appropriate to discuss briefly the financial information generally included in offering documents.

As part of the process of issuing a prospectus to raise capital, an enterprise typically must provide information about its financial position and results of operations to potential investors over and above that which it might provide to the normal users of its financial statements. In the interests of protecting the potential investor and to enable him to make the most informed decision possible about the risk involved, an additional burden of disclosure is placed on the issuer of the prospectus.

In North America, regulatory authorities concerned with the public offering of securities generally require the offering document to include audited financial statements which present a two year comparative balance sheet along with statements of income, retained earnings and changes in financial position for up to five years. This information serves to give the investor a historical earnings and funds flow trend. In addition, if the equity or debt offering is launched several months after the date of the most recent audited financial statements presented, interim financial information will be required to be included in the offering document. This information will usually be unaudited but it should be prepared using the same accounting policies as used in the audited financial statements. It is designed to give investors a current view of the financial position of the enterprise.

It may also be appropriate to provide pro forma financial statements in the prospectus. Pro forma financial statements are the historical financial statements adjusted to show the effect of a transaction or proposed transaction as if it had occurred previously. For example, if an enterprise intends to use the proceeds of an equity issue to retire existing bank indebtedness, or to acquire another enterprise, the pro forma financial statements would present the historical financial statements as though the proposed retirement or acquisition had taken place. This may assist the investor in understanding the nature and effect of the proposed use of the equity funds, and also give an indication of how the financial statements would look after the issue.

Finally, the presentation of future-oriented financial information, such as a forecast, in offering documents is increasingly recognized as both appropriate and relevant to investors. The presentation of such information will involve assumptions about certain future events, earnings and interest rates and normally will be required to be prepared in accordance with the accounting methods and principles followed in preparation of the historical financial statements. In order that investors do not place a higher degree of reliance on forecasts or other future financial information than is warranted, particular care is required in the preparation and disclosure of such information.

Summary

The focus in this section was directed towards important areas of presentation and disclosure for mining companies which have not been dealt with elsewhere in this chapter. In terms of overall pronouncements relating to presentation and disclosure in different countries, there is no doubt that the US rules and requirements are more comprehensive. While this results in less flexibility in presentation and disclosure in the US, particularly for SEC registrants, it does not mean that there is necessarily more detail in the financial statements of a typical US enterprise. On the contrary, financial statements of UK and Australian enterprises generally provide more detail regarding financial statement amounts. Some of this detail is included to comply with statutory requirements rather than pronouncements of the standard-setting bodies.

All of the areas discussed will be encountered frequently in mining companies. In addition, the nature and extent of activities by some mining companies will provide interesting presentation and disclosure issues for years to come.

APPENDIX I

COMPARISON OF INTERNATIONAL ACCOUNTING STANDARDS WITH ACCOUNTING RULES AND PRINCIPLES APPLICABLE IN CERTAIN MAJOR AREAS

The following comparison has been prepared in order to identify controversial issues between International Accounting Standards and local rules and principles in the stated countries. It has been restricted to situations, likely to have general application to mining companies, in which financial statements prepared in accordance with local rules and principles conflict with certain provisions of International Accounting Standards. International Accounting Standard No. 19, Accounting for Retirement Benefits in the Financial Statements of Employers, has general application to mining companies but is not included in this comparison because local rules and principles are under review in several of the stated countries. The following is a summary only and the full text of the original pronouncement should be referred to for a more thorough understanding of a standard.

References to Exposure Drafts are made in the absence of authoritative pronouncements. Exposure drafts precede the issuance of authoritative pronouncements and are subject to change in the approval process.

ACCOUNTING FOR MINERALS COMPANIES

International Accounting Standard		Accounting Principles Generally Accepted in the United States	Standard Accounting Practice in the United Kingdom	Accounting Standards in Australia	Accounting Principles Generally Accepted in Canada
Subject Matter	Controversial Issues				
1 Disclosure of Accounting Policies	IAS No. 1 requires presentation of comparative financial statements.	Presentation of comparative financial statements required for public companies (SEC regulations) and not required for private companies.	Presentation of comparative financial statements required.	Presentation of comparative financial statements required.	Presentation of comparative financial statements required.
2 Valuation and presentation of inventories in the context of the historical cost system	IAS No. 2 requires valuation of inventories at the lower of cost and net realizable value.	Require valuation of inventories at the lower of cost or current replacement cost provided current replacement cost is not in excess of net realizable value or below net realizable value reduced by the approximate normal profit margin. (In addition, U.S. GAAP permit valuation of certain inventories above cost.)	Valuation at the lower of cost and net realizable value required.	Valuation at the lower of cost and net realizable value required.	Valuation at the lower of cost and market is most common practice. (Significance of differences between IAS No. 2 and Canadian GAAP is presently under study by the Canadian Accounting Standards Committee.)
3 Consolidated Financial Statements	IAS No. 3 permits consolidation in the following cases:				
	. "holding company" owns majority of equity capital but less than half of voting stock.	. Consolidation not permitted.	. Consolidation not permitted.	. Consolidation permitted.	. Consolidation not permitted.
	. "holding company" has the power to control by statute or agreement with or without more than half of the equity interest.	. Consolidation permitted without ownership of more than half of the equity interest.	. Consolidation required if control is exercised by the power to nominate a majority of the board of directors.	. Consolidation required if control can be exercised. Control is a question of fact not prescribed through legal rules.	. Consolidation not permitted.
	IAS No. 3 requires disclosure of proportion of assets and liabilities to which different accounting principles have been applied, if they are included in a single balance sheet classification.	Disclosure of proportion of assets and liabilities to which different accounting principles have been applied, if they are included in a single balance sheet classification, not required.	Disclosure of proportion of assets and liabilities to which different accounting principles have been applied, if they are included in a single balance sheet classification, required.	Disclosure of proportion of assets and liabilities to which different accounting principles have been applied, if they are included in a single balance sheet classification, not required.	Disclosure of proportion of assets and liabilities to which different accounting principles have been applied, if they are included in a single balance sheet classification, not required.
	IAS No. 3 requires equity accounting for certain investment accounts.	Equity accounting required.	Equity accounting required.	Equity accounting required in supplementary financial information prepared on an equity basis.	Equity accounting required.

International Accounting Standard

Subject Matter	Controversial Issues	Accounting Principles Generally Accepted in the United States	Standard Accounting Practice in the United Kingdom	Accounting Standards in Australia	Accounting Principles Generally Accepted in Canada
4 Depreciation Accounting	IAS No. 4 requires disclosure of certain information including the useful lives or depreciation expense for the period and accumulated depreciation individually for each major class of depreciable asset.	Disclosure of such information not required individually for each major class of depreciable asset.	Disclosure required individually for each major class of depreciable asset. (In addition, disclosure is required of all movements during the year for individual classes of fixed assets.)	Disclosure of such information (except for estimated useful lives) required individually for each major class of depreciable asset.	Disclosure of such information not required individually for each major class of depreciable asset.
7 Statement of Changes in Financial Position	IAS No. 7 requires the presentation of a statement of changes in financial position.	Presentation of a statement of changes in financial position required.	Presentation of a statement of changes in financial position required.	Presentation of a statement of changes in financial position only required for companies with shares listed on an Australian stock exchange.	Presentation of a statement of changes in financial position required.
12 Accounting for Taxes on Income	IAS No. 12 provides that the full comprehensive method of tax allocation, i.e., deferred tax accounting, should normally be applied.	Comprehensive method of tax allocation, i.e., deferred tax accounting, required.	Provides that comprehensive method of tax allocation, i.e., deferred tax accounting, should normally be applied.	Comprehensive method of tax allocation, i.e., deferred tax accounting, required.	Comprehensive method of tax allocation, i.e., deferred tax accounting, required.
	IAS No. 12 permits that the partial allocation approach of deferred tax accounting may be applied when there is reasonable evidence that: . timing differences will not reverse for some considerable period (at least 3 years) and . there is no indication that the timing differences are likely to reverse after that period.	Partial allocation approach of deferred tax accounting not permitted.	Requires adoption of the partial allocation approach to deferred tax accounting, if criteria for the application of the partial allocation approach (same as under IAS No. 12) are met.	Partial allocation approach of deferred tax accounting not permitted.	Partial allocation approach of deferred tax accounting not permitted.
	IAS No. 12 provides that either the deferral or the liability method be used for the calculation of deferred taxes.	Deferral method required. The liability method is not acceptable.	Liability method presumed but by implication deferral method also allowed in limited situations.	Liability method required. Deferral method is not acceptable.	Deferral method required. The liability method is not acceptable.

ACCOUNTING FOR MINERALS COMPANIES

International Accounting Standard Subject Matter	Controversial Issues	Accounting Principles Generally Accepted in the United States	Standard Accounting Practice in the United Kingdom	Accounting Standards in Australia	Accounting Principles Generally Accepted in Canada
15 Information Reflecting the Effects of Changing Prices	IAS No. 15 requires disclosure on a supplementary basis for economically significant entities, of adjustments for the effects of changing prices on depreciation of fixed assets, cost of sales and monetary items and the overall effect of the adjustments on the results for the period.	Requires disclosure for large public companies. Requires disclosure of both current cost and constant dollar information.	SSAP No. 16 should give full compliance.	No requirement at present, however, a provisional standard has been issued.	Requires disclosure for large public companies. Requires disclosure of both current cost and constant dollar information with emphasis on current cost.
16 Accounting for Property, Plant and Equipment	IAS No. 16 allows for property, plant and equipment to be carried at an amount in excess of cost.	Property, plant and equipment is not allowed to be carried at an amount in excess of cost.	Property, plant and equipment may be revalued to an amount in excess of cost.	Property, plant and equipment may be revalued to an amount in excess of cost.	Property, plant and equipment may be revalued to an amount in excess of cost. Should not, however, occur in ordinary circumstances.
17 Accounting for Leases	IAS No. 17 requires that when a significant part of the lessor's business comprises operating leases, the lessor should disclose the amount of assets by each major class of asset. For sale and leaseback transactions, IAS No. 17 requires that: · for operating leases any profit or loss on the difference between sales proceeds and carrying value be recognized in income immediately, except that (a) any excess of fair value over sales proceeds, if compensated by future rentals at below market rates, or (b) any excess of sales proceeds over fair value, should be deferred and amortized over the period of expected use.	Requires that the cost of property on operating lease or held for leasing be disclosed by major class of property. For sale and leaseback transactions, FASB No. 13 requires that: · for operating leases any profit or loss on the sale be deferred and amortized in proportion to rental payments over the period of time the asset is expected to be used, except that where the fair value of the property is less than its undepreciated cost, a loss should be recognized immediately.	Exposure draft requires that operating leases should be shown by each major class of asset. For sale and leaseback transactions, an outstanding exposure draft requires that: · for operating leases any profit or loss on the difference between sales proceeds and carrying value should be recognized in income immediately, except that (a) any excess of fair value over sales proceeds if compensated by future rentals at below market rates, or (b) any excess of sales proceeds over fair value, should be deferred and amortized over the lease term.	Disclosure of operating leases by each major class of asset required. For sale and leaseback transactions, requires that: · for operating leases any profit or loss on the difference between sales price and carrying value should be recognized immediately, except that any difference between fair value and selling price should be deferred and amortized over the lease term.	Disclosure of cost of property held for leasing required. No requirement to show amount by each major class of asset. For sale and leaseback transactions, requires that: · for operating leases any profit or loss on the difference between sales proceeds and carrying amount be deferred and amortized over the lease term, except that (a) any excess of carrying amount over fair value be recognized as a loss immediately.

International Accounting Standard					
Subject Matter	Controversial Issues	Accounting Principles Generally Accepted in the United States	Standard Accounting Practice in the United Kingdom	Accounting Standards in Australia	Accounting Principles Generally Accepted in Canada
	. for finance leases a gain on the excess of sales proceeds over carrying amount should not be recognized as income immediately, but if recognized, it should be deferred and amortized over the lease term.	. for finance leases any gain or loss on the sale be deferred and amortized in proportion to the amortization of the leased asset, unless the fair value of the property is less than its undepreciated cost, in which case the loss should be recognized immediately.	. for finance leases any gain or loss on the difference between sales proceeds and carrying amount be deferred and amortized over the term of the lease except that any excess of carrying amount over fair value be recognized as a loss immediately.	. for finance leases any gain or loss on the sale should be deferred and amortized over the term of the lease. The amount of profit or loss to be deferred and amortized is calculated as the difference between selling price and fair value.	. for finance leases any gain or loss on the difference between sales proceeds and carrying amount be deferred and amortized in proportion to amortization of the leased assets (land straight-line) except that any excess of carrying amount over fair value be recognized as a loss immediately.
18 Revenue Recognition	IAS No. 18 requires that revenue arising in the ordinary course of business from the sale of goods or the rendering of services should be recognized when the terms have been effectively completed and it is reasonable to expect ultimate collection. IAS No. 18 requires that revenue arising from interest, royalties and dividends should only be recognized when no significant uncertainty exists as to measurement or collection.	U.S. principles for revenue recognition are based on ARB No. 43, APB Opinion 10 and FAS No. 48. These principles generally recognize revenue in a manner consistent with IAS No. 18.	Under the prudency concept, revenues are recognized only when realized in cash or in other assets, the ultimate cash realization of which can be assessed with reasonable certainty.	Matter not ruled upon.	Matter not ruled upon.
21 Accounting for the Effects of Changes in Foreign Exchange Rates					
(a) Translation of income statement items under the current rate method:	IAS No. 21 permits the average rate or the closing rate for the reporting period.	Average rate permitted in the absence of transaction date rate.	At an average rate or at the closing rate for the reporting period.	Matter not ruled upon. Predominant practice is at average rate.	Average rate permitted in the absence of transaction date rate.
(b) Treatment of exchange gains and losses on long-term monetary items:	Under IAS No. 21, such items are normally included in income statement of period but may be deferred and amortized to income over the period of the item.	Inclusion in income statement of period required, except for those gains and losses relating to certain hedging and intercompany transactions.	Generally included in income statement of period.	Matter not ruled upon. Predominant practice is to include in income statement of period.	Deferred and amortized to income over the period of the item.
(c) Treatment of non-monetary assets in hyper-inflationary economies:	Under IAS No. 21, preferable to adjust financial statements for the effects of changing prices before translation, but use of historical rates to translate such items is permitted.	Use of reporting currency as if it were the functional currency results in use of historical rates to "remeasure" such items.	Financial statements should be adjusted where possible to reflect current price levels before translation.	Matter not ruled upon.	Reporting currency considered appropriate unit of measure resulting in use of historical rates to translate such items.

ACCOUNTING FOR MINERALS COMPANIES

International Accounting Standard		Accounting Principles Generally Accepted in the United States	Standard Accounting Practice in the United Kingdom	Accounting Standards in Australia	Accounting Principles Generally Accepted in Canada
Subject Matter	Controversial Issues				
(d) Disclosure:	IAS No. 21 requires disclosure of the net exchange gain or loss in the income statement.	Requires disclosure of the net exchange gain or loss in the income statement.	Does not require disclosure of the net exchange gain or loss in the income statement.	Matter not ruled upon.	Requires disclosure of the net exchange gain or loss in the income statement.
	IAS No. 21 requires that the movement in reserves arising from translation adjustments be disclosed separately but does not have to be broken down.	Requires detailed explanation of the movements in the separate equity account.	Movement in reserves arising from translation adjustments must be disclosed separately aside to, or withdrawn from, reserves.	Companies code requires disclosure of any amount set aside to, or withdrawn from, reserves.	Requires detailed explanation of the movements in the separate equity account.
	IAS No. 21 requires disclosure of the method used for translating the financial statements of foreign operations and the procedure selected (i.e., closing or average rates) for translating the income statements of foreign entities.	Disclosure is encouraged.	Requires disclosure of accounting policies adopted. Non-exempt companies and groups should disclose the net exchange gains and losses on foreign currency borrowings identifying separately the amounts which have been offset in reserves and credited or charged to income.	Disclosure effectively required if the policy is significant (note that as there are no prescribed rules, alternatives are available).	
22 Accounting for Business Combinations					
(a) Use of the Pooling of Interests Method ("Pooling Method")	IAS No. 22 reserves the pooling method for a "uniting of interests" the prerequisites of which are: The shareholders of the combining enterprises achieve a continuing mutual sharing in the risks and benefits attaching to the combined enterprise, and (a) The basis of the transaction is principally an exchange of voting, common shares. (b) Effectively the whole of the net assets and operations of the combining enterprises are combined in one entity.	Pooling method only used for business combinations where ownership interests of two or more companies are united by an exchange of voting securities. Specific tests must also be met for the pooling method to be used.	Under Exposure Draft No. 31, specific conditions for the pooling method to be used must be met as follows: (a) Offer made to holders of both equity and voting shares of the offeree and has been approved by offeror shareholders. (b) Offer accepted by at least 90% of the holders of both equity and voting shares. (c) Consideration given for equity capital to be substantially in the form of equity capital and consideration given for voting non-equity capital to be substantially in the form of equity and/or voting non-equity capital.	Matter not ruled upon.	Pooling method reserved for business combination in which it is not possible to identify one of the parties as an acquirer.

219

International Accounting Standard		Accounting Principles Generally Accepted in the United States	Standard Accounting Practice in the United Kingdom	Accounting Standards in Australia	Accounting Principles Generally Accepted in Canada
Subject Matter	Controversial Issues				
(b) Application of the Purchase Method	IAS No. 22 requires that positive goodwill (i.e., cost exceeds fair value of assets acquired) be recognized in income on a systematic basis over its useful life or adjusted against equity.	Positive goodwill required to be recognized in income over useful life (maximum of 40 years). Adjustment to equity not permitted.	Under proposed SSAP No. 22, positive goodwill to be recognized in income on a systematic basis over useful life or adjusted against equity.	Under AAS No. 18, effective for periods ending on or after March 31, 1985, positive goodwill required to be recognized in income over useful life (maximum of 20 years). Adjustment to equity not permitted.	Positive goodwill required to be recognized in income over useful life (maximum of 40 years). Adjustment to equity not permitted.
	IAS No. 22 requires that negative goodwill (i.e., cost is less than fair value of assets acquired) be recognized in income on a systematic basis or allocated proportionately to depreciable non-monetary assets acquired.	Negative goodwill required to be allocated proportionately to non-current assets (except long-term investments in marketable securities). Excess remaining after reducing such assets to zero classified as deferred credit and amortized systematically to income.	Under proposed SSAP No. 22, negative goodwill to be credited directly to equity.	AAS No. 18 requires that negative goodwill be allocated proportionately to non-monetary assets. Excess remaining after reducing such assets to zero to be recognized in income.	Negative goodwill required to be allocated proportionately to non-monetary assets.
(c) Calculation of Minority Interest	IAS No. 22 prefers that minority interest be based on the fair values of assets acquired at the acquisition date (the "entity concept"). Alternatively it may be based on the carrying values of the assets and liabilities of the acquired subsidiary (the "parent company concept").	Minority interest can be calculated using either the entity or the parent company concept. Predominant practice is to use the parent company concept.	Minority interest required to be calculated using the entity concept.	Matter not ruled upon.	Minority interest is required to be calculated using the parent company concept. The use of the entity concept is not permitted.
(d) Realization of Benefits from Previously Un-Recognized Tax Loss Carry-Forwards	IAS No. 22 requires that the realization of such benefits be recognized in income unless goodwill arising on acquisition was adjusted against equity in which case it be adjusted against equity.	The realization of such benefits required to be adjusted against goodwill.	The realization of such benefits required to be treated as part of normal tax provision.	The realization of such benefits required to be adjusted against goodwill.	The realization of such benefits required to be treated as an extraordinary item of income when realized.
	IAS No. 22 requires that the carrying value of goodwill be reassessed to identify any content attributable to the benefits received which should be charged to income.				
23 Capitalization of Borrowing Costs	IAS No. 23 requires that a policy be established of either capitalizing or not capitalizing borrowing costs on assets that take a substantial time to be made ready for use or sale.	Capitalization of borrowing costs required if its effect, compared with the effect of expensing such costs, is material.	Matter not ruled upon.	Matter not ruled upon.	Matter not ruled upon.

APPENDIX II

FOREIGN CURRENCY TRANSLATION
-- A DISCUSSION OF METHODS AND APPROACHES

Accounting for Transactions Denominated in Foreign Currencies

General   During an accounting period, a company may enter into transactions which are denominated in foreign currencies, in other words, currencies other than the currency in which the company's accounts are maintained. The results of these transactions need to be translated into the currency in which the company maintains its accounts so that they may be reported in the financial statements, on a basis comparable with other transactions of the company.

Accounting Methods   Two approaches can be identified in accounting for transactions denominated in foreign currencies -- the "time of settlement" method and the "time of transaction" method. The "time of settlement" method is based on the premise that a transaction and its settlement are a single event so that any exchange difference arising from fluctuations in exchange rates that occur before settlement is accounted for as part of the transaction and not as a separate exchange gain or loss. The "time of transaction" method is based on the premise that a transaction and its settlement are separate events. The date of the sale or purchase establishes the exchange rate to be used in translation and, unless hedged, for example by a specific forward exchange contract, any fluctuation in the exchange rate between the date of the transaction and its settlement gives rise to a foreign exchange gain or loss which is accounted for separately. The "time of transaction" method generally represents standard accounting practice for foreign currency translation, although the UK, for example, also permits translation at the rate of exchange at which a transaction is contracted to be settled in the future.

Translation and Recording of Balances   For the translation and recording of transactions when the "time of transaction" method is used, transactions denominated in foreign currencies are translated and recorded in the company's accounting records at the date the transactions occurred using the exchange rates prevailing at that date. Subsequently, no adjustments are normally required to the amounts in local currency initially recorded for non-monetary assets, liabilities and income statement items resulting from foreign currency transactions. Non-monetary assets and liabilities are those that do not represent money or claims to money for example, plant, machinery and inventory. Once these items have been translated and recorded, they represent balances denominated in local currency. The amount of local currency established at the initial translation represents the cost of acquisition, unless the non-monetary assets are carried at market value rather than cost, in which case these items will be considered monetary in nature and an adjustment will be required to translate them at current rates. For monetary assets and liabilities subsequent adjustments may be required to the amounts in local currency initially recorded as a result of foreign currency transactions. Monetary assets and liabilities are those that represent money or claims to money such as cash, receivables and payables.

Translation of Unsettled Balances Resulting From Transactions   When transactions remain unsettled at a subsequent balance sheet date, the related outstanding receivables or payables still represent assets or liabilities denominated in foreign currencies which should be retranslated in order to provide the best available estimate of the amount that will ultimately be paid or received. If the retranslation results in local currency amounts differing from those originally recorded, adjustments need to be made in the accounting records. In most countries these receivables and payables are retranslated at exchange rates prevailing at the balance sheet date.

The Treatment of Exchange Gains and Losses on Transactions   Changes in exchange rates between the local currency and the foreign currencies in which transactions are denominated increase or decrease the amounts of local currency that will be received or paid upon settlement of the transaction. These increases or decreases represent exchange gains or losses from foreign currency transactions. When the initial recording of the transaction and settlement are within the same accounting period, the treatment of exchange gains and losses is relatively straightforward. These gains and losses are realized and therefore are generally recognized in the income statement. On the other hand, opinion is divided on the accounting treatment of unrealized exchange gains or losses arising from the translation of unsettled payables and receivables denominated in foreign currencies at a subsequent balance sheet date. Current practice around the world varies significantly with respect to the treatment of unrealized gains, with a greater degree of harmonization as to the treatment of unrealized losses.

The recognition in the income statement of exchange losses on unsettled transactions is either required or generally accepted practice in most countries. The treatment of exchange gains on unsettled transactions can differ between countries. In all countries where the above monetary/non-monetary distinction is made, the practice of recognizing gains on monetary items in the income statement tends to be the predominant practice. In some countries all gains are deferred on the grounds of prudence. A variation of this deferral approach is to recognize losses in the income statement after deducting previously deferred gains. In countries where the predominant practice is not to retranslate, unsettled transactions remain at historical exchange rates and therefore no gains are recognized until they are realized.

Accounting standards in the US, the UK and Canada
require that gains and losses on the translation
of short-term monetary assets and liabilities are
recognized in income in the period incurred. In
Australia, the recognition of gains and losses in
income in the period incurred is the predominant
practice. Under IAS No 21, all gains and losses
are required to be recognized in income for the
year except for certain inter-company items and
certain hedges. For long-term monetary items
Canada requires, and the International Accounting
Standard permits, the deferral and amortization
of gains and losses relating to such items. This
is also a method sometimes used in Australia.
Amortization of the exchange difference takes
place over the remaining period of the long-term
asset or liability.

Summary In summary, the "time of transaction"
method represents standard accounting practice
and current exchange rates are generally used for
the translation of monetary assets and liabilities. Unrealized exchange losses, arising from
the translation of unsettled balances denominated
in foreign currencies, are generally recognized
in the income statement, although in some countries long-term losses are deferred and amortized. Exchange gains arising from the translation of unsettled foreign currency denominated
balances are either recognized in current income,
deferred until realized or deferred and amortized
over the lives of the related non-current items.

Translation of Financial Statements of Foreign-
Based Entities

General A large number of companies conduct
foreign operations through foreign-based subsidiaries, divisions and affiliates. Normally,
these foreign-based entities maintain their books
and prepare financial statements in their local
currency, in accordance with the accounting principles and statutory requirements applicable in
the country of their operations. When consolidating the financial statements of foreign-based
entities with the financial statements of the
parent company two issues arise. The financial
statements of the foreign-based entity may have
to be restated in order to comply with accounting
principles applicable for the consolidated financial statements and they will also have to be
translated from their local currency into the
currency in which the consolidated financial
statements are expressed.

Restatement of Foreign Currency Financial
Statement Prior to Translation The main purpose
of consolidated financial statements is to
reflect the financial position and the results of
operations of a parent company and its subsidiaries and equity investees as if they were a
single entity. These financial statements can be
meaningful to the user only if the same broad
accounting principles have been followed in the
financial statements of all entities included in
the consolidation. Therefore, foreign currency
financial statements need to be restated in conformity with the accounting principles applicable
or the preparation of the consolidated financial
statements prior to translation.

Approaches to Translation and Translation Methods

Two different approaches to translation (and a
number of methods) can be identified. The differences arise from fundamentally different concepts of consolidation. The result is the use of
either one exchange rate, the current rate, or a
combination of current and historical rates for
the translation of all assets and liabilities
included in foreign currency financial statements. The current rate (otherwise known as the
closing rate) is the exchange rate prevailing at
the balance sheet date. The historical rate is
the exchange rate in effect at the date a specific transaction or event occurred. Under the
current rate approach, the current rate method of
translation is used. The current rate method
translates all balance sheet items, except for
equity, at the exchange rate in effect at the
balance sheet date. Income statement items are
either translated at the closing rate or at an
average rate for the reporting period.

The current rate approach is based on the concept
of the foreign operation as a separate unit from
the domestic operation. This approach assumes
groups of companies are composed of entities
operating independently but contributing to a
central fund of resources, including the parent
stockholders' funds. It may also assume that
assets used in local operations are largely financed out of local borrowings or earnings. In
short, stockholders of the parent company are
concerned primarily with the net investment in a
foreign operation. The objectives of the current
rate approach are to preserve the financial
results and relationships as measured and expressed in the respective "local" currency of a
foreign-based operation and to measure the effect
of rate changes on the net assets of foreign
operations, as it is the net investment in the
foreign operation that represents the investors'
total exposure to a rate exchange.

The current/historical rate approach is based on
the concept that a parent company and its foreign
entities are a single business undertaking.
Assets owned by a foreign operation are viewed as
indistinguishable from assets owned by the parent
company and should, therefore, be stated in the
same way as similar assets of the parent company. The objectives of this approach are to
measure and express in the currency of the ultimate reporting entity the financial results and
financial position of foreign-based operations as
the currency of the ultimate reporting entity is
considered the appropriate unit of measurement,
and to achieve in the translation process changes
in the unit of measurement without affecting
either the measurement bases for assets and liabilities or the timing of revenue and expense
recognition.

Under the current/historical rate approach there
are three methods of translation followed. The
monetary/non-monetary method generally translates

monetary assets and liabilities at the rate in effect at the balance sheet date and non-monetary assets and liabilities at historical rates. Assets and liabilities are monetary if they represent money or claims to money. All other items are considered non-monetary. Income statement items are normally translated at an average rate for the reporting period, except for cost of goods sold and depreciation and amortization charges related to non-monetary balance sheet items which are translated at historical rates.

The temporal method translates cash, receivables and payables and other assets and liabilities carried at current prices at the rate in effect at the balance sheet date. Assets and liabilities carried at past prices are translated at historical rates. Income statement items are normally translated at an average rate for the reporting period, except for cost of goods sold and depreciation and amortization charges related to balance sheet items carried at past prices, which are translated at historical rates. The current/non-current method generally translates current assets and liabilities at the rate in effect at the balance sheet date and non-current assets and liabilities at historical rates. Income statement items are normally translated at historical rates.

The current rate method and the current/historical rate approach are irreconcilable approaches to the translation of foreign currency financial statements. Current practice is divided with certain countries using both approaches depending on the circumstances of the foreign operations, the current rate approach being used for independent foreign operations and the current/historical rate approach for closely linked foreign operations. The new accounting standards issued in the US, the UK and Canada and IAS No 21 all recommend the use of both approaches depending on the particular circumstances of the foreign entity. In practice, the current rate method is likely to be used more frequently.

These developments reflect the trend of the standard-setting bodies toward a position that is based on econcomic reality. The conceptual basis for these standards is the economic effect of foreign currency fluctuations on a company's net foreign-based investment. This follows the unfortunate experiences of the US and Canada who withdrew previous standards which imposed foreign currency accounting practices based more on theory than economic reality and which led to widespread criticism. Current practice now takes account of foreign operations whose activities can be considered "covered" against the risk of a loss arising due to a change in exchange rates.

<u>The Treatment of Translation Gains and Losses</u>
Translation gains and losses result from translating the financial statements of a foreign entity using exchange rates that differ from those used to translate the foreign based entity's financial statements at the end of the previous period. The translation approach used, in other words the current rate method vs. the current/historical approach, should be the deciding factor in determining the treatment of translation gains and losses. The current rate approach, generally used when the foreign-based operation is considered independent or self-sustaining, assumes that the performance of that operation is best demonstrated by the financial results as measured and expressed in the respective foreign currency. Under this approach, exchange differences may result from many factors unrelated to the trading performance or financing arrangements of the foreign-based operation. Such exchange differences should therefore be appropriately excluded from income and taken directly to stockholders' interests.

The current/historical rate approach, generally used when the foreign-based operation is considered a direct extension of the operations of the investing company, assumes that the performance of that operation is best demonstrated by the financial results as reflected in currency of the ultimate reporting entity. Under this approach, financial statements should be translated in a manner that achieves the same effect as if all transactions of the foreign-based operation had been entered into by the investing company itself. Since, in this case, exchange differences are regarded as having a direct effect on the income and cash flow from operations of the investing company, they should be included in the income statement of the investing company.

In Canada, the US, the UK and under IAS No. 21 the above conceptual approach to the treatment of translation gains and losses has been formally adopted. The treatment depends on the translation approach used, which itself depends on the circumstances of the foreign entity. In Australia, where the current rate method is the predominant practice, all gains and losses are either recognized in income or are taken directly to equity.

Summary

To summarize, the present practice of translating foreign currency financial statements, the restatement of foreign currency financial statements prior to translation in accordance with accounting principles underlying the consolidated financial statements, is either required or is the predominant practice. There are two different and irreconcilable approaches to translating financial statements. These are the current rate approach and the current/historical rate approach. Use of these approaches differs from country to country. Under the current rate approach, the current rate method is applied and under the current/historical rate approach, the methods applied can be monetary/non-monetary, temporal and current/non-current. Canada, the US, the UK and the IAS all require the temporal method when an entity's foreign operations are integral to those of the parent company. Again, these accounting standards have introduced consistency between the translation approach used and the treatment of translation gains and losses. With respect to the latter, all gains and losses are recorded in income when the current/historical approach is used and as a component of equity when the current rate method is used.

APPENDIX III

ABBREVIATIONS USED IN THIS CHAPTER

AAS	Australian Statements of Accounting Standards
APB	Accounting Principles Board (predecessor of FASB) -- US
ARB	Accounting Research Bulletin (predecessor to APB opinions) -- US
CICA	Canadian Institute of Chartered Accountants
FAS	Financial Accounting Standard
FASB	Financial Accounting Standards Board -- US
IAS	International Accounting Standard
SSAP	Statement of Standard Accounting Practice -- UK

REFERENCES

Beresford, D.R., et al. Accounting for Income Taxes. 1983.

CICA. Analytical Review in a Research Study. 1983.

CICA. Financial Reporting for Non-Producing Mining Companies - A Research Study. 1967.

Ernst & Whinney. A Synopsis of International Accounting Standards and Exposure Drafts. 1982.

Ernst & Whinney. Foreign Currency Translation -- A Comparison of International Practice. 1983.

Mahon, J.J. Mahon's Industry Guide, Guide No. 14. Coal. 1980.

Mahon, J.J. Mahon's Industry Guide, Guide No. 15. Non-Ferrous Metals. 1980.

Ernst & Whinney publications can be obtained through any Ernst & Whinney office.

CHAPTER 4: ACCOUNTING FOR MINERALS COMPANIES

BIBLIOGRAPHY

Anon., 1977, "FASB--Board Issues Memorandum on Extractive Industries," Journal of Accountancy, Vol. 143, No. 2, Feb., p. 7.

Coe, T.L., 1983, "Accounting Practices in the Coal Industry: Some Unanswered Questions," Journal of Accountancy, Vol. 155, June, pp. 62-78.

Coffee, C.D., 1983, "Impact of Inconsistent Accounting Practices in the Coal Extraction Industry," Journal of Extractive Industries Accounting, Vol. 2, Spring, pp. 127-141.

Davies, B.J., 1980, "Practical Problems Preclude Meaningful Disclosure," Chartered Accountant in Australia, Vol. 50, Apr., pp. 19, 20, 23, 25.

Ferris, K.R., and Graham, B.R., 1982, "Amortization of Capitalized Costs in the Extractive Industries," Australian Accountant, Vol. 52, Aug., pp. 460-461.

McCarl, H.N., et al., 1980, "Non-Metallic Minerals," Industry Guides for Accountants, J.J. Mahon, Boston, Guide 16.

Ng, A.C., 1984, "Accounting for Pre-Production Costs of the Extractive Industries," Australian Accountant, Vol. 54, No. 3, Apr., pp. 187-190.

Swieringa, R.J., 1981, "Silver-Lined Bonds of Sunshine Mining," Accounting Review, Vol. 56, Jan., pp. 166-176.

Taylor, S., 1984, "Extractive Industries: Time for New Approach," Chartered Accountant in Australia, Vol. 54, Feb., pp. 45-48.

Whittred, G.P., 1978, "Accounting for the Extractive Industries: Use or Abuse of the Matching Principle? " Abacus (Australia), Vol. 14, No. 2, Dec., pp. 154-159.

Chapter 5

Minerals Industry Taxation

CHAPTER 5:   MINERALS INDUSTRY TAXATION

PREFACE

In contrast to Chapter 7 on "Operating Agreements," which will focus largely on taxation in the so-called "developing" countries, the seven papers in Chapter 5 discuss the tax laws as they affect mining in "developed," or perhaps more correctly, "capital exporting" countries.

Three of the papers give comparative data on taxation as it impacts mining within country, state and provincial jurisdictions. Each is based on an economic model so that the results are comparable. The first of these papers by Robert Davidoff, "A Comparison of Local, State and Federal Taxes in Eight U.S. States," illustrates how the potential profitability of certain mineral deposits can be greatly affected by the tax structure existing in a particular state.

In an analogous paper concerning "Current Trends in Canadian Mining Taxation," Robert Parsons describes the comparative tax situation for the eight Canadian provinces. In addition to providing a background on tax legislation in that country, the paper presents an overview of the events leading to the development of the present taxation regime in Canada.

A comparative analysis of the impact of tax laws on the rate of return on equity earned by mineral companies in thirteen capital exporting countries is the subject of a paper extracted from an earlier study by Coopers and Lybrand. Entitled "Comparison of International Taxation Regimes," the paper illustrates the different rates of return on equity expected from mining operations in the thirteen countries.

Three papers by authorities in the field of taxation focus on the interpretation of recent mineral tax legislation at the federal level in the United States. In "The Tax Structure of Foreign Mining Investment in the United States," del Castillo Nicasie discusses the advantages and disadvantages of partnerships, joint ventures, and the corporation as a vehicle for a foreign mineral entity doing business in the United States.

In his paper "Overview of United States Taxation of Mining Companies," Dennis McCarthy, cites the recent tax treatment afforded mining companies under the "Tax Equalization and Financial Responsibility Act of 1984." Among the issues covered are exploration and development expenditures, the handling of depreciation and depletion, investment tax credits, and sales and exchanges of mining property.

Similarly, John McCabe, in his paper on "Tax Planning Alternatives Available with the Use of Multiple Corporations in Mining Ventures," cites the comparative advantages of this approach under the new regulations.

The final paper in this chapter focuses on the impact of taxes on mining in Australia. In "The Crush on Mining Profits," Vivian Forbes, relates how mining has been the focus of special tax legislation in that country and gives a candid view of its impact.

Mark E. Emerson
Chapter Editor
Resource Exchange Corporation

5.1

A COMPARISION OF THE IMPACT OF LOCAL, STATE, AND FEDERAL
TAXES IN EIGHT U.S. STATES

Robert L. Davidoff

U.S. Bureau of Mines
Denver, Colorado

INTRODUCTION

The impact of taxation differs substantially from state to state in a complex fashion that depends on both the physical characteristics of a mineral deposit and on the price of the mineral commodity relative to its costs of production. A comparative analysis of the tax structures of eight states (Arizona, Colorado, Idaho, Montana, Nevada, New Mexico, Utah and Wisconsin) showed wide variations in rates of return for any given type of mineral property, with no state having either the most burdensome or least burdensome tax structure under all circumstances.

Three hypothetical mineral operations were analyzed: a producing open-pit copper property, a non-producing underground platinum operation, and a non-producing underground gold operation. None of them corresponds exactly to an existing property. Each of these property types is further defined to have ore feed grades at three different levels, leading to three potential profitability levels, referred to in this paper as "economic", "marginal", and "subeconomic". Thus the differential impact of state tax structures is measured for mineral properties of different types, as well as for different profitability levels. Montana is used as the base case.

This study does not address the topic of corporate structures or tax consolidation. There are no costs assumed external to the operation of the deposit (e.g. exploration to locate additional deposits).

Two types of analyses were performed for this study. The first calculates the total cumulative tax payments in each state using all applicable taxes (including U.S. federal income taxes and examines the resultant differences in discounted cash-flows rate of returns DCFROR's amoung the eight states. The second is a sensitivity analysis that examines each type of state tax separately (i.e. state income, property, and severance taxes) in order to isolate and examine the differences in tax payments that are due to the difference in rates and methods of computing specific types of taxes in each state.

The results of this study serve to illustrate the complex nature of interactions among various taxes. For example, property tax is a deduction in calculating severance tax in some state. Changes in property-tax and severance-tax payments both enter into the calculation of state and federal income-tax liability. This, in turn, could change the timing and usage of tax-loss carry forward amounts and investment tax credits. Additionally, the physical characteristics of a property, the method and costs at which the mineral will be extracted, and the profitability level at which each property operates will affect tax treatment.

Each mineral operation was evaluated at three different predetermined profitability levels, referred to using the terms economic, marginal and sub-economic. These three profitability levels are defined by the ore feed grades of the primary commodity that yield in constant terms, approximately an "economic" 18% DCFROR, "a marginal" 10% DCFROR, and a "sub-economic" 5% DCFROR, respectively, under the Montana tax structure. In the analyses, the mining and processing parameters for each of the three types of properties were assumed to be identical for each of the eight states. It was assumed that each operation produces at full operating capacity throughout its life. This implies the level of mineral commodity demand will support the inferred commodity price, such that each operation will be able to sell all its output.

The analyses performed for this study were done using a comprehensive economic evaluation simulator that is used in the Bureau of Mines, Division of Minerals Availability MAS program to perform DCFROR analyses. All economic evaluations performed upon these data (see appendix 2) were in constant January 1981 dollars. By-product prices used for the analyses were:

Gold....................$425.00/troy ounce
Palladium...............$200.00/troy ounce
Silver..................$ 10.00/troy ounce

Typical cash-flow calculation layout is given in Appendix 3.

The initial step for each evaluation is to define an "economic" profitability level scenario. An 18% DCFROR rate in Montana was considered the minimum sufficient to attract new capital to the mineral industry.

Then the price was sought to yield a 0% DCFROR (break-even) in Montana. This break-even rate is sufficient to cover the total cost of production but provides no positive return on the investment.

Next, holding the derived prices from the economic scenario constant, feed grade is varied to generate "marginal" and "sub-economic" scenarios in each state.

The primary commodity prices derived and then used throughout the analyses for cumulative taxes (including federal income taxes under the Montana tax structure) were:

	0%	18%
Copper price for the "economic" scenario	$0.77/lb	$0.83/lb
Platinum price for the "economic" scenario	$167/tr oz	$482/tr oz
Gold price for the "economic" scenario	$198/tr oz	$349/tr oz

RESULTS

For each of the mines analyzed, at each of the profitability levels, the states having the largest and smallest cumulative tax payments (including federal income taxes) are shown below:

Profitability Level	Price	Cumulative Tax Payments	
		Largest	Smallest

Producing Copper Mine

Economic	$0.77/lb	Montana	Nevada
	$0.83/lb	Montana	Nevada
Marginal	$0.77/lb	Montana	Nevada
	$0.83/lb	Montana	Nevada
Subeconomic	$0.77/lb	Montana	Idaho
	$0.83/lb	Montana	Nevada

Non-producing Platinum Operation

Economic	$167/tr oz	Arizona	Idaho
	482/tr oz	Utah	Nevada
Marginal	167/tr oz	Arizona	Idaho
	482/tr oz	Arizona	Nevada
Subeconomic	167/tr oz	Arizona	Idaho
	482/tr oz	Arizona	Idaho

Non-producing Gold Operation

Economic	$198/tr oz	Arizona	Idaho
	349/tr oz	Arizona	Nevada
Marginal	198/tr oz	Arizona	Idaho
	349/tr oz	Arizona	Nevada
Sub-economic	198/tr oz	Arizona	Idaho
	349/tr oz	Arizona	Nevada

SENSITIVITY ANALYSIS

Sensitivity analysis examines the taxes separately and in combination to see how the differences in rates and methods of computing tax liability lead to differences in tax payments and rates of return among states.

A similar methodology was used to determine the individual tax payments as was used to determine the cumulative tax payments. The ore feed grades that define the three profitability levels for each property type are identical to those used in calculating total cumulative tax payments. However, the price of the commodity (copper, platinum, or gold) was recalculated for each individual tax analysis. That is, the price necessary to attain a 0% or 18% DCFROR for the "economic" scenario was derived for each individual tax payment under the Montana tax structure, assuming the effective rates for all other taxes are zero. These derived prices were then used to determine the individual tax payments and the resultant DCFROR's, while holding price (and total revenues) constant within each state at each profitability level.

This disaggregated look at state taxes highlights how different states rely more heavily on different types of taxes for revenue. It also shows that individual tax payments change as the potential profitability level of a deposit changes, and gives some indications that the relative tax burden in a state can be high or low depending on the physical characteristics of a deposit.

Copper

The price per pound of copper necessary to attain a 0% and 18% DCFROR was calculated for the "economic" scenario under the Montana tax structure. These calculated prices were then used to determine DCFROR's and cumulative tax payments (including federal income taxes) for all the states for all levels of profitability. Figure 1 illustrates the DCFROR and cumulative tax payments for the eight state's tax structure analyzed for the "economic" scenarios. At both 0% and 18%, Nevada shows the highest DCFROR and Montana the lowest.

Figures 2 and 3 illustrate the same type of comparison as Figure 1 for the "marginal" and "sub-economic" scenarios, respectively. The data in Figures 1-3 show, as expected, that the DCFROR's of each state decrease as the ore grade and the associated level of profitability decline. For all three levels of profitability, the amount of cumulative taxes paid as a percentage of total revenues remains fairly constant. At a 0% DCFROR, the ratio (taxes paid to total revenues) lies between 0.8% and 2.6%. At an 18% DCFROR, the ratio (taxes paid to total revenues) stays in the boundaries of 2.4% to 4.1%.

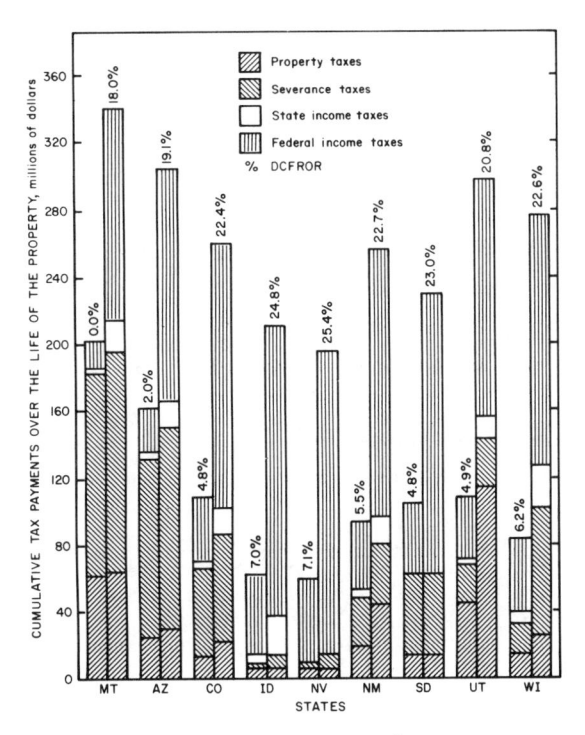

FIGURE 1. Individual and cumulative tax payments (including Federal income taxes) and the determined DCFROR for each state for the "economic copper property," using the copper price per pound necessary to attain a 0- and 18-percent DCFROR for the economic property under the Montana tax structure.

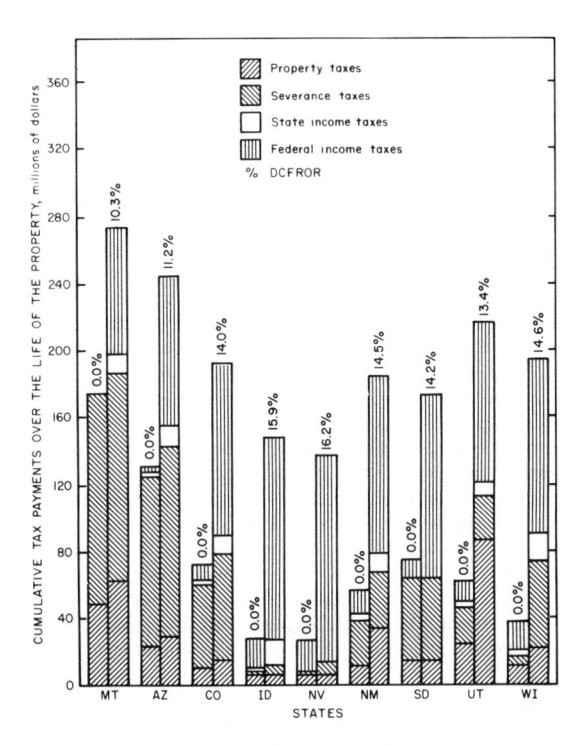

FIGURE 2. Individual and cumulative tax payments (including Federal income taxes) and the determined DCFROR for each state for the "marginal copper property," using the copper price per pound necessary to attain a 0- and 18-percent DCFROR for the economic property under the Montana tax structure.

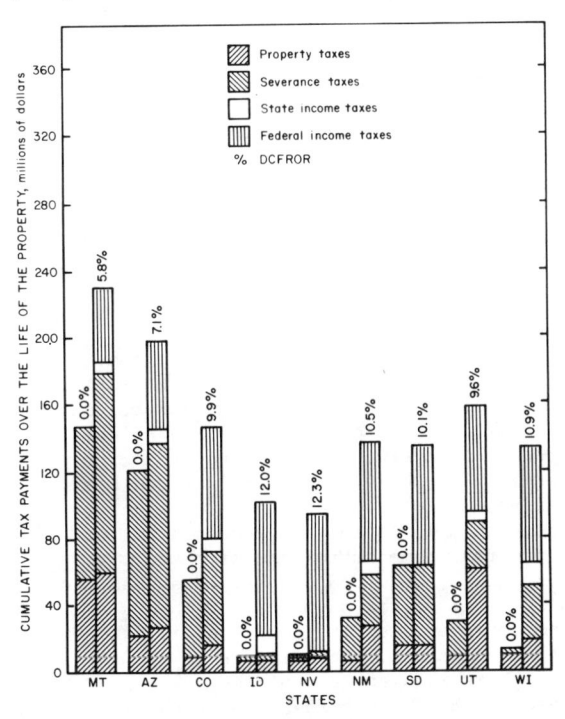

FIGURE 3. Individual and cumulative tax payments (including Federal income taxes) and the determined DCFROR for each state for the "subeconomic copper property," using the copper price per pound necessary to attain a 0- and 18-percent DCFROR for the economic property under the Montana tax structure.

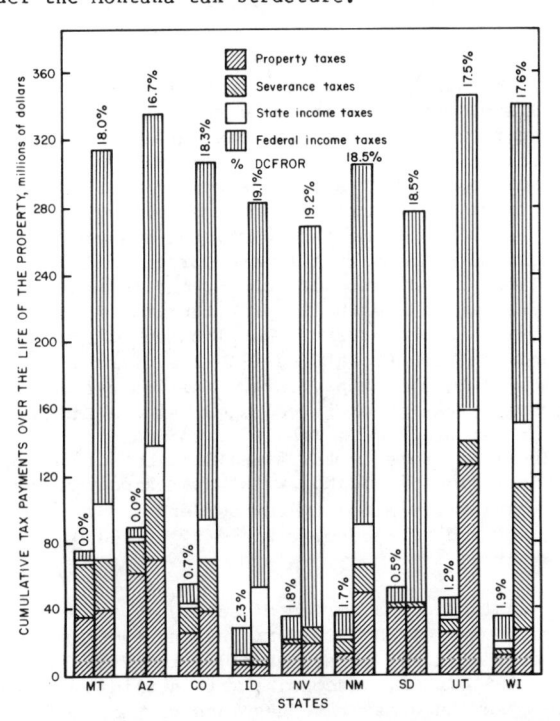

FIGURE 4. Individual and cumulative tax payments (including Federal income taxes) and the determined DCFROR for each state for the "economic platinum property," using the platinum price per troy ounce necessary to attain a 0- and 18-percent DCFROR for the economic property under the Montana tax structure.

Platinum

Figure 4 illustrates the DCFROR's and cumulative tax payments for the eight states' tax structures analyzed for the "economic" scenario. Nevada shows the highest DCFROR at the 18% level, while Idaho shows the highest DCFROR at the 0% level. Arizona shows the lowest DCFROR at both levels. Though Arizona's cumulative (undiscounted) tax payments in this instance are lower than either Utah's or Wisconsin's, the rate of return is lower because of a higher tax liability early in the life of the deposit. Arizona's Hoskold property tax is based on an estimate of the future worth of the deposit and payments begin as soon as the deposit's value can be established. Utah and Wisconsin have investment property taxes that do not generate payments until after development expenditures are made.

Figures 5 and 6 portray the same type of comparison as Figure 4, but are for the "marginal" and "sub-economic" scenarios, respectively. In these two cases Idaho has the highest DCFROR and Arizona remains the lowest DCFROR.

Taxes paid as a percentage of total revenues at breakeven (0%) lie within the ranges of 2.8%, to 8.8%, 3.1% to 10.1%, and 3.5% to 11.4% for the "economic", "marginal" and "sub-economic" levels of profitability, respectively. At 18% and the same three levels of profitability, the proportion of taxes paid to total revenues fall within the ranges of 15.4% to 19.8%, 19.7% to 25.3%, and 25.4% to 32.6%.

Comparing the burden of property tax only results of the study indicate that Arizona and Utah impose the largest property tax burdens on the proposed platinum operations (economic, marginal, and sub-economic). Arizona's combination of a Hoskold property tax and a high effective rate of 5.7% on the investment property tax has proven to be most burdensome at all three profitability levels at the 0% rate of return, and for the marginal and sub-economic profitability levels at the 18% rate of return. When looking at the "economic" scenario at the 18% rate of return, Utah has the largest tax payments. This is because of a property tax which is computed using a rate of 14% of net proceeds. The least burdensome stand-alone property tax overall at both rates of return resides in Idaho, with New Mexico and Utah also having low property tax burdens for the marginal and sub-economic property at the 0% rate of return.

Analyzing the burden of severance tax only at a breakeven (0%) DCFROR, Arizona has the largest tax burden at all three levels of profitability. At an 18% DCFROR, Wisconsin, with a sliding-scale rate based on gross revenues less certain deductions, has the largest burden at the "economic" profitability levels. Arizona remains the largest at the "marginal" and "sub-economic" level of profitability.

State income taxes have the same results for the platinum operation as they did for the copper property. Arizona, followed by Wisconsin, with their higher effective tax rates, have the largest state income tax burdens and lower DCFROR's.

In comparing the combined tax payments of property, severance, and state income taxes of the three platinum properties, excluding federal income taxes, at a 0% DCFROR, Arizona maintains the highest overall tax burden. At an 18% DCFROR, Arizona, Utah, and Wisconsin have the highest overall burden for the "economic" scenario, Arizona and Utah at the "marginal" scenario and Arizona at the "sub-economic" level.

Gold

Figure 7 illustrates the DCFROR and cumulative tax payments for the eight states' tax structures analyzed for the "economic" scenario. Nevada and Idaho, as in the analysis of the platinum property, show the highest DCFROR's. Arizona has the lowest, followed by Utah and Wisconsin.

Figures 8 and 9 portray the same type of comparison as did Figure 7, but the "marginal" and "sub-economic scenarios, respectively. In these cases also Nevada's and Idaho's overall tax structures are least burdensome, therefore most profitable. Whereas Arizona remains the least profitable, with the highest level of tax payments.

Taxes paid as a percentage of total revenues at the "economic" level of profitability fall in the range of 5.6% to 16.5% at the price that yields 0% and 16.9% to 23.0% at the price that yields 18%. For the "marginal" level, the ranges are 7.3% 21.5% at 0% and 22.0% to 30.0% at 18%. Finally, at the "sub-economic" level, the percentage distribution falls in the range of 8.6% to 25.5% and 26.1% to 35.5% for the 0% and 18% cases, respectively.

Comparing the burden of property tax only and severance tax only at a 0% DCFROR, Arizona imposes a larger property tax burden at all three levels of profitability of the proposed gold operation. Arizona's combination of Hoskold property tax and high effective rate of 5.7% on the investment property tax, leads to the highest level of property tax payments for the proposed (nonproducing) gold operations. Utah has the highest level of tax payments for the "economic" scenarios at the price yielding an 18% rate of return. This is due to a property tax with an rate of 14% of net proceeds. The least burdensome stand-alone property tax at the economic and marginal levels at both rates of return reside in Idaho. For the "sub-economic" level, Idaho is lowest at the 18% and New Mexico at 0%.

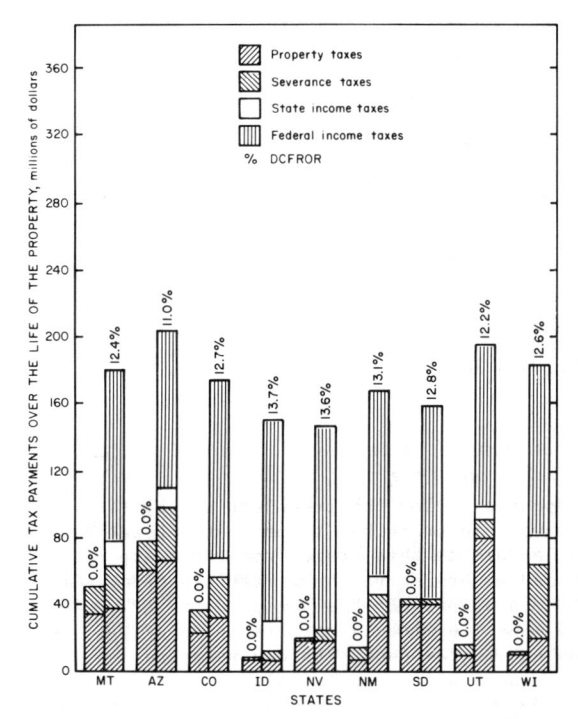

FIGURE 5. Individual and cumulative tax payments (including Federal income taxes) and the determined DCFROR for each state for the "marginal platinum property," using the platinum price per troy ounce necessary to attain a 0- and 18-percent DCFROR for the economic property under the Montana tax structure.

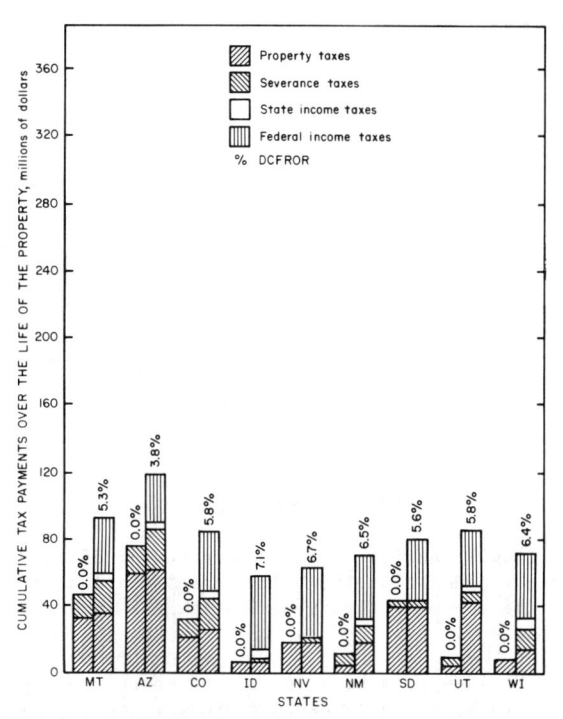

FIGURE 6. Individual and cumulative tax payments (including Federal income taxes) and the determined DCFROR for each state for the "subeconomic platinum property," using the platinum price per troy ounce necessary to attain a 0- and 18-percent DCFROR for the economic property under the Montana tax structure.

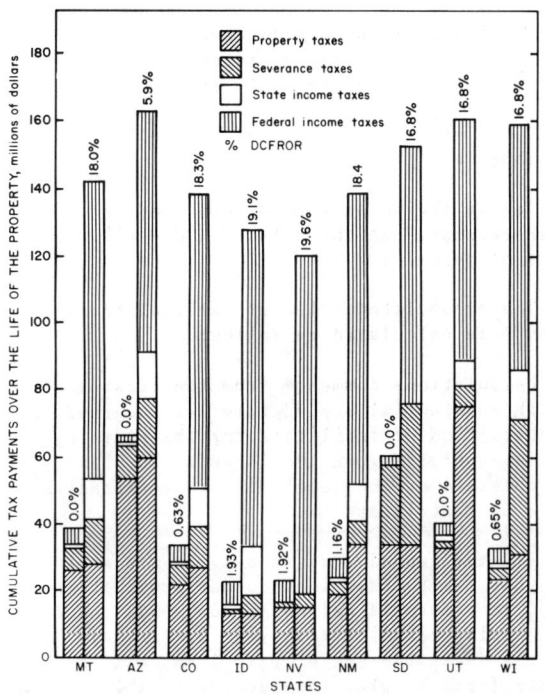

FIGURE 7. Individual and cumulative tax payments (including Federal income taxes) and the determined DCFROR for each state for the "economic gold property," using the gold price per troy ounce necessary to attain a 0- and 18-percent DCFROR for the economic property under the Montana tax structure.

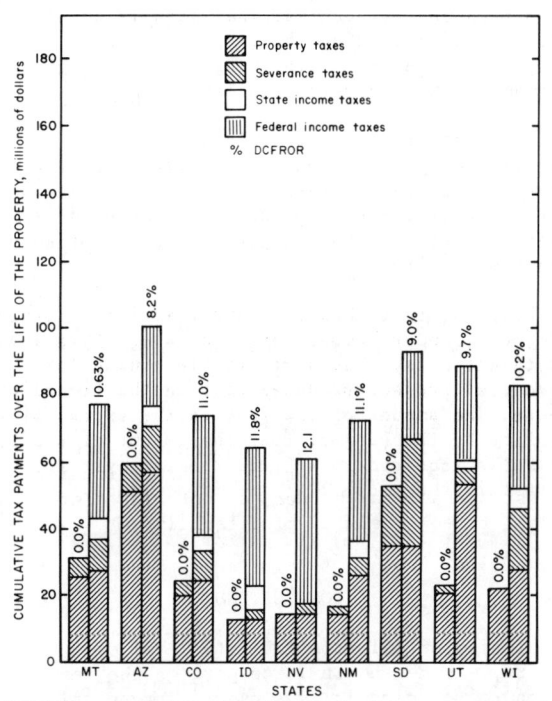

FIGURE 8. Individual and cumulative tax payments (including Federal income taxes) and the determined DCFROR for each state for the "marginal gold property," using the gold price per troy ounce necessary to attain a 0- and 18-percent DCFROR for the economic property under the Montana tax structure.

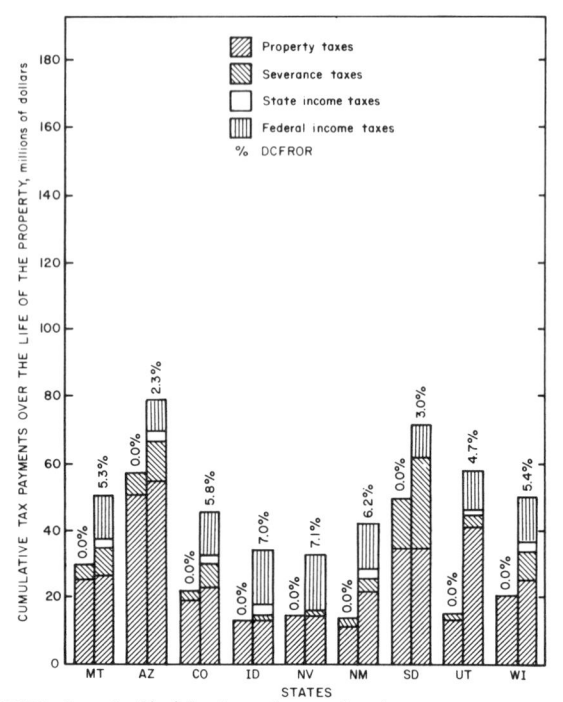

FIGURE 9. Individual and cumulative tax payments (including Federal income taxes) and the determined DCFROR for each state for the "subeconomic gold property," using the gold price per troy ounce necessary to attain a 0- and 18-percent DCFROR for the economic property under the Montana tax structure.

Analyzing the burden of severance tax only, at an 18% rate of return, Wisconsin, with a sliding-scale rate, again on has the largest tax burden at the "economic" profitability level. Arizona remains the largest at the "marginal" and "sub-economic" levels of profitability.

State income taxes have the same results for the gold operation as they did for the copper and platinum properties. Arizona and Wisconsin, with their higher effective tax rates, show larger state income tax burdens and therefore lower DCFROR's.

In comparing the combined tax payments of property, severance, and state income taxes of the three gold properties, Arizona again maintains the largest overall tax burden, at both rates of return and all three levels of profitability.

CONCLUSIONS

The potential profitability level of a mineral property may depend in large part on the tax structure to which it is subject. In general, progressive taxes based against gross revenues are much more detrimental to the economics of a mineral deposit than taxes based on net revenues. Net proceeds taxes with very high effective rates are also detrimental.

Property taxes are nearly independent of the profitability level of a mineral property and provide a relatively stable source of tax revenue to the states. Utah and New Mexico were exceptions to this general rule. Revenues decreased along with the expected profitability, since both states rely on net proceeds property taxes and allow deductions for all operating and transportation costs before computing tax liability. For properties that are currently non-producers, property taxes are the largest portion of the total state tax burden over the expected life of the mine in most cases.

Severance taxes are also less dependent on a mineral property's profitability level. Wisconsin, Idaho and Nevada were exceptions to this generalization. Severance tax revenues decreased along with profitability, since all three states base their tax on gross revenues less certain deductions (see Appendix 1). Severance taxes are generally a larger proportion of the state tax burden for producing properties.

State income taxes are highly dependent on a property's profitability level. However, Idaho was the only state analyzed that depended on income taxes for a substantial proportion of its revenue from mineral properties.

MINERAL INDUSTRY TAXATION

APPENDIX 1
SELECTED STATE TAXATION STRUCTURES

I. Severance taxes.
    A. Arizona.
       Tax based on value after milling: where the tax equals the rate times (commodity revenues less the sum of smelter operating costs, refinery operating costs and smelter to refinery transportation costs). The rate is 2.5%.

    B. Colorado.
       Tax based on value after mining: where the tax equals the rate times (commodity revenues less the sum of smelter operating costs, refinery operating costs, mill to smelter transportation costs, smelter to refinery transportation costs and a portion of the mill operating costs). The rate is 2.25%.

    C. Idaho.
       Tax based on gross revenues less certain deductions. The deductions are: all operating costs
         all transportation costs
         expensed exploration costs
         expensed development costs
         all amortization and depreciation present year's property tax royalty payments loan interest payments any and all depletion used

       The rate is 2.0%

    D. Montana.
       1. The Resource Indemnity Trust Tax. Tax based on value after mining (see Colorado). The rate is 0.5%.
       2. A sliding scale tax based on gross revenues less no deductions.

       The ranges and appropriate rates are:
         $0 - 100,000.........0.15%
         $100,000 - 250,000...0.58%
         $400,000 - 500,000...1.15%
         $250,000 - 400,000...0.86%
         greater than
         $500,000............1.44%

    E. Nevada.
       A net proceeds severance tax. Tax based on gross revenues less certain deductions. The deductions are:
         all operating costs
         all transportation costs
         all property taxes
         all royalty payments
         all amortization and depreciation
         all interest payments

       The rate is 1.3%.

    F. New Mexico
       Tax based on value after mining (see Colorado). The rate is 1.25%

    G. Utah.
       Tax based on value after mining (see Colorado).
       The rate is 1.0%

    H. Wisconsin.
       A sliding scale severance tax based on gross revenue less certain deductions. The deductions are:
         all operating costs
         all transportation costs
         expensed exploration costs
         expensed development costs
         all amortization and depreciation
         present year's property tax
         royalty payments
         loan interest payments

       The ranges and appropriate rates are:
         $0 - 100,000..................0.0%
         $100,000 - 4,000,000.........6.0%
         $4,000,000 - 10,000,000.....12.0%
         $10,000,000 - 20,000,000....16.0%
         $20,000,000 - 30,000,000....18.0%
         greater than 30,000,000.....20.0%

II. Property Taxes.

    A. Arizona.
       1. Hoskold Property Tax: A method of taxation in which the reserves in the ground are predicted and future worthed.
         Mill levy rate: 11.0%
         Assessment rate: 52.0%
         Redemption rate: 19.0%
         Interest rate: 8.0%

       2. Investment property tax
         Mill levy rate: 11.0%
         Assessment rate: 52.0%

    B. Colorado.
       1. Gross Proceeds Property Tax. Tax based on gross revenues less certain deductions. Deductions are:
           all operating costs
           all transportation costs
         Mill levy rate: 7.0%
         Assessment rate: 25.0%

       2. Investment Property Tax.
         Mill levy rate: 7.0%
         Assessment rate: 30.0%

    C. Idaho.
       1. Investment Property Tax.

Mill levy rate: 10.0%
Assessment rate: 20.0%

D. Montana.
1. Gross Proceeds Property Tax. Tax based on gross revenues less no deductions. Combined mill levy and assessment rate is: 0.66%

2. Investment Property Tax.
Mill levy rate: 22.0%
Assessment rate: 12.0%

E. Nevada.
1. Investment Property Tax.
Mill levy rate: 5.0%
Assessment rate: 35.0%

F. New Mexico.
1. A Net Proceeds Property Tax. A tax based on gross revenues less certain deductions. The deductions are:
   all operating costs
   all transportation costs
Mill levy rate: 5.0%
Assessment rate: 100.0%

2. Investment Property Tax.
Mill levy rate: 5.0%
Assessment rate: 33.0%

G. Utah.
1. A Net Proceeds Property Tax. A tax based on gross revenues less certain deductions. The deductions are:
   all operating costs
   all transportation costs
Mill levy rate: 14.0%
Assessment rate: 100.0%

2. Investment Property Tax.
Mill levy rate: 7.0%
Assessment rate: 22.0%

H. Wisconsin.
1. Hoskold Property Tax
Mill levy rate: 2.6%
Assessment rate: 100.0%
Redemption rate: 15.0%
Interest rate: 6.0%

2. Investment Property Tax.
Mill levy rate: 2.75%
Assessment rate: 100.0%

III. State Income Taxes

A. Arizona.
Gross revenues less certain deductions. Deductions are:
   all operating costs
   all transportation costs
   all amortization and depreciation
   federal taxes previous year
   severance taxes present year
   property taxes present year
   royalty payments
   loan interest payments
   any and all depletion used

0 - 2,000............. 0.0%
2,000 - 3,000......... 5.0%
3,000 - 4,000......... 6.5%
4,000 - 5,000......... 8.0%
5,000 - 6,000......... 9.0%
greater than 6,000.... 10.5%

B. Colorado.
Based on federal taxable income.
Rate is 5.0%

C. Idaho.
Based on federal taxable income.
Rate is 6.5%

D. Montana.
Based on federal taxable income.
rate is 6.75%

E. Nevada
No state income tax

F. New Mexico.
Based on federal taxable income.
Rate is 5.0%

G. Utah.
Based on federal taxable income.
Rate is 4.0%

H. Wisconsin.
Based on federal taxable income.
Rate is 7.9%

IV. Federal Income Taxes For All States.

Based on: Gross Revenue less:
   all operating costs
   all transportation costs
   expensed or amortized development
   expenses or amortized exploration
   loan interest payments
   depreciation
   royalty payments
   property tax
   statutory depletion
   severance tax
   state income tax
   tax loss carry
   Taxable Income

Federal tax rate used = sliding scale up to 46% Investment tax credits were used. Tax loss carry was used. Minimum federal tax was used.

APPENDIX 2

TYPICAL DATA FOR THE THREE GENERIC MINERAL
PROPERTIES USED FOR ANALYSIS

Copper:
    Preproduction years: 0
    Production years 29
    Average mine operating cost: $4.55/mt ore
    Average mill operating cost: $4.11/mt ore
    Annual tons of ore: 17,150,000
    By-products included: silver and gold
    Approx. capital investment over life + working
     capital: $101,318,000
    Total pounds recovered:
     8,823,250,000/8,366,541,00/8,112,691,000
     (economic /   marginal   / sub-economic)

Platinum:
    Preproduction years: 4
    Production years: 30
    Average mine operating cost: $40/mt ore
    Average mill operating cost: $22.00/mt ore
    Annual tons of ore: 380,000
    By-products included: paladium and gold
    Approx. capital investment over life + working
     capital: $87,720,000
    Total troy ounces recovered:
     2,287,000/1,525,000/1,287,000

Gold:
    Preproduction years: 2
    Production years: 12
    Average mine operating cost: $34/mt ore
    Average mill operating cost: $36.00/mt ore
    Annual tons of ore: 315,000
    By-products included: none
    Approx. capital investment over life + working
     capital: $124,140,000
    Total troy ounces recovered:
     1,913,000/1,473,000/1,243,000

APPENDIX 3

EXAMPLE OF TYPICAL ANNUAL CASH FLOW CALCULATIONS

Revenues
- less: mine operating costs
         mill operating costs
         smelter operating costs
         refinery operating costs
         all transportation costs
         expensed or amortized development
         expensed exploration loan interest
         payments depreciation royalty
         payments total property tax
             Before tax income

- less: depletion used
         severance tax
         tax loss carry
             State taxable income

- less: state income tax
             Federal taxable income

- less: federal income tax

- plus: tax adjustments
             Net income

- plus: depreciation
         depletion used
         deferred deductions

- less: equity investments

             Cash flow

5.2

CURRENT TRENDS IN CANADIAN MINING TAXATION

ROBERT B. PARSONS

PRICE WATERHOUSE, TORONTO

Abstract. Since 1978, new tax measures affecting the mining sector have been relatively favourable, evidencing a moderately enhanced government awareness and sensitivity towards the industry's concerns.

Today, although the industry continues to face a complex and sometimes capricious tax system, a wide range of tax provisions offer some excellent opportunities to the industry.

This paper examines the basic features of the present tax environment, discusses some of the more important changes of recent years, and comments on the outlook for mining taxation in Canada.

INTRODUCTION

Canada is well established as a major player on the international mining scene. Canada is the world's leading producer of nickel, zinc, potash, and asbestos, is the second largest producer of molybdenum, and in addition to these minerals, Canadian production of copper, gold, silver, and lead ranks high internationally. On a per capita basis Canada is first among major mineral producing countries in the world.

The importance of the mining industry to the Canadian economy will be appreciated when it is realized that over 80% of all production is exported, accounting for nearly 1/5 of the country's total exports. It is estimated that approximately 6% of Canada's total labour force owe their jobs, directly or indirectly, to the mining industry.

This paper examines the taxation of this important industry in Canada, in the context of the present political and economic environment, and comments on current trends in the taxation of the mining industry.

PRESENT ENVIRONMENT

With an area of almost 4 million square miles, Canada is the largest country in the Western Hemisphere and the second largest country in the world. A remarkable feature of this vast land is that most of its 25 million people live within 200 miles of the U.S. border.

Canada is a parliamentary democracy, divided into ten provinces and two territories. The division of powers between federal and provincial government is set out in the constitution, with considerable authority over certain matters being allocated to the provincial governments. The constitutional control over natural resources located within a province rests primarily with the province, although in various instances there is some degree of overlap with federal jurisdiction. The resulting interplay between federal and provincial authorities, and the occasional uncertainty over constitutional jurisdiction, have been the source of some inter-governmental conflict and consequent consternation for the mining industry, as indicated elsewhere in this paper.

Also, largely due to the importance of the respective roles of the various federal and provincial governments and to the difficulty which these governments seem to have in reaching a consensus on certain critical issues, Canada does not have an agreed upon national policy framework for the industry. Health, safety, environment, licensing and leasing arrangements, fiscal policies, and similar priorities can vary significantly among provinces.

The problems currently faced by the Canadian mining industry are typical of those encountered by producers in other major industrialized countries. The 1981/1983 worldwide recession, which marked the end of a decade of steadily increasing world mining capacity, was the direct reason for mine closings and temporary shutdowns. To date, soft prices continue to be the main concern in most sectors in the industry, as the world economies struggle to recover and a handful of minerally endowed developing nations relentlessly maximize production.

Emerging from this recession is a leaner Canadian mining industry. Major efforts of the past several years to improve productivity, in part through better technology, have achieved some success.

Other challenges facing the industry in the mid 1980's include an increased awareness of environmental matters, and the consistent threat of more government involvement in various facets of operations. Recent experiences suggest to the industry that the latter challenge could be the more serious one.

Taxation of the industry

Today, profits derived from most Canadian mining operations are generally subject to three different tiers of taxation. First, a federal income tax is levied on taxable income, generally at an effective rate of some 27%. In addition, the same taxable income is subject to a provincial income tax, normally at effective rates ranging from 4% to 12%, depending on the province in which the business operates. And finally, each of the producing provinces and territories imposes a mining tax, mineral royalty, or similar levy on some measure of production profits or revenues, as an economic rent in respect of the mineral resource which usually falls within the realm of provincial ownership. Since none of these taxes are deductible in computing the taxable profits or other base to which the other taxes are applied, an operator's burden in respect of a mining operation is the total amount of these three separate layers of tax. As a result, and because of a wide variety of other factors, including conflicting interjurisdictional policy objectives, the tax burden can in some cases become capricious.

Exhibit 1 depicts a comparative summary of the tax burden on a hypothetical new mine in various provinces. It is noted that this graph shows _effective_ tax rates: _marginal_ tax rates can be considerably higher. For example, the marginal tax rates in British Columbia, Ontario and Quebec can reach 60.5%, 66% and 61% respectively (Exhibit 3).

The income tax legislation, generally speaking, identifies three distinct stages of mining operations, and accords different treatment to each stage. The first two stages - being the actual extraction (or true mining) of the ore, and the processing of the ore to the prime metal stage - are regarded as mining activities for income tax purposes and are eligible for the various special rules (such as the resource allowance and the depletion allowance) which are accorded the industry. It is these two stages which are generally referred to as mining operations and which are addressed in this paper.

These processing operations (generally - concentrating, smelting, and refining) are often eligible for special incentive provisions in both the income tax and the provincial mining tax legislation.

The third recognized stage in the industry includes those activities which relate to processing beyond the prime metal stage, including semi-fabricating and fabricating operations. Activities of this stage are treated as manufacturing operations for income tax purposes and are not subject to provincial mining taxes or royalties. Manufacturing profits and assets are subject to a different set of rules than mining operations: these rules are not discussed in this paper.

The provincial mining taxes and royalties are not generally intended to be levied on profits derived from concentrating, smelting, or refining, the basic premise being that these provincial levies should be imposed on mining profits at the "pit's mouth". However, in the case of an integrated operation which includes, for example, a concentrator, the provincial mining tax or royalty is frequently calculated with reference to the total operating profit, with only a notional deduction being allowed by formula to reflect the processing profits included therein that might be attributable to the milling activities.

It is notable that the income tax and most provincial mining tax provisions distinguish between the preproduction and the production phases of mining operations, generally with expenditures and costs that are incurred during the preproduction period being accorded different treatment than the same expenditures and costs which are incurred during the production phase. Normally, the point of demarcation for these two phases is the date of commencement of commercial production. Although there is no definition of commercial production set out in any of the legislation (except in the Saskatchewan uranium royalty regulations), one rule of thumb which has evolved prescribes that the date of commencement of commercial production is the first day of the 90 day period throughout which the mine operated consistently at 60% of rated capacity. (In this case, capacity usually refers to throughput of the mill). Notwithstanding this generally accepted rule for determining the date of commencement of commercial production, there are clearly circumstances where the application of this rule would be inappropriate, and it is known that the taxation authorities have sometimes followed other approaches.

The income tax and provincial mining tax legislation also distinguish (although the distinction varies among jurisdictions) between mining conventional ores and mining industrial minerals. The tax treatment of these two different kinds of mining operations can be significantly different.

This paper deals with the taxation of conventional mining operations which would include base or precious metal deposits and coal deposits.

Exhibit 1

Summary of Effective Total Income and Mining Tax Rates on Hypothetical New Mine as a Percentage of Accounting Income

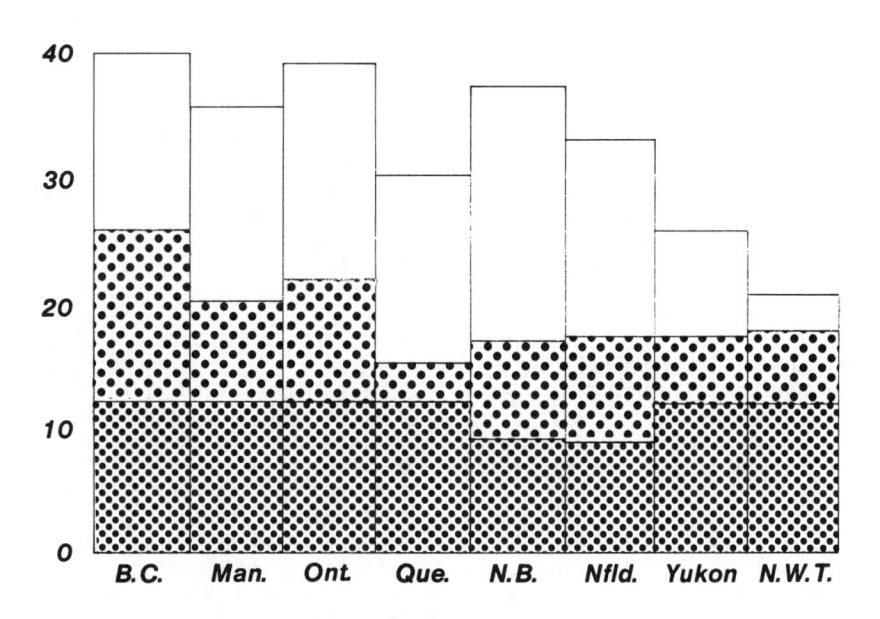

- Provincial Mining Tax
- Provincial Income Tax
- Federal Income Tax

Also, for income tax purposes, conventional mining operations include the mining of a bituminous sands deposit, oil sands deposit, and the mining of certain prescribed ores, such as asbestos, which would otherwise generally be referred to as industrial minerals.

Exhibit 2 illustrates the calculation of the total federal and provincial income and mining tax burden on a hypothetical Canadian mining operation over a thirteen year period, encompassing three years of preproduction development and ten years of production.

These calculations are based on the assumptions which are set out below. Unless otherwise indicated, all amounts are thousands of dollars.

The assumptions upon which the computer model is based include the following:

1. Reserves   25,000,000 tons

   Milling rate   2,500,000 tons/year
                   (700 tons/day)

   Grade   1% cu
                   .01 oz. au/ton
                   .06 oz. ag/ton

   Production   50,000,000 lbs. cu/year
                   25,000 oz. au/year
                   150,000 oz. ag/year

   Operating costs $16/ton milled (initially)

2. The capital costs that are incurred prior to the commencement of commercial production are as follows:

	Year 1	Year 2	Year 3
Purchase of mineral rights	1,000		
Exploration expense	4,000		
Development expense	5,000		
Cost of buildings, machinery, and equipment -			
Mining assets			10,000
Mill	10,000	10,000	10,000
Smelter	10,000	20,000	20,000

   No additional fixed assets are acquired following the commencement of production.

3. The interest rate throughout the period is 10%.

4. Product prices and annual operating costs increase at the rate of 8% per annum.

5. $1,000 is spent each year on off-site exploration.

6. The 3% inventory allowance is ignored.

7. Provincial capital taxes are ignored. These particular taxes, which are deductible for income tax purposes, vary as to both rate and basis among the provinces.

Income tax. The principal rules pertaining to the calculation of federal taxable income may be summarized in general terms as follows. It is noted that provincial income taxes are levied on federal taxable income allocable to a province under uniform (two factor) allocation rules followed by the federal and provincial governments. However, Ontario, Quebec, British Columbia, and Saskatchewan have in some cases particular rules for determining taxable income which differ in some respects from the federal rules.

Canadian exploration expense ("CEE"), as defined, includes:

(a) the cost of drilling, surveys, testing, and other expenses to determine the existence, location, extent or qualify of a mineral deposit, and

(b) expenses incurred prior to the start of production to bring a deposit into production, such as the cost of clearing, removing overburden, stripping, sinking a shaft, and constructing an adit or other underground entry.

A taxpayer may deduct its unclaimed CEE to the extent of income from any source. Any balance not deducted currently may be carried forward indefinitely for deduction in future years.

The cost of acquiring mineral rights, and the cost of acquiring other forms of ownership therein including royalty interests, constitute "Canadian development expense" ("CDE") for income tax purposes. A taxpayer's CDE is accounted for on a pool basis, with the unclaimed balance of the pool being reduced by the proceeds of disposition of such rights and interests. A taxpayer is entitled to an annual deduction on a declining balance basis equal to 30% of the unclaimed balance of his CDE.

The costs of buildings, machinery, and equipment used in mining operations ordinarily fall into either one of two "capital cost allowance" classes. Class 28 assets include certain buildings and equipment acquired prior to the commencement of commercial production of a new mine or that form part of a major expansion of an existing operation to increase capacity by at least 25%. Class 10 assets include buildings, machinery, and equipment acquired after the start of production.

The cost of Class 28 assets can be deducted in full against income from the new mine or major expansion. The combination of this feature together with the 100% deduction of CEE effectively means that all of the preproduction costs of a new mine can be recovered by the taxpayer before paying federal or provincial

MINERAL INDUSTRY TAXATION

Exhibit 2

ONTARIO MINING TAX MODEL

PREPARED BY PRICE WATERHOUSE

	YEAR 1	YEAR 2	YEAR 3	YEAR 4	YEAR 5	YEAR 6	YEAR 7	YEAR 8	YEAR 9	YEAR 10	YEAR 11	YEAR 12	YEAR 13	TOTAL
# UNITS PRODUCED-														
CU (LBS)	0	0	0	50000000	50000000	50000000	50000000	50000000	50000000	50000000	50000000	50000000	50000000	>>>>>>>
AU (OZ)	0	0	0	25000	25000	25000	25000	25000	25000	25000	25000	25000	25000	250000
AG (OZ)	0	0	0	150000	150000	150000	150000	150000	150000	150000	150000	150000	150000	1500000
PRICE PER UNIT-														
CU	0.00	0.00	0.00	1.00	1.08	1.17	1.26	1.36	1.47	1.59	1.71	1.85	2.00	
AU	0.00	0.00	0.00	500.00	540.00	583.20	629.86	680.24	734.66	793.44	856.91	925.47	999.50	
AG	0.00	0.00	0.00	8.00	8.64	9.33	10.08	10.88	11.75	12.69	13.71	14.81	15.99	
GROSS REVENUE-														
CU	0	0	0	50000	54000	58320	62986	68024	73466	79344	85691	92547	99950	724328
AU	0	0	0	12500	13500	14580	15746	17006	18367	19836	21423	23137	24988	181082
AG	0	0	0	1200	1296	1400	1512	1633	1763	1904	2057	2221	2399	17384
TOTAL	0	0	0	63700	68796	74300	80244	86663	93596	101084	109171	117905	127337	922794
OPERATING EXPENSES	0	0	0	40000	43200	46656	50388	54420	58773	63475	68553	74037	79960	579462
FEDERAL INCOME TAX														
NET REVENUE	0	0	0	23700	25596	27644	29855	32244	34823	37609	40618	43867	47376	343332
CCA-														
CLASS 10	0	0	0	0	0	0	0	0	0	0	0	0	0	0
CLASS 12	0	0	0	0	0	0	0	0	0	0	0	0	0	0
CLASS 28	0	0	0	0	7638	7432	0	16572	15816	16479	17063	0	0	81000
RESOURCE PROFITS	0	0	0	23700	17958	20212	29855	15672	19007	21130	23554	43867	47376	262331
RES ALLOWANCE	0	0	0	5925	4490	5053	7464	3918	4752	5282	5889	10967	11844	65583
CDE	0	0	0	300	210	147	103	72	50	35	25	17	12	972
CEE	0	0	0	6952	4048	1000	1000	1000	1000	1000	1000	1000	1000	19000
INT EXPENSE	3000	4800	8780	10523	9211	7578	5616	3274	2094	0	0	0	0	54874
RES PROFITS	-3000	-4800	-8780	0	0	6434	15673	7408	11111	14812	16641	31883	34520	121903
DEPLET ALLOWANCE	0	0	0	0	0	1609	3918	1852	2778	3703	4160	7971	7343	33333
INCOME	-3000	-4800	-8780	0	0	4826	11755	5556	8334	11109	12481	23912	27177	88569
PRIOR YRS LOSSES	0	0	0	0	0	4826	11754	0	0	0	0	0	0	16580
TAXABLE INCOME	-3000	-4800	-8780	0	0	0	0	5556	8334	11109	12481	23912	27177	71989
BASIC FEDERAL TAX	0	0	0	0	0	0	0	2000	3000	4000	4493	8608	9784	31886
INVESTMENT TAX CR	0	0	0	0	0	0	0	2000	3000	4000	0	0	0	9000
NET FEDERAL TAX	0	0	0	0	0	0	0	0	0	0	4493	8608	9784	22885

FINANCE FOR THE MINERALS INDUSTRY

ONTARIO MINING TAX MODEL
================================

PREPARED BY PRICE WATERHOUSE
================================

	YEAR 1	YEAR 2	YEAR 3	YEAR 4	YEAR 5	YEAR 6	YEAR 7	YEAR 8	YEAR 9	YEAR 10	YEAR 11	YEAR 12	YEAR 13	TOTAL
ONTARIO INCOME TAX														
NET REVENUE	0	0	0	23700	25596	27644	29855	32244	34823	37609	40618	43867	47376	343332
CCA-														
CLASS 10	0	0	0	0	0	0	0	0	0	0	0	0	0	0
CLASS 12	0	0	0	0	0	0	0	0	0	0	0	0	0	0
CLASS 28	0	0	0	2177	15385	19066	23240	23470	4662	-3000	-3999	0	0	81000
INTEREST EXPENSE	3000	4800	8780	10523	9211	7578	5616	3274	2094	0	0	0	0	54874
CDE	0	0	0	0	0	0	0	0	0	0	0	0	0	0
CEE	0	0	0	11000	1000	1000	1000	1000	1000	1000	1000	1000	1000	20000
RESOURCE PROFITS	-3000	-4800	-8780	0	0	0	0	4500	27068	39609	43617	42867	46376	187457
DEPLET ALLOWANCE	0	0	0	0	0	0	0	1500	9023	13203	14539	14289	15459	68012
INCOME	-3000	-4800	-8780	0	0	0	0	3000	18045	26406	29078	28578	30918	119445
PRIOR YEARS LOSSES	0	0	0	0	0	0	0	3000	13580	0	0	0	0	16580
TAXABLE INCOME	-3000	-4800	-8780	0	0	0	0	0	4465	26406	29078	28578	30918	102865
ONTARIO INCOME TAX	0	0	0	0	0	0	0	0	625	3697	4071	4001	4328	16722
ONTARIO MINING TAX														
NET REVENUE	0	0	0	23700	25596	27644	29855	32244	34823	37609	40618	43867	47376	343332
DEPRECIATION-														
15% CLASS	0	0	0	12000	12000	12000	12000	12000	12000	8000	0	0	0	80000
30% CLASS	0	0	0	0	0	0	0	0	0	0	0	0	0	0
100% CLASS	0	0	0	10000	0	0	0	0	0	0	0	0	0	10000
CEE	0	0	0	985	10015	1000	1000	1000	1000	1000	1000	1000	1000	19000
	0	0	0	715	3581	14644	16855	19244	21823	28609	39618	42867	46376	234332
PROCESSING ALLOWNC	0	0	0	465	2328	9518	10956	12508	12800	12800	12800	12800	12800	99775
TAXABLE PROFIT	0	0	0	250	1253	5125	5899	6735	9023	15809	26818	30067	33576	134557
MINING TAX PAYABLE	0	0	0	0	163	938	1092	1260	1717	3365	6458	7433	8485	30910
CASH FLOW														
NET REVENUE	0	0	0	23700	25596	27644	29855	32244	34823	37609	40618	43867	47376	343332
INTEREST EXPENSE	3000	4800	8780	10523	9211	7578	5616	3274	2094	0	0	0	0	54874
FEDERAL INCOME TAX	0	0	0	0	0	0	0	0	0	0	4493	8608	9784	22885
ONTARIO INCOME TAX	0	0	0	0	0	0	0	0	625	3697	4071	4001	4328	16722
ONTARIO MINING TAX	0	0	0	0	163	938	1092	1260	1717	3365	6458	7433	8485	30910
FIXED ASSETS - NET	20000	30000	40000	0	0	0	0	0	0	0	0	0	0	90000
EXPLORATION EXP	9000	0	0	1000	1000	1000	1000	1000	1000	1000	1000	1000	1000	19000
RESOURCE PROPERTY	1000	0	0	0	0	0	0	0	0	0	0	0	0	1000
	-33000	-34800	-48780	12177	15222	18128	22147	26710	29387	29547	24596	22825	23779	107940
OPENING CASH	0	-33000	-67800	-116580	-104403	-89181	-71052	-48905	-22195	7193	36740	61336	84161	0
CLOSING CASH	-33000	-67800	-116580	-104403	-89181	-71052	-48905	-22195	7193	36740	61336	84161	107940	107940

```
*******************************************************************
THIS MODEL IS INTENDED SOLELY TO ILLUSTRATE THE APPLICATION OF FEDERAL
AND PROVINCIAL INCOME AND MINING TAX PROVISIONS TO A "STAND-ALONE"
MINING OPERATION. THE CHARACTERISTICS OF THIS MINE MODEL ARE NOT IN-
TENDED TO REPRESENT THOSE OF AN ACTUAL MINE, AND ARE NOT NECESSARILY
INDICATIVE OF A TYPICAL MINING OPERATION.
*******************************************************************
```

income tax. (Some of the provincial mining tax regimes contain similar features so that, in those provinces, the capital cost of a new mine can be recovered before payment of either income taxes or mining tax.)

The cost of Class 10 assets is deductible at the rate of 30% per annum on a declining balance basis.

In addition to the deduction of CEE, CDE, and capital cost allowances for fixed assets, a taxpayer may be entitled to two special deductions which have no counterpart in the taxpayer's books of account - a "resource allowance" and the so-called "earned depletion allowance".

The resource allowance is intended to compensate for the non-deductibility of provincial mining taxes and royalties. This allowance represents a deduction in computing taxable income equal to 25% of mining profits as calculated before deducting CEE, interest expense and the earned depletion allowance: in some measure, it is a recognition that provincial mining taxes are not a deductible expense for income tax purposes.

The earned depletion allowance is an incentive provision designed primarily to encourage investment in exploration, new mines, and mineral processing facilities. The effect of this allowance is to provide a deduction equal to 4/3 of eligible costs. The taxpayer is entitled to deduct the lesser of the balance of his earned depletion base, and 25% of his resource profits as calculated after deducting the resource allowance, CEE, interest expense, and other related costs. The earned depletion base consists of $1 for every $3 spent on CEE, Class 28 assets, and certain processing assets - hence, the ability to deduct in total 4/3 of the actual eligible costs.

Another tax incentive which is of value to the mining industry is the ability to earn investment tax credits through expenditures on qualifying fixed assets and research and development. The investment tax credit, which can be offset against federal income tax otherwise payable, ranges from 7% to 50%, depending on the nature and location of the qualifying expenditure. Unused investment tax credits can be carried back three years and forward seven years. The amount of the credit reduces the deductible cost of the related expenditure.

In recognition of the fact that many corporations are often not able to claim their CEE, earned depletion allowance, and investment tax credits on a current basis, there are complex provisions in the legislation whereby the corporation can "flow through" such deductions and credits to investors for their utilization. These provisions can be extremely attractive in terms of raising capital which might otherwise be difficult to attract.

Mining tax. As indicated earlier and in Exhibit 1, provincial mining taxes or royalties can constitute a major component of an operator's total tax burden. The rates of mining tax differ considerably among the provinces, and reflect a wide spectrum of structures - from flat rates of tax (17.5% and 18% in British Columbia and Manitoba), to two stage taxes in New Brunswick and Newfoundland, and to progressive rate structures in Ontario, Quebec, the Yukon, and the Northwest Territories.

With some exceptions, the provincial mining taxes are generally imposed on some measure of operating profit - frequently involving a relatively simple calculation of gross receipts and operating expenses, including depreciation and amortization at specified rates. In order to restrict the tax base to mining profits, the provincial mining tax systems commonly include a processing allowance deduction which is intended to remove processing profits from the tax base. This allowance is generally calculated as a specified percentage of the cost of processing assets, with minima/maxima parameters for the allowance being established as percentages of taxable profit. Exhibit 4 summarizes the principal features of provincial mining taxes.

By means of various select provisions, including stepped-up processing allowances and investment allowances and credits which can vary significantly from province to province, some of the high statutory mining tax rates shown in Exhibit 3 can be substantially mitigated in certain circumstances to produce more acceptable levels of taxation.

The various income tax deductions and incentives, together with other similar provisions in the provincial mining tax regime, can have a highly favourable impact on the tax burden of an operating mine. Although nominal and marginal tax rates can reach severe heights (see Exhibit 3), more moderate actual effective tax rates can generally be experienced thanks in large part to these special incentive tax deductions and credits.

CURRENT TRENDS

Perhaps an appropriate way to describe the current state of Canadian mining taxation in terms of apparent trends would be to say that the various tax regimes are "on hold".

Compared to the wild melee which highlighted the mid-1970s, when the federal and provincial governments became embroiled in an unprecedented scramble to grab larger shares of the mining industry's cash flow, the period since 1978 has been one of relative calm.

From 1974 through 1977, the Canadian mining industry found itself playing the roll of the "ham in the sandwich", as the federal and provincial governments one after the other in

Exhibit 3

Summary of Federal and Provincial
Statutory Tax Rates (Note 1)
June, 1984

Province (Note 2)	Federal Income Tax	Provincial Income Tax	Provincial Mining Tax		Maximum Combined Marginal Rate (Note 3)
British Columbia	36%	16%		17.5%	60.5%
Manitoba	36%	15%		18 %	56.3%
Ontario	36%	14%	$ 0 - 250,000 250,000 - 1,000,000 1,000,000 - 10,000,000 10,000,000 - 20,000,000 over 20,000,000	0 % 15 % 20 % 25 % 30 %	66.3%
Quebec	36%	5.5%	$ 0 - 250,000 250,000 - 3,250,000 3,250,000 - 10,250,000 10,250,000 - 20,250,000 over 20,250,000	0 % 15 % 20 % 25 % 30 %	61.1%
New Brunswick	36%	15%	2% of net revenue plus 16% of net profit		55.9%
Newfoundland	36%	16%	15% of mine profit, plus 20% of excess of 20% of profit over royalties paid		59.0%
Yukon	36%	10%	$ 10,000 - 1,000,000 1,000,000 - 5,000,000 Every additional $5,000,000, tax rate increases by 1%	3 % 5 %	–
Northwest Territories	36%	10%	$ 10,000 - 1,000,000 1,000,000 - 5,000,000 Every additional $5,000,000, tax rate increases by 1% to maximum of 12%	3 % 5 %	46.5%

Notes

1. The above rates are the rates specified in the relevant legislation, and do not reflect any possible reductions on account of investment tax credits, resource allowances, and other provisions which may serve to reduce the taxpayer's effective tax rate.

2. Mineral production in the provinces of Alberta, Saskatchewan and Nova Scotia is generally subject to royalty rates and rules which may vary according to the mineral produced.

3. These rates reflect the 25% resource allowance deduction and automatic depletion deduction where applicable.

SUMMARY OF FEATURES OF PROVINCIAL MINING TAX LEGISLATION

Exhibit 4

	British Columbia	Manitoba	Ontario	Quebec	New Brunswick	Newfoundland	Yukon	Northwest Territories
Title of Statute	Mineral Resource Tax Act	The Metallic Minerals Royalty Act	The Mining Tax Act	Mining Duties Act	Metallic Minerals Tax Act	The Mining and Mineral Rights Tax Act	Yukon Quartz Mining Act	Territorial Lands Act
Tax Rate:								
Ontario								
$ 0 – 250,000			0%					
250,000 – 1,000,000			15%					
1,000,000 – 10,000,000			20%					
10,000,000 – 20,000,000			25%					
over – 20,000,000			30%					
Quebec								
$ 0 – 250,000				0%				
250,000 – 3,250,000				15%				
3,250,000 – 10,250,000				20%				
10,250,000 – 20,250,000				25%				
over – 20,250,000				30%				
Other provinces	17-1/2%	18%			2% on net revenue plus 16% on net profit in excess of $100,000	15% of mine profit plus 20% of excess of 20% of profit over royalties paid	$ 10 – 1,000,000 3% 1,000 – 5,000,000 5% Every additional $5,000,000, tax rate increases by 1%	$ 10 – 1,000,000 3% 1,000 – 5,000,000 5% Every additional $5,000,000, tax rate increases by 1% to maximum of 12%
Depreciation Mining Assets	30% declining balance (up to 100% of new mine income for new mine assets)	20% declining balance Depreciable balance reduced by investment tax credit claimed	30% straight-line basis (up to 100% of new mine income for new mine assets)	30% straight-line	5% minimum (no maximum) for new or expanded mine assets; other assets maximum 33-1/3%	10% straight-line	15% straight-line	15% straight-line
Processing Assets	Same as mining assets	Same as mining assets	15% straight-line basis	30% straight-line	5% minimum (no maximum)	Same as mining assets	Same as mining assets	Same as mining assets
Preproduction Expenses	100% deduction	Included in depreciable assets	100% deduction	100% deduction	5% minimum (no maximum)	Deductible over the life of the mine as estimated by the Minister	Nil	15% straight-line
Exploration Expenses	100% deduction	100% deduction	100% deduction	100% deduction	150% deduction	Ministerial discretion applies up to 100% deduction	100% deduction	100% deduction
Processing Allowance (% of Processing Assets):								
Concentrating	8%	10%	8%)	8%)	8%	8%	N/A	8%
Smelting	8%	10%	16%) (Note 1)	15%) (Note 2)	15%	8%		8%
Refining	8%	10%	20%)	15%)	15%	8%		8%
Northern Ontario –								
Refining			25%					
Semi-fabrication			30%					
	Based on original cost of processing assets	Based on original cost of depreciable processing assets, net of investment tax credit claimed	Based on original cost of depreciable processing assets	Based on original cost of depreciable processing assets	Based on original cost of depreciable processing assets	Based on original cost of depreciable processing assets	N/A	Based on original cost of depreciable processing assets

SUMMARY OF FEATURES OF PROVINCIAL MINING TAX LEGISLATION

	British Columbia	Manitoba	Ontario	Quebec	New Brunswick	Newfoundland	Yukon	Northwest Territories
Processing Allowance Minimum/Maximum % of Net Profit	15%/70%	15%/65%	15%/65%	15%/65%	15%/65%	no minimum/65% maximum	N/A	no minimum/65% maximum
Non-deductible Expenses	Royalties, depletion, cost of mining property (interest income offsets deductible interest expense)	Interest, royalties, depletion, cost of mining property	Interest, royalties, depletion, cost of mining property	Interest, royalties, depletion, cost of mining property	Interest, royalties, depletion, cost of mining property	Interest, royalties, depletion, cost of mining property	Interest, royalties, depletion, cost of mining property	Interest, royalties, depletion, cost of mining property
Special Features	A modified depletion allowance can be deducted, based on the federal concept of earned depletion, but eligible expenditures are restricted to certain expenditures incurred in B.C. after 1975.	Tax payable may be reduced by an investment tax credit equal to the lesser of: (i) 5% of prior year's undepreciated balance of certain assets and (ii) 50% of mining tax payable	Profits from certain operations may be taxed separately under graduated rate schedule: – new mines – significant expansion of existing mines – certain inactive mines reopened	Income may be averaged over a three-year period. 15% of losses incurred may be carried forward four years to reduce mining duties otherwise payable. Exploration and investment allowance: $1 allowance for every $3 of eligible expenditures. Eligible expenditures include Quebec exploration expenses and investments in new processing assets. Maximum allowance is 1/3 of mining profits before investment allowance.	New mine exempt from 2% royalty in first two years. In computing net revenues subject to 2% tax a milling and smelting allowance (8% of concentrating and 15% of smelting and refining assets) is deductible up to 25% of net revenue. In computing mine profit subject to 16% tax, a special deduction is allowed equal to 8% of the undepreciated cost of depreciable assets and preproduction development costs. The amount of 16% tax payable is reduced by 25% of eligible process research expenditures.	In computing mine profit subject to 15% tax, a deduction is allowed equal to the greater of 20% of profits (before this allowance) and non-Crown royalties paid.	In computing mine profit all taxes paid or payable upon mining, smelting or refining profits are deductible.	No royalties are payable in the first three years of production
Provincial Income Tax Rate (see Appendix K for features of provincial income tax legislation)	16%	15%	14%	5-1/2%	15%	16%	10%	10%

Notes:

1. Where an operator owns and operates smelting as well as concentrating assets, the 16% processing allowance applies to all processing assets. Similarly, the 20% processing allowance is applicable to refining assets as well as other processing assets used.
2. Where the operator does smelting or refining the 15% treatment allowance applies to all processing assets.

quick succession introduced entirely new sets of tax rules for the industry in an attempt to capture larger shares of the perceived "wind-fall profits" which record setting metal prices of the times were expected to generate.

It was not until 1978 that taxation authorities realized the error of their ways, and began to think about possible means to redress the jungle of taxes which strangled the industry.

The first step in achieving a rationalization of Canada's crazy patchwork quilt of mining taxes and policies was achieved with the release in November 1978 of a major federal-provincial report on the taxation of the industry. The report, prepared by a group of federal and provincial civil servants, set out a concise and informative discussion of the major taxation and incentive issues facing the mining industry, and put forward major recommendations for more rational and moderate taxation policies.

The report was prepared in response to a directive from the federal and provincial First Ministers in February 1978 to review the federal and provincial taxation of the resource areas. Although the report was more successful in identifying problems, rather than concrete solutions, and reflected some continuing differences of opinion between governments involved, it represented an important first step in a comprehensive review of Canada's resource tax policies.

The report of the senior civil servants set out the following conclusions about the level of taxation in the mining industry:

1. Despite the fact that the federal and provincial income tax burden on mining had risen in the 1970's, it was still lower than the income tax burden on other Canadian industries. (This comparison incidentally excluded any consideration of the burden of provincial mining taxes and royalties borne by the mining industry.)

2. The over-all tax level (including provincial mining tax and royalties) on mining, while generally substantially above that of Canada's manufacturing industries, was generally not out of line with those of other industries "keeping in mind that mining taxes and royalties are also, in part, payment to government for use of resources and are therefore similar to payments to private-sector owners rather than taxes".

3. International tax comparisons are difficult to make, and are not necessarily meaningful.

4. The contribution of the tax systems to the problems faced by the industry has been appreciably less than that of other factors, such as rising costs and depressed prices after 1974.

The report was considered briefly at the First Ministers' Conference on the Economy in November 1978, and was submitted to federal and provincial finance and resource ministers for further study and action. It represented a major new initiative both in federal-provincial co-operation and in a review of the fundamental features of mining taxation in Canada, and also disclosed something of the problems in dealing with the very issues that it raised. Although the federal and provincial officals did reach general agreement on some broad policies, they were unable to attain complete unanimity, and the discussions of several key issues (including the appropriate level of total resource taxation) reflect this disagreement.

Following the release of this report, a number of helpful changes were made to certain aspects of the taxation of mining operations. However, the basic framework for taxation which unfolded in the mid-70's remains intact.

At a time when the industry sensed that perhaps the storm clouds of taxation were beginning to clear, a deafening thunderclap sounded throughout Canada and abroad with the announcement in October 1980 of the federal government's National Energy Program ("NEP"). The NEP is a far-reaching program which substantially increased government involvement in virtually all aspects of the country's petroleum industry, and resulted in substantially higher tax burdens for petroleum producers and consumers.

With the pain of the 1970's federal/provincial conflict, and the resultant higher tax burdens, still fresh in their minds, Canada's mining men feared that their industry would be singled next for the kind of treatment thrust upon the oil and gas sector by the NEP. Despite continuing words of comfort from successive federal mines ministers that the energy sector represented a unique situation, concern and distrust lingered.

An important federal document released early in 1982 did much to allay the nagging fears of the mining industry. In March 1982 federal Mines Minister Judy Erola released a 157 page document entitled "Mineral Policy - A Discussion Paper". The discussion paper, which presents the federal government's assessment of the opportunities and constraints facing the Canadian mineral industry in the 1980's, was prepared within the context of three principal objectives - strengthening the resource base on which the industry rests, ensuring the adequacy of research activities needed to support development of the resource, and furthering the development of markets for the resource. Although the paper is successful in identifying the problems, constraints and opportunities of the industry, and offers some options and alternatives, it is almost totally devoid of immediate, practical recommendations.

With respect to taxation, the carefully worded federal paper indicates that consideration could be given to reducing the "disadvantages" faced by junior mining companies, and briefly discusses various suggestions to improve the existing tax rules, including: a system of exploration grants; tax credits; exemption from capital gains; and flow-through benefit provisions. The study devotes considerable space to the well-worn topic of comparative tax levels, and observes "taking all factors into consideration, the mining industry does not appear to be unfavourably treated compared to other sectors of the Canadian economy". The discussion paper is inconclusive regarding tax matters – the reader gets a sinking feeling that this important area has been dismissed with a shrug of the shoulders and a wave of the hand.

The reluctance of the mineral policy paper to make definitive recommendations for policy implementation is evident in the discussions of other important subjects. The study does an admirable job of identifying problems and issues in such areas as economic spin-offs, infrastructure, science and technology, employment, and the environment, then leaves it up to other federal bodies, other governments and the private sector to attend to required action.

The paper has been followed up by the federal government in respect of tax matters by the enactment of tax legislation designed to facilitate the financing of junior companies. The net effect of these changes is that, generally speaking, an investor in a mining company can, in certain circumstances, obtain an immediate 133% write-off for his cost of new shares issued by the company. For an investor residing in the province of Quebec, the write-off can be as high as 167%, resulting in a possible after-tax cost of 8¢ on the dollar for an investor in the top marginal tax bracket.

Overall, though, notwithstanding a sprinkling of generally favourable tax changes which have been introduced since 1978, the Canadian mining industry continues to face a highly complex, often confusing tax system. Despite acknowledgement by some government officials of the existence of certain potentially capricious provisions and other aberrations and inconsistencies in the legislation, little has been done in terms of improving the situation. Perhaps this is due in part to an evidently emerging school of thought in government and business circles which questions the importance of Canada's natural resource industries vis a vis the Canadian economy. Federal and provincial budgets in the past few years have tended to carry a common theme – that the thrust of Canada's future economic development will be tied to technological development and fostering the growth of basic manufacturing and high-tech industries. This shift in emphasis has been evolving over a period of years, but is becoming more pronounced as the private sector itself is slowly being persuaded that perhaps it should not automatically accept without question that Canada's future development will continue to be based on its natural resources.

With this kind of thinking beginning to permeate government and business planning, government finance ministers will naturally be somewhat reluctant to exhibit the greatest sympathy to the mining industry.

Another reason why such changes are unlikely to be forthcoming in the near future is that it seems to be exceedingly difficult for the various federal and provincial governments to reach a consensus on mineral policy matters. The mineral policy paper introduced by the federal government in March 1982 potentially faces a dusty death in government archives and libraries across the country if co-operation is not forthcoming from the provinces. And, neither federal nor provincial governments seem to be prepared to upset, with any serious effect, the equilibrium evidently in place in terms of federal/provincial resource sharing.

A couple of recent provincial budgets are of some interest, but would not seem to be indicative of any new or emerging trends in the taxation of the industry.

In its April, 1984 budget, the province of New Brunswick stunned the mining industry by announcing that it would replace its 16% flat rate mining tax with a progressive rate schedule encompassing rates of up to 30%, effective January 1, 1984. Only two months later, in June, 1984, having been bombarded by the industry for its proposed rate change, the province announced that it would defer the change and give it further study.

The province of Ontario in its May, 1984 budget announced that it would undertake a study of its mining tax regime. For some years, the province has been under fire for its progressive rate structure. However, it is understood to be the province's unofficial position that replacement of its progressive rate structure with a flat rate of mining tax would have to be accompanied by reductions in the province's processing allowance rates.

One emerging trend in Canadian taxation in general, which may have some bearing on mining taxation, involves a renewed clamour from the private sector of the economy for simplification of the tax rules. For the sake of equity and precision, Canada's income tax legislation has become unwieldy and complex in many areas, including that part of the legislation dealing with the extractive industries. The federal authorities apparently acknowledge this, and have set about to study ways of simplifying certain aspects of the Income Tax Act.

The first sign that the government may be taking tax simplification seriously came in the February 1984 budget, which included proposals to overhaul and simplify provisions relating to small business taxation. It is likely, though, that even if the mining tax provisions are to undergo any sort of simplification, the substance of the rules would remain the same.

For the first time since 1958, the federal Progressive Conservative party formed a majority government in September, 1984 following its resounding election victory. Despite the relatively pro-business character of the Conservatives, Canada's mining industry is not anticipating any monumental changes to mineral taxation policy. The industry might be affected, though, by those government policies which reflect a more receptive attitude toward foreign investment in Canada. Time will tell.

SUMMARY

Although existing Canadian mining taxation is not without its short-comings and pitfalls, the system overall seems to produce bearable burdens on most but not all mines. This is not to say that the tax rates for mining are either fair or reasonable: the tax system does contain anomalies and disincentives. But, at this moment, the real problem facing the industry is low mineral prices and a cost squeeze.

The ability of a new mine to recover its capital investment before payment of tax, the ability to flow through valuable tax write-offs to investors, and certain other provisions in the legislation compensate to some degree for relatively high statutory tax rates.

What the industry wants most in a tax sense is long-term stability in the rules, so that plans can be made with a greater degree of confidence that the premise for such plans will not be altered in the future. The mining industry would certainly like to be able to count on incentives which had some degree of permanence, rather than the ad hoc incentives which seem to have a habit of appearing and then vanishing as government policy goals and objectives change. The federal and provincial governments acknowledge this need for stability, and have undertaken without definitive promises to do their best to accomodate it.

The tax outlook for the Canadian mining industry appears to be one of calm. Although clouds loom on the horizon from past storms, these clouds seem to be moving slowly away. The federal and provincial governments, with some occasional exceptions, continue to try to convince the mining industry that the debacle of the mid-70's will not recur and that the policies of the NEP will not be extended to the mining industry or other sectors of the economy. However, industry observers and mining men who suffered through the 70's and who witnessed the sudden imposition of the NEP in 1980 continue to be wary.

Improved worldwide prices would solve a lot of what concerns Canada's mining industry today. It is hoped that firmer prices and the restoration of fair profit levels will not plant evil seeds in government minds.

5.3

OVERVIEW OF U.S. TAXATION OF MINING COMPANIES

Dennis J. McCarthy

Coopers & Lybrand, New York, New York

TAX TREATMENT OF EXPLORATION AND DEVELOPMENT EXPENSES

Exploration Expenditures

General Principles. Exploration expenditures are expenses incurred within the United States or the Outer Continental Shelf for the purpose of ascertaining the existence, location, extent or quality of any deposit of ore or other mineral. Section 617 of the Internal Revenue Code permits a taxpayer to elect to deduct these expenditures currently. In default of election, exploration expenditures must be capitalized and included in the basis of the mineral property for purposes of determining cost depletion.

Entitlement to Deduction. In order to qualify for current deduction, exploration expenditures must be incurred prior to the commencement of the development stage, i.e., when mineral deposits are shown to exist in sufficient quantity and quality to reasonably justify commercial exploitation. Generally, expenditures for geological and geophysical investigations, reconnaissance, surveying, testpitting, trenching, drilling, driving of exploration tunnels and adits and similar types of work constitute exploration expenditures if incurred prior to the commencement of development. However, expenditures to further delineate the extent and location of an existing commercially marketable deposit are not exploration expenditures.

Exploration expenditures may be incurred from within a producing mine to ascertain the existence of what appears to be a different ore deposit. Expenditures for property which is subject to an allowance for depreciation are not deductible under §617. However, depreciation properly allowable on assets used in exploration is an exploration expenditure.

Election to Deduct Exploration Expenditures. The election to deduct exploration expenditures is made by deducting these expenditures in the return for the first taxable year for which such treatment is desired. The election applies to all properties of the taxpayer. Once the election is made all exploration expenditures, in the year of election and thereafter, must be deducted, unless the election is revoked with the prior consent of the Commissioner.

The Tax Equity and Fiscal Responsibility Act of 1982 (TEFRA) made substantial changes in the treatment of exploration expenditures. For exploration expenses incurred after 1982 and before 1984 by a corporation, the amount currently deductible is limited to 85% of the expenditure. The remaining 15% must be capitalized and recovered over a five-year period. Although exploration costs capitalized under this provision are not part of the basis of the property for cost depletion purposes, investment credit may be claimed with respect to these expenditures.

Further changes have been added by the Tax Reform Act of 1984 with respect to expenditures made after 1984. The Act reduces the amount of otherwise currently deductible costs from 85% to 80%. Additionally, it limits the availability of the investment credit for capitalized mineral exploration costs to deposits located in the United States.

TEFRA also allows individuals a special method of recovery under which exploration expenditures may, at the election of the taxpayer, be amortized ratably over a 10-year period beginning with the year in which the costs are paid or incurred. If taxpayer does not use this method, the excess of the exploration expenditures actually deducted over the amount that would have been deducted if 10 year amortization had been elected is an item of tax preference subject to the alternative minimum tax.

Recapture of Exploration Expenditures. Exploration expenditures which have been deducted currently are included in income (recaptured) when and if the related property attains commercial production. The taxpayer may elect to recapture exploration expenses in either of two ways:

1. Include in gross income the entire amount of exploration expenditures previously deducted when the mine attains commercial production.

2. Forego depletion otherwise deductible with respect to the mine, to the extent of previously deducted and unrecaptured exploration expenditures.

This election may be made annually but all mines reaching the production stage during the year must be included in the election.

In general, miners elect to forego depletion allowances as it permits recapture to be spread over several years, in contrast to the inclusion in income method of recapture, in which the entire amount is recaptured in one year. Which method a taxpayer chooses depends on his income tax position. For example, a taxpayer with carryforward operating losses which might expire unused may well benefit from the current inclusion in income of previously deducted explorations costs.

Recapture on Disposition. Exploration expenditures are also recapturable if the property to which the expenditures relate is sold or assigned. This is accomplished by treating capital gain (taxed at 28 percent) from the sale or disposition of the mineral property, as ordinary income (taxed at 46 percent) to the extent of the exploration expenditures previously deducted but not recaptured. If a loss is recognized upon disposition there will be no recapture. In addition, there is no recapture upon a disposition of mineral property by gift, by transfer at death or in certain tax-free corporate or partnership transactions.

TEFRA provides that any exploration expenditures which are amortized over a 10-year period by individuals, or are capitalized by corporations and recovered over a five-year period, must be recaptured upon disposition of the property to which they relate but are not recaptured when the mine commences production.

Foreign Exploration Limitation. The current deduction for exploration expenditures made with respect to a deposit located outside the United States is limited to $400,000, reduced by the aggregate of all exploration expenditures (whether for foreign or domestic exploration) deducted or deferred by taxpayer during the preceding taxable years and all amounts (for domestic exploration) deducted during the current taxable year. Certain transferees of mineral property must, in determining the $400,000 limitation, take into account exploration expenses deducted by their transferor.

Development Costs

General Principles. Development expenditures are defined as costs incurred to prepare the mine for extraction which are incurred after the existence of ores in commercially marketable quantities has been disclosed. Thus, development expenditures must be incurred while the mine is in the development or production stage rather than the exploration stage.

A taxpayer can elect to deduct development expenditures currently or as the benefitted minerals are sold. The election only applies to that portion of the development expenditures which is in excess of the net receipts from the mine, received and accrued during the same taxable year and while the mine is in development. The current deduction or amortization of development expenditures, is taken into account in calculating the 50 percent of taxable income limitation for percentage depletion purposes. See Section II-C(2). Any deferred development expenses become a part of taxpayer's basis in the property, for all purposes other than cost depletion. This basis will be reduced as the deferred expenditures are amortized.

Pursuant to the provisions of TEFRA, for development expenditures incurred after December 31, 1982 in taxable years ending after that date, by a corporation, the amount currently deductible is limited to 85% of the development expenditures. The remaining 15% is capitalized (but not added to basis for purposes of determining cost depletion) and deducted over a five-year period. Investment tax credit is allowable on the 15% amount capitalized.

Further changes have been added by the Tax Reform Act of 1984 with respect to expenditures made after 1984. The Act reduces the amount of otherwise currently deductible costs from 85% to 80%. Additionally, it limits the availability of the investment credit for capitalized development costs to deposits located in the United States.

TEFRA also allows noncorporate taxpayers, a special method of recovery under which development expenditures may be amortized ratably over a 10-year period commencing with the year the costs are paid or incurred. If taxpayers do not use this method, the excess of the development expenditures actually deducted over the amount that would have been deducted if 10-year amortization had been elected, is an item of tax preference subject to alternative minimum tax.

Election to Defer Development Expenditures. In order to defer deduction of development expenditures, an election must be made. If no election is made, current deduction is required. The election to defer is made for each specific mine or deposit. Thus, taxpayer may defer development expenditures with respect to one mine and deduct them currently with respect to another. A new election must be made annually for each separate mine or deposit.

The decision to currently expense or defer development expenditures requires a clear understanding of the present value of various available tax benefits as well as insight into future business events. Some of the principal reasons for deferring the deduction for development expenditures include (a) the possible expiration of net operating losses, (b) limitation or expiration of carryovers relating to the investment tax and foreign tax credits, and (c) the desire to avoid the possible loss of the percentage depletion allowance due to the effect of a current deduction of development expenditures on the 50-percent-of-taxable income limitation.

Identification of Development Expenditures. Generally, development expenditures must be incurred for the purpose of preparing a mineral deposit for extraction. Thus, expenditures for sinking and securing a mine shaft that passes through an incompetent zone, including costs incurred for freezing the earth in the shaft area and for tubbing, cementing and grouting to secure the shaft from the intrusion of ground water are deductible as development expenses. Expenditures incurred to prepare a mine for production, by making the ore body accessible, such as costs incurred for driving shafts, drifts, cross cuts and for other production facilities to make the deposit accessible also qualify.

However, expenditures incurred in making an ore body accessible for production are not considered development expenditures if they take place on a continuing basis and if their amount is directly related to the rate of extraction of the mineral in the day-to-day mining cycle. Such costs are really in the nature of production costs. Thus, in the case of strip mining of coal, the cost of stripping overburden to expose the deposit for extraction was held not to be a development cost when removal was directly related to day-to-day mining operations.

Expenditures which are similar to exploration expenditures and which are incurred during development are considered development expenses. For example, expenditures for core drilling, incurred for the purpose of further delineating the extent and location of commercially marketable ore reserves, qualify as development expenditures, even though they are not incurred in the preparation of the mine for extraction. The purpose rather than the nature of the activity is the distinguishing factor between development and exploration expenditures.

Expenditures for acquiring or improving depreciable property do not qualify as development expenditures. However, to the extent that depreciable property is used in development, the depreciation is considered a development expenditure.

Carried-Interest Arrangements

If taxpayer agrees to pay exploration or development expenditures in connection with the acquisition of a fractional share of the working or operating interest, current deduction is available only to the extent of the fractional interest acquired. The remaining expenditures are considered part of the cost of the acquired interest and should be capitalized and recovered through depletion.

However, development expenditures incurred in connection with the acquisition of a fractional share of the working or operating interest may be deducted in full through the use of a carried interest. Under this arrangement, one party (the carrying party) may agree to incur development or exploration expenditures with respect to property owned by another party (the carried party) to earn an interest in that property. Generally the carrying party will retain the entire working interest in the property until his capital investment is recouped. After recoupment, a portion of the working interest reverts to the carried party and both parties share in profits, losses and capital expenditures.

The carrying party may elect current deduction of exploration and development expenses if he retains the entire working interest until recoupment. However, if the carried party participates in profits before the carrying party recovers his entire investment, the expenditures funded by the carrying party are capitalized as leasehold or property acquisition costs.

DEPRECIATION & DEPLETION

Depreciation

Accelerated Cost Recovery System.

GENERAL DISCUSSION: The Economic Recovery Tax Act of 1981 increased allowable rates of depreciation through the use of the Accelerated Cost Recovery System (ACRS). ACRS generally applies to tangible property, placed in service after 1980 and used in a trade or business or held for the production of income. ACRS replaces prior depreciation methods (with limited exceptions).

Under ACRS, depreciable property is categorized as either 3-year, 5-year or 10-year property, or 15-year real property, based upon the number of years necessary to fully depreciate an asset. It should be noted that the Tax Reform Act of 1984 has extended the ACRS write-off period for 15-year real property to 18 years where the property is placed in service after March 15, 1984. These "recovery periods" may be significantly shorter than the depreciation periods used under prior law. Generally, most mining equipment will fall into the five-year ACRS category.

ACRS does not differentiate between new or used property in the determination of applicable recovery periods or depreciation rates. The full cost of an asset may be depreciated without any reduction for salvage. Generally, under prior law, cost was diminished by salvage value, in determining allowable depreciation.

For all personal property and a limited class of real property, a half year's depreciation may be claimed in the year the asset is placed in service regardless of the month the asset was first used. No cost recovery deductions are permitted in the year of disposition. For most real property, the depreciation that may be claimed in the taxable year that the property is placed in service or disposed of is determined by the number of months during the year that the asset was in service.

ACRS was intended to apply only to assets acquired after 1980. Thus, "anti-churning" rules were provided to prevent taxpayers from qualifying property for ACRS that they or an affiliate owned or used during 1980. For property involved in a churning transaction, prior law and depreciation methods apply.

The table below specifies the applicable depreciation deductions under ACRS.

RECOVERY PERCENTAGES

Year of Ownership	Class of Investment		
	3-year	5-year	10-year
1	25	15	8
2	38	22	14
3	37	21	12
4		21	10
5		21	10
6			10
7			9
8			9
9			9
10			9
Total	100	100	100

The depreciation rates for 15-year and 18-year real property are accelerated write-offs based on the 175% declining balance method with an automatic switch to straight-line.

Election of Straight-Line Depreciation: A taxpayer who does not wish to use an accelerated method of depreciation with respect to recovery property may elect to use the straight-line method over the statutory recovery period or certain extended recovery periods.

Class of Property	Extended Recovery Period
3-year property	5 or 12 years
5-year property	12 or 25 years
10-year property	25 or 35 years
15-year property	35 or 45 years

Elections made with respect to personal property apply to all property in that class placed in service during the year of election. In the case of 15-year real property, the election is made on a property-by-property basis.

Units of Production Method. A taxpayer may, in appropriate circumstances, elect to apply the units of production method of cost recovery rather than ACRS. This method provides equal depreciation per unit of work performed by an asset. When the asset is acquired, the number of units of work which will be produced by the asset is estimated. The difference between the cost of the asset and salvage value, is divided by this estimate to determine a depreciation rate. For example, if a front-end loader costing $210,000 with a salvage value of $20,000 is estimated to have a working life of 19,000 hours, the depreciation rate is determined as follows:

$$\frac{\$190,000}{19,000} = \$10/hr$$

Thus, taxpayer may claim $10 of depreciation for each hour that the truck is used. When the truck has been depreciated to its salvage value, no further depreciation deductions are permitted.

Receding Face Doctrine: An expenditure which is incurred to maintain the present production of a mine because of the recession of the mine's working face may be deducted currently, even if it ordinarily would have to be capitalized and depreciated over the applicable recovery period. However, amortization is required if the expenditure increases the value of the mine, decreases production costs or restores property for which an allowance for exhaustion has been made. No special election or approval is required to apply the receding face doctrine.

Effects of Depreciation on Depletion: The use of ACRS will generally accelerate the recovery of capital costs but may decrease the percentage depletion allowance. Any increase in depreciation deductions resulting from the use of ACRS will reduce taxable income before depletion, thus reducing the "50% of taxable income" limitation on the depletion allowance. See Section II-C(2).

To alleviate this problem, taxpayer may wish to elect the straight-line or units of production method of cost recovery. Before this type of election is made, a discounted cash flow analysis should be undertaken comparing the benefit of the increased depletion deduction with the detriment of the slower depreciation deductions.

Depreciation Recapture: Generally, taxpayers are required to report gain on the sale of depreciable property as ordinary income to the extent of all or a portion of the depreciation deductions previously allowed with respect to the property. This applies even if the gain would otherwise be capital in nature.

Recapture of all depreciation deductions is required with respect to the following assets:

(1) Property which does not qualify for ACRS - all personal property.

(2) Recovery Property - personal property and 15-year nonresidential rental property, located within the United States which was not depreciated under the straight-line method.

However, for taxable years beginning after 1982, the amount of depreciation recaptured by a corporation, with respect to property in this class, is increased by 15% of the excess of (i) the amount that would be recapturable if the property were personal property over (ii) the amount of recapture, as computed above.

No recapture is required upon gifts, transfers at death and certain tax-free transfers.

Investment Tax Credit

General Discussion. A tax credit is allowed for investment in machinery, equipment and other depreciable tangible personal property which is recovery property or has a useful life of three years or more. Generally, for recovery property, the credit is equal to 10 percent of the cost of property with a recovery period of five years or longer and 6 percent of cost for property with a three-year recovery period.

For taxable years beginning after 1982, the amount of the credit which may be claimed in any taxable year is limited to the sum of (i) the lesser of the tax liability prior to application of the credit or $25,000, and (ii) 85 percent of the tax liability in excess of $25,000. Any investment tax credit which is not applied in the taxable year may be carried back three years and and forward fifteen years. Carryovers of investment credit are applied prior to using the current year's credit.

The maximum amount of "used" property cost with respect to which the credit may be claimed is $125,000 ($150,000 for taxable years beginning after 1987).

Generally, for property placed in service after December 31, 1982, the basis of assets that generate investment tax credit will be reduced by one-half of the amount of the credit. This reduction will affect both cost recovery deductions and gain or loss upon disposition of the asset. It can be avoided by electing, on a property-by-property basis, to reduce the credit by two percent; i.e., reducing the 10 percent credit to 8 percent, or the 6 percent credit to 4 percent.

Recapture of ITC. When property for which an investment tax credit was claimed is disposed of prior to the end of the life used for credit purposes, part or all of the credit is recaptured. The amount of recapture, which is a percentage of the initial credit, decreases with the passage of time. The amount recaptured is added to tax liability in the year of recapture.

Depletion

Nature of the Deduction. Although mining

companies are generally subject to the same tax rules as other entities, Congress has recognized the unique risks and the special nature of the mining industry by granting it a percentage depletion allowance. Percentage depletion permits the total deductions claimed over the life of a mine to exceed the total cost of the mine.

A taxpayer may claim a deduction for cost or percentage depletion, whichever is greater. This determination is made annually on a property-by-property basis. Percentage depletion is expressed as a percentage of the "gross income from the mine," the percentage varying with the type of mineral (Exhibit 2 specifies the depletion rates applicable to various minerals). The allowance for percentage depletion is limited to "50% of the taxable income from the property." Cost depletion is determined by amortizing the cost of the mineral property (with certain adjustments) over the total recoverable units of ore, as the ore is sold.

The deduction for depletion must be claimed when the mineral is sold rather than when it is extracted or shipped to an offsite storage facility. Depletion deductions reduce the adjusted basis of the mineral property. However basis is never reduced below zero.

Generally, only taxpayers possessing an economic interest in mineral property may claim percentage depletion. The "economic interest" concept requires: (1) an investment in minerals in place which legally entitles taxpayer to income from extraction of the mineral, and (2) that the mineral in place be the only source to which taxpayer can look for the return of his capital.

The requirement of an acquisition, "by investment", of minerals in place has been liberally construed. The investment may be acquired for cash, in exchange for other property, for services rendered, by gift, inheritance, or as a liquidating dividend. It is not necessary that the taxpayer incur a cost in the acquisition or retention of or that he have any tax basis in the investment. In fact, it is sufficient to show only that a clear capital interest exists.

Secondly, the interest acquired or retained must constitute an interest in "minerals in place". Generally speaking, there must exist a right to share in the minerals produced either by sharing in the proceeds from the disposition of such minerals or by taking production in kind. Taxpayer need not hold legal title to the minerals under state law.

Finally, in order to hold an economic interest, the minerals in place must be "the only source to which taxpayer can look for the return of his capital". This requires that taxpayer bear the ultimate risk of loss for a failure of production. In other words, he must depend entirely on the production of minerals for the return of his capital; there must be no alternative sources of recovery. For example, a contractual right to receive a definite sum of money within a definite time, with interim payments measured by production, is not an economic interest, because the right to share in the proceeds of production is incidental to the covenant to pay a certain sum.

Examples of economic interests include royalties, working interests, overriding royalties, net profits interests, and certain types of production payments.

Taxable Income from the Property. As noted above, the percentage depletion allowance cannot exceed 50% of "taxable income from the property". "Taxable income from the property" is defined as gross income from the property less allowable deductions (other than depletion) attributable to mining process.

An indirect cost, which is attributable to both mining and nonmining activities, must be apportioned on a reasonable and consistent basis in order to determine taxable income from the property. One acceptable method would be to use the relationship of the direct costs of each mining operation (to which the indirect cost relates) to total direct costs. However, other methods may be acceptable, depending on the particular facts and circumstances of a taxpayer's operations.

Some of the more common indirect costs to be considered in computing taxable income from the property are selling expenses, general and administrative overhead, state and local income taxes and foreign income taxes (if deducted rather than credited against U.S. income taxes).

Gross Income from Mining. As noted previously, percentage depletion is expressed as a percentage of the gross income from mining. Basically, gross income from mining is the value of minerals extracted from the mine after the application of specifically enumerated mining processes. The computation of gross income from mining has been a source of continual controversy for many mining companies. Two of the principal areas of controversy are:

    (i) the cutoff point (i.e., when do mining processes end and nonmining processes begin); and

(ii) the value of the mineral at the cutoff point.

Cutoff Point. In determining gross income from mining, taxpayer must compute the value of extracted minerals at the cutoff point. The cutoff point is that time when mining processes ends and nonmining processes begin. Congress has defined mining to include:

(i) Extraction of the ores or minerals from the ground

(ii) Transportation of ores from the point of extraction to the mill in which the defined mining treatment processes are applied. The transportation generally cannot exceed 50 miles unless the Commissioner of Internal Revenue finds that physical requirements dictate transportation beyond this distance

(iii) Certain enumerated mining treatment processes

It is important to note that a mining treatment process for one mineral may not be considered a mining process for another. Congress has defined the range of what are considered to be mining processes narrowly, and these do not necessarily include all the treatment processes normally associated with a particular mineral.

Method of Valuation. The second major area of conflict in determining gross income from mining is the manner of ascertaining the value of the mineral at the cutoff point. To the extent that minerals are sold to independent third parties at the cutoff point, the amount realized from that sale is generally the basis for determining gross income from mining. However, where the mineral is subject to nonmining processes prior to sale or is used as feed material in other operations, gross income from mining is computed by using a representative field or market price or by other appropriate method.

The representative field or market price method uses actual competitive sales of minerals at the cutoff point to determine an approximate price at which the mineral could have been sold in the market. If no representative price exists for a mineral of like kind or grade, a representative price for a like-kind mineral that is not of like grade may be used, with appropriate adjustments for difference in grade; however, those adjustments must be readily ascertainable.

If it is not possible to ascertain a representative field or market price, the proportionate profits method must be used unless either the Internal Revenue Service or the taxpayer determines a more appropriate method. One of the leading alternative methods is the representative schedule method, which uses representative charges for nonmining processes to work back from the refined mineral price to a mining cutoff price.

Generally, the mining company will attempt to use either a representative field or market price method, or representative schedule method, to determine gross income from mining in those instances where concentrates are not sold at the cutoff point. However, the IRS has been attacking the methods used by some companies and has attempted to apply the proportionate profits method. The proportionate profits method of determining gross income from mining is based on the theory that each dollar of cost is treated equally and is considered to earn an equal amount of profit. Gross income is computed by multiplying the sale value of the mineral by a fraction, whose numerator is equal to mining costs and whose denominator is equal to total mining and nonmining costs.

The application of percentage depletion is illustrated below:

(000s omitted)

Assumptions (U.S. copper mine):

Proceeds from sale of copper concentrates		$1,000
Gross Income from Mining		1,000
Cost of Mining and Milling	$ 600	
General, Selling and Administrative Expenses Attributable to Mining and Milling	200	800
Taxable Income from the Property (before deletion)		$ 200

Percentage Depletion Determination:
Lesser of (a) or (b)

(a) 15% of $1,000 $150

(b) 50% of $200 $100

    Percentage Depletion $100

Cost Depletion. The cost depletion allowance is computed by dividing the basis of the property by the total estimated number of recoverable units of ore to obtain a unit depletion rate, which is then applied to the units sold within the taxable year. Given the facts from the previous example, the cost depletion computation would be made in the following manner:

Basis of Property	$200,000
Total Ore Reserves	20,000,000 lbs. of copper
Depletion Rate	$.01/lb. of copper
Pounds of Copper Sold in the Year	600,000
Cost Depletion Allowance	$6,000

SALES AND EXCHANGES OF MINING PROPERTY

Classification of Property Interests

Mineral Interest. The term "mineral interest" refers to ownership of the minerals in place, without consideration of the surface ownership. The ownership of a mineral interest generally carries with it the right to make reasonable use of the surface to develop the property for production.

Operating Interest. The "operating interest", also called the "working interest", is the total mineral interest minus any nonoperating interests, such as royalty interests. It is burdened with the cost of developing and producing the minerals and thus generally receives a large share of the revenues from the property. Although the holder of the operating interest may subcontract the development and production work to others, he, nevertheless, ultimately bears the costs of exploration, development and operation.

In the usual mineral leasing arrangement, the mineral interest owner will retain a fractional interest (royalty interest) in the minerals that he previously owned in full, and convey the remaining interest (working interest) to another party who must develop and operate the property. The holder of the operating interest can sell all or a portion of that interest in order to spread the risk or obtain financing.

The share of mineral production accruing to the holder of the operating interest is taxed as ordinary income subject to depletion. Any amounts attributable to a nonoperating interest, such as a royalty interest, are excluded from the operator's gross income.

Royalty Interests. A royalty interest is a right to minerals in place that entitles its owner to a specified fraction, in kind or in value, of the total production from the property, free of the expenses of development and operation, except severance taxes. It is, in essence, a mineral interest stripped of the burdens and rights of developing the property.

The classic example of a royalty interest is the share of the gross value of production retained by a landowner upon the grant of a mineral lease, whereby the grantee is given the exclusive rights and burdens of development. A royalty interest may also be created by assignment, as when the owner of mineral rights conveys a fractional interest in the production from the property, reserving to himself all operating rights and burdens.

Amounts attributable to royalty interests are excluded from the operator's gross income. They are taxable to the royalty-holder as ordinary income subject to depletion.

Overriding Royalty and Net Profits Interests. An overriding royalty is very similar to the basic royalty interest, except that it is created out of the working interest and therefore its life is limited to that of the working interest. The holder of an overriding royalty is entitled to a specific fractional share of gross production and has no responsibility for developing or operating the property.

A net profits interest is similar to an overriding royalty in that it is created out of the working interest, has a similar duration, and bears no portion of the operating or development costs. However, the holder of a net profits interest is entitled to a share of the operator's net profit from operation of the property rather than a share of gross production.

Both net profits and overriding royalty interests are taxed in the same manner as royalty interests. Thus, amounts attributable to these interests are excluded from the operator's gross income. The royalty holder has ordinary income subject to depletion.

Production Payments. A production payment is a right to a specified share of the production from minerals in place (if, as and when produced), or the proceeds of production. It must have an expected economic life which is of shorter duration than the economic life of the related mineral property. Like a royalty interest, a production payment is a passive or nonoperating interest as long as the working interest owner fullfills certain agreed upon conditions. However, it differs from a royalty interest in that it is extintuished after the holder recoups his profit and capital from the mineral interest.

Carved-out production payments

    Nondevelopment: A production payment that is "sold or carved out" of a larger interest for reasons other than exploration and development is treated for tax purposes as a loan to the owner of the property, and not as an economic interest in minerals in place.

    Development carve-outs: A development carve-out (DCO) is a production payment whose proceeds are pledged and used by the mine owner for development or exploration. For tax purposes, the owner of the working interest is not required to recognize income at the time he assigns the DCO and receives payment therefor. Expenditures for development, which are funded by the sale of a DCO, are not recognized as development expenditures for tax purposes by the working interest owner. However, the proceeds realized from the sale of the mineral that are paid to the DCO owner are excluded from the taxable income of the working-interest owner, and the costs of mining these minerals are currently deductible. The DCO proceeds, in fact, must be used for developing the mineral property; to the extent these proceeds are not so used, a portion of the DCO loses its characterization and is treated as a loan.

    The DCO owner is considered to have made a capital investment. His tax basis in the DCO equals the cash paid and the adjusted basis of property exchanged. As payments are received by the holder of a DCO, a portion is treated as return of capital (usually cost depletion) and the balance is considered ordinary income.

    Retained production payments: A production payment retained on the transfer of a working interest will be treated as a purchase money mortgage loan in the burdened property.

DISTINGUISHING BETWEEN SALES AND LEASES

    The tax treatment of a transfer of a mineral interest frequently turns on whether the transaction is classified as a sale or lease. The gain recognized upon a sale of mineral property will often be taxed to the seller at capital gains rates (except for any recapture of exploration expenditures). However, if the property is leased, payments to the lessor will generally be taxed as ordinary income subject to depletion. On balance, a transferor of mineral property will usually prefer that the transaction be treated as a sale.

    In contrast, the transferee will normally prefer lease treatment. Upon a sale, payments by the buyer are included in the the basis of the mineral property and can only be recovered through cost depletion. However, these payments normally have little if any value, from a tax perspective, since percentage depletion is usually greater than cost depletion. In contrast payment made by a lessee are normally deductible (although they may decrease the lessee's gross income from mining, and thus, his percentage depletion allowance).

    Generally, if all or a portion of the operating interest in a mineral property is transferred and an economic interest in the property is retained which will extend over the life of the operating interest, the transaction will be treated as a lease. Thus, if A, the owner of the mineral rights (or a working interest) assigns a working interest to B, and retains a royalty, he has created a lease.

    Sale treatment will be accorded a transaction in which transferor gives up all of his rights in the mineral interest for a fixed payment which is not dependent upon production. If a fixed price is received in exchange for a fixed amount of minerals, which is less than the entire amount of the mineral in place, the transaction will also be treated as a sale. However, this type of transaction may be held to be a production payment.

    The assignment of any type of interest other than operating rights will generally not be treated as a lease. Thus, if the owner of a working interest assigns an overriding royalty or the owner of a royalty assigns a portion of the royalty, the transaction would not qualify as a lease. Sale treatment will be mandatory where the owner of a working interest assigns that interest, but retains an interest which does not qualify as an economic interest.

    In general, when the owner of a mineral interest assigns an undivided or divided interest therein, in exchange for cash or its equivalent, sale treatment will be mandatory. This applies even if only a portion of the working interest is assigned.

5.4

THE TAX STRUCTURE OF FOREIGN MINING INVESTMENTS IN THE UNITED STATES

Nicasio del Castillo

Coopers & Lybrand, New York, New York

INTRODUCTION

The tax structure of foreign mining investment into the United States has a significant impact on the financing and the profitability of the operation. For these reasons, it is critical for the foreign investor to develop and implement a well planned tax strategy that will enable its venture to:

(1) reduce its U.S. income tax liability

(2) lower U.S. withholding taxes when funds are transferred from the United States, and

(3) decrease or defer home country taxation on its U.S. earnings

ALTERNATIVE FORMS OF INVESTMENT

A threshold business and tax consideration is the form in which the investment is to be made. The most common alternatives include: forming or acquiring a U.S. mining corporation, directly establishing a branch mining operation, or entering into a joint venture or partnership with a U.S. mining company.

U.S. Corporations

The most common structure through which foreign mining companies engage in business in the United States is to form a U.S. mining corporation.

Forming a U.S. Corporation. If the investor decides to establish a new U.S. corporation the organization is relatively simple. No distinctions are made in the United States between "public" and "private" forms of incorporation. Minimum share capital is not usually required and share capital may consist of shares with no par value. Since corporate law is the responsibility of the individual states, the foreign investor must choose a state in which to incorporate.

If appreciated mining property is transferred to a corporation solely in exchange for stock or securities in that corporation, and the transferors have control of the corporation (defined as at least 80 percent of all voting shares and at least 80 percent of all other shares) immediately after the exchange, no gain or loss will be recognized. However, if the transferors receive cash or other mining properties in addition to stock, any resulting gain may, in certain situations, be taxed to the extent of the cash and the fair market value of other property received.

Acquiring an Existing U.S. Corporation. Acquiring an existing U.S. mining corporation is not as straight-forward as forming a corporation. This alternative may be accomplished through either a stock or an asset acquisition. Generally, with a stock acquisition the purchaser inherits all of the business and tax attributes of the former mining corporation. If the investor is able to acquire mining assets, however, the purchaser will usually be able to avoid any unknown or hidden liabilities that could be attached to an existing corporation.

Another principal advantage in acquiring assets is that the purchase price may be directly allocated to the acquired mines or other mineral-related assets with the result that the assets' depreciable base may be increased. Assuming the domestic seller and the foreign purchaser can agree, it is generally advisable to allocate the purchase price in the sales agreement to specific individual assets. From the foreign investor's standpoint, it is preferable to allocate the purchase price to depreciable assets such as buildings and equipment versus an allocation to mineral land holdings or intangibles such as goodwill, which are generally not depreciable.

It is not uncommon, however, for a seller mining company being acquired to insist for tax reasons that the investor purchase shares rather than assets. If the foreign investor acquires 80% or more of the total shares of the acquired mining corporation's stock within a 12 month period, then an election may be made to treat the stock purchase as a purchase of mine assets for tax purposes. When such an election is made, the domestic corporation's tax attributes are terminated and the depreciable tax cost of its mining assets are adjusted, as of the stock purchase date, to reflect the price that the foreign purchaser has paid for the stock. In effect, the domestic mining corporation is treated as if it had transferred all its assets to the foreign corporation in complete liquidation on the date of its stock purchase even though no actual liquidation is required or will have occurred.

Elections to treat a stock acquisition as an asset acquisition must be carefully planned because their success largely depends on how the purchase price of the acquired corporation's shares is allocated to the assets received. It is clearly advantageous to show high values for assets with short lives (e.g., ore stock or concentrates and fixed mine assets). Allocations to assets that cannot be amortized for tax purposes (e.g., costs of incorporation, goodwill, etc.) offer no benefit to the foreign purchaser.

Determining Federal Income Tax

The Internal Revenue Code generally treats any corporation as a separate taxable entity. Thus, the net economic results of mining operations are taxed to the corporation and do not pass through to shareholders. Income distributed to shareholders is generally subject to double taxation, once, at the corporate level when earned, and a second time, to the shareholders, upon distribution. Shareholders are usually not taxed on earnings which are retained by the corporation.

Taxable income is determined under U.S. tax accounting methods. The methods used will generally differ - in some respects at least - from the accounting principles used for preparing financial statements. A common reason for the disparity is that taxpayers can avail themselves of tax accounting rules that significantly reduce taxable income. Listed below (Table 1) is a comparison between the principal tax accounting methods used to compute a corporation's taxable income and the methods generally used for financial statement purposes.

TABLE 1

Comparison Between the Principal Tax Accounting Methods Used to Compute a Corporation's Taxable Income and the Methods Generally Used for Financial Statement Purposes

Item	Tax Return	Financial Statement
Depreciation		
Real property	18 years, accelerated depreciation (15 years for property placed in service before March 16, 1984)	Useful life, straight-line depreciation
Personal property	3, 5 or 10 years (depending on type of asset), accelerated depreciation	Useful life, straight-line depreciation
Pollution control equipment	5-year amortization	Useful life, straight-line depreciation
Mine exploration & development costs	May elect to expense as incurred	Capitalization and amortization over units produced
Valuation of inventories	May elect last-in, first-out (LIFO) regardless of flow	Generally, financials must conform to tax treatment if Lifo elected; some possibility of other valuation in non-U.S. consolidated financial statements
Interest and taxes during construction	10-year amortization for real property, otherwise deductible as accrued	Capitalization
Income on deferred payment sales	Installment method may be elected	Recognized in year sold
Long-term contracts	Income may be deferred until contract completed	Percentage-of-completion method of income recognition
Organization expenses	5-year amortization	Capitalization
Depletion	Percentage depletion (hard minerals)	Based on Cost

Once a U.S. Corporation's worldwide taxable income is determined, the following rates apply:

Taxable Income ($)	Tax on Column 1 ($)	% on Excess
0	0	15
25,000	4,250	18
50,000	9,250	30
75,000	16,750	40
100,000	26,750	46

The benefit of the lower bracketing is phased out when taxable income exceeds $1,000,000.

Two principal exceptions apply to the above general corporate income tax rates. First, long-term capital gains are taxed at a special preferential rate of 28%. On the other hand, capital losses may be used to offset capital gain, but may not be used to offset ordinary taxable income. Corporations may generally carry excess unused capital losses back to offset capital gains in its three prior taxable years and forward against future capital gains for a period of five years.

A second exception to the general corporate tax rates applies to gain or loss on the sale of depreciable personal property used by the mining company (e.g., machinery and equipment) and real property used in the company's operations which has been held for more that one year. Property falling into this category is to be separately aggregated. Any net gain, in excess of prior tax depreciation deductions, is capital in nature and any net loss is ordinary. Depreciation deductions (or the excess of accelerated over straight-line depreciation in the use of real property), to the extent of the gain are generally "recaptured", i.e. treated as ordinary income.

Exploration and Development Expenditures. With respect to exploration and development expenses incurred in tax years ending after 1982, the amount available for a current deduction is limited to 85% of the expenditure (80% for expenses incurred after 1984). The remaining 15% (or 20%) is capitalized and recovered using rates equivalent to the Accelerated Cost Recovery System (ACRS) table for five-year property. Accordingly, the capitalized exploration and development costs are not a part of a property's basis for depletion purposes. Also, the amortized portion of the capitalized exploration expenditures is recaptured only upon disposition of the property and is not recaptured when the mine commences production. Investment tax credit is available on the capitalized amounts provided the property is located in the United States.

Investment Tax Credits. For mining property placed in service after December 31, 1982, the basis of assets that generate investment tax credits will be reduced by one-half of these credits. This reduced basis will be used to determine both cost recovery deductions and gain or loss when the asset is sold or exchanged; except that if the investment credit is recaptured, one-half of the recapture amount is restored to basis. For leased equipment for which the lessee claims the credit, the lessee must include 50% of the ITC in income over the asset's ACRS period. The reduction in basis can be avoided for the regular investment tax credit on a property-by-property basis, by electing to reduce the credit by 2% (i.e., reducing the 10% credit to 8% for five-year property or the 6% credit to 4% for three-year property). This election is only available for the regular ITC. It should be noted that special transitional rules apply to assets constructed or acquired under contracts entered into after August 13, 1981 and effective on July 1, 1982, provided the assets are placed in service before 1986.

Depletion. Although mining companies are generally subject to the same tax rules as other taxable entities, Congress has recognized the unique risks and the special nature of the mining industry by granting it a percentage depletion allowance. Percentage depletion is an allowance expressed as a percentage of the "gross income from the property," the percentage varying with the type of mineral (see Exhibit II for depletion rates for the various minerals). Under the Tax Equity and Fiscal Responsibility Act of 1982 (TEFRA), the amount allowable to a corporation for percentage depletion with respect to iron ore and coal (including lignite) has been reduced. The allowance for percentage depletion is limited to 50% of the taxable income from the property. Depletion is allowed as a deduction when the mineral is sold rather than when it is extracted or shipped to an offsite storage facility. The taxpayer claims a deduction for cost or percentage depletion, whichever is greater with respect to each property and this determination is made annually. The percentage depletion allowance is not limited in amount to the taxpayer's basis in the property being mined, as is the case for cost depletion.

In addition to the above income tax rules, a 15% minimum tax is levied on certain items of income and expense that are regarded as tax-preference items. Common preference items for mining companies include the excess of percentage depletion over the property's tax basis and accelerated depreciation over straight-line depreciation for real property. For corporations, the minimum tax is imposed on the total tax preferences over the greater of: 1) $10,000 or 2) the income tax paid in the current year. Payment of the minimum tax is deferred if the benefit from the preference item is deferred (e.g., as a result of a net operating loss).

Consolidated Income Tax Returns. U.S. corporations that are 80% or more owned and controlled by a U.S. corporation may be included in a consolidated income tax return. This arrangement offers the advantage of offsetting losses from affiliated corporations against the profits of other affiliates in determining consolidated U.S. tax liability. Thus, mining companies may obtain a current benefit from certain preproduction losses and losses from other operations, assuming of course, that the consolidated group has sufficient taxable income to use the losses.

State and Local Income Taxes. State and local income taxes are also imposed on U.S. corporations. Despite the fact that these taxes are a deductible expense in computing taxable income for federal income tax purposes, they may be a significant cost that should be considered. Most states levy their respective income taxes based upon a corporation's federal taxable income with minor adjustments. Those corporations that are active in more than one state must usually allocate their state tax liability on the basis of a formula reflecting sales, property and payroll within the respective states where the corporation is conducting business.

California and a growing number of states impose the so called unitary tax. This method of taxation requires that a portion of the consolidated worldwide income of an affiliated group of companies be allocated to the state. This controversial system of taxation frequently produces anomalous results as a unitary tax state may impose a levy notwithstanding the fact that the subsidiary or division doing business in that particular state is not profitable.

FIRPTA. The Tax Reform Act of 1984 made major changes to the rules governing taxation of gains from sales of United States real property interests (USRPI) by foreign owners. Most importantly, the new law introduced withholding as the primary method of enforcement, instead of the extensive disclosure requirements of the Foreign Investment in Real Property Tax Act of 1980 (FIRPTA). As of January 1, 1985, purchasers of USRPI from foreign owners will generally be required to withhold the lesser of 10% of the selling price or the seller's maximum tax liability. The IRS will be required to make prompt determinations of seller's tax status upon request. With respect to reporting requirements, the 1984 Act generally requires disclosures only by the direct owners who would be liable for tax.

A USRPI includes a mine, well or other natural deposit as well as mineral royalty and personal property associated with the use of real property (including mining equipment). It also includes an interest in a U.S. corporation the value of whose USRPI constitutes more than half the value of all the company's real estate interests as well as its trade or business assets. Consequently, foreign investors owning stock in a U.S. mining company which meets the above criteria may be subject to the above rules. However, if the stock is traded on a securities exchange, the investor's interest will be a USRPI only if it equals more than 5% of the stock of the U.S. company.

Certain U.S. tax treaties which currently provide a tax exemption for gains on dispositions of capital assets (including USRPI) will continue to override FIRPTA until January 1, 1985. Renegotiation of a treaty prior to this date may further continue a treaty exemption for up to two years after the new treaty is signed.

Withholding Taxes. Dividends, interest, and royalties paid to foreign investors are generally subject to a U.S. withholding tax. There are no state or local withholding taxes on dividends, interest, or royalty payments. The statutory rate of withholding tax imposed by the Federal government is 30%, but that rate is frequently reduced by tax treaties. Income tax treaties between the United States and other countries reduce or eliminate the applicable withholding tax rates on dividends, interest and royalties. Exhibit I provides a summary of the withholding rates applicable to most countries through which foreign mining investment is likely to be made.

Dividend Planning. In many cases, paying dividends will result in little or no incremental home country income tax. As a general rule, tax planning in these cases is limited to a reduction of U.S. income and withholding tax. In those countries which do tax the dividend, no taxes are due if creditable U.S. taxes are equal to, or higher, than, the rate of taxation in the home country. If the U.S. taxes paid by a foreign owned subsidiary have been significantly decreased through tax accounting elections (e.g., accelerated depreciation, percentage depletion), then careful tax planning will frequently be required to minimize excess home country taxation.

To minimize the tax burden on the foreign investor, the home country tax implications of distributing the earnings from a U.S. corporation must be considered. While it is not possible to identify the precise foreign tax treatment of earnings distributed by a U.S. corporation engaged in mining operations without specific information, most countries apply one of the following rules:

° exempt net dividends from tax

° tax dividends received net of U.S. withholding tax

° tax gross dividends paid including amounts withheld in the U.S., and provide a credit for income and withholding tax paid on the dividend received.

If an intermediate holding company is not employed, a foreign investor from the countries listed in Table 2 would receive approximately the following amounts of net dividend on 100 U.S. dollars of pre-tax income.

Pretax income of U.S. corporation	100
State income tax	(6)
Federal income tax (46% of 100 less 6)	(43)
Gross dividend	51

TABLE 2

Overall Tax Burden on Dividends Paid by U.S. Corporation to a Foreign Corporate Shareholder

Investing Country	Gross Dividend	U.S. Withholding Tax*	Home-Country Tax Before Credit	Credit for U.S. Tax	Net Dividend
Canada	51	8	–	–	43
Germany	51	8	–	–	43
Italy	51	3	12	4	40
Japan	51	6	49	49	45
United Kingdom	51	3	52	52	48
France	51	3	–	–	48
Netherlands	51	3	–	–	48

* The withholding rates applied are the lowest available once certain ownership requirements have been met. The requirements vary by treaty from 10% to 95%.

Reducing Home Country Taxation

Common techniques to minimize potential home country taxation include: structuring the investment to permit tax free payment of intercorporate debt, retaining the earnings in the U.S. corporation, and interposing a foreign holding company between the foreign parent and its U.S. subsidiary.

By capitalizing a U.S. subsidiary with debt as well as equity (to the extent permitted under U.S. debt-equity principles) the interest due on the debt will be a tax-deductible expense to the U.S. subsidiary, thereby reducing its U.S. tax liability. In contrast, dividend payments are not a deductible business expense. The repayment of loan principal is a tax-free transaction. This distinction is even more important since the Tax Reform Act of 1984 eliminates the 30% withholding tax on portfolio interest paid by U.S. borrowers (including the United States) on the following obligations:

1) in registered form, or

2) in unregistered (i.e., bearer) form, provided the requirements for exemption under the TEFRA registration requirements are met.

However, foreign banks will remain subject to the 30% tax on loans made in the course of business as will interest paid to a foreign person who owns 10% or more of the obligor. This new rule will eliminate the 30% tax only for interest paid on obligations issued after July 18, 1984, the date of enactment. Thus, obligations outstanding as of that date will remain subject to the 30% tax.

Also, retention of earnings within a U.S. corporation will preclude further taxation as long as the earnings of the corporation are not accumulated beyond the reasonable business needs of the company. Since mining operations require substantial capital, mining companies have a great deal of flexibility in arguing that their earnings need to be accumulated.

Another possible planning technique would be to structure a U.S. investment through a holding company in a favorable tax jurisdiction. For example, dividends paid to a Dutch holding company are only subject to a 5% U.S. withholding tax and the Dutch tax authorities generally will issue advance rulings holding that dividend income from U.S. subsidiaries will be tax exempt provided the subsidiaries are operating companies. Other advantages of using a Dutch holding company include a 1% Dutch capital tax which may be minimized since there is no capital gains tax on the sale of stock in The Netherlands. By utilizing a holding company structure, the foreign investor will probably be able to redeploy its earnings derived from its U.S. mining operations without incurring incremental expense of home country taxation. This technique may be particularly useful to investors from high tax jurisdictions which do not have income tax treaties with the United States.

U.S. BRANCHES

Foreign mining companies can directly conduct their U.S. operations instead of operating through a U.S. corporation. Directly conducting activities in the U.S. in this manner is commonly referred to as a "branch." While the term "branch" is not used in the Internal Revenue Code, tax treaties between the United States and other nations include this term in the definition of "permanent establishment" which covers limited as well as extensive activities conducted through an office or other facility located in the United States.

Effectively Connected Business Income

The general rule is that active business income generated by a U.S. branch of a foreign business is taxed in the same manner and at the same rates in the case of as domestic corporations. Technically speaking, the income must be effectively connected with the conduct of a U.S. trade or business for this tax treatment to apply. Phrased in other words, the U.S. activities not carried on by the U.S. branch are usually not taxable under U.S. tax law.

Periodic Income

Periodic-type income not effectively connected with a U.S. business is subject to a flat 30% statutory rate of tax unless provided otherwise by a treaty. Periodic income generally includes passive investment types of income such as dividends, rents, royalties and interest. To the extent that periodic income is deemed to be effectively connected with a U.S. trade or business, that income will be taxed in the same manner as a domestic corporation would be taxed on that same income. The general rule is that, if periodic income is derived from assets used in or held for use in the conduct of a business or if the business activies significantly contribute to the realization of that income, then that income will be deemed to be effectively connected with a U.S. trade or business.

Home Country Taxation

U.S. branch income is frequently exempt from home country taxation through either the foreign country's domestic laws or tax treaties. If that is the case and U.S. branch income is not further taxed, a foreign investor may benefit from directly conducting its operations in this form. One potential benefit associated with directly investing through a branch is that U.S. losses, common among new mining operations, may, in the case of certain countries, be offset against the investor's home country taxable income. West Germany, for example1, permits branch losses to be deducted against a German corporation's taxable income. The benefit of this rule is limited by the fact that an equivalent amount of the branch's subsequent income usually must be taken into income when it is earned.

Those countries that do not exempt U.S. branch income typically do afford investing corporations relief by providing tax credits against home country taxation for U.S. taxes paid. Those same countries also usually permit U.S. losses to reduce the worldwide taxable income of the investing corporation which helps ease the financial burden of commencing a new mining operation.

PARTNERSHIPS

Since mining ventures involve, in many cases, significant risks, it is often advantageous for a company to spread them by encouraging others to invest. This enables a mining compnay to invest in several ventures on a reduced scale instead of investing in a single operation. As a result, noncorporate investment structures are frequently sought when less than an 80% interest in a venture is owned by one entity.

In its simplest form, a partnership permits a proportionate sharing of capital expenditures, income and expenses. For federal income tax purposes, the partnership is not taxed; instead, the partnership income or loss is passed through to each partner and is includible in each partner's federal income tax return. Most tax elections (e.g., the use of an accelerated method of depreciation and the deferral of development expenditures) are made by the partnership. Elections with respect to the deduction and recapture of exploration expenditures are made by the individual partners.

Partnership losses can be deducted by

the partner - frequently, a subsidiary of a parent corporation - to the extent of the partner's basis in the partnership. Generally, a partner's basis equals the capital contributed and a pro rata share of debt for which the partner is at risk.

The partnership form has been used where one party holds the rights to a mineral property but does not have adequate financial resources to exploit the property fully. Typically, a new investor would be invited into the venture to provide the necessary financing to develop the property and acquire the capital equipment. A tax feature unique to partnerships is the possibility of disproportionately allocating various income and expense items among the partners, provided these allocations have economic substance. Thus, development expenditures and depreciation on equipment financed by one partner can be allocated to that partner; other costs, revenue and expenses could be allocated on another basis. Alternatively, the partnership agreement may provide that the financing partner's capital investment be recouped from profits before any profit is shared among the other partners. Because of the flexibility of a partnership, it is an attractive vehicle for ventures where the parties have different financial situations or bring dissimilar assets to the partnership.

Sale of Partnership Interest

Generally, any gain realized upon the sale or exchange of a partnership interest will be capital in nature. However, to the extent of the selling partner's pro-rata share of partnership inventory which has substantially appreciated (i.e., inventory items whose fair market value is at least 120% of their basis to the partnership and which collectively make up more than 10% of the value of partnership's assets, other than money) and certain untaxed receivables, the gain will be ordinary, rather than capital. In addition, to the extent that the selling partner is relieved of a liability he is deemed to have received a distribution of cash, taxable, at capital gains rates, to the extent that it exceeds the basis of the partner's interest in the partnership. "Untaxed receivables" for this purpose, include the partners share of items which would be recaptured, if the related property was sold at fair market value.

Witholding Tax on Foreign Distributions

Distributions from a partnership are usually tax-free since the income effectively connected with a U.S. business has already been taxed to the partners; consequently no withholding tax is due on distributions to foreign partners. If a U.S. partnership has passive income not connected with a U.S. business, it must withhold tax from nonresident aliens and corporations on amounts of their distributive share which include such passive income.

Recognition as a Partnership for Tax Purposes

Notwithstanding the fact that foreign investors may have entered into a partnership agreement, the U.S. tax law will not automatically recognize an operation as a partnership. Generally speaking, there is a substantial risk that an operation will be treated as an "association" for tax purposes and taxed as a corporation if more than two or more of the following characteristics exist:

° centralized management

° limited liability

° continuity of life

° free transferability of interests.

An entity has continuity of life if the death, disability or withdrawal of any member of the organization will not cause its dissolution. Partnerships subject to one of the U.S. Uniform Partnership Acts generally lack this characteristic.

Centralized management means that the exclusive continuing power to make necessary management decisions is be concentrated in a managerial group, composed of less than all of the members, that has authority to act on behalf of the organization. A general partnership organized under one of the Uniform Partnership Acts generally does not have centralized management.

Limited liability exists if, under local law, there is no member of the organization who is personally liable for its debts. Personal liability means that the personal assets of a member may be used to satisfy claims if the asset of the organization are insufficient. In the case of a limited partnership, a general partner (who, under state law is generally considered to have unlimited liability) will be deemed, for federal tax purposes, to have limited liability, if he has no substantial assets, other than his interest in the partnership and acts as an agent of the limited partners.

Finally, free transferability of interests is present if a member is free to assign all of the attributes of his beneficial interest in the organization without the consent of other members. Thus, a member must be free to transfer not only his interest in the profits of the venture, but also his right to share in its control and assets. There is no free transferability if, under state law, such a transfer results in the dissolution of the old organization and the creation of a new organization.

JOINT VENTURES

Alternatively, a mineral joint venture is another non-corporate structure that permits the parties to jointly extract and produce minerals and to share costs jointly. Code section 761(a) affirmatively permits participants to elect not to be treated as partnership provided the operation does not engage in a profit-making activity. Theoretically, the participants are limited to receive only the minerals extracted or produced on their own account. This is the principal distinction between this form of operation versus a partnership which generally sells the mineral extracted or produced for the benefit of all partners. In restricted circumstances, however, joint venturers may appoint one party (or a third party) to sell the minerals without jeopardizing their election not to be treated as a partnership.

Since a true joint venture elects out of partnership treatment, each party can make his own tax elections with respect to his share of the venture's assets, expenses, and income generated from the mineral. This provides each participant with maximum flexibility to plan for his individual tax and financial needs.

Generally the election not to be treated as a partnership must be made and filed with the District Director. It is irrevocable as long as the venture remains qualified or until approval of revocation is secured from the Commissioner.

Although the "election out" may give investors more individual flexibility, it is not generally made when the investors have either contributed dissimilar assets or require profit-sharing arrangements that provide for non-pro-rata allocations. This is because the allocation of income or expenses between participants in a joint venture cannot differ from their ownership interests.

SUMMARY OF FACTORS TO CONSIDER IN STRUCTURING A MINING OPERATION

As noted earlier, the selection of the appropriate entity or entities is an important decision that can have significant effects. Tax considerations, as well as many other factors, play an important role in the selection process. In analyzing the choice of entity, the following factors should be considered:

Start-up Losses

If the venture is expected to incur losses during development and the participants would like to utilize these losses to reduce their taxable income, a partnership or a branch is generally a better choice than a corporation. Many countries permit foreign losses to offset home country taxable income. However, the benefit of this rule is usually limited by the fact that an equivalent amount of the branch's subsequent income must be taken into income when it is earned.

This is because a corporation is considered a separate taxable entity and generally does not pass losses through to its shareholders. Thus, early operating losses are usually not deductible by shareholders. They may, however, be used in later taxable years to offset venture income. In addition, if the venture is organized as a corporation, an 80% U.S. corporate shareholder may be able to take advantage of venture losses immediately by filing a consolidated return with other profitable affiliated corporations.

Earnings

If the venture expects to pay out a substantial portion of its earnings to participants, the use of a partnership or a branch operation, rather than a corporation, may in certain cases be preferable. Since a partnership is not a separate taxable entity, earnings are taxed a single time, to the partners. Similarly, branch income is subject to U.S. tax one time and the remittance of branch earnings is not subject to withholding tax.

In contrast, corporate profits are taxed twice, once to the corporation, when earned, and a second time, to the shareholders, upon distribution. However, the burden of double taxation may be mitigated to the extent that the venture retains earnings, since shareholders are not taxed on undistributed corporate earnings. This is an important concept as partners and foreign corporations with branch operations are in many cases taxed in their home countries on earnings, regardless of whether these earnings are distributed.

Allocation of Income and Losses

If participants desire to allocate items of income and loss among themselves in proportions that differ from their ownership shares, a partnership can provide maximum flexibility. Generally, the use of a partnership permits disproportionate allocations that have economic substance. For example, assume that one party holds mineral rights but does not have sufficient financial resources to exploit the property. If an investor is invited into the venture to provide financing for deveopment, the use of a partnership will permit the allocation to the investor of the development expenditures funded by him.

Liabilities

If a venture is exposed to significant contingent liabilities, a corporation has advantages over a partnership or a branch operation. Generally, a shareholder's liability is limited to the amount of money or other property that he has contributed to the corporation. In contrast, all general partners are personally liable for a partnerships's debts and obligations. Alternatively, a partner may be able to limit his liability by participating in a partnership through a newly formed wholly-owned subsidiary. When a branch operation is used, the liabilities of the foreign corporation are not limited to the branch's assets. Also, a branch operation may expose the foreign headquarter corporation's books and records to an examination by the Internal Revenue Service, whereas an examination on a local subsidiary is generally limited to the U.S. corporation's records.

Financing Needs

If it is expected that at some point the venture might make a public offering of securities, the use of a corporation is advantageous. Generally, public offerings of partnership interests are rarely made outside the tax-shelter field and involve a number of complications not present in a corporate offering. Also the presence of a U.S. corporation is more broadly accepted by the U.S. business community (e.g., customers, suppliers and creditors).

Mergers and Acquisitions

If it is expected that the venture may merge with or acquire another company, the merger may be completed more easily, more rapidly and with less expense if the business is organized in corporate form. Moreover, use of the corporate form will permit tax-free mergers and acquisitions, whereas the opportunity for tax-free transactions is more limited in the partnership context.

Transfers of Ownership

If it is expected that transfers of ownership will be frequent, the corporate form would again appear to be preferable. Generally, transfers of partnership interests or a branch operation are considerably more cumbersome than transfers of stock.

CONCLUSION

Numerous business and tax considerations are involved in the decision of how to structure foreign investment in the United States. When an investor has adequately identified its business and financial needs, then a well-planned tax strategy and investment structure may be developed and implemented.

EXHIBIT I

Withholding Tax Rates

The following table gives the basic provisions of how U.S. tax treaties apply to withholding on outward remittances of dividends, interest, and royalties. Under several treaties, the exemption or reduction in rate applies only if the recipient is subject to tax on such income in its country of residence; otherwise a 30% rate applies. So many exceptions and conditions apply, which are too numerous and detailed to list here, that the specific treaty must be reviewed in each case.

	Dividends		Interest*	Industrial Royalties
	To Parent	Other		
Nontreaty countries:	30	30	30	30
Treaty countries:				
Australia	15	15	30(1)	30 (1)
Austria	5	15	0	0
Belgium	15	15	15	0
Belgium Extensions (2)	15	15	15	0
Canada	10	15	15	10
Denmark	5	15	0	0
Egypt	5	5	12.5	15
Finland	5	15	0	0
France	5	15	10	5
Germany	15	15	0	0
Greece	30	30	0	0
Hungary	5	15	0	0
Iceland	5	15	0	0
Ireland	5	15	0	0
Italy	5	15	30	0
Japan	10	15	10	10

Jamaica	10	15	12.5	10
Republic of South Korea	10	15	12	15
Luxembourg	5	15	0	0
Malta	5	15	12.5	12.5
Morocco	10	15	15	10
Netherlands	5	15	0	0
Netherlands Antilles	5	15	0	0
New Zealand	5(4)	15	30 (3)	30 (3)
Norway	10	15	0	0
Pakistan	15	30	30	0
Philippines	20	25	15	15
Poland	5	15	0	10
Romania	10	10	10	15
Republic of South Africa	30	30	30	30
Sweden	5	15	0	0
Switzerland	5	15	5	0
Trinidad & Tobago	30	30	30	15
Union of Soviet Socialist Republics	30	30	0	0
United Kingdom	5	15	0	0
United Kingdom Extensions (5)	15	15	30	0

* As discussed in the text, the Tax Reform Act of 1984 eliminates the 30% withholding tax on portfolio interest paid by U.S. borrowers on obligations issued after July 18, 1984:

1) in registered form, and

2) in unregistered (i.e. bearer) form provided the requirements for exemption from the TEFRA registration requirements are met.

Notes:

(1) Rate will be reduced to 10% under a new treaty not yet in force.

(2) 1948 United States-Belgium treaty which applied to Burundi, Rwanda, and Zaire has been terminated effective January 1, 1984.

(3) Rate will be reduced to 10% under a new treaty not yet in force.

(4) Rate will be increased to 15% under a new treaty not yet in force.

(5) 1945 United States-United Kingdom treaty, which applied to the following U.K. overseas territories and former territories, has been terminated effective January 1, 1984:

Antigua-Baruda (Antigua) Malawi (Nyasaland)
Barbados   Montserrat
Belize (British Hounduras) St. Christopher, Nevis, and
                            Anguilla
Dominica   St. Lucia
Falkland Islands   St. Vincent
Gambia   Seychelles
Grenada   Sierra Leone
                          Zambia (Northern Rhodesia)

Exhibit II

Statutory Depletion Rates for Various Minerals*

1) 22% -
   A) sulphur and uranium; and

   B) if from deposits in the United states - anorthosite, clay, laterite and nephelinite syenite (to the extent that alumina and aluminum compounds are extracted therefrom), asbestos, bauxite, celestite, chromite, corundum, fluorspar, graphite, ilmenite, kyanite, mica, olivine, quartz crystals (radio grade), rutile, block steatite talc and zircon, and ores of the following metals: antimony, beryllium, bismuth, cadmium, cobalt, columbium, lead, lithium, manganese, mercury, molybdenum, nickel, platinum and platinum group metals, tantalum, thorium, tin, titanium, tungsten, vanadium and zinc.

2) 15% - if from deposits in the United States -
   A) gold, silver, copper and iron ore

   B) oil shale (except shale described in paragraph (5)).

3) 14% -
   A) metal mines (if paragraph (1)(B) or (2)(A) does not apply), rock asphalt and vermiculite

   B) if paragraph (1)(B), (5) or (6)(B) does not apply, ball clay, bentonite, china clay, sagger clay and clay used or sold for use for purposes dependent on its refractory properties.

4) 10% - asbestos (if paragraph (1)(B) does not apply), brucite, coal, litnite, perlite, sodium chloride and wollastonite.

5) 7-1/2% - clay and shal used or sold for use in the manufacture of sewer pipe or brick, and clay, shale, and slate used or sold for use as sintered or burned lightweight aggregates.

6) 5% –

   A) gravel, peat, pumice, sand, scoria, shale (except shale described in paragraph (2)(B) or (5)) and stone (except stone described in paragraph (7))

   B) clay used, or sold for use, in the manufacture of drainage and roofing tile, flower pots and kindred products

   C) if from brine wells – bromine, calcium chloride, magnesium chloride

7) 14% – all other minerals, including, but not limited to, aplite, barite, borax, calcium carbonates, diatomaceous earth, dolomite, feldspar, fullers earth, garnet, Gilsonite, granite, limestone, magnesite, magnesium carbonates, marble, mollusk shells (including clam shells and oyster shells), phosphate rock, potash, quartzite, slate soapstone, stone (used or sold for use by the mine owner or operator as dimension stone or ornamental stone), thenardite, tripoli, trona and (if paragraph (1)(B) does not apply) bauxite, flake graphite, fluorspar, lepidolite, mica, spodumene, and talc (including pyrophyllite), except that unless sold on bid in direct competition with a bona fide bid to sell a mineral listed in listed in paragraph (3), the percentage shall be 5% for any such other mineral (other than slate to which paragraph (5) applies) when used, or sold for use, by the mine owner or operator as riprap, ballast, road material, rubble, concrete aggregates, or for similar purposes. For purposes of this paragraph, the term "all other minerals" does not include –

   A) soil, sod, dirt, turf, water or mosses

   B) minerals from sea water, the air, or similar inexhaustible sources or

   C) oil and gas wells.

For purposes of this subsection, minerals (other than sodium chloride) extracted from brines pumped from a saline perennial lake within the United States shall not be considered minerals from an inexhaustible source.

5.5

COMPARISON OF INTERNATIONAL MINING TAX REGIMES

Coopers & Lybrand*

INTRODUCTION

From a tax viewpoint, United States mining companies are at a substantial disadvantage in the international mining area compared with companies in other major capital-exporting countries, according to a 1975 study by Coopers & Lybrand for the American Mining Congress. The study compared the relative tax position of a United States investor with investors from Belgium, Canada, France, Germany, Japan, the Netherlands, Switzerland, and the United Kingdom.

The comparison was based on average rates of return realizable by investor companies on four mine-model investments in twelve capital-importing countries. The countries in which these investments were assumed to be made are varied in their tax systems so that a wide range of possibilities was included in the analyses. Each mine model was standard so that the rates of return for each investment by each investor were affected solely by the respective tax systems of the capital-exporting and capital-importing countries.

UNITED STATES POSITION

The study found that the United States competitive position, in terms of the average rate of return on an overall basis, ranked next to last among the nine capital-exporting countries in the study. Further, the United States overall rate of return is more than 20% below that of the country (Japan) with the highest overall rate of return. On an average return-on-equity basis, the United States' position among the nine investor countries, for each of the mine investments studied, was: copper, eighth; iron ore, sixth; nickel, fifth; manganese, a tie for last place. Figure 1 shows the overall results realized by each capital-exporting country on the basis of average return on equity.

Figure 1
PERCENTAGE RELATIONSHIP OF RETURN ON EQUITY
Overall Average of Four Minerals

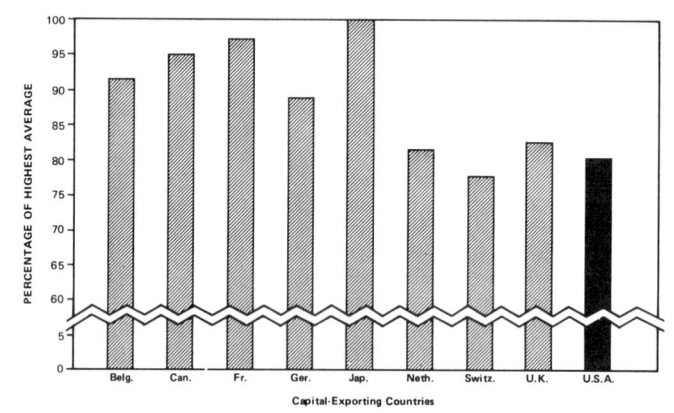

The frequency with which a given capital-exporting country attains the highest or next highest rate of return in a capital-importing country provides further insight and another basis for comparing the tax laws of the capital-exporting countries.

Of the situations compared in the study, the United States never attained the highest rate of return, and reached the second and third highest rate of return in only two instances. The tables showing the frequencies of rankings are given in Table 1 for each of the four mine models.

In summary, the United States mining company does not have a competitive advantage under the tax rules when competing for mineral concessions in other countries. The United States company, in fact, appears to be at a clear disadvantage.

The study also considered the effect of the following recent tax proposals on United States investors in international mining ventures:

---

*From The Coopers & Lybrand Newsletter, September 1975

Table 1

SUMMARY OF RANKINGS FOR INVESTMENTS
In Capital-Importing Countries
Based on Return on Equity

COPPER

Rank Based on Average Rank	Capital-Exporting Country	1st	2nd	3rd	4th	5th	6th	7th	8th	9th	Average Rank
1	JAPAN	3	2	—	1	2	—	—	—	—	2.6
2	BELGIUM	3	—	1	4	—	—	—	—	—	2.9
3	FRANCE	—	2	5	—	1	—	—	—	—	3.0
4	CANADA	1	2	—	—	—	2	2	—	—	4.6
5	GERMANY	1	2	1	—	—	1	1	1	1	4.8
6	NETHERLANDS	—	—	—	2	1	3	2	—	—	5.8
7	SWITZERLAND	—	—	—	2	1	1	—	2	2	6.8
8	U.K.	—	—	—	2	—	2	—	—	4	7.0
9	U.S.A.	—	—	—	1	—	1	1	5	—	7.1

IRON ORE

Rank Based on Average Rank	Capital-Exporting Country	1st	2nd	3rd	4th	5th	6th	7th	8th	9th	Average Rank
1	FRANCE	2	4	—	1	—	—	—	—	—	2.1
2	CANADA	4	—	—	—	—	—	2	—	—	3.1
2	JAPAN	2	1	1	1	2	—	—	—	—	3.1
4	BELGIUM	—	1	3	1	1	1	—	—	—	3.8
5	U.K.	—	1	—	—	4	—	1	—	1	5.6
6	NETHERLANDS	—	—	—	1	1	2	2	1	—	6.2
7	U.S.A.	—	—	—	1	1	1	3	1	—	6.4
8	GERMANY	—	—	2	—	—	1	2	2	6.7
9	SWITZERLAND	—	—	—	1	1	—	1	2	2	7.1

NICKEL

Rank Based on Average Rank	Capital-Exporting Country	1st	2nd	3rd	4th	5th	6th	7th	8th	9th	Average Rank
1	BELGIUM	1	—	3	3	—	—	—	—	—	3.1
2	CANADA	3	—	—	—	1	1	—	—	1	3.8
3	FRANCE	—	3	—	—	1	3	—	—	—	4.2
4	U.K.	1	3	—	—	—	—	1	1	1	4.4
5	U.S.A.	—	1	1	2	—	1	1	1	—	4.9
6	JAPAN	—	—	2	—	3	—	1	1	—	5.2
7	NETHERLANDS	—	—	1	1	2	—	3	—	—	5.4
8	GERMANY	2	—	—	—	—	—	—	1	4	6.6
8	SWITZERLAND	—	—	—	1	1	1	1	3	—	6.6

MANGANESE

Rank Based on Average Rank	Capital-Exporting Country	1st	2nd	3rd	4th	5th	6th	7th	8th	9th	Average Rank
1	FRANCE	1	3	2	—	—	—	—	—	—	2.2
2	JAPAN	1	3	2	—	—	—	—	—	—	2.3
3	BELGIUM	—	—	1	4	1	—	—	—	—	4.1
4	GERMANY	2	—	2	—	—	—	—	1	1	4.2
5	U.K.	2	—	—	—	—	—	1	1	2	5.9
6	NETHERLANDS	—	—	—	—	3	1	2	—	—	6.0
7	CANADA	—	—	—	1	—	2	2	1	—	6.4
7	SWITZERLAND	—	—	—	—	2	2	1	—	1	6.4
9	U.S.A.	—	—	—	—	1	2	2	1	7.6

- Foreign taxes paid as a deduction rather than as a credit

- Elimination of the per-country foreign tax credit limitation

- Recapture of tax benefits for foreign losses

- Exclusion of foreign income or loss from US taxable income

- Elimination of percentage depletion on foreign mineral income

Any one of these proposals, if adopted, would place the United States investor in last position on an overall basis. In many cases, moreover, such proposals would have a significantly adverse effect on the overall return on equity and would substantially worsen the existing United States competitive disadvantage.

The capital-importing countries included in the study, and the mineral investments assumed, are:

Australia	copper, iron ore, nickel manganese
Brazil	iron ore, manganese
Canada	copper, iron ore, nickel
Indonesia	coper, nickel
Iran	copper, manganese
Ireland	copper
Liberia	iron ore
Mexico	copper, manganese
New Caledonia	nickel
New Zealand	iron ore, nickel
Philippines	copper, iron ore, nickel, manganese
South Africa	copper, iron ore, nickel, manganese

ANALYTICAL METHODS

Return on equity (ROE) was the principal measure of the impact of the tax laws of the capital-importing and capital-exporting countries since most foreign mining investments involve significant debt financing. To eliminate the effect of different financing methods on ROE, constant debt:equity ratios and debt repayment terms were used for each comparative investment. Returns on investment (ROI), i.e., on total capital invested, were also computed in the actual study and showed standings similar to the ROE rankings. Computer programs were developed to reflect the tax laws and accounting rules for each capital-importing and capital-exporting country to compute ROE and ROI for each assumed investment in order to determine the tax elections that would produce the highest rate of return.

Many complex factors were analyzed in this study, including:

- The effect of doing business as a branch or foreign subsidiary in each capital-importing country,

- The effect of cash distributions of accounting principles at both the capital-importing and exporting-country level, and

- The tax systems of both capital-importing and exporting countries, including the tax elections available for optimizing cash flow.

The detailed hypothetical mine models used for the comparative analysis were based on statistics developed by members of the American Mining Congress. Coopers & Lybrand prepared a similar report to the American Mining Congress in March 1973. The current report concludes, in general, that the United States competitive position remains unchanged since the initial study. The model basics for the 1973 report are given in Tables 2, 3, and 4.

Table 2

COMPARISON OF AVERAGE RETURN ON INVESTMENT

(Percent)

	Fe	Cu	Ni	Mn
United States	9.6	11.6	8.6	8.0
Belgium	9.0	12.0	8.1	7.9
Canada	9.9	12.8	8.7	7.8
France	10.4	13.1	8.8	8.9
Germany	9.5	13.4	8.9	8.7
Japan	9.7	12.7	8.6	8.9
Netherlands	9.0	12.7	8.2	7.9
Switzerland	8.9	12.4	8.2	7.9
United Kingdom	10.2	11.6	8.4	8.1
Overall Average	9.6	12.5	8.5	8.2
US rank	5th	8th	4th	5th

Table 3

COMPARISON OF AVERAGE RETURN ON EQUITY

(Percent)

	Fe	Cu	Ni	Mn
United States	12.9	19.0	9.8	8.6
Belgium	11.7	21.0	8.8	8.6
Canada	15.7	23.1	11.4	8.4
France	14.9	24.3	10.2	11.4
Germany	14.0	26.2	11.3	11.0
Japan	13.2	25.0	10.0	10.9
Netherlands	10.8	20.5	8.8	8.3
Switzerland	10.4	19.9	8.8	8.3
United Kingdom	18.3	20.6	10.3	9.7
Overall Average	13.5	22.2	9.9	9.5
US rank	6th	Last	6th	5th

Table 4

BASIC COOPERS & LYBRAND MINE MODEL

(US$ millions unless otherwise stated)

	Fe	Cu	Ni	Mn
Total investment..	294	215.2	300.1	191.4
Interest during construction....	55.6	20.5	24.4	16.4
Debt (8%).........	220.5	166.6	208.2	147.6
Debt repayment (years).........	12	10	15	15
Equity............	73.5	48.6	91.9	43.8
Preproduction period (years)..	8	4	6	6
Average ore grade..	65%	0.99%	1.5%	47.5%
Mine life (years).	20	20	20	20
Est. annual gross revenue.........	105	64.2–101.5	70	45
Est. annual operating cost*.	47.8	29.5–43.8	27	17.2

*Exclusive of Depreciation

5.6

TAX PLANNING THROUGH THE USE OF MULTIPLE CORPORATIONS

JOHN J. McCABE, CPA

PARTNER, PRICE WATERHOUSE (NEW YORK)

INTRODUCTION

Over the years, Congress has written into the Internal Revenue Code various provisions aimed at lessening at least one financial burden faced by taxpayers in the mining industry - the federal income tax burden. These provisions include the election to deduct exploration expenses, the election to defer deduction of development expenses, and the percentage depletion deduction. Careful tax planning from the very beginning of a mining venture is essential in taking maximum advantage of these tax benefits. A major decision to be made is the choice of organizational structure. (For a more detailed discussion of this subject, see McCabe, John J., "Developing Structures for Investments in Mineral Property," New York University Institute on Federal Taxation, 42 NYU 24, 1984). One such structure is the use of multiple corporations, with mining activities conducted by one corporation and nonmining activities, such as manufacturing, financing and administration, conducted by the other corporation. This technique, which will be discussed in detail below, can be particularly advantageous in delaying recapture of previously deducted exploration expenses and in maximizing depletion deductions.

TAX BENEFITS IN MINING

Exploration

Subject to certain restrictions discussed below, exploration expenditures (i.e., expenditures paid or incurred during the taxable year for the purpose of ascertaining the existence, location, extent, or quality of any deposit of ore or other mineral, and paid or incurred before the beginning of the development state of the mine) are deductible at the election of the taxpayer (Section 617(a)).* Were it not for this option, exploration expenditures would be required to be capitalized and recovered through depletion, after production begins. Exploration of a mineral deposit can be very costly and may not result in commercial production for some time. If the taxpayer has income to be offset by the deductions, a mining venture can usually be structured in such a way as to allow the offset of mining losses against the other income. With today's high interest rates, this election can be quite valuable in present value terms.

Corporate and noncorporate taxpayers are subject to different restrictions on this tax benefit. Corporations may elect to deduct only 85 percent of eligible exploration expenditures (Section 291(b)). (The Tax Reform Act of 1984 reduced this to 80 percent effective for expenditures after December 31, 1984, in taxable years ending after such date). The remaining 15 percent (20 percent after December 31, 1984) is capitalized and recovered at the same rate as five-year ACRS property (Section 291(b)). Individual taxpayers are not subject to this 85 percent rule; however, any amount deducted in excess of ten-year straight-line amortization is a tax preference item for individuals (Section 57(a)(5)). Individuals may avoid the tax preference by electing to capitalize exploration expenditures and amortize them over ten years (Section 58(i)).

In order to prevent the taxpayer from, in effect, converting ordinary income into capital gain by selling property that has given rise to exploration expense deductions, the Code requires recapture of such deductions upon the occurrence of certain events (Section 617):

1. Upon disposition of the property, adjusted exploration expenditures are recaptured to the extent of gain realized on the disposition (Section 617(d)).

2. If the property is leased, the lessor is disallowed any depletion deduction until his adjusted exploration expenditures have been recaptured in the form of foregone depletion deductions (Section 617(c)).

*All section references are to the Internal Revenue Code of 1954, as amended, and to the Income Tax Regulations thereunder.

3. If the taxpayer retains the property and begins production, he can choose between foregoing depletion and taking into income immediately his adjusted exploration expenditures (Section 617(b)).

Under the last method, the recaptured expenses become part of the taxpayer's depletable basis. Normally, of course, the taxpayer will attempt to accelerate deductions and defer recapture to the maximum extent possible. The several alternatives available permit a great deal of flexibility in planning the mining project in such a way as to minimize income tax expense.

Development

Development expenditures are those paid or incurred during the taxable year for the development of a mine if paid or incurred after the existence of ores or minerals in commercially marketable quantities has been disclosed (Section 616(a)). While they are normally a deductible expense, an election is provided to defer development expenditures and deduct them ratably as the mineral is sold. Where development expenditures are deducted currently, however, they are subject to the same restrictions as exploration expenditures (i.e., the 85 percent rule for corporations, reduced to 80 percent by the Tax Reform Act of 1984, effective for expenditures after December 31, 1984, in taxable years ending after that date, and the tax preference rule for individuals). While an option to defer a deduction is normally less beneficial than an option to deduct, it does provide increased tax planning flexibility and under certain circumstances can result in permanent tax savings. For example, the ability to manage taxable income can help insure the utilization of expiring carryforwards of investment tax credits, foreign tax credits or net operating losses. In addition, it can be useful in avoiding the 50 percent of taxable income from the property limitation on percentage depletion, discussed below. There is no recapture of development expenditures.

Depletion

Section 611 of the Code provides for a deduction for cost depletion, to be calculated according to regulations prescribed by the Secretary. The method prescribed in the Regulations for calculating cost depletion is essentially the same as the units of production method of depreciation (Regulations Section 1.611-2(a)(1)). The deduction will be greater, the larger the taxpayer's basis and the greater the quantity of mineral removed during the year. The calculation will, of course, result in a zero deduction when basis has been fully recovered.

The Code also provides an alternative method of calculating depletion (Section 613). Percentage depletion is calculated by applying a percentage, determined by statute, to the taxpayer's gross income from mining, excluding any rents or royalties paid or incurred by the taxpayer with respect to the property. The allowable deduction is then limited to 50 percent of the taxable income from the property (TIFP) before any deduction for depletion.

In any year, the taxpayer is required to calculate depletion under both methods and to deduct the greater of cost or percentage depletion (Section 613(a)).

Several interesting and important characteristics of the percentage depletion deduction should be noted. First, in many situations, percentage depletion will exceed cost depletion, resulting in a relatively rapid recovery of basis. More importantly, since it is based entirely on gross income, no basis at all is required in order to claim the percentage depletion deduction. The taxpayer continues to be eligible for the deduction as long as he receives mining income from the property, even though his basis has been recovered. The lack of association between depletable basis and the percentage depletion deduction, in addition to its obvious advantages for the taxpayer, makes initial tax planning particularly important. For example, unlike many capital expenditures, where a large initial outlay will at least subsequently give rise to larger deductions (e.g., depreciation and amortization), it is possible that a large depletable basis will result in no tax benefit at all since percentage depletion deductions may exceed cost depletion deductions.

Finally, any deduction lost because of the 50 percent of taxable income limitation is lost permanently; it cannot be carried over or back to years in which the limitation is not exceeded.

From the preceding discussion, two chief conclusions can be drawn with regard to tax planning for maximizing percentage depletion:

1. Creation of depletable basis should be avoided under most circumstances, since no tax benefit may be derived from it.

2. Avoidance of the 50 percent of TIFP limitation may be accomplished by successful timing of income and deductions. The importance of this point may be illustrated by the following example:

	A	B
Gross income from the property	$400	$400
Deductions attributable to mining processes	290	320
Taxable income from the property	$110	$80
50 percent limitation	$55	$40
Statutory percentage of gross income from the property (e.g., 13 percent)	$52	$52
Percentage depletion allowable	$52	$40

Note that a 9.375 percent reduction in deductions attributable to mining, from $320 to $290, resulted in a 30 percent increase in percentage depletion allowable, from $40 to $52.

Tax benefit will usually be obtained in situations where the taxpayer is engaged in nonmining, as well as mining, activities and the increase in TIFP is the result of shifting expenses from the mining to the nonmining activity. When both activities are conducted within the structure of a single corporation, this will involve the apportionment of overhead and general administrative expenses. The Regulations require such apportionment, but do not require any particular method (Regulations Section 1.613-5(a)). Reasonableness and consistency are the main guidelines (Regulations Section 1.613-5(a), referring to Regulations Section 1.613-4(d)(2)). Permissible methods of apportionment include apportionment based on relative gross receipts or on direct expenses, with appropriate modifications for facts peculiar to the taxpayer. As will be illustrated below, the use of multiple corporations can help in avoiding the apportionment issue by segregating functions in separate corporations.

Section 631(c)

Where an owner of coal or domestic iron ore disposes of it, retaining an economic interest, his gain or loss will be considered Section 1231 gain or loss. His net gains for the taxable year will, therefore, be treated as capital gains, while net losses will be considered ordinary losses (Section 631(c); Regulations Section 1.631-3(a)(2); Section 1231(b)(2)). Prior to passage of the Tax Reform Act of 1984, the coal or iron ore was required to have been held for more than one year before the disposition. Effective with regard to property acquired after June 22, 1984, and before January 1, 1988, the holding period is reduced to six months (Section 631(c), as amended). The owner cannot be a co-adventurer, partner or principal in the mining of the coal or iron ore (Section 631(c); Regulations Section 1.631-3(b)(4)(i)). These terms have been said to imply a sharing of the risk and control of a mining venture so that income to the taxpayer is derived from direct operation of a mine rather than passive income as a lessor or sublessor retaining an economic interest (Revenue Ruling 73-33, 1973-1 C.B. 307).

Economic interest (which is also required in order for a taxpayer to be entitled to exploration, development or depletion expenditures) is defined by the IRS as follows:

An economic interest is possessed in every case in which the taxpayer has acquired by investment any interest in mineral in place... and secures, by any form of legal relationship, income derived from the extraction of the mineral...to which he must look for a return of his capital (Regulations Section 1.611-1(b)(1)).

In the past, disposals of coal involving transactions between related parties were eligible for Section 631(c) treatment. (Such transactions involving iron ore were specifically excluded by Section 631(c) and Regulations Section 1.631-3(e)(5)). The Tax Reform Act of 1984 amended Section 631(c) to accord coal the same treatment as iron ore, effective for royalties paid after September 30, 1985. Because of its limited applicability, this tax planning technique will be discussed only briefly below.

MULTIPLE CORPORATIONS TO MAXIMIZE TAX BENEFITS

The maximization of tax benefits through the use of multiple corporations requires the isolation of the actual mining of the mineral from all other corporate activities. This can be accomplished by causing one corporation (Parent) to acquire the mineral property (leasehold or fee) and then sublease it for an arm's-length royalty to a separate controlled corporation (Sub) that will actually mine the mineral. Under the right circumstances, this arrangement can reduce taxable income and income taxes for the consolidated group, without increasing cash expenses. The use of multiple corporations is intended to accomplish this in two ways. First, the total consolidated depletion deduction is increased. If Parent has acquired depletable basis in the property by, for example, paying the original owner a lease bonus, Parent will be entitled to a depletion deduction (most likely cost) against the royalty it receives from Sub. At the same time, Sub will be entitled to a percentage depletion deduction against its mining income. This is accomplished by structuring the corporation so that, of total consolidated income and expenses, Parent will recognize the maximum possible expense (this has no effect on Parent's cost depletion deduction) while Sub recognizes maximum possible income, maximizing its percentage depletion deduction and avoiding the 50 percent of TIFP limitation. For example, a potentially large expense such as interest can effectively be isolated in Parent without adversely impacting the 50 percent of TIFP limitation of Sub. Second, where Parent has exploration expenditures to recapture (see discussion above), they will be recaptured at a slower rate, against Parent's cost depletion, rather than against a single corporation's potentially greater percentage depletion.

It must be noted that "under the circumstances" is a key phrase here. Among the prerequisites for successful use of the multiple corporation technique are (1) a high depletable basis in the property, which potentially can give rise to large cost depletion deductions, (2) high gross income from mining which potentially can give rise to large percentage depletion deductions, (3) large expenses, i.e., interest that can be isolated at the Parent level, and (4) large exploration recapture exposure.

The multiple corporation technique can best be illustrated by example. In the first example, we assume that Parent has paid $500,000 for a leasehold interest in a mineral deposit with estimated recoverable reserves of 500,000 tons and has capitalized exploration costs of $500,000, giving Parent a depletable basis of $1,000,000. The original leasehold owner has retained a royalty, payable by Parent, of 10 percent of sales. The mineral sells for $50 per ton and is eligible for percentage depletion at the rate of 15 percent.

Parent subleases the property to Sub for a royalty of 15 percent of sales under terms that ensure that both Parent and Sub have an economic interest in the mineral. In order to avoid any interest expense in Sub, Parent will borrow any funds necessary and contribute these funds to the capital of Sub. Sub will actually mine the mineral. Sub's only significant asset consists of mining equipment which is five year ACRS property with a cost basis of $667,000. In the first year, 20,000 tons are sold. In the example, "adjusted gross income" means gross income from the property, adjusted by the royalty paid. Net income before percentage depletion equals taxable income from the property.

TABLE 1
MULTIPLE CORPORATIONS TO
INCREASE DEPLETION DEDUCTIONS

	Single Corporation	Multiple Corporations		
		Parent	Sub	Consolidated
Gross income:				
Mining	$1,000,000	-	$1,000,000	$1,000,000
Less - Royalty	(100,000)	($100,000)	(150,000)	(250,000)
Other	-	150,000	-	150,000
Adjusted gross income	900,000	50,000	850,000	900,000
Deductions:				
Depreciation	100,000	-	100,000	100,000
Interest	100,000	100,000	-	100,000
Labor	410,000	10,000	400,000	410,000
Cost depletion	-	40,000	-	40,000
Other	90,000	10,000	80,000	90,000
Net income before percentage depletion	200,000	(110,000)	270,000	160,000
Percentage depletion	100,000*	-	127,500	127,500
Taxable income	$ 100,000	($110,000)	$ 142,500	$ 32,500
Tax at 50%	$ 50,000	($ 55,000)	$ 71,250	$ 16,250

*Limited to 50% of TIFP

In this example, Parent receives ordinary income in the form of a royalty from Sub, and deducts the royalty it pays to the original leasehold owner. It also deducts interest, as well as labor and other expenses representing corporate overhead. Because of its high depletable basis, Parent's cost depletion deduction is very high, well in excess of the allowable percentage depletion, and this is the deduction Parent claims.

Sub, on the other hand, has no basis in the property and would calculate zero cost depletion. Its right to the percentage depletion deduction is in no way affected by Parent's cost depletion deduction. Under the facts of this example, a single corporation approach results in a partial loss of percentage depletion because the 50% of TIFP limitation is brought into play. Under the multiple corporation approach, however, many expenses are isolated in Parent, increasing Sub's taxable income sufficiently to avoid the limitation. Hence, the allowable percentage depletion increases even though gross income from mining, upon which the percentage depletion calculation is based, has decreased as a result of the royalty payment from Sub to Parent.

Prior to passage of the 1984 Tax Reform Act, the multiple corporation technique could also be applied to coal mining so that the consolidated group would obtain the benefits of both percentage depletion deductions (by Sub) and Section 631(c) (by Parent). In this scenario, rather than deduct cost depletion from its net royalties in arriving at ordinary taxable income, Parent would deduct "adjusted depletion basis" (calculated similarly to cost depletion, plus Section 272 expenses) from its net royalties. The resulting income would, under Section 631(c), be treated as Section 1231 gain, normally taxed at capital gains rates. Sub would still be entitled to percentage depletion. This technique has been eliminated effective for royalties paid after September 30, 1985 (Section 631(c) as amended).

A second example will illustrate the value of the multiple corporation technique in delaying recapture of deducted exploration expenditures. Under this scenario, Parent purchases the mineral interest as before, explores the property and elects to deduct all exploration expenditures under Section 617. When the mine is ready to begin production, Parent leases the property to Sub for an arm's-length royalty. All other assumptions are the same as in our first example.

TABLE 2
MULTIPLE CORPORATIONS TO
DELAY RECAPTURE OF EXPLORATION EXPENDITURES

	Single Corporation	Multiple Corporations		
		Parent	Sub	Consolidated
Gross income:				
Mining	$1,000,000	-	$1,000,000	$1,000,000
Less - Royalty	(100,000)	($100,000)	(150,000)	(250,000)
Other	-	150,000	-	(150,000)
Adjusted gross income	900,000	50,000	850,000	900,000
Deductions:				
Depreciation	100,000	-	100,000	100,000
Interest	100,000	100,000	-	100,000
Labor	410,000	10,000	400,000	410,000
Cost depletion	-	20,000	-	20,000
Other	90,000	10,000	80,000	90,000
Net income before percentage depletion	200,000	(90,000)	270,000	140,000
Percentage depletion	100,000*	-	127,500	127,500
	100,000	(90,000)	142,500	52,500
Exploration expenditures recaptured	100,000	20,000	-	20,000
Taxable income	$ 200,000	($ 70,000)	$ 142,500	$ 72,500
Tax at 50%	$ 100,000	($ 35,000)	$ 71,250	$ 36,250

*Limited to 50% of TIFP

As previously discussed, the exploration expenditures previously deducted are recaptured in the form of foregone depletion deductions. In the case of the single corporation, this is likely to mean the loss of large percentage depletion deductions in the early years of production. The multiple corporation technique reduces the amount of depletion that would otherwise be allowable to the recapturing corporation, thus reducing the amount recaptured in any year and deferring the total recapture, possibly over a period of many years. At the same time, this arrangement preserves the percentage depletion deduction for Sub.

CONCLUSION

In reality, of course, many factors, both tax and nontax-related, will enter into the choice of organizational structure for the exploration, development and extraction of a mineral deposit. These will include the need to minimize financing costs and operating expenses, as well as tax liability. Nevertheless, opportunities exist for increasing the after-tax return on an investment in mineral property by maximizing the benefits offered to the mining venturer by Congress through the Internal Revenue Code. It is often possible to take advantage of these opportunities without sacrificing other goals of the taxpayer, provided, of course, that the taxpayer is aware of them and alert to tax planning possibilities. The multiple corporation technique is one such possibility. By allowing a consolidated group to deduct at the same time both cost and percentage deletion, it can reduce taxable income and ensure that some tax benefit is derived from depletable basis. By delaying the recapture of deducted exploration expenditures and the resulting tax liability, it can increase the present value of the mineral investment. In any situation where the necessary prerequisites are likely to exist (i.e., high depletable basis and high gross income from mining), consideration of the multiple corporation technique should be included in tax planning for the mining venture.

5.7

THE CRUSH ON AUSTRALIAN MINING PROFITS

V. ("VIV") R. FORBES

MIM Holdings Limited
Brisbane, Queensland, Australia

INTRODUCTION

"The Liberal Party believes that the State's mineral resources belong to the people of Queensland and therefore it is essential that the state, on their behalf, obtains maximum benefit from the development of those resources." (Liberal Party, 1980 "Mineral Resources Policy" p.2). "Unless sound policies are implemented.... increased production of minerals will benefit only the already wealthy and the lucky few..... A resources tax is essential to ensure that above-normal profits..... are shared by the whole community.... Labor policies on taxation must ensure an adequate return from mining developments to the Australian people." (Labor Party Policy Paper, May 1982 "National Issues relating to rapid expansion of mineral production during the eighties". p.1, 37, 47).

Judging from these statements, key political parties in Australia believe that mining profits are lightly taxed and that shareholders get most of the benefits.

THE REALITY

The Government Share of Income statement in Table 1 attempts to measure the government share of the income generated by the "Big Five" Australian mining companies - Broken Hill Pty Ltd. (BHP), CRA Limited, Utah International, MIM Holdings Ltd., and Western Mining Corporation ("WMC"). These companies control over 70% of Australia's mining industry including all the most profitable mining operations. They dominate the mining and processing of iron and steel, aluminum, manganese, coal, oil and gas, copper, lead, zinc, silver, nickel, gold, ferro-alloys and salt in Australia. They are also major world exporters of most of these products. They are thus a representative sample of the profit-earning side of the Australian mining industry.

In their two most recent financial years these five companies had income before tax of A$6,980 million. The Government share of this income was A$6,285 million or 90%.

The total market capitalisation of the Big Five is about A$8,480 million. (December, 1982). Governments thus consumed the equivalent of 82% of their total value in just two years. This was taking place while the politicians drafted the above statements calling for an "adequate" share for the people.

Published with permission from Common Sense, Issue No.26, December 1982.

Table 1

The Government Share of Income ("GSI")
Statement for the Australian
Mining Industry*

	TOTAL FOR FIVE BIG MINERS FOR TWO YEARS		
	A$Million	%	Notes
SALES and other income	20,059		1
COST OF SALES (excluding taxes)			
Wages and Salaries	3,270		2
Goods and Services	7,673		3
Depreciation			
- historical cost	1,107		5
- current cost adjustment	722		5
Interest (net)	307		4
Total	13,079		
INCOME BEFORE ALL TAXES	6,980	100%	

GOVERNMENT SHARE OF INCOME

(a) Direct Taxes on Mining Companies

	Federal	State & Local	Foreign			Notes
Crude Oil Levy	2,881					4
Company Tax	1,262					4
Mineral Royalties		442				4
Payroll Tax		178				6
Sales Tax	162					7
Rail Freight Surcharge		136				8
Coal Export Duty	110					9
Foreign Taxes			25			10
Import Duty	23					11
Other Taxes	10	77				12
Less Subsidies	(21)					13
Sub-Total	4,427	833	25	5,285	76%	

(b) Taxes Levied on Others but originating as Mining Income

						Notes
PAYE Tax on Wages	874					14
Taxes on Dividends	126					15
Sub-Total				1,000	14%	
TOTAL GOVERNMENT SHARE OF INCOME	5,427	833	25	6,285	90%	
INCOME AFTER ALL TAXES				695	10%	
Less Income due to Minority Interests				140		5
NET INCOME LEFT FOR SHAREHOLDERS				555	8%	
DIVIDENDS Actually paid (net of tax)				576		16
NET CONSUMPTION OF CAPITAL				A$ 21 million		

* The Government Share of Income ("GSI") statement attempts to measure total government receipts from all sources which originated in mining income. It is producted by dissecting and reconstructing the audited profit and loss statements published by the Big Five Australian mining companies in their two most recent financial years. The companies and years covered are BHP - 1981 & 1982; CRA - 1980 & 1981; Utah - 1980 & 1981; MIM - 1981 & 1982 and WMC - 1981 & 1982. Some of these figures are based on estimates, but all are consistent with the published information examined. (Utah's operations were taken over by BHP in April 1984 - Editor).

Notes

1. Gross Sales before production levies; calculated from published figures for sales and oil levy and estimated coal levies.

2. Net Wages and salaries calculated by subtracting estimated PAYE tax from gross wages reported.

3. Calculated by subtracting wages and other items from reported cost of sales.

4. Reported in published accounts.

5. Estimated for Utah. All others published.

6. Estimated for BHP & WMC. All others published.

7. Estimated for all companies at 15% sales tax on 10% of goods purchased.

8. This figure for Utah alone. Not reported elsewhere.

9. $95 million reported by Utah. All others estimated.

10. PNG taxes paid by CRA Group.

11. CRA reports this figure. All others estimated.

12. Includes local rates and rent (A$32 million), pipeline fees (A$10 million), stamp duty (A$9 million), port and harbor surcharges (A$9 million), land tax (A$8 million) and red tape (A$19 million).

13. Research grants and export incentive and development grants. From government publications.

14. Calculated by taking account of average wages, tax scales, zone allowances and allowances for dependents.

15. Includes dividend withholding tax on foreign dividends and income tax at 30% on Australian dividends.

16. Gross dividends paid less estimated tax component.

SERVICES RECEIVED

What did the big miners get for the A$6 billion contribution to Australian governments? They should pay their share of general government services such as defence, law and order. In addition they get "free" advice and maps from the numerous state and federal departments devoted to minerals and energy.

Total federal defence spending is less than A$5 billion per year. The Big Five are thus paying a high price for free advice and maps. (Admittedly, BHP gets protection from steel imports. However this cost is not paid by governments - it is paid by consumers).

Their annual contribution to governments would pay the salaries of 175,000 public servants. This is generated by 120,000 workers employed by the Big Five. Each miner thus carries one and a half bureaucrats. Alternatively, the 175,000 public servant jobs provided by the taxes on the Big Five is equivalent to 43% of the federal public service.

TAXES ON JOBS

The gross payroll costs of the Big Five miners over the two years was A$4,322 million. However, the employees received only A$3,270 million. Payroll tax and PAYE tax quietly reaped over A$1 billion of unearned income for governments.

Presumably this was used to "create" jobs in the public sector. For every such paper shuffling job "created" in the public sector, another ore shovelling job was destroyed in the private sector. Any politician or union leader who talks about unemployment without mentioning this 25% fine for providing jobs is either stupid or dishonest.

FIGURE 1

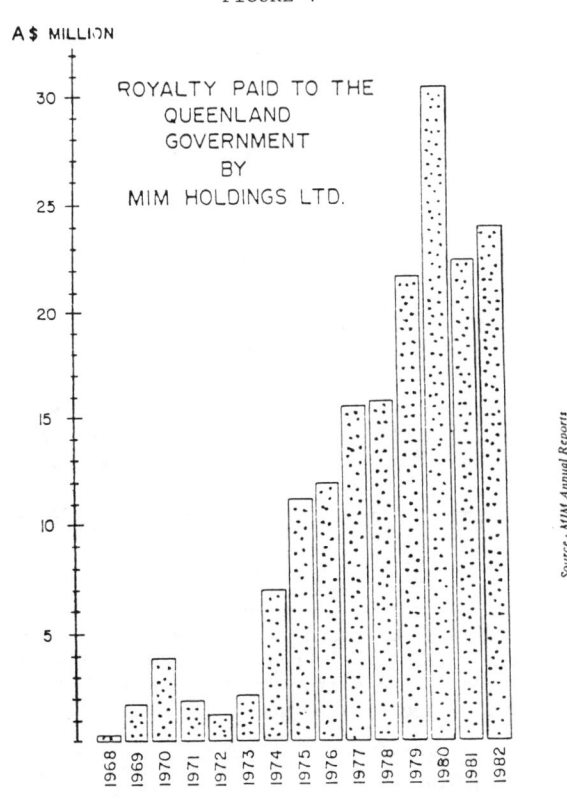

ROYALTIES

The lack of a real relationship between royalties and income further inflates the return to "the people". Figure 1 shows the royalties paid to state governments largely by the Mount Isa Mine. In the latest year the company paid royalties of A$24 million to governments while announcing losses of A$10 million to shareholders.

TAXES ON PRODUCTION

The biggest slice of the GSI is represented by taxes and levies on production. These include the crude oil levy (a complex and discriminatory tax whose cost falls largely on BHP and Esso Australia), the coal export duty (a complex and discriminatory tax designed to fall largely on Utah) and state mineral royalties (complex taxes levied on all mining companies irrespective of profitability). The effect of these is shown dramatically in Figure 2 below.

FIGURE 2

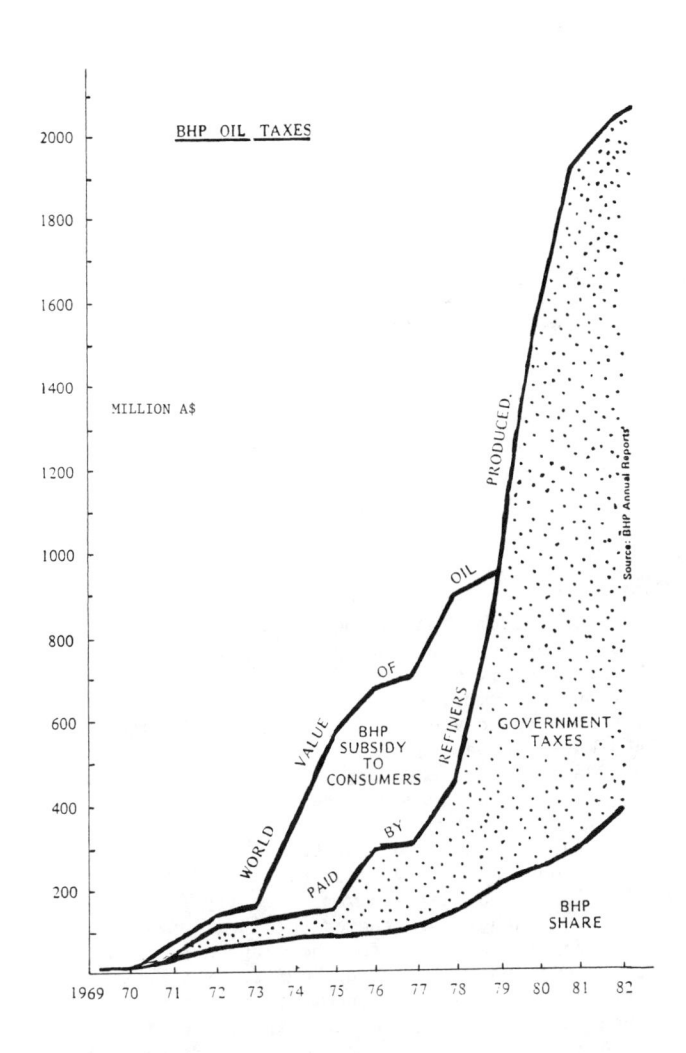

TAXES ON SUPPLIES

Consumption taxes tend to get hidden in the cost of many goods and services. They include sales tax at various rates on certain categories of goods and tariffs and quotas on imported goods. The estimated cost of these taxes shown in the GSI statement is only the direct component. However, the indirect, hidden and multiplied costs are considerable. For example, a mining company in outback Australia which decides to buy a vehicle for the mine manager may be faced by the following alternatives (illustrative figures only):

	Locally Made	Overseas Made	Cheapest Tax Free Price
	A$	A$	A$
Ex-Works Price	9,333	5,862	5,862
Sea Freight, say	-	100	100
Import Duty at 57.5%	-	3,371	-
Cost to Dealer	9,333	9,333	5,962
Dealers Margins (say 15%)	1,400	1,400	894
Rail Freight, say	100	100	100
Dealers Selling Price	10,833	10,833	6,956
Sales Tax at 20%	2,167	2,167	-
Cost to Miner	13,000	13,000	6,956

Taxes have thus increased the price of the vehicle from A$6,956 to A$13,000, an increase of 87%. Even if the miner chooses the locally made vehicle he effectively pays import duties because without duties he would have chosen the cheaper foreign vehicle. None of these hidden effects are included in the GSI statement. Moreover, because sales tax is levied on margins, freight and duties the effective sales tax on the imported vehicle is 37% of the ex-works price, not the nominal 20%.

DOUBLE TAXATION OF PROFITS

Individual shareholders are taxed twice on corporate profits. They paid corporate income tax before the dividends are declared and then pay personal income tax on what is left of the same income after dividends are received. Foreign shareholders pay corporate tax before dividends are declared and then pay 15% dividend with-holding tax before the dividends leave Australia.

Large Australian institutional shareholders do not pay tax on dividends received and small individual shareholders were recently given some relief. Larger more successful individual shareholders are penalised for their success. A non-tax-exempt shareholder in the 30% tax bracket in effect pays 46% of his income in company tax and then another 30% of what is left - a total burden of 62% or about twice his individual rate. With taxes at this level, individual shareholders are turning to more lucrative investments such as the legal horse betting, smuggling or buying tickets in the lottery.

This leaves the provision of equity capital to large unimaginative, safety-conscious financial institutions. Because individuals are becoming unwilling to invest their capital, companies are forced more and more into greater reliance on loans. This is producing a debt-laden economy which will be threatened with bankruptcy every time profits fall.

RED TAPE TAX

In 1978, the Bureau of Statistics carried out a survey of the cost incurred by small business in filling out compulsory forms from Australian government authorities such as the Bureau of Statistics, the tax department, corporate affairs, customs, trade, state and local governments. The cost for mining companies was about A$75 per employee per year.

Large companies are unlikely to spend a smaller proportion on red tape. In fact, they tend to be much more weak-kneed and conscientious with such requirements. They develop their own expensive bureaucracy who spend their time duelling honorably with the real bureaucrats. For the Big Five miners in our sample, the total red tape tax could be of the order of A$9 million per year.

MONETARY TAXES - INFLATION AND EXCHANGE CONTROLS

There is one fatal problem with unlimited democracy - no election has ever been won by promising the people less. They all want more, but promises cost money. Fulfilling just some of Australian political parties' pre-election promises has created chronic deficits.

There are only four ways in which governments can bridge these deficits:

Alternative	Result
1. Increase Taxes	Tax Payers Revolt
2. Cut Spending	Tax Consumers Revolt
3. Borrow Old Money	Interest Rates Increase
4. "Create" New Money	Currency Devalued

Alternatives 1 and 2 are the open, honest methods of bridging government deficits. However, because of past excesses, there are few policians willing to increase their use. Governments are thus increasingly forced into the more underhand financing alternatives:

Borrowing, which taxes the earnings of future generations

Money creation, which devalues the savings of past generations.

The combination of currency creation with exchange controls can be used as a savage tax on exporters. The domestic symptom of the fall in money value caused by creating new currency is general price inflation. The international symptom is devaluation of the currency compared with more stable currencies. The public has rightly concluded that devaluation of the currency is a sign of monetary mismanagement. Thus politicians try to hide the symptoms by imposing controls on wages, prices and exchange rates. This is like trying to cure a fever by smashing the thermometer. Devaluation does not cause price inflation - it is price inflation.

With free currency markets and no controls on wages or prices, there is a tendency for external currency devaluation to move in parallel with the fall in domestic purchasing power. However, in many countries including Australia, there was (as of December 1982) interference in currency markets without domestic price controls. (Australia abolished foreign exchange controls in late 1983 - Editor). Moreover, the interference in currency markets was not random. There was a conscious attempt to maintain an overvalued currency in order to hide some of the symptoms of monetary debasement.

The result causes a crush on mining profits. The overvalued exchange rate reduces foreign revenue while the uncontrolled domestic price inflation increases costs. The potential losses from this type of squeeze are shown in Figure 3. This process transfers wealth from exporters and savers to governments, to importers and to borrowers. This hidden tax is not included in the GSI statement.

Naturally not even governments can maintain an overvalued currency forever. Reality is periodically reasserted by a dramatic devaluation which, to be credible, is often overdone.

TAXING THE TRAINS

The Queensland government is the pace-setter in taxing the trains. Other Australian states are learning quickly. The taxes fall heaviest on the coal exporters. The Queensland government owns and operates all mineral railways in Queensland. Sugar mills are allowed to operate private railway lines, but no mining company is allowed to opt out of the state system (apart from underground mine trains, which fortunately were overlooked).

FIGURE 3

THE EFFECT OF
EXCHANGE RATES AND INFLATION
ON REVENUES FROM COPPER

The London Metal Exchange is the world barometer of copper prices. Australians selling on the LME receive the sterling price converted to Australian dollars at the official exchange rate.

Ⓐ Measures the loss of revenue caused by changes in the sterling - A$ exchange rate over the period.

Ⓑ Measures the loss of purchasing power of A$ receipts caused by domestic inflation over the period.

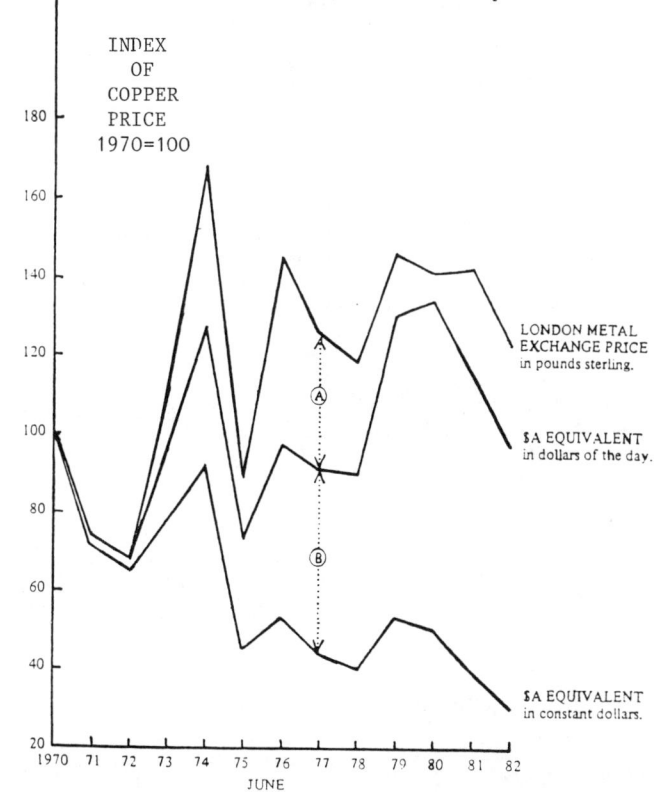

In addition, the nationalised railways have partial protection from competition from road hauliers, and all state governments are considering how to extend their legalised monopoly to pipelines and conveyors. (The Victorian Liberal Government recently imposed a pipeline tax of A$20 million per pipeline per year on Esso-BHP oil production from the Bass Strait). Armed with these advantages, the Queensland government has used its monopoly position to extract windfall returns from the coal companies.

For example, before getting approval to mine from the State government, every new coal project must agree to pay:

- all costs of railway construction
- all costs for locomotives and rolling stock
- all costs for loading and unloading facilities
- all future operating costs
- all escalation of future operating costs (plus some)
- a hefty profit surcharge which also escalates
- a future capital surcharge.

Despite all this, the railway is owned and operated by the government and the company has little say in waggon design, construction standards, operating procedures, manning levels or selection of rail routes.

Utah estimates that the rail profit surcharge is costing them over A$80 million per year and this is the only figure included in the GSI statement. Moreover, the figure excludes the cost of the capital tied up in railway assets.

This train tax will escalate swiftly as new coal projects reach production. It may create some jobs in the State government's bureaucracy, but it will strangle the creation of new jobs on the coal fields.

COMPULSORY CONTRIBUTIONS

A new method of taxing the miners is appearing in Queensland and to a lesser extent in New South Wales – compulsory contributions to infrastructure. Not only do new mining projects pay for much of the housing, roads, water, power, etc. used by their projects or their employees, but they are also being asked to pay for roads, libraries, swimming pools, sports facilities, and other amenities used largely by towns and people not connected with the mining project. In addition, it appears that such expenditure will not be a deduction for income tax.

TAXES ON PORTS

Port charges are another source of hidden unearned revenue for State governments. Hay Point in Queensland, Utah's export coal port, is a good example. This port was planned and constructed by Utah and its partners for their own use. All capital and operating costs have been paid by the companies and the port is operated by a Utah service company.

The charges levied on the port owners by State government authorities are (A$000):

Year Ending 30 June	1980	1981
Pilotage Fees	512	595
Harbor Dues	1,557	1,616
Conservancy Dues	737	805
Total	2,806	3,016

Source: Annual Report, 1981, Department of Harbours and Marine, Qld., p.20, 21.

The Pilotage Fees cover the costs of pilots for the coal carriers to and from the berth. With only 200 ships per year they could probably be done with just two pilots. It would seem hard to spend more than $300,000 per year on the service. From the point of view of the users, the rest is a tax-payment for which no related service is received.

Harbor Dues go into a fund which presumably will one day be spent on improvements to the harbor. In the meantime the companies are deprived of the use of the funds while the government invests them on the money market – more hidden taxes.

Conservancy Dues go into the State's consolidated revenue and are never seen again. What are they? The dictionary defines conservancy as "a board controlling a port". So users of Hay Point pay harbor dues and more harbor dues. For about A$3 million per year they get the value of two pilots and an unknown promise that harbor funds will be forthcoming.

It would take a specially-appointed Royal Commission to find out what happens at all the other ports of Australia, but the potential magnitude of this hidden tax can be gauged by the following list of ports in which mining companies are the chief users:-

Hay Point, Gladstone, Bowen, Townsville, Abbott Point, Weipa, Gove, Port Hedland, Dampier, Cape Lambert, Kwinana, Bunbury, Esperance, Whyalla, Yampi, Point Pirie,

Port Latta, King Island, Newcastle and Port Kembla.

CREEPING NATIONALISATION

The GSI statement indicates that governments have received 90% of the earnings of the Big Five miners in these two financial years. These are not unrepresentative years - they include years of record profit by BHP, CRA, Utah and WMC, as well as years of poor profits or losses by CRA and MIM. Moreover, there is no reason to suggest that this sample is unrepresentative of the whole mining and processing industry. Smaller companies are probably taxed less than BHP or Utah, but this is offset by all the loss-making exploration companies who also pay indirect taxes.

With the recent nationalisation of freehold coal properties in New South Wales, government ownership of Australian mineral deposits is now effectively 100%. Mining companies are merely tenants whose every decision is subject to approval by the government landlord. We have thus reached the startling position in Australia where governments own all mineral deposits, control all aspects of all mining operations, and receive 90% of the profits without contributing any capital - this amounts to virtual nationalisation of the industry by stealth.

What are the consequences of this? Firstly, the Big Five as a group are consuming capital - taxes, interest, and dividends paid exceed earnings from the operations. Secondly, the high rates of tax have removed much incentive for efficiency. Why struggle for efficient operations or for profitable diversification if 90% of the benefits go in taxes? Thirdly, the crush from all quarters on the return to the shareholders means that the chief beneficiaries of mining income are employees, suppliers, and governments. Faced with these alternatives, it is not hard to guess whose interest will become dominant in the eyes of management. Shareholders will tend to be forgotten.

CONCLUSIONS

The final grim consequence of the crush on mining profits will be a steady withdrawal of equity capital in the future. Politicians will talk of a capital strike as foreign investors sell up quietly and go home. Capital controls will be tightened as the same officials who once tried to prevent the entry of foreign capital now try to prevent its exit. The cost of capital will rise as individual shareholders choose private consumption and speculation rather than investment in public companies.

Debts will grow as mining companies struggle to replace plant out of devalued earnings. The heavy debt will magnify profit fluctuations. During the recurrent credit crises politicians will step in with conscripted capital from the Australian Industry Development Corporation, the Government Insurance Office and the Public Service Superannuation Fund. Almost no one will notice the steps to creeping nationalisation by direct and indirect taxation and levies.

CHAPTER 5: MINERALS INDUSTRY TAXATION

BIBLIOGRAPHY

Bennet, H.J., Thompson, J.G., Quiring, H.J., and Toland, J.E., 1970, "Financial Evaluation of Mineral Deposits Using Sensitivity and Probabilistic Analysis Methods," Information Circular 8495, U.S. Bureau of Mines, 82 pp.

Clement, G.K., Jr., Miller, R.L., Sievert, P.A., Avery, L., and Bennett, H., "Capital and Operating Cost Estimating System Manual for Mining and Beneficiation of Metallic and Nonmetallic Minerals Except Fossil Fuels in the United States and Canada," A Manual by Computer Services and Management Consultants, Inc. and Minerals Availability Field Office, U.S. Bureau of Mines, GPO Stock No. 024-004-02015-6, 149 pp.

Davidoff, R.L. and Hudelbrink, R.J., 1983, "Taxation and the Profitability of Mineral Operations in Seven Mountain States and Wisconsin: A Hypothetical Study," Mineral Issues, U.S. Bureau of Mines, May, 32 pp.

Johnson, E.E., and Bennett, H.J., 1968, "An Engineering and Economic Study of a Gold Mining Operation," Information Circular 8374, U.S. Bureau of Mines, 53 pp.

Kingston, G., 1977, "Reserve Classification of Identified Nonfuel Mineral Resources by the Bureau of Mines Minerals Availability System," Preprint, Mathematical Geology, Vol. 9, No. 3, 7 pp.

Chapter 6

Profit and Economic Rent

CHAPTER 6: PROFIT AND ECONOMIC RENT

PREFACE

Profitability in the minerals industry is largely sustained by its ability to extract "economic rent" from the wasting asset which is a mine. This chapter on "Profit and Economic Rent" describes how the two are linked from a theoretical as well as from a practical point of view.

In the first paper, "Economic Rent and Its Relationship to Finance," H.D. Drechsler introduces the concept of economic rent. Drawing upon material presented at SME short courses on mineral economics which he has conducted, Dr. Drechsler describes a hypothetical mining situation to illustrate this economic concept.

Peter Fletcher, in his paper, "Resource Rent Proposals," describes how the various state and federal governments in Australia and Papua, New Guinea have sought to improve their participation in the mining industry through royalties and other forms of economic rent. The paper explores some of the effects of these proposals on exploration and development of mineral resources in that country.

While governments in other countries have picked up on the concept of economic rent as well, economic theory does not specifically identify who is entitled to the profits that accrue from mineral resources. In the final paper concerning "Economic Rent Considerations in International Mineral Development Finance," John Hammes presents a banker's point of view. In today's economic environment markets for minerals and the availability of finance are key factors influencing the calculation and distribution of economic rent which may be claimed by various partners involved in mineral development in contrast to those that merely control the resource.

Certainly, an understanding of the concept of economic rent is a prerequisite for anyone dealing in the area of mineral development and operating agreements, the subject of the following chapter.

                                                        Simon D. Handelsman
                                                        Chapter Editor
                                                        Canadian Geoscience Corporation

6.1

ECONOMIC RENT AND ITS RELATIONSHIP TO FINANCE

Herbert D. Drechsler, Ph.D., P.Eng.

President, HDD Resource Consultants Ltd.

The objective of this paper is to identify the components of income above that necessary to keep a mine in production and relate those components to the profits of a mining firm. This is a discussion of "Economic Rent". Economic rent is the payment to the factors of production -- capital, labour, entreprenuer and the owner of the mineral deposit -- above the minimum necessary to keep all of these factors cooperating in the production of minerals from a particular deposit.

THE SOURCE OF ECONOMIC RENT IN THE MINING INDUSTRY

In the mining business each mine has a unique cost of production unrelated to any other mine. The cost of production is developed from many factors such as the technical characteristics of the deposit: depth of deposit, depth of overburden, areal extent, and type of deposit. Furthermore, there are locational factors which also alter cost of production. For example, identical mineral deposits located in Zambia or in the southwest U.S.A. would have different capital and operating costs. These differences occur because of different qualities of labour, ease of acquisition of operating supplies or investment capital and other similar factors.

To explain the source of rent, we shall use ideas developed by Alfred Marshall and Kenneth Boulding. Suppose a meteorite with special qualities exploded over earth and sent showers of small meteorites to the ground in various places. At some point in time, people learned of the special qualities of these meteorites and a demand arose and one mine was started. Suppose further that the cost of production to produce these meteorites was different for each deposit. Probably the lowest unit cost mine would commence production because if a higher unit cost mine started-up, eventually someone would discover a deposit that had a lower unit cost of production. The lower unit cost mine would take over the market and the high unit cost mine would cease production. The cost curve for the first mine would look like the following Figure 1.

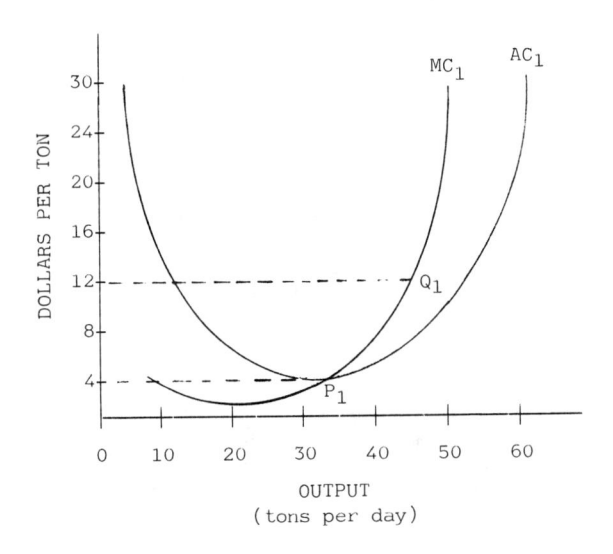

Figure 1.
Cost Curve For Mine A

The output for the first mine is on the horizontal axis with the cost per ton on the vertical axis. The curve labelled AC_1 is the average total cost of production over the range of possible production that could occur at the property. The total costs are composed of variable costs, which are dependent on output and fixed costs, which are dependent on time. For example, the cost of operating supplies would be a variable cost and normal profit would be a fixed cost.[1] The opportunity cost of the investment would also be part of the fixed costs.

In our example, all factors of production, are receiving the minimum necessary payment that will keep the mine operating. The average total cost curve is U-shaped because at zero output, the meteorite mine has no variable costs, only fixed costs which, on a unit basis, obviously are very high. To produce a small output of meteorites, the mining firm adds labour and supplies. These are components of variable costs. Since the mine has only a few employees and has inadequately utilized equipment, plant efficiency is low, the unit costs are high but less than with no output. As more labour and capital are added, output expands and cost becomes less until cost reaches some minimum point, P_1. As more labour and capital are added to the mine, efficiency begins to drop as workers interfere with each other; cost begins to rise because the proper amount of people and equipment are no longer utilized. Even if the mine becomes less efficient, output can still rise. Over the entire range of output possibilities, average total cost decreases at a decreasing rate, then becomes horizontal, then increases at an increasing rate. This is the familiar U-shaped curve shown in economic textbooks.

The line labelled MC_1 is the marginal cost of the average total cost. The marginal cost first declines, then becomes horizontal, then rises through the minimum average total cost at point P_1.

As demand for the special meteorites grows, price rises because initially there is no output. Price rises until it reaches $4 per ton, the minimum average total cost where Mine A commences production at an output of 28 tons per day. Let's assume 28 tons per day matches demand at a price of $4 per ton and our system is balanced.

If at some time demand should increase, then price would rise and output from the mine would increase to match that demand. Cost for each incremental increase in output would rise along the marginal cost curve, with output growing as shown on the horizontal axis. At a price of $12 per ton, the output would be at Q_1, or 37 tons per day. Let's assume that the minimum average cost of production of the next lowest unit cost mine, Mine B, is at $12 per ton. That mine would commence production at that price. Figure 2 is a cost ladder for four mining firms in our meteorite industry and that figure will show what happens when many firms enter the industry.

Using the same logic as previously discussed, we can see that for Mine A at a meteorite price of less than OH_1, nothing will be produced. At price OH_1, Mine A will begin production at output H_1P_1. As the price rises to K_1, the production of Mine A increases from H_1P_1 to K_1Q_1. However, at price OK_1, Mine B enters the industry because OK_1 is equal to the minimum average cost of Mine B. Mine B produces K_2Q_2 and the industry produces a total of $K_1Q_1 + K_2Q_2$ tons per day. If demand grows and price rises, the output of Mine A and Mine B increases following their marginal cost curves, until at price OL_1 the combined output is $L_1R_1 + L_2R_2$. However, at this price, Mine C comes into the industry and the total industry output increases by the quantity of L_3R_3. At price OM_1, the fourth Mine D, enters the industry and total industry output is $M_1S_1 + M_2S_2 + M_3S_3 + M_4S_4$. If price rose to ON_1, no new mine enters the industry but total output expands to include $N_1T_1 + N_2T_2 + N_3T_3 + N_4T_4$.

Economic rent is defined as any payment to a factor of production in excess of the minimum sum necessary to keep that factor from being used for some other purpose. In our system described above the factors include labour, capital, the mine owner, and the entreprenuer.

The mining industry has many firms each of which has a different cost of production, except as a coincidence. In this situation where there are low unit cost mines and high unit cost mines, it should be obvious that if the price were high enough to keep the high unit cost mine in production, that price is higher than that necessary to keep the low unit cost mine in production. The low unit cost mines are receiving more than the minimum necessary to keep operating. This means that the lower cost mines are receiving economic rent.

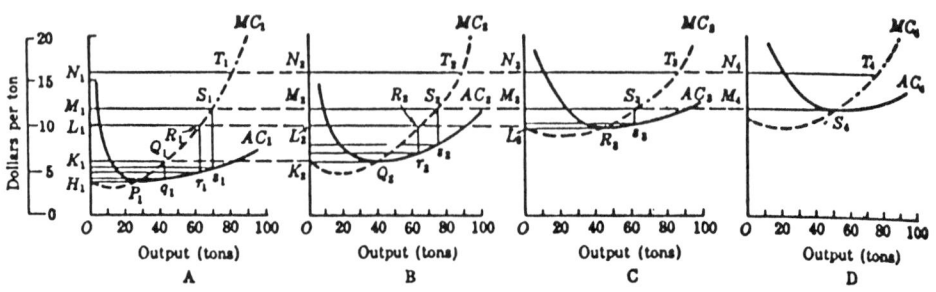

Figure 2.
Four Firm Cost Ladder

If a mining firm were to have zero net revenue, which occurs when all factors of production are receiving the minimum necessary to keep them working, the mine would continue operations. In Figure 2, when the price is as low as OH_1, even Mine A has a net revenue of zero. If the price rises to OK_1, economic rent forms for Mine A and is defined by the area $K_1Q_1 \times Q_1q_1$. As the price rises, all mines begin to gain economic rent, so that if price rises to OM_1, the economic rent in Mine A is $M_1S_1 \times S_1s_1$, in Mine B, is $M_2S_2 \times S_2s_2$, in Mine C is $M_3S_3 \times S_3s_3$, and in Mine D is nothing. The economic rent to each mine is different.

One thing becomes apparent, economic rent occurs because of the nature of each deposit, i.e., each deposit has a different unit cost of production. If all the cost curves for each mine were identical and price was equal to the minimum average cost, economic rent would be zero.

EXPLORATION AND ECONOMIC RENT

Most of the economic analysis of the evaluation of rent assumes that certain conditions will hold. Some of these conditions include: (1) the requirement of a competitive market for all factors of production; (2) the location of all deposits is known with certainty; and (3) the anticipated cost of production of all deposits is known before production commences. In mining, these last two assumptions are completely unrealistic except for pedagogic purposes. First, exploration processes to find mineral deposits are costly. In effect, the resource is renewed by a process of exploration. Second, expensive feasibility studies are required to identify and to estimate the cost of production. Third, mineral deposits are not automatically located in the optimum sequence of exploitation from lowest cost to highest cost deposits. Deposits are located in random sequence with some interspersed low unit cost and some high unit cost properties. The low unit cost properties enter the production stream to substitute for exhausted deposits or to add to supply as demand has grown. The high unit cost deposits, those with estimated cost of production higher than the price, are, in effect, placed into a mental mineral deposit inventory to await price rises that could occur as a result of the depletion process at other deposits under production, or changes in technology that would lower the anticipated cost of production.

We will assume that exploration for mineral deposits occurs because of the possibility of gaining normal profit and that anticipations of "luck" also stimulates systematic exploration. The "bonanza" concept is the hope of finding a deposit with relatively low cost of production that would develop economic rent that would accrue to the entreprenuer. The result of exploration success is a new deposit. The industry is always successful in satisfying the need for new deposits. Unfortunately, individual firms do not always find mineral deposits. Possibly, most companies fail in the search for mineral deposits.

Occasionally comment is made that more money has been used in the search for minerals than has been made in mining. This situation may be correct if subsidies have been paid to find minerals. In any event, funds for exploration must come from some source. These funds must be a part of the unit cost of production of the mining firm or part of the normal profit of the entrepreneur or both. A source of funds for exploration are funds that come from the economic rent available to low cost mining firms.

One argument, not tested, is that total economic rent in a mining industry only balances total exploration effort in that industry. That argument seems to make some rational sense, i.e., that total rent is consumed in the total exploration process. This argument assumes perfect information for all explorers throughout the world. However, if more funds are invested in exploration than returned as rent, then there is an inconsistency: there would be great losses, exploration would slow down, and the number of minable deposits would decrease. This would cause price to rise to increase the total available rent, until this system again went back into balance. In the reverse case, where too much economic rent was generated, exploration would expand, and may new deposits would be found. In this situation, price would fall until the economic rent decreased and the system would again go into balance.

There is a weakness in this argument that comes from the fallacy of division. What is true for the industry as a whole is not necessarily true for any particular firm. The amount of economic rent developed within the entire industry does not mean that all deposits accrue economic rent. A mining firm that finds an unusually low cost deposit, may receive large amounts of economic rent. Most of the economic rent in the industry may go to that firm with the remaining deposits obtaining little rent.

MINING FINANCE

In North American finance, discussion usually revolves about companies that have share capital and are traded on some stock exchange. However, when considering the entire world, most mining firms are either privately or governmentally owned and are not traded over a stock exchange. Economic rent accrues to mineral deposits because of the inherent nature of the deposit and is not related to ownership structure. Therefore, the discussion of economic rent applies to any type of ownership structure in the control of a mining firm.

Funds flow into and out of a mining firm as follows:

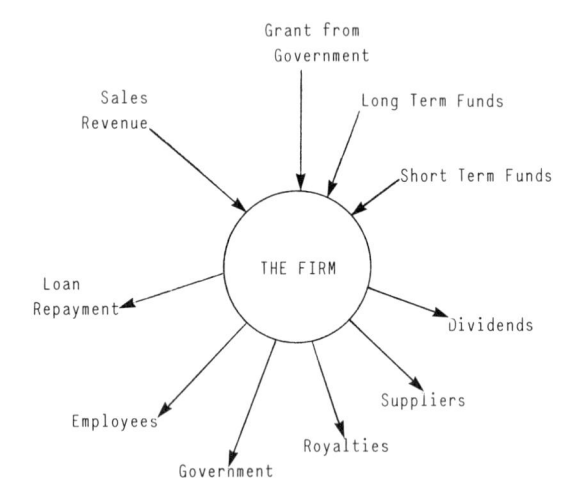

Figure 3.
Funds Flow in a Mining Firm

The figure shows that there are several sources of funds. For example, revenue from sales of concentrate, metal or ore; grants from government; and long and short term loans. In some countries such as Zambia where the copper mines are controlled by the government, the revenue to mines is a budgetary-political decision made by the leadership of the country. In some countries where the mines are also owned and controlled by government such as in Chile, the sales revenue flows directly to the mining firm and income is taxed by government.

Revenue from sales may contain economic rent if the economic costs are less than that revenue. Some of the revenue from gifts, grants, loans or other such sources may also contain some element of economic rent. This occurs because economic rent exceeds that sum which is necessary to maintain the industry. However, to encourage an industry to grow, government may provide surplus payments to producers and this revenue can be considered economic rent. It is the amount greater than that necessary to maintain the industry. In Canada, surplus payment to gold mines after World War II helped sustain an uneconomical gold mining industry. Economic rent may occur inherently from any source of revenue if it provides a surplus payment.

The cash entering the firm is reduced through payment to employees, loan payments, taxes, and payments to suppliers and to the owners of the mine as royalties or dividends. The normal payments to all these groups are the minimum necessary to keep them working to support the mine and its ancillary plant. If the employees are paid less than that necessary to keep them working they will leave. This argument is satisfactory in the absence of any union because of the assumption of labour mobility. This means that if the employees don't accept the wages from the operation, they are free to find jobs elsewhere and leave. Obviously that situation does not hold.

Furthermore, because we don't know where new deposits of minerals are likely to be found, the owner may have difficulty in finding a new property and may not want to leave the existing property even if the revenue is low.

If any income accrues to the various groups that receive income from the mine, these groups can also negotiate to receive a portion of the economic rent if such an amount exists. However, it is extremely difficult to correctly estimate the amount of rent that may exist in the revenue stream of the company.

WHO GETS ECONOMIC RENT

The inputs to the production process are the capital, the employer, the entrepreneur or managers, and the mineral deposit. Our economy has limited quantities of capital, strong labour unions and relatively scarce experienced employees, scarce entrepreneurial management talent, and few low cost ore bodies. The battle for the largest share of economic rent, if any exists, is severe. As long as each input gets its normal income, all these factors continue to work. If the price rises and economic rent enters the income stream, the owner of the deposit will usually try to increase their income. If the government runs the deposit, after taxes will increase. Labour unions will attempt to negotiate for higher wages and if the firm has large quantities of economic rent available, the employees will certainly receive a wage increase composed primarily of economic rent. The entreprenuer or manager will certainly act to increase their income and finally the owner of the plant and equipment will want to increase the dividend return to them. The division of the economic rent then becomes an exercise in political power rather than an economic argument. The final result of the division cannot be easily forecast.

The result of this struggle of the division of economic rent is relatively easy to forecast. First, the economic rent that is available at a higher cyclical price will cause permanent division of the net income stream. This division will exist when lower cyclical prices exist and no rent is available. The normal profit will then be split and eventually one or the other of the groups which depend on their normal income will receive less than that necessary to keep them working. That group will leave and the mine will cease production.

Second, government, through various tax methods such as royalties and income taxes, often will attempt to scalp off the economic rent without stopping the production process. This action may not reduce the income to the employees and management, but could reduce the income to the capital provider and the supplier of consumable goods. In this situation, the mines continue to operate, but as the capital and supply bases deteriorate, the production rate drops. Eventually the mines cease production.

Economic rent is a supplier of profit in excess of that necessary to maintain the mine. The key problem of economic rent is its existance and the desire of many groups to scalp that stream of income from the mine operation. In the effort to compete for economic rent often more than ten percent of the profit is removed from the firm. The result is eventual decay of the mining plant and reduction of operations. There is no good way to estimate the quantity of economic rent that any one deposit can accrue. This situation occurs because of uncertainty of future prices and costs, and the ultimate size of the mineral deposit.

FOOTNOTE

1. Normal profit is the minimum profit necessary to keep the entrepreneur engaged to organize the capital, labour and ore body and produce from the mine.

6.2

RESOURCE RENT TAX PROPOSALS IN AUSTRALIA

PETER H. FLETCHER

PRICE WATERHOUSE, SYDNEY, AUSTRALIA

THE ECONOMIC THEORY

It will assist in the understanding of the Resource Rent Tax ("RRT") proposals in Australia if the economic theory behind the tax is briefly explained.

The idea goes back to the concept of rent put forward by the nineteenth century economist David Ricardo.

His analysis was broadly as follows:

A tenant who farms land which is just on the margin of economic viability cannot afford to pay his landlord any rent, because the value of the crop produced by such marginal land is equal to the cost of producing the crop, plus the cost of keeping the tenant alive. There is nothing left out of which to pay rent to landlord.

A tenant who farms good land can afford to pay his landlord rent -- but the maximum amount of rent that he can be asked to pay is the excess of the value of the crop produced by the good land over the cost of producing the crop and of keeping the tenant alive.

Adapting this theory to the taxation of mining projects, modern economists who favour this system of taxation say that a profitable mining (or resource) project produces rent in the Ricardian sense. The amount of the rent is the excess of the profits by the project over the amount of profit which is required to attract the taxpayer to invest in the project in the first place.

This explains why the tax is called "Resource Rent Tax".

PAPUA NEW GUINEA PRECEDENT

Although Papua New Guinea ("PNG") is now independent of Australia, there are still close links between the two countries, and strong similarities between their respective tax regimes.

PNG introduced an Additional Profits Tax ("APT") in 1978. In many respects, Australia's RRT proposals follow the precedent set by PNG's APT. However, in practical terms, not a great deal can be learned at this stage from PNG's experiences since, according to the writer's understanding, no PNG mining project has yet paid any APT. The Bougainville mine is not liable to APT, being subject to tax under its own separate Act, which contains a different formula.

Outline of PNG Additional Profits Tax.

PNG has two versions of Additional Profits Tax ("APT") - one applying to general mining, the other to petroleum mining. The two versions are very similar. The general mining version is described below.

The tax applies on a licence by licence basis. This means that losses on Licence A cannot be offset against profits on Licence B. Expenditure relating to two or more licences, is apportioned "in such manner as the Chief Controller thinks reasonable".

Tax is levied on the "accumulated value of net cash receipts" at the rate of $(70 - n)\%$ where n=the rate of income tax.

The accumulated value of net cash receipts ("av") is calculated as follows:

Calculations commence from the first year in which substantial expenditure was incurred on the construction of facilities for the carrying out of mining operations on the licence. For that year the av consists of the total allowable expenditure to date on that lease.

For the second year, the av is the av at the end of year one indexed in accordance with the statutory formula set out below, plus the total allowable expenditure in year 2. For year 3, the av is the av at the end of year 2 adjusted by the index factors, plus the total allowable expenditure in year 3.

This process is repeated year after year. When the licence starts to earn revenue, the revenue is deducted from total allowable expenditure for that year. If revenue exceeds total allowable expenditure for the year, the excess revenue is deducted from the indexed av at the close of the preceding year.

APT becomes payable when and if the revenue in a particular year is sufficiently large to eliminate the indexed av, and the tax rate is applied to the excess of revenue over the indexed av.

Once APT is paid for a particular year, the av at the beginning of the following year is deemed to be zero. APT is therefore payable on the net cash receipts of that following year.

Taxable revenue consists (broadly) of:

Assessable income from mining operations, but excluding interest income; and,
Proceeds of sale of mining rights or information.

Allowable expenditure consists (broadly) of:

Operating expenses, but excluding depreciation, interest and prior year losses;
Capital expenditure on mining facilities (deductible 100% in the year in which it is incurred);
Exploration expenditure;
Income Tax

All items of income and expense are included only insofar as they relate to that particular licence.

The indexation adjustments are given by the following formula:

$$\frac{F}{E} \times (100\% + R) \times av$$

F = US dollar/Kina exchange rate for current year
E = US dollar/Kina exchange rate for previous year
R = 20% or, if the taxpayer so elects,
    the US Prime Rate + 12%

As noted above, the fundamental principles of Australia's RRT proposals are very similar to those of PNG's APT, although naturally there are important differences in detail.

HISTORY OF RRT PROPOSALS IN AUSTRALIA

The Australian Labor Party embraced the idea of RRT at its Federal Conference in 1977. RRT was still a part of official Labor policy on its election to the Federal Government in 1983.

The Government therefore published a Discussion Paper in December 1983. This Paper announced the Government's intention to introduce RRT with effect from 1 July 1984.

RRT was initially to appply to the petroleum sector only. It would cover both onshore and offshore fields; both existing fields and fields yet to be discovered. The Discussion Paper indicated in outline the form which RRT was to take in Australia, and invited comments.

Once successfully introduced in the petroleum sector, it was envisaged that RRT would subsequently be extended to cover other mining sector, coal and uranium being among the principal candidates for early selection for this purpose.

Industry reaction was far from favourable and the Government has now substantially modified its proposals, as outlined below.

Existing Mining Tax Regime in Australia

Before discussing the details of the latest RRT proposals, it is perhaps appropriate to explain briefly the existing tax regime as it applies to mining companies in Australia.

The onshore area of Australia is divided into six States, plus the Northern Territory and the Australian Capital Territory. Each State and the Northern Territory has jurisdiction over the granting of mining licences within its borders, and up to the three-mile territorial sea limit. The Federal Government has jurisdiction over the Australian Capital Territory and offshore areas beyond the three-mile limit.

In most cases, mining projects in Australia are subject to a royalty or excise levied by the State or Federal Government responsible for granting the licence, plus income tax levied by the Federal Government.

In the case of petroleum mining, the typical rate of State and Northern Territory royalties is 10%, based on production.

The Federal Government charges a crude oil excise levy on "old" oil - i.e.oil produced from fields discovered on or before 17 September 1975. This levy is charged on a sliding scale based on annual production from a field or area, the top rate being in excess of 80%. The crude oil levy on Bass Strait production represents a major source of Australian Federal Government revenue.

"New" offshore oil is not subject to any royalty or levy under present legislation.

In addition, income from mining operations in Australia is normally subject to Federal income tax at 46% (or 51% if the taxpayer is a company which is not resident in Australia for income tax purposes). The exception is income from gold mining in Australia which is exempt from Federal income taxes.

There are no State income taxes at present in Australia.

CURRENT RRT PROPOSALS

As of the date of writing, the principal policy elements of the RRT arrangements are set out in a Treasury Press Release, based on a Statement by the Treasurer the Hon. P.J. Keating, MP and the Minister for Resources and Energy, Senator the Hon. Peter Walsh on 27 June 1984 (referred to hereafter as the "June 1984 Statement").

RRT is to apply with effect from 1 July, 1984 in relation to offshore areas where the Commonwealth's Petroleum (Submerged Lands) Act applies, other than in specified areas which will continue to be subject to excise and royalty arrangements. RRT will not be applied in respect of oil revenues within the jurisdiction of any of the States or the Northern Territory.

The existing crude oil levy on "old" oil will continue to apply.

As of the date of writing (August 1984), there is some uncertainty as to the tax regime which will apply to some projects which are already in the pipeline. At one stage it was proposed that a new excise regime would be introduced on a basis similar to that applying to "old" oil, but at lower rates. However, Senator Walsh has recently canvassed the idea of applying a version of RRT to proposed new developments in Bass Strait, (principally the Bream field).

The unfortunate result will be that there will be a variety of different Federal tax regimes for offshore oil plus different royalty arrangements in each State and Territory.

In its June 1984 Statement the Federal Government repeats its intention to approach the States with a view to replacing State royalties with RRT at a later stage. Taxpayers can be forgiven for a degree of scepticism as to whether the States would be willing to relinquish their rights to royalties.

The June 1984 Statement contains no mention of any extension of RRT to other sectors of the mining industry besides petroleum. Whilst this cannot be ruled out as a possibility for the future, it appears to represent a low-priority item in Government thinking at present.

As of August 1984, no RRT legislation has yet been published even in draft form. The only official written information on RRT is contained in the various Press Releases issued by the Government. (For references, see end of paper).

Details of RRT Rules

RRT is to be levied on a project-by-project basis. The rules for determining a project will be specified in the legislation but the basic principles are that:-

- the project will represent an integrated investment and could include a number of proximate fields. Broadly, the boundaries of an integrated investment will comprise a production licence area and treatment and other facilities and operations outside that area which are integral to the production of a 'marketable' petroleum product;

- the taxable output of a project (that is, the 'marketable' petroleum product) will be treated as 'marketable' for assessment purposes at the first point in the production process at which it is saleable commercially, even though an actual sale may not have taken place at that point;

- if no sale takes place at that point, or where a non-arm's length sale occurs, an income value will be attributed to the product at the RRT assessment point;

- project boundaries for RRT assessment will not extend beyond the petroleum production stage to downstream activities such as refineries and facilities for transporting 'marketable' products. This means that neither expenditure on downstream activities, nor value added to products through those activities, will be taken into account in calculating liability for RRT;

- the scope of project expenditure and income to be taken into account will encompass certain infrastructure where this is integral to the production of a 'marketable' product, including social infrastructure (eg housing and associatred facilities of the kind that qualify for deduction under the petroleum mining provisions of the income tax law) provided principally for employees of the project and their dependants, and office buildings situated at or proximate to the site of petroleum operations; and

- expenses not directly related to the project will be excluded. For example, where an entity has diverse interests, only one of which is a project assessable for RRT, only those costs incurred at its head office which are solely attributable to the RRT project will be deductible for RRT purposes. Clearly identified expenditures, such as project engineering design costs carried out in the head office therefore would be deductible, even though this might involve an apportionment of some employees' wage costs between time spent on that activity and the remainder spent on other activities not directly associated with an RRT project. General overhead costs incurred at head office would not, however, be deductible.

A project's liability for RRT will remain unaffected by changes in ownership or implementation of farm-in agreements. The project participants will be assessed on the basis of eligible project expenditure and receipts. A deduction will not be available for any cash payments made for entering a joint venture nor will such a payment be assessable to the vendor for RRT purposes.

In the case of a full acquisition of a company's interest in a project, the purchaser will be entitled to claim deductions for expenditure not already recouped by the vendor for RRT purposes. With partial acquisitions, however, deductions will accrue to the party actually incurring the eligible expenditure.

RRT is to be deductible for income tax purposes (as royalties, or crude oil levy are), but income tax is not deductible for RRT purposes.

It is interesting to note that under the PNG system, petroleum income tax is deductible for APT purposes - this is the opposite of the proposed Australian system. The theory behind RRT would appear to require that RRT should be the only tax applicable to a resource project - but it seems that theoretical purity must give way to political reality.

Calculation of RRT

The proposed rate of RRT is 40%, with a threshold rate of return equal to the long term bond rate plus 15%.

The basic principle behind the tax is that allowable expenditures are accumulated, and carried forward compounding annually at the threshold rate of return. Taxable revenue is deducted from allowable expenditure for this purpose.

When taxable revenue exceeds the accumulated, compounded value of allowable expenditures, RRT is payable on the excess.

The concept is the same as that adopted for PNG's APT, with the exception that there is no adjustment for movements in the US$/local currency exchange rate.

To illustrate the procedure involve in RRT calculations, assume a project with the following expenditure/income profile:

Year	Capital Expenditure $ million	Operating Expenditure $ million	Income $ million
1	100	Nil	Nil
2	100	Nil	Nil
3	100	20	200
4	Nil	50	500
5	Nil	50	500

It is assumed that the long term bond rate is 15%, so that the threshold rate of return is 30%. The RRT computation would proceed as follows:

```
                                              $ million
Year 1   Capital expenditure                      100
         Accumulated net expenditure c/fwd        100

Year 2   Accumulated net expenditure b/fwd        100
         Add: 30% threshold rate of return         30
         Add: Year 2 capital expenditure          100
         Accumulated net expenditure c/fwd        230

Year 3   Accumulated net expenditure b/fwd        230
         Add: 30% threshold rate of return         69
         Add: Year 3 capital expenditure          100
         Add: Year 3 operating expenditure         20
         Deduct: Year 3 income                   (200)
         Accumulated net expenditure c/fwd        219

Year 4   Accumulated net expenditure b/fwd        219
         Add: 30% threshold rate of return         66
         Add: Year 4 operating expenditure         50
         Deduct: Year 4 income                   (500)
         Net Income subject to RRT               (165)

         RRT payable at 40%                        66

Year 5   Year 5 income                            500
         Deduct: Year 5 operating exp.           ( 50)
         Net income subject to RRT                450

         RRT payable at 40%                       180
```

The following points should be noted:

1. Capital expenditure is deductible in full in the year in which it is incurred, in the same way as operating expenditure.

2. The threshold rate of return is applied each year to the accumulated net expenditure brought forward from the end of the previous year. Exploration expenditure incurred in the 5 year period prior to the granting of the first production licence may be compounded year by year at the threshold rate. Expenditure prior to that 5 year period will be compounded year by year at a rate equal to the GDP deflator, and deducted after all expenditure subject to the threshold rate has been deducted.

3. Once the accumulated net expenditure is exhausted, the threshold rate compounding claim a refund of RRT previously paid - but this issue is yet to be resolved.

4. There are of course many other uncertainties over the details of RRT such as the treatment of exchange fluctuations.

The June 1984 Statement contains a fairly full list of what expenditures will be allowable and what income will be taxable for RRT purposes. The main items are summarised below - but it should be borne in mind that in all cases the income and expenditure must relate to the project in question.

Income for RRT Purposes

The following items will be taxable.

1. Receipts from sale of crude oil, condensate, natural gas, LPG etc.

2. In vertically integrated operations, e.g. where the taxpayer sells refined products rather than crude oil, the taxpayer will be deemed for RRT purposes to have sold his petroleum production at market value at the point where it first reaches a commercially marketable state.

3. Proceeds of sale, lease or hire of assets in respect of which an RRT deduction has been allowed. (It is not clear whether sale proceeds would be limited to original cost, in cases where proceeds exceed cost).

4. Insurance recoveries received on project property for which an RRT deduction has been allowed.

Interest, private override royalty income and other non-project related income will not be subject to RRT.

Allowable Deductions for RRT Purposes

The following items of expenditure will be deductible:

1. Exploration expenditure on geological, geophysical and geochemical surveys, exploration drilling and appraisal drilling.

2. Capital expenditure on offshore production platforms, drilling plant and equipment and overheads at the wellhead.

3. Capital expenditure on pipelines/tankers for transporting petroleum to the shore.

4. Capital expenditure on initial treatment plant.

5. Operating costs relating to the production, transportation to shore, and initial treatment of petroleum.

Interest, private override royalty payments and non-projects related expenses will not be deductible for RRT purposes.

The Government has announced that a system of provisional instalments of RRT will be incorporated in the RRT legislation to reduce delays in RRT payments, but no RRT payable in respect of any income year will be collected in advance of that year.

PRINCIPAL UNRESOLVED POINTS IN RELATION TO RRT

The Government has not yet determined the treatment to be accorded to end of project life losses. For example, the cost of removing an offshore platform, cleaning-up, sealing wells etc. may result in substantial losses for RRT purposes at the end of the life of a project. The taxpayer may be permitted to carry such losses back and borrowings, inclusion of RRT as a creditable tax for the purposes of double taxation relief, but perhaps the last major uncertainty which needs to be mentioned here is whether RRT will be extended to onshore petroleum projects, and other sectors of the mining industry.

REFERENCES

The main sources on RRT are the following three Australian Federal Government press Releases:

1. Discussion Paper on Reserve Rent Tax in the Petroleum Sector, December 1983.

2. Outline of a "Greenfields" Resource Rent Tax in the Petroleum Sector, April 1984.

3. Resource Rent Tax on "Greenfields" Offshore Petroleum Projects, 27 June 1984.

6.3

ECONOMIC RENT CONSIDERATIONS IN INTERNATIONAL MINERAL DEVELOPMENT FINANCE

JOHN K. HAMMES

Vice President
Citibank, New York

INTRODUCTION

From the point of view of the consumer, the cost of mineral commodities might be viewed as the total price industry pays for mine output. Similarly, the mining company engaged in the on-going business of finding, developing and operating new mines is concerned very much today with the distribution of all the revenue paid for mine output and not just with its direct or controllable cash costs. In other words, what will the labor force, the supplier of energy, explosives, and reagents, and the supplier of government services, management, capital, and the ore reserves receive for their contribution to production? A significant amount might be categorized as what economic theory classifies as economic rent.

Mining companies are finding it increasingly difficult to negotiate concession or work contracts to develop ore reserves in lesser developed countries, just as they find themselves accused of taking excess profits in their own countries. The principal difficulty in reaching mutually satisfactory agreements in many cases is the difficulty in establishing an acceptable distribution of income, in particular the distribution of economic rent. (See Chapter 7). Further, it is logical that the decisions and policies of the interested parties be they government or industry should be and will be influenced by their competitive position relative to determining the distribution of economic rent. The raising of finance for a mining project today may be inhibited by too biased an extraction of economic rent by entities other than the risk takers, the mining companies.

ECONOMIC RENT

A British economist (Ricardo D. McGraw-Hill, 1955) first proposed the concept of economic rent (See Drechsler, H.D. earlier in this Chapter.) Ricardo pointed out that no rent would be paid until land became scarce. Ricardo also stressed the distinction that the price of farm produce was not higher because of the rent paid on farm land but that the rent or price of more fertile farm land was higher because the price of produce is higher.

This led to the concept that economic rent is not a cost and that it is therefore not price determining, but is itself price determined. Economic rent may be viewed as a cost if cost is equated with the concept of payment or distribution of income to any factor used in production. However, if it is a criterion that a payment must be made to the supplier of the factor of production, then economic rent is not truly a cost because it is not necessarily distributed or "paid" to the supplier.

Consider the commodity A which occurs in three grades: A1, A2, and A3. Any grade may be used as a substitute for another grade in all of the industrial uses of A. However, in industries X and Y the different grades have different productivities. Table 1 shows the relative productivities of each grade in industries X and Y, and in all other industries $\lessgtr$.

TABLE 1

RELATIVE PRODUCTIVITIES

COMMODITY GRADE	Industry	X	Y	$\lessgtr$
	A1	10	4	2
	A2	5	3	2
	A3	2	2	2

Based on a paper presented to the SME Fall Meeting, September 1974.

Examining this table we see that in industry X, grade A1 produces 10 units of output for each 2 units produced by a similar quantity grade A3 and grade A2 produces 5 units of output for each 2 units produced by a similar quantity of grade A3. We might therefore conclude that, for uses in industry X, A1 is worth five times the value of A3 and twice the value of A2.

What, in fact, does economic theory say regarding the actual prices which industry will pay for the various grades of A. Consider that the basic price/value of A in most of its industrial uses $\leq$, is $1.00, and assume that the demand for A1 in industries X and Y is less than the total supply of A1. What will the price of A1 be? The table of relative productivities tells us that industry X would pay up to $2.00 for A1 but would then pay $1.00 for A2 before paying more than $2.00 for A1. Industry Y would similarly pay up to $1.33 for grade A1. The economic theory then tells us the price actually paid by industries X and Y will lie in the range between $1.00 and the upper limits we have established. The actual price is not determinitive and will depend on the bargaining position and strategy of the industry and of the suppliers of A1. On the one hand, industry X knows that a price of $1.01 is a better price for A1 than can be obtained by suppliers in most other markets, whereas the suppliers know that A1 is worth much more than A3 for uses in industry X. Before looking further at who gets this economic rent, other implications of the relative productivity table are worth examining.

Suppose there are 1000 units of A1 available and the needs of industry X are 500 units and of industry Y are 700 units, what will the price of A1 be? In this case, where industry needs exceeds supply, industries X and Y will bid up the price so that industry X pays just enough over $1.33 to obtain all 500 units it wants and industry Y obtains the remaining 500 units at $1.33 and buys A2 for the remainder of its production requirements.

Similarly, assume there are 1000 units of A1 available but that the needs in industry X is 2000 units and in industry Y is 3000 units. The pricing will be similar to that in the previous example as long as there is sufficient A2 to meet the additional needs of industry X. However, if A2 is in relatively short supply, its price will be bid up and the price of A1 will increase owing to its partial dependence on the price of A2.

These examples serve to illustrate the two essential elements of economic rent. Economic rent came about for two reasons: 1) the particular input or, grades of input have different productivities in industrial application, and 2) the input or grade is scarce in supply (Samuelson, P.A., 1955).

RENT IN THE MINERAL INDUSTRIES

If one considers the two essential elements of economic rent, it is quite clear that the concept is very much applicable to the mineral industries. Variations in the tenor, metallurgical characteristics, location and attitude of ore mineralization satisfy the requirement that the inputs have varying productivities. Secondly, ore mineralization can be regarded as scarce. The more important concern is the question of what factors in production receive the economic rent and just how does one determine economic rent.

There is no clear answer to either of these questions. No formula or other quantitative analytical procedure provides a solution. Earlier in the discussion of input pricing it was stated that the suppliers know that A1 is worth much more than A3 for uses in industry X. This statement implies relatively complete and accurate knowledge of industry economics by the involved parties. This is one of the problem areas in economic analysis. The assumption that all of the facts are known is not true. If economic rent for a particular mine represents the difference between the marginal cost of the highest-cost production of all mines required to meet industry needs and the costs of that mine's production, how readily calculable is economic rent?

Another problem unique to the mineral industry and affecting our concept of cost versus rent is the cost of finding new mineral deposits. Herfindahl has pointed out (Herfindahl, O.C., 1961) that the long-run cost of minerals is broader than that of mining cost and that the cost of exploration and evaluation must be covered in the price of a mineral to make that industry viable. This concept leads to the conclusion that good mines must provide a portion of the funds spent on exploration activities that are unsuccessful.

One other area of difficulty in our differentiation between cost and economic rent is the economist's concept of profit. A "normal profit" in economic theory is the interest or return on capital. As labor and supplies are "paid for", so is there a cost associated with the provision of capital and the financing thereof, and this normal profit is not a distribution of economic rent. It is as much a cost of production in the economic sense as are the wages paid labor and the payments for supplies. Just as there are problems with allocation of the long-run finding costs for minerals, there are areas of disagreement as to what a normal profit is. What appears as an attractive or excessive accounting rate of return in the eyes of the administration of a lesser-developed-country during the third year of a mine's life, which happens to coincide with very high metal prices, might not be so attractive when the mining company views the overall long-term discounted-cash-flow rate-of-return on the venture. Such considerations as varying risk (See Chapter 9) and opportunity cost affect the interested parties concept of a normal return. (See Chapter 3).

THE DISTRIBUTION OF ECONOMIC RENT

The fact that economic rent is a difficult concept to put numbers to in the real world does not prevent its existence and, in the case of the minerals industry does not prevent it from being a significant part of the total revenue from mine production. What are the economic forces that influence the distribution of rent? How will the distribution depend on the bargaining position and strategy of the firms in the industry and of the suppliers of the input including the financiers?

The concept of economic theory provides a starting point. Much of the theory in microeconomics has developed to explain the behavior of the firm under conditions of perfect competition.

But in the real world, it is difficult, if not impossible, to point to an industry which operates under the conditions of perfect competition. Open entry under perfect competition implies that new mining companies can enter into and exit from the production of mined products at will. But what of the problems of ore reserve availability, technical know-how, finance etc? Alternatively, can a firm obtain all these elements necessary to become a mineral producer, and at what speed? For these reasons, economists have theorized (Fellner, E., 1960) on the firm's behavior under conditions varying from those of perfect competition.

It is difficult to determine the difficulty or ease of entry into each mineral industry. Nevertheless, one can make the qualitative judgement that the greater control of entry in the hands of an interested party the more likely that party will receive the greatest distribution of rent. For example, the small exploration company which has found an interesting prospect but which has done little to develop the property and which brings little capital, production capability or marketing know-how, etc. to the situation is likely to turn the property over to a larger mining company for a relatively small carried interest.

In another example, it seems a logical generalization that a mining company financing the development of a new iron mine in a lesser-developed-country would receive a larger part of the economic rent than would the developer of a copper mine in that country. There are fewer iron-mining companies bidding for properties and iron mining is more of a marketing, transportation, and finance business than many other mineral industries. Not only is the provision of capital important because it plays a very significant role as a control on entry and is therefore an important consideration in this rather theoretical concept of the distribution of rent, but it is also important because international mineral development is indeed being impeded by the inability of mining projects to obtain finance.

CAPITAL IN NEW MINE DEVELOPMENT

It is an acknowledged fact that the capital cost per new unit of mine production capacity is increasing at a distressing rate. A principal external factor inhibiting finance for the minerals industry is turmoil in the international financial markets. Increases in foreign exchange flows to the OPEC countries to pay for energy, and floating exchange rates and the failure of some banks, e.g. owing to foreign exchange losses, have resulted in a drastic cutback in the availability of credit and particularly, the availability of long-term eurocurrency borrowings which are a principal source of finance for mining projects. The long construction times and the resultant lengthy grace periods and long payouts on mining loans have always been distasteful to lenders because of their concern with liquidity, and the recent decrease in the appetite for long-term loans has hit hard at a number of mining projects looking for financing. In addition, the high interest-rate environment of the past few years coupled with the increasingly capital-intensive nature of mine development has resulted in more projects becoming interest-rate sensitive with the result that the economic returns to the mining company are not commensurate with the risk.

Inherent factors contributing to difficulty in mine finance are the many risks associated with the industry, and a lender's unwillingness to finance new development in the face of the uncertainty of achieving forecast cash flows. A lender is always concerned not just that a new project is going to be technically and economically viable, but also with the assurance that the loan will be repaid before a substantial portion of the funds are distributed to the other interested parties. This ability to allocate economic rent to the burden of financing a minerals project is a key reason for the development of project financing new minerals developments, particularly at a high percentage of debt, nowadays at the 70-80% levels (See Chapter 11). This can be contrasted to a manufacturing venture which usually carries a much lower share of economic rent.

If the financial structure is an equitable arrangement for all the parties and the economic rent is not squeezed out of the development, then a lender feels more confident and the mining company and its host government will devote the necessary resources to making the mine a continuing success.

SUMMARY AND CONCLUSIONS

Two related problems of great importance in the pattern of international mineral development are the provision of capital for new mine development (the subject of this book) and the equitable sharing of the wealth produced from new mines. (See Chapter 5 and 7).

The provision of capital or financing will be aided by the negotiation of equitable sharing of the economic rent in a minerals development. The negotiation of concession or other development agreements is complicated by diverse opinions as to what constitutes a fair return on capital and by the difficulties in agreeing on the disposition of those portions of "venture profit" that we have described in economic theory or the long run finding cost and economic rent. While the latter concept is a difficult one to apply in business negotiations, the application of the theory in assessing one's financial position or in formulating a bargaining strategy has merit.

CHAPTER 6: PROFIT AND ECONOMIC RENT

BIBLIOGRAPHY

Boulding, K.E., 1966, <u>Economic Analysis</u>, Harper and Rowe, New York.

Carlisle, D., 1954, "The Economics of a Fund Resource: Mining," <u>American Economic Review</u>, Vol. 44, Sept., pp. 545-616.

Fellner, W., 1960, "Competition Among the Few," A.M. Kelley.

Gray, L.C., 1914, "Rent Under the Assumption of Exhaustibility," <u>Quarterly Journal of Economics</u>, Harvard University Press, Cambridge, MA, May.

Helliwell, J., 1975, "Economic Rent From Mineral Resources: Concepts and Perceptions, Collections and Conflicts," <u>CIM Directory</u>, pp. 101-105.

Herfindahl, O.C., 1961, "The Long-Run Cost of Minerals," <u>Three Studies in Mineral Economics</u>, Resources for the Future, Inc.

Kierans, E., 1973, "Report on Natural Resources Policy in Manitoba," Dept. of Economics, McGill University.

Owen, B.E. and Kops, W.J., 1979, "The Impact of Policy Change on Decisions in the Mineral Industry," Centre for Resource Studies, Queen's University, Kingston, Ontario.

Ricardo, D., 1955, "On Rent," <u>Readings in Economics</u>, McGraw-Hill, New York.

Samuelson, P.A., 1955, Chapters 20 and 27, <u>Economics</u>, McGraw-Hill, New York.

Stonier and Hague, 1964, <u>A Textbook of Economic Theory</u>, 3rd ed., John Wiley & Sons, New York, pp. 273-298.

Wilson, W.G., 1975, "Economic Rent-What Is It And Does It Exist?" <u>CIM Directory</u>, pp. 114-117.

Chapter 7

Impact of Mineral Development and Operating Agreements on Financing

CHAPTER 7: IMPACT OF MINERAL DEVELOPMENT AND OPERATING AGREEMENTS ON FINANCING

PREFACE

The papers in this chapter discuss various aspects of relations between governments, primarily in the developing countries, and international mining companies. In the paper on "Mineral Exploration and Development Agreements," Wolfgang Gluschke provides an overview from the perspective of the United Nations. He introduces the reader to the different types of mineral development agreements and the areas covered by these contracts, focusing on the stages of exploration and development, corporate structure, and taxation.

To illustrate some of the issues which may impact the negotiating process, Pedro Lizaur, in "Sharing Risks and Rewards in International Contract Negotiations," describes a hypothetical natural resource project sponsored by a foreign firm, based on four alternative model contracts. The paper discusses the various criteria which may be employed by parties evaluating prospective projects and the impact which alternative contracts have on the standards of evaluation. From this understanding a negotiating process can be constructed to address the specific interests of both the foreign private party and a government entity.

Following Lizaur, two contributions discuss recent contractual agreements in a more specific context. In "Alternatives for Countries' Mining Finance in Recent Mineral Development Agreements," Steven Zorn evaluates the relative importance of the principal contractual mechanisms impacting mineral financing which have evolved in the late 1970s and early 1980s and explains their contribution to the development of mineral projects during those years. Thomas Walde's paper gives the perspective of the legal advisor concerning "Three Recent Mineral Development Agreements in South America." This region has seen a great variety of contracts ranging from general to highly specific. The examples given by the author are representative of the region's diversity and have significance and application beyond the specific cases mentioned.

The last article looks at the question of mineral agreements and finance from the point of view of an international engineer-banker. In the paper entitled "Mineral Industry Response to Mineral Operating Agreements," Kurt Pralle provides candid comments concerning developments in this sphere during the past two decades.

                                                                Wolfgang O. Gluschke
                                                                Chapter Editor
                                                                United Nations

7.1

MINERAL EXPLORATION AND DEVELOPMENT AGREEMENTS: AN OVERVIEW

Wolfgang O. Gluschke

Natural Resources and Energy Division
United Nations
New York, NY

INTRODUCTION

Virtually all countries have general legislation covering most aspects of mining and mineral processing, including investment and tax laws, safety and health regulation, and specific rules applying to exploration reporting procedures. Yet most developing countries negotiate individual agreements with foreign investors for major mineral projects, in particular when all or part of the expected output is exported. Such agreements are known under several different names, as for example concession agreement; technical assistance agreement or service contract; management agreement; production sharing contract; joint venture; or contract of work. The objectives and conditions governing such contracts are as diverse as these titles are.

While a mining project is only one among a large mumber of similar undertakings in industrial countries, in most developing countries such projects usually have a dominating position within the country's economy. The general legislation is often inadequate and insufficiently detailed to cover all aspects of large projects with extensive foreign inputs and international markets. Therefore, the need for specific arrangements. Specific arrangements may also become necessary due to the complexity of markets for mineral products, the large number of participants - equity partners, suppliers, financial institutions, buyers, etc. - or economic and political conditions in the host country and fiscal regulations in the home country of the investors.

Most agreements are concluded between governments, or government agencies, in developing countries and private industry headquartered in developed market economies - Canada and the United States of America in North America, Western European countries, Japan, South Africa and Australia - or, in some instances, with state-owned firms in these countries. There are fewer agreements between developing countries and the centrally planned economy countries - Eastern Europe and the USSR - which cannot be considered here because of absence of adequate information.

Agreements can be concluded at any of the different stages from exploration through closing of a mine. While most contracts are negotiated at the time when a potentially viable ore body has been delineated and investigated sufficiently to anticipate a realistic time for commencement of production, other agreements start with "grass-roots" exploration, during operation of a mine, smelter or refinery or, exceptionally, after a temporary shutdown. While several major provisions are included in all types of agreements, others are covered in much more detail (or not at all) in some contracts reflecting the specific work programme of each project. While it is realized that subdivision and classification of complex contractual relations is necessarily arbitrary to a certain extent, such grouping is useful in practical terms, especially for the non-specialist. The following discussion, therefore, will start with a description of the major forms of contractual relation - concession, investment contract, joint venture, management agreement, service and production sharing contracts. Major issues covered in all, or almost all, agreements will then be dealt with. These include stages of development, corporate structure and control, fiscal arrangements, infrastructure development and environmental law, and dispute settlement and renegotiation.

FORMS OF CONTRACT RELATIONS

There has been marked change in the type of agreement entered into during the 1960s (and before) and the early 1970s and the period thereafter. While straight "concessions" - granting of exclusive rights to foreign investors over a certain deposit or areas with little restriction or well-defined obligations - were the rule during the first half of this century, more complex agreements have evolved since. Some of these cover every conceivable aspect, foreseen or only possible, and define rights and obligations of all partners to the last detail. Over the years, agreements have without doubt become more beneficial for the developing host countries, in part as a result of well-publicized nationalizations, as for example in Chile, and in part due to changed attitudes in all countries towards the legitimate interest of the raw materials producing developing countries(1).

Concession

The mineral exploration or development concession, granting exclusive rights to the concessionnaire for long periods of time and often over a large area, has been the traditional form of relationship between private industry and host governments for many decades. More recently, however, agreement has evolved to the effect that such concessions are to the disadvantage of the developing countries since they "excluded the host government from participating in the ownership, control and operation of the undertaking". Concessions were generally granted for very long periods of time, in some instances up to 100 years, and fees or rents were often low. Companies were seldom required to surrender parts (or all) of the concession area if they failed to undertake exploration or other development work. In the view of developing countries, such concession "created an enclave status for the transnational corporation fortified by a regime or economic and legal arrangements so formidable and pervasive that it overtly challenged the sovereignty of the host government over its natural resources"(2).

Foreign Investment Agreement

After the change of government in Chile in 1973, this country concluded a series of foreign investment contracts with mining companies, among them Compania Minera San Jose, a subsidiary of St. Joe Minerals of the United States, Foote Mineral Company (subsidiary of Newmont Mining Corporation), and Noranda Mines Limited. St. Joe meanwhile has completed development of the El Indio gold-copper-silver mine at an investment cost of some US$250 million; Foote is developing lithium deposits in the Atacama Desert in the northern part of the country; and Noranda was interested in the development of the Andacollo copper deposit though had to withdraw in 1980 because of lack of financing.

In these and other agreements, the Chilean Foreign Investment Committee and the respective foreign firm agree on financial and fiscal conditions under which future investment can take place; little coverage, however, is extended to question of timing, operational levels, sales, and regional impact. The agreements generally run for a period of 30 years and set limits on the time until which the investment will have to be completed, as for example eight years from the date of signing the agreement. The St. Joe Agreement states that "San Jose shall have the right freely to determine the level of its production, the method and timing of its shipments, and the volume of its sales. The right to determine the level of production shall include the right to suspend production entirely"(3).

The agreements define the "Authorized Capital Investment" (e.g. US$23 million for Foote, US$100 million for St. Joe, and US$350 million for Noranda) to be made by firms incorporated in Chile, either with or without government participation. In the case of St. Joe, the "Authorized Capital Investment may be carried out by San Jose or by any corporation, partnership or other legal entities". Noranda's equity share was 51% and Foote's share of the Sociedad Chilena de Litio Ltda. is 55% initially while the Corporacion de Fomento de la Producion, a state-owned firm, holds 45%; the agreement anticipates that "the company may in the future be converted into a sociedad anonima (Corporation)". The contracts define in detail the guarantees of the Chilean government and regulations concerning foreign exchange, accounting procedures, and levels of taxation. The companies have the right of free access to foreign exchange markets; retention abroad of export proceeds; remittance abroad of interest amortization payments for loans or suppliers' credits, capital investment following the sale or liquidation, and profits earned from operations; and to maintain a foreign currency account in Chile. Past expenditures are part of the Authorized Capital Investment.

The government guarantees a maximum tax of not more than 49.5%. The effective tax rate of 49.5% "shall be maintained at a constant level by increasing or decreasing the rate of the additional tax by an amount required to make the total tax burden on the net income...equal to 49.5% thereof"(4). Operating losses can be carried forward for up to 5 years. "An applicable income tax rate of 49.5% shall not, however, prevent San Jose from utilizing any tax credit which may be available under Chilean laws at any time". The company can take advantage of future, lower, tax rates, if it so desires, since "San Jose may elect at any time to have it's net income taxed at the same rates which prevail for Chilean nationals, in which case San Jose shall irrevocably forfeit the guarantee of an invariable rate of income taxation". The company can choose between normal and accelerated depreciation of capital assets until the start of commercial production.

Joint Venture

The most common form of contractual relations is that of a joint venture between the government - or a government agency - and the foreign investor(s). Both sides have an equity position, with either party having a majority of shares. The maximum participation allowed for foreign investors may be determined by general legislation, as for example in Mexico (49%), or the distribution of equity is being negotiated separately in each agreement. The government's equity can consist of fully paid-in capital; it can be in lieu of other inputs, e.g. the value of the ore body or infrastructure; or the government can receive "free equity" or can purchase equity on concessionary terms. A number of contracts give the government the option of purchasing shares at a later stage at a predetermined price. Obligations and rights of the parties are not necessarily divided according to the relative equity participation. Joint ventures are the most interesting and complex contractual arrangements. A detailed discussion of joint venture arrangements is outside the scope of this paper, however.

Management Contracts

Management contracts specify services to be provided by a company or companies for a set fee, a percentage of sales receipts from the commencement of production, or a combination of fixed fees and commissions. In a strict sense, the company has no equity position, but such agreements are not very common. Examples are the "Technical Assistance Agreement" between the Sar Chesmeh Copper Mining Company, fully owned by the Government of Iran, and Anacanda-Iran, a wholly-owned subsidiary of Anaconda Company (which at that time was an independent firm and not yet a division of Atlantic Richfield (ARCO), and the Service and Technical and Administrative Agreements between Rosaria Dominica S.A. and Rosario Resources Corporation, the first in force in 1979 and 1980 following the forced sale of Rosario Dominicana to the government of the Dominican Republic, and the second running for three years from the beginning of 1981.

More common are management agreements for projects in which the managing company also has an equity interest though the contracts can be formally separate. Among the several examples only some better known can be mentioned here: Ghana, after acquiring a majority ownership of various diamond and gold mines, concluded management agreements with Consolidated African Selection Trust (Akwatia diamond mine) and Ashanti Goldfields Corporation, a subsidiary of Lonrho Limited (United Kingdom), in 1973; an "Administrative Agreement" between Cerro Colorado Copper Corporation, Panama, in which Texasgulf Inc, held 20% of equity until it sold this equity to RTZ in 1980, and Texasgulf as Administrator was concluded in February 1976, at the same time as the "Association Agreement"; and the Colombian agreements with Exxon (Cerrejon coal project) and a group of three companies (Hanna Mining, Standard Oil of California, and Billiton, a member of the Royal Dutch Shell group) formed to develop, jointly with the government, the Cerro Matoso nickel deposit, appoint the foreign partners as operator (technical management). In many cases, the foreign partner, or one of the partners, acts as sales agent and receives a predetermined commission for these services. For example, Billiton receives 6% of revenue of nickel sales from the Cerro Matoso project though the company also agreed to purchase the total output for a period of 15 years.

"Pure" management contracts – agreements without equity participation by the managing firm – are usually possible only in those developing countries which possess large capital resources, for example from oil exports as in the case of Iran, or where an existing operation assures income from which the contractor can be paid. The Technical and Administrative Agreement between Rosario Dominicana and Rosario Resources sets the fee at 1% of proceeds from sales before taxes, though not less than $1 million and not more than $1.75 million annually, derived from the Pueblo Viejo gold and silver mine. Services to be rendered include geological surveys and exploration, feasibility studies in the vicinity of the existing mine, engineering and economic planning, supervision of present operations, and assistance in marketing, negotiations with buyers, and transportation. The fees will be adjusted automatically every year in accordance with the United States Consumer Price Index of the preceding year.

The Sar Chesmeh Copper Mining Company (SC) was to pay a monthly fee fixed for each year until commencement of commercial production and a percentage of the gross sales value thereafter; this payment was 2% of the sales value for the first three years and would decline progressively until it reaches 0.2% in the tenth and last year of the agreement. In addition, Anaconda-Iran would be reimbursed for salaries, wages and fringe benefits for personnel on secondment (most of the staff would be on secondment according to the agreement), for the cost of an administrative office, and for various other costs.

In the case of the Cerro Colorado copper project, the Administration Agreement specifies the following fees to be paid to Texasgulf:

(a) during the evaluation period (including the feasibility study) a flat fee of US$500,000, to be capitalized and counted towards Texasgulf's equity contribution if the project goes ahead, or deferred until termination after completion of the feasibility study;

(b) 1.25% of design and construction costs (the US$500,000 mentioned above to be credited to these payments), to be paid during the design and construction period;

(c) a percentage of 1.5% of gross sales (2.5% for sulphuric acid, if produced) for the first 5 years of production, gradually declining to 0.75% at the end of the contract period (fifteenth year of commercial production); and

(d) reimbursement of costs related to personnel services, fringe benefits for employees, materials, sales costs, and other expenses.

Service Contract and Production Sharing

While the terms "management agreement" and "service contract" are often used for similar arrangements, the term "service contract" is here meant to refer to cases where the foreign investor provides all the funds for early stages of exploration and evaluation and carries the total risk, without any contribution by the government. If such activities are successful, the government acquires shares without payment ("free equity") or on concessional terms. In case of a production sharing agreement, the government's benefits are "in kind", or the government can choose between a certain percentage of shares and delivery of part of the project's output.

While service and production sharing contracts, as defined here, are common for oil and natural gas exploration and development agreements, they are rare in the "hard mineral" field; only one major contract between the Government of Indonesia through the state coal company PN Batubara, with Shell Mijnbouw, concluded in 1975, can be mentioned here. In this contract, production costs were calculated as a fixed percentage of total output ("cost coal"), with the remaining production to be divided between the government and the foreign investor. Thus, benefits accruing to the foreign company very much depend on the efficiency of operations and the price of coal realized by the project. There is, in addition, a "windfall profits" tax of 20% if the price of coal rises above US$25 per ton(5).

MAJOR ISSUES

Even though contractual relations can have various different forms, several issues are dealt with in all or most agreements in a comparable manner. These are usually areas of special concern for either the government of the respective host country or the companies investing in that country, or both. While some are of a practical nature and relate to the distribution of benefits or obligations, others are politically sensitive. Since individual contracts can differ substantially the following discussion relies to a large extent on examples of stipulations of specific contracts rather than broad generalizations. It is believed, however, that the given examples are fairly representative of a majority of agreements entered into during the last 10 to 15 years.

Staged Development

In many recent agreements, activities from initial exploration to production are subdivided into separate stages, and both the government and the foreign investors can in most instances withdraw from further involvement in the project without penalty at specialized intervals. Very detailed provisions in this respect are contained in the agreements between the government of Panama and Texasgulf of 1977 concerning the Cerro Colorado Copper project and the "Contract of Work" between Indonesia and RTZ, also of 1977, which covers all stages from exploration through production. In some cases, as many as five stages are distinguished: general survey, exploration, feasibility study, construction, and production.

Virtually all agreements which include exploration define the area in which exploration is to be carried out; specify which minerals or metals are included or excluded; set a time schedule for different stages of exploration; define the percentage of the original area to be surrendered at specified stages; determine rental fees or minimum expenditure requirements; regulate the role of the government or government agencies; and define reporting procedures. In most instances the government assumes the role of regulator but refrains from active participation during the period of exploration, though not necessarily at later stages, e.g. during development and production.

The contracts discussed here in some detail as examples are the Concession Agreement between the Government of Paraguay and the Anschutz Corporation of Denver, Colorado, USA, concluded in late 1975, and the "Contract of Work" between Indonesia and P.T. Rio Tinto Indonesia (subsidiary of Rio Tinto Zinc, Ltd. (RTZ) UK) and Conzinc Rio Tinto of Australia, Ltd., itself a subsidiary of RTZ of 1977. For Indonesia, this was the first such contract after many years and represents the so-called "third generation" of agreements.

The Anschutz contract covers the entire eastern part of Paraguay (east of the Paraguay river) and virtually all minerals and metals (with the exception of hydrocarbons). The exploration period is seven years, to be divided into two stages of three and four years each; an extension for another two years is possible if so desired by the concessionaire(6). During the first period, the company has to spend a minimum of US$300,000 and double this amount during the second period; an additional expenditure of US$500,000 is required should the company exercise the option of a two-year extension. The company can at any time terminate work without penalty if results are negative. At the end of the first three years 25% of the initial area has to be surrendered, another 25% after seven years, and an additional 25% after the first year of an extension.

A much more detailed sequence of activities is required from RTZ in Indonesia. Not later than six months after the signing of the "Contract of Work", the company has to start a "General Survey Period" of not more than 12 months; minimum expenditures during this time will not be less than US$20 per square kilometer (this amounts to US$400,000 if we assume that the initial area is about 20,000 square kilometres). During this and all subsequent periods the company can relinquish at any time any area on the "grounds that the continuation ... is no longer a commercially feasible or practical proposition". The General survey period is to be followed by an "Exploration Period" of not more than 36 months. "The programme of exploration shall as appropriate involve detailed geology, geophysics, geochemistry, sampling, pitting and drilling". The company has to spend at least US$150 per square kilometre; "the computation of such spending is based on the size of the Contract Area on the commencement of the Exploration Period. Following the Exploration Period, "the company shall commence studies to determine the feasibility of commercially developing the deposit or deposits in question. The company will be allowed a period of 12 months to complete such studies and to select and delineate the area in which the company shall commence operation". During these periods, the company has to surrender increasing percentages of the initial area until it is reduced to not more than 25% at the end of the Exploration Period. The following construction period is limited to 30 months, and the production period is to extend over 30 years. Commencement of operation should start not more than 8 years "since the initiation of the general survey. Upon request by the company, the Government will consider an extension of the operating phase".

The Cerro Colorado agreement allows 24 months for the feasibility study, sets a period of three months for both partners to decide whether to continue or not, and a 5 year construction period. However, neither the feasibility study period nor the time allowed for a decision were long enough, and both periods were extended.

Staged development provisions allow both sides to reconsider the advantages and disadvantages of a project after additional information has become available or the market situation has changed substantially. In most instances, a right of first refusal is granted the partner which withdraws if the project is continued with another partner, and the original participant will be reimbursed for the expenses needed for the feasibility study(7).

Corporate Structure and Control

Virtually all contracts contain provisions on the corporate structure of the joint venture, the rights and obligation of the partners, financing, the amount of capital to be invested, the distribution and issue of shares, the formation of management or technical committees and the appointment of an administrator or managing company. Such provisions are the most important since they determine the distribution of obligations and the share of benefits going to the partners. In rare instances, the contract may be signed by the government and a company which promises to attempt to find additional share holders. For example, the Ok Tedi agreement beween the Government of Papua New Guinea and Dampier Mining Company Limited (Damco), a wholly owned subsidiary of Broken Hill Proprietary Co. Limited of Australia, states that "Damco shall actively and diligently seek partners who may be prepared to take equity interest in the Project...". In 1981 a new company, Ok Tedi Limited was formed, and additional share holders are Amoco Minerals (subsidiary of Standard Oil of Ohio) with 30%, Kupferexplorationsgesellschaft (a group of German companies) with 20%, and the Government with 20%; the Government had the option of acquiring 20% of equity in the original contract.

An example of a joint venture between the government of a developing country and a number of companies is the Cerro Matoso nickel project in Colombia (figure 1). After many changes during exploration and negotiations, the Government now holds 45%, CONIQUEL (a subsidiary of Hanna Mining and SOCAL, both in the United States) 20%, and Billiton, a subsidiary of the Royal Dutch Shell Group, 35%. Though its equity share is only 20%, CONIQUEL has been appointed as the manager of the project, and Billiton is responsible for the commercialization of the total output.

Equity participation by the different partners is different in almost all contracts, and can change over the life of the agreement. In Chile, for example, the companies can have a minority

Figure 1. Structure of the Cerro Matoso Joint venture (1979)

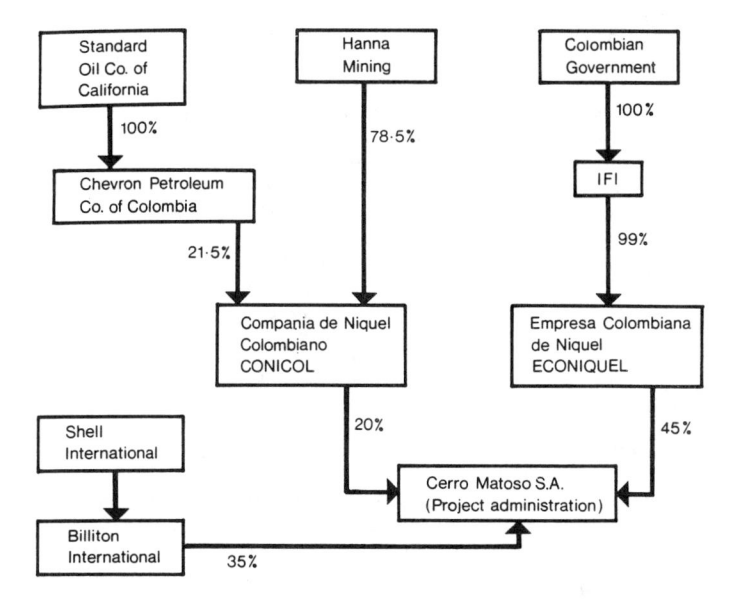

Source: Jaimes, Gilberto F., "Evolution of the Nickel Contract of Cerro Matoso, Colombia", Legal and Institutional Arrangements in Minerals Development, pp. 199-205, Mining Journal Books, London, 1982.

position (Noranda with 49%) or majority participation (Foote with 55%); the Panamanian Government initially held 80% of the Cerro Colorado project though this was reduced to 51% after RTZ replaced Texasgulf; Papua New Guinea has no equity in the Bougainville Copper Mine and 20% in the Ok Tedi project; the Indonesian Government has no equity interest in P.T Rio Tinto Indonesia initially but RTZ is required to offer not less than 5% of the issued shares each year, starting with the year after commercial production commences, to the Government or to Indonesian nationals until "not less than 51% of the total shares issued shall have been offered and either issued to or purchased by Indonesians". At least 30% of issued shares are to be offered to the Government.

Such subsequent increase in the ownership of projects is becoming more common, and in most cases where such "phase-out" is part of an agreement, the price to be paid for such shares is fixed. In the Indonesian agreement, the price is the higher of two alternatives: the depreciated book value of the total investment (including reinvested funds) or a multiple of six and two thirds of earnings (profits after corporate taxes averaged over the preceeding 5 years). Texasgulf has to sell all shares to the Panamanian Government after 20 years at a price calculated as the sum of retained earnings and 8 times average annual net earnings.

In most cases the Government pays for its share of equity; fairly frequently, however, it either receives "free equity" or the foreign investors advance capital for the Government's participation, to be paid back during production out of revenues ("carried interest"). Examples of "free equity" are Botswana's 50% ownership, together with DeBeers Consolidated Mines, in DeBeers Botswana Mining Company which exploits the Orapa diamond deposit at an annual capacity of 4.5 million carats(8); 50% participation of the Liberian Government in iron ore mines; or the US$30 million credited to the Panamanian Government in the Cerro Colorado Agreement as a compensation for the value of the ore deposits and other privileges. The term "free equity" is often inaccurately used. In most cases, it is "equity issued as consideration for non-cash contributions to the venture"(9). In Liberia, for example, it is in lieu of tax payments.

Provisions on finance and provision of the required capital are often not very detailed. Where the government provides all the funds, a management contract, as discussed earlier, is the normal arrangement. In all other cases, the foreign partners provide a part or all of the capital investment needed before production can start. In some contracts, it is simply stated that financing is the responsibility of the companies or that "the parties shall have joint responsibility for and shall cooperate in arranging for all such financing by way of loans and other financial obligations, and shall employ all reasonable efforts to this end"(10). The Government may provide financing for infrastructure requirements - railway or roads, housing, schools etc. - in order to maintain control over these parts which are directly linked to the country's economy in general; Governments also often have access to concessionary funding for such infrastructure from international organizations (e.g. the World Bank and regional development banks) or bilateral assistance institutions.

Control of operations is often exercized through management, executive or technical committees which have to ratify all major decisions such as selection of contractors, changes in the project's parameters, sales contracts, issuance of additional shares, election of officers, or change of the terms of incorporation. The composition of such committees may or may not reflect the relative participation in the joint venture; for example, Cerro Colorado's Executive Committee has 5 members; two from the Government, two from the minority share holder (Texasgulf), and one from the Administrator (also Texasgulf). The Colombian Government, though with only 45% of the shares, nevertheless has a veto right over major decisions in the Cerro Matoso nickel project.

Actual control is not necessarily in the hands of the majority share holder but is often regulated through specific provisions in the contract. Developing countries often complain that the international mining companies control most projects, irrespective of contractual arrangements, because of their predominance in international markets, in finance and know-how.

Taxation

While mining companies in developed countries operate under legislation generally applicable to corporate activities, virtually all developing countries negotiate taxation and other fiscal arrangements specifically for each project. Taxation provisions are generally complex and lengthy, and a large number of types of payments to be made to the federal or district governments are known. The most important of these are royalties, income tax (or corporation tax), dividend or dividend remittance tax, import and export duties, land rent, fees and charges for services by the government, and excise taxes.

Royalties. Royalties are in most cases payments at a fixed amount or percentage for each ton produced or exported, or more often, a fixed or variable percentage of the value of production. The value of production has to be defined, for example at the mine mouth, F.O.B., or at some other point. Usually, the value of production is calculated on the basis of reference prices such as the London Metal Exchange, quotation in trade journals (Metal Bulletin or Metals Week), or other mutually agreed pricing mechanism. If the product is a concentrate, costs of smelting and refining are deductible, as are other costs, such as transportation, normal losses, insurance, etc. One example of such a straight forward royalty is the payment of 1.25% of the F.O.B. revenue to the Papua New Guinea Government by the Bougainville and Ok Tedi projects.

But royalty levels can be determined according to various other formulae. Brown and Faber identify 25 different types of royalties and quasi-royalties, including variable percentage of sales, per ton, or of profits; revenue and production sharing (without equity participation); and posting of "artificial" prices for tax purposes(11). According to the authors, a "mineral royalty

(i) is a payment to the owner of mineral rights for the right to extract and sell the ore;

(ii) is a charge upon production;

(iii) should be variable with the value of the ore being mined;

(iv) should be so designed that the increase in the value of the ore accrues mainly to the owners."(12).

Royalty payments are in almost all cases deductible from revenue before calculation of income tax and are therefore treated like any other cost (i.e. "expensed"). Exceptionally, royalties are not deductible or, also not common, are credited against income tax payments.

Some agreements set payments at a fixed percentage of the metal ingot price. According to the agreement between the Jamaican Government and

Kaiser Bauxite of 1975, the company pays, in addition to 0.50 Jamaican cents per long ton bauxite, a "production levy" of 7.5% of the average realized price of primary aluminum metal received by Aluminum Company of America, Kaiser Aluminum and Reynolds Metals Company computed on a three-company arithmetical average...".

Since royalties have to be paid irrespective of the profitability of the company, even when the project operates at a loss, invariably royalties are generally at levels of a few percent. Exceptions are relatively high royalties in Jamaica (as mentioned above), Colombia (8% for Cerro Matoso), and on diamonds in a number of commonwealth countries(13). Royalties can also be different for unprocessed and processed products, with lower levels on the latter if the host country wants to encourage processing within its own territory.

Income Tax. Almost all countries impose income or corporation taxes, to be calculated on the basis of revenues less production and other costs, such as depreciation and amortization, royalties, interests, and other allowances. Tax rates differ though in most countries are between 25 and 50%. For example, corporate tax is 25% in Paraguay, one of the lowest, and not more than 49.5% in Chile; in the latter country, the Government guarantees that, even if general tax rates are higher at a future time, the companies' tax would remain at that level if the company elects not to be taxed at the general rate (as mentioned previously). Such guarantees, for the duration of the contract or a certain period of time, are not uncommon. The Ok Tedi agreement with Dampier sets the income tax at 35% during the "investment recovery period" even if general tax levels will be higher in the future. "Tax holidays" - periods during which no income tax is due - were included in many earlier contracts but appear to be less common in more recent agreements. Two examples are the agreements between the Indonesian Government and Freeport Indonesia (FI) for the Ertzberg copper project, concluded in 1967, and the original contract for the development of the Bougainville copper-gold mine, also of 1967. The Indonesian agreement states that FI "shall not have liabilities for income tax purposes ... until the experation of 36 calender months after commencement of the Operating Period... Thereafter, FI shall pay income taxes... The rate of such income taxes for the first 84 calender months next following such 36-month period shall be 35%, and the rate of such income taxes thereafter for the remaining period of this Agreement shall be 41.75%".

The original Bougainville agreement also provided for a "tax holiday" which was abolished in the renegotiated contract in 1974: "The income of the Company shall be exempt from income tax for a period commencing on the date on which the Company first enters into commercial production of copper concentrates under this Agreement and ending on the last day of the period of three years next following that date...". The renegotiated agreement entered into in 1974 provided for maintaining that tax-free period from the commencement of commercial production (April 1972) to the end of 1973 but introduced an income tax retroactively from the beginning of 1974.

While many agreements in one way or the other tie the effective income tax rate applicable to specific projects to taxation levels in force in the host country generally, there are often provisions which "freeze" such tax rates or which attempt to inhibit a change of rates during the contract period. Generally there are two different approaches: some countries promise not to change tax levels except in cases where legislation specifically refers to a project, and others require that the tax levels agreed upon in any agreement be subject to general taxation in force at any one time. The Ok Tedi contract, for example, states that "except where the contrary intention appears, either expressly or by implication, the provisions of the Income Tax Act which are not inconsistant with the provisions of this Agreement are applicable to the Company, and the Company shall pay income tax at the normal rates at which companies are liable to pay income tax in Papua New Guinea". The Cerro Colorado Agreement required that "Cerro Colorado SA shall be subject to the payment of income tax to the Nation ... on the basis of a fixed rate of 50% of its net taxable income".

Effective tax levels depend to a large extent on accounting procedures and depreciation schedules. In addition to "tax holidays", there are several techniques designed to reduce tax levels during the initial period of production so that the foreign investors can repay loans and recover invested capital. The most important are a lower tax rate (as discussed before) and accelerated depreciation. An example of accelerated depreciation is provided by the Ok Tedi agreement which allows 25% depreciation of initial capital expenditures if the "target income level" is not achieved during the first four years of commercial production. For all practical purposes, this provision reduces income tax to a level where recovery of the invested capital is likely within four years after startup of production.

Another device reducing income taxes is a provision for loss carry-forward. For example, in the previously mentioned contract between the Indonesian Government and Freeport Indonesia losses could be applied to the subsequent four years. A very exceptional provision is contained in one contract which stipulates that "in computing taxable income of the Company for each accounting period, the net operating loss incurred for the accounting periods commenced within ten years prior to the beginning date of each accounting period shall be deductable from the taxable income of each accounting period". Another agreement requires that "a net loss for any taxable year shall be carried over and deducted in the succeeding taxable year or years until such loss shall have been fully absorbed against net taxable income for such succeeding year or years, provided, however, that in any such succeeding year, the amount of the prior year's loss carried over and deducted shall not exceed 25% of the amount which would be taxable income for such preceeding year...". These examples refer to rather generous allowances, but comparable, though less generous, carry-forward provisions are common in recent agreements and have in part replaced "tax holiday" provisions.

Accelerated depreciation is probably the least "visible" of tax concessions granted to foreign investors, and the least objectionable politically from the viewpoint of the host country. As mentioned before, the Chilean agreements provide for generous depreciation schedules, and the Ok Tedi contract provides for higher depreciation rates if the "target income level" is not achieved (as discussed before).

Some recent contracts, in particular in Asia, introduce an "additional profits tax" to be paid in addition to normal income tax in case exceptionally high profits are realized. The introduction of such a tax is based on the conviction that profits above a level necessary to attract the investment by foreign investors should mostly go the the host country. Additional profits tax schemes exist for projects in Indonesia, Botswana, Papua New Guinea and other countries. As an example, the formula used in the latter country shall be used here though only the general principle of this complicated arrangement can be illustrated.

"The Company shall pay Additional Profits Tax ... if the accumulated value of Net Cash Receipts in any Tax Year is positive ... when calculated in accordance with the following formula:

Additional Profits Tax = A x (.70 - T) where:

A = accumulated value of Net Cash Receipts and
T = the normal percentage rate of company income tax in Papua New Guinea".

<u>Other Financial Provisions</u>. As mentioned previously, several other taxes or fees are included in mineral development agreements, such as land rents, customs fees, import and export duties, stamp fees, and dividend withholding tax. Most important of these in terms of amount to be paid to the Government are the latter. Usually, dividends remitted to share holders outside the host country are subject to a levy of 10-15%, though not all agreements include such tax. Most agreements waive many of the less important fees, including export and import duties. However, preferential treatment is given to local suppliers in several cases, and an import tax may be payable if comparable local products are available.

Since interest payments are virtually always deductible as costs, there are often limitations on the level of debt in relation to equity. For example, a debt-equity ratio of 60:40 may be allowed, or the debt may be limited to three times equity. Most agreements disallow debt to be provided by equity partners, but some older agreements, as for example in West Africa, still permit loans to be provided by affiliates or subsidiaries.

Most contracts give the foreign investor the right to maintain foreign exchange accounts either with the national banks in the country or, more often, abroad. Such foreign exchange is needed to pay for materials and services, service debt payments, or pay dividends to the companies.

Many agreements contain lengthy provisions defining gross income, net income, deductions allowable for tax calculation, and other important fiscal regulations. Gross income is relatively easy to determine for mineral products for which terminal markets - commodity exchanges - exist but very difficult for products that are sold to affiliates, such as bauxite, alumina or iron ore. In such cases of "transfer pricing" artificial reference prices may be used or revenue may be determined on the basis of posted metal prices, as is the case for some bauxite transactions (as previously discussed).

General Considerations

Tax provisions and accounting regulations are very complex and it is completely impossible to cover this subject in the space available here. There is also often a close relationship between provisions of individual contracts and the general legislation of the host country of the project and the host countries of the companies. Taxes paid in the country where the project is located may be deductible from income tax in the country where the company has its headquarters, but not royalties. Double taxation agreements exist between some countries so that some of the tax issues need not be negotiated separately.

The relative share of benefits that each partner receives from a project depends on the total of all arrangements and not on any one provision alone. Therefore, the different taxes and fees, corporate structure, lengths of contracts, and permissible changes have to be seen as a whole package and not separately. A "free equity" of 50%, given in lieu of taxes, may not be the most advantageous deal for a developing country.

The benefits to be derived from mineral projects also depend on the nature of the individual project. "Bonanzas", such as the Bougainville copper mine and diamond mines in Botswana, make it possible that the Government receives a major share of total economic rent while at the same time allowing the investors a substantle return on invested capital. Real problems in determining the benefits to accrue to the country and the foreign companies develop for marginal, or somewhat profitable projects, and these projects are the majority. Difficulties also arise when conditions turn out to be different from what was anticipated when the agreement was signed. For example, the Bougainville copper mine opened at a time of high prices for copper and was so profitable that the Government insisted on a renegotiation of the original contract, and the company concurred reluctantly. On the other hand, production costs at the Exmibal nickel mine in Guatemala, jointly owned by INCO (80%) and Hanna Mining, have been very high due to steeply rising oil prices while prices for nickel have been depressed for years.

The result is that the operation ran at a loss, and there were disputes over royalty and tax payments between the Government and the companies.

The level and timing of taxation and other financial arrangements determine the relative distribution of benefits derived from a mineral project. There appear to be two main currents of thought on the most appropriate type of agreement between developing countries and international mining firms. Several countries, in particular in Latin America and in Africa, rely on equity participation, either in a majority position or at least a strong minority position, combined with corresponding control over major decisions in regard to investment, major project components, marketing and timing, On the other hand, there are countries which rely more on taxation and other financial provisions, including such devices as the additional profits tax, to secure a major share of benefits without necessarily insisting on substantive equity participation.

Recent developments have convinced many developing countries that they can capture a major share of a project's rent. These countries believe that they should get most of a project's rent above a level which allows the company a "fair return on investment", or what the Government considers a fair return on investment.

Finally, one has to emphasize the importance of stability. Mining projects take many years for exploration, feasibility study and construction. During those years the bargaining position of the host country increases until during production the foreign investors, since the capital investment has been made, depend on the co-operation of the country. Stability in tax matters is probably more important than the actual level of taxation.

INFRASTRUCTURE DEVELOPMENT AND ENVIRONMENTAL PROTECTION

Mining projects in developing countries require considerable investment in infrastructure since such projects are often a far distance from centres of population, available sources of energy and water, and ports and transportation facilities. Investment in such infrastructure - roads, railways, bridges, ports, pipelines, airports, power and water supply systems, public welfare facilities (schools, hospitals, etc.), offices, warehouses, communication facilities - can at times be higher than in the mine itself. One example of very high infrastructure costs is the Cerrejon coal project in Colombia: a railway of some 160 km and a port to handle ships of up to 150,000 tons capacity will require US$830 million, and housing for a community of about 10,000, US$315 million(14), out of a total investment of US$3.2 billion.

Infrastructure development is a sensitive issue since it relates directly to Government activities in other sectors. Facilities created for a mining project will often be used by other projects or the general public, and a formula has to be found to allocate costs - both for investment and operation - and eventual fees to be paid for the use of facilities among these different users. Almost always the Government will insist on some form of control over infrastructure and insist on approval rights before construction begins.

Usually, the Government will be credited for any contribution made to the project. The Government may have better access to concessionary finance from international lending institutions, as for example the World Bank or regional development banks. For example, the Cerro Colorado agreement states that "the Nation or the Cerro Colorado Mining Development Corporation may finance the construction of highways, port facilities and other items of infrastructure through the use of such intergovernment or foreign development funds as they may obtain... The cost of such construction, excluding interest and other finance charges, shall be entered on the books of Cerro Colorado S.A., as capital paid in by the Cerro Colorado Mining Development Corporation".

For the Selebi-Pikwe nickel-copper project in Botswana, the Government financed "a power plant, a dam, a pipeline, roads, an airport, rail lines, and a town with all its related facilities. Funding for the infrastructure was obtained from the World Bank, United States Agency for International Development (USAID), the Canadian International Development Agency (CIDA), and the Botswana Government"(15). The loans are about US$70 million and are Government obligations, to be paid out of revenue from the project, though backed by "take or pay" provisions.

Environmental protection measures, though not foreseen in all agreements, are becoming more common and may be dealt with in considerable detail in more recent contracts. Provisions are rather general in most cases, as for example in the Guatemalan nickel agreement: "Exmibal assumes the obligation to invest in special installations and take the necessary measures to avoid contamination of water and air..."and" to rehabilitate mined areas within a reasonable period, in order to restore them to a level of economic activity equal to or better than the existing level before the areas were mined out". The RTZ agreement with Indonesia stipulates that "the Company shall conduct its operations under this Agreement in such a manner as not to harm the human environment and shall utilize the best mining industry practices to protect natural resources against unnecessary damage, to prevent pollution and harmful emissions into the environment in its operations and to dispose of waste materials in a manner consistent with good waste disposal practices. The Company shall otherwise conform with the relevant environmental protection laws and regulations which may from time to time be in force in Indonesia".

OTHER ISSUES RELATED TO FINANCE

As discussed before, several contracts do not define conditions on which financing for a project will be assured. However, there are financing implications in almost all provisions of agreements with developing countries, even those which on first examination would not appear to have an influence on these questions. In industrialized market economy countries, legislation governing environmental protection and health and safety aspects, for example, impose heavy costs on mining companies, but these are not always covered in agreements with developing countries.

A major factor in deciding the return to equity to an investor is the ratio between equity and loans. A high percentage of debt is considered desirable by most international companies on the grounds that this assures a high return if the project is successful. On the other hand, developing countries attempt to limit the loan-to-equity ratio because they are afraid of high interest payments which would have preference over return to equity investments. Another issue addressed in many agreements is the use of loans provided by equity investors. Basically, companies with an equity interest should not be allowed to provide loan finance since this would allow guaranteed income and dilute the risk exposure of equity investment. However, some countries allow such debt financing, but usually limit the percentage share of such loans provided by equity partners.

Loans coming from equity partners also are controversial because interest payments would continue even when the project would operate at a loss. This means that the foreign investor would continue to receive income while, at the same time, the host government would receive no income or would be forced to accept losses if overall receipts are not sufficient to cover costs.

There are other provisions in many agreements which, though they do not directly relate to issues of finance, are nevertheless important for the overall success of a particular project. These include foreign exchange regulations, marketing arrangements, and most importantly accounting procedures. The latter issue is addressed in a previous chapter in this publication.

During the last two or three years, not many new agreements have been concluded between developing counties and foreign investors because of the overall economic difficulties and the apparent oversupply of most minerals and metals. Therefore, it is difficult to ascertain whether new forms of contractual relationships are emerging or whether the types of agreements typical of the 1970's will continue to dominate the relations between the international mining companies and the developing countries during the 1980's.

FOOTNOTES

1. Detailed information on forms of agreements and past developments can be found in Smith, David and Louis Wells, _Negotiating Third World Mineral Agreements: Promise as Prologue_, Ballinger, Cambridge, Massachusetts, 1975; Mikesell, Raymond F., _Foreign Investment in the Petroleum and Mineral Industries: Case Study in Investor-Host Country Relations_, Johns Hopkins, Baltimore, 1971; Asante, Samuel K.B., 'Restructuring Transnational Mineral Agreements', _American Journal of International Law_, vol. 73,m 1979; and Kirchner, Christian, et al., _Mining Ventures in Developing Countries_, Kluwer-Deventer, Netherlands, 1979.

2. Annotations in this paragraph are taken from Asante, Samuel K.B., 'Restructuring Transnational Mineral Agreements', _American Journal of International Law_, vol. 73, 1979.

3. Agreement between the Compania Minera San Jose, Inc. and the Chilean Foreign Investment Committee, 1977, para. 4.14.

4. Quotations in this and the following paragraph are from the St. Joe Agreement (footnote 3).

5. Hutabarat, Hamonangan, 'Comparison of Terms of Agreements for Hard-Rock Minerals in Indonesia', _Legal and Institutional Arrangements in Minerals Development_, United Nations, Mining Journal Limited, London, 1982, p. 206-215.

6. Cardenas, Emilio, 'Mine Investment Legislation: A Latin American Outlook', _Legal and Institutional Arrangements in Minerals Development_, United Nations, Mining Journal Books, London, 1982, p. 68-84.

7. Zorn, Stephen A., 'New Development in Third World Mining Ageements', _Natural Resources Forum_, vol. 1, No. 3, 1977, p. 239-250.

8. Johnson, Charles J., 'Minerals Objectives, Policies and Strategies in Botswana - Analysis and Lessons', _Natural Resources Forum_, vol. 5, No. 4, October 1981, pp. 347-367.

9. A detailed discussion can be found in Freeman, Peter, 'Ambiguities in the Operational Meanings of 'Free Equity' and the Additional Profits Tax', paper presented at the Conference on "Mineral Policies to Achieve Development Objectives", East-West Center, Honolulu, 9-13 June, 1980.

10. Cerro Colorado Association Agreement of 1976.

11. Brown, Roland and Michael Faber, Some Policy and Legal Issues Affecting Mining Legislation and Agreements in African Commonwealth Countries, Commonwealth Secretariat, London, 1977, 107 pp.

12. Brown and Faber, op. cit., p. 59.

13. Brown and Faber, op. cit., p. 54.

14. _Financial Times_, London, 8 October 1981.

15. Tibone, M.C., 'Renegotiating the Selebi-Phikwe Contract in Botswana', _Legal and Institutional Arrangements in Minerals Development_, United Nations, Mining Journal Books, London, 1982, p. 193-198.

ADDITIONAL REFERENCES

Harkin, Daniel A., "Systematic Mineral Exploration", _Natural Resources Forum_, vol. 1,

No. 1, October 1976, p.29-37.

Carman, John, "United Nations Mineral Exploration Activities: 1960-1976", Natural Resources Forum, vol. 1, No. 4, July 1977, p.317-336.

United Nations Secretariat, "Two Decades in Mineral Resources Development", Paper 6, "Mining Engineering in the Developing Countries: the Role of the United Nations", Paper 7, and "Mineral Engineering in the Developing World: the Role of the United Nations", Paper 8, Institution of Mining and Metallurgy Symposium on National and International Management of Mineral Resources, London, 1980.

United Nations Secretariat, "Two Decades of Mineral Resources Development: The Role of the United Nations', Natural Resources Forum, vol. 5, No. 1, January 1981, pp. 15-30.

Lewis, Alvin, "The United Nations in Mineral Development", Engineering and Mining Journal, January 1983, p. 68-70.

Trabat, Thomas J., Brazil's State-Owned Enterprises, Cambridge University Press, New York, 1983, 294 pp.

Moment, Samuel, The Pricing of Bauxite from Principal Exporting Countries, 1974-1978, United Nations Industrial Development Organization, Vienna, 1980.

Hellawell, R. and D. Wallace, Negotiating Foreign Investments, A Manual for the Third World, 2 vols., 1982.

7.2

SHARING RISKS AND REWARDS IN INTERNATIONAL CONTRACT NEGOTIATIONS

PEDRO L. LIZAUR

CROCKER NATIONAL BANK

INTRODUCTION

The degree of success of an association between parties in establishing a new venture ultimately depends on three factors:

a. the venture's underlying economic strength, which is based on the complementarity of the comparative advantages brought by each party to the negotiating table;

b. the ability to reach an agreement which satisfies each party's objectives and constraints; and,

c. the capacity to design and implement a funding structure which is consistent with the project economics.

A thorough assessment of the venture's profitability and associated risks under alternative technical and economic scenarios is the basis for a sound "go" or "no go" decision. In-depth evaluation of the risks and rewards to each of the parties involved under alternative legal and financial structure must be the foundation for a long lasting "how to go" decision.

The allocation of risks and rewards should be consistent with the objectives and constraints of all parties and the sharing mechanisms, established in the contract, should be flexible to assure the long-term stability of the agreement. Rapidly changing economic conditions during the last decade have had a dramatic impact on the envisaged allocation of risks and rewards between host countries and private investors.

The effects of these changes, and the fact that the relative strengths of the parties involved may shift during the life of a project (e.g. before and after major capital investments have been made), have forced the renegotiation of a multitude of contracts. Therefore, the present and future stability of an economic agreement would depend not only on the originally designed provisions, but also on provisions which allow for the automatic adjustment of rewards to project sponsors under a variety of future economic situations.

A pre-condition for the allocation and quantification of risks and rewards is the definition of standards of measurement used by each of the parties. This is critical in international negotiations when one of the parties represents the national interests of a country. The understanding and use of different standards for the measurement of rewards for risk taken between a government and private enterprise facilitate the negotiation of a satisfactory agreement. This is possible because certain trade-offs between the negotiating parties may have different relative values.

STANDARDS OF MEASUREMENT OF REWARDS FOR RISK TAKEN

The negotiating process takes a step forward when the parties are able to focus on their interests, explore different alternative solutions and then base results on a set of objective standards. Unfortunately, the parties frequently put more effort into "staking out" and defending positions, than into understanding their intrinsic interests. Moreover, when one party, such as a governemnt owned corporation, represents a country's national interests, the negotiation process increases in complexity. Private firms focus on commercial profitability, using discounted cash flow techniques such as NPV, IRR, or ROE. In contrast, the government is also concerned with the project's impact on the local socio-economic environment.

Without entering into a discussion of alternative economic development theories and approaches, there exist several possible measures that could be pragmatically utilized by each party in evaluating the socio-economic impact of a project.

The more complete measure is the National Profitability (NP). The NP is a discounted cash flow procedure similar in structure to IRR or NPV but in which the project's financial cash inflows and outflows and discount rates are adjusted to take into consideration the local conditions. The NP is calculated on an unlevered basis and all taxes and duties are removed from the project cash outflows. The adjustments made by the use of "shadow" prices, "border" prices and "social" discount rates represent an attempt by the government to estimate social opportunity costs for project inputs and outputs. The determination of appropriate shadow pricing levels is a complex process which requires value judgements on socio-economic issues vis-a-vis the country's overall development program.

Another comprehensive measure is the National Wealth Effect (NWE). It quantifies the project rewards to the government for risks taken. The NWE is based on the country's social cash flow, defined as the flow of the country's commitment of resources to (outflows) and benefits from (inflows) the venture, and calculated at the social discount rate which reflects the opportunity cost of capital in investments elsewhere in the economy. The NWE is most effectively used in structuring the venture during the final negotiation stage.

Measures of more limited scope, the Net Foreign Exchange Effect and Employment Effect Ratio, are widely used since they do not require shadow pricing adjustments. The former reflects the venture's contribution to the country's stock of foreign exchange reserves, while the latter is the ratio between the venture's investment cost and the total number of man-years of employment created by the venture.

Assuming that the project design has been agreed upon and that it meets the required National Profitability, the parties involved negotiate "how to go" about implementing the project. The design of the ownership and financial structure, based on the allocation of risks and rewards under alternative technical and economic scenarios, is the subject of the final negotiation. At this stage, the parties should search for an efficient allocation of rewards.

An allocation is considered efficient if it falls in the Negotiations Efficiency Frontier (NEF). The NEF is here defined as the locus of different agreements such that an increase in NWE cannot be achieved without a decrease in financial profitability. Point M (Plot 1) represents an efficient allocation under the economic scenario defined by price P_1. In addition, changes could be introduced in the parameters of alternative model contracts making the parties indifferent between them. However, when changes in the economic scenario occur, model contracts reallocate rewards differently among the parties. This is illustrated by point M (Plot 1) under price P_1, and by points X, Y, Z under prices P_0 and P_3.

Plot 1
FOREIGN INVESTOR ROE

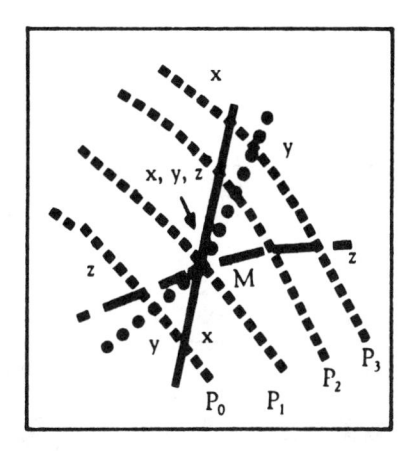

Gov't NWE

QUANTIFYING REWARDS FOR RISK TAKEN

To illustrate the discussion, we will utilize a hypothetical natural resource project sponsored by a foreign firm and four alternative model contracts. The basic project design is described in Table 1. It is assumed that this design has been discussed and accepted by all parties. For the sake of simplicity, this example assumes no inflation.

Table 1
PROJECT DESIGN FOR A NATURAL RESOURCE VENTURE

- Life of the venture 25 years starting on signature date.
- Final production capacity 5 million units, to be reached 4 years after starting operations.
- Operating costs at full capacity US$18 per unit.
- Construction period 4 years.
- Capital costs US$200 million ($5MM in year 1, $20MM in year 2, $125MM in year 3, $50MM in year 4).
- Pre-investment expenditures by Government $10MM. (These expenditures are only taken into consideration for computing the government economic returns).

We assume the government takes a 10% equity participation in the project with its equity payments advanced at no interest by the foreign concern. This advance is then recouped through government dividends redirected to the foreign investor. In addition, the four model contracts (A through D) have different clauses regarding royalties and taxes which are simplified proxies for common agreements in natural resource ventures.

Contract -A- (Table 2) simulates a concessionary agreement with the host country receiving royalty payments at a fixed rate. The royalty would typically reflect the estimated value of the ore in the ground under an assumed set of economic, technical, and geological conditions.

Table 2
MODEL CONTRACT -A-

- A royalty valued as 10% of gross revenues;
- 8 year income tax holiday; and,
- 22.5% income tax rate for years 9 through 13, and 45% thereafter.

Contract -B- (Table 3) represents a simplified production sharing agreement with royalty payments determined by a non-linear function linked to accumulated production. Although the increase in royalty payments is not directly related to the venture's profitability, the non-linear function is intended to increase government rewards when real output price increases are foreseen.

Table 3
MODEL CONTRACT -B-

- Royalty payment of 10% of gross revenues increasing to 15% when accumulated production reaches 10 million units, and to 20% after accumulated production has reached 25 million units;
- Income tax as for contract -A- including the 8 year tax holiday.

Model contracts -C- and -D- (Tables 4 and 5) reflect non-linear profit sharing linked to pre-agreed levels of venture profitability as measured by IRR (after tax). A royalty payment of 10% on gross revenues has been maintained reflecting the estimated value of the ore underground. In contract -C- the government's reward is increased by additional income taxes linked to the project IRR. Contract -D- represents a net profit sharing, with the government receiving after tax "preferred" dividends linked to pre-agreed levels of project profitability.

Table 4
MODEL CONTRACT -C-

- A royalty payment of 10% of gross revenues.
- Income tax rate of zero up to an achieved project IRR of 10%, 30% between a project IRR bigger than 10% and smaller than 30%, and income tax bracket of 50% thereafter.

Table 5
MODEL CONTRACT -D-

- Royalty payment of 10% of gross revenues.
- Income tax brackets and tax holiday as for contracts -A- and -B-.
- The government receives preferred dividends (P.D.) according to the following schedule:
  --Zero P.D. if project IRR $\leq$ 10%;
  --20% P.D. when 10% $\leq$ project IRR $\leq$ 15%;
  --30% P.D. when 15% $\leq$ project IRR $\leq$ 20%;
  --40% P.D. when 20% $\leq$ project IRR $\leq$ 30%; and,
  --60% P.D. when project IRR $\geq$ 30%.
- Common dividends to shareholders are distributed after payment, if any, of preferred dividends to the government.

The allocation of rewards between the government and the foreign investor is shown in Plot 2 for the four model contracts under consideration. The foreign investor's profitability, measured by its return on equity (ROE), and the NWE to the government using its NPV with a social discount factor of 10%, is shown under different project life horizons.

Plot 2
FOREIGN INVESTOR ROE

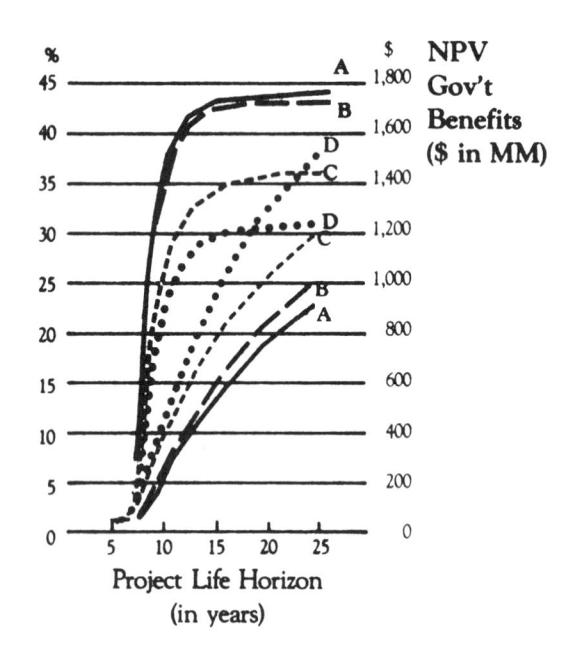

None of the proposed model contracts have the negative feature of fixing the ROE to the foreign investor. The non-linear models allow for smaller ROE's rates of growth when the project's economic life horizon is extended.

From the point of view of a foreign investor, the value of the return on equity over the life of the project (25 years in this example) is important, but also the shape of the ROE curves when the ROE is computed over a variable project life horizon as shown in Plot 2. Other things equal, curves showing the steepest slopes in early years would be preferred. They indicate that the foreign investor would be able to recoup its investment and make an acceptable return in the shortest possible period. In this regard, each of the four model contracts would allow the investor to achieve a 15% real return within less than one year of each other. The increase in political risk associated with lengthening by six months the minimum life of the project required to reach an adequate return might be perceived as minimal. From the host country point of view, however, the most vital measure of profitability at the stage of final negotiations is NWE calculated over the whole life of the project. By definition the government is not exposed to political risk. Consequently, under "ceteris pribus" conditions, it should be willing to trade lower returns in the early years for higher economic returns over the life of the project.

A quick glance at Plot 2 also reveals the models' different allocation of rewards to both parties. These differences in profitability become more obvious when the project life horizon is extended to 25 years. From the government's standpoint, model -D- is preferred to -C-, -C- is preferred to -B- and -B- is preferred to -A-. The reverse being also true from the foreign investor standpoint. This resulted from the additional compensation to the government for assuming different risks that were introduced from model contract -A- to model contract -D-.

Assuming that the fixed royalty payment reflects the value of the undeveloped natural resource, the benefits accruing to the government under model contract -A- represent payment for the right of doing business in the country plus the value of the undeveloped natural resource without any allowance for risk taken. Model contract -B-, however, recognizes the government's risk in predetermining the value of the natural resource over the life of the project -- resource risk. The rewards to the government under model contract -B- should, therefore, always be larger than those under contract -A-. On the other hand, we should expect smaller returns than under model contracts -C- and -D- since the government only takes the risk for an undervalued resource increasing its compensation independently of the financial performance of the project.

The allocation of higher rewards to the government by models -C- and -D- reflect the partial and/or total absorption by the government of the resource risk, the market risk, the technical risk and the geological risk. Lower market prices than expected, unforeseen geological complications and unexpected technical problems would negatively affect the rate of return of the project, therefore, delaying the activation of the trigger mechanisms for higher government participation built into the models.

The economic impact on government rewards attributable to its partial absorption of technical and geological risks under model contract -D- is illustrated in Plot 3. Higher than expected operating costs not only reduce the foreign investor's ROE but dramatically affect the NPV of government benefits over the life of the project.

Plot 3
FOREIGN INVESTOR ROE

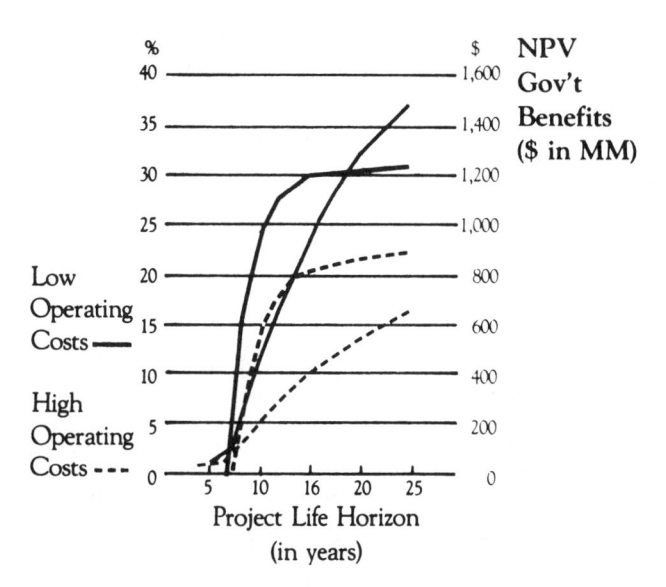

The behavior of the four model contracts under changing economic conditions (e.g. high or low unit prices) is shown in Plot 4. The allocation of additional rewards attributable to higher unit prices is also accomplished in relation to the risks taken by both parties. The slopes of the lines A-A, B-B, C-C, and D-D represent the sharing of additional rewards. Obviously, the government shares a larger portion of additional rewards under models C and D reflecting the additional risks it has taken.

Plot 4
FOREIGN INVESTOR ROE

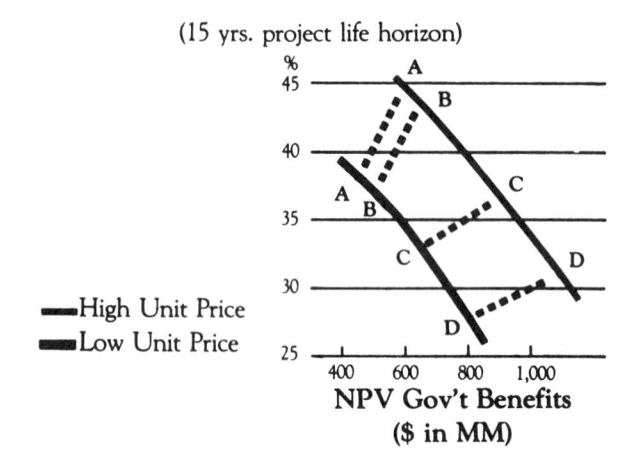

(15 yrs. project life horizon)

— High Unit Price
— Low Unit Price

NPV Gov't Benefits
($ in MM)

CONCLUSION

It is widely accepted that an understanding of the specific interests and constraints of each party engaged in international negotiations not only unburdens the negotiating process, but also facilitates reaching a fair agreement. However, rapidly changing economic conditions and the desire of host country investors, public or private, to be more actively involved in projects originally sponsored by foreign enterprises requires the additional effort of quantifying rewards for risks taken under alternative scenarios. The understanding and the use of different standards of measurement by a foreign private party and a government entity allow for the construction of an agreement that will meet their specific interests. In addition, the design of appropriate mechanisms to automatically allocate rewards for risks taken under a variety of technical, geological and economic scenarios would be the foundation for a sound and long lasting association.

7.3

ALTERNATIVES FOR DEVELOPING COUNTRY MINING FINANCE

Stephen Zorn

Tanzer Natural Resources Associates
New York, New York

INTRODUCTION

The process of financing mining projects in developing countries has changed greatly in the past 15 years. New methods of financing, notably the use of syndicated Eurocurrency project loans, have appeared, and direct equity financing by transnational mining corporations has become less significant. These changes are of particular importance to developing countries which have many mineral deposits still awaiting development, or even discovery. If major new mines are to be financed, host governments will need to obtain more sophisticated information than has hitherto been widely available as to the various sources of finance currently available and as to possibilities for innovative approaches to financing, such as the combination of several different kinds of financing in a single package.

This paper reviews the major sources of mining finance potentially available to developing countries and highlights those problems in obtaining adequate financing that developing country producers have begun to experience in the last few years and that are likely to persist through the 1980s. Section I surveys the major types of mining finance; section II reviews specific financing problems, including (a) the withdrawal of the transnational corporations from developing countries; (b) inflation and the size of new projects; (c) developing country indebtedness; and (d) competition among projects in different countries.

SOURCES OF DEVELOPING COUNTRY MINING CAPITAL[1/]

There is a wide range of financing mechanisms which are potentially available to developing countries with economically viable mining projects. Moreover, these financing mechanisms have not been fully exploited by developing countries, and more aggressive pursuit of financing by the governments of these countries may well reveal greater possibilities in respect of any particular project than may initially have appeared. However, virtually all the funds available from these sources are available only to projects which meet the standards of viability acceptable to lenders and others in the business community. If a project does not, for example, promise to generate a discounted cash flow rate of return of at least 15 per cent on total funds employed, or if it does not appear likely to generate sufficient cash flow in the early years of production to cover debt service requirements with a comfortable margin in reserve, most of the sources of financing will not be available. Only where a government has its own financial surplus (as in the case, for example, of the industrialization program of the Saudi Arabian government) can it effectively undertake projects which do not satisfy the definition of economic viability that is accepted at any given time by the worldwide financial community.

Internal Sources In Developing Countries

Internal financing capacity in developing countries is restricted to a relatively small number of countries - typically those large enough to support a well-rounded industrial sector, in which mining is only one of a number of dynamic industries. While there are large mining companies in certain other countries, the ability of such companies to finance new projects (or even, in some cases, to generate the cash necessary to maintain existing levels of production) is highly problematical.

Within developing countries, the possible sources of internally-generated mining finance are the following:

(1) state-owned enterprises;
(2) private capital;
(3) the banking system, and in particular those financing agencies that have a special involvement with the mining sector; and
(4) direct budgetary appropriations.

State-Owned Mining Companies   Four state-owned mining and industrial companies in the Third World rank among the 500 largest corporations outside the United States.2/ The three are the Zambian Industrial and Mining Corporation, Gécamines of Zaire, Codelco-Chile, and Companhia Vale do Rio Doce (CVRD) of Brazil.

In many other countries, however, the prospect for the financing of mining projects through state enterprises appears to be limited, for reasons both of financial resources and of government policy. Small countries do not have the capital required to fund a major state mining enterprise, even if government policy favored such a step. In many other countries, government policy has favored private-sector development.

Thus, while state enterprises are likely to be participants in mining projects, it does not appear likely that they will be major sources of funds for such projects and especially not for the very large projects which are becoming typical of the copper, nickel, iron ore, and bauxite-aluminum industries.

Even where state mining enterprises exist, they are often subject to the effects of unpredictable price and revenue fluctuations. The effect of this price instability is particularly marked in the case of copper, where the recent price depression has visibly affected the abilities of two very large state enterprises, Zimco in Zambia and Gécamines in Zaire, to finance expansion programmes. Those state companies that are producers of minerals with more stable and predictable prices (for example, CVRD, with its concentration on iron ore and aluminum) may be better able to make long-term financial commitments.

External sources of finance, in deciding whether to lend funds, apply roughly the same criteria to state-owned mining enterprises that they do to the transnational mining corporations. In addition, commercial banks typically require fairly rapid payback of loans, and charge substantial premiums in interest rates on loans to developing countries' state enterprises. This more expensive private finance may well offset any advantage that state mining enterprises gain by virtue of their access to loan funds from international lending agencies like the World Bank.

State mining enterprises are also subject to the same pressures as private companies regarding payment of their operating surplus to the central government for general revenue purposes. If anything, the ability of state enterprises to retain a share of their surplus sufficient for further investment needs may be less than that of privately owned companies, since the directors of a state enterprise will often be government officials concerned with other financial needs in addition to those of the mining sector, and hence more willing to respond to government pressure to hand over the surplus to the Treasury.

Private Domestic Capital   The number of developing countries in which private domestic capital has become a significant factor in mining development is extremely limited. Only India, the Philippines, Malaysia, Mexico and Brazil can be said to have a mining industry in which a significant proportion of total production capacity is owned by private domestic capital.

The likelihood that private domestic capital will be a factor in mining development is a function of the broader economic structure of any particular country. Relatively few of the developing countries have highly developed private capital markets or a broad spectrum of industrial activity under the control of private domestic capital. Where private domestic capital does exist, moreover, it is often invested not in industrial production (including mining) but rather in trade or property, or agriculture.

While other patterns may exist in different developing countries, the example of the Philippines as well as a similar pattern in Mexico3/ suggests that a number of countries may possess the potential for the development of certain mines by private capital, but that such development will probably tend to be limited to relatively small-scale projects. (Whether developing country governments wish to encourage -- or to permit -- private development is, of course, another matter.) The large mineral developments (150,000 ton copper mines, 10 million-ton iron ore mines or 200,000 ton aluminum smelters) appear too large for private capital in most developing countries to play a significant role.

National Mineral Development Banks   In several developing countries, specialized financial institutions exist with the express purpose of fostering mineral development. Examples of such institutions include the Banco Minero de Peru, the Companhia de Pesquisas de Recursos Minerais in Brazil (CPRM), and the Comision de Fomento Minero in Mexico.4/

The functions of these institutions include (a) financing of grassroots exploration and feasibility studies; (b) providing loans or equity investment for national enterprises (private or public) involved in mineral development; and (c) providing technical assistance for small-scale mining.

By centralizing technical and financial assistance, agencies such as those just discussed should be of significant assistance in promoting the financing of new mining capacity. Development of such facilities might be useful as a means of promoting small-scale mining projects.

Budgetary Appropriations for Mining   If private capital is not available for mining development, or if a political decision to limit the private sector has been made, then any local capital investment will be made either through a state enterprise, or directly by the government itself, by way of budgetary appropriations. Thus, if the government of a developing country wants to have significant mineral development, and if the options of development by existing state enterprises or by private capital are not available, the government has the choice of either having the project developed by foreign capital or of somehow finding government funds for the project.

In view of the very large cost of major mineral projects, a budgetary appropriation would mean that the government might not have sufficient budget funds for a variety of other projects, including schools, roads, agricultural development, etc. And, in view of the willingness of foreign capital (either through the mining companies or the banks) to invest in mining projects, while they will usually not invest in social projects, a government may be reluctant to commit its limited funds to mining.

Direct Foreign Investment

Historically (especially before the mid-1960s) most of the capital required for mining projects in developing countries has been provided as the result of direct foreign investment. For the period up until 1960, informed estimates are that close to 90 per cent of total capital requirements for developing country mining were met either by private equity or by internal generation of cash from ongoing operations.[5] Good examples of major mining regions developed in this way are the Zambian Copperbelt and the Chilean copper industry. In each case, initial equity stakes were built up by reinvestment of surpluses over a considerable period of years.

In the post-war period, the trans national mining companies continued to be the major force in such developments as the new Pacific Rim porphyry copper mines, though in this period non-equity external finance had already assumed a more important role, through project financing of developments such as Bougainville Copper in Papua New Guinea or Cuajone in Peru.

In the cases of bauxite and iron ore, projects in developing countries have tended to be undertaken as part of the worldwide integrated operations of transnational corporations. The Caribbean bauxite industry (as well as some bauxite projects in West Africa), and the West African iron ore industry were largely the result of this form of financing.

As mining projects continue to increase in cost, however, and as host-country governments continue to seek greater direct involvement, usually through equity holdings, it is becoming increasingly unlikely that the financing of projects through equity investment by transnational corporations in the context of their worldwide integrated operations will be of major significance for developing countries, although the technique continues to be used for projects within some of the industrial countries. There appear to be two reasons for this trend.

First, the increased concern of developing countries about sovereignty over their natural resources, resulting in substantial government equity positions in many developing country mining projects and in nationalization of some foreign-owned ventures, has given the mining companies a perception of higher risk levels in developing countries. The result has been that exploration efforts and investment decisions have tended to focus on the supposedly "safe" mineral-rich countries of the industrialized world, notably Canada, Australia, the United States and South Africa. Secondly, even if the transnational mining companies were willing to put substantial funds into the developing countries, the size of individual projects has grown, while the companies' own financial ability is in most cases significantly less than it was a decade earlier.

One response of both the developing countries and the mining companies to the emergence of sovereignty over resources as an area of concern has been an attempt to separate mining companies' provision of technical skills from the equity risk. This separation can occur through the use of joint ventures, in which the transnational mining company may have only a minor equity stake but acts as operator of the project, responsible for planning and execution, or through service contracts, where the mining company takes no risk at all.

This new pattern may be an adequate solution to mining companies' increasing fears of the political risk in developing countries, but it offers no solution to the countries' financing problem. The major commercial banks often insist that a mining company, with technical qualifications that the banks consider adequate for the project, have a substantial equity stake; the banks appear to consider this sort of equity involvement necessary to insure the mining company's full commitment to the success of the project. Secondly, many mining companies themselves consider the return on a service or management contract inadequate, compared to the potential return on a successful equity investment. Traditionally, mining companies have looked for very high returns on their investment in successful projects to balance the inevitable losses on unsuccessful exploration, producing an acceptable overall rate of return for the mining company as a whole. In most cases, the service contract arrangement does not include the possibility of these very high returns.

Transnational mining companies have reduced their equity involvement in the developing countries not only because of perceived political risks but also, and perhaps more importantly, because the companies' own financial capabilities have been reduced. A squeeze on companies' cash flow in recent years has made it difficult for them to raise money in their traditional capital markets.

The effect of rapid inflation during the 1970s, combined with the historic-cost basis of calculating capital allowances in most countries' tax laws, severely restricted the mining companies' ability to generate surplus cash from their ongoing operations. For some minerals, notably copper, this trend has been reinforced by a severe depression in market prices, to levels that in real terms are as low as any since the Depression of the 1930s. Most mining companies in the industrialized nations also had to meet heavy capital expenditure commitments for pollution control requirements on their existing plants, further weakening their financial position. As a result, debt-to-equity ratios for many mining companies have increased to the point where it is difficult for them to obtain further funds in the financial markets of the industrial countries. Thus developing countries will have to look increasingly to other sources for high-risk exploration funds, and will be able to involve mining companies only when the existence or potential of a commercially profitable deposit has been proven.

Project Financing

In the 1960s project financing became established as the major technique for funding developing-country mining projects. Project financing has become a major funding source worldwide. By the 1970s, it had become the major technique for funding projects in the centrally planned economies of Eastern Europe as well.

Project financing emerged because capital costs for new projects were growing beyond the internal cash generating and borrowing capacity of existing mining firms; this was especially true of financing needed to develop low-grade orebodies in remote areas -- primarily in the developing countries -- where substantial amounts were required for basic infrastructure as well as for the mining facilities themselves. Project financing was encouraged in the late 1960s and early 1970s by a growing concern around the world with the question of possible exhaustion of resources. This produced a favorable public exposure for those banks willing to lend for mining projects. Finally, the rapid expansion of the Eurodollar market, and the movement of hundreds of banks into Eurodollar financing, which was relatively free of government regulation, created a favorable climate for increased lending.

Without attempting a precise definition of project financing, we can say that it contains at least one of the following elements:

(1) a basic reliance on the anticipated cash flow of the project itself to repay the debt, in contrast to reliance on the over-all credit of the project sponsors;

(2) a matching of several different sources of finance to the needs of the project, rather than reliance on a single source of funds; and

(3) a sharing of the risks involved, usually by syndicating the project loan among banks on as wide a basis as possible.

From the point of view of the transnational mining corporations, project financing is normally more expensive than direct borrowing on the strength of the parent company's balance sheet (or, in the case of government-sponsored projects in developing countries or in the centrally planned economies, on the strength of the government's own general credit rating).

The use of greater amounts of loan finance makes possible a higher return on the direct equity investment; if the overall rate of return on a project is, for example, 15 per cent, but two-thirds of the project capital is in loan funds at 10 per cent, then the return on equity might be in the order of 25 per cent. Along with the opportunity for higher returns, of course, went greater risks, since the loan funds had first call on available cash, but this risk did not prevent extensive use of project financing in the late 1960s and early 1970s.

Project loans have traditionally been structured on a floating-rate basis, at a premium of 1/2 to 1 1/2 per cent above the London interbank offering rate (LIBOR). Interest rates are adjusted fairly frequently to reflect changes in the cost of money to the banks themselves.

Firm long-term sales contracts for a major share of the output of the project have often been a precondition for commercial bank credits to projects. In many cases, these contracts may not have been strictly necessary to guarantee a market for the products; refined copper, lead zinc or tin, for example, could always be sold on the London Metal Exchange. But the advance sales commitments do serve the purpose of reassuring the bankers that funds will be generated to repay the loans. And, in the cases of products that are less readily marketable through the international metal exchanges (bauxite, iron ore, or copper concentrates, for example), a sales contract may be necessary to guarantee a market.

In addition to facilitating commercial bank financing, the sales contracts have often been combined with consumer credits. Japan, in particular, has been a major source of funding through consumer credits for mining projects. Project financing arrangements also typically include an important measure of supplier credits, supported by guarantees issued by the governments of industrialized nations.

For reasons to be discussed below, project financing is not a panacea for all problems of mining finance. In particular, the current debt crisis of developing countries has barred all but the most creditworthy of them from Eurocurrency markets for the time being. Project financing may have ceased to be a dominant mode of finance.

Export Finance and Supplier Credits

Export credits from the companies and/or countries which supply mining equipment have recently become a major feature of mining finance, although there are a few somewhat older projects in which supplier credits and export financing were of primary importance. It has, in fact, become common for the export finance component of a project to be the first to be negotiated, because of advantageous financial terms typically associated with such financing.

Export credits are available from all of the OECD countries. Usually, they involve fairly long repayment periods and fixed interest rates. In addition, export credit schemes normally involve a government guarantee mechanism, sometimes as part of the same system that insures equity investments in developing countries and sometimes through a separate government agency.

Interest rates usually vary over a relatively narrow range. In the US Eximbank scheme, for example, rates range from 8 per cent for loans of up to six years to a maximum of 9 per cent for loans of 14 years.

Generally, credits are limited to five years for "wealthy" countries, but 10 years or more for the developing countries.

Despite the attractiveness of the export finance schemes, several problems prevent their full application to the entire range of developing country mining projects.

First, the system is relatively rigid. Export credits for terms longer than five years are governed by the Berne Convention, which requires inflexible and rapid commencement of repayment. Thus if technical problems reduce expected cash flow in the early years of a project's life, the inflexibility of the export credit arrangements may act to increase the financial pressure on the project

Secondly, export credits have a limited application to the mining phase of projects. Typically, a large open-pit mining project will incur only 45 per cent of its costs in imported capital equiment and services that qualify for export credits. If only a portion of this 45 per cent can be financed through the export credit institutions, this may mean that only 35-40 per cent of the total cost can be met through this mechanism. The proportion of imported goods and services is much higher for smelter and refinery projects.

Thirdly, export credits are obviously not applicable for projects in those countries with sizeable domestic mining equipment industries of their own. Mexico, Brazil and the Republic of Korea are current examples of such countries, and other developing nations have ambitions, expressed in their development loans, to build mining equipment industries. Thus, over time, the scope for export credit financing in the developing countries may decrease.

Fourthly, export credits are always dependent on the guarantee of the exporting country government. In the case of developing countries which are judged uncreditworthy by the guarantee institutions in the developed nations, no guarantee will be provided, even if a particular mining project is otherwise financially viable. Similarly, there have been instances in which developing countries were not granted guarantees because of political differences.

Fifthly, export credit financing commitments have tended to grow more rapidly than the capital base of the export finance institutions. In the United States, for example, Eximbank loan commitments increased by 300 per cent in the 1970s, while the agency's capital and reserves increased by only 25 per cent. While this increasing loan-to-capital ratio is a general feature of the Western banking system over the past decade, it appears to be especially marked in the case of the export financing agencies, raising the question of whether these agencies are adequately capitalized.

Most of the major industrial countries have also established some form of investment guarantee scheme covering, inter alia, mining investments in developing countries by companies from the industrial nations. Since the schemes cover both equity and loan investment, they facilitate the procurement of finance and the assembling of complete financial packages for mining projects in developing countries.

The agreements normally do not provide automatic protection against nationalization, but generally the host country is required to undertake to provide fair compensation without undue delay (of course, exactly what constitutes fair compensation is often a matter of strenuous debate, and occasionally for arbitration). The investment agreements can either be in a very detailed form, dealing with a number of specific issues (as in the case of those negotiated by the Federal Republic of Germany), or more general, containing only an expression of good intentions and establishing a procedural mechanism for resolving disputes (as in the case of those negociated by the USA).

The coverage under the schemes is subject to some limitations. For example, each of the schemes has a limit on the total amount of coverage that can be provided worldwide, determined by the extent to which the sponsoring government is prepared to underwrite the coverage. In addition, most schemes limit coverage in any one country; OPIC, for example, does not permit more than 10 per cent of its outstanding coverage to be in any single country.

From the point of view of both host countries and mining companies and lenders, there are several additional drawbacks to the insurance schemes as they apply to mining projects. For example, many of the national schemes are simply too small to cope with the massive investments required in new mining projects. The schemes are intended to operate on a more or less commercial basis; they have in fact managed to do so, and have even reduced premium rates over time, but only by rejecting some large projects that posed potential threats to their solvency. Even OPIC, by far the largest of the insurance programs, was severely embarrassed by the insured losses suffered by US mining companies in Chile. Until the settlement of the OPIC claim against the Chilean government, this coverage had led OPIC's auditors to question the adequacy of the organization's financial reserves. And, since the Chilean nationalization in 1971, OPIC officials have been reluctant to undertake high-risk coverage of similar large mining projects.

Consumer Credits, Production Payments and Compensation Trade Arrangements

Consumer Credits   Several industrialized countries which are heavily dependent on imports of foreign minerals, notably Japan and the Federal Republic of Germany, have made mine development funds available for projects to supply their mineral processing industries.

In Germany the government agency, Kreditanstalt für Wiederaufbau, has been the most significant agency for extending finance tied to long-term supply agreements. Since the early 1960s, KFW has participated in financing a number of mining projects in developing countries as well as Canada and Australia, with the aim of assuring long-term supplies for German industry. Such loans are usually extended for 10-15 years, at fixed interest rates.

Japanese interest in financing mine development tied to supplies has been equally active. Much of the recent development of the porphyry copper mines in the Pacific Rim region has involved Japanese finance, and most of the Pacific Rim (Papua New Guinea, Philippines, British Columbia, Chile and Peru) new production has gone to Japanese consumers.

Japan has also been involved in financing mining development through the Overseas Mineral Resources Development Corporation (OMRD), a government enterprise that participates with the major Japanese mining and trading companies on a 50-50 basis in new mining projects.

The current outlook for consumer credits, however, is somewhat clouded because of decreasing concern in the major importing countries over security of minerals supply. It is not clear at present what degree of continued support agencies like KFW and OMRD will receive in the near future for further supply-linked projects.

Production Payments   A financing device related to consumer credits is the production payment, under which banks advance funds against the security of the minerals in the ground, and recover the funds as the minerals are produced and sold. This arrangement has frequently been used in North American mining development.

There are, however, two significant obstacles to widespread use of production payment financing in developing-country projects. First, the banks that extend this kind of financing prefer to make it available for products whose prices are likely to be fairly stable, or at least are not likely to decline. Thus, production payments have been widely used in coal and oil development, but only infrequently for non-ferrous metals.

A second obstacle to the use of production payments in developing countries is that such arrangements normally involve a legal claim by the lenders on the minerals in the ground, as security for the loan. Such a claim raises serious constitutional, legal and political problems for a state which asserts, as most developing countries do, that the minerals in the ground are the property of the state and cannot be alienated until they are produced.

Compensation Trade Arrangements   6/   One way to avoid the difficulties of production payments financing, while still linking the financing of mining projects to future production, is through "compensation trade," a form of long-term financing which involves the acquisition of plant, equipment and modern technology by the recipient country (usually through a state enterprise) from a foreign supplier on the basis of foreign bank loans and credits, and the eventual repayment of the loans and credits through the sale of the resulting products to the same foreign supplier, pursuant to a long-term purchase contract.

The concept of compensation trade was pioneered by the centrally planned economies of Eastern Europe and the USSR in the 1960s and 1970s, and is now being applied with increasing frequency around the world, particularly for large-scale energy and raw material exploration and development projects.

The principles which apply in structuring compensation arrangements in large-scale resource development projects are basically similar, whether the project is in coal, oil and gas, timber, ammonia or other products. Since compensation arrangements in large-scale development projects are necessarily complex and long-term undertakings, they tend to be worked out by the parties over a series of distinct phases moving towards project implementation. Typically, these phases are as follows:

(a) Joint Preliminary Feasibility Study
(b) General Agreement
(c) Arrangement of Bank Financing
(d) Conclusion of Implementing Contracts
   (i) Supply of Equipment, licences, etc.
   (ii) Long-term Counterpurchase of Product

The Joint Preliminary Feasibility Study usually entails exchanges of visits and consultations between the parties on the basis of which evaluations are made as regards the technical and economic viability of the contemplated undertaking. During this initial phase, the identities of the respective parties in the development project are normally defined. On the side of the foreign supplier, particularly in the very large and diversified projects, a consortium of companies may be created which together bring to the project the ranges of technologies as well as financing and marketing capabilities required for the overall realization.

The General Agreement is the key document for purposes of structuring a prospective resource development project on a compensation basis. The main objectives of the General Agreement are:

(a) to define the basic compensation concept to be applied;
(b) to establish an indicative maximum project budget;
(c) to propose an approach for the obtaining of necessary investment financing;
(d) to agree on a method of pricing applicable to the long-term counter-purchase of resultant product;
(e) to establish a time frame for project implementation; and
(f) to identify the contracting parties and modus operandi for the negotiating of implementing contracts.

In a legal sense, the General Agreement is more in the nature of a memorandum of understanding than an enforceable contractual obligation. On the side of the recipient country, it is often signed at the governmental (i.e., ministerial) level, without any contractual waiver of sovereign immunity. Moreover, it typically contains a clause expressly precluding either litigation or arbitration of any disputes which may arise thereunder. All such disputes are to be resolved solely by negotiation between the parties. Consequently, the General Agreement is intended merely as a framework within which the respective parties may proceed to take the necessary steps towards concrete contractual implementation of the project.

The General Agreement does not constitute a contractual commitment either as to the foreign party's supply or the compensatory counter-purchases. Both commitments are usually the subject of separate contracts between the foreign supplier or members of the foreign consortium, on the one hand, and appropriate foreign trade organizations or enterprises of the recipient country on the other, once financing arrangements have been obtained and contract terms and conditions are negotiated. Compensation arrangements are normally not simple barters. The foreign supplier will be paid in cash for its delivery of equipment, through the bank financing, and will later pay cash upon purchases of resultant products, with an ultimate balance of payments either in equilibrium or in favor of the recipient country.

Another key issue which is invariably covered in the General Agreement is the financing of the investment costs of the project. The General Agreement will normally define a global project budget for financing purposes and will, to a greater or lesser extent, place on the foreign supplier the responsibility of assisting in arranging the necessary financing through his government finance facilities or private commercial banks. Where a consortium of suppliers is involved or where it is contemplated that equipment will be supplied from various countries, it may be necessary or desirable to approach a number of government credit authorities. Typically, foreign financing will cover only 80 per cent or 90 per cent of project costs, the balance representing risk capital on the part of the recipient country.

The General Agreement also usually attempts to deal with the difficult issue of pricing of products for the long-term counterpurchase agreement. Generally speaking, prices are set by reference to international market prices for products of similar quality at time of shipment. The problem then is how to ascertain the international market price. One rather common approach is to agree that prices will be determined by the parties on a CIF port of delivery basis for each six month (or appropriate) period. Where prices are set on a FOB basis, it is usually provided that differences in freight rates will also be taken into account.

In the case of certain products, the price determination may be facilitated by the existence of a recognized international index or market quotation. A minimum price for product is sometimes provided in General Agreements, with the understanding that if the world market falls below such minimum level, there has been a fundamental change of circumstances requiring the parties to meet and renegotiate a new equitable arrangement under the new circumstances.

On the basis of the General Agreement, the parties will next proceed to arrange for the necessary long-term bank financing. Frequently, for major projects, the foreign supplier's government may already have expressed its willingness to support the project by providing government credits and credit guarantees at preferential rates.

While a preliminary commitment on financing is sought prior to contract negotiations, the actual financing agreement is normally concluded only after the supply contract has been signed. The ultimate availability of financing, particularly in very large scale projects such as coal or oil exploration and development, will depend on the evaluation of the economic viability of such projects.

Multilateral Financial Institutions

The World Bank and the regional development banks have not been major factors in mining finance; less than 2 per cent of their loans in the 1970s were for mineral development and processing. None the less, the importance of these international agencies is greater than the loan figures would indicate. First, public international financial agency loans are concentrated in those mining investments that occur in the developing countries. Secondly, the funds tend to be focused on a few key projects, which can be of great importance to the economies of particular developing countries. Thirdly, international agency funds often serve as catalysts for much larger loans and investments from other sources. Thus it has been estimated that, in the past 25 years, direct World Bank funds accounted for only 1-1 1/2 per cent of mineral sector investment in the developing countries, but that those projects in which the Bank played a part accounted for 6-8 per cent of such investment, taking into account total loans and investments in all projects where the Bank was involved.7/ In these projects, the Bank funds, on average, accounted for 25 per cent of total project costs; the remainder was met by other lenders and equity investors.

At present levels of funding, the World Bank can be expected to be directly involved in about one-quarter of all major developing country projects. Lending at this level would appear to be of considerable significance for developing country mineral industries. The basic financial problem for developing country projects does not appear to be an overall shortage of investment funds, but rather specific problems associated with packaging these funds in a way consistent with the development of the mining projects. In particular, the higher debt levels in major new projects often create strains in the early years of project life between lenders, equity investors and the host government, all of whom are anxious to obtain rapid and substantial returns from the project. Where the use of Bank funds permits a more orderly organization of debt service, or where it attracts additional finance on advantageous terms, the World Bank (or regional development bank) funds may have an importance greater than their numerical value would indicate.

In addition to the World Bank/IDA lending, the International Finance Corporation has also played a significant role in smaller mining projects. For the five years beginning in 1977, the level of IFC investment in mining projects was in the order of $50-75 million annually, which was spread over projects where the total investment, including the IFC share, reached totals of $400-500 million per year.

Thus planned levels of international agency involvement in developing country mining finance, while not very large in comparison either to total lending by these agencies or to total investment in the minerals sector, may continue to be sufficient to have a marked impact on several developing countries.

In addition to providing large amounts of finance for major mining projects, international financial agencies have at least three other significant roles in relation to developing country mining projects. These roles are:

(a) acting as catalysts for the involvement of other participants;
(b) providing technical assistance in the early stages of project development; and
(c) undertaking general surveys that can lead to the identification of viable mining projects.

Non-Traditional Sources of Finance

Petrodollars   The high profitability of oil companies in recent years and their large cash flows have led a number of these companies to become involved in non-fuel mineral projects. This involvement has occurred in two ways. First, a number of oil companies have set up new mining subsidiaries and entered upon all phases of mineral exploration, evaluation, and production. Examples of this kind of involvement include Exxon, which has focused its mineral exploration effort in North and South America, and Amoco Minerals, a division of Standard Oil of Indiana, which has taken substantial equity positions in several projects, including Tenke-Fungurume copper mine in Zaire, where Amoco held a 28 per cent equity share, and the Ok Tedi copper prospect in Papua New Guinea.

The second kind of oil company involvement has been through the purchase of operating mining companies, their assets or of their equity. The most significant cases are the purchase of Billiton by Shell Oil, of Kennecott by Sohio (itself a subsidiary of British Petroleum), and of Anaconda by Arco.

Oil companies setting up their own mineral divisions or injecting sizeable amounts of new capital into older mining and metal companies can provide a new source of capital for mining projects. In addition, the greater familiarity of oil company managements with operations in developing countries, as compared to that of mining company managements, may lead to a less restrictive view of potential investments in developing countries, less hampered by perceptions of risk.

On the other hand, oil companies have traditionally looked for very high rates of return on their successful projects, to compensate for the large amounts spent in unsuccessful exploration; this history of seeking high return may deter some investments in marginal projects, and appears to have contributed to these companies' recent dissatisfaction with their mining investments.

A possible parallel to the funds invested in metal mining by the oil companies may arise from the surpluses generated by some oil-producing countries. As yet, however, the evidence that petrodollars will make a large, identifiable contribution to the needs of developing country mining finance is uncertain at best.

In large part, the role of petrodollars has already been discussed, since much of the liquidity in the Eurocurrency market is provided by deposits of surpluses from oil-producing nations. In particular, the very largest banks -- those that tend to be the lead banks in mining project finance -- are by far the largest recipients of petrodollar deposits.

There have been a few cases in which Arab states have made funds available for specific mining projects. Two of the most significant are the $100 million loan from the Libyan Arab Bank to the Gécamines copper expansion project in Zaire and the 1977 grant of $100 million from the Government of Saudi Arabia to the Sultanate of Oman for development of a copper project in the latter state. Other identifiable injections of petrodollars into mining projects include deposits of Arab funds with Japanese banks, which then made equivalent loans to the Tenke-Fungurume copper project in Zaire, and the purchase of a share in the equity of Lonrho by Arab interests.

In general, Arab investment policy has been conservative and has avoided mining projects because of their high risk factor. By far the more significant impact of petrodollars will be through the liquidity that they give to the Eurocurrency market, and through more general assistance to developing country governments (either bilaterally or through contributions to international agencies like the World Bank/IDA or the IMF) that allow those governments to include mining projects in ongoing overall development plans. The recent decline in oil prices, moreover, appears likely to reduce the availability of this sources of finance.

<u>Insurance Companies</u>   Like the oil companies, insurance companies in many of the industrial countries are major sources of capital. The stable and predictable financial structure of insurance companies (as opposed to commercial banks whose deposits are for relatively short terms) allows the insurance companies to consider very long-term lending. Such lending can be advantageous to mining projects, which carry a heavy debt service burden in the early years of production.

Insurance company finance has been widely used in the US mining industry for many years, particularly in the iron-ore sector. In developing countries, there are two major mining projects that have relied on insurance company financing. These are the Ertsberg (Gunung Bijih) copper project in Indonesia and the Falconbridge Dominicana nickel mine in the Dominican Republic.

These findings illustrate both the advantages and disadvantages of insurance company financing for mining projects in developing countries. The major advantage, clearly is the long term of the loans. On the other hand, interest rates tend to be high, and the security and guarantee arrangements demanded by the insurance companies are stringent. Moreover, insurance company funds have thus far been available only to projects in which all the equity was held by private companies, and not by the host governments. It remains to be seen whether insurance companies would be willing to advance funds based on undertakings of a host government, in the absence of transnational mining company guarantees.

FINANCING PROBLEMS IN THE 1980s

A number of specific problems have recently affected developing countries which had sought funds to develop deposits. These problems appear likely to persist through the 1980s. The most important of them are the following:

(1) the withdrawal of transnational mining companies from developing countries;
(2) inflation and the size of new mining projects; and
(3) developing countries' balance-of-payments deficits and indebtedness.

Withdrawal of Transnational Companies

The heavy concentration of Western mining company exploration and development spending in a few industrialized countries has already been noted. In recent years, as much as 80 per cent of these companies' spending has been limited to the U.S., Canada, Australia and South Africa. The reasons for this withdrawal from the Third World are well known; they include the increasing assertions of sovereignty by developing country governments as well as the increasingly weak and over-extended financial position of many of the major mining companies.

One response to the inability of individual companies to take on major projects has been the formation of international consortia, bringing together companies from several different industrial countries. By raising the threat of diplomatic and economic reprisals from several countries, such consortia inhibit nationalistic action by host governments.8/ Even the use of this device, however, has not kept mining companies as active in developing countries as some governments would wish. And, while funds for development, and even for exploration, may be available from a number of sources outside the mining companies, those firms do possess a substantial amount of technical and managerial expertise which may be more difficult than money for a developing country to obtain.

Inflation and the Size of Projects

Inflation, to the extent that it is not fully anticipated in the original financing of a mining project, imposes an additional financing burden, and this has to be met either through further borrowing or through additional commitments of shareholders' or governement funds. If further borrowing is required, capital expendiure may have to be delayed while the new financing is arranged, and this delay may lead to a further increase in costs, producing a spiral of rising project debt and higher capital spending.

Inflation introduces an element of uncertainty into mining ventures, even if metal prices move with the inflation rate in the long term. For any single mining project, prices and costs may well not move together in the short or medium term. A mine coming onstream when prices and costs are not moving together may face a severe cash flow problem.

A related financing problem for developing country projects is their great absolute size in many cases, and the huge amounts of finance required. The increasing financial requirements are the result of a large increase in per-ton real investment costs in the early 1970s, of the growing capital intensities in mining, and of the scale economies which can be reaped by large size operations. The financial requirements become particularly heavy where substantial amounts of infrastructure are involved.

The sheer size of these projects has made the assembling of finance difficult. No single financial institution or mining company is normally prepared to put such very large amounts at risk on a single project. Hence the trend towards development of projects by consortia of companies and towards the formation of banking consortia to spread the financing risk. But the need for these elaborate arrangements increases the time required for putting together financial packages for projects and makes it more likely that there will be a slip in the financing process somewhere before the project is completed.

Finally, the large absolute size of projects contributes to the detrimental impact of inflation. When total project costs were much lower than they are today, potential cost overruns were small enough in absolute terms to be accounted for in the banks' assessment of a company's or a government's financial ability to bring a project to completion. The very large cost of current projects, however, combined with high and unpredictable inflation rates, makes the entire process of assessing the ability of a project to generate the cash flow required to repay its financing much more uncertain than it was 15 or 20 years ago.

From the point of view of smaller developing countries, the large absolute size of many modern mining projects may mean that the host country will have to subordinate virtually all its other development objectives to accommodate the financial, infrastructure, and manpower needs of a big mining undertaking, and that the host country's overall credit rating and ability to borrow for other purposes will be affected by the large financial exposure that the banks may have in the mining project. Finally, the large size of projects tends to produce competition among host countries for projects, since not all can go forward simultaneously. The usual result of this kind of competition is a weakening of the host country's bargaining position vis-à-vis the transnational mining corporation when such matters as taxation, royalties, and equity structure are being negotiated.

Developing Country Indebtedness

The overall problem of Third World debt has received considerable attention. Since Mexico's suspension of debt payments in August, 1982, hardly a month has gone by without a new debt crisis in some developing country.

A result of the debt crisis is that lending for projects in many countries has been cut off entirely, because the commercial banks have determined that the country has reached an overall debt limit.

From the point of view of the developing country governments, the recent increase in debt has created a situation in which proposed mining projects will be seen as competing with other potential uses for loan funds; exteral finance will not be available for all desired projects.

CONCLUSIONS

In the mid-1970s, the outlook for financing mining and mineral processing projects in developing countries appeared relatively optimistic. Western countries' concern with assuring secure supplies of industrial raw materials, combined with the greatly increased liquidity of world financial markets, led to a situation in which developing countries were being courted by consumers and investors. More recently, however, the emergence of what appears to be chronic over-supply in such key markets as copper and iron ore, together with the deepening debt crisis of the developing countries, has reversed this optimistic outlook.

In the 1980s, it appears likely that few large-scale base metal projects in developing countries will be financed on any basis other than massive support from the host government (as, for example, in the case of Carajas in Brazil). The major commercial banks have all but given up on mining project finance, and the transnational mining companies have concentrated their own (limited) efforts in a very few countries, primarily those that are already industrialized. The danger, from the point of view of the developing countries, in this situation, is that countries will begin destructive, cut-throat competition to attract what few remaining investment funds are available, thus denying themselves future revenues from any projects that are successfully brought into production. Evidence of this sort of competition is already available, in the increasing use of such incentives as tax holidays and exemptions.

A more productive long-term strategy for many developing countries might well be to focus on the develoment of their mineral resources for use in national or regional industrialization (thus echoing the way in which the mining industry developed in the United States), and giving up the increasingly unlikely prospects of succeeding in improving their status in Western mineral and financial markets. Such a strategy, however, depends on a far greater degree of cooperation among developing countries than has yet been apparent.

NOTES

1. A discussion in this section draws on M. Radetzki and S. Zorn, _Financing Mining Projects in Developing Countries_ (London, Mining Journal Books Ltd., 1979), pp. 32-127.

2. _Fortune_, August 20, 1984.

3. Radetzki and Zorn, _op. cit._, pp. 46-47.

4. For details of the Mexican and Brazilian cases, see _ibid._, pp. 48-51.

5. _Minerals: Salient Issues_, Report of the Secretary-General to the Fifth Session of the Committee on Natural Resources, May 1977 (UN Document E/C.7/68).

6. The discussion in this section relies heavily on J.M. Herzfeld, "Compensation Arrangements in Coal and Other Resource Development Projects", (mimeo), report prepared for the UN Centre on Transnational Corporations, April 1980.

7. R. Bossen and B. Varon, _The Mining Industry and the Developing Countries_ (Washington, 1977) p. 210.

8. See T.H. Moran, "Transnational Strategies of Protection and Defense by Multinational Corporations: Spreading the Risk and Raising the Cost for Nationalization in Natural Resources", _International Organization_, Vol. 30 (1973), pp. 273 et seq.

7.4

THREE RECENT MINERAL DEVELOPMENT AGREEMENTS IN SOUTH AMERICA

Thomas W. Wälde

Natural Resources and Energy Division
Department of Technical Co-operation for Development
United Nations, New York

BACKGROUND

Foreign investment has been a major factor in bringing about the substantial mining industries of Chile, Peru, Bolivia, Venezuela, Brasil and Guyana. But economic nationalism and sentiment against foreign domination of vital national industries grew in the 1960s and most major mining projects were transferred into state ownership. Among the state mining enterprises formed were CODELCO - Chile, MINERO-PERU/CENTROMIN - Peru, COMIBOL - Bolivia (in 1952), Hierro-PERU, GUY-BAU/BIDCO - Guyana, etc. New investment during that time was discouraged because it was subject to the restrictive conditions of the investment regulations of the Andean-Pact, a grouping of Chile, Bolivia, Peru, Ecuador, Venezuela and Colombia. The philisophy was that mineral resources were to be developed exclusively by national enterprises. Essential foreign inputs had to be acquired "depackaged" as separate purchases of equipment and technical assistance services.

As a result of the worldwide economic recession which hit the mining industries in 1980-1982, many projects that were viable in 1970 were not developed by the new state enterprises due to lack of capital, managerial and technical abilities and due to problems with marketing. As a consequence of the long-term decline of metal prices, governments experienced difficulty in getting investment organized through national or foreign enterprises and have had to face up to the trade-offs between national control and active mineral development. A change has since taken place in governmental policies. The 1981 Peruvian mining law, the 1982 mining law of Uruguay, the 1979/1980 mineral promotion laws of Argentina, all witness the gradual shifting of emphasis from restriction to promotion. The investment rules of the Andean Pact have been progressively softened through national regulation and today are more or less irrelevant. Nationalizations and forced renegotiations have not occurred in the last few years. Countries have gone to great lengths (without much success) to re-attract foreign investment on terms that accommodate the political sensitivities of the South American countries and the practical requirements of investors and financiers.

For this analysis, three contracts have been selected which reflect this accommodation in three South American countries, where mineral production plays a major role in the national economy. These contracts reflect both the attitudes and traditions of the individual countries, but also mark the shift in mineral investment policies that has taken place.

EL INDIO AGREEMENT

The 1977 investment agreement between Chile and St. Joe Minerals marked a dramatic relationship between Chile and foreign mining investors. In Chile, mining of copper has played a major role in the country's history. Large scale copper mining was initiated and developed by Kennecott and Anaconda. The nationalization of these two companies under the Allende government in July 1971 started a wave of nationalizations of foreign-held mining and petroleum operations since then in many parts of the world. The military government supported the new Chilean state mining companies (CODELCO and ENAMI), but it also attempted to accelerate mineral development by promoting foreign private investment, thus reversing the nationalistic policies followed by the Allende and the preceding Frei government.

The new policy by the military government is expressed in Decreto-Ley 600 (1974) which governs foreign investment in Chile. Decreto-Ley 600 is

---

NOTE: Much of the information relied upon in this report is based on advisory missions carried out since 1981 by the writer as United Nations interregional adviser on mineral development legislation or by DTCD mineral development legislation consultants to Colombia, Guyana, Costa Rica, Honduras, Guatemala, Dominican Republic, Haïti, Bolivia, Surinam and Argentina. The opinions expressed, however, are exclusively those of the author.

different from comparable investment legislation of the time, notably Decision 24 of the Andean Pact, that imposed restrictions and obligations on foreign investment. Is not surprising, therefore, that Chile left the Andean Pact shortly after passing its 1974 Foreign Investment Law. Most mineral development/investment agreements combine obligations and performance requirements, including special mining taxes, with guarantees and some tax incentives. The Chilean investment law, on the other hand, is rather similar to the investment laws and investment agreements that have been passed in a number of French-speaking African countries ("Convention d'etablissement"). In principle, the law authorizes the government to grant foreign investors a large number of guarantees and tax and related exemptions. These are embodied in the form of an "agreement", instead of a simple administrative grant, to increase the expectation of stability. Most agreements concluded at the same period differ relatively little from each other.

The 1974 investment law was not viewed favourably by all members of the ruling group in Chile, because it discriminated in favor of foreign investors through the grant of rights that were unavailable to national investors. Some of the most far reaching privileges (e.g. the 30 years stabilization guarantee) were reduced in the 1977 modification of the law.

The main features of the Chilean foreign investment laws, operative through foreign investment agreements are:

- maximum tax guarantees
- repatriation of capital
- accelerated depreciation
- loss carry-forward provisions
- foreign exchange privileges

In addition, the military government split up the operative and regulatory role of CODELCO. CODELCO is in charge of production and marketing of copper in Chile. The Comision Chilena del Cobre, on the other hand, regulates and supervises copper production companies.

Chile has concluded a number of agreements under the 1974 foreign investment law. The 1975 agreement with Metallgesellschaft in Toqui started the foreign investment in Chile's mineral resources. In 1977, agreements were concluded with Noranda (a 51/49% joint venture between Noranda and the Chilean state enterprise ENAMI to develop the Andacollo copper deposit), with Superior Oil/Falconbridge/McIntyre Mines (a 51/49% joint venture with the state enterprise CODELCO to develop the Quebrada Blanca copper deposit), with Foote Minerals (a Newmont Mining subsidiary), a 55/45% joint venture with the state development corporation CORFO to develop lithium in the Atacama desert, with Nippon Mining (to develop the Cerro Colorado copper deposit), and finally with Compania Minera San Jose, Inc., a 100% controlled subsidiary of St. Joe Minerals to develop the El Indio gold/copper deposit. In 1978, EXXON Minerals purchased the La Disputada copper mine from the state enterprise ENAMI and concluded a foreign investment agreement. In 1979, ANACONDA concluded a foreign investment agreement for exploration and development of the Los Pelambres deposits. In 1982, Getty Oil and Utah International signed an agreement authorising investment in copper mining up to US$1.5 billion in La Escondida copper deposit.

Due to low copper prices since, the commercial viability of many of these copper projects is not clear. In 1983 Noranda withdrew from the Andacollo project due to financing difficulties and was reimbursed for its exploration expenditures. Nippon Mining abandoned the Cerro Colorado deposit. Metallgesellschaft withdrew from the Toqui project. Anaconda also stopped work in Los Pelambres due to the high cost of the project. The lithium project is going forward and EXXON has continued to invest in exploration, however. It has been reported that among the major mining projects initiated in the 1970s the El Indio project is the only one that showed an acceptable rate of return.

The El Indio deposit is located at about 4,000 m above sea level in the Chilean Andes. The deposit contains gold, silver and copper. It was exploited for decades on a small scale until it attracted the attention of St. Joe Minerals in 1974. St. Joe purchased mining rights from private owners through Compania Minera San Jose, which now owns 80.6% of El Indio; the rest belongs to private Chilean owners. Intensive exploration started in September 1976 and accelerated in 1977 after the investment agreement was signed. By 1978, the orebody had been delineated as having two kinds of ores: direct shipping ores (49,222 mt, with 345 gr/mt of gold, 145 gr/mt of silver and 2.60% copper) and ores to be processed in the plant,(3.1 million mt, with 12 gr/mt of gold, 144 gr/mt silver, and 3.52% copper). The decision to develop the mine was taken in late 1978 and in late 1981 the first commercial shipments started. 1982 operations yielded sales of ca. 12,000 kg of gold, 19,900 kg of silver and 9,700 mt of copper. St. Joe has reportedly spent over US$214 million in developing the deposit.

CERREJON COAL AGREEMENT

After extensive, and ultimately fruitless, discussions between the Government of Colombia and Peabody International to develop the coal deposit of Cerrejon, Block A (Central Cerrejon), covering about 38,000 ha, the government granted the mining rights to the state petroleum company (ECOPETROL) in 1976. ECOPETROL put the project up for international bidding on the basis of a Contrato de Asociacion. EXXON was selected and established INTERCOR as its project subsidiary. The mining rights were subsequently transferred to CARBOCOL, a newly established state enterprise for coal development. After the conclusion of the agreement in 1976, intensive exploration took place, resulting in the declaration of a commercial discovery in 1980. An investment decision was made and by 1981, project development had started (e.g. roads, airstrip, railway, port facilities, socio-economic impact study). Total investment was estimated, in 1979, at more than US$1.2 billion and is likely to be substantially higher. Commercial production is expected to start by 1985, and to last for at least 23 years with annual production of 15 million tons. The project is one of the major coal development projects worldwide and the largest investment project in Colombia.

CARBOCOL obtained its mining rights under Mining Law No. 60 of 1967 and No. 20 of 1969 as amended by Decree No. 1275 of 1970 and Colombia's petroleum code. The investment agreement, embodied in the Contract of Association, required approval under applicable investment regulations. Foreign exchange privileges (in particular the right to transfer proceeds obtained from the sale of INTERCOR's assets and INTERCOR's net profits) are granted under Resolution 23 of December 16 of 1976 of Colombia's National Council of Economic and Social Policy, CONPES (Cf. Annex).

GUYANA/COGEMA URANIUM AGREEMENT
(1982)

Guyana, a South-American Caribbean country, is an important producer of bauxite. Bauxite operations, taken over between 1972-1974 from foreign investors, have since been run by the Guyanese state companies (GUY-BAU/GUY-MINE/BIDCO). Due to technical, financial and managerial problems, and also due to the economic recession and the erosion of Guyana's strong position as a producer of calcined bauxite, production and revenues has fallen considerably and foreign assistance has been sought to improve the performance of Guyanese bauxite mining operations. Apart from bauxite, there is a lively activity of small- and middle-scale miners extracting gold, mostly by dredging operations from placer deposits. Most of the gold is smuggled out, to avoid having to sell it to government authorities for overvalued Guyanese currency.

Guyana has been trying for years to diversify its mineral production by encouraging gold, manganese, petroleum and uranium development, primarily by foreign companies. As there were indications of interesting uranium deposits, talks were held around 1980 with Uranerzbergbau, a F.R. German uranium company. This brought no result due to different concepts about revenue-sharing. In 1979, a non-exclusive prospecting license was granted to COGEMA, a subsidiary of CEA, the French government's nuclear power agency, for uranium prospection. Under the terms of the agreement, COGEMA was entitled to prospect for 3 years for uranium; it could then apply for an exclusive exploration right under a contract to be negotiated. The prospecting agreement was rather ambiguous since it was not clear what type of agreement was envisaged and what would happen in case of disagreement on the terms of such agreement.

Negotiations between COGEMA and Ministry of Mines and Energy and the Guyana Geology and Mines Commission led to an agreement in February 1982. Both sides were pressed to accommodation. The Guyana government was anxious to obtain foreign investment and to demonstrate its acceptance by foreign mining companies; in addition, the exploration of uranium was to be conducted in an area of Guyana that is claimed by Venezuela. Getting French state involvement was hence seen as an asset in Guyana's dispute with Venezuela. COGEMA, on the other hand, was seeking to diversify its sources of supply of uranium for France's very energetic and far-reaching nuclear power programme, up to now mainly dependent on Gabon, Niger and Namibia. Subsidized by government, it was involved in a major strategy to open up new sources of supply, resulting in COGEMA's participation in exploration in Zambia, Colombia, Australia, Indonesia, Canada and other countries.

Guyana was assisted by the United Nations and a consultant from the Commonwealth Secretariat. A recent uranium agreement with Tanzania with Uranerzbergbau played the role of setting an example. Though the uranium price was falling dramatically and most forecasts predicted considerably oversupply for the future, COGEMA has continued exploration in Guyana and has spent about US$20 million to date on the basis of the agreement. The agreement is based on the Guyana mining law (of British origin) and applicable tax legislation. Compared with Colombia and Chile, the Guyanese bargaining team had much more leeway to negotiate and was less constrained by applicable mining, investment, company, customs or tax regulations. The smaller size of Guyana and a tradition of British law might be factors that have encouraged such comparative flexibility.

TYPE OF CONTRACT

El Indio

In contrast to the Colombian and Guyanese agreements, the 1977 Chile/St. Joe El Indio contract is an "investment agreement" exclusively between the state, acting through its investment commission, and St. Joe as investor applying for investment incentives and guarantees under applicable investment legislation. Most issues otherwise regulated in mineral development agreements (joint venture, revenue-sharing, including special taxes, royalties, fees, etc) between governments and state enterprises and foreign investors are absent. The investment law is used primarily as a device to attract investment in non-CODELCO projects, often with private Chilean partners, through incentives and guarantees. While other issues are notably absent, the incentives and guarantees granted are ample, detailed and comprehensive, to provide assurances against repetition of the nationalizations under the previous government.

The main issues in the negotiation with San Jose were the question of tax stabilization, depreciation/amortization rates, import duty exemptions, foreign exchange privileges, non-interference by government in marketing and production, other tax exemptions, debt/equity ratios, and arbitration issues. A major issue was the scope of foreign exchange privileges. The history of foreign investment in Latin America is fraught with difficulties of foreign exchange restrictions. Countries tend to appropriate all foreign exchange proceeds from mining projects, while foreign investors tend to use foreign exchange proceeds to service debt, to repatriate profits and to protect the investment against nationalisation. In addition, exchange rates with an artificially high value for domestic currency are sometimes imposed resulting in the imposition of quasi-taxes on foreign investment.

The foreign exchange privileges of the agreement have to be seen in light of such experiences. St. Joe may retain export proceeds abroad, open foreign currency accounts in Chile and have free access to foreign exchange for carrying out the investment. These privileges are meant to ensure that the investor may service its debt and obtain profits in foreign exchange.

Another important issue was the exemption from taxes and the limitation from new taxes. These guarantees and incentives were borne out of the experience of investors with new taxes and other quasi-fiscal levies (e.g. the Jamaican bauxite levy imposed in 1974) that disrupt the financial system that existed at the time of the investment decision. Renegotiation of fiscal arrangements and imposition of new taxes, levies and other additional government participations of a fiscal character are a prominent feature of government/investor relationships in the 1970s. St. Joe and the other companies were insisting on getting as many and as comprehensive guarantees from Chile as was possible. Accordingly, the government promised that it will not levy taxes on the sale of assets by St. Joe. This is possibly in reference to the Allende government's claim that "excessive profits" be deducted from compensation payable after the nationalisation of Kennecott and Anaconda in 1971. The maximum tax burden is 49.5% on earned income. Given that Chile does not use production based royalties, export taxes or similar levies, combined with the promise not to impose any new taxes or comparable fiscal levies, this promise (including the corporate income, the dividend withholding and a housing tax) effectively restricts government taxation to ca. 50% of taxable income. In contrast, most other mineral producers (including Canada, Australia and South Africa), in addition to a general income tax, dividends withholding taxes and property taxes, impose production based royalties. In copper this is between 1 and 5 % of the value of metal contained in ore.

Recently special mining taxes have been imposed in some countries on additional windfall profits exceeding a stipulated rate of return on investment. The Chilean method of avoiding special mining taxes is certainly an incentive of considerable power. It was probably based on the philosophy prevailing in Chile at that time which favoured effective incentives for an aggressive investment promotion policy.

Other fiscal exemptions relate to interest paid on foreign loans. Some countries levy a tax on foreign loans, which makes financing charges a substantial burden both for the investor and for the income-collecting tax authorities. St. Joe is also granted an exemption from import duties, provided products of comparable quality are not available at comparable terms in Chile.

A salient feature of the agreement is the freezing of income tax and customs duty rates for the term of the agreement, i.e. for 30 years and the guarantee that no additional taxes will be imposed in the future, except if St. Joe opts for a subsequent, generally applicable tax system out of its own volition. This freezing clause (authorized by the 1974 act and reduced under the 1977 act to 10-years) is infrequent, as it ties down the state's sovereign power of taxation for an inordinately long period. Few countries would go so far as to oblige themselves and subsequent governments for such a long period. In a number of countries (e.g. Jamaica), courts have held that a government can not tie the hands of the following governments and arguments are raised to the effect that any provision to that effect is counter to the basic principle of permanent sovereignty over natural resources. For example, the recent arbitral award in Kuweit v. AMINOIL attributes only a relative effect to a contractual stabilization clause. A previous award (LIAMCO v. Libya) recognizes the state's right to abrogate such stabilization clauses by nationalization with compensation. A close reading of the El Indio agreement, however, reveals that Chile may have reserved its right to abrogate the guarantees given, albeit under the condition of compensation for damages. Art. 4.26 of the agreement provides for "payment of full and adequate compensation for any injury or damage to San Jose caused directly or indirectly thereby" (i.e. if the guarantees under the agreement are impaired, attenuated or abrogated). Presumably, Chile would not be able under international law to abrogate these provisions for compensation, even if it may be entitled to abrogate the guarantees mentioned above.

Another guarantee of considerable importance relates to the freedom of the investor to determine production and marketing policies without interference from the government. Again, this clause (which is frequent in African "conventions d'etablissement") is borne out of the experience of mining companies with governments that impose minimum production targets even in times of unprofitable operations. Their objective is to maintain foreign exchange earnings or mining employment, and to sell below market prices to domestic consumers in the framework of intergovernmental arrangements. St. Joe's position is further bolstered by the provision that St. Joe "shall have the right to sell and export subject to such laws and regulations as are in effect at the date of this Agreement". This provision recognizes the supervisory powers of the Comision Chilena del Cobre based on applicable law in 1977 (e.g. its right to check if sales to affiliated enterprises conform to world market conditions), and the rules of the board of the Chilean Copper Commission on fixing the Chilean producer price to the LME price. It exempts St. Joe from subsequent, and more intensive government intervention, as for example the imposition of OPEC-style "posted prices".

Cerrejon

The 1976 Colombia/EXXON Cerrejon coal agreement is a "contract of association". It has been used by ECOPETROL since 1959 for large operations with foreign investors in petroleum. Up to 1980, 62 contracts of that type were signed. The contract of association is a non-corporate joint venture in which the Colombian state enterprise holds the mining right, the foreign partner assumes the risk of exploration and the partners share in both project development expenditures and project revenues. It differs both from the form of the equity joint venture, used in Colombia for the 1970 Cerro Matoso nickel agreement and from operations contracts used by ECOMINAS for smaller mining operations, principally concerning emeralds, managed and financed by private mining companies. There, operations contracts mean that the state enterprises leases the area, viz. the mining rights, against financial participation based on the prospective, real value of production. Other major projects (e.g. the Cerrejon Block A coal deposit) are operated directly by CARBOCOL on the basis of service and turnkey contracts with engineering companies.

Guyana Uranium

The 1982 Guyana/Cogema uranium contract provides that in the event of commercial discovery and development a joint stock company will be set up. In contrast to Chile, where there is no state participation in the El Indio project, both Colombia and Guyana use a joint venture model. While the Colombian joint venture is a mere contractual joint venture, without any reliance on the models of company law, Guyana uses the joint stock company to constitute the form of cooperation. In both Guyana and Colombia, as will be seen, the foreign partner assumes the role of the manager, with supervision and policy-making delegated to joint organs of cooperation. Thus, the system of a separate management contract (as in the 1976/1980 Panama Cerro Colorado agreements) is integrated into the organisation of the joint venture itself.

ORGANISATION, MANAGEMENT AND CONTROL

El Indio

The Chile/St. Joe contract leaves corporate organisation to the private companies forming a joint venture. Both Colombia and Guyana have established a delicate structure of management by the foreign company, while supervision and policy-making rests with a joint organs. Carefully inserted checks and balances bring about the necessary compromise between the government's interest in control, participation and eventual full national control, and the companies'and investors'requirement of full management during the exploration and extending into the production phase.

Cerrejon

In the Colombian case, the "Operator", i.e. INTERCOR, will be set up as an independent company and undertake construction according to an agreed work programme. Expenses will be charged to a joint account. Three years are allotted for the construction period, with possibilities for an extension. The exploitation period starts with the first shipment of coal to seagoing vessels and lasts until the end of the contract's term (30 years). It is not envisaged that a jointly owned corporation or a partnership will be set up. Parties have no power of attorney for each other and no joint liability exists. However, contrary to such denials of corporate joint-ventureship, the 50/50% participation in a joint, high-risk and long-term coal development project requires a quasi-corporate organization to allow the partners to achieve their community of interest. It is therefore interesting to observe how INTERCOR and CARBOCOL have constructed a non-corporate institutional network for joint project cooperation.

INTERCOR is the "Operator", and is entitled to appoint the "manager" of the joint project. The appointment of the "manager" will require consultations with CARBOCOL. Operations will be subject to an "executive committee" who will supervise construction and operations; evaluate the operator's performance, approve work, investment and expenditure programmes, approve expenditures and contracts of a large scale, appoint auditors and supervise the management of the joint account. The executive committee consists of representatives of CARBOCOL and INTERCOR in equal number. Its decision requires unanimity. It is to meet four to five times a year and is empowered to establish sub-committees. The executive committee is the main body for organizing project cooperation. In case the unanimity rule prevents a decision, recourse will be had to the chief executives of both partners and eventually the contract's machinery for dispute settlement is to break the deadlock. This three-tiered process of decision-making may become rather heavy, but it can be expected that the expectation of long and protracted negotiations will motivate the partners on the executive committee to seek a solution themselves, thereby avoiding the escalation to high-level consultations and ultimately arbitration and litigation.

The executive committee supervises the operator. The operator is free to appoint personnel, but has to comply with general guidelines of the executive committee.

The joint account is the main instrument to organize the financial relationship between the parties. The joint account is to be set up if commercially exploitable deposits are discovered. This provision reflects the risk-character of the agreement. The foreign partner bears the exploration risk. A full partnership with a sharing of costs, benefits and risks only takes effect if the exploration results in commercially exploitable deposits. Through the joint account Colombia fully participates, if not in the exploration risk, in the risk related to the development of the mine and the marketing of the output. Also, it has to secure financing and thereby fully assume the risks of financing, e.g. increasing interests rates in spite of a possible shortfall in revenues and cost overruns. The Joint Account is characterized by a 50/50% distribution of expenses and benefits. All expenses are charged to the joint account. If a party is in default with respect to its mandatory contribution to the joint account the other party will receive interest at commercial rates on its positive position in the account. As expenses will be incurred in Colombian pesos and US dollars, special rules are provided for currency conversion. If CARBOCOL is in default with its obligations to contribute, INTERCOR can use its royalty payment obligations to cover the default and equalize the joint account positions of both parties.

Guyana Uranium

In Guyana, the government is unable to contribute risk capital to uranium development due to difficulties concerning availability of foreign exchange and is unlikely to find commercial financing. Given its history of nationalizations and its professed socialism, it was anxious to command the mining venture. The resulting compromise is unique. Guyana obtains a 25% free equity in the project protected from capital increases and against any obligation to secure shareholders loan guarantees. In addition, Guyana will control the board of directors of the prospective joint operating company, irrespective of its 25% equity share. Finally, Guyana retains an option to acquire up to 51%, either when the joint operating company is set up, at a price reflecting COGEMA's paid-up capital, or at commercial rates assessed by independent valuers.

In exchange, COGEMA insisted that it retain full management powers and be protected by unanimity rules on the board of directors in all questions concerning major decisions. Additional checks and balances consist of a "Guyanization" of the positions of chief financial controller, 5 years after commissioning of the mill, and general manager 10 years after commissioning of the mill. Other control devices are a joint marketing committee to monitor long-term sales contracts, a dispute-settlement method in case of pricing disputes, and elaborate rules on allowable expenditures. The Guyana contract is rather unique in that COGEMA accepted Guyanese majority irrespective of Guyanese free equity on the basis of extensive management powers and unanimity requirements for COGEMA.

FINANCIAL REGIME

El Indio

In the Chile - St. Joe agreement, fiscal instruments are incorporated from general tax legislation and frozen. This promise relates to a maximum tax burden of 49.5% on earned income. Chile does not use production-based royalties or similar levies, export taxes etc. and is not to impose any new taxes or comparable fiscal levies. This promise (including the corporate income, the dividend withholding and a housing tax) effectively restricts government taxation to ca. 50% of taxable income.

The financial impact of these fiscal incentives is bolstered by accelerated depreciation of fixed assets and by a low 5-year amortization of preproduction expenses. These provisions are frequent in modern mining agreements. Some agreements concluded under the impact of double-digit inflation even adjust preproduction expenses by reference to an inflation or cost-of-capital index.

Cerrejon

In the Colombia/EXXON agreement, the fiscal regime has to be seen in light of the fact that CARBOCOL, the Colombian state enterprise, assumes 50% of project development expenditures, and thereby a heavy, and rather unusual financing burden and risk for a state mining enterprise. The revenues of the Colombian state are generated by fiscal instruments and by the participation due to CARBOCOL as partner of the association contract. The Colombian state receives income primarily through regular income taxes (at 52%). A recent analysis (Zorn, "Coal and Uranium Investment Agreements", Natural Resources Forum, Vol. 6, (1982), p. 350) assumes that, apart from "participation revenue" (cf. infra) no income taxes are due. This reading of the contract ignores Art. 16.4 and 30 of the agreement indicating that regular income tax is paid under applicable income tax law. As the contract regulates only the INTERCOR /CARBOCOL cooperation, and as it does not constitute a waiver of generally applicable legislation, the fact that the contract itself does not provide details on general income tax can in no way be used to assume that income tax is not payable.

In Colombia, as in other countries with a strong legal tradition, the government can not waive or modify the applicability of general legislation except if expressly authorized by law to do so. This reading is confirmed by an analysis of the contract by the Colombian National Planning Department (1981, op. cit., at 2.2.1).

The second method of generating government income is through the three-tier revenue-sharing system between INTERCOR and CARBOCOL as partners of the contractual joint venture, CARBOCOL receives (1) royalties; (2) a share of the coal produced; and (3) participation revenue, a kind of an additional tax on profits exceeding a basic rate of return on investment.

Coal production is shared between both partners at the rate of 50/50. In addition, INTERCOR has to provide CARBOCOL a 15% royalty on its share. A part of the royalty goes to provincial and municipal authorities under Colombian law. CARBOCOL may take the royalty either in cash or in kind. In both cases, the royalty is calculated on the mine-mouth value or volume, with transportation and storage charges for delivery to the export terminals to be accounted for separately.

A very interesting fiscal instrument is the "participation revenue" due CARBOCOL. INTERCOR income (not including US $-indexed depreciation) exceeding a rate of return of 35% on accumulated total investment is subject to the tax/participation mechanism at rates that rise progressively in proportion to the absolute amount of pre-tax (but after-royalty) income exceeding the 35% rate-of-return level (Cf. Art. 16.2). This mechanism has to be seen in the light of the fact that INTERCOR registered all of its capital contribution as equity investment so that there will be no deduction of interest on INTERCOR's capital contribution when calculating income subject to taxation and the participation mechanism. An evaluation of the financial terms conducted for the Colombian National Planning Department (para. 2.2.1) comes to the conclusion that the participation mechanism, in terms of net present value, has a financial impact equivalent to additional royalties at a rate, depending on the prices for thermal coal, of between 2 and 7%. The threshold 35% rate of return for imposing the additional tax is considered high, particularly when the average rate of return earned by US mining companies on equity investment is ca. 15%. While a 35% threshold is unusually high, unlike additional profit taxes imposed elsewhere (e.g. Guyana, Tanzania, Papua New Guinea), the Colombian participation mechanism is imposed on a total investment figure that is not adjusted by reference to an inflation index. Hence, as inflation increases the nominal value of revenues, but not the nominal value of the investment base, the 35% target will be easier to reach. Also, since INTERCOR does not deduct interest when calculating its revenues for tax purposes, revenues are increased which makes it easier to reach the 35% threshold. Finally, while financial experts always emphasize that additional profits taxes should be calculated by using a rate of return on equity and not on total investment. In actual practice, it is difficult to find any contract or mineral tax regulation using that approach .

Considered as a whole, the financial regime of the contract based on a 57.5/42.5 (royalty taken into account) production-sharing, the 52% income tax and the participation mechanism imposed on pre-tax profits is more favorable than is claimed by commentators. Compared to more recent coal contracts (e.g. the 1981 agreements between Indonesia and US coal companies) with lower royalties, lower income taxes and without an additional profits tax, Colombia may not have made a bad deal. Such evaluation holds true even if it is considered that Colombia has to contribute 50% of the expenses for developing and operating the coal deposit. The investor has to assume exploration at his own risk and is not reimbursed for this and other pre-production expenses (as is often done in exploration/ development agreements) except by amortization of such pre-production expenditures. By sharing expenditures and production, CARBOCOL is more heavily involved both in the commercial and technical risk (and the benefits) of the Cerrejon project than Indonesia. In that country the government's role is restricted to taxation without participation in the venture with its own risk capital.

Guyana Uranium

In the Guyana/Cogema agreement, the philosophy is that Guyana should not be exposed to foreign-exchange liabilities and the fiscal regime should be responsive to the ups and downs of the volatile uranium market. Guyana's 25% free equity is protected against an increase of capital and mandatory shareholders guarantees or loans. Guyana is only obliged to provide advances through goods, services and products or cash in Guyana currency. A debt/equity ratio of 4:1 (in Chile: 85:15) is agreed upon.

Royalties are payable at a rate of 3% of the sales value of minerals. The rate will increase to 5 % if the ratio of the total annual production costs including operating costs, interests, amortization, depreciation and the 3% royalty to annual sales value is less than 85%. By this device, Guyana increases early cash-flow for the company and obtains the higher royalty only at a later stage, when market conditions and operating costs allow it. COGEMA will pay income and corporation tax at a rate of 45% and obtains accelerated depreciation/amortization (5 x 20%), exemption from customs duties and a 7-year loss carry-forward. The prospective joint company will also be entitled to an exploration allowance for future exploration outside the mining area up to 10% of total capital expenditure in the first five years, and 5% thereafter. A "special mining royalty" is established which triggers an additional 25% tax on income exceeding an internal rate of return on total investment of

10% plus either, at the company's option, the US inflation rate or the US domestic corporate borrowing rate. Other methods of government revenue-sharing consist in a 15% withholding tax on dividends and the government's free 25% equity.

Foreign exchange rights are established to meet foreign exchange obligations and to pay amounts due to COGEMA. These rights are guaranteed by the Bank of Guyana.

LEGAL STATUS

In all three cases, the legal status of a contract between the government and a foreign company is difficult to ascertain, as South American countries have always emphasized the Calvo doctrine which subjects such agreements to national law and jurisdiction. Investors have tried to secure some protection from subsequent government interference by having access to international law and arbitral tribunals.

El Indio

The Chile/St. Joe contract provides for stabilization ("freezing") of the law valid at the time of conclusion of the contract for the duration of the agreement. Surprisingly, given the investment promotion philosophy of Chile at that time, the agreement follows the Calvo-principles and omits any reference to international arbitration (e.g. ICSID, ICC or external ad hoc arbitration, e.g. with UNCITRAL rules).

Cerrejon

The Colombia/EXXON contract is subject only to Colombian law and to the jurisdiction of Colombian courts. INTERCOR waives diplomatic protection except in the case of a narrowly defined denial of justice. Colombia, therefore, has closely followed the Calvo principles which have often been denounced by foreign companies as inimical to investment security. However, their impact in Colombia may be lessened and their acceptability by EXXON may be increased because Colombia has enjoyed an unusual legalistic tradition and independence of the judiciary which may make the unilateral change of contractual terms by way of legislative action less likely than in other countries.

While the agreement does not provide for international arbitration on investment disputes, preferred by foreign investors, it contains a special procedure for "technical disputes". They are to be submitted to a three expert panel. If both parties can not agree, the chairman will be appointed by the managing board of the Colombian Association of Engineers, and in the case of accounting disputes, by the Central Board of the Bogota Accountants. Legal, technical and non-technical disputes are exclusively within the jurisdiction of Colombian courts.

The investor's concern for stability of terms of the agreement are evidenced by several clauses. Art. 16.4 seeks to stabilize the impact of future changes of the general income tax law in Colombia. While, true to the nature of the agreement creating a legal obligation only between INTERCOR and CARBOCOL, but not with the Colombian government as such, the article does not attempt to "freeze" the tax legislation applicable in 1976, but rather provides that the financial impact of a change of general income tax rules on INTERCOR will be offset by a corresponding modification of the "participation revenue" due to CARBOCOL. In other words, CARBOCOL will have to reduce its claims against INTERCOR, if INTERCOR is taxed by the Colombian state more heavily than assumed in 1976. This method of stabilizing the fiscal regime is less questionable than traditional "freezing clauses" which were open to attack on the ground that a sovereign state can not tie its hands by contract with a private enterprise. However, the clause does not cover all contingencies. For example, one could envisage an increase or imposition of taxes on mineral production that exceeded the amount payable under the participation mechanism. In that case, INTERCOR would be hard put to claim compensation from CARBOCOL.

Art. 39 declares the foreign exchange regime determined by the National Council of Economic and Social Policy (Cf. supra) to be one of the "basic conditions" of the contract and the agreement incorporates the full text of the pertinent decision. The agreement does not provide for a procedure or sanction if that foreign exchange regime is modified. Presumably, INTERCOR would be entitled to claim breach of contract, terminate the agreement and ask for compensation.

Guyana Uranium

The 1982 Guyana/Cogema uranium agreement witnesses the difficulty of accommodating national sovereignty versus investment stability concerns. Apart from a procedure of expert determination in case of pricing disputes for uranium exports, disputes will be subject to a conciliation procedure between the Minister and the chief executive of COGEMA. Otherwise, the chief executive of the Inter-American Development Bank may appoint a conciliator who can make a nonbinding recommendation. Otherwise, the dispute will be settled by arbitration on the basis of the UNCITRAL arbitration rules. The appointing authority will be the Secretary-General of the Commonwealth Secretariat or the Secretary General of the International Court of Justice. Other technical disputes will be determined by experts using the ICSID's rules on fact-finding, to be appointed by the appointing authority for arbitration (cf. supra).

The agreement is governed by the law of Guyana. However, the terms of the agreement are to prevail over general law. Tax rules applicable at the conclusion of the contract are frozen for the duration of the contract.

CONCLUSIONS

The three agreements analysed in detail with respect to form of contract, management and control, financial regime and legal status illustrate the new mechanisms of contractual accommodation between governments and investors in the wake of the wave of foreign investment restriction and nationalizations of the early 1970s. In all three cases, economic nationalism is clearly present and the governments take pains to uphold national sovereignty. The governments take equal pains to accommodate legitimate concerns of the investor and financiers related to the specific nature of the project. The specific situation in each country has strongly influenced the outcome of negotiations. While Chile provides guarantees on all questions where mining investors have been severely affected by previous government actions, Colombia's relatively reduced indebtedness in 1976 made it more ready to assume 50% of the financing burden of the project. Political sensitivities seem to have made the Colombian government shun full legal association with the foreign partner in the form of a joint venture company. Guyana's strong interest in obtaining uranium exploration in areas disputed with Venezuela made it willing to grant a large exclusive exploration right. The contract strongly emphasizes the importance for Guyana to have a strong visible representation in any future uranium company. Also, Guyana's foreign exchange problems are reflected by its unwillingness to expose itself to foreign exchange debt. Financial revenues are emphasized, although in a very flexible form, responsive to investor needs for initial cash flow. In toto, the three contracts illustrate that the developments of the 1970s may have made it more challenging to achieve sophisticated mineral agreements. Nevertheless, these agreements illustrate that when both sides share an interest in investment, there is ample room for developing stable, but flexible contractual mechanisms that accommodate the main concerns of both partners even in view of the changes in the economic environment.

References

Asante, S.K.B. and Stockmayer, Albrecht, 1982, "Evolution of development contracts: The issue of effective control," Legal and Institutional Arrangements in Minerals Development, UN/DSE, London: Mining Journal Ltd , 1982, pp. 53-6420.

Asante, S.K.B., 1979, "Restructuring Transnational Mineral Agreements," American Journal of International Law, Volume 73 (1979), p. 335 and "Stability of Contractual Relations in the Transnational Investment Process," International and Comparative Law Quarterly, No. 28 (1979) p. 401;

Bartels, Martin et al., 1981, Bibliography on Transnational Law of Natural Resources (Frankfurt: Kluwer, 1981)

Bender, Matthew, 1984, Natural Resources Forum, Vol. 8, No.3, 1984, pp. 279-289.

Bosson, Rex and Varon, Bension, 1977, The Mining Industry and the Developing Countries (Oxford: Oxford University Press, 1977)

Cormick, Roger, 1983, "Legal Issues in Project Finance," Journal of Energy and Natural Resources Law, Vol. 1, No. 1, 1983, pp. 21-44

Dobozi, Istvan, 1983, "Mineral development: Co-operation between socialist countries and developing countries", Natural Resources Forum, Vol.7, October 1983, pp. 339-351

Eschenlauer, Arthur, 1984, "The New Twist in Energy Project Financing," International Energy Law, International Bar Association, Section on Energy and Natural Resources Law.

Fischer, Peter and Wälde, Thomas, "International Concessions and Related Instruments," Oceana, Vol. I-V, 1981 et. seq.

Garnaut, Ross and Ross, Anthony Clunies, 1975, "Uncertainty, risk aversion and the taxation of natural resources projects," Economic Journal, No. 85,(1975) p. 272.

Gillis, Malcolm et al., 1980, Tax and Investment Policies for Hard Minerals: Public and Multinatinal Enterprises in Indonesia, (Cambridge: Ballinger, 1980)

Gillis, Malcolm, for example in Public-enterprise finance: "Toward Synthesis," Public enterprises in LDCs, (Cambridge: 1982) and Economics of Development, (New York: Norton, 1983), cf. also supra, FN 11.

Gillis, Malcolm, especially Tax and Investment Policies for Hard Minerals: Public and Multinational Enterprises in Indonesia (Cambridge: Ballinger, 1980).

Legoux, Pierre, 1980, "Report on Mining Legislation in African Countries," ECA document E/CN.13 of 6 June 1980

McGill, Stewart, 1983, "Project Financing Applied to the Ok Tedi Mine - A Government Perspective," Natural Resources Forum, Vol. 7, No. 2, April 1983, pp. 115-131

Mikdashi, Zuhayr, 1976, The International Politics of Natural Resources, (Ithaca: Cornell University Press, 1976).

Mikesell, Raymond, 1984, Foreign Investment in Mining Projects, (Cambridge: Oelgeschlager, Gunn and Hain, 1983) with a chapter on Botswana summarized in: Natural Resources Forum, Vol. 8, No.3, 1984, pp. 279-289.

Moran, Theodore, 1974, Copper in Chile, (Princeton)

Prast, W.G. and Lax, Howard, 1983, "Political Risk as a Variable in TNC Decision-Making," Natural Resources Forum, Vol.6, No. 2, April 1982, pp. 183-199.

Radetzki, Marian and Zorn, Stephen, 1979, Financing Mining Projects in Developing Countries, (London: Mining Journal Ltd., 1979)

Smith, David and Wells, Louis, 1975, Negotiating Third World Mineral Agreements, 1975.

Stobart, Christopher, 1984, "Effect of Government Involvement on the Economics of the Base-Metals Industry," Natural Resources Forum, Vol. 8, No. 3, July 1984, pp. 259-267.

Stockmayer, Albrecht and Suratgar, David, 1982, Contributions on project financing arrangements in mining projects in: UN/DSE, "Legal and Institutional Arrangements in Minerals Devlopment," Mining Journal, (1982).

Stockmayer, Albrecht, Projekt-Finanzierung und Kreditsicherung, (Frankfurt: Kluwer, 1982) (English translation expected) summarized in "Legal and institutional arrangements in minerals development", UN/DSE, supra note 18.

Suratgar, David, 1982, "International Project Finance," Natural Resources Forum, Vol. 6, No. 2, 1982, pp. 113-127

Vernon, Raymond, 1981, "State-owned Enterprises in Latin American Exports," Quarterly Review of Economics and Business, Vol. 21, No. 2 (Summer, 1981) pp. 97, 104.

Wälde, Thomas, 1982, "Permanent Sovereignty over Natural Resources: Recent Developments in the Minerals Sector," Natural Resources Forum, Vol. 7, No. 3, 1983, pp. 239-253.

"Third World Mineral Development: Current Issues," Columbia Journal of World Business, Vol. XIX, Spring 1984, p. 27

Ibidem, "Third World Mineral Development: Recent Literature," CTC Reporter, 1984, p. 52

Ibidem, "Lifting the Veil from Transnational Mineral Contracts, A Review of Recent Literature," Natural Resources Forum, Vol. 1, April 1977.

Ibidem, "Revision of Transnational Investment Agreements," Laws of the Americas, Vol. 10, (1978) p. 265 and: UN/DSE, "Legal and Institutional Arrangements for Minerals Development," Op. Cit. supra note 18, 24 pp. 166-285.

United Nations, "State Petroleum Enterprises in Developing Countries," (New York: Pergamon, 1980).

Woakes, Michael and Carman, John ed., 1983, AGID Guide to Mineral Resources Development, and State of Montana, Handbook for Small Mining Enterprises, (State of Montana: March 1976).

Zorn, Stephen, "New Developments in Third World Mining Agreements," Natural Resources Forum, Vol.1, No. 3, April 1977, pp. 239-251

Zorn, Stephen, "Coal and Uranium Investment Agreements in the Third World," Natural Resources Forum, Vol.6, No.4, October 1982, pp. 345-359

United Nations, The Future of Small-Scale Mining, (New York: McGraw Hill, 1980).

United Nations, Small-Scale Mining in the Developing Countries, 1972, Sales No. E.72.II.A.4

Zakaryia, Hasan, "Impact of Changing Circumstances on the Revision of Petroleum Contracts," MEES 12 (July 11, 1969), Supplement.

7.5

THE EFFECT OF HOST GOVERNMENT ATTITUDE UPON
FOREIGN INVESTMENT IN MINING

G.E. PRALLE

Citibank, N.A.
New York, NY

INTRODUCTION

The decade from 1969 to 1979 saw the crest of a wave of investments by companies and investors, primarily from the major industrial nations, in the metals and minerals industries of the so-called lesser developed countries. This was brought about by two major considerations: 1). The higher ore grades and therefore the higher profit potential of the mineral deposits in the third world countries, and 2). The desire, particularly by the European countries and by Japan, to ensure a safe supply of raw materials not available in their own countries.

The decade began with the start of the Toquepala mine in Peru and the major investments in the Bougainville mine in Papua New Guinea and the Ertsberg mine in Indonesia. It ended with the start-up of Southern Peru Copper Company (SPCC) Cuajone mine in Peru and Inco's Soroako mine in Indonesia and the investments in several major alumina or aluminum projects in Australia and Brazil.

Concurrent with this huge flow of funds into the mineral resources industry of the developing world was the build-up of a psychological fear in the host countries of exploitation/domination by the foreign investors. This resulted in an increasing trend towards: 1). Nationalization of existing mineral ventures, and 2). Tougher legal/fiscal frameworks for new ones.

NATIONALIZATION

Nationalization appeared in its earliest form as "Mexicanization" of the petroleum industry in Mexico in the 1930's, followed later by Mexicanization of the mining industry in general. This was enacted with the Mexican Mining Law of 1961 which required that 51% of the equity be owned by Mexican nationals. Later in the decade the equity share for foreign investors was reduced to 33 1/3% for the developers of the Frasch sulphur operations in the Isthumus of Tehuantepec. However, the law was not fully implemented in the copper industry until late 1970 with the purchase of 51% of the ownership in the Cananea Mining Company.

The trend toward nationalization of the copper industry progressed to Chile where in 1967 under the Government of Eduardo Frei the major North American copper companies were induced to sell a one-half interest in their Chilean mines to the State. Likewise, in 1970, "Zambianization" took place, when the two major operating companies Roan Selection Trust (Amax) and Zamanglo (Anglo American) were restructured into Roan Consolidated Mines (RCM) and Nchanga Consolidated Mines (NCCM) providing the government with 51% of the equity ownership. Whilst the earlier forms of nationalization in general constituted a fairly priced buy-out (El Teniente, Chile for US$80 million, RCM for US$117.8 million, NCCM for US$178.7 million), the total nationalization of the Chilean mining companies in 1970 under Salvador Allende did not.

Later, "Western" countries joined this trend and on October 29, 1973, Mr. Gough Whitlam proclaimed that "overseas capital must continue... in partnership with Australian capital" and "in some special energy cases we have an objective of 100% Australian ownership". In 1974 Mr. Pierre Trudeau stated: "The Liberal Party of Canada sets as an objective that new resources projects in the natural resource field should have at least 50% and preferably 60% Canadian equity ownership".

FOREIGN INVESTMENT

The situation for the foreign investor in the worldwide resource industry rapidly deteriorated from then on. New features appeared in mineral exploitation and operating agreements, centering not only on local government/local companies ownership but also on increased taxation. The new tax rate (mineral tax, royalty, and income tax) for the semi-government companies NCCM and RCM in Zambia became an effective 73%. Likewise the much publicized and discussed renegotiated Bougainville agreement brought dramatic increases in the total "take" by the PNG Government.

More innovative forms of indirect taxation made their appearance and were added to the normal tax burdens of the mineral industry which already suffered from above average levies. One such form was the requirement that the mining company sell its output to a marketing company owned by the host Government to "ensure that the highest return be obtained". Among the first of such marketing agencies was Metal Marketing Corporation of Zambia Ltd. (Memaco) and Minero Peru Comercial S.A. (Minpeco). Memaco collects a commission from the producers and Minpeco progressed one step further in requiring that most of the blister copper produced by SPCC's Cuajone mine be refined in the copper refinery of Minpeco's affiliate Minero Peru, where refining charges per pound have consistently exceeded those levied upon the SPCC sister operation Toquepala by foreign refiners. More recently we have seen the formation of the Minerals Marketing Board in 1982 in Zimbabwe which totally controls all sales functions in that country.

The true effect of those happenings in the 1960's and early to mid 1970's was largely obscured by the strong general expansion of the worldwide mining industry. Numerous new ventures in copper, nickel, iron ore, bauxite, coal and other minerals were initiated based primarily on the assumption of impending shortages brought about by high growth rates in consumption as well as the generally accepted myths concerning unending geometric progression in demand for raw materials promalgated by the Club of Rome and others. Yet even in those heydays of foreign investment, reaction to the Whitlam pronouncements was swift and brutal: A massive exodus of exploration teams and their budgets occurred in such an unmistakable way that the following Liberal Government saw fit to moderate the rules to their present form. These involve a process of "naturalization" through increased Australian ownership brought about not in a precipitous manner but over time and in a logical businesslike way.

Similarly, when in 1974 the new British Columbia Premier David Barrett initiated the rather abnormal NDP taxation measures, which were directed against all mining companies, and which resulted in some instances in a taxation rate of over 100%, the effect upon exploration expenditures and new capital investment in the Province were truly dramatic. Exploration expenditures alone dropped from US$53 million in 1970 to US$20 million in 1974, despite the fact that 1974 was an all time peak metal price year.

Similiarly, in Zambia, a country in which mining accounts for the largest part of GNP, and well over 90% of foreign exchange income, foreign investment in the mining industry totally disappeared with the exception of some relatively minor exploration for uranium by AGIP of Italy and Power Reactor Corp. of Japan. Likewise, in neighboring Zimbabwe, undoubtedly an exceptionally mineral rich country, there has been no foreign investment except some expansion of Union Carbide's chromium operation. Following postnationalization of Union Miniere, Zaire experienced some modest sized investments in minerals such as the Musoshi Mine and the Kinseza Mine sponsored by Japanese interests. But the giant US$950 million Tenke Fungurume project was abandoned after expenditures exceeding US$200 million.

At the same time, the Philippines, which has had no nationalizations or particularly anti-business or anti-foreign investment attitudes, continued to attract substantial investments and financings from the United States and other industrial nations with major flows of funds into such companies as Atlas, Marinduque, Benguet, Marcopper, and others. (In mid-1984 President Marcos reiterated that foreign capital was most welcome, and complete foreign ownership of a natural resource company is indeed permissable.)

These observations find support in the accompanying compilation by World Mining, September, 1977, which lists seventy-five major copper deposits awaiting development. Since that publication, only six of these deposits have been brought into operation. Three are in Canada; Valley Copper, Highmont, and Sam Goosly; two are in the Philippines; Basay and Dizon, and one, Monyawa, is in Burma. Several other deposits not included in the compilation have also been developed. These include Amacan (North Davao) in the Philippines, Lo Aguirre in Chile, and H-W in Canada. Deposits listed in the compilation that are being developed include Tintay in Peru and Teutonic Bore in Australia. Escondida in Chile, which contains 545 million tonnes assaying 2% copper, and Olympic Dam/Roxby Downs, Australia, containing 400 million tonnes of 1.0% copper plus uranium and gold, are also approaching development.

PROJECT DEFERRAL

By the late 1970's an over-supply position manifested itself with long term implications in many areas of the metals and minerals industry, particularly in copper, nickel, molybdenum, iron ore, and bauxite, where most of the capital expenditures had taken place. The net result was a noticeable fall-off in foreign investment in the field of natural resources. In fact, most of the super projects then pending, such as Cerro Colorado (copper) in Panama, Gag Island (nickel) in Indonesia, Pachon (copper) in Argentina, Jabiluka (uranium) in Australia, and others were "deferred". Those mega projects which did in fact go ahead, such as the La Caridad copper mine in Mexico, and the Carajas iron ore project in Brazil, did so only because of massive local government sponsorship/ownership/commitment.

The reason for these and other project deferrals is that in the late 1970's the mineral industry, and with it the foreign investment situation, underwent two very distinct policy changes: 1). The mega projects were replaced with small to at best moderate sized projects, including incremental projects, and 2). Precious metal projects replaced base metal projects in priority.

In both cases the objectives were the same: 1). To reduce political risk through lower visibility; 2). To reduce financial risk through smaller investments, better timing, and more controlled technology; 3). Reduce market risk through smaller output and more attractive commodities

Precious metal mines, which lend themselves to small investments with precise scheduling and proven technology, proliferated, particularly in Australia, Canada, and the United States, but also in Mexico, Peru, and Chile. The Golden Sunlight gold mine in Montana came on stream for under US$50 million and the Real de Angeles silver mine in Mexico for US$170 million as compared to the La Caridad expenditures which to date total over US$1.1 billion.

Meanwhile, some host Governments had come to realize that: 1). The foreign investor was rapidly becoming a rarity, putting his money where it was safest; 2). The value of the indigenous mineral to the world at large was far less than previously thought, and 3). The mining business was a marginal business best left to professionals.

NEW PROJECT INCENTIVES

To reverse the strong decline in foreign investment in the late 1970's some countries began to actively woo new entrants into the natural resources and metals business. The Republic of Ireland, which has had a long history of pursuit of foreign capital and the employment which results from it, stepped up its campaign and in 1978 landed the US$558 million Aughinish Alumina project (later to actually cost well over US$1.2 billion) through the "extras" offered, i.e. a tax holiday until 1990, 8 million in cash grants, a 29.2 million debt guarantee, interest subsidies, training grants, freedom from import and export duties, etc.

Likewise, the Province of Quebec, through and in association with its provincial utility Hydro Quebec and it's investment company Societe Generale de Financement, managed not only to cause Quebec-based Alcan Aluminum to expand significantly, but also to attract a consortium headed by Pechiney of France to build a new US$1.2 billion, 230,000 tons per year aluminum smelter based on a very favourable 25-year electric power contract, full financing participation in proportion to ownership, and infrastructural help. Furthermore, discussions were begun which may lead to a second such project headed by Kaiser Aluminum.

As we enter the mid-1980's Chile stands out in its efforts to attract foreign investments in its huge reservoir of undeveloped mineral deposits. The decade-long hiatus caused particularly by the events under Allende continued well into the years of the pro-business/pro-foreign investment government of General Pinochet and only ended when it became apparent that the new attitude appeared to be durable. During the recent several years the following frame work has evolved through formalized decree laws or through case by case negotiation:

* Full foreign ownership is permissible

* The State will not interfere in the planning process

* Government will assist in or carry most of the infrastructure burdens

* The investor can do his own product marketing

* To provide a fast return most plant and equipment can be depreciated over one to three years

* Profits can be remitted out of the country

Whilst foreign investment dropped in many countries and essentially disappeared in others, this above policy has had a number of major successes for Chile including:

* Massive exploration expenditures by North American, European, and Japanese mining companies

* An expansion of US$25 million by the Hochschild-owned Mantos Blancos copper mining company in 1979

* A second expansion of US$75 million by the same group in 1984

* The US$250 million installation of a mine/concentrator/roaster at El Indio by St. Joe Minerals Company

* The approaching construction of the Cerro Colorado copper mine by Rio Algom Ltd., estimated to cost US$255 million

* The approaching construction of the US$1.2 billion Escondida project now owned by Broken Hill Proprietary Ltd. and Getty Minerals.

At the same time, Argentina, which shares with Chile the exceptionally favourable geology of that part of the Andean mountain chain, has not been able to attract a single investment into that area from abroad.

If we attempt to trace the movements of foreign capital we see it follow channels where it is welcomed and assisted and see it abandon areas where it is oppressed. Perhaps this is the main reason why the United States leads all other nations in having attracted foreign investment in its natural resource and metals industry. The Commerce Department in May 1984 published data indicating that total Japanese direct investment in the United States exceeded US$10 billion for the first time in 1983, a significant portion thereof being in metals and mining. European investment exceeds this, and even developing countries have investments in the US. Although rules and regulations in the United States do not particularly favor either the natural resource industry or the foreign investor therein, the rules are fixed, the legal framework is stable, and the capital in the outflow are entirely free.

Known World Copper Deposits Awaiting

UNITED STATES

Coastal Mining Company (Getty/Hanna)
West Casa Grande, Arizona. 250,000,000 tons (500,000,000?) averaging 1.0 percent copper before dilution.

Kennecott Copper Corporation
Safford, Arizona. 500,000,000 tons (+) of 0.5 percent copper.
Ladysmith, Wisconsin, 4,000,000 tons of ore averaging 4.8 percent copper.
Spar Lake, Montana. 43,500,000 tons of ore averaging 0.74 percent copper and 1.54 ounces silver per ton (Asarco conducting feasibility study).

Inspiration Consolidated Copper Company
Joe Bush property, Arizona. 90,000,000 tons 0.7 percent copper.

Newmont Mining Corporation
Vekol Hills, Arizona. 100,000,000 tons averaging 0.56 percent copper.

Cities Service Company
Miami East, Arizona. 50,000,000 tons averaging 1.95 percent copper.

Continental Oil Company
Florence, Arizona. 800,000,000 tons averaging 0.4 percent copper.

Kerr McGee Corporation
Red Mountain, Patagonia, Arizona. No ore reserve figures published.

Exxon
Crandon, Wisconsin. Two discoveries of massive sulphide copper/zinc ore. No ore reserve figures published.

Phelps Dodge Corporation
Safford, Arizona. 400,000,000 tons ore averaging 0.72 percent copper.
Copper Basin, Arizona. 159,000,000 tons ore averaging 0.55 percent copper and 0.02 percent molybdenum.

Arco (Anaconda)
Ann Mason property, Nevada. A potential of 495,000,000 tons of copper bearing material with average grade of 0.40 percent total copper.
Bear property, Nevada. Indications of 500,000,000 tons copper bearing material averaging 0.40 percent total copper.

Anamax Mining Company
East Helvetia, Arizona. 211,000,000 tons copper bearing material averaging 0.56 percent total copper plus 22,000,000 tons grading 0.55 percent acid soluble copper and 2 miles away, 24,000,000 tons of 0.75 percent oxide-sulphide.

Ponce Mining Company (Puerto Rico)
Amax/Kennecott Nearby but separate ore bodies totaling 244,000,000 tons averaging 0.73 percent copper.

CANADA

Bethlehem Copper Corporation
J-A Zone: Highland Valley area near Ashcroft, British Columbia. 286,000,000 tons averaging 0.43 percent copper and 0.017 percent molybdenum.
Maggie: 35 miles northwest of J-A Zone. 200,000,000 tons averaging 0.4 percent copper equivalent.

Cominco Limited
Valley Copper Mines (81 percent-owned). Lake Zone; Highland Valley area near Ashcroft, British Columbia, 800,000,000 tons averaging 0.48 percent copper.

Highmont Mining Corporation Limited (Teck 45 percent)
Highmont property: Highland Valley area, British Columbia, 140,000,000 tons averaging 0.40 percent copper equivalent.

Liard Copper Mines Limited
Schaft Creek property, British Columbia, 294,000,000 tons averaging 0.40 percent copper and 0.036 percent molybdenite.

Stikine Copper Limited (Kennecott, Hudson Bay)
Stikine River area, British Columbia, 138,000,000 tons of + 1.0 percent copper.

Equity Mining Corporation
Sam Goosly property near Houston, British Columbia, 43,500,000 tons grading 0.33 percent copper, 2.78 ounces silver, and 0.026 ounces gold per ton.

Granby Mining Company Limited
Huckleberry project near Smithers, British Columbia, 85,000,000 tons averaging 0.401 percent copper.

Noranda Mines Limited
Coldstream River Area, Revelstoke, British Columbia, 3,200,000 tons grading 4.49 percent copper, 3.24 percent zinc, and 0.6 ounce silver per ton before dilution.

Silver Standard Mines Limited
Minto area, near Carmacks, Yukon Territory (Asarco 50 percent). 4,700,000 tons averaging 1.78 percent copper.

Placer Development Limited
Berg prospect, near Houston, British Columbia, 400,000,000 tons averaging 0.40 percent copper and 0.05 percent molybdenum.

Texasgulf Inc.
Izok Lake, Northwest Territory.

MEXICO

Hudson Bay Mining and Smelting Company, Limited.
La Verde deposit, Gabriel Zamara, Michoacan. 81,300,000 tons grading 0.699 percent plus silver and gold values.

Tormex Mining Developers Limited (Penoles)
Santo Tomas prospect, northern Sinaloa. 150,000,000 tons averaging 0.5 percent recoverable copper.

Cobre de Sonora, S. A. de C. V.
Santa Rosa and **Pilares** deposits, Sonora. 143,000,000 tons 0.83 percent copper with silver and molybdenum values.

Industrial Minera Mexico
El Arco. Baja California. 630,000,000 tons 0.6 percent copper.

PANAMA

Texasgulf Inc.
Cerro Colorado. 1,000,000,000 tons (+) of 0.6 percent (+) copper.

Cobre Panama. (Japanese consortium)
Petaquilla prospect. 300,000,000 tons 0.6 percent copper.

ECUADOR

Caucha Direccion General de Geologia Minas (DGGM) is prospecting and reassessing this porphyry copper deposit discovered some years ago. At that time ore reserves were estimated at 55,000,000 tons averaging less than 0.5 percent copper and prospect was considered uneconomical. Work also being done at **San Miguel, San Bartolome,** and **Angas** prospects.

ARGENTINA

Cia Minera Aguilar S.A. (St. Joe)
El Pachon. San Juan Province. 870,000,000 tons grading 0.59 percent copper and 0.015 percent molybdenum.

YMAD (state-owned)
La Alumbrera, Catamarca Province. 220,000,000 tons grading 0.5 percent copper and 0.0225 ounce of gold per ton.

PERU

Centromin
Toromocho Ore reserves estimated at 330,000,000 tons averaging 0.766 percent copper, 0.2 percent molybdenum and 0.28 ounce of silver per ton.

Mineroperu
Santa Rosa. Sulphide ore body underlying and adjacent to Cerro Verde begins production 1977. Santa Rosa ore

Source: World Mining, 1977, Sept., pp. 110-111.

Favorable Conditions For Development

reserves estimated at (1,200,000,000 tons?) averaging 0.55 percent copper.
Quellevaco Ore reserves estimated at 450,000,000 tons averaging 0.8 percent copper.
Michiquillay Ore reserves estimated at 460,000,000 tons averaging 0.75 percent copper.
Tintaya Ore reserves estimated at 44,000,000 tons sulphides averaging 2.0 percent copper and 11,000,000 tons of oxides averaging 2.2 percent copper.
Chalcobamba and **Ferrobamba**. Relatively less explored than some other major deposits. Ore reserves to date put at 33,000,000 tons averaging 2.0 to 3.0 percent copper.
Berenguela. Ore reserves estimated at 17,600,000 tons averaging 1.26 percent copper and 4.27 ounces of silver per ton.

Mitsubishi/Homestake
Pashpap Ore reserves estimated at 53,000,000 tons averaging 0.86 percent copper and 0.5 percent (?) molybdenum.

CHILE

Enami (?)-Noranda
Andacolla. 280 miles north of Santiago. 330,000,000 tons 0.7 percent copper.

Codelco.
El Abra 29 miles north of Chuquicamata. 550,000,000 (+) tons of 1.0 percent sulphide and oxide ore.
Quebrada Blanca North of El Abra. 165,000,000 tons of 1 percent.
Pampa Norte. North of Chuquiccamata. 440,000,000 tons averaging 0.8 percent copper (oxide).

Los Pelambris North of Andacollo. 440,000,000 tons averaging 0.8 percent copper.

Cerro Colorado. Near Peruvian border. 110,000,000 tons of 1.2 percent copper.

PAPUA NEW GUINEA

Broken Hill Proprietary Limited (PNG government)
OK Tedi. 250,000,000 tons 0.85 percent copper.

MIM Holdings Limited
Frieda River prospect. 366,000,000 tons averaging 0.45 percent copper.

Triako Mines-Buka Minerals-Kennco Explorations
Yandera prospect. 338,000,000 tons of 0.42 percent copper.

AUSTRALIA

Western Mining Company
Andamooka region (Olympic Dam) South Australia. Significant prospect reported.

AMAX Inc. (Consortium)
Golden Grove, near Yagloo, Western Australia. 11,000,000 tons 3.5 percent copper.

Western Mining Corporation
Teutonic Bore, north of Kalgoorlie, Western Australia. (Western Selcast and Mount Isa 60:40). Significant copper/zinc discovery.

PAKISTAN

Fort Saindak in Chagai district of Baluchistan. Three ore bodies. East body 261,000,000 tons average grade 0.37 percent copper. South body 55,000,000 tons average grade 0.47 percent copper. North body 20,000,000 tons 0.45 percent copper. Marginal values of molybdenum present in East and South deposits. Feasibility study was begun April 1976.

MONGOLIA

Erdenetiyn-Ovoo. Extensive copper/molybdenum deposit reported. Tonnage and grade not given. However, a plant with annual capacity of 16,000,000 tons of ore is reported under construction with a 1978 completion date.

ZAIRE

SMTF
Tenke-Fungurume deposits contain reserves estimated at 56,000,000 tons mixed ores averaging 5.7 percent copper and 0.44 pecent cobalt. Development postponed.

ZAMBIA

Lumwana deposit. 220,000,000 tons of 1 percent copper bearing material indicated.

Kansanshi. 10,000,000 tons/ore averaging 3.06 percent copper. Development postponed.

PHILIPPINES

Lepanto Consolidated
Hinobaan project. Ore reserves 100,000,000 tons averaging 0.50 percent copper

Benguet Consolidated
Tawi-Tawi project, Benguet province. Ore reserves 165,000,000 tons averaging 0.39 percent copper.

In addition to the deposits listed above by George Munroe, the editors of WORLD MINING add the following deposits compiled from records in the WORLD MINING Resource Research Department.

UNITED STATES

Amax, Inc.
Kirwin. Small porphyry, estimated 70,000,000 tons of 0.75 percent.

Arco (Anaconda)
Heddleston. 93,000,000 tons at 0.48 percent.
Deep Continental North, 497,600,000 0.49 percent copper.
Deep Low Grade, 820,000,000 tons of 0.74 percent total copper.
Hall, 54,000,000 tons of 0.46.

Placer Amex Inc.
Lights Creek. 350,000,000 tons of 0.34 percent.

Ownership unknown.
Washington state.
Mazama, 470,000,000 of 0.37 percent.
Sampson, 100,000,000 tons of 0.5 percent.
North Fork, 44,000,000 tons of 0.5 percent copper.

BULGARIA

Elatsite, being developed for production. Reserves reported at 250,000,000 tons of 0.45 percent copper.

BURMA

Moniwa, 66,000,000 tons of 0.8 percent.

PHILIPPINES

Marcopper Mining Corporation
Toledo tailing ore body. 200,000,000 tons assaying about 0.57 percent.

C.D.C.P. Mining Corporation
Basay, 220,000,000 tons assaying 0.40 percent. An open pit mine planned.

Benguet Consolidated Inc.
Dizon, 100,000,000 tons assaying 0.42 percent.

OUTLOOK

In the future we will surely see continued competition for the limited pool of foreign investment funds. Countries will realize that their natural resources are not unique in the world but, on the contrary, must compete for oversupplied markets. Mining companies no longer have unrealistic expectations of either high ore grades, high product prices, or unusual profit potential in foreign countries. Investors will commit their funds only where business conditions clearly indicate that the investment is welcome and where a reasonable return can be achieved. This includes an equitable fiscal treatment comparable to other industries, equal and business-like financial participation by local organizations when this is mandated, freedom of foreign exhange flows and profit remittance, and full participation by the host government in general infrastructural costs.

It is probable that these conditions will exist in a very limited number of countries. These fortunate few are likely to attract the lion's share of the private investment available for mining in the future.

CHAPTER 7: IMPACT OF MINERAL DEVELOPMENT AND OPERATING AGREEMENTS ON FINANCING

BIBLIOGRAPHY

Asante, S.K.B. and Stockmayer, A., 1982, "Evolution of Development Contracts: The Issue of Effective Control," Legal and Institutional Arrangements in Minerals Development, Mining Journal Ltd., UN/DSE, London, pp. 53-6420.

Asante, S.K.B., 1979, "Restructuring Transnational Mineral Agreements," American Journal of International Law, Vol. 73, p. 335 and "Stability of Contractual Relations in the Transnational Investment Process," International and Comparative Law Quarterly, No. 28, p. 401.

Bartels, M. et al., 1981, Bibliography on Transnational Law of Natural Resources, Kluwer, Frankfurt.

Bender, M., 1984, Natural Resources Forum, Vol. 8, No. 3, pp. 279-289.

Bosson, R. and Varon, B., 1977, The Mining Industry and the Developing Countries, Oxford University Press.

Carman, J., 1977, "United Nations Mineral Exploration Activities: 1960-1976," Natural Resources Forum, Vol. 1, No. 4, July, pp. 317-336.

Cormick, R., 1983, "Legal Issues in Project Finance," Journal of Energy and Natural Resources Law, Vol. 1, No. 1, pp. 21-44.

Dobozi, I., 1983, "Mineral Development: Cooperation Between Socialist Countries and Developing Countries," Natural Resources Forum, Vol. 7, Oct., pp. 339-351.

Eschenlauer, A., 1984, "The New Twist in Energy Project Financing," International Energy Law, International Bar Association, Section on Energy and Natural Resources Law.

Fischer, P. and Walde, T., 1981, "International Concessions and Related Instruments," Oceana, Vols. I-V, 1981 et seq.

Garnaut, R. and Ross, A. C., 1975, "Uncertainty, Risk Aversion and the Taxation of Natural Resources Projects," Economic Journal, No. 85, p. 272.

Gillis, M. et al., 1980, Tax and Investment Policies for Hard Minerals: Public and Multinational Enterprises in Indonesia, Ballinger, Cambridge.

Gillis, M., 1982, "Toward Synthesis," Public Enterprises in LDCs, Cambridge.

Gillis, M., 1983, Economics of Development, Norton, New York.

Gillis, M., 1980, Tax and Investment Policies for Hard Minerals: Public and Multinational Enterprises in Indonesia, Ballinger, Cambridge.

Harkin, D. A., 1976, "Systematic Mineral Exploration," Natural Resources Forum, Vol. 1, No. 1, Oct., pp. 29-37.

Hellawell, R. and Wallace D., 1982, Negotiating Foreign Investments, A Manual for the Third World, 2 vols.

Legoux, P., 1980, "Report on Mining Legislation in African Countries," ECA document E/CN.13 of 6 June.

Lewis, A., 1983, "The United Nations in Mineral Development," Engineering and Mining Journal, Jan., pp. 68-70.

McGill, S., 1983, "Project Financing Applied to the Ok Tedi Mine - A Government Perspective," Natural Resources Forum, Vol. 7, No. 2, Apr., pp. 115-131.

Mikdashi, Z., 1976, The International Politics of Natural Resources, Cornell University Press, Ithaca.

Mikesell, R., 1984, Foreign Investment in Mining Projects, Oelgeschlager, Gunn and Hain, Cambridge, with a chapter on Botswana summarized in Natural Resources Forum, Vol. 8, No. 3, pp. 279-289.

Moment, S., 1980, The Pricing of Bauxite from Principal Exporting Countries, 1974-1978, United Nations Industrial Development Organization, Vienna.

Moran, T., 1974, Copper in Chile, Princeton.

Prast, W. G. and Lax, H., 1983, "Political Risk as a Variable in TNC Decision-Making," Natural Resources Forum, Vol. 6, No. 2, Apr., pp. 183-199.

Radetzki, M. and Zorn S., 1979, Financing Mining Projects in Developing Countries, Mining Journal Ltd., London.

Smith, D. and Wells, L., 1975, Negotiating Third World Mineral Agreements.

Stobart, C., 1984, "Effect of Government Involvement on the Economics of the Base-Metals Industry," Natural Resources Forum, Vol. 8, No. 3, July, pp. 259-267.

Stockmayer, A. and Suratgar, D., 1982, Contributions on project financing arrangements in mining projects in UN/DSE, "Legal and Institutional Arrangements in Minerals Development," Mining Journal.

Stockmayer, A., 1982, Projekt-Finanzierung und Kreditsicherung, Kluwer, Frankfurt (English translation expected) summarized in "Legal and Institutional Arrangements in Minerals Development," UN/DSE.

Suratgar, D., 1982, "International Project Finance," Natural Resources Forum, Vol. 6, No. 2, pp. 113-127.

Trabat, T. J., 1983, Brazil's State-Owned Enterprises, Cambridge University Press, New York, 294 pp.

United Nations Secretariat, 1980, "Two Decades in Mineral Resources Development," Paper 6, "Mining Engineering in the Developing Countries: the Role of the United Nations," Paper 7, and "Mineral Engineering in the Developing World: the Role of the United Nations," Paper 8, Symposium on National and International Management of Mineral Resources, Institution of Mining and Metallurgy, London.

United Nations Secretariat, 1981, "Two Decades of Mineral Resources Development: The Role of the United Nations", Natural Resources Forum, Vol. 5, No. 1, Jan. 1, pp. 15-30.

Vernon, R., 1981, "State-owned Enterprises in Latin American Exports," Quarterly Review of Economics and Business, Vol. 21, No. 2, Summer, pp. 97, 104.

Walde, T., 1982, "Permanent Sovereignty over Natural Resources: Recent Developments in the Minerals Sector," Natural Resources Forum, Vol. 7, No. 3, pp. 239-253.

Walde, T., 1984, "Third World Mineral Development: Current Issues," Columbia Journal of World Business, Vol. XIX, Spring, p. 27.

Walde, T., 1984, "Third World Mineral Development: Recent Literature," CTC Reporter, p. 52.

Walde, T., 1977, "Lifting the Veil from Transnational Mineral Contracts, A Review of Recent Literature," Natural Resources Forum, Vol. 1, Apr.

Walde, T., 1978, "Revision of Transnational Investment Agreements," Laws of the Americas, Vol. 10, p. 265 and UN/DSE, "Legal and Institutional Arrangements for Minerals Development," pp. 166-285.

United Nations, 1980, "State Petroleum Enterprises in Developing Countries," Pergamon, New York.

Woakes, M. and Carman, J., eds., 1983, AGID Guide to Mineral Resources Development, and State of Montana, Handbook for Small Mining Enterprises, State of Montana, March, 1976.

Zorn, S., 1977, "New Developments in Third World Mining Agreements," Natural Resources Forum, Vol. 1, No. 3, Apr., pp. 239-251.

Zorn, S., 1982, "Coal and Uranium Investment Agreements in the Third World," Natural Resources Forum, Vol. 6, No. 4, Oct., pp. 345-359.

United Nations, 1980, The Future of Small-Scale Mining, McGraw Hill, New York.

United Nations, 1972, Small-Scale Mining in the Developing Countries, Sales No. E.72.II.A.4.

Zakaryia, H., 1969, "Impact of Changing Circumstances on the Revision of Petroleum Contracts," MEES 12, July 11, Supplement.

Chapter 8

Strategic and Financial Planning

CHAPTER 8: STRATEGIC AND FINANCIAL PLANNING

PREFACE

Every firm, whether or not in the minerals business, has a general business and financial strategy. This strategy may be fashioned either explicitly through a formalized planning process within the firm, or it may have evolved implicitly by default through the activities of the various corporate functional departments. Increasingly, firms are coming to realize that a coordinated, formalized strategic planning process facilitates the development of an optimum overall business strategy and increases a firm's chances for financial success.

Six papers are included in this chapter in order to review various key aspects of strategic and financial planning. R. A. Arnold's paper, "The Evolution of the Planning Process in the U.S.," introduces the subject by reviewing the evolution of formal business planning processes and systems in response to the increasing complexity of the business environment.

Certainly fundamental to any successful financial and business strategy is knowing who and what you are, who's there with you, and how things may change in the future. A meaningful business strategy requires a realistic and sophisticated view of competitive advantages in one's businesses. Such competitive knowledge enables one to best position one's company in order to maximize the value of those capabilities that distinguish its businesses from those of its competitors. L. Kovisars' paper, "Competitive Analysis," discusses approaches to gathering, constructing and analyzing competitive cost information.

In "Assessing Strategies for Natural Resource Companies," by B. C. Castleman, portfolio analysis techniques are reviewed as effective tools to help the planner evaluate strategies for optimizing corporate wealth. Specifically, the development of a portfolio of investments in different projects where competitive advantage can be achieved and maintained is the essence of strategic management. This paper describes an approach to defining the critical elements of a successful strategy and gives examples of the application of these analytical tools to the various mining and extractive businesses.

The development of a strategic plan is of little value, however, unless it is executed. At its best, an unexecuted plan leads to confusion and time wasted, and at its worst it could lead to cynicism and a loss of credibility.

Generally for a mineral producer, the execution of a strategic plan involves the reinvestment of corporate funds into the acquisition and/or development of mineral properties. Ideally, a firm would discover and develop new properties on its own accord in order to meet its growth plans. However, firms are relying increasingly upon the acquisition and divestiture of properties and operations in order to meet their strategic objectives. H. J. Sandri's paper, "Mineral Industry Acquisition Analysis," reviews specific reasons and criteria for mineral property acquisitions, and discusses appropriate techniques for the valuation of mineral assets.

Whereas the above papers deal with the overall development and execution of corporate strategy, "Financial Management of Diversified Companies," by P. J. Maxworthy, deals specifically with the issues of financial policy and strategy. The final paper, "Long-Range Planning of a Copper Mining Company," by F. Buttazzoni and J. Munita, presents one specific company's view of its own business planning process and discusses the benefits derived from this approach.

The premise of this chapter is that formal strategic and financial planning activities have an important role in the optimization of corporate value and financial well-being. The contributing authors are gratefully acknowledged for their insightful elaboration upon this theme. As chapter editor, I have included a bibliography at the end of this chapter for the reader interested in additional information on the subject of strategic and financial planning.

William R. Bush
Chapter Editor
Director, Business Development
Duval Corporation

8.1

THE EVOLUTION OF THE PLANNING PROCESS IN THE U. S.

Richard A. Arnold

Pennzoil Company
Houston, Texas

ABSTRACT

The evolution of planning processes in the United States from the early 1900's is briefly examined. From one year operating budgets, planning processes have evolved into sophisticated analyses of alternative long-term future strategies and the management issues and practices which relate to their implementation. Increasing complexity within the business environment has led to numerous planning methodologies and extensive use of computers to analyze operating and financial information. All of these developments, however, have not supplanted the real need for decision-makers to continue to give important weight to the intuitive or soft data inputs into the planning process.

INTRODUCTION

The purpose of this paper is to trace the historical development of planning processes and techniques in order to better understand current practices and future developments in strategic and financial planning. First, the development of planning and control processes from early in the century to the present will be reviewed. Next, some observations about the current "state of the planning process" will be presented; followed by concluding thoughts concerning the future of the planning process.

For this discussion the word "process" will be used rather than the word "system" - which perhaps is the more common or accepted form. One definition of process is "something going on", while a definition for system is "an organized or established procedure". I'm sure that we can all agree that something has been going on in planning. However, there may be some disagreement about how organized we are or how established any planning system is. To some of the management in my company, system implies formal, rigid, standard, controlled, etc. In spite of this, control is not the primary or perhaps even secondary objective of strategic planning activities at Pennzoil, as it is in some companies.

THE DEVELOPMENT OF PLANNING IN THE U.S.

Planning and control processes began with the development of accounting systems in the early 1900's which enabled organizations to develop policy and procedure manuals, annual budgets and suitable operating and financial controls. The emphasis was on the short-term and control measures were financially oriented. The establishment of other performance control measures led, after World War II, to Management by Objective (MBO) programs, which for the most part were short-term oriented and linked to a compensation system. Meanwhile, the U.S. economy had progressed from satisfaction of basic demand in the early 1900's to rapid growth, then deceleration during the 1930's, to internationalization, technology proliferation and saturation of growth in some industries in the 1950-1980 period. Global competition, stagnation, a resurgence in technological developments, apparent limits on growth for basic industries and the like have characterized the 1980's. Changes in markets, technologies, government regulations, economic expectations and the world competitive structure have brought us from a stable or incrementally changing environment to one of unexpected discontinuities and unpredictable surprises.

With credit to SRI International, Table 1 is an attempt to put the evolution of planning processes relative to time and the business environment into some perspective.

Table 1
Evolution of General Management Systems

Time	1900	1930	1950	1980
Strategic Information Perspective / Environmental Challenges	Satisfaction of basic demand / Acceleration of growth	Response to customer preferences / Deceleration of growth	Internationalization / Technology proliferation / Saturation of growth	Global competition / Stagflation / Socio-political pressures / Limits on growth
Time Perspective / Change				
Past — Stable / Incremental	POLICY AND PROCEDURE MANUALS • Financial Control		CONTROL	
↓ Future — Incremental		• Budgeting	• Management by Objective • Long-Range Planning	
Familiar discontinuities (threat/opportunity)	PLANNING		• Strategic Planning	
Novel discontinuities			• Strategic Management	
"Weak signals"			• Issue Management	
Unpredictable surprises	FLEXIBILITY		• Surprise Management	

Source: SRI, *Managing Strategic Surprise*, H. Igor Ansoff

In the 1950's, planning processes began to evolve from the basic annual budget and MBO control systems. Following are some of the terms used to describe major planning processes, and they are listed roughly in their order of development:
o Long-Range Planning
o Strategic Planning
o Strategic Management
o Issue Management
o Surprise Management

MAJOR PLANNING PROCESSES

Long-range planning, one of the earliest planning processes adopted by many companies and still widely used, is an extrapolation from the present and short-term budget into the future using macroeconomic models or other correlation factors. This process is useful for core businesses where market changes are incremental rather than drastic, the competitive climate is relatively stable, and new business additions are a minor part of the planning horizon. This planning process seeks to answer the question "If all goes well about as we expect under normal circumstances, where will we be in 19XX?" Unfortunately, much of our business environment does not fit with the conditions necessary to have a reasonable chance of having this type of planning process work successfully most of the time.

In the sixties and seventies, three established trends continued and changed the nature of planning. First, the widening use of computers brought increased computational power, favoring increased use of econometric models, sensitivity analysis, Monte Carlo simulation and other mathematically intensive activities. Second, in this phase the planning horizon was extended out further (to the 5-20 years range) and more intensive efforts were made to monitor the total business environment. Third, efforts began to concentrate on internal analysis of mission, competitive position, strengths and weaknesses, opportunities and threats leading to a process which included mission, objectives, goals, strategies and implementation programs. Strategic planning, as many know it, had begun to emerge, championed by such companies as General Electric and IBM.

As staffs wrestled with the masses of data poured out in the increasingly complex long-range plans, the problems of interpretation and implementation moved to the fore. Strenuous efforts were made to make sense of a mountain of data. The response, in many cases, was to develop a formal strategic planning process.

During this period, the use of various mathematical or strategic reasoning techniques emerged such as a correlation of relative market share, capital intensity, value added, quality, customer concentration and other factors affecting profitability or maximization of cash flow. A great deal of research emerged from the work at General Electric, the PIMS program and other places.

In my opinion, this work made a significant contribution to our understanding of how many businesses function and why some are more profitable than others. However, the rules of the game as developed in this research do not apply to all businesses (exceptions may be natural resource and many distribution type companies) and do not explain all of the factors which affect profitability or the attractiveness of a specific business. Where applicable, a relatively high degree of correlation to profitability has been claimed (70% - 80%). But, we must understand that 1) the conditions which may have dictated success in the past may not apply in the future, 2) sometimes going counter to the majority is the best strategy and 3) hard work, superior management, and simply luck and timing can counterbalance many apparent strengths of the competition.

In multi-business companies, <u>portfolio management</u> became an essential part of the <u>strategic management process</u>. Businesses were arrayed by star, cash cow, problem child and dog categories (or similar terms). Total corporate strategy became the optimization of future cash flows by the allocation of capital resources to specific businesses and the adoption of different growth, contain, harvest and divest strategies for business segments. Market share, life cycle analysis, market characteristics and growth, cost position and other positioning factors became essential elements of strategic analysis and management. A further step was ascertaining whether businesses were moving toward another strategic position and what steps were needed to be taken to change their course (if that was determined to be desirable). Cash flow analysis and projections were the key quantitative end products of the process.

In many circles, this form of strategic planning and strategic management is presently viewed with a high degree of skepticism and disenchantment, perhaps for some of the following reasons:

o The planning process, or decision-making exercise, depends on the right inputs. Faulty or overly simplistic data even with good analysis can lead to bad decisions.
o Much of the analysis depends on the assumption that market factors will not change, i.e. a good industry environment will hold into the future. This is often a faulty assumption.
o Competitive changes, which often completely undermine a strategic position, are not anticipated.
o The concept that businesses with common positioning characteristics should follow the same strategies was carried too far in some cases.

Much of the preceding material leads to a very important point which many companies began to address in the 1970's or even before. We live in an age of discontinuity and surprise. Projecting or forecasting a single set of external and internal conditions and establishing objectives, goals, strategies and long-lasting investment programs based only on these assumptions can be disastrous. History has demonstrated that even the most elaborate forecasting models, or the most highly regarded experts, can be wrong and recently, it seems, often have been wrong.

<u>Strategic alternatives planning</u> assumed greater importance as the results of too rigid, too generalized, one course planning became apparent. Basically, this process compares an array of reasonably possible alternative future environments against alternative internal strategies. The objective is to test internal strategies against external events to determine what limited number of strategies or strategy would enable the company to fare the best or perhaps even survive in the future. Contingency or alternative plans can be developed to be put into action should external (and perhaps some internal trends) indicate that a different environment is beginning to develop. The important feature about this planning process is the thinking process that management has to go through to anticipate things that can go wrong as well as things that can go better and then to develop a workable strategy. Further, early warning signs can be developed to indicate that a business or industry is starting to take a different path than anticipated by the highest probability case. Initiation of an alternative strategy then may be warranted.

<u>Issue management</u>, a variation of strategic management, is an attempt to develop by a top down-bottom up process the selection, analysis and implementation of programs dealing with key strategic issues affecting the future of the corporation. Priorities are developed. Often, task forces are formed to deal with the analysis, planning and execution of projects. In some companies, key issues have been those government, social, legislative, economic situations on which a company feels it can bring influence to bear; those large and external issues which will influence a company's future and, therefore, which are of great strategic importance.

<u>Surprise management</u> has been put into practice in recent years by a few companies when none of the planning and management processes in place seemed to provide an answer or an adequate response to a large surprise. The event or events came as a complete surprise; existing plans, even contingency plans, were not applicable; the problems

created by the event were not familiar; an action plan was immediately needed; and perhaps company or institutional survival was at stake.

Procedures to deal with such a situation had to be developed including non-standard communication responses, development of information handling centers, division of responsibility, quick-acting task forces, and novel approaches to new problems. Obviously, as in any emergency or crisis situation, procedures developed before the real crisis and some practice beforehand are advised.

PLANNING ANALYSIS TECHNIQUES

Various techniques are used to project, analyze, position, value, examine alternatives and otherwise manipulate data in all major planning processes. There is not time in this paper to discuss each. As stated by Ernest C. Miller in his book <u>Advanced Techniques For Strategic Planning</u>, AMA Research Study 104, American Management Association, 18 techniques were used in an AMA survey on the subject. They are listed below:
1. Simulation
2. Linear programming
3. Correlational analysis
4. Mathematical models
5. Statistical decision making
6. Decision trees
7. Bayesian analysis
8. Exponential smoothing
9. Delphi technique
10. Econometric methods
11. Game theory
12. PERT
13. Critical path method
14. Scenarios
15. Dynamic programming
16. Input/output analysis
17. Risk analysis
18. Present worth

SOME OBSERVATIONS

Most companies which are engaged in strategic planning employ a combination of the principal planning processes, although only a relative few are heavily involved in issue and surprise planning management. There are many companies still at the budgeting and long-range planning stage.

Following are some general observations that I have relative to the current state of the planning process:
- o In as complicated an arena as the business world, planning processes too often seek simple answers, although this is frequently mandated by circumstances or lack of reliable information or research.
- o Too much emphasis has been placed on the computational aspect of planning when great detail is not warranted in many circumstances.
- o Too much emphasis is placed on the accuracy (to be proved later) of a single set of numbers, or even several sets of numbers, when the important aspect of any planning process is the thought process (if done correctly) that management is forced to go through.
- o The appropriate balance between top down-bottom up planning in multi-divisional companies, while sought, is not often achieved.
- o A very common problem in most U.S. corporations is ensuring the proper use of planning staffs by maintaining close relationships with both the C.E.O. and operating management.
- o Many U.S. corporations still give lip service to long-range, strategic and financial planning when in reality the focus is too often short-term.

The head planner of most U.S. companies is the chief executive officer. The companies commonly recognized as successful in the planning world use a combination of hard data and systems (with good research) to back up the intuitive thinking (soft data) of the C.E.O. and key management personnel. Corporate and divisional management must maintain an active, leading role in the planning process. Too little or too much reliance on planning staffs and hard quantitative data can be equally harmful. The proper blend between qualitative and quantitative analysis and strategic thinking is the desired objective of any current and future planning process. If there is a current trend in planning, I believe it is back toward a more optimum balance between hard and soft data analytical approaches rather than the recent emphasis on hard, detailed quantitative analysis.

In conclusion, the following statement summarizes my philosophy:

PLAN FOR THE FUTURE
BUT
PREPARE FOR THE UNANTICIPATED

8.2

COMPETITIVE ANALYSIS

Leons Kovisars
Director, Mining and Metals

Chase Econometrics,
Bala Cynwyd, Pa. 19004

INTRODUCTION

Competitive analysis is the collection and evaluation of data on a product, a product line or on an entire industry. Much of the focus of competitive analysis centers on production volume and the price targets of the competitors. In actual practice capacity may be used in place of production volume. Similarly, unit production costs and individual producing units -- mines, mills, smelters, refineries -- may be the targets of the analysis.

The purpose of competitive analysis is to provide information for strategic planning. The analysis will locate the position of the company in the industry and define the competitive margin -- or lack of a margin -- between the company and other producers. The analysis can show the current situation and may be expanded to examine the outlook for the future. The forecast option is particularly valuable because it will show the opportunities for new project development or the expansion of current production. Similarly, the forecast will show the potential threat from new entrants, and may provide early warning to reconsider operations that may become uncompetitive in the future.

Production cost analysis is a clearly understood means to evaluate producers and estimate the outlook for their profitability. The cumulative cost for the industry outlines the supply structure and defines the marginal cost of production used in developing price forecast as explained in the subsequent section of this paper.

Competitive analysis is particularly important to the mining industry, notably to metal producers. The final product is typically a purified metal conforming to standards, and interchangable with the output of other producers, i.e. a fungible commodity. Brand loyalty is questionable in the mineral industry. Consumers will show a preference for specific brands, but this preference will be limited to a premium of several percent over the general market price. Price and assurance of supply are the keys to marketing success. Assurance of supply depends on geopolitical considerations and production capabilities. Price will be closely related to cost of production and a thorough competitive analysis should describe the cost structure of the industry. This snapshot will locate one's own position and provide an immediate and readily understandable evaluation of the hazards and opportunities inherent in this position. In the following section we shall discuss elements, methods and sources for competitive evaluation.

ELEMENTS OF COMPETITIVE ANALYSIS

The practice of systematic collection and analysis of data on competition is hardly a new concept; see, for example, Wall (1) or Porter (2). However, most discussions have been in the context of conventional industrial or manufacturing industries, with a focus on product lines and brand marketing. These considerations are not irrelevant to mining but they do have only a limited application. The key to success in the mining industry is minimizing the cost of production. However, this simple concept becomes complex when we consider relevant costs and the planning horizon.

Production costs include the factors consumed in production, namely labor, fuel, energy, materials and supplies. The cost also includes purchased and contracted services such as freight, smelting and refining, sampling and sales required to convert ore to the finished metal. Royalties and ad valorem taxes paid on gross production are direct expenses, as is the cost of ad-ministration required by the production units. The sum of these outflows is the total cash cost, the expense that has to be paid in cash in order to continue production.

Indirect costs incurred by a producer will commonly include allocated headquarters expenses, as well as the capital charges associated with equity investment and debt. Profit is not generally considered an indirect cost even though it represents the cost of using the owner's capital; similarly, income taxes represent the government's participation in the profits of the enterprise. Interest payments are considered as a direct expense in some cost studies, but many competitive evaluations would consider interest as a indirect cost related to fixed capital. Treatment of indirect costs should vary in a competitive analysis. The **total cash cost** (cash breakeven cost) is the relevant figure for comparing the existing producers. The **total cost**, which includes interest and all other capital charges, is significant in analyzing long-term viability of the company, as well as the decision to develop new production capacity, but is less important when comparing the costs of production of established producers. Economic theory focuses on relevant costs which will be different for projects at different stages of development.

Cash costs of production afford a readily understandable means for comparing different mines. Cash costs are applicable to a long-run operating decision, but they offer only limited insight for the near-term shutdown decision. The analysis of this operational decision requires the separation of variable cash from fixed cash costs; the latter may include such items as property taxes, insurance, purchases under "take or pay" contracts, and similar non-discretionary costs. Some labor costs are also fixed, and may include the "care and maintenance" staff expense, unemployment compensation, unfunded vested pension liabilities, salaries for contract labor in remote operations, and similar costs.

The cash expenses of government-controlled mines are largely internal to the country and paid in national currency. Such expense would include much of the labor and some of the other production costs for example, the costs of grinding media purchased from a national steel company. Such expenses can be regarded as fixed for the purpose of deriving a production decision.

In summary, the relevant figures for a production decision will change with the length of the planning horizon. The near-term variable costs impact the decision to operate rather than place the mine on a suspended or standby condition. A somewhat longer horizon for an operating mine would consider the total cash cost as the relevant cost. For a project in the planning or development stage the total cost, including taxes and profit, is the relevant figure for determining the economic viability of the project.

METHODS FOR COMPETITIVE ANALYSIS

The above discussion implies that production costs will be standardized and comprehensive to allow comparisons of equivalent data for the total industry. In practice, however, data on production costs are usually scarce and rarely comparable. Production costs are sensitive business information and most companies prefer to keep this data private. When cost figures are released it is usually difficult to identify what elements are included, and disaggregation into cost components is rarely possible.

It is clear that because of the above, the industry cost structure cannot be defined by straight-forward compilation. The alternatives are:

- Detailed analysis of financial and production data
- Standardized engineering cost estimation

FINANCIAL AND PRODUCTION ANALYSIS

Public companies are required to file financial reports which may contain a wealth of data on costs incurred in prouction and sales. Local and national regulatory agencies commonly require a parallel reporting of non-financial data on the actual production activities. Much of this production data is also reported by the producers in their financial statements--annual reports, 8K, 10K, 10Q and similar reports. A careful compilation and analysis of these reports can provide a very good estimate for the cost of production at specific operations

The data gathering and analysis is clearly a tedious job requiring resources that few mining companies care to expend. This approach relies upon data of variable depth and breadth of coverage and reporting. Industry analyses which rely on the evaluation of a company's published financial reports will usually be limited to the major producing mines, and a coverage of 75 to 85 percent of the industry-wide output will be typical for a "comprehensive" study using this approach.

ENGINEERING COST ESTIMATION

Engineering cost estimation abandons the attempt to extract detailed data from financial aggregates. Cost estimation will focus on the physical materials balance for each producing unit -- mine, mill, smelter, refinery. The approach is in effect project evaluation applied to the group of projects constituting the industry. The method is very flexible and can produce reasonably good cost estimates with only limited information. The basic objective of engineering cost evaluation are to:

- Determine the units of input required per unit of output (labor-hours per tonne ore milled; liters of diesel fuel per tonne hauled, etc.)

- Convert these physical input units to cash equivalents using the cost of the inputs at the work site (pesos per man-hour; US$ per liter of diesel fuel, etc.)

- Convert the local cash equivalents to standard currency — typically US-dollars — using the appropriate exchange rate

The depth of estimation can vary from rough "ballpark" estimates of dollars per tonne ore mined and milled, to detailed estimates of labor, energy and material costs by specific cost centers. For example, engineering cost analysis may differentiate the cost of production stoping in full-gravity block caving into the cost for labor, fuel for support equipment, explosives, replacement parts, drill bits and electricity. The production stoping cost center will be one of the ten to thirty operating areas specifically analyzed for the mining stage. Similar detail can be developed for concentration (milling) and for further processing. A practical exposition of this approach is detailed in Capital and Operating Cost Estimating System Handbook (3); a summary report of the results of such a set of detailed calculations is shown in Table 1.

Application of Cost Estimation

The author has applied the engineering cost evaluation method for estimation of mining/milling costs at a wide variety of mines and concentrators. Most of this work has been with copper mines but it has also been applied to nickel, molybdenum, uranium and precious metal mines. The method only considers cash costs and does not include provisions for depreciation, income taxes, incentives such as depletion or tax credits, or other non-cash costs. Expenses incurred outside the mine, or of a nature that are difficult to allocate on a per tonne basis are separately evaluated as "other expenses". Such expenses may include interest costs and allocated corporate expense. The general computational flow is to determine the annual:

- Tonnes of ore produced
- Tonnes of metal recovered

then calculate the annual total and variable production costs by major cost element for unit operations (mine, mill, support). The cost elements include:

- Labor
- Materials and supplies
- Energy and fuel

Adding the cash costs of transportation, smelting and refining and sales to the production cost for the mine and mill determined above yields the total cash cost for that particular mine. All costs and credits may be computed as periodic, i.e. annual, daily, etc., sums, allowing subsequent grouping into several cost classes -- mine/mill cash costs, total cash costs, variable cash costs — and restatement into unit costs eg., cost per tonne ore produced, cost per pound mineral recovered. Byproduct/coproduct credits may be subtracted to determine the net cash cost or the costs may be prorated to the individual mineral products.

Several details of the calculations on an annualized basis include:

- **Annual Ore Production:** determined by multiplying the effective mill capacity by the operating rate (tonnes per day x days per year)

- **Annual Metal Production:** metal grade x recovery x annual ore production. Grade changes may be included in forecasts of future production.

- **Labor Cost:** the annual labor cost is the product ration of the mine), and the annual labor cost per employee. The labor cost includes normal wages, as well as all additional expenses related to employment. These expenses would include bonuses, payroll taxes, fringe benefits, housing and medical costs; in sum, all the personnel-related expenses incurred by the operating unit. Note that this method of computation includes much of what is generally labeled as "general and administrative" costs, or "overhead."

- **Materials and Supplies (M&S) Cost:** the cost of materials and supplies is defined to include all expendable items consumed in the production and processing of ore, but excludes fuel, power, and utilities services. The usage is determined by engineering estimation of the M&S requirements for individual unit operations at mines and mills. Such unit operations include drill and blast, haulage (by specific method), hoisting, ventilation, grinding, flotation, etc. The usage is converted to local cost at local rates.

- **Cost of Energy, Fuel, and Utilities:** the cost of these consumed items is calculated similarly to the M&S cost determined above. Electric energy and fuel consumption (diesel, oil, gas) should be estimated for unit operations and converted to cost at local rates.

- **Property Tax and Insurance:** this is the local property tax, and not income or gross revenue taxes. The estimation of this tax may be based on a percentage of the undepreciated fixed plant value. The percentage varies from 1 to 3 percent; 2 percent is representative of many producers.

- **Royalty:** the royalty and ad valorem cost may be determined by applying an individual royalty rate at each mine to the net value of all metal production. Export taxes based on percentage of the value may be included in this calculation.

- **Transportation:** separate estimates for truck, rail and sea haulage should be determined for each mine using the individual haulage distances and estimated tonne per mile charges.

- **Smelting and Refining Cost:** Smelting and refining costs may be different for producers with integrated plants compared to those using custom treatment. Cost estimates should consider concentrate grade, smelter recovery, and treatment charges.

- **Sales Cost:** the cost of sales may be determined as a share of gross revenues, or as a fixed charge per unit of mineral produced.

- **Other Cash Costs:** other costs necessary for the operation of each mine unit are included in this cost element as applicable.

Cost calculation may be made on the basis of constant or fixed values for physical inputs or on a current value or inflated cost basis. The second approach allows the use of differing cost escalations for key components such as labor, energy and materials in each country.

Individual calculations should be made for each production center or mine/mill complex. By-product revenue allocation or cost prorationing will reduce the total to yield the annual net cash cost for the producing center. Dividing this figure by the annual production yields the cash cost per unit of metal produced. It is often desirable to convert from local calculations made on a current value basis to a final global comparison on a constant value basis using the appropriate deflator for expected inflation.

Coproduct And Byproduct Credits

The "right way" to allocate coproduct and by-product costs, income, and credits is often a hotly debated topic. The incremental revenue approach allocates all mining costs to the production of the primary mineral. The cost for the primary mineral also includes all other costs up to the by-product/co-product recovery circuit. The incremental income from by- product/coproduct recovery is calculated as:

Income = revenue - incremental cost incurred in metal recovery

This income is applied as a credit to the primary metal cost calculation. The credit is a cash item and thus reduces the cash cost for a producer.

The allocation of all costs to the primary mineral can produce considerable distortions in cost evaluation. Relatively small changes in the by-/co-product market prices will cause large changes in the apparent production cost for the primary mineral. A common alternative approach allocates the costs and credits on the basis of relative revenue derived from the total operation. This alternative approach prorates costs to all of the mineral products. The proration basis may be the volume or tonnage of the individual mineral products, but more commonly it will be the share of revenue contributed to the operation.

Effective versus Nominal Or Nameplate Capacity

The effective production capacity (tonnes of mineral per year) may be determined on the basis of historic production for each mine, including actual mine and mill production capacities, annual operating schedules, ore grades, and recoveries. The effective capacity should be based on normal sustainable production and would be equivalent to "typical" production recorded for a specific mine/mill. This figure should not be confused with "nominal" or "nameplate" capacity; the effective capacity commonly ranges 90-95 percent of the nominal capacity. In the absence of grade declines, the effective capacity is particularly dependent on the operating schedule, which may vary from 365 days per year for some centers to less than 200 days for problem producers.

Disruptions caused by wars, natural disasters, labor disputes, and other factors should not be included in the capacity determination. These disruptions are usually temporary and their inclusion in cost calculations for a particular year would tend to distort estimates of the normal costs of production.

Computation And Data Sources

Data is the key to the engineering cost method. The widespread availability of powerful, low-cost computers has reduced the previously formidable engineering calculations to routine programs. Several mining engineering companies, for example, Wright Engineers, Ltd., Vancouver, have developed standard in-house cost estimation systems that are available for use by their clients. These cost estimating programs are not usually available for commercial purchase. MET Research Corp. has developed a series of capital and operating cost programs which have been made available to clients; other consultants may have similar products in the near future.

The data for engineering cost evaluation must include considerable detail on the individual projects:

- Mining and concentration methods and equipment used
- Ore grades and recoveries;
- Waste to ore ratios
- Capacity and production levels
- Mining depths and haulage distances
- Labor force
- Smelter and refinery agreements

Much of this data is available in the standard mineral industry journals and publications, or may be included in public as well or private databases. Principal information sources would include:

- American Metal Market
- Engineering and Mining Journal
- Metals Week
- Mining Bulletin
- Mining Engineering
- Mining Journal/Magazine
- Skilling's Mining Review

Several of these journals publish regular summaries of mineral investment activity, which provide good outlines for many of the mining projects. In addition, a number of national and local publications are excellent sources of mineral investment information, for example:

Newsletter of the Chamber of Mines (Philippines)
Northern Miner (Canada)
Revista Mineria (Peru)

Private reviews commonly provide excellent detail that is not usually available in public sources. Most mining companies maintain extensive files on mineral projects, and such data may be expected to provide the core of the information for the project. Consultants will commonly have information on many of the significant projects for the industry in which they specialize. This data may include actual costs for specific projects, which will provide benchmarks for cost evaluations.

Data Presentation

Detailed competitive analyses will produce enormous volumes of data. Numerous options are available for tabular and graphic data presentation. Perhaps the most comprehensive is the supply curve --the graph of industry capacity versus cost. For example, the worldwide cash costs of producing copper can be plotted as the tonnages of copper that will be produced at increasing cost levels. This presentation of the copper production capacity results in a cumulative cost function which shows the tonnage of copper that can be produced at or below a given cash cost level. If, for instance, the lowest cost mine produces 100,000 tpy at a cost of 23 cents per pound, and the next lowest cost mine produces 150,000 tpy at 25 cents per pound, the next 100,000 tpy at 29 cents per pound, then the cumulative cost schedule is as follows:

Cash Cost (US cents per pound)	Production Capacity (tonnes per year copper)
0 — 22.90	0
23 — 24.9	100,000
25 — 28.9	250,000
29 — -----	350,000
etc. etc.	

An example of such a supply curve is shown in Figure 1. The example uses prorated cash costs as the vertical axis and the cumulative production capacity as the horizontal axis (the actual numerical values have been deleted). The value of this graph for defining the structure of the industry and the competitive position of any specific mine is immediately apparent.

The extention of the engineering cost evaluation to future operations allows consideration of:

.Cost changes at current operations
.Competition from new production capacity
.Depletion of reserves or production suspensions at current operations

Recalculation of industry production costs for future operations at each mine may include the effects of grade changes, depletion and change in by-product values. New mines may be included based on the available estimates for the completion of projects currently under development or in the planning stage. Costs for these projects may be estimated on the same basis as that used for the current producers. The more tenuous projects should be regarded as general estimates of new project development, and are very likely to be subject to major revisions in the future.

EVALUATING IMPACTS OF STRUCTURAL CHANGES ON COMPETITION

The decade of the 1970's brought about large changes in mineral production costs, rearranging the competitive positions of many producers and major producing regions. Costs increased for all producers but the differing impacts of energy and labor price escalations scrambled the positions of producers in the industry cost rankings. An extraordinary combination of economic and political events occurred during the 1970's:

.The **energy shocks** of 1973 and 1978 gave a competitive advantage to producers in regions with lower energy costs, such as the United States, and accelerated the expansion of more efficient treatment processes such as flash smelters.

.**High inflation** impacted all producers, but particularly those in high-cost labor areas.

.**Exchange rate** movements, including devaluation and revaluation of currencies.

.**Restrictive anti-pollution legislation** in most industrial nations has led to sharply increased smelting and refining charges for material treated in these areas.

The impact of these factors caused costs to double in the United States copper industry between 1972 and 1977, even though the general inflation during the same period increased less than 60 percent. Similar increases in production costs were experienced by many copper producers outside the United States, as well as by the producers of other metals and minerals. By early 1984 inflation had abated in the industrial nations and even in most less developed countries. The impact of variable inflation on individual mineral projects and competitive positions within the industry can be readily evaluated using engineering cost evaluation.

COMPETITIVE ANALYSIS AND MARKET PRICE FORECASTING

A common approach to evaluating prices in the future focuses on the price required to encourage investment in new production capacity. This "target price" is defined to be equal to the total cost-- cash cost plus capital charges -- of the future marginal producer. This approach requires the definition of the geological, mining and financing conditions of the future high- cost producer. The validity of this "target pricing" approach is questionable, but it is worth examining the method applied to a well-studied metal -- copper.
Copper deposits in North American — Canada and United States -- have been commonly used as models of the highest cost producers. The recent focus of development activity has shifted to Latin America, suggesting that a marginal deposit in a country such as Chile, may be a more realistic model for the target pricing evaluation.

The operating costs for the model copper producer will be assumed to be comparable to the costs incurred by similar mines and processing plants currently in production. Capital investments for a mine/mill complex sited in North America in the early 1980's are estimated to be about US$8500 per daily tonne of ore milled. Comparable projects in Chile would exceed US$11,000 per daily tonne due to higher infrastructure costs. Investment for a new smelter/refinery would be about US$5500 per annual tonne copper treated.

We will consider two projects, one in North America and one in Chile. The project parameters will be comparable except that the grade for the Chilean deposit will be 0.85 percent copper versus 0.50 percent in North America.

Additional differences between the North American and Chilean models are:

- The higher grade and recovery in Chile boosts copper output 1.7 times

- Higher concentrate grade in Chile will reduce smelting costs and capital investment

- Loan terms are expected to be similar -- two-thirds of financeable investment for a twelve year term -- but the interest rate will be 12 percent in Chile vs 11 percent in North America. We should also expect that the loan term may be shorter for the Chile example-- say, 8 to 10 years versus 12 years.

- The owners will require a higher rate of return as a risk premium for the Chilean project -- 17 percent vs 15 percent.

The estimates shown in Table 2 are only generalized approximations, but they are comparable in rigor to the estimates used in other "target pricing" models, and should demonstrate this application of engineering cost evaluation to competitive analysis.

The totals in Table 2 indicate that a potential investor in a marginal Chilean project will require reasonable assurance that the long-run copper price will average at least $1.70 per pound; the comparable North American model will require almost $2.00 per pound. Two conclusions are apparent:

- If the "target price" concept is correct then copper prices will have to triple from the levels shown in 1983-1984

- If the concept is invalid, then these marginal projects have a very good chance of being subeconomic. We do not intend to argue the merits of either alternative, but the impoverished state of the copper industry, in the early 1980's is emphasized by the desperate straits of the new projects with their heavy debt loads. Clearly the planners expected financial success, but global economics, technological changes, and competition yielded an outcome radically different from the planning targets.

An alternative approach to price forecasting will seek to develop the industry supply curve (cost curve) for the future periods. Engineering cost evaluations can be developed for the individual projects expected to be operating in the future-- including the marginal projects expected to be developed. The future price will be determined by the marginal producer, i.e., the last mine required to satisfy the future level of demand for that mineral. Clearly, this approach requires more data and effort than the use of the "target price:" however, the basic component of analysis is the same: competitive evaluation using engineering cost evaluation.

REFERENCES CITED

1. Wall, Jerry L., 1974, "What the competition is doing: your need to know.", Harvard Business Review, November/December, p.22-166.

2. Porter, Michael E., 1980, Competitive Strategy: Techniques for Analyzing Industries and Competitors, The Free Press, New York

3. STRAAM Engineers, Inc., 1979, Capital and Operating Cost Estimating System Handbook, US Bureau of Mines Contract J0255026

Table 1. Example of Production Cost Estimation
All Costs in Constant Mid-1983 US$

PRODUCTION SUMMARY

million tonnes ore	4.0
million pounds molybdenum	8.9
thousand tonnes molybdenum	4.0
tonnes copper	0.0

COST SUMMARY

	Million $/year	Ore + Waste $/ton	Ore $/ton	Mo $/lb
MINING				
labor	10.44	1.12	2.62	1.17
energy	3.18	0.34	0.80	0.36
material & supplies	6.02	0.64	1.51	0.68
subtotal	19.64	2.10	4.94	2.21
MILLING				
labor	5.93	NA	1.49	0.67
energy	6.08	NA	1.53	0.68
material & supplies	5.02	NA	1.26	0.56
subtotal	17.03	NA	4.28	1.91
OTHER COSTS				
freight	0.64	NA	NA	0.07
roasting	2.81	NA	NA	0.32
sales & other	1.60	NA	NA	0.18
TOTAL CASH COST	41.72	NA	NA	4.69
byproduct credits	0.00	NA	NA	0.00
NET CAST COST	41.72	NA	NA	4.69

Table 2. Engineering Estimation of Copper Production Costs.

	Cents per Pound Copper (1982 US$)	
Cost item	North America	Chile
CASH COSTS		
mining and milling	52	40
smelting and refining	20	16
freight	6	8
other	4	5
subtotal	82	69
CAPITAL CHARGES		
depreciation and amortization	24	20
interest	8	7
other	4	1
income taxes	31	29
return on equity (profit)	47	44
subtotal	114	101
TOTAL COST	196	170

1. COPPER PRODUCTION COSTS

cents/pound copper

tonnes copper

8.3

ASSESSING STRATEGIES FOR
NATURAL RESOURCE COMPANIES

by

Breaux B. Castleman
Vice President
Booz, Allen & Hamilton Inc.
Houston, Texas

The ultimate test of a business strategy is whether it achieves the owner's objectives. Most shareholders are interested in growth and profits, and share values on stock exchanges are very often closely correlated with profit growth and profitability measured as return on equity (ROE). Strategic management, then, involves managing investments or resource allocations to optimize long-term growth and real return on equity, within reasonable limits of risk.

Figure 1 shows a plot of the five year, weighted average ROE of a number of communications companies against their stock market valuations in 1983, measured as a ratio of market value to book value. The principal explanation of the differences in the relative valuations of the individual stocks is historical ROE, weighted toward the most recent years. Further analysis demonstrates the added importance of growth in explaining the differences in stock market to book ratios.

Research conducted by Booz Allen and others in recent years suggests that the relationships between ROE and market to book multiples hold true in industry after industry. Furthermore, we have found that in assessing business strategies for companies in these industries, the analyses should focus on whether proposed strategies are likely to lead to consistent long-term growth in real ROE, without excessive leverage.

Our research also reveals, however, that these approaches to strategy assessment are not consistently applicable in the natural resource extractive industries. Figure 2 shows a similar comparison of market to book values against five year average ROE for a group of nine mining companies in 1983. Market value of the shares shows little correlation with ROE.

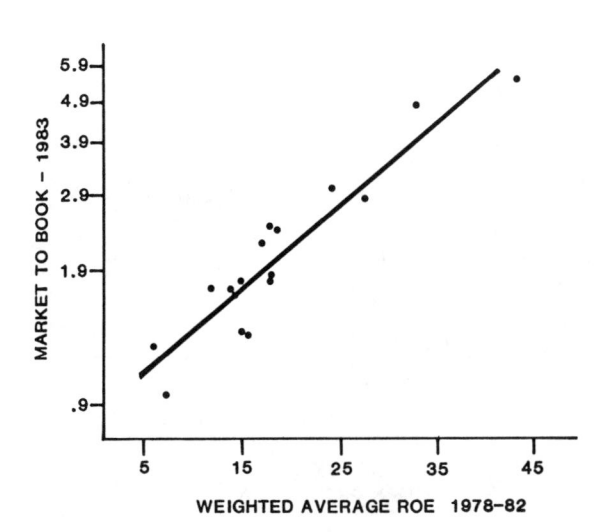

FIGURE 1. COMMUNICATIONS INDUSTRY

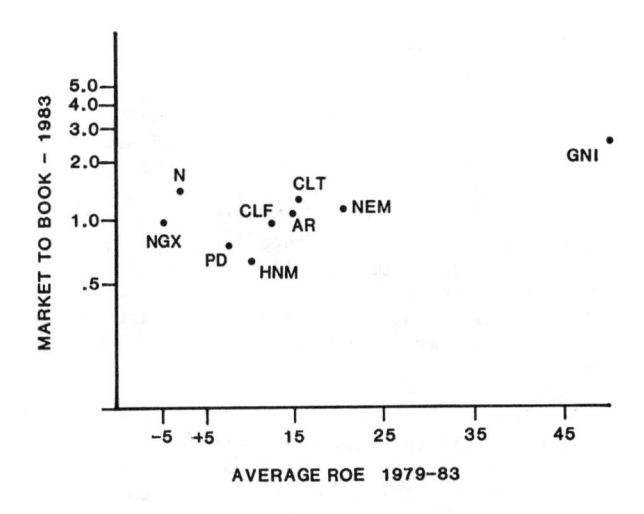

FIGURE 2. MINING INDUSTRY

Analysis of the data for other extractive industries indicates a similar pattern; that is, low correlation between share values and historical ROE. This suggests that strategies for natural resource companies should be developed and evaluated differently from those of industrial companies. Why this is so and which elements in the environment and in the company are significant in strategy assessment in the natural resource industries are explored below.

GLOBAL COMMODITIES AND SHAREHOLDER VALUE

A quick scan of the business environment in which extractive industries operate reveals four important factors which influence profit, performance and value.

First, most natural resources are true commodities. With but minor exceptions, it is almost impossible to achieve strategic advantage through differentiating the product. Nor is it possible to add real value to the product at the extraction stage since, by definition, the product is a natural resource. The consequences are that profitability in the industry as a whole is tightly linked to movements of the market price of the individual mineral. As in other industries, profits of most similar companies tend to rise and fall together, but this is much more pronounced in natural resource businesses.

Second, most extractive industries are fully mature. The rapid expansion of industrial capacity has slowed to a crawl. This means slow or no growth in minerals demand which is driven by real growth in the manufacturing sectors of the economies of industrialized and industrializing nations.

Third, extractive industries are global in scope with numerous participants, many of whom are government owned companies. During recent times, capacity to produce has increased faster than demand in many natural resource markets. Moreover, some government controlled companies have national objectives (to generate foreign exchange or employ their citizens) that result in continued capacity expansion and production at uneconomic price levels. This situation describes a "perfectly competitive" global market, some of whose participants behave "irrationally" for political purposes.

Fourth, natural resources are increasingly viewed as national treasure by the countries where deposits occur. This passion for ownership, combined with modern economic theory and confiscatory taxes, tends to limit profitability during periods of prosperity and increase costs during periods of economic distress.

The combination of these four elements defines a business environment of slow growth with rising supply in the form of excess capacity, and price fluctuations tied to general industrial cycles. As a consequence, long-term downward pressure on prices for most commodity minerals is moderated infrequently by upturns in the business cycle or by concerted action to restrict supply. Moreover, the industrial dynamics and political tax pressures in global extractive industries are such that the negative aspects of the situation are amplified.

If this assessment of the situation is accurate, one would expect to find that mining shares tend to sell at a discount to other, more attractive industries. A quick comparison of the market to book ratios in Figures 1 and 2 confirms that. Further, one observes that the opportunity to achieve high, sustained market to book ratios is greater in the fast growing, high profit telecommunications industry where market to book ratios of 2.5 to 6.0 are not unusual.

Finally, one would expect that when investors expect an upturn in the global business cycle they anticipate real increases in most commodity prices. Given the extractive industry dynamics referred to above, one would expect efficient capital markets to anticipate improved performance and to bid up the value of mining company shares. In fact, Figure 3 shows that during the recovery in the U.S. economy in 1983, the market to book ratios of most mining company shares in our sample increased handsomely, though real ROE's improved only marginally or not at all.

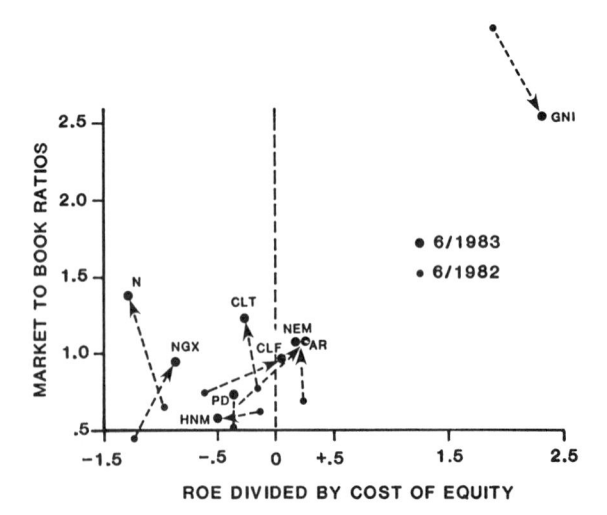

FIGURE 3. MINING INDUSTRY - SHARE PRICE DYNAMICS

The importance of these macro economic effects should not be minimized in assessing viable strategies for natural resource companies. It is extremely difficult to devise realistic strategies to increase shareholder wealth under these conditions.

VALUE CREATION DILEMMA

Unlike most industrial companies, extractive industry companies create value through the discovery and exploitation of natural resources. If such a company has demonstrated a capability to explore for and consistently increase its reserves of minerals, it will be perceived as capable of growth and value creation.

This same company must then demonstrate the ability to profitably convert those reserves to cash either by sale in place or by development. The more successful an individual company in these two activities, the more likely it is to receive a relatively higher market valuation. Great Northern Mining's (GNI) position on Figure 2 is based on investor expectations that GNI can continue to discover new reserves and earn profits at the 45% ROE level. Figure 3 indicates a slight decline in investor confidence between 1982 and 1983 as to whether GNI can continue its performance. This loss of confidence in GNI resulted in a drop in its market to book ratio despite its own increasing profitability and rising industry expectations.

This phenomenon of share values tracking closely with discovery rates or opportunities for discovery of new reserves is particularly pronounced in smaller, nonintegrated mining and oil companies.

Interestingly, when the mining industry has a succession of profitable years, the companies tend to be evaluated more on the basis of comparative returns on equity. Figure 4 shows the market to book versus five year average ROE for our mining group for the relatively prosperous 1976-80 period.

Achieving a consistent growth rate with adequate returns over long periods is very difficult when the inherent profitability of the industry is low. The more low return assets one obtains, the more difficult it is to grow and maintain margins.

Faced with this dilemma, some extraction companies, notably oil companies, find themselves resorting to radical restructuring. They spin off producing properties and exploration activities in an effort to realize stock market values for their profitable, growing businesses while minimizing the impact of low return refining and marketing activities.

Such companies as Transco Energy and KN Energy have successfully spun off exploration and development activities, to be followed by sharp increases in the combined share values. Those companies which failed to restructure rapidly have found their share prices doubling as they were acquired (Gulf Oil).

The need to restructure and downsize an enterprise creates a second dilemma for the management of a natural resource company because of the enormous capital requirements of the business. Still, this type of restructuring is driven by external market forces and its strategic criticality must be assessed before turning to operating strategies.

CRITERIA FOR ASSESSING RESOURCE ALLOCATION STRATEGIES

Having achieved an appropriate overall structure, natural resource companies must develop operating strategies that will optimize growth, profit and risk over long periods of time. Strategic management in this situation involves allocating resources to high profit, high growth mineral exploration and development projects.

In minerals exploration, strategy involves allocating resources to a portfolio of exploration opportunities that, when combined, produce predictable cash flows and profits for reinvestment. For example, in petroleum exploration, the basic competitive arena is a hydrocarbon producing province which, for simplicity sake, can be termed a basin. Some such basins are inherently more attractive than others because of success rates, size of finds, costs per barrel, and flow rates. Figure 5 shows an assessment of gas producing basins. Those basins with large initial potential, shallow depths and high exploration success are deemed as most attractive.

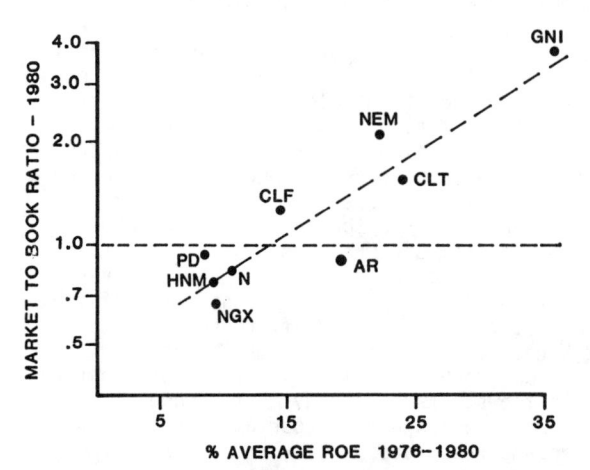

FIGURE 4. MINING COMPANIES - 1980

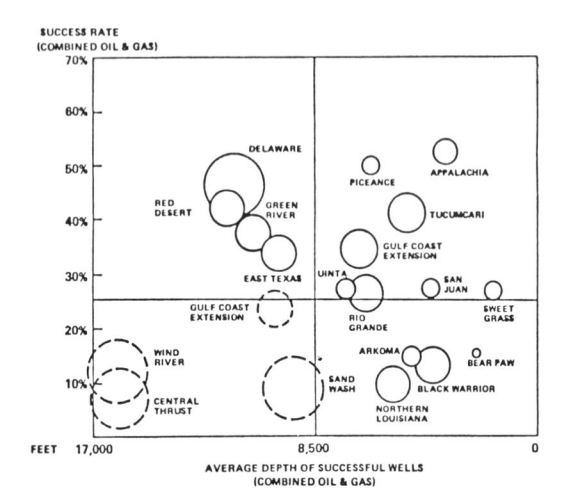

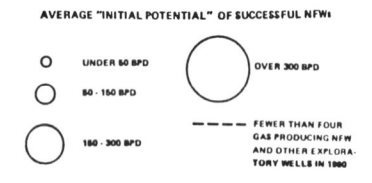

FIGURE 5. PETROLEUM BASIN ASSESSMENT

This analysis of a company's current and prospective basins can be used to assess its profit potential. Of course, this is simply another way of saying that higher grade ore bodies and larger mines reduce the cost of extraction. In other natural resource industries, this analysis can be done by assessing each major existing project and property for its economic potential. Comparisons should be made with industry averages and likely competitor performance. Figure 6 resulted from such an analysis. It shows rapidly declining cost per unit of uranium extracted relative to ore grade.

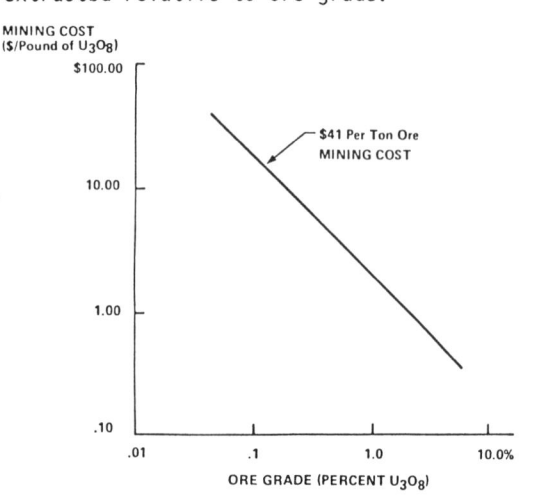

FIGURE 6. EFFECT OF ORE GRADE ON MINING COST

Furthermore, economies of scale favor the company mining a comparable grade deposit of larger size. Figure 7 demonstrates that higher capacity salt mines in the U.S. in 1980 achieved vastly greater productivity than smaller mines.

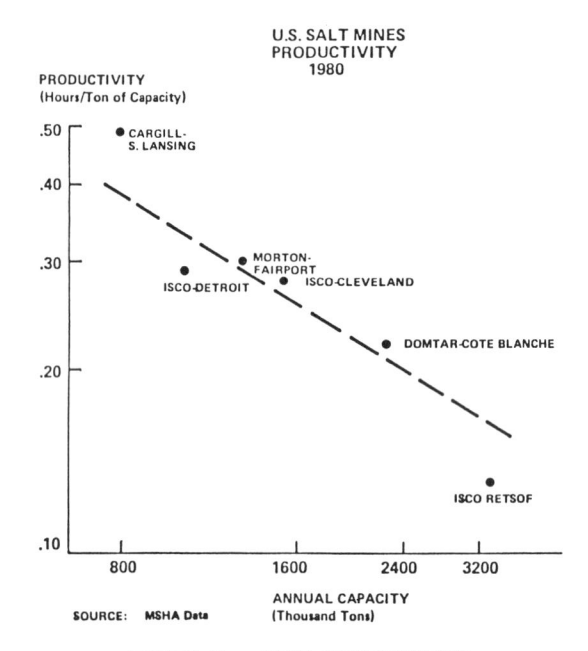

FIGURE 7. MINE PRODUCTIVITY COMPARISONS - 1980

These kinds of economic assessments of a company's undeveloped and developed holdings are desirable to begin to determine potential operating economics.

These kinds of assessments can be done for oil or gas or other minerals provinces and projects using a variety of techniques. The objective of the analysis is to determine whether a company is exploring in provinces where success will be translated into comparatively superior profits.

Further, it is important to know the profit and cash flow profile of individual basin programs in order to develop a rational resource allocation strategy. For example, Figure 8 shows a simplified comparative cash flow profile for a large offshore new field discovery and a smaller onshore find. The offshore development requires the building of an expensive platform and produces negative cash flows in the hundreds of millions of dollars over eight to ten years. Only a consortium of the largest companies could afford such an undertaking. By contrast, the smaller onshore prospect requires only a few years of negative cash flows. Each company must choose prospects which are appropriate for its own size and resources.

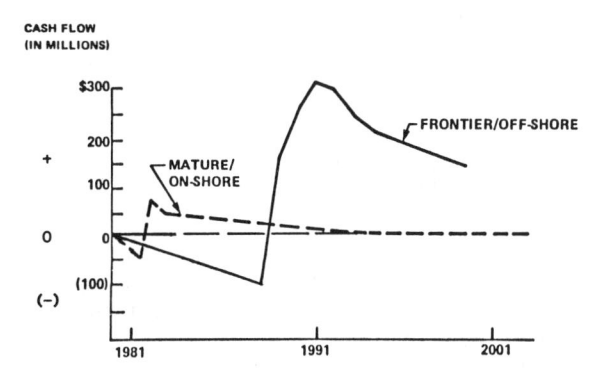

FIGURE 8. COMPARISON OF CASH FLOW PROFILES

The sum of all the individual wells determines the cash profile of the basin. Having selected basins which contain attractive prospects with proper cash flow characteristics, the next criteria for allocating resources involves balancing the timing of the overall funds flows of all the basins.

If an exploration company works exclusively in frontier areas looking for large deposits with long pay backs, it will quickly exhaust its investable funds. This can be avoided by balancing frontier investments with those in more mature, positive cash flow basins. Figure 9 shows the results of years of insightful investment in basins in various stages of the petroleum life cycle. The combined cash flow characteristics of basins produce a balanced portfolio.

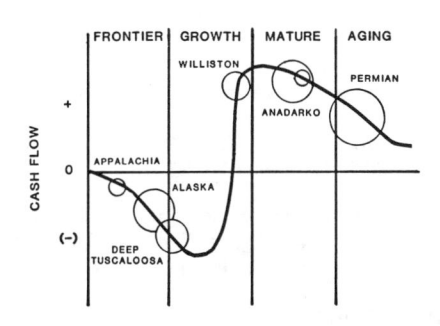

FIGURE 9. BALANCING THE PORTFOLIO OF BASINS

Then, having selected basins with the appropriate financial characteristics, the exploration company must evaluate its ability to achieve exploration success. In the oil exploration business, the strengths required vary according to the basin. Figure 10 summarizes and contrasts typical criteria for assessing exploration strengths in frontier and aging basins.

	FRONTIER	AGING
UNDEVELOPED LEASEHOLDS	●	●
GEOPHYSICAL INTERPRETATION	●	◐
"CREATIVE" GEOLOGISTS	●	○
GEOLOGISTS EXPERIENCED IN BASIN	○	●
CRITICAL MASS OF TALENT	●	●

FIGURE 10. CRITICAL STRENGTHS CRITERIA

Our experience suggests that a successful petroleum exploration company must have an inventory of low cost leaseholds; the ability to interpret geophysical intelligence; creative geologists in frontier areas where there is no control; geologists with extensive experience in mature areas where a premium results for knowing the area; and a critical mass of exploration and operating talent that permits the company to bring more "horsepower" to bear in evaluating opportunities quickly.

These criteria for assessing corporate strengths and weaknesses by basin will vary depending upon the particular natural resource being sought. Conducting this type of analysis, however, is important to determining one's ability to achieve a low cost position in the resource.

With these assessments of the basins and the company's competitive position completed, it only remains to produce an investment policy matrix. Figure 11 provides an illustration of such a matrix for a typical petroleum exploration company. Obviously, investment policy in this mythical company would favor attractive prospects in frontier areas (with high growth potential) where the company has a strong competitive position (Appalachia). Maintenance investments must be made in the mature Permian Basin to continue to produce cash flows for frontier plays.

This company should probably improve its cost position in Alaska and probably divest its holdings in the Anadarko to fund more attractive opportunities elsewhere.

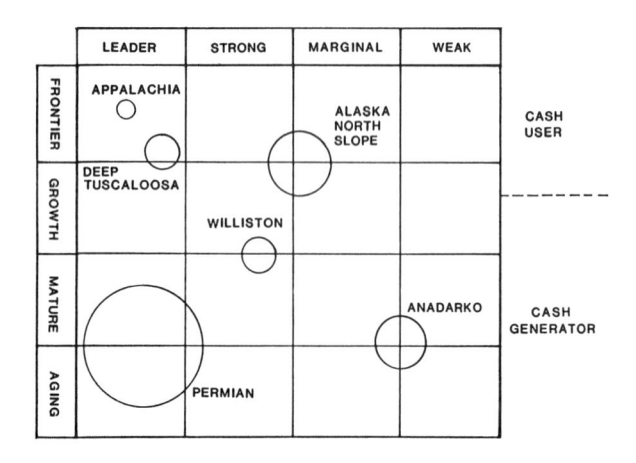

FIGURE 11. EXPLORATION COMPANY ILLUSTRATIVE POLICY MATRIX

Booz Allen has found that conducting this type of structural assessment provides valuable insights into the strategic health of natural resource enterprises. After conducting such an analysis and making decisions regarding the portfolio of resource basins and their roles in the company's future, management typically turns its attention to improving operations and lowering costs of production. These types of investments should also be evaluated in the context of the investment policy matrix. The wisdom of these investments should be evaluated based on their ability to improve competitive position in an individual basin.

Corporate investments in exploration and development technology and mining process technology and equipment are all designed to improve a company's productivity and, hence, its overall cost production. A favorable cost position permits a natural resource producer to limit losses in periods of depressed commodity prices and maximize profit in periods of prosperity. Since such a producer is at the mercy of powerful market forces, he must necessarily focus his strategic management on achieving a low cost position in selected mineral provinces--while continuing to grow. In this manner, the company is most likely to optimize shareholder wealth.

SUMMARY

Successful strategies for natural resource extractive companies should be designed to increase shareholder wealth. This is most likely to be done by achieving a combination of growth in reserves and high sustained returns on equity.

Because the natural resource industries produce true commodities under nearly "perfectly competitive" market conditions, there is little opportunity to differentiate one's product and achieve superior price realization. Moreover, the highly cyclical and slowly growing demand for commodity minerals virtually ensures periods of low or no profitability for all but the lowest cost producer.

Under these conditions, strategy assessment concepts call for identification of those critical factors that are most likely to lead to establishing low cost reserve positions. An example of analytical tools for use in evaluating petroleum exploration strategies was given. These included procedures for understanding

- Basin Economic Characteristics
- Project or Program Cash Flows
- Basin Life Cycles and Funds Flow Characteristics
- Factors Leading to Successful Exploration
- Resource Allocation Tools

Similar techniques are used in other extractive industries. All are designed to optimize resource allocation to achieve the twin strategic objectives of consistent long-term growth and profitability.

NOTES

1) Stock symbols used on exhibits

   AR - ASARCO
   CLF - Cleveland Cliffs
   CLT - Cominco
   GNI - Great Northern
   HNM - Hanna Mining
   N   - Inco
   NEM - Newmont Mining
   NGX - Northgate Exploration
   PD - Phelps Dodge

2) Cost of Equity calculation

   COE = 90 day TBill rate + stock beta (7.6 equity premium)

8.4

MINERAL INDUSTRY ACQUISITION ANALYSIS

Henry J. Sandri, Jr.

Burlington Northern Inc.

ACQUISITION JUSTIFICATION

There are many reasons why companies consider acquisitions as a means to expand their operations. Some wish to reduce the impact of business cycles on their businesses. Others seek firms within the same industry for expansion and increased market share. Still others diversify in order to avoid takeovers. Whatever the reasons or justification, an evaluation of an acquisition candidate should be systematic and thorough.

To many financial managers the concept of a systematic acquisition evaluation indicates the need for a regular set of evaluation criteria. While these are available from numerous sources, one basic parameter frequently overlooked is the application of these criteria to an industry which is different from the acquirer. This becomes particularly important when evaluating firms within the mineral industry. The mineral industry is significantly different from other industries to require different valuing techniques.

This paper will present a number of reasons why an acquisition candidate in the mineral industry should be evaluated differently from other acquisitions, including by other mineral firms. Additionally, this paper will review the most appropriate means of valuing a mineral firm.

Firms considering the acquisition of a mineral firm should understand the nature of the industry before ever making a tender offer. Some of the general considerations include the following:

° Level of Capital Intensity: In the mid-1980's the start up cost for a gold mine has been between 20-200 million dollars and 500-1,500 million dollars for a bauxite operation.

° Global Nature of Competition: Except for certain industrial minerals, most minerals and metals are traded worldwide and many on national exchanges, such as COMEX.

° Start-Up Time: Many mining operations in developed and developing countries have long lead times, 7 to 10 years in some cases, prior to the start-up of production.

° Long Term Nature of Investment: For most mineral operations payback usually occurs at least 7 years after production has started. Additionally, many mining operations will continue producing for 80-100 years.

° High Level of Risk: The number of factors that a mining firm cannot control such as grade, volume, world supply and demand, plus the long payback periods tend to classify mineral firms at a higher than average risk compared to other investments.

° Cyclical Nature: The mineral industry is cyclical in nature, experiencing long troughs and explosive peaks, thus making timing a critical issue.

° Labor Intensive: Even with the large capital requirements needed to operate a mine, most mines are labor intensive in proportion to their capital equipment.

° Location: Obviously, the nature of this industry requires operators to work in difficult locations, frequently far from civilized localities, in poor weather conditions and without proper supply and distribution infrastructure.

One of the most important considerations is an understanding of asset replacement and asset value. Unlike manufacturing concerns, firms in the mineral industry are working toward the elimination, through depletion, of their asset base. Product flexibility is not available to a mining operation. A mineral firm basically has two options as its resource base is depleted: 1) replace the asset with new production, or 2) deplete the asset and close down operation, as in the South African gold mining model referred to later in this paper. An understanding of the asset replacement needs will substantially benefit firms who seek mineral acquisitions.

The concept of asset depletion is not applicable to most industries, especially manufacturing firms. Manufactured products follow typical life cycles which result in the eventual maturation and elimination of a product. Upon reaching the maturity stage of the product, the manufacturing firm can either find a new product, expand to new markets or shut down operations. The life cycle for the product is driven by its demand and its ease of substitutability. If during the life of the product, raw material supplies become scarce or unavailable, the manufacturer typically seeks alternate sources of supply or substitute materials. On the contrary, mining firms must seek new operations or shut down when their supply of material is depleted. This concept of needed replacement of the mineral asset base is reflected in the depletion allowance applicable to mineral extraction firms.

In order to prolong the eventual shut down requirements faced by firms, acquisition and exploration are used as replacement vehicles. Both are accepted methods of asset replacement as well as asset portfolio building. As with manufacturing companies, large mining firms tend toward the acquisition route while smaller concerns employ the exploration approach (large manufacturing firms buy a higher proportion of new products than invent them, just the reverse of smaller firms). In general, less than 25 percent of all new mining properties are developed by the original exploration company.

Reviewing the number of large and small acquisitions in the mineral industry in the late 1970's and early 1980's, the largest proportion have been made by mineral firms, followed by companies in the natural resource business other than mining (energy, forestry, real estate) and lastly followed by non-resource firms. The asset replacement need and/or the understanding of this concept, primarily by the other resource firms, has probably been a prime motivator in these acquisitions.

There are three general categories of reasons for making mineral industry acquisitions. In some circumstances the reasons for an acquisition can fall under more than one category. The reasons exemplifying these categories have been pared down from the normal list in order to reflect or highlight why mineral acquisitions should be reviewed with a different attitude than other acquisitions. The three categories are property replacement, business, and managerial in nature.

Property Replacement Reasons

These reasons support acquisitions that directly benefit the buyer by extending the life of their current operations through acquiring similar type properties. Additionally, firms seeking to add properties to a mineral asset portfolio tend to use these motives. The following list contains frequently cited reasons for property replacement acquisitions.

° Extend corporate life

° Extend operational longevity

° Replace depleated assets

° Secure inventory of properties

° Increase operating portfolio of properties

Business Reasons

Business reasons tend to support acquisitions that directly benefit the buyer's business with enhanced corporate earnings. This type of acquisition is expected to provide growth, increase profitability and, potentially, solve problems of the buyer and seller. Frequently these motives support the synergistic potential of the organizations and it is generally contemplated that the newly acquired business will operate at least as well as previously. In many cases both of the parties expect that the resulting entity will be stronger than both of the original organizations taken separately. The following business reasons is typically considered in a mineral acquisition.

° Extend an existing product line

° Acquire a broader market

° Better control supply

° Secure necessary assets

° Obtain key personnel

° Combine operations and take advantages of economies of scale

° Acquire cheap undervalued assets

° Speculate on price plays

Management Reasons

This third category revolves around the motivational factors for a company or individuals and how they will benefit from an acquisition. In numerous cases the acquisitions are prestige related and are not based on sound financial principles. Many of these acquisitions are predatory and do not fit with current lines of business. In some instances the actions by a few participants within one industry can lead to a near merger frenzy, defying all business rational. Examples of management reasons for mineral acquisitions include:

° Executive status, compensation, commitment or other reasons benefitting the executive management and board

- Financial transfers such as moving capital from one country to another or liquidating acquired profits

- Conglomerating or diversifying for the sake of doing so

- Deter potential takeovers by pursuing acquisitions

The acquisition of a mineral firm can add to the portfolio of an acquiring firm in numerous ways, even if it does not have any mineral industry background. The more common reasons supporting the concept of a mineral industry acquisition, other than asset replacement, are listed below.

- Speculation: Due to the cyclical nature of the industry, firms may acquire mineral assets as a diversified investment with the idea of turning these assets as prices reach cyclical highs (numerous mining properties were sold at exorbitant profits when gold and silver exploded in price in 1979 and 1980).

- Return on Investment: Mineral company acquisitions, when made economically, should provide a 7-12 year payout for the purchase including premium payments over existing market values.

- Diversification: Despite the current down market, mineral firms offer asset diversification for both mineral and non-mineral firms.

- Strategic Positioning: Increases in world tension, the potential for bankruptcy in numerous third world nations, and increases in strategic stockpile purchases could lead to higher than average returns for strategic mineral investments.

- Security: It is generally a safer investment to acquire domestic assets compared to foreign assets or certain other domestic investments.

- Entry Cost: Acquiring a mineral firm can be cheaper than starting in a new industry. The biggest problems facing new investors in the mineral industry are the high cost of reserve acquisition, lack of industry experience and an inability to alter production goals to changing markets.

- Taxes: Due to the depletion allowance and high capital equipment needs, resulting in high depreciation, mineral firms may benefit a corporate tax position better than other industrial income streams or tax shelters.

ASSET VALUATION

Any and all of the reasons above can be used to justify an acquisition; however, none may be strong enough to compel a company to acquire another operation. Even the asset depletion reason requires a firm to determine that it will not follow the South African gold mining model, which closes down operation as it depletes a mine. Assuming a firm has determined that a mineral company is an acceptable acquisition target, the next problem arises. The problem is how to properly evaluate a mineral firm. In general, there are numerous methods of evaluating an acquisition candidate. These may include some type of book value analysis, a discounted cash flow approach or asset valuation to name a few. While the methods chosen usually reflect the confidence and historical use of the evaluator, some methods are more applicable than others. For example, the concept of book value may be very applicable for manufacturing operations, but it doesn't work well on natural resource assets that may have been acquired 30 to 50 years previously. The following three valuation methods are recommended for evaluating a mineral firm. While each of these has negative implications, they tend to better represent the value of a mineral firm than other methods.

- Discounted Income: This method discounts future net income of the operations. This method is more appropriate for mill and refinery complexes than the resources of mining operations. It is most appropriate for overseas operations where ownership and stability may be a problem (it may also be appropriate for certain restricted domestic properties, such as Alaskan operations).

- Discounted Cash Flow: This method expands on the discounted income model by including depreciation, depletion and cash flow items annually generated. This system is most appropriate for large industrial complexes such as steel mills. However, it does not properly consider the value of mineral reserves beyond the time period used for evaluation.

- Asset Valuation: This method calculates the present value of mineral reserves, fixed assets and investments less long term debt. This method is most appropriate for mineral firms primarily involved in mining activities.

A chart of the applicable use of these techniques against various mineral types is presented in Table I, following this page.

Discounted Income Model

This approach relies almost exclusively on the income statements of the company under review. The income statements must be taken for the next 15 years of operations and all costs, profits, margins, etc. must be forecast over that period. The net income from these forecasted income statements are discounted at an appropriate discount factor in order to determine the present value. When properly done, each item of cost and pricing should be forecasted based on it's future expected value, not only the inflation rate. Another major consideration is economic changes that may

Table 1. Mine Type and Applicable Evaluation Method

Mine Type	Discounted Income	Discounted Cash Flow	Asset Valuation
Precious			
Mines	Poor	Fair	Good
Mills	Fair	Good	Good
Base			
Mines	Poor	Fair	Good
Mills	Fair	Good	Fair
Ferrous			
Mines	Poor	Fair	Good
Mills	Fair	Good	Poor
Strategic			
Mines	Poor	Fair	Good
Mills	Fair	Good	Poor
Industrial			
Mines	Poor	Fair	Good
Mills	Fair	Good	Good
Uranium			
Mines	Poor	Fair	Good
Mills	Fair	Good	Good
Coal			
Mines	Poor	Fair	Good
Foreign			
Mines	Good	Fair	Fair
Mills	Good	Fair	Fair

impact the output of the operation. Additionally, equipment replacement assumptions can dramatically alter an operation's net income.

This method is most applicable for operations and companies where control is shared or in question. For example, the Discounted Income approach is appropriate for the following:

° All foreign operations,

° Minority equity positions in domestic operations,

° Partnerships and joint ventures, and

° Annuities, production payments, or royalty income.

Discounted Cash Flow Model

The traditional Discounted Cash Flow approach is also an acceptable method for valuing mineral companies, especially those lacking large mineral reserves, or those which are considered downstream processors, such as steel companies. However, for mining companies or firms owning large mineral reserves, this approach provides no method to evaluate the value, in either current or future terms, of the mineral reserve base. In order to use the Discounted Cash Flow approach, an analyst should modify the system to somehow include a reserve valuation.

Asset Valuation Model

This approach basically compiles the value of the asset base of the mineral firm less its debts. Because of the asset replacement requirements and a need to understand mineral assets, this method is considered one of the best in determining the value of the reserves. In order to carry out an Asset Valuation calculation, an analyst must understand the differences between proven, probable and potential ore reserves. This is because the Asset Valuation technique, while acknowledging the importance of probable and potential reserves, is based on proven ore reserves.

Proven ore is ore which has had tonnages and grades computed from detailed sampling. The ore tonnages have been calculated from drill holes, outcrops and surface and underground operations in such a manner as to block out ore bodies on at least four sides. The resulting ore body should be well defined in geological and grade content as well as physical size and shape. Additionally, the ore body must be able to be mined by the current means employed. If an ore body is well delineated but mining infrastructure is not available for production the ore should not be considered proven. The valuation of proven reserves should be based on the current extraction and recovery methods. For example, if a precious metal mine is recovering only 60 percent of the metal available compared to similar operations extracting over 85 percent, the valuation recovery factor should be 0.6 not 0.85 (however, the acquiring firm should consider the 0.15 differential in recovery potential as a bonanza after acquisition if it plans to upgrade the recovery process). Additionally, if an ore body contains currently sub-economic grade material that, with reasonable assumptions on supply, demand and price, will shortly increase to ore grade material, the sub-economic material should be defined as proven (a probability factor may be placed on the sub-economic ore to reflect this price speculation).

Probable ore is ore which has been measured and evaluated but not at the degree of certainty associated with proven ore. One common estimate of probable ore is that its dimensions have been blocked out on three sides. Frequently a definite grade can be assigned to the ore, however, proper sampling has not been carried out for confirmation purposes. Usually this ore can be reclassified proven by additional orebody delineation or by improving the existing mining or extracting systems. Probable ore reserves are not used in calculating Asset Value because in their current state significant questions remain as to grade and tonnage. Improperly adding probable ore to proven ore reserves for Asset Valuation reasons may impose severe negative consequences. For example, the cost of delineation may be extremely prohibitive; the results may be marginal or negative, thus shortening the life of a mining operation; the extraction required may be more expensive, lowering the margin of the newly proven ore; proven ore expansion may increase taxes without increasing production. However, an evaluation of probable ore reserves serves to illuminate potential pitfalls or bonanzas for acquiring firms.

Possible ore is the third category of mineral reserve. Typically, quantitative estimates are based on an estimated extension of the ore body. This inferred extension of the ore body may be based on similar geologic structures as well as adjacent mine workings. Since grades and tonnages are not available for possible ore, this category is not used in Asset Valuation.

Even though the Asset Valuation method uses only proven ore in it's calculation, some modification may be required because mining firms tend to underestimate their reserves for various reasons, some of which follow:

° In some western states, state and local taxes are calculated on the size of the proven mineral reserves.

° The cost of reserve delineation is quite high and discourages the building of a large proven reserve base beyond a 10 year horizon.

° Changing economic conditions will alter reserve calculations to a degree that mineral wastes once thought to be worthless are now high-grade materials.

° Large reserves are not usually reflected in stock prices and a large reserve position increases the potential of an acquisition.

The following steps are required in order to perform an Asset Valuation of a mineral firm.

1. Calculate the total proven ore reserves for each metal and mineral. This calculation should include grade, volume, dilution factor, rate of extraction and escalated prices for the next 15 years. Mineral prices do not tend to move at the rate of inflation, some move faster and other slower and frequently on a cyclical basis. The inflation rate should not be used unless other estimates are not available.

2. Calculate the annual profit per mineral and metal. This is done by determining the cost to extract each unit of mineral and subtracting this from the estimated price to determine profit. The per unit profit is then multiplied times total volume to determine total profit. The costs for extraction and processing should be inflated, but at an appropriate rate for the mining industry.

3. Calculate the present value of the mineral assets. In determining a discount factor, three approaches commonly used are:

   ° Liberal Discount: Assuming the rate of inflation is lower than the prime rate, use the current and future rate of inflation.

   ° Traditional Discount: Use the prime rate or approximately "prime + 1".

   ° Conservative Discount: Use your current and future rate of reinvestment. In economic terms, this is the true opportunity cost of capital.

4. Calculate the present value of the fixed assets of the entity being acquired and add this to the mineral asset value.

5. Calculate the present value of listed securities and discount value of unlisted securities. Calculate a discounted present value of foreign operations over a 5 to 10 year period and add all these values to the totals in step 4. The five to ten year time frame for foreign operations is chosen as a

moderate to low probability of takeovers. A high probability should be given no value. Additionally, foreign operations should be discounted at a faster rate than domestic operations because of the general lack of overseas control.

6. Calculate the total long-term debt and other liabilities and deduct these values from present value total calculated in step 5. This results in the net asset value of the corporation. In order to determine the net asset value per share, divide the net asset value by the number of shares outstanding.

The resulting figure is the asset value per share of the potential candidate. In general the asset value per share is approximately 1.5 X to 3 X the actual share value. One final calculation is needed to determine the market value from the asset value. This involves an estimate of the cost of capital and the minimum acceptable rate of return on new projects.

Most corporations use the cost of capital as a primary measure of the rate of return on an investment. As previously mentioned this cost or discount rate should reflect the liberal, traditional or conservative position of the firm making the analysis. At the present time 14-15 percent appears to be a good estimate for the cost of capital for most large U.S. firms. The second figure to determine is the hurdle rate or minimum acceptable rate of return on new projects. For firms in the mining industry who have successfully run operations similar to the one being reviewed, the hurdle rate is usually several percentage points above the cost of capital. This might be as high as 18-22 percent at the present time. For firms unfamiliar with the operations of the company being reviewed or not associated with mining, higher percentage values should be added. For example, it is not unreasonable for a firm to view it's first mining venture in terms of a 25-30 percent return; 15 percentage points equaling the cost of capital, 5-7 percentage points exemplifying the risk of the venture, and the remaining percentage points added as the risk of not understanding the business. Market Value is then computed based on the following formula:

$$MV = AV \times \frac{DR}{(MR = AR)}$$

Where: MV = Market Value

AV = Asset Value

DR = Discount Rate or Cost of Capital

MR = Minimum Rate of Return (basically DR + mining risk)

AR = Additional Risk (lower for mineral firms than non-mineral firms)

The Market Value rate calculated provides a suggested mid-range market price per share for a mineral firm. Table II illustrates an example of an Asset Valuation for a mining firm.

Table II. Asset Value and Market Value Calculations

BONANZA MINING COMPANY

Asset Value Per Share Calculation

Proven Reserves	Discounted (15%) Asset Value Per Reserve
Gold (2.1 MM oz)	$13.24
Silver (1.4 MM oz)	0.17
Lead (2.1 MM tons)	8.15
Zinc (0.7 MM tons)	6.34
Copper (1.7 MM tons)	2.19
Steam Coal (144.9 MM tons)	16.40
Metallurgical Coal (33.1 MM tons)	5.62
Other Domestic Operations	4.03
Foreign Operations & Securities	2.11
Current Fixed Assets	7.04
Long Term Debt	(8.73)
Asset Value Per Share (11.8 MM shares)	$56.56

Market Value Per Share Calculation

Discount Rate: 15%   Minimum Acceptable Return: 18%

Additional Risk: 0% $MV = AV \times \frac{DR}{(MR + AR)}$

$MV = \underline{\$47.13} = \$56.56 \times \frac{15}{18}$

In the example, it was assumed that a mineral firm was evaluating Bonanza Company and therefore, the additional risk (AR) was non-existent. The market value of Bonanza is $47 in this example. If the real market price of Bonanza were less than $47, the company is extremely overvalued relative to the market and/or the required acquisition return of the evaluating company.

CONCLUSION

The mineral industry is different enough from other industries to require a different type of analysis for an acquisition. This basic difference stems from the large asset base associated with a mining firm and its need for asset replacement. One method of acquisition evaluation that does properly assess a mineral firm is Asset Valuation. This method allows for an evaluation of the productive asset base for a selected number of years. While the Asset Valuation method is not foolproof, it works better than other traditional approaches. In any event, a mineral acquisition analysis should always include an asset evaluation.

REFERENCES

Anders, G: Gramm, W.P.; Maurice, S.C.; and Smithson, C.W.; The Economics of Mineral Extraction, New York: Prager Publishing, 1980.

Bailey, P.A.; "The Problem of Converting Resources to Reserves," Mining Engineering, 28 (1976): 27-37

Banks, F.E.; "A Note on Some Theoretical Issues of Resource Depletion," Journal of Economic Theory, 9 (1974): 238-44.

Bing, G.; Corporate Acquisitions, New York: Gulf Publishing Company, 1980.

_____; Corporate Divestment, Houston: Gulf Publishing Company, 1980.

Cameron, J.I.; "Investment Theory land Mineral Investment Practice," Presented at the Meetings of the Council of Economics of the AIME, Washington, D.C., 1977.

Coopers and Lybrand; Checking Into an Acquisition Candidate, Coopers and Lybrand Printing, 1982.

Cummins, A.B., editor; Mining Engineering Handbook, Vol. 2, New York: AIME, 1973.

Herfindahl, O.C.; l"The Process of Investment in Mineral Industries," in D.B. Brooks (editor), Resource Economics, Baltimore: John Hopkins University Press, 1974.

Ise, J.; "The Theory of Value as Applied to Natural Resources," American Economic Review, 64 (1974): 284-91.

Park, W.R. and Moillie, J.B.; Strategic Analysis for Venture Evaluation, New York: Van Nostrand Reinhold Company, 1982.

Peterson, F.M.; "A Theory of Mining and Exploring for Exhaustible Resources," Mimeograph, College Park: University of Maryland, 1975.

Salter, M.S., and Weinhold, W.A.; Diversification Through Acquisition, New York: The Free Press, 1979.

8.5

FINANCIAL MANAGEMENT OF DIVERSIFIED COMPANIES

by

Peter J. Maxworthy
Peko-Wallsend Ltd.
Group Treasurer
Sydney, Australia

INTRODUCTION

There is no hard and fast rule on whether functions within a company, diversified or otherwise, should be strongly centralized or decentralized. In all diverse organizations, there are certain basic functions which remain common (or central) to each element of the business, viz:

    Personnel Management
    Cash Management
    Taxation Management
    Legal
    Planning
    Accounting Standards

It would follow that each of these common factors requires some supervision at a central level to ensure that overall group objectives are met. It is true that other functions including:

    Industrial Relations
    Marketing and Distribution
    Production
    Research and Development
    Purchasing
    Management Accounting

may also be common to each element of the organization. However, there is far less commonality in the degree to which these functions are central to the efficient performance and financial well-being of the business.

Role of Operating Units

The resources of operating centers may be viewed as a bank of funds, the nature of which can be changed from time to time, but the employment of which must provide for ongoing increases in the net worth of the total enterprise expressed in terms of market values. Operating units are required to continually review products, methods, and opportunities and to monitor avenues for incremental investment--all with a view to attaining the necessary additive net present value of future cash flows of the enterprise.

Operating units should be permitted sufficient autonomy to enable the best use to be made of the skills and experience of the people it employs in order to produce a return appropriate to the value of assets employed.

Role of Head Office

The head office will typically comprise a specialist group--the leaner the better if one strongly adopts the decentralist approach--often acting as Service Centers but also performing a control function in the corporate sense.

The head office function is a cost center. It may be instrumental in developing strategies for increasing revenues or reducing costs on a group scale, but it does not generate revenue. Every dollar spent on this function is an extra dollar to be generated by the operating centers in order to achieve an increase in corporate wealth.

TREASURY MANAGEMENT

Treasury management is concerned with the provision of finance, cash-flow management, and risk analysis. Each decision taken within the enterprise has a cash implication. It follows that a well-integrated central Treasury function can, in all likelihood, produce economic benefits.

The degree to which the function should be centralized depends largely upon the unique nature of the operating units it is servicing. Levels of ownership, the calibre of unit operating managers, and the nature and geographic distribution of the business are all factors determining the degree to which the Treasury function is centralized.

In the case of Peko-Wallsend, the Treasury Department is small, comprising four people. While divisional cash-flow management is largely decentralized, Treasury is responsible for the daily management of funds surplus to the immediate needs of operating units. Treasury also operates as central banker and provides funds to meet both long- and short-term demands for cash by operating units.

Short-term placement and borrowings, long-term funding and the determination as to how those funds are denominated falls under the Treasury umbrella as does the assessment and management of risks associated with commodity prices, exchange rates, interest rates, etc.

Treasury management therefore has implications for the profit and loss account by its effect on:

    Interest expense and income
    Foreign exchange variations--exposures related to income and expense items and short- to medium-term loan account denominated in other than one's domestic currency (in this case, Australian dollars--"A$")
    Revenue-stream risk assessment and management
    Tax expense

and on the balance sheet via:

    Movements in assets and liabilities-- denominated in other than Australian dollars
    The capital structure
    Debt mix
    Working capital levels
    Capital expenditures

Throughout, Treasury's primary concern is the implication of each of the above factors for the net present value of future corporate cash flows.

Cash Management

The object of cash management is to shorten the receivables cycle, to lengthen the payables cycle, and to ensure that no liquid assets remain idle. Where cost benefits arise from (i) economies of scale and (ii) matching processes, then these factors should be exploited to the maximum.

One of the problems associated with managing cash is the time delay in becoming aware of inefficiencies. Bankers are steadily developing computer-based systems to improve the timeliness of reporting cash balances and making cash transfers and consolidations. Besides reducing the company's requirement for short-term funds, these developments assist the Treasurer in his role as the company's central banker.

Management of Financial Risks

In order to graphically illustrate the issues with which a diversified mining company's Treasury Department must come to terms, a hypothetical enterprise has been constructed, ABC Company, the financial attributes of which are shown in Tables 1 and 2. It produces two commodities, both with pricing denominated in U.S. dollars ("US$").

TABLE 1
ABC Company
Typical Profit & Loss

Revenue Stream

Commodity 1 - US$/tonne	$200
Tonnes sold	500,000
Commodity 2 - US$/kg	$ 50
Kilograms sold	1,500,000
Exchange rate	0.9000
Total Revenue	$194,444

Variable Costs

Commodity 1 - A$/tonne	$100
Commodity 2 - A$/kg	$ 20
Total Variable Costs	$ 80,000
Fixed Costs	$ 60,000
Profit before Interest and Tax (P.B.I.T.)	$ 54,444
Less:	
Interest Expense	7,500
Tax Expense	21,594
NET PROFIT AFTER TAX (P.A.T.)	$ 25,350

TABLE 2
ABC Company
Typical Cash Flow

P.B.I.T.	54,444
Dep'n (net of rep capex)	4,000
Cash to Service External Obligations	58,444
Less:	
Tax	21,594
Debt Service	12,500
Equity Service	20,000
Cash Available for Additional Debt Service	4,350

Present Balance Sheet

Shareholders' Funds	200,000
Debt	50,000
	250,000
Fixed Assets	225,000
Working Capital	25,000
	250,000
Debt/Equity Ratio	25.00%
Return to Equity	12.67%

Commodity 1

This product is sold to customers situated in the Southeast Asian Basin. The company has a strong competitive advantage over other major world suppliers of Commodity 1 because of its relatively close proximity to the market and excellent record in meeting delivery requirements. Experience indicates that even during recessionary periods the company is able to sell every tonne of product it can produce.

The price of the commodity is, however, set on the world market and is very volatile. There is limited ability to sell the commodity on a long- or even medium-term contract basis.

Statistical analysis of historical price trends shows a high standard deviation of plus or minus 20% from the median price measured over time. The product has a variable cost of about 45% of the current selling price expressed in Australian dollars (A$). In its own right, Commodity 1 attracts a high level of fixed cost and therefore has a relatively high break-even selling price.

Commodity 2

This is also sold to industrial users situated in the Southeast Asian Basin. The users of this product are more widely spread so while some customer commonality exists for the two products, a wider customer base is serviced for Commodity 2.

The commodity is in plentiful supply, but it is customary to sell on a long-term supply contract basis subject to plus or minus 10% volume variations. Although the company has been subjected to volume cuts of more than 10% under long-term supply contracts, it has been able to make up the shortfall with spot sales because of the market diversity.

The standard deviation for volume of annual sales has been shown to be plus or minus 8%. The product is priced in world markets and statistical analysis of historical price trends shows a standard deviation of plus or minus 10% from the median over time. Variable costs attributable to each unit of production represent about 35% of current price measured in terms of A$. On a stand-alone basis, Commodity 2 also has a relatively high level of fixed costs. It too has a high break-even selling price.

History. The company has experienced violent fluctuations in earnings due to the high volatility in the price of its only product, Commodity 1. It recognized that if it merged with a producer of Commodity 2, it could achieve an overall saving in fixed costs attributable to both commodities, resulting in lower break-even prices.

On closer examination of the decision to acquire the Commodity 2 company and with the benefit of hindsight, it has been discovered that additional benefits accrued when the two commodities are put together. Statistical analysis shows:

1. Price movements for Commodity 1 lag price movements for Commodity 2 such that when the price for Commodity 1 is peaking, the price for Commodity 2 is bottoming. Price movements are negatively correlated.

2. Volume movements for Commodity 2 closely follow price movements for Commodity 1. Heavy volumes tend to be delivered when the price for Commodity 2 is low. Volume movement is positively correlated to the price of Commodity 1 but negatively correlated to the price of Commodity 2.

3. Price movements of Commodity 1 tend to follow closely the performance of the US$. As the US$ weakens, the price of Commodity 1 strengthens--a further negative correlation.

4. The A$/US$ relationship has a standard deviation of plus or minus 8% from the median exchange rate.

Figure 1 (supported by Tables 3 and 4) illustrates the effect of these correlations.

Without volume changes, a greater overall revenue stream is generated when the price of Commodity 1 is at its lowest point (see Figure 2). This reflects the effect of the exchange rate (movements in which relate to the total revenue stream) and the price of Commodity 2 being negatively correlated to Commodity 1.

In practice it would be appropriate to attach probabilities to each of these events and to the likely degree of correlation. For the purposes of this exercise and in the interests of simplicity, it has been assumed that correlations (whether positive or negative) are direct. Although over-simplified, this example does illustrate the need to assess individual risks in a global sense rather than in isolation. The Treasurer of the ABC Company may be covering the company's US$ foreign exchange exposure earnestly, believing he or she is reducing risk. In fact, the company has been denied one of the benefits of risk diversification by removing a natural hedge.

Within Peko, Treasury "buys" those risks from the operating units for which there are liquid futures markets and actively trades these positions. However, this is done within a general framework which permits the department to act in the knowledge that by trading, it may well be adding to overall corporate risk.

Capital Structure

Once the likely range has been determined for the projected revenue stream, one should track through the effects to cash flow (see Figure 3).

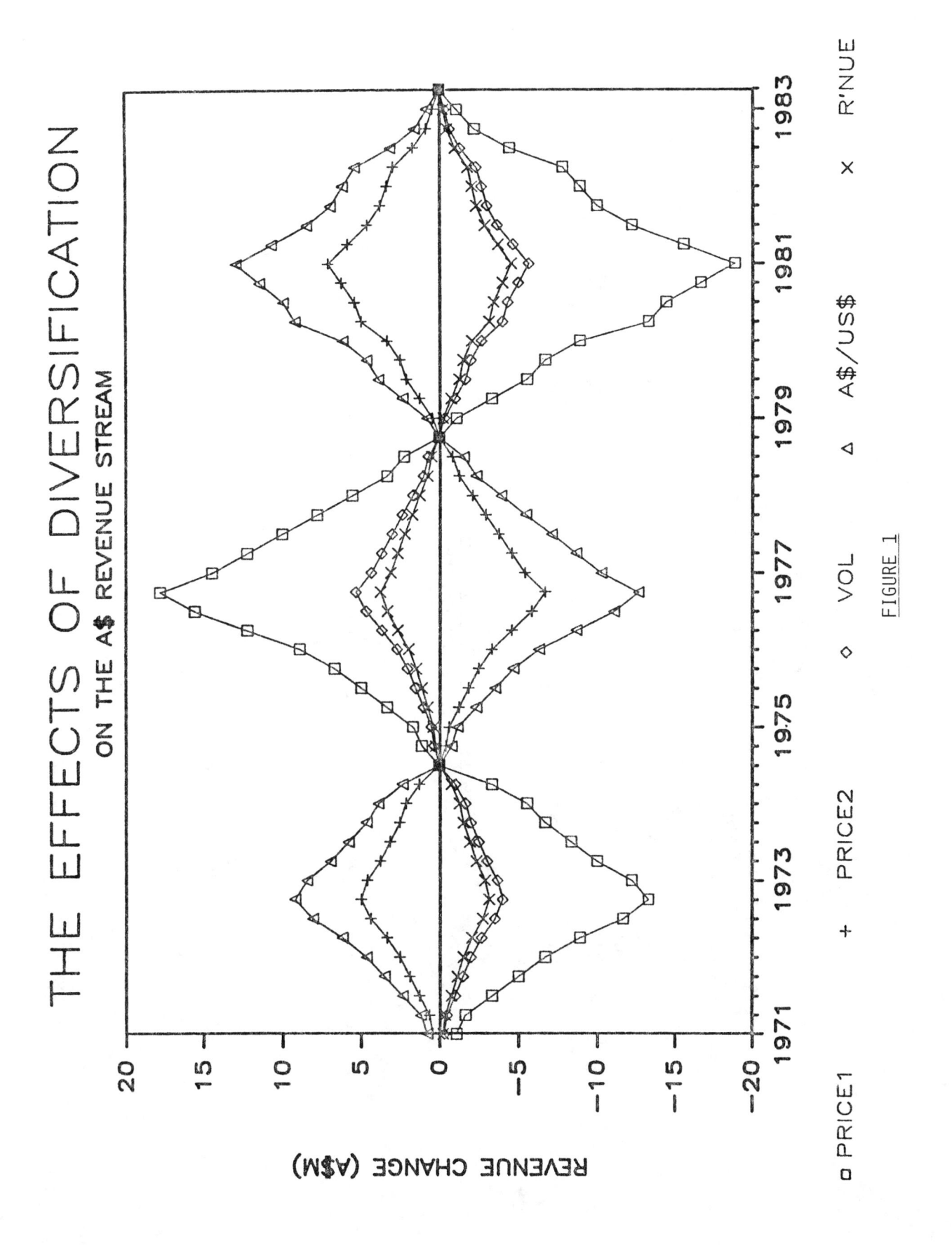

THE EFFECTS OF DIVERSIFICATION
ON THE A$ REVENUE STREAM

FIGURE 1

□ PRICE1   + PRICE2   ◇ VOL   △ A$/US$ × R'NUE

TABLE 3
Effect of Changes On Revenue Stream

Commodity 1 US$ Price % Change	Commodity 2 US$ Price % Change	Commodity 2 Volume % Change	Exchange Rate % Change	Commodity 1 Price A$m	Commodity 2 Price A$m	Commodity 2 Volume A$m	Exchange Rate A$m	Revenue Stream A$m
-20.00	10.00	-8.00	8.00	-22.2	8.3	-6.7	15.1	-5.4
-15.00	7.50	-6.00	6.00	-16.7	6.3	-5.0	11.4	-4.0
-10.00	5.00	-4.00	4.00	-11.1	4.2	-3.3	7.7	-2.6
-5.00	2.50	-2.00	2.00	-5.6	2.1	-1.7	3.9	-1.3
0.00	0.00	0.00	0.00	0.0	0.0	0.0	0.0	0.0
5.00	-2.50	2.00	-2.00	5.6	-2.1	1.7	-3.9	1.2
10.00	-5.00	4.00	-4.00	11.1	-4.2	3.3	-7.9	2.4
15.00	-7.50	6.00	-6.00	16.7	-6.3	5.0	-11.9	3.5
20.00	-10.00	8.00	-8.00	22.2	-8.3	6.7	-15.9	4.6

TABLE 4
Aggregate Effect of Changes on Cash Available
To Service New Debt
(A$ Millions)

Revenue Stream	Variable Costs	Tax Expense	Cash for New Debt	Cash for External Obligations	Tax Expense	Debt Service	Equity Service	Cash for New Debt
-5.4	2.4	1.4	-1.6	55.4	20.2	12.5	20.0	2.7
-4.0	1.8	1.0	-1.2	56.3	20.6	12.5	20.0	3.2
-2.6	1.2	0.6	-0.8	57.0	20.9	12.5	20.0	3.6
-1.3	0.6	0.3	-0.4	57.8	21.3	12.5	20.0	4.0
0.0*	0.0	0.0	0.0	58.4	21.6	12.5	20.0	4.4
1.2	-0.6	-0.3	0.3	59.1	21.9	12.5	20.0	4.7
2.4	-1.2	-0.6	0.7	59.6	22.1	12.5	20.0	5.0
3.5	-1.8	-0.8	0.9	60.2	22.4	12.5	20.0	5.3
4.6	-2.4	-1.0	1.2	60.7	22.6	12.5	20.0	5.6

*Current level of sales.

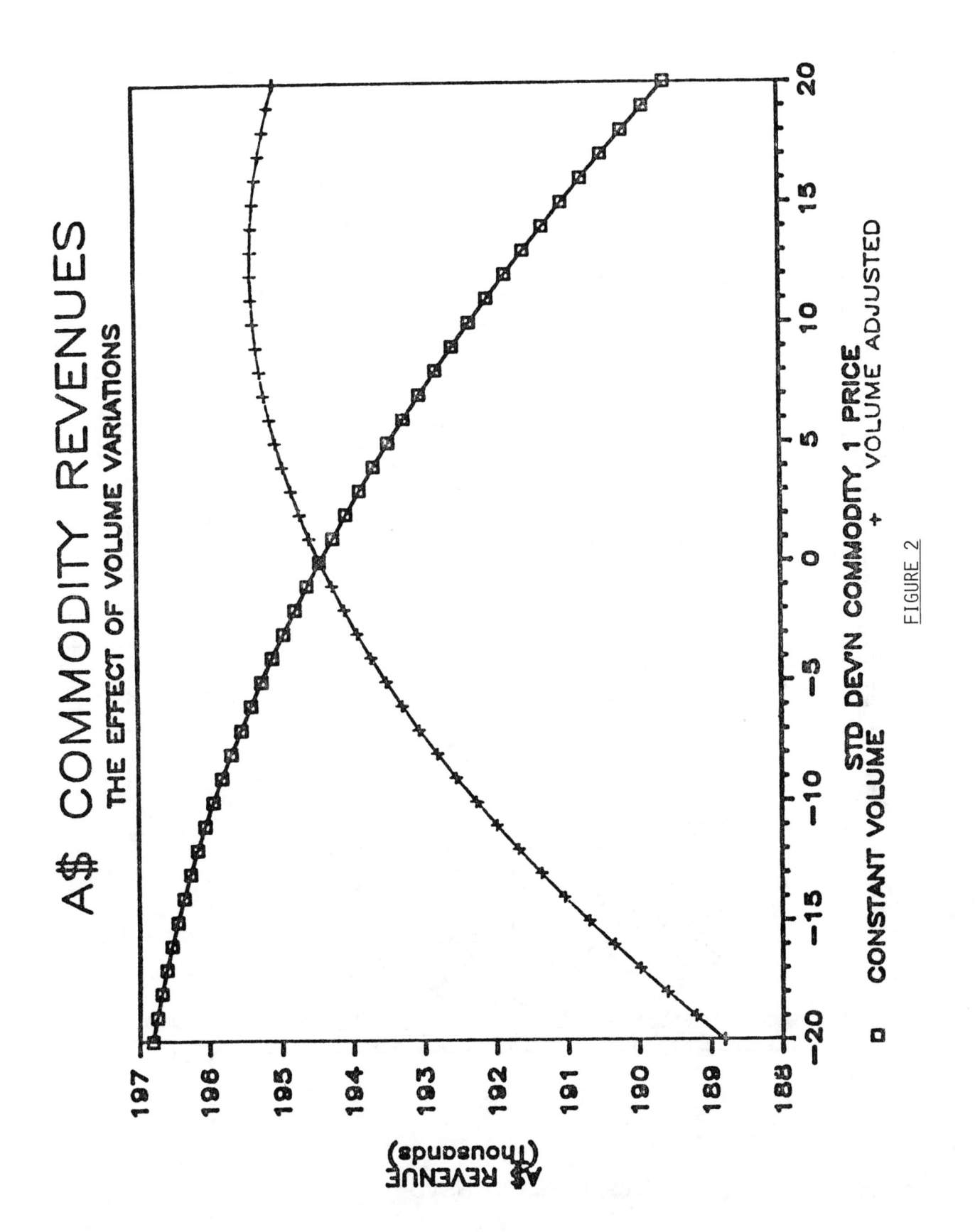

A$ COMMODITY REVENUES
THE EFFECT OF VOLUME VARIATIONS

FIGURE 2

CASH AVAILABLE
TO MEET EXTERNAL OBLIGATIONS

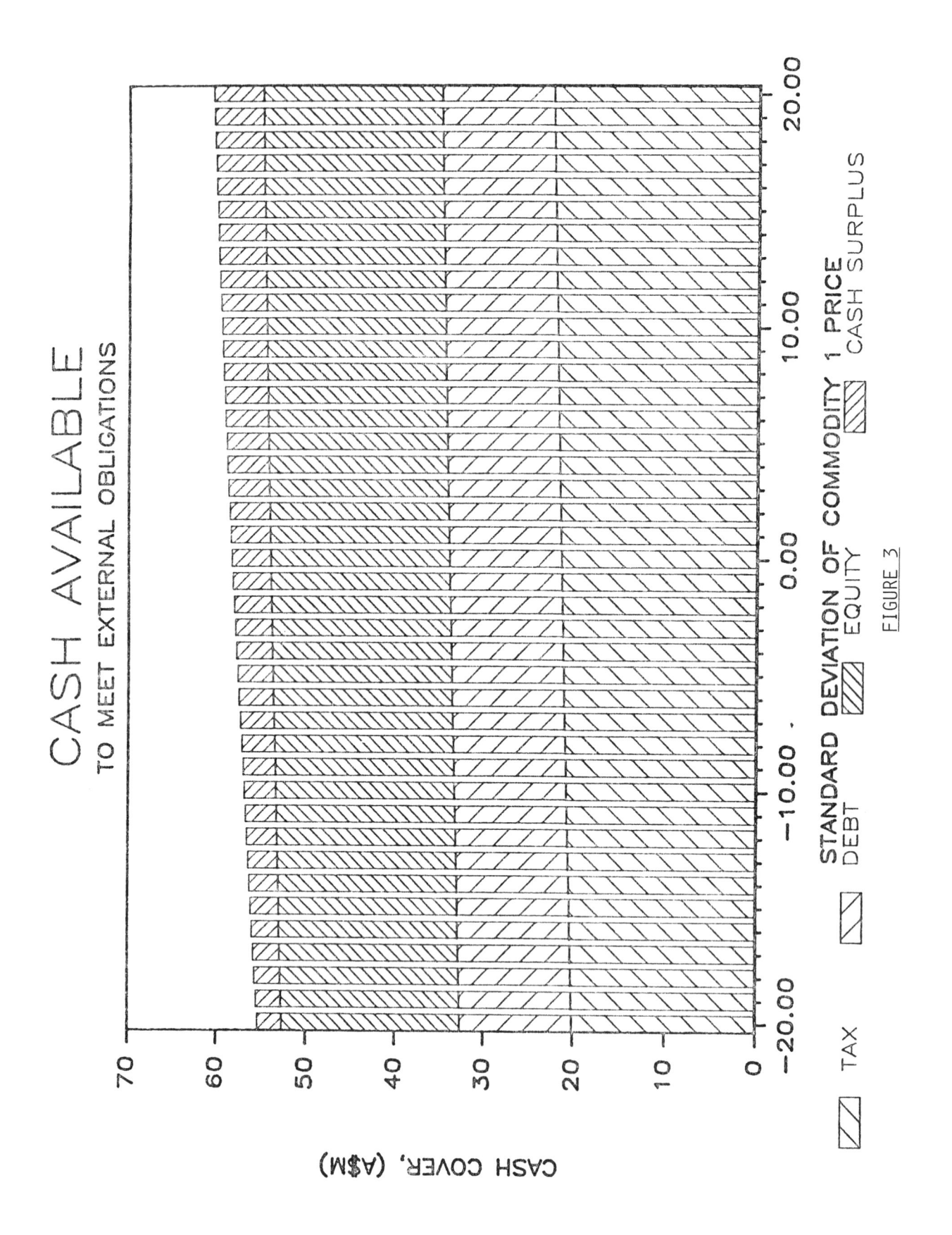

FIGURE 3

STANDARD DEVIATION OF COMMODITY 1 PRICE

Legend: TAX, DEBT, EQUITY, CASH SURPLUS

In establishing the ABC Company's typical cash-flow characteristics, the following assumptions and observations have been made:

Prices for Commodities 1 and 2 are currently running close to the historical trend line.

Fixed operating costs ($60 million per annum) include depreciation charges of $10 million per annum.

Tax is paid in the year in which it is incurred.

Replacement capital is required to be met at the rate of $6 million per annum.

Debt/equity ratio is pitched at 25% by the Board of Directors.

Debt attracts an interest rate of 15% per annum.

Debt principal is repayable over 10 years.

It is considered that equity participants should be rewarded with dividend payments calculated at the rate of 10% per annum.

At the current level of sales, operating costs (both fixed and variable) and debt/equity, the enterprise will generate cash totalling $58.4 million to service tax, debt and equity (see Tables 1 and 2). Tax payment ($21.6 million), debt service ($5 million set aside for repayment of principal and $7.5 million in pre-tax interest), and equity service ($20 million for dividends) use $54.1 million of this cash generation. This leaves a surplus of $4.4 million. The table reflecting sensitivities (refer to Table 4) shows that this cash surplus during the worst period in a cycle will be reduced by $1.6 million to approximately $2.7 million assuming the correlations are maintained.

Debt Level

In principal, it is well established that debt is less expensive than equity funding. The higher the level of debt, the greater are the returns to shareholders measured in terms of dividends on the one hand and incremental share price on the other. The price shareholders pay for the promise of increased returns is increased risk. Increased debt adds fixed cost (interest after tax) to be covered by the operating units and creates a priority on cash (principal and interest payments) which would otherwise be available to shareholders.

However, if the objective is to increase corporate wealth, then clearly one should opt for the level of debt which meets the following criteria:

Improves returns to shareholders

Is capable of being serviced in the bad times as well as the good

Provides an acceptable level of comfort to shareholders (i.e. does not adversely affect the historical price/earnings ratio for share prices)

Is acceptable to the organization's lenders

Figure 4 shows the effect on return on shareholders' funds by increasing the ratio of debt to equity. At zero debt, there is no benefit gained by equity holders from tax shelter on interest cost. At a ratio of 66.7%, the return to equity has increased from 11.76% to 14.2%, an increase of around 35%.

But it should be remembered that in the bad times, the cash available to cover new debt dropped by around $1.6 million in the extreme case. From Figure 5 (refer to Table 5), it can

TABLE 5
Debt/Equity Mix
(A$ Millions)

Debt	Equity	Ratio	Cash for External Obligations	Tax Payments	Debt Service	Equity Service	Cash for Additional Debt	Return to Equity
0	250,000	0.00%	58,444	25,044	0	25,000	8,400	11.76%
20,000	230,000	8.70%	58,444	23,664	5,000	23,000	6,780	12.08%
40,000	210,000	19.05%	58,444	22,284	10,000	21,000	5,160	12.46%
60,000	190,000	31.58%	58,444	20,904	15,000	19,000	3,540	12.92%
80,000	170,000	47.06%	58,444	19,524	20,000	17,000	1,920	13.48%
100,000	150,000	66.67%	58,444	18,144	25,000	15,000	300	14.20%
120,000	130,000	92.31%	58,444	16,764	30,000	13,000	-1,320	15.14%
140,000	110,000	127.27%	58,444	15,384	35,000	11,000	-2,940	16.42%
160,000	90,000	177.78%	58,444	14,004	40,000	9,000	-4,560	18.27%
180,000	70,000	257.14%	58,444	12,624	45,000	7,000	-6,180	21.17%
200,000	50,000	400.00%	58,444	11,244	50,000	5,000	-7,800	26.40%

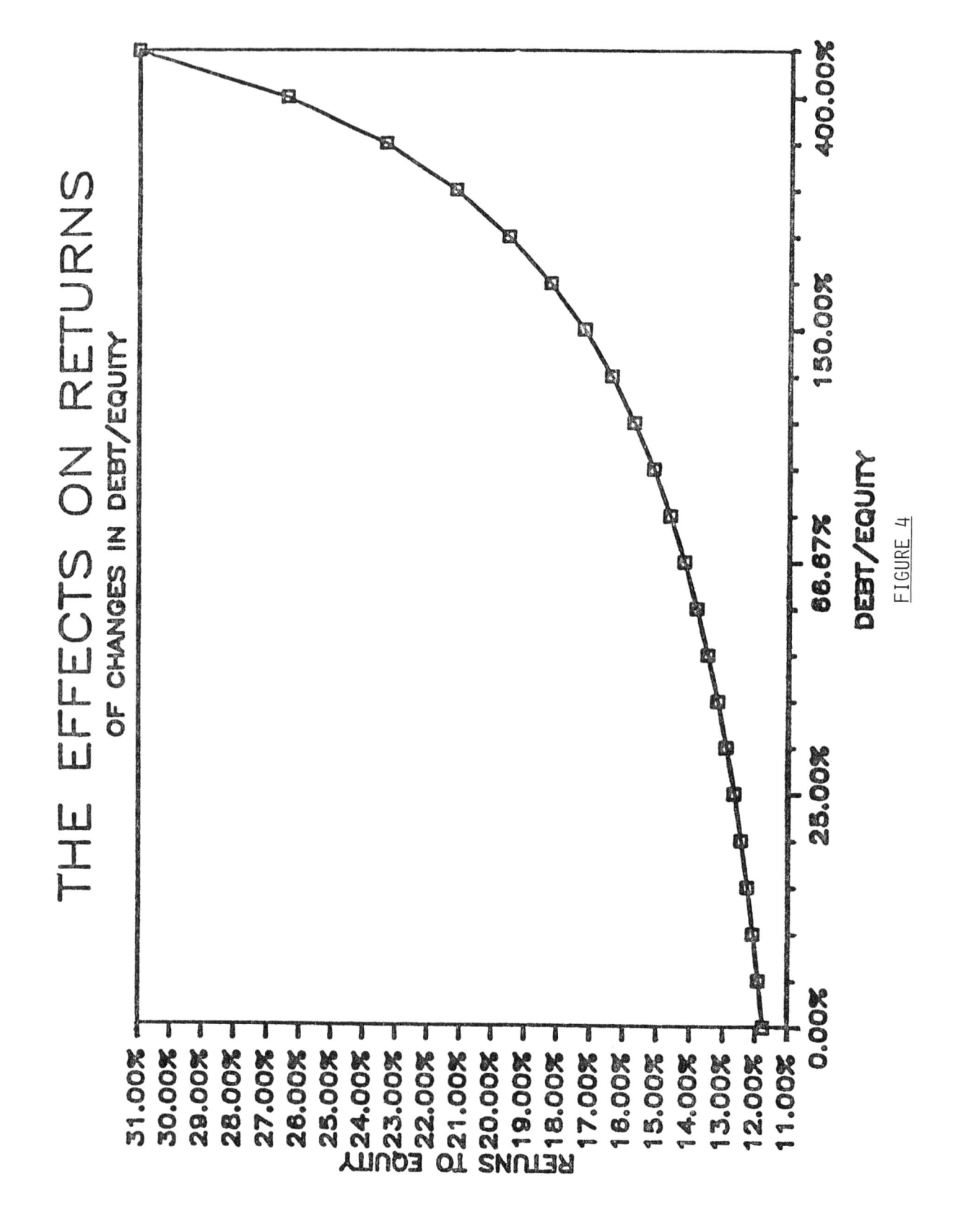

THE EFFECTS ON RETURNS
OF CHANGES IN DEBT/EQUITY

FIGURE 4

CASH AVAILABLE TO SERVICE
EXTERNAL OBLIGATIONS

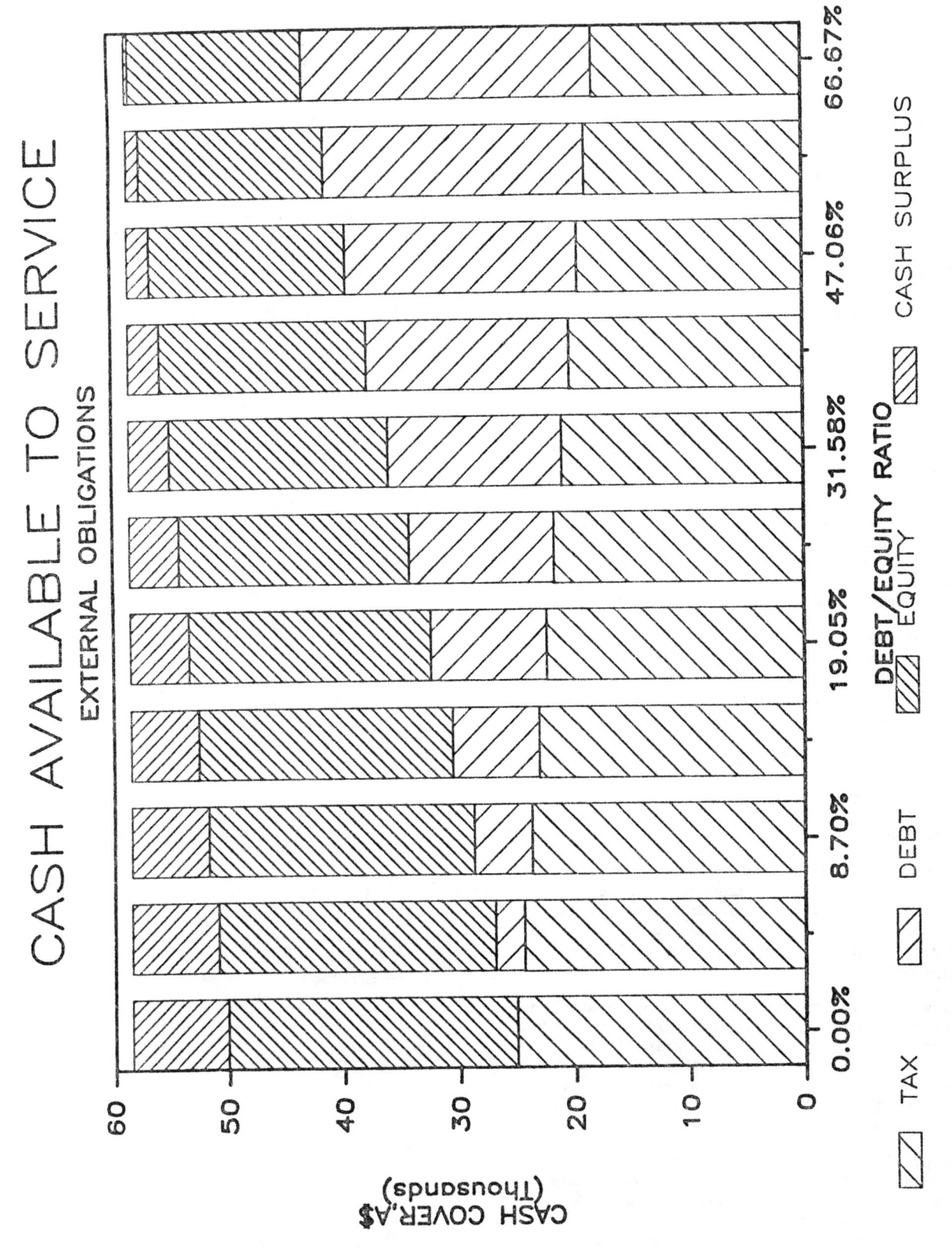

FIGURE 5

Legend: TAX | DEBT | EQUITY | CASH SURPLUS

DEBT/EQUITY RATIO: 0.00%, 8.70%, 19.05%, 31.58%, 47.06%, 66.67%

Y-axis: CASH COVER, $ (Thousands), 0–60

be seen that at debt/equity of 100% there is negative cash available for new debt. Given further that there is an inherent down side of $1.6 million, this level of debt could not be recommended.

From Table 5, an appropriate debt/equity level appears to be around 45-50%. This means the company could increase its borrowings by around $35 million with comfort. In the event Directors are prepared to modify dividend policy during the bad times, a ratio of 65-70% would not be unthinkable.

The subject of capital structure (see Chapter 2 of this book) is a matter which needs to be regularly addressed and updated in the light of current economic and political circumstances. It should be varied from time to time to reflect new trends, given that historical relationships are not always reliable guides to the future.

This method of determining an appropriate debt/equity mix does, however, offer an alternative to the traditional approach of relying on external advisers and "gut feel."

Debt Currency Mix. It is necessary to depart from the above worked example since there are no overseas assets in the ABC Company. The establishment of a satisfactory currency mix of debt exposure is essentially a matching process. It is appropriate to closely match assets denominated in an external currency with liabilities which are similar in amount and denomination. Provided the match is maintained over time, this will have the effect of eliminating balance sheet or translation exposure to foreign currency fluctuations.

Similarly, it will be necessary to determine the economic life of those assets which are not funded by equity capital. A stream of long-term debt should be put in place to cover remaining fixed assets.

Short- to medium-term debt can be used to fund working capital requirements. It must be recognized, however, that working capital has an element of permanency. Short-term funds, by their very nature, may not always be available to the company.

Security. Whether borrowings should be secured or unsecured depends largely on the history of the company. In a diverse organization it is preferable to arrange all borrowings on an unsecured basis rather than provide security on a group of assets which may be geographically widespread.

The provision of security also has the effect of tying up assets. For example, it may be that a warehousing site is to be moved in the interests of efficiency and that that site has been used as security for a long-term, low-interest mortgage borrowing. It has been my experience that mortgage lenders will not accept replacement securities. They will require that the loan to which the security refers be repaid and that a fresh loan be arranged in respect of the alternative security. This may come at an inopportune time from an interest-rate viewpoint.

Where all borrowings are contracted on an unsecured basis, the company will be required to undertake to remain within certain restrictive financial covenants and to refrain from denigrating the lenders' position, in terms of priority on cash and any future granting of security, as it was perceived at the time the loan was put in place. It should be noted, however, that where all loans are the subject of covenants and negative pledges, should the company default on any one of its loans, there will probably be cross-default clauses which will cause all loans to be in default. In a highly geared company, this may well result in an unwelcome loss of control.

This paper is condensed from a presentation made by the author at the 1983 Australian Corporate Finance Conference.

8.6

LONG-RANGE PLANNING IN A COPPER MINING COMPANY

FERNANDO A. BUTTAZZONI AND JORGE C. MUNITA

CODELCO-CHILE

INTRODUCTION

To provide a background for discussing long-range financial and operational planning in the mining industry, we would like to begin by describing the role of planning in Codelco. Having taken part in Codelco's planning process since its inception in 1976, we believe that useful conclusions may be drawn for the experience we shall relate.

Although Codelco is a state-owned copper producer with an important role in the Chilean economy, and is consequently very much at the core of national economic planning, the following discussion is not related to government planning but to planning in an operationally and commercially autonomous business concern with all the characteristics of a private enterprise. The only way in which Codelco differs from private business is that its financial successes or failures have a more direct impact on the country's ability to participate in international trade, to meet its international financial commitments and to achieve economic and social growth.

Thus, what follows is a presentation of the conceptual features of the planning process at Codelco and a description of the current state of planning in the company. It is understood that certain conceptual components are basic to any formal strategic planning process, and that the components of Codelco's planning process are therefore generally applicable in similar companies.

First, the mission and scope of operation of the company must be defined, and goals must be set to achieve leadership in the business activities in which it engages. Having set this framework, the company must constantly scrutinize those internal elements and conditions that are essential to fulfill its mission and achieve leadership. In Codelco this means considering the actions which can best serve to define future capacities, taking into account declining ore grades and determining optimal production levels; watching over the productivity and skills of its labor force, and; keeping abreast of technological developments in non-ferrous mining and processing. If the company is proficient in these three areas, it will be cost efficient and therefore competitive.

Codelco must also watch over the business environment in which it operates. The company sells concentrate, blister and fire-refined copper, wirebar, cathodes and wire rod and must study the behavior of the market to attune its product mix to the demand existing for these various commodities in the international market and the copper consuming industries in general.

By examining the internal and external environment in which the company conducts its production and commercial operations, it is able to diagnose its strengths and weaknesses, to identify the hazards it must overcome to stay ahead in the copper trade, and to take full advantage of business opportunities as they arise.

CHARACTERISTICS OF CODELCO-CHILE

Although our subject is Codelco, readers should keep in mind that there are many other copper producing operations in Chile, all in the private sector, and many valuable deposits that may be exploited in the future. Codelco is the only public sector copper producer, although there is one other state enterprise that operates smelting and refining facilities principally to serve a large number of very small producers of copper concentrate.

In the mid-1980's, Codelco is the largest copper mining company of the world and the largest producer and exporter of copper with yearly sales totalling close to US$ 1.8 billion in 1983. Moreover, it is a highly profitable copper company at a time when the world copper industry as a whole has reported losses for several years.

Codelco-Chile is a fairly new organization. The history of the four mining operations it manages is much longer: El Teniente was opened in 1912, Chuquicamata in 1915, Salvador in 1959 -- with installations of the old Potrerillos mine dating from 1927 -- and Andina in 1970. Having had separate owners, the management systems of the four operations were entirely different and were superseded by the nationalization of the four mines in 1971. Some time elapsed before an appropriate managerial structure could be devised to ensure the efficient operation of these important national economic resources. Finally, in 1976 Codelco-Chile was established as the legal owner and operator of the four operations. Since that year the four mines have operated as divisions of the company, with their activities coordinated from corporate headquarters in Santiago, where certain centralized functions, such as sales and finance, are performed or supervised. In accordance with the economic philosophy prevailing in Chile, the company is bound by the rules and regulations that apply to business in general:

- no government subsidies are promised or received
- income and business taxes are paid
- profitable operations are expected.

In 1983 Codelco's four mines produced a total of 1,012,055 metric tons of copper and 15,264 metric tons of molybdenum, as well as fair amounts of dore metal and anode slimes containing silver and gold. The Chuquicamata and El Teniente Divisions contributed 55 percent and 30 percent, respectively, of the toal copper produced, and the two smaller operations, Salvador and Andina, 9 percent and 6 percent, respectively. For comparison, Chuquicamata produces at about the same level as all of Zambia, while output at El Teniente well exceeds the totals for Australia or the Philippines.

The following table combines the production figures, sales income and profits of Codelco, and total transfers of funds to the National Treasury (profits plus taxes) during the eight years the company has been in existence.

Table 1: Codelco - Copper production, income from sales, profits, and payments to the National Treasury, 1976-1983

Year	Copper production (metric tons '000)	Total sales (US$ million)	Net profits (US$ million)	To Treasury (US$ million)
1976	854.1	1,268.0	178.6	423.2
1977	892.7	1,182.2	159.0	393.8
1978	876.5	1,264.3	232.8	363.1
1979	910.2	2,071.4	467.3	871.2
1980	904.5	2,280.8	401.3	1,006.6
1981	893.6	1,741.1	97.3	480.3
1982	1,032.9	2,659.7	160.6	523.9
1983	1,012.1	1,774.1	220.7	678.5

It will be noted that Codelco has increased copper production in the past years. Current output is even more notable compared with copper production when the four mines were nationalized, at which time their combined output was 571,000 metric tons (1971). This success may be attributed to good management, but that is too broad a statement. It is more accurate to say that production has increased because there have been large mine and processing capacity expansions; technological improvements have been made, and the operations have seldom been disrupted by strikes. However, the best reason for the growth of Codelco's copper output is that the company has available remarkably large ore reserves of good quality. The planning process is geared to support the orderly development of all of Codelco's assets, from the orebodies, through the equipment, to our employees. Let us now examine the operating characteristics and history of Codelco.

Reserves

The identified reserves of the four Codelco mines presently amount to 112.6 million metric tons of contained fine copper, and total Chilean reserves presently stand at 145.9 million metric tons. The next largest reserves of copper of any one country are those of the United States, and they are estimated at 94.6 million metric tons. The reserves of Zambia and Zaire added together do not exceed 42.6 million metric tons. Hence it is evident that Codelco has more copper in its four deposits than what is available in entire continents. Moreover, several major ore bodies have been identified in Chile in recent years, so that Chilean ore reserves are growing, whereas in the rest of the world few new mines worthy of attention have been found, and none as large as those in Chile. These reserves are distributed as follows:

	Identified reserves Million mt fine Cu	Useful life of deposit-years
Chuquicamata	44.9	64
Salvador	3.2	30
Andina	16.9	220
El Teniente	47.6	125
Total	112.6	

Even if Salvador is becoming depleted, its reserves are still higher than those of many new mines that are now being considered for exploitation in other parts of the world.

As for the Andina deposit -- until now the smallest Codelco operation -- ongoing geological studies are revealing that it has enormous potential and may easily be as large as El Teniente. In fact, one satellite ore body, Sur Sur, was brought on stream during 1983 as an open pit operation with initial ore grades ranging between 2 and 3 percent -- quite an improvement over the ores grading 1 percent copper that were being mined at Andina. This specific sector is expected to contain 30 million metric tons of ore with an average grade of 1.5 percent, and under it is a still larger ore body, grading 1.2 percent copper, which will have to be exploited using underground mining methods when the open pit is exhausted in about ten years.

The readers of this paper are probably well informed about Chuquicamata and El Teniente, for they are an important part of the history of large-scale copper mining and will continue to play a leading role in the industry for many years to come. Although the ore grades in the two deposits are steadily dropping, our long-range plans show that cut-off grades at the mines by the year 2000 will be well above the current head grades at typical North American mines. These high copper grades provide Chile a crucial natural advantage in copper mining. The quantity of copper recovered per ton of ore mined and milled in Chile is about three times higher than that in the United States. The effect of ore grade on costs is compounded throughout the production process since less energy and fuel are needed, less materials are used and smelting costs are lower because of a much higher copper grade in the concentrate.

Processing and utility capacities

Development planning has played an important role in Codelco, leading to successive mine and concentrator expansions involving considerable financial resources and resulting in production increases in spite of deteriorating ore grades in the deposits. The largest growth of production was achieved before the market depression started in 1981, and the capacity improvements made at that time have helped the company to weather the copper crisis successfully. It will be noted that production in 1982 and 1983 was stable, and no production increases were made in 1984. Among other important considerations in planning the operations, attention has been given to balancing production capacities, developing the most marketable product mix, introducing technological improvements to up-grade installations and lower production costs, and to careful allocation of scarce investment resources.

The following table illustrates the developments achieved during the past eight years:

	1976	1984
Ore extraction (10^6 mtpy of mineral)	53.6	80.5
Overburden removal (10^6 mtpy of mineral)	30.0	62.0
Concentration (10^6 mtpy of mineral)	51.5	80.0
Vat Leaching (10^6 mtpy of mineral)	6.0	6.0
Smelting (10^3 mtpy of concentrate)	1,800.0	2,270.0
Electrolytic refining (10^3 mtpy of copper)	460.0	490.0
Fire-refining (10^3 mtpy of copper)	117.0	140.0
Electrowinning (10^3 mtpy of copper)	48.0	48.0
Installed power generation capacity (MW)	414.0	442.6
Water supply (liters per second)	5,500.0	7,110.0

Cost reductions

Codelco's cost reduction programs were planned and went into effect long before the price of copper began to drop in the early 1980's; therefore, the prolonged price crisis has hurt the company considerably less than those competitors who were caught with costly, obsolescent plants.

Production cost surveys carried out by prestigious international experts, involving mine-by-mine cost comparisons, conclude that Codelco is clearly the world's lowest-cost producer of copper. The main reasons for this are high labor productivity and, as mentioned earlier, high ore grades.

In the area of labor productivity, the following table illustrates the improvements achieved:

Table 2. Productivity in Codelco - Metric tons of fine copper per year per man

Year	Mine and concentrator		Smelter and refinery	
	Per Direct Employees	Per Total Employees	Per Direct Employees	Per Total Employees
1976	76	38	144	59
1977	80	40	164	66
1978	86	42	179	74
1979	80	39	192	79
1980	81	40	200	83
1981	82	43	183	81
1982	97	47	303	146
1983	96	48	291	142

These figures, of course, reflect the numerous technological improvements introduced in all phases of the operations, including automation and improved processes, which have made it possible to reduce the overall labor force from 30,948 in the mid-70's to 25,657 at the end of 1983. One interesting development is that the professional staff of Codelco is fairly stable, whereas the number of workers with special skills is increasing.

Production Costs, the Key to Profitability

Between 1976 and the end of 1983, Codelco invested over US$ 1,500 million in the ongoing development program, and the funds allocated in 1984 for this purpose totalled US$ 346.1 million. The investments now being made will enhance Codelco's relative efficiency, or comparative advantages, in copper mining. The natural advantage that the company enjoys, that is, the quality and quantity of Chilean copper minerals, and those advantages that have been painstakingly developed, i.e., efficient, technologically up-to-date installations, high productivity, and economies of scale, have all combined to lower Codelco's production costs. And low production costs are the only key to profitability given the historical variability of prices and the uncertainty inherent in the future.

LONG-RANGE PLANNING SYSTEMS

Basic elements of planning

There are many ways of structuring a planning process to meet the particular needs of a business. However, there are certain essential features that are common to all, for example, an organized framework of procedures and systems must be established. The planning process must be designed in such a way that it gradually embraces all sectors of the organization, and --

most important of all -- it must have the full support and confidence of the company's top management, for planning is an instrument for decision making and it is worthless if it is not used as such.

Planning is a continuing process with an advancing time horizon. The nature of the business dictates the length of the planning horizon. In the case of the mining industry, in which the execution of development projects takes time and the resulting works are expected to have a long useful life, a period from twenty to twenty-five years is frequently found appropriate. The fundamental characteristics of an advancing time horizon is that when the current year ends, it disappears from the planning horizon and a new final year is incorporated. This rotation results in a general yearly review and updating of all the company's plans. This planning flexibility enables the company to cope with a constantly changing business environment.

Planning performs a variety of functions. There is strategic planning, which defines the objectives of the company and the general courses of action required to attain those objectives. A more elaborate definition is that strategic planning determines the future business attitudes of the company with regard to product-mix and markets, profitability, growth, technological innovation, management-labor relations and its external relationships. Then there is tactical planning, which generally refers to short-term courses of action to implement longer-term strategies. A final example is project planning, a function that is closely related to operations planning and capital investments, but it can also refer to functions that can be analyzed and developed independently, such as marketing, R&D, financing, and others.

Planning analyzes the company internally and also the environmental conditions in which it operates, including the general business and economic climate, political events, technological developments, market behavior and legal constraints. These analyses are helpful in identifying a company's strengths and weaknesses, or to alert the company to outside events that can represent threats or opportunities.

Figure No. 1 illustrates how the time horizon advances year by year and also how strategic and operations planning interact with project planning.

Structuring a planning process

Before planning can commence, a company must know where it wants to go -- it must have clear, attainable objectives. In business, the objectives are likely to be economic, for instance, maximization of return on investments, cost minimization, product diversification, or others. These objectives must be such that they can be understood and shared by all sectors of the organization, because all are functionally interrelated, interactive parts of a whole, and lack of planning in one sector lessens the overall effectiveness and validity of the planning effort.

Once the objectives have been set, a framework of procedures and systems can be revised which translate the company's business objectives into viable plans. Figure No. 2 shows the elements of a planning structure, in which the determination of objectives is the first stage of the process.

It can be seen that planning requires a large body of internal and external information. Internally, the company must have a thorough knowledge of its resources, and externally, a good understanding of the constraints and opportunities that exist. All this information must be analyzed and organized to serve the purposes of planning, that is, the formulation and evaluation of technically and financially feasible plans to achieve the company objectives.

There are, of course, alternate ways towards an objective, depending on a range of complex internal and external factors. For example, a mining company may have all the mineral resources needed for growth, it may have at its disposal the best available technologies, highly skilled personnel, a well-developed infrastructure, and ample capital resources. But market conditions may be negative: demand for the products of the company may be static or threatened by substitute products; perhaps the company will be unable to produce at competitive costs at a time of low market prices; perhaps the company's competitors are promoting protectionist measures that will curtail its markets.

Financial considerations are also important. External funding, interest rates, inflation rates, the attitude of the shareholders towards the re-investment of capital funds, are among the constraints that must be taken into account.

The data gathering and evaluating processes then lead to the formulation of alternative courses of action to materialize the company objectives. These alternatives consist of series of interrelated projects to be carried out according to a coordinated schedule to achieve specific capacity increases, or operational and other improvements. After they have been evaluated exhaustively, the combination of development alternatives that can best serve the business interests of the company is recommended to the decision makers for approval as the following year's development plan. The formulation, analysis, and selection of projects; the structuring and evaluation of alternatives; the complementary financial and market analyses, and the selection of the development plan must be accomplished within a definite timeframe -- the planning cycle -- and according to procedures that are generally specified to the last detail in the planning guidelines of the company.

The planning cycle

The planning cycle, of which an example is shown in Figure 3, establishes the planning activities

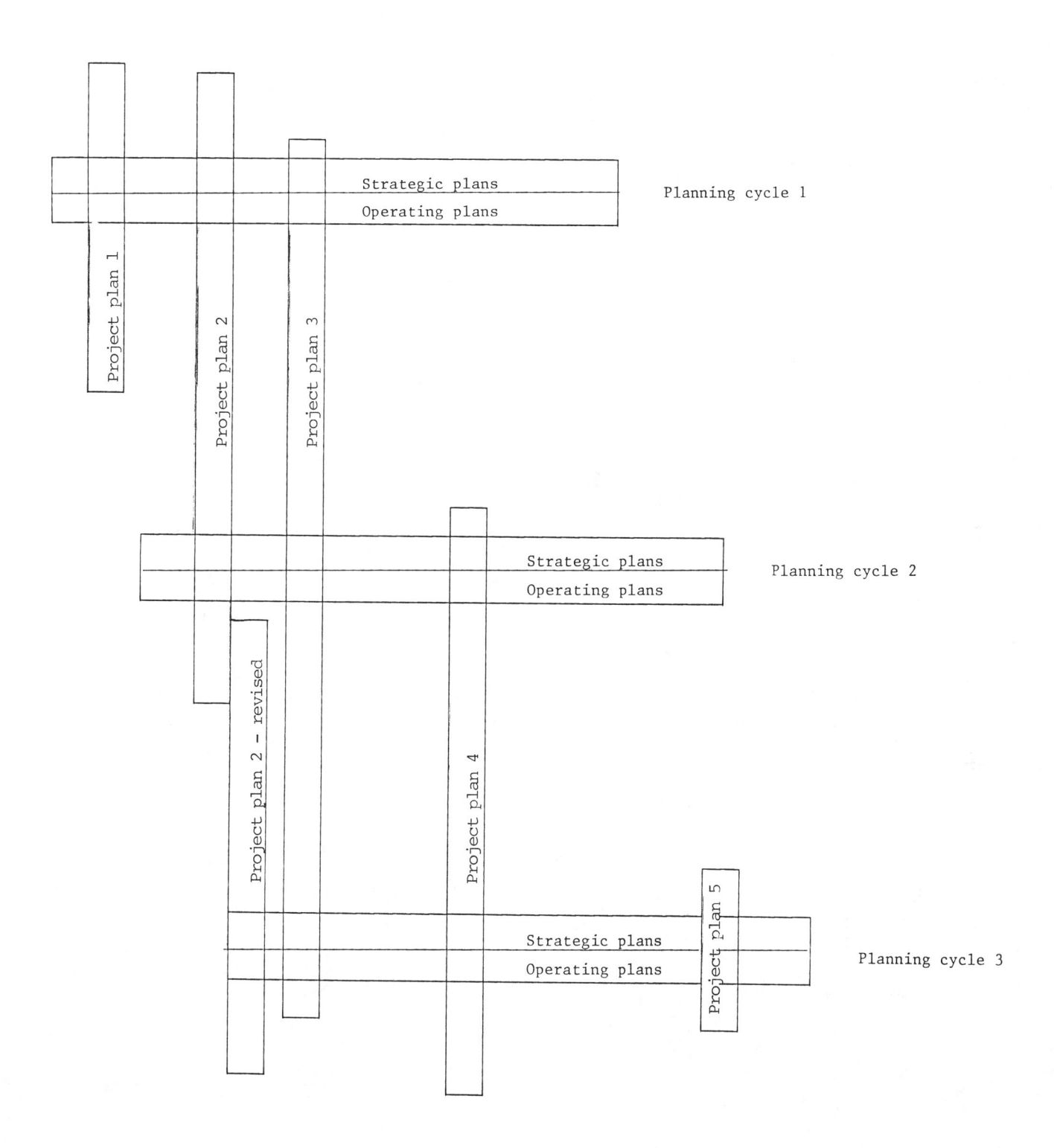

Figure 1. <u>Relations of project plans to long-range plans</u>

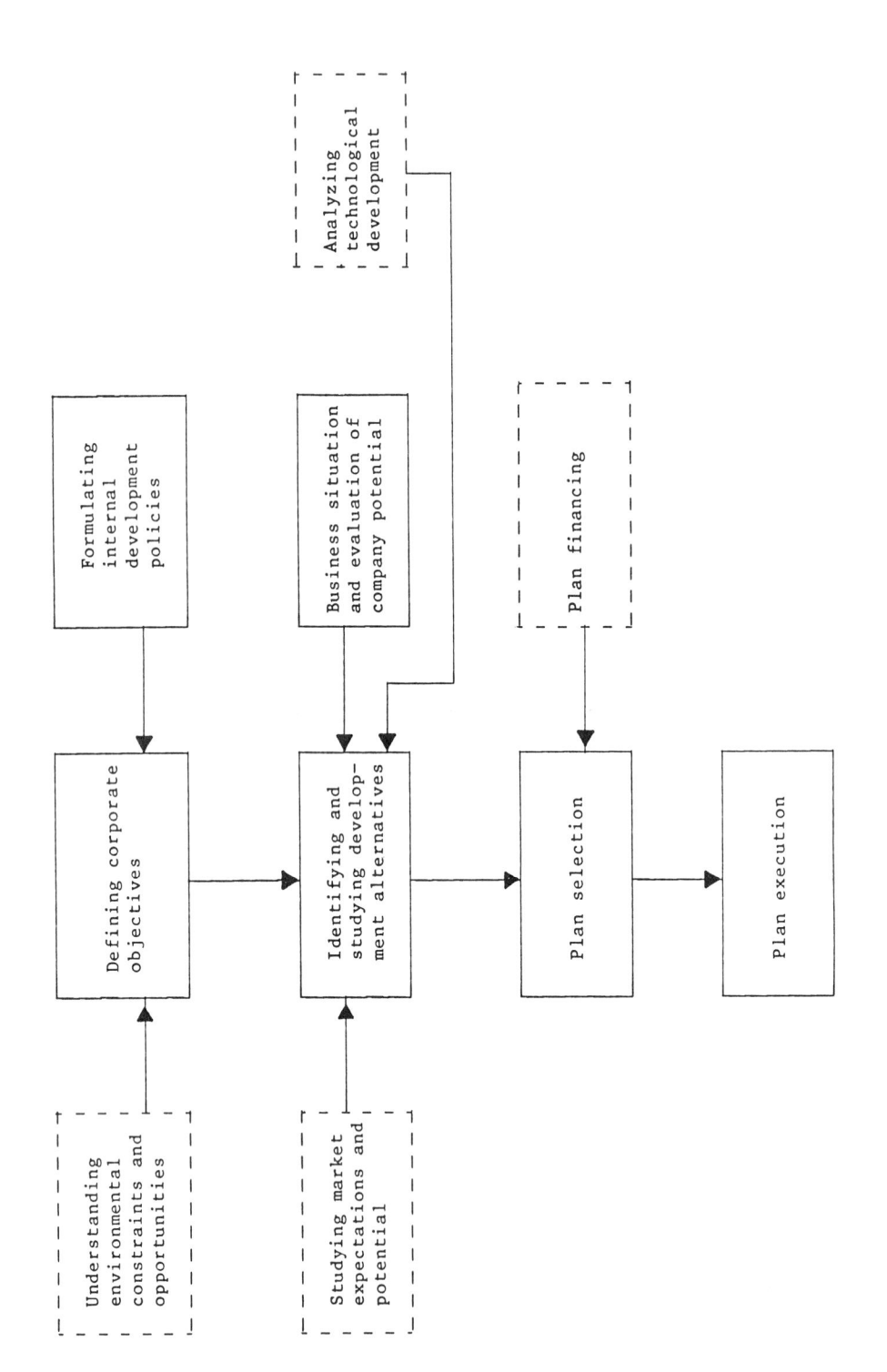

Figure 2. Structuring a planning process – Elements involved

Note: Broken line identifies external factors

STRATEGIC, FINANCIAL PLANNING

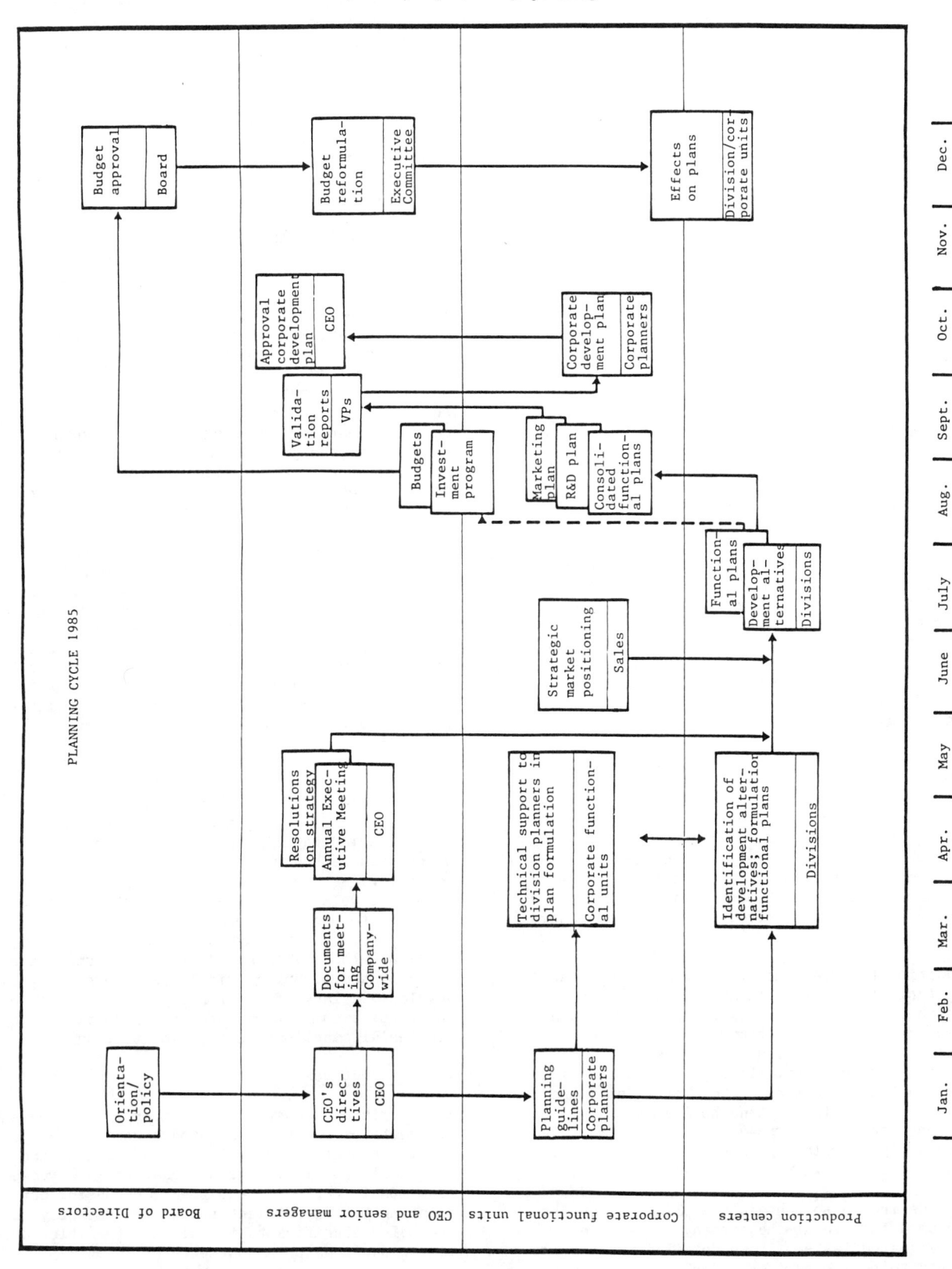

Figure 3. *Example of planning cycle*

at four different levels of responsibility and sequence in which the established planning tasks are carried out during the cycle. It may be seen that there is a two-way exchange of information going on at all times during the scheduled cycle. As in the case of the planning horizon, the length of the cycle will vary according to the nature of the business activity involved, and also according to the rate at which changes in the business environment occur. In mining, the cycle generally is a year long, and within that timeframe, a strict schedule of activities is set up, allowing a precise period of time for the completion of each task. Moreover, since the process is repetitive, the cycle, or schedule, does not vary significantly from one year to another and becomes a familiar routine with all participants knowing what they are expected to do, how it is done, and when it must be completed so that the ensuing planning task can be undertaken.

As the figure shows, at the start of the year, top management issues its directives, restating the company's objectives and policies, thus outlining in general terms the work to be carried out. At the same time, the corporate planning nucleus prepares and distributes the planning guidelines to the senior executives and planners involved in the process. This includes participants in the production centers where development options are identified and formulated and also those functional units that are incorporated in the process and are required to formulate supporting plans to improve specific functions, such as marketing, research and development, client relations, community development, and others.

Planning guidelines

In Codelco, the mechanics of the planning process are explained in the planning guidelines. Together with the directives of the chief executive officer, this is one of basic managerial and planning documents of the company. These guidelines, which were relatively simple instructions when planning was first undertaken in the company, are representative of the way the Codelco planning process has evolved over time and become increasingly complex.

In addition to establishing the scope and general characteristics of the planning process, the guidelines determine the new planning cycle and provide detailed instructions on data gathering routines, the preparation of reports, the formulation of development alternatives at the divisions, as the production centers of Codelco are called, and functional plans both at the division and corporate levels. Special attention is given to the analyses relating to strategic positioning in the markets in which the company operates.

The guidelines also contain the planning formats to be used throughout the company to carry out the designated tasks. These detailed formats are expected to be error-proof. For example, the divisions are required to report on measured, indicated, and inferred copper reserves and resources, indicating tonnages, copper and molybdenum ore grades, in accordance with the terminology of the U.S. Bureau of Mines and the U.S. Geological Survey. These reports are supported by up-dated reports on drilling.

Statistical data is also required on power consumption according to the different kinds of fuels used, on power generation and purchases and sales of energy and on the industrial and drinking water used in power generation, copper production processes and in services.

In order to formulate the development alternatives, full information is required with regard to potential production based on existing reserves. This includes computerized mine schedules with the following parameters: operating days per year, mining rates of ore and waste, stripping ratios, cut-off grade, copper and molybdenum head grades, and fine copper and molybdenum content. Also included are milling schedules for sulphide ore, indicating operating days per year, ore milling rates, copper and molybdenum head grades, copper and molybdenum recovery percentages, expected copper and molybdenum grades in the concentrates, and fine copper and molybdenum content. Equally complete data on extraction and beneficiation of oxide ores are required. Reports on smelting schedules include present and future smelting capacities, days of operation per year, copper grade in concentrates, recovery percentages, tonnages of blister and fire-refined copper produced. The refinery schedules include days of operation per year, anode loads and purified solutions processed electrolytically, efficiency rates and total cathode output. Data on casting is also requested, although this activity is highly dependent on market variations.

With the above technical data it is possible to forecast the detailed overall marketable production of each division, and the company as a whole, including all the electro-refined and electrowon cathodes, blister and fire-refined copper, wire bars, copper concentrate, and copper forms, as well as the by-products that Codelco can ship to world markets.

Since all the divisions and corporate units are bound by the planning guidelines, the company has developed a large fund of homogeneous, reliable information that it can use in a wide variety of managerial functions in addition to planning.

The alternatives

In the mining industry, any number of development alternatives may be formulated based on the available mineral resources, technologies, and technical capacity. The alternatives can range from merely replacing equipment or phasing out an existing but inefficient production capacity to proposing significant, technically feasible capacity expansions or adjustments.

The formulation of alternatives at the operating level generally does not contemplate marketing of financial constraints in designing future scenarios, unless such limitations are real and are bound to effect the company no matter what plan is eventually approved.

A variety of good development alternatives will lend flexibility to the decision making process and lead to the approval of a development plan. Among the alternatives it is always important to include a base case to be used as a reference in measuring the merits of the various development options being considered. The base case consists of the best minimal course of action open to the company in the event that none of the development alternatives formulated are approved, either because of a lack of investment funds or because none of the alternatives proposed turn out to be significantly better than the base case.

In mining the natural characteristics of the mineral deposit involved are the prime consideration in defining a base case. If the ore grades in the deposit are comparatively low and the costs of production high then there is room for doubt that the mining venture will remain profitable. Should this occur, the course of action to be contemplated in the base case could be the immediate closure or a programmed phasing out of the operation. However, if the situation is not entirely negative, the base case should consider the possibility of continuing the operation at the level of activity requiring the lowest investment.

In the case of a competitive mining operation with low production costs, based on high grade ores, and where there is no doubt that the operation will continue, the base case should consider what level of exploitation is appropriate to obtain the highest return without further investment. Since closing the mine obviously is not an option in this situation, the base case will probably propose maintaining current ore extraction levels or the prevailing output of fine metal. There will be intermediate options, depending on the estimated efficiency of each and the lowest amount of capital the owners are willing to invest or reinvest in the operation.

The base case must represent a realistic option, that is to say, it must be the best of all the minimal feasible alternatives. Otherwise, if a financially unfavorable plan is adopted, other slightly more favorable plans may appear profitable by comparison.

As we indicated earlier, the information relating to a development plan consists principally of the investments involved, complete production data, and detailed information on operating costs. Figure 4 presents an example of a development alternative which is compared against a base case scenario.

The estimated investment is the sum of the capital costs of all the projects combined in a given development alternative intended to improve the level of operation of the business.

Future output of each product and by- or co-product is estimated, taking into account the technical improvements to be made, the expected output and efficiency of the installations that are contemplated and the capacity increases programmed at the different stages of the production process. With the results of these analyses, it becomes possible to estimate the associated future operating costs.

One important premise is that the implementation schedule first outlined for each development alternative under consideration is the "best" possible, and that any change of timing in the execution and start-up of projects has a negative effect on the economic efficiency of the respective plan. Due to this, each alternative must be thoroughly tested by trial and error to obtain the best final version.

Participation of functional units in the planning process

Once the formulation of a development alternative has been completed, it must be technically validated by the pertinent corporate functional unit, generally the engineering department in the case of mining companies. This entails reviewing the assumptions used to define the associated efficiency and productivity increases to be achieved; the technical feasibility of the interrelated projects proposed; the cost criteria applied, and the general coherence and homogeneity of the alternative. The reviewing unit also verifies the correct application of the planning procedures and methods devised at corporate level for use throughout the company.

Following the technical validation of the development alternative, an economic evaluation from the corporate standpoint takes place. Although at this stage all the projects and the alternative as a whole will have been carefully evaluated by the production center planners, the corporate review is considered necessary to:

i) ascertain that a uniform evaluation criterion has been applied at all stages of the process, and

ii) that the alternatives can be submitted to subsequent comparative analyses, or that they can be combined and recombined to define the overall corporate development plan. This flexibility is required to adjust the plan to the financial limitations that frequently arise in development planning. The analyses must therefore provide for possible downward adjustments if the funds allocated for the plan are insufficient to carry it out in all its parts.

Investment profile
$ million

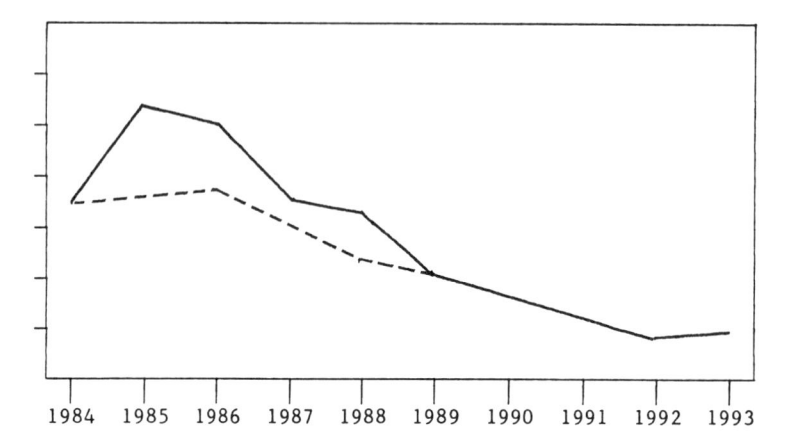

Production profile
000' tonnes

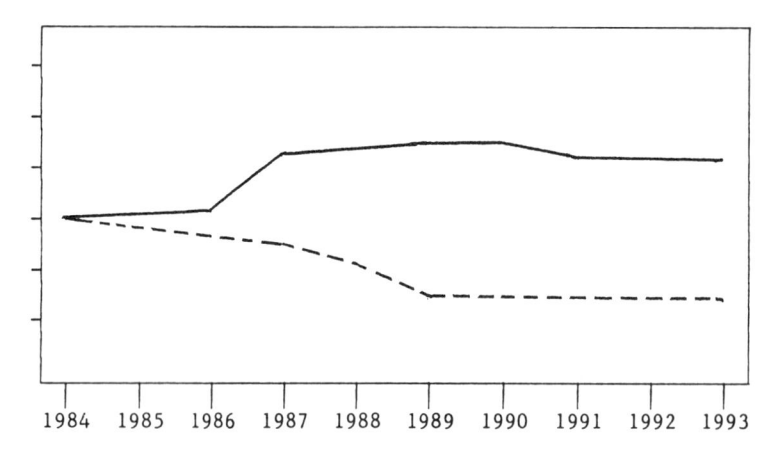

Operating costs profile
¢/lb

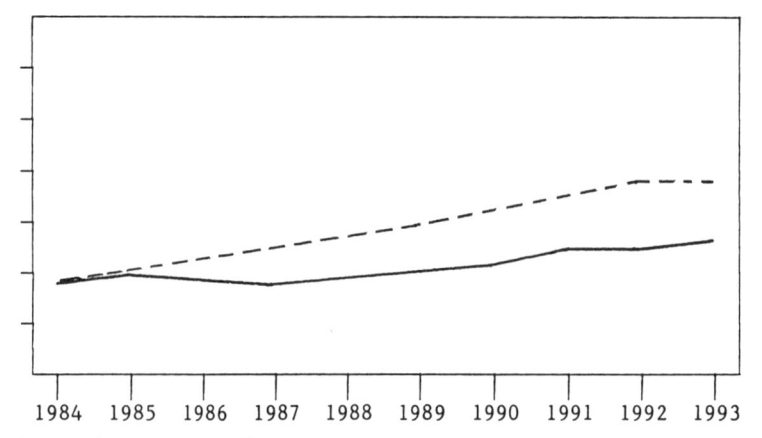

Figure 4. <u>Presentation of development alternative and base case</u>

Development alternative being considered - unbroken line
Base case - broken line

The analytical process is then taken up by the corporate financial and marketing units which have no counterparts in the production centers. The role of the marketing analysts is to consider how the aggregated output increases provided for under the alternatives match future market expectations. Among the aspects that must be considered in this connection are the proposed product mix and production increases; the possible effect on the market of increasing supplies; the continuing validity of the assumptions on prices, premiums and discounts on sales used in formulating the alternatives, and the possibility of obtaining a more flexible product mix by means of outside custom treatment of raw materials. These market conditions must be studied in the light of strategic positioning studies conducted by the company of each of the markets where it is active and in connection with each of the products it has to offer.

When the economic viability of a plan has been established, financial analyses are undertaken to determine the investment funds it requires and whether these funds can be obtained. This is done by projecting future results of the plan in relation to a pre-established scenario, defining the sources of funds, and determining how the funds are to be applied.

To establish the sources of funds, it is necessary to first clearly formulate a profit retention policy. This makes it possible to compare the advantages of a range of financing alternatives based on external loans. Whatever the resulting amount of credit may be, the cash flow projection should leave no doubt that repayment of the debt in normal terms will be possible. The company's debt capacity is determined on the basis of a maximum debt-equity ratio.

Future price and risk

Forecasting is necessarily used in evaluating plans. Some forecasts are comparatively easy to make, as, for instance, those relating to physical variables defined by applying known technical relationships, such as the metal content of concentrates -- which can be estimated from the relevant ore characteristics, the applicable processing technology, and the scale of the installations.

Other forecasts are elusive because they refer to variables that present erratic patterns of behavior and also because they are beyond any company's control. In mining, one such variable is the future price of the commodity. In the case of copper, the economic results of an operation are extremely price sensitive. In spite, however, of the high degree of uncertainty and subjectivity inherent in estimating future prices, these forecasts must still be made and applied in order to be able to analyze future cash flow patterns, the feasibility of servicing obligations to be incurred, and the profit expectations of a given plan. Recognizing the difficulties involved, the price forecast will consist of the best possible approximation to the results expected from the plan after evaluating all the other variables. On the one hand, the risk of forecasting an overly optimistic future price is obvious; on the other hand, a pessimistic price forecast can result in the rejection of ultimately worthwhile projects or plans. Therefore, one useful complement to an economic analysis of this type is to define an equilibrium price to measure the sensitivity to future prices of the plan. The equilibrium price is actually the future constant price required to obtain a rate of profitability at least equal to the minimum acceptable rate of return on the investment to be made. Thus, a typical decision company managers must face is to define the minimum price level that will allow them to recommend implementation of the plan. This decision reflects the degree of risk that management is willing to accept for the company.

In mining, the natural characteristics of the ore body will often give an indication of the risk involved, but risk will never be eliminated entirely, being, as it is, inherent in any type of business.

FINANCIAL CONSIDERATIONS

There are two distinct producer groups in world copper mining, with sharply contrasting features. One group is composed of companies that may be financially poor but are the owners of highly profitable ore deposits, while the other is made up of generally more wealthy companies operating comparatively low-grade mines. The first group includes most of the mining operations in developing countries, whether publicly or privately owned, and the second is represented by the leading copper producers in the industrial countries.

The current economic recession has impacted all sectors of the world copper industry with extreme severity. Certainly both producer groups have had to face financial hardships, but the point we would like to emphasize is that the causes of their respective difficulties are radically different.

Many of the low-grade deposits have been operated at a loss for many months, and in some cases, for years, and others have been closed temporarily or permanently by their owners. These operators are struggling to survive until conditions improve for the industry. It is not that they do not have access to financial resources, for these companies can resort to equity investments, bond issues, and equipment leasing, but their effort is to encourage lenders by matching risk to reward expectations.

In contrast, it has become evident that the operators of high-grade mineral deposits are in a much better position to continue operating even if adverse market conditions last much longer. Nevertheless, although these businesses are economically sound, they are highly vulnerable to financial constraints resulting largely from the generalized recession. The recession has had dire effects on the developing economies, which are now finding it extremely difficult to service their foreign financial commitments. Thus, the only sources of capital funds available to these companies are short-term lines of credit and long-term foreign loans, the latter of which could easily be arranged with private banks of the industrialized countries until recently, but are now more difficult to obtain.

In sum, the copper mining industry is in a period of transition in which painful adjustments shall be made. However, in the long term, and assuming minimal protectionist constraints, a different copper industry will emerge in which only cost-efficient companies, capable of offering competitively priced products, will find space.

Our experience in Codelco has proved that planning is the best instrument available to cope with the periodical ups and downs of the copper trade, and particularly with the far-reaching changes that are foreseen in metal consumption patterns.

CHAPTER 8: STRATEGIC AND FINANCIAL PLANNING

BIBLIOGRAPHY

Albert, K. J., ed., 1980, "Top Management Strategy, Planning, and Control," Handbook of Business Problem Solving, McGraw-Hill, New York.

Ansoff, F. H., 1965, Corporate Strategy, McGraw-Hill, New York.

Bierman, H., Jr., 1980, Strategic Financial Planning: A Manager's Guide to Improving Profit Performance, The Free Press, New York.

Bing, G., 1980, Corporate Acquisitions, Gulf Publishing Company, Houston, Tex.

Bing, G., 1980, Corporate Divestment, Gulf Publishing Company, Houston, Tex.

Bradley, J. W. and Korn, P. H., 1981, Acquisition and Corporate Development: A Contemporary Perspective for the Manager, Lexington Books, Lexington, Mass.

Buzzell, R. D., 1983, "Is Vertical Integration Profitable," Harvard Business Review, Jan.-Feb., pp. 92-102.

Coopers and Lybrand, 1982, Checking Into an Acquisition Candidate, Coopers and Lybrand Printing.

Fogg, C. D., 1976, "New Business Planning: The Acquisition Process," Industrial Marketing Management, Vol. 5, pp. 95-113.

Fruhan, W. E., Jr., 1979, Financial Strategy, Richard D. Irwin, Homewood, Ill.

Gourgues, H. W., Jr., 1983, Financial Planning Handbook, New York Institute of Finance, New York.

Hall, W. K., 1980, "Survival Strategies in a Hostile Environment," Harvard Business Review, Sept.-Oct., pp. 75-85.

Hamermesh, R. G. and Silk, S. B., 1979, "How to Compete in Stagnant Industries," Harvard Business Review, Sept.-Oct., pp. 161-168.

Harrigan, K. R. and Porter, M. E., 1983, "Endgame Strategies for Declining Industries," Harvard Business Review, July-Aug., pp. 111-120.

Harvard Business Review, 1983, Financial Management, Wiley, New York.

Harvey, J. L. and Newgarden, A. M., ed., 1969, Management Guides to Mergers and Acquisitions, Wiley-Interscience, New York.

Hilton, P., 1970, Planning Corporate Growth and Diversification, McGraw-Hill, New York.

Ise, J., 1974, "The Theory of Value Applied to Natural Resources," American Economic Review, Vol. 64, pp. 284-291.

Miller, E. C., 1971, Advanced Techniques for Strategic Planning, AMA Research Study 104, American Management Association, New York.

Mullins, D. W., Jr., 1982, "Does the Capital Asset Pricing Model Work," Harvard Business Review, Jan.-Feb., pp. 105-114.

Navin, T. R., 1978, Copper Mining and Management, University of Arizona Press, Tucson, Arizona.

Porter, M. E., 1980, Competitive Strategy: Techniques for Analyzing Industries and Competitors, The Free Press, New York.

Rothschild, W. E., 1976, Putting It All Together: A Guide to Strategic Thinking, AMACOM, New York.

Rothschild, W. E., 1979, Strategic Alternatives: Selection, Development and Implementation, AMACOM, New York.

Salter, M. S. and Weinhold, W. A., 1978, "Diversification Via Acquisition: Creating Value," Harvard Business Review, July-Aug., pp. 166-176.

Salter, M. S. and Weinhold, W. A., 1979, Diversification Through Acquisition, The Free Press, New York.

Sammon, W. L., Kurland, M. A. and Spitalnic, R., 1984, Business Competitor Intelligence: Methods of Collecting, Organizing, and Using Information, Ronald Press, New York.

Singhvi, S. and Jain, S., 1974, Planning for Corporate Growth, Planning Executives Institute, Oxford, Ohio.

Steiner, G. A., 1979, Strategic Planning: What Every Manager Must Know, The Free Press, New York.

STRAAM Engineers, Inc., 1979, Capital and Operating Cost Estimating System Handbook, U.S. Bureau of Mines Contract J0255026.

Uyterhoeven, H. E. R., Ackerman, R. W., and Rosenblum, J. W., 1977, Strategy and Organization: Text and Cases in General Management, Irwin, Homewood, Ill.

Vancil, R. F., ed., 1970, Financial Executive's Handbook, Dow Jones, Irwin, Homewood, Ill.

VanHorne, J. C., 1983, Financial Management and Policy, Prentice-Hall, Englewood Cliffs, NJ.

Wall, J. L., 1974, "What the Competition is Doing: Your Need to Know," Harvard Business Review, Nov.-Dec., pp. 22-166.

Yavitz, B. and Newman, W. H., 1982, Strategy in Action: The Execution, Politics, and Payoff of Business Planning, The Free Press, New York.

Chapter 9

Risk Analysis for Mine Finance

CHAPTER 9: RISK ANALYSIS FOR MINE FINANCE

PREFACE

For most financial institutions the key to lending money is not just to find projects that need cash, but to determine which projects have satisfactory risk profiles so they can repay the loans advanced. Companies funding a project must also consider the risks to potential profitability of the project before allocating development capital. To properly evaluate viability, the specific project risks must be identified and an educated decision made as to the acceptability of these risks. To help define the project risks associated with mineral ventures and to suggest ways to minimize these risks, eleven papers have been included in this chapter. "Analysis of Risk Sharing" by C. Richard Tinsley outlines the different systems of risk analysis and defines 14 key risks which must be considered in financing mineral ventures, particularly "project financings" where the risks are shared among the company and the lender.

Reserves are the fundamental basis of any mining project. Operating costs and profit margins are based on the volume and grade of ore produced. In "Understanding the Risks in Coal Reserve Estimates" by James E. McNulty, volume or quantity determinations are discussed focusing on the qualifications, experience and objectivity of the estimator. The risk of ore grade or quality are discussed in great detail, and are viewed as equally important as the risk of ore quantity or volume. McNulty suggests that those risks can be minimized by careful selection of the testing facilities and supervisor.

Once a mineral deposit has been identified and accurately delineated, the next question which must addressed is: Can the mineral be extracted profitably? A discussion of the risks which a company entering a mining venture must consider is presented in C. Allen Born's paper "Operating Risk." Emphasis is not limited only to cost risks, such as development and replacement costs, operating costs and financial costs, but Born suggests that a key issue is whether the producer will be capable of meeting various industry operating averages in ore production.

In "Operational Risk Assessment in Mining Enterprises," by Peter J. Szabo, five important areas for operating cost control are discussed: management, labor, equipment availability, maintenance, and climate. Each of these can cause the demise of an otherwise ordinary project. The financial analyst should have a knowledge of the project and environment (labor pool, climate, operator's experience, etc.) in order to evaluate the acceptability of the inherent risks.

With international projects additional risk of interest rate fluctuations and currency movements are imposed on financing. In "Hedging Interest Rate and Exchange Rate Risks" James L. Poole describes how interest rate risks can be offset with the use of equity financing, hedging of financial futures, or interest-rate swaps. Poole also discusses how exchange rate risks may be minimized by forward markets, hedging on the future markets, or borrowing in the currency that the product will be sold.

A change in the political situation within the host country can cause major variations in the profitibility of a mining project. "A Systematic Approach to Political Risk Analysis" by William D. Coplin and Michael K. O'Leary outlines the development of matrix system for a political risk analysis, called the Prince System. Some interesting tables are also presented on the risk of various world producers of minerals. In "Country Risk Analysis," by John R. Stuermer and Peter Allen, the ability and willingness of a country to repay external debt is discussed. Two examples of this are Chile and Mexico. Potential factors affecting country risks which are discussed include the effectiveness of government control, social unrest, as well as external economic and political pressures.

In order to promote economic development of the less-developed countries, the U.S. government established the Overseas Private Investment Corporation (OPIC) to insure some political uncertainties. B. Thomas Mansbach's paper "OPIC Insurance Programs for the Mining Sector" discusses coverage of four basic types of political risk: currency, inconvertibility, expropriation, loss or interference with operations caused by war or revolution.

In "Feasibility Studies and Other Pre-Project Estimates," Grover R. Castle selected 18 actual mine financings since 1965 to analyze one of the key risks in a project development: completion. This paper also reviews the actual performance of these projects compared with feasibility study estimates and concludes that "there is a very high probability that any new project shall run into problems and even after construction has been completed, there are plenty of risks that can cause problems."

Generally, construction starts before a project is funded. It is important to monitor costs during construction because of capital spending limits and to be assured that the money is well spent. "Cost Overrun Risk and How It Was Minimized During Construction of the Mt. Gunnison Mine" by Les P. Haldane describes a successful construction project for a Colorado coal mine. The final paper in the chapter by Charles Berry, "Conventional and Non-Conventional Risks Insurance for Mining Projects," describes how risks insured under conventional policies relate to the lenders' interests. Berry also discusses trends in unconventional insurance cover available from the private insurance market.

An understanding of the risks influences the judgement about a mineral project's financial rewards. It is encouraging to see the authors provide guidance and solutions for risk evaluation and mitigation in addition to describing the high degree of risk which characterizes the operations of the minerals industry.

John C. Robison III
Chapter Editor
First National Bank of Chicago

9.1

ANALYSIS OF RISK SHARING

C. Richard Tinsley

European Banking Company Limited
London, England

INTRODUCTION

The economic analysis (Chapter 3), the engineering studies (Chapter 10), the credit structure (and the consequential funding sources) - Chapter 11, and the overall feasibility structure for any venture are all directed to determining one major decision element: How much and how certain is the reward for taking these risks? Risk assessment is a surprisingly weak area of analysis and is usually addressed in a checklist fashion. It is much preferable to deal within a risk system so that the inevitable tradeoffs among risks are correctly made.

DEFINITION

The first step toward risk definition must be to ascertain all of the risks to which a project or a company is exposed; otherwise one will remain ignorant of the possibility of a financially crippling loss until it occurs. (Carter and Crockford, 1974). However, simply stating what "risk" is can lead to debate. For example, risk could be (adapted from Vaughan, 1982):

- the chance of a loss
- uncertainty of loss
- possibility of loss
- uncertainty
- dispersion of actual from expected results
- probability of an outcome different from the one expected
- possibility of an occurrence of an undesirable contingency
- a condition in which a possibility of a loss exists.

For the minerals industry financier, the last definition is probably the most appropriate. However, from a mineral company's point of view the definition might well include the possibility of not achieving the expected financial return.

RISK CLASSIFICATION

Risk identification will be readily agreed as the all-important first step. There are five major "branches" of risk classification. These are:

- Statistical
- Modellers
- Checklist
- Project Financing
- Insurance

The Statistical set discovered probability theory and any risk discussion soon deteriorates into a mathematical jungle (Megill, 1977; Hartman, 1976; Reutlinger, 1970; and Pouliquen, 1970). Their use of terminology tells its own story: coin toss (binomial distribution), gamblers' ruin, delphi (oracle) techniques, monte-carlo simulations, random walks, etcetera. These techniques inadequately represent the judgements necessary in an increasingly complicated and interrelated world.

The risk Modellers have turned to the computer to handle these inter-relationships but soon find barriers because of "simultaneity". Certainly the computer has its uses to conduct various analyses (marketed as "sensitivity" analyses) but very often one finds that this involves changing one or two variables at a time, all other things being equal. Unfortunately, in a modern and complex world all other things do not remain equal. The statisticians will, no doubt, jump to their own defence by citing this

---

Based on presentations to the SME Minerals Resource Management Committee Short Course, "Project Financing in the Development of Mineral Resources", SME - AIME Annual Meeting, Dallas, February 13-14, 1982 and the Canadian Institute of Mining and Metallurgy's Second Mineral Economics Symposium "Risk Management in Mining", Vancouver, Canada. November 21-23, 1982.

very point as the reason probabilistic tools are necessary. But the inter-relationship of risks is difficult to quantify and represent mathematically. Translation : specifying the covariances tends to be the stumbling block in practice. (Squire and van der Tak, 1975).

The Checklist champions rely on a list of key points to audit and, q.e.d., the risks are covered. Incredibly, for some of these, there is not even a discussion of what risk might be. (Nevitt, 2nd Edition). This checklist approach is particularly rampant in the feasibility study business and is a major reason why less than 10% of the 185 feasibility studies presented to me for loan applications have provided satisfactory raw material on which to base a decision. (Tinsley, 1982). Only three of these studies contained a risk discussion. Checklists are useful but should not be the only technique applied. Their popularity springs, I believe, from over-specialisation of team members. At least, the insurance industry recognises: "It is impossible to (develop) a complete (check) list" but then goes on the note: "In any case it is far better for a responsible official to prepare his own questionnaire" - another checklist! (Carter and Crockford, 1974).

The Project-Finance community discovered risk classification early in its career and has generally done the best job in defining the various risk categories and the methods of analysis applied to structuring their financing documentation. Appendix 1 gives a compilation of 11 of the better discussions by various bankers or their engineers. It is disturbing to realise that there are glaring omissions from this tabulation, most notably, environmental risk.

Finally, the Insurance community treats risk quite differently from all of the above. Their attention is generally focused on asset loss/reinstatement (See Berry's paper in Chapter 9) and the industry is fond of distinguishing between "pure" and "speculative" risk wherein pure risk involves only the possibility of a loss. In general, pure risks are the only ones that can be insured whereas speculative risks (where there is also the possibility of a gain) are uninsurable. This is changing fast, however, as insurance underwriters begin to discover that project "speculative-risk" insurance often rates a higher premium. Thus far, this new area is called "business interruption" or "specialty risks" insurance, which looks increasingly like project finance. In fact, these underwriters admit "they're acting as investment bankers and venture capitalists, dealing with risks the (insurance) industry considers uninsurable". (Le Roux) However, a dictum of insurance is that the portfolio should comprise a relatively large number of independent (pure) risks. Otherwise claims tend to come in bunches. Besides keying off actuarial tables based on historical data, Insurers - like the Modellers - run afoul of the overlap of risks.

RISK SHARING BY PROJECT FINANCING

The main attraction of project financing is the ability to pass certain risks to the lenders, although this has a perceived cost (risk premium) and lengthens the negotiating process. (Tinsley, 1980). It is possible to use project financing to isolate a specific project from the ongoing business of the company and thereby not jeopardize the debt capacity of the organization. For example, a very large, new mining project, which is bigger than the existing total business of the company, would be an ideal candidate for project financing. Another example of the successful application of project financing is where more than one mine is being developed at the same time. One or two of the mines may be project financed. Yet a further example of the suitability of project financing to the mining industry is where a project financing is arranged from an existing mine to generate acquisition funds or even develop another mine. A project financing could first be arranged from one mine to develop a second and, a few years later, from the first two mines to develop a third mine.

Misconceptions

A bank will undertake to absorb the consequences of certain risks and whenever it agrees to do so without further recourse to the other assets or financial strength of the sponsors, the finance is often called non-recourse financing. There is no such thing as totally non-recourse financing. Many types of project financings are "limited recourse" where the limits on the recourse vary from project to project. But some types of project financing can have full recourse to the sponsors.

A second misconception is that limited-recourse financings are, by definition, always off-balance sheet. There is a great deal of confusion between the risk-sharing nature of project financing (described here) and the balance sheet treatment which is a matter for accountants. Some varieties of project financing are recorded on the balance sheet but usually not as a long-term debt item (important); others are simply outlined in the notes accompanying the financial statements; and a few project financings are totally off-balance sheet and do not even appear in the notes.

RISK CATEGORIES

The categories of risk sharing in a minerals project financing are as follows:

Within the Company's Control
Operating: Technical
             Cost
             Management
Participant
Engineering
Completion

Outside the Company's Control
Reserve
Market
Infrastructure
Environmental
Political
Force Majeure: Temporary
                 Permanent
Foreign Exchange

Within the Bank's Control
Syndication
Funding
Legal (?)

The terms "Commercial", "Project", or "Economic" risk are too imprecise to be useful. (See Appendix 1 for other Risk Classifications.)

CASHFLOW CATEGORIES

A classification of the impact of each project risk on cashflow predictability is as follows:

Cash Flow Calculation	Main Risk Impact Thereon
Tons Recovered	Reserve, Force Majeure, Technical
Times: Prices Received	Market, Infrastructure
Equals: Gross Proceeds	
Less: Royalties, Fees	Political
Operating Costs	Cost, Management, Environmental
Interest Expense	Funding, Syndication
Depreciation/Amortization	Completion
Overhead	Participation
Taxes	Political, Legal
Equals: Net After Tax	
Plus: Depreciation/Amortization	Completion
Less: Capital Expenditures	Engineering
Loan Repayment	Debt/Equity Ratio, Funding, Foreign Exchange

The following is a summary of the various components of risk and the general rules applied to each.

WITHIN THE COMPANY'S CONTROL

Operating Risk

(This is also known as the production risk).

The operating risk has three interrelated components: technical, management, and costs.

The ability to economically extract the reserve depends on the engineering, experience, and quality of staff applied to the project.

Technical Component: As a general rule if:

(a)   Known and proven technology is used;

(b)   The facilities are projected to remain technologically competitive; and
(c)   Plant life is twice loan life;

then the banks will take the Technical Risk. If an untried process is incorporated into a new mining project, some banks may require a process warranty. Project financing is seldom applied to new technology. Even though the Technical Risk has been passed to the banks, continuing covenants will be required to ensure that the mining company applies generally accepted practices and obeys the applicable laws including environmental regulations. (This is known as the "prudent operator" clause).

Cost Risk Component: A key element in the cash flow projections, which are the basis of assessing project financings, is the operating cost projection. This Cost Risk component is sometimes partly absorbed by the banks by means of an economic test. However, the size of a project financing loan will depend on the scope of the economic test, if any. The selection of the stream of proceeds will also influence the size of the loan. For example, if gross proceeds are dedicated to loan repayment - common in a US oil project financing - then the company is taking the operating cost and tax risk. However, the economic test may override the dedication of proceeds to the project financing.

One way banks evaluate Cost Risk in the absence of adequate sales contract coverage (Market Risk) is to examine the position of the mine on the "cost curve" relative to all the other producers of the given commodity. This is regularly done on a formal basis for operating and total costs in metal-mine financing and is undertaken implicitly for coal-mine project financing. As a general rule, a mine in the lower half or lower third of the cost curve will be readily able to pass the Cost Risk to the banks.

Management Component: The experience of the management in applying the given technology is crucial when considering the projected operating performance of a project. The availability of a sufficiently trained workforce may also need close examination in some locations. The banks may seek an employment contract or "key-man" clause to ensure the continued involvement of the company and of strategic individuals. A technology-supply agreement may be needed to cover this risk component.

It is difficult to define the dividing line separating these three components of Operating Risks, but all are generally under the control of the sponsors and are easier to absorb if the sponsors have a successful history of building and operating similar projects.

Participant Risk

(This is also known as the credit or sponsor risk)

The stature of the other companies in the

project may have an impact on the project financing especially if such participants are weak financially or technically. If there is a weak or inexperienced participant in a joint venture, the lenders may require gross guarantees.

It is very important to study the joint venture agreement (JVA) since its structure may seriously inhibit project financing. In general, assignment of the joint venturer's unencumbered interest should be possible without consent for the purposes of financing. The cancellation, abandonment, and force majeure clauses may be crucial to the project finance document, which, after all, has its roots in the JVA itself. Close scrutiny of the JVA will concentrate on provisions for forfeiture, dilution of interest, and compulsory contributions by the other joint venturers. (Ladbury, 1979).

Engineering Risk

(Also known as design risk)

This poorly understood risk is often counted as part of Completion Risk since engineering and design flaws become quickly noticeable as the project encounters difficulty in construction or startup. The risk here revolves around the poor quality of the engineer/design work which can also have a crippling impact upon the cashflow stream well after the (excess) capital has been spent pre-completion to counter the hardware problem.

This risk can arise from poor professional advice or the selection of an inappropriate or inexperienced firm for the technology or location involved. (See Berry's paper in this Chapter for some insurance aspects of this risk.)

Completion Risk

(Also called the construction, development, or cost overrun risk).

Broadly speaking, a lender expects loan proceeds to be spent on building a project which is on time, at budget, and is capable of producing sufficient cash flow after completion to repay the loan comfortably. The Completion Risk is usually not taken by banks in project lending and financial support is necessary prior to completion.

The usual format is to structure an objective completion test which, when satisfied, signals that the pre-completion supports and undertakings are released and the project has commenced its limited-recourse status. The completion test is usually not the same as the engineering/construction completion used by the mine construction contractor.

There are two main types of completion tests. One calls for construction to be finished by a certain date and the loan is immediately due if completion has not occurred by that fixed date. A more common test is a performance completion test which might have all or some of the following types of components:

a. Continuous operation for A consecutive months;
b. [90%] of designed production achieved and delivered;
c. B cubic metres of stripping at C dilution;
d. Acceptable process plant recoveries/availability;
e. On-specification product(s); and
f. Achieve a defined operating cost per tonne.

Completion tests are becoming more complex as banks become more experienced in sizing up the acceptable levels of risk and as companies realise the possibilities of risk shedding while limiting the impact on their balance sheets. Additional completion test categories now seen include the following:

a. Sales completion test;
b. Reserve life greater than D years;
c. Net present value of net cash flows greater than [150%] of loan outstanding;
d. Financial covenants on working capital, debt:equity, net worth, etc. are satisfied; and
e. Minimum ore grades.

The implication of a company accepting the Completion Risk is that cost overruns must be met by the company and, in the case of completion not being attained by a certain date, not achieving a project financing at all and the loan becomes a corporate credit.

OUTSIDE THE COMPANY'S CONTROL

Reserve Risk

(This may also be known as the reservoir or supply risk).

The banks can almost always accept the reserve risk and the estimation of reserves has become a key element in banks' evaluation of a mine project financing. Most banks will only accept proven reserves. The link of the reserves to the mining plan is especially important for underground mines where poor mining practices can isolate some of the reserves.

As a general rule, mineable reserves should be twice as large as the reserves to be mined during the loan repayment period or, put another way, the loan repayment period cannot be more than half the mine's life. If reserves are believed to be insufficient, the banks may ask for a reserve warranty.

Market Risk

(Sometimes called the sales or price risk).

Market Risk occurs, for example, when the sales price falls; market share drops (perhaps due to a shift in freight rates or due to a new entry by a lower-cost producer); demand for a product ceases; or sales are lost due to deteriorating quality of the project's output. In certain circumstances, the risk arises where sales contracts are cancellable after a period of below-par deliveries. This is one of the most critical areas of risk absorption by the banks.

The best coverage of Market Risk comes from long-term offtake contracts extending beyond the end of the loan life, although sales contracts are not often feasible or desirable for certain commodities. These contracts should be with a reliable consumer, adequately cover escalation and, preferably, have floor prices. As a rule, contracts with three-year price reopeners will be regarded as only three-year contracts even though the deal is for a "volume" or "evergreen" contract. Contracts will be evaluated with the minimum tonnage option and after discounting for other delivery conditions.

Long-term sales contracts are more prevalent in certain commodities such as steam coal although it is usually preferable not to have mines fully contracted as it may be more beneficial to encourage a miner to retain some tonnage for spot sales to capitalise on short-term price fluctuations.

Special techniques have been developed to handle shorter-term sales contracts, although it would be preferable for some tonnage to be sold under a longer-term contract to minimally cover the debt. The competitive position of a specific commodity to other product substitutes in the market is examined in detail and a technical link with a consumer is deemed highly desirable. The availability of alternate markets (e.g. the sale of coking coal as steam coal) is regarded as a positive contribution to minimising Market Risk.

Sometimes a sales completion test will be required as well as the performance completion test (See Completion Risk above) before the banks assume the marketing risk. As a general rule of thumb, the banks aim for the net present value of the sales proceeds net of operating costs to be at least 1.5 times the amount of loan outstanding. Sales completion is sometimes written as an ongoing obligation where the sales contracts are very short.

Infrastructure Risk

(Also called the transportation risk).

Transportation is a very important component of many mineral projects and banks will need to be assured that the chosen infrastructure will remain technologically and economically competitive over time, especially in comparison with other existing or potential production centers. The materials handling required to deliver or export the production can sometimes equal or exceed the mine's operating and capital costs. Moreover, port capacity may become the limiting factor for remote projects.

If the sales price is not on an f.o.b. mine basis, the ability to pass through transportation costs in any sales arrangements means that the producer absorbs this risk. In some locations, the government builds and operates the rail and port, but one must be careful that the operator is not "taxed" through higher freight rates or user fees. In general, the banks will wish to review closely the transportation studies and, perhaps, to inspect the existing facilities or the proposed sites. Independent certification will assist the bank's absorption of this risk component.

Environmental Risk

The cost and market impacts caused by environmental pressures on minerals projects is well described by Simon Strauss in Chapter 1. This risk must be addressed for mines in all locations of the world and very often the project financing cannot proceed without favorable assurance of environmental compliance from the local regulatory bodies with which the sponsors must deal.

This risk category can also arise due to the location of the project e.g. near towns or highways or in proximity to wilderness, heritage, native reserve, or scenic areas. This is not simply a sub-set of political risk but a distinct risk aspect of minerals projects today.

Political Risk

(Actually a mix of risk categories such as nationalisation, currency inconvertibility, regulatory, and tax risks).

Each bank has a continuing review process of country risk based on an evaluation of the political and economic outlook. Banks are then willing to take some degree of the political risk for certain countries. Loans to sovereign borrowers in the country establish the market "price" for such risk in the country itself. Some of the most sophisticated petroleum project financings being arranged today are almost solely for political-risk sharing.

In the event banks decide they cannot absorb Political Risk, then some of the risks may be covered by government-sponsored export-credit agencies; organisations like the Overseas Private Investment Corporation (OPIC) for US lenders; or through insurance from groups such as Lloyds of London or private insurance sources. Some surprisingly advantageous Political Risk insurance is now being written by large insurance groups; however, the amount is quite small in relation to the capital required for a new mining project. If the project financing parallels a World Bank or export-credit financing, then some (but not complete) comfort may be derived from the potential political leverage from these government agencies.

The risk of currency inconvertibility is usually due to arbitrary government action caused by serious balance-of-payments problems, which should have been identified in the banks' economic risk analysis process. A mechanism will usually be set up to accrue the local currency in the expectation that remittance will be permitted at a future date.

Other more difficult areas which must be judged on a case-by-case basis concern the following:

1. The development agreement;
2. War and insurrection;
3. Tax and ownership changes (creeping nationalisation);
4. Borrowing restraints (e.g. variable deposit requirements); and
5. Non-government political activists (unions, environmentalists, terrorists).

The development agreement with the government can be subject to direct change or cancellation, or to the exercise of other governmental powers (e.g. export controls). It is not uncommon for project lenders to seek a direct guarantee from the government not to subvert the development agreement.

In the case of war and insurrection, the banks may, for example, share the residual risk after a pre-agreed level of damage has been done. Guarantee mechanisms would otherwise be required in certain countries.

The minerals industry is a wide-open target for creeping nationalisation through a noose of continually changing tax or royalty regimes. Banks will try to anticipate the impact of tax changes on the ability of the project to stand on its own and may not accept more than a defined portion of this risk.

The most difficult aspect of Political Risk assessment is the impact of non-government political activists on a mining project. It is quite common to see additional loan repayment time made available to a project where strikes are expected. In fact, force majeure coverage may be very important in such an instance.

Force Majeure Risk

Force majeure concerns events outside of the control of either the borrower or the lender. Some element of political risk may be included in the definition of force majeure.

Lenders can usually absorb temporary Force Majeure Risk during both the construction and operating phases of the mine, due, for example, to a strike or an unavoidable delay in the delivery of a critical equipment part. But lenders are naturally cautious about permanent force majeure situations which are irreversible. Many force majeure events coincide with the other risk categories such as technical risk (water inflow) or political risk (government regulations), but lenders can often accept some of the permanent Force Majeure Risk through project financing. The expectation of such risk will lead to a requirement for a flexible transaction.

Business interruption insurance is becoming more prevalent for the minerals/metals industries and is another way, favorably viewed by the banks in a project financing, to cover Force Majeure Risk. However, it is not yet available in amounts exceeding about $ 40 million.

Foreign Exchange Risk

Besides currency inconvertibility (a political risk), foreign exchange exposure can occur when project revenues differ from the currency of the project loan. Lenders usually avoid Foreign Exchange Risk in project financing.

The best hedge is to match the loan currency to the (underlying) currency in which the price of the product is set. A further step is to match equipment purchases to the sales revenue currency. However, if a large loan is required, the available loan currencies may be limited, for example, to Eurodollars.

WITHIN THE BANK'S CONTROL

Syndication Risk

(May be called the financing or underwriting risk).

Once the terms and conditions of the project financing have been negotiated and documented, the actual financing has some risks which should be understood. Of course, if the loan is only given by one bank then this risk category does not apply. However, by virtue of the usually large amounts involved, most projects involve a syndicate of banks which has its own conventions.

The lead managers (1-3 banks) carry the negotiating responsibility with the borrower and the participating banks. The agent bank (usually one of the lead managers) arranges the documentation and loan disbursements.

The lead managers then arrange a syndicate of banks, first choosing the managers (2-6 banks) and the co-managers (3-10 banks) whose ranking depends on their amounts of the loan; and finally the participating banks.

The types of bank syndicates vary from fully underwritten or club deals among the lead managers to a best-efforts commitment to raise money over and above the amount committed by the lead managers. Generally, a fully-underwritten financing requires a slightly higher interest margin than a best-efforts financing. If it is decided to arrange a project financing in a number of sequential borrowings, then the borrower takes the risk that the interest margin may increase.

Funding Risk

Most mine project financings are funded on a floating-rate basis due to the necessity for great flexibility in drawdowns and repayment. If interest rates escalate uncontrollably, the available cashflow can be correspondingly reduced and may not be sufficient to repay principal when due.

Banks usually accept this risk indirectly in a project financing. Fixed-rate project financings are rarely used and are usually only made available in co-financings from government or quasi-government entities such as export-credit agencies or the World Bank.

Legal Risk

The burden of legal documentation (which in the end apportions the risks among lenders, insurers, governments, buyers, and sponsors) usually rests on the banks and its advisers. There is some risk that professional advisers will create risks in the documents which can affect the tax position, tenure, security, enforcement and other attributes so heavily negotiated in the risk-sharing process embodied in the project financing in the first place. Second opinions, judgement, and experienced staff and advisers are perhaps the only way to mitigate this final risk category.

RISK ALLOCATION

A break down of the general risk absorption by the various parties in a project financing is summarised below:

Risk Category	Absorbed by:	If "Risky" by:
Reserve	Bank	Reserve Warranty
Operating:		
Technical	Bank	Company, Warranty
Cost	Bank	
Management	Bank	Contract
Infrastructure	Bank, Consumer	Consumer
Environmental	Bank	Company
Market	Bank, Sales Contract.	Buy-back
Political	Bank	Insurers, Government
Force Majeure	Bank	Insurers, Company
Foreign Exchange	Bank, Company	Hedging
Funding	Bank	Swaps, Hedging
Participant	Bank	JVA, Contract
Engineering	Bank	Insurers
Completion	Company	Constructor
Syndication	Bank (underwritten)	Company (best efforts)
Legal	Bank	Insurers

CONCLUSION

A full understanding of the risks and the general rules applied by project-finance bankers can result in remarkably good loan agreements without too much difficulty. An early dialogue with the banks will serve to validate the company's expectations of risk sharing in a project financing. A firm understanding of these objectives should be established, perhaps with the assistance of a financial adviser.

As a business strategy, project financing protects the company from having its existence threatened or its financing capability eroded due to the failure of a particular project. This is especially important when a company is developing a large new mine or plant or is bringing a number of projects into production concurrently.

REFERENCES

1. Carter, R.L. and Crockford G.N. (Editors), Handbook of Risk Management, Kluwer-Harrap Handbooks 1974 (updated).

2. Hartman, D.G., "Foreign Investment and Finance with Risk", The Quarterly Journal of Economics, May 1976, pp 213-232.

3. Ladbury, R.A., "Limited Recourse Financing", The Australian Mining and Petroleum Law Association. Third Annual Conference, Perth, 17-19 May, 1979.

4. Le Roux, M., "Specialty Risk : Insurance for the Future", Institutional Investor (Special Advertising Section).

5. Megill, R.E., An Introduction to Risk Analysis, Petroleum Publishing Company, 1977.

6. Nevitt, P.K., Project Financing, AMR International Inc., Second Edition (updated).

7. Pouliquen, L.Y., Risk Analysis in Project Appraisal, World Bank Staff Occasional Papers, Number 11, 1970.

8. Reutlinger S., Techniques for Project Appraisal under Uncertainty, World Bank Staff Occasional Papers, Number 10, 1970.

9. Squire L. and van der Tak, H.G., Economic Analysis of Projects, World Bank, John Hopkins Press, 1975, p 45.

10. Tinsley, C.R., "Mine Project Financing", Proc. Second Australian Coal Conference, Surfers Paradise, Queensland, Australia, April 1980.

11. Tinsley, C.R., "Project Financing - What The Banks Will Want". Proc. Metals Economics Group Seminar, Denver, February 4-5, 1982.

12. Vaughan, E.J., Fundamentals of Risk and Insurance, J. Wiley, 1982.

APPENDIX 1

Different markets use separate languages when attempting to define risks. The bulk of this paper is drawn from the language used to project finance minerals developments. Table 1 shows how 11 different banks or bankers assessed risk prior to the publication of this book. These are real differences in response to as well as perception of risk between bankers.

For comparison, Table 2 gives the view of risk using the language of the insurance industry (Carter and Crockford, 1974). Again the focus of the insurers is on physical losses or pure risks as distinct from speculative or business risks which are deemed outside of the insurance realm and more to within the purview of management sciences.

Table 3 has been developed by the author as another way of classifying risks according to their causes, namely: human, government, natural, impersonal. This properly breaks up force majeure risk which is often lumped together into one category. These classifications are useful in risk trade off.

Table 1
BANKERS' RISK ASSESSMENTS

SOURCE DATE	CHEM. 1975	ANZ 1978	C.R. TINSLEY 1980	IRVING TRUST 1980	CREDIT LYON. 1980	EUROPEAN 1981	CROCKER 1981	US BANK 1981	CHASE MAN. 1982	AUSTRALIAN 1982	BofA 1984
RESERVE	X	X	X	X	X	X	X	X	X	X	X
TECHNICAL	X	X	X	X	X	X	X	X	X	X	X
COST	X	X	X		X		X	X		X	
MANAGEMENT				X		X					X
INFRASTRUCTURE		X				X				X	
MARKET	X	X	X	X	X	X	X	X	X	X	X
POLITICAL	X	X	X		X	X	X	X	X	X	X
FORCE MAJEURE	X		X		X	X			X		X
FOR. EXCHANGE	X	X	X				X			X	X
PARTICIPANT		X	X		X	X				X	
COMPLETION	X	X	X	X	X	X	X	X	X	X	X
SYNDICATION			X								

Table 2

INSURANCE INDUSTRY

Pure Risks (Risk Management)   Losses

Social   Property
Nature   Liability
Personal   Personnel
Technical   Financial
                                     Business
                                       Interruption

Business Risks (Management Sciences)

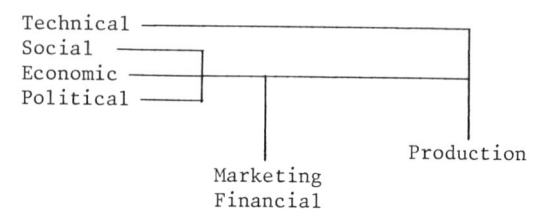

Technical ─────┐
Social ────────┤
Economic ──────┤
Political ─────┘
              │   │
              │   Production
              Marketing
              Financial

Table 3

FOUR FACTORS

Human

    Operating
    Force Majeure (strikes)
    Completion
    Syndication
    Legal

Government

    Infrastructure
    Political
    Force Majeure (blockades, exchange controls)

Nature

    Reserves
    Force Majeure (floods)
    Environmental

Impersonal

    Market
    Foreign Exchange
    Participant
    Funding

9.2

UNDERSTANDING THE RISKS IN COAL RESERVE ESTIMATES

by

Mr. James E. McNulty

Vice President
Paul Weir Company
Chicago, Illinois 60606

Abstract. Coal reserve estimates are prepared using a simple equation: the product of area, bed thickness, density and recovery factors. Each of these elements involves a degree of risk based on certain assumptions. Reserve area, thickness and density rely on certain physical data whose accuracy, reliability and sufficiency may be problematical. Estimates of recovery factors for mining and coal preparation are often subjective. Most reserve estimates assume, sometimes incorrectly, that the subject reserves will be completely extracted. The principle that certain coal resources are reserves, if they can be considered economically recoverable at the time of determination, needs to be reexamined in the light of recent market cycles. Reserve estimating requires the exercise of judgments, the validity of which reflect the qualifications, experience and objectivity of the estimator.

INTRODUCTION

Reserve estimates are necessary for the evaluation of the financial model of a mining project, just as reserves are necessary for production. Coal reserve estimates are prepared using a simple equation: the product of area, coal thickness, density and recovery. If, in a general way, risk is exposure to danger, then in a financial sense, it can be exposure to failure as a result of inadequate, unreliable or inaccurate data upon which decisions are made. Each element in the coal reserve equation involves judgments which create different types and degrees of exposure.

AREA

Coal reserve estimates are usually made on maps of individual coal beds although, increasingly, reserve estimates are being generated by computer methods so that a map in the conventional sense is not necessary to the operation.

The conventional documentation for a coal reserve estimate, however generated, would include a map or maps showing extent of coal occurrence and the points of data which support the estimate. In most reserve estimating, area is determined from the coal bed map using a plane polar planimeter or electronic digitizer.

Unless an estimate is totally fraudulent, the area used in the estimate does not introduce errors, which would substantially affect the accuracy of the results. The subject reserve area usually represents someone's reasonable interpretation of reliable subsurface data.

However, in mapping of surface minable coal, the size of reserve areas are affected by the accuracy of topographic mapping. In most surface mining projects, large scale topographic mapping is generated for a number of activities ranging from reserve studies to mine design. Outcrops, subcrops, coverlines, reserve and overburden quantities developed from such maps have a much greater accuracy than those derived from 7-1/2 minute quadrangle mapping.

Quadrangle maps usually have a topographic contour interval of 6.1 meters (20 feet). According to national map accuracy standards, the elevation of 90 percent of the points on these maps would be within one-half the contour interval and the balance within a full contour interval. Depending upon the degree of slope (spacing between contour lines) estimate of minable areas and overburden quantities may not be within 20 percent of actual quantities.

Photographically enlarging government topographic quadrangle maps does not increase their accuracy.

How the outcrop (or subcrop) of a surface minable area is determined also can affect the accuracy of the determination of minable area and consequently the reserve estimate. Natural outcrops where coal is exposed are rare. The shallow extent of a coal bed is usually covered by soil or other unconsolidated surface material. Also, coal close to the surface is usually weathered, resulting in a deterioration in calorific value and other desirable qualities. Ideally, the outcrop or subcrop should be determined by closely spaced drilling. At the other end of the scale would be an outcrop projected from a crossplot of a coal bed structure map and a 7-1/2 minute topographic quad.

The sizes of reserve areas of underground mines are not much affected by the accuracy of topographic mapping and quadrangle maps are frequently used in deep mining reserve evaluations.

COAL THICKNESS

The second element of the equation is the thickness of the coal bed. Thickness is the most important attribute of a coal bed because of its effect on the economics of recovery (Weir and McNulty, 1981). In understanding a reserve estimate, it is important to establish how thickness is determined and how it is applied.

Although calculations are involved, coal reserves are estimated rather than calculated. It is much like taking a public opinion poll and like a poll, the accuracy of the result is dependent upon the size and distribution of the sample population.

In order to establish the thickness of a coal bed or beds in a reserve area, a certain density of data is required. What that density is cannot be simply stated. Data, however, must be sufficient in quality, quantity and distribution to describe the variability in bed thickness.

Properly faced up, an outcrop presents the best opportunity for the measurement and description of coal beds. Ribs and faces in active mines are also good, but conditions of observation are always less than ideal in underground mines because of the limitations of artificial light.

Drilling is the traditional method of prospecting for coal at distances from outcrops. Historically, the diamond core drill has been the most extensively used tool in coal exploration. Cores of coal, properly recovered, enable accurate measurements and descriptions, as well as material for chemical analysis (McNulty and Ball, 1977).

Although rotary drilling cannot provide the detail to be had from cores, it is a faster and cheaper method of obtaining some types of data. It can, under certain conditions, be done using compressed air as the circulating medium (within certain depth limits), thereby minimizing concerns of water sources and water haulage. Since rotary drilling is widely used in water supply development, such rigs have almost a universal availability.

However, rotary drilling by itself is a most unreliable method of subsurface data gathering. The interpretation or log of the results of a rotary hole relies on the identification of cuttings and variations in drilling speed and time. In the case of lost circulation, there are no cuttings to identify. Boundaries of coal beds are difficult to recognize, a problem which increases with decreasing bed thickness.

Since accurate measurements of coal bed thickness cannot be made by rotary drilling, the accuracy of any reserve estimate relying solely on that type exploration has to be suspect.

Rotary drilling aided by downhole geophysical logging, however, is well regarded as an accurate method of identifying coal in the subsurface and determining coal bed thickness. Over the past 20 years, methods and equipment have been developed specifically for the logging of coal. Geophysical logging is now a standard exploration procedure in core holes as well as rotary holes.

In order to maximize the value of this tool, it is important to utilize up-to-date equipment and to have experienced personnel to interpret the results.

After the quality of thickness data, it is important to evaluate their adequacy on the basis of density and distribution. Investigators usually try to classify estimates by the qualitative characterization of data density with the terms "Proved" and "Measured."

In a U.S. Bureau of Mines publication (Dowd, et al, 1950), measured coal was defined as "coal for which tonnage is computed from dimensions revealed in outcrops, trenches, mine workings, and drill holes. The points of observation and measurement are so closely spaced, and the thickness and extent of the coal are so well defined that the computed tonnage is judged to be accurate within 20 percent or less of the true tonnage. The limits of accuracy should be stated. Although the spacing of the points of observation necesary to demonstrate continuity will vary in different regions according to the habit of the coal beds, the points of observation are, in general, of the order of 1/2 mile (0.8 km) apart. The outer limit of a block of measured coal, therefore, shall be of the order of 1/4 mile (0.4 km) from the last point of positive information (that is, roughly one-half the distance between points of observation)."

This is essentially the same definition used by Averitt (1969) in developing and classifying a national coal resource estimate.

Later, the USGS (1976) and USBM adopted a more streamlined definition for measured: "Coal for which estimates of the rank, quality and quantity have been computed, within a margin of error of less than 20 percent from sample analyses and measurements from closely spaced and geologically well known sample sites."

Wood, et al (1983) in a section titled "Glossary of Coal Classification and Supplementary Terms" define measured as, "The highest degree of geologic assurance. Estimates of quantity are computed partly from dimensions revealed in outcrops, trenches, workings and drill holes and partly by projection of thickness, sample, and geologic data not exceeding a specified distance and depth. Rank is calculated from the results of detailed sampling that may be located at some distance from this type of resource and may be on the same or other coal beds. The sites for thickness measurements are so closely spaced and the geologic character so well defined that the average thickness, areal extent, size, shape, and depth of coal beds are well established. <u>However, a single measurement can be used to classify nearby coal as measured</u>....." (Emphasis added).

It is difficult to understand how a single point can establish the "areal extent, size and shape" of a coal deposit.

In the same document (Wood, et al, 1983) under the heading of "Specific Criteria," measured is described as, "Accessed and virgin coal that lies within a radius of 1/4 mile (0.4 km) of a point of thickness of coal measurement..."

Noticeably absent from this last definition are the phrases "so closely spaced" and "so well defined" of earlier definitions.

The term "Proved" as it is used in reserve estimates tends to rely more on the judgment of the estimator than specific distances from points of observation. There isn't anything magic about 0.4 km (1/4 mile). Some coal beds are quite variable in thickness within much shorter distances. Some exhibit rather consistent thicknesses for much greater distances. It is important to make a judgment of the adequacy of data density on the basis of the interpretation of all available geologic data, including the apparent variability in thickness indicated by those data.

Some consideration must also be given to minimum coal thickness in reserve evaluations. In underground mines in the U.S., the height of equipment must be about 30 cm (12 inches) less than the minimum height of the seam being worked, without taking roof or floor, as well. This limits seam thickness minable with most mining equipment to about 90 cm (36 inches). High quality coal is being surface mined down to 18 inches and less in thickness. However, where coal is thin, larger areas must be mined. Haulage and reclamation costs are greater and a higher percentage of coal is lost in loading.

DENSITY

The third element in the reserve factor is density and like area, is usually not a cause for concern in a coal reserve estimate.

Since coal is a heterogeneous solid, the densities used are averages rather than absolutes. In most estimates, general values based on rank of coal are used rather than project or site specific values. This is because of the difficulty of accurately measuring in place density and to a lesser degree, with variability.

Like any other laboratory determination, the validity of the result of a specific gravity test depends upon the sample: its condition and what it represents. Obviously, only full seam samples should be used because individual coal materials have a wide range of densities. The condition of the sample should, as closely as possible, approximate that of the in situ condition.

It is very difficult to get representative values from the testing of small ($+50$ mm) cores. Even a core that is in reasonably good condition will break into a number of short segments. The material lost at breaks between plies and from core surface pitting is sufficient to affect the results of the determinations. Losses of material and moisture due to sample handling can also affect results.

Larger size samples are preferable to smaller, but collection of these is invariably more expensive.

Large diameter drill cores ($+150$ mm) provide good samples because the greater core strength results in fewer losses due to breakage and longer core segments. Losses due to surface pitting relative to total core volume are substantially smaller.

Vertical variability of density within a coal bed can be addressed by good core recovery and total bed testing. Horizontal variability can only be determined by a sufficient number of tests.

Channel samples, block samples or column samples collected from nearby mines, exploration adits or fresh faces excavated at outcrops can also provide good specific gravity determinations. These types of sites, however, are usually expensive and, consequently, fewer in number than core holes in a typical exploration program. The distribution of such openings is controlled by topography, so the distribution of sample points is not always sufficient to represent the geometry of the subject reserves. At moderate depths, large diameter cores are the most economical solution.

Where the density of coal in place is not known, it is customary to use values that have wide acceptance in the technical community. For reserve purposes, and in the absence of specific data, the practice is to use density factors based on coal rank as used by the USGS. These are:

Bituminous coal 1.324 tons per cubic meter
(1800 st per acre-foot)

Subbituminous coal 1.306 tons per cubic meter
(1775 st per acre-foot)

Lignite 1.288 tons per cubic meter
(1750 st per acre-foot)

In U.S. practice, reserves are usually estimated using one of these factors and no attempt is made to determine the actual density of the coal in the ground. There are two reasons for this. The first, as previously noted, is the difficulty in accurately determining, with confidence, the density of coal in place. The second has to do with the net effect of the range of possibilities. The difference between the factors for lignite and subbituminous coal is 1.4 percent, as is the difference between subbituminous and bituminous coal. The difference between the factor for lignite and that for bituminous coal is 2.8 percent.

Standard values should not be used where coal bed measurements include substantial quantities of high gravity, noncoal material, which occurs usually as partings of shale or clay, especially if the areal distribution is irregular. Thickness of partings is usually subtracted from the total coal bed thickness before the appropriate density factor is applied.

There are cases in the western and southwestern U.S. in coals of Cretaceous age and younger, where thin partings are too numerous to be removed economically either by selective mining or mechanical beneficiation. Coal and included partings are mined and burned in thermal electric generating stations. Deposits such as these may have an in place coal density of 1.398 tons per cubic meter (1900 st per acre-foot).

Some producing companies use a density factor of 1.282 tons per cubic meter (1742 st per acre-foot) for their reserve inventories. This is a conservative value.

The range of all of the values noted herein is from 1.282 tons per cubic meter (1742 st per acre-foot) to 1.398 tons per cubic meter (1900 st per acre-foot). The larger of these is only 9 percent greater than the smaller. Considering the errors that can be introduced by measurement of map areas, coal bed thickness determinations and the application of recovery factors to coal in place, density factors used in the reserve equation are usually not critical to the accuracy of reserve estimates.

RECOVERY FACTORS

Having found the product of area times coal thickness times density, the estimator has arrived at a quantity which represents the amount of coal in the ground or the "coal-in-place" value. In order to determine the quantity of coal likely to end up in trucks, trains or ships, the coal-in-place estimate must be reduced by a factor, which represents losses expected to be sustained in coal production and, where required, beneficiation or coal preparation.

Coal Preparation Recovery Factors

Risks that are associated with the application of recovery factors to coal preparation processes are less than those for mining recoveries, because they usually are based on the results of testing done by an independent laboratory, and those results and the methods by which they were generated can be evaluated.

Erroneous estimates, however, can be obtained by the misapplication of test results.

Coal preparation, necessary and/or desirable in order to meet certain markets, usually involves the removal of heavier materials in the process stream in order to reduce ash content and thereby enhance calorific value. In many instances, sulfur is reduced as well. In general, the quantity of material that is recovered as product is inversely proportional to the enhancement of quality over that of the raw feed. In the development of a project, a determination is made as to where the loss in material is balanced by the acceptability of the product.

These determinations are made using the results of screen analyses and/or washability tests. The pitfall here is that commercial scale preparation plants do not operate with 100 percent efficiency. The recoveries of product coal through a washery may be only 90 percent of the theoretical values determined in the laboratory. They may be even less if inappropriate equipment is installed in an operation, so that it must operate at reduced efficiency in order to produce the quality required by a particular economic model (Byrom, 1984). In such situations, production from the reserves goes to refuse disposal rather than coal markets.

Mining Recovery Factors

The mining recovery factor represents the greatest exposure of all the elements of a coal reserve estimate, because of its subjective nature. The other elements are constructive factors in that they combine to create a quantity of a certain magnitude. The fact that the mining recovery factor reduces the quantity of available resources creates an atmosphere that promotes the generation of optimism instead of pessimism, sometimes at the expense of realism.

Coal mines rarely recover all of their assigned reserves. There are usually more economically attractive alternatives than the recovery of reserves which are too deep or distant from central facilities and at the far reaches of power, haulage and ventilation systems, which may be old, obsolete or in poor repair.

The geometry of reserves may also have an adverse effect on total recovery. Irregular shaped deposits cannot be effectively worked where maximizing recovery would result in poor machine application or availability.

Geologic factors, which greatly influence the recoverability of coal, are often difficult to quantify in terms of anticipated losses. Where possible, it is very helpful to evaluate the experience of nearby mines having the same or similar conditions.

Changes in mining plans or interruptions of mining operations usually result in losses of coal reserves.

Also, having an effect on overall recovery is what is called the illegal fraction (Kaiser, et al, 1980). This includes populated areas, highways, pipelines, railroads, rivers and reservoirs that could be impaired by mining or its attendant activities. Although the statutory and regulatory restrictions that may be imposed on an operation may be difficult to assess, their affect on reserve recovery must be addressed.

Averitt (1969), in consideration of a national resource estimate, used a recovery factor of 50 percent of coal in place. This was based in part on a study (Lowrie, 1968) of some 200 mines which produced more than half of the country's underground production in 1963. The average recovery of those mines was 57 percent within a range of from 29 to 91 percent. All were under less than 300 meters (1,000 feet) of cover. The higher recovery from surface mining was considered to have a minimal effect because of the small proportion of surface minable resources in the total.

For partial extraction room-and-pillar mining, where pillars are left in place for support of the surface, mining recovery may vary from 60 percent under very shallow cover to as little as 35 to 40 percent at depths of 300 meters (1,000 feet) or more. Second mining or "pulling pillars" has approached 90 percent recovery, but this is exceptional and a more general overall average might be 65 percent. With longwall mining, although probably 95 percent of the coal is recovered within the longwall panel itself, the overall mining recovery will depend upon the recovery in the development entries as well. Overall recoveries may vary from 60 to 70 percent (Weir and McNulty, 1981).

Mining recovery at surface mines is somewhat dependent upon coal bed thickness since losses in thin beds will be proportionately higher than in thick beds. For thin beds, the recovery may be 80 percent, compared with 90 percent or more for thick coals.

In assessing overall recovery, it is necessary to consider not just the "mined" area but the "mined over" area. The "mined" area is that portion of a coal reserve that is within the perimeter of coal recovery operations. The "mined over" area includes the "mined" area as well as all areas of available reserves left unmined for whatever reason. If the difference between the two areas is great, the effect on overall recovery will be dramatic.

In most reserve estimates, the mining recovery factor used is based on the judgement of the estimator reflecting experience under similar conditions. If detailed life of mine projections have been made, more refined estimates of mining recovery can be applied. However, most reserve estimates are generated prior to the stage of project development when detailed projections are available.

ECONOMICS AT THE TIME OF DETERMINATION

Reserves have, in the past, been defined as quantities that can be economically extracted or produced at the time of determination. For the most part, this is an excellent principle, especially since it removes the opportunity to speculate about the movement of coal prices. This has worked very well for many years when coal prices were steady or slowly rising. However, in the decade of the 70's, the steep rise in coal prices contributed to the perception of a continued rise in prices. Examples abound where certain deposits were reserves under the economic test of the late 70's and subeconomic resources today. To benefit from this lesson, we need to examine recoverability and price sensitivity relative to loan terms and payback periods.

SUMMARY

The most subjective element of a coal reserve estimate is the mining recovery factor. In analyzing the estimate, this factor deserves the most scrutiny. The exposure that can be created by inadequate, inaccurate or poorly interpreted thickness data ranks second. Designation of reserve areas, especially in surface mining, can be a problem, but the density conversion factor rarely is.

Financial institutions, and others, assessing adequacy of coal reserves attract risk which is related to the expertise of in-house personnel or independent consultants performing reserve studies (Pouch, 1980). All such studies require judgment, and the qualifications and experience of the estimator are all important. Too often, reserve estimates which are relied upon are not objective, or realistic, and resulting quantities are either exaggerated or diminished.

REFERENCES

Averitt, P., 1969, "Coal Resources of the United States, January 1, 1967," Bulletin 1275, U.S. Geological Survey, 116 pp.

Byrom, R., 1984, Private Communication.

Dowd, J.J., Turnbull, L.A., Toenges, A.L., Cooper, H.M., Abernethy, R.F., Reynolds, D.A. and Crentz, W.L., 1950, "Estimate of Known Recoverable Reserves of Coking Coal in Indiana County, Pennsylvania," Report of Investigations 4757, U.S. Bureau of Mines, 22 pp.

Kaiser, W.R., Ayers, W.B., Jr. and LaBrie, L.W., 1980, "Lignite Resources in Texas," Report of Investigations No. 104, The University of Texas at Austin, Bureau of Economic Geology, 52 pp.

Lowrie, R.L., 1968, "Recovery Percentage of Bituminous Coal Deposits of the United States," Report of Investigations No. 7109, U.S. Bureau of Mines, 19 pp.

McNulty, J.E. and Ball, C.G., 1977, "Coal Exploration," in Energy Technology Handbook, ed. by Douglas M. Considine, McGraw Hill, New York, pp. 1-37 – 1-47.

Pouch, R.C., 1980, "What You and Your Banker Must Know About Project Financing," Coal Mining and Processing, Vol. 17, No. 1, pp. 52-56.

U.S. Geological Survey, 1976, "Coal Resource Classification System of the U.S. Bureau of Mines and U.S. Geological Survey," Bulletin 1450-B, U.S. Geological Survey, 7 pp.

Weir, J.P. and McNulty, J.E., 1981, "Effects of Physical Characteristics of Coal Occurrence and Current Mining Methods on Economic Recoverability," Geological Society of America Bulletin, Part I, Vol. 92, pp. 558-562.

Wood, G.H., Jr., Kehn, T.M., Carter, M.D. and Culbertson, W.C., 1983, "Coal Resource Classification System of the U.S. Geological Survey," Circular 891, U.S. Geological Survey, 65 pp.

9.3

Operating Risk
(Improving the Odds in a High Risk Industry)

C. Allen Born

Placer Development Company
Vancouver, B. C.

INTRODUCTION

As the title of this paper suggests, the author does not believe that risk can be managed. Rather, the objective is to identify the risk which will be present in a given venture and assess the level of that risk which is acceptable. More often than not, the risk which is present is subject to little or no modification. After identifying and measuring the risk a corporation will decide to accept it and proceed in such a way that it will be less exposed, or it will decide not to proceed at all. The exercise for a corporation is different only in degree from everyday judgements made by average people in average situations. However, that word "degree" means that it is considerably more complex on the corporate level.

There is good reason for the additional complexity of corporate risk evaluation. The variables influencing a company's decision can be more numerous and the penalties for miscalculation much greater than those normally encountered by an individual.

It should be understood that no analysis of risk, certainly not one as brief as the following, can do more than touch on some of the high points.

Before examining the sorts of risk which affect the development and operation of specific mines, let us consider the broader aspect of the term with which mining companies, and all other business corporations, must deal. Companies must determine their long-term objectives and strategies within the context of a constantly changing world. To plot their direction they must attempt to read the future by seeking answers to questions such as:
How will society evolve?
Which businesses will thrive?
In which of these should we place our emphasis?

If a company hopes to be around five or ten years from now it must make the attempt to answer such questions in order to judge the impact of changing patterns in the evolving business environment. Information is available to allow the lessons of the past to be projected on the future. We have data on birth rates, economic booms and busts, patterns of conflict within and between nations, drought and other forces of nature as well as individual events which cannot be predicted, but which over time, do seem to occur in regular cycles.

Ultimately the question "where will we be and what will we be doing a decade from now?" must be answered by one or more senior people with responsibility for the company's future. The compilation and interpretation of hard data by competent support staff, can only be used up to a certain point in determining future corporate strategy since there will be alternative possible futures and a high degree of risk will always remain. At this point those elusive qualities of experience, instinct, and - "gut feeling" - become important factors in choosing a long-term corporate direction.

The goals of a company have a significant bearing on its attitude towards risk. This in turn dictates the "hurdle rate of return on investment" which a project decision-maker will require. The hurdle rate is the minimum-discount-percentage-rate applied to forecast cash flows to give a present value of zero to the All-Equity-Funded Investor.

The feasibility study for a project is a fundamental tool in the management of risk. An experienced team has an established and well-understood set of ground rules for the preparation and assembly of a feasibility study. The necessary degree of conservation is built in at all levels of study data and, if the project is technically feasible, the hurdle rate becomes a catch-all measure of risk in a base-case project cash flow.

Now, we can focus on the management of risk in the development and operation of a mine. Before a mining company attempts to analyze those risks it must assess what can be termed its "inside" risks. The company must determine the extent to which it is completent to develop and operate a proposed mine. Questions which it must ask about itself:
- Do we have experience with this material?
- If not, can we acquire it?
- Is the proposed mining method one that we know?
- Is the company organized to handle the increased load?
- If not, what changes are indicated?
- Is new technology involved?
- If yes, can we develop it?

When the company is satisfied that it can manage and minimize the internal risks it can turn to the analysis of the outside risks attendant on the project.

This paper is confined to the analysis of the risks confronting a mineral developer and producer in Canada which enjoys a reputation as a "safe" or "low risk" area for investment. Whether this evaluation will continue to be representative after the following brief survey of risk in Canada, remains to be seen. However, I think you will agree that a description of mining risks in Canada is a good base from which to calculate the greater risk generally associated with mining investment in many other countries. If mining is a risky business in Canada, how much more is it elsewhere?

There are three areas of major risk with which a company must deal in developing and operating a mine. These are:
(1) market prices and demand
(2) costs of capital, operation and finance
(3) taxation and regulation

MARKET RISKS

The price of metals is the single most important factor in determining a mine's profitability. Virtually any mine development can be justified if the price is high enough. Conversely, if the price falls low enough, almost any mine can become a loss-making proposition. The estimation of metal prices and sales volume is extremely difficult since prices tend to fluctuate to extremes and are generally outside the control of companies. The risk associated with price forecasts, therefore, is quite high.

How can we reduce this risk? Economic and technical trends combined with production costs and the costs of competitive materials give only a fair indication of future price trends. As a general guide, however, the cost of competitive substitutes often sets an upper limit, scrap costs set a lower limit and production costs indicate an expected price. The economic structure of the particular metal or mineral usually gives a good indication of price volatility. The structural analysis should include the length of the supply chain from mine to end consumer, the relative number of buyers and sellers at each step, the pricing methods and the amount of producer discipline required to adjust production to consumption and demand.

Price expectations should be adjusted in line with the cyclical nature of economic activity since prices can fall below full production costs for extended periods during part of each cycle. Most miners are familiar with the business cycle of 3-6 years, but there is also the capital investment cycle, with particular attention to the third phase in which economics run out of liquid savings. The resulting economic contraction usually tends to be severe as recent experience indicates.

COST RISKS

The second major area of risk concerns the estimating of capital, operating and financial costs.

Three approaches are useful in limiting the risk of over-runs in capital costs. First, have the senior design engineers and operating superintendents review the detailed estimates which will become the basis for their budget responsibility. Construction cost indices are available for cross-checks and the advice of construction firms and equipment manufacturers is also very helpful. However, some error will always be present because such important variables as weather and labour, cannot be forecast accurately. Second, use the initial design period to look for less expensive alternatives and to assure close coordination of the whole design and construction team. Finally, whenever possible, time development for the trough in a normal business cycle of 3-6 years. In this period excess construction capacity tends to keep prices competitive and delays to a minimum.

Operating Costs Operating Cost estimates should be treated in a similar fashion. Experienced operators should review the detailed operating cost estimates. Labour, supply and maintenance indices should be used where available as a cross-check. Visits to similar operations are also very useful.

Probably the best long-term security for a potential mine is a favourable competitive cost of production. While many factors bear directly on competitive costs, the most important are usually the grade and recovery expected from a deposit, as well as sufficient tonnage to carry the venture through a number of cycles, providing an opportunity to recover investment and earn a profit. The risk of operating cost estimates is that production will be limited by lower grades, greater dilution, higher grinding indices and worse recovery than expected. Adequate drilling, bulk sampling and laboratory testing can minimize and quantify this risk. On many disseminated deposits, statistics can be used to determine when the drilling is adequate.

While current competitive costs must be considered, they should also be projected for the first several years of expected production to include possible new mines, relative cost trends, and all forms of relevant taxes.

Good timing is again a factor in commencing production at the beginning of a recovery so that the probability of high prices during the critical payback period, is increased. However, the mine must be financially viable at the lowest price expected during the recession or favourable forward prices should be established before construction starts, especially if this is required in order to obtain financing on reasonable terms.

Financial costs   Financing costs include interest and exchange rates. These are currently very unstable and difficult to estimate accurately since they involve prediction of central bank behaviour.

Unstable interest and exchange rates, combined with periodic bouts of inflation, have significantly raised the risks on all capital investment projects. Increased government impediments to natural resource investment and trade flows have made the situation worse, which brings us to the third and last major area of risk to be discussed here.

TAXATION AND REGULATION RISKS

The suggestion that government poses major risks in terms of taxation and regulation is not new although the ingenuity through which complex problems are made more complex, is worthy of admiration. Taxation in this context means any cost imposed or influenced by government, including duties, royalties, and export levies. It is common to think of taxes in terms of income taxes and mining royalties, but there are also indirect levies such as those on sales, excise duty and workers' compensation. The possibility of changes in direct or indirect taxes is the risk to be quantified. We are trying to measure what taxes and regulations will impact on production costs and what changes will affect the balance of profit left for the mine developer. The Canadian Income Tax Act of 1972, is an example of a drastic change in taxes. In a single move it raised the effective tax rates, made taxes payable earlier in a mine's life, created a feeling of uncertainty in the industry, and by taxing capital gains, reduced the availability of risk capital so critical to mine development.

As mine development is a long-term investment, both the current level and future trend of taxation should be evaluated. The best indicators are past trends in the ratio of total government spending to gross national product, as well as current trends in government revenues and expenses. Another method is regular polling of attitudes in the civil service, among political leaders and in the general electorate.

An example of a current trend with negative implications for international companies may be found in California which proposes to tax resident corporations on the basis of their world income. It seems fairly obvious that the state is about to lose a number of companies as they move to nearby jurisdictions which continue to offer realistic corporate tax laws.

There are at least three other areas in which government represents risk for mine development and operation:
   Fiscal and monetary policy - this exerts a crucial influence on the level of inflation which has severely eroded mining profits during the last decade. The result has been a declining amount of capital for investment in our industry.
   Nationalization - through acquisition of operating mines, government establishes a lever to manipulate segments of the industry.
   Regulation - this refers both to rules and constraints imposed on the industry through legislation or through "soft regulation" such as guidelines, procedures and policies - sometimes not even in written form. These are being used with increasing frequency by Canadian Government agencies.

In summary, the major categories of risk in developing and operating mines concern price forecasts, cost estimates and taxation. While many risks can be reduced with careful preparation, others have to be accepted as part of the mining environment. Nevertheless, they can and should be measured. Probably the best overall methods of reducing project risks are careful evaluation by experienced teams and the timing of construction and operation within economic cycles.

9.4

OPERATIONAL RISK ASSESSMENT OF MINING ENTERPRISES

Peter J. Szabo

BJ Leonard & Co., Denver, Colorado

INTRODUCTION

The time period from the detailed feasibility study to the post investment audity usually ranges from three to eight years depending upon the nature of the mining project considered. It is the only after the post investment audit that we know the adequacy of the assumptions made concerning operational risk assessment in the detailed feasibility or acquisition economics.

The post investment audit results from numerous mining projects can best be described by a verse by John Greenleaf Whittier's Maaud Muller. It reads
    "For all sad words of tongue or pen the saddest are these 'It might have been!'"

Dissecting the negative results of numerous mining projects we find faulty assumptions concerning markets, geology, capital investment and operational risk assessment. Although in hindsight many of these assumptions could have been considered impossible or impractice to have forecast with any degree of certainty, we find that refined assumptions concerning operational risk assessment could have been made with only a modicum of additional "Due Diligence".

Anyone who invests in a mining property should employ a much higher discount rate of future cash flow, (usually in the minimum range of 25%-30%), than he would expect from such safe investments as Treasury Bills or high grade Utility Bonds. The nature of this steep discount translates to a large uncertainty concerning the timing and actualization of assumptions contained in the detailed feasibility study. We must remember that when a project gets beyond 20 years, any contribution to the value of the project of cash flow is minimal due to the nature of the discounting process. Thus for each full or partial day lost in the development or early operation stage of a project, the financial impact will be greater than similar lost days later on in the project.

Adequate operational risk assissment is particularly critical when bidding on competitive tenders for coal deposits in remote locations such as the Bowen Basin in Queensland, Australia. Because of the market attractiveness of these coal deposits for supplying end users in the Far East, Middle East and Europe, state governemnts require competitive tender biddings via the Authority to Prospect process. Front end bonuses and super royalty payments must be commited before the actual mining conditions are known. Time to submit firm offers from the actual request for proposal vary from 90 to 120 days. Thus any operational risk assessment must be detailed, realistic and reflective of the actual project.

When one searches the literature, one finds limited qualitative data on operational risk assessment in valuation papers. Usually this facet of valuation takes a secondary role to such areas as reserves, markets, and environmental considerations. When one comes to the question of operational risk assessment more often than not one finds the use of "rules-of-thumb", such as tons per man shift, average production, average seam height, etc., are used to forecast results on which banks and other financial institutions are required to make investment decisions. Because of the extremely competitive environment of the 1970's and 1980's numerous financial institutions approved loans without adequate "Due Diligence", and are currently paying a steep price in terms of non performing loans and charge offs.

One must always remember that each operation is unique with its own characteristic operations environment. This environment encompases such factors as management, labor, climate, availability of equipment, maintenance of equipment etc. A systems approach must be employed in assessing any mining project, i.e. Each part of the mine to market chain can effect a different part of the overall system. Only when such impacts taken individually are assessed for their cumulative effect on the operation can we adequately assess the impact of operational risk. Such sophisticated techniques as regression analysis must be used in forecasting operational risk.

The one cardinal rule in operational risk assessment is that the evaluator must mine and market the project on paper. By using mine feasibility studies of the project concerned as well as deriving assumptions from surrounding operations, quantification of risk parameters can be given some credibility.

This paper will focus on five key components of operational risk assessment, i.e. management, labor, equipment availability, maintenance, and climate. Each will be treated briefly identifying those salient features that numerous times are not found in many feasibility studies used for financing purposes.

Management

Often times the nuances of the management team are ignored when assessing mining projects or acquisitions. Yet the project evaluator must always remember that in this business probably more than any other business actual management practices can make or break a project. This is especially true in the area of acquisitions as numerous major oil companies found out the brutal reality of failure to account for differing corporate cultures of oil and mining companies in the 1970's and 1980's. The evaluator must remember especially in acquisitions that the acquired company's corporate culture has directly influenced its profitability. The "Big Oil" or acquiring company's corporate culture on the mining company must be realistically assessed in light of additional overhead costs, morale considerations, and synergistic effects.

In Eastern Kentucky several oil companies purchased coal companies and immediately severed traditional banking and legal relationships and imposed "Big Oil" corporate staff structure on local personnel. The resulting increased land costs from irate citizens of the community contributed to increased mining costs especially strip mining costs. Adequate coordination and supressed oil company egoes could have resulted in expected profitability instead of excuses to upper management for poor performance.

On new mining projects a whole different set of operational problems regarding management availability must be assessed. Numerous mining projects are started at the height of a boom. New companies especially "Big Oil" companies have had to scramble to assemble management teams to get their projects on stream. In their overzealousness to get part of the action, a President or Vice President of acquisitions will be hired without a thorough evaluation of his capabilities. This key player will be the nucleus of the whole mining team. Unfortunately during boom times top talent is rarely available and so the new mining upstart usually settles for a second or thrid string quarterback to assemble a team who again in boom times are usually second stringers. Executive search firms have made a handsome return in filling organizations with their present staff. Is it any wonder that when the results of the post investment audit are in that large inefficiencies in management are noted. A thorough analysis of the management situation before entering a project or acquisition could have avoided or at least accounted for anticipated inefficiencies in the projections.

In dealing with major projects in remote locations, the question of timing of acquiring adequate staff must be thoroughly analyzed. The lack of an open pit foremen or maintenance superintendant can result in adverse impacts that normally many evaluators would not consider when reviewing the "Big Picture".

We must always ask ourselves has management allowed in their assessment of a project adequate time for formation of the required project team. If the management has not accounted adequately for this then the adjustment of the development schedule must be adjusted.

Labor

Labor is a critical factor of the operational risk component. Indeed in certain countries such as Australia a less than thorough assessment of the labor situation would be foolhardy in determining project viability. For even preliminary feasibility purposes state or even district analysis of the labor situation should be formulated. As an example using average man days worked in the United States would not be representative of the actual excellent labor conditions in many coal fields of the west or locally in the western part of Eastern Kentucky. On the other hand the extremely poor labor conditions historically experienced in District 17 of the United Mine Workers of West Virginia would not be representative of the national labor force in coal mining operations. This varing state of labor situations stresses the importance of local assessment of labor conditions.

One should not assume that the assessment of the labor situation ends at the mine site. On the contrary in countries such as Australia one must scrutinize the entire mine to port chain to determine the adequacy of assumptions concerning effective man days. Australia is notorious for poor labor conditions with numerous wild cat strikes a common occurence. Since the union ethic is so strong in that country even a small union's pickets will not be crossed by other union's members. A common practice of labor slowdowns although keeping the actual days worked constant can result in the actual effective man days worked being significantly diminished.

Strikes anywhere along the mine to port chain can result in excessive stockpiles of ore or coal that can ultimately shut down operations that do not have adequate storage capacity at the mine. This is yet another example of how each component of the operations must be analyzed on a system wide basis. Failure to account for actual effective man days worked will result in an underestimate of labor costs and therefore falsely inflate the projects estimated value.

Equipment Availability

Equipment availability is often treated lightly if at all in the preliminary study. Many times the actual realization that specific equipment needed for the project will not be available for a project will not be realized until the detailed feasibility study or actual development has commenced. This is especially true when projects are, or acquisitions are commenced during boom times when the temptation to get in on the action overrides prudent engineering requirements. Often times companies even though assessing equipment availability with various manufacturers fail to assess what the competition might be doing in the way of acquisitions or project expansion. A classic example occurred during the scarce and several companies found their profits squeezed as a result of not being able to obtain sufficient dozers, front end loaders, augers, or other capital equipment. A not to uncommon sight during the 1973-1974 coal boom was the actual highjacking of mining equipment in Eastern Kentucky. For those companies that needed the equipment and couldn't acquire it the lessons will be well remembered.

Equipment availability at fovorable prices can be a definite asset to an otherwise noneconomic project. Shortly after the coal boom of 1973-1974 was over, manufacturers of mining equipment were making excellent deals on various pieces of equipment because of excess inventory and order cancallations. For those shrewd enough to detect this window in mining capital equipment availability and were prepared to exploit it significant returns were theirs for the taking.

Maintenance

Especially in the area of mining acquisitions the need to evaluate the acquired company's maintenance program can not be overemphasized. Many "Big Oil" companies once the decision was made to enter the coal business acquired Appalachian coal producers with only the slightest analysis of how the maintenance program or lack thereof contributed to the profitability of a mining company. Only through detailed analysis of a company's maintenance record can we hope to put in proper prospective whether historical financials are reflective of what future operations will produce. Should the maintenance program have been deficient the new owner will have to pay in the way of increased production or maintenance costs or increased capital equipment costs for replacement capital equipment. Again it must be remembered that maintenance can not be considered as an isolated operational risk but must be considered in light of the overall mining system. Inadequate maintenance effects the number of effective man days worked which in turn impacts project costs.

Climate

Climactic impact on mining operations can have significant effect upon project profitability. In estimating such parameters as actual effective man days worked, efficiency of equipment and highwall stability the climactic effects must be scrutinized carefully. Many evaluators will at least in a general way consider climactic impacts in their preliminary feasibility, however a detailed assessment is critical especially when operating in non temperate zones. Numerous oil companies after acquiring Appalachain coal companies were shocked to find out how severe Appalachian winters and springs could be on accurate assessment of effective man days worked and equipment maintenance.

Similarly the harsh winters and subsequent spring run-off in the Rocky Mountain region effected mining operations and transportation networks in more than general way. Particularly noteworthy was the earth slide which occurred on April 14, 1983 near Thistle, Utah. This natural disaster that severed the main line of the Rio Grande Railroad through the Rocky Mountains stopped shipments of coal over that railroad for selected producers in

Colorado and Utah for 81 days. This interruption had a significant impact on coal producer's profitability.

When operating out of the temperate zone adequate climactic assessment is even more critical especially in the arctic or tropical environment. What we have been talking about are normal or limited variations from normal weather patterns. However when evaluating major projects in the tropics or adjacent zones, the impact of catastrophic climactic events such as El Nino can not be ignored. El Nino, which is a reference to the Christ child, is a warm current that normally occurs off the coast of Peru around Christmas time. Depending on the severity of this event topical weather can be effected significantly. The years 1982-1983 will go down in history as one of the most memorable for catastrophic climactic impacts on mining projects. Developers of such notable mining projects as Ok Tedi In Papua , New Guinea will remember the lessons of El Nino well. Failure to account for this catastrophic climactic event which resulted in the worst local drought in 50 years forced project developers to abandon use of the Fly River main supply route to the Ok Tedi project when the river became impassable. For five months aircraft instead of barges had to be utilized for supplies until the normal tropical rains returned and allowed the barges to float off the sandbars.

Although El Nino events have been known as far back as 1726 and on average occur once every four years, the probability of the severity of such events must be assessed and contingency plans developed with associated costs before commiting to any mining project. Mine developers can not be lulled into a flase sense of security by assuming that normal weather in the tropics, or for that matter any place else, will continue. Mother Nature has a way of always supprising you when you least expect it.

CONCLUSION

Mining projects are too fraught with risks to discount the all important parameter of operational risk. By careful detailed analysis of operational risk many a project which was thought viable should have been relegated to the inactive project file. Financiers should not be quick to jump into projects without adequate operational risk assessment, regardless of the perceived value of the reserves or customer relationships. The lessons of past failed mining projects are all too numerous to be ignored. As the famous philosopher George Santayana once wrote "Those who cannot remember the past are condemned to repeat it."

References

Anon, 1983, "Adios, Maybe to El Nino," Time, Vol. 122, No. 12, Sept. 19, pp. 67.

Anon, 1984, "News From Bechtel," News Release From the Bechtel Group Inc.

Anon, 1983, "Papua New Guinea," Mining Annual Review.

Anon, 1983, "Ok Tedi Will Go On Stream in 1984, Despite Strikes, Jungle Cimate," Engineering & Mining Journal, Vol. 184, No. 8, Aug., pp. 15.

Anon, 1983, "Thistle, Utah; We've Tunneled Through Disaster to Continue Our Commitment to Your Transportation Needs," Rio Grande The Action Railroad, July.

Bennewitz, G. A., 1984, "Current Status of Western Railroads Relating to Transportation of Export Coal," paper presented at the 4th U.S.-Japan Coal Conference, San Francisco, CA, May 21.

Chave, L., 1982, "The Acquisitions That Haven't Paid Off," New York Times, Vol. 131, No. 45, 266, Mar. 28.

Cane, M., 1983, "Oceanographic Events During El Nino," Science, Vol. 222, No. 4629, Dec., pp. 1189-1194.

Fuller, D. B., 1983, "Oil Firms Hit Dry Hole by Diversifying," Denver Post, Oct. 30, reprinted from the Los Angeles Times.

Garcia, O., Anderson, M., 1983, "The Activity in the Tropical Pacific During the Last Three ENSO Events (El Nino Southern Oscillation)," paper presented at the 8th Annual Climactic Diagnostic Workshop, Toronto, Canada, Oct.

Rasmusson, E., Wallace, J., 1983, "Meteorological Aspects of El Nino Southern Oscillation," Science, Vol. 222, Dec.

Shao, M., 1983, "Metal Fatigue, Exxon's Mining Unit Finds it Tough Going After 16 Years in the Field," Wall Street Journal, Jan. 16.

Szabo, P. J., 1983, "Successful Mining Acquisitions Can They Really Be Made," paper presented at the Denver Section Meeting, Society of Mining Engineers of AIME, Oct.

9.5

DEALING WITH INTEREST RATE AND EXCHANGE RATE RISKS

JAMES L. POOLE

INTRODUCTION

Companies in the mining industry are subjected at times to currency exchange rate risks and interest rate risks. The former occurs any time a firm deals in more than one currency. The latter occurs anytime a firm is financing itself with borrowings that have a floating interest rate (although a fixed rate can be a problem at times).

FOREIGN EXCHANGE RISKS

There are two general reasons why exchange rate risks (FX risk) can occur: First, a firm may be building a mine or plant in another country where financing for the plan is denominated in funds other than the country hosting the plant. For example, if a Canadian firm is building a smelter in another country which is being financed with Canadian dollars, the C$100 million for labor, which must be paid in the local currency, has subjected the firm to a C$100 million FX exposure risk. If after the cost estimates were made the local currency appreciates 30% against the Canadian dollar, the cost of labor has increased to C$130 million. (Of course, the local currency could also devalue during this period making the labor cost component less than estimated.)

The other reason exchange rate risk arises is due to the fact that the costs of production in one country producing minerals may be incurred in one currency while the production is sold in another currency, such as coal being exported from Canada to Japan under contracts denominated in Yen. In this example, the Canadian firm is exporting coal to Japan and being paid in yen in 30 days at a contract price of 100 Y per ton. Furthermore, assume that the mining firm anticipates a depreciation of the yen against the Canadian dollar of 5% in the next 30 days. Therefore, the mining firm stands to lose 5% of the value of its shipment in a month's time not including carrying costs.

In both cases, the mining company risks an adverse change in the exchange rate, that is, the currency the company holds may decrease in value relative to the other currency.

There are a number of ways to mitigate the exposure to foreign exchange which can be utilized by the mining company. Optimally, a company should try to match all their inflows and all their outflows in the same currency. For example, if a Canadian firm has mining costs and debt service costs both denominated in Canadian dollars then 100% of their receipts would need to be in Canadian dollars. If a Canadian firm has 70% of its cash outflows in Canadian dollars and 30% of its cash outflow in American dollars in the form of a loan amortization then their receipts would need to be 70% in Canadian dollars and 30% in U.S. dollars. This general principle of matching cash receipts and cash costs in the same currency works fine in a perfect world but what of the real world when these cannot be matched? In actual practice, however, the financial markets have developed a number of ways in which foreign exchange risks can be managed or reduced.

FORWARD MARKETS

One method is to use the forward markets which are made on a world-wide basis by the commerical banks for foreign exchange. If a firm wanted to lock in the current exchange rate for the receipt of a foreign currency at some point in the future, they could do so by contracting on the forward market with a bank selling that foreign currency at the current rate at the future date. For example, assume that a Canadian firm is exporting coal to Japan which would pay yen in 90 days. If a firm wanted to lock in today's exchange rate they could contract with a bank to sell on the forward market yen in 90 days. When the yen was received the contract would be executed by selling the yen to the bank and receiving the previously agreed upon number of Canadian dollars. Generally speaking, the forward market can be used to sell forward about 12 months. The costs of selling (or buying) forward

Based on a paper presented to the 2nd CIM Mineral Economics Symposium, Vancouver, B.C., November 1982

contracts more than 12 months in the future becomes expensive.

A second way that a company can circumvent exchange rate risk is by buying contracts in the financial futures market. Using the previous example of a Canadian company receiving yen in 90 days, the firm could go through a broker and purchase a contract on the futures exchange to sell yen in 90 days. When the 90 days are up and the yen are received, the company would sell the contract and any loss that one received on receipt of the yen would be offset by the corresponding gain in the futures contract and vice versa. This is little more than the classical hedging approach. The disadvantages of using the futures market as opposed to the forward market is that future contracts have a specific settlement date, at a specific amount and a specific exchange rate. These are not negotiable as in a forward contract.

BORROWINGS

A third way to avoid exchange rate risk is available through borrowings. Again taking the example of yen being received in 90 days and assuming that the mining company prefers the existing exchange rate, the firm could borrow the yen and immediately convert them to Canadian dollars. When the yen was received in 90 days the loan would be repaid. The only thing that the mining firm must be sure of is that the interest charges associated with borrowing are built into the price of commodity being sold.

CURRENCY EXCHANGES

A fourth method to reduce foreign exchange risk is the currency swap. This transaction, which is usually arranged by a bank, is very complex, is relatively long term in duration (i.e., 5 years), expensive and is relatively difficult to establish. The simplest form of currency swap is shown in Figure 1. Its establishment, operation and reexchange at maturity is shown in Figure 2. This type of currency swap can be off balance sheet with the exchange fee often having favorable tax treatments.

Figure 1 *

CURRENCY SWAPS

Simplest Form - £/$

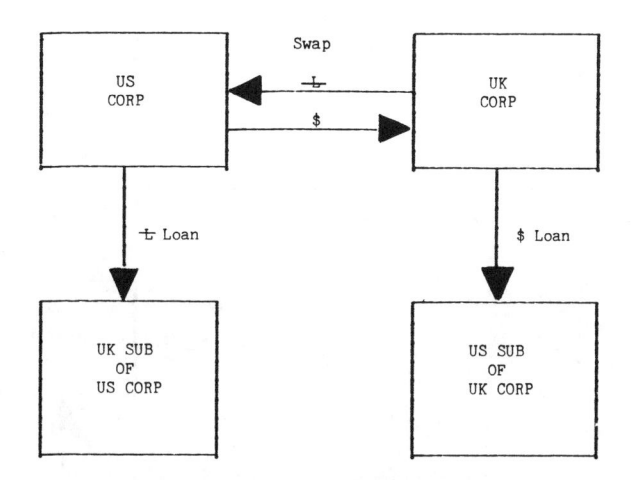

```
              Swap
  ┌─────┐    ←─£──    ┌─────┐
  │ US  │             │ UK  │
  │CORP │    ──$─→    │CORP │
  └──┬──┘             └──┬──┘
     │ £ Loan            │ $ Loan
     ▼                   ▼
  ┌─────┐             ┌─────┐
  │UK SUB│            │US SUB│
  │  OF  │            │  OF  │
  │US CORP│           │UK CORP│
  └─────┘             └─────┘
```

* The flows reverse at the end of the agreement at the same exchange rate as on the initial exchange date.

Figure 2

INITIAL EXCHANGE

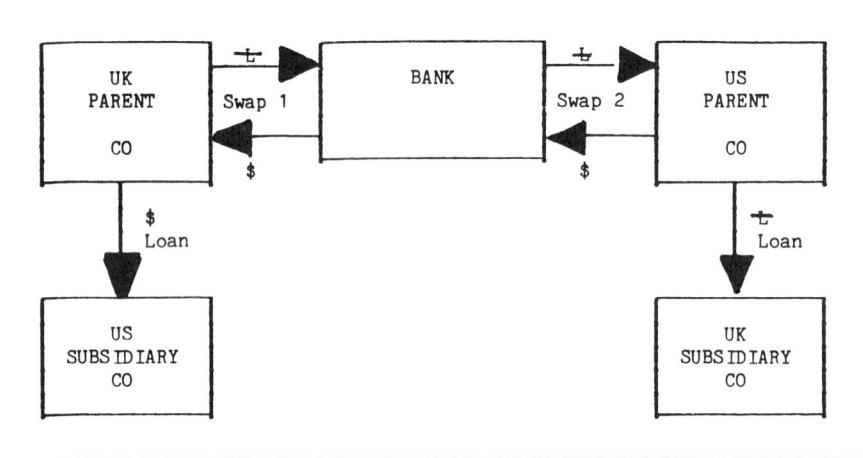

ANNUAL EXCHANGE FEE PAYMENT

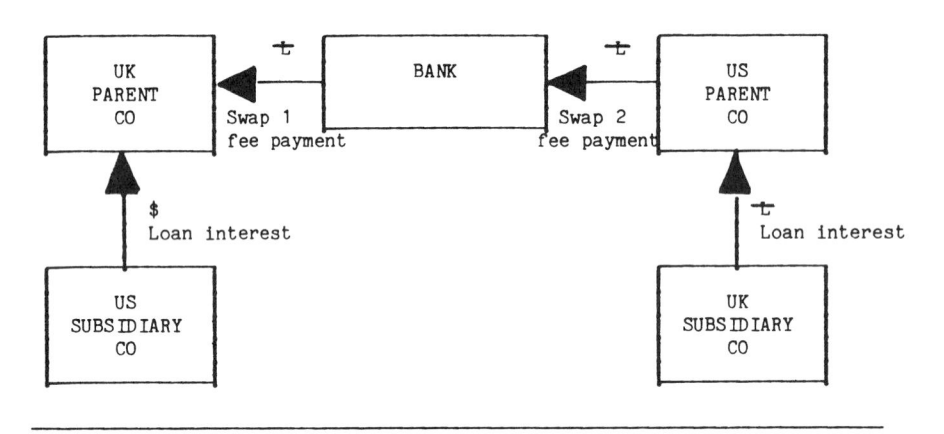

RE-EXCHANGE AT MATURITY

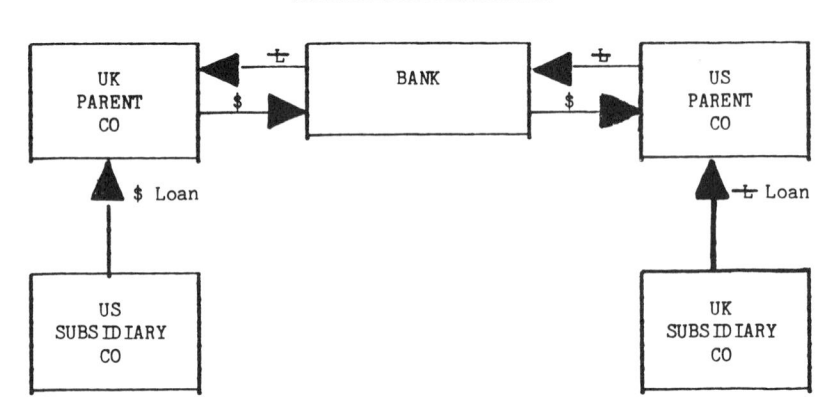

Another variation on currency hedging can be via back-to-back and parallel loan structures, often with adverse balance sheet and income effects and generally more complex security and documentation than swaps. Figure 3 shows the underlying transaction.

Figure 3

STRUCTURE OF BACK-TO-BACK LOANS

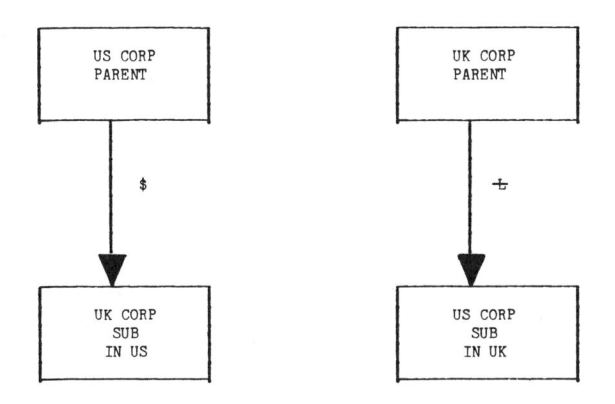

US$/£

```
┌─────────────┐              ┌─────────────┐
│  US CORP    │              │  UK CORP    │
│  PARENT     │              │  PARENT     │
└──────┬──────┘              └──────┬──────┘
       │ $                          │ £
       ▼                            ▼
┌─────────────┐              ┌─────────────┐
│  UK CORP    │              │  US CORP    │
│  SUB        │              │  SUB        │
│  IN US      │              │  IN UK      │
└─────────────┘              └─────────────┘
```

In summary, the mining company has four financial tools at its disposal to help dampen the effects of foreign exchange risk. These tools are: 1) the forward markets at the banks; 2) the use of hedging on the financial futures market; 3) borrowing in the currency and converting it to the desired currency and then repaying the loan upon receipt of the payment; and 4) currency exchange or parallel loan structures.

INTEREST RATE RISKS

Interest rate risks occur when a firm uses borrowed funds and pays a floating interest rate on those funds (such a risk can also occur if a firm borrows on a fixed rate basis and the interest rate drops). There are three ways to circumvent or to dampen the effects of interest rate risks. The first way is to use equity financing, however, that has the cost of the dilution of ownership and potentially higher capital costs of dividends. The second way is with hedging of financial futures and the third way is through interest rate swaps.

FINANCIAL FUTURES

Interest rate hedging works in a very similar fashion to exchange rate hedging. If a mining company anticipates that interest rates will increase, the firm can go into the financial futures market and hedge on the interest rate. This is done by first determining which of the interest rate futures (i.e., Treasury Bills or Bank CD's) correlates best with the interest that the company is paying on the loan. For example, assuming that a company is paying a Eurodollar based rate, the company would find that the change in the Eurodollar rate correlates very well with changes in CD rates. If the firm has a U.S. prime rate or Canadian prime rate based financing they would be able to use either T-Bills or CD's but the hedge would not be perfect (called cross-hedge risk). Nonetheless, assuming a firm is interested in locking in the existing interest rate for the next 12 months the firm can go through a broker and purchase financial future contracts.

Generally, the settlement date of the contract should correspond with the interest rate payments or interest option exercise dates. As with exchange rate risks, any loss that one receives because of an increase in interest rates on the loan is compensated for by the profit gained from selling the contract and vice versa. The use of financial futures to hedge interest rates can go out to about 18 to 24 months. However, they work best from 6 to 18 months and are least expensive from 1-12 months.

INTEREST-RATE SWAPS

The interest-rate swap is very similar in principle to hedging on the financial futures market, but a swap is more complex from a documentation standpoint and it requires a bank to operate as middleman. However, interest rate swaps can be done for much larger amounts of money and for a longer period of time than hedging. In principle an interest-rate swap links a company that pays a floating rate, called a "fixed-rate taker," to a firm that pays a fixed-interest rate, called a "floating-rate taker".

Specifically, a fixed-rate taker, such as a mining company, contracts with a commerical bank to pay a fixed rate to the bank. This rate is in turn passed along to the floating-rate taker. On the other side, a floating-rate taker contracts with the same bank to pay to the bank a floating rate which is then in turn paid to the mining firm.

An example of this can be seen in Figure 4 below. Firm A, which is the fixed-rate taker, has a loan with Bank B which carries a floating rate. Firm B has fixed-rate borrowings usually from a public debt market such as the Eurobond market or the U.S. bond market. Firm B contracts with the Bank in the middle (Bank A) to pay Bank A some multiple (usually less than 100%) of the floating rate that Firm A pays. Likewise Firm A contracts with Bank A to pay some multiple (usually less than 100%) of the fixed rate that Bank B pays to the debt market. In this way Firm A is assured of getting its multiple of its floating debt requirement to pay Bank B and in return it pays a multiple of the fixed rate portion of Firm B's debt.

Figure 4

INTEREST-RATE SWAP

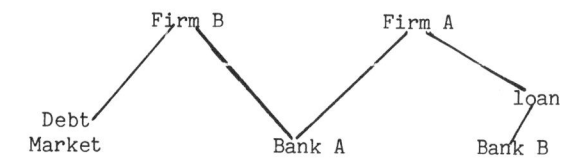

```
        Firm B           Firm A
        /   \           /      \
       /     \         /        \
      /       \       /          \ Loan
   Debt        \     /            \
   Market      Bank A            Bank B
```

In general, interest rate swaps are usually done for loans in the $25 to $75 million category and can be done for periods of 2 to 7 years. A series of credit decisions must be made in every interest rate swap. Bank A must determine the creditworthiness of Firm A and Firm B to assure that they will pay their contractual requirements of the interest. Also, the Firms A and B must determine the creditworthiness of Bank A to be assured that the interest payments will be paid.

In summary, it is possible through the use of either equity financing, financial futures hedging or interest-rate swaps to circumvent or dampen some of the effects of interest rate risks that borrowers experience in debt financing. However, it should be noted that the methods to circumvent interest rate risks are relatively complex and should be used by the more financially sophisticated firm, usually with the help of an investment bank or a commerical bank.

CONCLUSION

In conclusion, today's financial environment subjects the mining industry to many risks which it must evaluate and strive to reduce in an effort to maximize earnings. Two of these risks that the mining industry faces are caused by floating interest rates and floating exchange rates. It should be noted, however, that the financial methods to hedge these risks are primarily meant for the more financially sophisticated firms and should only be used in their initial stages with the help of a merchant bank or a commercial bank.

9.6

A SYSTEMATIC APPROACH TO POLITICAL RISK ANALYSIS

William D. Coplin & Michael K. O'Leary

Maxwell School, Syracuse University and Frost and Sullivan, Inc.

ABSTRACT

Risks from political instability and government policies restricting equity ownership, local operations and transfers of payments affect the profitability of foreign mining ventures. More than seat-of-the-pants judgments and old-boy networks are required to anticipate these risks and to take actions appropriate to offsetting them. A system for making political forecasts, called the Prince System, and a series of charts relaying risks to a production of key minerals has been developed.

INTRODUCTION

Firms with extractive operations outside their home country do not have to be told that political stability and government decisions play a big role in the success of their operations. The takeover of copper firms in Chile during the Allende administration in 1971 and the recent decision by Venezuela to reduce Reynolds Metals equity share in a smelting operation from 50% to 28% are two examples of hundreds of events that have determined the profitability of extractive operations over the past decade. Both risks and opportunities for international firms seeking to produce and market raw materials are heavily affected by political conditions. While news of political events travel much faster than in the past and host governments are more clever in shaping the risks as well as the opportunities, politics has always been part of the mining business.

Political risk analysis, the assessment of macro-political changes and specific governmental decisions on the operations of extractive industries needs to become as systematic and integrated into corporate decision-making as economic analysis has become over the past two decades. Since events in Iran, however, there has been the steady growth of a set of tools to make political forecasts and integrate those forecasts into business decision-making on a systematic basis.

The reason for the emphasis on a systematic approach can be found in recent failures of political risk analysis. On the one hand, relying on traditional sources such as government and bank officials in either the home or host country or corporate staff - local or headquarters - in almost all cases produces a bias in favour of the status quo. The failure to forecast the collapse of the Shah of Iran in 1978-79 was a result of this bias. On the other hand, relying on newspapers and published academic sources runs the risk of overemphasizing the forces of change or at the very least exposing oneself to so many conflicting opinions that making a political risk forecast becomes itself a full-time job.

Finally, there is a need to be able to make comparisons across projects and across countries. Each corporation needs to develop a rating system that allows for such comparisons but at the same time gains the legitimacy and acceptance that is needed for it to be used. Rating systems aggregate information required at higher levels of decision-making but their use is ultimately determined by the care with which the specific pieces of information and analysis are developed and integrated.

The last section of the paper will provide some displays relating our political risk ratings to mineral productions in countries producing aluminum, bauxite, cadmium, copper, gold, lead, nickel, silver, tin, and zinc. These tables will demonstrate the utility of a systematic approach which yields uniform ratings.

SYSTEMATIC POLITICAL RISK ASSESSMENTS

Two basic tasks have to be completed in creating a systematic approach to political risk assessment within any extractive firm. The first task is the process of identifying and assessing the importance of critical political factors that may cause risk to the company's operations. The second task is forecasting the likelihood that each of these factors will take place.

Identifying and Weighting Political Factors

Each extractive industry faces somewhat different risks to its business operation, and frequently the risks vary from country to country.

Therefore, the most important initial step is to identify each type of major political event or condition that could affect the industry's operation. The first factor is the question of "regime stability". Each analysis has to begin with the question of how long the current regime will remain in power and under what conditions it might change. The question looks at both superficial regime changes such as the election of a new government in a two-party state where there is little difference between the two parties and major regime change such as the revolution in Iran in 1979. It may be equally important to identify the most likely alternatives to the current regime in order to make contacts with the next government.

The next factor to consider is the current level of "political turmoil" within the system, as well as what it will be in the future under the present regime as well as likely alternative regimes. Political turmoil includes political strikes, labour violence, terrorist activities, public demonstrations, riots and mob violence, guerrilla warfare, civil, war and international warfare. Turmoil should be analyzed in terms of whether or not it is directed at extractive operations or will indirectly affect such operations as a result of disruptions in labour supply, transportation and communications or international commercial activities.

The next set of factors to be assessed fall in the general category of "restrictions on international business." A general evaluation of the orientation of the country's major political actors toward outside investment could be conducted in order to assess the potential for actions resulting from nationalistic or anti-private enterprise sentiment. In addition, the current and future levels of restrictions with respect to the following five categories needs to be explored:

- Restrictions on foreign equity ownership
- Requirements for use of national or local products
- Taxation discrimination
- Restrictions on repatriating capital and profits
- Exchange restrictions

Another general area of analysis has to do with labor quality and costs, which figure heavily in the business viability of any product.

Finally, we should emphasize that what we have discussed above concerns political risks international business faces because of their international character. All foreign operations potentially face the same risks as their domestic counterparts. Changes in tax laws, environmental restrictions or decisions about infrastructure developments (e.g. transportation and communication) need to be considered in a thorough analysis of political factors shaping international business.

In order to prepare a systematic analysis for one's own operations, one will need to construct a list that begins with the basic items already mentioned. These include: Regime Change, Political Turmoil, Foreign Equity Restrictions, Local Operation Restrictions, Taxation Discrimination, Repatriation Restrictions, Foreign Exchange Restrictions, Labor Costs. Then add other items important to the operations. Relevant staff members should collectively produce such a list as a first step in developing the rating system.

Forecasting Events and Conditions

Most corporations have neither the staff time nor expertise to make the forecasts. However, in order to develop that capability or to intelligently use outside forecasting services or consultants, it is necessary that the users of political risk analysis become at least minimally familiar with various approaches to forecasting.

Most forecasting is done by using relatively few approaches. Table I lists those approaches, their current use, and their strengths and weaknesses. As the table suggests, country specialists have to be used not only as a source of information but also a source of analysis. Except for the little used statistical and computer simulation approaches, all approaches rely on the perceptions of experts. It is important to develop techniques in which a panel of experts can be surveyed and their views integrated. An eclectic approach that combines closed-choice questions, descriptive qualitative analysis, and an analysis model like the Prince Model is necessary. By getting specialists to express their views in different formats, it helps to improve the coherence and reliability of their analysis.

Three criteria in selecting specialists are: objectivity; access to current, reliable and accurate information; and analytical ability. A local plant manager, a bureaucrat from the host country, a banker, or a political opponent to a current government, all have reasons to be biased. In contrast, professional scholars or researchers in government or the private sector who have reputations as objective analysts are more likely to conform to expectations of objectivity.

Access to relevant information is knowledge of the key political actors in the system, current issues and economic conditions. The access has to be based on both knowledge of published and unpublished printed information as well as personal contact with government officials, other political actors and other experts in the field.

Finally, the country specialist must have well-developed analytical capabilities. In addition, the specialist needs to have an imagination which will allow anticipation of alternatives that would not be considered through complete reliance on simple straight line projections. The willingness and ability to come up with alternatives and to speculate on the implications of those alternatives for international business is a key analytical capability necessary for solid political-risk analysis.

It is difficult to find country specialists who satisfy all three criteria. Objectivity and analytical ability are frequently possessed by academics who often tend to have access to the core of political conditions. Nationals of the country tend to have good access but have difficulty in providing objective and analytical discussions. For these reasons, it is necessary to be on continuous lookout for country specialists who have both perspective and information. Furthermore, it is important to keep tabs on the analysis and forecasts such specialists provide. Checking their own internal consistency and the degree to which their forecasts turn out to be accurate is vital to developing a first-rate political risk analysis capability.

STANDARDIZED RISK RATINGS

The approaches discussed in the previous section can also be used to produce a rating system in which risks can be compared across countries or across projects. To illustrate the utility of a systematic risk rating system, we conclude this paper by presenting a series of tables that describe how levels of production for ten metals are related to the levels of political risk within the producer countries. In each table box, countries are listed in descending order of production.

The political risk ratings are then cross-tabulated with levels of production as provided by the 1980 edition of the Statistics Yearbook published by the American Bureau of Mines. By looking at the intercorrelation between the political risk rating and the levels of production one can see the degree to which supplies of metals are at risk from political factors.

To provide a brief analysis of the relationship between political risks and the aggregate production of minerals for the countries studied, we have indicated in a row percentage the breakdown between percentage of production and the three risk categories. Over 80% of the bauxite, gold, nickel, silver, and tin are produced in countries where political risks are either medium or high. Over 20% of the bauxite, tin, and zinc is produced in countries in which political risks are considered high. These findings illustrate not only the usefulness of general risks ratings for evaluating the degree to which metal production is vulnerable to political risks but also that political risk is very important to businesses that depend upon minerals. Even if those businesses use domestically-based sources of supply, political risk will affect market prices and ultimately on the decisions that have to be made. A rating system similar to the one described in this paper can be an important tool in planning and decision-making for the minerals industry.

Table 1

Approaches to Forecasting	Current use	Strengths/Weaknesses
Qualtitative description and analysis	Extensively used by in-house staffs, consultants and publications	Provides rich detail. Time-consuming to read and completely dependent on writer's information and analysis. Comparisons between countries almost impossible.
Closed-choice questions such as what is the likelihood of of a given regime change	Used by some commercial forecasters and part of in-house checklists	Quick to read. Can be used to survey a panel of experts. Subject to writer's information and analysis. Forces too much simplification.
Statistical or computer simulation models based on quantitative data	Used primarily by accademics	Current quantitative data not available. Usually poor predictor on political events.
Analytical model which directs to make explicit assumptions and provide specified information	Used by Frost and Sullivan and by several in-house staffs	Allows for quick comparison of writer's experts political assumption. Supports comparisons between countries, and over time. Completely dependent upon writer's information.

Table 2

ALUMINUM LEVEL OF PRODUCTION (000's of short tons)

	Very low (0-95)	Low (95-250)	Medium (250-420)	High (420-5,150)	Row %
LOW	Taiwan Cameroon	Austria	U.K Italy Netherlands	U.S.A. West Germany Norway	64.4
MEDIUM	Sweden Mexico Turkey South Korea	India New Zealand Greece Egypt South Africa	Venezuela Australia	Japan Canada France Spain	26.2
HIGH	Iran	Yugoslavia Argentina	China Brazil Romania		9.3

Sources: Frost & Sullivan World Political Risk Forecasts, August 1982; American Bureau of Metals Statistics Yearbook, 1980.
The above countries accounted for 80% of total world production in 1980.

Table 3

BAUXITE LEVEL OF PRODUCTION (000's of short tons)

Risk	Very Low (0-150)	Low (150-1,500)	Medium (1,500-3,000)	High (3,000-32,000)	Row %
LOW	Italy		U.S.A.		2.7
MEDIUM	Turkey Spain Pakistan	Indonesia Malaysia	France India	Australia Jamaica Greece	80.1
HIGH		Romania Dominican Rep.	China	Brazil Yugoslavia	17.2

Sources: Frost & Sullivan Political Risk Forecasts, August 1982; American Bureau of Metals Statistics Yearbook, 1980.
The above countries accounted for 65.7% of total world production in 1980.

Table 4

CADMIUM LEVEL OF PRODUCTION (000's of pounds)

Risk	Very Low (0-250)	Low (250-650)	Medium (650-1,800)	High (1,800-5,000)	Row %
LOW	Austria	Norway	Italy Finland Netherlands U.K.	U.S.A. Belgium West Germany	62.9
MEDIUM	India South Korea Turkey Zambia	Zaire Peru	France Mexico Spain	Japan Canada Australia	32.6
HIGH	Romania	China Yugoslavia Algeria			4.5

Sources: Frost & Sullivan World Political Risk Forecasts, August 1982; American Bureau of Metals Statistics Yearbook, 1980.
The above countries accounted for 77.1% of total world production in 1980.

Table 5

COPPER LEVEL OF PRODUCTION (000's of short tons)

	Very Low (0-1.5)	Low (1.5-32)	Medium (32-185)	High (185-1,300)	Row %
LOW	West Germany Taiwan Italy	Ireland	Finland Norway	United States	21.5
MEDIUM	Ecuador South Korea France Israel	Malaysia Turkey India Morocco Portugal	Mexico Indonesia Spain Japan Sweden	Chile Canada Zambia Zaire Peru Philippines Australia South Africa	72.6
HIGH	Algeria Argentina	Zimbabwe Bolivia Iran	China Yugoslavia Romania		5.9

Sources: Frost & Sullivan World Political Risk Forecasts, August 1982; American Bureau of Metals Statistics Yearbook, 1980.
The above countries accounted for 76.9% of total world production in 1980.

Table 6

GOLD LEVEL OF PRODUCTION (000's of short tons)

	Very Low (0-15)	Low (15-70)	Medium (70-250)	High (250-22,000)	Row %
LOW	Taiwan West Germany	Finland		United States	3.5
MEDIUM	Portugal Zambia New Zealand Malaysia Ecuador	Indonesia France South Korea Venezuela Costa Rica	Mexico Peru Chile Japan Spain India Zaire Sweden	South Africa Canada Philippines Australia Colombia	87.7
HIGH	Argentina El Salvador	Nicaragua Romania Bolivia	China Yugoslavia	Brazil Zimbabwe Dominican Rep.	8.8

Sources: Frost & Sullivan World Political Risk Forecasts, August 1982; American Bureau of Metals Statistics Yearbook, 1980.
The above countries accounted for 75.5% of total world production in 1980.

Table 7

LEAD LEVEL OF PRODUCTION (000's of short tons)

	Very Low (0-5)	Low (5-20)	Medium (20-90)	High (90-650)	Row %
LOW	Austria Norway U.K. Finland	Italy	Ireland West Germany	United States	27.4
MEDIUM	Philippines Chile Ecuador Colombia	Thailand Zambia South Korea India Tunisia Turkey	Spain Sweden Japan France	Australia Canada Peru Mexico Morocco South Africa	57.4
HIGH	Algeria	Bolivia Iran	Romania Argentina Brazil	China Yugoslavia	15.2

Sources: Frost & Sullivan World Political Risk Forecasts, August 1982; American Bureau of Metals Statistics Yearbook, 1980.
The above countries accounted for 73.1% of total world production in 1980.

Table 8

NICKEL LEVEL OF PRODUCTION (000's of short tons)

	Very Low (0-1.8)	Low (1.8-3.6)	Medium (3.6-35)	High (35-215)	Row %
LOW	Norway	Finland	United States		4.4
MEDIUM	Morocco Mexico		South Africa Greece	Canada Australia Philippines Indonesia	86.5
HIGH	Dominican Rep.	China Brazil	Zimbabwe		9.0

Sources: Frost & Sullivan World Political Risk Forecasts, August 1982; American Bureau of Metals Statistics Yearbook, 1980.

Table 9

SILVER LEVEL OF PRODUCTION (000's of troy ounces)

	Very Low (0-400)	Low (400-1,700)	Medium (1,700-5,400)	High (5,400-51,000)	Row %
LOW	Taiwan	Italy Ireland Finland West Germany		U.S.A.	16.4
MEDIUM	India Tunisia Colombia Ecuador Portugal New Zealand	Zambia Indonesia	Zaire Spain France South Korea Morocco Philippines Greece	Mexico Peru Canada Australia Japan Chile Sweden South Africa	75.7
HIGH	Nicaragua El Salvador Algeria	Dominican Rep. Romania Zimbabwe Brazil	Yugoslavia China Argentina	Bolivia	7.9

Sources: Frost & Sullivan World Political Forecasts, August 1982; American Bureau of Metals Statistics Yearbook, 1980.
The above countries accounted for 76.3% of total world production in 1980.

Table 10

TIN LEVEL OF PRODUCTION (metric tons)

	Very Low (0-225.5)	Low (225-850)	Medium (850-9,000)	High (9,000-62,000)	Row %
LOW	United States Cameroon		United Kingdom		1.9
MEDIUM	Mexico South Korea Zambia	Peru Japan Spain Portugal Canada	Zaire South Africa	Malaysia Thailand Indonesia Australia	76.8
HIGH	Argentina	Brazil	Bolivia Nigeria Zimbabwe		21.2

Sources: Frost & Sullivan World Political Risk Forecasts, August 1982; American Bureau of Metal Statistics Yearbook, 1980.
The above countries accounted for 94.2% of total world production of 1980.

Table 11

ZINC LEVEL OF PRODUCTION (000's of short tons)

	Very Low (0-21)	Low (21-55)	Medium (55-175)	High 175-1,200	Row %
LOW	Austria	Norway	West Germany Italy Finland	United States Ireland	24.1
MEDIUM	Morocco Philippines Tunisia Chile Thailand Ecuador Colombia	France Turkey Greece India Zambia	Zaire South Africa South Korea	Canada Peru Australia Mexico Japan Spain Sweden	65.0
HIGH	Algeria	Romania Argentina Iran	China Yugoslavia Brazil Bolivia		11.0

Sources: Frost & Sullivan World Political Risk Forecasts, August 1982; American Bureau of Metals Statistics Yearbook, 1980.
The above countries accounted for 73.3% of total world production in 1980.

APPENDIX I

The Prince Model:
Probabilities and Decision Structure Charts

The Prince Model is used to generate probability scores for the most likely regime and for political turmoil, restrictions on international business and trade restrictions. It is also used to construct the Decision Structure Charts used to illustrate the political forces underlying these probabilities. This appendix describes and explains how the Prince Model is used to generate probabilities and how to interpret the Decision Structure Charts that are derived from the Prince calculations.

The Prince Model

The Prince Model has been developed by the authors from many years of refinement and application. It is useful in a wide variety of political situations, and is one of the few approaches to political forecasting that is both systematic and relevant to decision-makers. It provides a means for the rigorous analysis of information, producing calculations that result in forecasts of the probability of specific political outcomes. At the same time, it uses the invaluable and unique expertise of specialists - both to supply information for the model and to adjust the calculations produced by the model according to qualitative, subjective knowledge about a given country.

Data for the Model

The first step in using the Prince Model to obtain information for a country forecast is to conduct a survey of the team of expert analysts (at least three for each country). Each team member answers a questionnaire that includes several "Prince Charts", which are used to record the positions of major individuals, groups and institutions on a particular action that could affect international business in the country. In completing the charts, experts are first asked to list at least seven actors that are able to influence each action during the next eighteen months. The actors selected may be individuals, groups or ministries within the government, opponents of the government, and individuals or groups in the society, such as business, unions, or ethnic organizations. Actors may also include foreign individuals or institutions, such as the International Monetary Fund, or other governments.

The experts then indicate on the Prince Chart their estimates of the position of each actor listed, according to the four categories listed below:

1. <u>Orientation:</u> The current general attitude of the actor toward the action. The actor's orientation is classified into one of three categories: Support (S), Neutrality (N), or Opposition (O)

2. <u>Certainty:</u> The firmness of the actor's orientation. For group actors, certainty is a function of the extent to which there is consensus among the actor's membership in supporting or opposing the action. Certainty is measured on a scale ranging from 1 (little or no certainty) to 5 (extremely high certainty).

3. <u>Power:</u> The degree to which the actor can exert influence, directly or indirectly, in support of or in opposition to the action relative to all other actors. The bases of an actor's power, as well as the ways in which this power may be exercised, are varied. Power may be based on such factors as group size, wealth, physical resources, institutional authority, prestige, and political skill. Power is measured on a scale ranging from 1 (little or no power) to 5 (extremely high power).

4. <u>Salience:</u> The importance the actor attaches to supporting or opposing the action relative to all other actions with which that actor is concerned. Salience is measured on a scale ranging from 1 (little or no importance) to 5 (extremely high importance).

The Prince Charts obtained from each of the team members are analyzed to measure the degree of consensus among the experts. If the team members differ significantly on estimates of an actor's position, additional information is sought. After obtaining a clear consensus among the experts, the individual charts are then combined into a single set of estimates. Figure 1 is an example of a Prince Chart showing the combined judgment of the experts on the positions of actors supporting and opposing the Marcos regime in the Philippines (in June 1980, when it was compiled).

Decision Structure Charts

The information in the Prince Chart is used to create a Decision Structure Chart for each risk factor. Decision Structure Charts visually portray the forces supporting and opposing each factor, presenting an over-all picture of the probable action concerning the factor.

On the chart, each actor is placed along the vertical axis according to whether it supports, is neutral toward, or opposes the action, and by the certainty of the actor's support or opposition. The actors are located along the horizontal axis according to their importance in determining the outcome of a decision as measured by their power (ability to influence the decision) and salience (the likelihood that an actor will use its power to influence the decision). Thus, the position of the actors in relation to the two axes indicates their roles in determining the action.

Actors located in the upper right corner of the chart are those that exert the maximum weight in support of the action. Actors in the lower right corner of the chart are those that exert the maximum weight against the action. Actors located in the middle of the vertical axis are either undecided or likely to shift positions. Actors located in the upper left are strong supporters of the action, but carry little weight in determining its outcome. Actors in the lower left are opponents with little influence.

Figure 2 is the Decision Structure Chart based on the Prince Chart for the Philippines (Fig. 1). The two charts show significant differences among the actors in the Philippines concerning a regime change. One extreme position is that of Ferdinand Marcos, located in the upper right corner of the Decision Structure Chart, not only adamantly in support of his regime but also having the capability to exert maximum weight to support his orientation. A different extreme position is that of Raul S. Manglapus, located in the lower left corner of the Decision Structure Chart, opposed to the regime, yet with little ability to bring about change.

The political analysis can best be understood by examining the Decision Structure Chart in conjunction with the narrative analysis provided for each estimate. The narrative appearing with the chart describes the factors underlying the positions and the importance of the actors, possible coalition changes, other actors that may become important, and the impact of the changes on international business.

Probability Calculations

The estimates of the probability of the occurrence of an action in the next eighteen months come from the country specialists' analysis of the actor's positions. Each actor is assigned a total score in each Prince Chart. This "Prince Score" is the product of the firmness of the actor's orientation to the action (certainty), the actor's ability to influence the outcome of the action (power), and the importance of the action to the actor (salience). The score has a positive sign if the actor supports the action and a negative sign if the actor opposes the action. Figure 3 presents these actor-by-actor scores for support of the Marcos regime.

FIGURE 1

PRINCE CHART ON MARCOS REGIME CONTINUATION

CURRENT ESTIMATES

ORIENTATION	CERTAINTY, POWER, SALIENCE	
+ Supports	1 Little/None	4 High
O Neutral	2 Slight	5 Extremely High
— Opposes	3 Moderate	

ACTOR	ORIENTATION	CERTAINTY	POWER	SALIENCE		PRINCE SCORE
Benigno Aquino	— x	5 x	2 x	5 =	—	50
Armed Forces	+ x	4 x	3 x	4 =	+	48
Cardinal Sin	— x	5 x	1 x	5 =	—	25
Christian Left	— x	5 x	1 x	3 =	—	15
Economic Elite	+ x	4 x	3 x	2 =	+	24
Moderate Opposition	— x	2 x	2 x	3 =	—	12
Labor	— x	3 x	1 x	3 =	—	9
Raul S. Manglapus	— x	5 x	1 x	3 =	—	15
Ferdinand Marcos	+ x	5 x	5 x	5 =	+	125
Moro National Liberation Front	— x	3 x	2 x	2 =	—	12
New People's Army	— x	5 x	2 x	2 =	—	20
Cabinet	+ x	5 x	2 x	4 =	+	40
Overseas Opposition	— x	2 x	1 x	3 =	—	6

FIGURE 2

DECISION STRUCTURE CHART
MARCOS REGIME CONTINUATION

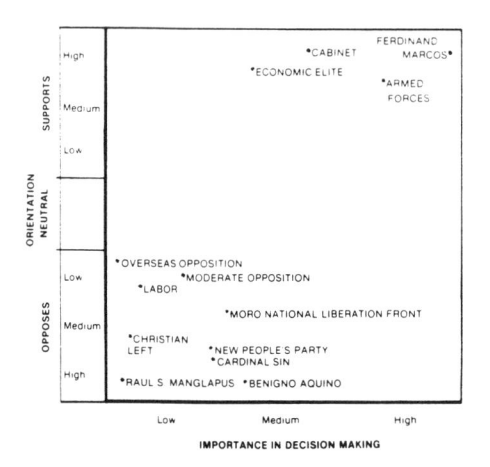

[Chart: Orientation (Supports High/Medium/Low, Neutral, Opposes Low/Medium/High) vs. Importance in Decision Making (Low/Medium/High)]

Supports:
- High: CABINET, FERDINAND MARCOS
- Medium: ECONOMIC ELITE, ARMED FORCES

Opposes:
- Low: OVERSEAS OPPOSITION, MODERATE OPPOSITION, LABOR
- Medium: MORO NATIONAL LIBERATION FRONT
- High: CHRISTIAN LEFT, NEW PEOPLE'S PARTY, CARDINAL SIN, RAUL S. MANGLAPUS, BENIGNO AQUINO

FIGURE 3

SCORES FOR SUPPORT OF MARCOS REGIME

ACTOR	ORIENTATION	CERTAINTY	POWER	SALIENCE		PRINCE SCORE
Benigno Aquino	−	× 5	× 2	× 5 =	−	50
Armed Forces	+	× 4	× 3	× 4 =	+	48
Cardinal Sin	−	× 5	× 1	× 5 =	−	25
Christian Left	−	× 5	× 1	× 3 =	−	15
Economic Elite	+	× 4	× 3	× 2 =	+	24
Moderate Opposition	−	× 2	× 2	× 3 =	−	12
Labor	−	× 3	× 1	× 3 =	−	9
Raul S. Manglapus	−	× 5	× 1	× 3 =	−	15
Ferdinand Marcos	+	× 5	× 5	× 5 =	+	125
Moro National Liberation Front	−	× 3	× 2	× 2 =	−	12
New People's Army	−	× 5	× 2	× 2 =	−	20
Cabinet	+	× 5	× 2	× 4 =	+	40
Overseas Opposition	−	× 2	× 1	× 3 =	−	6

PROBABILITY EQUATION:

$$\frac{\Sigma W_s}{\Sigma W_t} = \frac{\text{Sum of Total Positives Actor Scores (including 1/2 of each Neutral Actor Score)}}{\text{Sum of Absolute Value of all Actor Scores (including Neutral Actors)}} = \frac{237}{401}$$

Probability = 59%

The major opponent of the regime, Benigno Aquino, is assigned a total of − 50, while Ferdinand Marcos is assigned a total score of + 125. These scores reflect the actors' opposition or support for the regime and also their respective importance in either opposing or supporting the regime. This Prince Chart indicates that Ferdinand Marcos' ability to support his regime is stronger than Benigno Aquino's ability to help bring about a regime change. (Editor's Note: Aquino was assassinated in 1983)

The chart also shows the large number of opponents of the Marcos regime. The ability of these opponents, each relatively weak, to coalesce and increase their strength is what constitutes a danger to the regime.

As Figure 3 shows, in the Philippines example the total weight in the political system affecting the support for the rgime was 401. This is the denominator in the probability equation. The numerator in the probability equation, the sum of the positive actor scores, was 237. The quotient, the probability of the Marcos regime's remaining in power for the next eighteen months (following June 1980) was 59%. The model can be applied to forecasting the likelihood of regime stability, political turmoil, restrictions on international business, and trade restrictions.

Appendix I is adapted from User's Notes to World Political Risk Forecasts published by Frost and Sullivan, 1982.

Based on a paper presented at the 2nd CIM Minerals Economics Committee Symposium, "Risk Management in Mining", November 1982, Vancouver, B.C.

REFERENCE

William D Coplin and Michael K O'Leary, Everyman's PRINCE: A Guide to Understanding Your Political Problems, North Scituate, Mass: Duxbury Publishing Company, 1976. Revised Edition

9.7

COUNTRY RISK ANALYSIS

JOHN R. STUERMER AND PETER ALLEN

FIRST NATIONAL BANK OF CHICAGO

COUNTRY RISK ANALYSIS

A company operating in a foreign country assumes all the risks that it would at home. However, beyond these, it assumes risks that arise from the unique political economic, financial conditions of the foreign country which it operates. These risks are known as "country risks". The importance of these risks has grown appreciably in the recent past (particularly in the aftermath of Iran's revolution in the late 1970s and Mexico's debt rescheduling of 1982-83), and as a result, country risk analysis has become an increasingly useful management tool.

Country risk analysis seeks to assess the probability that countries will service their aggregate foreign liabilities on time, as well as the probability that country conditions could inhibit the debt service capability of private sector debtors. This involves assessment of both the <u>ability</u> and the <u>willingness</u> of countries to meet their external liabilities in a timely fashion. "Ability" is linked economic and financial variables: principally, a country's economic structure, foreign exchange generating capacity and debt service burden. On the other hand, "willingness" is tied to political factors: mainly the economic/financial orientation of government policy and the stability of economic decision making.

Ability: Economic, Financial Conditions and Prospects

Judgement of a country's ability to meet foreign financial obligations begins with a detailed study of a country's external balance sheet which compares the country's foreign liabilities and assets. Table 1 below presents country balance sheets for Mexico, Korea and Indonesia.

TABLE 1

The External Balance Sheet, 1983
(billions of dollars)

	Indonesia	Mexico	Korea
Foreign Liabilities By Source:	31.1	87.8	40.5
Bank	13.1	75.5	26.3
Non-bank	18.0	12.1	14.2
Foreign Reserves	9.3	4.7	7.0

The large magnitude of Mexico's external debt (nearly $90 billion), compared to foreign assets of approximately $5 billion, illustrates the weakness of this country's external payments position, and the main reason that it has been unable to meet interest and principal payments and been forced to reschedule its foreign debt. On the other hand, the relationship between Korea's external liabilities and assets ($50.5 billion and $7 billion, respectively), is more favorable and helps explain why Korea has been able to meet debt service payments and maintain its access to the international capital markets. Indonesia has even a more favorable relationship between its foreign assets and liabilities, $31 billion and $9 billion, respectively, and its debt is mostly lower interest rate, non-bank debt, making for a less onerous debt service burden.

After the balance sheet, the analysis turns to a comparison of the country's foreign exchange earning capacity and its prospective financing requirements, that is, the country's future earnings stream versus foreign currency needs for debt service and imports. This comparison entails a detailed forecast of the balance of payments, focusing on the current account balance (the balance of the country's trade in both goods and services). The current account balance provides a measure of a country's ability (if in surplus) or inability (if in deficit) to save foreign exchange (or "hard currency").

The current account positions of Mexico, Korea and Indonesia between 1982 and 1984 are illustrated in Table 2. Mexico dramatically reversed its current account from a $3.8 billion deficit in 1982 to a $5.5 billion surplus in 1983, a change that was driven by severe import cutbacks and export promotion required by debt rescheduling. Korea and Indonesia also improved their current account positions in 1983 (by about $1 billion dollar each), but had to make relatively modest adjustments in policy because past policies limited the size of their respective foreign debt burdens and maintained their access to international financial markets. These improvements by Korea and Indonesia are forecast to be sustained in 1984 and beyond, while Mexico's newly created current account surplus is expected to decrease in 1984 and to revert into deficit thereafter. This means that Mexico will probably remain unable to meet fully its debt service requirement at least until the 1990s.

TABLE 2

Current Account Trends
(billions of dollars)

	1982	1983	1984
Mexico	-4.9	5.5	3.8
Korea	-2.6	-1.7	-1.7
Indonesia	-6.5	-5.7	-1.7

As these examples show, the most important element of current account forecasts is an evaluation of government economic policy. This involves consideration of the size and composition of public sector spending and the budgetary deficit, the growth of credit and money aggregates, the appropriateness of interest and exchange rate policies, the extent of subsidies and other price distortions, and the formulation of wage policy. If economic policy is well managed, then a country's balance of payments should remain relatively stable and problems caused by exogenous shocks or cyclical declines in export demand should be transitory.

Balance of payments projections are based upon a central forecast of the world economic environment. The major variables are: interest rates, commodity prices, industrial country, oil prices, growth, etc. Since the world economy is perhaps the most powerful influence on individual country creditworthiness, an important aspect of country risk analysis is to test country sensitivity to alternative assumptions/forecasts about external conditions.

The final element in the assessment of a country's ability to meet its foreign obligations is the country's access to the capital markets: will the country be able to raise all the money implied by the combination of domestic economic policies and likely external conditions? This leads to an effort to identify financing sources and a judgment about the availability of credit. If credit is not available in sufficient quantities, economic policies will have to be adjusted or payments problems will begin piling up. This judgment is crucial to the overall assessment.

Willingness: Political Conditions and Prospects

Evaluation of a country's willingness to meet foreign liabilities focusses on the stability of economic policy. This begins with an analysis of a country's political structure, customs and institutions in an effort to determine how major decisions related to economic management are taken. Analysis of this process extends into an assessment of a government's susceptibility to change, its the capacity to manage conflict among competing groups in society, and its dependence on key personalities and vulnerability to disruption from external or internal sources. Particular emphasis is placed upon indentifying whether a workable mechanism for leadership succession is in place.

An important part of the analysis is an assessment of the strength of political opposition, legal or illegal. This entails evaluation of political parties, labor unions, military organizations, religious groups and other important actors as well as an understanding of their sources of support and relative political impact. It likewise includes an assessment of the probability of significant political change. This can occur within the political system—a shift between political parties in legislature—or outside the system—through revolution, coup, or other dramatic changes. It is difficult to project such discontinuities, but countries that are more susceptible to radical change than others can in some cases be identified and isolated before a crisis.

Additional political judgments focus on three general areas: the effectiveness of government control, the potential for social unrest, and the impact of external factors. Effectiveness of control refers to the government's ability to formulate and implement economic and social policy as well as the capacity to resolve political conflicts. One important element of this is the quality of the political leadership succession process; in this context, predictability and orderliness are crucial.

The potential for social unrest is associated with the distribution of income, cultural and historical tendencies, the extent of political dissatisfaction, the availability of processes to express dissatisfaction, and the prevalence of violence in a society. Finally, external factors include consideration of the external security threats that a country may face, its relations with the U.S. and other major powers, as well as its membership in regional security associations and organizations. The analysis of individual country geopolitical conditions must be placed in a regional or international context. This has both political and economic dimensions.

CONCLUSION

Country risk analysis is designed to develop informed judgements on the present and future creditworthiness of countries. While the objective is simple and straightforward, the process is very complex, involving a multitude of variables that can differ widely from one country to another. Nevertheless, country risk analysis is a vital tool of senior bank managment at a time of significant uncertainty in the world financial system.

9.8

OPIC Insurance Programs for the Mining Sector

B. Thomas Mansbach

Overseas Private Investment Corporation

INTRODUCTION

The Overseas Private Investment Corporation ("OPIC") is a U.S. Government agency mandated to promote the economic development of the less developed countries ("LDCs") by assisting U.S. private investment in the Third World. Generally, investment decisions turn on the optimism or pessimism of the private sector's financial analysis of a project, with market price of product a major unknown in a very chancy set of projections. Discussions with American mineral companies, however, show that, more often than not, in developing country projects it is the political risks of investing in mineral exploration and extraction that pose the most serious deterrent to investment. In several of the projects described here, investments would not have been made if OPIC had not been involved providing support to the project sponsors and their funding sources.

Clearly it is not OPIC's business to promote U.S. investment in a project which is not commercially viable, but OPIC's programs can neutralize the political uncertainties which act as disincentives to investment in otherwise commercially viable projects. That is to say, OPIC's programs are designed to reinforce market decisions and not distort them.

Although OPIC is an independent agency of the U.S. Government, it is managed by a Board of Directors, the majority of whom come from the private sector. This, coupled with the statutory requirement that it be self-sustaining and operate with due regard for the principles of prudent risk management, requires that it operate on a business-like basis. Although established as a development agency, to the business public OPIC is both an insurance company and a lending institution, with two major programs which are available to U.S. investors in the minerals sector: its political risk insurance program and its finance program. Each has been specially tailored to the meet needs of the mining industry.

The insurance program provides coverage against four basic types of political risk: Currency Inconvertibility, Expropriation, loss caused by War, Revolution and Insurrection and Interference with Operations as a result of an act of War, Revolution or Insurrection. Coverage against the risks of Civil Strife can be added to both War, Revolution and Insurrection coverage and Interference with Operations.

Inconvertibility coverage provides protection against a deterioration in an investor's ability to convert local currency into U.S. dollars. This deterioration can be the result of either active or passive delays. Active delays consist of a change in exchange laws or regulations that leads to inconvertibility delays of 30 days or more. For example, if the exchange regulations at the time of contract execution allow an investor to transfer earnings equal to 18% of its registered capital per annum and the government of the host country later changes the law to allow conversion of only 12%, the insured would convert the 12% transferable under the new exchange regulations with the government foreign exchange authorities, delivering the remaining 6% to OPIC for conversion at 99% of the then prevailing rate of exchange. (OPIC does not cover devaluation risks.) Upon delivery of the local currency to OPIC, normally at the U.S. Embassy in the project country, the insured's account in the U.S. would be credited with dollars.

Delays can also result from what is termed passive blockage, which includes administrative delays of more than 60 days (sometimes longer) caused by the bureaucratic tie-ups or pigeonholding of transfer requests, as well as delays suffered simply because the government of the host country does not have sufficient foreign exchange. The transfer procedure is the same as with active blockage – the local currency is delivered to OPIC in the project country and OPIC provides dollars in the United States at the then prevailing rate of exchange.

The second coverage -- Expropriation -- protects not only against the loss of the investment due to the well-known risks of expropriation, nationalization or confiscation by the host government, but also against

"Creeping Expropriation" -- actions by the host government which, although not a seizure per se, prevent the investor from constructing or operating the insured project.

Under Expropriation, OPIC also offers mining companies coverage for special contractual relationships entered into with the government, such as those for the provision of water or electricity. Thus, if an investment in an aluminum project were insured with OPIC and that company had an agreement with the government to purchase water or power at a specified rate and the rate was increased to a level which materially adversely affected the return on the investment project, such an act could be covered by OPIC.

The advantages of OPIC coverage are not solely financial, especially in cases of expropriation. Frequently, OPIC's mere presence is sufficient to lessen the possibility of seizure or significant change in an investor's project agreement. In other instances, OPIC has used the maturation period required under expropriation coverage to work with the parties in reaching a settlement.

War, Revolution, Insurrection coverage (hereinafter referred to as "W/R/I" coverage) protects against damage to tangible property as a result of such activities. Examples here are obvious. If on-site equipment is damaged, destroyed or seized and retained because of war, revolution or insurrection, OPIC will pay to the insured the lesser of the item's original cost or the cost to repair or replace it. This includes major damage to mine shafts if damaged or destroyed by a covered event.

The fourth and final coverage is Interference with Operations (hereinafter referred to as "I/W/O" coverage). This cover protects against an interruption in operations because of war, revolution or insurrection, whether or not such acts have caused damage to physical assets. Thus, if the fighting in an area has not resulted in damage to project assets but has nonetheless rendered it impossible or unreasonably hazardous to continue to operate there for six months, OPIC would pay the costs of maintaining the insured investment during the period the project is closed. The insured, in turn, would transfer to OPIC the project securities evidencing his investment, which OPIC would hold for five years. If at any time during that five-year period the insured feels he could resume project operations, he simply returns the insurance compensation to OPIC, without interest, and OPIC returns the project securities to the insured. By the same token, if at any time during that period OPIC believes that the insured can resume operations, OPIC can "put" the project securities to the insured in exchange for a reconveyance of the insurance compensation. After five years, however, all bets are off -- if the situation has not calmed down, the insured keeps the money and OPIC owns the securities evidencing the insured investment. Coverage against Civil Strife -- terrorism, sabotage and the like, can be added to either War, Revolution, or Insurrection or Interference with Operations. If added to W/R/I, it would protect the project assets against physical damage as a result of Civil Strife. Adding Civil Strife coverage to I/W/O coverage protects the insured against project closure as a result of Civil Strife. Civil Strife coverage is available only in conjunction with W/R/I or I/W/O coverage; it cannot be purchased separately and it does not cover Civil Strife where the principal motivation is a student or labor issue.

OPIC's insurance coverages are designed to neutralize the political uncertainties associated with an investment, so that the risks attendant in operating in the developing countries are comparable to those of operating in the U.S. or elsewhere in the developed world. Inconvertibility provides a transfer function and protects cash flow. Expropriation protects net worth. War, Revolution or Insurrection protects physical assets and Interference with Operations provides bridge financing, which is of particular importance for smaller companies with a disproportionate amount of their assets tied up in the Third World countries.

The insurance is available to eligible investors as defined by OPIC's implementing legislation (the Foreign Assistance Act of 1969, as amended) and include: (a) U.S. citizens, (b) corporations, partnerships and other business organizations substantially beneficially owned (i.e. more than 50%) by U.S. citizens, and (c) foreign subsidiaries at least 95% owned by U.S. citizens or corporations, as defined above. Where a company is less than 50% U.S.-owned and the foreign shareholders can secure participation by similar government agencies in their own countries (MITI, ECGD, COFACE), OPIC can step in to provide coverage for the U.S. percentage interest.

Insurance is available for any amount of U.S. interest in a project as long as the investor's contribution is to a new project or for an expansion of an existing project. OPIC will insure most types of investments including those made pursuant to concession or production-sharing agreements, long-term leases, loans of equipment, take-or-pay contracts or risk-sharing contracts, as well as traditional debt and equity investments.

Generally, coverage is provided for 90% of the initial investment plus an additional 180-450% coverage for retained earnings, up to a maximum insured amount of $100-125 million. The term of coverage generally corresponds to the construction period plus ten years, but is in no case greater than 20 years.

In addition to the equity coverage just described, OPIC offers coverage to institutional lenders to mining projects.

Here the coverage is designed to protect lenders against interruptions in the payment stream of principal or interest because of Inconvertibility, Expropriation, W/R/I or I/W/O. If the default on a payment continues for a period of three months in the case of Expropriation (one month for a subsequent consecutive default) or one month in the case or W/R/I, the lender is compensated for the amount of the insured portion of the defaulted installment (both principal and interest). If the loan is by a U.S. financial institution, coverage is provided for 100% of the loan amount, up to a maximum insured amount of $100-125 million. Coverage generally runs the full term of the loan plus a grace period.

OPIC's coverage tends to be imaginative and flexible. The agency has worked with commercial banks, international financial institutions and supplier credit programs to structure coverage that best protects the investments of the mining companies and their lenders. A few years ago, several banks used OPIC's institutional lenders cover to allay concerns arising more from risks in the countries surrounding the host country than within the host country itself. The principal concern was that regional turmoil would cut the rail links to the host country, interrupting imports of supplies and exports of product resulting in a default on project loans. OPIC provided coverage to the banks against events that might affect rail lines and lead to an interference with the operations of the project. What OPIC, in essence, covered were not solely events within the country itself but within the region. While OPIC does not, as a general rule, cover risks on a regional basis, in this case the regional risk was acceptable and was covered under the contract. Thus if a project sponsor has a need that does not seem to be covered in OPIC existing programs, alternative structures can be developed that will allow OPIC to provide the required assistance.

In a second case, while the lending banks were satisfied with the ability of the project to pay out based on projected market conditions and the operation of the project under the project agreements as originally negotiated, they had nonetheless viewed the venture with skepticism, mainly because of concern over the political stability of the host government and a possible forced renegotiation of the basic project agreements. When the project sponsors approached OPIC, they were hard pressed to find a bank willing to take those risks. OPIC facilitated the financing by insuring the agreements as originally negotiated, so that any material change in the agreements (including the joint venture agreement, the operating agreement, the offshore trust agreement, the marketing agreement and the technical service agreement) that resulted in a default in an installment of debt, would constitute a compensable act.

In other cases, OPIC has not only provided equity coverage to the project sponsor itself but also to thousands of portfolio shareholders who participated in separate offerings to buy shares in the project.

OPIC's insurance is used often as a financing tool -- not just as an insurance contract. Frequently, even after the proposed project financing has been structured to minimize the risk -- using a consortia of experienced lenders from several different countries, getting the support and participation of the World Bank and its affiliate, the International Finance Corporation -- certain significant political uncertainties remain. In those instances OPIC participation can be the lynchpin to bringing in the banks, to making the deal feasible from a financial and political viewpoint.

OPIC's experience in the worldwide mining sector is diverse. To date, we have written $3.9 billion of political risk coverage to 30 projects in 25 countries.

In addition to political risk insurance, OPIC also has a finance program under which it provides medium to long-term financing (5-15 years) to projects sponsored by U.S. companies. Since 1970, OPIC has lent $235 million to seven mining projects in seven countries. Direct loans are made to projects sponsored by small businesses[1]/ and guaranteed loans funded by U.S. financial institutions are made to large projects. OPIC funding is often provided on a project finance basis - without recourse to the U.S. sponsor - and can range from $150,000 to as much as $50 million. All-in lending costs for OPIC finance programs, while not concessionary, are attractive. Because the cost of money on an OPIC-guaranteed loan -- which is in essence government paper because it is backed by the full faith and credit of the U.S. Government -- is less than conventional loans, the all-in rate on the funds (which includes both a commercial and political risk guaranty) approximates the commercial rates for medium and long-term money. Moreover, the longer term and flexible grace periods offered under the program often allow OPIC to complement other shorter-term lenders in a financial package for a project.

Because OPIC finance acts as a project lender, funding -- as distinguished from OPIC's political risk insurance -- in the minerals area is limited to projects in the development/production stage. A strong U.S. interest, usually defined as a significant equity position on the part of the U.S.

---

[1]/By "small business", OPIC means a company with sales of less than $118 million and a net worth of less than $44 million.

sponsor, is a requirement for OPIC's financial support. The agency looks both to the commercial viability of the project and to the managerial and technical expertise of the investor when evaluating requests for financing.

OPIC financing has been used successfully in conjunction with export credit financing, development bank lending, and commercial finance. In the latter case, OPIC is often willing to take lengthier maturities and guarantee fixed rate obligations. Co-financing is most often the norm in large mining ventures, since no single source can provide the massive credits needed to launch projects of such magnitude. In these cases OPIC, as a senior lender, is willing to share the collateral security of the project with other senior lenders. For example, this approach was used in funding a $750 million dollar gold and copper mining venture in the remote jungles of a newly emerged Asian nation. In that case, commercial loans were structured over five years, and the OPIC facility, the export credits and the development bank loans over eight years. OPIC has also used its maximum guaranty coverage in several other mining projects in the Caribbean, Africa, and Asia, often in conjunction with OPIC's political risk insurance program.

Finally, in addition to the finance and insurance programs, OPIC offers several pre-investment services including the Investor Information Service (a clearing house of U.S. government information of interest to potential investors in the developing countries), Feasibility Study Funding, Opportunity Bank (a project/sponsor matching service) and Investment Missions, which may be useful to U.S. investors looking for mining opportunities in the developing nations.

For additional information on OPIC, phone or write:

Information Officer
Overseas Private Investment Corporation
1129 20th Street, N.W.
Washington, D.C. 20527
(800) 424-6742

9.9

FEASIBILITY STUDIES AND OTHER PRE-PROJECT ESTIMATES.
HOW RELIABLE ARE THEY?

Mr. Grover R. Castle
Vice President, Chemical Bank

New York, New York

INTRODUCTION

Banks are being asked to assume more and more of the project risk in connection with project financings and, as a result, they are seeking to become more sophisticated in the way that they approach these kinds of financings. For example, a large number of banks that are active with the mining companies have established mining groups which typically will include mining engineers, economists, and bankers, in order to specialize their service to the mining industry. These banks have geared up these special service staffs so that they will be better equipped in their lending activities to analyze, measure and evaluate the project risk. The hope is that a team of these specialists will be able to predict the potential success or failure of these new "green fields" mining ventures that the banks are asked to finance. This paper shall focus on this subject and try to determine just how successful these specialists can hope to become as they analyze these various projects that are brought to them for financing.

The approach that shall be taken in this paper shall be to examine how successful the mining companies themselves have been in the past with this same problem. These mining companies should be the most sophisticated in their approach to the analysis of the potential for success of a new mine because they have had the most experience in evaluating these projects. After all, who knows more about developing a mine than the companies that have developed similar mines in the past?

For the purposes of this paper, the writer has selected a sample of 18 mines that have been developed over the years and measured the actual performance of each of these sample ventures against the original projections. All of these cases have involved the development of large projects where the sponsors or partners in the venture are substantial and sophisticated companies and where the feasibility studies have been prepared by engineering firms that are considered very professional in the industry or alternatively where the feasibility has been done by a comparable in-house staff. The sample cases have been chosen from a period which starts in 1965 and continues through 1981 with the cases distributed over that period as indicated below:

Approximate Date of Commencement Of Development	Number Of Cases
1965 through 1969	5
1970's	8
1980 - 1981	5
	18

The types of mines (or mineral being mined) selected are briefly summarized below:

Mineral	Number Of Projects
Iron Ore	5
Copper	3
Nickel	3
Coal - Metallurgical	2
- Steam	1
Uranium	2
Bauxite	1
Phosphate Rock	1
	18

Fourteen of the projects are open pit mines, three are underground mines and one mine starts on the surface and gradually goes underground. Seven of the mines are in the the far east with the remaining three in either South or Central America.

The following four specific questions were examined for each project:

1. Was the project constructed on time, and if not, why not?

2. Did the project experience any overrun costs (i.e. did the development of the project cost more than originally projected) and if so what caused the overrun?

3. Was the mine ultimately able to achieve the design production level and if not, why not?

4. Has the actual cash flow generated by the project measured up to the original forecast and if not what is the cause of the shortfall?

Only four (22%) of these projects experienced no problems in the four categories mentioned above and fourteen (78%) of the projects had problems. Eleven of these fourteen experienced trouble in all four of the above categories. Thirteen (72%) of these cases had severe problems and five of the thirteen have (to date) been unable to generate a positive cash flow. Two of the projects (numbers 9 and 11) have been actually closed down and moth-balled because the problems were so severe. The results of this survey are summarized in the Summary Table on the next page.

The Summary Table shown does not reveal any obvious patterns except that all of the copper projects and all of the nickel projects had serious problems. The cases with problems tend to be fairly equally distributed over the period from 1965 to 1981. Furthermore, except for this concentration of problems in these nickel and copper cases there are no other concentrations of problems in any particular mineral as can be seen from the Chart that follows:

Type Of Mineral	Total Number Of Projects	Number With No Problems	Number With Minor Problems	Number With Severe Problems
Iron Ore	5	2	1	2
Copper	3	-	-	3
Nickel	3	-	-	3
Coal				
Metallurgical	2	1	-	1
Steam	1	-	-	1
Uranium	2	1	-	1
Bauxite	1	-	-	1
Phosphate Rock	1	-	-	1
	18	4	1	13

In the next section of this paper, each of these questions is discussed separately.

Construction Delays

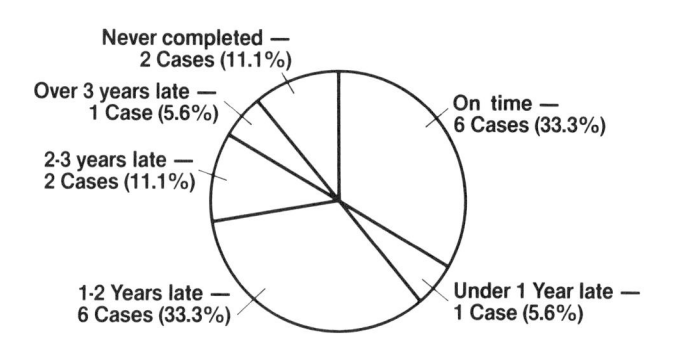

Pie chart:
- Never completed — 2 Cases (11.1%)
- Over 3 years late — 1 Case (5.6%)
- 2-3 years late — 2 Cases (11.1%)
- 1-2 Years late — 6 Cases (33.3%)
- On time — 6 Cases (33.3%)
- Under 1 Year late — 1 Case (5.6%)

CONSTRUCTION DELAYS

Twelve (66.7%) of the cases experienced Construction Delays and five (28% of all the cases) of these twelve were more than two years late. Two (11%) of the projects were never completed.

The most common cause (7 cases) of these delays involved some aspect of the design of the mine. These design problems are listed below:

1. Unsuitable equipment was ordered for the mine.

2. Corrosion required replacement of a kiln.

3. Unanticipated water in mine made equipment and mine plan unsuitable.

4. The hostile environment in jungle location made construction more difficult.

5. Cave in.

6. New technology caused start-up delays.

7. Problems with pollution abatement equipment.

8. Metallurgical recovery or quality problems.

9. Equipment quality not satisfactory.

Three other projects were delayed by strikes and in two cases labor productivity was below original estimates. Depressed commodity prices were responsible for the delay in two of the projects and at one mine the delay was caused by a delay in the delivery of equipment.

RISK ANALYSIS FOR MINE FINANCE

SUMMARY TABLE

Analysis Of Eighteen Actual Mining Projects Included In Survey

Project Number	Date (Start of Development)	Mineral	Type Of Mine	Location	Construction Delay	Cost Overruns	Achieve Design Production Level	Achieve Projected Cash Flow
1	1965	Iron Ore	Open Pit	Far East	On Time	None	Achieved On Time	Achieved On Time
2	1965	Iron Ore	Open Pit	Far East	4 Years Late	9%	4 Years Late	First six years didn't Cover Debt Service
3	1967	Iron Ore	Open Pit	Far East	On Time	None	Achieved On Time	Achieved On Time
4	1968	Copper, Molybdenum, Silver	Open Pit	North America	2 Years Late	64%	2 Years Late	Unprofitable For 2 Years Maturities Extended
5	1969	Coking Coal	Open Pit	North America	3 Years Late	50%*	3 Years Late	Unprofitable for 3 Years
6	1970	Copper	Open Pit	Far East	On Time	41%	2 Years Late	Achieved On Time
7	1971	Iron Ore	Open Pit	Far East	On Time	None	1 Year Late	Lower than Expected but Covered Debt Service
8	1972	Iron Ore	Open Pit	North America	14 Months Late	37%	14 Months Late	Close to Projections
9	1973	Nickel	Open Pit	South or Central America	2-1/2 Years Late	87%	Never Achieved	Never Earned Cash Flow
10	1973	Nickel Cobalt	Open Pit	Far East	One Year Late*	57%	Never Achieved	Never Earned Cash Flow
11	1974	Uranium	Underground	North America	Never Achieved	50%*	Never Achieved	Never Earned Cash Flow
12	1975	Steam Coal	Underground	North America	2 Years Late	10%	Never Achieved	Never Achieved Projection
13	1979	Nickel	Open Pit	South or Central America	16 Months Late	59%	Never Achieved	Never Earned Cash Flow
14	1980	Metallurgical Coal	Underground	North America	On Time	None	Achieved On Time	Achieved On Time
15	1980	Gold, Silver Copper	Open Pit and Underground	South or Central America	About 1 Year	100%	15 Months Late	Achieved On Time
16	1980	Bauxite	Open Pit	Far East	5 Months Late	34%	5 Months Late To Date	Never Achieved Projection
17	1980	Uranium	Open Pit	Far East	On Time	None	Achieved On Time	Achieved On Time
18	1981	Phospate Rock	Open Pit	North America	Never Achieved	None	2 Years Late	Never Earned Cash Flow

*Estimate

Overruns

- No overrun Completed on budget — 6 Cases (33.3%)
- Overrun less than 20% — 2 Cases (11.1%)
- Overrun from 20% to 50% of original cost — 5 Cases (27.8%)
- Overrun of more than 50% — 5 Cases (27.8%)

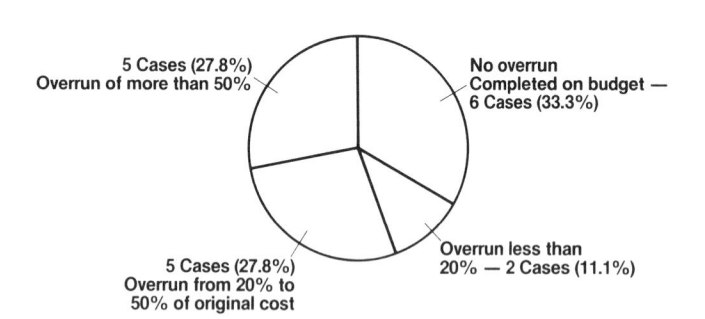

OVERRUNS

Two thirds (12) of the cases experienced overruns and in ten (56% of all the cases) of these twelve, the amount of the overrun was more than 20% of the original cost.

Again design problems were the most prevalent cause of overruns with this cause accounting for the problem in seven of the overrun cases and all of the same causes listed (under Completion Delays) above (except for item 9) were a factor in the cause of these overruns. In one other case an overrun was caused by a problem in the design of a conveyor belt that could not handle the full capacity of the project. Four of the projects cost more due to inflation and in three of the projects higher interest costs contributed to an overrun. The purchase of additional equipment caused an overrun in two cases and one of the projects was effected by each of the following causes:

1. Labor problems.
2. Low productivity.
3. Depressed commodity prices.

Actual Production Vs. Budget

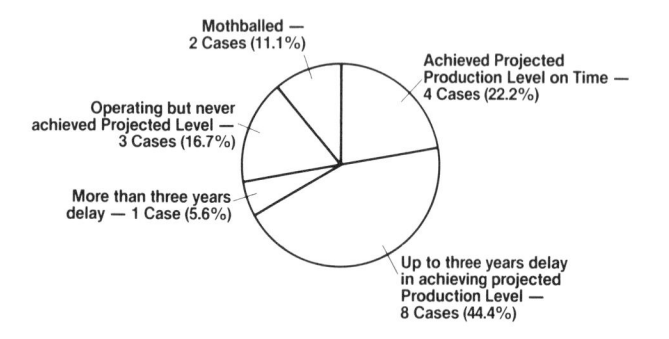

- Achieved Projected Production Level on Time — 4 Cases (22.2%)
- Up to three years delay in achieving projected Production Level — 8 Cases (44.4%)
- More than three years delay — 1 Case (5.6%)
- Operating but never achieved Projected Level — 3 Cases (16.7%)
- Mothballed — 2 Cases (11.1%)

PRODUCTION LEVEL

Fourteen (78%) of the sample cases had difficulties in achieving design production levels and in six (33% of all the cases) of these fourteen mines the production problems continued for more than three years.

Many of the same design problems which caused the Delays in Construction and the Overruns also resulted in actual production falling below the original budget. Five of the projects had unsatisfactory production levels due to Design problems and these were caused by the reasons listed as items 1, 7, 8 and 9 under Construction Delays. Another design problem that impaired a project's ability to produce at expected levels was the design of a conveyer belt that could not handle full capacity.

Depressed commodity prices was the reason why five other cases did not achieve design production levels and at one mine the buyer under a long-term purchase contract purchased less than the minimum provided in the contract. In three other projects the production levels were effected by strikes and in two cases low labor productivity contributed to the failure of the project to achieve design production levels on schedule.

Actual Cash Flow Vs. Budget

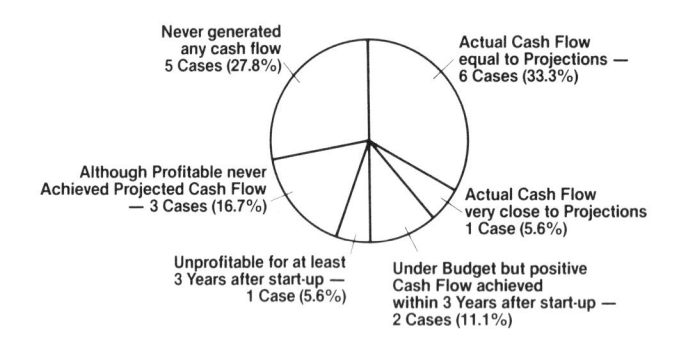

- Actual Cash Flow equal to Projections — 6 Cases (33.3%)
- Actual Cash Flow very close to Projections — 1 Case (5.6%)
- Under Budget but positive Cash Flow achieved within 3 Years after start-up — 2 Cases (11.1%)
- Unprofitable for at least 3 Years after start-up — 1 Case (5.6%)
- Although Profitable never Achieved Projected Cash Flow — 3 Cases (16.7%)
- Never generated any cash flow — 5 Cases (27.8%)

CASH FLOW LEVEL

The actual cash flow in twelve (67%) of the projects were below the projections. In one of these cases the project didn't cover debt service during the first six years of operation and in another case the project was unprofitable for 3 years. In three (17% of all the cases) of the cases the project never achieved budgeted levels and in five (28% of all cases) of the mines the operation has never generated a positive cash flow.

Depressed commodity prices was the cause of this earnings shortfall in six (including all three nickel cases) out of the twelve cases that were under budget. In four other cases the cause of the miscalculation was that operating costs exceeded projected costs and in two of these four the increased operating cost was primarily related to the cost of energy. Low labor productivity was a factor in connection with two of the projects and in one other case, the cause of the earnings problem was due to the use of equipment that was unsuitable for the mine.

CONCLUSION

The obvious conclusion from this survey is that there is a very high probability that any new project shall run into problems and even after construction has been completed, there are plenty of risks that can cause problems. Furthermore, a high percentage of these problems are serious enough to cause concern as to whether the project will be able to pay out the financing. The most common causes of these problems are related to the design of the project and depressed commodity prices. If the largest most sophisticated mining companies are experiencing problems of this magnitude, then it would seem that the banks should be cautious in their approach to project financings. It is interesting to note that despite the gloomy picture descibed in this paper no banks have lost money (as of the date of this paper) in connection with any of the project financings covering these eighteen sample cases. The main reason why the banks financing these projects have not taken a loss is that in all of these cases the sponsors have provided completion undertakings and or sufficient equity which have forced the sponsors to stick with these projects and overcome the problems that developed.

9.10

COST OVERRUN RISK AND HOW IT WAS MINIMIZED
DURING CONSTRUCTION OF THE MT. GUNNISON MINE

Mr. L. P. (Les) Haldane

ARCO Oil and Gas Company
Cost and Scheduling Supervisor
Anchorage, Alaska

ABSTRACT

Building a small underground coal mine is a risky endeavor, especially in the recent economic environment. ARCO's Mt. Gunnison No. 1 Mine, located in western Colorado, was designed and built as a small expandable speculation mine with initial production to be sold on the spot market, so it was very important to minimize capital exposure. This mine, operated by West Elk Coal Company, a subsidiary of Anaconda Minerals, was successfully built under budget at $35 million, and has been operating successfully since completion of construction in late 1982.

The initial planning phase of a project is extremely important for testing the economic and technical feasibility of a proposal. The construction phase, by contrast, carries with it significant cost and quality risks which can relate to long-term economic and operating problems and therefore directly affect the viability of the project. This paper discusses several different aspects of the cost overrun risk involved in building an underground coal mine, and how attempts were made to minimize these risks during construction of the Mt. Gunnison Mine. Good people, construction planning, and execution are essential factors in assuring a quality, cost-effective project, with corresponding risk minimization during the construction phase and long term favorable implications for the operating phase.

INTRODUCTION

Purpose

The purpose of this paper is to discuss risk in terms of building a small underground coal mine, and how this risk was successfully minimized during construction of the Mt. Gunnison Mine. It is meant to represent a practical approach to effective construction management for reducing cost overrun risk.

Scope

There are generally three broad phases to a mining project: the planning phase, the construction phase, and the operating phase. Each of these phases has its associated risk.

The planning stage of a project is critical. This is where the economic and technical feasibility of the project are tested, and financing and detailed design are completed. The biggest risk during this phase is that the reserve is not all that it is thought to be in terms of size, quality and location.

The operating phase also has risks in that the operating costs (labor, materials, overhead) are higher than anticipated, or the market is such that prices or demand were not what was anticipated.

FIGURE I
MT. GUNNISON COAL PROPERTY

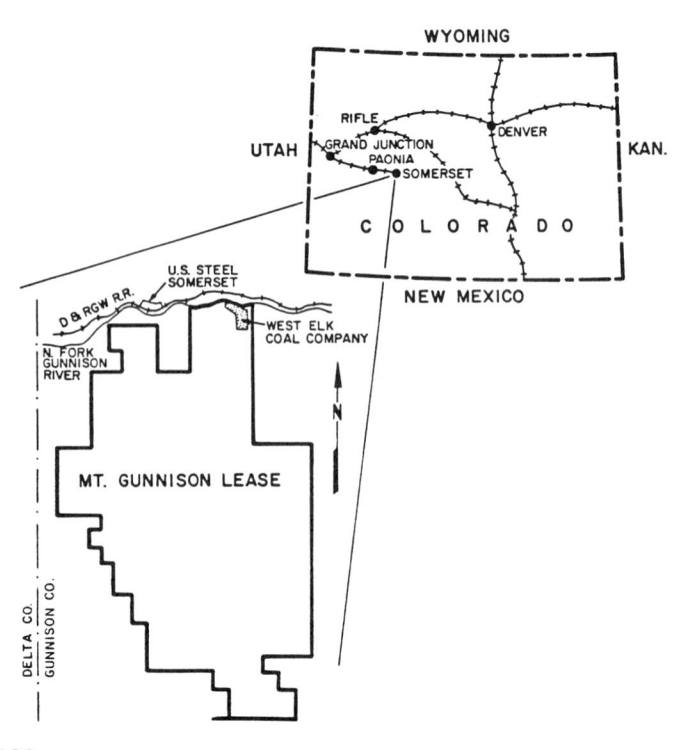

However, this paper will focus on the risks during the construction phase. This stage is one of the most vulnerable to risk. Ineffective planning, organization and control can yield both cost overruns and long-term operating problems, and therefore make a mining project very unprofitable. Good people, construction planning and execution can minimize this risk.

Project Description - Mt. Gunnison No. 1 Mine

The Mt. Gunnison Mine is located in western Colorado in the rugged West Elk Mountain range about 130 km (80 miles) southeast of Grand Junction and 16 km (10 miles) east of Paonia in Gunnison County. Figure 1 shows the location of this property.

The mine is located in a steep valley cut by the North Fork of the Gunnison River. The mine shares the valley with the river, a road (Colorado Highway 133), a railroad (Denver & Rio Grande), a 44 Kv power line, and several other coal mines. The geologic setting is one of several sand and shale formations intermingled with several coal seams. The portal bench of the mine is located near an old landslide area.

The project directive for this mine was to design and build a small coal mine readily expandable to 1.3 million tonnes per year (1.4 million tons per year), and not to preclude further expansion to 2.5 million tonnes per year (2.8 million tons per year). Since this was to be a speculative mine with the small initial tonnage to be sold on the spot market, it was imperative that capital exposure be minimized. Any item that was not required for the operation of the small mine was to be deferred until justified by future coal sales.

The surface construction consisted of earthwork (roads, runoff retention ponds, portal bench, and facility pads), a coal handling system (conveyor belts, breaker, silo and loadout facility), buildings (warehouse, administrative office, mine office and changehouse) and several support facilities such as a substation, a railroad spur, and sewer and water treatment plants. Figure 2 shows a plot plan of the surface facilities.

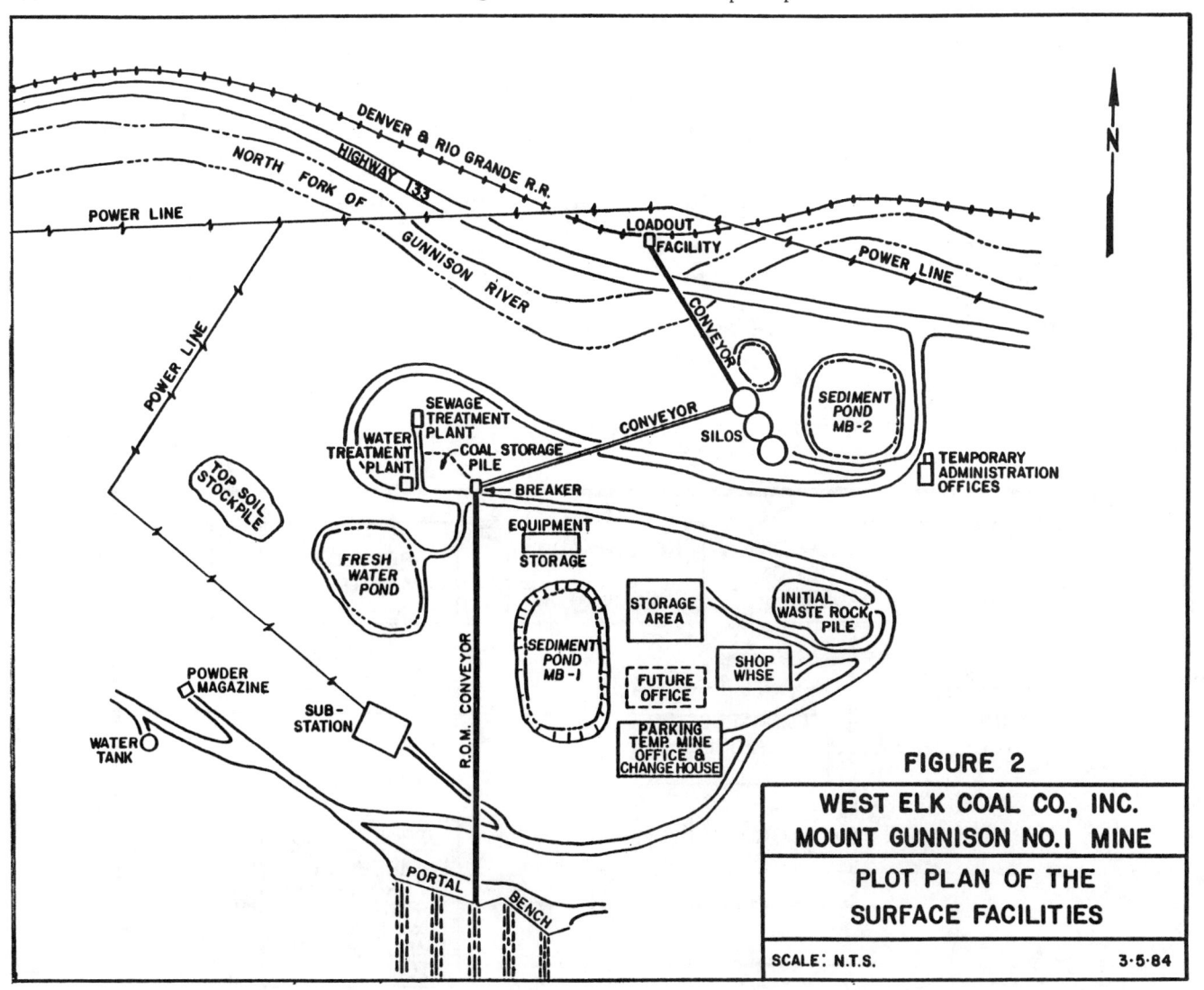

FIGURE 2
WEST ELK COAL CO., INC.
MOUNT GUNNISON NO. 1 MINE
PLOT PLAN OF THE SURFACE FACILITIES
SCALE: N.T.S.   3-5-84

The underground construction consisted of driving five lined tunnels approximately 30 meters (100 feet) in length horizontally to the coal. The mine was designed as a room and pillar mine using highly automated state-of-the-art equipment including continuous miners and shuttle cars.

Project Milestones

The major milestones of the project are listed below:

June '79 Basic Engineering complete.
Aug. '79 Started permitting process.
Dec. '79 Project funding approved.
June '80 Surface design contract awarded.
Dec. '80 Coal Handling System contract awarded.
Apr. '81 Earthwork contract awarded.
May '81   Project funding revision approved.
Aug. '81 Start of construction.
Sept.'81 Landslide movement identified.
Nov. '81 Slide stabilized with compacted buttress.
Jan. '82 Substation energized.
Jan. '82 Production started.
Apr. '82 Rock anchors completed and slide further stabilized.
Apr. '82 Silo slip completed.
Aug. '82 Water treatment facilities completed.
Sept.'82 Foundation and steel work for coal handling system completed.
Oct. '82 Sewage Treatment Facilities completed.
Oct. '82 Entire Coal Handling System Operational.
Oct. '82 Asphalting of access road completed.
Nov. '82 Sealing of access road completed.
Nov. '82 Portal Bench completed.
Dec. '82 Construction essentially complete.

PEOPLE

Project Management

The first major factor in minimizing cost over-run risk in the construction phase is to have a good project management team, both in terms of organization and in qualifications/ability. Without good people, the planning and execution of the project will suffer, with corresponding detrimental impacts on cost and quality.

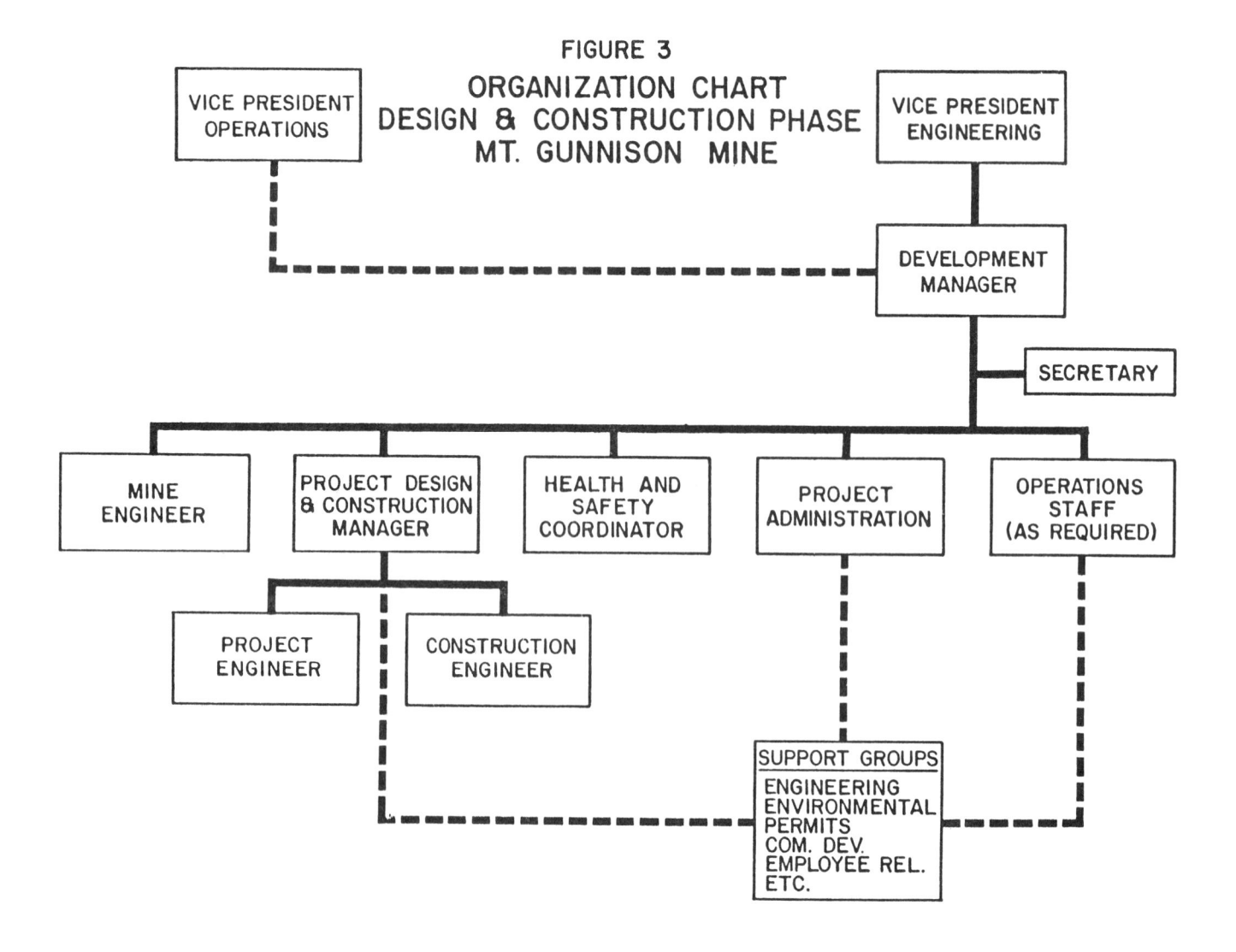

FIGURE 3
ORGANIZATION CHART
DESIGN & CONSTRUCTION PHASE
MT. GUNNISON MINE

Organization

The project management team on the Mt. Gunnison project was organized as a small on-site group (see Figure 3). It was set up such that the owner (in this case ARCO) acted as the general contractor. This allows much better control over the job.

Even though the project team was small, it was able to draw a vast amount of administrative/ engineering support from the home office in Denver. This allowed the team to work efficiently on the job without people "bumping into each other", while at the same time drawing the proper amount of expertise and support to get the job done.

Teamwork was stressed in this organization to allow good communication and cooperation, but everyone also had clearly defined objectives, authorities and responsibilities. These factors added to the efficiency of the team because everyone knew what the other person was doing.

Qualifications/Ability

Although often difficult to always obtain, it is also important to have people on the project team that have good qualifications and abilities. This involves having people with technical competence, flexibility, good communication skills, an action orientation, and hopefully with some historical involvement with the project. These qualifications and abilities were apparent on the Mt. Gunnison project team. This was very helpful in addressing problems that were encountered during construction and thus minimized the risk of a cost overrun. For example, the initial earthwork around the old landslide area caused some gradual ground movement that threatened the entire portal bench area and the access road. With lots of help, the project team was able to identify the problem, review the alternatives, and quickly implement corrective action that stopped the movement and possibly saved several million dollars in rework.

Technical competence and flexibility allows members of the project team to efficiently handle a variety of functions during construction. Every member of the team at some point might be required to perform the various functions of construction such as project control, quality control, safety and engineering.

Communication skills are important because there is constant interaction with the construction contractors, management, regulatory agencies, and support groups. The importance of communication will be discussed further in a later section of this paper.

People who are action oriented, hard working, with a "can do" spirit are invaluable on construction projects. As problems arise, members of the team must be willing to meet the challenges, take risks and make decisions so that the project does not flounder as costs increase.

Finally, people who have had some historical involvement with the project during the feasibility and initial planning stages are also important because they are familiar with how and why decisions were made. This avoids the problem of "reinventing the wheel" or making some costly change because some important factor was overlooked.

PLANNING

Construction Planning

Construction planning is another key factor in minimizing cost overrun risk because it involves making strategies to minimize costs at the front end and identifying potential problems before they happen. Good project planning involves setting a sound contracting strategy, establishing a project budget and schedule, and constant interaction with "outside" interests.

Contracting Strategy

Once the scope of work has been clearly identified, discrete work packages can be segmented to allow the work to go to specialty contractors. For example, ARCO was able to go to separate contractors for the coal handling system, earthwork, tunneling and sewer/water treatment plants on this project. This allowed ARCO to include contractors with expertise in these areas on each of the bidders lists. ARCO also tried to chose small, local, quality contractors who were "hungry" for work. Contractors with familiarity and expertise in certain types of work reduces the risk of a cost overrun.

A detailed scope of work also allows the work to go out for lump sum competitive bidding, which further decreases the cost overrun risk. Of course, the contractor is going to have to include some of his risk in the bid, but as long as the work scope is well defined and the bids are truly competitive, then this additional cost should be minimal. This requires good design packages with detailed specifications, and "tight" contract language which lays out reporting requirements, schedule requirements, personnel, scope change procedures, and required interfaces with the owner and other contractors.

Another factor that might be considered a part of contracting strategy is the use of non-union labor. The Mt. Gunnison project was fortunate in that local skilled non-union labor was readily available. Non-union laborers can handle a variety of tasks and thus can be used more efficiently; and with non-union labor the problem of labor strife is minimized. This was especially important on the Mt. Gunnison project because the mine was to be operated with non-union labor. These factors combined can reduce the risk of a cost overrun.

Project Budget and Schedule

The establishment of a project budget and schedule is another important planning tool because "targets" are set up and problem areas are identified. The budget should be broken down into as much detail as possible to correspond with the work packages described earlier. The budget should be based on the known scope of work, but there should also be an allowance (or contingency) for unknowns. For example, the work to correct the unexpected landslide problem during construction of the Mt. Gunnison Mine was covered by contingency.

The project schedule should be closely related to the budget and should also be broken down into as much detail as possible to correspond to the work packages. This master schedule serves to tie all the work together and to identify problem areas such as equipment, material or labor shortages. All of the constraints, such as weather, permits, work-arounds, labor, material and equipment should be considered. On this project, a CPM scheduling technique was used successfully to identify critical paths. A master schedule was invaluable on this project in integrating the work of some 20 contractors.

Interaction With "Outside" Interests

The final factor in good construction planning is constant interaction/communication with, and sensitivity to, the needs of the community, permitting and environmental agencies (this project required 28 different permits), operations, upper management, and the various contractors. Of course, this interaction is important not only prior to construction, but also during construction.

Interaction with these groups is essential for smooth planning, construction and reduced cost overrun risk. A delay in a permit, for example, could shut down construction for a period of time, with subsequent increased costs for demobilization and then remobilization. Good working relations with other home office support groups is also important (for example: accounting to get the bills paid, contract administration for contract and related services, and employee relations to handle personnel needs). Legal, land, labor relations and purchasing help may also be required.

Constant interaction of the construction group with the operations group is important because they are the "customer". When the construction group leaves, the operations group will have to make the mine work. This was especially critical on this project because production at the mine started six months after the start of construction. Naturally, it is important to start producing as soon as possible, but this causes interface problems as construction proceeds. Good communication to review common work areas and resolve schedule conflicts is essential for uninterrupted construction.

Both informal and formal meetings on a daily/weekly/monthly basis with all the above mentioned groups on the Mt. Gunnison project was of great help in keeping the lines of communication open. People in the project team should be named as single point of contact coordinators for operations, community development/public relations, environmental/permitting and safety (MSHA/OSHA) interfaces to promote good construction planning. On the Mt. Gunnison project, for example, the mine engineer was appointed the single point of contact between the project team and operations.

EXECUTION

Construction Execution

Good construction execution is the final major factor in minimizing the risk of cost overruns. This involves not just getting the work done, but getting it done in a controlled manner with proper support. The risk here, of course, is that if the work is not done in a controlled manner or without proper support, then the costs will be out of control, or the quality of the project will be such that it causes long-term operating problems during production. This means that proper project control, quality control, and engineering support are required.

Project Control

Project control involves the activities of cost control, scheduling, and estimating. Cost control means setting up a budget for each contractor prior to construction and controlling against it to avoid the "it costs what it costs" syndrome. Cost control involves more than just checking physical progress, signing invoices and collecting costs. It involves control over "change orders" for increases or decreases in scope, cost benefit analysis to constantly look for ways to do the job cheaper, and also good forecasting of all known and possible work to provide advance warning for potential overruns. Cost overrun risk decreases as the project nears completion, so contingencies can be reduced.

Cost control is less important for lump sum jobs versus cost plus jobs, but careful attention must still be paid to change orders. Change orders should be limited to cost effective or safety/regulatory considerations. "Cosmetic" or convenience changes should be kept to a minimum, especially on a speculative project such as this mine where the project directive calls for keeping capital exposure to a minimum.

As mentioned earlier, scheduling is also a very important part of planning and controlling a job. As with a budget, each contractor should establish a schedule within the framework of the master project schedule and should be controlled against it. The contractor must be very involved in the scheduling process to heighten his awareness of problems or constraints. Schedule control is especially important on lump sum jobs because the contractor may want to keep his work force light to minimize his costs.

Estimating is another essential activity in project control. Estimate review of contractor budgets is most important. Change orders must be reviewed for accuracy and reasonableness, and alternatives for new work must be reviewed for cost effectiveness.

On the Mt. Gunnison project, the construction engineer and project engineer performed these project control duties with support from the home office.

Quality Control

Quality control goes hand in hand with project control in minimizing the risk of cost overruns on a project. Rework always costs extra money, and the old saying "the bitterness of poor quality still remains after the sweetness of low cost is forgotten" always holds true. Operations must have a mine that they can live with for many years. Quality control requires constant monitoring of the contractor's work by the project management group. Again, this function was handled for the most part by the design and construction team on the Mt. Gunnison project. The construction staff should monitor quality and safety related activities to ensure that the contractor is fulfilling his responsibility to achieve quality and safety in construction and that construction is being accomplished in accordance with drawings and specifications. Formal sign-off criteria and procedures must be established between the contractor and the project team, and then passed on to the operations group for turnover of care, custody and control.

Engineering Support

Engineering support is also an important part of construction execution. Engineering support is required to review and approve change orders, provide redesign support for design changes/errors, and provide design fixes or work-arounds for unforeseen topographic/geologic anomolies (such as the landslide problem). Engineering support also plays an important part in quality control. The project team can call in the experts as required to help in this function.

CONCLUSION

Good people, planning and execution are key factors in minimizing the risk of cost overruns in a construction project, and certainly played a major role in the success of the Mt. Gunnison project.

9.11

CONVENTIONAL AND NONCONVENTIONAL RISKS
INSURANCE FOR MINING PROJECTS

Charles Berry
Berry Palmer & Lyle Limited
London, England

ABSTRACT

Conventional and nonconventional insurance is examined from the point of view of a financier of a major mining and minerals project. The paper reviews the elements of a conventional insurance program; explains why such a program is not fully responsive to the project financier's needs; examines some of the new insurance coverages that can supplement a traditional program; and emphasises that these insurance markets are still in the early stages of their development with consequent effects on capacity (insured amount) and to some extent price (insurance premium).

INTRODUCTION

"Project finance" does not refer to lending to creditworthy entities in respect of a project or to export-credit related financings. In both cases the lender's risk assessment remains primarily one of credit evaluation based on the financial statements of the borrower or the guarantor (usually a country in the case of government export-credit programs). Rather "project finance" refers to lending direct to a project where the lenders only source of recovering the loan is from the revenue stream generated by successful operations. It is in this context that the subject of insurance becomes particularly relevant to the lender, and particularly difficult.

Since the evaluation of risks in the context of mine finance is the subject of this chapter, it is appropriate to consider insurance in the context of project finance (described further in Chapter 11) to fully analyse the capabilities of the insurance industry to handle the risks involved.

CONVENTIONAL INSURANCE

Conventional insurance and conventional insurance thinking is about physical loss and damage. An "all risks" insurance policy covers all risks of loss or damage to physical property; a liability policy covers liabilities for physical loss or damage to another person or property; even a consequential loss policy is geared to consequential losses following physical damage. The insurance industry's thoughts are so geared to concepts of accidents and physical loss that often it talks of providing an "all risk" policy in the context of a project financing without qualifying this by saying it means all risks to physical property. This does not include all of the risks which a minerals project faces as described by the other authors in this chapter.

An insurance policy geared to physical loss or damage to property falls far short of offering complete protection to financing institutions providing project finance. This does not mean, however, that the physical loss coverages are unimportant. On the contrary, it remains the basis of the project insurances. Two developments have in recent years improved such programs from the point of view of lending institutions, namely the advent of owner-controlled insurances and of improved techniques for protecting the lender's interest.

Owner Control

Traditionally where a project had several contractors and suppliers at work during the construction phase, each separate contractor was responsible for its own insurances. Even though contracts might specify consistently what each contractor should insure, such a decentralized system has several disadvantages, such as

inconsistency of coverage for different elements of the project

over-insurance, or double insurance incurring unnecessary premium cost

under-insurance, or gaps in the insurance cover

arguments between different insurers over responsibility for a loss leading to delay and extra expense

payment of claims to contractors rather than to project owners and their financiers.

Under an owner-controlled insurance program, a project avoids many of these pitfalls. The owner arranges and controls a single insurance program which covers all the works during construction and operating phases, and all equipment incorporated in the project from the time it is shipped to the project. The project has better cover, negotiated usually with a single underwriter or set of underwriters, and the project brings its buying power into play by securing lower overall premiums.

The Lender's Interest

A well-thought-out, owner-controlled insurance program may well not be enough to protect the lenders even against losses caused by conventionally insured, physical-damage claims. Two particular problems remain. The first is one of breach of the policy terms and conditions by the insured owner or project sponsor. An insurance policy is conditional and many of the conditions may impose obligations on the owner which the lenders cannot be sure will be fulfilled. If they are breached, a claim may be denied and lenders are left high and dry. The second problem is one of the lenders access to claims proceeds.

Increasingly conventional project insurances allow the lending institutions interests as a financier of a project to be recognised in the cover. A variety of techniques may be used to improve the lender's control of the insurances and access to claim proceeds, e.g.

loss-payee clauses, which nominate the lenders as the recipient of any claims moneys (perhaps above a certain limit)

assignment of policies, where the lenders are assigned the rights of the policy, as opposed to merely the proceeds

joint insureds, where the lenders become joint policy holders

warranty wavers, which allow for payments of claims to the lenders in certain circumstances, despite the fact that the insurers could deny liability to the owners as policy holders because of breach of condition or warranty in the policy

banker's or lender's interest policies, designed as another solution to the problem that the lenders are at risk to the legitimate denial of a claim by insurers due to the policy holders breach of policy conditions.

WHY CONVENTIONAL "ALL RISKS" INSURANCES FAIL TO COVER ALL RISKS

A well-planned conventional program still leaves a lender with many risks some of which may be transferable through insurance. The limitations of conventional insurances, geared to physical damage, are well illustrated by considering the design risk on a project. Assume that in order for a particular project to be viable, raw material has to be processed at site, and the processing plant is designed to produce a certain quantity and quality of processed material. If the processing plant fails to meet its design requirements the project may fail. We are told the conventional insurances include cover for "faulty design". What is actually covered?

Figure 1 illustrates the types of losses which may flow from design failures. As can be seen such conventional insurance may only cover the cost of repairing and replacing equipment damaged as a

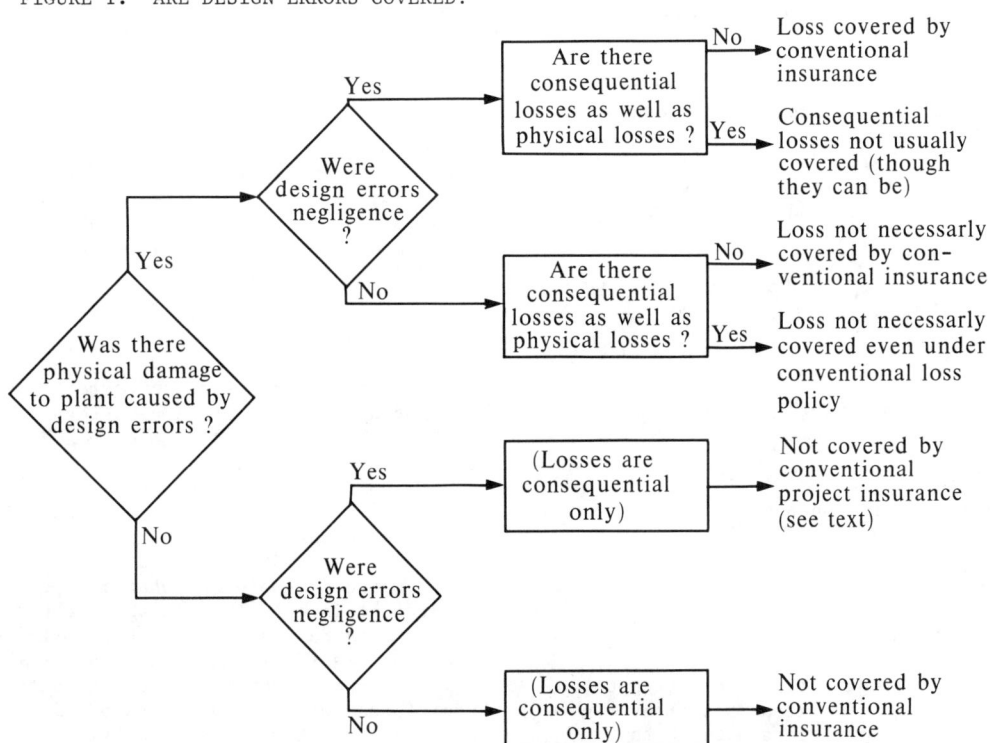

FIGURE 1. ARE DESIGN ERRORS COVERED?

result of faulty design. There could well be doubt as to coverage unless negligence can be shown. Consequential losses flowing from even negligently caused physical damage, for example the delays and extra interest costs, are not necessarily covered. Cover for consequential losses are usually the subject of a separate policy which may or may not be available and taken out, as an extension to the conventional program.

Where there is no physical damage, for example where the processing plant will simply not run to designed specification, conventional insurances do not apply. Where negligence can be proved against a designer or contractor, the owner and the lender may be able to pursue damages and may ultimately benefit from the negligent party's professional indemnity insurance. Indeed in certain circumstances an owner may arrange a project-based professional indemnity cover for all participants. But again even where this is available it is an indirect route to insurers; the first claim rests on the courts.

The above illustrates that conventional insurance, geared to physical damage, is well suited to asset-based lending, but falls very much short of the mark when applied to project lending. Where lending is secured by physical assets, conventional insurance bolsters the lenders fundamental assumption, that the asset will be available if needed.

However, a key tenet underlying lending on a project-finance basis is that the mining project produces a revenue stream sufficient to service its debts; not simply that the project assets remain physically intact, although where these assets form the main part of the second line of defense, the project's security package in a legal sense. Physical loss or damage could halt the revenue stream, but so could many other eventualities which do not involve damage. In these circumstances, there is little comfort for the lenders in the physical fabric of the project. The project equipment and infrastructure becomes simply so much scrap, lying idly in some remote location. The availability of the crucial revenue stream in sufficient quantity to repay the lenders may depend on key assumptions concerning

the availability of the mineral reserves in sufficient quantity/quality

the costs of extraction

the availability of a market (tonnage)

price

the continuing co-operation and blessing of a host government

interest rates

performance by key players

A well-structured project financing attempts to minimise these risks to the extent possible. In particular, the project designers attempt to achieve a secure lending structure by building a legal framework around the project intended to cover key risks, for example

concession and/or joint venture agreements designed to reinforce host government co-operation

sales contracts designed to reduce the market risk

contracts to secure the performance of key players

But such a framework remains unlikely to eliminate these risks. The political risk is not eliminated by an agreement with a host government: a new government may abrogate the deal. The sales contract reduces the market risk by turning the risk into one of performance by a specified buyer or one of certain defined causes of force majeure which may relieve the obligation to purchase. And emerging from this legal framework are certain areas of identifiable and unacceptable risks not involving physical damage for which insurance is necessary.

UNCONVENTIONAL INSURANCE

Risks to owners and project lenders arising from these key risk categories has led to a demand for unconventional insurance products. The insurance market has responded significantly in certain areas, with little enthusiasm in others. Below are listed some areas of non-physical damage risk with an indication of what is possible in today's insurance market. It is a changing market and worth checking with professionals in the market, including the rapidly emerging private insurance market, on current conditions and premiums.

Design Risk

In the example of the processing plant which simply does not perform or fails to meet design criteria, performance guarantee insurance can be used to insure this risk. Known also as efficacy insurance, there is a small but growing market in the USA and London.

Political Risk

Political risk insurance is perhaps the most developed area of the unconventional insurance market with substantial capacity and growing competition between underwriters in the USA, London and to some extent other parts of Europe. A variety of covers are available essentially designed to guarantee that foreign governments will honor their commitments to a particular venture. These markets supplement what may be available from government investment insurance schemes such as OPIC in the US and government export-credit agencies such as FCIA/Exiusbank. Commercial-insurers political-risks cover has been used to provide additional security to owners of mining projects and their lenders since the early 1970s.

Performance Risk on Key Parties

There is evidence in the credit insurance market that worthwhile insolvency consequential-loss covers may be developed. Such covers would provide welcome additional protection to projects and their lenders against the risk of insolvency of key players where the balance sheet of the key player is quite sufficient to support the burden of work the key players themselves perform, but completely inadequate to support the level of bonding that would be required to cover against the possible consequences of their failure.

Shortfall of Mineral Deposits

Covers have been placed in London and probably in the US to insure investors in oil wells against a shortfall in the expected volume of recoverable oil. There may well be circumstances where there is a measure of uncertainty as to the volume of economically recoverable mineral deposits. Such a policy could be of comfort to lenders.

Loss of Market

There is no insurance market which directly insures against simple loss of market, and nor is there any expectation that one will develop. But where the project designers have attempted to avoid or freeze the market risk by entering into long-term take-or-pay contracts, political- and credit-risk insurance may be adaptable to insure two risk areas which must always remain, namely the risk of the Buyer (or supplier) reneging on their obligations and the risk of force majeure events of a political or other nature, preventing the contract working.

Price, Fx and interest rates

Rumours that there are insurance markets available to cover directly price movements, foreign exchange rates, interest rates, and cost escalation are false. This is likely to remain the case. (Editor's note: Other financing tools are available to mitigate these risks such as interest-rate swaps, foreign-exchange and commodity-price hedges and options, and fixed-price turnkey construction contracts and certain types of management agreements.) This means that it remains impossible to obtain insurance cover where the cause of loss is stated as one of these events. The rumours start because, for example, credit insurance policies may involve running some indirect exchange fluctuations risk where they agree to indemnify for the cost of closing out a forward exchange position where the buyer defaults on his payment obligation. Though this involves indirectly covering "exchange fluctuations" the cause of loss is still the insurable risk of buyer default. Another example is the case in which a physical damage policy covers the current cost of replacement. This is in a sense insuring against inflation.

PRIVATE INSURANCE

The above are a few specifics. More generally the private insurance markets facilities for financially oriented insurances are illustrated by Figure 2. Traditionally financiers have tended to look to government insurance agencies to find, for example, political risk insurance but the private market can have several advantages

policy terms and wordings adapted for each separate risk

willingness to innovate

speed of response

risk related rating

absense of restrictions on sourcing of goods and other eligibility criteria

However, the market for these types of unconventional cover are in their early stages of development and two attributes are, therefore, notable. The first is that these types of cover tend to be produced as a result of specific requests for a particular situation rather than sold as a ready-made package. This is to be expected, but means that the owners and financiers must often initiate the demand for such covers. Secondly, the capacity for a new type of cover in the insurance market takes a long time to develop. Thus, for many types of cover capacity must be measured in millions or tens of millions, though in the political risk field capacities in the hundreds of millions are not unheard of.

The above illustrate insurers involvement in unconventional insurances, far removed from the usual concern with loss or damage to physical assets. However, physical damage covers may also be innovative and unconventional when applied to project financing. Examples of this are conventional perils insurance geared specifically to lenders needs such as Startup Delay Insurance and Business Interruption Insurance, where insurers indemnify for the extra interest costs following physical damage losses or accidents.

CONCLUSION

Unconventional insurances are emerging which can complement both the company's and the financier's provision of funding for a new minerals development. Although it is not possible to allay all risks, conventional and nonconventional risks coverage should be examined and the market tested in detail either before arranging finance or contemporaneously with the assembly of investment and loan finance.

FIGURE 2. APPLICATIONS OF FINANCIAL INSURANCE PRODUCTS.

TYPE OF FINANCIAL INSURANCE / FINANCIAL RISKS ASSOCIATED WITH	Commercial Credit Insurance	Export Credit Insurance	Government Default	Currency Inconvertibility	Termination of Contract due to: - Buyer Repudiation - Embargo - War, Civil War, etc. - Other Force Majeure	On-Demand Bond, Unfair Calling	Supplier Default	Technical Non-Performance Indemnity	Non-Ratification Indemnity	Recourse Indemnity	Confiscation, Expropriation and Nationalisation	Creeping Expropriation	Deprivation	Residual Value Insurance
TRADE	•	•	•	•		•	•							
MAJOR EXPORTS	•	•	•	•	•	•	•		•	•				
INTERNATIONAL PROJECTS		•	•	•	•	•	•	•	•	•	•	•	•	
FOREIGN INVESTMENTS				•							•	•	•	
LEASES	•	•	•	•	•						•		•	•
SOVEREIGN LOANS			•	•										

CHAPTER 9: RISK ANALYSIS FOR MINE FINANCE

BIBLIOGRAPHY

Barratt, D.J., 1982, "Performance Risk," Proceedings, Second CIM Mineral Economics Committee Symposium, Risk Management in Mining, Nov. 21-23, Vancouver, British Columbia.

Bauer, H.F., 1982, "Market Risk Definition," Proceedings, Second CIM Mineral Economics Committee Symposium, Risk Management in Mining, Nov. 21-23, Vancouver, British Columbia.

Baylis, K.E., 1982, "Coal Market Risks," Proceedings, Second CIM Mineral Economics Committee Symposium, Risk Management in Mining, Nov. 21-23, Vancouver, British Columbia.

Betterley, R.S., 1977, "Unearthing Risks: The Risk Identification Committee," Weekly Underwriter, Apr. 23, p. 3,13.

Brant, A.A., 1982, "Exploration Risk-Fringe Factors," Proceedings, Second CIM Mineral Economics Committee Symposium, Risk Management in Mining, Nov. 21-23, Vancouver, British Columbia.

Burn, R.G., 1982, "Exploration Risk," Proceedings, Second CIM Mineral Economics Committee Symposium, Risk Management in Mining, Nov. 21-23, Vancouver, British Columbia.

Cameron, P., 1982, "Labour Risks," Proceedings, Second CIM Mineral Economics Committee Symposium, Risk Management in Mining, Nov. 21-23, Vancouver, British Columbia.

Carter, R.L., ed., 1973, Handbook of Insurance, Kluwer-Harrap Handbooks.

Carter, R.L. and Crockford, G.N., eds., 1974, Handbook of Risk Management, Kluwer-Harrap Handbooks.

David, M., 1982, "Ore Reserves Risk," Proceedings, Second CIM Mineral Economics Committee Symposium, Risk Management in Mining, Nov. 21-23, Vancouver, British Columbia.

Denenberg, H.S., Eilers, R.D., Melone, J.J. and Zelten, R.A., 1974, Risk and Insurance, Prentice Hall.

Fujisaki, A., 1983, "Managing the Risk of Product Demand," CIM Bulletin, Vol. 76, No. 858, Oct.

Hammes, J.K., 1982, "Financial Risk Definition," Proceedings, Second CIM Mineral Economics Committee Symposium, Risk Management in Mining, Nov. 21-23, Vancouver, British Columbia.

Hartman, D.G., 1976, "Foreign Investment and Finance with Risk," The Quarterly Journal of Economics, May, pp. 213-232.

Hendrick, K.C., 1983, "Managing Metal Price Risks," CIM Bulletin, Vol. 76, No. 858, Oct.

Krembs, J.A. and Perkins, J.G., 1981, "Business Interruption Interdependency: The Hidden Exposure," Risk Management, Nov., pp. 12-24.

Le Roux, M., "Specialty Risk: Insurance for the Future," Institutional Investor (Special Advertising Section).

Lowder, S.J., 1982, "Risk Management: Key to Profitability," The Chartered Accountant in Australia, June, pp. 48-51.

McClelland, R.H., 1982, "Political Risk Definition," Proceedings, Second CIM Mineral Economics Committee Symposium, Risk Management in Mining, Nov. 21-23, Vancouver, British Columbia.

Megill, R.E., 1977, An Introduction to Risk Analysis, Petroleum Publishing Co.

Miller, R.G., 1981, "Risk and Uncertainty," Foresight, Sept.

Nevitt, P.K., Project Financing, 2nd ed., AMR International Inc.

Ongpin, J.V., 1983, "Risk Management Strategy in Mining," CIM Bulletin, Vol. 76, No. 858, Oct.

Pouliquen, L.Y., 1970, "Risk Analysis in Project Appraisal," World Bank Staff Occasional Papers, No. 11.

Reutlinger, S., 1970, "Techniques for Project Appraisal Under Uncertainty," World Bank Staff Occasional Papers, No. 10.

Schreiber, H.W., and Emerson, M.E., 1984, North American Hardrock Gold Deposits: An Analysis of the Discovery Costs and Cash Flow Potential," Engineering & Mining Journal, Oct., p. 50.

Scott, A.T., 1967, "The Theory of the Mine Under Conditions of Certainty," Extractive Resources and Taxation, Mason Gaffney, ed., The University of Wisconsin Press, pp. 25-60.

Somm, A.F., 1982, "Transportation, Smelting and Refining Charges," Proceedings, Second CIM Mineral Economics Committee Symposium, Risk Management in Mining, Nov. 21-23, Vancouver, British Columbia.

Spencer H. and Temple C., 1975, "Project Risk Management," Foresight, Nov., pp. 3-34.

Squire, L. and van der Tak, H.G., 1975, Economic Analysis of Projects, World Bank, John Hopkins Press, p. 45.

Tinsley, C.R., 1982, "Project Financing – What the Banks Will Want," Metals Economics Group Seminar, Denver, Feb. 4-5.

Tinsley, C.R., 1982, "Risk Trade-Off," Proceedings, Second CIM Mineral Economics Committee Symposium, Risk Management in Mining, Nov. 21-23, Vancouver, British Columbia.

Vaughan, E.J., 1982, Fundamentals of Risk and Insurance, J. Wiley, New York.

Chapter 10

Information Requirements

CHAPTER 10: INFORMATION REQUIREMENTS

PREFACE

Finance is often regarded as possessing mystical attributes, yet it seems to hold the key to the development of mining projects. Finance for the minerals industry is not a mysterious business. The information required at one state of financing a project can be drastically different from another phase and each source of funding has its own conventions and idiosyncrasies. The authors in Chapter 11 have also described their approaches and requirements for information, supplementing this chapter.

Seven papers are presented in this Chapter. The first four papers and the last two papers give details about project finance which demands the most thorough assembly of information. The objective of project finance is that the project stand on its own once built; therefore, it requires the most comprehensive review of feasibility.

Mark Emerson's paper, "Independent Engineering Information for Project Financing," describes the process of data gathering and activities of the independent engineers. Not all project financings require this review, although Norman Gibbs and John Stroka, in their paper "What Bankers Look for in Project Loan Applications," state that some lenders indeed "prefer that technical and cost elements in a mine feasibility study be prepared, or verified, by an independent consultant of repute." Hans Schreiber's paper, "The Role of the Independent Consulting Firm in Project Financing," goes through areas of verification and includes aspects that are useful in developing negotiations with lenders. Tom Bispham's paper, "Technical and Financial Elements of a Mining Project Loan Request: Preparing a Complete Information Memorandum," provides a summary of the information requirements for the "Bankable Document." Not a document which one puts in a bank vault; it is, rather, the input required to develop the concise document required for a loan syndication among a group of banks.

Gary Thomason's paper, "Understanding the Loan Approval Process," reveals that different information is required at each layer of the bank's approval process. Richard Tinsley further expands on this topic in his paper, "Banking Needs for Project Development Financing," rejecting the notion that the Bankable Document is always appropriate. While checklists are useful, the complexity in present day minerals project analysis, marketing, and financing defies the industry's collective skills, especially given the rate of change of the world's economies.

Approaching the stock market has its own conventions, many of which are regulated as stock exchanges seek to provide adequate and accurate disclosure to prospective investors. Affleck and Stevenson's paper, "Information Requirements for Equity Issues" describes this information; for example, the information for an equity prospectus. These requirements are so stiff, particularly in the US, that many smaller companies prefer to raise money from limited investor groups through private placements which do not require such vigorous disclosures.

At the other end of the scale are the insurance industry placements in Europe. Insurance brokers typically prepare one or two-page summaries of projects to obtain cover from underwriters. European investors are also prepared to proceed with little hard data, particularly for gold-backed instruments. In the case of commercial paper or bond issues, the information needed is more descriptive business and financial disclosure rather than tonnes and tire wear.

Many financial institutions have in-house capabilities to develop information, opinions, and presentations about minerals ventures. Included are banks, the World Bank/IFC, insurance companies, and certain leasing specialists whose officers can become rattled if not allowed to examine all the information themselves.

Finally, financial information is best understood when presented with proper accounting conventions and in the format which the recipient will use to make his financial assessment. Chapter 4 presented basic accounting principles applied to minerals projects around the world. The development of on-going financial information gives invaluable feedback to the company and its financiers.

The necessary information is a function of the level and type of finance being raised. Since the disclosures required vary, it is wise to ask for advice early, then review comparable information packages before making a presentation to the financial marketplace.

C. Richard Tinsley
Chapter Editor
European Banking Company Limited

10.1

INDEPENDENT ENGINEERING INFORMATION
FOR PROJECT FINANCING

Mark E. Emerson

Resource Exchange Corp.
New York, New York

INTRODUCTION

A long time has passed since geologists encountered encouraging mineralization in their regional exploration program for base metals deposits. Detailed drilling during the following two field seasons served to delineate the configuration of the deposit. Various statistical techniques were employed to provide confidence levels regarding the extent of the reserves and to categorize them according to whether proved, indicated or inferred. At this juncture, it appears that the company has an orebody. Time elapsed? - three years; money spent? - $2.5 million. Much remains to be done before the orebody becomes a mine.

The company urgently needs to develop new orebodies to replace its existing mines which are being depleted. It therefore has expended sizable amounts in exploration, the life blood of the company. A staff of specialists is maintained to evaluate new projects and to conduct feasibility studies.

The scale of new mining projects today is beyond the resources of most mining companies. Bank financing, therefore, is often vital if the company is to survive. The following scenario describes how a company goes about evaluating mine projects and what is needed to meet lender's criteria in securing the needed financing.

MINE PLANNING

Now that the company engineers have taken over the project, their concern is no longer with finding ore but in developing it. An exploration shaft has been put down and a tunnel driven into the orebody to verify, on a large scale, what has been seen so far only in pencil-like drill cores. A large tonnage bulk sample is mined and shipped off to an independent metallurgical process laboratory for beneficiation testing. The budget is expended to cover engineering studies. Preliminary capital and operating costs are developed in-house and profitability studies are made based on assumptions concerning the markets for the main product and the indicated gold, silver, and other by-products. Time elapsed? - four years; cost to date? - $3.5-$4.5 million.

The staff prepares its own pre-feasibility study. After review, the Executive Committee authorizes management to proceed with full-scale studies. The task is to make preliminary designs and to develop definitive capital and operating cost estimates. Specialized consultants will be engaged to study slope stability for the open pit and rock strength if underground recovery proves more feasible. If it is near the sea, an oceanographic report might be commissioned to see if the tailings can be disposed of off shore. Environmental impact studies are commissioned from a firm familiar with government requirements and regulations in this important area. The company subscribes for an expensive multi-client study of the supply-demand-price outlook for the product to support its own long-range price forecast.

Now all the numbers are available to make a definitive appraisal of the value of the project. Cash flow analyses and sensitivity studies follow. Reams of computer printouts are generated. Time elapsed: five years. Amount spent? - up to $9 million, depending on the size, complexity and attractiveness of the project. Marginal projects always take longer.

Finally, an engineering and construction (E&C) company is given the task of integrating and compiling all the data into a feasibility study. The results of dozens of reports, memoranda, computer printouts, and charts are boiled down to one final summary report of a few hundred pages together with conclusions and recommendations.

After thoughtful review at a board meeting, the directors authorize management to proceed with detailed engineering work - provided that

---

Presented as part of an SME Minerals Resource Management Committee Short Course, "Project Financing in the Development of Mineral Resources," SME-AIME Annual Meeting, Chicago, Illinois, Feb. 21-22, 1981.

financing is available. In approaching the commercial banks, it will be necessary to reveal all of the confidential information on the project. The object is commonly to obtain the maximum amount of financing with the least equity in the project and with minimal conditions on the parent firm. It may be one of the company's objectives to have the financing done off balance sheet. If the project can stand on its own two feet, financially, the company will preserve its existing assets unencumbered by the new financial commitment. This is one of the main attractions of project financing to mining companies.

Just as in financing a home, debt service in a mine venture is the single largest cash commitment made. Unlike home financing, which may consist of level payments for a period of 25 or 30 years, the loan on a mining project is rarely for more than 10 years and payments usually decline over time, with the largest share of the principal and interest paid in the early years following startup. (Interest payments are highest at startup.)

With long experience that startup difficulties are the rule rather than the exception and that metal and mineral prices are hightly cyclical and largely unpredictable, the company perceives itself as highly vulnerable to its financing commitment. Thus, it seeks to moderate the terms of the loan and to share the risk with others as needed. It will seek to make its completion commitment to the lender as flexible as possible. Often the company will bring in a joint venture partner in the form of another mining company or a buyer of the product. The object is to reduce its risk and to make the financing package more attractive to the lender and, hopefully, the terms more palatable to the borrower.

FINANCING - A DIFFERENT PERSPECTIVE

Non-recourse financing is a rare situation. Whether it is an insurance company, investment banker, or commercial bank, the lending institution is anxious to lend long-term money to good projects. A good project, according to one institution, is one that will proceed on schedule, with no overruns and offers a high probability of returning all borrowed funds plus interest in a relatively short period (7 to 10 years). Since good projects are hard to find the competition among lenders is fierce. Whether your project will be in this enviable position depends on how well you have done your homework for the feasibility study.

Unless the sponsor of the project (the borrower) understands the lender's criteria he is unlikely to receive the funds sought, at least not unconditionally. The lending institution has a requirement for a general base of information and it has only a limited time to devote to its appraisal. The company has lavished millions of dollars on the project and devoted untold man hours of effort -- not only from those of your in-house staff, but those of the many service organizations you have engaged.

On the other hand, even the most active lender in the area of project finance may have but one technical person (an engineer) on its staff. The backlog of projects needing his or her attention is such that he can devote no more than one day, perhaps two days for a large project, deciding whether or not yours merits further consideration.

At this juncture, the quality of the feasibility presentation is all-important. Superfluous information will only detract and inadequate information delay your financing. What the engineer seeks is assurance that his bank will be free of all technical risks if it undertakes the financing of the project.

The reasons for the banker's concern are apparent if one places oneself at his or her desk. He has no control over the location of the drill holes. He did not select the engineering and construction company. Having a small staff, the time that can be devoted to the project is miniscule compared to that of the sponsor and his support group. More important is the fact that, lacking equity, the lender obtains only a modest return on his money. If he cannot share in the rewards of a good project why should he share in its risks?

THE FEASIBILITY STUDY

Technical and Economic Requirements

If done properly, the project's feasibility study will be a synopsis of all the work to date, with an analysis of the cash generated based on volume and market price expectations. The bank's engineer looks to it for backup information in five essential areas:

The Exploration Report

Here should be given all the background needed to evaluate the reserves. The lender wants to know what kind of drilling was undertaken, when and in what detail it was done.

He is not so much interested in total reserves as in the tonnage that is proven, indicated and inferred at different cutoff grades and recoverable within the deposit location. Having reviewed and visited many other mines, the bank's engineer will draw comparisons to rate this deposit in terms of size, grade, and its overburden to ore ratio. The confidence in the reserve estimates is important. In a non-recourse financing the lender will want proven reserves to exceed that necessary to support operations over the life of the funding by a factor of two. Proven reserves must be adequate for the life of the funding; about ten years for a large project and six years for a small project where corporate guarantees are involved.

INFORMATION REQUIREMENTS

Mine Plan

If the reserves meet the test, the bank engineer will go on to the mine plan. From initiation to production is a period of expenditure without any income during these years when the value of a dollar is the highest (present value). Backup consulting reports on geotechnical aspects are a necessity at this point. The engineer is interested in the equipment specified; its make, size, replacements, etc. He will look carefully at the environmental problems of waste disposal and surface disruption. Too often he has seen cash generation delayed for years before an environmental impact statement was approved.

Metallurgical Processing

As a mining engineer or geologist (the two most common professionals on bank staffs), the banker cannot be expected to be an expert on milling and processing. He will have more confidence in the data if a reliable E&C company has placed its seal of approval on the preliminary design and on the capital and operating costs estimates. He will also be more comfortable with this outside opinion because engineering and construction firms tend to be closer to technological developments in the industry because of their wide exposure to client work. The bank's engineer is familiar enough with process flow diagrams to know when "tried and true" technology is being proposed vs. new untested innovations. The latter carry inherent technological risk and will likely require a strong commitment by the sponsor if he is to receive the desired financing. The engineer will also look carefully at material balances in the smelting and refining stages if this is part of the project. Energy intensive procedures whese days require a commitment of a low-cost energy source to be attractive.

Capital and Operating Costs

These are generally prepared by an engineering company. An acceptable level of confidence in the data will need to be in the order to +/- 20% at this stage. The banker will also look for sufficient detail on capital costs to make comparisons with other projects. He will want contingencies and escalation to be specified.

To the banker, a high capital cost may be warranted by low operating costs. He will want this to be given in terms of both cash costs (the break-even costs) and total costs (which include the proposed debt service). Since the expenditures will be made over a period of time, inflation must be considered. Historically, costs have been shown in constant dollars so that the reader can make his own adjustments for inflation. The tendency these days is to use current dollars in the analysis since not all items of cost are affected by inflation equally, for example - taxation.

The location of the project weighs heavily in the banker's appraisal of the timing of cash income. Remote areas requiring staging for construction and the need for new town sites, power facilities, et al, are far more apt to fall far behind schedule than a project in, say, Arizona. The lender has no control over the construction scheduling and will assume none of the risk in this area. He will look to the sponsor for a guarantee or other forms of credit support until completiion when the loan becomes limited-recourse financing to the sponsor.

The Market Study

There are two main factors that influence revenues for a mining project. The volume of production and the price received for the primary product and its by-products. The sponsor or the purchaser is in a position to control the first; but unless a captive market is involved, or product sold on long-term contract, or there is some way for the producer to differentiate the product in the marketplace, neither the sponsor or the lender will be in a position to influence the price. Invariably, the economics of mineral projects are more affected by a change in the price received for the product than in an equal change, percentage-wise, in the capital or operating costs.

Considering its importance, it is surprising how often time is lavished on certain details of cost or operation and the market aspects are glossed over. Granted, markets are much more difficult to quantify but a carefully designed market plan, with assumptions spelled out and uncertainties set forth, will go a long way toward satisfying the banker.

Evaluation of the Feasibility Study

Although many professional individuals and reliable engineering and consulting firms contributed to the feasibility study, not everyone is without bias. It is just as hard for a borrower to be objective about a pet project as, say, an engineering firm that expects to make the detailed drawings and direct construction, for which it will pocket a percentage of the total costs. Thus, the need for independent verification arises.

This is the role for firms to whom banks turn for a independent audit of the data in the feasibility study and for comment concerning the assumption upon which the financial projections are based. A stamp of approval by a reputable consulting firm is tantamount often to having a "bankable" report.

Unlike most bankers, the independent consulting firm will look at the supporting documents behind the feasibility study. These will be requisitioned and reviewed with attention focused upon those critical areas which can affect profitability of the project. If this is to be an open-pit mine, a specialist in open-pit mining will study the supporting reports, paying careful attention to the drill-hole spacing, the stripping ratios indicated, and configuration of the pit designed, the changes in grade with depth, the haulage distances involved, the equipment specified, the spares, and replacement schedules, etc. After making a sensitivity analysis, he may see that the project could benefit from a higher or from a lower production level. In that case, his advice may be that the sponsor do such an analysis. Similarly, a specialist in beneficiation will critically examine the technology selected. He will pay particular attention to the backup documents by the outside firm that did the test work on the bulk sample. He will consider the alternatives and perhaps recommend other techniques to recover primary and secondary products.

The consulting firm generally assigns a manager to the project who will select a team of specialists as needed. The project manager's job is to integrate data from all of the specialists into one concise evaluation report. The time for such an endeavor is variable but anything over six to eight weeks may be considered a luxury at this stage.

Estimates of Collateral Value

Since banks are making other large loans, they need backing of some kind. If the borrower is not willing to mortgage its own assets, then the banker must look to the project itself. The reserves in the ground and the extraction facilities being installed have value only to the extent that after all cash operating costs are paid, including royalties and taxes (the landowner and usually the Government having first and second call), there is money available to pay off the debt under a wide range of conditions. Knowing that the projections of cash available are just projections, and that they depend on many variables, the banker will want a cushion of at least 1-1/2 times the loan for cash thrown off by the project.

Financial Benchmark

While there are many means to evaluate a project, the universally accepted test is "cash flow." This is nothing more than a determination of the generation of cash by the project period by period. Items such as depletion and depreciation are merely means of reducing taxes and are added back in calculating the cash at the bottom line. Less important than the total cash flow over the life of the loan or the venture is the timing of receipts. Obviously, a dollar received now is more valuable than one received 10 years in the future. The banker is interested in lending money only to projects that are fully self-supporting. He looks at the cash flows on a project basis; that is, 100% equity basis and at various levels of leverage. The reason for looking at it on a project basis is that the banker wants to be sure the venture can carry itself in case the sponsor defaults, in which case the lender ends up owning the project since the loan is paid from the project's assets -- not the sponsor's.

PROJECT EVALUATION

Unlike the banker, who is primarily interested in getting his money back safely with interest, the sponsor is interested in making a profit. It is true that the "carrot" in exploration is largely the discovery of a bonanza. However, boards of directors often will settle for something more modest in the way of a return, provided the project meets certain financial and non-financial needs of the firm. Discounting of future cash flows is a standard procedure for determining the return on investment. If the return is adequate "on paper" on a 100% equity basis, the company will seek to enhance that return by using the bank's money to increase its return through debt levere. The risks in this are great because the leverage can just as readily magnify a loss. Miners are aware that there is no reward without risk. They have always risen to the challenge.

INFORMATION REQUIREMENTS

SUPPORTING DOCUMENTS FOR A REPORT ON
THE FEASIBILITY OF AN OPEN PIT COPPER MINE

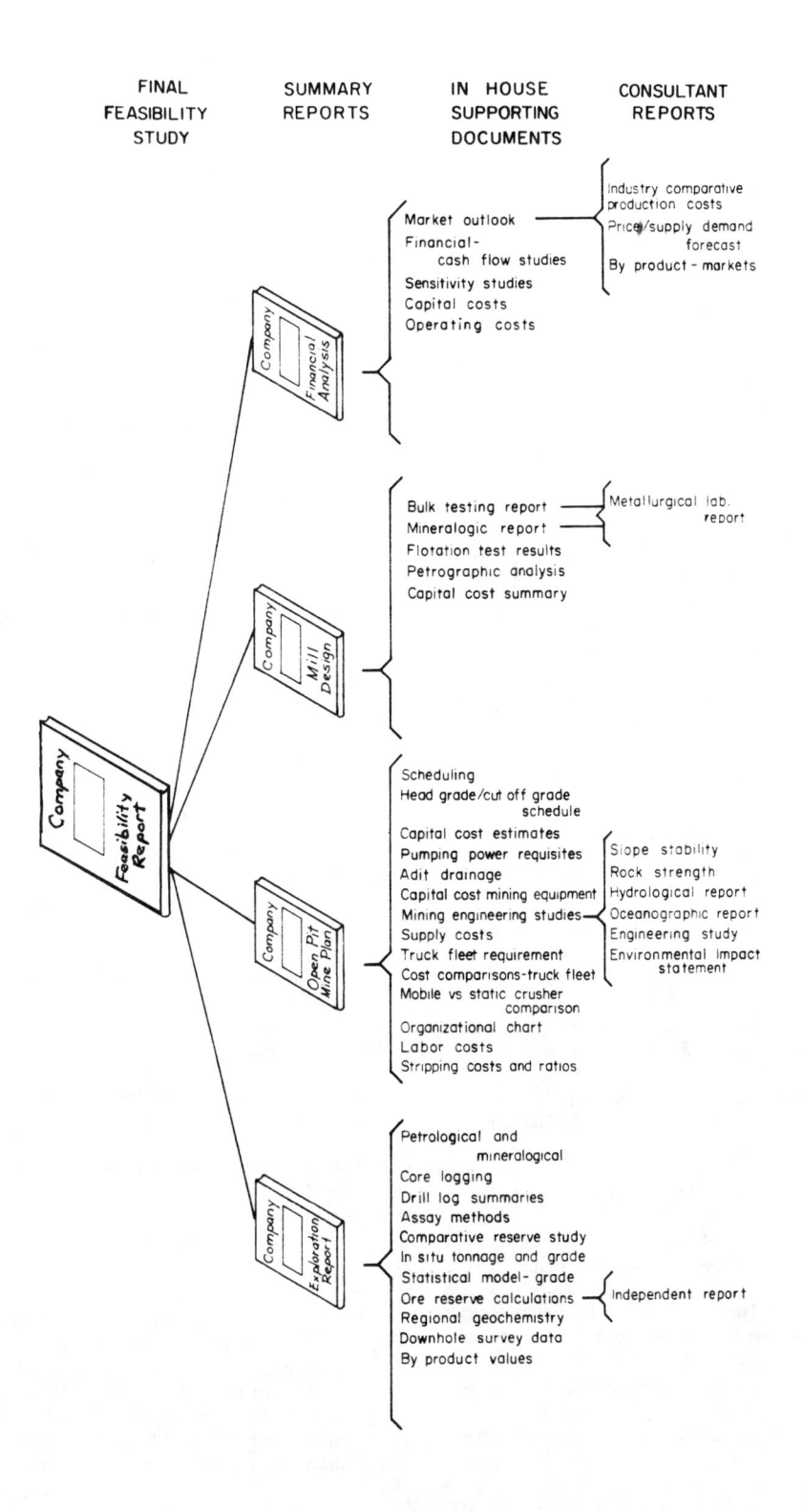

FINAL FEASIBILITY STUDY	SUMMARY REPORTS	IN HOUSE SUPPORTING DOCUMENTS	CONSULTANT REPORTS
Company Feasibility Report	Company Financial Analysis	Market outlook Financial-cash flow studies Sensitivity studies Capital costs Operating costs	Industry comparative production costs Price/supply demand forecast By product-markets
	Company Mill Design	Bulk testing report Mineralogic report Flotation test results Petrographic analysis Capital cost summary	Metallurgical lab. report
	Company Open Pit Mine Plan	Scheduling Head grade/cut off grade schedule Capital cost estimates Pumping power requisites Adit drainage Capital cost mining equipment Mining engineering studies Supply costs Truck fleet requirement Cost comparisons-truck fleet Mobile vs static crusher comparison Organizational chart Labor costs Stripping costs and ratios	Slope stability Rock strength Hydrological report Oceanographic report Engineering study Environmental impact statement
	Company Exploration Report	Petrological and mineralogical Core logging Drill log summaries Assay methods Comparative reserve study In situ tonnage and grade Statistical model-grade Ore reserve calculations Regional geochemistry Downhole survey data By product values	Independent report

10.2

THE ROLE OF THE INDEPENDENT CONSULTING FIRM IN PROJECT FINANCING

Hans W. Schreiber

President, Behre Dolbear & Company, Inc., New York, New York

INTRODUCTION

At the end of the day, the decision by the sponsor to proceed with a project and to seek financing, or the decision by the financier to grant financing are made on a judgmental basis. Cash flow projections, sensitivity and risk analysis, the innumerable ratios applied by the financier, the so-called feasibility study, etc. are all only tools, and the degree to which they affect decisions is as highly varied as are the personality differences and business conduct and outlooks of the persons who render the decision of "go or no-go". The independent consultant, then, is expected to provide an additional increment of credulity and/or reliability to the "tools" which others will use in reaching a decision; the consultant is really not in a position to <u>directly recommend</u> a "go or no-go".

PROJECT FINANCE INGREDIENTS

The project finance situation has basically three ingredients:

- The project;
- The inherent risks attendant with project failure;
- The parties.

The project, to be viable for financing, will have certain essential characteristics, namely:

1. Clear title or rights to the mineral;
2. Proven definition/establishment of mineral quantity;
3. Financial viability;
4. Physical access consistent with financial viability;
5. A market for the product at a price and for the quantity consistent with financial viability;
6. Transportation/shipping from the project facility to the market at a cost consistent with financial viability;
7. Availability and demonstrated expertise of the developer of the facility;
8. Demonstrated expertise of the operator of the facility;
9. Production processes do not involve unproven new technology;
10. Availability of management personnel;
11. The property and the facilities have value (and thus can be used as collateral);
12. Government approvals/support, as required, will have been obtained;
13. The political environment is essentially friendly and stable;
14. The owner(s) of the project have the financial strength to make equity contributions consistent with total required investment capital and project risk;
15. Backing by credit support other than that of the financier (i.e. guarantees);

The inherent risks of failure are (Nevitt, 2nd Edition);

1. Completion delays with consequent delays in the generation of revenue
2. Cost overruns
3. Technical failure
4. Financial failure of contractors
5. Government interference
6. Uninsured losses
7. Increased price or shortages of raw materials, supplies
8. Technical obsolescence
9. Loss of competitive position in the market place
10. Expropriation
11. Inadequate/poor management

The parties directly or indirectly integral to the project financing are:

- The mineral rights owner
- The developer
- The operator
- The equity investor

Based on a presentation to the SME Minerals Resource Management Committee Short Course on Project Finance in the Development of Mineral Resources and on "Overview of and Special Considerations Regarding Mine Project Financing from the Viewpoint of the Independent Consultant" to the XVI Convention of Mining Engineers, Lima, Peru, November 1982.

- The purchaser of the product
- The lender of loan capital
- The "independent" consultant

In most situations:

1. The mineral rights owner and the operator become one;
2. The operator and the equity investor become one, although non-operating equity investors are common, e.g. the purchasers of the product.

CONSULTANT'S ROLE

The activities of the independent consultant are given by Emerson in the previous paper. The independent consultant's role, is, in a sense, defined by the extent to which the consultant, whether he is acting for the sponsor or the financier, can contribute to or enhance reliability in the characteristics of a viable project or to the extent the consultant can minimize, uncover and define the risks attendant to project failure. The consultant's influence (along with the influence of other parties to the project) is perhaps best generalized and summarized in matrix form as shown in Figure 1.

Figure 1

IMPACT ON ANALYSIS BY THE PARTIES INVOLVED IN THE EVALUATION OF A PROJECT FOR PROJECT FINANCING

Characteristics	Owner/operator	Developer	Equity investor	Product purchaser	Loan Financier	Consultant	Risks of Failure
government approval/support; friendly/stable political/environment	minimal to moderate	none to minimal	minimal	moderate	moderate to great	none to minimal	government interference; expropriation
clear title	great	none	none	none	none	moderate to great	
proven definition of mineral quantity	great	none	none	none	none	great	
financial viability; market/market price; transportation/shipping	moderate to great	none	highly variable	great	none	great	materials and supplies shortages and price increases; loss of competitive position
expertise of developer; expertise of operator; process technology	moderate to great	great	none to moderate	none	none	moderate to great	completion delays, cost overruns, contractor financial failure, technical failure and obsolescence
management	great	none	moderate to great	minimal to moderate	great	moderate to great	poor management

The consultant has influence through:

1. Review of feasibility studies/reports produced by the owner/operator or by producing his own feasibility report;

2. Independence - i.e. capability to deliver an outside opinion, an opinion with the least bias;

The consultant exerts his role vis-a-vis financier by:

1. Making statements about:

   a) Technical viability
   b) Economic viability
   c) Reserve confirmation

2. Setting valuations

3. Analyzing the soundness and reliability of the financier's back-up position in the event of project failure

4. Relating profit optimization to loan repayment requirements

5. Continued monitoring during the project development and during production until the loan is amortized.

This can sometimes lead to the consultant exercising influence on the financier (on behalf of the sponsor) regarding:

1. In the event of negotiations leading to project financing

   a) the total outstanding period of a loan;

   b) the payback period of a loan;

   c) the setting of the first year of loan principal payback;

2. Gaining relief from the financier on behalf of the sponsor(s) regarding financial commitments once development or production are in progress and events have proved contrary to predictions based on apparently sound practice;

3. Representing a non-technical member of a joint venture or a consortium-syndicate as a technical advisor on specific aspects of a project and/or on the commercial aspect of the project in its entirety.

FINANCIAL INFORMATION

Cash flow projections and reports need to be prepared and the independent consultant can assist in validating the figures presented.

The owners, in viewing the cash flow, are particularly interested in a number of items: (For methodology, see also Schenck, Chapter 3.)

1. the discounted cash flow, rate of return (DCF-ROR), which reflects the true interest earned on the equity capital taking into account the time value of money (usually at least 15%);

2. the pay-back period (often not longer than 3 to 4 years after production begins);

3. the percentage of reserves required to repay the equity capital (usually not greater than 20 to 25%);

4. the ratio of net revenue to net operating profit (usually not less than 1.5 and 2.0 to 1);

5. net income (before or after tax) per participant or per share.

The lender(s) to the project, in viewing the cash flow, are also particularly interested in a number of items; which, however, are quite different from those which interest the owners. The lenders are particularly interested in loan coverage and in sensitivities and, thus, look carefully at:

1. the ratio of the life of the proven reserves to the total loan period (usually at least 2 to 1);

2. the ratio of the mine project's total project (net) cash flow to the loan principal (often at least 2 to 1, thus an effective 3 to 1);

3. the ratio of the loan principal to the effective equity capital (currently not greater than 4 to 1);

4. the ratio of each year's operating cash flow to the principal repayment over the years the loan is to be amortized (usually not less than 1.25 to 1 and often much greater);

5. the ratio of the gross operating profit to the interest expense, cumulatively and annually, over the years the loan is not completely amortized (usually not less than 2.0 to 1 and more often around 3.0 to 1).

CRITICAL TECHNICAL ASPECTS

There are a number of technical or "engineering" aspects which generally are universal to most mine projects and are critical to their projections of financial performance. Thus, these aspects receive particular attention in the engineering consultant's review. They are:

1. In-situ reserve estimation and categorization

   a) density of data points in keeping with nature of deposit;

   b) sampling technique and exposure to a favorable or unfavorable bias in the establishment of grade (salting);

   c) consideration of high unit value - low bulk effect on sampling results (nugget effect);

   d) application consistency of "rules" to calculations.

2. Run-of-mine or mill-feed reserve estimation (the effect of dilution)

   a) relationship of ore to host rock;

   b) relationship of host rock to mining method;

   c) relationship of mining method to rock testing.

3. Mining

   a) cost relationships of labor, materials and supplies;

   b) comparative productivity;

   c) special, site-related hazards;

   d) adequacy of planning.

4. Processing

   a) weight and rate processing relationship between pilot plant testing and envisioned full production;

   b) representativeness of the bulk sample(s) used for pilot plant testing to the ore deposit and the ore types;

   c) calculations of metallurgical balance and recoveries;

   d) the degree to which the process or parts of the process are tried and proven or are innovative;

   e) the avaiability and quantity adequacy of "support resources"; e.g. water, power, chemicals, reagents, etc.;

   f) transportation.

5. Marketing

   a) the acceptability and appeal of the product(s) to the general market;

   b) the degree to which the product(s) must conform to an unusual specification (in terms of size or content criteria) to meet a particular or speciality market;

   c) the ease with which the market can replace or substitute the subject products;

   d) the nature of the sales contract(s)

ESTABLISHING ENGINEERING INTEGRITY

The manner or methodology by which a project's engineering integrity is established is varied, and is a function of:

1. the past performance of the owner, the similarity of the project to be financed with the owner's past and present business,

INFORMATION REQUIREMENTS

and the apparent in-house technical skills of the owner;

2. the size of the mine project and its complexity (ore types, mining method, mill circuits, types of concentrates, etc.);

3. the ratio of equity capital to loan capital;

4. the lender's past experience with mine projects of similar nature, the degree of his desire to make a loan, and his knowledge of the particular industry or commodity (there is competition among commercial lenders to provide funds for attractive projects;

5. the "nature' of the owner - e.g. whether a large coporation or a syndicate group composed of private individuals - and his likely commitment to the project.

The basic methodologies (or paths) are illustrated in Figure 2 and range from an in-house feasibility study fundamentally accepted as is by the lender, (1), to a feasibility study undertaken by an engineering and construction firm (with input from the owner and from independent consultant specialists) vetted by both the lender's in-house technical expertise and by independent consultant specialists engaged by the bank (whose fees and expenses are eventually paid by the owner, (6).

SENSITIVITY AND RISK ASSUMPTION

The fact that many lenders request at least two cash flow scenarios - a base case situation and a "most likely" situation - is tantamount to an initial sensitivity determination (see Chapter 3). "Sensitivity" is simply a determination of the degree or extent any one factor or any combination of factors which bear on financial performance (be it profit or cash flow or rate-of-return) actually affect financial performance. The following are some of the factors which bear on financial performance in general decreasing order of sensitivity.

1. price;
2. grade;
3. recovery;
4. operating cost;
5. funding costs;
6. utilization of capacity;
7. capital cost;
8. delays in commencement of production;
9. period over which capital is infused;
10. tonnage (when increased).

(Editor: See Chapter 3 and also Castle, Chapter 9)

With an adequate number of case runs to provide data points, both the actual sensitivity and relative sensitivity of each of the variables can be graphed to produce a sensitivity "rose" (Figure

Fig. 2

SCHEMATIC PRESENTATION of ENTITIES (Not Always) INVOLVED and RELATIONSHIPS APPLIED in ESTABLISHING FEASIBILITY STUDY INTEGRITY

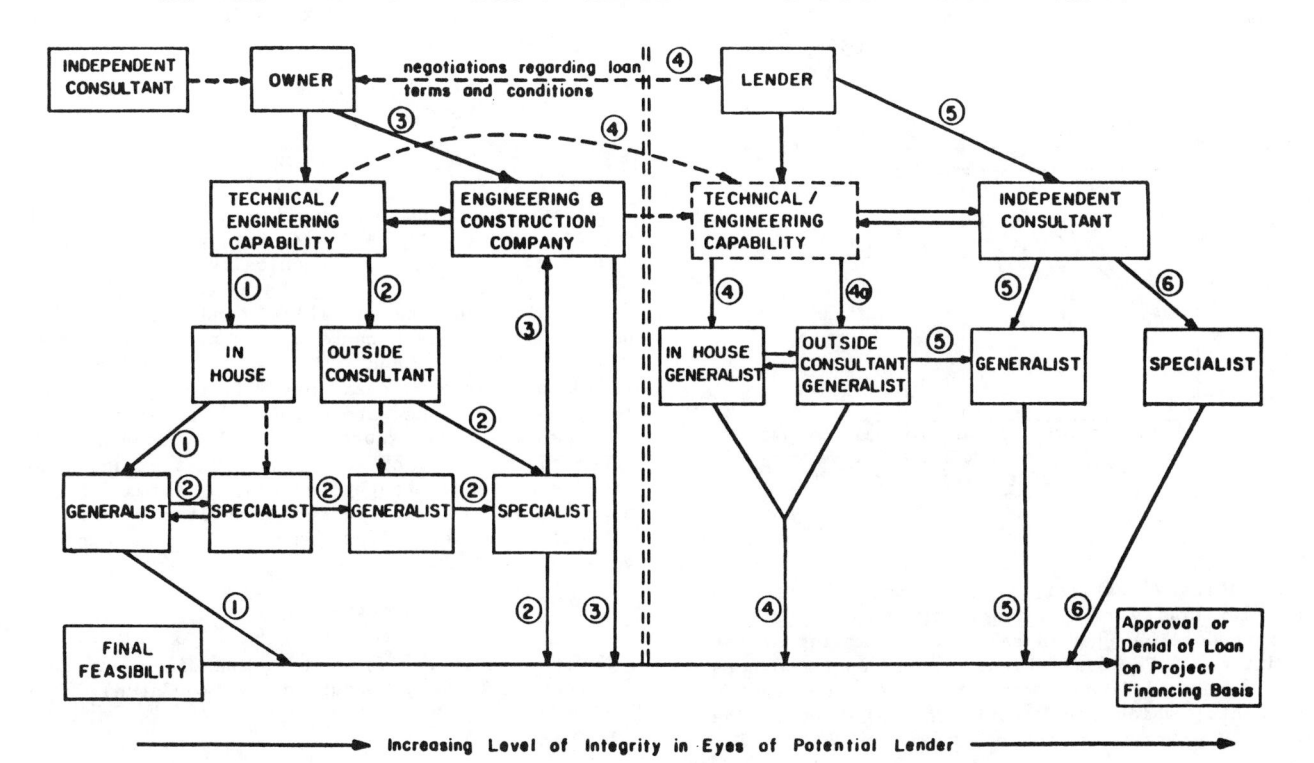

⟶ Increasing Level of Integrity in Eyes of Potential Lender ⟶

3). The steeper the line, the greater the sensitivity of the variable being scrutinized. While the sensitivity of most variables tends to be a straight line function, a few – such as tonnage and capacity utilization – tend not to be.

It should be noted that tonnage quickly can become the most sensitive factor if lower tonnages than the feasibility study are encountered. Generally, the sensitivity of a given variable decreases under favorable conditions because of increasing effective tax rates as margins tend to improve.

Sensitivity from the downside perspective allows the lender to "see" the quantitative difference in factors which underlie the "most likely" situation (for which the lender may reasonably expect realization of agreed repayment and interest schedules) vis-a-vis the factors which underlie the base case situation (at which point the lender may reasonably expect either repayment or interest schedules or both to come under pressure. The quantitative difference (the sensitivity) allows the lender to make the decision whether or not to lend based on his judgement of the acceptability and probability of experiencing this risk sensitivity.

Figure 3

SENSITIVITY "ROSE"

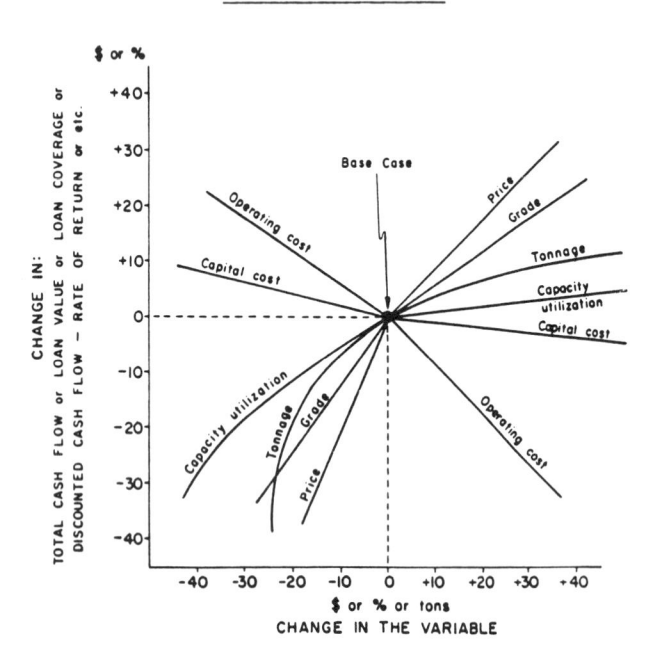

LOAN COVENANTS

As indicated earlier, project financing reduces the flexibility of the owners of a mining project to introduce changes in response to varying market conditions. The limitations imposed on the owners are usually defined by affirmative and negative covenants, which are negotiable depending on the overall attractiveness of the project. The independent consultant can often provide valuable input to these negotiations. The more important affirmative covenants are:

1. completion tests, both as regards utilization of capacity, and quantity and quality of product (see Tinsley, Chapter 9);

2. maintenance of sales contract(s);

3. proper and agreed application and utilization of borrowed funds;

4. maintenance of an agreed level of working capital;

5. maintenance of agreed insurance coverage;

6. maintenance of the lender's security position and subordination of all other debt;

The more important negative covenants are:

1. no disposing of assets;

2. no merging or consolidation of the owner;

3. no mortgages, charges, or liens to be incurred;

4. no pre-payment of any recognised debts or obligations unless also to the project financier on a pari-passu basis;

5. no changes in the scope or the business of the project;

6. no payment of dividends before retirement of loan, except from "excess income' as defined and agreed;

7. no decrease below an agreed and defined level of net worth.

CONCLUSION

It is important to realize that banks will compete among one another for projects which appear financially and technically sound. Thus, owners of attractive projects – those which are financially and technically sound – are in a position to approach banks with confidence and not with a "hat-in-hand" atitude. Owners may reasonably expect that any of the terms and conditions of the above listed covenants are negotiable. The approach to the lenders with independent verification and analysis and the broad experience brought to strategy formation and negotiation by independent consultants can be effective and efficient from a time and expense point of view.

REFERENCE

Nevitt, P.K., Project Financing, A.M.R. International Inc., Second Edition (undated)

10.3

WHAT BANKERS LOOK FOR IN PROJECT LOAN APPLICATIONS

Norman J. Gibbs and John Sroka

ANZ Banking Group
Melbourne, Australia

INTRODUCTION

At the point a company decides to begin mine development and wishes to convince lending institutions that the proposed operation will return their borrowed funds, plus interest, over the terms of the loan, communications between lender and borrower are crucial. The importance of the mine feasibility presentation should not be underplayed, especially as new projects continue to increase in size, complexity, and cost. While the content and emphasis of a report to the lender will vary according to the particular project and type of debt financing sought, there is a general base of information upon which all such studies are built.

THE PROJECT'S TECHNICAL ASPECTS

Here, the object is to satisfy the lender that he will not be unduly exposed to technical risks as a result of the physical scope of the project.

Reserves

The feasibility report submitted to financial institutions should include an adequate description of the geologic setting, mineralogy, ore types, and reserves broken down into such recognized categories as proven/probable/possible or measured/indicated/inferred. Rather than a bald statement of reserves, lenders prefer to have detailed information on:

- History of exploration

- Drilling, both percussion and core, with core recovery, drill hole spacing, analyses, and continuity of mineralization

- Bulk sampling, with core assay comparisons

- Overburden, with stripping ratios

- Specific methodology of reserve calculation, and

- Geologic reserves vs. mineable reserves.

As a rule of thumb, a lender would prefer to see that reserves in the proven class alone are sufficient to support operations at the design rates until the borrowed funds are repaid. (See also Schreiber above in this Chapter.)

Mining

The rationale for adoption of a particular mining technique should be presented as well as descriptions and justifications for:

- Preproduction development

- Preliminary mine plan, with forecasts of grades and material movements

- Geotechnical test work which influenced decisions

- Mining recoveries and efficiencies

- Major equipment, with type, model, rating life, spares, replacement schedules, utilization, and productivity

- Manpower requirements and work schedules

- Design assumptions

- Ore stockpiling and blending

- Environmental requirements, and

- Waste disposal

Processing

If a new ore body is to be developed or a new process used, it is important to provide details of test work, scale up, and extrapolation factors. A basic process description with a summary of the studies which justify the intended processing route should be included with the following data:

Adapted from a paper "Feasibility Studies for Financing Purposes", presented at the Australasian Institute of Mining and Metallurgy Conference, North Queensland, September 1978 and published in SME's <u>Mining Engineering</u> in December 1978.

Process flow diagrams and plant layout

Major equipment list, with ratings and capacities

Product and co-product specifications and tolerances

Manpower requirements and work schedules

Fuel, power, and consumables usage factors, and

Environmental considerations.

Lenders are sensitive to design parameters derived from test work on samples, since miscalculations can lead to process deficiencies at start-up and, consequently, protracted delays and equipment modification. The net effect could be a disillusioned lender who is asked to provide additional funds and/or rearrange the original debt servicing requirements.

Infrastructure

The lender needs to be satisfied that adequate provision has been made, especially at projects in remote locations, for items such as:

Water/power supplies and services

Maintenance workshops

Transport facilities, and

Housing

Construction

In this area, lenders are wary that projects may take longer and cost more to construct than original projections. (See Castle, Chapter 9.) The sponsor should include in his feasibility study a presentation on how he intends to carry out the engineering, procurement, construction, and management (EPCM) functions necessary to take the plan from drawing board to reality. This will include details of:

Responsibility for EPCM functions, including qualifications of the in-house staff and contractors/consultants

Construction schedule with allowances for lost time due to weather and industrial disputes

Manpower requirements

Construction camp and facilities, and

Labor agreements.

COSTING INFORMATION

As mining companies look to larger operations for economies of scale, they also place greater dependence upon lenders for larger amounts of debt funding. Consequently, the costing aspect - both capital and operating - in a mine feasibility report is of fundamental importance to the potential lender.

Capital Costs

The lender requires sufficient background information and detail to establish a confidence rating in the estimate, and to allow comparisons of the project cost proposals with similar elements at operating mines. The project sponsor should also indicate how cost control, monitoring, and cost-trend forecasting will be instituted.

The typical framework of a base cost estimate may consist of the following major components:

Direct costs, such as equipment, bulk materials, and the labor for their installation

Distributable costs, which cover the job-site services provided in support of the direct cost components

Home office costs for off-site services, and

Gross margin, including an allowance for home office overhead and profit margin

It is essential that definitions of estimating terminology be included, as well as the basis and methodology for establishing the estimate. Descriptions should include:

Date of estimate

Summary of costs into categories consistent with the report's "Technical" section

Nature and quantity of quotations

Sources of major equipment

Allocation of freight, sales tax, landing costs, and duties

Installation basis, subcontract/direct hire

Construction labor productivity, working schedules, and cost assumptions on rates and benefits, and

Timing of the cost expenditure pattern.

Assuming that the estimate is presented in the format of a base cost at a certain date, to which are added allowances for contingencies and escalation, then the project sponsor should describe his methodology for the allowances.

There is frequently a misunderstanding associated with the term "contingency". The dictionary meaning is misleading since it indicates that a cost item may or may not be incurred, depending upon some chance occurrence. In the estimating context there is no doubt that such a cost shall be incurred. Contingency refers to those costs which are regularly encountered but cannot be tied down at the study stage.

To estimate escalation, the sponsor company should, at the very least, split the capital cost into labor and material components for which inflation indices are readily available. There are two areas which cause estimating difficulties: (1) Occasionally, project sponsors do not apply escalation factors to contingency costs; and (2) project sponsors sometimes do not account for the full time lag between the date the estimate was prepared and the most current project commencement date.

Lenders are also interested in a statement defining the accuracy of the estimate, which allows them to assess the impact of overrun on the project's overall cash flow. The accuracy of an initial "order-of-magnitude" cost estimate would range from around +/-25% to +/-15%. A preliminary estimate where many more engineering man-hours were consumed would have a typical accuracy ranging between +/-15% and +/-5%. For a definitive estimate where perhaps 40% of the total design engineering man-hours had been expended and firm bids received, the estimate would be expected to have an accuracy of +/-5% to +/-3%. The tolerance band in any particular estimate need not be symmetrical, e.g., to say "-5% to +8%" is perfectly acceptable.

Operating Costs

Correct estimation of operating costs can be of more ultimate importance to the long-term success of a project than the initial capital costs. A typical approach for presenting operating costs is to use the same cost centers that were established for the capital cost estimate, with estimates prepared on the basis of labor, materials, and service requirements. Descriptions should include:

Methodology of cost estimation

Specification of costs into fixed and variable components

Labor cost rates by craft

Manpower requirements

Allowances for training, absenteeism, and turnover

Equipment running cost assumptions

Materials, power, fuel, and consumables usage factors and unit costs

Freight, royalty, and selling expenses

Overhead cost estimates

Inflation rate trend studies for major components, and

Effects on future costs due to expected changes in operational factors such as deeper mining and longer haul distances.

MARKETING

Just as important as costing assumptions are the project sponsor's marketing assumptions, for they determine the income forecasts. Although marketing studies are inherently less precise than costing studies, the lender needs to be satisfied that the assumptions are well-founded and conservative. An overview of the total market scene should be provided with every feasibility study, even if the product is to be sold on a long-term contract basis.

Market Overview

In a presentation of the total product marketing scene, the following information should be included:

Industry structure

Supply demand-relationships, historical and projected

Determinants of demand

Pricing trends

Basis of competition

Substitutes, possible technological changes, and

Reserves

Specific market aspects

Price and volume are the two market elements that account for a project's income. The lender would prefer to see most of the output committed in advance under a long-term sales contract extending beyond the term of the proposed loan. To support a project's revenue forecast, the following types of information should be provided:

Selling arrangement - spot or term contracts

Price forecasts and justifications

Price basis - c.i.f/f.o.b., negotiated, or tied to a market price

Currency exchange risk

Volume forecast and justifications

Number and size of buyers

If a term contract - details of buyers, currency, length of term, volume and maximum-minimum options, stockpile requirements, quality requirements and penalty provisions, protective clauses, maintenance of competitive pricing, and renegotiation options

Government regulations on export approval, tariff protection, and price levels, and

Specific marketing advantages/disadvantages of the particular project under study.

ECONOMIC VIABILITY

Of primary importance to a lender is the ability of a proposed project to generate sufficient income to meet operating costs, tax payments, and debt servicing obligations under a reasonably wide range of conditions. The main test of overall economic viability, from a lender's point of view, is through cashflow analysis.

A base case cashflow forecast that extends at least through the life of the proposed loans - and preferably for some additional years - should be prepared under the costing and marketing assumptions justified elsewhere in the report. The forecasting periods should be semiannual during the construction and project start-up, then annually thereafter. A typical cashflow format would show:

Construction/development expenditures

Equity fund contributions

Loan fund drawdowns

Receipts from sales

Operating costs, segregated by category

Capital additions and replacements

Movement in working capital

Taxation and royalty payments

Scheduled debt service payments

Dividend payments, and

Overall cash surplus and deficit position

Lenders generally judge the cashflow forecast on the basis of debt service coverage, e.g. a common requirement is for the surplus cash (before debt service) in each year to be at least one-and-a-half-times the proposed debt service for that year.

In this section of the report, sensitivity studies to identify the key criteria affecting the economic viability of the project should also be presented. The effects of inflation and price increases on capital and operating costs should be investigated, as well as the effects of:

Time delays in achieving design output

Changes in ore grade

Lower-than-design plant output

Changes in debt-equity ratios, and

Currency changes.

OTHER FACTORS

In addition to the technical, costing, marketing and cashflow aspects, there are other types of information which form a part of the feasibility report to lenders:

Project sponsor

Lenders require a description of the sponsor's activities and scope of operation, plus historical and current financial performance. Sources of funds, both equity and loan, for the capital cost of the project should be submitted with especially detailed information on the equity funds.

If more than one sponsor is involved, additional information concerning the joint venture agreement is required: each party's equity contribution, financial responsibilities, technical input, and provision of management manpower, plus the joint venture's legal and operational/organizational structure.

Independent verification

The feasibility report should identify those aspects that were investigated and prepared by the sponsor's own organization, as differentiated from work carried out by external independent consultants. A lender who is accustomed to reviewing financial statements that require statutory independent audit would prefer that technical and cost elements in a mine feasibility study be prepared, or verified, by an independent consultant of repute. (See Emerson and Schreiber above in this Chapter.)

Statutory requirements

Project approvals from government, local through federal, that have been sought and granted should be submitted. Although ideally these should be settled before a sponsor looks for loan funds, it is sometimes not practical, and in these cases the sponsor should report the status of approvals still pending.

Labor relations

A presentation on labor relations pertinent to the project, the steps taken by the sponsor to minimize industrial disputes, and the allowances provided for such disputes in the planning and cost areas should be included with the report.

Future implications

The lender is concerned with fast-paced changes that could affect the medium and long-term prospects for the project. A relatively common concern is the possibility of product substitution, or obsolescence. However, the sponsor should also consider other potentially serious consequences. For instance, where government approvals have been given, they can sometimes be substantially revised or even totally rescinded through a change in government policy. Risks concerning such currently sensitive areas as environment, foreign ownership, royalties, tariff protection, and export need to be assessed. The floating of major currencies in recent years has added a further complication in this category of risk.

CONCLUSION

A potential lender needs to be satisfied that the proposed project will not only repay the borrowed funds, but also meet other financial commitments essential to the maintenance of that source of cash flow. In assessing the risks associated with the loan, a lender will sometimes need to know almost as much about the proposed operation as the sponsor. It is considered that the most appropriate method of providing a lender with the information on which to make a lending decision is through the submission of a comprehensive feasibility study. In practice, this would involve the sponsor in little additional effort since he should have made a similar presentation for his own investment evaluation purposes.

10.4

TECHNICAL AND FINANCIAL ELEMENTS OF A MINING PROJECT LOAN REQUEST:
PREPARING A COMPLETE INFORMATION MEMORANDUM

Thomas P. Bispham

Dames & Moore,
New York, New York.

INTRODUCTION

From the lender's point of view, the greatest period of risk in a project financing occurs during the construction phase. Therefore, most projects are supported during this phase by completion guarantees supplied by some creditworthy entity; for example, the project sponsors in the case of a joint venture, such as that of W.R. Grace, Hanna Mining, and Liberty Capital; or by a strong contract with a utility or foreign importer/investor, such as the way North American Coal Corporation has structured many of its transactions.

THE PROJECT LOAN REQUEST

A project financing is a more complex form of term loan than one using an ongoing corporation's balance sheet as a credit base and, therefore, a potential lender will subject the loan request to a more rigorous analysis. To support this effort, the financial institution will require a more detailed information memorandum to evaluate the credit and technical risks of a project - especially since much of the necessary information is not otherwise available. (See also Gibbs and Sroka paper above in this Chapter)

A well-prepared information memorandum will impress prospective lenders with the planning ability and general competence of management. Also, the process of preparing the memorandum will help management understand the true extent of the opportunities and limitations of its venture.

Appendix A outlines the topics which comprise a complete information memorandum for a loan request. Of some interest to those seeking to sell a currently-operating mining venture, a similar information memorandum should be prepared i.e., covering the same topics discussed here, for the investment banker to use with potential purchasers. For that purpose, though, one might not want to prepare quite as much detail as for a lender.

Appendix B outlines the elements of a reserve evaluation and feasibility study, often called the "Bankable Document." Usually, these studies are prepared by independent engineering consultants for use by potential borrowers and sellers. (Potential purchasers will often contract for a similar study for their use in evaluating an acquisition.)

ELEMENTS OF A LOAN INFORMATION MEMORANDUM

1. Outline: The information memorandum should begin with a one-page outline covering such topics as:

   a. The borrower's name, address, and phone number.
   b. Principal contact at the borrower.
   c. Amount of the proposed financing.
   d. Desired pricing and structure of the proposed financing. This should include a description of the amount and source of equity in the proposed financing.
   e. Use of the proceeds and timing.

2. Summary: The most important information contained in the memorandum should be summarized in two pages.

Note - If the borrower is a recently-formed single-purpose company, then elements 3, 4, and 5 will apply to the guarantors of completion and to the project sponsors.

3. The Borrower: This section is a brief biography of the borrower including such topics as date and state of incorporation, founder(s), lines of business, etc.

4. Capitalization: The borrower's current debt and equity position, including leases and trade debt, with amounts, terms, names, and addresses enumerated. Equity owners, and percentage ownership, should be identified.

5. Historical Financial Statements: If available, a minimum of five year's worth of complete historical financial statements

Based on a paper presented at Coal Conference and Expo VI (Sponsored by Coal Age Magazine), Louisville, Kentucky, October 1981.

should be provided. Breakdowns of sales and cost categories and explanations of any abrupt changes in accountants or in accounts (such as a loss or a dramatic increase in receivables unaccompanied by a concomitant sales increase) should be provided.

6. Independent Reserve Evaluation and Feasibility Study (The Bankable Document):
Where a project involves the development of coal reserves, or the construction of a facility dependent on production from coal reserves, a lender usually will require verification of those reserves by an independent technical expert.

The lender will want to see maps describing the geology and leases or land ownership status, showing drill holes and/or sample points. The lender will be concerned with the quantity of measured coal reserves (call "proven" by many bankers) and whether they will have a sufficient mineable life to more than "cover" the life of the proposed financing.

The feasibility study should confirm that the project can be completed according to the technical specifications at the projected cost. The lender will be interested in the plans for mining and cleaning the coal plus the operator's ability to achieve the estimated mining, cleaning, and transportation costs, as they relate to the overall economics of the project.

Often, since most financial institutions do not wish to be subjected to the market risk, a major element in evaluating the economics of a coal project is the sales contract. Essential details of the sales contract should be provided:

a. How strong is the purchaser?
b. Does the term of the contract extend beyond the term of the proposed loan?
c. What percent of design capacity is covered by the contract? Are the economic projections based on contract quantities?
d. What quality requirements are in the contract? Can the mining operation meet them easily?
e. What are the Force Majeure provisions?
f. Under what conditions can the buyer refuse to take coal under the contract?
g. Does the contract sales price compare favorably with that used in the economic projections? Is there price escalation to cover operating cost increases?

Lenders will also expect a discussion of the status of the project's permits and other regulatory issues. Delays caused by litigation and/or regulatory agency actions can destroy the economic viability of a project.

Most lenders prefer that the feasibility study and economic projections be prepared by a well recognized and independent engineering consultant whose opinion they respect. (See Emerson and Schreiber above in this Chapter.)

Often, a review of historical, as well as future markets, is appropriate in order to develop additional cash flow forecasts to result in three "cases:" Optimistic, Pessimistic, and Most-Likely. These "sensitivity analyses" might use a lower/higher sales price per ton and/or demonstrate the effect of an arbitrary increase in costs (for example, interest or labor).

Most consultants will enter all of the relevant financial data into a computer program making such sensitivity analyses much easier to generate, with the product not only in a more usable form, but with fewer arithmetic errors.

The project loan request is much more than an engineering "package" with a cover letter. It is an opportunity for you to tell your company's story; that is, one opportunity per financial institution.

Appendix A

Elements of a Complete Information Memorandum

(These are general topic headings used when preparing an information memorandum for companies who are seeking either a mining project loan or to sell an on-going mining operation.)

1. An outline of the Basic Proposition (e.g. Loan request or terms of sale)

2. A short summary of the Memorandum's most important information

3. A description of the borrower and its history. (This, and elements 4 and 5 also apply to guarantors of completion for loans.)

4. A detailed description of the borrower's current capitalization

5. A five-year set of historical financial statements

6. An independent reserve evaluation and feasibility study, the "Bankable Document." (See Appendix B for a more detailed description.)

Appendix B

Elements of the Reserve Evaluation and Feasibility Study: "The Bankable Document"

1. Land status, with maps

2. Geology, with maps showing drill holes and/or sample points.

3. Coal reserves showing measured, indicated, and inferred reserves, plus a discussion of the parameters and techniques used

4. Mining method and equipment details, with estimates of capital expenditure requirements and operating and labor costs to achieve optimum production rates over

project life

5. Mining projections and production schedules

6. Coal preparation method and equipment with similar cost detail

7. Status of permits and other regulatory issues

8. Market considerations, including contract details, competition, and transportation costs

9. A detailed financial forecast to cover the period for which the loan will be outstanding (or, in the case of a potential sale, at least five years). This will include all the financial data developed during the feasibility study. In many instances, especially those involving a financing, three forecasts should be presented; Optimistic, Pessimistic, and Most-Likely (See also Arne in Chapter 10).

10.5

INFORMATION REQUIREMENTS FOR EQUITY ISSUES

E. L. Affleck, F.C.A.
Director, Filings
Superintendent of Brokers Office
Vancouver, British Columbia

W. G. Stevenson, P.ENG.
Mining Engineer Advisor
Superintendent of Brokers Office
Vancouver, British Columbia

ABSTRACT

In both Canada and the United States, a company making a public offering of securities must comply with regulations which have been passed by Federal, State and/or Provincial Legislatures.

These regulations usually require the distribution of an acceptable prospectus which provides essential background information about the issuer, the project and the share offering. When the public offering has been accepted, the unlisted company may then seek a listing of its shares for trading on a Stock Exchange. Once listed, the company will be subject to rules of the Stock Exchange.

This paper will deal with the specific information required for inclusion into a prospectus being submitted for a resource company and particularly those parts of concern to mining engineers and geologists.

MINE FINANCING IN BRITISH COLUMBIA THROUGH THE VANCOUVER STOCK EXCHANGE

INTRODUCTION

When the American Institute of Mining and Metallurgy was formed some 115 years ago the money necessary to find, develop, and to bring a natural resource property into production was only a fraction of that required today. Funds for exploration development and production were provided by grub stakes from local merchants, wealthy individuals, syndicates and the large and successful mining companies. Only a small part of the financing for mining ventures was provided by the sale of stock. This means of financing was much less prevalent then than it is today.

This change from individual and small group financing to financing by thousands of individuals through the facilities of stock exchanges has been gradual but persistent.

We may presume that a "let the buyer beware" attitude was as prevalent 115 years ago as it is today but a great many more individuals are involved today, and some of these individuals are presumed to be less sophisticated than the grub stakers of 115 years ago. Along with the change from a few investors financing a mining project to thousands of individuals participating in financing has come gradual governmental intervention which has been designed to protect the thousands of investors who are involved in the financing.

Legislative agencies, congressional committees, royal commissions, panel discussions and judicial processes have contributed to the acts and regulations which have been drawn up and passed by state, provincial and federal governments to protect the investor.

In the process of enacting this protective legislation the goals have never been clearly defined. There have consistently been two schools of thought: first, there are those who believe that if the buyer is provided with full, true and plain disclosure no matter what the venture entails, the offering should be allowed to proceed. Second, there are those who believe the regulatory bodies should assess the venture and if it appears the investor is unlikely to recover anything, the venture offering should not be allowed to proceed.

Because of these two schools of thought, tensions and conflicts have developed, concessions have been reached and the resulting legislation has often not clearly spelled out the duties and responsibilities for the administration of the acts and regulations.

In addition to federal, state and provincial regulatory bodies, a number of self-governing associations of securities brokers and dealers are also involved in the securities regulatory process. Most prominent among these self-governing associations are the stock exchanges, which in North America regulate not only much of the secondary trading (i.e. trading in previously issued shares) but also much of the equity financing (i.e. successive offerings of shares from treasury) carried out by listed companies.

The Vancouver Stock Exchange is constituted under a British Columbia statute, but in order to maintain operations, it must continue to be "recognized" under the British Columbia SECURITIES ACT, a statute which is administered by the Superintendent of Brokers, a British Columbia Government Official. Subject to appropriate appeal procedures, the Superintendent in empowered to withdraw "recognition" where he deems it to be in the public interest to do so.

This paper will be chiefly concerned with the financing of mining ventures through public offerings. Over the past 15 years Vancouver has become one of the most active centres for raising venture capital for resource development in North America. This paper will analyze the process for funding a company through to the offices of the Superintendent of Brokers of British Columbia and the Vancouver Stock Exchange.

Two stages of development must be completed before a company may apply to list its securities on the Vancouver Stock Exchange:

1. The company must qualify for a certificate of incorporation and must raise "seed capital" through private financing. A company incorporating under the British Columbia COMPANY ACT must satisfy the B.C. Registrar of Companies (a British Columbia Government Official) that the relevant parts of the COMPANY ACT have been complied with in order to obtain a certificate from him.

2. The company must then make a distribution of its shares through an offering to the public. If a public subscription is to be carried out in British Columbia, the requirements of the British Columbia SECURITIES ACT (administered as previously indicated by the Superintendent of Brokers) must be complied with. These requirements are discussed under the headings "TECHNICAL REPORTS" and "PUBLIC FINANCING".

Once a public distribution of shares has been accomplished, the company may apply for a listing with a stock exchange. Most of the companies applying to list shares on the Vancouver Stock Exchange have effected a public distribution through a share subscription in British Columbia. Having secured a listing, not only may trading of previously distributed shares proceed, but the company may also carry out subsequent public financings through offering additional distributions of treasury shares through the exchange. The Vancouver Stock Exchange requires that a "Statement of Material Facts" (similar to the prospectus described under the "PUBLIC FINANCING" heading) be submitted for approval before any additional offering of treasury shares may be made through the exchange.

A company which has completed one public distribution of shares but which has not obtained a listing on an exchange may engage in further public financing by making subsequent public offerings of treasury shares in compliance with securities regulation. An unlisted company, however, may find itself at a disadvantage in seeking further public financing through treasury share offerings since it is unable to offer the investor the degree of liquidity afforded by the stock exchange market.

The salient features of public financing in the U.S.A. have been attached as Appendix "A".

Throughout this article reference has been (will be) made to various statutes, regulations and forms which comprise the body of securities regulations in the U.S.A. and Canada. Reference to the source for the regulations is attached as Appendix "B".

PRIVATE FINANCING

It has long been recognized that the discovery and the early stages of exploration and development are much less expensive to fund than are the latter stages leading up to production which require large capital investments. It is also recognized that the latter stages of a viable mining project are easier to finance through bank loans, lines of credit and mortgages than are the earlier and riskier stages.

The financing of a mining venture today typically begins with a modest cash outlay by a founding group who assume a high risk that their outlay will not be recovered. A private or "non-reporting" company is usually incorporated, and to raise the capital required to start up the enterprise, the founders subscribe cash for low-price, high risk shares. In British Columbia, as long as the solicitation of subscriptions is confined to persons and entities having a close association with the principals, the company may claim "private" status. It need concern itself with securities regulation only to the extent of taking care that treasury shares are issued for consideration which would be acceptable to regulatory authority at such time as the company seeks to "go public".

A salient and early transaction usually involves the sale to the private company by its principals of an unproven, natural resource property. For many years in British Columbia, principals were permitted to acquire 750,000 shares and cash reimbursement equal to their out-of-pocket outlay as consideration for vending an unproven property into the company. Such shares were escrowed pending the development of the property. This policy was recently changed and now principals are prohibited from taking any "up front" share consideration for vending in unproven property but in order to afford such principals a measure of control over the affairs of the company and an incentive, they are permitted instead to purchase 750,000 shares for cash at 1 cent per share.

The work carried on by a private company at this early stage usually involves commissioning a mining engineer to report on the mining property or properties which the company has acquired or plans to acquire.

An astutely managed company should find a technical report written by an independent and qualified person to be of immediate assistance in raising exploration money. The recommendations for the initial program of exploration and development to be carried out on a property may be implemented using proceeds from the private subscription. Generally, the greater the progress made on exploratory work prior to going public, the lesser the risk to which the public subscriber is exposed. As a consequence of reduced risk it should be possible to set a higher price per share for the public subscriber.

TECHNICAL REPORTS

When the company elects to convert from a private to a public company, one of the requirements is the compilation of a technical report(s) on the company's natural resource properties for submission to the Superintendent of British Columbia for review and approval. The author of the report must be independent of the company and considered qualified by the Superintendent of Brokers. The content of such a report is not prescribed in detail by the SECURITIES ACT or SECURITIES ACT REGULATIONS of British Columbia, but a specifications as to the content may be obtained by consulting NATIONAL POLICY STATEMENT 2A, a set of guidelines adopted by all securities jurisdictions in Canada, and LOCAL POLICY STATEMENT #3-01, a supplementary set promulgated by the Superintendent of Brokers of British Columbia.

The following topics should be covered in the technical report:

- Introduction
- Property Description
- Location and Access
- History of past exploration, development and production
- Geology and Mineralization
- Reserves, tonnage, grade and classification
- Conclusions
- Recommendations and estimated costs for implementing the recommended program

PUBLIC FINANCING

At such time as a company seeks to "go public" a prospectus must also be submitted to with the Superintendent of Brokers of British Columbia. When the company seeks to explore and develop a mining property through such a public issue the prospectus filing must be accompanied by supplementary documentation, foremost of which is the technical report(s) referred to previously and a set of audited financial statements. The prospectus and the supporting documentation are reviewed by staff of the Superintendent's office and when the revisions indicated by the office have been carried out the prospectus is receipted. Since no order may be accepted from a subscriber until a definitive copy of the prospectus has been delivered to him, receipting of the prospectus is required before active solicitation of public subscribers may begin.

CONTENTS OF A PROSPECTUS

The typical prospectus is a narrative document designed to provide background about the issuer, its principals, its past and current operations as well as a detailed description of the proposed use of the funds being solicited from public subscribers. Provided the prospectus discloses sufficient information on the topical items set out in the SECURITIES ACT REGULATIONS (Form 10), the order of presentation is left to the discretion of the issuer, except where some specific location such as the cover page is prescribed. The following items are required to be incorporated in the prospectus:

Distribution Spread

(i.e. the portion of the share price taken by a registered dealer as compensation for participating in the sale of shares.)

The price per unit of the securities offered to the public, the gross proceeds arising from a full subscription to the securities offered, the per unit and aggregate amount to be allocated from the offering as compensation to registrants participating in the offering is provided. The net per unit and net aggregate amount to be received from the subscription by the issuer must be prominently disclosed on the face page of the prospectus.

Plan of Distribution

There are three means by which the securities offering is to be made to the public. First: a firm underwriting agreement or a stand-by underwriting may be negotiated with a registered investment dealer (who in effect guarantees the sale of the complete offering). Second: a dealer may simply agree to use his best efforts in selling the securities, in which case if a set of minimum subscription is not achieved within a stated offering period the entire offering is cancelled and all subscriptions are refunded. Third: the company seeking financing may register as a security issuer and may sell its own securities.

Use of Proceeds

This is one of the more salient elements. While a reasonable amount of proceeds may be designated for unallocated working capital, the majority of the proceeds should be allocated to asset acquisitions and program expenditures which have been planned in some detail. If the issuer seeks to use subscription proceeds to pay off old debts incurred for purposes bearing no relationship to the forecast exploration and development programs, the Superintendent has the option of denying the offering. Where insufficient working capital is on hand to pay off old debts, a private subscription or an offering to existing shareholders is deemed to be a more appropriate method of funding.

Sales for Other Than Cash

If any securities are concurrently being offered for other than cash, details of such an offering must be disclosed.

Share and Loan Capital Structure

With the information disclosed, a prospective investor should be able to determine the prices at which shares have been issued for cash and the number of such shares issued, the "distribution spread" on any of such issued, the number of shares issued to date for other than cash, and the total number of shares which will be outstanding at the conclusion of the public offering.

The number of issued shares in the hands of principals, is disclosed in the prospectus under the information about "insiders".

If the issuer or any subsidiary of the issuer has issued debt securities, details are provided.

Since the issuer may make subsequent offering of shares to the public from treasury after the offering covered by the prospectus has been concluded, the information on the number of securities which will be outstanding at the conclusion of the prospectus offering may be of transitory interest. However, if an overwhelming number of shares has already been issued in proportion to the number of the shares offered on the prospectus, presumably the investor can draw some conclusion about the degree of control and interest already in the hands of insiders etc., and the prospects of a per share return on investment.

Name and Incorporation of the Issuer

The jurisdiction, date of incorporation of the issuer, address of head office and registered office is provided. Where the company has lain dormant for some time, the date on which the company recommenced activities should be provided.

Description of Business and Property of the Issuer

Extracted from the engineers report.

Incorporation within One Year - Preliminary Expenses

A company, which has been incorporated for less than one year prior to the date of the balance sheet contained in the prospectus is drawn up, must provide a statement of the amount or estimated amount of administrative and development expenses including the amount already expended and the estimated future expenditures.

Promoters

The SECURITIES ACT provides a lengthy definition of a "promoter". Essentially a promoter is an individual or entity who acting alone or in conjunction with others takes the initiative in founding, organizing or reorganizing the company which is seeking funding, and/or who has received or who will receive substantial share consideration for services and property contributed to the issuer company.

Details of securities which promoters have acquired from the issuer and everything which promoters have vended or contributed to the issuer should be disclosed.

This is one of the units of "insider information" which will enable the prospective investor to determine the extent to which promoters have benefitted or are likely to benefit from their association with the issuer at the expense of the investor.

Pending Legal Proceedings

An investor will be able to gain from this disclosure whether legal proceedings would effect his investment.

Issuance of Shares

A description of the rights and obligations for each class of shares being offered.

Issuance of Obligations

A clear description of the rights, security and obligations will be provided for each debt security.

Issuance of Other Securities

Details of the rights, warrants and obligations which are attached to the securities will be provided.

Dividend Record

The amount of dividends or other distributions paid per share in each fiscal year by the issuer during its last five years preceding the date of the prospectus should be provided.

INFORMATION REQUIREMENTS

Directors and Officers

This is one of the more salient items of insider information required and will provide the names and home addresses of all directors and officers of the issuer, the positions and offices held with the issuer by each, and the principal occupations within the five preceding years of each director and officer.

The Superintendent usually presses to have the relevant track record as well as the qualifications and experience of each director and officer disclosed, and the association of such individuals with other companies which may be competing with the issuer for properties or for public funding.

Remuneration of Directors and Senior Officers

This item will provide information as to salaries paid to these insiders as a group, and the pension benefits paid or to be paid under management contracts to companies controlled by these insiders.

Disclosure in this and the succeeding three items will show the extent to which these insiders have already benefitted and stand to benefit from their association with the issuer.

Options to Purchase Securities

Detailed disclosure should be provided as to options to purchase securities from the issuer or any of its subsidiaries which are:
(a) held or proposed to be held by all directors and senior officers of the company of any subsidiary of the issuer as a group,
(b) held or proposed to be held by all other employees of the issuer and any subsidiary of the issuer as a group,
(c) held or proposed to be held by any other person or company, naming each such person or company.

Escrowed Securities

Shares received in consideration for vending of property, or for shares purchased at 1 cent per share by principals will be disclosed.

Principal Holders of Securities

Each individual or company currently holding more than 10% of any class of shares will be disclosed.

In addition the percentage of issued shares of each class currently held by all directors and senior officers as a group will be provided.

Prior Sales

This item shows the prices and the number of securities of the class offered in the prospectus have been sold within 12 months prior to the date of the prospectus.

Interest of Management and Others in Material Transactions

This item will disclose any insider interest in material transactions not previously required by any other item.

Auditors, Transfer Agents and Registrars

Name and location of each of the above.

Material Contracts and Other Material Facts

These last three items should provide a "catch-all" for any previously undisclosed facts which might have a bearing on the decision of an investor whether to subscribe to the offering.

CONCLUSION

The information units required to be disclosed in a prospectus have been discussed in the order in which they appear in Form 10 of the SECURITIES ACT REGULATIONS. A moment's reflection will indicate that there are five main categories of information required:

(a) Information on securities authorized, issued and outstanding, and the consideration received by the issuing company (issuer) from securities already issued. This enables the investor to compare the price asked of him with the consideration heretofore (and possibly hereafter) to be received by the issuer for its securities.

(b) Background information on the activities of the issuer to date including properties currently held, previously abandoned, etc.

(c) Information about insiders, and the extent to which they have heretofore benefited from participation in the issuer, as well as the extent to which they are likely to benefit in future.

(d) Information on the use to be made of the proceeds of the proposed issue as well as proceeds already raised and available to the issuer.

(e) Other information likely to influence the decision of a prospective investor.

An appropriately drawn prospectus will not relieve the prospective investor from making a decision whether or not to invest in a particular mining venture, but it should provide him with all information relevant to a decision.

It is greatly to the interest of those who have a long-term stake in mining ventures that investor confidence in the risk capital market be preserved through adequate disclosure of information to prospective investors by means of prospectuses, technical reports and financial statements. Provided the individual investor can make some realistic appraisal of the potential risks and potential rewards relating to a speculative venture, he is likely to continue to provide the risk capital which is so sorely require by mining mining ventures in their pre-producing stages.

APPENDIX "A"

The Situation in the U.S.A.

A company seeking to offer securities to the public in the United States of America will usually be required to register under the (Federal) SECURITIES ACT with the U.S. Securities and Exchange Commission. A prospectus will be filed with the S.E.C., and the adequacy and accuracy of the information included in the prospectus will be reviewed by the S.E.C. before the document is approved for distribution to prospective investors. The Commission has the authority to exempt offerings not exceeding $5,000,000 from certain registration and disclosure requirements.

In addition to complying with federal securities law, a company must comply with registration and disclosure requirements under the securities laws of each of the states in which the securities are to be offered. As has been the case in Canada, measures have been taken to bring about considerable uniformity in state law and the minimize the overlapping of conflicting federal and state requirements. Furthermore, securities listed on the New York and American Stock Exchanges are exempt from much of the state requirements.

If the offering and distribution of securities is placed in the hands of a member(s) of the National Association of Securities Dealers, the offering document will also have to be reviewed by the NASD Committee on Corporate Financing to determine whether the compensation afforded the underwriter is fair and reasonable.

Securities and Exchange Commission Registration Forms, eg. Form S-1, S-2, S-3 etc. prescribe the contents of various types of prospectuses and the manner in which information is to be disclosed to prospective investors. As in British Columbia the information required is expected to be presented in a narrative form. A junior natural resource company would likely use Form S-18 which specifically requires disclosure of information relating to each property of a mining company, its location, means of access, form of title, history of operations and work done. Disclosure of proven (measured) and probable (indicated) reserves is allowed.

Form S-18 requires sundry maps and other supplemental information to accompany the prospectus filing of a mining company issuer. The issuer must furnish to S.E.C. staff a list of every material engineering, geological or metallurgical report concerning a property including governmental reports, which are known and available to the issuer. S.E.C. staff then reviews the list, identifies those reports which it already has on file, and requests the remainder to be submitted.

While with the consent of a mining engineer his report may become an integral part of a prospectus filing, his involvement in the filing process is not quite so clearly set out as it is in the policy statements of the British Columbia jurisdiction.

APPENDIX "B"

References

The handiest comprehensive indexed reference source for U.S. Federal Securities regulation is probably the C.C.H. FEDERAL SECURITIES LAW REPORTER looseleaf service which contains laws, regulations, forms, rulings and decisions pertaining to Federal regulation of securities.

The C.C.H. CANADIAN SECURITIES LAW REPORTER provides a similarly comprehensive indexed reference service for both Federal and Provincial regulation of securities in Canada.

Addresses for C.C.H. publications:

U.S.A.

(a) Commerce Clearing House, Inc.
420 Lexington Avenue
New York, NY
10017

(b) Commerce Clearing House, Inc.
4025 W. Peterson Avenue
Chicago, Illinois
60646

(c) Commerce Clearing House, Inc.
425 - 13th Street, N.W.
Washington, D.C.
20004

Canada
C.C.H. Canadian Limited
6 Garamond Court
Don Mills, Ontario
M3C 1Z5

10.6

UNDERSTANDING THE LOAN APPROVAL PROCESS

Gary P. Thomason

Republic Bank of Dallas
Dallas, Texas

INTRODUCTION

One may have heard about how various projects were financed or certain companies were successful in obtaining a bank loan, but there are many more projects and companies who fail to get the money they need. At least a portion of these failures, and a large percentage of the frustrations connected with even the successful financings, are a result of not understanding the process of obtaining money. Recent years have seen a proliferation of debt financing sources available to finance minerals operations (see the next Chapter 11), but these represent only a minor portion of the minerals financing market. The fundamental source of borrowed money to many minerals related companies remains the banks. Therefore, understanding the "Loan Approval Process" can easily mean the difference between getting the money or not.

LOAN APPROVAL PROCESS

The initial stages of the loan approval process should begin long before the bankers, and the organizations they represent, enter upon the scene. The loan approval can be broken down into six stages:

1) Generation
2) Preparation
3) Presentation
4) Evaluation
5) Approval
6) Documentation

As you can see in Figure 1, the bank is only directly involved in the second half of the process. The perils in obtaining financing for the smaller minerals venture are primarily in the first three stages, not the last three. The minerals venture making application for a loan can, in almost every case, improve its chances of success by carefully planning and preparing itself and the information that will be required of it in the loan approval process.

Figure 1

AREAS OF ACTIVE PARTICIPATION
IN THE LOAN APPROVAL PROCESS

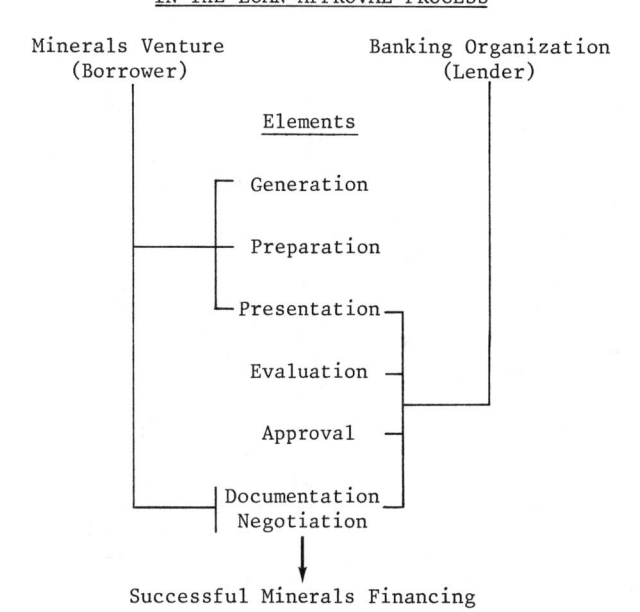

Minerals Venture (Borrower)   Banking Organization (Lender)

Elements
- Generation
- Preparation
- Presentation
- Evaluation
- Approval
- Documentation
Negotiation

↓

Successful Minerals Financing

Generation

This first step in the process is deciding on the need for a loan. (Earlier papers in this Book and in Chapter 11 provide some guidance.) It is surprising how many ventures confuse the need for money with the need for a loan.

Suffice to say that the equity investor risks both the timing of the yield on his capital and, ultimately, his capital. The lender, on the other hand, advances temporary funds which he expects to be repaid on a specific schedule with a fixed

---

Based on a paper presented at the SME-AIME Annual Meeting Atlanta, Georgia, March 6-10, 1983.

return. The limited nature of the lender's return is the driving force behind his need to quantify and limit the risks he is willing and able to assume.

Preparation

The materials needed to prepare for presentation to a bank are the same items used in deciding that a project was viable.

All financial institutions will evaluate a loan request by addressing, in one way or another, the age-old "Five Cs" of credit:

- Character of Management
- Financial Capacity of the Venture
- Collateral
- Financial Condition of the Venture
- Condition of the Industry

Knowledge of how the banker will approach the evaluation of a loan proposal can assist in shaping the presentation to address the key questions and concerns.

Character of Management: This area encompasses the organization of the company, its owners (stockholder, partner, parent company), officers and their backgrounds. The bank will have to be satisfied that management can successfully organize and run a mining operation and has in-house the necessary technical expertise, besides the legal and financial capability. A lawyer and an accountant, who know both mining and investment, are absolute necessities in properly organizing a minerals venture. Their expertise will be a key component in winning the confidence of investors and bankers alike. It is worth the effort to choose the right people so that you can benefit from their previous mining, underwriting and loan negotiating experience. One last item: no matter how glowingly management is presented, the banker will want to meet personally with key people to develop his or her own impressions and, possibly, do independent background checks.

Financial Capacity of the Venture: Financial Capacity is most easily defined as the certainty of cash flow. The cash flow proforma should be a year-by-year projection of the venture's income and expenses, including capital expenditures. In creating proforma statements, it is wise to include a most-likely scenario and a breakeven level at which the venture can meet its projected debt service, both principal and interest.

Collateral:

This will be the bulkiest section of the presentation. Of primary importance as collateral in a mineral venture are the mineral reserves, the land, and the ownership thereof. These two items are inevitably assigned to the bank by way of security. Therefore, the quality of the reserve calculations should be of the highest order. It is not an "absolute must" to have a reserve study done by a reputable outside firm, but it is strongly recommended. A good consultant's report on reserves, the mining and processing methodology, and the estimated capital and operating costs can be considered to be a key piece of collateral in itself. A feasibility report, although the expense to produce such a study can be burdensome to many smaller enterprises, can provide most of the essential data from which proformas, marketing plans, mine permitting and many other essential presentation features are derived.

Additional collateral support will ultimately come from the mine development - the bricks, mortar, equipment and plant for which the prospective loan proceeds are to be used.

Financial Condition of the Venture:

This section is probably the simplest to present as it is basically the balance sheet of the venture, whether project or company. It represents the existing creditworthiness and the credit record of the entity to be financed or which may be required to provide a guarantee to the new financing. Items such as explanatory footnotes or statements of personal or corporate financial condition are particularly important where guarantees are involved.

Industry Condition:

There seems to be a lack of concern in a large portion of loan applications to this critical factor. It is absolutely critical to obtaining a loan that one not only understand, but is able to communicate to the prospective lender the economics of the end product. The importance of a venture's competitive advantage within the industry cannot be overstated. Whether generated by efficiency of production, quality of reserves or product, proximity to market or that rare bird, a firm purchase contract, the project competitive advantage can make a project attractive even in the worst of times in any industry.

This expansion on the "Five Cs" of credit should provide a start on the journey to "thinking like a banker" and, hopefully, has helped provide a backdrop to the next step - Presentation.

Presentation

In this area, the individuality of banks and bankers become important. A good beginning point is to remember that to whom one makes the presentation may be as important as the presentation itself. For all but the smallest and simplest of transactions, it is recommended to consider several banks who have recognized expertise in the mining and minerals area. From this group one should begin to reduce the number of banks to a manageable subset by getting recommendations from your local banker, others in the industry, or any other source available to you.

It is important not to confine oneself to any one bank at the presentation stage because comparability and good old competition between the banks can be to your advantage. Most importantly, do not feel trapped into using the bank around the corner; the vast majority of banks are too small to have specialized minerals staffs and just because a bank is in your back yard does not mean it understands the problems.

Armed with the presentation, call the lending officers at the chosen few banks; give a brief outline of your proposal; and, if there is interest, send a more complete package. No effort should be spared to make the material interesting, clear, and comprehensive. For example, do not send hand-written notes, photocopies of colored maps where the colors are not shown, nor materials with typing errors. The more sophisticated the audience, the more professional the presentation should be.

How the information is set forth should be guided by what will happen when the proposal hits the bank. A lending officer, not necessarily the one originally called, will review the package. The best one can hope for in the two to four hours he will spend in the initial review of your package is that it stands out from the one or two others he or she may be reviewing that day. Many bankers may not fully understand all of the technicalities of the package, but they can recognize a professional and detailed job.

The next step is also extremely important. The banks which respond to the package positively will either visit you or have you visit them. It is an advantage to have the bankers come to you, both psychologically and to provide an early opportunity for them to physically review the property. At this interview, while the primary purpose is to answer questions and expand on your packaged material, one should spend time evaluating the character of the bank and the lending officer with whom you are visiting. The quality of the lending officer on your account will be the single most important aspect of getting a satisfactory loan approved within the organization. Therefore, your final choice should be based not only on the expertise and reputation of the respective institution, but the capability of the account officer as well.

In summary, having narrowed the prospective banks to the two or three where there is interest and the character of the bank and of the lending officer meets your criteria, the loan approval process leaves your direct control and becomes the responsibility of the lending officer.

Evaluation

A banker will review the "Five Cs" of credit although the order of priority may be different than listed above. The lending officer should call with questions. No presentation can be so comprehensive and so clear that no questions are asked. A commitment should be made to keep accurate information flowing freely.

The lending officer should begin interacting with ideas on loan structure, bank criteria, and general ideas and advice. Use the banker as a resource. See if the ideas make sense to the project and incorporate them if they do. If you have selected the bank properly, the lender may have seen several similar projects from which he can draw parallels and forsee potential problem areas.

The lending officer should begin to dictate what additional information he requires and in return be willing to commit to a time frame for giving you a final answer. At this point, the lending officer must complete his evaluation and "sally forth" within the bank as your advocate for loan approval.

Approval

Approval of a loan can take from a few minutes to several weeks and the process varies from organization to organization. The one constant is the lending officer; he is responsible for selling your venture to his management. Very few banks allow their lending officers to unilaterally make loans, both on the "two heads are better than one" theory, and as a control mechanism. The lending officer has taken your presentation, the information he has requested, his perceptions, and a potential loan structure and put together a recommendation memorandum which he feels will be attractive to his management. The approval of this package may take many different forms:

1) One senior bank official's signature.
2) Several senior official's signatures.
3) Presentation to a loan committee.
4) Even presentation to the bank's board of directors.

Two points need to be reinforced. The lending officer is now your advocate; he is selling to his management. Therefore, the capability of the officer and the quality of the information provided to him are crucial to success.

The loan approval does not occur in a vacuum, the approving officials are balancing various alternative uses of the institution's funds every time they look at a loan package. The fact that one bank declines to provide a loan does not mean it is not a good, bankable deal to other banking organizations. Several turndowns, on the other hand, from knowledgeable organizations, coupled with their reasoning, should cause serious reconsideration of either the way the loan request is being presented or of the validity of the loan itself. Assuming that the lending officer's own approval memorandum appears attractive to the approving authorities, they will make what changes in structure they feel are needed and set the pricing of the loan.

To avoid disappointment, a borrower should always be aware that until the loan documents are signed, loan structure is open to change. Broad parameters discussed with the officer will be negotiated and refined with the approving authority. When the bank's commitment letter arrives at your desk, it is the culmination of the bank's internal process and should closely represent what you and your banking officer have discussed.

Documentation

One is, optimistically, two thirds of the way there when the commitment letter arrives. Many people do not consider documentation negotiation as part of the approval process. But the commitment letter does not have to be accepted out of hand. A change in the terms of the commitment letter is by no means impossible - nothing is written in stone. One should understand, however, that different banking organizations give their officers varying degrees of latitude in working with the commitment terms. In some organizations documentation, loan covenants, and pricing to a degree, are left to the lending officer's discretion. Other organizations will require the lending officer to return to the approving authority for each minor change you might request. Therefore, it is not productive to ask for a lot of changes which do not really make an important difference. Even if you agree in principle, getting the final loan document in a form you can live with is not always easy.

Do ask for changes where a well-thought-out, logical reason can be presented for being unwilling to agree to certain terms. Lending officers have been known, from time to time, to place too many and too restrictive loan covenants on a borrower. In discussing a bank's commitment letter all things are open to questions, however, hopefully the relationship with the lending officer has grown so that both understand when there is no room left to maneuver.

An attorney should review the documents, both for legalities and to make recommendations on terms and conditions. However, it will ultimately be the principal's responsibility to negotiate the business aspects of the loan. Banking organizations do not act capriciously. They do, however, try to protect their interests with every possible device at their disposal. The banks and lending officers know they have certain powers over borrowers. Unlike the case in other contract negotiations, many borrowers feel they have few, if any, alternatives but to take what the bankers dish out. In my opinion, one needs to develop a negotiating strategy to obtain a loan agreement that meets the requirements without putting a stranglehold on the venture. The broad parameters of such a strategy should be:

1) Think like a banker and identify the bank's objectives;

2) Meet the objectives that are reasonable;

3) Provide alternatives, or sound reasoning for elimination of those objectives that are unreasonable or unnecessary.

Negotiation is the process of compromising on important points for the purpose of reaching a mutually satisfactory solution. Remember, the banker has a lot of time and effort invested at this point, he will not be obstinate nor unyielding to change without reason. Be careful. Probe responses and always read between the lines; knowledge of what areas the banker can relax will often give direction on which trade-offs can be made to get the type of arrangement desired. One last point: banks, like any other organization, react to competition, and in many cases, it may influence some banks to propose more lenient terms in order to obtain the business. I do not suggest that all ventures, particularly smaller or grass-roots ventures, attempt to use the "bid" approach to obtaining loan commitments, but larger companies looking for project financing may find this a useful technique. In any case, the more information on hand on the type of covenants various banks have used in similar situations, the better armed the negotiators are to recognize when the requirements of a bank are unduly and perhaps unnecessarily restrictive.

SUMMARY

To "think like a banker" means one can be prepared to get the loan in the most efficient manner. By being one step ahead of the process, by knowing what is needed, one can be one of those companies or projects that everyone is always hearing about.

The loan approval process is often needlessly mystified by both those who have the money and those who need it. There is always money for good projects, but many potentially good projects never get off the ground because they lack the preparation and presentation which allows the financial community to appropriately analyse their attractiveness.

REFERENCES

Arne, Kenneth G., "Basic Concepts of Mine Financing," Montana Mining Association Annual Meeting, May, 1981.

Bispham, Thomas P., "Coal Mining Project Financings: Lenders' and Borrowers' Considerations," Dames & Moore Engineering Bulletin/59, April, 1982.

Brooks, William A., Hursh, Dale S., "Acquiring Capital Funds for Coal Mining," Mining Engineering, September, 1982.

McGinty, Thomas E., "Project Organisation and Finance," The Cleveland-Cliffs Iron Company, Cleveland, Ohio, 1981.

USDA Forest Service, "Anatomy of a Mine from Prospect to Production," U.S. Department of Agriculture, Ogden, Utah, 1977.

Whitney, J.W. and Whitney, R.E., "Investment and Risk Analysis in the Minerals Industry," Whitney & Whitney, Inc., Reno, Nevada, 1981.

Young, William L. "Small Mine Financing in Canada: One Person's Experience," Mining Engineering, June, 1982.

10.7

BANKING NEEDS FOR PROJECT DEVELOPMENT FINANCING

C. R. Tinsley

European Banking Company Limited
London, England

ABSTRACT

A strictly limited number of specialist project finance banks exists, with the necessary specialist personnel able to address project cost-competitiveness, cost and commodity forecasting, joint venture negotiations, and completion and sales undertakings. Different techniques of internal bank appraisal are applied, with the usual equity investment project valuation criteria being seldom useful for loan assessment purposes.

Company attention paid to the preparation of a sensitivity structured bankable presentation can prove most effective for securing development project finance. Substantial scope is possible for savings of time and money.

INTRODUCTION

A minerals project today can, and very often must, be financed on the basis of 75-90 per cent debt. Hanna Mining (Beal, 1971) cited 55 per cent debt as high for the original Iron Ore Company of Canada development in 1951. In 1973, a major North American bank (English, 1973) preferred 50 per cent debt although "we have stretched bank loans to 60/75 per cent of costs". The trend toward higher debt levels has become even more noticeable as banks continue to refine their lending techniques such as the specialised "project-finance" varieties.

Banks are becoming involved at progressively earlier stages in the project evaluation period to ensure that the evaluation process will provide the raw material which will satisfy the project lender yet retain the flexibility so necessary for minerals development financing.

DIFFERENT APPROVAL PROCESSES

One Australian development received visits from more than 200 banks. The company's treasury staff became quite adept at determining whether the visiting bank was a tourist, a scout, an officer, or a general. Not only does the level and authority of the individual bank's representative vary considerably but the banks' approval processes can be quite different. In order to assess each bank's needs, it is advisable early on to determine the approval process and final authority for loan approval.

The following list gives the main approval variations.

Progressive Authority

The bank will have established progressive loan-size commitments that can be undertaken by individuals or committee. A large bank can usually commit up to $10 million at a committee level, but experience with project commitments suggests that most require board approval.

Loan Committee

No matter what the size of the financing is, a presentation has to be made to a credit committee which usually meets once a week. Committee members, used to balance sheet analysis, either rubber stamp a project-finance presentation due to its complexity or send it back to the drawing board.

Matrix

The bank may have established matrix management whereby the financing must be simultaneously approved, for example, by the mining department, the multinational accounts department (depending on the company involved), and the international banking department (depending on the country involved). In these circumstances, a project loan approval may have more to do with the bank's relationship with the company than with the merits of the project.

Based on a presentation to the Aus.I.M.M. Sydney Branch, Project Development Symposium, November 1983

Sequential

The project must first be approved by the mining department. Then it is presented elsewhere (for example to a regional headquarters in Manila or Hong Kong for an offshore bank) before it even gets to head office.

In any of these variations, the potential for blockage en route is very high unless the various parts of the bank have been well prepared and the loan originator is given the time and opportunity to develop a first-class presentation. In this respect, a project loan application can sink or swim on the professionalism and credibility within the bank of the individual concerned (Thomason, 1983). If the loan originator does not properly understand the minerals industry, an endeavor should be made to have the loan originated further up the approval process or with the active involvement of the bank's mining staff. A good way to accomplish this is to arrange for a site visit for the bank's mining engineer prior to the final loan application being presented.

No self-respecting bank in this area operates without its own engineers although some European banks are fond of using the engineering talents either of outside consultants or the mining organisations in which they or their governments have investments (Sarmet, 1980). Most banks use their engineers purely as consultants and, however knowledgeable and personable the engineer may be, one must ascertain whether the engineer has any authority in the credit structuring or approval process other than a technical veto. The engineer's input is, of course, a vital part of the approval process and only a well-organised feasibility study permits evaluation in the relatively short timeframe the engineer may have for any given project evaluation.

There should be no mystery surrounding the bank's loan evaluation process. The banks look at the same information as the company, often in the same order.

1. reserves
2. mine design
3. process testwork
4. project sequencing/optimisation
5. capital costs
6. operating costs
7. cashflow projections

The sole numerical difference is the overlay of various loan amounts and structures on the cashflow projections usually directed at measuring the relationship of projected surplus cashflows to the debt service requirements in each year. (See Schreiber's paper in this Chapter). Banks seldom use net present value ("NPV"), rate of return, or payback criteria in minerals finance evaluations. Tax optimisation cases are run once the percentage debt has been established.

Sensitivity analysis is an area for wide divergence of opinion as each bank may make a different judgement on the importance of a particular variable. A knowledge of these criteria may only be derived from experience as most banks are reluctant to open up their many rules of thumb or internal threshold levels for fear of introducing further matters of negotiation into an already complex area of risk judgement and trade-off.

The provision of cashflow projections in the format actually used by the bank not only saves times for the banker/engineer but can allow the devining of some of these cashflow criteria. An offer to run sensitivity analyses (assuming a good loan routine is built into the computer model) may also reveal the sensitive areas being examined by the bank, an aspect which may be useful in subsequent negotiations (Tinsley, 1982a). Since most project finance applications have only one opportunity for review per bank, these time-savings for the reviewer can also reflect well on the company's financial professionalism.

THE PROJECT EVALUATION PROCESS

Exploration is usually funded by tax-shelter money. Once a discovery is made, the situation usually requires venture/corporate funds (Arne, 1981) through to the feasibility study and the investment decision itself. Banks rarely finance feasibility studies or exploration expenses although many governments, through their export-credit or aid agencies, do finance feasibility studies in other countries (generally-lesser developed nations).

However, one should inform the banks directly or through the mining press whenever a feasibility study is commissioned so that they can begin to anticipate the development and discuss it when visiting with the executive or treasury staffs. Experienced banks recognise that the lead time is long for a minerals development and can thus gather information relevant to the market, regional, or competitive positions of the expected development.

When a banker visits a company, the minutes of the meeting are usually written up for internal communication purposes and some "priming" about a new project is certain to be noted and will maintain the bank's attention to both the project and the company. Some companies use the project as "bait" or as a "reward" to banks who do them favors in the corporate debt or banking service arena. This practice can lead to difficulties since the project loan can become too tightly identified with the corporate debt exposure of the bank in question which can effectively defeat the stand-alone objective of project finance. There should be "horses for courses" (Wightman, 1983), with the project-finance banks best able to provide the most flexible finance structure. The community of banks with sound project finance capability is actually very small - at most 30 banks for a fairly simple financing and perhaps 15 banks for a complex package. The number has recently declined as banks reassess their project exposure and their problems with massive sovereign debt reschedulings worldwide.

Some companies decide to have a bank as an advisor during feasibility study stage to ensure early input from the majority source of funds. The feasibility study cost and duration can often be substantially reduced, for example by tailoring the scope of studies to include coverage of areas of concern for the banks. A bank's assistance at this early stage can also focus attention on the competitive position of the project in world terms, a discipline often difficult for a project finance team of purely technical staff wrestling with sieve sizes and pump diameters. It is also not readily admitted that the experienced banker may be able to make contributions in the following areas.

The Commodity Outlook

This can be important for the metals, some of which cannot be financed today e.g. nickel and molybdenum. Banks may also have long-standing experience with competitive analysis to rank projects.

Joint Venture Agreement

No amount of drafting expertise in the loan documentation can overcome deficiencies in the joint venture structure or state agreements. For example, cross charging or cross-default can prohibit certain advantageous lending mechanisms.

Completion

The obligations of the company to build the development on time and at budget may influence plant design toward reliability, proven technology, modularisation, and even standby circuit/stopes/benches. The choice of the engineer-design-construct ("EC") company can also influence the bank's views on the need for recourse to the corporate balance sheet rather than the project on its own.

Sales Contracts

In some coal and industrial minerals sales contracts, the minimum sales contract criteria was specified by the bank which, in some instances, required a new sales practice for that industry (Tinsley, 1982b).

The Bankable Documents

The bank can advise which reports are required and in what manner to present and summarise them. No matter how these are presented, the loan initiators within the bank will have to do a new writeup to summarise the project and the proposed financing/risk absorption, partly as an exercise to display internally that the project has been thoroughly reviewed.

The fabled Bankable Document, a single volume, has faded into the annals of history as the banks refine their loan structures and must therefore closely assess not only the project analysis itself but the process of analysis, the experience of the people and the strength of the project management team to be involved. The banks should be presented with supporting documentation. Emerson (1981) gives a good guide to the outside, independent consulting reports commonly expected. (See this Chapter).

Experience has shown that up to 85 per cent of the loan application material initially submitted for bank financing do not fully satisfy the banks' review process. This highlights the fact that many companies and EDC firms do not fully understand what the banks require. A new cover sheet on the feasibility report is seldom adequate.

Large Australian mining companies such as BHP (Flew, 1983), CRA (Wightman, 1983), CSR (Willis, 1983), and Western Mining Corporation (Crook, 1982) all state that they approach the banks either late in the project evaluation cycle or indeed after the feasibility study has been agreed by the board as a basis for investment.

For specific projects or joint-venture situations, it has become the fashion in Australia to appoint a financial adviser to fine tune the financing amount, sourcing, and costs, and even to bargain with the banks. This often reduced the financing process to a call for tenders and selecting either a

1. lead managing bank (partially underwritten finance)

2. lead management group (usually fully underwritten finance)

3. club of banks (where the best parts of the various tenders are massaged into a package accepted by a fairly large number of banks)

The merits of each technique needs to be carefully considered. For example, the club approach may require iterative internal presentations within a bank often trying to overcome risk elements that were unacceptable on the previous run. The practitioners of these approaches not surprisingly caution against the timetable delays caused by the extended finance negotiations. A 1/8 per cent shaving of margin can never be compensated by the, say 12 per cent, increase in capital costs irreversibly caused by "delay escalation" caused by the negotiations to squeeze out the 1/8 per cent.

Mental arithmetic points to a $1-million-a-month increase in a $100-million project's costs compared with a $500,000 saving of margin over the total finance life. Some companies like Placer Development in Canada and Exxon for Colombian coal have apparently decided not to arrange the finance prior to completion, thereby avoiding any delay from finance negotiations. Few mining companies have the strength to adopt this policy.

THE BANKABLE PRESENTATION

The composition and means of presentations are well covered elsewhere in this Chapter.

Although checklists are a useful drill for airline pilots and feasibility study engineers, the presentation to the banks should not follow that format. A better approach is to ensure that the risk areas assessed by the banks are fully and succinctly addressed. The bank can devote a maximum of 10 days to two weeks for a complex project credit review which will include the time allotted for any site visit. The early dialogue recommended above and cashflow formatting can double the effective analysis period. Simply providing three copies of all material to each bank can encourage the bank to divide the analytical workload or to conduct concurrent reviews e.g. in a sequential approval process. Extra copies of background studies expeditiously despatched upon request reflects well on the perception of the sponsor's planning and preparedness.

It is helpful if the reviewer is given a list of contact names should questions arise in particular areas. Banks appreciate openness and any attempt to hide information will be spotted immediately by the experienced reviewer with an immediate loss of credibility in the data being provided.

Also the qualifications and track record of consultants and key management staff can inspire confidence in an organisation's analytical competence and management experience to avoid/handle the difficulties that inevitably occur in the minerals business.

An understanding of the people and their professional and managerial strengths and weaknesses is the most difficult part of the assessment of a project development's viability. Professionals would like to rely on the reduction of the project to numbers as proof positive of competitive strength and economic viability. It is not numbers that develop a project. It is people.

An early understanding of what information the banks' personnel require has the potential for substantial savings in time and money both during and after the feasibility study stage. Early dialogue can improve the acceptance of a project as a "bankable" proposition which can be readily approved.

REFERENCES

Arne, K.G.,"Basic Concepts of Mine Financing", Presented at the Montana Mining Association Annual Meeting, May 6-8, 1981

Beal, R.E., "Multi-Layer Financing Packages for Major Mining Projects," Presented at the 1971 Mining Show of the American Mining Congress, Las Vegas, Nevada, October 11-14, 1971

Crook, S., "Corporate Financing: Lenders and Borrowers", The Australian Accountant, November 1982, pp. 618-620, 622

Emerson, M.E., "Role of the Engineer in Project Financing,"Presented at Short Course on Project Financing in the Development of Mineral Resrouces, S.M.E. - A.I.M.E. Annual Meeting, Chicago, Illinois, February 21-22, 1981

English, L.A., "The Changing Nature of Mine Finance in Canada,"Presented at the A.I.M.E. Annual Meeting, Chicago, February 26, 1973

Flew, R.J., "Analysing Project Viability: A Corporate Approach," Presented at a Conference on Project Finance in Australia, Sydney N.S.W. March 23-25, 1983

Sarmet, M., 1980. International Project Financing, Banque, No. 392, February 1980

Thomason, G.P., "Understanding the Loan Approval Process," S.M.E. - A.I.M.E. Preprint No. 83-59, Annual Meeting, Atlanta, Georgia, March 1983

Tinsley, C.R., 1982a. "Project Financing - What the Banks Will Want, "Presented to a Metals Economics Group Conference, Mine Development in the 80's : Changing Considerations in the Feasibility Study, Denver, Colorado, February 4-5, 1982

Tinsley, C.R. 1982b. "Industrial Minerals Sales Contracts from a Banker's Perspective," Presented at the Fifth International Industrial Minerals Congress, Madrid, Spain, April 28, 1982

Wightman, K., "A Borrower's Approach to Project Finance," Presented at a Conference on Project Finance in Australia, Sydney, NSW, March 23-25 1983

Willis, J., "Structuring the Corporate Financial Package," Presented at a Conference on Project Finance in Australia, Sydney, NSW, March 23-25, 1983

Chapter 11

Sources of Finance

CHAPTER 11: SOURCES OF FINANCE

PREFACE

Chapter 11 gives details on the forms of equity finance and presents guidelines on the criteria for bank and non-bank institutional financing. A healthy minerals company or a strong new project (either exploration or development) can use a broad range of financing tools. Rather than dashing to the local stock exchange or friendly banker, armed with the best-looking presentation (after studying Chapter 10), each individual or enterprise needs to pause and identify its financing objectives and the ultimate goals of financial return or corporate control.

This chapter was written by practitioners in each funding sector who have also included guidelines, approach techniques, and information requirements to supplement Chapter 10. Despite the diversity, common elements of risk vs. reward pervade the thinking about each funding source. Nevertheless, a single source of funding is not usually sufficient for all stages of a project or for all the requirements of an operating company. The 20 papers in the chapter strongly reflect the diversity of minerals industry finance.

Tomek Ulatowski's paper, "Sources of Funding for Mineral Projects," introduces a wide spectrum of financing sources for international mining projects. Ken Arne, in his paper "Basic Sources of Mine Financing," explains the relationship of the funding source to the risk phase of the project's development. Arne also describes funding sources to be found in the US. Herb Drechsler completes the overview in his paper "Mining Finance" by reviewing different financing types in reference to ASARCO's financial statements. The Case Studies at the end of Drechsler's paper show how financial commitment letters are laid out.

Chapter 9 outlined the risks the "pure" project financier must face. Richard Tinsely has discussed this aspect in his paper "Project Financing Supports and Structuring," highlighting the emerging trend toward financial packaging which includes capital market sources of funding such as commercial paper, floating-rate notes and leasing. Gary Castle's classic "Project Financing - Guidelines for the Commercial Banker" remains as true now as when first published in 1975. His analysis of project financing problems is instructive and sobering. Kyle Wightman's paper "Project Finance - Does it exist in the Mining Industry: CRA's Experience" arises from hands-on experience within a large mining company as it financed a series of projects in its $1 billion program to raise capital.

Four papers follow, presenting views on raising equity for the minerals industry. Ed Cruft's paper "How to Finance Mineral Prospects" describes a strategy for a junior mining company raising funds for exploration, acquisitions and expansions from carefully selected sources of funding tailored to each activity. Mike Chender's company publishes a newsletter on mine development from which an article "Financing the Juniors" discusses financing obtained from trading companies. John Brock speaks of the difficulties encountered when an exploration company promotes prospects, particularly to producing minerals companies. His paper "Operating a Promotional Exploration Company," advises that geological knowledge is necessary during the presentation and negotiation of prospects. Fred Brooks' "Financing the Development of Small Mining Projects - an Operator's Viewpoint" sets down considerations for negotiation when the small company wishes to retain operatorship. Each author recommends that multiple talents and flexibility are assets in negotiating with financiers.

Jim Boettcher's paper "Venture Capital Financing" presents an existing view from another leading edge in prospect development. Brian Gorval describes the means of "Public Financing as a Source of Funding for the Canadian Mineral Industry" from the perspective of the flourishing Vancouver Stock Exchange which has experienced a boom in minerals company listings and in raising capital for mineral ventures.

Chris Baiz focuses attention on "Financing the Industrial Minerals Industry" whose marketing aspects are difficult to analyse. Some of the problems encountered in this part of the minerals industry are also addressed in Richard Tinsley's paper "Special Considerations in Project Finance for Industrial Minerals."

As mentioned above, financial packaging is used even for small project development financings. Victoria Yablonsky, Bob Gillham and Gary Castle have provided a very up-to-the-minute look at how the US commerical paper markets were used within project finance "risk" structures to achieve lower funding costs for two major Australian coal-mine financings in 1983 and 1984. This is evidence that a wider range of bank, currency, and capital markets are becoming available for medium and large-sized projects.

Export-credit programs are another layer of funding officially supported by various governments to support their own national exports. Rides' paper "Worldwide Export Credit Programs" and the supplement on the Export Import Bank of the US provide a perspective on this avenue of funding which, although usually low cost, is hard to arrange expeditiously. Hartsell Cash describes supra-national financing sources in "World Bank Financing," a traditional source of debt which has been vital to many developments in the less credit-worthy countries. The supplement on the International Finance Corporation, an agency of the World Bank, discusses the equity activities of this international agency which has a close involvement with many minerals developments worldwide. Co-financing with any of these agencies is becoming important to countries rescheduling national or private debt to mitigate political risk.

Leasing has been the mainstay of the coal industry. Rod Ravenscroft's paper looks at "Leveraged Leasing for the Minerals Industry" in major minerals producing countries and provides guidelines on the evaluation of leveraged leasing as a funding source.

The chapter closes with a paper on utilizing metal trading companies and commodities exchanges as a means of funding production. This type of financing has increased in popularity as the industry turns its attention to precious metals worldwide and Reg Rowe and Simon Handelsman's paper, "Using Gold as a Financing Mechanism," illustrates an approach for offering low-cost funds to a gold producer.

With some 30-40 different sources from which to seek funding, how can the individual or company make the right selection? The providers of funds are no less human or warm than the people in the minerals industry and will be able to provide valuable guidance if approached correctly and in a timely manner. Knowledge of the different sources of funding will either eliminate (and thus save the time of preparation and approach) or open new vistas for funding the minerals industry from the gleam in the geologist's eye to the time when the energy or minerals are ultimately produced.

C. Richard Tinsley
Chapter Editor
European Banking Company Limited

11.1

SOURCES OF FUNDING FOR MINERAL PROJECTS

TOMEK ULATOWSKI

CROCKER NATIONAL BANK

This presentation discusses the necessary ingredients for the creation of an acceptable credit structure, i.e., the structure that enables the borrower to attract the funding from different groups of lenders. As the amount of debt in the capitalization of new projects increases, it becomes quite critical that the mining venture sponsors obtain funding in a manner which maximizes the project economics.

Attention normally is focused on the inherent operating cost characteristics of a venture such as the cost of mining, milling, transportation, smelting, and refining. The projected costs are derived from lengthy feasibility studies often taking millions of dollars and years to prepare. For a variety of reasons as the amount of equity directly provided by the sponsors declines, the balance must be made up from external credit sources. As a result, once the project becomes operational, debt service requirements assume crucial importance. For most new ventures, in the early years of production, the total amount of periodic debt amortization charges together with the projected interest expenses equals or exceeds the total operating cost. Utmost care must be taken, therefore, to reducing these costs which will then increase the chances for the implementation of the project. Project sponsors must carefully evaluate different funding sources in terms of their potential economic impact and, concurrently, evaluate alternative credit structures in order to be able to attract the selected funding vehicles.

In this presentation I will discuss credit sources in the order of their relative appeal; i.e., from the standpoint of maximizing rates of return to the project owners. The tenor of the credit, that is the period from signing a loan agreement to the time of last repayment under the loan agreement, is undoubtedly the most important variable in evaluating the attractiveness of debt sources. Therefore, I will begin by discussing debt sources with the longest maturities and will end with a description of credits with relatively shorter tenors. The availability of different funding possibilities will naturally depend on the geographic location of the proposed venture, the creditworthiness of the borrower and the sponsors, the project economics, and finally the actual liquidity conditions in the capital markets at the time that the funds are committed.

U.S. LONG-TERM DEBT MARKET - PUBLIC OFFERINGS AND PRIVATE PLACEMENTS

The long-lived nature of most mining projects dictates that the majority of the mine financing be raised from long-term sources. In today's world capital markets U.S. debt issues probably carry the longest maturities available. The maturities can range from 5 to 40 years. Various types of long-term corporate debt instruments can be differentiated from one another by their terms, seniority as to claims against the assets of the issuer, legal form, and whether issued privately or through a public offering. The terms of these debt issues vary widely. For example, the in-terest rate which is typically fixed at the time of issuance, is determined by the credit quality of the issuer, the maturity, the average life, the seniority of the debt, the quality of control covenants or type of assets pledged as security, and other exogenous factors such as: economic policy, general level of interest rates and cur-rent level of financing activity.

The entire principal amount of the debt may be due and payable at maturity (known as a "bullet" maturity) or, more typically, through periodic repayments in equal or staggered amounts. Almost all such debt issues allow for prepayment at the option of the issuer, but require varying amounts of repayment penalties. Disbursements are made over a relatively short period after signing of the debt documentation which may not coincide with the actual funding needs of the project.

Financial covenants can be quite stringent, covering areas such as the issuer's financial leverage, working capital, payment of dividends and other distributions, merger and acquisition activities, and limitations on the disposition of property.

Long-term debt issues may also be distinguished by the seniority of their claims against the assets of the issuer in the event of liquidation. Secured bonds, for example, have a claim against specific real and personal property. If these claims are not satisfied from the proceeds of the sale of such property, they will rank on equal footing with other senior unsecured debt for the balance of the claim. Secured bonds often have fewer, if any, covenants compared to unsecured issues.

Next in security are senior unsecured debt instruments. These instruments have a senior first claim against the unencumbered assets of the issuer. Often referred to as debentures, these issues do not have the protection afforded by a lien on specific property. Thus, the lenders often require a more comprehensive set of financial covenants.

Subordinated issues are junior to all senior creditors. Generally this means that before any payments against principal are made to subordinated creditors in a liquidation, the senior creditors must be satisfied in full. Since the subordinated issues are more risky than the senior debt, the overall yield to the investor is usually higher. The increased yield may take the form of a higher interest rate or an "equity kicker" associated with the issue in the form of warrants or through a conversion feature into common stock.

Public Offerings - Taxable Securities

Perhaps the most familiar vehicle is the underwritten offer of taxable securities. These offerings are registered with the Securities and Exchange Commission (SEC). Some of the advantages of this financing vehicle are: fixed interest rates at relatively low levels, long maturities (maturities out to 30 years and beyond are common), and relatively few restrictive covenants. The disadvantages include: relative inflexibility in drawdowns, difficulties in explaining a project financing within the confines of an SEC approved prospectus, significant prepayment penalties, and the high cost of doing an underwritten offering including underwriters' discount, registration fees, printing and legal expenses.

The securities issued in these marketplaces include senior debt, subordinated debt, preferred stock and common stock. The senior debt may be either secured or unsecured. The subordinated debt and preferred stock may or may not be convertible, and any of the securities may be issued with warrants attached. In 1983 over $54.6 billion was raised in the taxable securities markets in 643 issues, compared to $57.2 billion raised in 642 issues in 1982.

Private Placements - Taxable Securities

In 1983, twice as many financings were done privately as were done publicly although private placements of securities generally tend to be substantially smaller than securities sold in the public markets. The advantages of a private placement are: confidentialiity, ease of describing a complicated transaction to a limited number of sophisticated institutional investors, flexibility in timing of drawdowns, and relatively lower issuing expenses. The disadvantages include: generally slightly higher interest rates, more restricted covenants, significant prepayment penalties, and slightly shorter final maturities and average lives.

Tax Exempt Securities

In 1983 issuers raised $123.0 billion through tax exempt public financings of which $27.0 billion was in the form of revenue bonds. In 1982 revenue bonds amounted to $19.0 billion of a total of $87.2 billion issued in the tax-exempt securities market. These securities are backed principally by the revenue generated by the project being funded. Among the assets which can be financed by revenue bonds are pollution control equipment and dock and wharf facilities.

General Comment

For practical purposes, the U.S. long-term capital market is essentially limited to domestic high quality credit issuers. For a new mining venture to obtain funding from the above sources it would require very strong support from creditworthy sponsors. Although, in theory, it may be possible for a foreign borrower to obtain funds in the U.S. long-term capital market, limitations such as the 1% foreign basket criteria imposed on insurance companies (which are often principal investors in such issues) may make it exceedingly difficult.

OFF-SHORE LONG-TERM FINANCING

This category covers conventional offerings of securities done outside of the United States, principally in Europe.

Eurobond Market

The Eurobond market is an international capital market which companies and countries use to raise capital denominated in a variety of currencies. U.S. dollar issues represent, however, the majority of the capital raised (see Table No.1). Since this market began functioning in 1963, it has been used to raise over $200 billion. In 1982 and 1983 approximately $43.6 billion and $44.2 billion was raised in the Eurobond market, respectively.

SOURCES OF FINANCE

Table No. 1
THE EUROBOND MARKET BREAKDOWN IN 1983*

Currencies	No. of Issues	Amount (US$ million)	1983%	vs. 1982%
US Dollar	292	$35,152.58	79.50	82.10
Deutsche Mark	79	4,129.05	9.33	9.90
ECU	44	1,713.53	3.88	1.60
Sterling	23	1,564.70	3.54	1.80
Canadian $	26	1,074.84	2.43	3.00
Yen	4	295.85	0.67	1.10
Australian $	8	215.41	0.49	0
Norwegian Kr.	2	40.96	0.09	0.10
Kuwaiti Dinar	1	17.08	0.04	0.40
New Zealand $	2	15.15	0.03	0
	481	$44,219.15	100.00	100.00

Type of Issues

Straight Issues	346	26,499.51	59.93	67.40
Floating Rate Notes	71	12,872.54	29.11	23.20
Convertible/ Shares	32	1,673.33	3.78	2.90
Warrant/Shares	30	3,019.77	6.83	1.30
Zero Coupons	2	154.00	0.35	4.70
Exchange Options	-	-	-	0.50
	481	$44,219.15	100.00	100.00

Regional Breakdown

Europe	240	24,352.35	55.07	35.90
USA	73	6,355.71	14.37	32.10
Japan	62	4,301.47	9.73	5.00
Canada	45	3,914.95	8.85	16.00
Australia/ New Zealand	21	1,455.44	3.30	2.90
Other countries	40	3,839.23	8.68	8.10
	481	$44,219.15	100.00	100.00

* Source: Agefi International Financing Review

The Eurobond market generally provides unsecured fixed or floating rate financings with maturities ranging between 5 and 15 years. The amounts raised for individual issues are generally smaller than in the United States and vary between $10 to $200 million.

The Eurobond market is currently unregulated. Issues are never made directly to the public. This market is restricted to companies which are well known and very creditworthy. In this respect, the Eurobond market is more demanding than that of the United States bond market. Interest rates have been somewhat higher than those required for a U.S. public offering. Issuing expenses, particularly the underwriting discounts, are typically larger than for comparable U.S. issues. Tax considerations, principally withholding taxes, pose severe problems for U.S. corporations which do not have offshore operations. However, it is relatively easy to set up a nassau financing operation if the cost savings justify the expense.

Euro Currency Private Placements

In the Eurobond market, private placements describe practically any type of financing which is not an underwritten public offering. Institutional investors have played an increasing role in the last few years in the Eurobond segment of the private placement market. They include offshore institutions, or institutions located in countries with large resources (such as OPEC members) whose local markets do not provide adequate investment opportunities. For these investors the dollar is the principal currency of denomination, but because OPEC countries are more concerned with currency fluctuations, private placement commitments from these sources will tend to have shorter maturities.

Individuals continue to play a significant role in the private placement of Eurobond financings; however, this source of funds has been more volatile, principally due to currency expectations. Placements are often bought, in the first place, only because of the lack of a suitable alternative, and investors want the flexibility to switch out of them with minimum loss in light of subsequent currency developments. This explains the normally shorter maturities, the small denominations, and (because of the costs associated with bond distribution), a commission structure close to that of public offerings.

Private placements in the Eurobond market have most of the characteristics of a broadly distributed public offering, except for the absence of a large underwriting group, and, in many cases, a listing.

Other Markets

Recently, alternatives to the Eurobond market have begun to develop in Japan and in few of the more developed Asian countries. Access to these markets is tightly controlled and restricted to extremely recognizable names. In addition to the qualities which make a borrower acceptable, the same tax problems for United States issuers are present in these non-European markets as in the Eurobond market.

INTERNATIONAL LENDING AGENCIES

Since World War II, international lending agencies have played a major role in channeling considerable amounts of loan capital to developing countries. Most of the funds have been used to augment the countries' industrial infrastructure and to support major investments in agriculture and transportation. The most important lending agency in this context is the International Bank for Reconstruction and Development Bank (World Bank or IBRD), Inter-American Development Bank (IDB) and its equivalents in Asia and Africa which have also played significant roles in providing long-term financing to many projects. International lending agencies generally will provide financing for ventures with a strong local government involvement and in most instances with a government guarantee. Since mining projects are thought to be relatively attractive in terms of their overall economic appeal, the involvement of international agencies in most cases would be limited, since these ventures can be financed on a more commercial basis by international and domestic banking institutions. Still, the support of the international lending agencies may be sought in circumstances where major infrastructure is required which could not be directly supported by a mining venture.

Loans by international lending institutions frequently are extended with tenors of up to 30 years, may be at fixed interest rates, and are available in a variety of internationally accepted currencies. The international agencies generally make their loans for certain specific aspects of an overall project following a lengthy evaluation and analysis. In the 1970's, two new lending concepts emerged, referred to as "co-financing" and "complementary financing", whereby the international lending agency financing is tied to commercial bank financing. Basically, co-financing and complementary financing combine two sources of capital, provided by commercial bank syndicates and international lending agencies, into one package. These vehicles offer advantages for both the borrower and the lender. In co-financing done with the World Bank, the commercial bank loans are linked through optional cross-default clauses with the World Bank acting as a channel for disbursements and repayments if the commercial bank so desires. In complementary financings, done recently on a number of occasions with the IDB, a commercial bank syndicate participates in the IDB's own loan and receives all the guaranties and support that the IDB obtains. In World Bank co-financing, commercial banks and the World Bank each prepare its own loan documentation with the borrower. In complementary financing, as structured by the IDB, there is only one set of loan documents signed by the borrower, that of the IDB.

An international organization which has been quite important in developing mining projects in less developing countries is the International Finance Corporation (IFC). The IFC was established in 1956 as an affiliate of the World Bank. The World Bank, the IDB, and other official international lending agencies work more directly with the governments. The IFC's role, on the other hand, is to support private enterprises in the developing member countries. IFC investments, either in the form of equity or debt, do not require government guaranties but must hold out the prospect of earning a profit, and must benefit the economy of the country in which the project is situated. The IFC's role is to mobilize private capital in the country concerned as well as loan funds from international capital markets. Its involvement in terms of ownership is normally less than 25% and only rarely is the IFC the single largest shareholder. The ultimate goal of the IFC is to sell its investment once the project is successful.

The high degree of professionalism, the agency's reputation, and its international character often make it quite desirable to have the IFC become involved with a mining project to benefit from the advice it gives and capital which it may provide.

SUPPLIER AND/OR BUYER FOREIGN GOVERNMENT SUPPORTED CREDITS

As a means of supporting the export potential of a country's industrial and manufacturing sectors, most developed countries, as well as an increasing number of developing nations, provide local exporters with financial assistance by making available credits and guaranties to foreign buyers on concessionary terms. Export equipment financing, as it is often known, is a potentially important source of funding for any mining project since the financing terms are often considerably superior to the alternatives available in the marketplace.

The size of this credit market is very considerable; to illustrate, long term (five years or longer) international export credit insurance and guaranty authorizations extended by France, Germany, Japan, the United Kingdom and the United States amounted over US$16.7 billion in 1983. Credits to the mining industry accounted for roughly 12% or US$2 billion of this total. Official financial support for U.S. exports through the activities of the Export-Import Bank of the United States (Eximbank) in 1983 amounted to US$8.5 billion in export credit insurance and guaranty authorizations, and US$0.8 billion in medium and long term credits (down from a high-point of US$5.4 billion in 1981). The performance and lending practices of the OECD countries in supporting exports are governed in principle by the International Arrangement on Officially Supported Export Credits. This arrangement establishes the guidelines for the extension of credit, but is not legally enforceable. Its successful operation depends upon the "good faith" and reciprocity exercised by the 20 participating nations. This cooperation, however, has been breaking down recently and borrowers have been able to extract more attractive terms than officially agreed upon.

As a general rule, a cash down payment of 15% is required. Presently, the repayment terms may not exceed 8-1/2 years for relatively wealthy and intermediate countries, and 10 years for relatively poor countries. Minimum interest rates vary depending on the repayment period and the development category of the country currently, ranging from 10% to 12% for over 5-year credits. In the Arrangement there is no agreement as to the currencies in which the loans can be made. This leaves considerable elbow room for extension of credit, de facto, on even more concessionary terms.

The mechanics of export financing are straightforward. Basically there are two types of credit takers: either the buyer of the product is the direct borrower and therefore the term "buyer credit", or the supplier becomes the obligor, giving rise to the term "supplier credit". The majority of credits, however, are arranged with the foreign buyer being the borrower. Typically, all equipment and services provided by a country to a project qualify for export financing. Also, some of the local costs can be financed by the export credit institution. When a country does not produce a certain component necessary to complement the exported equipment, the export credit agency could also finance the cost of the foreign acquisition of that component. The percentage of foreign purchases which can be financed by the export credit agency varies from country to country. In one famous case the Export Credit Guaranty Department of the United Kingdom offered to finance Pan American's acquisition of Lockheed L-1011 airframes, even though the airframes were manufactured in the United States. Britain did this as she was exporting Rolls Royce jet engines, notwithstanding the fact that the engines represented only 20% of the purchase price of each aircraft with the total contract value amounting to $500 million.

Very often the down payment required in export credits is financed by commercial banks in conjunction with the export agency. In the case of the U.S. Eximbank, it is possible to have just one loan agreement, although the terms between different credit tranches may differ.

In addition to the direct extension of export credits, credit agencies provide considerable insurance protection covering political as well as commercial risks. The beneficiaries of the insurance policies are typically the commercial banks or other senior lenders. In the United States, the Eximbank and the Foreign Credit Insurance Association (FCIA) can grant insurance coverage of up to 90% of the credit in return for an individually negotiated premium fee ranging from 1% to 6% per annum.

Export credit agencies have played -- and will continue to play -- a major role in lending to mining ventures around the world. For the sponsors of the project, it is important to recognize that these credits are attractive sources of funds. The terms of such loans vary from country to country. In making equipment purchase decisions, the project sponsors should carefully evaluate both the normal technological and cost criteria and the financing possibilities. When equipment of similar quality is produced by a number of countries or when engineering construction services can be obtained from different geographically located sources, it is possible that the sponsors will be able to extract better purchase terms simply by knowing about the opportunities elsewhere. It has been my experience that a comprehensive and knowledgeable approach to acquisition questions, which also takes into account the financing aspects, indeed can be very beneficial to the project sponsors.

LEASE FINANCING

In general, a lease transaction occurs when the possession, use or control of equipment or facility has been transferred from one party (the lessor) to another party (the lessee) while the lawful title of that equipment or facility remains with the lessor.

Direct Leases

The lessor owns the equipment during and at the end of a lease term. The lessor takes the depreciation deduction and the lessee may deduct the full lease payments for tax purposes. The investment tax credit may be taken by the leasing company or, by agreement, by the lessee. Most of the tax benefits claimed by leasing companies are passed through to the lessee in the form of lower rental rates.

Capital Lease ("on balance sheet")

A capital lease is generally treated as if equipment or facility were owned by the lessee. A lease is considered a capital lease by the lessee if it meets any of the following criteria:

1) The lease transfers ownership of the property to the lessee by the end of the lease term.

2) The lease contains an option to purchase the property at below market price.

3) The lease term is equal to 75% or more of the economic life of the new property.

4) The present value of the rentals and other minimal payments is equal to 90% or more of the fair market value of the leased property less any related investment tax credit retained by the lessor.

Operating Lease ("off balance sheet")

Generally any lease not considered a capital lease is an operating lease. Lease payments are expensed in income statement.

Leverage Leases

Similar to direct leases, however, in a direct lease the lessor provides 100% of the capital from its own funds, whereas in a leverage lease the lessor becomes owner of the equipment by providing only a percentage (20% to 40%) of the required capital. The remaining capital in the leverage lease is borrowed from institutional investors on a nonrecourse basis to the lessor. Typically, leverage leases are secured by a first lien on the equipment, the assignment of the lease and the assignment of the lease payments.

Leasing can be an attractive alternative for funding some of the costs of a mining project. Effectively it is possible to obtain long-term financing by way of leasing at a lower cost than if a direct loan was raised, so long as the lessee is unable to use all the tax benefits. In return for trading the various tax benefits to a third party, the lessee receives cash credits which reduce after-tax funding costs. The cash credits are realized through a rental or a lease payment obligation which is significantly below the conventional debt service alternative.

LOCAL MEDIUM-TERM INSTITUTIONAL LENDERS

Although no capital market can match the depth and the resilence of the U.S. capital market, in a number of countries it is possible at times to secure relatively longer term credits from local private or government supported inter-bank institutions. For example, the Australian Resources Development Bank (ARDB) (owned by trading banks), and the Australian Industrial Development Corporation (AIDC), which is government owned, have provided funding for mining projects with maturities in excess of those available in the commercial bank market. The participation of such institutions can be very helpful in the development of new mining ventures. Funding that they provide may be used to build the infrastructure or even the production facilities. In certain countries, such funding can be obtained only in local currency but in others it is also possible to borrow in U.S. dollars or another desirable currency. The financial resources of these lending institutions generally, however, are not sufficient to provide all the funds needed from external sources. Similarly, some of the funding, but in local currency, can sometimes be obtained from certain private institutions in that country, such as life insurance companies and pension funds. Again, however, the total amounts available in most instances are rather small.

INTERNATIONAL COMMERCIAL BANKS

An international commercial bank syndicate probably is the most flexible source of funding for any major project. The primary disadvantage of utilizing this source, however, is the relatively short tenor of the loans. In today's market typical tenors would not exceed much more than 10 years from the time of entering into a firm lending commitment to the time of last repayment. If we were to consider a 3 - 4 year construction period, a project would have to repay the loans in 6 to 7 years. Under ordinary circumstances and high leverage this may be difficult to accomplish. Therefore, in order to enhance the economic viability of the project, the amount of commercial bank loans should only be fixed after a determination is made as to the maximum availability of longer term credits.

Commercial bank loan syndicates are based either on Eurodollar pricing terms, i.e., with a spread over the London Interbank Offered Rate (LIBOR), or on U.S. or other country prime rate pricing. Commercial bank credits, in most instances, therefore, have floating interest rates fixed in the Eurodollar market for periods of up to six months. The spread over LIBOR depends on the quality of the borrower and the market conditions. In today's environment this spread may vary from 1/2% to 3%.

Most credits for international mining projects are arranged in the Eurodollar market with pricing, therefore, tied to LIBOR. U.S. domestic borrowers tend to borrow on a U.S. prime rate basis. In certain rare instances fixed rate credit may be obtained from commercial banks. However, because of the banks' liability structure, they are most reluctant to enter into such fixed interest term loan commitments.

Commercial bank financing either Eurodollar or domestic, is typically arranged by a group of managing banks and agented by one leading bank. Project lending loan transactions can be very complicated and require lengthy documentation. In addition to the interest rate, the borrower is expected to pay management fees, particularly in the Eurodollar transactions, of up to 2% of the total loan on signing of the loan agreement and also a commitment fee of up to 1% per annum on the undisbursed funds. Commercial bank term loans normally, however, do not contain prepayment penalties in case of refinancing or earlier repayment.

Commercial banks stand ready to provide adequate funding for most mining projects on the condition that these projects are of a self-liquidating nature and can withstand the vagaries of fluctuations in commodity prices. It is very advisable for the project sponsors to collaborate with commercial banks very early in the project's development since these institutions often have not only commercial bank funding capability but also they can assist the sponsors in the investigation of other more attractive sources of funds.

11.2

BASIC SOURCES OF MINE FINANCING

Kenneth G. Arne

Vice President and General Manager
Paul Lime, Tucson, Arizona

INTRODUCTION

The real problem is finding the "appropriate" financing sources by matching the risk and reward expectations of the various classes of investors and lenders.

Generally, lenders who advance funds temporarily are interested in assuming only limited risk and conversely expect to earn only a limited reward. The equity owners on the other hand, are expected to take most of the risk of the enterprise, but should receive a reward commensurate with that risk. In particular they will receive the benefit of the company's increase in net worth over time.

MINE DEVELOPMENT CYCLE

Consider the development cycle of a mining project beginning with rank wildcat exploration (Figure 1)

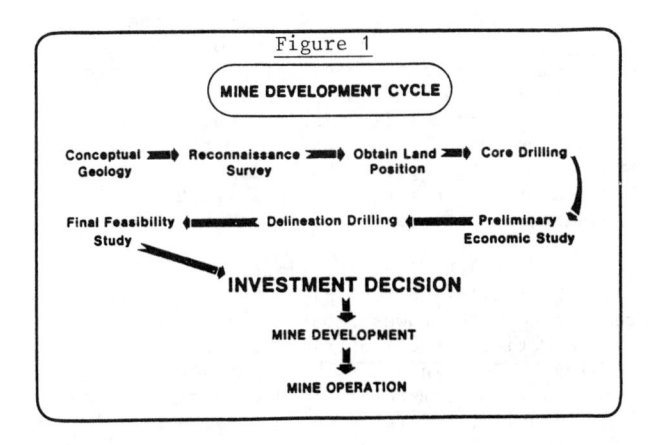

Figure 1
MINE DEVELOPMENT CYCLE

Just as different professionals must be utilized at different stages, so must different funding sources be used at different points in the pipeline. Figure 2 shows the general sources of funding:

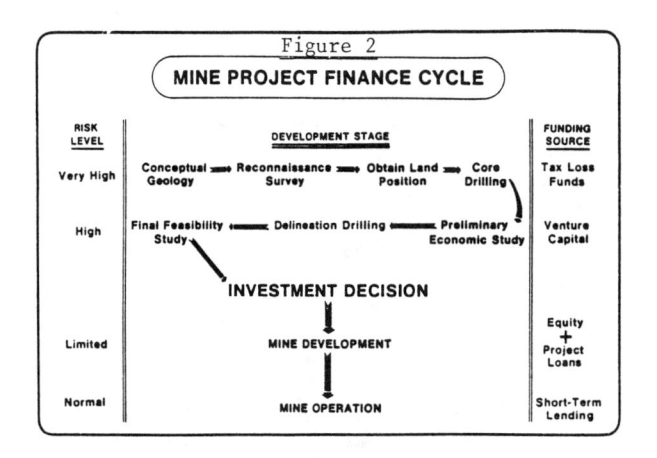

Figure 2
MINE PROJECT FINANCE CYCLE

This concept can perhaps be better understood if we compare the risk with the capital requirements. Suppose each area within a triangle is proportional to the risk taken (see Figure 3):

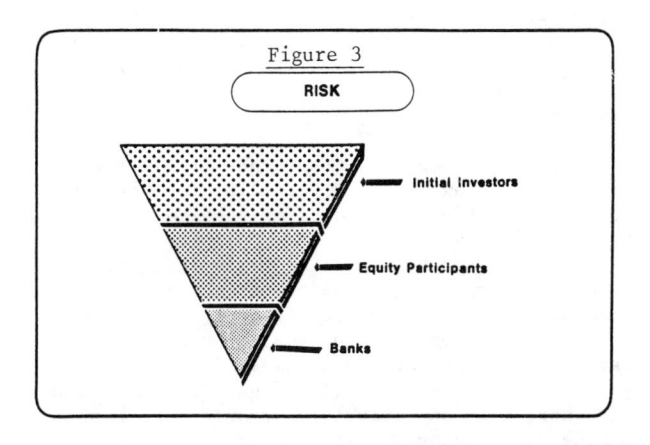

Figure 3
RISK

Based on a paper "Basic Concepts of Mine Financing" presented at the Montana Mining Association Annual Meeting, May 6-8, 1981

527

Risk

Obviously the bulk of the risk falls on the initial investors who begin the development cycle, with a substantial but lesser amount falling on the second stage equity participants, and almost none on the banks. Now compare this with the amount of funding available from these sources to bring the mine on stream. The initial investors may risk losing their entire bet, but it should, relative to the total cost of the mine, be a modest bet, since generally the exploration costs of a mine project are small compared to the concrete and steel expenditures for mine and plant construction. The second stage equity participants will invest more funds, but at the proving up stage should be exposed to lower risks than those assumed by the initial investors. Finally, the bankers, who take a very small risk, are capable of providing most of the funds for mine development (See Figure 4).

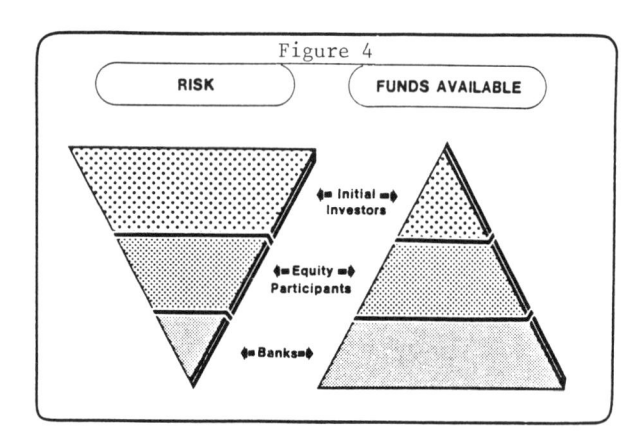

Figure 4

RISK FUNDS AVAILABLE

← Initial →
Investors

← Equity →
Participants

← Banks →

By definition, risk is the probability that the desired outcome will not be achieved. Most of us would agree that the preliminary stages of exploration are very risky indeed, given that the desired outcome is to establish a profitable mine. (See Brock's paper in this Chapter.)

Tax-Loss Funds

Tax-Loss Funds is money that if not invested in mineral exploration, drilling for oil, or some other activity, would simply be paid over to the friendly government tax collector. Corporations and high bracket individuals do generate income against which expenditures for mineral exploration can be offset for tax calculation purposes. A few public ventures have been set up for mining as limited partnerships much the same as oil and gas drilling funds and there are many privately organized limited-partnership ventures. (See Cruft's paper in this Chapter.) A word of caution is needed. Competent legal and tax accounting advice is an absolute necessity to properly organize these ventures, and unless you personally know some wealthy individuals, you will need to interest an underwriter in the project in order to find individuals willing to invest in the mine.

Proving up a mine

The second financing stage, which is labelled as High Risk, is essentially the "proving up" portion of the mine development cycle. At this point the geologic risk is small, since the deposit has been drilled. However, uncertainties in ore grade, capital costs, mine operating costs, mill processing scheme, and metal market outlook still remain.

At this stage, the general lack of certainty pretty much precludes lenders from advancing funds so the search must continue for additional equity funding to finance the delineation and feasibility study work. (For further details, see Brock's paper in this Chapter.)

An alternative is to seek out a private venture-capital company. These are firms, or in some cases joint ventures, organized for the purpose of making investments in small, unknown firms which usually are in a new technology or have some other special feature which could be the basis for a rapid increase in stock value. Typically a venture capital company expects its profit to come from stock price appreciation after the new technology has proven to be a market success and the company's stock is actively traded by the investing public. Long term ownership of the stock or control of the company are not generally objectives of the venture capital companies. Venture capital companies are usually staffed by professional investment managers, so a complete business plan backed by solid technical detail is important in the discussions with these firms. (See Boettcher's paper later in this Chapter.)

Debt Financing

Once the delineation work is finished, the mining and metallurgical designs complete, the environmental permits in place, and the economic studies indicate that the project is both technically and financially viable, debt financing can be considered. At this point the risks have been reduced to a point that a lender, with his limited profit objective and his limited risk exposure, can become interested in the project.

Despite the fact that interest rates are near their all-time real high, debt financing remains both the lowest cost (since interest is tax deductible) and most flexible source of funding. Debt in one form or another can be obtained from commercial finance companies, leasing companies, federal government agencies, local or state development authorities, commercial banks, and insurance companies. Each of these classes of lenders has its own specific type of deal in which it is interested.

Commercial-finance companies tend to work on smaller and higher risk credits, for example, financing a single bulldozer or dragline. These are, to use a trade term, "collateral-based" lenders. That is, the finance company looks solely to the value of the single piece of equipment as its security, regardless of the overall financial situation of the borrower. Thus, retrievability and alternate uses of the machinery elsewhere are of prime importance.

Leasing Companies rent equipment rather than money, though in many cases this distinction is artificial. Leasing companies will finance equipment from individual items through complete plants under a variety of schemes. Tax considerations for the sponsor company are important here, since many of these deals hinge on who can take best advantage of the investment tax credit, accelerated depreciation, and residual value of the equipment. (See also Ravenscroft's paper in this Chapter.) (One must admit to be fascinated by the possibilities explained by the leasing people. It seems that whatever deal can be concocted by the mind of man has been put into a lease arrangement.)

Federal agencies also have a hand in the debt business, usually to promote a specific program deemed to be in the nation's overall best interest. The US Small Business Administration is probably known to you in its efforts to promote business enterprises; US Farmers Home Administration will back many types of industrial activity that will aid rural areas; and the Department of Energy has a program for small underground coal miners. In many cases the government agency does not directly lend money, but rather guarantees the loan, so that the local banker actually handles the loan request and disburses the funds. The current administration in Washington, D.C. has announced its intent to cut back funding of some of these programs, so US Federal Government guarantees may become less important than before.

State and local political subdivisions or development authorities in the USA can serve as intermediary agencies to enable a private company to obtain funds on favorable terms. These arrangements typically start with a local government unit, a city or county, for example, organizing a special "industrial development district" which becomes a legal subdivision of a state with the same status that municipalties enjoy. Bonds issued by state or local governments are exempt from US Federal income tax, and, within limits, bonds issued by an industrial development district are treated as municipal bonds for tax purposes. Because of their tax-exempt status, municipal bonds generally carry lower interest rates than do comparable private bonds.

A private company will arrange for a local industrial development district to acquire a plant site and construct a plant. The funds for this are provided by the sale of bonds of the industrial development district which in fact lends its name only to the project, the financial support normally comes from the sponsor company with no obligation on the general taxpayers. The company then enters into a long-term lease for the facilities with the annual lease payment going to amortize the district's bonds.

Commercial banks can provide funds to a mining company under a variety of arrangements. Originally, bank lending was for short-term working capital needs, financing the build-up of a seasonal inventory for example. During this century, term lending to established companies has become common. Term lending means loans for several years duration (6-10 years) for the purposes of financing permanent investments in plant and equipment. Even more recently the concept of the project financing has become common with commercial bankers. The idea of project financing is that the lender looks only to the new venture to repay the loan and has only limited recourse to the sponsor company. Rather than relying upon a strong balance sheet to provide assurance that the loan will be repaid, the lender will carefully examine the operation itself to determine if it is technically and economically viable since the cashflow from the project will pay interest and principal on the loan. (See Chapter 9 and 10 for more detail.)

Beyond the lending function, commercial banks also offer services which should be noted. Payroll preparation, handling of accounts receivable, money-market investments, corporate trust activities such as payment of dividends and shareholder record keeping, financial consulting, and foreign exchange transactions are some of the day-to-day treasury activities which can be obtained at commercial banks.

Insurance companies receive large amounts of money on a regular basis from the premium payments of their policyholders. These funds must be invested somewhere (remember that a dollar not put to work is a wasted dollar), and one place these funds can be invested is in direct loans to business. In the trade these are called private placements since the deal is done with a small group of investor companies and not offered to the general public through Security and Exchange Commission procedures. Private placements with insurance companies offer some advantages vis-a-vis commercial banks: a fixed rate of interest can often be negotiated, and a term longer than the 6-10 year range for commercial banks is generally attractive to the insurance companies. The other side of the coin is that insurance companies will deal only with companies of the highest credit standing and generally will not take project risk.

11.3

MINING FINANCE

Herbert D. Drechsler

The University of British Columbia
Vancouver, British Columbia, Canada

ABSTRACT

Mining Finance is the interrelated aspects of "determining the total amount of funds to employ in a mining firm; allocating these funds efficiently to various assets; and, obtaining the best mix of financing in relation to the overall valuation of the firm." The first section is an overview of the financing problem through examination and discussion of the Balance Sheet and Income Statement. The next section is a discussion of the nature of the criteria developed in feasibility study. The third section describes type of capital funds, and the final section describes case studies of mining finance.

INTRODUCTION

A prime assumption is that the mining firm wants to maximize the market price of the firm's common shares. For government-owned firms, the objective is to maximize the value of the remaining mine reserves. These objectives are broader than the usual assumption that the firm is owned by "shareholders" and more specific than profit maximization.

Major political changes have occurred in the countries which have mines so that it is no longer a certainty that the firm has shareholders and the shares are traded on a stock exchange. Many firms are owned by government and the ownership is not subject to trading. As discussed above it is assumed that these government-controlled firms have the same objectives as traditional firms. However, financial management is necessary whether or not the firms are for profit or not-for-profit oriented. Therefore this paper is applicable to all types of mining companies.

Much of the following material has been developed from work done by Peter Lusztig and Bernhard Schwab, <u>Managerial Financial in a Canadian Setting</u> (Toronto: Butterworths, 1977); Ezra Solomon, <u>The Theory of Financial Management</u>. (New York: Columbia University Press, 1963); and V. Fred Weston and Eugene F. Brigham, <u>Essentials of Managerial Finance</u>, Third Edition (Ninsdale, Illinois: Dryden Press, 1974)

THE ANNUAL REPORT

This paper will first identify elements of the firms annual report using ASARCO's statements as an example. This is done so that in the discussion on funds acquisition the reader will understand how funds are both an asset and a liability to the firm. Second is a discussion of the decision criteria used in typical feasibility studies. These criteria help evaluate the financing decision. Following that section, there will be described the types of capital acquisition funds which are used to purchase assets with which to earn income. And the fourth section will have a series of case studies with which to illustrate the varieties of long run funds for new mine or expansion investments, and some constraints.

Mining is a business and accounting is the language of business. This language has a special vocabulary which is best shown in corporate annual reports. (See Chapter 4). Tables 1 and 2 are the principle financial statements shown in the ASARCO, Inc., 1976 annual report. Table 1 is the consolidated Balance Sheet and Table 2 is the Statement of Earnings. Accountants and financial managers often recommend reading an annual report in the following manner:

1. Read the Auditor's Report. If any unusual accounting practice has been done, or if the auditor does not completely agree with the accounting methods of the firm, this will be shown in the report.

2. Read the Notes to Financial Statements. The notes to the accounting statements explain in great detail the underlying principles used in the preparation of the key financial statements.

3. Read the Balance Sheet, Statement of Earnings, and Statement of Changes in Financial Position, and form an opinion of the firm.

4) Read the Statement by the President and

Based on a paper published in SME's <u>Mineral Processing Plant Design</u>, 2nd ed., 1980.

Chairman of the firm. See if this statement coincides with the previously made judgement of the firm.

Table 1A
ASARCO INCORPORATED
Consolidated Balance Sheet
As of December 31
($000,000)

Assets	1976	1975
Current Assets		
Cash	$ 30.4	$ 34.6
Accounts and Notes Receivable	104.3	102.7
Inventories	187.3	246.5
Materials and Supplies	56.1	56.9
Prepaid Expenses	1.6	1.2
Total Current Assets	379.7	441.9
Investments	422.5	378.8
Property		
Buildings and Equipment	794.7	742.7
Mineral land	146.2	142.3
Other	31.4	29.7
Total Property	972.3	914.7
Less: Accumulated Depreciation and Depletion	326.4	288.0
Net Property	645.9	626.6
Funds Committed to Construction	69.5	–
Other Assets	26.0	27.1
Total Assets	$1,543.7	$1,474.6

Table 1B
ASARCO INCORPORATED
Consolidated Balance Sheet
As of December 31
($000,000)

Liabilities	1976	1975
Current Liabilities		
Accounts and Notes Payable		
Bank Loans	$ 6.2	$ 6.6
Trade	63.7	87.6
Other	24.1	23.1
Accrued Liabilities		
Salaries and Wages	5.2	5.8
Taxes on Income	15.9	10.4
Other Taxes	10.4	9.6
Current Portion of:		
Long Term Debt	5.0	8.6
Other	23.0	22.1
Total Current Liabilities	153.5	173.8
Long Term Debt	400.4	341.9
Deferred Credits	87.0	73.5
Other	45.4	51.3

Equity	1976	1975
Preferred Stock		
Authorized – 10,000,000 shares @ no par value		
Common Stock		
Authorized – 40,000,000 shares @ no par value		
Issued – 27,678,223 shares	339.2	339.2
Retained Earnings	533.9	510.3
Other	6.7	6.7
Less: Treasury Stock	22.4	22.2
Total Equity	857.4	834.0
Total Liabilities and Equity	$1,543.7	$1,474.5

Table 2
ASARCO INCORPORATED
Consolidated Statement of Earnings
for the Year Ended 1976 and 1975
(000's)

	1976	1975
Sales of Products and Services	$1,103.7	$1,004.6
Cost of Products and Services	961.0	898.3
Income From Products and Services	142.8	106.3
Equity in Earnings of Nonconsolidated Associated Companies	30.7	27.8
Other Income	8.7	5.7
Closing Perth Amboy Refinery		(20.5)
Income Before Deductions	182.1	119.3
Deductions		
Research Expense	4.4	5.3
Exploration Expense	7.2	7.1
Selling and Administrative Expenses	30.8	30.4
Corporate Taxes and Interest	1.3	2.8
Depreciation and Depletion	50.7	36.5
Interest on Debt	33.1	21.9
Total Deductions	127.9	103.9
Earnings Before Income Tax	54.2	15.4
Taxes on Income	11.9	(10.0)
Net Earnings	$ 42.3	$ 25.9
Net Earnings per Share	$ 1.58	$ 0.95
Cash Dividends Per Share	$ 0.70	$ 1.05

BALANCE SHEET

The balance sheet is a statement of a firm's financial position on one specified date. It records the firm's assets which are the resources which are expected to earn income, as well as its liabilities which are the economic obligations of the firm, and shareholders' equity which is the source of some of the assets. The name "balance sheet" is derived from the fact that the total of liabilities and equity equals assets.

In the ASARCO Notes to Financial Statements are shown the significant accounting policies.

They include statements on the principles of consolidation, inventories, property, depletion and depreciation, taxes or income, copper surcharges, exploration, mine development and retirement plans. As an example, the policy on inventories is:

> Metals at primary smelters and refineries and at secondary metal plants are at last-in, first-out cost (LIFO). Metals at mines are at first-in, first-out (FIFO). Inventories are written down to market, if lower than cost. Primary metals sold at firm prices for future delivery are included at the sales price.

ASSETS

Assets are the income-generating portion of the balance sheet. Assets are divided into smaller categories to indicate their sources. In the case of ASARCO, the classifications are:

1. Current assets $30,431,000: These are cash plus assets that can be quickly converted to cash. In the example current assets include cash and short-term deposits, or marketable securities which are a short-term investment of funds not immediately needed.

2. Accounts Receivable $104,359,000: These are funds due from customers. This figure often includes an allowance for doubtful accounts.

3. Inventories $187,318,000: This figure includes four basic items. First are the metal inventories of smelters, refineries, and secondary metal plants. Second is the provisional cost of metals purchased for which a price has not been fixed. Third are sales prices for metals sold at firm contracts for future delivery, and fourth are mine ore inventories. Therefore this figure includes both metal, concentrates, and ore, only a part of which can be quickly converted to cash.

4. Materials and Supplies $56,103,000: This is the average cost value of normal operating supplies for all plants and offices.

5. Prepaid Expenses $1,588,000: These are short term prepayments to suppliers of operating materials.

6. Total Current Assets $379,799,000: This is the sum of the above listed assets.

7. Investments $422,481,000: This figure shows the value of ASARCO's ownership in other companies. The value is determined by the total of original cost plus the companies share in accumulated retained earnings, less dividends to ASARCO. Some of ASARCO's investments include M.I.M. Holdings Limited, Southern Peru Copper Corporation, and Revere Copper and Brass Incorporated. Investments which are intended to be converted to cash during an operating cycle are included in current assets.

8. Property $972,339,000: This figure includes tangible assets such as buildings and equipment, mineral lands, land other than mineral land, and automobiles. The value shown in the Balance Sheet is original cost or current market if it is less than cost. Maintenance and repairs are charged to earnings as they occur. These assets are depreciated at various rates, but, in general, assets at plants are depreciated on a straight line basis except those assets placed into service at primary smelters and refineries during the period 1961 and 1970 which are depreciated by accelerated methods. Depreciation and depletion at mines are generally computed on the ore reserve method.

9. Other categories: This includes funds committed to construction $69,525,000 and other assets $26,003,000. Sometimes in some Annual Reports under Other Assets is an item for deferred mine development. This account is for major development which is capitalized rather than amortized currently as an expense.

LIABILITIES

Liabilities are the firm's financial obligations to organizations or individuals outside the firm. They are listed in the balance sheet and are divided into two categories:

1. Current liabilities: these are liabilities which fall due during the coming year (or the operating cycle if longer than a year). They are accounts payable $94,088,000. Accrued liabilities (such as wages, salaries, interest) $31,419,000, and the current payments of the long term debt and reserves for plant designs. Current liabilities may also include notes payable. The total current liabilities of ASARCO are $153,547,000. The reserves listed represent the anticipated costs to be incurred after 1977 associated with the closing of the Baltimore and Perth Amboy copper refineries and the Amarillo zinc plant.

A major consideration is that reserves are not set-aside cash. They are accounting charges. The only location of cash in the firm's accounting statements is under the cash account on the assets side of the balance sheet.

2. Long-term liabilities $532,804,000: these are liabilities which fall due beyond one year (or the operating cycle). This includes the long-term debt of the firm, and other long-term accounts payable or reserves. Long term borrowing of the

firm for the purpose of developing a mine would fall into this account.

SHAREHOLDERS EQUITY

The final category is stockholders' equity, $857,387,000. Equity is the excess of total assets over liabilities. In this example it is divided into preferred stock (unused); common stock of 40,000,000 shares of which 27,678,223 are issued and other capital including retained earnings. Retained earnings are not cash. They are accounting earnings which are developed from the consolidated income statement and kept within the firm.

INCOME STATEMENTS

An income statement is a record of the income and cost flows of a firm within a period of time. The net income is calculated as the excess revenue over expenses and taxes, and reflects changes in shareholders' equity. Net income also is a guide to the firm's debt-paying capacity and its ability to produce long-run earnings and dividends.

Revenue: The first figure on an income statement, $1,103,737 is revenue. It is the gross increase of assets that represents the total sales value of products delivered and services rendered. Revenue can be recognized at many points in the business process: when orders are secured, when the goods are manufactured, when they are delivered or when payment is received. Most accountants use the point of sale as the time to record revenue. For instance, at Craigmont Mines Limited in 1976:

> Concentrate revenue is recognized at the time of sale. However, copper concentrate sales not finalized at year end have been recorded at prices estimated to be in effect on finalization dates.

> Not included in the revenue are proceeds from borrowings, investments by owners, or adjustment of revenue from previous accounting periods.

EXPENSES

Expenses are deducted from revenue and are essentially costs incurred for receiving revenue. They represent a negative effect on stockholders' equity.

Cost of Production Services: $960,981. This expense in the total of direct supplies and labor necessary to produce the goods or services for sale. Indirect costs are generally shown below in Deductions.

In the Income before Deduction ASARCO labels a special Cost, Unusual Items, at a negative $20,500. In fact this is an increase in income. This account comes from a reserve account which anticipated the closing of the Perth Amboy Copper Smelter.

Deductions: $127,922. Total deductions includes the following items: Research Expense $4,416; Exploration Expense $7,185; Selling and Administrative Expenses $30,817; Corporate Taxes and Interest $1,331 (This item is not federal income tax);

Depreciation and Depletions (To be discussed further in this chapter) $50,667; and Interest on Debt $33,106.

Income Tax $11,865. Income tax is calculated from the tax statement which is not a part of the annual report. The annual report shows accounts on a "book" basis as compared with a tax basis for the government.

Depreciation Methods: Depreciation and depletion are accounting processes that convert assets into expenses. The major methods of calculating depreciation are:

a. Straight-Line

$$\text{Annual Depreciation Expense} = \frac{\text{Cost of Asset} - \text{Estimated Scrap Value}}{\text{Estimated Useful Life}}$$

b. Sum of Years Digits Q

$$\text{Annual Depreciation Expense} = \frac{1}{\sum_{Q=1}^{X} Q} (X - A)E$$

where: X = Estimated life of asset
A = Year of depreciation
E = Asset cost minus estimated scrap value (net cost)

c. Declining Balance

A uniform rate of depreciation (e.g. 25% per annum) is applied to the declining net cost of the asset.

d. Unit of production

The net cost of the asset is divided by the number of hours, or tons, or whatever production unit is relevant. This unit of production depreciation rate is then multiplied by the number of units the machine was run or tons produced from the mine during the year. This determines the depreciation of the machine or depletion of the mine in that year.

NATURE OF DECISION CRITERIA FROM THE FEASIBILITY STUDY

The methods outlined in this section are used to rank investment proposals (see also Cole's paper in Chapter 3). Since the amount of money available for projects is not infinite, it is necessary to assess the economic worth of one investment over another. Each proposal can then be accepted or rejected on the basis of its "hurdle rate" which is the cut-off rate for decision. In all the following methods, the decision is "yes, invest" or "no, don't invest". There are no "maybe" decisions. Obviously there is no decision maker who would base an investment of hundreds of millions of dollars on a single economic criteria. There are numerous other factors to consider such as the risk propensity of the decision maker, the historic strategy and decision process of the firm, the political interrelationship between members of the decision groups, and the concern of the firm with the external world.

Internal Rate of Return Method

The internal rate of return is the rate of discount which, applied to the cash flows for investment, will yield a net present value of zero. This method calculates the discount rate which equates the present value of expected returns to the initial investment outlay. The equation for determining this discount rate r is:

$$\sum_{t=0}^{n} \frac{C_t}{(1+r)^t} = 0$$

where C_t is the cash inflow or outflow and n is the number of years the flow is expected.

For example: If the initial outlay of $20,000 is expected to yield a cash inflow of $6,000 per year for the next 3 years the equation would be:

$$20,000 = \frac{6,000}{(1+r)} + \frac{6,000}{(1+r)^2} + \frac{6,000}{(1+r)^3}$$

Computer programmes are available to solve for r. Manual solutions require trial and error methods for solution. In the example above if 10 percent is first substituted for r, the result is $19,894.80. If 9 percent is then substituted for r, the result is $20,250.35 thus straddling the answer. By interpolation it is possible to find the discount rate that will equal $20,000; i.e. 9.7 percent. The internal rate of return is then compared with a required rate of return which is the firm's weighted average cost of capital in the case of unlimited capital availability. In the usual case where capital constraints are binding, the hurdle-rate must be raised with opportunity cost the relevant criteria against which new investments are measured. In numerous conversations with managers of mining firms, the author has found that 15 percent in real terms is the common discount rate used for decision purposes.

Net Present Value Method

The net present value is the sum of all future cash flows generated by a project with each cash flow discounted to the present. The discount rate used is the firm's weighted cost of capital or the more usual opportunity cost of capital as illustrated above. The equation used for calculation is:

$$\text{Net Present Value} = \sum_{t=0}^{n} \frac{C_t}{(1+r)^t}$$

where C_t is the cash flow, n is the number of years the flow is expected and r is the rate of discount.

For example: If the cash flow is $3,000 per year for the next 5 years after an initial outlay of $10,000 at a required rate of return of 9 percent, the Net Present Value is:

$$\text{Net Present Value} = -10,000 + \frac{3,000}{(1+.09)} + \frac{3,000}{(1+.09)^2}$$
$$+ \frac{3,000}{(1+.09)^3} + \frac{3,000}{(1+.09)^4} + \frac{3,000}{(1+.09)^5}$$
$$= 1668.96$$

The net present value number indicates the excess or short fall in cash flows in present value monetary terms, once financing charges at a rate of 9 percent were met. A positive cash flow indicates that a project is acceptable while a negative cash flow indicates rejection of the investment.

Payback Method

The payback method is the period of time required for the expected after-tax cash flows to recover the initial investment.

For example: If an investment of $20,000 has an annual cash inflow of $6,000 per year, the payback period would be:

$$\text{Payback period} = \frac{\$20,000}{\$6,000} = 3.3 \text{ years}$$

Projects are ranked with quicker payback being preferred.

This method has several drawbacks. It does not take into account the time if cash outflows directly into the investment and it ignores returns beyond the payback period. By ignoring proposals with longer payback periods it dismisses projects such as those involved in introducing and developing new products or tapping new markets. Payback may be used as a tool for evaluating risk. The longer the payback period is, the greater the risk and the less liquid the investment.

Average Annual Rate of Return Method

This rate of return is calculated as a ratio of the average annual profits after taxes to the average book value of the investment over the life of the project.

For example: Assume the average annual profit over a five-year period is $3,000. The average net investment of a $10,000 project is the difference between the beginning book value and the ending book value. The mine plant usually has no value at the end of the mine life, then the average book value of the investment is:

$(\frac{\$10,000 - 0}{2})$, or $5,000 the

Average Annual Rate of Return = $\frac{\$3,000}{\$5,000}$ X 100% = 60 percent.

Projects are ranked according to the average annual rate of return with higher percentages preferred. Managers can set a hurdle rate at any acceptable level. If the average rate of return is higher than the required percentage, the project is accepted, otherwise it is rejected. The method ignores the time flow of the profits; it it based on accounting profits rather than cash flows; and the hurdle rate is not consistent with managers.

In normative terms, the internal rate of return and the net present value are conceptually better decision rules. They both include the time value of money and the pattern of cash inflows and outflows. The net present value is a theoretically better method for decision-making because it is consistent for an investment proposal.

The payback period and the average annual rate of return methods both have usefulness to the manager. They give some indication of the risk potential of the investment project and give non time-oriented decision rules.

The investment decision criteria give management an assessment of the impact of new projects or the financial statements. As such, they are dealing with forecasting the effect of present decisions. Managers utilize as many tools as possible to aid in increasing the accuracy of the forecast. This section has not dealt with uncertainty as an input in decision criteria. Managers do not make decisions in isolation from the internal and external environment of the firm, therefore the decision methods of this section must be considered interrelated with non-financial project selection criteria.

MAJOR SOURCES OF FUNDS

In Figure 1, the classification is first by period of time between raising the monies and final repayment in the case of debt. Two periods of time are considered. Short-term financing is defined as debt maturing in one or two years or less. Long-term financing is defined as debt beyond two or more years. The time period is of an arbitrary nature, and there is much interlap between the two categories.

The firm has two prime alternative sources of financing. There are internal and external sources. The internal source of funds is the net cash inflow through sales of goods and services. This inflow is composed of retained earnings and the sum of the non-cash reductions of income before calculation of income tax. Recall from an earlier section that depreciation and depletion are not cash accounts but reflect costs of operating plant and equipment and recognition of the mine as a wasting asset. For tax purposes, these accounting costs are applied to operating income to calculate net taxable income. In the sense that depreciation and depletion are accounting costs, in the calculation, monies equivalent to these costs may be considered as a source of internal financing.

The internally generated sources of funds are adequate for providing smaller amounts of capital for day-to-day operations. This includes capital for financing the accounts receivable, inventories, and using cash to pay the day-to-day bills and wages. For long-term investments, these sources are inadequate. Most large mines are underinvested.

Figure 1

SOURCES OF FUNDS FOR A MINING FIRM

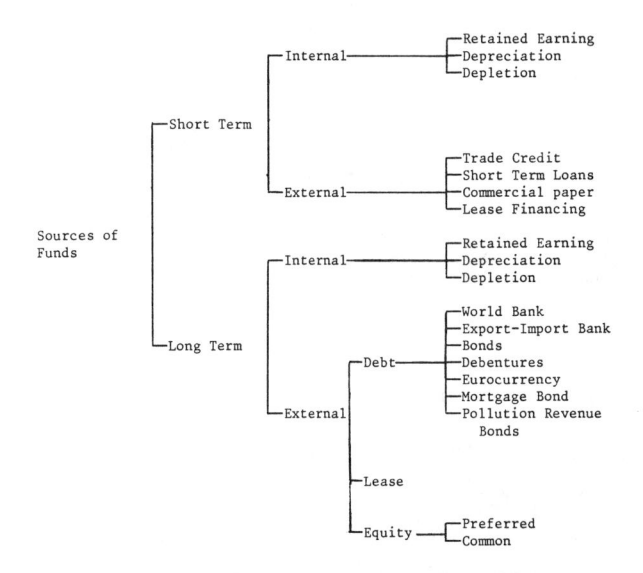

Sources of Funds
- Short Term
  - Internal
    - Retained Earning
    - Depreciation
    - Depletion
  - External
    - Trade Credit
    - Short Term Loans
    - Commercial paper
    - Lease Financing
- Long Term
  - Internal
    - Retained Earning
    - Depreciation
    - Depletion
  - External
    - Debt
      - World Bank
      - Export-Import Bank
      - Bonds
      - Debentures
      - Eurocurrency
      - Mortgage Bond
      - Pollution Revenue Bonds
    - Lease
    - Equity
      - Preferred
      - Common

Source: Adopted from Peter Lusztig; Bernhard Schwab, *Managerial Finance in a Canadian Setting* (Toronto, Butterworths, 1977) p. 209.

The full extent of the mine is unknown except on the day the mine is depleted. The size of the mining and processing plant is a function of the ore reserve size among other elements. The final ore reserve size is unknown on the day the plant is first constructed, and it is initially built smaller than optimum. As more ore reserve is located more plant is added. Investments are also needed to replace equipment destroyed in accidents or when just worn out. This investment maintains mines and processing plants at constant output with no change in cost.

Investment is needed to reduce the cost of operations through technological advance or efficiencies developed by substituting new equipment for personnel. Finally, investment is needed to produce more output as demand for mineral products grows.

Mining firms are not providing enough internally generated funds to support mining and processing operations. In the period between 1961 and 1970 and between 1971 and 1975 many firms made substantial increases in their debt positions. This increase is shown in Table 3. For the nine mining firms the debt/equity ratio was 0.03 in 1961, and 0.20 in 1970, and was 0.37 in 1975. The book value of the mining firms in the fifteen-year period of time increased 2500 percent. (See also Strauss, Chapter 1). For long term purpose, these firms borrowed heavily from banks and other locations rather than sell more shares in the financial markets. This is a strong indication that debt financing is important in the financial structure of the firm.

Table 3

DEBT-EQUITY OF SOME MINING COMPANIES

	1961		1975	
	Book Value	Debt	Book Value	Debt
Copper Range	$ 51,605	$ 36,496	$ 105,284	$ 31,270
Johns-Manville	283,649	-	580,512	186,322
Phelps Dodge	394,528	-	893,263	522,509
Int. Nickel	569,764	-	1,484,353	611,236
Newmont	112,000	-	648,331	251,437
ASARCO	381,306	-	833,961	341,934
Kennecott	748,478	6,540	1,410,423	406,387
Anaconda	814,200	65,000	1,210,918	238,279
AMAX	262,600	15,300	1,365,170	543,200
TOTALS	$3,618,129	$ 123,336	$8,532,215	$3,132,574

Debt-Equity ratio   1961 - 0.03   in 1970 - 0.20   in 1975 - 0.37

Source: Annual Reports, Moody's Industrial.

TRADE CREDIT

Trade credit is a form of short-term financing. When goods are delivered and not paid for at the point of delivery, trade credit exists between the supplier and the purchaser. For example in the ASARCO Balance Sheet is shown Accounts Payable of $94,000,000. These are supplier credits to the firm. Trade credits are one of the major sources of short term financing. Trade credits are convenient and easy for mining firms to acquire. As long as the buyer maintains his credit standing, suppliers are willing to provide the financing. These credits are useful because they remain available during periods of monetary tightness or other periods when funds from other short-term sources are scarce. The prime form of trade credit is open account. Under this system the supplier issues an invoice listing the item, its price, the total amount due and the credit terms. It is called open because the buyer does not sign a formal recognition of debt. The supplier's evidence of the debt is simply an accounting book entry and the purchase and shipping orders. The great use of trade credit is evidence of the trust between buyers and sellers.

SHORT-TERM LOANS

Short-term loans from banks are like trade credit in that they usually mature within a year. They are not regarded as permanent financing. Banks often stipulate that firms be out of debt for a short period each year as a guarantee that short-term loans are not static.

Short-term loans can be either secured or unsecured. Unsecured loans are those which do not require collateral. They can take the form of a line of credit in which a maximum amount that can be borrowed is specified. Loans beneath this amount are readily accepted, although lines of credit represent no legal commitment on the bank's behalf. In order to ensure a legal commitment from the bank a revolving-credit agreement must be arranged. In this agreement the bank must lend to a firm up to a maximum specified amount. Another form of unsecured short-term loan is the transaction loan which is borrowing for a single purpose. When the purpose is fulfilled or completed, the loan is repaid. Often banks require that firms maintain a minimum average balance of 15 to 20 percent of the outstanding loan in a chequing account. For the bank this is a form of security since it increases the liquidity of the firm and makes funds available if the loan payments default.

If the bank requires even more security, it asks for collateral. This security is used to guarantee that the loan is paid. If the obligations of the loan are not met, the collateral is sold and the lender receives his funds. Accounts receivable and inventory are assets which a firm may use as collateral. When using accounts receivable as security, the bank or finance company accepts those with good credit ratings. Because of the chance of fraud, the percentage amount advanced upon the face value would probably be low. Inventories are rated according to their marketability. Highly marketable inventories would receive a high percentage of their market value, whereas a highly specialized piece of equipment with low marketability would not be readily accepted as collateral. Collateral becomes necessary when the firm's credit is new or their credit rating is low. Another form of collateral is outside assets such as stocks or bonds. The percentage amount lenders are willing to pay on the market value of bonds is usually high. In addition, bonds often require compensating balances which are a requirement that the borrower maintain a fixed percentage of the loan as an interest-free deposit in the bank. Also bonds sometimes require standby fees on lines of credit. These fees are charged on the portion of the loan not borrowed and act as a restraint on the extension of lines of credit.

An example of this type of loan is the short term debt of Hudson Bay Mining and Smelting Company, Ltd., as shown in the prospectus for $50,000,000, 10½% Sinking Fund Debentures Due 1995 of December 9, 1975.

> The Company has a line of credit with Canadian chartered banks and also has borrowing privileges with an associated company ($7,959,000). The loans from these sources are at interest rates equal to the prime rate in the case of the parent company and at normal commercial rates in the case of subsidiaries. The maximum outstanding amount under these arrangements for the eighteen months ended June, 1975, was $11,009,000. The approximate average outstanding balance for the period was $2,500,000. The average rate of interest on outstanding loans at June 30, 1975, was 9.8% and the approximate average rate throughout the eighteen months, 12.4%. There are no compensating balance requirements under these loan arrangements.
>
> The outstanding notes payable ($17,180,000) are for terms of between 30 days and six months and bear an average interest rate of approximately 7.4%. To support its commercial paper operations, the Company has a standby line of credit in addition to its open line of credit. The cost of this standby line of credit is 1/2% on the amount of the line.

COMMERCIAL PAPER

Commercial paper is a short-term, unsecured promissory note issued by major companies such as Hudson Bay Mining and Smelting Co. Limited, as above and sold to banks, insurance companies, pension funds and other business firms. Phelps Dodge Corporation lists $72,178,000 of commercial paper in the 1976 Annual Report. It is used as a supplement to bank credit and as an alternate source of funds during periods of tight money. It may mature in anywhere from 24 hours to 365 days, although most notes are 30, 60 and 90 days. Advantages of commercial paper include such items as: the interest rate is often one-quarter of one percent to 2 percent below the prime rate; there is prestige associated with the acceptance of a firm's commercial paper; speed of fund acquisition; and advice from the commercial paper dealers.

Another example of short-term or medium-term borrowing is illustrated in the 1976 Phelps Dodge Annual Report:

> The corporation has entered into a revolving credit and term loan agreement with various banks which permits the corporation to borrow at any time through September 1979 up to $150,000,000 at 116% of the agent bank's prime rate of interest. During the revolving period a commitment fee is payable on the unused portion of the bank's commitment at an annual rate of 1/2 of 1% and an availability fee in lieu of compensating balance is payable at an annual rate equal to 9% of the agent bank's prime rate of interest. Borrowings under the agreement which are outstanding at the end of the revolving period bear additional interest increasing from 3/8 to 5/8 of 1% per annum and are repayable from April 1981 through October 1984. There have been no borrowings under this agreement to date. (Commercial-paper finance is also covered by Yablonsky, Gillham and Castle later in this Chapter 11).

LONG-TERM DEBT

Long-term debt is a debt that matures after one year and often has a term of several years or more. It usually has a final repayment date as well as periodic interest payments. Collateral in the form of stocks, bonds, machinery or equipment, or the mining property including ore reserves is usually requested on these loans. Repayment is usually on a regular basis, for example, every quarter or semi-annually. Prepayment is accompanied by a penalty. The interest rate may be higher than for a short-term loan because of inflation and risk considerations and can be either a fixed rate or a rate that is variable with the bank's prime rate. As with short-term loans, the borrower may be requested to sustain a minimum average balance. Other requests from the bankers may be that net working capital be over some minimum, that changes in management be approved by the lender, and that periodic financial statements be submitted.

A typical example of a long-term note is illustrated in the 1976 ASARCO annual report. These are the 7 7/8% notes due April 1, 1994. Payable in annual instalments of $4.1 million commencing in 1978. The notes to this debt indicate:

> The note agreement contains convenants relating to maintenance of certain financial ratios and limiting long-term debt and declaration of cash dividends. Under the dividend limitations at December 31, 1976, $114.9 million of retained earnings were available for payment of cash dividends. However, in making dividend payments, consolidated current assets must be maintained at not less that 175% of adjusted current liabilities which amounted to $148.5 million at December 31, 1976.

In this case, ASARCO will pay an interest charge of $782.50 per annum for every $1,000 of face value of debt authorized. The balance sheet limitations control the amount of debt ASARCO may borrow.

There are many ways of providing lenders security for loans. For instance, in the 1975 Teck Corporation Limited annual report is shown:

Current bank loans aggregating $8,737,000 and the $4,100,000 term bank loan are secured by certain accounts receivable, marketable securities and investments.

The $10,803,000 term bank loan is secured by a collateral fixed and floating charge debenture on the company's portion of the assets of the Newfoundland Zinc Project of approximately $11,500,000 plus specific assignment of concentrate accounts receivable and concentrate inventories from the project of approximately $1,350,000.

The mortgage loan payable is secured by the company's 45% interest in the Highmont mineral properties.

SINKING FUND DEBENTURES

A debenture is an unsecured bond. Debentures are unsecured in the sense that they have no claim to any specific property but they are backed by the earning power of the firm. One restriction of a debenture often seen is that the firm shall not pledge its assets to other creditors. There are subordinated debentures which are also unsecured. Upon liquidation, these creditors have even less claim to the assets of the firm, although they are ahead of preferred and common stockholders.

Retirement of bonds or debentures can occur in several ways. First, debt can be retired by payment of face value at maturity; second, by periodic repayment of principal over the life of the debt; third, by conversion if the security is convertible; and fourth, by calling the debt before maturity if there is a call feature in the debt.

Borrowers usually require firms to provide for a sinking fund provision. This provision typically requires the firm to buy and retire a portion of the bonds issue each year. The bonds to be retired may be drawn in a lottery or the funds from the sinking fund may be used to buy bonds on the open market. An example of this are the Falconbridge Nickel Mines Limited 7.75% Sinking Fund debentures maturing February 1991:

> No portion of the principal is due on the 7.75% debentures until 1977 when the company is required to commence sinking fund repayment sufficient to retire $1,250,000 principal amount of debentures in each of the years 1977 to 1990. (From the 1976 Annual Report).

In the area of secured bonds there are mortgage bonds which place a claim upon a specific asset. Creditors usually make sure that the market value of the collateral exceeds the value of bond issue because if corporation defaults and the creditor must foreclose, he sells the collateral to recover the amount of the bond issue. If it is not fully recovered, the bondholder becomes a general creditor for the remaining amount. Falconbridge Dominicana, C. por A. also has 9½% mortgage bonds for company housing at the Dominican property, repayable monthly to 1993, in Dominican Republic currency.

Mortgage bonds may place severe limitation upon the borrowing power of the firm as well as requiring collateral. In the case of Falconbridge Dominicana:

> All loans are secured by a first mortgage on the assets of Falcondo (The Dominican property) and by a contract under which Falconbridge Nickel Mines Limited has agreed to buy all ferronickel of commercial value produced by Falcondo. Falconbridge is also obligated to provide 60% of all the funds required by Falcondo to enable it to meet its operating costs and debt service obligations in the event receipts from the sale of ferronickel produced by Falcondo and other receipts are insufficient for that purpose. (From the 1976 Annual Report).

EURODOLLAR BONDS

Eurocurrency is a deposit liability, denominated in a currency banked in a country outside of the country of origin. Eurodollars are USA dollars deposited in a bank outside of the United States. Other examples are French francs in London or d-marks in France.

The sources of Eurocurrency are foreign governments or business organizations who want to hold currency outside the country of origin and banks with foreign currency outside the country of origin. The demand for Eurocurrency comes from individuals, government and organizations that need funds. Phelps Dodge in 1975 had $25,000,000 or 7½% Eurodollar Notes due in 1977.

MISCELLANEOUS BONDS FINANCING

Below are listed several alternative methods of financing which are found in annual reports. One method becoming more popular is the Air Quality Control Obligation. These are bonds issued by the town in which the mining firm has operations. The proceeds from the bonds are used for pollution control purposes. Repayment is made by the company by agreement with the town. A typical bond of this type is illustrated in the 1975 Kennecott Annual Report.

> On May 16, 1975, the town of Hurley, New Mexico issued a 7 3/8% Pollution Control Revenue Bond in the amount of $21,000,000 payable in 20 equal quarterly instalments commencing August 1, 1980 and ending May 1, 1985. Proceeds of the bond issue were made to the Company for the purpose of financing certain pollution control facilities at Kennecott's Chino Mines Division.

Purchasers of mineral products often supply funds for the equipment and development of mining properties. For instance, the Atlas Consolidated Mining and Development Corporation 1976 Annual

Report shows:

Pesos 55,406,218; Loan from Mitsubishi Metal Corporation (MMC) with interest at the rate of 8 1/6% per annum, payable through deductions from the proceeds of concentrate shipments, collateralized by a standby letter of credit of US$2,000,000 which is automatically reinstated every year for a period of 10 years to 1980, in favour of MMC.

PUBLIC SOURCES OF LONG-TERM CREDIT

In the United States, the Export-Import Bank (Eximbank) is a federal agency specializing in foreign lending. This bank aids export trade through long-term direct capital loans for US goods or services, rebinding credits, an insurance and guarantee program for US exporters and guarantees for foreign financial institutions. An example is at the Feni-Rudnici nickel property in Yugoslavia. On May 30, 1974, the Bank authorized a $38.7 million limit loan at 6% per annum and two financial guarantees for $51.6 million. The loans were to buy US goods and technical services to develop an ore body, a smelter, and processing facilities.

In another case Kennecott in 1964 financed the enlargement of the El Teniente copper mine in Chile through sale of 51 percent of the corporate equity to the government and by a loan from Eximbank. The loan was to be repaid over a ten to fifteen year period and was to be unconditionally guaranteed by the government of Chile. This situation aided Kennecott in gaining compensation when the mine was nationalized.

In Canada, the Export Development Corporation (EDC) provides similar financing functions as does the Eximbank, i.e. guarantee loans, finance exporting, and insure overseas investments. For example, EDC provided $20.75 million and $57.25 million in financing the sales of Canadian goods and services to nickel mines in Guatemala and Indonesia. The loans were made on a commercial basis. The Export-Import Bank of Japan provides direct loans to exporters upon recommendation of the exporters commercial bank. Other governments such as the United Kingdom, France, Germany and Italy also provide direct loan for exported goods and services.

Several international agencies provide some financing, primarily for infrastructure in mining areas. This includes the World Bank Group composed of the International Bank for Reconstruction and Development (IBRD), the International Development Association (IDA), and the International Finance Corporation (IFC). The loans are generally made to governments for the purpose of developing infrastructure such as roads, power-generating facilities, etc.

LEASE FINANCING

One means of obtaining the use of buildings, land and equipment is leasing. In leasing, the lessor retains title to the facilities but makes them available to a lessee in return for periodic lease payments. The lease usually is limited to a certain number of years by the economic life of the physical asset with the lessor carrying the risk of obsolescence. The lessor gains the depreciation allowance, rental income, and any residual value remaining upon termination of the lease. The lessee gains the equipment and converts the cost from capital cost to expense, for the advantages.

The two main kinds of leasing are sale and leaseback, and direct leasing. Sale and leaseback occurs when a firm sells an asset it owns to an outside party, such as a financial institution, and then leases the asset from the outside party. The kind of financial institution involved in this kind of arrangement varies with the asset to be sold and leased back. Insurance companies usually deal with real estate; commercial banks and specialized leasing companies deal with equipment.

Direct leasing involves simply renting equipment and maybe service leases or financial leases. (See also Caldon in Chapter 3). Service leases provide the equipment as well as its maintenance. This type of lease is cancellable, that is, the lessee can return the equipment and cancel the lease before the full term of the lease has expired. Financial leases often do not include maintenance and cannot be cancelled in midstream. The lessor gains the full cost of the asset plus a financial return. Usually the firm arranges the purchase and delivery of the equipment and also arranges for the financial institution to buy the equipment. The firm and the financial institution form a lease agreement on the equipment. (Leveraged leasing is covered by Ravenscroft later in this Chapter 11.)

COMMON STOCK

The common stockholders collectively own a firm. They are residual owners in that they are entitled to the earnings and assets of the firm only after debt and preferred stock dividends have been paid. Common stock can be considered a perpetual loan which has no maturity date. In addition, the company has no legal obligations to pay dividends to common stockholders. Common stockholders have various rights in relation to the proceedings of the firm. They have the right to vote, inspect financial records (although this privilege is limited), elect the directors, and can authorize the issue of common stock, preferred stock, bonds, debentures and other securities. Voting privileges can be given to another person by proxy.

Common stockholders may sell their shares to another individual and receive the funds. This is usually done through a stockholder through a stock market. The only change this entails is

a new name on the stock certificate. There is
no change in the capital or financial status of
the company. However, if a company offers new
common stock, the common stockholders must be
offered enough shares to maintain their proportionate ownership in the firm.

Common stock may be considered authorized
or issued. In the ASARCO 1976 Annual Report
40,000,000 shares without par value are authorized
with 27,678,223 shares issued. Authorization
indicates the maximum number of shares the
company can issue. If the company amends its
charter, it can authorize more shares. Increasing
the authorization requires the assent of the
shareholders. Unissued shares permit the management flexibility to grant stock options, pursue
mergers, or raise additional capital on short
notice.

Issued or outstanding shares are that
portion of the authorized shares which have been
sold or issued by the corporation and held outside of the firm.

Company shares can be with or without par
value. ASARCO's shares are without par value
while Newmont shares have a par value of $1.60.
Par value is a stated monetary face value. Par
value simply provides information showing the
lowest possible investment behind the company's
historical offering of shares. It has no value
today and can confuse investors because a
particular value for the stock can be incorrectly
assumed. The value of a common stock is its
market price.

There can be more than one class of common
stock in a company. For instance, the International Nickel Company of Canada has Class A
and Class B common shares.

> The two classes are interconvertible at
> any time and are similar in all respects,
> including dividend rights, except that
> dividends on Class B shares may be declared
> payable out of "1971 capital surplus on
> hand" as defined by the Income Tax Act of
> Canada.

Some classes of shares may have different numbers
of votes. For example in the 1975 Teck Corporation Limited annual report indicates:

> The Class "A" shares carry the right to
> 100 votes per share and the Class "B"
> shares carry the right to 1 vote per share;
> in all other reports the Class "A" and "B"
> shares rank equally.

Common shares can be used to raise funds.
New mining firms often must resort to selling
common stocks to acquire funds because the
company has insufficient credit ability to raise
capital from other sources. In general, some
factors to consider in selling common shares to
acquire capital are:

1. Funds received are made available permanently.
2. The firm has no obligation to make
   dividend payments.
3. During buoyant stock market periods,
   common shares may be easy to sell at
   a relatively high price.
4. The credit standby of the firm is
   enhanced by the increase in equity.
5. Exiting shareholders may have their
   control diminished when new shares
   are sold.

(See also Gorval's paper later in this Chapter 11.)

PREFERRED STOCK

This is a debt that lies between common
stock and bond or debenture holding. Preferred
stockholders rank ahead of common shareholders
but behind the lenders in claims on the firms
assets. No dividends are paid to common shareholders until the preferred shareholders' dividends are paid. Preferred stock has voting
privileges, limited to special circumstances such
as when a period of time has passed without their
dividend being paid.

There are advantages to the firm in using
preferred stock to attain funds. Unlike debt,
where lack of payments has serious consequences,
there are no legal repercussions if dividends
are not paid. To preferred shareholders, however,
preferred stock often has provisions for carrying
unpaid dividends forward from one year to the
next. Before dividends are paid to common
stockholders, arrears from preferred stock must
be paid. A rare feature sometimes included in
preferred stock is called participation. In
this case the preferred stock is paid its stated
dividend and common stock is paid dividends up to
this amount. Any excess income remaining is
split between common and preferred stockholders.
Preferred stock can be retired by using the call
feature, a sinking fund deposit, or converting
the preferred stock into common stock.

The main advantage of preferred stock over
debt is that there is no legal obligation to pay
dividends. When cash becomes inadequate this
can be an important difference. Preferred stock
also increases the equity base from the view
point of creditors and therefore helps the
financial condition.

Newmont Mining Corporation illustrates
some complexities in preferred shares. In the
1976 Annual Report is shown:

> Preferred stock, par value $5.00;
> authorized 5,000,000 shares; issuable
> in series; Series A, $4.50 Cumulative
> Convertible (Liquidating value $100
> per share).

Preferred shares have a par value which is
meaningful. In the above case normally the
liquidating value would be $5.00 but it is stated
that this value is $100. The annual dividend is
$4.50. The cumulative feature is a protective

device - all past preferred dividends must be paid before any dividends on common stock are paid. Further is the Newmont report:

> Each share of Series A, $4.50 Cumulative Convertible Preferred Stock is convertible into 3 1/8 shares of common stock, and is callable for redemption in whole or in part at any time on at least 30 days notice at a price commencing with $104.50 per share on January 1, 1974 and declining annually thereafter by $0.50 to $100 a share on and after January 1, 1983.

CASE STUDIES: CAPITAL ACQUISITION

This section will illustrate with examples the varying means used to acquire investment funds for mining firms. In some inventories the actual names of the mining firm and the capital lending agent have been changed.

Case Study One

Introduction: The Borrower, X Corporation is the largest mining company in the Republic of Y. It is engaged in the mining of copper, ore and the processing of the ore into concentrate for export.

As a result of significant new copper discoveries, X has planned a substantial expansion of its existing operations. This overall expansion program is referred to herein as the "main-stage program", consisting of:

1. Development of an open pit mine to include a 35,000 short-tons-per-day (stpd) copper concentrator (total cost, approximately $100 million).

2. Construction of a copper smelter and refinery (total cost, approximately $200 million).

The "first-stage program" to be undertaken in advance of the expansion project above, consists of the purchase of machinery and equipment at an approximate cost of $30 million. A significant amount of this equipment has already been placed on order due to long delivery lead times. The "first-stage program" is to be financed through $30 million in Eurocurrency commercial bank loans.

This financing will eventually become an integral part of the main stage financing, which will be undertaken when economic and financial conditions permit. It is anticipated that many of the banks participating in the "first-stage" financing will also join in the main-stage financing. If the $300 million main-stage program does not go forward as anticipated, the equipment purchase with "first-stage" financing will serve as normal replacements for equipment in the Borrowers existing operations.

Summary of Loan

Borrower: X Corporation

Amount: US$30,000,000

Purpose: To finance the purchase of equipment needed for the development of the open-pit mine.

Latest Availability: June 30, 1975

Final Maturity: January 1, 1980

Repayment Schedule: Principal will be repayable in eight equal semi-annual instalments with payments commencing July 1, 1976.

Interest Rate: 2 1/4% per annum over the arithmetic mean (rounded upward to the nearest whole multiple of 1/8 of 1%) of the 6-month London interbank offered rate for relevant deposits as quoted in "Z" banks London office. Interest will be payable quarterly in arrears from the date of first drawdown.

Commitment Fee: 3/4% per annum of the undrawn portion payable quarterly in arrears, commencing March 3, 1975.

Taxes: All payments of principal and interest are to be free of host country withholding or other taxes (except "Z" banks' income taxes).

Reserve Requirement: The Borrower will reimburse any additional costs incurred by the lenders as a result of the imposition of any reserve requirements in respect of the loan or resulting from imposition of any interest equalization tax or other tax.

Minimum Participation: US$1 million for any one Bank.

Governing Law: The Loan will be governed and construed in accordance with the Laws of the United Kingdom.

Other Conditions

a. The making or continuation of the loan will be conditional upon the availability of funds in the offshore currency market.

b. The Borrower is required to maintain a minimum working capital ratio of 150%. In addition there will be certain restrictions on borrowing, on the mortgaging of property whether now owned or hereafter required, the paying of cash dividends, and the acquisition of the Borrowers Capital stock.

c. Appropriate covenants relating to the activities of the Borrower to the end that should the "main-stage program" not materialize or be significantly delayed, the first-stage program will be controlled so as not to impair the financial capability of the Borrower to service the first-stage financing.

d. Managing banks must succeed in syndicating 100% of the total financing required for the first-stage development program.

e. There shall be no material adverse changes affecting the Borrower prior to signing the loan agreement.

f. The Borrower shall have obtained all the

necessary approvals from the host country government or its agencies as required.

g. There shall exist no event of default under any loans or obligations.

Case Study Two

The following case study is from a Canadian bank in a recent (1975-76) loan which provided funds for capital investment and working capital.

Summary of Loan

Borrower: LMN Company

Amount: Loan #1 US$38,000,000 project term loan
Loan #2 US/Cdn$4,000,000 working capital

Purpose: To finance construction and development of mine and mill complex. To finance accounts receivable and inventory.

Availability: Drawdowns under the project term loan can be made until December 30, 1978.

Rates of Interest: Bank's US Base Rate plus 1 1/4% payable monthly on the basis of a 360-day year or selected maturities of London Interbank rate plus 1 1/2%, interest payable quarterly.

For Cdn dollars, the Bank's Prime Rate plus 3/4% interest payable monthly. For US dollars, the Bank's US Base Rate plus 3/4% interest payable monthly.

Standby Fee: To October 31, 1976, 1/4% per annum; thereafter 1/2% per annum on undrawn portion of loan #1.

Mandatory Repayments: Eighteen equal quarterly repayments of $2,111,111 commencing the earlier of September 30, 1979 or six months following completion date.

Following completion date, mandatory repayments may be deferred during periods of force majeure of up to six months, provided such deferred repayments are made in four equal quarterly instalments following the end of the force majeure period.

Final Maturity: December 31, 1983.

Mandatory Prepayments: In any fiscal year, excess cash flow (to be defined as cash generated above mandatory repayments and capital expenditures) is to be dedicated to prepayments on the Bank loan, payment of amounts outstanding under the deficiency guarantee and interest on any income debentures on a basis to be determined.

Security: Loan Agreement containing the normal positive and negative covenants (including but not limited to) ownership, operation of the project, maintenance of specific working capital level, prohibition against additional borrowings and pledging of security to others.

Other terms include:
- A first fixed and floating charge debenture on the mine-mill and mineral property.
- Assignment of proceeds to be received under the sales contract with the smelter.
- Assignment of Insurance Policy normal for operations similar in scope.
- Unconditional deficiency guarantee for up to US$10,000,000 from the smelter, deficiency guarantee to reduce pro-rata as repayments on the term loan are made.
- Controls on concentrate inventories and General Assignment of Book Debts.

Conditions Precedent: Infusion by PQR Company (co-venturer) or other party acceptable to the Bank of US$12,000,000 into the project (excluding present deferred development costs of approximately $4 million).

The borrower shall have entered into a long-term sales contract with the smelter with terms and conditions as outlined in the Technical & Economic Summary dated February 1976. (Not Included).

LMN Company and PQR Company have entered into a fixed-price contract for construction-development of the project on the approximate terms and conditions outlined in the PQR letter. The contract is to call for the completion of the mine-mill as per design specifications and satisfactory operations on a continuous basis at acceptable product quantity and quality levels on a basis to be agreed with the Bank.

The Loan Agreement and all security and supporting documents shall have been finalized and registered where necessary.

The project shall have been reviewed by the Bank's consultant and the Technical & Economic Summary shall have been confirmed to the Bank's satisfaction.

Other Terms and Conditions: Legal costs and all out-of-pocket expenses of the lender and its consultant related to the preparation of documentation to be for the account of the Borrower.

Case Study Three

The following is taken from the Financial Post, May 22, 1976:

Investment Group Tries New Wrinkle for Mine Financing

Mining exploration men for the past few years have been trying to evolve new methods of raising exploration money, faced with tight restrictions put in the way of such efforts by stock exchange and securities commissions which are proponents of the old ways of financing the penny-mine business.

The novel concept, the brainchild of the company director, attempts to reduce risks in the highly speculative mineral exploration field by mixing debt and equity to increase protection for investors.

Funds Reinvested

"We're breaking with traditional forms of mining investment in Canada to provide an investment vehicle combining safety of principal with the speculative appeal of former small exploration companies." The Director says, "Here's how the five-year plan works":

A new company, First Mineral Explorations, Inc., will offer $10.5 million of noninterest bearing collateral trust bonds, and two million common shares, in 10,000 units of one $1,000 bond and 200 shares. The price per unit: $1,050.

The $10 million capital (after underwriting and other costs) will be invested in short-term government and government-guaranteed securities as well as bank and trust company deposits.

The interest on the investment, estimated at $1 million per year, will be used for mineral exploration.

At the end of the five-year period, the bonds are redeemable.

The investors' return will be reflected in the value of the common shares, depending on the company's exploration success.

The bonds and common shares will be transferable only in units, not separately, until a date set by the directors, not before June 30, 1978, not later than June 30, 1980.

The draft prospectus notes that "in the opinion of the directors", the $1 million-per-year income "is adequate to enable the company to carry out a diversified, well-balanced exploration program".

It continues:

"It is anticipated that specific projects may be undertaken alone or jointly with others on a joint venture basis, and that exploration work may be done on the property of other mining companies in consideration for shares of such other companies and under agreements whereunder the write-offs for tax purposes will be available to the company rather than the company which owns the property."

Preliminary Prospectus

"As the bonds offered by this prospectus will have the benefit of a pledge of securities which will ensure their retirement in full at maturity, there will be no covenants in the trust deed securing the bonds, which would inhibit the company from borrowing moneys or otherwise financing itself for the purposes of acquiring directly or indirectly interests in properties thought to be capable of economic production and of bringing any such properties into production".

Case Study Four

The Tara lead and zinc project: (Ireland)

Financing of the Tara lead-zinc project at Navan, 30 miles north of Dublin, was arranged in two parts: US$100 million in bank loans and US$30 from the Canadian government's Export Development Corporation (EDC).

The $100 million portion was arranged by Toronto Dominion Bank and three Irish banks, Allied Irish Investment Bank, Bank of Ireland, and the Ulster Investment Bank. (The latter three provided 30% of the bank loan).

In addition, Toronto Dominion Bank Investments (UIC) Ltd. (owned by the TD bank), Citibank N.A., First National Bank of Chicago, Bank of Nova Scotia, Commerce International Trust Ltd. (owned by Canadian Imperial Bank of Commerce), and Banque Canadienne Nationale (Europe) - a subsidiary of Bank Canadiene Nationale - participated in the consortium.

These banks and buyer credit loans have been guaranteed unconditionally by Tara Exploration and Development Company which owns 75% interest in the Tara Mines Limited. The Irish government owns 25% of Tara mines. There is a standby facility of up to $15 million in the case of overruns to be provided by Bank of Nova Scotia and the TD and Commerce subsidiaries.

Case Study Five

From the New York Times, June 16, 1976

US Steel

The nation's largest steel-maker faced a dilemma in its financing decision. On the one hand, its ratio of debt to equity had risen to 32 percent - considered high for an AA-rated company - making debt inadvisable.

On the other hand its stock is selling below "book" or net asset value - $52.38 compared to $59 - making equity equally unattractive. Book value is the theoretical worth of the company, based on balance sheet figures.

Even though the market value of the company can vary widely analysts still worry about stock's being sold below book value because technically that means new stockholders get a piece of the company at cut-rate prices.

Pressed for capital to finance $800 million in planned expenditures this year, the steelmaker settled on a hybrid: convertible debentures - which come close to avoiding the pitfalls of both equity and debt.

Strictly speaking, the issue is debt - the company promises to repay $400 million to lenders on July 1, 2001 while making semiannual interest payments until then. The wrinkle is that debenture holders can turn their certificates in for stock at a set price, expected to be 10 to 12 percent above the stock price on the date when the debentures are sold.

In other words, if the stock market booms and the price of US Steel soars, debenture holders will have a chance to reap substantial capital gains by converting.

For this option, buyers of convertible issues will settle for a return expected to be 6 percent on the steel issue - markedly reducing the company's interest costs. But the major plus for US Steel in offering convertibles at a conversion price close to book value is that debt-to-equity ratio will drop without the dilutive effects of selling stock below book.

Financial experts expect the issue to sell easily - particularly to public pension funds, which are limited by law as to how much common stock they can hold.

Case Study Six

As the final case study, a more complete project financing scheme will be presented. Much of the data has been developed by Mr. Grover R. Castle, Vice President, Petroleum and Mineral Division, Chemical Bank, New York. This document includes an addendum to provide for political risk.

Proposed 3 1/4 Year Standby Eurodollar Commitment
Followed by a 3 1/2 Year Eurodollar Term Loan
to Canadian Coal Company

Amount: $75,000,000 US Dollars

Purpose: Canadian Coal Company (Coal Company) has been formed to construct and develop a new deep coal mine and preparation plant (Mine) in Canada. The Partners (Partners) in the Coal Company are listed below:

Partners	Percentage Ownership
Canadian Steel Company	35%
Canadian Mining Company	25%
Japanese Trading Company	20%
Provincial Government	20%
	100%

The Mine is estimated to have proved but undeveloped reserves exceeding 90 million tons of low-sulphur medium-volatile recoverable washed coal and the total projected cost of developing the mine is $100,000,000. The Proposed Loan shall finance 75% of the cost of the Mine and the balance of the cost shall be supplied by pro rata (in proportion to Percentage Ownership) equity investments by the Partners. Initial production of the Mine is expected to start in 1980 and the eventual design capacity of the Mine is planned at 3,000,000 tons per year. The coal is expected to be sold to Japanese utilities under long term contracts.

Standby Period: From September 30, 1977 through December 31, 1980.

Standby Fee: 1/2 of 1% per annum on the average daily unused portion payable at the end of each interest period.

Management Fee: To be determined.

Borrowings: Prior to each Borrowing the Banks shall receive a certificate indicating that the proceeds of the Borrowing shall be used for the development of the mine. Also at the time of each Borrowing, the Partners shall invest cash into the Coal Company which shall be used for the development of the Mine and which shall be in an amount equal to one third of the amount of the Borrowing.

Interest Rate: (X) percent over the three, six or nine month cost of funds for Eurodollars payable at the end of each quarter.

Amortization of Term Loan: Commencing March 31, 1981, the Proposed Loan shall mature in 14 quarterly instalments through June 30, 1984. The first seven instalments shall each equal 3.57% of the original amount of the Proposed Loan and the next six instalments shall each equal 10.71% of the original amount of the Proposed Loan. Assuming the full $75,000,000 is borrowed then the first seven instalments shall each equal $2,678,571, the next six instalments shall each equal $8,035,714, and the final instalment shall equal $8,035,719.

Alternate Interest Rate: If any Bank is unable to quote rates to prime banks in the London Interbank Market for offering of US Dollars, then such Bank shall no longer be obligated to make advances unless such Bank (in its sole discretion) and the Coal Company agree as to what interest rate shall apply to such advance.

Taxes and Governmental Action: All payments by the Coal Company (principal, interest, fees and reimbursement of expenses) will be made in US Dollars free and clear of any taxes, levies, imposts and duties (except for taxes on the overall profits of the Banks). The Coal Company shall pay additional interest to compensate for any such taxes.

If any change in applicable law or regulation subjects any Bank's Loan to any US interest equalization tax or any other tax, levy, impost or duty on loans to foreign borrowers, or changes the basis of taxation of payments to any Bank of principal, interest or fees payable (except for tax on overall profits of such Bank) or imposes or modifies its reserve requirements or imposes other conditions, to increase the cost of making or maintaining the Loan, the Coal Company shall pay additional interest to compensate for the resultant increase in the cost of making or maintaining the Loan.

In the event of any of the foregoing, the Coal Company shall be entitled to repay forthwith all or any part of the Proposed Loan then outstanding with accrued interest thereon and the funding loss, if any, whereupon that Bank shall be discharged from liability to make any further loans.

Prepayments: In whole or in part at the end of any interest period at the Coal Company's option without penalty. All Prepayments shall be applied against the Proposed Loan in the inverse order of maturity.

Collateral: The Proposed Loan shall be secured by a Real Estate Mortgage (Mortgage) on the Mine and Floating Charge (Floating Charge) on production from the Mine. The Debenture providing for the Mortgage and the Floating Charge shall also provide for an Assignment of the Sales Contracts (as hereinafter defined).

Other Conditions of Lending: Usual, plus prior to any borrowing hereunder:

1. All Governmental approvals and clearances shall be obtained.

2. The Banks shall be satisfied with an analysis of the Feasibility Report prepared by Canadian Engineering Company. This would include an inspection of the basic data and the Mine site and the Banks shall be satisfied that the Coal Company owns at the Mine in excess of 90 million tons of low sulphur medium volatile recoverable clean coal. The Banks shall also be satisfied with the feasibility of developing the Mine for $100,000,000.

3. The Banks shall be satisfied with the terms of the License (License), authorizing the Coal Company to exploit the coal reserves in the Mine and all consents required under the License for the Mortgage shall have been obtained from the Province of Y.

Undertakings by the Partners: The Partners shall:

A. Complete (Completion) the Mine prior to March 31, 1981. Completion shall include:
   1. The operation of the Mine for a continuous period of 120 days and during that period:
      a) the Mine shall have produced 950,000 tons of coal in accordance with a predetermined set of specifications;
      b) the cost of operating the Mine shall not exceed a level which shall be $23 per ton less than the escalated Sales Price provided in the Sales Contracts (hereinafter defined) and the cost of transporting the coal shall not exceed a level which shall be $28 per ton less than the escalated Sales Price in the Sales Contract.
   2. A requirement that all costs of Completing the Mine shall have been paid for by either the Proposed Loan or Partners Contributions.
   3. A requirement that on the date of Completion, no environmental suits shall be threatened or in process.
   4. A requirement that on the date of Completion, Sales Contracts (Sales Contracts) covering the sale of at least 85% of design capacity shall be in full force and effect. The purchasers under such Sales Contracts shall be acceptable to the Banks and the terms of the Sales Contract shall conform to the specifications set out below:

Term: At least 15 years
Price: The Base Price shall be $46 per ton and there shall be provision in the sales contract for escalation above the Base Price for all cost (of operating the Mine) increases subsequent to the date of the Loan Agreement. All payments under the Sales Contract shall be in US or Canadian Dollars.
Quality: The Sales Contract may include a penalty on the Sales Price if the coal does not meet the following minimum standards (average of each shipload):

Moisture	7.0 maximum
Inherent moisture, air dry basis	2.0% maximum
Volatile matter, air dry basis	25% maximum
Ash, air-dry basis	10% maximum
Sulphur, air-dry basis	0.5% maximum
BTU, air-dry basis	13,500 minimum
FSI	7.5 maximum

Force Majeure and Events of Default: Usual provisions in normal commercial contracts, but a purchaser shall not be permitted to cancel a Sales Contract in the event that coal can be purchased at a lower price from another seller.

   5. A requirement that on the date of Completion the Coal Company shall have Insurance (Insurance) on the Mine in amounts, covering risks and with underwriters all of which are satisfactory to the Banks. Such Insurance shall have a loss payable clause in favor of the Banks.

B. Represent and warrant that they own the License to exploit the coal reserves in the Mine and such license is in full force and effect; warrant that on the date of Completion all governmental clearances have been obtained; and warrant that on the date of Completion no amendments to the License have been signed since the copy that has been furnished to the Banks.

All undertakings by the Partners shall be on a several basis in proportion to their Ownership Percentage.

Covenants of the Coal Company:

A. The Coal Company shall:
   1. Maintain existence, pay taxes, royalties and liabilities when due, maintain the Mine in good repair, and defend the Mine against all claims.
   2. Operate the Mine in a good and workmanlike manner.
   3. Maintain Insurance on the Mine.

4. Furnish annual audited, quarterly interim unaudited statements and quarterly production reports.
5. Permit representatives of the Banks to inspect the Mine.
6. Give written notice of any claim or demand effecting title to the Mine or any adverse actions against such title.
7. Complete the Mine prior to March 31, 1981.
8. Pay all closing costs in connection with the Proposed Loan.
9. Maintain working capital of at least $1,800,000 at all times.
10. Maintain the License in full force and effect.

B. The Coal Company shall not:
1. Incur, create or permit to exist any indebtedness for borrowed money, except the Proposed Loan.
2. Incur, create or permit to exist any mortgage, lien or any encumbrance (including CSA's, purchase money mortgages and other title retention agreements and including any production payments) on any of its assets, whether now owned or hereafter acquired, except for the usual exceptions and the Collateral provided for herein.
3. Guarantee, endorse or otherwise in any way become contingently liable for any indebtedness of any other party (including take or pay or working capital maintenance agreements).
4. Incur or become liable for the payment of rent.
5. Acquire all or a substantial part of the assets or merge or consolidate with any other entity.
6. Sell all or a substantial part of its assets.
7. Make any loans or advances or investments in any other entity.
8. Make any distributions to the Partners or retire any of the partnership capital except that the Coal Company may make distributions out of 50% of Excess Cash Flow earned on a cumulative basis subsequent to the Completion of the Mine. Excess Cash Flow shall be defined as the amount by which the sum of net income plus depreciation plus all non cash charges earned in any year exceeds $24,000,000.
9. Amend the License or the Sales Contracts.
10. Engage in any business other than the operation of the Mine.
11. Grant any Section 88 lien to any Canadian bank.

Events of Default: In the event:

1. Of non payment of principal or interest when due.
2. Of misrepresentation
3. Of voluntary or involuntary bankruptcy of the Coal Company or any of the Partners (prior to Completion only as to the Partners).
4. Of cross acceleration.
5. The Coal Company is no longer owned by the Partners in proportion to their Percentage Ownership.
6. Of final judgment against the Coal Company.
7. Of default under any Sales Contract.
8. Of default under any other provision of the Loan Agreement.
9. The License is no longer in full force and effect.
10. The Mine produces less than 1,275,000 tons on an average basis in any six month period after Completion.

Subject to approval of Counsel for the Banks as to all legal details.

Canadian Coal Company Venture
(Modifications to Financial Plan in Order to provide for assumption of Political Risk)

Other Conditions of Lending: The Banks shall receive a letter from the Host Country addressed to the Banks confirming that:

a) there shall be no change in the provisions of the License (including those relating to taxes) during the payout of the Proposed Loan;
b) the Mine shall be given its fair share of goods and services at a fair cost;
c) the Host Country is aware of the terms of the Proposed Loan and shall not take any action that will have an adverse effect on the payout of the Proposed Loan.

Undertakings by the Partners: The Partners shall:

A. Complete (Completion) the Mine prior to March 31, 1981. Completion shall include:
1. The operation of the Mine for a continuous period of 120 days and during that period:
   a) the Mine shall have produced 950,000 tons of coal in accordance with a predetermined set of specifications.
   b) the cost of operating the Mine shall not exceed a level which shall be $23 per ton less than the escalated Sales Price provided in the Sales Contracts (hereinafter defined) and the cost of transporting the coal shall not exceed a level which shall be $28 per ton less than the escalated Sales Price in the Sales Contract.
2. A requirement that all costs of Completing the Mine shall have been paid for by either the Proposed Loan or Partners Contributions.
3. A requirement that on the date of Completion, no environmental suits shall be threatened or in process.
4. A requirement that on the date of Completion, Sales Contracts (Sales Contracts) covering the sale of at

least 85% of design capacity shall be in full force and effect. The purchasers under such Sales Contracts shall be acceptable to the Banks and the terms of the Sales Contracts shall conform to the specification set out below:

Term: At least 15 years
Price: The Base Price shall be $46 per ton and there shall be provision in the Sales Contract for escalation above the Base Price for all cost (of operating the Mine) increases subsequent to the date of the Loan Agreement. All payments under the Sales Contract shall be in US or Canadian Dollars.
Quality: The Sales Contracts may include a penalty on the Sales Price if the coal does not meet the following minimum standards (average of each shipload):

Moisture	7.0 maximum
Inherent moisture, air dry basis	2.0% maximum
Volatile matter, air dry basis	2.5% maximum
Ash, air-dry basis	10% maximum
Sulphur, air-dry basis	.5% maximum
BTU, air-dry basis	13,500 minimum
FSI	7.5 maximum

Force Majeure and Events of Default: Usual provisions in normal commercial contracts, but a purchaser shall not be permitted to cancel a Sales Contract in the event that coal can be purchased at a lower price from another seller.

5. A requirement that on the date of Completion the Coal Company shall have Insurance (Insurance) on the Mine in amounts, covering risks and with underwriters all of which are satisfactory to the Banks. Such Insurance shall have a loss payable clause in favor of the Banks.

B. Represent and warrant that they own the License to exploit the coal reserves in the Mine and such License is in full force and effect; warrant that on the date of Completion all governmental clearances have been obtained; and warrant that on the date of Completion no amendments to the License have been signed since the copy that has been furnished to the Banks.

In the event of a Political Event the Partners shall no longer be obligated to perform the Undertakings described above during the continuance of such Political Event. However, if such Political Event shall subsequently be corrected then the Undertakings hereunder shall henceforth be reinstated.

Political Event shall be defined as:
A. Any action by the Host Country (including taxation or legislation) which does not constitute the due exercise of rights under the License and which has the following consequences:
   (i) expropriation or nationlization;
   (ii) preventing the Coal Company from Completing the Mine;
   (iii) preventing the Partners from exercising their fundamental rights.

B. War, Insurrection or civil strife which prevents the Coal Company from Completing the Mine.

C. Failure of the Host Country or its Central Bank to convert local currency into US Dollars in order to permit the Coal Company to make the applications on the Proposed Loan that are provided for under the amortization section of this Outline.

All undertakings by the Partners shall be on a several basis in proportion to their Ownership Percentage and the Undertakings by the Host Country shall include a waiver of sovereign immunity.

Events of Default   In the event:

9. Of a Political Event.

SELECTED REFERENCES

Castle, Grover R., "Project Financing - Guidelines for the Commercial Banker", Journal of Commercial Bank Lending, April 1975, pp. 14-30.

Chender, Michael, "Financing New Production in the Copper Industry Calls for New Approaches", Mining Engineering, December 1976, pp 20-23.

Daniels, J.D., et al., International Business: Environment and Operations (Reading, Mass: Addison-Wesley, 1976).

Drechsler, Herbert D., "A Source of Capital : Multi-Nation Investment", Proceedings of the Council of Economics, AIME, Annual Meeting, San Francisco, California, 1972.

Drechsler, H. and Schwab, B., "Economic Evaluation of Mining Ventures with Capital Constraints: Interrelationships between Net Present Value, Payback, and Regret Due to Investment Delay", Proceedings of the IX World Mining Congress, III-2, May 1976, Dusseldorf, Germany.

Drechsler, H. and Stephenson, J., "The Effect of Inflation on the Evaluation of Mines", Canadian Mining and Metallurgical Bulletin, February 1977, pp. 76-82.

Gentry, Donald W., "Buy-or-Lease Decisions for Capital Equipment", Mining Engineering, September 1975, pp 39-43.

Horngren, Charles T., Accounting for Management Control, third edition, (Englewood Cliffs: Prentice-Hall, Inc. 1975).

Mackenzie, Brian W., "Evaluating the Economics of Mine Development", Canadian Mining Journal, December 1970.

Lusztig, Peter and Schwab, Bernhard, Managerial Finance in a Canadian Setting, (Toronto: Butterworth and Company, 1973).

Solomon, Ezra, The Theory of Financial Management (New York: Columbia University Press, 1963).

Ulatowski, Tomec, Frohling, Edward S, and Lewis, F. Milton, "Delaying Debt During Early Development: A New Approach to Mine Financing", Engineering and Mining Journal, May, 1977, pp 65-70.

Van Horne, James C., Financial Management and Policy (Englewood Cliffs, N.J.: Prentice-Hall, Inc. 1971).

Wearly, William R., "Implications of Escalating Capital and Development Costs", Mining Congress Journal, June 1976, pp 33-40.

Weingartner, M., "Some New Views on the Payback Period and Capital Budgeting Decision", Management Science, August 1979, pp B-594 to B-607.

Wells, Howard M., "Decision Under Uncertainty", Journal of the South African Institute of Mining and Metallurgy, April 1976, pp 375-382.

Weston, J. Fred and Brigham, Eugene F., Essentials of Managerial Finance, third edition (Hinsdale, Ill: The Dryden Press, 1974)

11.4

PROJECT FINANCE SUPPORTS AND STRUCTURING

C. Richard Tinsley

European Banking Company Limited
London, England

INTRODUCTION

Project financing has been defined by various authors in this book. It is founded on reliance upon the project's own future cash flows and secondarily on the collateral value of the asset base developed. It is chiefly provided by the banking community who, after an assessment of the risks involved (See Chapter 9) analyse the company's presentation (See Chapter 10) in order to structure a financing package often at a leverage of up to 80% debt. This paper describes the various support and structuring techniques used to delimit and abate the risk components perceived from the banks' experience on project development especially for the period before cash flow is generated (pre completion).

ADVANTAGES

The advantages of project financing extend beyond the offloading of risks to banks. Among the advantages are the following:

1. Longer loan maturity than conventional, unsecured financing;

2. Loan disbursement and repayment tailored to specific mine;

3. Ability of a joint venture without a "track record" to raise financing;

4. Obtain financing where further direct borrowing (corporate credit) limited by existing indentures;

5. Possibility of arranging finance off-balance sheets;

6. Leverage return on equity;

7. Obtain tax savings with certain project financing structures; and

8. Obtain flexible loan repayment conditions.

It can also be said that from the point of view of the company, project financing has certain disadvantages which can be summarised as follows:

1. Perceived higher cost (risk premium) compared with straight corporate credit;

2. More documentation; and

3. Need to negotiate risk-sharing aspects.

The main attraction of project financing is the ability to pass certain risks on to the lenders within a flexibly funded transaction. (For a more detailed description of the specific risks see the author's paper at the beginning of Chapter 9. The same risk system is followed here.)

The typical project financing has corporate support prior to completion and will usually carefully control the resultant cashflow after completion to ensure the bank does not unnecessarily assume further risks of an equity nature.

WITHIN THE COMPANY'S CONTROL

Operating Risk

This is broken into three interrelated segments for ease of description.

Technical Component: This must be specifically addressed if new technology or unproven adaptations are to be used. The banks will have scrutinized the sampling and bulk testing work, perhaps grilling the geological, mineral processing, metallurgical, and mining engineering staff or consultants directly. A number of supports are available such as:

TECHNOLOGY GUARANTEE: Either the company itself or the process developer will guarantee that the technology will work for a period of, say, six months to two years after startup. If not, then financial penalties will apply although these are rarely sufficient to repay the loan. Many process developers are weak financially and may have little substance to back such a guarantee.

TECHNOLOGY MANAGEMENT: This may be seen where proprietary technology developments are to be provided to a project. The aluminium producers favor this as a means of control and to obtain an extra return from their R&D and investment in the new venture. The management agreement reassures the banks that new technological developments will continue to be made available to the project so that it can maintain its technological competitiveness.

TECHNOLOGY INSURANCE: In some cases, insurance cover may be raised from unconventional or even conventional insurance markets. (See Berry's paper, Chapter 9.) This can be expected to be difficult for brand-new technologies.

QUALITY ASSURANCE: An equipment supplier or technology licenser may guarantee the resultant product's specifications which to a large part can cover the technical component of operating risk. This is often seen in the industrial minerals industry.

FLEET ASSURANCE: An equipment provider or lessor may guarantee to replace individual items found faulty from a fleet "in the neighborhood." This may be a little difficult for a dragline in, say, Thailand although it may be easy for a piece of dredging equipment there.

ALTERNATE SOURCING: If the technology does not work out then the company undertakes to provide ore from elsewhere (if the mine) or treat the ore elsewhere (if the plant) with no impact on cashflows. This is rarely seen.

BUSINESS INTERRUPTION INSURANCE which takes effect upon startup can provide useful protection of the assets or the loan in the event of lost income arising from an accident. If a total loss, the banks will often want to be the loss payee under this type of insurance policy as well as the all-risks general insurance policy.

Cost Component: This is often confused with market risk but it is a key component in measuring the robustness of a project when facing market risk. The usual measure here is the cumulative production - cost curve, if properly prepared gives both operating and total costs and show the changes expected over time. (Cost curve analysis has not done too well in factoring in dramatic currency changes in the last few years.) The major supports are as follows:

SALES CONTRACT: Escalation provisions are the chief tool to cover cost risk and have many formulations in descending order of support:

Base Price plus actual escalation
               formula escalation
               floor price (escalation)
Fixed Prices with inbuilt escalation (but pre-

Market Prices linked to others; to the cheapest alternative; to a published index/price.

Hopefully, the minerals producer can keep any productivity gain but this is sometimes caught in the formula. Additional bias in sales contract components to cover the risk in cost elements, e.g. energy, can help here.

COST GUARANTEE: Sometimes suppliers to the industry will guarantee cost components either of consumables or even resulting from their plant and equipment. A guaranteed reagent consumption or a low-cost hydro power cost are examples here. Sometimes, advantageous transfer pricing can provide a support here within a large company. Fixed treatment and refining charges or low concentrate penalties may provide such support.

COST WAIVERS: A government will sometimes waive taxes or royalties for a period of time, often at least until a project loan has been retired or payback has been achieved, particularly if the project is a generator of export earnings. (This has, however, an element of political risk should a regime change or the law or decree be repealed). The company itself may waive management charges, royalties, dividend rights, or interest charges for a specified period. The capital component of infrastructure charges may be held flat or waived until debt service is retired.

ECONOMIC TEST: Should operating costs continually exceed cash inflow, say for one or two years, then some project financings (usually with rock-solid reserve positions) permit the operator to close the mine on the condition that if this condition is reversed, then the situation will be studied by an independent engineer and, if a reopening recommendation is made, then the mine will be required to be restarted.

Management Component: Besides market risk, this is perhaps the most intangible yet vital component to the operation's cash flows which are the very core of project financing. Besides high skills and a proven track record, lenders may seek comfort from:

MANAGEMENT AGREEMENTS whereby (a) the key staff are encouraged to stay with the development or (b) the sponsor agrees to put whatever management talent it has or can find to the task of building and operating the project. Bonuses or share options are the most usual incentives besides cash. This type of agreement is common where the company either has experience with the technology or regional conditions.

KEY-MAN INSURANCE: This may literally insure the life of a few key people to a level equal to the debt outstanding should one or more of these individuals die.

LABOR CONTRACTS may be sufficiently firm and reliable to underpin this risk component. This can generally only occur in a mature industry. Texasgulf's decision to put all of its Kidd Creek, Ontario, staff on salary is a good example here.

SOURCES OF FINANCE

TRAINING AGREEMENTS are often part of a technology supply agreement and provides useful support. There may even be a commitment to supply the executive staff for a period of years after startup. Some training agreements imposed by governments to enforce greater participation by their nationals in the industry, may increase the management risk eventually especially where tribal rivalries are hard to control.

Participant Risk

The credit standing, technical and managerial quality of a participant may call for special structures in a project financing. The most common problem facing the project financies is a financially weak sponsor. A financially strong company chooses to project finance less often since it can access other funding sources on a corporate basis often at a lower cost and with less hassle compared with documenting the complex risk-sharing mechanisms and supports in a project financing. Some large companies feel that they cannot "take a walk" from a mine that is project financed since it has their name associated with it; some banks depend on these companies feeling that way; and consequently the credit basis is essentially the corporation and not just the project.

The main support structures to handle participant risk in a project financing are as follows:

Joint Venture Agreement: This document often defines the joint venturers interrelationships among themselves which, if there is an inbalance in creditworthiness, can add to the overall credit support to the project financing. Conversely, sometimes these provisions such as cross-charging or compulsory contributions/dilutions can interfere with the project financing.

Contingent Financial Support: This can be provided to meet the Completion risk or may provide on going support to the financing such as working capital maintenance agreements or contingent equity underwriting agreements or guarantees from a strong parent.

Financial Ratios: The credit status of the participant may be preserved at least until completion through a set of financial ratios such as a current ratio of 1.2 to 1.0; debt to tangible net worth not greater than 2.0 to 1.0; minimum tangible net worth; maximum amount of long-term debt. Certain subordination structures may also be necessary to control intercompany loans, charges, dividends, and legal security.

Off-Balance-Sheet: The credit risk of some highly leveraged participants may be further stage managed through the use of off-balance-sheet structures which usually require corporate techniques to reduce ownership below 50% (rather than financing techniques). Accounting standards around the world are moving toward putting most financial obligations on the balance sheet and even counting it as long-term debt. The capitalisation of lease obligations on balance sheet is a good example here for a financing tool that used to have this source of funds off-balance-sheet.

Cross-collateralization and cross-default clauses can gather in support from other project participants or even other non-project activities of the sponsors.

Engineering

Flaws in capital budgeting or poor design estimations are only coverable by outsiders under:

Insurance: A professional indemnity insurance policy may cover such items as errors and omissions but usually nowhere near the level of cover commensurate with the project finance loan size. Other unconventional insurance policies can be obtained (See Berry, Chapter 9) for design errors at a more realistic cover level.

Independent Certification whereby an independent authority (hopefully with a good credit standing) will warrant that it agrees with the project studies on the basis of its due diligence exercise.

Completion Risk

This is the one risk the banks most fear to take. In most project financings, this is certain to be the risk the banks do not take. Numerous devices have been structured to package this risk more tolerably and the author can identify at least six major varieties of completion support.

Completion Guarantee: This requires the loan to be repaid by the company rather than the project's cash flows either in full or on an agreed repayment schedule if the mine is not completed by a certain date. It is seldom used in project financing today.

Completion Undertaking: The sponsors undertake to put in however much money it takes to make the project meet the objective completion test (refer to the author's paper in Chapter 9). The completion test will, in all likelihood, pick up other risk categories as well. Therefore, the completion test is achieving support for other risk areas. In a disaster situation then there may be recourse to the sponsors for the whole loan.

Overrun Undertaking: In this case the sponsors agree to provide only the overrun amounts above the pre-agreed debt and equity financing total.

Equity and Debt Subscription: In the event of a cost overrun or a delay forcing capitalization of interest, both the banks and the company agree to contribute further funds, either pro-rata or in leap-frogging tranches, to meet the overrun with a ceiling on the banks' exposure at some stage.

Standby Facility: These may be contingent underwriting facilities provided to cover cost overruns perhaps from strong parent or related companies.

For example, interest may only be permitted to be capitalised to a given cumulative dollar ceiling beyond which the sponsors must pay interest when due.

Default Agreement: This is a variation of a completion guarantee except for a "stand-still" period after all the base financing has been provided, say 90 days within which talks proceed between the banks and the sponsors to hopefully reach agreement on how to finance the increment to achieve project completion. The commitment of both sides to proceed may be halted early if estimates show the project will come in above the financed budget.

The definition of the completion test and these completion support agreements is often the most testy in project finance negotiations. Once completion has been achieved the project commences its limited-recourse or sometimes non-recourse status and the banks can no longer go back to the sponsors and their other activities for payments under the project finance agreement.

Insurance: This can cover completion risk either through contract works policies or startup delay policies which latter may also cover the capitalised interest bill caused by a delay.

Turnkey Contract: The turnkey construction contract may be framed in such a way as to provide the necessary support for the completion aspect of a project financing. However, most incorporate project testing periods which may be acceptable at measuring physical completion but are not long enough to cover the performance completion test in a project-finance document. The various types of equipment or technical guarantees referred to under the technical risks section above can also be integrated with this turnkey support.

OUTSIDE THE COMPANY'S CONTROL

Reserve Risk

The usual rule of thumb is that proven reserve life needs to be more than twice the loan life. One of the difficulties here is the definition of what is "proven" or of a lesser reserve category. An independent reserve report is usually required.

Should reserves fall short of the required tonnage at the right classification, the following can be used:

Reserve Undertaking: If periodically or at loan maturity, reserves do not achieve a pre-agreed standard then the bank may have recourse to the sponsors for all or a portion of the project loan outstanding at the time, under an undertaking document by which the sponsor warranted that the reserves would meet the standard.

Reserve Proveup: The company must convert tonnages from a lower classification to a "proven" standard each year, before a certain date or even startup. These usually must be certified by an independent consultant. In not, recourse to the sponsor's other cashflow and assets is required or further collateral has to be provided to the bank.

Reserve Depletion Protection: If expanded reserves are not proven up (or tonnage is lost because of poor mining practices or an accident) then the debt repayment schedule will be faster. This support structure is widely used in oil and gas financings keyed to a net present value (NPV) of reserves according to the "decline curve" of production and field depletion. To an extent, this NPV test also covers operating as well as market risk.

Reserve Collateral: Until reserves reach the "proven" standard, additional collateral is held by the bank and is progressively released as the loan principal reduces or tonnage meets the standard. Such collateral could be access to other reserves and their cashflows, cash, shares, or guarantees.

Reserve Weighting: Banks hate to do it, but discounted probable, possible (inferred, indicated) reserve tonnages may be added to the proven tonnage to make up the loan life coverage required by the bank.

Reserve Insurance: This has been used in the oil and gas industry (See Berry's paper in Chapter 9). There is no reason why the minerals industry, which usually has a quantum leap in better reserve definition than oil and gas reservoirs, cannot obtain insurance coverage.

Market Risk

This key risk in project financing is the most difficult to cover and usually has the weakest support overall within the project's collection of risks. Supports and structures applicable here include:

Sales Contracts which, besides covering the cost component of Operating Risk (See above), provide support for market risk but true "take-or-pay" contracts are now rare in the minerals industry. The most common variety is "take-and-pay" although, as markets world-wide have become weaker in recessionary times, this has unfortunately shifted to "take-or-breach." The predilection for equity or fairness clauses in oriental purchase contracts, while seeming to balance the commercial reality between supplier and buyer, are too frequently a cause for price reopeners.

A sales contract should preferably be longer than the loan life (unusual to achieve) and should have protection for the supplier against contract cancellation and force majeure. Some contracts may provide short-term credit support through cash on presenting "clean" shipping documents, perhaps with such payments backed up by a bond.

Merchant Financing: If a metals-trading company is providing money and an offtake contract

SOURCES OF FINANCE

(usually to secure an exclusivity on the sales agency for the project), then this support can be useful provided the metal-trading-company's market position and credit standing is sound. This route often covers volume offtake but seldom offers solid long-term price protection.

Consumer Financing: If the consumers are providing debt or equity (or both) plus a contract, then this ties in their support to the project. (See the Ranger uranium case study in Chapter 12.) It does, however, provide the consumer with too much insight into operating costs and market strategies such that too high a percentage of this form of tied support can be hazardous.

Buy-back Clauses: If the product remains unsold, sometimes a buy-back arrangement may be structured at least to accommodate debt service repayments. This can also come in the form of a market displacement guarantee whereby the producer agrees to reduce production at an existing plant in favor of the new producer to allow the shortfall to be placed within the existing customer base.

Advanced Sales: The money required for development can sometimes be entirely from an advanced sale, provided tax on such a large block of income can be properly handled. This has been popular in the uranium industry and to a much lesser extent in the aluminium industry.

Throughput Agreements are fairly simple support tools that have developed more strongly in the oil and gas industries particularly for common-user facilities such as pipelines. The users covenant to put sufficient material through the asset or entity involved to generate sufficient throughput revenue (through tolling charges or handling fees) to retire the debt for the asset/entity. Severe constraints on impossibility of performance and force majeure ensure that the cashflow eventuates more or less on a "hell-or-high-water" basis. This bite or recourse aspect is the primary reason why so few are seen in minerals industry financings.

Infrastructure Risk

This is a large component of many mining projects since discoveries are most often in remote areas. The types of supports available here are:

F.O.B. Sales Contracts may put the risk of infrastructure upon the buyer if it is phrased as at the mine gate or well head. This can, however, have a sting in that the purchaser may choose not to arrange to collect the product e.g. as the Japanese have been doing by delaying or refusing to send coal carriers to various ports around the world. If a mining company has no say in the infrastructure chain, it risks technical absolesence, for example, if rail lines are not maintained or ports not deepened to keep up with the growth in bulk carrier size.

Pooled Infrastructure Agreements are popular in some mining districts particularly for water supply, power, telecommunications, the townsite and port. Often a new entrant has to rebate earlier participants contributions but in many cases a new company benefits from the costs of the existing infrastructure. Companies have a harder time getting together on overland transportation.

Government Commitments: The government may commit to fund, develop, and maintain the requisite infrastructure. This should ensure their support for the success of the venture. However, it does present the government with an easy way to levy indirect taxes through freight rates and handling charges. If matters are not progressing well on the project's economic performance or governmental relations, it may be tempting for government to seek control or a shareholding using the infrastructure card as the point of pressure.

Environmental Risk

This is coming to the forefront of risks to be covered in a project financing. Although the environmental impact measurement is requiring high-tech equipment and understanding, some support mechanisms have been developed, such as:

Rehabilitation Guarantee: This is a pool of funds or a guarantee provided to the local authorities and is established at a level to rehabilitate the area more or less should the mine/plant have to shut down for any reason. (See Ranger uranium case study in Chapter 12 for one of the first and largest such guarantee which was eventually provided by the project-finance banks.) This type of rehabilitation bonding on a smaller scale has been caused by surface-mining regulations in particular. State authorities are also increasingly demanding in this area.

Environmental Management: This may be provided voluntarily or by regulation whereby alternate operating strategies can be adopted to bring operations into compliance in any one day or season.

Rehabilitation Waiver: Some states waive environmental requirements in areas where mining is benign or where the area is beyond reprieve. Such a waiver may need to be additionally supported by letters of comfort or commitment from the ministries concerned or perhaps within the development act, decree, or agreement with government.

Environmental Warranty: This transfers the risk to the company by giving the banks recourse should environmental problems endanger cashflow and, in turn, debt service.

Political Risk

This risk is one that the banks in their wisdom felt most equipped to handle, that is until the LDC debt mountain started to cast a shadow over loan portfolios everywhere. Nonetheless this is a prime motivation to involve project financing since it brings a wider constituency to the forefront of a government's thinking on a project. Some of the main structures and supports are listed below:

Development Agreement: This may be phrased in strong language with external arbitration provisions. It will never, however, control an intransigent sovereign power. The emergency provisions within a country, when invoked, can usually overwhelm the development agreement.

Insurance: Some political risk agencies such as OPIC in the US, Lloyds of London, and various private insurers (See Chapter 9) will provide very effective insurance against currency inconvertibility, creeping nationalisation, unfair calling of performance bonds, and the like. Usually a delay of six months is built in to ensure that the government action is not a temporary squeeze.

War and Insurrection Residual: In some tender political locations, banks have been willing to absorb some of this risk due to damage caused to the mine or plant from such a reason. For example, the first $10 million in damage may be accepted by the banks. This is still a rarity.

Tax Indemnification: In almost all project financings, the banks bear the brunt of subsequent tax changes (short of "creeping nationlisation"). However, should tax rates begin to jeopardize debt service, then there may be recourse for the increment of taxes. It may also trigger a more rapid armotization of the loan. Imposition of additional interest withholding taxes is most commonly borne by the sponsors.

Offshore Payment Agent: One route to provide temporary or accruing support, especially against currency inconvertibility, is to ensure that all foreign exchange is kept offshore, possibly in a secured account, and debt service is made from this account. Due to the debt rescheduling restraints so pervasive in LDC countries, this is an increasingly popular route in export-oriented project financings.

Currency Inconvertibility Agreements usually provide for the accrual and local investment of blocked currency with a view to repatriation at a later date. This may also require the sponsor to make good either through a currency swap or perhaps an asset swap or access to additional collateral such as a guarantee. This is a form of Foreign Exchange risk.

Cofinancings: The provision of a project financing in parallel with a national or supranational body loan can be strong protection to the project lenders. The theory goes that a government will generally exempt such financings from freezes, embargoes, moratoria, or reschedulings since it cannot afford to lose the support of that national or supranational body in the greater scheme of its national debt. It is not a foolproof method, but is a powerful support in a project financing.

Local National Participation: An effort to include local nationals in a project as equity investors and local banks as lenders can forestall adverse government actions. This may involve the construction of different classes of shareholders and debt.

Force Majeure

This is one of the risks which seems to affect almost every mining project at some stage of its development or operation. Supports can come from:

Insurance can cover acts of nature but is less able to cope with acts of man or acts of government, for example the pervasive was risks exclusions in most insurance policies. Business interruption insurance perhaps applies more readily here than under the technical component of Operating risk since, if negligence or management causes can be proven, most insurance policies would not pay up.

Deferral provisions in the credit agreement can provide for automatic deferral of principal payments due to poor or no cashflows caused by force majeure. There may be a limit in the deferral amount or extension period; equally there may be a "claw-back" mechanism to prior cash distributions to the sponsors should a shortfall occur; or there may be a balance provided with prior debt service paid as expected or ahead of time.

Foreign Exchange

This risk arises when cash flow is emerging in one currency and debt service required in another currency. However, the underlying commodity price may provide a natural hedge e.g. a gold project with a US$ project financing.

Hedging: Various techniques provide for relatively short-term foreign exchange hedging either with banks or with currency futures contracts on the IMM in Chicago.

Swaps: Longer-term currency hedges can be arranged through swap transactions either on a back-to-back basis or through parallel loans. These are unusual within project financing structures.

Barter: Various elements of countertrade can be utilised to exchange capital from different currencies for the hardware to build the project and the repayment for such plant and equipment. This is not always a perfect hedge since the accounts typically have to be established in one or another currency.

WITHIN THE BANK'S CONTROL

Syndication Risk

This risk arises when the banking syndicate is being arranged. It is covered by:

Underwriting Agreement: This provides that the lead bank(s) will provide all of the project financing required whether or not other banks are willing to join the bank syndicate later and participate in the risk.

Broad Syndication: Addition of banks who have (1) relationships with the purchasers (2) experience in project financings, (3) experience of operating in the particular region, and (4) represent a diverse set of international banks can give support to a project financing. The second point is important as there is nothing worse than a panicking banker trying to address a waiver down the line without that banker having a commercial and experiential background on the matter in question.

Funding Risk

This risk that interest rates will get out of hand is difficult to cover off through supports and guarantees, although some techniques can be applied. It may be recalled that most project financings are funded with floating-rate money due to the repayment flexibility required for a sound structure.

Interest Make-up Agreement: The parent/sponsor may agree to bear all interest expense in excess of, say, 15% p.a. Sometimes, this excess may be recaptured from future cashflows or the sponsor may have built up a paper "credit" when interest rates were below this level.

Interest Protection Agreement: The banks may also agree to put a ceiling (cap) on interest rates if they float above a pre-agreed level. Again this could usually be recaptured when interest rates returned below the agreed level. This is not popular with banks after some major money-center banks got into trouble with big portfolios of this type of loan in the late 1970s.

Hedging: Some hedge mechanisms are available through forward interest-rate markets e.g. the IMM in Chicago, the LIFFE in London, and in some measure should longer-term Libor funding periods be permitted (beyond the normal three and six-month interest periods). These range in term from a few days to 18 months on the exchanges to possibly a few years for Libor.

Swaps: Interest-rate swaps, i.e. floating to fixed, can be layered into a project financing although it is unlikely to be on the full amount because of the penalties should a swap have to be broken. Swap length is usually one to seven years. (See Poole's paper in Chapter 9.)

Alternate Funding: The option to lower-cost sources of funds can, of course, help reduce this risk. Gold funding and financial instruments such as commercial-paper (covered elsewhere in this Chapter) provide flexible alternatives whereas export credits or leasing tend to have rigid repayment terms viz one cannot readily stop leasing and then re-lease the equipment a few months later. These are accessed via letters of credit or guarantees; therefore, a bank/consortium credit is substituted for the project risk in the eyes of the other funding sources.

Supplier Credits: Sometimes suppliers have access to subsidized local funding which enable them to offer favorable financing terms to the mining project. Some suppliers may use their own lower-cost funding pool or may grant concessionary financing terms in order to clinch an order.

Legal Risk

The risk of professional advisers creating unworkable, faulty, or unenforceable documentary structures are difficult to cover but some supports are available such as:

Title Insurance: This can sometimes be effective to cover deficiency in the title to tenure of the mine site. Title searches should be done professionally.

Legal Opinions: This element of the project finance documentation needs to be thoroughly scrutinized well before loan signing to ensure the lawyers have not embodied serious caveats into their opinions. A second opinion or jurisprudential statements, e.g. from a Queen's Counsel, may be well worth the expense. International joint ventures can be tricky from a cross-border and language standpoint.

SECURITY

The security requirements for a project financing are extensive and usually include:

- a mortgage or first charge (fixed and floating)
- rights to the mining titles, leases, licenses, or tenements
- assignment of the sales contracts
- priority/subordination of joint venture agreement elements
- assignment of the management, technology, or operating agreements
- cash deposit collateral accounts
- foreign paying-agency agreements
- assignment of insurance contract

Conventional production loans can have all of the above whereas production payments in the US may only have a mortgage over the production payment itself. A production payment is a variation of a forward purchase arrangement which has accepted contractual and accounting criteria in some jurisdictions. It was a popular technique for US coal projects until tax benefits were eliminated in 1969.

DEBT : EQUITY RATIO

A continuum of ratios is possible and the maximum amount of debt will finally be determined from sensitivity analysis. In general, the higher the percentage debt the tighter the project finance structure, supports, undertakings, and covenants. For example, a solid layer of supports may be required to assure on-going operating costs are soundly covered. Tougher sales contract clauses will be required in high debt situations.

TAX EFFICIENCY

A major component of project finance structuring, especially for multiparty arrangements (See Cockburn's paper in Chapter 2) is the taxation impact. If a company is in a tax-loss position for the foreseeable future, it should examine the structures which ensure that it does not forego any tax deductions and that it either lowers its deductions from the project's development or passes them on to others for value. A lease is an obvious candidate for study in this context. Trusts can also have the desired effect. Both trusts and leases are often expensive to establish because of the thickness of the documentation. Preference capital can also be tax effective in some jurisdictions.

JOINT VEHICLES

A joint borrowing or sales company structure may be desired by parties of generally equal credit standings. This can also achieve an off-balance-sheet financing effect if there are more than two parties and any party is below 50% ownership.

THIRD-PARTY GUARANTORS

Some governments or strong multinationals may be prepared to issue guarantees for some of the project's financing. Export-credit agencies when in the supplier credit mode provide such a guarantee and they may require a government or corporate guarantee when providing a buyer credit.

Some of these supports are viewed as particularly strong if they have a maturity longer than the expected project financing. Sometimes the banks' own later maturities are guaranteed by the government agency or a supranational body.

CONCLUSIONS

Selection of the support structures in a project financing is heavily driven by the risks to be covered and the balance sheet and tax objectives of the sponsor(s). The large number of supports described above should not become a checklist since project financing requires sophisticated judgements to arrive at the trade-off in these various supports to achieve both the lender's and the borrower's objectives and risk comfort levels. Yet the complexity of supports demonstrates the flexibility that has needed to be developed in order to make this source of funds flow.

Project financings are live documents i.e. one can expect changing conditions to have to be accommodated. The fewer loan documentary waivers or amendments required on an on-going basis is a good measure of the skill with which to documentation was stitched together.

Just to bring all these complexities together, a term sheet follows that has been handed around the minerals project finance community in recent years. It is understood to have been developed in 1977.

Mr. Uno Grande
The State Mining Co.,
Ruritania.

Dear Uno,

Please pardon me for writing this on the back of an envelope. It's all I had on the plane back to London.

As I said, we and the bank syndicate can let you have the US$100,000,000 for your big new hole in the ground.

This is how we see the deal: -

1. Send us a telex

    You can have the money any time up to 12 months from now. Just send us a telex. In good time please, say, five banking days. Big round amounts only, we don't deal in peanuts. We will each chip in our bit and no more.

2. We've got shareholders too

    Pay us back our money. Eight equal lots, one every six months starting 30 months from now.

3. You want out

    If you want to pay back early, that's fine, but you must call us up 30 days ahead and pay us 1/4% consolation fee. Big rounded amounts only. Early pay-back means a shorter deal.

4. Milk of human kindness

    As I explained to you, we unfortunately have to charge for this money. I congratulate you on beating us down to 1% over the London Interbank Offered Rate (LIBOR). Charity runs in our blood.

    We will fix the interest rate every three or six months at your choice (five banking days again, please). You pay the interest at the end of each period. I hope you took on board my explanation of how we work out interest periods, the 11.00 a.m. (London time) routine, etc. Remember? Anyway, just leave the mechanics to us as we always do it.

    If you don't pay on the nail, we can add an

extra 1% to the usual rate till you pay up.

5. **That's your problem**

   I know it's very difficult for you to understand, but we don't carry the $100,000,000 around in our pockets. We have to get it elsewhere in London. If we can't, naturally we'll get round a table with you and talk about other ways and means. But if we don't see eye to eye after, say, 30 days, you pay us back. And that's the end. It's too bad that you may not be able to get the money either, that's your problem.

6. **The Taxman**

   You pay us in spendable dollars of the US of A. at our New York agent in Clearing House Funds. And we want the full amount, i.e. you pay the taxman and top up our money.

7. **Extras**

   So far, the authorities have left us alone. Cost to us is cost to you plus spread. But we have to face facts. Some central bank, taxman or other like person may decide to poke his nose into our business. The deal could become more pricey for one of us. Reserves, different taxes, that sort of thing. If that happens, you pay us the extra. We will tell you how much and you can't argue. But if any of us banks ups the cost, then you can take him out.

8. **We don't want to go to jail**

   If our side of the deal runs foul of the law, no more money from the bank affected and you take him out straightaway. Plus the unwinding costs.

9. **The paperwork**

   You can't have any money until your directors, the central bank and our lawyers have given us their O.K. the way we like it.

10. **Promises, promises**

    You promise us:
    i) Your company is there in good shape;
    ii) Your company can do this deal and you, Uno, can sign;
    iii) It's all legal;
    iv) The authorities have given their thumbs-up;
    v) No mistakes in your last financials; Things haven't got worse since then;
    vi) Nobody's suing you for big money;
    vii) You are sticking by the terms of your other deals;
    viii) The fact-sheet we sent round about you sets it out like it is.

11. **Do's and Don'ts**

    a) Don't put your assets in hock;
    b) If the balloon goes up, we get equal pay-out with your other deals;
    c) Send us your fiscals within 90 days of year-end;
    d) Send us other info when we ask for it;
    e) Dig the hole asap.

12. **The plug**

    Our money back straightaway and not another cent if:
    i) You don't comply;
    ii) You have told us a lie;
    iii) You don't stick by the terms of your other deals;
    iv) You go bust;
    v) You vanish;
    vi) Your other creditors move in;
    vii) We don't like the way things are going for you financially;
    viii) Your hole in the ground doesn't get dug like you said or fills up with water; etc.

13. **No stabs in the back**

    Uno, this bit is between us and the banks.

    You, colleagues in the syndicate, appoint us as your leader to run this deal. We are delighted to be of assistance and value your esteemed confidence. But, just to avoid any unpleasant misunderstandings, we have to make some things clear. It's every man for himself. We don't have to tell you what we know: you check it out yourselves. If we have slipped from the very highest standards of veracity in order to get you into this deal, keep your eyes open next time. We can believe everything the lawyers or anybody else tells us. We can do other deals with the borrower and pocket the profit. If it's between us and you, we can look after No.1. Naturally, we will do what most of you want within reason, but if we foul up, no liability. Sorry.

14. **Boiler-plate**

    You can mostly skip this part, Uno, since it's the boiler-plate.

    a) You will pay us our out-of-pockets, including the lawyers. I much enjoyed eating out in Ruritania at your expense;
    b) We could lose money if you don't pay when we said. You will see us whole, especially for the unwinds;
    c) You pay the stamps;
    d) If we turn a blind eye once, it doesn't mean we'll do so next time;
    e) We don't have to write it all out here - we can still throw the book at you;
    f) We can give other banks a slice of the action any time. We can switch to our other offices;

g) I'm a lousy linguist and I don't speak Ruritanian, beautiful language though it is. Please help out with translations;

h) If the judge gives us dinarios, etc. you make up the difference;

i) If you don't pay, we can grab any money you lodge with us.

15. The Rules

I was most touched by your patriotism, Uno, but you must appreciate that if we play by your Ruritanian rules, His Most Majestic Excellency The Sun King of Ruritania can change the rules in the middle of the game. So, if you don't mind, we'll keep to the English rule-book.

16. The Judge

a) English judge to sort out any problems. Or New York. Or anywhere else we care to name. We can send the invite c/o your offices in London and New York. Don't say it's inconvenient.

b) I have to speak to metaphors here. If you park your car on a yellow line, we can give you a ticket. And tow your car away. Even if it's marked CD.

Assuming you like this deal, Uno, please say so.

Yours hopefully,

Joe Y. Zed

Joe Y. Zed
....
Moneybank

It's O.K. by me.

Uno Grande

Uno Grande
....
State Mining Co.

It's O.K. by us.

	US$000,000
Moneybank	20
Manybanks	20
Muslimbank	20
Moltobanco	10
Magnifiquebanque	10
Misyomobank	10
Meanbank	8
Meanestbank	2

11.5

PROJECT FINANCING - GUIDELINES FOR THE COMMERCIAL BANKER

Grover R. Castle

Vice President, Chemical Bank,
New York, New York

INTRODUCTION

"Project Financing" or "Project Loans" have become increasingly popular in recent years, and commercial banks are being asked to consider more and more of such financing. However, the analysis of this type of loan is quite different from the analysis of a loan to a financially sound corporation based on the company's balance sheet. For this reason Chemical Bank did a study of project loans in an attempt to be more sophisticated in our approach to them. As part of this study, we examined 29 project financings for which information was available. In this article I shall describe project loans and set forth the guidelines we developed, as an aid to credit officers who might have occasion to analyse this type of financing.

DEFINITION OF "PROJECT FINANCING" OR "PROJECT LOAN"

The terms "project financing" and "project loan" commonly describe the various methods that banks and institutional lenders use to finance the construction of new projects on a basis whereby payout is anticipated from the revenue stream generated by the project. A project financing often involves a loan to a new entity formed specifically to own the project.

The analysis of a project loan proposal is quite different from the analysis of other term-loan proposals since it involves a credit decision based on the review of a projection or forecast rather than of historical earnings and/or it involves the reliance on contractual obligations of third parties.

Lenders are asked to assume the added risks associated with project loans because:

1. Project financing may permit a company to obtain additional leverage. It is possible for companies to arrange financing on a project basis that could not be arranged as a direct borrowing. For example, a company may not be permitted to borrow more money directly, due to a limitation in a bond indenture. However, a project financing can sometimes be negotiated on a basis not prohibited by the indenture.

2. Often the sponsors of, or the stockholders in, the project are seeking an off-balance-sheet financing. For example, 11 of the 29 projects in our study were arranged on an off-balance-sheet basis and one showed on the balance sheet as a deferred income item. An off-balance-sheet financing, for the purpose of this article, is a financing which does not show as a direct liability on the balance sheet of the sponsors of or stockholders in a project.

3. Projects are often located in a foreign country. The financing is arranged so that the lenders assume the political risk on the loan because many companies rightly believe that this arrangement affords a degree of insulation from political risk not otherwise obtainable.

4. Many projects are jointly owned by several sponsors. This form of financing lends itself well to a situation where a new entity is formed to own a project and do the borrowing.

Project loans can be used to finance a variety of activities including development of iron-ore deposits and associated pelletizing plants; development of coal reserves; construction of pipelines, alumina plants, fertilizer plants and refineries; etc. The specific areas of analysis can be quite different, depending upon the type of project being financed.

It is important, therefore, to classify the proposal according to the type of project, and also by the nature of the financing - non-recourse financing, financing covered by a guarantee of completion and financing supported by an undertaking lasting for the life of the financing. I will discuss each of these types in more detail later, but first, I should like to make some

Copyright 1981 by Robert Morris Associates. Reprinted with permission from Classics in Commercial Bank Lending, (Philadelphia: Robert Morris Associates, 1981) pp. 242-258.

general comments about the risks involved in new projects.

THE PROJECT RISK

Experience indicates that there is a high probability that any new project will experience some form of trouble. For example, of the 29 projects in the study, only 17 can be evaluated from a performance standpoint. The other 12 either have not matured to the point where they can be evaluated or information on performance is insufficient. Of the 17 that could be evaluated, 14 (82%) have run into some form of trouble. Analysis showed, the most frequent types of problems encountered were:

Cost overruns

This proved to be the most common difficulty associated with new projects and, in my opinion, the greatest risk. Of the 17 projects in my sample, 12 (71%) experienced cost overruns. These overruns ranged from 300% to 5% of original cost; 8 projects (47%) had overruns of more than 20%. I believe that the problem is actually worse than these statistics indicate because the projects in the sample are all mature and were completed before the current inflation and materials shortages. These factors have aggravated and will continue to aggravate the cost-overrun problem.

Completion delays

Another problem probably as common as cost overruns is inability to complete a project within the time span originally anticipated. Of the 17 projects evaluated, 10 (59%) experienced completion delays. The delays ranged from as long as 20 months (ignoring an extreme case where one project was abandoned) to as short a time as a month, and 5 projects (29%) were delayed for 6 months or longer. The average construction period in the projects that were delayed was about 33 months.

Actual annual cash flows not equal to original projections

Of the 17 mature projects 6 (35%) did not generate the cash flow originally projected. Reasons for the variances included interference by a foreign government, underestimation of operating costs, low productivity of foreign labor, miscalculation of the characteristics of an ore reserve, more frequent equipment replacements than originally projected, and changes in parities of currencies. (A foreign project with sales contracts in US dollars will suffer if there is a dollar devaluation.)

Market problems

A project may be initiated to meet a legitimate demand for a particular product in short supply but, by the time completion occurs, the shortage has been overcome by competing projects. As a result, the selling price of the product drops below what was originally projected. This proved to be a problem in one project of the sample.

Reserve miscalculations

Obviously if the project involves development of a mineral deposit and if the payback is dependent solely upon minerals in the ground, then reserve evaluation is critical. A miscalculation of the reserves became a problem in one project.

Political risk

Overseas projects involve added risks because foreign governments may take action that will affect the projects. In addition to the usual political risks such as expropriation, war and inconvertibility, a number of more subtle actions may also cause trouble. Some of these are:

- Increased taxes or royalties.
- Requiring equipment to be purchased within the host country.
- Imposition of import duties on raw materials needed for the project.
- Delays in completion caused by the time needed to obtain import licenses or required parts.

One project in my sample experienced foreign government interference.

Project inefficiency

Obviously the most extreme case of inefficiency is one in which the process does not work at all, and I know of only one such situation. However, an inefficient process can have a serious effect on overall economics, and can also adversely affect repayment of a loan dependent, for example, on ore reserves. In such a case, a larger than anticipated percentage of ore could be used up because of refining inefficiencies.

Of the 17 matured projects in the study, 9 (53%) had what I would describe as severe trouble (2 ended in bankruptcy and 6 others did not generate enough cash flow during some period of the financing to cover payments of principal). This, I believe, is a high incidence of trouble.

Despite the dismal picture that I have just described, in only one of the 17 projects did the lenders take a loss, and then they lost money only after the project sponsor went bankrupt.

The lenders fared well in spite of the statistics because they were able to shift many, if not all, the risks to others. If all the risk is not going to be shifted, then it is important to analyse carefully every potential area of weakness in any new project to be sure that you are covered when and if trouble occurs. For this reason, I believe that the analysis of a project loan requires a high degree of sophistication in more than one area and is best accomplished by a team effort. For example, a project involving the development of a mineral reserve in a foreign country should require:

1. A project financing specialist to negotiate the financial terms and help structure the deal.

2. A qualified engineer to appraise the mineral reserve.

3. An international specialist to analyse country risk.

THE THREE TYPES OF PROJECT FINANCING

Nonrecourse project financings

In this case the lenders can look only to the project for a payout since there are no backup undertakings from the stockholders. This type of financing is quite rare because of the high degree of risk involved, and in fact, I could only find three cases in the study that would properly fall into this category. This method of financing should probably be confined to projects where no new technology is involved or where special circumstances (such as a high percentage of equity) reduce the risk.

The analysis of any nonrecourse project financing should consider the following points:

1. Equity: Probably the first and most important question in any nonrecourse project financing is the amount of equity that should be required. There is no established rule of thumb as to the precise equity required since it should vary with the nature of the risk. However, I wish to emphasize that a large equity investment is of vital importance, although it may gradually become less important as the undertakings from the stockholders become stronger.

The average equity in the three nonrecourse project financings in my sample was 59% (computed as a percentage of original estimated cost) and ranged from a high of 93% to a low of 35%. In the project with a 93% equity, outside collateral was contributed to the project to support the financing and the 93% was computed by assigning a value to this collateral.

Those projects in the study that were supported by a guarantee of completion had an average equity of 35%, with a range from 67% to 18%. The projects supported by an undertaking lasting for the life of the financing had an average equity of 27%, with a range from a 67% to 2 1/2%.

While the amount of equity is determined primarily by the type of financing and the type of project, other factors also enter into its calculation. These factors include the financial strength of the stockholders, the coverage "for the life of the reserves" (discussed later), the type of geological formation, the geological characteristics of the minerals, etc.

2. Feasibility study and economic projection: A feasibility study and economic projection should always be required if for no other reason than to force the sponsors to think through all the problems involved. The feasibility study should confirm that the project can be completed to meet technical specifications at the estimated cost. The economic projection should forecast the amount of production, sales, operating costs and earnings that will be generated over the life of the project.

It is preferable that both the feasibility study and the economic projection be prepared by a well recognized and independent engineering firm, but it is not always possible to insist on this. If an independent study is not available, I usually attempt to verify the figures to the extent that it is possible. This might be done by checkings or by comparisons with other similar projects. The analysis of the project should include a computation of coverage on an annual basis.

ANNUAL COVERAGE measures the ability of the project to cover the required payments in each year. This coverage is computed by dividing cash flow in each year by the amount of required debt service in that year. It is also advisable to do an average coverage of the debt service during the payout of the loan to get an idea of how the coverage looks for the entire term. This average can be arrived at by simply averaging all the annual coverages.

Consideration should be given as to whether the cost figures and market assumptions used in the forecast are reasonable. A review of historical as well as future markets might be appropriate to determine whether to revise the forecast using a lower price per unit and/or arbitrarily assuming increases in costs. Furthermore, if the project is located in a foreign country it might be appropriate to see what would happen to the coverages if the host country increased taxes or if there were a devaluation of either the dollar or the local currency.

It might also be useful to know what the breakeven level is on a project. The average coverage during the payout of the financings in the study ranged from 1.10X to 4.77X. I do not like to see a coverage of less than 1.50X because, as proved to be true of the projects analysed in the study, there is too great a chance that maturities will need to be extended.

3. Reserve analysis: With projects involving the development of a mineral reserve or the construction of a facility such as a pipeline dependent on mineral reserves, it is customary to require verification of the reserves to be developed or to be dedicated to the pipeline by an independent technical expert. The only exception to this rule would be if the documents include an undertaking by financially responsible third parties to indemnify the lenders if the reserves are not as extensive as originally estimated.

Our bank has the technical staff and is equipped to "analyse the reserve risk" (i.e., the risk that the oil or other mineral is not as extensive as originally estimated), without requiring an outside independent study, except in large projects which would require a large engineering department to do the study. However, we prefer to simply verify an independent study. Such a verification involves starting with an independent reserve report and reviewing the calculations that have been made and the methods used by the outside consultant. Verification of reserves is important in order to be certain that, by our standards, there is a coverage for the payout on the financing during the life of the reserves.

THE LIFE OF THE RESERVE COVERAGE is different from the annual coverage and average coverage discussed earlier. Chemical Bank's minimum criterion is that our analysis must indicate that, at the time the loan is projected to pay out, at least half the reserves and at least one half the future net revenue are still remaining.

Let's take an example of a mineral deposit with 45 million tons of ore reserves and a project financing projected to pay out in its eighth year. The projection indicates that 17,350,000 tons of ore will be consumed during the first eight years, which amounts to 39% of the total 45 million tons. Furthermore, the projections indicate that a total of $343,990,000 of future net revenue (cash flow) will be earned during the life of the reserves and only $133,568,000 (39%) of the total future net revenue will be used during the first eight years. This project fits within our standards because more than 50% (61%, in this case) of the reserves and future net revenue are remaining at payout. This coverage for the life of the reserves is critical because all oil and gas or mining proposals involve wasting assets that, once depleted, are gone forever. Thereafter the project is no longer viable after the reserves are used up.

"PROVED PRODUCING" is an important consideration in mineral projects. We are reluctant to make loans against mineral properties except in cases where the reserves are thus classified. Obviously a rigid policy in this regard would not permit financing the development of any new mineral reserves without a guarantee or some other third-party backup. We commonly relax this rule in the case of hard-mineral projects provided we receive a guarantee of completion (which will be discussed later) and provided further that such reserves, though nonproducing, are classified as proved.

We feel that it is acceptable to loan against proved nonproducing hard-mineral reserves on this basis because it is relatively easy to determine with a high degree of accuracy the extent and composition of an ore body. The more difficult part of a hard-mineral project lies in producing a concentrate or refined product that meets design specifications as to quality and quantity, and producing the concentrate within estimated cost limits.

PROVED NONPRODUCING PETROLEUM RESERVES in a few cases have been financed without a guarantee or other backup, but each of these exceptions has been reviewed very carefully. I am reluctant to take the development risk in a project loan based on potential oil or gas reserves because it is much harder to define the true extent of petroleum reserves than it is of hard minerals. On the other hand, the products produced can be sold more easily and with almost no processing, and normally producing costs are a less significant part of sales.

To summarize: when a project loan involves taking the so-called "reserve risk," I would not suggest a nonrecourse project financing unless the reserves are classified as proved producing. This effectively means that such financing is not really possible in a situation where the proceeds of the financing would be used for development of the properties. If the financing is backed by a guarantee of completion and the reserves are classified as proved nonproducing, it is not unusual for the lenders to take the reserve risk in a project loan if a hard-mineral reserve is being developed. It is possible - but more difficult - to arrange such a financing if the mineral is oil and gas.

CLASSIFICATION OF A MINERAL DEPOSIT as proved requires that a well or a number of wells must have been drilled to define an oil and gas property, or core holes trenches or shafts, etc. in the case of a hard-mineral deposit. A banker not technically trained and/or without experience should not attempt to classify a mineral deposit, but should rely on experts. The type and nature of the mineral deposit dictate the extent of exploration necessary to define the mineral deposit.

To classify oil and gas accumulations as proved nonproducing requires the drilling of enough test wells to verify the magnitude of the reserves and that quantifiable reserves can be recovered. Typically, this would include an accumulation in a "thick pay section" with a rather uncomplicated trapping mechanism (i.e., a consistent indication of oil or gas without faulting). In the case of hard minerals, there should be core holes enough, not only to outline a hard-mineral deposit, but also to give good data as to the analysis of the ore. The number of core holes obviously depends upon the geology and can vary from as much as 600-foot centers to a 5-meter grid.

4. Market: The lenders must be satisfied through either market studies or long-term sales contracts that a market is available for the project output. In all the nonrecourse financings in the study the lenders were protected by a dedication of a substantial percentage of the output under minimum price sales contracts with an acceptable party, under terms running until well after the final payout.

Nonrecourse financing analyses should include a careful review of the terms of the contract and particular attention should be paid to the following features:

- Obligor: How strong is the purchaser?
- Term: Does it extend beyond payout?
- Quantity: What percent of design capacity is covered by the contract? Are the projections based on contract quantities?
- Quality: What quality requirements are in the contract?
- Force majeure provisions.
- Default provisions: Under what conditions can the buyer refuse to take under the contract?
- Price: The sales price should compare favorably with the price used in the economic projections. Also it is preferable to have price escalation to cover increases in operating costs.

5. Political or country risk: If the project is located outside the United States and if the financing is structured so that the lenders assume the political risk (i.e., any undertakings from the stockholders or others do not indemnify the lenders against war, expropriation, inconvertibility, changes in taxes, etc.), the analysis of the proposal takes on a whole new dimension. A decision has to be made as to whether the lender wishes to assume this political risk, and this would presumably involve a country study by the territorial officer in the international division of the bank.

Assuming that a bank is willing to take the political risk in the foreign country where the project is to be located and further that the financing is structured in such a way that the lenders take the political risk, then the following steps ought to be taken:

Demand a copy of the concession and review it looking for provisions relating to:

- Term: If the concession runs for only 15 years, then the computation of the coverage for the life of the reserves should include only reserves projected to be produced during the first 15 years.
- Taxes and/or royalties that need to be paid to the host country.
- Basis for cancellation or default.
- Required development schedule.
- Host-country participation in ownership.
- Host-country right to take output.
- Transfer out of the host country of proceeds of sales of output from the project.

Require a letter from the government of the host country addressed to the bank and stating:

1. That there shall be no change in the provisions of the concession (including those relating to taxes) during the payout of the financing.

2. That the project shall be given its fair share of goods and services within the host country.

3. That the host country is aware of the terms of the financing and that the host country shall not take any action that will have an adverse effect on the payout of the loan.

Obviously, the requirements in any such letter - in fact the actual requirement of such a letter - basically depend upon the confidence that the lender has in the host country. In eight of the projects used in the study, the lenders took the political risk and in only two of those did the lenders obtain such a letter. In one case, the letter took the form of a virtual guarantee of payment by the host government. Five of the projects were in countries considered politically stable. In one case no letter was required because nationals from the host country owned a fairly large equity interest in the project.

Require a letter from the central bank of the host country guaranteeing the transfer out of that country of proceeds of sale in convertible currencies in amounts needed to cover the payments on the project's debts. This should be a standard and routine requirement in some form in connection with any project financing in which the lenders are taking the political risk.

6. Environmental considerations and governmental clearance: The lenders must be satisfied that all governmental approvals have been obtained and that there is no risk or only minimal risk of environmental suits. Environmental actions can delay or halt construction of new projects.

For example, through November 1972, Con Edison had spent over $21 million on its Storm King project for a pumped-storage hydroelectric plant at Cornwall, New York., and the project has been held up since 1963 by environmental suits. In May 1973, Con Edison estimated that the cost of the project has been increased by $292 million as a direct result of this delay. Another well publicized example is the TAPS pipeline from the North Slope of Alaska to Valdez. This involved a five-year fight over environmental issues before Congress authorized construction.

Obviously a lender would not wish to advance funds on a nonrecourse basis before all clearances have been obtained. An analysis of the environmental issues should include the following:

Determination of the aspects of the project requiring government clearances, such as air emission, emission into streams or water bodies, diversion of streams, disposition of waste or noise emission.

Determination of the particular governmental body that has jurisdiction over each aspect of the project - state, Federal or local.

Examination of the procedural aspects of the clearance.

- Nature of approval: The Environmental Protection Administration and many states often require filing for approval a written impact statement (prepared either by the sponsors or by the governmental authority) on new projects. Other agencies may require only oral hearings or a petition.
- Timing of approval: Some authorities require approval in stages with an initial approval based on the plans, followed by intermediate approvals based on the implementation of these plans and a final approval based on actual operation of the project.

Study of new legislation. Be alert to current and impending legislation trends. The environmental aspect is a new and popular problem with new legislation being enacted every day.

7. Supply contracts: The lenders should be satisfied that all feedstock and power required by the plant or project are available at costs which do not exceed those in the economic projections. It is preferable that both feedstock and power be covered by long-term fixed-price contracts, and these should be analyzed just like a sales contract.

8. Builder: The builder of the project is always important and particularly so when no guarantee of completion is available. The analyst will wish to make a thorough check of the builder's reputation and performance. Proper analysis of a nonrecourse project financing should also include a careful review of the construction contract to examine such provisions as:

- Is it a fixed price?
- Is there a requirement that the project be completed prior to a fixed date?
- Are there penalties for delay?

9. Title: Title is a particularly important consideration in projects involving reserves of oil, gas and minerals or the construction of a pipeline and should be considered in connection with any new project. The lenders should be satisfied that the project has "good and marketable" title to the land on which the project is to be located and the mineral reserve to be mined. The most common way to satisfy this requirement is with a title warranty from a financially responsible party, but in a nonrecourse financing this might be accomplished by title opinions from an independent law firm. It is also possible to obtain title insurance from a title insurance company.

10. Insurance: During those periods when the lenders are not protected by undertakings, the project should be protected by insurance with a loss-payable clause in favor of the lenders. Some obvious risks to be considered are: builders' risk, property damage, comprehensive general liability, workmen's compensation and employers' liability, automobile liability, boiler and machinery coverage and, where appropriate, excess (umbrella) coverage. This is a very specialised field and it is best to seek technical advice from qualified specialists.

11. Operator: Careful consideration should be given to the operator's financial strength as well as demonstrated ability to operate the project. A review of the operating agreement is an integral part of the analysis of a project financing.

Project financing covered by a guarantee of completion

Many of the more common risks can be eliminated from a project financing if the lenders negotiate a guarantee of completion whereby the project stockholders agree to complete the project. After completion, the lenders are essentially on their own, relying only on sales contracts, supply contracts and the project itself.

Because the lenders are on their own after completion, all eleven points discussed above under nonrecourse project financing are also pertinent here, except that title can usually be covered by a warranty rather than title opinions or title insurance.

The form of the guarantee of completion is very important. Ideally it should cover the following elements:

1. Completion date and overrun costs: The obligation of the guarantor to complete by a certain date and to cover all overrun costs in the form of equity should be stated. Of the 11 projects in the study where financing was covered by a guarantee of completion, 6 (55%) required completion by a certain date usually somewhat later than the projected completion, and 8 (73%) specifically obligated the guarantors to cover all overruns with equity.

Occasionally the completion undertakings will only be a simple covenant by the stockholder to complete without a date. However, in such a case if the project is delayed due to governmental action or an environmental suit and the stockholders decide to fight these actions in the courts, the lenders would probably be forced to stand by during this period with no recourse to the guarantors.

2. Definition of completion: Completion should be defined as the time when the project shall have operated for a period (to be negotiated) according to the specifications in the feasibility study and during that period have produced a specified quantity of output which meets a specified quality test at a cost no greater than indicated in the economic projections.

If a refining process is involved, the guarantee might also require a certain degree of efficiency. For example, the definition of completion of a plant might require that it operate for 90 consecutive days and during that period produce 100,000 barrels a day (90% of design capacity) of product which meets a predetermined specification (usually the specifications in a sales contract) with an operating cost per barrel of no more than 50¢ per barrel. One completion guarantee that I am familiar with contained the points mentioned above plus a requirement that, during the initial consecutive three-month period of operation, the project earn 80% of the amount of cash flow set forth in the economic projections.

However, only 6 (55%) of the 11 projects in the study sample contain any operating test at all and only 2 (18%) contained a cash-flow test in the definition of completion.

The definition of completion sometimes contains a requirement that the project have a minimum amount of working capital on the date of completion. Two (18%) of the 11 study projects contained such a provision.

3. <u>Maturities prior to completion</u>: Of the 11 completion guarantees in the study, 5 (45%) contain a continuing obligation of the guarantors to cover any maturities on the loan prior to defined completion; one provides the guarantors with an option of purchasing the notes. In the absence of this type of provision, the lenders could be forced to use stockholders for damages and, to prove damages, the lenders might be required to go through a foreclosure proceeding to establish the amount of their loss.

Three of the five undertakings mentioned contain an obligation of the stockholders to pay off the loan if the project is not completed prior to a certain date well beyond the expected completion date.

4. <u>Acceleration provisions</u>: If the loan accelerates prior to completion, it is preferable to require that the obligation to pay under the completion guarantee also accelerates. However, this is difficult to negotiate; in fact, only 3 (27%) of the 11 guarantees among the study projects contain such a provision. In many situations the obligation to meet the maturities is confined to the original maturities, regardless of acceleration against the borrower.

5. <u>Bankruptcy or default</u>: In the event the obligor under the completion guarantee goes into bankruptcy or defaults under another agreement, it is preferable to have acceleration of the obligation under the completion guarantee regardless of the status of the project. This feature is difficult to obtain with large triple A guarantors, and only 2 (18%) of the 11 guarantees in the sample contain such a feature.

6. <u>Financial covenants</u>: It is possible to include financial covenants restricting the guarantors in the completion guarantee. This is not a common feature and was included in only one of the completion guarantees in the study.

7. <u>Environmental clearances</u>: The definition of completion can also require that all environmental clearances have been obtained before completion occurs. This is most important in domestic projects, and more important with some projects than with others. Only 2 (18%) of the 11 guarantees in the study contained such a provision.

8. <u>Prohibition of sale of interest</u>: The completion guarantee can also require that the guarantors will not sell their interest in the project. This is a fairly common requirement. Of the guarantees contained reviewed 8 (73%) included such a provision, and one of the three exceptions provided for a default in the loan agreement (against the borrower) if the stock was no longer owned by the sponsors.

9. <u>Title warranty</u>: Whenever appropriate, the completion guarantee can also contain a warranty of title to the reserves of oil, gas or minerals being developed.

As you can see from this discussion all completion undertakings are different. The final product is the result of a variety of factors such as the complexity of the project, the amount of lender's confidence in the sponsors, the type of project and how well the lenders negotiate the terms of the proposal.

Project financing supported by an undertaking for the life of the financing

It is possible to shift all or substantially all the risk to the sponsors of the project through various types of undertakings which go far beyond the obligation to complete. In these cases, it is still contemplated that the financing will pay out from the project's cash flow, but, if the cash flow doesn't cover debt service, the lenders have recourse against the sponsors. Most of these undertakings make unnecessary the eleven points discussed under nonrecourse project financing. Some forms of these undertakings are:

> Minimum working capital undertaking or working capital maintenance agreement - Under this type of arrangement, the sponsors agree to maintain at all times a minimum working capital in the borrower to assure the lenders that the borrower shall have sufficient funds to meet maturities on financing. If working capital as defined falls below a specified level, the agreement requires stockholders to make subordinated advances (usually on a pro rata basis) in amounts sufficient to increase working capital to the required level. This type of undertaking was used in three projects in the sample studied. In two of them the financings were off balance sheet, and the third has not been taken down as yet.

> Cost company arrangement - Under a cost company arrangement, the stockholders are unconditionally obligated under an operating agreement assigned to the lenders to provide the cost company with their pro rata share of all amounts needed to pay operating costs

for the project plus principal and interest on the project's borrowings as they become due. Funds are furnished by the stockholders in proportion to their stock ownership. The arrangement is usually off balance sheet for the stockholder.

This arrangement is common in connection with iron-ore projects. In a typical case, each stockholder received from the project a pro rata share of the ore produced. The cost company does not sell any ore and has no net income, and each stockholder includes in income statements a pro rata share of the cost company's production and operating expenses.

Throughput arrangement - Pipeline projects are often supported by throughput agreements assigning amounts due the pipeline. Each stockholder is severally obligated to ship or cause to be shipped through the pipeline, in proportion to stock ownership, liquid hydrocarbons in a total amount which will provide sufficient cash to pay all the pipeline's expenses and liabilities as they come due.

If for any reason, including failure to complete the line and cessation of operation, the pipeline has insufficient cash to pay all expenses and liabilities, the stockholders shall contribute on a several basis an amount sufficient to augment such cash deficiency to meet all liabilities and expenses.

So long as there is sufficient cash from other shippers to pay all expenses and liabilities, no stockholder will be obligated to ship its full requirement as described above. Throughput-agreement financing is usually off balance sheet.

Tolling contracts - Alumina and other plants often obtain financings based on the assignment of a tolling contract whereby the participants in the projects are unconditionally obligated to pay their pro rata share of amounts needed to meet the principal and interest on financing. A tolling contract is very similar to the so-called "take or pay contract" whereby the purchaser is obligated to pay for purchases regardless of whether the seller is able to deliver or not. The tolling contract is usually off balance sheet.

GENERAL COMMENTS ON STRUCTURING OF PROJECT FINANCINGS

Interest rates

Usually lenders are able to demand a higher interest rate on project financings than on direct loans made to the sponsors. A nonrecourse project financing would be expected to carry a higher rate than a financing backed by a guarantee of completion. A financing backed by an undertaking for the life of the financing would carry the lowest rate of the three types of loans. A comparison of the rates charged in the projects studied bears this out, as shown below:

Type of Project	Average Rate*
Domestic Dollar Financings	
Nonrecourse	16.67%
Covered by Guaranty of Completion	15.79%
Covered by Continuing Undertaking	15.53%
Eurodollar Financings	
Nonrecourse	None
Covered by Guaranty of Completion	12.55%
Covered by Continuing Undertaking	12.25%

*All rates computed using the following assumptions:

1. Prime assumed to be 12% on the date the loans were negotiated for the life of loan on domestic dollar financings.

2. London Interbank rate assumed to be 11% on the date the loans were negotiated for all Eurodollar financings.

3. All rates are computed on a no-balance basis.

4. All computations are calculated without provision for reserve requirement.

5. All fixed-rate financings are related to prime on the date the loan was negotiated. The average rate is computed by using the same absolute spread over a 12% prime as the fixed rate was over prime on the date the loan was negotiated.

Maturity

It is important that the maturity schedule be set up so that it fits within projected earnings. For example, in some projects the projected cash flow will not support equal payments over the term of the loan, and, therefore, a balloon payment is sometimes set at the final maturity, or the payments might start out low percentage-wise and gradually increase in later years. When there is a balloon payment or when the later maturities are weighted more heavily than early payments, it is customary to provide for an earnings recapture clause (sometimes called a mandatory prepayment provision) whereby a percentage of cash flow exceeding a certain level is applied against the loan in the inverse order of maturity to reduce the balloon payment. Five of the projects in the sample had such clauses.

Often a flexible maturity schedule provides that payments commence three months after completion, whenever that occurs, with a requirement that these maturities start no later than a certain date regardless of the date of completion.

With all loans where the lenders take the

reserve risk, it is preferable that the financing pay out through a dedication of a percentage of gross revenue with a minimum payment schedule rather than through fixed maturities. For example, where total future net revenue is approximately 57% of total ore value (gross revenue or sales) we would attempt to get a dedication of 55% of gross revenue (ore value) commencing with completion of the project. This dedicated revenue would be applied against the financing first to interest and then to principal on a monthly or quarterly basis.

This type of payout preserves the coverage in the life of the reserve test even if the oil, gas or other mineral is extracted faster than originally projected. If the mineral is extracted faster than originally expected and the financing pays out with equal quarterly maturities, then obviously fewer reserves would remain at final maturity than anticipated.

It is also possible to provide that if the dedicated percentage does not pay down the loan in accordance with a minimum schedule, then the percentage dedicated increases to the extent necessary to meet that schedule.

Covenants

Generally speaking, the loan agreement confines the borrower's activities to the operation of the project, which means that a full set of financial covenants are enforced against the borrower.

Events of default

In addition to the usual events of default relating to the borrower, it is customary to have an event of default if:

- The project is not completed by a certain date.
- There is a default under a supply or sales contract or other key contracts such as tolling contracts or throughput agreements.
- If any of the stockholders become bankrupt, breach a loan agreement or dispose of their stock in the project. This is appropriate where there is an undertaking that runs for the life of the loan. It is also appropriate prior to completion where there is a guarantee of completion. It is often difficult to negotiate this provision if the project is jointly owned by several major companies, and it is sometimes acceptable to have acceleration of only a defaulting stockholder's pro rata share of the loan.

While the above types of default are customary, there are cases where the only defaults relate to the borrower and there is no default if the plant is not completed by a certain date.

Collateral

The collateral usually includes an assignment of any supply contract, sales contract and other key contracts such as tolling contracts or throughput agreements, and often also includes a mortgage on the project plus a pledge of the stock of the borrower. Furthermore, it is often appropriate for the lenders to be named as beneficiary under any insurance coverage on the project.

Editor's Note: This article first appeared in The Journal of COMMERCIAL BANK LENDING, April 1975. See also Grover Castle's paper, "Feasibility Studies and Other Pre-Project Estimates. How Reliable Are They?" in Chapter 9 of this Book.

11.6

PROJECT FINANCE - DOES IT EXIST IN THE
MINING INDUSTRY : CRA'S EXPERIENCE

Kyle Wightman

CRA Limited
Treasurer - Projects
Melbourne, Australia

INTRODUCTION

CRA Limited is the largest mining group in Australia by value of sales and the second largest company in Australia in terms of share market capitalisation. In the past 25 years, CRA, its predecessors and affiliates have been involved in a number of mining and further processing developments in Australia and elsewhere. Many of these have been financed using project financing techniques. These include Comalco's development of the Weipa bauxite deposits in Northern Queensland in the late 1950's, the Hamersley iron ore developments in the Pilbara region of Western Australia in the mid 1960's, the Queensland Alumina Limited alumina refinery in Gladstone, Queensland, at about the same time, and the Bougainville copper project in 1969/70.

In the past three years CRA has again utilised project financing techniques to fund its Tarong steaming coal project in Queensland, the Blair Athol steaming coal project also in that State and CRA's interest in the Argyle diamond project in the Kimberley region of Western Austrlia.

PROJECT FINANCE DEFINED

Nevitt Definition

Nevitt's is the most frequently quoted definition -

"A financing of a particular economic unit in which a lender is satisfied to look initially to the cash flows and earnings of the economic unit as the source of funds from which a loan will be repaid and to the assets of the economic unit as collateral for the loan."

Risk and Balance Sheet Treatment

It is important to focus in more detail on two aspects of project financing, risk and balance sheet treatment. The assessment of the various elements of risk in a project and the sharing of that risk between sponsors and lenders is the most critical feature. At one end of the spectrum is full absorption of risk by the sponsor(s). This can be described as "on risk" or "on credit" or "full recourse" to the sponsor(s) and is effectively a corporate financing. At the other end of the spectrum the lenders absorb the whole risk in a non-recourse financing which is "off risk" or "off credit" for the sponsor.

The other aspect is the sponsor's balance sheet treatment. If the balance sheet of the economic unit undertaking the project and the financing need not be consolidated into the balance sheet of the sponsor (because ownership does not exceed 50%) then the financing may be described as off-balance sheet. If consolidation is required then the financing is described as on-balance sheet.

Most project financings involve some recourse to the sponsor, i.e. they are limited recourse. Very few are non-recourse or off credit to the sponsor. Limited recourse financings may be on-balance sheet although many may also be off-balance sheet insofar as consolidation is concerned. However, sponsor support may very well be a contingent liability to the sponsor and require disclosure by way of note to the balance sheet should consolidation not be required.

CORPORATE FINANCE CONTRASTED

A financing which is on credit or full recourse to the sponsor, whether it is directly borrowed or guaranteed by that sponsor, is a corporate financing. Lenders will tend to base their credit assessment on a historical assessment of the sponsor's activities and its financial results rather than an assessment of the project or a detailed assessment of the future activities and prospects of the sponsor.

PROJECT FINANCE - A BORROWER'S PERCEPTION

Although CRA could choose to arrange all finance on a corporate basis (except perhaps for the very largest projects or acquisitions), project financing techniques provide a number of specific advantages which CRA has noted and which should be carefully considered by a project sponsor.

Advantages

Risk Reduction. Project financing reduces risk for the project sponsor.

LENDER RECOURSE LIMITED: Because lenders' recourse is limited beyond project assets and cash flow to full or partial sponsor support in relation to specific risk events or amounts, sponsor risk is reduced. This means that, in the event of project difficulty, a lender's ability to attach, take control or wind up the other business activities of the sponsor is restricted or prohibited.

REPAYMENT DEFERRALS: Project financing can achieve loan repayment deferrals (automatic debt rescheduling) if project cash flow is insufficient to meet targeted loan repayments.

STRONG NEGOTIATING POSITION IF DIFFICULTIES ARISE: Project financing puts the project sponsor in a strong negotiating position in the event of project difficulties resulting from economic recession, disruption to markets or other reasons. The more limited the support provided to lenders by a sponsor, the stronger the position of the sponsor when dealing with lenders in the event of the financed project encountering difficulties. In this situation the sponsor may have the option to "turn over the keys" to the lenders or abandon the troubled project. On the other hand, the project sponsor could use the option of abandonment to strengthen its negotiating position with the lenders to obtain any desired further rescheduling of the project borrowings.

LIMITS CROSS DEFAULT: Project financing can limit cross default through to other borrowings of the sponsor. CRA's corporate loans are all inter-related; a default in one loan can trigger defaults in all of the other loans. Defaults under project finance loans, however, can be limited or excluded from affecting CRA's other corporate loans.

REDUCED POLITICAL RISK: Project financing can reduce political risk in the case of offshore projects.

THIRD PARTY AUDIT: Project financing provides project sponsors with a third party audit. The process of presenting a project to potential lenders for objective review, as a basis for loan offers, can provide a useful additional perspective to the decision-making process.

Preserves Sponsor's Borrowing Capacity. When lenders and rating agencies appraise the credit-worthiness of a company, they discount the existing debt burden according to the extent to which it is non-recourse. In other words, they tend to view such borrowings favorably. This then has two effects.

- Firstly, it preserves the sponsor's borrowing capacity to a greater extent than would result from undertaking all borrowings on a corporate or full recourse basis.

- Secondly, the terms and conditions of future corporate borrowings will be improved as a result of this preserved borrowing capacity.

Extends Bank Lending Limits. All lenders have internal policies which establish a limit on the amount of money that can be lent to any specific credit or borrower. Because limited recourse financing establishes a credit entity (i.e. THE PROJECT), separate from the sponsor, banks can choose to lend more money to fund activities of a large borrower than they normally would if all projects of that borrower were funded as corporate borrowings. In this way access to borrowings from lenders with existing relationships with the sponsor can be expanded without the need to necessarily expand the list of lenders.

Allows Higher Gearing. Detailed scrutiny of a project by lenders gives them confidence to provide a higher level of gearing. This is evidenced by the accepted practice of 70%-75% gearing in projects as opposed to corporate gearing of less than 50% for most mining companies.

Longer Loan Maturities. The more common five-to seven-year loan period for corporate borrowings can be extended to twelve or more years with project financing as the repayment schedule is tailored specifically to the project cash flow.

Disadvantages

The disadvantages of project financing are:

- complexity
- additional time and effort to implement and administer
- additional lending margins and fees applying to this incremental financing
- the provision of security over project assets to lenders
- additional legal expenses and possibly financial advisory fees
- a reduction in the level of confidentiality applying to the proposed new project.

ARE THE BENEFITS REAL OR ILLUSORY?

Risk Reduction is Real

The sharing of risk with lenders provides a sponsor with options in the event a project runs into difficulties. The limiting of the extent of lender recourse to the sponsor could ensure the survival of the sponsor under circumstances where the project is sufficiently large and other activities of the sponsor are in difficulty at the same time. The burden of a corporate loan for the same amount under similar circumstances could bring the sponsor down or increase its vulnerability to takeover.

Off-Balance Sheet Techniques may be Illusory

Although off-balance sheet techniques may keep project debt away from specific inclusion in the liability side of a sponsor's balance sheet, contingent liabilities in relation to sponsor supports will require disclosure.

Sophisticated lenders and rating agencies will look beyond the balance sheet treatment of project debt to the recourse or credit aspects of such debt as far as the sponsor is concerned.

However, because of imperfections in the market place in terms of disclosure requirements in various countries and differing levels of sophistication in various debt and equity markets, appearances by way of off-balance sheet techniques can be of benefit to sponsors in their future capital raisings.

Would a Sponsor Abandon a Project?

Project lenders will tend to commit only to those projects which have sponsors who are regarded as least likely to ever abandon or walk away.

Equally, sponsors who fall into this category may themselves be of the view that they would never walk away from a project to which they commit, particularly in the mining industry with the impact that may have on relationships with Governments in the country and state concerned. They may also be of the view that they would never leave lenders to one of their projects to suffer any loss.

However, over the last few years there has been an increasing preparedness on the part of project lenders, in loan documentation, to facilitate the avoidance of cross default to other sponsor corporate indebtedness, to allow project abandonment under defined circumstances and to spell out very clearly the limitations on lenders access to other assets and cash flow of the sponsors should project cash flow and assets be insufficient to pay interest and repay debt.

Under these circumstances and recognising the rapidly changing world in which we live, responsible sponsors could decide to walk away from projects, particularly where corporate survival was at stake.

CRA Approach

CRA's three recent major projects, Tarong, Blair Athol and Argyle have all been financed on a limited recourse, on-balance sheet basis. In each case, corporate borrowings could have been undertaken. However, each was considered on a case-by-case basis and the project approach was taken for risk reduction reasons and to preserve CRA's borrowing capacity thereby reducing the cost of future corporate borrowings.

Even though the borrowings for CRA's interest in these projects will be on-balance sheet for CRA, care is being taken in the balance sheet to highlight their limited recourse nature.

Care has also been taken in CRA's most recent consolidated balance sheet to highlight that portion of the debt of a 58.77% owned consortium company, New Zealand Aluminium Smelters Limited (NZAS), which is supported by non-CRA members of the consortium. The support is provided by take-or-pay contracts for the shares of the product of that aluminium smelter to which the non-CRA consortium members are entitled.

CRA also arranged finance in 1980/81 for a small project outside Australia, in which it has a minority interest, effectively on a non-recourse basis. The only undertaking made by CRA was to maintain its shareholding for a minimum period, which was less than the loan period. The minority interest meant that the debt was off balance sheet for CRA.

The highlighting of both on-balance sheet categories of debt, i.e. the limited recourse debt for Tarong, Blair Athol and Argyle, and the NZAS debt which is supported by non-CRA members of the NZAS consortium, enables future lenders and rating agencies to focus on the true nature of these borrowings and discount them accordingly in their credit assessment of CRA.

Tables 1 and 2 below show extracts from CRA's balance sheet at 31 December 1983. Table 1 shows the non-current liabilities section of the balance sheet and Table 2 shows notes 10(a) and 10(b) which explain the two categories of debt described above.

TABLE 1

CRA LIMITED - EXTRACT FROM BALANCE SHEET

Conventional consolidation				Equity consolidation	
1983	1982		Notes	1983	1982
(thousands of dollars)				(thousands of dollars)	
		Liabilities			
		Maturing later than one year -			
223 995	74 025	Limited recourse debt - secured	10(a)	223 995	74 025
		Loans supported by Companies other than			
63 127		the Group - secured	10(b)	63 127	
3 140		- unsecured		3 140	
226 110	354 460	Other loans - secured	10(c)	226 110	354 460
1 008 979	916 168	- unsecured		1 008 979	916 168
106 794	114 409	Provisions	11	106 794	114 409
1 632 145	1 459 062			1 632 145	1 459 062

SOURCES OF FINANCE

TABLE 2

CRA LIMITED

EXTRACT FROM NOTES TO THE ACCOUNTS

Liabilities maturing later than one year – Loans	Currency	Repayable between	Interest rate %	Consolidated Principal outstanding at the year end	
				1983	1982
				(thousands of dollars)	
(a) LIMITED RECOURSE DEBT					
Syndicate of Australian Banks (KCC) (limited recourse over Tarong project)	Aust./US dollars	1986/92	variable	82 070	45 196
Syndicate of International Banks (KCC) (limited recourse over Blair Athol project)	Aust./US dollars	1985/90	variable	141 925	28 829
				223 995	74 025
(b) LOANS – Supported by Companies other than the Group					
The participants in New Zealand Aluminium Smelters (NZAS), Comalco 58.72%, Showa 20.64%, Sumitomo 20.64%, are each separately responsible under long term contractual arrangements for the servicing of their respective shares of certain loans made to NZAS. The undermentioned loans to NZAS are those for which participants other than Comalco have undertaken contractual obligations to provide, directly or indirectly, funds to meet debt servicing.					
SECURED					
Debenture loans (NZAS)	US dollars	1984/2050	7.86 to 7.98	19 343	
Debenture loans (NZAS)	NZ dollars	1984/2006	7.71 to 7.72	4 681	
Debenture loans (NZAS)	Yen	1984/92	7.18 to 8.12	45 933	
				69 957	
UNSECURED					
Other loans (NZAS)	Yen	1984/93	7.86 to 7.98	3 918	
Total				73 875	
Less: Instalments due within 12 months (Note 12)				7 608	
				66 267	

The A$224 million limited recourse debt includes Tarong and Blair Athol debt drawn to the balance sheet date. With Argyle drawdowns now underway maximum borrowings under the three facilities could reach approximately A$860 million when these projects are completed.

RECENT CRA PROJECT FINANCINGS

Tarong

Construction of the first stage of the Tarong steaming coal project in Queensland has recently been completed. The project is located about 150 km North-West of Brisbane and is dedicated to the supply of coal to the 1,400-megawatt Tarong Power Station, owned by the Queensland Electricity Generating Board (QEGB) which is under construction adjacent to the coal deposit. The first of our 350 megawatt generating sets completed in this power station has just commenced supplying base-load power into the Queensland grid.

The coal project is 100% owned by CRA and has a particularly bankable sales contract with the QEGB, a state government statutory authority. It covers a period of 15 years and provides a base price plus escalation clauses and minimum contractural tonnages.

A limited-recourse project financing in Australian dollars totalling approximately A$260 million on an 80:20 debt:equity ratio was arranged in 1981-82 with Westpac (then Bank of New South Wales) as lead bank. Because the market risk was minimal it was a fairly straightforward task to convince banks to accept not only that risk but also the reserve, production and political risks. Recourse to CRA was limited to a completion guarantee and a recapture of prior cash flow up to an agreed limit.

Exceptions to the negative pledge in CRA corporate borrowing arrangements allow security to be given over new project assets. This enabled security over the Tarong mining lease and other project assets to be provided to the Tarong lenders.

However, the circumstances under which lenders can exercise their security have been limited as far as possible in the Tarong loan documentation by defining many events beyond the control of the project manager, including an inability to meet scheduled loan repayments, as special prepayment events rather than events of default. The occurrence of a special prepayment event does not allow lenders to exercise their security whereas an event of default does. Although CRA's most recent corporate borrowing arrangements provide for exceptions to their cross default clauses in relation to the exercise of security by project lenders, previous arrangements which are still outstanding do not. Accordingly, the exercise of security by lenders could cause cross default to CRA corporate debt and it is therefore important for CRA to limit the extent to which such rights arise for its project lenders.

If a special prepayment event occurs, Tarong lenders can either take a transfer of the project in full settlement of the outstanding loan obligations or instruct the project manager to operate and/or dispose of the project for the benefit of lenders. This latter mechanism is termed the self-receivership provision which provides lenders with a mechanism to effectively take charge, operate and dispose of project assets, i.e. powers analogous to those available through the exercise of their security, while at the same time precluding them from formally exercising that security.

The Tarong project provided an ideal opportunity for CRA to undertake a limited-recourse project financing with litle difficulty and with relatively low incremental costs and expenses.

Blair Athol

Construction of the Blair Athol steaming coal project in Central Queensland, near Clermont in the Bowen Basin, has also recently been completed. This project is one of the most competitive of all the large-scale steaming coal mines developed or planned in Australia and is based on export sales to the Japanese market.

The project is structured as a joint venture in which CRA has a 50.22% interest and management control. Atlantic Richfield (ARCO) has a 15.39% interest, and Australian companies, ACI and Bundaberg Sugar, have each taken up a 12.195% interest. Sales contracts with Japanese power utilities which provide minimum contractual tonnages for a 15 year period, together with a base price and escalation clauses for the first five years, underpin the development. The Japanese customers also took up a 10% interest in the joint venture.

A limited-recourse project financing was arranged by CRA in 1981-82 with the assistance of The First Boston Corporation as financial advisor. Initially debt was arranged on a 70:30 debt:equity ratio for the full project to facilitate cohesion among joint venturers and to assist incoming Australian joint venturers, required under Australia's foreign investment regulations, in financing their interest in the project.

Ultimately, the foreign joint venturers, i.e. ARCO and the Japanese customers, elected not to participate in the financing package. Accordingly, the final debt package covered 70% of the cost of the 75% of the project held by CRA and the two incoming Australian joint venturers. The amount was US$450 million. Bank of Montreal was appointed agent bank after offers were sought and received on a club basis.

At the time this financing was undertaken coal risk in Australia was attractive to banks. Accordingly, fine margins were achieved by the borrowers with recourse beyond the project assets and cash flow being limited to an assignment by the borrowing joint venturers of the completion undertakings given by project joint venturers to one another and a guarantee of recapture of prior cash flow up to an agreed limit. Lenders were prepared to accept the market, reserve, production, foreign exchange and political risks.

As in Tarong, security over project assets was provided to lenders, subject, however, to limiting the events pursuant to which security can be exercised and the potential for cross default to other CRA debt by use of the special prepayment event mechanism. In addition, the Blair Athol loan arrangement had an abandonment clause which permitted the joint venturers to abandon the project under certain specified adverse circumstances.

The Blair Athol financing took some time and effort to put into place, due largely to the changes in the joint venture participations and the decision by the foreign joint venturers not to participate in the financing package. However, the borrowing joint venturers are satisfied with the result. The involvement of a significant number of banks in the project, and particularly Japanese banks, is comforting under the changed circumstances now prevailing in the coal industry with customers seeking to reduce both price and tonnages specified in sales contracts.

Argyle

Construction of the Argyle diamond project in the Kimberley region of northern Western Australia commenced in December 1983 and is scheduled for completion in the last quarter of 1985.

The Argyle diamond mine will be the largest in the world by volume of production. It will increase the world supply of diamonds by some 50% in volume or carat terms, although this will represent only a 4% increase in supply in value terms due to the low average quality of Argyle diamonds. In addition the diamond industry is extremely complex and secretive. Diamonds are not a homogenous product. No two diamonds are the same and value varies over an enormously wide range depending on size and quality.

To further increase the financing challenge the project is structured as a joint venture in which CRA has a 56.8% interest and management rights.

Ashton Mining Limited has 38.2% and Northern Mining Corporation NL 5.0%. All joint venturers were new to the diamond industry and certainly were not equal in terms of financial strength and expertise.

However, on the positive side the processing technology and the broad principles of the recovery techniques are not new. They have been proven in existing diamond operations including the existing smaller scale alluvial diamond treatment plant at Argyle.

Marketing is the key variable in the project and much attention has been given to this aspect by the joint venturers, their financial advisors and ultimately the lenders to each joint venturer. So-called bankable sales contracts in the familiar coal industry style are not available, although CRA and Ashton Mining have entered into a sales contract with the Central Selling Organisation for the majority of their share of production for the period to the end of the first five years of production from large scale mining operations.

CRA arranged a limited-recourse project financing in 1983/84 totalling US$235 milion to finance its interest in the project on a 75:25 debt:equity ratio. The other joint venturers also arranged financing for their respective interests in the project at the same time, as agreement was not reached to finance on a co-ordinated basis.

The considerable learning curve facing anyone involved with this project was the major problem. This was due to the scarcity of accessible information about this secretive industry. The interesting phenomenon was that the more we learned about the project and the diamond industry, the more comfortable we became with both. This phenomenon was experienced in turn by CRA and the other joint venturers, their financial advisors and, ultimately, the banks.

CRA's financial advisor was again The First Boston Corporation and banks were approached to submit offers on either a club basis or a fully underwritten basis. Initial reactions were mixed, with the most positive responses coming from those banks who had been looking at the project for a long period, perhaps through earlier involvement with the other joint venturers. A group of four banks who submitted an underwritten offer won the mandate. The group comprised Westpac, also as agent bank, Continental, Societe Generale and Toronto-Dominion. Shortly thereafter they were joined by Fuji and Dresdner to make a group of six Lead Managers and underwriters. Fuji and Dresdner had earlier submitted very positive club style responses to the initial invitation. The subsequent syndication of the facilities by the Westpac group was a great success. This reflected the increasing level of knowledge and comfort among the potential lenders about the industry and the project.

Once again, in this financing, recourse beyond the project assets and cash flow to CRA as sponsor was limited to completion and recapture of prior cash flow up to an agreed limit. However, this time the completion risk was covered by CRA guaranteeing completion. Lenders were once again prepared to accept the market, reserve, production, foreign exchange and political risks even though this was a new commodity for the project lending market and the world was in the midst of a severe recession.

The provision of second ranking security to lenders (subject to a first ranking crosscharge in favour of non-defaulting joint venturers) was a feature but subject again to the special prepayment event mechanism to limit cross default to other CRA debt. An abandonment provision was also a feature of this financing.

POST MORTEM

Is the time, effort and cost worthwhile? We do ask this question from time to time in CRA and opinions differ.

Most doubts seem to be expressed when the accounts are rolling in for legal and financial advisory fees and travel expenses, during the establishment of the loan arrangements. It is often forgotten that these costs can be fairly minor when put in perspective, i.e. amortised and expressed as an annual percentage over the period and amount of the loan.

What is comforting about undertaking and completing a limited recourse project financing is the knowledge that other institutions, the banks, have considered very carefully your new project proposal, have endorsed it, have agreed to take on specific risks associated with the project and have committed to advance their funds on that basis. The resultant reduction in sponsor risk and options available to the sponsor under adverse circumstances are meaningful results which CRA, on balance, has regarded as worthwhile.

In this rapidly changing world in which we live, the sharing of risk and the preservation of corporate credit or borrowing capacity are becoming more and more critical to corporate survival and growth. Limited recourse project financing techniques have a role in achieving these objectives.

REFERENCES

Nevitt, Peter K., Undated, An Overview of Project Financing, Project Financing, Second Edition, AMR International, Inc., pp. 1.

CRA Limited, Annual Report 1983, Melbourne, Australia pp. 37 and 44.

11.7

HOW TO FINANCE MINERAL PROSPECTS

Edgar F. Cruft, Chairman and President

Nord Resources Corporation
Dayton, Ohio

INTRODUCTION

It is sometimes said that "mines are made, and not found." I rather doubt that the exploration geologist would be overly sympathetic to that statement, and, of course, like most one-liners it is only partly true. Access to different financial markets are important at the mineral prospect stage. The modern mine financier has a much more complex, and interesting, job than the financiers of an earlier era, for example, the Canadian financiers of the 40's and 50's who had one simple set of options; i.e. float a company, promote it, and sell an equity stock issue, and then promote it some more.

QUALITIES OF A FINANCIER

Because technical people are seldom taught financing in their basic training, and because they are usually not involved in it until at quite a senior level in their careers, the whole field often acquires a certain mystique. The engineering professions have too often believed financing is the province of the accountant or financial executive and just as we in the technical world are often too quick to bristle if the latter try to tread on our domain, the opposite can just as easily be true. Unfortunately, this is another example of a compartmentalized individual, and that is one thing the successful financier cannot be.

However, to be a successful financier one does not need extensive training or experience in accounting or finance. In fact it can sometimes be a disadvantage to get too buried in working only from the standpoint of discounted cash-flow analyses, ROI's, etc. It must be stressed, however, that one had better know how these analyses are made, and be knowledgeable in understanding a balance sheet, and have a good understanding of financial terminology. The difference between a simple project-type debt financing with a commercial bank and a more inventive and creative financial package comes from an intuition and an experience relating to risk-reward factors. The most important element is "common-sense", based on knowledge and experience.

The goal is structure a financial package that fits the project, and in so doing it is important to recognize that it must fit within the limitations of the group supplying the money. I find insensitivity to recognize these constraints the most common mistake made in the field. Put yourself in the other person's place and understand his sources of funds and financing limitations.

The engineer has an advantage in financing, as he or she has (or should have) a better idea than anyone of the value of the product, be it the mine, prospect, or company i.e. the intangible or potential worth which cannot be put on a pro-forma. He also knows, or should know, the risks involved. Two projects might look the same on paper, but a good financier must have a sense of when to take whatever deal he can get, and when to hold out based on his evaluation of future developments.

It is essential to have a good sense of commodity values and trends, trends affecting the general economy (money supply, interest rates, effects of energy costs, etc.), and political considerations. He must also have a good understanding of the science and art of taxation as deals can often be financed using the government as a silent and unwitting partner by proper use of the tax laws.

WHEN TO SEEK FINANCE

The first critical decision to make in financing a mineral prospect is "when". At the exploration level, when it is only a dream in the geologist's eye, the alternatives are more limited than when it is a fully-engineered project.

The "when" decision depends on how much in internal funds the company has to play with, and how desirable it is to invite financial participation to spread the risk. Assuming that decision is not taken out of lack of money, no one can categorically say when to do it. There is

---

From a paper presented to the SME-AIME Annual Meeting, New Orleans, February 1979

no formula; it is a business decision; and the influencing factors are very individual. We all know large companies that have set themselves back many years, or almost gone under, by attempting to self-finance huge projects that have gone belly-up. Large companies can sometimes survive that; small ones seldom can.

PARTNERSHIPS

The oil companies have used tax-shelter financing very successfully at the earliest exploration stage. Some 60 smaller independent oil companies raised $572 million in publicly registered drilling funds from U.S. investors in 1978, not including sizeable sums in private funds or funds from foreign sources. In this technique the promoting company forms a limited partnership and induces a large number (for public programs) or a small number of wealthy investors (for private programs) to drill oil wells for which they obtain a high deduction from income for tax purposes in the current year. Assuming they hit, they will over the years receive an income from oil and gas operations.

Most of the oil and gas programs are, however, very conservative and tend to drill many more development wells than exploration wells. Their overall track record relating to return on investment is quite poor, and their success is mostly because of marked increase in oil and gas prices than clever exploration. Nevertheless, the concept is a good one for investor and explorer.

The technique has not been effectively applied to mineral exploration (with some legitimate and illegitimate exceptions), because all the money normally has to be spent in the year in which it is raised, with no commitments to fund for later years. This is seldom a problem for oil and gas drilling, but in mineral exploration, particularly at the grass-roots level, a program needs staying-power, i.e. 2-4 years. Nord has been successful in using these tax-shelter programs, but always with wealthy private investors who will make a contractual or best-efforts agreement to continue funding for more than one year. As an example, we recently did a program for basic exploration in Australia and New Guinea with investors who committed for two years. (See Appendix 1).

In mineral exploration the pay-out is also different to oil and gas. While we, at Nord, are working on several interesting projects involving uranium, chrome, nickel, and cobalt, it is obvious they will require a further capital investment of several hundred million dollars and 5-7 years to bring a mine to production. The investors neither want to wait that time nor incur that huge expense, and they make the assumption, (reasonably justifies in view of our history) that we will be able to obtain project financing and get an initial return from another party by spin-off of part of the interest.

FINANCE FROM A MINING COMPANY

This brings us to the next level of prospect development, at which an ore-body has been fairly well defined, but extensive further test-work (confirmation drilling, mining feasibility, and metallurgical testing) is required.

Nord's approach to date has normally been to bring in a major company in the mineral area as a partner. An example would involve the Sierra Leone rutile deposit where we spent $500,000 in the early 1970's and brought Bethlehem Steel in as a major partner in what ended up as a $60 million project. We jointly obtained $20 million of project financing from the US Exim Bank of Washington, D.C. and Nord received $2.25 million from Bethlehem in 1975, mostly as a cash purchase, but with some as a share against future project revenues. Bethlehem also financed on Nord's behalf 15% of our equity interest in the project (at 1% over prime) with no Nord guarantees on repayment. The property goes on stream this month, at a time when the rutile future looks good, after a couple of years of depressed prices.

While today we could have retained a larger interest in the project as a stronger company able to bear a greater financial burden, this kind of financing for a young, growing company has many advantages. It gives a good front-end cash payment well in excess of investment, essential to continued operations or to reimburse original partnership investors, and provides equity interest at no further cost and with no <u>further company guarantee</u>. I stress this latter aspect of non-recourse financing as an essential way for a young company to go. (Note more recent developments given in Appendix 2.)

OFF-BALANCE-SHEET FINANCE

If one loads up the company with debt on its balance sheet at an early stage, it restricts the ability to grow because of an intolerable debt burden. Further growth must await pay-out from projects before starting new ventures.

Almost all Nord's financing has been "off-balance-sheet" type, with recourse only to a subsidiary company formed for the particular project. In fact, the only recourse or "balance-sheet" financing we have (in 1979) is a $2.5 million loan from Amex Bank Ltd., of London, a merchant banking group. This was used for general corporate use and developments, and partially as a "grub-stake" for our initial entry in Australia and the South Pacific. (See Appendix 3 for changes since 1979.)

EUROPEAN SOURCES

What are the advantages (and disadvantages) of European financial sources? Well, at the prospect development or pre-feasibility level there are several advantages.

In the United States the banking laws

preclude most American banks from taking equity positions in financings. European and most other foreign banks have no such limitations. This means that when the U.S. commercial banks make a loan to a project, particularly on a non-recourse basis, they have to absorb the risk by taking higher interest rates, charging loan fees, requiring stiffer terms, and/or requiring compensating balances. Of course, if they are loaning against the corporate credit of a very large company this does not apply.

The European merchant banker will make loans on a much more agreeable financial basis with a lower initial financial burden, which is important to a new project. One does not get something for nothing, however, and he will take equity participation either directly or as a function of profits. They might well take an equity in the participation in the parent company through stock or warrants, and to do this it helps if you are a public company.

Our initial financings were done in London with such a group, i.e. Rothschild Intercontinental Bank (which later became assimilated into Amex Bank Ltd., a subsidiary of American Express). We have done equity financing with a number of European institutions. In fact over twenty groups are stockholders now in our company.

The center for much of this financing is London which has access to wide sources of capital from around the world. Interestingly enough we have just completed some institutional financing which involved Scottish groups who are aggressive in investing in North American natural resources. Some French banks also are very aggressive and competitive in exotic financings requiring a mixture of debt and equity. The Germans and Swiss tend to be more conventional, in my opinion, but will do imaginative straight project-type financings on a large scale.

ACQUISITION FINANCE

Part of Nord's interest is in growing through acquisition of producing mineral properties, and financing these can be almost identical to financing a project which is ready to go into full scale production. As an example we recently acquired the Georgia kaolin paper coater operations of a major U.S. company.

We purchased the property for $2.5 million and expanded filter capacity by new equipment costing approximately $1 million. This $3.5 million acquisition was done by a major U.S. commercial bank, which loaned 80% against the project in a subsidiary company (again as a non-recourse note, keeping the parent balance sheet clean). In this instance we obtained an equity partner from New York who put up 20% and obtained certain accelerated tax benefits. The bottom line was that while we had a tremendous commitment in terms of management and credibility, our total financial liability on the project was zero. Luckily, we have completed one year of operations with an operating profit (before debt service) of about three times that of the previous owners. We intend to do more on this formula, but, as in all things, the first one has to come out right to do it again. (See Appendix 4 for the status as at the 1983 Annual Report date.)

OUTLOOK

Less and less money will be available, particularly at the middle or interium stages of prospect development. This is largely because of huge amounts of capital being held in hands which are traditionally disinclined to invest in the risk areas, and even less in mineral projects. By this I mean the OPEC nations, particularly the Arab bloc. Perhaps they will change their investment style. Imaginative financial packaging will be more important for the traditional non-energy minerals sector.

Appendix 1

NORD-HIGHLANDS MINERAL VENTURE-1
(including Nord Australex Limited Partnership)

The Limited Partnership (Australex) was formed to continue the exploration and preliminary development work started by the Company in Australia, Papua New Guinea, the Fiji Islands and other areas in the South Pacific.

The profits and losses of the Partnership are allocated to the partners in proportion to their respective interests in the Partnership, except that the profits and losses are allocated 99% to the limited partners and 1% to the Company until the earlier of: (a) the capital contributions of the limited partners are reduced to zero; or (b) there is cash flow as defined in the Partnership agreement. The cash flow of the Partnership will be allocated entirely to the limited partners until they have received an amount equal to their capital contribution. The cash flow will then be allocated to the Company until it has recovered its initial capital contribution. Thereafter, the cash flow will be allocated in proportion to the partners' respective interests in the Partnership.

The limited partners have agreed that all administrative, geological and technical services necessary to conduct partnership activities will be performed by the Company.

On 1 November, 1979, Australex transferred all of its mineral rights carried at $404,410 to Nord-Highlands Mineral Venture-1. This Joint Venture was formed by Nord (including Australex) and Highlands (including Hicor Mineral Exploration Series-1) to engage generally in mineral exploration and development activities.

The Joint Venture agreement, which terminates 31 December, 2019, provides that Highlands shall be charged 100% of the Venture's costs and shall receive varying interests in the Venture's revenues (75% or 100%, depending on the prospect) until it has received distributions in cash or in kind that equal its capital contribution (payout). Australex will be charged with all pre-venture acquisition, development and testing costs.

When payout is complete, Nord Highlands share in costs and revenues equally on certain prospects and 2/3 Highlands and 1/3 Nord on others. Australex shares only in the costs and revenues associated with mineral interests in any country in which the Partnership had previously commenced mineral exploration (Australia, Papua New Guinea and the Fiji Islands).

Highlands is required to contribute additional capital to the Joint Venture through March, 1986, based on a percentage of limited partners' gross subscriptions in Hicor Mineral Exploration Series-1, to preserve its revenue interest.

Nord, as manager, receives no fee. However, Nord is being reimbursed for all direct expenses and also a portion of its overhead associated with operations of the Joint Venture.

Source: Nord Resources Corporation, 1983 Annual Report; Notes.

Appendix 2

ACQUISITION OF FOREIGN OPERATIONS

In December, 1982, the Company attained 100% (by acquiring the remaining 85% interest outstanding) ownership of Sierra Rutile Limited (SRL), a company engaged in mining and producing rutile, with operations in Sierra Leone, Africa.

In connection with the purchase of SRL, the seller forgave an advance of $750,000 previously made to the Company, which was to have been paid to the seller solely out of the Company's portion of cash flow from SRL (based on its original 15% interest in SRL). This amount has been reflected as an extra-ordinary gain in 1982 operations. In addition, the seller assigned to the Company $22,400,000 of amounts owed to the seller by SRL. This amount has been eliminated upon applying the purchase method of accounting to the acquisition.

At 31 December, 1982, the Company accounted for its investment ($692,000) at cost due to economic instability and political and other uncertainties. As a result of negotiations with the government of Sierra Leone during 1983, a number of these uncertainties have been removed and the Company has consolidated the operations of SRL in the fourth quarter of 1983. In addition, under the present agreement with the government of Sierra Leone, the funds of SRL may be retained in any country. The balance sheet at 31 December, 1982 has been reclassified to give effect to the consolidation of SRL in 1983. Subsequent to 30 June, 1992, the government of Sierra Leone may, at its option acquire 47% of the shares of SRL at a purchase price of 47% of its book value.

Under a marketing agreement between SRL and a third party, advances which bear interest at 110% of prime rate are being made at the rate of $170 per ton to SRL for each ton of rutile produced from 1 November, 1982 through 31 December 1984, to a maximum of 110,000 tons produced in the period. These advances are secured by rutile inventory with a carrying value of approximately $3,000,000 at 31 December 1983. It is the Company's opinion that SRL can finance its own operations internally considering the advances persuant to the above marketing agreement and, if required, short-term advances from the Company.

SIERRA RUTILE LIMITED (SRL)

The amounts due under the credit agreement and the note payable are non-interest bearing, except for $5,250,000 due under the credit agreement, which bears interest at $7\frac{1}{2}$% per year. Payment of interest on the $5,250,000 has been deferred through 31 December 1984. Interest on the above amounts has been imputed at 11% and the resulting discount is being charged to expense over the term of the agreements. The face amount of the debt and the net present value at 31 December 1983 is as follows:

	Face Amount	Net Present Value
Credit agreement...	$6,462,000	$4,704,942
Note payable.......	7,500,000	4,327,092
	$13,962,000	$9,032,034

Payments under these agreements will be in semi-annual installments beginning June, 1985. The total semi-annual payments will be in decreasing amounts from $945,000 (June, 1985) to $750,000 (June, 1994), with a final payment of $108,000 due January, 1995.

The amount due under the financing agreement of $16,706,613 is non-interest bearing (no interest was imputed on this debt as there are no fixed and determinable payment dates or amounts). The amount is due in 2005 with payments required as early as January, 1986 in the event certain cash flows (as defined in the financing agreement) are realized.

Substantially all the assets of SRL with a carrying amount of $26,700,000 are pledged as collateral under the above debt.

Source: Nord-Resources Corporation, 1983 Annual Report; Notes.

Appendix 3

In connection with previous bank loan agreements with Amex Bank Ltd (Amex), warrants were issued to purchase share of the Company's Common Stock. During 1981, Amex exercised its option to require the Company to purchase for $370,000 a warrant for 319,796 shares of the Company's Common Stock. During 1982 Amex exercised a warrant to purchase 158,046 shares at an aggregate price of $777,523, reduced by 68,920 shares, exercisable at $7,02 per share, and 8,954 shares, exercisable at $4.01 per share, are outstanding at 31 December 1983. These warrants expire on 8 January, 1987 or earlier if certain conditions are met.

Source: Nord Resources Corporation, 1983 Annual Report; Notes.

Appendix 4

NORD KAOLIN COMPANY (NKC)

Prior to 1982, NKC entered into a bank loan agreement to fund the expansion of its kaolin facility. Payments on the balance of the loan are to be in quarterly installments of $200,000 through 31 December 1985, thereafter quarterly installments of $500,000 through September 1987 and the remaining $4,566,000 due 31 December 1987. The bank has the option to significantly accelerate the debt repayment schedule if NKC does not obtain financing to replace this debt by 29 September 1984. Interest is payable quarterly at prime rate plus $1\frac{1}{2}\%$ ($12\frac{1}{2}\%$ at 31 December 1983).

In connection with this loan agreement, the bank has also committed $2,715,000, in the form of a letter of credit, to support pollution control Industrial Development Revenue Bonds privately placed with an institutional investor. The bonds bear interest at 10% and are due in annual principal payments of $271,500 beginning 1 January 1992. Interest is payable semi-annually in January and July. The fee for the letter of credit is equal to 1% of the letter of credit amount per year.

NKC also pays an amount of interest, designated as a loan fee, equal to 10% of NKC's income before taxes (as defined) until the cumulative amount of the loan fee payments shall be $1,100,000 and 5% thereafter until such cumulative amount shall be $1,350,000 and $2\frac{1}{2}\%$ thereafter until the cumulative amount is $1,500,000. Through 31 December 1983, $347,000 of loan fee has been paid or accrued. The loan agreement also provides that until the cumulative amount of the loan fee payments aggregates $500,000, should the income before taxes be less than $5,000,000 for any year, the loan fee for such year shall be increased by 1% for each $100,000 or fraction thereof that income before taxes is less than $5,000,000 to a maximum of 15%.

Substantially all the assets of NKC with a carrying amount of $22,300,000 are pledged as collateral under the above debt. The bank loan agreement, as amended in April 1984, and effective as of 1 January, 1984, contains certain restrictive covenants which include, among others, a specified net worth, a specified current ratio and limits on payments to the partners, additional debt, capital expenditures and lease commitments. The loan agreement also restricts NKC from transferring any funds to Nord in the form of cash dividends, loans or advances. The amount of such restricted net assets at 31 December 1983 was $4,240,000, which represents Nord's share of the net assets of NKC.

11.8

FINANCING THE JUNIORS

Michael Chender

Metals Economics Group
Boulder, Colorado

INTRODUCTION

The project financing described elsewhere in this book is available to major mining companies on the basis of a strong balance sheet, a significant equity position in a project of proven feasibility, and a demonstrated ability to bring ore deposits into production. "Junior" mining companies, which are most active in Canada, Australia, the US, and South Africa, by definition do not meet these criteria. In many cases, junior companies will need to raise money simply to cover the costs of acquiring and exploring potential deposits. In what follows we will look primarily at junior financing practices in the US and Canada, although the same general principles will apply in Australia and South Africa.

The principal vehicles for junior mine company financing are limited partnerships, the private placement of securities and debentures, and public offering of securities, and property deals with stronger companies, in the form of leases, earn-ins and co-ventures. The cost (in terms of dollars or equity) and, to some extent, the applicability of using these difficult approaches varies with the stage of development of the property and with the corresponding degree of risk to the investor. Much junior company fund raising is, in fact, "interim" financing in that it allows the company to improve its properties and develop its assets to the stage where more favorable financing arrangements may become available.

Companies will therefore often use several of the approaches described above during the development life of a specific property.

LIMITED PARTNERSHIPS

Limited partnerships and syndications are a popular way of structuring the initial private financing of mine developments because of their attractive tax consequences for the well-to-do investor. Costs of exploration and development can in many countries be passed through as deductions to the limited partners, who are thus able to gamble on a possible windfall at little real cost. In addition to private Canadian funds, much German money flowed into Canadian exploration ventures on this basis, but stopped in 1982 due to economic conditions in West Germany.

In the US, recent regulations have, in the great majority of cases, taken away the ability to "shelter" income in mining limited partnerships. Deductions taken by the individual investor cannot exceed the amount of money he actually has at risk. Nevertheless, the allowable deductions are still attractive and some of the major mining companies are now beginning to look at limited partnerships as a vehicle for attracting additional exploration funding. Ranchers Exploration and Development has successfully put together several drilling funds on this basis.

THE PRIVATE PLACEMENT

Initial funding for the acquisition and preliminary exploration of prospects is often accomplished through a private placement of unregistered securities or, where it is not a problem to carry debt on the statement, by a private placement sale of debentures. A fledgling company wishing to do a public offering may well be advised by an underwriter to accomplish a small private placement as a precondition of the larger offering. In this way, the sale of the registered securities is not inhibited by the company starting out with a debt-ridden balance sheet and no demonstrated ability to secure the additional financing generally necessary beyond the proceeds of that offering.

In the US, the securities offered in a private placement are not registered with the Securities and Exchange Commission (S.E.C.) but are restricted as to their resale, and could be further restricted by a underwriting agreement

---

Based on an article published in Mine Development Monthly, May 1981. Reprinted with permission of Metals Economics Group, Denver.

executed after their sale. Because they carry a high degree of risk and restriction, they are offered only (by regulation) to investors possessing "sophistication" and "adequate means," who are theoretically able to hold the securities for an indefinite time.

In the event that a subsequent public offering is successful, the privately-held securities may be registered for sale (after a waiting period mandated by local securities regulations and perhaps by the underwriting agreement), at which time the return to the initial investors and the corresponding costs to the junior may be quite substantial.

Although the private placement can be done at any time, when substantial development funds are required a more broad-based financing vehicle may be more approproate.

PENNY STOCK MARKETS

Certainly the most visible and controversial source of financing junior company activity is through the public offering of company securities on the various over-the-counter markets and the so-called "penny" stock exchanges.

Canada's Vancouver exchange is the world's most active in listing junior mining companies, with C$217 million raised and close to 200 new companies listed in 1980 - over 90% of which were oil and gas and/or mining concerns. Author's update: In 1983 close to 300 companies listed in Vancouver and over 30 in the US. Vancouver set new records in that year for money raised, largely as a result of the Hemlo play.) In the US, the rapidly growing Denver over-the-counter market produced underwritings of US $280 million in oil and gas and mining issues in 1980, a figure expected to reach $350 million in 1981. Mining issues are less common in Denver (and in the US penny stock markets in general) with nine mining companies newly listed in 1980.

Active markets for junior mining company offerings also exist in Australia and South Africa.

The fact that financing is available at all to companies of marginal financial strength, little demonstrated ability to conduct the business in which they are involved, and often only fragmentary evidence that the subject of that business even exists at the proposed location, testifies to the lustre of metals and their lure of windfall profits in the public imagination.

To some in the mining industry, the penny stock markets seem rife with dishonest schemes perpetrated upon a gullible and greedy public by stockbrokers and underwriters who know nothing about the mining business, thereby making "legitimate" venture capital more difficult to raise. This is more likely the exception than the rule. Nevertheless, the relative ease or difficulty of raising money on these markets often has more to do with the general mood of the market at the time and less with the merits of a particular offering. Currently, the Denver market is hot; local feeling is that any mining issue that is even vaguely promising can sell. Of course, all this may change in six months, and even a company that has properties with merit might not be able to find a substantial underwriting.

COSTS OF A PUBLIC OFFERING

In the US, a typical underwriting agreement by which a public offering is effected can be expected to net the mining company about 85% of the proceeds from the sale of stock. Of the remaining 15%, 10% will go to the underwriters in the form of commissions and approximately 5% will go towards paying the accountable (legal, accounting, printing, mailing, etc.) and nonaccountable (usually fixed at 2% or 2-1/2%) expenses incurred by the underwriter in connection with the offering.

In addition, both the purchasers of common stock and the underwriters are given the opportunity to profit from a rise in the market by being issued warrants, at nominal cost, to purchase additional shares at low, fixed prices. These warrants are exercisable only over a given time period after the offering. The purchaser of the shares often gets one warrant for every two shares, while the underwriters usually get warrants for additional shares equal to 10% of the issue.

The implications of this arrangement must be considered by the company in terms of securing future financing. If use of the development funds raised in the public offering should result in even mildly promising indications, it would not be uncommon for the value of a stock to double or even triple. For instance, in the Equity Gold example below, let us say that one of their gold properties started to show promise in the third year after the offering. At that time the underwriter would have the option to purchase 600,000 shares of the company at a price of $0.605/share by exercising his warrants. If the stock, initially issued at $0.50/share would then be trading at $1.25/share, the exercise of the warrants would yield the company $363,000, rather than the $750,000 in gross proceeds that an offer of 600,000 new shares would theoretically net. Exercise of the shareholder's warrants for 3 million shares at $0.75, if we use the same figure of $1.25/share, would net the company $2,250,000 rather than the $3,750,000 that would be realized from an issuance of a like number of new shares.

The costs of a public offering in Vancouver are higher than in the US, with underwriters taking a commission of 25% on new issues at C$0.50/share or under, and 20% on issues valued at between $0.51 and $1.00. Commissions gradually decrease on a sliding scale to 10% for issues at above $5.00/share.

In the US, however, original stockholders cannot sell their shares for at least two years after their company goes public. Stockholders of companies going public on the Vancouver exchange can sell up to 25% of their shares after only ninety days. The remainder can be sold in increments of up to 25% at a time, over successive ninety-day periods

Equity Gold Offering

To further illustrate some more of the principles described above, we can look at the offering done by Equity Gold Inc., a Colorado firm, which was organized in early 1980 and made an over-the-counter offering of shares in September. The company brought together several people with major company mine-development experience and offered three leased properties that had previously been mined for gold and silver as proposed exploration targets. Although fairly extensive historical documentation existed, the two principal properties had not had a thorough on-site inspection because of flooding and cave-ins. The third, a small gravel prospect, had trenching and assaying done by the previous lessee.

Through the previous May the company had incurred expenses of $223,000 for the purchase of property and equipment and for preliminary operating costs, which it covered through the private placement of stock. The public offering was for three-million units at $1 per unit. (Each unit contained two shares and a warrant to purchase an additional share at any time in the next three years for $0.75.) The stock has appreciated strongly and is today selling in the neighborhood of $1/share.

The underwriters were given warrants at the nominal cost of $100 to purchase up to 600,000 shares starting after 13 months. Purchases could be made at a price of $0.535/share in the second year, $0.57 in the third year, $0.605 in the fourth year and $0.64 in the fifth year.

The six-million shares offered represent only 21% of Equity Gold's stock. Even if all the warrants were to be exercised, the company would only be giving away 32% of its equity.

The initial anticipated net proceeds of the offering were to be distributed as follows: $650,000 in one case and $630,000 in the other for the rehabilitation and exploration of the two principal prospects; $50,000 for the evaluation of the third; $9,000 to repay a loan; and $1,071,000 in "uncommitted" funds to be used for cost overruns, further exploration, or acquisition of mineral properties as the company would see fit. In fact, in February Equity Gold acquired an old silver mine in Leadville, Colorado.

Should any of its prospects appear to merit further development after the initially budgeted exploration work is completed, it is doubtful that the remaining funds will suffice. Equity Gold will then be looking to a further stage of financing. (The S.E.C. subsequently brought an injunction against trading in Equity stock. The company was found to have given misleading information in an attempt to manipulate its stock price.)

The purchase of junior mining company securities is for the most part entered into by unsophisticated investors with limited resources. Generally speaking, junior companies doing public offerings should present several properties that exploration funds will be devoted to, so as to offer the investor some leverage in the light of the great odds against the ultimate viability of any relatively untested prospect.

Alternatively, a substantial percentage of the proceeds from the offering may be dedicated to funding the acquisition of new properties for exploration, in which case the public (and underwriters) will need to be convinced that the company's management is particularly capable to undertake this successfully.

The high capital costs of mine development and mill infrastructure construction are more appropriately borne by some form of organization involving mining companies of sufficient expertise and financial strength to complete the undertaking.

PROPERTY DEALS

The most common vehicle for raising funds for the exploration and development to the production stage of mining properties held by juniors is the property deal. A junior mining company comes to an agreement with another, usually more senior company to split the equity or eventual proceeds from production in exchange for work and/or cash invested in the property. The property deal allows the junior company to recover its acquisition and preliminary exploration expenditures, retain a significant participation in the property if it should desire to do so, and not deplete its often very limited capital resources in the pursuit of one deposit.

The two most typical forms of property deals, broadly speaking, are leases and earn-ins (also commonly referred to as option agreements).

Lessor Retains Royalty

A lease agreement will usually call for the lessee to make regular rental payments and to commit to spend a certain amount of money in exploring the property within a given time. Should the property go to production the lessee will be obligated to make a royalty payment to the lessor based on some measure of production income. The contract will frequently give the lessor an option to convert that royalty at some point to a working interest; or, as described below, it may give the lessee an option to buy out the lessor's royalty interest. If neither option exists, or is not exercised, the lease will revert to the lessor after a specified period of time.

A simple example is provided by a 1978 lease of mining claims owned by Amalgamated Larder Mines to Kerr Addison. The property, consisting of gold-bearing claims with several old worked-out shafts in the Larder Lake, Ontario area, is located adjacent to Kerr Addison's producing gold mine.

Amalgamated Larder-Kerr Addison Lease

Obligations of Kerr Addison to Amalgamated Larder were as follows:

	PAYMENTS	WORK COMMITMENTS
Initial	C$ 5,000	
by 9/1/79	additional $ 5,000	5,000 feet of diamond drilling
by 9/1/80	$ 10,000	10,000 feet (cumulative) of diamond drilling
by 9/1/81	$ 30,000	
by 9/1/82	$ 50,000	$400,000 spent (including drilling)
by 9/1/83	$100,000	$600,000 (cumulative)
by 9/1/84	$800,000	$1,000,000 (cumulative)

Up until September 1, 1984, Kerr Addison had the option of purchasing the claims from Amalgamated Larder for $1 million. It may surrender its option at any time and will lose it by not meeting the above schedule.

Royalty payable to Amalgamated Larder is the greater, in a given year, of 10% of net profits of $0.40 per ton of ore mined. In addition, if Kerr Addison purchases the property, it is obligated to pay Amalgamated Larder $100,000 per year from the date of purchase until commercial mining begins as "advance royalties," against the royalty payments mandated above. Finally, Kerr Addison has the option within ten years of purchase to pay $3 million and terminate its royalty obligations.

The initial fee and subsequent rental payments for the first several years are relatively low. Both rentals and work commitment balloon after several years, as is typical, at which time it is expected that the lessee will have ascertained whether or not he has a property that merits development. This represents a compromise between the lessor's desire to see his property developed and the lessee's wish to take as little unnecessary risk as possible. As it happened, because it soon encountered promising indications of a new gold structure at depth, Kerr had already spent over $400,000 on the property by the end of April 1980.

Teck-Grenoble Earn-In

Under an earn-in agreement, the company holding the prospect gives another company the right to develop a vested interest in the property by spending a fixed amount on exploration or development. The property holder will in this case maintain an equity, rather than royalty interest.

A typical earn-in agreement is represented by a contract recently negotiated by Teck Corp. with Grenoble Energy, regarding the development of Grenoble's gold-silver-copper property in the Greenwood district of British Columbia. This property had seen a fair amount of work, including the driving of several adits, before Teck came into the picture. Teck currently estimates potential reserves at the mine in the area of 500,000 st.

Grenoble granted Teck the right to earn a 60% interest in the property and 300,000 shares of Grenoble, by spending up to C$1 million in exploring the property by March 1, 1984. If it decides to put the property into production, Teck will arrange all necessary financing, most likely as a loan against the property. This would put both parties at risk for the financing, to the extent of their interest in the property. If a production decision is made, Teck will also receive one share of Grenoble for every $5.00 spent up to a maximum of an additional one-million shares. This would give it an interest of approximately one third in Grenoble.

In an earn-in deal of another level of magnitude, Silver King has granted Gulf Oil the right to earn a 51% interest in its properties in Nevada's Ward mining district by the expenditure of $10.5 million in exploration and development over a seven-year period. Gulf's commitments are much higher, but in this case, Silver King had done considerable previous drilling to prove 11 to 12 million st of reserves worth about $80/st at today's prices, before making the deal with Gulf.

A similar case is a recent production earn-in on a Canadian gold mine. Cumo Resources is putting up the funding required (up to C$3 million) to bring Consolidated Louanna Gold Mine's O'Sullivan Lake property into production at 140 tpd. After Cumo's expenses plus interest are paid back out of 75% of net profits, it will retain a 25% interest in the property, with Consolidated Louanna holding 75%. This seems like a rather favorable deal for Consolidated Louanna, which will realize 75% of the profit probably without putting up any funding. But before completing the deal, Consolidated Louanna had already developed 113,000 tons of probable and possible reserves grading 0.30 oz/st gold and completed a feasibility study projecting a profit of C$36/st at a gold price of C$500/oz.

An earn-in or option will be attractive in cases where the company holding the property can support a working interest that may require it to provide funds. Then the interest retained is "carried" -- without obligations to support expenses -- a company's general position becomes similar to that of a royalty holder, except insofar as its equity holding may give it a voice in management of the project.

As anyone with experience in the financing of small ventures is well aware, the variations possible on the basic types of agreements described above are limited only by the imagination.

The choice and negotiation of a financing vehicle at a given time in a project's development will obviously be a function of a great number of considerations, including corporate philosophy and goals, size and known feasibility of the project, amount of dollars already invested, tax implications, existing obligations on the projects, debt position, cash needs and the like.

However, in practice the financing method may often be dictated by circumstances rather than choice. Some companies will undertake the time and expense of a public offering because despite trying, they are unable to interest a major company in taking an option on their claims. Others, thinking in terms of a public offering, may be directly approached by eager corporate suitors offering generous terms in order to consolidate the claims in the area in which the junior company property is located.

BACHELOR LAKE GOLD MINES' OFFERING

An interesting case is that of Quebec Sturgeon River Mines, a Toronto-based company, which did a $10.5 million public offering (3,000,000 shares at $3.50/share) in June 1980, to fund the development to production of its Bachelor Lake Gold Mine at Lesueur Township, Quebec.

As mentioned before, production financing on this scale through a public offering is rather unusual. (Sonora Mining recently topped previous records with a C$40 million offering on the Toronto Stock Exchange for production financing of a California gold property.) The Bachelor Lake deposit had been known since 1946 and about $10 million in total had already been spent on it by the time of the offering, with Quebec Sturgeon River's costs in acquiring, exploring and developing the property, totalling over $4 million. Quebec Sturgeon River had been negotiating bank financing for the property some ten years ago, but that was dropped when a change in Canada's tax-credit regulations led to a modification of bank terms. The company then considered an offering of debentures but rejected it as too costly. Joint Ventures were ruled out as Quebec Sturgeon River found that the majors it approached wanted majority stock ownership and control, positions the company was unwilling to give up.

The company ultimately found an underwriter willing to take on the large amount they needed. Quebec Sturgeon River was left with 55% of its stock after the offering and, important to them, a debt-free position. As for the costs of the financing, since the company planned to spend the proceeds over a twelve-month period, costs were essentially paid for by the interest accruing on the unspent portion of the $10.5 million over the course of the year.

CONCLUSION

The existing deals in the marketplace are worth studying for their guidance on the way the numerous financing options are being tied together for junior mining companies. Tax-shelter structures, farmins/outs, and public offerings are the most common vehicles used by the junior companies for raising finance.

11.9

FUNDING A PROMOTIONAL EXPLORATION COMPANY

JOHN S. BROCK

WELCOME NORTH MINES, LTD., VANCOUVER, B.C.

INTRODUCTION

The Oxford Dictionary says that the word "promoter" has usually been used in an opprobrious sense since at least 1876. This popular image of mining promoters is of cigar-chomping shysters and Brooklyn-Bridge vendors, improving their standard of living with the aid of funds raised through the sale of stock to little old ladies. It is unfortunate that mining promoters are considered crooks until they can prove their integrity and honesty by finding a mine.

A common rule of thumb states that for every 1,000 prospects, 100 come to be drilled, and one becomes a mine. However, it is also a statistic that there have been mining promotions, where money raised has gone elsewhere than into the ground; Junior exploration companies can be handy vehicles for unscrupulous financial activity. One must remember though that every stock exchange has had its share of sound mining promotions that, because of the one in 1,000 odds, have failed for legitimate reasons.

The promoter must be able to rationalize the worth of a mineral discovery which, on an industry-average basis, probably cost more to find than the discounted value of its net return, which return is not an assured percentage of invested capital, and furthermore, is ultimately from a wasting asset. So, it is easy to understand why the promoter must constantly justify the credibility of his operation to his grubstakers during the course of the "five phases of exploration," those being (1) euphoria, (2) disillusionment, (3) search for the guilty, (4) punishment for the innocent, and (5) distinction for the uninvolved. The promoter with the exploration organization that does not have the benefit of positive cash flow, by definition for the sake of this article, works for a promotional outfit, a Junior Company, that must raise its exploration budget from a variety of sources.

Promotion, as discussed here, will not entirely refer to the formation and operation of a joint-stock company but rather to the overall methods of raising risk capital for mineral exploration ventures from outside sources.

THE NECESSITY OF PROMOTION

In Canada, the publicly financed Junior Company has had an enviable record of mine-finding success. In the Province of British Columbia alone, the discovery of fifteen out of twenty-four mines has been as a result of the activity of the Junior Company (B. C. & Yukon Chamber of Mines, 1977). In the Province of Ontario, one research study indicated that between 1951 and 1974 the 'Juniors' spent only 28 percent of the total exploration funds but were responsible for 62 percent of all economic discoveries (Freyman, 1978).

The reasons for the success of the Juniors are tied to a variety of factors, the main one being personal initiative based on incentive. In other words, the principals, or their employees have 'a piece of the action.' The smaller companies can operate faster and more efficiently than their larger counterparts. The Junior Company is also destined to take some bigger risks, the 'long shots,' many of which efforts by virtue of the Junior's perseverance coupled with a touch of 'ore is where you find it' philosophy, result in major discoveries.

The individuals, syndicates, private and publicly financed Junior Companies, because of high costs of exploration and development, cannot hope to finance most mineral prospects all the way to production. Exceptions do exist, those being the lower capital cost, small, high-grade deposits.

Private investors and publicly financed Junior companies, therefore, can usually afford to finance most exploration through the initial drill-testing stage. The major resource company will enter a mining deal with a Junior Company usually by way of option or joint venture agreement, and will finance to production to earn an equity in the prospect or the Junior Company or both. The conventional lending institutions, banks and government agencies, will usually only participate through loaning a portion of capital

---
Based on a presentation to the SME-AIME Annual Meeting, New Orleans, February 21, 1979. Updated by the author in March 1984

costs financing for production. In summary, the sources of capital sought depend on the maturity of the exploration project, and the promoter will mix or switch from one source to the other as the project advances in accordance with the amount of dollars required and the best financial leverage obtainable for various levels of risk present. (See Carraghian's paper in Chapter 2.)

THE INGREDIENTS FOR A PROMOTION

Ideally, the promotional company must have expertise -the promoters should be geologists or prospectors, with good business instincts. A knowledge of where to find the mine goes hand-in-hand with knowing where to find the money. A key reason for the promoter to be a geologist is that, in the case of selling to other resource companies, he or she will initially be dealing with a geologist on the other side of the table. As negotiations progress, he may also, if not able to immediately draw on outside expertise, have to wear the hats of an accountant, minerals economist, and lawyer. (See also Cruft's paper in this Chapter.)

The prospecting ability and geological expertise of the promotional company's personnel are crucial - they are the ones who must evaluate and discover or acquire the mineral prospects on which the company's fortunes are hung. A previous track record of success and exploration-program operation can also help, especially if the promoter is also going to act as operator of the exploration project. The mining community is small, even on a global scale, and reputation is important.

The promoter must anticipate commodity cycles so that he can prepare for demand. It is odd that the Boards of Directors of most major resources companies don't explore for their favorite commodity until the price is high. The promoter should be ready for this corporate aberration with a portfolio of good prospects acquired during metal-market slumps.

The geologist-promoter must be in tune with commodity demand, metallogenic fads and today's popular geological models. Whether it be base metals or precious metals, porphyry deposits, carbonate-hosted deposits, shale or clastic-hosted skarn deposits, or volcanogenic massive sulphides - all are subject to cyclical levels of enthusiasm. The promoter must be ready for the enthusiasts; they can be found at any mining convention or symposium.

The promoter should also look for the individual prospector or small company that does not have the expertise or capabilities to finance and be prepared to act as a scout or middleman. The promoter can also act in the same capacity for a large mining company that wants to get involved in mineral exploration for a new commodity or in a new geographic area.

PROMOTER AS MARKET MAKER

There are no real regulatory provisions in Canada for either defining or recognizing the role of the Market-Maker within the Securities Act or the Income Tax Act. In the United States, these people are recognized and called Specialists, and fullfil somewhat the same function, although they may also serve as Brokers.

The Market-Maker recognizes attractive resource situations that require funding and, for a consideration, usually settled by way of shares, will assist by acting as a liaison in arranging the Broker financing, creating a market through promotion and maintaining an orderly market by regulating buy and sell orders through one or more Brokers. An orderly market is important when the track record of a company is examined by potential future investors. The job requires public-relations expertise and proper dissemination of news: something for which most management of Resource Companies have no time or ability.

When over-regulation by the Toronto Stock Exchange and the Ontario Securities Commission sent the Bay-Street Promoter* west to Vancouver, it was much to the discomfort of the local regulatory bodies. Many of these Promoters are still with us today, are active and have exhibited a tremendous track record insofar as greatly assisting in the making of new discoveries, one example being Murray Pezim of Hemlo gold fame. (See Chapter 12.)

Good Market-Makers are a special breed and Vancouver doesn't have enough to support all the Junior issues. The Brokerage Houses, in the future, could fill the gap by creating recognized, in-house Specialists.

SOURCES OF RISK CAPITAL

THE PUBLIC

Public financing through the sale of shares of Junior mining and exploration companies has usually been considered to be the most effective and popular way to mobilize private and venture risk capital for early stages of exploration and development. However, in spite of the fact that the contributions of the Juniors to mine-finding cannot be argued against, many North American securities commissions, tax legislators and other 'protechtionist' bodies have since "Comstock Lode" days legislated and regulated the publicly financed Junior stock company out of business. Many Junior companies in the United States were the first to go; eastern Canadian Juniors have been on the wane. The Vancouver Stock Exchange and the Government of the Province of British Columbia must be commended for apparently realizing the importance of raising risk capital from the public in natural resource exploration and development. (See Gorval's paper in this chapter.)

---

* Bay Street, the location of the Toronto Stock Exchange

Public financing allows the promoter a chance to capitalize on his efforts through a market for his own shares, although regulations restricting the sale of the promoter's shares have made it so difficult to realize his incentive that other sources of capital are often more appealing (Lau, 1979).

The promoter using public funds raised through the stock market should be aware of some of the following points:

1) The capital required to finance the high cost of exploration and the resultant dilution of shareholder's equity through repeated public financings eventually leads most public stock companies on a route to self-destruction through over-capitalization or dilution. At some point in the company's growth the shareholder will benefit more from financing of its mineral properties through joint venture and option agreements with second parties, which will spread the risk and reduce equity in specific properties, but will preserve equity in the company's share capital.

2) When financing publicly, the promoter should plan to finance for the total amount of funds required to complete the project. Too many times the publicly financed company grinds to a halt, in debt, half way through a project that cannot be completed. Refinancing at that stage can be impossible.

3) The promoter, once in charge of public funds, may find that the regulatory bodies can restrict him from certain types of exploration deals. For instance, public companies require more disclosure of information, thereby limiting confidentiality of exploration results. Restrictions may be placed on deals that involve transfer of stock to the participating partner.

4) More is required in the way of administrative procedure to satisfy the regulatory bureaucracy. At times the stock exchange, the shareholder of the public company and, indeed, stock brokers may request and subsequently receive information dealing with results of exploration, which because of its preliminary nature could be subject to misinterpretation.

5) Publicly traded companies can be vulnerable to market manipulation and takeover by outside interests against the best interests and wishes of the company's management.

6) When financing through the public, the promoter must choose a broker who realizes he must compromise the best interests of the company and of the public. The company must be satisfied that it is receiving fair dollar value for the sale of its capital and the shareholder must be given a decent value for his risk dollar (Buckland, 1979).

7) The promoter must prepare for a public financing with a prospectus that does not put all the shareholders' eggs in one basket. The risk of financing the exploration of one unproven mineral prospect is too great. The promoter therefore has to have a longer-term plan for his shareholders - other projects, acquisitions, and exploration programs. Most promoters will have to deal with public financings more than once during their working life - credibility should therefore be established and maintained.

8) The sophisticated speculative investor is becoming increasingly interested in good management. Therefore, beware those Juniors with part-time, unqualified boards of directors, who make decisions on behalf of their shareholders on the basis of opinions and recommendations by third parties who are not held accountable. The ideal management core consists of capable full-time employees responsible to a strong board of directors. Equal emphasis must be placed on those two philosphically opposed departments - exploration and administration. The exploration department to be responsible for choice of commodity to be searched for, acquisition and evaluation of work plans. The administration department to be responsible for budget, accounting, land records, legal and reporting requirements.

The benefits of public financing, where allowed, through the sale of stock are multifold. It is a relatively fast and simple way to raise risk capital, it allows the public to participate and, yes, gamble, on a discretionary basis on a country's resource development. (It is interesting to watch many government regulators protecting the public from the speculative mining stock promoter by regulating the promoter out of business, while at the same time promoting government-sponsored lotteries by way of the prime-time television commercials. The only television coverage allowed to mining promoters is via the investigative reporters of the Sunday night magazine shows.)

Broker: The role of Broker is often confused with that of the Market-Maker or Promoter. Perhaps, in the past, the Broker-Promoter were one and the same. But today, the chief role of the Broker is as an underwriter, the provider of funds. His research seeks out the best vehicles in need of financing. Innovative methods of financing are considered, such as a combination of equity and allowance of deductibility of Canadian Exploration Expense. The Broker must be prepared on a success-contingent basis to follow through with further financings as a particular project matures. The Broker, these days, is also arranging joint venture participation between resource company clients. Some Vancouver Brokers have anticipated that the capital requirements of local companies are more than can be provided locally and, therefore, are expanding through branch offices to other parts of Canada and overseas. Recently-seen investor participation through the Vancouver Stock Exchange is about 20 percent American, 50 percent Canadian, and 30 percent European.

Some Brokers estimate that 70% of the total funds raised went to oil and gas, or about C$600 million. Over the next three years, it is estimated that at least C$800 million will be raised for Junior Mining. Vancouver's lead underwriters now know that much of that oil and gas money was not well-spent (no pun intended!). A similar situation could arise with respect to the spoke of financings of gold heap-leach prospects in the United States.)

The Broker must also make his client money, therefore necessitating the eventual sale of shares for the turnover of invested capital. This is a necessary part of the financing equation if the Broker is to maintain a pool of funds for continued underwritings. At this point, it is more the role of the "Market-Maker" to ensure that an orderly market is maintained.

THE CORPORATE SOURCE

The corporate source for risk capital represents the largest amount of available capital. Typical corporate sources for mineral exploration funding are the producing mining companies (the Majors), venture-capital groups, and metal brokers.

A key advantage is that one is dealing with one entity only. There is usually no direct and significant involvement with brokers, stock exchanges, and securities commissions. The deal is normally structured with optional provisions for financing to completion, whereas public financing is generally done in success-contingent stages. The corporate promotion usually involves giving up equity in the property rather than in the company as in a stock deal. The advantages and disadvantages of dealing with corporate sources hinge on their financing, operating, and management capabilities and the promoter's negotiating ability.

Research on the corporation's requirements and capabilities is a prerequisite. Some corporations that have ready cash may lack operating capabilities. Some may not be capable of financing senior stages of development. Realize, though, that mining and exploration do not necessarily go hand-in-hand. Some exploration is funded because of other priorities, such as tax reasons, or the obtaining of captive sources of metals. It can save the promoter a few steps to first deal with an operating company that is hungry for new reserves of its favorite commodity. (See also Cruft in this chapter.)

Oil companies traditionally pay higher 'front-end' payments. Mining companies have operating expertise. Metal brokers may offer better royalty schemes.

Some corporations have disbanded their own exploration efforts and maintain exploration departments that serve only as scouting agents for property acquisitions. Some have given up entirely and just buy their way in at feasibility study stage, thus leaving the pioneering work to the Juniors and independents (Lau, 1979).

The promoter must be ready with counters for the reasons for rejection, many of which are unrealistic. It is interesting to collect major company advertisements for mining properties and then match them to 'turn-'em-down' letters for properties that were later explored and developed by others, for example:

> "If you get excited, we get excited.", but the letter of rejection said: "It doesn't fit our corporate objectives."

> "Give us a call and let's talk.", says another, but when you call they say: "No budget."

> "We go anywhere.", claims one major international organization, but when they get there it's "Too remote."

> "We participate in and finance mineral resource development.", the response when queried: "Talk to us after drilling it yourself."

> One other giant says that "We are prepared to examine, option...", but when it comes time to arrange for the field trip: "If we find ourselves in the area we may take a look."

The field examination is, in many cases, done by a staff geologist who usually has less geological experience than the exploration manager or vice president who will make the final decision to acquire the prospect. Unfortunately, if the staff geologist's vested interest is only the monthly pay cheque, he may not want to risk job security by recommending your prospect, and therefore becomes known to the promoter as a 'turn-'em-down artist'. If the staff geologist makes the "safe" decision to turn your property down, senior levels of his management may never hear of it. One should make sure they do.

Unless obliged to, promoters should try not to deal with the same company twice unless they have terminated the previous deal. The same participant in all the promoter's ventures can stretch the participant's budget and staff capabilities and will naturally divert attention to the most seemingly favorable prospect. The promoter must also remember to deal with his neighbor first; by virtue of his presence in the area he has already established his priorities.

FINANCING ARRANGEMENTS

The promoter must realize that most deals with major corporations are so bad for the Junior company that publicly traded shares of the small company invariably suffer a collapse in price when the terms are announced. A few of the many points to negotiate hard for are cash option payments, work commitments, earning level, and carried interest.

Cash Option Payments: It's surprising but not uncommon to see the reluctance of many companies to pay 'front-end' money. They explain that they would rather see it go 'into the ground,' while at the same time pretending not to realize that the front-end payment usually represents a partial recoupment of costs that have already gone into the ground, and accordingly, have substantially lowered the risk. The promoter must argue that the cost of finding and exploring the prospect has been done at less expense by the Junior company than could have been done by the major. Front-end payments should be kept low, but not so low as to fail to recoup a good portion of the risk money expended to find the prospect. The promoter might have to be prepared to lower the amount of continuing optional cash payments in order to allow room to negotiate for a higher retained equity in the property, yet at the same time he must ensure that the cash option payments are high enough to provide an incentive for the participant to carry out a reasonable and continuing work program.

Work Commitments: Work commitments are called for so that the property doesn't 'sit on the shelf', but should not be so onerous that they increase the chance of the participant writing off the property at an early stage and premature termination of the agreement. A call for diamond-drill commitments prior to completion of geological assessment can have a detrimental effect. The work commitments should be structured in accordance with the participant's willingness to allot budgets from year to year. Work commitments should specify both dollar amounts that can be cumulative from year to year, as well as a work plan. A contrary theory offered by some promoters emphasizes no work commitments in lieu of very attractive cash option payments.

Earning Level: For tax reasons; the participant may want to earn his interest immediately. If such is the case, he must be obliged to maintain his interest by performing work commitments and possibly making cash option payments. The property will not be very marketable the next time around if the original participant has walked away and taken an interest with him, therefore the deal should be structured to allow his relinquishment of interest in total if the agreement is terminated. The promoter should leave himself an option to participate by way of 'back-in' provisions, preferably without a penalty, to a working interest at as late a stage in the development of the property as possible. Although his chances of financing the working interest are remote, that option to maintain a working interest may be dealt to a third party. The net discounted worth of the working interest as compared to the worth of a carried interest or royalty must always be kept in mind prior to making a decision to 'back-in'. It is surprising how many minority partners elect to back in to a working interest when they could have kept a more valuable non-assessable or carried interest.

Carried Interests: Net Smelter Return Royalties or Mine Value Royalties are better than today's Net Profits Interests. There is a nasty tendency of major companies these days to 'prime-plus' the original, simple Net Profits Interest to death. The major rapidly assumes the role of a banker with exhorbitant interest charges on all cost categories, to the extent that the net worth of the Net Profits Interest is worthless. The Mining Industry sorely lacks a model form for Definition of Net Profits and Royalty Interests, the petroleum and coal industry has such standard forms. The promoter should ensure that buy-outs of carried interests are at his option.

RECENT DEVELOPMENTS

Most major mining houses, because of exploration budget restraints experienced during the 1981-1983 recession, pulled back from participation with Junior companies during the early phases of exploration. This has in part re-defined the role of the Major as a corporate source of financing.

Some of the more imaginative larger mining corporations are now emerging as acquisitors in share positions of Junior Companies. Examples are Noranda, with Golden Sceptre and Goliath; Tech, with International Corona; and Placer, with Skyline. All of those deals could pave the way to the possibility of eventual control of the Junior and its holdings by way of take-down of shares.

A few Major companies, who in order to keep well-qualified staff employed, are considering the formation of consulting divisions mainly to provide services to the Junior mining companies - an excellent method of maintaining exposure to the current exploration scene.

Some Majors, recognizing public financing as an alternative to being inactive, are forming Juniors. Some for the purpose of continuing their exploration efforts. Amax's newly formed Canamex will, for example, equity finance its exploration program with Canadian public funds.

It is logical to assume that the most successful Juniors will eventually look to the Majors for expertise in the matters of feasibility, production, marketing, and support for debt financing. It is also logical for the management of most Junior Companies to welcome new management to assume the routine of production matters. At this point, most true explorationists are glad to go on to a new grass-roots venture.

PROMOTER AS OPERATOR

A promoter has generally attained his relatively independent and entrepreneurial position through a certain degree of non-conformity to the 'motherhood' principles of the larger corporation. If required to report to that corporation in accordance with their restrictive operating procedures, he may encounter problems.

Admittedly, the promoter-operator may wish to act as operator of a venture which emphasizes his geological capabilities and expertise and may provide income through a management fee to help in defraying overhead costs. The promoter must be prepared to evaluate his joint venture participant's capabilities as operator, especially if the deal involves the promoter's right to back in to a minority working interest position. Having to match dollars with a major corporate partner who has assumed control and the position of operator can be an unpleasant and costly experience for the small independent who does not easily conform to big company policy.

GOVERNMENT SOURCES

Many governments these days don't need to be promoted - they just move in through partial or total nationalization, through an increase in royalties or tax levies, or through the right to elect to participate in exploration through their own Minerals Corporations. In 1974, five out of ten of Canada's provinces entered the mining business by one or more of these procedures; many have now backed off after realizing the damage done.

In spite of the fact that there is no justification for the government to be in exploration or mining, the promoter can, perhaps, justify his dealing with them secure in the knowledge that he will probably save the taxpayer a few dollars. Some have found it advantageous to promote government agencies by selling expertise and a personal mine-finding incentive, which qualifications the civil servant may not possess.

Promotion of a government requires a demonstration that a curtailment in resource development will not help the economy and that export of raw natural resources that will help balance-of-payment problems would be useful. Governments should also be taught that foreign risk capital for resources development is a gift.

The promoter must consider tax funds as the best form of government participation in mineral exploration and development, whereby the individual spending exploration dollars can claim one hundred percent or more of those expenditures against personal income. Tax funds have worked well in the oil and gas industry where the prospects for a more immediate return on investment are quicker than in the mining game. Tax funds have a place in the Junior Company financing scheme, where investors get the benefit of the write-off and are bonused with stock, flow-through shares, in the capital of the company. This formula, which has been slow to catch on until recently in Canada, on a simplistic basis allows the Junior company an underwriting at half the equity dilution, the shareholder gets the tax benefit and can have some liquidity on the other 50 cents of his dollar through owning a marketable security.

CONCLUSION

The incentives may differ from year to year but the Mining Promoter will always have to identify the problems of mine finding, and work around them, and convince someone somewhere to expend time, effort and money to discover and produce minerals. J. Paul Getty said: "Plainly a business must supply a need. In so doing it must give value for value received. The value received - or the price of the service rendered or product sold - must be fair, low enough to be within the buyer's ability to pay, yet high enough to give the business a reasonable profit." That statement summarizes the financing operations of a true promoter.

REFERENCES

British Columbia & Yukon Chamber of Mines, "Public Financing of Mineral Prospects in British Columbia," Public Finance Committee Submission to Ministry of Consumer & Corporate Affairs, Vancouver, B.C., June 1977.

Buckland, C.C., "The Role of the Broker in Public Mine and Exploration Financing," Address to 67th Annual General Meeting, B.C. and Yukon Chamber of Mines, January 19, 1979.

Freyman, A.J.. "The Role of Smaller Enterprises in the Canadian Mineral Industry with a Focus on Ontario," Report, Ontario Ministry of Natural Resources 1978.

Lau, J.T., Private and Public Mine Financing, Address to 67th Annual General Meeting, B.C. and Yukon Chamber of Mines, January 19, 1979.

11.10

FINANCING THE DEVELOPMENT OF SMALL MINING PROJECTS - AN OPERATOR'S VIEWPOINT

Fred H. Brooks, President
Brooks Minerals, Inc.

Lakewood, Colorado

INTRODUCTION

The toughest job for any mining company, large or small, is to locate and identify a property which it feels has the potential for development and which can be tied up through location, lease or option under terms within their means. In the case of a small company, this precludes grass-roots exploration with its attendant high costs and the extended time required to develop a prospect to the point of being a project. The small mining company, therefore must limit its search to known properties which, for various reasons, have been shut down or were never thoroughly developed at the time of discovery. These types of prospects are located either through literature search, familiarity with many mining areas in general, or through contacts and submittals from a variety of sources.

In order to develop a marketable presentation to any investor, the basic requirements are the same:

- the property must demonstrate the potential to provide a satisfactory return on investment

- the evaluation study must be thorough and professionally developed.

In order to promote any mining property, it is first necessary to know what you have.

INFORMATION DEVELOPMENT

We normally begin with a detailed literature search of the district, assembling all available data on the basic geology, production records, and old reports of this or neighboring properties and any other information which can be found. This will aid in the general understanding of the basic geology and hopefully allow a fairly detailed knowledge of the specific occurrence. With this information one can begin to speculate on the reserve potential and develop some high and low figures for estimating purposes as to mining method, production rate, metallurgy and begin developing estimated capital and operating costs. Cashflow runs are developed for varying tonnages (high and low), estimated high and low grades of metal and at differing market prices for the product.

FINANCE CONTACTS

Brooks Minerals, from its inception, has been an operating company and we have been able to assemble a small but experienced operating staff. For this reason we try to put together projects and joint ventures where we can be the operator and maintain a certain degree of control.

Whether due to familiarity with many colleagues in the mining industry or due to unfamiliarity with the banking houses and securities markets, the approach to raising finance has generally been to contact the established mining companies.

This requires a familiarity with the corporate philosophies of these companies with respect to their development goals and financial capabilities. A common mistake is to shotgun the proposal to many companies of different sizes and goals in hopes that one might be interested. It is more beneficial, when possible, to tailor a proposal to each company, stressing the positive points where the specific proposal best meets their interests.

One company may be more interested in a development program with fairly certain and conservative return potential. The property may well contain both possibilities but a lack of understanding of the individual company's method of operation could spoil the deal. In another instance, one company may look at a property as a moderate income producer over an extended life, yet a second may elect to high-grade the deposit over a shorter period in order to provide early cash flow or take advantage of a favorable market situation.

MAKING THE DEAL

There are as many types of deals as there

are properties and people to run them. Here again, a thorough feasibility and cost study will tell just what the potential value is and enable one to drive a realistically hard bargain, knowing what profits there will be to split up. An old adage in negotiating agreements which is never out-dated is, "The best agreement is one in which both parties can make money". The investing mining company expects and is entitled to a reasonable rate of return. The profit potential above that rate of return is the area for negotiation.

A company may negotiate backing for a finder's fee and royalty interest; stay in for a net-profits participation; or if it has the capability for operation, hold out for an operator's interest with participation in profits.

This latter approach limits which mining companies can be approached. It also requires a high degree of confidence in the small mining company's ability to perform satisfactorily. For this reason a highly qualified nucleus of staff people and preferably a moderatley good track record is a prerequisite. The small mining operator needs to demonstrate the ability to plan the mine development, mill construction where applicable and operate the property more efficiently than the funding company.

This sounds tougher than it actually is since, the efficient small mine operator can generally outperform a large company in operating a small, labor intensive operation in the following ways:

- Overhead is almost always lower

- Small companies can often, but not always, get through the bureaucratic jungle with less resistance and red tape

- The small company manager, generally with a vested interest, provides close supervision and often tighter cost controls

- The small operator's staff can generally perform a multitude of engineering and operating disciplines.

CONCLUSION

Over the past several decades, the major mining companies have been gravitating toward larger, lower-grade and highly-mechanized operations. With the recent dramatic rise in precious metal prices, many of the older small vein mines are being re-evaluated. These types of deposits do not lend themselves to high tonnage, mechanized production methods. This is the principal reason why the small mine operator has regained his position in the industry and why larger mining companies, who in the past would seldom consider being the non-operator with a junior partner, have come to reconsider their position.

11.11

VENTURE CAPITAL FOR THE MINING INDUSTRY

JAMES H. BOETTCHER

PARTNER, VICKERS BOETTCHER MASSY-GREENE PARTNERS

INTRODUCTION

There are many and varied sources of finance available to the mining industry for exploration, development and/or the operation of mining projects and companies. These sources include equity from major as well as "penny" stock exchanges, precious metal mutual funds, private investor groups (from "grubstakes" to tax-oriented limited partnerships) and banks. The variety is representative of the fact that the world's capital markets are segmented and stratified according to risk-reward characteristics. This interrelationship is summarized in Exhibit 1 and 2 which relate (very approximately) various sources of finance to a typical mining project's life cycle and risk-return characteristics. The objective of this paper is to introduce, as it applies to the mining industry, a somewhat recent phenomenon represented by one of these capital markets segments referred to as "venture capital."

Venture capital investing per se, which today is a multibillion dollars per annum investment business, goes back as far as Queen Isabela's backing of Columbus' "wild ideas." The real institutionalization of the venture capital process really began in earnest, however, after World War II. Today over 700 private venture capital sources (and more government related ones) exist providing funding to a range of industries, but typically related to the "high-tech" business. The table to the right shows industries financed by venture capital funds between 1980 and 1982. Conspicuous by its absence is, of course, mining which is not a great surprise since typical mining projects up until recently were large, required long lead times and were undertaken by established mining houses. Also, as most of you can appreciate, <u>successful</u> investing in the mining business requires a great deal of experience and understanding of the industry which most venture capital funds do not have.

TARGETED RETURN

The rate of return objective of most venture capital funds is, according to the legalized version, to realize a return which is "substantially in excess of that obtainable through conventional equity investments."

One may say that a common objective of venture capital funds is to realize a 20% to 30% compound annual internal rate of return on capital over their life which is often ten years.

Venture Capital 1980 - 1982
<u>Disbursements by Industry</u>

	Percent of Total Number of Investments		
	1982	1981	1980
Communications	11%	12%	11%
Computer Related	38	29	26
Other Electronics Related	13	13	9
Genetic Engineering	3	5	8
Medical/Health Related	8	7	9
Energy Related	5	6	20
Consumer Related	6	5	4
Industrial Automation	4	6	3
Industrial Products	4	6	2
Other	8	11	8
TOTAL:	100%	100%	100%

Source: <u>Venture Capital Journal</u>

This objective has been recently more feasible in the mining sector due to the development of smaller precious metal projects which cost less and are quicker to generate cash flow than the behemoth projects of the 1970s. In addition to precious metals, venture capitalists have invested or looked at a variety of other mining projects including ones as prosaic as a high margin "kitty litter" operation, sand-gravel and other industrial minerals operations, the leveraged buyouts of coal mines or mineral divisions of major energy or mining corporations, and even in in-situ copper or gold leaching ventures. A typical investment is in the range of $100,000 to $3,000,000 although some can go as high as $5,000,000. This implies a project size up to, say, $30,000,000 if a venture capital fund does it on its own. Larger projects can be done if the equity portion of the investment is syndicated with other investors.

SOURCES OF FINANCE

VENTURE CAPITAL vs OTHER SOURCES

There are several characteristics that differentiate venture capital funds from other sources of mining finance. One of the main differences is venture capital's principal source of funding which is primarily large institutional investors such as bank trust departments, insurance companies, university endowments, and corporate pension as well as Taft Hartley plans. These sources typically invest as limited partners in a blind pool which is managed by one or more general partners. What is particularly exciting about this approach to natural resource investing is that such venture capital funds represent intermediary vehicles between the multibillion dollar pools of capital and the natural resource industry which is a tremendous user of capital. In order to keep this bridge between investment opportunities and capital sources open, these funds must develop solid credibility and a track record showing consistent real rates of return which meet the objectives outlined above.

In addition to its sources of capital, a second important differentiation between venture capital and other sources of mine finance is highlighted in a definition of venture capital by Stanley E. Pratt, probably the leading chronicler of the venture capital industry, as:

"...participating investors seeking to <u>add value</u> through ongoing long-term oriented involvement with a continuing business involvement." (emphasis added)

Venture capital investors are usually looking for situations where they can "add value" in a way that complements the skills, resources and capabilities of a venture's sponsors or promoters (hereafter referred to as "the Principals") so that all parties make more money. The extent of a venture capital investor's involvement varies from case to case but has to be based on a mutual feeling that value is being added so the investor's involvement is not seen as unwanted meddling. One reason the venture capital investor can add value is that he usually has extensive industry and/or financial experience.

Ways a venture capital investor can be of assistance and add value include, to mention a few:

- Brainstorming with the Principals on an overall business development strategy, beginning where they are and looking ahead to help make sure that all decisions are made with an eye toward arranging the future sources of financing that will facilitate the timely development of a project;

- Helping develop and, if necessary, write a comprehensive business plan and strategy for the project or company's development providing inputs, especially with regard to financial assumptions, as well as acting as a sounding board often in the "devil's advocate" role, to test the reality of the business plan's goals, objectives and assumptions;

- Identifying key reports or studies, as well as their content, that need to be done to assure the project's viability and secure financing. Also, suggesting top consultants who could undertake such work;

- Helping the consultants to produce the studies or reports as required to get their work in a form most useful to the financial community;

- Helping to identify and recruit key people to actually be brought into the company as appropriate (we often find that third party can be helpful to not only identify key members of the management team, but also to discuss with them, and often their families, the prospects for a company in terms of other oppotunities the key management person may be considering);

- Helping identify, structure and negotiate alternative sources of funding for the project including banks, mutual funds, resource companies, penny stock exchanges, private institutions and major stock exchanges both in the U.S. and abroad; and

- Helping to structure an entrepreneurial environment in which all work as a team for the good of a project -- this could include, for example, the development of management equity earn-in programs.

Very often, especially in the early days of a small company or project's life, this type of assistance is difficult to buy and can be instrumental in determining the success of the venture. The venture capitalist's patience, perspective and fortitude of having been through many similar situations before can be an invaluable ingredient provided it is delivered in an attitude of cooperation and mutuality of interests.

STRUCTURING AND PRICING

To realize their ambitious objectives, venture capital firms usually have specific preferences for the types of deals they seek in terms of its stage of development which can, in turn, be related to risk. These varied stages are, again, shown in Exhibit I with their counterparts to the mining industry. The greater the perceived risk, the greater must be the expected potential return. As can be seen by the dashed line in Exhibit I, our firm's objective for the deployment of funds in terms of risk-reward relationship is roughly as follows:

% of Capital Invested	Classification of Financing	Approximate Expected Rate of Return
40%	Fourth Stage (Mezzanine or Bridge)	25% - 35%
40%	2nd or 3rd Stage	30% - 45%
20%	Seed Capital	100% - 300%

These guidelines will accordingly influence how selected deals are evaluated, structured and priced. The expected rate of return in the 20% to 30% range is arrived at differently within each risk/return category.

A number of factors affect how an investment is structured and priced. These include:

- How much money are the Principals putting up relative to the total funds initially required?

- What is the total amount of equity funding required to launch the venture? (How much additional dilution over the years will be necessary to keep the business moving according to its stated business plan?)

- What is the expected time required for the project or company to proceed through the various stages including exploration, feasibility, engineering and design, construction, and realize full operation? When might the first positive cash flow begin and how long thereafter to projected payback?

- At what stage in the project of company's development are its Principals coming in to request funding?

- Is the investment opportunity a one-project opportunity (i.e. which will be developed and mined to completion) or is it a multi-project ongoing company? If it is the latter, how attractive will the company and its product be in the stock market, and what kind of price/earnings multiple will it be able to command now and in the future when it is decided to take the company public or sell to or merge with another company?

- What is the potential upside for the investment and what options exist, from our perspective, for liquidating it in the future? (Most venture capital investors tend to take a 3- to 7-year time frame for their investing, although some have a 10-year partnership. Consequently they must see a way to liquidate an investment within that time frame, for example, by selling assets by receiving dividends or interest, by taking a company public or by selling it to another larger company.)

- What is the potential downside, i.e. how much of the total investment could be lost if the project does not go according to plan? What are the probabilities that the project will not go as planned? Are there any sources of additional or secondary collateral or alternative payment?

- What specific risks is the investor really being asked to assume (e.g., reserve risk, completion risk, market risk, metallurgical risk, price risk)? As in any project financing, the various risks are segregated and, if the financial architect involved does his work properly, laid off and allocated to those best able to accept them.

The process of structuring and pricing an investment usually begins with an operating statement and cash flow forecast. These are possible where one is looking at a leveraged buyout of a division or operating mine, a venture capital investment in a growing or multi-project company or start-up project where a reasonable amount of drilling has been performed but perhaps no plant or mill has yet been built. The venture capital investor would work over the assumptions very carefully in conjunction with the Principals and perhaps consultants retained by the venture capital firm. The result of such efforts would be to develop a base case, or expected cash flow projection for the project as well as a good feeling as to the risk associated with the projections. By the time this is done, one would also have a good feeling from the Principals involved regarding their objectives, constraints and priorities. Based on these factors and on the project's characteristics as discussed above, an investment would be structured in one or some combination of the following elements:

Common Stock: If an investment were in, for example, all common stock, a highly simplified way of pricing it would be to solve the following equation for the percentage shares the fund would require to receive its targeted return:

$$\frac{(FNDINV) \times 100}{(PVFNCF)} = \% \, SH$$

Where:

PVFNCF = Present Value of the venture's Future Net Cash Flow (available to shareholders - discounted at the Expected Rate of Return)

FNDINV = Fund Investment

% SH = Percentage of shares or ownership of venture required to give Expected Rate of Return

The PVFNCF could by itself be quite complicated if it were decided to treat the investment as a "stock market" as opposed to cash flow, play. The table shown in Exhibit 3 was, for example, one used to decide on the likelihood of realizing the 30% return over a 3-year period using a convertible subordinated note with a coupon of 10% for an investment in a company with an initial value of $12.5 million. An explanation of how the table works is given in the Exhibit. It indicates that to realize a 30% return the company's value must increase from $12.5 million in the first year to $22.5 million by the end of the third year. A range of earnings that would be required to realize this valuation assuming alternative stock market price earnings ratios is given. If it is assumed that stocks for this type of mining venture are selling in the range of 12 times earnings at the end of year 3, the venture would, for example, have to be earning $1.873 millions to provide a 30% return. One would then see where the venture was today and what its game plan was for the future to increase earnings before deciding whether or not there was a realistic chance of this occurring.

Loan With Some Equity Kicker: A basic loan, either secured or unsecured could be made in a variety of ways again depending on the circumstances of both the project and the objectives of its Principals. The loan could, for example, be at a high or low fixed rate or a floating rate (e.g. two points below prime). Interest could be payable currently, it could be accrued or capitalized, or there could be no interest at all with all consideration coming out of some form of production payment or royalties. Equity kickers can take a variety of forms from a net smelter return percentage or net profits interest to some form of production payment tied either to a rate of return or the recovery of some fixed amount of money or stock options or warrants which would be evaluated much as the common stock described above. The venture capital investor would again work from the base case cash flow forecasts to decide what combination of loan repayment, interest rate and equity kicker would give the desired risk adjusted expected rate of return.

Combination Equity/Debt Vehicle: This category includes such vehicles as subordinated convertible notes or convertible preferred shares in conjunction with stock options or warrants. One interesting transaction like this called for the purchase of 10% notes which were convertible, at the investor's option, within a 3-year period from the date of purchase, into either 10% of a common stock of the company or 10% joint venture working interest in a specific project.

Thus the structure and pricing of a "venture capital" investment for a mining project does not follow any set procedures and is the result of a variety of interrelating factors. Luckily, through the creative use of options or contingencies, one is usually able to structure deals that both the investor, the perennial "skeptics," and the project's Principals, the perennial "optimists," can live with. Loan agreements or preferred stock indentures give management the flexibility to run a company as long as things go reasonably well but also give investors the opportunity to exert more influence if problems develop. This is not all bad as we have found some discipline is good for young companies especially in the financial control and planning areas.

PREPARING THE APPROACH

Before approaching the venture capital firm interested in mining opportunities, Principals should think about a few questions regarding their own situations.

1. **Is venture capital really the type of funding that I need?** As mentioned in the Intoduction, capital markets are segmented according to the risk-reward characteristics of the various providers of funds. It is therefore useful to differentiate venture capital according to its value-added approach. It is not necessarily inexpensive but it should be added that public equity offerings are probably equally as expensive if not more so. On the other hand, certain types of private placements or tax-oriented limited partnerships, not to mention bank funding if it is available, may be much more attractive from a cost standpoint.

2. **What is unique or attractive about my particular situation?** Venture capitalists in general see a large number of proposals, and since there are relatively few venture capitalists active in the mining area, the problem facing a would-be seeker of venture capital funds is that he must get a venture capitalist's attention as quickly as possible in order to get a good hearing. This can be done in a number of ways but the question of what is unique or particularly attractive about a mining project company should be carefully thought through. This could include proprietary patent or process rights, a unique and exciting geologic find, well written long-term sales contracts, a particularly strong management team with complementary skills, a particularly strong and impressive board of directors, etc.

3. **Have I invested the money I had available wisely?** People trying to get projects off the ground or buy out existing companies or divisions are individuals usually running on a very limited budget. In order to approach any provider of potential funding, including venture capitalists, it is in the best interest of the individuals to spend such limited resources wisely. This could include, for example, on an orebody that has "good smells" but not much systematic sampling, a few strategically placed drill holes which would give a good indication of the reserve potential. On a disseminated or placer deposit, it could include having a <u>reputable</u> geologist take independently verifiable samples, even if only with a pick up and back hoe, which were assayed by a <u>reputable</u> laboratory. Of course one also runs the risk of killing a project with such a sampling but that will be necessary sooner or later. If it is a new or unique mineral process, money could be wisely spent, for example, performing a marketing survey to identify clearly and dramatically what the potential applicability of the process is in terms of dollars and cents growth opportunities.

4. **If everything works out what is the maximum upside potential for the project?** This may not be possible for certain types of exploratory efforts but in most instances some projections can be done with credibility. Careful thought should go into the projections, for this type of exercise can be a two-edged sword. A simple but professional assessment of a project's potential on one page could get an investor's attention. Alternatively, four or five pages based on a few notes from old mine workings extrapolating out using $600/oz gold without deducting operating costs may only indicate the Principals bringing in the proposal are either naive or too promotion oriented.

5. **What is the depth of management team's knowledge and experience as applicable to the project?** If it's appropriate for a given project, putting together an impressive, professional management team can be very helpful in terms of getting a venture capitalist's attention. The areas, of course, that need to be addressed either through reputable consulting arrangements or in-house management include geology, metallurgy, mine engineering and operations, finance and legal not to mention general management skills such as leadership, management by objective capabilities, discipline and a public relations flair. Some of these skills and expertise can be brought in by a qualified outside board of directors that have indicated the willingness to take an active participation in certain stages of the company's or project's life cycle.

A clear representation of the status in this management area is important for a venture capitalist as it will partly determine how "labor intensive" an investment in a venture is likely to be. If the management team, directors and consultants assembled already are in fact quite capable, with minimum input from a venture capitalist, to proceed with the successful development of a project, a venture capitalist may structure and price the deal less "rigorously." He would feel comfortable that an excessive amount of time would not be required on his part to monitor the investment and would thus be available for other investments. On the other hand, if he felt that the basic resources were attractive but that a good deal of his time would be required to make the project happen in a successful way working carefully with the Principals, he may price the deal with more of a risk premium and require certain additional covenants or conditions in the documentation.

6. **What are the major strengths, weaknesses and risks of your venture?** This is kind of a catchall but it may be useful, prior to presenting your proposal to play "devil's advocate" with several members of the management team based on some of the comments in papers such as this one.

7. **Is a business plan or at least a good presentation developed?** As mentioned earlier, venture capitalists are quite prepared to help entrepreneurs develop project concepts or business plans. What one should have prepared before seeking venture capital funding really depends therefore on the stage at which the venture capitalist is approached. At whatever stage it is, however, it does pay to have thought through carefully, and to have summarized and documented all relevant facts and information available regarding a project in an orderly and concise manner to make as favorable as possible an initial impression on the venture capitalist. For example, on one occasion we actually liked the basic underlying elements of a project but lost interest in it after a very turgid and confused initial presentation by a team of people which we felt, no matter how much money was spent, could not make the project happen successfully.

VENTURE CAPITALISTS IN THE NATURAL RESOURCES

Two good sources for identifying venture capital firms are Capital Publishing Corporation's 1983 <u>Guide to Venture Capital Sources</u> and A. David Silver's <u>Who's Who in Venture Capital</u>, published in 1984 by Ronald Press. One major U.S. fund in formation at the time of writing not included in either book is the VenturesTrident Limited Partnership of Boston. Its focus will be principally on precious metals projects.

CONCLUSION

The sponsors or promoters of mining projects of any kind should be aware of as many sources of capital for their project as possible. Venture capital, according to the definition used in this paper, is a distinct source that is quite exciting for the smaller mining opportunities. This is true because of the flexibility with regard to structuring of investments and the "value-added" approach and considerable assistance that venture capitalists are interested and quite willing to provide to project developers. Venture capital is also exciting due to the fact that funds are, in many cases, provided through venture capital partnerships from large pools of institutional capital. Such funds should have a significant impact on the availability of attractively priced funding for the small- and medium-sized mineral projects and companies for some time to come.

SOURCES OF FINANCE

Exhibit 1

General Interrelationship
Perceived Risks — Required Expected Return
Compared with Stages of Business or Project Development (1)

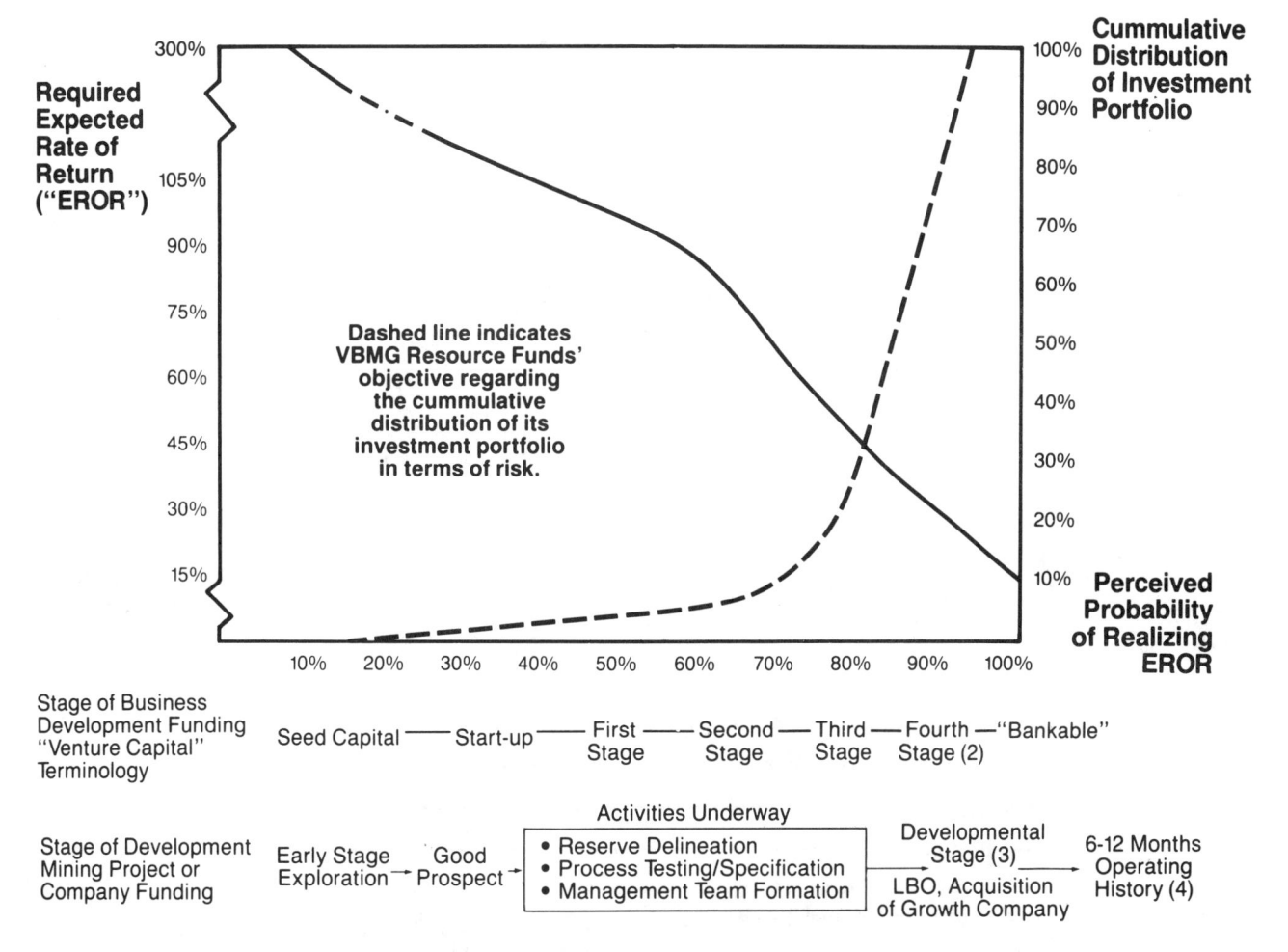

Required Expected Rate of Return ("EROR"): 300%, 105%, 90%, 75%, 60%, 45%, 30%, 15%

Cummulative Distribution of Investment Portfolio: 100%, 90%, 80%, 70%, 60%, 50%, 40%, 30%, 20%, 10%

Dashed line indicates VBMG Resource Funds' objective regarding the cummulative distribution of its investment portfolio in terms of risk.

Perceived Probability of Realizing EROR: 10% – 100%

Stage of Business Development Funding "Venture Capital" Terminology: Seed Capital — Start-up — First Stage — Second Stage — Third Stage — Fourth Stage (2) — "Bankable"

Stage of Development Mining Project or Company Funding: Early Stage Exploration → Good Prospect → [Activities Underway: • Reserve Delineation • Process Testing/Specification • Management Team Formation] → Developmental Stage (3), LBO, Acquisition of Growth Company → 6-12 Months Operating History (4)

Notes

(1) The diagram depicts general risk-reward relationships and how they (very approximately in the mind of the author) correlate with various stages of mining project or company funding. The diagram is highly simplified to illustrate certain points and could, for example, be expanded to include how different risks (e.g., loss of capital) would affect its shape.

(2) Fourth stage also is referred to as "mezzanine" or "bridge" financing.

(3) Considerable latitude exists in the definition of a "developmental stage" project. In this context it is assumed to be one where the present value generated by cash flow from proven reserves is projected to equal approximately 150% of the project's cost and satisfactory process and metallurgical solutions have been reached.

(4) A "bankable" mining project here is one capable of receiving a loan secured only by the assets of the project where the lender looks only to the cashflows from it for repayment. Bankers must "expect" these loans will be repaid on schedule going into them, thus the 100% Perceived Probability of Realizing EROR.

Vickers Boettcher Massy-Greene

Exhibit 2

General Interrelationship
Perceived Risks — Required Expected Return
Compared with Potential Mining Funding Sources

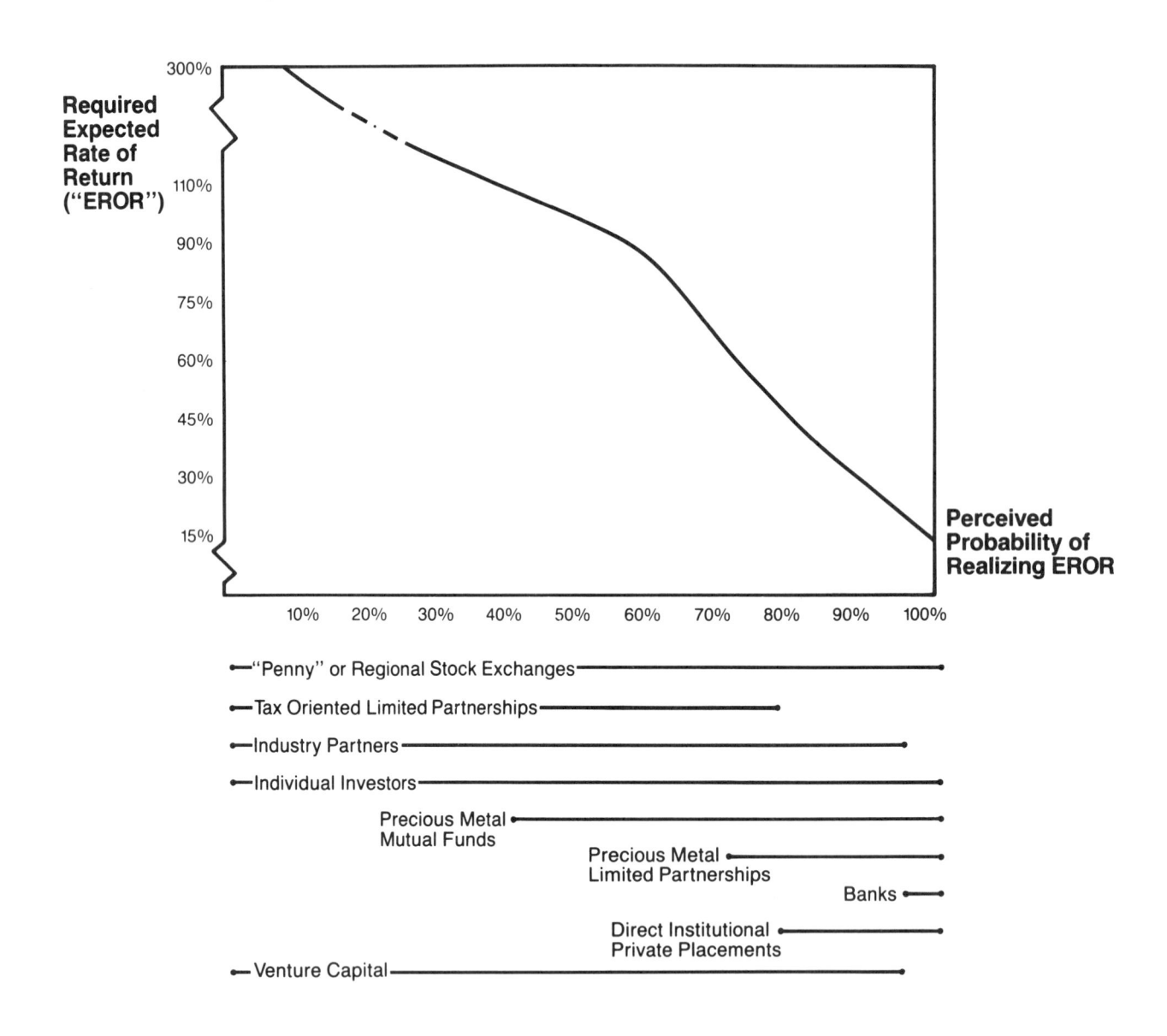

Required Expected Rate of Return ("EROR") (y-axis: 15%, 30%, 45%, 60%, 75%, 90%, 110%, 300%)

Perceived Probability of Realizing EROR (x-axis: 10% to 100%)

Funding sources by approximate range:
- "Penny" or Regional Stock Exchanges (~10% to 100%)
- Tax Oriented Limited Partnerships (~10% to 75%)
- Industry Partners (~10% to 90%)
- Individual Investors (~10% to 100%)
- Precious Metal Mutual Funds (~35% to 90%)
- Precious Metal Limited Partnerships (~60% to 100%)
- Banks (~90% to 100%)
- Direct Institutional Private Placements (~60% to 100%)
- Venture Capital (~10% to 90%)

Note Sources of funds for mining companies or projects are indicated according to approximate risk-reward relationships reflecting the stratification and segmentation of capital markets. The particular interests and requirements of each institution within a given "source" category will vary widely.

Vickers Boettcher Massy-Greene

EXHIBIT 3

Common Stock Valuation Example

RETURN	FV (3rd Yr)	P/E=4	P/E=6	P/E=8	P/E=10	EARNINGS P/E=12	P/E=14	P/E=16	P/E=18	P/E=20
10.00%	12500	3125	2083	1563	1250	1042	893	781	694	625
12.50%	13560	3390	2260	1695	1356	1130	969	848	753	678
15.00%	14670	3668	2445	1834	1467	1223	1048	917	815	734
17.50%	15833	3958	2639	1979	1583	1319	1131	990	880	792
20.00%	17050	4263	2848	2131	1705	1421	1218	1066	947	853
22.50%	18321	4580	3054	2290	1832	1527	1309	1145	1018	916
25.00%	19648	4912	3275	2456	1965	1637	1403	1228	1092	982
27.50%	21033	5258	3506	2629	2103	1753	1502	1315	1169	1052
30.00%	22475	5619	3746	2809	2248	1873	1605	1405	1249	1124
32.50%	23977	5994	3996	2997	2398	1998	1713	1499	1332	1199
35.00%	25539	6385	4257	3192	2554	2128	1824	1596	1419	1277
40.00%	28850	7213	4808	3606	2885	2404	2061	1803	1603	1443
50.00%	36250	9063	6042	4531	3625	3021	2589	2266	2014	1813
75.00%	59727	14932	9955	7466	5973	4977	4266	3733	3318	2986
100.00%	91250	22813	15208	11406	9125	7604	6518	5703	5069	4563

The far left column ("Return") represents the return on the Fund's invested capital over three years based on the Future Values ("FV") of the Company at the end of the third year shown in the second column. The value assumed at the time of investment was $12,500,000. Because the investment vehicle was a subordinated convertible note paying 10% interest, the return to the Fund over these years would be 10% if the FV remained $12,500,000 as shown above. The table itself indicates what the third year earnings would have to be, given a range of possible prices/earnings ratios, to give the FV's corresponding to a range of returns on the Fund's investment.

VICKERS BOETTCHER MASSY-GREENE

11.12

PUBLIC FINANCING AS A SOURCE OF FUNDING FOR THE CANADIAN MINERAL INDUSTRY

Brian J. Gorval, M.B.A.   and   Robert L. Kemeny, B.Sc., A.R.S.M., P.Eng.

Vice-President, HDD Resource Consultants Ltd.   President, Anglo Canadian Mining Corporation

INTRODUCTION

Financing, or providing adequate capital at low cost for developing and bringing a natural resource property into production, is a fundamental requirement for profitable operation. Under the present economic environment the cost of financing any project is gaining more importance than ever.

The fundamental keys for success in mining are "the Four M's": Mineral, Market, Management and Money. In recent years, with capital and operating costs skyrocketing, increased emphasis on obtaining adequate and low cost capital funds for bringing in a prospect is essential.

For a small mining company there are limited sources for raising funds. In most cases these companies have neither a substantial asset base, nor any cash flow, and they are deemed as risky investments. The issue of shares of capital stock of such a company is probably the most common method of raising funds via the public purse. Alternatively, debt could be used and a source of such capital might be private investors or financial institutions. The company which intends to raise funds from the investing public by way of the sale of equity, subsequently lists those shares for trading on a stock exchange.

Going and staying public commits a company to a multitude of complex laws, rules and reporting requirements. The governing bodies are the watch dogs for the public. The largest of such an organization of this kind in the world is the Securities and Exchange Commission (SEC) in the United States.

In Canada, the Securities Laws come from within provincial jurisdiction. Most provinces have their own Securities Act, and Securities Commission. It is the Commission which administers the Act, and hence controls the distribution and trading of securities within the province. Extra provincial registration is required for trading and distribution in more than one province.

In British Columbia, prior to a public offering, the approval of the Superintendent of Brokers (SOB) and of the VSE must be obtained.

THE VANCOUVER AND TORONTO STOCK EXCHANGES

There are five stock exchanges in Canada which supply a trading forum in which mining companies may operate. The stock exchanges are located in Montreal, Toronto, Winnipeg, Calgary, and Vancouver. Of the five stock exchanges, the Vancouver Stock Exchange (VSE) and the Toronto Stock Exchange (TSE) are the most heavily used for resource-based issues and are the focus of this paper. The objective of this section is to provide a brief background on the VSE and TSE.

The Vancouver Stock Exchange

The VSE is Canada's predominant stock exchange for financing resource-based, junior issues. Since 1907, the VSE has acted as a source of funds for mining, exploration and energy companies throughout Canada and the world. Currently, there are over 1400 companies listed on the VSE. During 1983, over 3.1 billion shares traded at a value exceeding CDN$3.96 million.

The VSE has developed a reputation as one of the few exchanges where a speculative resource venture may obtain a listing and subsequent financing in a relatively short time, generally five to ten months.

The SOB, the Investment Dealer's Association of Canada (IDA), and the VSE all regulate trading at the Exchange. Any irregularities in a company's stock dealings are investigated and trading may be halted by the VSE until a satisfactory explanation has been received from the suspended company. Companies can be suspended from trading or delisted for various reasons. One reason is misrepresentation of facts in documentation submitted to the SOB, VSE, or to the public, through news releases.

All applications made to the VSE for listing shares must be sponsored by a member of the VSE. The procedures to take a company public and obtain a listing on the VSE are fairly standard. Potential investors may obtain the listing requirements by contacting the VSE at the Stock Exchange Tower, P.O. Box 10333, 609 Granville Street, Vancouver, British Columbia, Canada, V7Y 1H1. Telephone: (604) 689-3334.

Trading Volume - VSE. Table 1 (Financings on the VSE) shows the number of venture capital financings that took place on the VSE over the period 1980 to 1983. The money was raised mostly by mining and oil exploration companies.

TABLE 1
FINANCINGS ON THE VSE[1]

Year	Number	Value ($ Million)
1980	494	193
1981	305	190
1982	110	46
1983[2]	182	139

1. Includes all funds raised through Prospectuses (new public company), Statement of Material Facts (established public company), underwritings, options, best effort and unit offering agency agreements and right offerings, but does not include private placements, etc.
2. First seven (7) months only.

The table shows that, over the period 1980 to 1982, the number of financings decreased. Financings in 1982 were down to 22 percent of the level of 1980 financings. Financings on the VSE began to drop in the early summer of 1981. The number of new financings dropped rapidly during the rest of the summer and collapsed during the last four months of 1981 and the first seven months of 1982, corresponding to the depths of the recession. New financings started to increase in the last five months of 1982. The table shows that the number of financings in the first seven months of 1983 have exceeded total financings for 1982 by 65 percent, indicating a strong recovery in the venture capital market.

Value of Shares Traded - VSE. Table 1 also shows the value of VSE financings for the period 1980 to 1983. The table shows that the value of financings exhibited a similar trend as the number of financings. In 1982, the amount of money raised for all venture financings dropped significantly to $46 million from $193 million in 1980, but increased substantially in the first seven months of 1983.

The Toronto Stock Exchange

The TSE is more heavily used by major companies for raising money than is the VSE. Many of the older, established mining companies, such as INCO and Noranda, are listed on the TSE.

The TSE's regulations have been designed to encourage senior stock issues rather than junior companies from using the Exchange because of previously bad publicity resulting from stock scandals. Recently, the TSE revised its listings standards to attract junior companies. A company which has never undertaken a public financing will be allowed, once conditionally approved for listing on the Exchange, to use an offering prospectus distribution, through the facilities of the Exchange for its initial public offering. How much of the venture capital market the TSE may attract, will depend on the final rules governing new issues, and the time required in Toronto to bring out a new issue. The VSE's rules and SOB regulations are still more specialized to encourage junior issues than those of the TSE. Consequently, the VSE will likely continue to be the preferred market for junior resource companies fund-raising.

Applications for listing shares on the TSE must be sponsored by a member of the TSE. Investors may obtain the listing requirements by contacting the TSE at the Toronto Stock Exchange, 2 First Canadian Place, Exchange Tower, Toronto, Ontario, M5X 1J2. Telephone: (416) 947-4670.

Trading Volume - TSE. Table 2 (Financings on the TSE) presents estimates of the number of financings of junior mining companies on the TSE for the years 1980 to 1983. Excluded from this table are financings by large mining companies, such as INCO and Cominco, companies with operating mines and those companies with large oil and gas interests relative to their mining interests.

TABLE 2
FINANCINGS ON THE TSE[1]

Year	Number	Value ($ Million)
1980	15	12
1981	7	9
1982	5	11
1983[2]	23 (18)	72 (58)

1. Includes funds raised through public offerings, fixed price offerings, Statement of Material Facts and rights, but does not include private placements. Excluded, are the large established mining companies, companies with operating mines and those companies with large oil and gas interests relative to their mining interest.
2. First seven (7) months only.

The table shows a similar annual trend in the number of financings as seen on the VSE over this time period. From 1980 to 1982, there was a decline in the number of financings, while in 1983 there was a strong increase. Some of this increase in 1983 is likely due to the less restrictive listing rules applicable to junior resource companies.

Value of Shares Traded - TSE. Table 2 also shows the value of financings of junior mining companies on the TSE for the period 1980 to 1983. The table shows that the value of financings changed little from 1980 to 1982. However, there was a large increase in value in 1983. The large growth in the value of financings in 1983 may be due in part to the general economic recovery and relaxation of listing requirements for junior mining companies.

PROCEDURES TO PUBLICLY FINANCE A JUNIOR MINING COMPANY

The objective of this section is to illustrate and describe the various steps that a junior mining company would take in order to raise funds through a public offering on the VSE. Figure 1 (Flowchart of Events for a Public Offering for a Mining Company) illustrates the various stages that take place from the formation of the private company, up to the listing and trading of the public company.

The first step is to incorporate a company. The company must have at least three directors, the majority of whom should be Canadian residents, and at least one, a resident of British Columbia.

This first stage in the history of the company which is "going public", is the private stage, because during this period, no public trading of the company's shares is permitted. Shareholders at the private stage must be members of the family, friends, relations, or close business associates of the management team of the company. Such shareholders cannot dispose of their shares to the general public, and any future transaction of such shares are to be regulated by the SOB and VSE. The private stage is a very important phase of the process because the share distribution at this stage will determine the ability of the management of the company to maintain effective control once it is public and trading on the VSE.

Once the company has been incorporated, three types of share transactions may occur at the private stage: (1) for promoters; (2) for property; and (3) for seed capital.

Promoters Transaction

The success of a company primarily depends on the owners or principals who are often called the "Promoters". In order to provide principals of the company with voting control, the SOB, under the current rules, allows 750,000 common shares to be issued by the company, at a price of one cent per share. This block of stock is distributed among the promoters, directors and officers, and serves as an incentive for all individuals of the management team. This block of shares is called "the Escrow Share Block". It has the same voting rights as all other common shares, but cannot be disposed of without the approval of the SOB. These shares are released from escrow from time to time by the regulatory authorities, if the company undertakes, in a satisfactory manner, the exploration and development of its resource properties and conducts its business according to the rules and regulations of the regulatory bodies. If these shares are not completely released for any reason over a ten year period, then the balance will be deemed cancelled by the regulatory authorities.

Property Transaction Shares

The SOB has established two categories of natural resource properties. They are: (1) a property without proven, measured reserves, called an "Indeterminate Value" (IV) property; and (2) a property on which sufficient work has been carried out and which has proven, measured reserves called a "Determinate Value" (DV) property.

A maximum of 200,000 shares may be issued prior to production for the acquisition of an IV property. Usually the property is acquired under an option agreement, and no more than 25,000 shares may be issued on signing the option agreement. All future deliveries of blocks of shares, should be dependent upon and contingent to favourable exploration results to avoid issuing shares for a worthless property. The vendor may receive an additional block of shares, and be reimbursed for previous expenditures on the optioned property under special circumstances which include significant exploration results.

The company may issue a block of shares which will depend on the present value of the property, if proven reserves have been established on a DV property. The shares will have to be priced the same as the price per share of the first public offering. Shares issued for this class of property are usually free from any trading restrictions although they might be required to be pooled or escrowed, depending on the SOB's policy.

Properties of both the above categories may be acquired primarily in the following ways: (1) from a prospector, company or previous operator; (2) management can grubstake a prospector; and (3) the directors of the company can vend in or personally stake a property.

Seed Capital Shares

During the early stages of the company, management has to determine the required amount of seed money necessary to maintain the company prior to its public offering. Once the amount of money has

SOURCES OF FINANCE

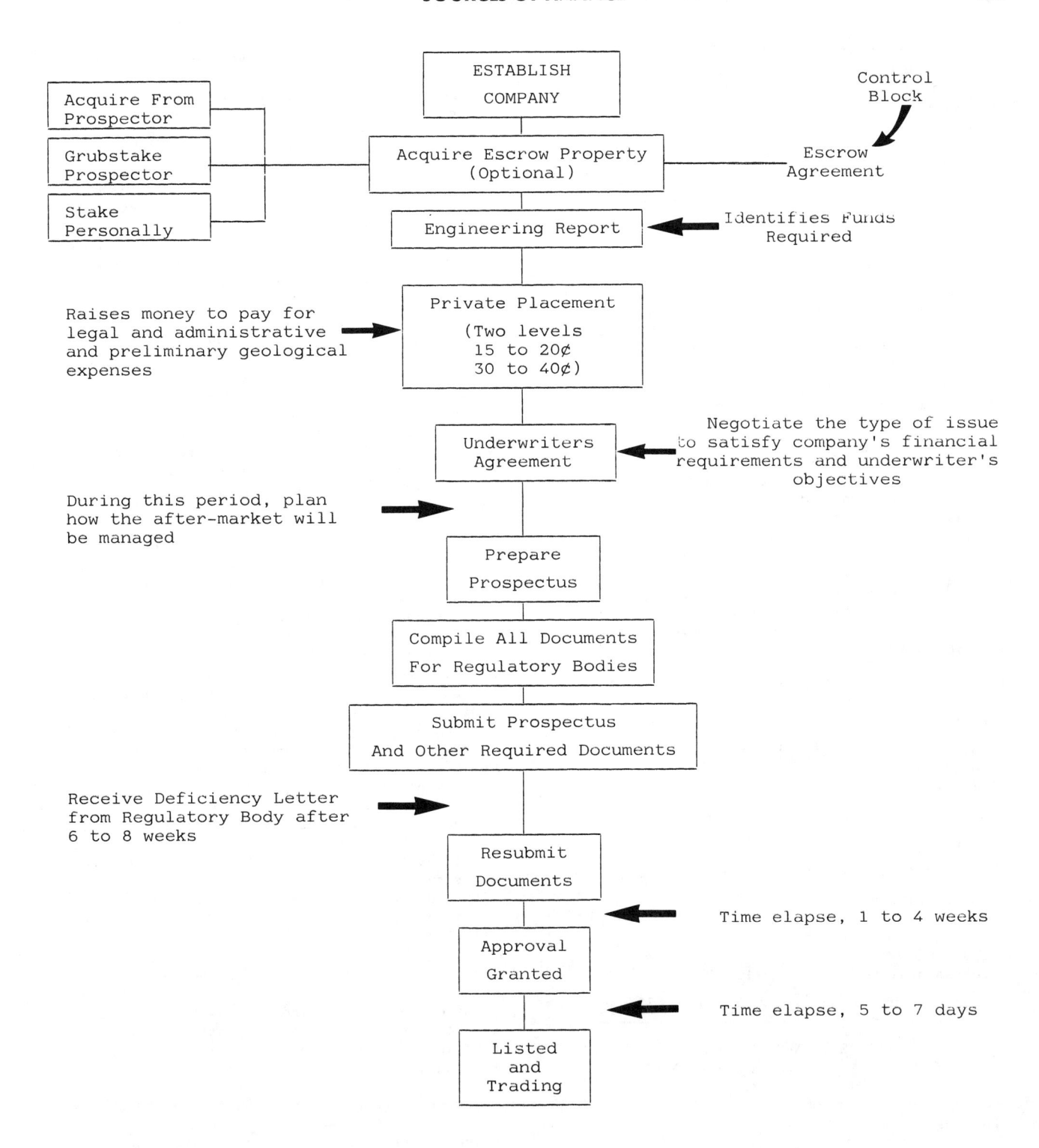

FIGURE 1. FLOWCHART OF EVENTS FOR A PUBLIC OFFERING FOR A MINING COMPANY

been identified, management moves to raise this money through a private placement. The private placement of the seed capital shares is generally conducted in two stages. The first level of financing is a share issue generally totalling about 400,000 - 800,000 shares, priced at a legal minimum of fifteen cents per share. The second private placement generally comprises a placement of thirty cent to forty cent shares totalling about 200,000 - 400,000 shares.

Depending on the price of shares offered to the public, the seed capital shares may be required to be "pooled" or held for a period of time, and will be released for trading only in quarterly intervals after the shares are listed on the VSE. At the present time, seed capital shares sold to private investors at a price equal to or exceeding fifty percent of the first public offering price require no pooling.

The company should raise at least $75,000 through the sale of seed capital shares to be funded for preliminary legal, administrative, consulting and exploration expenses. Figure 2 (ABC Resources Corporation - Proforma Use of Funds) illustrates how funds raised at a private stage are used. Under current policies, the company at the private stage must have raised at least $20,000 by selling seed capital shares, and must spend a minimum $15,000 on its property before the SOB will except the company's key filing document, the "Prospectus".

FUNDS FROM PRIVATE PLACEMENT		$ 75,000
USE OF FUNDS:		
Geochemical sampling and geological surveying - Gold Property	$ 20,000	
Assaying	2,000	
Property Payments	10,000	
Metallurgical Studies	2,000	
Administration	9,000	
Travel	6,000	
Legal and Professional Fees	10,000	
Corporate Purposes	16,000	75,000

FIGURE 2. ABC RESOURCES CORPORATION - PROFORMA USE OF FUNDS

The Prospectus

A prospectus is a detailed legal document which must be filed with and approved by the SOB. The form and content of a prospectus is established by the SOB. The prospectus contains audited financial statements, technical reports and a summary of the key management, properties and financial structure of the company. A detailed description of the information required for inclusion with a prospectus is provided by E. L. Affleck and W. G. Stevenson in Chapter 10.

The technical report or engineering report referred to above, must include the most relevant facts related to the property which include location, geology, work programs with budget and recommendations. All reports must be prepared by qualified experts in good standing and members of the Association of Professional Engineers, or a Fellow in the Geological Association of Canada. The cost of the report will vary depending on the location of the property, but commonly ranges between CDN$2,000 to CDN$3,000 for a property located in Canada.

During the preparation of the prospectus, the company's lawyers should also be preparing the other necessary documents required by regulatory bodies. In British Columbia, the SOB requires several documents. Some of these documents include: audited financial statements, underwriting agreements, engineering reports, title opinions, certified copies of all material contracts, escrow agreements, and pooling agreements.

The SOB reviews the various documents, which generally for the VSE takes a period of six to eight weeks, and produces a "Deficiency Letter" which is sent to the company and its lawyer. The deficiency letter indicates the various changes or deficiencies, if any, in the company's submission to the SOB. It is then the responsibility of the company and its lawyer to correct the deficiencies and resubmit the documents which indicate that such deficiencies have been corrected. The SOB then reviews the new submission and after a period of one to four weeks another deficiency letter is made to the company or approval is granted.

Once the prospectus has been accepted by the SOB, the shares to be issued by the company are qualified for public distribution. The public offering of the shares is called "Primary Distribution". There are two major types of public offering: (1) underwriting; and (2) agency offering.

Listing Procedures on the Vancouver Stock Exchange

After approval of the SOB has been granted, the company can make an application to the VSE to have its securities conditionally listed on the Exchange.

Under which ever type of offering the shares are going to be distributed with the public, the company must comply with the listing requirements of the VSE. The main requirements are the following:

1. The applicant company must have management expertise in the field of the activities of the company.

2. Be sponsored for listing by a member of the VSE.

3. Have an acceptable Transfer Agent in Vancouver, British Columbia.
4. Have completed a public offering (Primary Distribution) of at least 250,000 of its shares.
5. Have a minimum 150 public shareholders, exclusive of any of the applicant companies, insiders, directors and officers, principals and promoters.
6. Have a minimum working capital of $30,000 unallocated funds, above the budgeted funds for exploration and development work.

Once the minimum requirements are met by the company, the following documents have to be forwarded to the VSE and SOB.

1. List of Escrow shareholders.
2. Complete list of all shareholders.
3. Updated audited financial statement (no more than 120 days old).
4. The name and address of the Trust Company to be used.
5. Pooling, Escrow and Underwriters Agreements.

All the above documentation will be reviewed and vetted by the regulatory authorities, and if they are satisfied, the shares of the company are authorized for primary distribution.

The initial distribution of the securities of the applicant company can then be made on the trading floor of the Exchange, by the underwriter or agent on a day (offering day), within 90 days of the conditional listing of the applicant company's securities. Different fees have to be paid to the VSE, details of which are shown in Figure 3 (Vancouver Stock Exchange Schedule of Fees), as of May 1984.

The final stage of the procedure takes place once the offering through the Exchange has been completed. The company is required to file a declaration that no material changes from the disclosed facts in the prospectus have taken place, and all distribution and financial requirements by the regulatory bodies have been satisfactorily met, then the shares of the company will be "fully listed, posted and called" for trading on the VSE.

TYPES OF SHARE OFFERINGS

The objective of this section is to describe the three main types of share offerings used to raise money on the VSE.

Unit Offerings

The most popular financing issues used today are unit or warrant deals. In these issues, the investor receives a common share and a number of warrants, usually one or two, to buy a further share six months later at a higher price. In some cases, if only one warrant is involved initially, these units have a piggyback feature which enables the exerciser of the first warrant to receive another warrant good for another six months.

These deals were an invention of necessity in less exciting markets when the public and the brokers needed an incentive as an inducement to complete a deal. Unit financings are still very popular with the investor and are becoming a very valuable tool for the issuer. Although these financings may create overdilution, they virtually assure in "up" markets, a continuity of financing to the issuer without going through the tedious and costly process of submitting and vetting a new statement of material fact. This process can cost between CDN$5,000 and CDN$15,000, and take anywhere from three months to a year to complete.

The following is a typical example of a unit offering for the VSE.

> "The Issuer by its Agents hereby offers (the "Offering") to the public Eight Hundred Thousand (800,000) units (the "Units"), each Unit consisting of one (1) common share and one (1) Series "A" Share Purchase Warrant. The Offering will be made within the Offering Period, being a period of fifteen (15) days after the issuance of a receipt for this Prospectus (the "Effective Date") by the Superintendent of Brokers of the Province of British Columbia (the "Superintendent"). The offering price of the Units shall be Fifty Cents (50¢) per unit."

Shares Only Issue

A common method of financing junior mining companies is the straight shares issue with an option to the underwriting broker. This type of financing was prevalent in the 70's when markets were performing well. However, today these issues are less common because of the difficult economic times, large number of financings, and innovative financing techniques such as flow-through shares. They are generally only done by companies who absolutely want to minimize dilution.

The best efforts financings, which also minimize dilution, are basically only a product of shakey markets, as there is no guarantee of placement by the underwriting broker. The following is a typical example of a straight share issue on a best efforts basis.

> "The Company, by this Prospectus, offers to sell to the public in British Columbia, Three Hundred Thousand (300,000) of its common shares as fully paid and non-assessable at a price of Thirty Cents (30¢) per share subject to a commission in respect of some or all of such shares not exceeding Three Cents (3¢) per share. These shares will be sold on a best efforts basis through registered securities dealers who will receive a commission. The

VANCOUVER STOCK EXCHANGE

Schedule of Fees

Application for Listing (See Notes)	
- Industrial and Resource Sections	$3,600 - $15,000
- Development Section	$2,200
Substitutional Listing or transfer from Development Section	$1,375 - $15,000
Annual Sustaining Fees (See Notes)	
- Non-exempt companies	$925
- Exempt companies	$1,260
- Each Additional Class of Listed Security	$120
Underwriting Agreement (including option)	$540
Fixed Price Offering	$540
Fixed Price Offering with one series of Warrants	$975
- Second series of Warrants	$335
Secondary or Shareholder Offering	$390
Best Efforts Financing	$540
Private Placement	$775
Amalgamation, Merger, Takeover Bid	$825
Rights Offering	$540
Name Change	$335
Alteration in Capital	$335
Listing New Series or Class of Equity (See Notes)	$440 - $15,000
Creation of New Series of Debt	$540
Property Acquisition Agreement	$540
Management and/or Employee Stock Option or Stock Purchase Plan	$335
Amendments to Agreements and/or Statements of Material Facts	$285
General Corporate Material Change	$390
Transfers Within Escrow - Each Transferor	$115
Processing*	$230 - $2,500
Printing Fee - Filing Statements and Listing Applications	$155
- Amendments	$55

EXECUTED DOCUMENTS AND FULL FEE MUST ACCOMPANY ALL SUBMISSIONS

Effective January 1, 1984

*****PROCESSING FEE**

If an inordinate amount of time is required to process submissions from a listed company, or a submission is rejected, a fee will be charged based on the amount of time spent on the submission.

NOTES

INITIAL LISTING FEE

The fee will be calculated on the basis of the total issued capital of all classes to be listed: $2,200 on the first million shares, and $525 for each additional million shares or part thereof. The minimum listing fee shall be $3,600 unless the shares of the company are to be or have been previously, listed on the Development Section. The maximum listing fee shall be $15,000 regardless of the number of issued shares or classes of stock. The fee for listings on the Development Section shall be $2,200.

TRANSFERS FROM DEVELOPMENT SECTION

The fee shall be calculated as to $1,375 on the first four million shares listed, and $525 for each additional million shares, or part thereof, to a maximum of $15,000.

SUBSEQUENT LISTINGS

In the event that any listed company issues additional shares of any class of stock already listed, or lists a new series or class of equity capital on the Exchange, the minimum fee shall be $440 and the maximum fee $15,000. The fee shall be calculated on the basis of $100 for each 100,000 shares up to one million shares and $250 for each additional million shares, or part thereof, up to the $15,000 maximum.

Neither shares issued under employee stock option agreements, nor shares issued upon exercise of share purchase warrants will be subject to the additional listing fee, except those shares issued upon exercise of share purchase warrants with a term of less than one year from date of issue. Shares issued as a result of a private placement or a Statement of Material Facts, will also be exempt from additional listing fees.

ANNUAL SUSTAINING FEES

All companies exempt from filing requirements shall pay an annual sustaining fee of $1,260. Exempt companies are not required to submit a filing statement or annual questionnaire with the Exchange, but file by letter in the event of change in capitalization, financing, merger, etc. In such cases, the normal filing fee is applicable.

Non-exempt companies shall pay a $925 annual sustaining fee.

All companies will be charged an additional $120 annual sustaining fee on each class of shares listed in addition to the main listing. This additional sustaining fee applies primarily to preferred share issues, and warrant issues having a term in excess of one year.

FIGURE 3. VANCOUVER STOCK EXCHANGE SCHEDULE OF FEES

Company reserves the right to accept or reject applications for these shares in whole or in part."

Flow-Through Shares

Flow-through shares are a new method of financing available to junior mining companies. Under recent amendments to the Canadian Income Tax Act and changes in the Federal budget, mining companies doing exploration not now in a taxpaying situation and not expecting to be within the next several years may issue tax-sheltering securities.

A typical flow-through share offering would be made as follows:

"This Offering is designed for individuals and corporations who can take advantage of the income tax deductions through the financing of Canadian corporations involved in exploration and development of mining properties.

A unit purchased from the Offering will incur $25,000 of expenses which constitute Canadian Exploration Expenses (CEE) and Canadian Development Expenses (CDE) to earn 35,000 common shares of ABC Corporation.

The Offering is a minimum of 20 units (700,000 shares) and a maximum of 60 units (2,100,000 shares). Each unit earns 35,000 common shares of ABC Corporation. The shares will be issued in bearer form by July 1, 1985, and will be held in trust until January 1, 1986.

Each unit of 35,000 common shares will be earned as follows:
a) $6,250 upon subscription
b) $6,250 on November 15, 1984
c) $6,250 on January 15, 1985
d) $6,250 on March 15, 1985."

In a flow-through share agreement, an investor agrees to incur <u>Canadian Exploration Expenses</u> (CEE) or <u>Canadian Development Expenses</u> (CDE) solely as consideration for shares, other than "prescribed shares", of the capital stock of the company.[1] These shares have become known as "flow-through" shares because the deduction from income for these expenditures incurred, which is normally available only to the company, flows through as a deduction from income available to the investor. In addition, the April 19, 1983, Budget relaxed the rules for the deduction of <u>Earned Depletion</u> (ED), which increases the attractiveness of such agreements to the mining investor.[2]

The investor is able to claim a deduction from income in the year that the CEE or CDE are incurred. The investor is also able to claim a deduction from any source of income of the ED earned in the year. When the investor sells the flow-through shares the disposition is treated as a capital disposition, provided the investor is not a dealer or trader in securities. For income tax purposes, the flow-through shares have a cost base of nil.

<u>Financing Exploration or Development Costs</u>. The use of flow-through shares can help finance the cost of Canadian exploration or development programs without creating any additional liabilities for the company. The corporation is not entitled to deduct the costs of the exploration or development programs as these costs flow through to the investor, but it benefits from the exploration or development programs carried out.

Mining companies going the flow-through route receive more money than they would from an issue of regular common shares – up to a 50 percent premium in some cases because they can price the issuer higher due to tax benefits. They also acquire the funds at a significantly lower after-tax cost.

The premium allows them to obtain additional government incentive grants because the company is able to spend more money than otherwise would have been available.

From the investor's viewpoint, the advantages are that he will be entitled to a deduction from income for the amount invested and he will also own shares of the capital stock of the company, which can be sold after the exploration or development costs are incurred. In addition, the investor may obtain capital appreciation by holding the shares as an investment.

Existing shareholders may experience some dilution in the value of their shares unless they take part in the flow-through share agreement to the extent of their share ownership in the company, or the flow-through shares are issued at a sufficient premium to compensate for the surrender of the tax deductions transferred to the investors.

<u>Tax Effect Of Investment</u>. The following example considers the financial impact on a Canadian investor in a 50 percent tax bracket with sufficient other income to claim all deductions at that level. The investor spends $100 on shares that qualify as a CEE. All exploration spending earns depletion which is used by the investor.

Figure 4 (Flow-Through Share Tax Deduction) shows the calculation for arriving at the after-tax cost of investment. In the following example, the investor's after-tax cost of investment is $33, because of the tax benefit received from deducting CEE and ED.

Deduction for CEE Incurred	$100
Deduction for Earned Depletion	$ 33
Total Deductions	$133
Tax Saving at the 50% Rate	$ 67
Amount Expended	$100
Less Tax Savings	67
After-Tax Cost of Investment	$ 33

FIGURE 4. FLOW-THROUGH
SHARE TAX DEDUCTION

<u>Investor Liability</u>. Despite the benefit of flow-through shares, the investor should be aware of a potential liability. Because flow-through shares financing requires the issuer to act as agent, investors, as principals, could conceivably be responsible for liability arising from the exploration expenditures.

The issuer must carry insurance covering itself and its investors and provide a further indemnity to cover all risks related to making expenditures. Details of the insurance and indemnity provided must be printed in the prospectus.

An investor should make sure that the issuer has the insurance coverage to protect the investor, otherwise the investor is at risk. Ultimately, it is up to the investor to weigh the tax benefits of the investment against the liability problem.

ROLE OF UNDERWRITERS

In the case of a public offering, new securities are made available for sale to the general public resident in British Columbia. While a few corporations may undertake to retail their own securities, they generally lack the expertise and required facilities for floating large public issues. Therefore, the services of an investment dealer are engaged to underwrite the issue. Underwriting refers to the purchase of a new offering of securities by the investment dealer. Purchases take place at a set price and on a particular date. Once the underwriting is contracted the success of the offering from the issuer's standpoint is assured, unless it is a best efforts offering. The underwriter meanwhile arranges to retail the securities to investors at a higher price than was paid the issuer with the spread providing the underwriter compensation.

The reputation of the underwriters and principals of the company are all important to the success of public financing. The underwriter not only provides marketing services and advice on timing, type, and terms of the issue, but also assumes all risks associated with the offering to which it would otherwise have been born by the issuer. In effect, the underwriter provides his expertise to assist the mining company in raising venture capital.

The primary approach to underwriting involves a firm commitment under which the investment dealer purchases the entire issue. Exchange for the risk assumed through the purchase, as well as the management and subsequent selling effort the underwriter is compensated by the spread between the price at which the security is to be sold to the public and the amount actually paid to the issuing corporation. In Vancouver, this spread can go as high as 25 percent of the issue price (the price to the public).

An investment dealer will not always agree to underwrite any company. Generally, they look for companies with "steak and sizzle". The term implies the following: that the company has at least one property which is a good exploration target, but has no glamour - the steak; and the company has a property which may be pure grass-roots exploration, however, it has mystery, people will get excited about it - the sizzle. The underwriters have less difficulty placing a company with sizzle than one without. Also, an after-market is easier to maintain if the company has some mystery to it.

Common with mining issues are underwriters' options. The underwriter receives a series of options in blocks of unissued shares in exchange for a firm commitment underwriting of a set amount. This usual form of compensation, gives the underwriter the right to buy a set number of shares at predetermined prices which are above the original issue price and rising over time.

In contrast to a firm commitment underwriting is the agency offering on a best efforts distribution basis. In such cases, the investment dealer undertakes to do little more than attempt to retail the securities for the issuer and he receives a commission on sales made. Where the firm seeking funds has a doubtful future, certain dealers may be willing to handle its issues but not guarantee their sale. As the fees are likely to be less, best efforts distribution may also be favoured by an issuer when it is felt that the quality of offering is such that they require no underwriting.

Figure 5 (Examples of Underwriters Compensation) shows the various ways that underwriters are usually compensated for their efforts.

Underwriter's Discount: underwriter may receive by way of discount from the prevailing market price:

Market Price Per Share	Maximum Discount
- up to $0.50.........	25%
- $0.51 to $1.00......	20%
- $1.01 to $2.00......	18%
- $2.01 to $5.00......	15%
- over $5.00..........	10%

and if shares are sold to public at price greater than underwriting price plus twice the appropriate discount, excess must be paid to company.

Agent's Commission: agent may receive, by way of commission, 6% of gross proceeds where all shares are not sold or 7-1/2% if all shares are sold.

Underwriter's Option: underwriter may be granted option to purchase company's treasury shares; cannot exceed the number of shares underwritten; exercise price must exceed underwriting price by not less than 25 cents per share; if underwriting price exceeds $1.00 per share, no more than 300,000 shares can be optioned at one price; option exercise period(s) cannot exceed nine months from date of underwriting.

Agent's Warrant: agent who guarantees sales of all shares may be granted warrant to purchase company's treasury shares; cannot exceed 50% of the shares involved in the offering; exercise price must not exceed prescribed increments; warrant exercise period cannot exceed six months from date of offering.

FIGURE 5. EXAMPLES OF UNDERWRITERS COMPENSATION

ROLE OF PROMOTERS

The promoters of a junior mining company are the directors of the company. Their objectives are primarily to secure sufficient funds to enable the company to operate and to acquire properties and participate in business ventures of merit to enhance the development and growth of the company.

Promotional Firms

It is quite common for junior mining companies to hire the services of an advertising or public relations firm. This firm then has several responsibilities. First, it would design the various promotional literature that the company would need including news release forms, business cards, stationery, annual reports, quarterly reports, etc. Second, it would provide a list of names of potential investors (stock brokers, investment dealers) to send news releases to in order to develop a market for the stock. Third, it would arrange various seminars and meetings with potential investors (sometimes called "dog and pony shows") for officers of the company to attend to describe the progress and operations of the company. Fourth, the firm would work in conjunction with the management of the company to prepare news releases and other public information which is being made available to potential investors. The fees for these types of services commonly run between CDN$3,000 and CDN$5,000 per month with an initial start-up fee of about CDN$8,000 to CDN$12,000.

Public Relations Consultant

Another common action for a junior mining company to take is to hire a public relations consultant. This individual is then responsible for assisting the company in providing information to potential buyers of the stock in order to maintain an orderly market for the company's shares. This consultant generally becomes an employee or officer of the company with a contract renewable on a six month basis. This individual will do some of the jobs that a promotional firm would do, such as arranging for seminars and meetings, and identifying groups of potential investors. However, the consultant generally does not become involved in designing corporate literature.

The objective of promotion in general is to provide information to investors so that they can make an investment decision about the company. The investor has to decide if he can make money by buying the stock of the company. The investor can make money if he buys low and sells high.

Figure 6 (An Example of Price Movements for a Publicly Traded Junior Company) illustrates the behaviour of the price of a public stock over time. The figure shows that the company came out with an underwriting at a price of fifty cents per share. The objective of promotion is to provide information such that demand for the company stock increases to the point where the price of the stock increases to a level above fifty cents per share, for example one dollar per share. This enables the people who had faith in the company and bought the shares at fifty cents to make a profit. Over time the price of the stock will drop because of the behavioural actions of investors. However, management will continue to provide information about the company's investments to keep investors interested in the company. These investors may purchase company shares again which again will raise the price of the stock to some higher level. This action repeats itself throughout the life of the company.

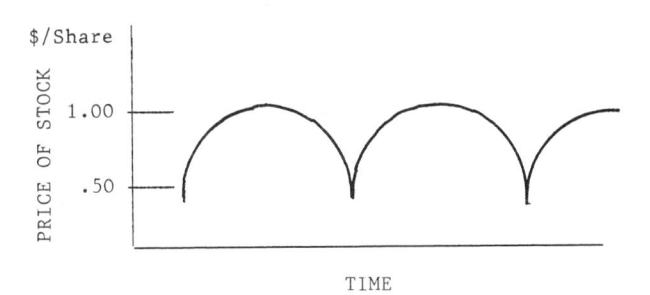

FIGURE 6. AN EXAMPLE OF PRICE MOVEMENTS FOR A PUBLICLY TRADED JUNIOR COMPANY

The services of these individuals commonly costs about CDN$2,500 to CDN$3,000 per month, plus expenses, plus a number of stock options. Stock options usually total about 100,000 shares.

ROLE OF LEGAL COUNSEL, ENGINEERS AND AUDITORS

The role of legal counsel is to prepare the necessary documentation, such as the prospectus, give title opinions, prepare escrow agreements, underwriter agreements, and to satisfy the requirements of the regulatory bodies. Legal counsel will also assist the management to undertake title searches on all the property acquisitions of the company, as well as act on behalf of the company regarding any legal action that is required.

Technical reports on natural resource properties in Canada should be prepared by a mining or geological engineer who is a member in good standing of an appropriate association of professional engineers. In case of an out-of-province, out-of-country property, the author of the report should be a member of an acceptable professional association. The author of the report, called "Qualifying Report", should include a Letter of Consent that the report can be used for a public offering, and also a declaration of what the author's interest, if any, is in the company, or in the property. A short summary of the professional background of the author is also required to be included in the report.

The auditor who will review the financial statements of the company should be a chartered accountant and a member of an Association of Chartered Accountants.

From the initial incorporation of a private company, to the point where the company is listed and trading as a public company, the cost of professional services (legal counsel, engineering consultants, auditors) ranges between CDN$25,000 and CDN$35,000, depending on the complexity of the public issue.

CONCLUSION

The VSE and, to some extent, the TSE, provide excellent vehicles for raising venture capital for Canadian junior mining companies.

The Security Commission and other regulatory agencies exercise a quasi judicial role in controlling the creation of a public company, the distribution and trading of its securities, as well as all of their business transactions in the future.

Anyone who is interested in pursuing the avenue of financing through a public vehicle should be prepared for the following:

1. To spend initially approximately CDN$75,000 to incorporate and administer a public mining company in Canada, under the constant surveillance of the regulatory bodies. Additional funds are required when the company becomes a going concern.

2. Prepare the legal, technical and financial documentation necessary to take a company public.

3. Be prepared to work on the task dilligently for a period of four to twelve months before listing and trading is accomplished.

Managing a public company puts additional responsibilities on the shoulders of its directors and principals. Regular reporting to the shareholders, annual meetings and full disclosure on all business transactions is required. The risk of a take over, and losing control over the company, is always there.

FOOTNOTES

1. The acquisition cost of a mineral property are included in the Cumulative Canadian Development Expense pool. A taxpayer may deduct up to 30 percent of the balance of the pool against income. All acquisition costs are included in the same pool (not segregated by property). All Crown grants related to CDE are also credited to the pool.

    Included in Canadian Exploration Expenses are expenses incurred in Canada for the purpose of determining the existence, location, extent or quality of a mineral resource including expenses incurred in the course of:
    - prospecting
    - carrying out geological, geophysical or geochemical surveys.
    - drilling by rotary, diamond, percussion or other methods.
    - trenching, digging test pits and preliminary sampling.
    - clearing, removing overburden and stripping.
    - sinking a mine shaft, constructing an adit or other underground entry.

    Cumulative Canadian Exploration Expenses are deductible up to 100 percent in any year. As with CDE, Crown grants related to CEE are credited to the pool and credit balances are included in income.

2. Depletion is earned on the basis of $1 for each $3 expended on, basically, CEE and certain tangible assets (which would not be of interest to a flow-through investor). CEE created from certain capitalized overhead expenses does not qualify.

    Earned depletion is deductible subject to two limits: balance in the pool (i.e. the $1 for every $3 of qualifying expenditures); and 25 percent of the taxpayers income after deducting CDE and CEE.

REFERENCES

The Toronto Stock Exchange, Information and Media Services, April, 1984.

The Vancouver Stock Exchange, Public Relations Department, April, 1984.

"The Vancouver Stock Exchange and Mineral Exploration in the 1980's", Mining Review, Volume 3, Number 6, November/December, 1983, p. 10.

11.13

COMMERCIAL PAPER:
AN INNOVATIVE SOURCE OF FINANCING FOR MINING PROJECTS

VICTORIA YABLONSKY, ROBERT GILLHAM, AND GROVER R. CASTLE

CHEMICAL BANK, NEW YORK, NEW YORK

INTRODUCTION

The commercial paper market, unique to the United States, is a direct exchange of funds between high-grade borrowers and large lenders; because it eliminates both the bank's role as middleman and also some significant transaction costs, commercial paper is the lowest cost short term instrument in the world. For this reason, virtually all borrowers would like to enter the market, but only issuers with the highest rating from Moody's and Standard & Poor can attract purchasers at the best rates.

COMMERCIAL PAPER

The commercial paper market, in existence for over 100 years, has traditionally been used for short-term financings, with most paper maturing within 30-45 days. Under the U. S. Securities Act of 1933, (section 3(a)3) industrial paper can avoid registration only if it matures within 270 days and is used for current transactions. It was first used to finance the short-term seasonal working capital needs of manufacturing companies. Since World War II, commercial paper use has grown steadily, serving in the 1950's as the market in which finance companies financed the explosion in consumer credit purchases of heavy consumer durables.

LETTER OF CREDIT BACKING

As far as is known, the use of letters of credit to support energy-related commercial paper began about 13 years ago with an arrangement to finance the nuclear fuel for Con Edison's Indian Point 3 Plant. Goldman, Sachs was the architect of this financing and Chemical Bank provided the letter of credit support. After an early burst of interest, this type of financing wobbled severely in the difficult credit markets of the mid-seventies before regaining serious momentum in 1976 and 1977.

A mining project would not normally qualify as a borrower in the commercial paper market because of its credit standing and because its large capital needs require long term financing. However, lately, commercial paper has begun to take the place of a bank term loan in limited recourse project financings. L/C backed commercial paper took this new and startling turn in 1982 when it was used as part of the $700 million financing for the development of the Newlands-Collinsville coal mines of Mount Isa Mines Ltd. in Australia. This was the first use of L/C backed commercial paper in a long-term limited recourse project financing. First Boston acted as financial advisor and, again, Chemical Bank provided sole letter of credit support of $155 million. The $785 million BHP financing for the acquisition of the Queensland (Australia) coal assets of Utah International from General Electric Company followed the next year; there we acted with two other Front Banks to provide $270 million of L/C support.

These two important transactions indicate that not only has commercial paper become an integral part of a long-term financing package, but also that it is no longer the exclusive domain of the best known U. S. companies. Lesser-rated companies and a multitude of foreign borrowers have gained access to this "exclusive club." How has this become possible?

OPERATION OF AN L/C

An instrument of the commercial banks, the "direct pay" letter of credit, has enabled lesser-rated companies to issue commercial paper in their own names. The letter of credit allows the issuer to substitute the credit rating of the supporting bank for its own. The L/C in effect allows a lesser-rated company to rent the bank's credit rating. This upgrading is possible because the bank is unconditionally obliged to pay the noteholders at maturity and then turn to the Issuer for reimbursement. The Issuer commonly "rolls over" a new short-term issue and reimburses the supporting bank from the sale of the new issue until cash flows from the project allow the borrower to reduce the level of paper outstanding. The commitment for an L/C facility, up to 12 years in the case of the Broken Hill transaction, governs the life of the commercial paper tranche of the financing. The term of the L/C commitment thus has the power to extend the ordinary 30-45 day life of a commerical paper note through repeated "rollovers" to a 12-year period. Moreover, in the event that an illiquid market prevents rollover, or if at any time the Issuer's cash flows prove inadequate for scheduled reductions in outstanding notes, or even in the event of permanent inability to pay through bankruptcy, the L/C issuing bank is nonetheless obliged to make payment on maturing notes under its commitment, and this irrevocable obligation is what makes the L/C such a powerful tool. The L/C issuing bank assumes project risk and default risk; the noteholders do not. It is this certainty that the banks will pay which enables the project to substitute the bank's credit rating for its own.

Moreover, the Securities Act specifically exempts from registration paper which is in effect guaranteed by a bank through its L/C's, without requiring the current transaction tests which industrial paper must pass. Thus both the commercial paper and instruments guaranteed by a bank may be used for long-term investments and still be exempt from registration.

ADVANTAGES TO COMPANIES

The banks, then, served several important functions in the Mount Isa and BHP transactions. First it made available one of the United States' most attractive funding sources to a lesser-rated foreign issuer borrowing, moreover, on a project rather than on a corporate basis. Chemical Bank, as credit bank, took project risk, while the ultimate purchasers of commercial paper enjoyed a highly rated name. The ultimate funding cost is therefore more attractive.

Since they characteristically have huge capital needs, Mount Isa's and BHP's ability to include an additional capital market among their funding sources and to obtain advantageous commercial paper rates can strongly enhance their viability. And the commercial paper market, currently at $213 bn, is one of the largest capital markets in the world.

Yet another cost advantage accrued from the use of L/C-backed commercial paper in both the BHP and the Mount Isa transactions. Australian tax law requires the borrower to pay a 10% withholding tax on interest paid to a non-resident lender. Since in most instances lenders are unwilling to absorb this tax, the borrower must then "gross up" the interest paid to the lender by an amount sufficient to make the lender whole on an after-tax basis. This of course increases the project's interest expense. Certain instruments, however, are exempt from Australian withholding tax, among them "bearer debentures" such as commercial paper, which are issued by a resident, are sold outside of Australia and are meant for wide public distribution. Thus no withholding tax was payable on the Commercial Paper tranches in the BHP and Mount Isa deals.

BACKUP STRUCTURE

This structure, in addition to cost advantages, provided funding flexibility. Suppose, for instance, the scenario occurs where the Commercial Paper market dries up for any reason, and through no inner fault of its own, the project cannot sustain its needed level of paper borrowing through rollovers. To deal with this possibility, the Mount Isa transaction was structured to include a contingent Revolving Credit Loan, which could be drawn down in the event maturing paper could not be rolled over. Similarly, in BHP it was agreed to provide interim financing in such a situation.

Fitting a Commercial Paper tranche into a limited-recourse project financing is not without its legal complexities. As mentioned earlier, an essential step in the process is to obtain a top rating from the two major U. S. rating services. To satisfy the agencies that the L/C backup for Mount Isa Mines and BHP in fact worked as an iron-clad commitment, a number of legal issues had to be addressed which relate to the supporting bank's attempt to honor its L/C in the event of bankruptcy.

In a bankruptcy, any payments made within 90 days before insolvency is declared can be reclaimed by the bankruptcy trustee under the notion that the beneficiary is receiving a preferential payment, i.e., reducing the pool of funds to which all creditors are entitled. If the project went bankrupt, every noteholder paid during the preceding 90 days could be subject to a preference attack. The essence of the L/C mechanism is its "direct pay" function. If Mount Isa or the borrowers in BHP, through their Finance companies, paid noteholders directly, payments could be subject to a preference recall. However, under a direct pay L/C, the noteholder presents his note to the bank, who pays it under the L/C commitment. The note is paid from bank funds, not the Finance Company's funds. The bank is then reimbursed subsequently, either by funds transfered from the Finance Companies or by proceeds of the rollover of Commercial Paper. The bank thus interposes itself between Noteholder and project and alone is exposed to preference recall.

The BHP transactions was especially complex because it involved a flock of underlying borrowers. If any borrower failed to perform its obligation to remit funds to the Finance Company to pay maturing paper, the resulting default by the Finance Company could bring down the entire commercial paper program and endanger the fully performing borrowers. The banks' L/C's, however, functioned to provide assurance to all borrowers that even if one borrower could not produce his required payments, noteholders would still be paid under the L/C, and hence all other borrowers could continue to issue commercial paper through the Finance Company, which could continue to be viable.

Only a handfull of banks do innovative type of financing such as these, and within this group, few will do unsecured financings, because the credit structure can be a very fragile one. In order to satisfy the commercial paper rating agencies, or more properly, their lawyers, the banking camel has been forced to go through the eye of the needle, emerging on the other side shorn of virtually any chances of repayment until months after a credit problem develops. A bank's decision to engage in these two transactions, therefore, was based on extremely sound economics.

ADVANTAGES TO BANKS

Some of the reasons why banks are both interested in this innovation and rightfully cautious will be obvious, others will be less so. A major advantage of this rental of a bank's quite high rating to another party (and it should be noted that not all banks enjoy the highest ratings) is the fee income generated by these transactions. Certain bankers have been heard to comment favorably on the boost these types of facilities can give their bank's ROA, as it is generally assumed no sustained bank borrowing will ever take place. Additionally, banks have enjoyed the commercial-paper dealing fee as well.

The methodology behind the L/C facility, however, recalls the princess who kissed a frog, thereby transforming the latter into a prince. The problem a bank faces in the use of L/C's to support commercial paper is: how many frogs can it kiss without risk of being turned into a frog itself as a result of its excess?

Bank's penchant for renting out their credit rating for individual transactions can have serious repercussions for the bank's funding needs as a whole. Since banks and bank holding companies are daily selling billions of their own CD's and commercial paper, at what point in time does the proliferation of commercial paper supported by a particular bank begin to affect the ability of the bank to fund its own liability position at the lowest cost? An excessive amount of our paper in the market would increase supply and lower demand for our paper, leading eventually to a higher borrowing cost. And, a decision to commit the bank's valuable L/C space to a single piece of business for 12 years will use up our availability in this market for a long time to come.

Moreover, banks must consider the accounting implications of L/C's. These, of course, are disclosed as a contingent liability. Their sheer size, rather than quality, which is not ascertainable, could give pause to securities analysts as well as to the regulatory agencies. Over the past four years, standby L/C's issued by U. S. banks have grown at a compound rate of 32% and total more than $65 bn (March, 1983) as compared with $27.9 bn at the end of 1979.

CONCLUSION

The bottom line for a mining project is, of course, what are the savings experienced by using letter-of-credit backed commercial paper instead of straight term-loan financing? The money saved, not to mention the name exposure in the U. S. capital market that a foreign corporation can achieve this way, is considerable. The LIBOR/Commercial Paper pricing differential (before calculation of a variety of fees applicable to both instruments) was 128 basis points in 1982, and 78 basis points in 1983, and this differential is expected to continue. Had the BHP transaction been in place from 1982, for example, approximately $5.5 million would have been saved over the last two years by using commercial paper rather than term loan financing. As for the Australian Withholding Tax, an additional $7.2 million would have been saved (prior to other tax considerations of the participating banks), for a total of something under $13 million over two years. When comparable savings are projected out over a 12 year financing life, this is a not inconsiderable saving.

11.14

FINANCING THE INDUSTRIAL MINERALS INDUSTRY

Christian F. Baiz III

Manufacturers Hanover Trust Co.
New York, New York

INTRODUCTION

Industrial minerals can have similar financing requirements to those needs of metaliferrous and coal mining projects. One of the outstanding differences is the understanding of the many different markets that these industrial minerals represent. There are in excess of 30 principal industrial minerals ranging in terms of annual production from the largest quantity such as sand and gravel, to possibly the smallest quantity of production of a legitimate industrial mineral which may be defined as industrial and rough gem diamond. (See Tables 1 and 2)

Project finance techniques, analytical methods, and information requirements are covered elsewhere in this Chapter and Chapter 10. Some comments on equity finance aspects follow.

EQUITY AMOUNT

The amount of equity in a financing transaction provides a cushion to the lender in case the project runs into trouble. Generally speaking, the greater the amount of risk in a project, the more equity a lender will require. The amount of cash equity in a project financing transaction is between 25 and 40 percent. The amount of leverage (debt/equity) is negotiable based on such factors as the strength and extent of the sponsor's undertakings, of the project's economics, and the term of a strong sales contract. Naturally, a borrower seeks to obtain a loan with as little equity required as can be negotiated.

Zero percent equity is usually associated with a take-or-pay contract (a hell-or-high-water unconditional commitment under which the purchaser must pay for the commodity as though it were shipped regardless of whether it is actually mined or shipped) with a financially strong corporate entity i.e., the parent company or an end user. As a take-or-pay contract is tantamount to a guarantee of payment, the banker views the credit basis of the loan as being other than the borrower's mining operation and therefore will analyze the creditworthiness of the institution purchasing the commodity on the basis of its corporate balance sheet. Do not confuse a take-or-pay contract with a take-if-tendered or take-and-pay contract.

Another means in the United States of obtaining zero percent equity is the carved out production payment financing, typical of coal mine financing. This involves the sale, or conveyance, by the owner of the mineral property of a production payment to a nominally capitalized company. The nominally capitalized company is formed solely to buy the production payment and usually borrows 100 percent of the purchase price from a lender.

Equity amounts of 10 to 20 percent generally are found in transactions which employ techniques such as an advance sale or a subordinated lease/loan. These agreements are viewed as quasi equity by most banks for several reasons. First, the proceeds represent a cash infusion early in the project's life - usually prior to the bank lending any funds. Thus, the repayment schedule is "behind" that of the bank debt. In the case of an advance sale, repayment is in the form of a long-term supply contract with favorable terms, usually priced close to the mining operation's costs.

There is also usually a requirement that all costs of the project be paid for by the loan or equity, often making the sponsors financially responsible for any cost overruns. Cost overruns are a major risk in any project, large or small, and are often due to unforeseen factors such as wildcat strikes, a piece of faulty equipment, or any event which delays completion, e.g., remote location, weather, etc...

The question of how overruns are covered can be negotiated. For example, in addition to the original debt/equity ratio of the project, there can also be a cost overrun ratio that will

---

Based on a presentation at the First International SME-AIME Fall Meeting, Honolulu, Hawaii, Sept. 4-9, 1982.

Table 1

SELECTED INDUSTRIAL MINERALS
GROSS AVERAGE VALUES IN 1981
(US$ per short ton)

Diamonds, Rough Gem @ $170/ct	$771,115,000
Diamonds, Ind'l @ $5.34,imp't'd	26,700,000
@ $1.50, CSO	7,500,000
Asbestos, f.o.b. mine, avg.	355
Boron, Granulated Pentahydrate	185
Fluorspar, Acid-spar	165-175
Met'l Pellets, 70% eff.CaF2	110
Diatomite	152
Kaolin, Georgia	25-175
Sulfur, Elemental	110
Potash	64
Portland Cement	54
Feldspar	51
Lime	46
Barite, Mine avg.	33
Phosphate Rock, 68-70% BPL	29
Rock Salt	16
Stone Crushed	3.43
Sand & Gravel	3.10

Table 2

INDUSTRIAL MINERALS
WORLD & US PRODUCTION
1981
(in million short tons)

COMMODITY	WORLD	U.S.
Cement	970	74
Stone, Crushed	2,800	900
Phosphate	143	57
Lime	135	19
Sulfur	57	12
Sodium Carbonate	8.537	8.350
Boron Minerals	1.144	.750
Bromine (lbs)	706.3	431.8
Gypsum	80.6	12.0
Diatomite	1.682	.707
Sodium Sulfate	2.261	.595
Barite	8.300	2.400
Asbestos (mt)	4.981	.081
Feldspar	3.650	.690
Fluorspar	5.050	.115
Diamonds (Gem & Industrial)	10	-0-

establish how much of the cost overrun the sponsor will be liable for providing funds to meet the completion of the project.

If the project sponsors do not wish to be subject to completion guarantees, it may be possible to negotiate a position that the sponsors will guarantee the payment of interest throughout the life of the loan. In that way, even if there were a problem with the project achieving completion, the bank's loan would always be "current" or paid-up on interest. Ultimately, it could be reasoned that the sponsors would tire of paying the interest and would then pay up the entire loan.

MARKET RISK

This one aspect makes lending to the development of industrial minerals projects most interesting due to the breadth of industrial mineral commodities that might be project financed. Here, one recognizes a divergence in analysis of a specific mining project away from the more basic commodities such as coal, copper, or bauxite/alumina.

Historically, industrial minerals have proven to be more profitable than other business functions of conglomerates participating in this side of the mining industry. For companies producing both metals and industrial minerals, the latter will consistently return a high margin on sales without experiencing the precipitous cyclicality which metals must tolerate. Rather than market prices as the prime mover to profitability of industrial minerals, unlike metals, other key factors impact these commodities. Examination of companies' operations will show that:

-- The greatest potential for profit is in specialty physical minerals, particularly pigments or process aids, rather than chemical or physical mineral commodities.

-- The potential lies in markets that are segmented either by application or geography, often relatively small with a...

-- Diffuse pattern of many customers, none dominant, and few likely to enter the business.

-- The best situation is one where large economic deposits are owned by only a few companies serving the particular market, (Some diamond examples are given in Table 3), and a...

-- Relatively high investment to enter the business.

-- The enterprise should also require end-use knowhow in major applications,

-- High technical service, and...

-- Good distribution (direct or through a distributor network) to consuming industries.

Table 3

DIAMOND MINING PROJECTS

(million)

Location	Western Africa	Western Africa	South-East Asia
Project cost	US$50	US$63.2	US$345
Project cost/ Anticipated annual production of carats	US$208	US$53	US$16
Type of deposit	Kimberlite	Alluvials	Alluvials and Kimberlite
Reserves	7.7 tonnes proven, probable, possible	1.2 cts proven	Alluvials: 3t Kimberlite: 150t
Average grade	0.46 cts/t	0.15 to 1.15 cts/m3	Alluv. 0.4-5 ct/t Kimb. 3.5-9 ct/t
Quality	Not given, however, deposits have historically yielded 40% gem and 60% industrial	90% gem and near-gem; 10% industrial	Alluvials: 85% near-gem and industrial Kimberlite: 90% near-gem and industrial
Anticipated annual production	0.24 cts/year	0.2 cts/year	Alluvials: 2 cts/year Kimberlite: 20 cts/year
Mining method	Underground	Draglines & trucks	Alluvials: N/A Kimberlite: open cut
Estimated start-up date	mid-1983?	mid-1983	Alluvials: mid to late 1982 Kimberlite: 1985/6
Valuation average	Not available	US$204-231/ct	US$7.83-US$18.62 /ct (subject to dispute)

These factors would produce a business with limited present and future competition, insignificant captive production, and one where the products can be differentiated and sold to many diverse customers on the basis of performance, service and convenience, but not on price.

The general outlook for industrial minerals consumption is for continued growth if one assumes a continually growing industrial society. Such a growth in demand is likely to be steady and far less cyclical than for metals.
With some exceptions, e.g., structural minerals and metallurgical fluxes, most industrial minerals are consumed in such non-durable goods as fertilizer, paper, and paint, or in the processing of oil, food, and beverages.

CONCLUSION

In conclusion, by the diversity of uses and the volumes of production of the various industrial minerals, it is immediately recognized that this business is large and complex. It can be highly profitable and given the right conditions should be easily financed within the private sector.

REFERENCES

Anonymous, "Industrial Minerals, Dollars in Dirt".

Baiz, C.F., III, "The Private Sector Role in Private Financing of the Mineral Industry".

Bispham, T.P., "1982, Coal Mining Project Financings: Lenders' and Borrowers' Considerations", in ENGINEERING BULLETIN No. 59. Los Angeles: Dames & Moore.

U.S. Bureau of Mines, "Mineral Commodity Summaries - 1982".

11.15

SPECIAL CONSIDERATIONS IN PROJECT FINANCE
FOR THE INDUSTRIAL MINERALS INDUSTRY

C. Richard Tinsley

European Banking Group
London, England

INTRODUCTION

Documentary complications arise from the risk apportionment in project financing which generally means that once the project is up and running and has satisfied the lender's completion test, the loan agreement specifically limits the lender's recourse to the project and its assets. It is the purpose of this paper to describe the special considerations which apply to industrial minerals projects and explore the reasons why so few project financings are negotiated in this industry.

SIZE

Small projects do not warrant the relatively high costs of negotiating and documenting project financing. The extent of the documentation is directly proportionate to either (a) the degree of risk apportionment, or (b) the amount of control given to lawyers to negotiate the documents. Although one would like to believe that project-finance documentation could be off-the-shelf, it is almost always necessary to tailor the documents to the mine development itself and the individual desires of the borrower. Accordingly, the documentation costs and initial expenses are seldom less than $40,000, and for complex projects can reach $400,000. In addition, banks often charge an up-front fee to partly compensate them for their staff time and partly to increase their yield on what is deemed to be a "riskier" loan asset.

Another factor is the generally lower debt:equity ratio when bankers are uncertain of the accuracy of the cash flow projections. The ratio of the amount of cash flow left over after debt service to debt service itself is a key tool for project finance (debt service coverage). Bankers will require a higher coverage ratio for industrial minerals projects because of the relatively higher uncertainties in the following categories:

Prices
Markets
Geology/Quality
Transportation
Environmental

This, in turn, means that lower debt:equity ratios are the norm for new industrial minerals projects.

INDUSTRY STRUCTURE

It is hardly worthwhile to generalise about a field as diverse as industrial minerals, but there are four other structural reasons why project financing has not been widely applied.

1. The most obvious example is the predominance of small-sized operations, largely a function of transportation economics. Geological conditions also contribute to their dispersion.

2. Often, the product is only a small part of a company's business in contrast to many petroleum, metals, or coal companies whose sole business is the production of the raw material. The construction industry is a good example here for cement, gypsum, crushed stone, sand and gravel, and the drilling-mud services business for barite and bentonite. A company is unlikely to jeopardise its main business through the failure of a project financing on one of its key inputs. The corrollary is that the mined input is often regarded as a cost centre and not run on the sort of economic basis suitable for project financing.

3. There may be a reluctance to reveal key aspects of an industrial minerals venture to an outsider for purely competitive reasons. Unlike the other minerals industries which readily share technical information at plant visits and

---

Based on a presentation at the 112th AIME Annual Meeting, Atlanta, Georgia, March 7-10, 1983

meetings, such as the AIME, the industrial minerals sector is the most secretive in my experience. Understandably, process technology is often critical to the competitive position of an industrial minerals project which, after all, rarely produces one fungible product but a product range (quality) which is usually determined by the process. The best examples here are diatomite, ground calcium carbonate, synthetic rutile/slags, and, to a lesser extent, kaolin.

4. In other instances, the market outlook is of strategic importance and even though bankers are very reliable at adhering to matters of confidentiality, the information may be so sensitive that it cannot be revealed and thus, bank financing may be precluded. Until recently, good examples were diamonds and lithium.

For a project financing to get off the ground, the company will need to allow the bankers full access to all information to assist the process of risk analysis and its conversion into the cash flow projections.

PRICES

Price projections for industrial minerals are the most difficult consideration. First of all, what is the "price"? As can be seen on the accompanying graphs, the chief published source, the monthly Industrial Minerals from London, has wide price ranges and list prices which do not enable the determination of any marker price increases in the projection (Mayberry, 1981) which is clearly inadequate for commodities with prices that can and do swing widely according to the business cycle and external events. It is a fallacy to regard industrial minerals prices as somehow being more stable than metals prices with the consequent protection of the mine investor's yield.

As will be evident in the next section on Markets, the price projection is often related to the carve out of market share which is usually transportation or import-substitution oriented. Shifts in ocean freight costs, which can make up half of the delivered price, are even more difficult to predict.

The company must, therefore, present a very sophisticated analysis of these factors to get banks to believe the projections, or else must secure price protection via sales arrangements. The latter is rare for industrial minerals. (Tinsley, 1982)

MARKETS

The banker capable of project financing will be following many commodity prices simultaneously, and some may even have a long-run view on key commodities such as copper, coal, nickel, or even diamonds. But I know few bankers who can give their view of the feldspar market, for example. Therefore, thorough market studies are a prerequisite to any industrial minerals project financings. Some would say it is the key input for an industrial minerals venture. (McVey, 1981).

This market study should analyse the producers just as extensively as the consumers. It should also examine substitution and technological changes especially for the consumers. If possible, this study should be validated by an independent consultant, although such experienced consultants are few and far between in the industrial minerals industry.

There is also too little analysis of marketing practices and the entry strategy for the new development. Industrial minerals often involve a fairly slow market growth based on trial shipments, often at a discount. Entry at a discount may also raise customers' expectations of future discounts. There is also extensive brand loyalty in the industry which reinforces customers' natural conservatism on purchasing policy, again since the industrial minerals input is often a small cost component in the final product.

GEOLOGY/QUALITY

As significant as processing is to most industrial minerals venturers, the most profitable ventures invariably start off with good quality reserves in a good location. However, it is not sufficient to simply estimate the reserve tonnage as it relates to the mining plan. It is more important to demonstrate that the grade can be maintained to yield proper quality products on a consistent basis. Customer loyalty rests on consistent delivery, quality, and, of course, price, and there is nothing worse than a faltering introduction of a product into the marketplace because of inferior product quality (McVey), which can often stem from the reserve base.

In the studies I have seen, the reserve estimate is too often left up to the geologist who may not have a full appreciation of the important quality characteristics of the product, or the process. Accordingly, vigorous bulk sampling, pilot-plant programmes, and test marketing will need to validate the producability of the reserves.

I have also seen too few analyses of the geological parameters with which competitors are dealing (Mayberry), or the prospects for discoveries. As a general comment, industrial minerals cannot withstand the sort of exploration budgets thrown at finding metal mines. There is, then, a significant risk that future (inadvertent?) discoveries could alter the competitive position of the venture.

TRANSPORTATION

Few industrial minerals projects can support a new infrastructure, except perhaps port modifications. Too often, the mining industry assumes that the risks inherent in the outside world end at the mine gate. This aspect is

FINANCE FOR THE MINERALS INDUSTRY

AUSTRALIAN RUTILE CONCENTRATES
(min. 95% TiO₂, bulk, FOB/FID)

SOURCE: INDUSTRIAL MINERALS

DIATOMITE, U.S., FLUX – CALCINED
(DELIVERED U.K.)

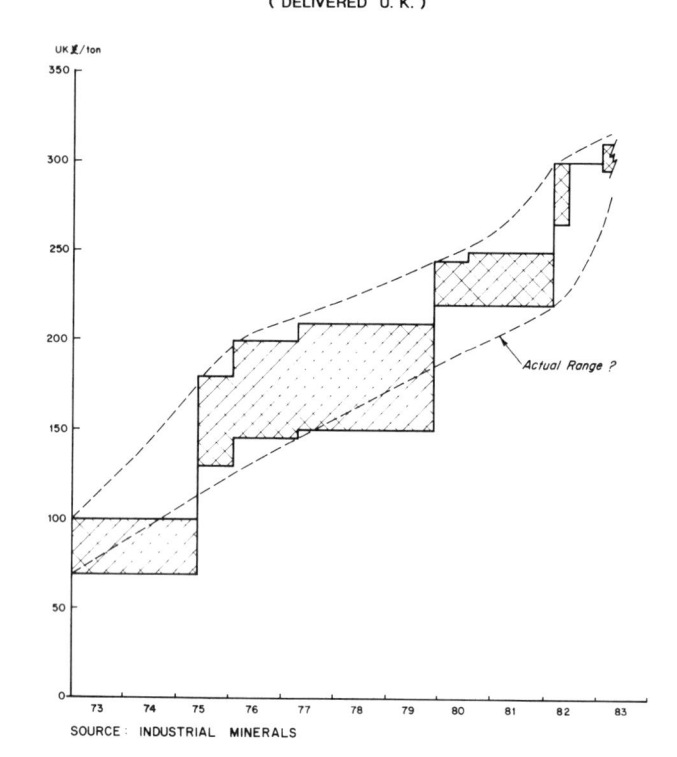

SOURCE: INDUSTRIAL MINERALS

FLORIDA PHOSPHATE, LAND PEBBLE
(70 – 72% BPL, ROM, dry basis, underground, bulk, ex-mine average)

SOURCE: INDUSTRIAL MINERALS

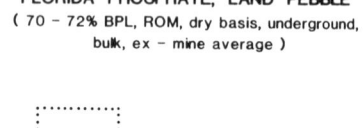

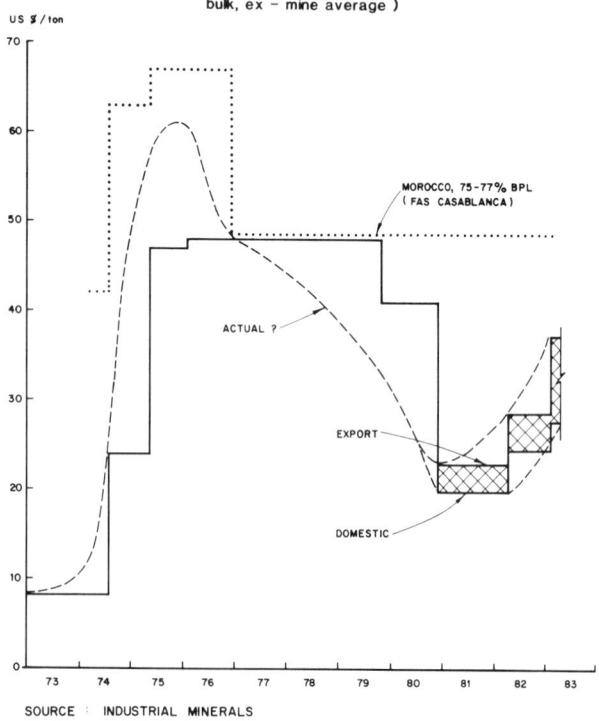

KAOLIN, COATING
(FOB U.K.)

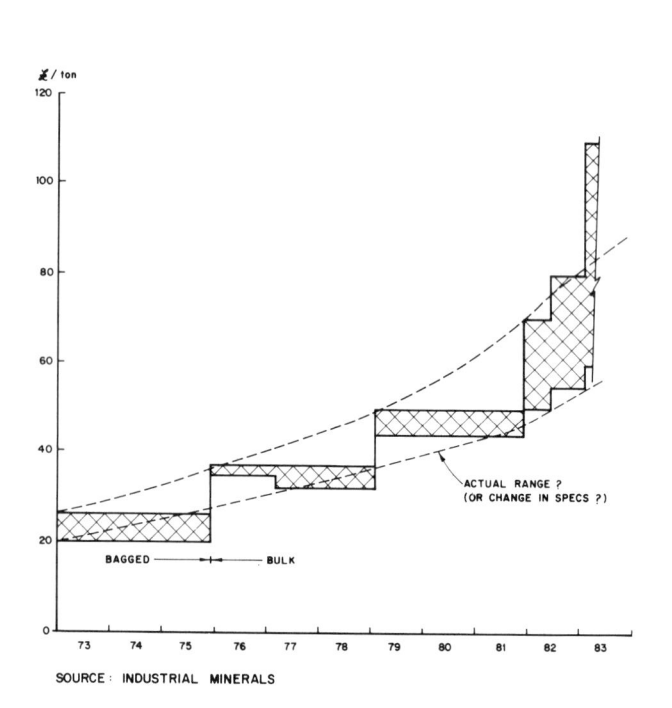

SOURCE: INDUSTRIAL MINERALS

often poorly handled in most loan applications. There is a temptation to believe that all other producers will be similarly affected by any transportation changes. (The same is true about changes in energy costs.)

ENVIRONMENTAL

The industrial minerals business is only now coming under significant pressure from environmental factors. The recent situations with asbestosis and kaolinosis speak for themselves. A review of the consuming industries' environmental risks is, however, lacking in most project finance applications. This may also relate to the changes in purchasing policies due to environmental considerations, e.g. sulphate pigment plants for ilmenite and synthetic versus natural soda ash.

CONCLUSIONS

Project financing for industrial minerals developments is often precluded by the small size of the financing. Simply determining the price projections and market position for a new venture is much more difficult than for other commodities which are higher value, fungible, and with readily known market prices or sales contract coverage.

The industrial minerals business is highly dependent on reserves and location, but quality and transportation economics are too often handled in a cavalier manner in loan applications. To a great extent, this is evidence that the company seeking a project finance loan has not understood the scope of studies required to analyse an industrial-minerals project financing. The project finance banker should be one of the first experts consulted, rather than the last to be contacted to get the capital through that "magic" but poorly understood "bankable document." It is preferable to see all the studies as well as an incisive summary which become the bankable documents.

REFERENCES

McVey, H., "Industrial Mineral Marketing - Logic and Illogic", SME-AIME Fall Meeting, Denver, Colorado, November 18-20, 1981, Preprint No. 81-331.

Mayberry, R.C., "Phosphate Reserves, Supply and Demand -- Southeastern Atlantic Coastal States 1980-2000 A.D., SME-AIME Fall Meeting, Denver, Colorado, November 18-20, 1981, Preprint No. 81-409, p.11.

Tinsley, C.R., "Industrial Minerals Sales Contracts from a Banker's Perspective", 5th International Industrial Minerals Conference, Madrid, Spain, April 28, 1982.

11.16

TRADE FINANCING

SUPPLEMENT

INTRODUCTION

Trade financing is a particularly important component of short-term financing for a minerals company since errors, extra costs, or payment delays/defaults can easily wipe out the profit margin on a particular shipment or can cause problems when arranging purchases of e.g. reagents. It is required to make any minerals operation function and does not have to be cross-border to require proper management.

Sources of working capital are as follows:

1. Extended credit terms from suppliers

2. Inventory loans from banks, finance companies

3. Accounts receivable financing or factoring

4. Credit lines or overdraft facilities from banks.

TRADE CREDITS

These are provided under the familiar terms "net 10 days" or payment due by the "10th day of the following month." Thirty days for payment is usual but longer periods can be negotiated with regular suppliers. Shipment may be made on open account with specified periods for settlement. In the mining industry, a custom smelter gains trade credit by paying for concentrates three months after the month of shipment.

INVENTORY LOANS

Various forms of inventory finance are provided via "floor-plan" or warehouse financing. A set percentage, usually 60%-80%, of a good-quality, readily saleable inventory item (e.g. bullion, concentrates, ingots) will be provided by a bank or finance company which will usually have security over these products and has a right to inspect the goods from time to time.

Legal security will be achieved by warehouse receipts, floating liens, chattel mortgages, trust receipts, or through collateral management service companies. The latter certify control over stock or distribution outlets and often carry insurance against fraud. Costs range around 1.5% - 2.0% p.a. or more plus some fees to cover out-of-pockets or other costs.

ACCOUNTS RECEIVABLES

Accounts receivable financing is based on the company's generation of satisfactory receivables which are continuously assigned to a bank or finance company ("factoring"). Because of the risks in being able to collect receivables, such as payments to a smelter for its blister copper shipments to a refinery, this form of financing costs as much as 3% - 5% of the debt amount purchased plus an additional charge for processing paperwork.

Receivables can be purchased with or without recourse and on a notification or non-notification basis to the client's customer to, respectively, pay the invoice directly to the bank or not.

LETTERS OF CREDIT

Letters of Credit ("L/C's") are a common feature of international trade finance whereby a bank, at the request of the buyer, issues a written undertaking to an exporter to pay a specified sum of money within a set time on presentation of previously specified documents. This undertaking means that the exporter is no longer relying on the credit standing and integrity of the foreign importer. A dispute about shipment does not otherwise affect the obligation to pay upon presentation of a "clean" set of documents. These documents include the bill of lading and other documentation, often on standard "form" bases.

There are a number of conventions and techniques in L/C's such as irrevocable/revocable L/C's, time L/C's, confirmation/advising L/C's and transferable L/C's. These are tools to adjust for the varying risks that the importer will not pay. Other routes where one is comfortable with an importer include documentary sight bills (D/P bill), documentary term bills (D/A bill), and, of course, open account.

ACCEPTANCES

Once an import/export has been satisfactorily completed ("accepted"), a payment will be made in a given period. This acceptance may then be traded in the market at a discount and the exporter gets his money earlier. This has become a multibillion component of international trade payments. Rates fluctuate daily.

A FORFAIT

The developing forfaiting market is described in Mike Bradbury's paper in this Chapter. It covers the purchase of bills of exchange (out to five years) by the forfaiting bank without recourse to the exporter. These bills usually bear the guarantee of the importer's bank. The forfaiter collects the amount of the bills as they mature from the importer. A forfait fees range up to 4% p.a with the variability heavily dependent on the "quality" of the importer's country risk.

CREDIT INSURANCE

As mentioned in the Supplement on North American Export Credit Programs in this Chapter, many nations have credit insurance entities besides private insurance (see Berry's paper, Chapter 9). These often contain comprehensive insurance against such risks as political risks (currency inconvertibility, for example) or commercial risks (e.g. bankruptcy of the importers). A host of alternatives exists and one should check with a major bank or insurance agent experienced with international payments to quote rates and advise on the alternatives.

HEDGING

Currency exchange risks for trade finance are usually not covered by the swap and parallel exchanges covered by Poole in Chapter 9. The money market or forward-exchange market provide the means to take forward cover for this risk through, respectively, either borrowing the foreign currency (converting and investing the proceeds) or having a forward contract to sell the foreign currency to lock in the exchange rate. Complications such as exchange controls, interest-rate differentials, or excessive coverage costs have to be watched carefully since by blindly hedging, one may end up with increased risk.

CONCLUSIONS

Although routine, attention to trade financing is important because of the loss of margin or delay costs inherent in making a payment. Many techniques are available and the minerals company treasurer should not simply rely on its suppliers and purchasers to arrange this financing but should check the costs, methods and currency exposures and decide to control this important financing cost directly. The techniques are evolving quickly; pricing changes almost daily; and it is worthwhile to seek advice from bankers, factoring houses, and finance companies on prevailing costs and methods.

REFERENCES

Baughn, W.H., Walker, C.E. (eds.) The Bankers Handbook, Revised Edition, Dow Jones - Irwin, 1978.

Bruce, R., McKern, B., Pollard, I., Handbook of Australian Corporate Finance, Butterworths, 1984.

Rodriquez, R.M., Carter, E.E., International Finance Management, Second Edition, Prentice-Hall, 1979.

Tinsley, C.R., "Mine Finance", Chapter in Mineral Industry Costs, Northwest Mining Association, 1981.

11.17

FORFAITING EXPORT FINANCE IN THE FREE MARKET

Michael Bradbury

Prudential-Bache Trade Finance Ltd.
London, England

INTRODUCTION

A major change in financing exports in recent years has been the decreased importance of subsidized export credits, following the commitment of the OECD countries to eliminate the subsidy element. Subsidized export credit rates are now often more expensive than market rates as the financing for exporters has decreased in volume and importance.

The forfait market, in recognition of this decreasing importance, provides a rapidly expanding source of nonrecource finance for exporters. The following is a step by step guide to what it is, how it works, its benefits and problems.

WHAT IS FORFAITING?

Forfaiting is not new. It has been used in Europe for about 25 years to provide export finance for both capital goods and commodities without recourse to the exporter. Although long the preserve of exporters in Germany, and more lately Italy, forfaiting is now used increasingly by exporters in Great Britain, the United States, Canada and Scandinavia. It is a fairly simple operation with easy straight-forward documentation.

The transaction involves the purchase by the forfaiting bank of a bill of exchange or a series of bills without recourse to the exporter. These bills will have been drawn by the exporter, accepted by the importer, and will usually bear the unconditional guarantee of the importer's bank. They are normally drawn payable at intervals during the credit period agreed between importer and exporter and are purchased at a discount by the forfaiter. The exporter thus obtains his cash payment and the importer receives his credit period. The forfaiter collects the amount of the bills as they mature directly from the importer.

Credit periods vary, depending upon the type of goods and the country and commercial risk of the importer. The period is generally between one and five years with six monthly maturities. The usual forfait currencies are US dollars, Deutschmarks and Swiss francs and others, such as sterling, French francs, and the Scandinavian currencies are being used more frequently.

CHOOSING FORFAITING

Most types of financing require the continued involvement of the exporter in the financial side of the transaction during the credit period. Forfaiting has certain comparative advantages. In particular, it is possible to forfait 100 percent of the risk without recourse to the exporter, unlike government sponsored export finance schemes, where some retention of the risk by the exporter is required.

In addition, exports from country sources do not present problems and forfaiting can also cover trade within the EEC. Other benefits for the exporter are:

(a) The exporter's balance sheet is improved; debts are turned into cash thus improving the liquidity of the company and enhancing its cash flow. And there are no contingent liabilities for bills of exchange.

(b) If the sales contract is in a foreign currency forfaiting removes the risk to the exporter of adverse movements in foreign exchange during the credit period.

(c) Fluctuations in interest rates during the credit period can be ignored since in forfaiting the transaction is usually done on fixed interest rate terms on a discount basis.

(d) The political and transfer risks are taken by the forfaiter.

(e) The commercial risk is taken by the forfaiter.

(f) The costs of administration and collection etc. during the credit period are removed.

SOURCES OF FINANCE

(g) The exporter's normal bank limits or lines of credit are freed for other purposes.

(h) Forfaiting is a very simple transaction with a minimum of documentation. Agreement to deal with the transaction, including future commitments and a fixed interest rate quotation, can be given by the forfaiter in a matter of hours.

There are occasions when forfaiting is not the most effective way of financing exports. The importer may not be willing or able to provide the bank guarantee which the forfaiter usually requires. This could be because credit lines with his bank are fully used, or he may not want to pay the guarantee commission to be charged by the guaranteeing bank. Commitment periods in forfaiting are not generally as long as OECD subsidized schemes, and will not be as long as the exporter requires in some cases. Most forfaiters, however, will be happy to give commitments to buy transactions on good risk countries for up to one year in advance, which is adequate for the majority of contracts. A further disadvantage may arise when exports are priced in a currency which has high interest rates such as the French franc. In these cases forfaiting can be expensive.

On balance, however, the pros for forfaiting far outweigh the the cons, and given a wider exporter education, its use will grow more quickly over the next few years, unless the next round of negotiations conducted in the OECD over the future of the consensus rates results in a large reduction in those rates. This must be considered unlikely, given the opposition from the US and British governments to further official subsidies and the 1983 commitment by OECD countries to eliminate the subsidiary element in export credits.

COSTS AND CHARGES

The cost of forfaiting varies depending on the level of interest rates for the particular currency at the time of the forfaiter's commitment, and on the forfaiter's assessment of the credit risks relating to the importing country and to the guaranteeing bank. The interest cost is made up of:

(a) a charge for using the money and covering the interest rate risk; this in effect is the forfaiter's refinancing cost and is usually based on the cost of funds in the Euromarkets.

(b) a charge for covering the political and transfer risks and the commercial risk attaching to the guarantor-the margin. This varies from country to country and from guarantor to guarantor.

There are two ways in which forfaiters calculate (a), above. Some price on the basis of the LIBOR (London Interbank Rate) rate applicable to the average life of the transaction; e.g. on a five year transaction repayable by 10 six monthly installments the average life of the transaction is 2 1/4 years and the LIBOR rate for this period would be used. The method becoming more frequently used, particularly in London, is to take a weighted average of the 10 LIBOR rates applicable to the 10 maturity dates.

The addition of (a) plus (b), above, gives the yield, known as the discount or yield compounded semi annually (DYCSA). As most forfaiting transactions are made by deducting the interest by way of a discount immediately on purchase, this yield must be translated into a discount rate, which is the usual method of quotation made by the forfaiter to the exporter. For a typical forfait series of bills this is a complex calculation and is usually done by computer. Forfaiters may also publish simple tables showing approximate conversions.

Additional costs, which are also part of the calculation are: (a) days of grace, and (b) a commitment fee.

Days of grace are an additional number of days of interest charged by the forfaiter and reflect the number of day's delay normally experienced with payments made from the debtor country. These range from none up to, say, 20 on some countries which banks are prepared to forfait.

COMMITMENT FEE

A commitment fee is payable when an exporter accepts a firm offer of purchase by the bank in advance of the delivery of the paper. Banks will commit to purchase paper, often at a fixed rate of interest some months in advance of the paper being available for discounting. The maximum commitment period will vary depending on the country of the importer and the currency concerned, but could be up to 12 months. The fee is usually expressed either as a percentage per annum or as a per mille per month and is calculated on the face value of the paper to be forfaited. The details of the bank's commitments are usually set out in a commitment letter which is binding on the bank and the exporter. If for any reason the paper is not delivered, the commitment fee is still payable by the exporter.

TYPICAL TRANSACTION

As an example, take an export of a piece of mining equipment (or a certain tonnage of concentrates) in the USA and sold by the US company to a German buyer with shipment to be made in June 1983 (the dates coincide with a real transaction and take in all calculations). The contract price for a cash sale is US$740,000 but the buyer requires five year credit terms with repayments to be made in 10 six monthly instalments. He will pay the interest costs and will provide the guarantee (or aval) of XYZ Bank, which is acceptable to the forfaiter.

On April 4, 1983 the exporter approaches the forfaiter who confirms that he is willing to quote for a forfaiting proposition covering the export to be completed three months time bearing the aval of XYZ Bank. The forfaiter is happy to calculate the amount of the promissory notes (so that after discounting the exporter will receive $740,000) and quote a straight discount rate of 9 7/8 per cent per annum (pa). He also wishes to charge two days' grace and a commitment fee of 1 per cent pa. This commitment fee in the forfaiter's charge for agreeing on April 1 to purchase, at a fixed discount rate, paper which will not be available until after shipment is made three months later. He will stipulate an expiry date for his commitment, i.e. that the paper must be in his hands before, say, July 29, 1983. This will allow sufficient time after the export has taken place, for the exporter to obtain the promissory notes duly avalised/ guaranteed and to present them for discounting. For ease of the following calculations, however, we assume that both shipment and discounting take place on June 30, 1983.

The usual method of calculation is to divide the principal into equal amounts and then to calculate the interest on the declining balance. If all the interest costs are to be passed on to the importer it will be necessary to know the rate of interest equivalent to the straight discount rate quoted by the forfaiter. In our example, a straight discount rate of 9 7/8% p.a. is the equivalent of a give formula (showing how 9 7/8 pa = 12.3840 pa) yield of 12.3840% per cent pa. The contract price of $740,000 is divided into 10 equal instalments of $74,000.

Interest is then calculated at the applicable interest rate (12.384 per cent pa) on the exact number of days between, say, date of shipment, June 30, 1983 and the due dates of the six monthly promissory notes. Interest may also be calculated from the invoice date, date of acceptance, etc. depending upon what is agreed between exporter and importer. (See TABLE 1)

CALCULATION OF AMOUNT

Since, in accordance with normal practice in the forfait market, the discount is calculated on the basis of the actual number of days outstanding on a base of 360 a day year (365/360), it is better for the exporter to calculate the interest element in the bills in the same way. In the example cited the exporter obtains promissory notes for a total face value of US$996,086.96 which reflects his invoice price plus the interest.

Due Date	Capital	No. of Days	Interest	Face Value of Prom.Notes
02.01.85	74,000	186	47,348.09	121,348.09
02.07.85	74,000	182	41,696.86	115,696.86
31.12.85	74,000	182	37,063.99	111,063.88
01.07.86	74,000	182	32,430.89	106,430.89
31.12.86	74,000	183	27,950.65	101,950.65
30.06.87	74,000	181	23,037.64	97,037.64
31.12.87	74,000	184	18,735.59	92,735.59
30.06.88	74,000	181	13,822.59	87,822.59
31.12.88	74,000	184	9,367.79	83,367.79
30.06.89	74,000	182	4,632.98	78,632.98
	740,000	1,827		996,086.96

FORMULA: $I = \dfrac{p \times t \times r}{360 \times 100}$ *Dates adjusted for Saturdays and Sundays

Where I = amount of interest

P = principal amount outstanding (on a reducing balance basis)

t = time i.e. number of days

r = interest rate to be charged

If the promissory notes issued by the importer in favor of the exporter are presented to the forfaiter on June 30, 1983 the calculations apply as follows:

$$\dfrac{FV \times t}{100} = N \text{ and } \dfrac{N \times d}{360} = D \text{ and } FB - D = PV$$

Where FV = face value of the promissory note(s)

t = time in number of days i.e. days from date of discount to maturity adjusted for Saturday, Sunday and days of grace

N = "discount numbers"

d = discount rate in % p.a.

D = amount of discount

PV = present value, i.e. after deducting discount from face amount of promissory note(s)

(FV) Face Value of Bills (US$)	(t) Days	(N) Discount Number	(d) Discount (US$)	(PV) Net Proceeds (US$)
121,348.09	188*	228,134.41	6,257.85	740,000.02
115,696.86	370	428,078.38	11,742.43	103,954.43
111,063.88	552	613,072.62	16,816.92	94,246.96
106,430.89	734	781,202.73	21,428.82	85,002.07
101,950.65	917	934,887.46	25,644.48	76,306.17
97,037.64	1098	1,065,473.29	29,226.52	67,811.12
92,735.59	1282	1,118,870.26	32,611.37	60,124.22
87,822.59	1463	1,284,844.49	35,244.00	52,578.59
83,367.79	1647	1,373,067.50	37,664.00	45,703.79
78,632.98	1829	1,438,197.20	39,450.55	39,182.43
996,086.96		9,335,838.34**	256,086.94	740,000.02

* i.e. 30 June 1983 - 21 January 1984 plus 2 days' grace

** It is not necessary to calculate the discount and net proceeds for each individual note. The calculation can be made on the series as a whole e.g.:

$D = \dfrac{N \times d}{360} = \dfrac{9,335,828.34 \times 9.875}{360} = \$256,08$

PV = FV - d =   996,086.96
              - 256,086.94
              US$740,000.00

In this example the exporter obtains immediate cash net proceeds of $740,000 and the importer pays a total cost of $996,086.90 and receives credit over five years.

In addition the exporter will have to pay the agreed commitment fee to the forfaiter for the period starting on the date the forfaiter committed to buy the paper at a fixed rate of discount until the actual date of discount. This amounts to $2,490.22 at 1 per cent pa for the period between April 1, and June 30, on the face value of the bills in the example, on a 365/360 day basis.

If the exporter and importer agree to invoice in Deutschmarks or Swiss francs instead of US dollars the figures would reflect the lower interest costs in those currencies, as follows:

Contract price US$740,000

@2.41 = DM 1,783,400   @2.06 = SW.Fcs. 1,524,400

Discount rate, DM: 6 7/8%   Sw.Fcs: 5 5/58%

Total bills, DM: 2,182,566.30 Sw.Fcs: 1,795,538.79

If the importer can be persuaded to take such a currency risk then the lower interest costs will make the exporter's price more competitive.

The exporter will only have a currency exposure from the date he signs the commercial contract until the date of discounting of the promissory notes and this may be covered by a forward exchange contract in the usual way.

DOCUMENTATION

Forfaiting documentation is simple and straightforward. There is usually a short facility letter, similar to the specimen, below, which is issued by the forfaiter and confirmed by the exporter, setting out the terms and conditions of the transactions.

A.N. Exporter Ltd.
Dear Sirs,

Forfaiting Transaction No.	Without Recourse purchase of promissory notes

On the basis of information which you have given to us (which is detailed below) and subject to receipt of satisfactory documentation to confirm our purchase and your sale, without recourse to you, of the following bills of exchange in the form of the enclosed specimen:

Amount in the effective currency of (country)	United States dollars nine hundred and ninety six thousand and eighty six and 96/100.
Evidenced by: promissory notes (details)	promissory notes in the international form dated 30 June 1984
Acceptor/Issuer:	Deutsche Importer AG
Avalor:	XYZ Bank AG, Frankfurt
Maturity:	10 semi-annual instalments first maturity 2 Jan. 1985 - US$121,348.09; last maturity 30 June 1989 - US$78,632.98
Domicile of payment:	Aval bank
Terms:	9 7/8% p.a. straight discount calculated on the basis of a 365/360 days year plus 2 days' grace
Special Instructions	Promissory notes to bear appropriate German Stamp duty. Signature of issuer to be confirmed by his bankers.

(Place and date of issue   London 30 June 1984   US$ 121,348.09

On   2 January 1985   for value received pay against this bill of exchange

to the order of   A.N. Exporter Ltd.   (?)

One hundred and twentyone thousand three hundred and forty-eight and 09/100

effective payment to be made in   United States Dollars

free and clear of any deductions, taxes, imposts, collection charges, levies or duties present or future of any. This bill of exchange is payable at XYZ Bank A.G., Frankfurt

Drawn on   Deutsche Importer A.G.,   per pro A.G. Exporters(?)

             47 Kaiserstrasse,

             Frankfurt,

             Federal Republic of Germany.   Director

Delivery of Documents
to us: by 29 July 1983

We confirm that we waive our right of recourse against you as drawer of these bills and will endeavor to obtain a similar undertaking from any subsequent purchaser from us.

This transaction shall be construed and interpreted in accordance with the provisions of English law and shall take effect as if it were made in England. You agree to submit to the non-exclusive jurisdiction of the Supreme Court of Judicature in London and appoint _____ as your irrevocable agents for the acceptance of service in respect of any proceedings in connection with this transaction.*

You have given us the following information in the course of negotiations and we are buying on this basis.

1. The transaction underlying the issue of the bills is the export of mining equipment from U.S.A. to W. Germany.

The obligations of the importer under your agreement with him are not dependent upon performance of the goods or any other conditions which require satisfaction.

2. You will not take any action which might affect the validity of the bills.

3. This transaction constituted valid and legal obligation on your part and it is within your power to enter into this transaction.

4. All necessary export and import licences in connection with the underlying transaction have been properly obtained.

5. All necessary approvals for the admittance of the foreign exchange upon the due dates by the importer have been obtained.

6. Payment of the bills will be made free of any tax, impost, levy or duty present or future.

*Note: U.K. registered companies do not need appoint agent.

For Forfaiting Bank Limited

Authorized Signatory

Memorandum of Acceptance

We hereby accept the above terms and conditions

For and on behalf of

A.N. Exporter Ltd.

           Director(s)

Dated:

(1) To ensure that the payments are made in the currency specified in the bill or note, the effective currency clause is always included whenever the instruments are denominated in a currency of a country that differs from the place of payment.

(2) To ensure that the forfaiter resolves the full face down amount of the bill or note, as he has no recourse to the exporter or prior holders for deductions made in the importer's country, the clause "free of any taxes, imposts, levies, duties, or collection charges present or future of any nature" is included in the wording of the bill or note.

The bills will also bear an "aval" (or guarantee). This is the forfaiter's preferred form of security of payment of the bill or note. It is recognized in most countries as an unconditional and irrevocable guarantee to pay on the due date and it also takes on the inherent abstract nature of transferability of the bill or notes on which it appears.

For an "aval" to be acceptable to a forfaiter, the bank avalising must be acceptable in terms of risk. This avalor must be an internationally recognized bank which is not only for the security of the forfaiter purchasing the paper from the exporter but also for others in case the paper is sold into the secondary market.

The aval may be placed on the face of the bill or promissory note or on the reverse of the instrument.

Sometimes a formal guarantee is issued instead of an aval, particularly in some countries where the aval is not a recognized form of guarantee under the local laws. The guarantee wording varies: occasionally it is written on the "bill of exchange" but more usually it takes the form of a separate document. Alternatively the security for payment of the bill may take the form of a blank endorsement by the guarantor. The wording of the separate guarantee varies but the main points to be included are that it should be (a) irrevocable, (b) unconditional, (c) divisible, and (d) assignable.

The exporter usually obtains the documentation for forfaiting through his bank's documentary collection system. He sends the bills of exchange with his shipping documents with instructions for them to be released to the importer only against acceptance of the bills and the aval or guarantee of the named bank.

When these are returned to the exporter they are then ready to be forfaited and the forfaiter has only to satisfy himself with the authenticity of the bills and the genuineness and capacity of the signatures. He will normally ask for a confirmation from the bankers of the drawer and drawee of respect of the signatures. The signatures of the avalising bank will often be held on file by the forfaiter. Once any other conditions which were mentioned in the facility letter have been satisfied, the discounting then takes place, as described earlier.

Transactions may be so large that they are syndicated between two or more banks, either by splitting up the bills between them and discounting separate branches, or by one bank holding the bills and issuing participation certificates to the other funding banks.

Another alternative is for one bank to fund the whole transaction but to reduce its risk on the particular transaction by taking guarantees from other banks.

When the transaction has been completed, the forfaiter has the choice of keeping the transaction in his portfolio or selling it as a package to another bank, thereby creating a secondary market in the paper.

The claims by the exporter on the importer must be in an easily understandable and negotiable form; therefore bills of exchange and promissory notes are the instruments most often used. For readers familiar with the normal type of bills and notes employed in international trade, some difference will be apparent in the bills and notes used in forfaiting.

11.18

EUROPEAN EXPORT CREDIT PROGRAMMES

EDWIN A. RIDES

EXECUTIVE DIRECTOR - FIRST INTERSTATE LIMITED

HISTORICAL ORIGIN OF THE EXPORT CREDIT AGENCIES

Almost since the inception of international commerce sellers, having manufactured and/or sold and shipped goods to an overseas buyer, have been obliged to accept the risk that they will not be paid for one reason or another. There are many reasons why payment may not be forthcoming, for example, the goods may be lost at sea, or rejected by the buyer or the buyer may simply not be creditworthy. These risks were brought into sharp relief as international trading became more sophisticated and the practice of giving credit to buyers gathered momentum. Inevitably enterprises sprang up which offered some mitigation of this exposure, accepting houses and later confirming houses and banks were prepared to assume acceptable credit risks on behalf of the seller, usually by discounting trade paper, and for many years the market ticked over reasonably satisfactorily. However, after the 1914-18 War international trade was at an extremely low ebb and its inherent risks were accentuated to a point where they were often commercially unacceptable.

Yet a healthy and vigorous international market was indispensable to recovery of the world economy and a number of governments, led by the United Kingdom, perceived a need to offer their exporters a comprehensive form of insurance, particularly to facilitate the extension of credit to foreign buyers. In the United Kingdom the Export Credits Guarantee Department (ECGD) was set up to encourage trade through the provision of export credit insurance. Over the ensuing years others followed suit and now most of the developed trading nations and many developing nations have similar agencies.

PRESENT DAY ROLE OF THE EXPORT CREDIT AGENCIES (ECAs)

Traditionally then the ECA's role was to encourage trade through the provision to exporters of insurance against the risks of non-payment by overseas buyers of their goods and services. However, during the 1970's the manufacturing industries of most of the developed nations found themselves facing severe competition for business, particularly in the developing world. As a result the more aggressive ECA's began to introduce ancillary forms of cover designed to reduce exporters' risks outside of the credit field. For instance, most of the European agencies now have a form of cover against foreign exchange risks where contracts are denominated in currencies other than that of the exporter's country. Most of the agencies also provide cover to exporters against the risk of unfair calling of bid, advance payment, performance bonds etc and in certain countries, where high inflation rates were endemic, a form of cover against cost escalation. In certain circumstances it is also possible for an exporter to obtain cover against losses arising as a result of the involvency of a partner in a consortium or joint venture.

These are all forms of insurance but, also during the 1970's, many of the national export credit programmes took a major step away from their original raison d'etre by facilitating the provision of preferential financing terms for the acquisition of capital equipment by an overseas buyer. This new line of business quickly became an important ingredient of major export financing packages and, as we shall see later, turned out to be extremely costly for the countries providing it.

The Insurance Function

The insurance function, particularly of the so called political risks inherent in international financing operations, is now almost taken for granted by the exporting community worldwide. Over the years it has played a significant part in facilitating and encouraging international trade. In most cases the "political risks" are borne by the state, although a number of private insurance companies are now offering cover for these risks.

The principal ECA's in Western Europe and the forms of insurance they offer are shown in the Exhibit at the end of this account.

<u>Credit Insurance</u> Inevitably there are differences in the detail of the ECA's programmes but they all provide essentially the same cover against the credit risks involved in exporting. Typically, exporters receive insurance for between 90 and 95 per cent of principal and contractual interest associated with a particular transaction and most of the agencies also provide insurance or guarantees to banks who may have purchased the exporters' receivables or who have furnished credit directly to the importer. Such guarantees are generally on even more favourable terms and in the UK, Italy and Spain, for instance, can be for as much as 100 per cent of princpal and interest.

For exporters, the cover would typically include:

1. Commercial risks:-

   (i)   insolvency of the buyer;

   (ii) the buyer's failure to pay within six months of due date for goods which he has accepted;

   (iii) the buyer's refusal to take up goods which have been despatched to him (where not caused or excused by the exporter's failure to properly perform his part of the contract);

   (iv) war and certain other events preventing performance of the contract provided that the event is not one normally insured with commercial insurers;

   (v)   cancellation or non-renewal of an export licence or the prohibition or restriction on export of goods by law. This is covered only where the pre-shipment risks are insured.

2. Political risks:-

   (i)   a general moratorium on external debt decreed by the government of the buyer's country or of a third country through which payment must be made;

   (ii) any other action by the government of the buyer's country which prevents performance of the contract in whole or in part;

   (iii) political events, economic difficulties, legislative or administrative measures arising overseas which prevent or delay the transfer of payments or deposits made in respect of the contract;

   (iv) legal discharge of a debt (not being legal discharge under the terms of the contract) in a foreign currency, which results in a shortfall at the date of transfer.

The banks may enjoy either an unconditional guarantee or an improved form of insurance from most of the European agencies. This has proved to be a major inducement to banks to participate in this type of business even in circumstances where the risks would otherwise be unacceptable to them.

<u>Complementary Forms of Insurance</u> Over the years, in response to demand from their exporters, most of the ECAs have supplemented their credit insurance services with a range of complementary facilities. These facilities were not available in a convenient form in the private insurance market and, coupled with the credit insurance, they have proved invaluable in supporting the exporting potential of the nations concerned.

BID AND PERFORMANCE BOND COVER Typically this type of cover takes two forms,

Counter guarantee facilities to banks issuing bonds on behalf of exporters who are not fully recourseworthy, and

Cover for the exporter against unfair calling of bonds by the buyer. Most ECAs offer this facility only for contracts with public buyers, but it has been in great demand. There has been a number of cases of capricious calling of bonds and practices, particularly in buyers markets like the Middle East, have become heavily loaded against the exporter who feels the need for protection in a more flexible form than that offered by the surety companies.

EXCHANGE RISK COVER This type of facility is made available to exporters invoicing in foreign currency. It take two forms; (1) a limited indemnity, intended for use with the forward exchange market against exchange losses incurred by the exporter as a result of payment default by the overseas buyer, and (2) full exchange risk cover against any loss suffered by the exporter in converting his foreign currency receipts into local currency.

These facilities are usually reciprocal, that is, the exporter is insured against losses but pays any foreign currency gains to the ECA. There is normally a first loss of up to 3% which has to be borne by the exporter and the indemnity is confined to major convertible currency.

INSURANCE FOR OVERSEAS INVESTMENTS This facility is particularly pertinent where a mining company, for instance, is making a significant investment in a developing country where there is a risk of arbitrary political action, e.g. expropriation or nationalisation of assets, or where restrictions may be imposed on the remittance of profits earned by the investment in the host country. Many of the major ECAs run schemes of this sort and they have played a significant part in encouraging overseas investment.

The Financial Support Function
(Capital Goods only)

During the mid to late 1970's most of the European programmes provided support for financing capital goods exports at preferential interest rates. Some high interest rate countries, among them the UK, France and Italy, provided significant subsidies to their exporters by this means. Finance was made available to overseas buyers by the commercial banks through buyers credit facilities at rates which were significantly below market and the difference was borne by the appointed agencies of the governments concerned. In the UK this was done by ECGD, in France through the BFCE and the Bank of France, in Italy by Mediocredito Centrale and now in Spain by Instituto de Credito Oficial (ICO). Other agencies also provided support on more or less generous terms (see Exhibit).

Initially there was a free for all between the major exporting nations which led to mutually destructive competition on financing terms in order to secure major capital equipment orders. Not only was this extremely expensive for the governments concerned (in the UK, for instance, it was costing £500 m per annum at its peak) but it was self-defeating and ultimately the only beneficiary was the buyer. Fortunately for the tax payer in these countries, the governments of OECD member nations met with a view to bringing some sort of order to the export credit scene and in 1977/78 agreement was reached between them on minimum interest rates, maximum credit periods and minimum down payments which would be applied to future capital equipment export credits. This agreement, which was known as the "Consensus", also prescribed that loan repayments had to be in equal semi-annual instalments and capitalisation of in interest during the construction or pre-commissioning period was permitted only exceptionally.

Although the Consensus was not binding it was generally adhered to by member nations, but where it was breached by one country provision was made for other countries competing for a particular piece of business to match the offending offer after appropriate consultation. The Consensus conferred some order on the market but the minimum interest rates were pitched at levels which still required significant subsidies. In the early 1980's resistance to the formula from the United States and certain low interest rate countries gathered momentum and the minimum interest rates (which were reviewed annually under the Consensus) were progressively increased to bring them nearer to the then current market rates.

Even then the annual rate reviews were invariably an occasion for protracted haggling between the member countries and in 1983 a formula was evolved which provided for automatic adjustment of the Consensus rates. The current arrangement is that there is a matrix of interest rates applicable to financing exports from high interest rate countries which is adjusted automatically at six monthly intervals (January and July each year), in line with movements (as opposed to the actual rates) occurring in the previous six months in government bond rates for the five SDR currencies, i.e. US Dollars, Pounds Sterling, Deutschemarks, Yen and French Francs. Currently the matrix is as follows:-

	Category 1 (relatively rich countries)	Category 2 (intermediate countries)	Category 3 (relatively poor countries)
Minimum Interest Rate	% p.a. fixed	% p.a. fixed	% p.a. fixed
For credits between 2-5 years inclusive	13.35	11.55	10.70
For credits over 5 years	13.60	11.90	10.70
Maximum credit period	5 yrs	8 1/2 yrs	10 yrs
Minimum Downpayment	20%	15%	15%

In October 1984 the high interest rate currencies (HIRCs) included the US$ and £ sterling, and several of the ECAs are prepared to support financing in these currencies at these matrix rates. Other HIRCs such as the French Franc, Lira and the Peseta are normally available only in respect of exports from France, Italy and Spain respectively and at rates pitched somewhat above the Consensus minima.

Major convertible currencies with market interest rates which fall below matrix levels are known as low interest rate currencies (LIRCs). At present they include the Yen, Deutschemark, Swiss Franc, Dutch Florin, and the Austrian Schilling. The LIRC Consensus rates are determined by reference, not to the SDR rate, but to government borrowing rates for the currency in question plus a margin of 20 basis points. They are revised monthly and the following table shows the rates extant in October 1984.

Currency		LIRC Rate
Sterling		13.55%
US Dollars		14.58%
Deutschemark	(i)	8.72% (SDRs 40m or less)
	(ii)	9.52% (SDRs over 40m)
Dutch Florins		10.65%
Swiss Francs	(i)	7.25% for periods of up to 8 years from the point at which the fixed interest rate is committed.
	(ii)	7.5% for periods in excess of 8 years but not more than 10 years from the point at which the fixed interest rate is committed.
Canadian Dollars		13.56%
Finnmark	(i)	11.65% up to 5 years credit
	(ii)	11.9% 5-8 1/2 years credit
Austrian Schilling	(i)	9.325% 2-5 years credit
	(ii)	9.075% 5-8 years credit
	(iii)	9.2% over 8 years credit
Yen		8.0%

NOTES

(i) In contrast to other LIRCs the fixed interest rate for officially supported Swiss Franc loans may not be committed prior to date of contract.

(ii) The LIRC rates for the Dutch Florin Swiss Franc and Japanese Yen may be revised in the course of any monthly period.

It is open to any exporter who is prepared to invoice his goods in these currencies to seek support from his national ECA where banks are unable to arrange finance on more favourable terms. However, the LIRC rates are comparatively high and in practice the banks have been able to dispense with the support of the ECAs in a number of recent transactions involving these currencies.

Foreign Currency Financing An important development, pioneered by ECGD in the late 1970's, was the provision of interest rate support for transactions invoiced and financed in foreign currencies, i.e. other than the domestic currency of the exporting country. Because of the salutary effect it has on the exporting countries' balance of payments, ECGD's lead was followed by others including Mediocredito in Italy and, more recently, by the BFCE in France and ICO in Spain. Under these schemes support is normally available only for major convertible currencies and, in practice, the bulk of the business transacted has been denominated in US Dollars, although Deutschemarks, Swiss Francs and, very exceptionally, Canadian Dollars and Singapore Dollars have been supported by the ECA's where it has been deemed necessary to win a major piece of business.

Mixed Credits Most of the European countries operate arrangements for linking conventional export credits with soft loans which constitute a part of their international aid budget. Although the allocation of soft loans to individual projects is made by the aid agency in the exporting country the ECA's usually have a hand in sponsoring a mixed credit and some countries (notably France and Japan) were quick to use this device to secure important orders particularly among the less developed nations.

APPLICATION TO THE MINING INDUSTRY

By now the uninitiated might well be asking what relevance all this has to the mining industry. The answer is that during the export credit bonanza of the 1970's several major mining projects benefited from support from the ECAs and some of them would never have been started without it. Through the provision of guarantees and indemnities against the political and commercial risks inherent in this type of transaction, the ECAs enabled finance to be raised which would not otherwise have been available. The key, of course, was for the project sponsor to purchase capital equipment (and associated services) from the countries of ECA's who were prepared to accept the exposure. This sometimes resulted in orders being spread among a large number of countries in order to optimise the export credit input.

The ECAs financial support function was another major inducement to bring them into a transaction. Subsidised interest rates comprised an important ingredient in any well structured package; the somewhat rigid repayment terms associated with such financing being outweighed by the financial advantage. An ideal repayment profile would typically comprise development bank funds for longer maturities, export credits for medium term maturities and commercial lending for the balance. Under the terms of the OECD Consensus it was even possible to include in the export credit an element of finance for local costs on the same preferential terms as were made available for imported capital goods and services.

Of course the usefulness of official export credits does not stop with commissioning of a mine and associated processing plant. There is usually a continuing need for plant and equipment in the shape of replacements, spares and updating technology during the life of the mine. Worn out plant and equipment and obsolescent processes create an ongoing need for specially tailored financial facilities and the ECAs have had an important part to play in meeting this requirement.

RECENT DEVELOPMENTS ON THE EXPORT CREDIT SCENE

Together with the international development banks and aid agencies the more aggressive ECA's have been prepared to accept exposures in countries which have been foresaken by the commercial banks. In these circumstances it is perhaps not surprising that, almost without exception, they have sustained heavy losses over the past three to four years. Indeed without state backing several of them would have been bankrupt by now.

In an attempt to stem their losses they have been obliged to restrict cover (or even "come off cover") for many developing countries. The restrictions may take the form of strict limits to the amount of business they will entertain in a particular market, curtailment of the percentage of cover and term of the credit and the imposition of significant increases in insurance premiums. All this has made it much more difficult to mount major projects in developing countries. The greater part of Latin America, Africa, and even some parts of the Middle and Far East are now out of court. Of course Australia, North America and most of Western Europe remain persona grata from the ECA standpoint as are Algeria, India, Malaysia and The Peoples' Republic of China; several of which are of potential interest to the mining industry.

The developing countries still gain a considerable advantage from accessing the Category 2 and Category 3 interest rates which remain substantially below market for such countries. Where the national export credit agencies have no appetite for business in these countries it is sometimes possible to arrange some protection through commercial insurers, for example, the American Insurance Group (AIG) and Lloyds of London. Where commercial banks and the international development banks are participating in a major financing, it may also be possible to persuade some of the more flexible agencies to accept the project risk. This is most likely to be a plausible approach where they can be insulated from the country risk through long term take or pay contracts the income from which is paid into offshore accounts and applied first to service the commercial and export credit loans. (Examples of this are to be found in the Cuajone copper project in Peru and the ill starred Tenke Fungarume project in Zaire).

In the developed markets, on the other hand, the consensus rates are not, at present, likely to be attractive to the better quality borrowers. As consensus rates tend to follow, but lag some months behind, movement in the market rates, they are attractive when market rates are rising and relatively unattractive when they are falling. However, it is important to note that the Consensus rate prevailing at the time the ECA commits to support the financing is applied for the duration of the loan. Thus it is possible for the buyer, through judicious anticipation of market trends, to "lock in" at a favourable rate particularly when market Consensus rates are in the lower quartile of the interest rate cycle.

However, the conclusion is inescapable that the ECA's financial support programs have been seriously devalued by the recent changes to the Consensus. Fortunately "necessity is the mother of invention" and this old adageholds true even in the field of export credit. Over the past year or so a number of innovative ideas have emerged for combining export credit insurance with market based financing instruments, the aim being to produce funds on more favourable terms than would otherwise be available to an overseas buyer. For instance the export credit guarantee, which is effectively a sovereign commitment to honour the debt from the country of the exporter, should produce funds from the private placement market, in most of the major convertible currencies at very fine rates. This could be particularly advantageous to borrowers with second and third tier credit ratings.

Other variations on the theme include export credit/leasing facilities and hybrid export credit/currency and interest rate swaps. Variations of this sort must be tailored to suit the transaction. Of course, they will not always be appropriate, but the object is to extract as much legitimate advantage as possible from the export credit programmes, development agencies and the money markets. These are skills which the banking fraternity is rapidly developing and sponsors of major mining projects should be sure to explore all these possibilities with their banks before concluding their financial arrangements.

SOURCES OF FINANCE

| COUNTRY | AGENCY | INSURANCE/GUARANTEES |||||| FUNDING ||
| | | CREDIT RISK || EXCHANGE RISK | OVERSEAS INVESTMENT | PERFORMANCE ETC. BONDS || LOCAL CURRENCY | FOREIGN CURRENCY |
		EXPORTERS	BANKS			UNFAIR CALLING	BANK G'TEES		
Belgium	OND COPRAMEX	X	X	X		X		X(1)	X
France	COFACE BFCE	X	X	X	X	X	X	X	X
Germany	HERMES AKA	X	X	X		X	X	X(2)	
Great Britain	ECGD	X	X	X(3)	X	X	X	X	X
Italy	SACE MEDIOCREDITO CENTRALE	X	X	X	X	X(4)	X	X	X
Netherlands	NCM EFA	X	X	X	X	X		X(5)	
Spain	CESCE ICO BANCO EXTERIOR	X	X	X				X	X
Sweden	EKN SEK	X	X	X		X		X	X
Switzerland	ERG	X	X			X			

NOTES

(1) COPRAMEX - does not subsidise export credits but it is possible to obtain subsidy on a case by case basis from the IRG and agency of the Belgian State.

(2) AKA - provides funds at a fine market rate but does not subsidise from public funds.

(3) Limited to exchange losses stemming from buyers default.

(4) Limited availability.

(5) Not generally available - applied selectively.

For more detailed information on the range of services of the ECAs refer to Chase Guide to Government Export Credit Agencies.

REFERENCES

The Chase Manhattan Bank, N.A. 1984, <u>Chase Guide to Government Export Credit Agencies</u>

11.19

NORTH AMERICAN EXPORT CREDIT PROGRAMS

SUPPLEMENT

INTRODUCTION

Besides the European export credit agencies covered in Ted Rides' paper in this Chapter, other agencies are also active (see Appendix) but the most notable for mining projects are The Export-Import Bank of the United States ("US Exim") and the Export Development Corporation ("EDC") of Canada. Mexico's Fomex is basically a short-term trade-finance fund administered by the Central Bank.

US Exim

This independent agency can provide loans, guarantees, and insurance. Like all export-credit agencies, its mandate is to support the export of goods and services from the US as a means of competing against other countries' products. If a piece of mining equipment can only be made in the US, the US Exim may very well not provide any financial assistance. (It is important not to place the order for any item for which export credit is sought because, once ordered, the situation is no longer competitive in US Exim's eyes.)

Direct Credits and financial guarantees ("buyer credits") are commonly in amounts greater than $5 million and for terms longer than six years. An example here is the $48.5 million provided to PT International Nickel Indonesia (see Robert de Gavre's paper in Chapter 12).

US Exim usually requires a 15% downpayment and, therefore, finances up to 85% of the goods; less if the US content is lower. The dollar funding can be US prime, London Interbank Offered Rate (LIBOR), or other sources. Exim can fund in either or both of the following ways:

1. Under a US bank Letter of Credit ("L/C")

2. Reimbursement to the buyer/borrower via a US bank account.

US Exim will, without charge, issue a preliminary commitment. It will, however, need to approve the various credit agreements and prefers to tie other parallel financings into one loan agreement. US Exim's total commitments are large, up to $58 billion in aggregate under various programs.

US Exim has been the world's leader in medium- to long-term minerals project financings/ co-financings but suffers from (i) aggressive foreign export-credit competition whereby other nations mix in soft-loans or aid programs ("mixed credits") to increase the percentage financed and lower the interest margin and (ii) an imbalance in its own cost of funds.

A common US Exim package would be as follows:

15% Banks under US Exim guarantee, floating rate over prime or LIBOR.

35% US Exim funding, fixed rate

35% Pefco funding, fixed rate

85% Drawn down pro-rata under US bank L/C

15% Cash down payment (may have been financed within a project/corporate credit)

100%

Repayment usually begins six months after planned start-up in equal semi-annual instalments, with the banks usually paid out first, Pefco next, and US Exim last.

Medium-term credits ("supplier credits") are issued for US equipment exports for terms up to five years, again for 85% of the equipment costs although the exporter is expected to bear at least 10% out of that 85% for its own risk. The buyer is required to be a first-class credit such as some governments or highly credit-worthy companies. Otherwise a bank L/C would be required. (This may be wrapped into an overall project-finance package for a minerals development. The project-finance banks provide an L/C to US Exim instead of funding that portion themselves.)

US Exim guarantees 100% of the political risk and either 85% or 95% of the commercial risk for the US exporter.

<u>Discount loans</u> are a means for US Exim to get US banks to fund exports under a one-time buy-back or discount arrangement. US Exim commits to fund the loan at any time at 1% below the US bank's usual price for funds, e.g. prime, but not less than the US Federal Reserve discount rate. The 85% financing criterion applies.

US Exim guarantees or other government insurance is optional. The 1% p.a. limit on interest spread means that other fees, costs, and expenses are baked into the front-end fee.

<u>Engineering Multiple Program</u> is a new vehicle to enable US engineering companies to offer finance for 85% of feasibility studies and other engineering services.

<u>Pefco</u>, the acronym for the Private Export Funding Corporation, is owned by 54 banks, 7 corporations, and an investment bank. It can only fund under a US Exim guarantee of principal and interest. It usually does so at fixed interest rates. A commitment fee of 0.5% - 1% is charged on the unutilised portion of a Pefco commitment.

<u>FCIA</u> is the initials for the Foreign Credit Insurance Association, which is composed of some 50 US insurance companies. It insures payments for US exports against political and commercial risks. The main difference between Pefco and US Exim are:

1. Insurance is issued direct to the exporters who can raise finance by assigning the insurance policy's proceeds to a bank (plus providing a keep-harmless covenant to that bank.)

2. US Exim relies on the credit judgement of the financing bank, whereas FCIA does its own credit assessment under slightly different credit criteria.

Premium costs vary with the situation, percentage coverage, and the deductible policy at the time but vary from 0.15% - 0.9% for short-term (up to 180 days) and 0.7% - 4.7% for medium-term (up to five years) depending on the buyer and payment terms.

EDC

This Canadian Crown Corporation can issue export credit insurance, guarantees, and loans.

<u>Loans</u>: EDC can make supplier-credit loans on fixed and floating-rate bases to buyers of Canadian goods and services. In international bidding, excess amounts can be raised if it can be shown to provide benefits to Canada. EDC is quite flexible as to what constitutes "Canadian content."

EDC likes to work with other financing entities by way of parallel loans, complimentary financings, or participations in EDC loans. The large credit provided to the Tintaya copper project in Peru is an excellent example of EDC's role in this area. This loan was made up of (US$):

$85 million	EDC Supplier Credit - 15 years
15	EDC Buyer Credit (for local costs) - 15 years
<u>155</u>	<u>Bank syndicate - 10 years</u>
US$215 million	Guaranteed by the Republic of Peru

Most of EDC's loans are in US$ but fixed-rate loans are available in Canadian $, deutschemarks, Swiss francs, and yen. Floating-rates are sometimes available also in French francs, Kuwaiti dinar, and pounds sterling.

EDC will work with the Canadian International Development Agency to provide mixed credits for infrastructure, e.g. complementing an EDC project financing, but only to match foreign competition.

<u>Guarantees</u> are provided to banks or other lenders who finance exports of goods and services as follows:

1. Under insurance by EDC, non-recourse supplier finance.

2. Performance security, e.g. against premature collection against a L/C for example.

3. Bid-bond guarantees.

4. Loan guarantees.

Under an EDC guarantee, funding usually costs 0.5% above LIBOR.

<u>Export-credit insurance</u> is provided for 90% of political and commercial risk for short-term (to 180 days) and medium-term (to five years) exports under a wide variety of programs.

<u>Forfaiting</u> or note-purchase facilities are also provided by EDC for credit periods of two to five years. EDC's letter of interest specifies the form of promissory note and bank guarantee required. A commitment fee is payable if a formal EDC offer is accepted. Upon signature of the contract and satisfaction of the contract conditions precedent, the notes are discounted or purchased by EDC at a predetermined discount rate and held by EDC until maturity without further recourse to the supplier.

CONCLUSION

Export credits from North American agencies are worthwhile considering as part of a project-financing package or as a base load of equipment-related financings for a mine expansion or for equipment replacements. It may be a key component is accessing political and commercial risk insurance for a minerals development in a "risky" or uncreditworthy country. Both US Exim and EDC are experienced with minerals project financings worldwide.

REFERENCES

Chase Manhattan Bank, "Export Credits", chapter in Trade Banking Group handbook, 1984

Dunn, A. and Knight, M., Export Finance, Euromoney Publications, London 1982.

Financial Times, "Export Credit Service", regular updating service, 1984.

APPENDIX

MAIN EXPORT-CREDIT ACRONYMS

Country	Agency/Insurance Entity
US	Eximbank, Pefco
Canada	EDC
U.K.	ECGD
France	BFCE, Coface
Germany	KfW, AKA, Hermes
Italy	Mediorcredito, SACE
Japan	Eximbank, EID

OTHERS

Country	Agency/Insurance Entity
Belgium	OND, Creditexport
Brazil	Finex, IRB
Mexico	Fomex
Singapore	ECICS
Austria	OKB, EFPS
Denmark	DEF, EKR
Finland	VTL, FEC
Hong Kong	HKECIC
Israel	IFTRIC
Korea	KExim
Norway	EF, GIEK
Sweden	EKN, SEK
Switzerland	ERG
Australia	EFIC
Iceland	ECF
India	ECGC
Ireland	ICI
Malaysia	MECI
Netherlands	NCM
Pakistan	ECGS
Portugal	Cosec
New Zealand	EXGO
S. Africa	IDC, CGIC
Spain	CESCE
China-Taiwan	Exim
Venezuela	Finexpo

11.20

WORLD BANK GROUP FINANCING

L. Hartsell Cash

Mining Adviser, World Bank
Washington, D.C.

INTRODUCTION

Created in 1944 to help rebuild those economies, principally in Europe, which were seriously damaged or destroyed during the Second World War, the World Bank--or to use its correct name, the International Bank for Reconstruction and Development--began its operational phase by making a loan to the Government of France to accomplish two tasks: (i) to import coal for space heating purposes; and (ii) to buy new machinery and equipment to help rehabilitate its coal mines. Once the thrust of reconstruction in postwar Europe was largely accomplished, the World Bank Group (for today the World Bank has three separate lending windows) turned to mobilizing funds to improve living standards in developing countries, which remains its primary goal today. Over this period of time, the World Bank Group has continued to assist the world mining industry in various ways and, since 1955 the World Bank Group has financed mining or mineral processing projects with loans/credits totalling $1.5 billion.

The World Bank Group consists of two institutions--the World Bank itself and the International Finance Corporation (IFC), which was founded as a separate institution in 1956 and three separate lending windows--the World Bank, the International Development Association (IDA), and IFC. The World Bank loans are provided at commercially-related interest rates but have grace periods usually of up to 3 to 5 years and repayment periods ranging from around 15 to 20 years. Since the World Bank raises its operating funds on the international capital markets in varying currencies and in varying interest rates, it relends a basket of currencies at floating interest rates which are fixed every six months to reflect the Bank's own borrowing cost. At the present time, the interest rate is just below 10% per year and in addition, a one-time front-end fee of 1/4 of 1% is charged. Also, a commitment fee of 3/4% per year is charged on the undisbursed amount of the loan itself. For loans to industrial and mining projects, it is normal to charge an additional guarantee fee of 1 to 1.5% in addition to the annual interest rate. This guarantee fee is payable to the government of the country in which the project is carried out since, by its charter, the World Bank is required to have the guarantee of the host government for each loan made in a specific country. By far the greatest source of lending funds within the World Bank Group is that of the World Bank itself and for the fiscal year ending June 30, 1984, the World Bank approved loans totalling $12 billion.

The second lending window within the World Bank Group is IDA which performs the same functions as the Bank itself and has the same staff. Nevertheless, its loans (called credits) are made on much softer terms and are made to the poorest nations only. IDA funds are subscribed by the wealthier countries, usually as grants, and these are relent as credits to those countries with the lowest per capita income. IDA credits are made for 50 year periods, including 10 years of grace, to the government itself. The credits are interest free, but there is a service charge of 3/4% per year on the amount withdrawn and outstanding to defray the administrative costs of IDA. Credits approved during fiscal 1984 totalled $3.5 billion.

The third lending window is IFC. While IFC is a separate corporation and operates quite separately from the World Bank and has its own staff and regulations, it works very closely with the World Bank on its overall approach to the mining sector and to a particular country. IFC's mission is essentially to supply venture capital and to stimulate development of the local private sector in capital markets as well as to promote the international flow of private capital to developing countries. Unlike the World Bank and IDA, the IFC does not lend directly to governments nor does it require a government guarantee or underwriting of its loans. IFC lending is at a fixed interest rate which is currently about 14% per year. IFC finances all types of commercial enterprises, both public and private, and IFC can make investments direct in equity and can organize and lead loan syndications. These last two functions are not carried out by the World Bank itself. Total loans and investments approved in Fiscal Year 1984 equalled $700 million.

It is the goal of both the World Bank and IFC to mobilize as much outside capital as possible to help finance its projects and thus the size of their loans in relation to total project costs normally run between 15 and 30%. In smaller projects or projects where it may simply be impossible to mobilize outside private capital, their participation may be somewhat higher.

The procurement requirements of the two institutions differ as well. It is the practice of both the World Bank and IDA to require that their loans or credits be expended to purchase equipment procured under international competitive bidding. This differs from IFC which can expend its funds for any types of equipment provided they are purchased at economic prices and fit the requirements of the project. In the case of each of these three lending windows, however, the goods procured must come from a World Bank member country or Switzerland, or Taiwan (China) if they are to be financed by the World Bank Group.

TYPES OF PROJECTS

The World Bank participates in the financing of three basic types of investments: (i) the development of greenfield installations and expansion of existing mining operations, (ii) the restructuring and rehabilitation of ongoing mining/processing operations, and (iii) exploration and pre-investment work as well as technical assistance.

First, World Bank financing for mine development of greenfield installations or mine/plant expansion has been the traditional involvement of Bank financing demands of the sixties and seventies when mineral output in the developing countries more than doubled. The World Bank is often approached to finance solely a portion of an infrastructure component associated with mining ventures. As the viability of such infrastructure investment depends primarily on the viability and management of the mining component, the Bank--as a matter of policy, seeks a direct involvement in the mining component before considering financing of any associated infrastructure investment. It is recognized however that, in selected instances, there may be sound reasons for the Bank to finance infrastructure, a power plant for instance, without an investment in the mine itself.

The second type of project is what we call rehabilitation or restructuring investments. Such operations address cases where either low metal prices or financial difficulties force mining companies to trim and rationalize their operations, with the result that the efficiencies of their mines are lowered and operating costs increase. In such cases there is often a role for the World Bank Group to help developing countries focus on the source of the problems, help them obtain needed technical assistance, and perhaps take some difficult decisions to help resolve their problems. These rehabilitation efforts may be directed toward removing processing bottlenecks, strengthening and clarifying management's role, replacing old equipment, providing critically-needed spare parts, or decreasing production costs.

The third area of involvement by the World Bank is the financing of exploration, feasibility studies, engineering work and technical assistance. In recent years this has primarily involved exploration and feasibility work on coal deposits to ascertain the viability of mine-mouth power plants. On a selective basis, however, the Bank is willing to consider financing feasibility work for non-fuel minerals depending on the likelihood that a viable project will emerge. This assistance would be provided in order to attract foreign equity and loan capital and to improve the developing country's ability to negotiate realistic agreements with foreign investors. The Bank also finances technical assistance as required to help governments assess their country's mineral prospects and determine the most appropriate way of developing the sector or a particular mine.

ECONOMIC PRIORITY AND VIABILITY

World Bank lending takes place within the context of the Bank's overall economic and sector work. The Bank undertakes economic work and analysis on a regular basis for each of its borrowing member countries. For each country, the Bank prepares every one to two years an economic report which provides the basis for the Bank's own decision-making on financial and technical assistance. This report also serves as a planning tool for governments and as a basic document for other national and international lending agencies. The report describes and analyzes the country's overall economic situation, creditworthiness and principal problems and outlines a general order of development policies and sectoral priorities.

The economic report is generally supplemented by more detailed analysis of individual sectors and the relationships between them, which helps to clarify the relative importance of alternative projects in achieving the country's development goals. The general economic work as well as detailed sector work, permits a continuing dialogue between the Bank and each borrowing country, a better assessment of available resources (technical, human and financial), and a better definition of each country's development objectives. The Bank safety standards. In the case of large export projects this will include the associated railroad, port and other facilities. This appraisal includes a review of the project cost estimates including contingencies--for both capital and operating costs. Where feasibility work has not been undertaken or is considered inadequate, the Bank is prepared to consider assistance in financing such work where considered justified by project prospects.

A critical part of the technical preparation of a project is the project management and monitoring procedures. Successful project implementation depends on sound institutional arrangements, good procedures and capable, experienced staff. Along with adequate technical design and specification, a strong and capable management is essential for a viable project. Management must be able to adapt to changes in the market, technology and the country's political situation and should be given the autonomy to operate independently of direct or indirect government intervention.

FINANCIAL VIABILITY

The project must also be financially viable and should not require subsidies to survive. The financial viability of the project is assessed in terms of several related measures including expected profitability, cost competitiveness with other producers, financial strength of project financial structure including completion arrangements and sales contracts. While the Bank Articles of Agreement require that a government guarantee be given on all IBRD loans, the Bank basically looks to the soundness of the project and its sponsors, and to project completion agreements rather than the government guarantee to ensure the loan is repaid. The expected financial rate of return of the project is estimated taking into account the market and other risks of the project and the project's cost competitiveness with existing producers. The financial position of the sponsor is also examined since the sponsor must be in a position to cover the early years of negative cash flow until the project reaches full production and positive cash flow. The cash flow of the project should be sufficient for debt servicing, as would be appropriate in any commercial lending.

Accordingly, the basic financial structure of the project (i.e., the relative proportion of debt and equity financing) must be sound. The project agreements will typically include guarantees from the project sponsors or other entities to ensure the provision of additional funds to meet cost overruns and close any financing gap. The basic project structure and agreements should provide reasonable risk sharing between the parties involved. If project management is being undertaken by an outside group the Bank likes to see that group with a substantial equity share in the project, so that their earnings from the project are closely linked to the projects performance. Finally, marketing contracts are also desirable especially for bulk export products such as coal, iron ore and bauxite even though such contracts may have only limited enforceability in practice.

OWNERSHIP AND MANAGEMENT

The World Bank will finance projects in the public, private or mixed sectors provided there is sound and experienced management to plan and implement the project. There should be a clear delineation of the duties and responsibilities of the Board of Directors and of day-to-day management. Experienced and capable personnel should be available to head the principal departments--operations, sales, finance, planning, engineering and administration.

In cases where the principal sponsor is a foreign corporation, it is essential that there be a suitable local partner-- either public or private--with a financial stake in the operation and with a defined role. It is highly desirable that this be an active partner with local business experience and, hopefully, good contracts within the local banking and government groups.

Attention will be paid to plans for attracting and training local personnel in all phases of the operation and, where necessary, suitable training should be established. If training is not available locally for highly technical positions, a plan for training abroad should be formulated.

MARKET AND MARKETING

It is essential that there be a ready market for the mine's output. Sponsors are expected to have carried out sufficient market research to clearly establish the market for production. The Bank's appraisal of commercial prospects will entail full review of market prospects, the details of the marketing plan or sales arrangements, and the organization necessary to effectuate such marketing efforts. In most cases it will be necessary to enter into sales contracts for a substantial portion of the expected output.

SPLIT OF BENEFITS

In cases where a foreign entrepreneur proposes to open a new mine with related infrastructure in a developing country, the Bank analyzes the project and the related legal arrangements to determine whether they are likely to lead to a stable long-term collaboration which, in terms of both risks and benefits, is fair to both parties. This will entail review of financial projections under a wide range of scenarios so that an equitable arrangement can be achieved.

COFINANCING

Both the World Bank and IFC will assist the project sponsor in securing outside financing from other international financial institutions, bilateral sources, commercial banks or equipment suppliers. In large, complex projects it may be desirable to employ an investment advisor to assist with this task, and the World Bank Group will work closely with them in completing the financing plan.

Other lenders have found that such cofinancing offers several advantages. For official agencies cofinancing is an efficient way of combining professional expertise. Export credit agencies can be satisfied that their resources will be applied to well-conceived and appraised projects and commercial lenders can be assured that loan proceeds will be used for the intended purposes. More generally, the risks attached to lending in developing countries can be reduced through cofinancing. The arrangements for association with the World Bank reduce the risk that the borrower may not repay even if the project goes well.

Borrowers also find cofinancing advantageous. In particular, it gives them greater assurance that high-priority investments can be financed on the best terms available. In addition, cofinancing can improve their access to each source of funding--official, export credit, and commercial.

In an effort to make its cofinancing effort more effective, the World Bank has recently begun to use three new cofinancing instruments--direct participation in the later maturities of commercial loans, guaranteeing of the later maturities of commercial loans, and taking contingent participation in the final maturity of a commercial loan designed with a fixed level of installments combining floating interest rates and variable-principal repayments. It is hoped that these will enable the Bank to attract more cofinancing to its mineral projects in the future.

11.21

INTERNATIONAL FINANCE CORPORATION

SUPPLEMENT

INTRODUCTION

The International Finance Corporation (IFC) is affiliated with the International Bank for Reconstruction and Development (World Bank) but operates with a separate staff and funding. It supplements the activities of the Bank by providing, to private business, equity and loans without government guarantees in whatever form and combination most suited to the project. IFC frequently complements the Bank's activities by utilizing, for example, the Bank's public sector infrastructure activities to support IFC's private sector projects. In the case of local financial institutions especially, IFC collaborates with the Bank by providing technical and financial assistance.

IFC is the world's largest multilateral organization which finances private sector business in developing countries. IFC's fundamental purpose is to promote the economic development of its developing member countries through the support of the private sector. Of the 124 member countries of IFC, more than 100 are developing countries.

Since it began operations in 1956, IFC has been associated with more than 2,000 companies and financial institutions in supporting over 700 business ventures in more than 80 countries. The total capital costs of these projects exceeds $26,000 million. In the 1984 fiscal year, IFC approved 62 projects in 37 countries with a total project cost of $2,473 million.

IFC will make investments only if sufficient capital cannot be obtained on reasonable terms from other sources. IFC often serves as the catalyst for a project by encouraging other investors—from inside and outside the host country—to make their own equity investments along with the local sponsor for a particular project.

IFC'S SPECIAL ROLE

The significance of IFC's role extends beyond just its financing and technical assistance: IFC's capability to raise investor confidence in a project and to help all the participants to arrive at mutually satisfactory investment arrangements is its distinctive character.

Direct Investments

IFC invests in all types of projects both large and small. Heavy industrial ventures not only supply essential materials and equipment for other domestic enterprises, but also earn foreign exchange through exports. In recent years, increased emphasis has been placed on agriculture and agribusiness products, aquaculture, food and food processing, and other renewable and non-renewable resources such as metals and energy.

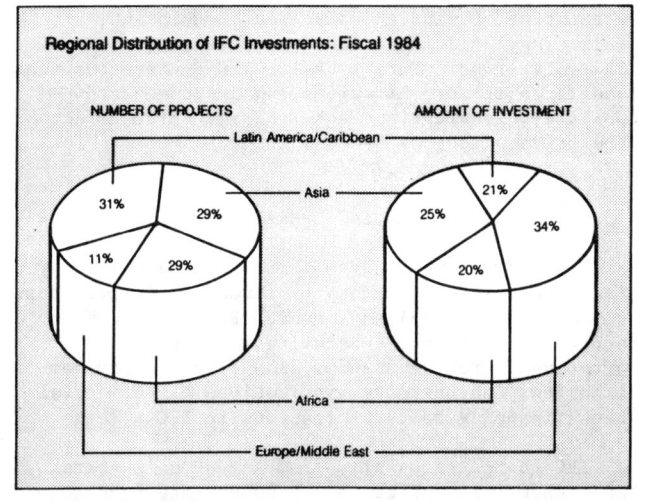

Regional Distribution of IFC Investments: Fiscal 1984

NUMBER OF PROJECTS: Latin America/Caribbean 31%, Asia 29%, Europe/Middle East 29%, Africa 11%

AMOUNT OF INVESTMENT: Latin America/Caribbean 21%, Asia 25%, Africa 34%, Europe/Middle East 20%

Raising Capital

An overall objective of IFC is to assist entrepreneurs and others in the private sector raise the needed capital for business ventures. IFC does not compete with, nor seek to replace private initiative and capital. Rather, IFC helps to raise funds in order to undertake projects that would otherwise be held back by the lack of adequate funding.

IFC can do this in several ways: IFC can invest in equity and make loans for its own account. But beyond that, IFC can reach out into the international capital markets in search of additional capital. Because of IFC's 25 years experience and record, outside investors who might otherwise be deterred from investing in a project are often encouraged to do so in parallel with IFC or by taking up syndicated portions of IFC investments. For every dollar of IFC participation, for example, five and one-half dollars were invested by others.

IFC helps local investors, who might be unfamiliar or inexperienced in dealing with the international financial community, find and utilize sources of capital that otherwise might not be possible. In this way, IFC can be a bridge between the international capital markets and the local businessmen in a developing country.

Technical Assistance

The Corporation has a broad mandate to help its member countries and individual enterprises by providing technical assistance based on its long experience and accumulated know-how. The Corporation's international staff--in which 66 countries are represented--has accumulated much financial, legal and technical expertise relevant not only to investment opportunities but more so in finding the solutions to investment problems in the developing world. IFC can also draw on the World Bank's wide experience, as well as on its own contacts with financial institutions and development agencies.

IFC frequently provides specific technical assistance while appraising and monitoring individual projects. This help may involve advice on accounting, financial management, equipment, marketing, administration or any other aspects of the enterprise. More broadly, IFC advises business and financial organizations and governments on policies, institutions and programs which affect the private sector.

FINANCING OPERATIONS

The Corporation's versatility and adaptability is evident in the variety of financing it provides. Depending on a project's need, IFC can invest in equity, make loans, underwrite securities offerings, provide standby financing--in any combination, and organize syndications of commercial bank financing as participations in IFC's loans.

IFC loans are usually from seven to 12 years and, when held for its own account, are made at fixed rates. In mid 1984, IFC commenced offering floating rates as well.

The syndicated portion of IFC's financing is normally made at floating rates. Loans are usually denominated in U.S. dollars but can be made in other currencies depending on the project's requirements.

IFC tailors the nature and amounts of its financing to the project's needs. In keeping with its private sector orientation, IFC neither requires nor accepts government guarantees on its investments. It does, however, assure itself that the government has no objection to the project or to IFC's involvement in it. IFC also seeks to ensure that foreign exchange will be available to service IFC's loans and other repatriation requirements.

Although the Corporation emphasizes the private sector, it may invest in "mixed" government/private enterprise projects when they contribute to private sector development.

Capital Resources

IFC's capital, subscribed by 124 member countries, was approximately $544 million as of June 30, 1983. The IFC Board approved a capital increase of $640 million in mid 1984. The resources available to IFC are supported by accumulated earnings of $230 million, by repayments and by sales to others of IFC investments. The Corporation can borrow from the World Bank and other sources up to four times the amount of its unimpaired subscribed capital and accumulated earnings -- about $2,990 million. A new five-year program announced in mid-1984 projects IFC investing $7,000 million in $30,000 million worth of private sector ventures in developing countries around the world.

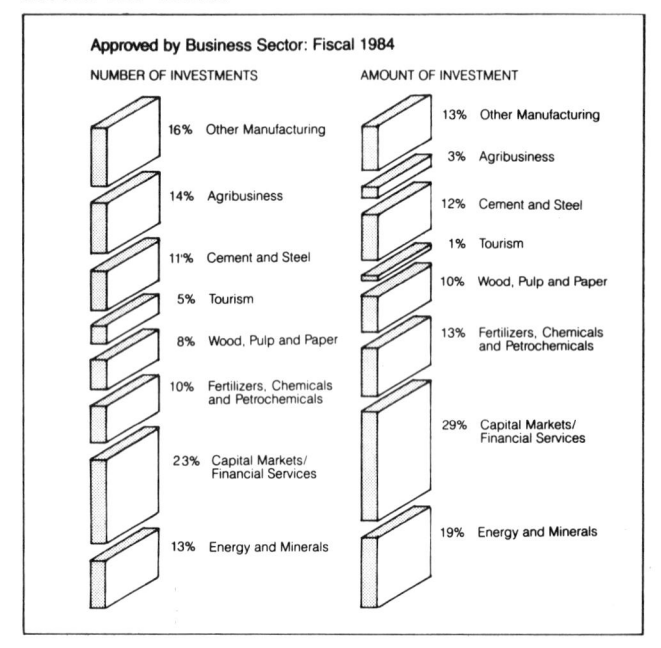

Approved by Business Sector: Fiscal 1984

NUMBER OF INVESTMENTS	AMOUNT OF INVESTMENT
16% Other Manufacturing	13% Other Manufacturing
14% Agribusiness	3% Agribusiness
11% Cement and Steel	12% Cement and Steel
5% Tourism	1% Tourism
8% Wood, Pulp and Paper	10% Wood, Pulp and Paper
10% Fertilizers, Chemicals and Petrochemicals	13% Fertilizers, Chemicals and Petrochemicals
23% Capital Markets/Financial Services	29% Capital Markets/Financial Services
13% Energy and Minerals	19% Energy and Minerals

PROJECT PROPOSALS

Requests for IFC assistance come from a wide variety of sources: it is often from an entrepreneur in a developing country who wants to establish a new business or expand an existing enterprise. It may be from a company that IFC has already assisted and wants to expand further or enter a new field.

Companies and financial institutions in the industrial countries also approach IFC with their proposals for joint ventures with developing country sponsors. On some occasions, a government may suggest that IFC look into a particular sector it wants developed by private investors.

There is no standard form of application for IFC

financing. However, certain preliminary information is needed to enable IFC to decide whether an investment proposal warrants consideration:

--a brief description of the proposed project and the technology involved;

--the estimated total cost and amount of financing needed, with an indication of the proportion to be provided by the sponsors;

--its legal status and financial history;

--information on cost and availability of raw materials and other inputs;

--the market envisaged, potential competition and export opportunities.

Guidelines

IFC will assist a project only if there is a benefit to the economy of the host country, if there is a prospect that the venture will earn a profit, if there is provision for immediate or eventual local participation and if the host government does not object.

IFC never invests alone; IFC expects to supplement and mobilize private capital, not replace it. IFC will make investments in a wide variety of sectors -- manufacturing, mining, tourism, agribusiness, financial institutions and others but will not invest in such activities as land speculation, trading companies and luxury goods.

IFC invests for its own account generally no more than 25 percent of the project cost. Normally its investments are between $1 million and $50 million.

Source: IFC Basic Information pamphlet and 1984 Annual Report.

HOW IFC WORKS WITH COMMERCIAL BANKS

IFC has cooperated with financial institutions, primarily commercial banks, since the early 1960s. The most active program involves the syndication of participations -- that is, an arrangement whereby banks are offered "shares" or "participations" in an IFC loan with the banks taking the same credit risks, on a pro rata basis, as IFC. Under this technique, the loan agreement is signed by IFC and the borrower; a participation agreement by IFC and the participating commercial banks. Unlike a normal Eurocurrency credit agreement, the participating bank is not a direct lender -- it has no direct contractual arrangement with the borrower.

Loan Agreement

The IFC loan agreement normally provides for two portions, one for the account of IFC with terms, including fixed interest rates, and the second for the account of participating banks with normal market conditions. These conditions generally include a floating interest rate, usually tied to the six month London Interbank Offered Rate (LIBOR), together with a spread, commitment fee and front end participation fee which are negotiated in each instance. The maturities run out a minimum of five to seven years -- occasionally longer -- depending on the market's preference for a particular country. Often the IFC portion of the loan will comprise maturities longer than those applicable to the commercial banks. If the loan is secured, for example by a mortgage, and should it become necessary to foreclose (and fortunately IFC has not had much experience in this connection), the participants would share on a pro-rata basis in the proceeds resulting from any such foreclosure proceedings. The commercial bank participation is otherwise without recourse to IFC.

IFC, as the lender of record, will administer the loan. Its responsibilities include sending out billings, collecting payments, promptly distributing to participants their pro-rata shares, overseeing the progress of the project and administering the loan documentation. The various legal rights of the participants are set forth in the documentation, and, at a minimum, there is always close and continuing consultation between IFC and the respective participants in the administration of the loan.

This participation program is important to IFC, both in its role as a catalyst in furthering the flow of funds from capital exporting to developing countries and also in assisting projects which would otherwise be too large for IFC to consider for its own account. To illustrate this point, out of total cumulative commitments of $4.5 billion since IFC's establishment in 1956 through fiscal year 1983, participations placed with financial institutions totaled $1.7 billion, or approximately 38 percent. The annual volume of participations has increased significantly in recent years, from $80 million in fiscal year 1976 against $419 million authorized last year.

Why has IFC's participation and syndication business been successful? Essentially because the banks see a convenient way of participating in a profitable, expanding market -- namely, corporate project lending on a selective basis in developing countries. IFC employs substantial staff and financial resources in appraising projects prior to the time the Corporation itself decides to commit and a commercial bank is asked to participate -- and IFC's reports are available to prospective participants. Participating banks are aware that these appraisal practices represent a key factor explaining IFC's satisfactory loss ratio.

Loan Syndication

IFC will organize syndications in its loans in various ways. In most instances, IFC will identify prospective participants based on its own experience in the market, in addition to existing relationships between the borrower and commercial banks. Initially, IFC informally outlines the project to a bank and, if interest is shown, follows this up with a copy of its Appraisal Report, a Term Sheet identifying the principal conditions of the loan, including pricing, together with a firm offer. In most cases, IFC would manage the transaction. In a few other cases, especially where the amount of funds to be raised from commercial banks is substantial, the borrower, with IFC's advice, may request one or more major banks to use their strength, again in consultation with IFC, to syndicate the IFC participation in the loan. Once the financial package is arranged, IFC would administer the loan. An example of this arrangement was the recent syndication with 15 Japanese banks in support of the El Dikheila steel project in Egypt. Fuji Bank acted as lead manager, Bank of Tokyo and Sanwa Bank as managers and Dai-Ichi Kangyo Bank and the Industrial Bank of Japan as co-managers.

Other Means of Cooperation

IFC works with commercial banks in other ways. For example, IFC and banks may independently, but in parallel, finance a particular project (with separate documentation, not via the participation agreements described below).

In other instances, a portion of the financial plan might include, in addition to the IFC financing, credits provided with the support of a national export guarantee organization, such as the Export-Import Bank of Japan. Again, close coordination would be in each lender's interest.

Also, on a few occasions, banks will work with IFC in connection with equity or quasi-equity investments, including the purchase of subordinated, convertible debentures.

One additional area of collaboration is of prime importance to IFC -- namely, the referral by commercial banks to IFC of projects on behalf of their clients in order to obtain financing for a portion of the funds required. IFC attempts to make clear that its relationships with commercial banks do not always involve IFC taking the initiative vis-a-vis the banks, i.e., it is not a one-way street situation.

CONCLUSION

"Co-financing" and other corresponding programs are receiving increasing attention -- both from the major international agencies and also a growing number of internationally minded commercial banks. A result may well be that a larger portion of future project lending to developing countries will be carried out by banks in partnership with international institutions such as the IFC. It is in the interest of commercial banks to be familiar with the specific features of the available programs, so as to be able to benefit should the right project present itself for joint collaboration.

Source: IFC Pamphlet

11.22

LEVERAGED LEASING FOR THE MINERALS INDUSTRY

Rodney G. Ravenscroft

Lease Underwriting Limited, Sydney, Australia

Leasing is not a new phenomenon. Experts have identified its origins as early as biblical times. It is a method of finance that has grown rapidly and is now used in most countries, including those of South East Asia, South America, the Middle East and even the People's Republic of China.

Internationally, the leasing world divides fairly easily into two groups and this is determined, by and large, according to whether a country has Anglo-Saxon origins or not. Most of the Anglo-Saxon countries have seen tax-oriented leasing become a sophisticated means of financing the acquisition of large and costly items of equipment and large projects.

Countries such as France, Spain and Japan have a highly developed leasing industry but it is not tax-oriented. This paper will not address leasing of this kind.

LEASING - THE CONCEPT

Wide acceptance of leasing as an alternative for financing capital equipment has grown with the recognition by the users of equipment that it is use and not ownership that generates profits. Whilst this seems obvious, memories of repossession of assets during the Depression and attitudes towards leasing as "last resort financing" have raised strong philosophical resistance that has only recently begun to be overturned.

A financial lease is a commitment by way of contract between a user of equipment (lessee) and a financial owner (lessor) under which the lessee makes a series of payments in return for use of the equipment. The lessee is acquiring economic value associated with ownership but not ownership of the equipment itself - that remains with the lessor.

A lease is a method of financing the use of an asset; other methods finance the assets themselves.

It is the taxation aspects that then become important. The lessee is giving up some or all of the taxation benefits of ownership such as depreciation and investment tax credits. The lessor claims these benefits, as he is entitled to do as owner, and takes them into consideration when determining the level of lease payments that are necessary to generate his required rate of return. These lease payments are, of course, lower than they would have been without those taxation benefits. The lease payments themselves are fully tax deductible.

This is tax shelter leasing. The lessor is acquiring taxation deductions that are applied fully and immediately against his other taxable income so as to reduce the amount of tax he is to pay to the revenue authorities.

It is this concept that makes financial leasing work.

Governments provide incentives by way of tax concessions to encourage capital investment. Governments presume that prudent capital investment will offer employment opportunities, generate sales and streamline economic activity. These, in turn, will lead to the collection of government revenue.

It makes no difference at all whether these incentives are claimed by a commercial or industrial owner/user or by a financial owner who makes equipment available for use by industry. Rather, the issue is to ensure that if an incentive is available, it is used so as to achieve its desired results.

No greater or lesser incentive is available for leasing and leasing does not misuse the taxation aspects of equipment ownership.

The critical issue is whether the full value

of these incentives is reflected in lease payments to the advantage of the lessee.

One of the most efficient vehicles to ensure that this does occur is the leveraged lease.

LEVERAGED LEASING

Leveraged leasing developed "to satisfy a need for lease financing of especially large capital equipment projects with economic lives of 10 to 25 years".[1]

Leveraged leasing allows a company the use of an asset at a cost usually significantly below the rate at which it could obtain conventional finance such as a commercial loan.

The three fundamental aspects of leveraged leasing have already been identified above. Leveraged leasing

- is tax oriented financing - that is, leveraged leasing is applicable in situations where the assets to be used entitle their owner to taxation deductions;

- has application in especially large financings - but none in small transactions; and

- provides a long-term form of finance.

In a leveraged lease transaction the lessor provides only a portion of the asset cost from his own resources, generally in the order 15-30% of cost, and claims tax benefits on the entire equipment cost. The balance of required funds is borrowed from institutional lenders specifically for the project and in accordance with the credit standing of the lessee or the project to be financed.

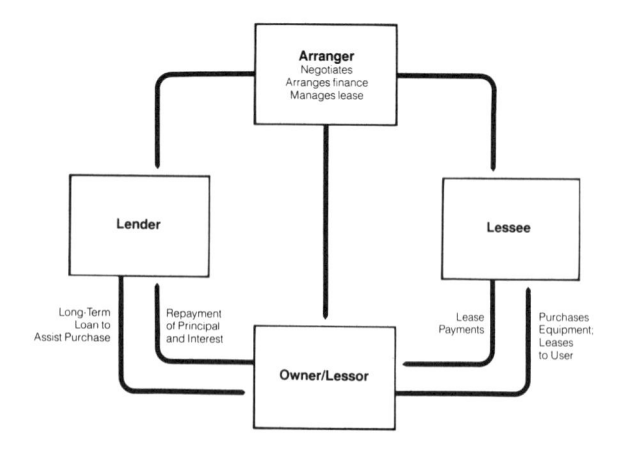

FIGURE 1: SIMPLE LEVERAGED LEASE STRUCTURE

Typical lessors are banks, finance companies and merchant or investment banks. Typical lenders are life insurance companies, pension or superannuation funds and, sometimes, banks.

A lessor can participate in his own right or, if the transaction is of sufficient size or risk, the lessor position can be syndicated amongst a number of organisations who participate through a partnership or a trust.

The lessor normally borrows funds from the lender on a non-recourse basis so as to emphasise reliance of lenders solely on the lessee. Non-recourse and limited recourse debt is well known in the financing of mining industry assets, in particular in project finance where the use to which funds are put may not give rise to a readily realisable asset. Lenders have to satisfy themselves as to the viability of the project and its likely cash flows, etc. and do not take a charge over other assets of the borrower.

This aspect of leveraged leasing also distinguishes the technique further from traditional leasing. In the more common forms of leasing, banks and leasing companies borrow funds from the public, from institutions or elsewhere based on their own credit, to enable them to acquire a myriad of assets for lease to their many, many clients. In a leveraged lease, the credit standing of the lessor does not impact the rate of interest at which debt funds are raised for the transaction.

A lessee's alternative methods of funding high-cost assets or projects are by debt from banks or institutions whether by debenture or otherwise, an equity raising by ordinary or preference shares, some form of innovative finance (including leveraged leasing) or a combination of these.

The determining issue as to which alternative is successful will be cost and other financial considerations (some of which are considered later).

Before this, however, a brief review of the financiers' positions should enable a more complete understanding of leveraged leasing.

THE LESSOR'S POSITION

Whilst there are many variations to the form of analysis used by lessors, all are basically similar. All forms are lengthy and complex and rely on the use of computer models.

The starting point is to calculate the annual after-tax cash flows that arise in the transaction. This is straight forward:

Taxable Income = lease payments received −
                 interest on debt −
                 depreciation
Taxable Income x tax rate = tax saved or
                            paid

This is then an input in the monthly cash flow analysis

Inflow − Outflows = net cash flow

Inflows are rentals, residual value and tax savings (if any, including investment tax credit). Outflows are debt amortisation (principal and interest) tax payments and the lessor investments.

The resulting series of cash flows can be solved to determine the yield on the lessor's investment.

Leveraged lease transactions generate taxation savings in the early years (because deductions are greater than rentals) and taxation payments in the latter years. For the lessor this is a form of tax deferment. Indeed, as a general rule, the longer the payment of taxes can be deferred, the more efficient will be the lease structure.

The consequence of these tax payments is that there are negative monthly cash flows later in the transaction. The analysis therefore includes a sinking fund to accumulate sufficient of early cash flows to provide for the later taxation payments. An earning rate applies to monthly sinking fund balances, but at a conservative, a supposedly "effortless" rate.

The lessor's after-tax cash flows for yield analysis are therefore those after establishment of the sinking fund (only those cash flows that are not in the sinking fund). There is more than one way of assessing the yield analysis [2] because of the existence of negative monthly cash flows or future liabilities. The most commonly accepted approach, the multiple investment sinking fund method, is premised on negative cash flows representing a future investment when there remain sufficient further positive cash flows to enable a return to be generated. If this is not so, then the negative cash flow is a liability and must be provided for.

Computer models arrange the cash flows so as to generate the greatest advantage in the transaction. The variables that give greatest scope here are those that enable different amortisation patterns with debt.

The analysis is relatively complex. Seemingly trivial variations in assumptions can result in quite significant changes in lessor yield.

THE LESSEE'S POSITION

It is important that a lessee has at least a basic understanding of the lessor analysis and certainly a detailed understanding of the assumptions the lessor has assumed for the yield calculation. This is necessary, amongst other reasons, because of the growing inclination of lessors to protect themselves by indemnity from many of the assumptions used to establish the yield.

The lessor's real risks are

. that the lessee will default, and in this case the lessor's recovery rights are subrogated to lenders;

. that his own assessable income will fall to a level at which he is unable to fully and immediately use the taxation losses that result from the leveraged lease investment; and

. the tax rate will change so as to either diminish tax losses or increase tax payments − both of which will reduce the yield from this investment. Of course opposite movements will improve the yield.

Lessors appear willing to accept the first two risks, are less willing to accept the tax rate risk, and are generally totally unwilling to accept any other change from the original basis on which they make the investment.

It is therefore critical that a lessee understands and agrees with the assumptions from which the lease payments are derived and has a good grasp of possible changes and the impact that these will have on the payments that he is to make under the contract.

THE LENDER'S POSITION

Lenders provide most of the funds in a leveraged lease. Many lenders and types of lender can participate, each for a different term and at varying rates of interest − if that best suits both the transaction and the lender.

Every form of debt available to a lessee can be incorporated within a leveraged lease. In particular, multi-currency loans, export-import bank loans and supplier or buyer credit facilities have been used.

When the lender assesses the prospect of a loan, the issues to be considered are the credit standing of the lessee, the term of the loan, the rate of amortisation and, sometimes, the likely future value of the asset being financed. These loans are specific and, in a case of lessee default, lenders can gain immediate access to the asset financed for realisation of part or all of the debt outstanding. If there is a remaining shortfall a lender then ranks as an unsecured creditor of the lessee.

The ability to take immediate action rather then sheltering under the complicated security umbrella of a general charge over a lessee's assets (often the position in a direct loan) is usually seen as a worthwhile advantage by lenders.

The Mining Industry

The creation of economic growth is a major goal of each nation and six basic categories of resource are necessary - natural resources, capital, labour, and technological, managerial and entrepreneural skills. Growth is dependent on international factors and the theories of resource differentials and comparative advantage are fundamental in this. They say, put very simply, that basic economic forces work towards out-flows from the countries in which resources are in relatively abundant supply into those countries where the supply is relatively poor and that such a movement is beneficial to both countries.

Of course nations erect barriers to hamper these flows when they see short-term and, often, political advantage in doing so.

The mining industry often attracts special consideration. This is because mining is a unique and important form of economic activity. It is unique in that it is "concerned entirely with the exploration of exhaustible resources. The amount of a particular mineral in the earth's crust may be large, but it is limited". [3]

Mining must always be considered separately from manufacturing industry because whilst the latter is constructive and involves the employment of labour in creative activity, mining is concerned only with taking out, using, depleting, in other words exploiting limited quantities of natural resources.

A manufacturing corporation is building and investing; something new is being brought into or built into the local economy. The risk factor, viewed most broadly, is reduced for in todays world of industrial development, establishment of a facility that contributes in a positive way to a nation's development is often guaranteed to succeed. Such an organisation can grow with the nation, expand in those areas of fastest and newest geographic growth, change its product as necessary, improve the quality and in these ways fulfill its own growth requirements.

Mining, on the other hand, is a business fundamentally concerned with an ore body which is fixed in location, unalterable as to quality and limited as to quantity. The problem of finding and defining reserves and planning extraction involves many, many risks and the process is usually very expensive.

Ore bodies have more often than not been found in remote locations where no infrastructure exists. Products are sold on world markets at prices fixed by economic and, increasingly, political considerations.

Revenues can and do fluctuate wildly and unpredictably.

Mining is concerned with a "non-creatable" and a "non-crop" product. Companies involved in mining are working with a wasting asset. Mines run out, quality deteriorates, processes become more expensive at deeper levels.

THE ATTITUDE OF GOVERNMENTS

Governments and corporations often take different views, even philosophically, of the mining business. Corporations view the product as a natural resource, an international commodity. Corporations are interested in reliability of supply and it makes little difference whether tin is from Malaysia or Bolivia, lead is from Yugoslavia or USA, or tungsten from China or Portugal.

Governments have a responsibility to provide a climate conducive to attracting investment funds and to encourage development. However they have regard for national resources and most have some anxiety about exploitation, particularly foreign exploitation. A typical sentiment is expressed by the Peruvian Government:

> the mines are the exclusive and irrevocable property of the state, and the latter as only owner thereof, may grant to individual parties the right to explore and exploit certain limited extensions under the conditions stipulated by law.

Further, to give some comfort,

> No conscientious person, no country, no government or any justice tribunal could approve the stripping of the national resources of a generous nation that has always granted and shall continue to give the security offered by its laws to foreign investors, who have come, are coming and shall come to live and work honestly in our land. [4]

The very nature of the mining industry makes

it a difficult one to be assured of on-going, even growing profitability. Governments often seem intent by their nationalistic, political interests to make life even more difficult for the industry. To Governments, mining is concerned with a national resource.

THE ATTITUDE OF THE INDUSTRY

In spite of the differences in attitude, the nature of mining and many problems, great mining empires have been built and great success has been realised.

The mining industry has been fortunate to have had many men with great vision and real entrepreneurial spirit. But most of these have been geologists, metalurgists and engineers - the captains of this particular ship.

In our own corporate experience this spirit does not often extend throughout a mining corporation, particularly to areas like finance. Indeed it could be easy to generalise and say that financial people in the mining industry are renowned for their conservatism.

It is common for mining houses to adopt the traditional stance of owning assets, even though the company might often have significant unrecognised tax losses or have a low average rate of taxation.

But there are some signs of change.

Leveraged leasing is innovative and it can be employed to improve the viability of a project. Yet it is not a common or popular method of financing in the industry but there are certainly examples of its growing use as a means of financing.

The significant mining countries of the Western world are Australia, Brazil, Canada, South Africa and U.S.A. Most of these are Anglo Saxon countries whose system of law best suits leveraged leasing.

Experience of United Kingdom

The most Anglo-Saxon of these countries has to be the United Kingdom.

Classic leveraged leases have not really been arranged in the U.K. although the U.K. probably has the most active tax-oriented leasing market outside U.S.A.

The Equipment Leasing Association makes up approximately 90% of the finance lease market in the U.K. They publish annual statistics. Business written in calendar 1983 and published in March 1984 is set out in Figure 2. Extractive industry assets financed by lease are either oil exploration equipment or included in the "plant and machinery" category.

Until recently almost all depreciable equipment has qualified for a 100% first year allowance. The entire capital cost could be claimed as a taxation deduction in the year equipment was installed. It is difficult for a traditional leveraged lease to find application in this situation. The Finance Act 1982 began to change this. Amongst other things the first year allowance was altered to more conventional depreciation for assets used by offshore lessees. However the Budget presented in March 1984 announced the progressive abolition of 100% allowances for all assets and its replacement with more conventional depreciation. This is certain to lead to the emergence of traditional leveraged leases in the U.K.

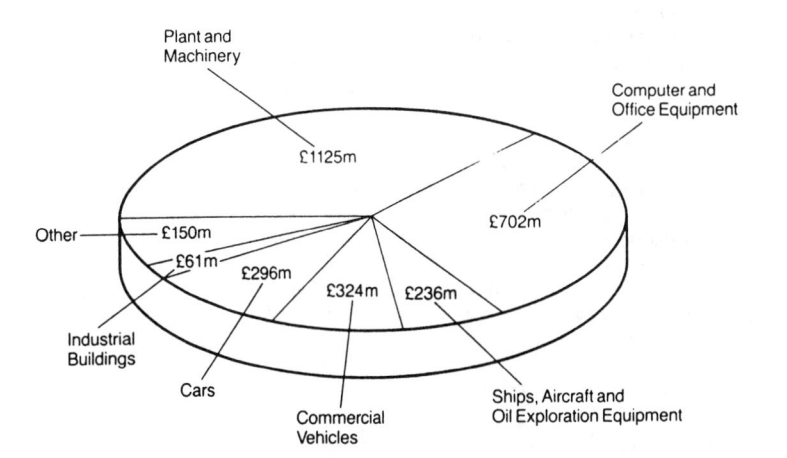

FIGURE 2: ELA ASSETS BY TYPE - 1983
Source: Leasing Digest March 1984

Many oil rigs, particularly those for use in the North Sea, have been financed by U.K. leases. Perhaps the most memorable is the transaction awarded to Lloyd's Leasing Limited in December 1981. This was to finance an advanced semi-submersible drilling rig for St. Vincent Drilling Limited, a joint venture between Ben Odeco and British National Oil Corporation (BNOC). The rig is to be sub-chartered to BNOC and used to develop some of BNOC's more remote licensed areas.

£90 million was to be drawn over the three year construction period with Lloyd's as the sole lessor, but with 85% of the total cost provided as subsidised rate finance obtained under the U.K. Shipowners scheme at 8.75% p.a. The term is 6 1/2 years from commissioning and the lease rate to St. Vincent was said to be between 5% and 6% p.a. Rentals did not become payable until 1984 and any cost escalations during construction will be incorporated into the rig cost. St. Vincent has the right to sub-charter the rig if necessary and also has options to renew the lease.

The transaction offers almost total flexibility to the lessee and at a very attractive cost.

Debt funds in U.K. leases must be raised on a full recourse basis. For some time there has been lobbying for Inland Revenue to allow non-recourse funding but lessors have been able to achieve attractive lease results on thecurrent basis - as in the St. Vincent lease.

The biggest mining organisation in the U.K. is the National Coal Board. As a Nationalised Industry the NCB is statutorily liable to corporation tax and can lease. Unfortunately the guidelines for leasing by Nationalised Industry allow them only to lease equipment that is not specific to their own industry and to lease from lessors who are in the business of providing credit as commercial lessors. As a consequence there is very little equipment and even fewer lessors who qualify.

According to Tom Clark at Lloyd's International Leasing Limited, whilst a couple of draglines have been leased to the coal industry in the U.K., they have received no publicity. The mining industry is effectively not a candidate for leasing because of the nationalised structure of the bulk of the industry and the guidelines developed by Government.

United States of America

Leveraged leasing started in U.S.A.

The aggressive free enterprise market in that country accepts leasing and rental at all levels. Their system also diverts from the conventional British idea (being closer to that in many Continental countries) and allows the title and rights to minerals to rest with the land owner on whose property a deposit exists.

There are three characteristics of a leveraged lease that must be met in U.S.A. and that distinguish this kind of financing:

1. lessors are required to structure each transaction so as to produce an overall profit and to have a reasonable cash flow,

2. lessors must maintain at least a 20% investment in leased assets throughout the lease term (although this investment may be in the future residual value),

3. lessors must accept the risk of equipment ownership and cannot offset this with insurance or other protection against the lessee.

Leveraged leasing is so well developed in USA that it has quickly adapted to the many legislative and regulatory changes over the last few years. Jargon such as safe-harbour, grandfather, TBT, ERTA, TEFRA abound and the abundance of advisors, lawyers, lobbyists and so on indicate that the product is a mature one indeed. At the time of writing this paper there is yet another change and, although legislation has been passed, its applications will not be known until certain Treasury Regulations are promulgated. However, none of this can hide or detract from the sheer size of the leveraged lease market in U.S.A. or the volume of transactions that are put in place each year.

Yet in 1981, according to U.S. Treasury figures, only $514 million in assets were financed for companys in the mining industry - about 3% of the value of all assets leased that year.

Industries in U.S.A. that most commonly use leasing are lumber and paper, railroads, chemicals, manufacturing of primary metals, the airlines and utilities.

But the mining industry has used leveraged leasing in U.S.A. almost since the inception of this funding alternative.

Pacific Power & Light's Bridger Coal, Black Thunder Coal Mining, Pittston Coal and Westmoreland Coal were all involved in the early days. Draglines were funded over terms of up to 20 years, dump trucks for 7 years and long-wall miners for up to 10 years. Jersey Miniere a Gulf & Western joint venture, has recently financed zinc operation costing $30 million for about 15 years.

But there have recently been some spectacular transactions.

Colorado Ute recently concluded a leveraged lease arranged for them by Merrill Lynch. This transaction funded equipment costing approximately $350 million for coal mining operations. The equipment comprises conveyors, coal cars, and mining equipment.

However, Basin Electric is still drawing funds at the time of writing in their $1.1 billion lease of mining equipment.

In the U.S.A. mining equipment costing tens of millions of dollars, hundreds of millions of dollars and even thousands of millions of dollars (billions) have been financed by leveraged lease.

This does not, of itself, legitimise the funding alternative but does demonstrate its application in major transactions and in carefully analysed situations.

Canada

Leveraged leasing is not arranged in Canada. Under current practice the cost of equipment for depreciation purposes (capital cost allowance or CCA in Canadian terms) can be no greater than the amount for which the lessor is "at risk". When debt is arranged on a non-recourse basis, without liability to a lessor, the lessor's only risk is his contribution to the assets. For this reason he can only claim depreciation on this small portion of equipment cost - which is a very clear disincentive for leveraged leasing and the result is that none is arranged on a conventional basis.

In fact very little mining equipment is leased in Canada. Statistics Canada has produced data showing that in 1980, 1981 and 1982 the percentage of new equipment purchases financed by all kinds of lease in the mining and petroleum sector amounted to 3.68%, 2.06% and 1% respectively. At December 1982 there were only 106 individual leases for this kind of equipment.

South Africa

South Africa has a well developed leveraged lease market and it is a country where mining equipment is prominent on a list of equipment leased. Leasing (including hire purchase) has been the fastest growing area of bank advances for several years, outstripping conventional and traditional bank advances.

Tax legislation governing mining allowances is extremely complex but, for the most part, the effect is a 100% deduction of all capital expenditure in the year in which it is incurred. However, mining allowances can only be offset against income derived from mining operations. Strictly speaking, the allowances arising out of one mining operation may not be deducted from income from another mining operation. The enforcement of provisions of this kind makes leasing an inaccessible form of finance for many major mining projects, except under extremely contrived situations.

However, conventional depreciation (wear and tear allowances) that specifically apply to manufacturing industry can sometimes by used in mining situations. The sensitivity of the Finance Authorities in South Africa to growing deficits has recently seen a tightening of regulations. Changes to legislation now require that tax allowances arising from the leasing of assets such as aircraft and productive equipment may only be offset against leasing income. Because of the involvement of financial institutions in large leasing transactions over the past few years, the amount of leasing tax base that is available for the absorption of leasing allowances is now limited, and this has reduced the scope for the leasing of expensive items of mining equipment.

In recent years suspensive sales (hire purchase or credit sales) have become the most effective form of non-equity finance used by the mining sector (but specifically not for gold or diamond mining). The efficiency of this approach arose from the ability of mining companies to capitalise suspensive sale finance charges into the cost of the asset and then write off the total cost immediately. This, of course, is the exact opposite of the very strong financial condition of many of the South African mining companies.

So whilst leasing is well developed and well used in South Africa, particularly in the mining industry, the opportunity for leveraged leasing is presently restricted because of legislation.

Australia

As in South Africa, leveraged leasing has become a significant financial alternative used in the funding mix by companies in Australia. The first asset financed this way in Australia was a copper smelter. As the copper price fell in 1975, the smelter was closed and put on a care-and-maintenance basis. A few years later the smelter was re-commissioned using a new and costly separation process. The new equipment and re-commissioning costs were also financed by another leveraged lease. Both transactions, of course, had to be cross referenced in documentation but the lessee's flexibility to finance and, more importantly, to use (or not use) the equipment as he and only he decided was unimpeded.

The term of the initial lease for this mining house was 15 years and the add-on lease was designed to match its terms. The lessee further protected his options with rights to terminate the financing at any time after eight years of the contract had run or to extend for up to another fourteen years.

The cost of debt funds was approximately 11% when this lease was written and the lease payments equated to a cost of about 7 1/2% p.a.

Leveraged leasing's easiest to understand and perhaps most common application is in truck and shovel operations. It is an application such as this that can highlight one of the important features of leveraged leasing.

The useful life of trucks is not likely to be more than four years in a testing mining site environment but the useful life of a big shovel is more than likely some longer term.

It is common therefore in truck and shovel operations to have the transaction packaged with two different terms - one appropriate for each kind of asset.

There is an alternative view that says you

should seek a lease term that is as long as possible, because of the much lower annual lease charges that result from longer terms. This viewpoint was developed in the U.S. aircraft industry, where is became common to seek a lease term for as long as 30 years, well beyond the period that the airline intended to use the asset.

We have found that financiers are often prepared to oblige in meeting a lessee's commercial desires such as this - but not for a term longer than the likely useful life of the equipment.

So, again, in a truck and shovel operation the most attractive financial result is likely to occur with two lease terms - one to suit each kind of equipment.

Quite a few draglines have been financed by leveraged lease in Australia as have washeries, preparation plants, and underground drilling equipment (such as long wall miners). The peripheral activities such as infrastructure, transport (railroad equipment), stacking and reclaiming units and port facilities have also been financed this way.

My firm has recently been involved with a major coal miner. The cost estimate to develop a new steaming and coking coal deposit was $250 million and the company decided on a form of project finance so as to keep the funding off its balance sheet and hopefully to achieve cost-effective financing.

As often happens, the company's financial advisers determined that the date for commissioning the first Marion dragline, which cost $30 million, was not opportune to put project finance in place. The project had no revenues at that time, however whilst other operations were profitable, the client was not paying tax at the full rate.

The client, as is appropriate in a context such as this, decided to finance the dragline by leveraged lease but wanted to ensure that the lenders would be temporary lenders only. It was proposed they either be replaced or the terms renegotiated when the project finance was put in place - in the most pessimistic view no longer than three years from dragline commissioning.

This strategy had a number of benefits:

. Net cash flow could be realised at the earliest possible date,

. Taxation incentives were optimised,

. The underlying financing term was designed to assist net cash flow. The lease term is 12 years,

. Flexibility for the ultimate project was totally retained.

In fact the lease was put in place within 28 days from decision to proceed to signing of documentation. Lenders are international banks with funding in United States dollars, the currency of the sales contracts. Lessors are Australian financial institutions as they must be to get access to Australian tax incentives.

Now, fifteen months later, the project is about to be financed in full. A major preparation plant, vehicles and sundry expenditure will also be financed by a leveraged lease.

Further sales contracts are now in place and all the lenders will be co-ordinated in an endeavour to obtain a better interest rate.

The security will be the project itself. No guarantees (other than a completion guarantee) will be provided.

There are many similar examples of Australian leveraged leases but, in mining, coal is the most significant application.

Similar tax legislation to that in South Africa restricts mining industry deductions to taxpayers and, indeed, income that is derived from mining activities. However this has proved less restrictive in the Australian environment.

THE BENEFITS OF LEASING

Advocates of the leasing business include some exceptional salesmen. There are certainly publications available that list pages and pages of the advantages of leasing.

One sets out twenty seven reasons for leasing rather than buying equipment and develops the argument to such an extent that, to be against leasing is akin to adopting a position against motherhood.

Most national accounting bodies offer rigorous critiques on leasing and the following comment is typical. One says "The advocates of leasing are quick to repeat the often quoted advantages of leasing. Some of these advantages, however, are illusory, and must be considered carefully with offsetting disadvantages. From the point of view of the cost of leasing only a careful evaluation using the most appropriate method will give the true picture. Nevertheless certain subjective or non-quantifiable considerations may need to be taken into account before a final decision can be made regarding the desirability or otherwise of a certain leasing proposal".[5]

There certainly are advantages but, by and large, if a cost benefit cannot be demonstrated, it can be difficult to substantiate the cause of leasing in any particular situation.

Important benefits relevant to leveraged leasing other than cost can be the provision of 100 percent financing (which can include installation and other "soft" costs), as a form of consortium finance leveraged leasing can introduce new and alternative funding sources, lease payments can often be structured to more accurately reflect the revenues that are generated by the equipment financed.

AN EXAMPLE

Table 1 sets out the assumptions and data of an example. The assumed equipment cost is $1000 and the equipment depreciates on a straight line basis at 20% p.a. or may be depreciated on the 150% declining balance method.

The corporate rate of taxation is 46% and a prior year taxpayer convention is used.

The equipment is commissioned and paid for in July 1984 and a funding term of 10 years is required. The estimated market value of the equipment after 10 years is 10% of its original cost.

A leveraged lease with an 10 year term and a 10 percent residual value is structured. The lessor's desired return on his investment is 20 percent p.a. pre-tax. The interest rate on lenders' funds (the lessor's borrowing) is 15.5 percent p.a.

Given the assumptions about the equipment and the prevailing interest rates, a leveraged lease can be structured with a lessee interest rate over 300 percentage points below the lessor borrowing rate.

The interest rate from 19 semi-annual payments in arrears each of $83.98 and one final payment of $183.98 (including the value of the residual that is guaranteed) is 12.47% p.a.

The equipment is funded 76.1202 percent with debt and 23.8798 percent with lessor contribution. The loan is repaid in a non-level way so as to best suit the structure.

TABLE 1: EXAMPLE ASSUMPTIONS

A. Leased Asset

Cost:	$1,000
Depreciation Rate:	30% p.a. diminishing value
	20% p.a. straight line
Installation and Funding Date:	July 1984
Funding Term:	10 years
Market Value at end of Funding Term:	10% of cost

B. Lease Structure

Term:	10 years, commencing July 1984
Residual Value:	10%
Lessor Funding:	76.1202% debt
	23.8798% equity
Interest Rate on Loan:	15.5% p.a.
Lessor Yield:	20% p.a.
Rental Payments:	20 semi-annual in arrears payments each equal to $83.98
Loan Repayments:	20 semi-annual payments comprising principal and interest as follows:
	11 x $84.13
	1 x $45.60
	2 x $54.32
	2 x $53.44
	2 x $50.89
	1 x $84.13
	1 x $58.12

C. Borrow Alternative Assumptions

Funding:	100% of cost. Loan drawn down July 1984
Interest Rate:	15.5% p.a.
Repayments:	20 equal payments in arrears with a balloon payment of $100 at end
Depreciation Method:	Diminishing value

Table 2 sets out the lessor's annual tax returns for each year of the transaction. The tax statement comprises assessable lease rentals and deductible interest and depreciation. The lessor is in a total taxation paying position the financing produces an aggregate taxable income or contribution to revenue.

In the first five years of the lessor's investment the taxation deductions are greater than the rent received and the lessor has a tax loss by virtue of the investment. All subsequent years generate taxation payments.

This is tax deferral.

This is the heart of leveraged leasing. This method of financing relies on both the existence of significant taxation benefits that can be transferred to a lessor and the ability to defer ultimate payment of taxes. It is virtually impossible to produce a lease interest rate lower than the interest rate on lenders funds (a negative spread) if the lease term is very short, say three years.

In this example the lessor relies to only a

very small extent on retained cash to generate his return. By and large the return is generated by non-cash items - in effect the use of taxation losses that occur in the early years.

The rentals are determined so as to repay only the lenders funds which comprise 76 percent of the asset cost.

It is because the lessor relies on this that rentals can be so low and a negative spread is produced. The lessee is very clearly being passed back the benefits or incentives to investment through rentals reduced below the cost of lenders funds.

TABLE 2: LESSOR TAX STATEMENT

Tax Year	1 Lease Payts & Residual	2 Interest	3 Deprec	4 Taxable Income 1-(2+3)	5 Tax Loss/ Pay't
1984	70	49	125	-104	-48
1985	168	113	263	-207	-95
1986	168	104	200	-136	-62
1987	168	93	200	-125	-58
1988	168	81	200	-113	-52
1989	168	67	13	88	41
1990	168	53	0	115	53
1991	168	45	0	123	57
1992	168	35	0	133	61
1993	168	24	0	144	66
1994	198	6	0	192	88
Total	1780	669	1000	110	51

Table 3 sets out the annual cash flows of the lessor. This comprises lease rentals less debt amortisation plus any tax saved or less any tax paid. The resulting net cash flow is available to pay back the investment together with the yield (or profit).

Whilst in many leveraged leases the lessor would recover his investment if there was no tax deferral, that cash flow would be unlikely to produce an acceptable return.

This is not tax avoidance - simply a transfer of those benefits that are available to any owner/user of capital goods to a financial owner. Nothing new has been added. The financial transaction has been structured in accordance with the tax and legal requirements, as permitted, so as to produce an attractive cost to the user of equipment.

COST COMPARISON

If a lessee is not earning assessable income or if the lessee has significant carry forward taxation losses then a lease versus borrow analysis based on cost competitiveness should be relatively easy. Nevertheless it is necessary to carry out a proper analysis - and this requires a comparison of after-tax cash flows of leasing and ownership financed by borrowing.

It is important to make the comparison a meaningful one and to compare like with like. Both financing terms should be the same, the pattern of loan amortisation should be the same as the pattern of lease payments, and consideration should be given to the end of the financing term. Most common when residual values are guaranteed is to assume in the lease case that the residual value is paid, title to the asset passes to the lessee then the equipment is sold to realise the same amount as the residual value. In the ownership alternative the equipment is sold, also for proceeds equal to the residual value.

TABLE 3: LESSOR ANNUAL CASH FLOW

Tax Year	Invt	Lease Rent & Res	Princ & Int	Tax Liab	After Tax Cash Flows	Retn on Invt	Invt Balc
1984	239				-239		247
1985		168	168	-72	72	32	199
1986		168	168	-119	119	16	97
1987		168	168	-46	46	9	60
1988		168	168	-55	55	4	9
1989		168	168	-49	49	0	-40
1990		168	130	87	-49	0	8
1991		168	109	59	0	0	7
1992		168	107	59	2	0	4
1993		168	101	64	3	0	1
1994		268	142	69	57	1	-57
1995			99	-99	99	1	42
1996			-44		44	3	-
Total	239	1780	1429	52	60	66	

Table 4 sets out the tax result of the alternatives. Table 5 derives the after-tax cash flows of leasing and ownership. This shows that leveraged leasing is cost competitive with ownership of equipment financed by borrowing.

TABLE 4: TAX RESULTS OF LEASE & BORROW COMPARED

	Borrow Alternative					Lease Alternative		
Tax Year	Sale Inc	Int on Loan	Depn at 30%	Reduc Taxab Inc	Tax Sav 46%	Lease Payts	Reduct Taxab Income	Tax Sav 46%
1985		153	300	453	209	168	168	77
1986		147	210	357	164	168	168	77
1987		139	147	286	131	168	168	77
1988		130	103	233	107	168	168	77
1989		119	72	191	88	168	168	77
1990		107	50	157	72	168	168	77
1991		93	35	128	59	168	168	77
1992		76	25	101	46	168	168	77
1993		57	17	74	34	168	168	77
1994		34	41	-25	-12	168	168	77
Total	100	1055	1000	1955	898	1680	1680	770

Table 6 contains the discounted net present value when the discount rate used is the after-tax equivalent of the lessee's marginal cost of debt. The result, again, demonstrates a benefit to leasing. If the lessee was not a current taxpayer the benefit would have been much greater.

Leveraged leasing can be a most advantageous form of financing. Its use requires 3 pre-conditions:
. the existence of equipment with tax benefits of substance
. medium to long term finance
. a high cost of equipment

TABLE 5: AFTER TAX CASH OUTFLOWS OF LEASE & BORROW COMPARED

	Borrow Alternative				Lease Alternative		
Tax Year	Sale Inc	Princ & Int	Tax Saving	Net Cash Outf	Lease Rent	Tax Saving	Net Cash Outf
1985		195	0	-195	168	0	-168
1986		195	209	14	168	77	-91
1987		195	164	-31	168	77	-91
1988		195	131	-64	168	77	-91
1989		195	107	-88	168	77	-91
1990		195	88	-107	168	77	-91
1991		195	72	-123	168	77	-91
1992		195	59	-136	168	77	-91
1993		195	46	-149	168	77	-91
1994	100	295	34	-161	168	77	-91
1995		0	-12	-12	0	77	77
Total	100	2050	898	1052	1680	770	910

TABLE 6: SUMMARY OF CASH OUTFLOWS PER $1000

	Total Net Cash Outflow	Total Discounted Net Cash Outflow
Lease	$ 910	$647
Borrow	$1052	$664
Difference	$-142	$-17

If these are not present then leveraged leasing can be an expensive even unfortunate financing alternative.

SOURCES OF LEVERAGED LEASING

Financial markets continue to become more complex. New financial products, structures and innovations proliferate. The speed of these developments makes it difficult for the staff of large corporations to stay abreast. Today we rely more and more on specialists and leveraged leasing is no exception.

Leveraged leasing is best arranged by people who are experts in the field. Such people are found with investment and merchant banks, in banks or with specialist leveraged lease firms. Wherever, the expertise should be able to be substantiated with specific transactions and with personal references from Treasurers who have experienced the packager's ability at first hand.

When it comes to putting a lease in place it is important to remember that all issues relating to the equipment itself are determined solely by the lessee. He nominates a lease life usually to fit the economic life of the equipment. He chooses the equipment and is the only party in a lease transaction to liaise with manufacturers or suppliers and builders.

A leasing company or packager arranges the leveraged lease. Often the packager takes no direct stake in the transaction but negotiates with the lessee, arranges the lessor and lender participants, and manages the lease throughout its life or arranges for a trustee to manage the transaction.

In return, the packager is paid a fee by the lessor. This is recovered by the lessor together with his principal investment.

A vital pre-requisite to becoming involved as lessee in a leveraged lease is to determine who is the person responsible for the financing aspects of the project and to ensure he stays involved in the action. From the outset, adopt a philosophy that even though you have contracted the packaging of the lease, it is your transaction. Don't hesitate to ask questions - remember that the packager has a clear and unambiguous responsibility to the lessee.

Step 1 is to call for proposals. In every market there are a number of competent and professional firms who can put the lease together but my advice would be to deal only with a firm that has been recommended to you.

Naturally you will be basing your selection of packager/manager, largely on who can provide the lowest lease cost to your organisation but, equally importantly, a packager must have a demonstrated ability to successfully arrange and manage the transaction.

Step 2 is to make your choice from the proposals received. A method of evaluation is described above. If the bid is on a "best endeavours" rather than a firm basis, have the successful packager confirm that he will produce the lessor and lender funds as set out in his proposal. I suggest that the packager be given a very short time to do so. He must also confirm that the lease as proposed can be achieved. If the packager can't do that, one could assume that his original proposal was based more on hope than reality, and it's time to look for a new packager.

Step Number 3 can be undertaken while you are waiting for the confirmation from the packager. This requires checking the taxation aspects with experienced tax advisers. Focus on the tax structure of the lease, check whether it involves any tax indemnities, and get advice on the likelihood and cost of the indemnities being invoked.

This may vary the final cost to the lessee, in total, quite substantially. Sometimes it will be commercially viable to provide indemnities as the potential benefits will outweigh the cost. In addition, the lessee's fall back or worst case position will be an eminently acceptable one.

Step 4, the preparation of documents, is also a vital one for the lessee. Care must be taken at a commerical as well as a legal level so that the documents reflect the lessee's position clearly and accurately. The packager will be able to give experienced advice on the attitude of the different parties, but high-quality and independent legal opinion must also be obtained.

LEVERAGED LEASING TOMORROW

Leveraged leasing is now a mature product. But it is more than just a financial product. It is an idea - a concept - that has very broad application.

Leveraged leasing as a concept looks at an application that requires innovative approaches to taxation, cash flows and, sometimes, security in financing. It is a form of engineering where the different flows are designed to suit the unique circumstances of a commercial problem. This has particular relevance in the mining industry.

Governments offer tax incentives to attract particular industries, to encourage the replacement and modernization of equipment, and to assist development of capital projects. Mining industry costs can be determined in advance with a high degree of certainty. However revenues fluctuate widely with movements in product price and volume. Leveraged leasing has particular relevance in this situation. The mining industry is all about taking risks but minimising risks.

Leveraged leasing is one approach to financing that can match financial innovation with technical innovation to enhance project viability to improve financial returns.

Notes:

The author is appreciative of assistance from Zoe Katzen - Nedbank in South Africa, David Murray - Saturn Lease Underwriting Limited in U.K., and Gary Civello - U.S. Lease Financing Inc. in U.S.A.

1. First Chicago Leasing Corporation Leveraged Leasing - a new alternative financing 1973 p5

2. See Van Ewyk, O "Leveraged Leasing - What Makes it Profitable for the lessor" in Lease or Buy Decision Centre for Professional Development, South Melbourne, Australia

3. Barger H & Schurr S H The Mining Industries 1899 - 1939 p 249

4. Government of Peru Petroleum in Peru pp 17-18, 23

5. Van Ewyk op.cit p12

BIBLIOGRAPHY

World Leasing Yearbook 1983 and World Leasing Yearbook 1984 Hawkins Publishers Limited, London

Van Horne, James C 1971 Financial Management and Policy Prentice Hall

First Chicago Leasing Corporation, 1973 Leveraged Leasing: a new alternative in Financing

Manufacturers Hanover Leasing Corporation Leveraged Leasing: A new choice for the user of capital equipment

Smith P R "A Straightforward Approach to Leveraged Leasing" in Journal of Commercial Bank Lending July 1973

Van Ewyk, O "Leveraged Leasing - What Makes it Profitable for the Lessor" in Lease or Buy Decision, Centre for Professional Development, South Melbourne, Australia

Bruce R, McKern B and Pollard I, 1983 Handbook of Australian Corporate Finance Butterworths

Glynn JJ "Equipment Leasing" in Certified Accountant April 1979

11.23

USING A GOLD LOAN AS A FINANCING MECHANISM: CASE STUDIES

Reg M. M. Rowe, C.A.
President, Lodestar Energy Inc.
Vancouver, British Columbia

Simon D. Handelsman, P. Eng.
President, Canadian Geoscience Corporation
Vancouver, British Columbia

ABSTRACT

The mechanism of using gold as the basis for a loan was suggested to the treasurer of a mining company operating two gold heap leach mines at a time of high interest rates. Due to the seasonal nature of cash flows from operations, and the relatively low cost of funds this scheme was attractive. The paper describes the agreement with the bank, gives an example of the transactions, and describes other types of gold loans.

INTRODUCTION

At a heap-leach-gold mine, operating costs are expended in advance of any metal revenues. For a year-round operation this will tend to settle down into a constant flow of income and expenditures, financed by an adequate working capital.

However, many heap leach mining operations, especially in the western United States, can operate their leaching circuits only for part of the year, due to snow and freezing conditions. In such cases most of the expenses come early in the season, and the revenue from metal sales comes up to several months later at the end of, or after the close of, the leaching season. The on-going expenses of offices, salaries and wages, and mining have to be financed ahead of sales revenue (without considering forward sales or other price protection schemes).

In 1981 the mining company's treasurer was approached by the company banker with the suggestion of a gold loan. The concept is fairly simple. Bullion held in a vault does not generate any income, but generates the costs of storage and insurance, etc. The metal value appreciates or depreciates with the movement in the gold price.

The bank implements the loan in the following fashion:

1. The mining company (the "Buyer") enters into a Gold Purchase Agreement with the Bank. (Appendix 1).

2. Under this agreement the Bank delivers an amount of gold to the Buyer, and charges the cost of the gold to the Buyer's loan account.

3. The Buyer immediately sells the gold to the Bank, and the proceeds are credited to, say, the Buyer's current account.

4. The Buyer pays interest to the Bank on a monthly basis at the agreed rate.

When the Buyer repays the loan:

5. The Buyer may repay the loan either by delivering the same quantity of gold as taken down in transaction (2); or

6. The Buyer may make payment to the Bank an amount equal to the quantity of gold taken down in transaction (2) multiplied by the effective gold price at the time of repayment.

TERMS

At the time of the "loan" the current bank rate was 17.5% and the mining had term deposits which were earning historically high interest payments. The rate charged on the "loan" was 6%. On an interest rate basis, the "loan" was more attractive than cashing out the term deposits.

The terms and conditions are contained in Appendix 1. From a review of this document, it is clear (Item 13) that the loan has similar terms to a demand loan.

RISK ASPECTS

There are essentially two principal risks.

Production Risk

There is a risk to the Bank that due to operating or other conditions the mine is not able to produce gold. In this particular situation, the mining company had sufficient assets in the form of term deposits to make the likelihood of non-repayment minimal, due to the demand aspect of the loan.

Although the mine had a short production operating record, it had 10 years of proven reserves, and had demonstrated the ability to produce gold. The mine operators were confident that they could produce more than 40,000 ounces of gold in a year.

Price Risk

All of the price risk is taken by the Buyer. This may be offset by a hedging program. In this case it was handled by watching the gold market, and anticipating the need to repay the loan if the price moved up beyond a particular level.

ECONOMIC CONSIDERATIONS

The following example was used to assess the risks:

Borrow 10,000 ounces @ $500.00/oz

a) in November Gold is $400/oz

   repay either 10,000 ounces or $4 million.

b) in November Gold is $600/oz

   repay either 10,000 ounces or $6 million.

By the time a decision was reached the loan was taken down at about $425/oz.

GOLD LOAN AS A LEASE

Another example using gold as a financing mechanism is illustrated by the following example.

An operating gold mine purchases gold from a bank -- in this case the transaction is actually structured as a lease rather than a straight loan, although the terms and conditions are similar.

The mine purchases (leases) gold from the bank.

The mine had previously locked in its revenue from its next 18 months production by selling forward in the normal hedging market, say 80,000 ounces at $420/oz ($33.6 million).

Currently (at the time of the loan) the gold price is $350.00/oz -- so the mine did a smart thing, and has a paper profit of $70/oz.

The corporation has term loans and a revolving credit, with a debt of, say, $60 million -- at (then) current interest rates of 14%.

The company now liquidates its forward position (futures contract) for a $70/oz profit ($5.6 million).

The mine has obtained the future profit of $70.00 per ounce from the option contract, and has an interest savings on the loan, which is now 2% instead of 14%. Also there is an interest savings, as part of the loan is paid down.

The mine makes repayment to the bank of the gold from mine production, instead of making physical deliveries under the option contract.

Clearly, the company is no worse off than it was before in respect to price risk; and it has reduced the cost of funds substantially.

Revolving Credit

A more complicated version of this arrangement may be developed if the borrower wants to operate the gold loan option so that he may go in and out of the transactions in cash or gold on a revolving basis. A reason that the borrower may require the agreements structured as a revolving credit is to ensure liquidity.

Suppose the borrower takes down $28 million representing the 80,000 ounces at $350/oz. If the price of gold goes to $500/oz then his loan liability is $40 million. Thus the bank, because of the credit regulations, has to be sure that the borrower has the strength to support this larger amount.

In this example the commitment is to lend say, six months from now, and the lender (or leasor) does not know what the demand on funds (or physical gold) may be then.

Unfortunately, the principals in these cases prefer not to be identified. The acknowledgments are anonymous.

APPENDIX 1

GOLD PURCHASE AGREEMENT

AGREEMENT made as of the -- day of ---, 19--, between PARTICULAR BANK (the "Bank") and A MINING COMPANY (the "Buyer").

WHEREAS the Buyer proposes to purchase gold bullion from the Bank from time to time and the parties desire to set forth the terms and conditions applicable to each such purchase.

WITNESSETH that for valuable consideration it is agreed as follows:

1. In this agreement all references to ounces are to fine ounces troy weight.

2. Each purchase of gold by the Buyer from the Bank shall be initiated by a purchase order in substantially the form attached hereto delivered to the ----- Branch of the Bank in and signed by duly authorized officers of the Buyer, and the terms and conditions of this agreement shall apply to each such purchase.

3. Gold purchased hereunder shall be in the form of gold bars having a minimum of 995 parts of

fine gold per 1,000 parts.

4. Delivery of gold purchased hereunder shall be made at the ---- of the Bank in ---- and shall be effected by delivery to a bonded carrier selected by the Buyer and approved by the Bank, whereupon title to and the property in such gold shall pass to the Buyer, and the Buyer shall assume all risk of damage to or loss or destruction of such gold from any cause whatsoever.

5. The Buyer promises and agrees to pay to the Bank at the said Branch the purchase price of gold purchased hereunder following demand therefore by the Bank from time to time.

6. The purchase price shall be payable in lawful currency of the United States of America ("U.S. funds") and shall be the amount determined by multiplying the total number of Unpaid Ounces at the time payment is actually made by the Buyer by the Price in effect at such time. As used in this agreement "Unpaid ounces" at a particular time means the total number of ounces of gold purchased hereunder up to and including such time less (i) the number of ounces, if any, for which payment has therefore been made pursuant to paragraph 5 or 7 hereof, and (ii) the number of ounces, if any, for which credit has therefore been given pursuant to paragraph 8 hereof, and the "Price" at a particular time means the Bank's offer price per ounce of gold bullion in effect at such time as conclusively determined by the Gold Trading Department of the Bank at its Head Office.

7. The Buyer may at any time make payment for all Unpaid Ounces or from time to time make payment for one or more Unpaid Ounces by paying to the Bank a sum of U.S. funds equal to the amount determined by multiplying the number of Unpaid Ounces for which payment is being made by the Price in effect at the time of such payment by the Buyer.

8. The Buyer may at its option from time to time delivery to the Bank at ---- gold of the same quality as gold purchased hereunder in such quantitites (not exceeding the then number of Unpaid Ounces) as the Buyer may see fit. Gold so delivered by the Buyer to the Bank shall be subject to examination by the Bank in order to verify fineness and weight, and the Bank may refuse to accept any gold which is not of the same quality as that purchased hereunder. For purposes of determining the number of Unpaid Ounces at any time, the Buyer shall be given credit for each ounce of gold so delivered by the Buyer and accepted by the Bank at or prior to such time. Nothing herein contained shall obligate the Buyer to deliver gold to the Bank.

9. The Buyer promises and agrees to pay to the Bank in U.S. funds on the last business day of each month interest on the amount of the purchase price from time to time outstanding, as well after as before default, until paid. Such interest shall be calculated daily and shall be at a rate per annum of ---- subject to change at the discretion of the Bank, provided that if and whenever the rate is changed the Bank will notify the Buyer of such change at least 30 days in advance. For the purpose of computing interest hereunder the purchase price per ounce shall be determined each day and shall be determined each day and shall be equal to the afternoon London fixing price per ounce on such day, and the purchase price on any day on which there is no afternoon London fixing shall be the most recent afternoon London fixing price immediately prior to such day.

10. The purchase price in respect of each purchase of gold hereunder shall be deemed to be outstanding upon delivery of such gold to the Buyer in accordance with paragraph 4 hereof.

11. The Buyer shall be responsible for and shall indemnify the Bank against all taxes and other levies applicable to gold purchased hereunder.

12. This agreement shall become effective upon acceptance by the Bank of the initial purchase order of the Buyer, provided that nothing herein contained shall obligate the Buyer to give further purchase orders to the Bank or obligate the Bank to accept any such purchase.

13. This agreement may be terminated on a date to be specified in written notice given by either party to the other at least thirty days prior to the date so specified; provided however that termination of this agreement shall not affect any obligation which had accrued or was accruing hereunder prior to the effective date of termination so specified. Any notice, demand or other communication hereunder shall be in writing and shall be hand delivered or mailed by registered mail during any period when normal postal services prevail, addressed, if to the Buyer, to ----

or if to the Bank to ----

or at such other address as the party to receive the same shall have furnished to the other party, and shall be deemed to have been given if hand delivered upon receipt by the addressee, or if mailed on the second business day following the date of mailing.

14. This agreement shall enure to the benefit of and be binding upon the respective successors and assigns of the parties hereto, provided that this agreement may not be assigned by the Buyer without the written consent of the Bank.

IN WITNESS WHEREOF the parties hereto have executed this agreement.

------------ (Buyer)
by

------------ (Bank)
by

APPENDIX 2

FORM OF PURCHASE ORDER AND RECEIPT

PURCHASE ORDER to Bank,

Date   19   Number

Quantity of gold bullion

Delivery date

This purchase order is given pursuant to the Gold Purchase Agreement between the undersigned and the Bank dated

ACCEPTED --------- 19

------------------------------(Buyer)

------------------------------(Bank)

by

RECEIPT is hereby acknowledged of the delivery of the gold bullion covered by the above purchase order.

19

CHAPTER 11 : SOURCES OF FINANCE

BIBLIOGRAPHY

Brooks, W.A. and Hursh, D.S., "Acquiring Capital Funds for Coal Mining," Mining Engineering, September 1982, pp 1332-1335.

Bruce, R. et al, ed. Handbook of Australian Corporate Finance, Butterworths, 1983.

Castle, G.R., "Term Lending - A Guide to Negotiating Term Loan Covenants and Other Financial Restrictions," Journal of Commercial Bank Lending, November 1980, pp 26-39.

Castle, G.R., "The Role of the Petroleum Engineer in the Project Financing Team," Trans. SPE 9571, 1981 Economics & Evaluation Symposium of the Society of Petroleum Engineers of AIME, Dallas, February 25-27, 1981.

Curtis, R.R., "Debt Financing," Coal Economics and Taxation Symposium, Coal Assn. of Canada, Saskatchewan, 1978, pp 27-33.

Emerson, C., Project Financing, The Financial Times Business Enterprises Limited, London, 1983.

Haworth, G.R. and Aimone, T.J., "How Major Mines will be Financed in the Future," Mining Engineering, Vol. 30, No.9, September 1978, pp 1292-1298.

Hobbs, W., "Financing the Aluminium Industry in the 1980s," First International Aluminium Conference, Metal Bulletin, Madrid, Spain, September 29, 1980.

International Project Finance, Section in The Banker, December 1977, pp 45-71.

Ladbury, R.A., "Limited Recourse Financing," The Australian Mining and Petroleum Law Association, Third Annual Conference, Perth, 17-19 May, 1979.

Lees, F.A., International Banking & Finance, MacMillan Press, 1975.

McKechnie, G. (ed), Energy Finance, Euromoney Publications, London, January 1983.

Mintz, F., "Reserve Based Financing - Specific Requirements and Alternatives, " Trans SPE 9578, Economics and Evaluation Symposium of SPE of AIME, Dallas, February 25-27, 1981, pp 79-89.

Nevitt, P., Project Financing, Euromoney Publications/AMR International, 4th Ed., 1983.

Radetzki, M. and Zorn S., Financing Mining Projects in Developing Countries, United Nations Study, Mining Journal Books, London, 1979.

Rodriguez, R.M. and Carter, E.E., International Financial Management, 2nd Ed., Prentice-Hall, 1979.

Summers, K.R., "Lending to the Coal Industry," Journal of Commercial Bank Lending, Vol. 61, No.3, November 1978, pp 8-18.

Tinsley, C.R., "Mine Finance," Mineral Industry Costs, Northwest Mining Association, 1981, pp 213-220.

Ulatowski, T., "Financing of Direct Reduction Development," SME - AIME preprint No. 78-K-353, Fall Meeting, Florida, September 11-13, 1978.

Wang, P., "Leveraged Buy-out Financing of a Coal Mine," Mergers and Acquisitions, Fall 1977, pp 34-45.

Wanless R.M., Finance for Mine Management, Chapman and Hall, 1983.

Wilson, W.W., "Coal Mine Development Financing During the Next Decade," Professional Lease Management Inc. Second Coal Conference, New Orleans, March 10, 1976.

Zimmerman, C.Z., "An Approach to Writing Loan Agreement Covenants," Journal of Commercial Bank Lending, November 1975, pp 2-17.

Chapter 12

Case Studies

CHAPTER 12: CASE STUDIES

PREFACE

This chapter presents case studies which illustrate the wide variety of international mine financings completed over the past decade.

The range of minerals involved includes precious metals (gold and silver), base metals (copper, lead and zinc), ferroalloys (nickel), an industrial mineral, and fuels (coal and uranium).

The case studies have been selected from a diverse geographical context, including projects in Canada (five), Mexico (three), Pacific rim countries (three), and Australia and the United States (one each). In addition, one paper describes how some companies have developed flexible strategies for survival in today's recessionary environment by sourcing funds on a global basis.

In preparing this chapter the editors sought to include case histories detailing a broad range of situations involving mineral finance. Some were updated from material presented at previous MRM Committee short courses and sessions dealing with international finance. The post-installation appraisals are instructive by explaining why some projects have failed to meet the financiers expectations and why others have been particularly successful.

Certain of the case studies focus specific attention on particular risks or issues, including sovereign or country risk, recessionary financing, junior company financings, project financings, and financing under conditions of severe cost inflation. The projects range in size from small mine acquisitions to multiparty financings in excess of US$1.0 billion.

Chapter 12 is the largest chapter in this book, principally because the cases are detailed with considerable background and support information presented. In most cases the authors have sought to provide a step-by-step behind-the-scenes view of the financing as it developed.

This chapter should be most instructive to the reader, as it incorporates the concepts and situations presented throughout the book into a comprehensive whole, the "successfully financed mineral project."

                                                                            Simon D. Handelsman
                                                                             Chapter Editor
                                                                             Canadian Geoscience Corporation

12.1

FINANCING THREE GOLDS

PETER A. ALLEN

President, Lac Minerals Ltd.
Toronto, Ontario

INTRODUCTION

Lac Minerals Ltd., a significant North American producer of gold and other metals for over three decades, recently completed the financing for its third discovery in an eight year period. From 1975 to 1982, three important gold deposits in Quebec and Ontario: namely Bousquet, Doyon and Hemlo, were located and tested by company geologists. When current development plans are complete, Lac will have injected $400 million into these deposits.

A discussion of these projects and their financing reveals some extraordinary challenges faced by top mining management. Responses to these challenges developed by Lac management may interest executives planning to proceed with new North American gold or base metal mines.

MINE DE BOUSQUET

In 1976, we explored a property containing a large low grade surface gold deposit. The geologial setting was exciting and we tested recent rethinking about the volcanigenic genesis of gold and base metal occurrences. These ideas helped us to persist in drilling a nearby conductor and 600,000 ton, 0.22 ounce/ton orebody was outlined. With gold at $Cdn.150 per ounce the feasibility was approved and work started February 1978.

Availability of a company custom mill 35 miles away reduced our immediate capital needs to $8.5 million.

First of all, understanding the geological significance of results at Bousquet was critical. Mine and surface exploration geologists at Lac are continually and closely working together to better understand the nature of gold deposits in the Shield. This management emphasis on both pure and mine geology meant that a much larger ore reserve could confidently be predicted. (A proposal to quickly develop and ship the ore to a nearby company mine was discarded.) Instead of designing just for the initial plan of 500 tons per day we therefore included headframe, shaft and hoisting capacity for 1,700 tons and provided a compartment for future shaft deepening. These prebuilt additions were welcomed within two years.

The type of collaboration whereby surface and mine geologists make study and field projects of each other's problems sets the foundation for this predictive ability. At Lac, the traditional responsibility for and supervision of the mine geologist by the mine manager is modified by this collaboration. This adds interest and scope to the geologist's work and lets him think beyond production requirements. Besides such management considerations, dedicated, professional geologists with integrity and lots of practical experience are needed to plan successful developments like Bousquet.

Second, projecting future bullion revenue was critical and our experience selling production for future delivery gave us a way to ensure that loan repayment at term would not be jeopardized by falling metal prices. Output from a single mine is often reduced or curtailed for numerous reasons, so "selling forward" is done best and with least risk by companies like Lac with more than one or two producing locations for that metal. At first, a forward marketing program was considered unwise, speculative and unorthodox by analysts and other producers. Declining markets in the early 1980s prolonged gold project capital recoveries however, and critics were better able to appreciate the usefulness of such a tool. The notion that gold miners shouldn't sell forward is rooted in the belief that long-run gold prices will rise as modern social theory implementation in democracies inevitably erodes currency values. Selling against this trend goes against this industry belief or "ethic" and is at odds with the company shareholder who expects the gold price to rise not fall. Nonetheless, an efficient tool is available to the gold producer (not to most other metal producers) to quickly disengage from adverse price trends or to lock in spot prices instantaneously for two years or more of profits.

Gold had recently retreated from U.S.$200 to $104 and now had reversed to $150. Lac had been forced to close several of its producers during the late 1960s because central bank-controlled gold prices failed to match inflationary pressures. Fortunately, the company survived this hardship because of its unwillingness to finance development with debt. We now, therefore, sold enough gold forward at $150, so that if the price dropped to $104 the loan would still be repaid on schedule.

Knowing as well that the loan could easily be repaid from Lac's healthy cash flow and treasury, limited recourse five year term floating rate bank debt was chosen. This kind of funding was only recently available for gold projects. Back in 1970, a new pattern for mine financing was signalled by a $74 million floating rate bank loan for Gibraltar Mine's porphyry copper development. Although this wasn't approved so much on the excellent merit of the property as on the parent company's guarantee, a trend was nonetheless started towards much more easy, flexible credit. Both domestic and international credit markets were becoming structurally transformed and the resultant supermarket-type competition for corporate finance spilled over into our own high-risk resource industry which traditionally has financed with a proportionately large equity component and smaller long-term fixed rate obligations. Consequently, today, the larger North American mining companies are staggering under huge debt burdens which threaten their survival if metal prices remain depressed.

Lack of sophisticated attention by lenders towards such topics as geology, metallurgy and metal markets often has led to defaults during the last decade. Banks have experienced a 'steep learning curve' on the risks of financing small companies with a single mine or large ones dependant on buoyant metal prices to repay debt. Inevitably, stricter credit has resulted. One large Canadian bank stopped financing gold projects after 1982 altogether.

Fortunately at Bousquet, escalating production and interest rate costs during 1978-80 were balanced by much higher gold prices. These in turn, however, dealt us a paper loss on forward sales when a seven month completion delay occurred. Make up deliveries from other company operations helped, but completion on time becomes especially important during inflation in costs or prices if metal is presold. In spite of poor ground conditions, however, the excellent continuity of the ore allowed us to double production and the loan was repaid in two years.

In summary, then, mine development must have as its foundation sound geological understanding. The technical and design work during feasibility must be done by well qualified, experienced people.

The essence of this work must be appreciated by the executives planning the financing for correct risk evaluation in the rapidly changing market environment. Calculus theory and the computer are two useful guides in exploring possible outcomes but there is no substitute for the executive's understanding of the productive capability and technical limitations of the orebody. Armed with this understanding, he can plan cash flows necessary in future years to discharge capital and service charges and to continue the ongoing development which is always more expensive and urgent than planned. Escalating feasibility metal prices should only be used with great caution and reliance on discount cash flow and various rate of return yardsticks distance the planner from reality somewhat and are better used for comparisons. The size of projected annual cash flows, especially during the first few years of production, must be carefully scrutinized for unrealistic assumptions and unreliability. Likelihood and ability of increasing production within a couple of years should be very high and allowed for in design. This is the pre-eminent safety factor which lets management protect healthy margins in the face of adversity and is as important to the creditor as to the producer. Conservative safety factors are as applicable in financings as they are in engineering design.

Today, five years after startup, Bousquet production has tripled and ore reserves have increased ten times. The mine cash flow last year was $25 million and further expansion is planned, demonstrating that the planning begun at feasibility never really stops. It should include not only early repayment of external financing but a healthy enough cash generating potential to internally finance substantial expansion.

LA MINE DOYEN

Following the Bousquet discovery, a 51% interest was acquired in an adjacent property with a known submarginal underground gold deposit. Overburden had been assumed to be up to 150 feet thick in places and our geologists felt that a $6,000 verification program was needed. The overburden was found to be much shallower, permitting a drill program to test a 300 foot subsurface horizon. The resultant rich open pit orebody discovery will produce well over a million ounces of gold when completed in 1988. Again the geological environment was excellent with a zone initially 1,000 feet long dipping 65 degrees with mining widths of 100 feet, permitting an eventual pit design target depth of 300 feet on a final waste-ore ratio of 6:1. The discovery during 1977 led to a feasibility plan being approved in September 1979 by the joint venture partners under Lac's management. Another unused Lac mill 30 miles away was available to handle 2,000 tons per day, permitting a fast and extremely low capital entry of $6.7 million. Equipment was leased, substantial contractor involvement was planned including trucking to the custom mill, and the project was generally expedited into production by March 1980. These steps were taken to permit internal financing, to maximize cash flow in the early years, to make an early determination of the orebody's characteristics, and to take advantage of exceptionally high metal prices. Our development of Doyon was greatly facilitated by both the experience and cash flow gained from the earlier Bousquet project. The ore characteristics, however, differed from Bousquet's and initially our consulting metallurgist advised roasting, which would have substantially delayed the project, raised its cost four times and possibly invalidated it environmentally. The Company's long operating experience on various ore types helped to overcome this problem and the addition of large quantities of lime permitted high recoveries in an extended but reasonable retention time using a conventional cyanide circuit.

Today, La Mine Doyon remains a relatively low cost producer like Bousquet, mainly by constantly increasing the scale of operations which has kept unit costs below $250 Cdn. per ounce. Fifty ton trucks are now used with ore and waste removal exceeding 30,000 tons per day on stripping ratios of 8:1, making it certainly one of the largest open pit gold producers in North America.

These two highly successful operations helped to multiply Lac's cash flow ten times and permitted much higher exploration and project development. Geologically, it honed up and reinforced our technical concepts about syngenetic gold deposits, improving our ability to recognize favourable suites of rock types during exploration. It did not, however, improve the balance sheet of the Company a great deal as we continued to follow the usual path of most miners in plowing back all available earnings after tax and dividends.

HEMLO, WILLIAMS PROJECT

In mid 1982, our exploration team located a significant surface zone grading 0.2 ounces per ton over a strike length of 500 feet open to the west and at depth, with widths of about 60 feet and dipping 65 degrees. Approximately three million tons of ore, including over a million tons open pit indicated an obviously economic deposit. More important, the geological environment to the west and at depth on the Company's property appeared to be excellent.

A 1982 year-end consolidation and simplifying reorganization broadened interest in Lac and increased the price of its shares. By May 1983 prevailing market conditions and wide-spread investment dealer support permitted a treasury issue of two million common shares for $52 million. Subsequently, the orebody expanded by drilling to over forty million tons of slightly lower grade and in late summer a feasibility for pit and underground was started for review by April 1984.

Technical and economic considerations on the project were gradually satisfied. Company cash flow which had averaged $60 million per year for three years was expected to continue, more than satisfying internal project needs of $40 million a year. Moreover, forward sold gold for 1984 guaranteed the year's budget eliminating this unknown.

By February 1984, we realized that the project would cost about $250 million and that internal funds including cash flow projections still left a $100 million gap by mid-1985 using current gold prices of $375 U.S. Management, therefore elected to raise the funds before the formal feasibility completion. We felt it important to take advantage of favourable gold, bond and stock markets and were sceptical of these continuing short term. Recent history demonstrates that there are quite short 'windows' during which successful new gold related issues can be floated. Although gold markets themselves have developed enormous depth and liquidity, they are very sensitive to many other conditions - interest rates, oil prices, industrial production, inflation, currency movements, international debt, banking crises, military conflicts, and normal gold supply/demand factors. In short, we felt a successful gold related financing would be more a matter of availability and timing than price. Windows for other straight issues such as preferred shares or debentures had similarly become shorter and less predictable.

In February, therefore, the Company began gathering prospectus information and explored finance questions with dealers and institutions. The Company's long standing conservative bias towards financing mine development principally with equity and keeping debt in easily manageable limits at all times because of the inherent risks of the business, tended to provide for an eventual equity component to the financing, although less than a year had passed since the last one. Straight and convertible preferreds were considered and were rejected as being more expensive to carry than straight common. (Dilution was not considered to be serious since the issue of one million shares, say, would only be approximately three and a half percent dilution.) Straight debt was available in Canada and Europe but with a prime rate of 10.5% and expected to rise slightly during the year, a less expensive instrument was sought.

Recent European interest in gold related securities allowed us to work closely with our principal underwriter to design what has become a highly successful, innovative and bell-weather issue of 8% Euro-debentures with a gold warrant attahed. (This low coupon was more than 550 points under the market.) The debenture is unsecured except for standard covenants and the warrant permits a conversion into gold at U.S.$460 any time during the next five years. ($460 was a 15% premium over gold at the time of issue but if the coupon had been at the market of about 13-5/8%, the premium would have worked out to about 47%.) Concern was expressed at the time about the conversion price of gold and the premium over the market. The purchasers need not have worried. The issue was heavily over-subscribed at the time of pricing and has been one of the few recent debenture issues to remain at par in spite of subsequent weak bond, gold and other markets. The lesson here is that the unusually long call on gold of five years has made the warrant price relatively insensitive to gold price fluctuations. Inadvertently, Lac has become the world's largest option writer and consequently, the warrant has become a closely watched standard in the gold futures market.

This Euro-issue was deployed first to take advantage of a window in gold markets and active, stable European bond markets. The issue was not sized larger than $50 million U.S. for three reasons: first, to keep a tight market; second, to limit the amount of bullion for possible warrant conversion should the price rise above U.S.$460 towards the end of the five year period, (much prior conversion is not contemplated because of the usual market premium which will make it more attractive for vendors to sell in the marketplace rather than convert); third, from a longer run perspective it was considered prudent not to over expose the Company to foreign institutional lenders who perhaps lacked an intimate knowledge and caring relationship with the issuer.

This then has turned out to be one of the few ideal investment instruments for both the issuer and purchaser. The Company will save very large amounts of interest over expected prevailing rates and should gold prices eventually rise above U.S.$460, it is expected that both productive levels and cash flow will be substantially much larger by proportion than the obligation to make available a small portion of its output. The creditor should continue to enjoy the benefits outlined of both a two way bet and a stable market, should he continue to hold the unit. Purchasers of the stripped warrant will own perhaps the only such call currently available, which undoubtedly will provide flexibility to their portfolios, and buyers of the stripped debenture will enjoy the usual capital gain, a high effective rate and in some cases, foreign exchange advantages. This type of multi-faceted instrument is characteristic of today's structurally transformed credit markets. More diverse financial institutions with much wider investment needs are today competing for corporate issues both as agents and principals. New liquid international markets continue to open up and broaden and changing currency markets add to their complexity. The freeing of the bullion price and the subsequent fascinating development of its worldwide market has played a significant part in this restructuring of financial markets.

This point was emphasised by our Euro-issue, and continued demand for high quality gold-related debt or even equity instruments should persist. That no other issues have been presented at time of this writing confirms that, outside of South Africa, there are very few sound multi-mine gold producing companies that could issue an obligation of comparable quality.

A few weeks later, the Company further enhanced the debenture with a one million common share issue, which all together raised $95 million Cdn., putting the treasury at $175 million. Together with expected internal cash flow and the improved cash generation from this working capital position itself, the Company should now be able to finance its projects for the next three years.

SUMMARY

Mine financing requires concentration and a variety of skills on the part of the Chief Executive Officers and his staff. The CEO may well be advised to guide the process personally to avoid misunderstandings, delays and long meetings. Ideally, his own experience or that of his staff should include a thorough understanding derived from experience on geological, construction, market and mining matters. He must be able to realistically visualize what is most likely to be produced and at what cost. His analysis will have to include contingent run-outs of many possible but unhappy eventualities so that risk may be truly evaluated, and the security correctly designed.

The dual objectives of a successful financing are a happy purchaser, and production profits as planned. Thus the executive widens his stewardship to include both a bigger company and a broader investment following. He, therefore, must exercise more leadership, honour additional obligations, fulfil new promises.

Financing well ahead of requirement is a good idea given the shortened wave-length of today's markets. Metal prices on the other hand often stay depressed for unpredictably long periods, and are subject to rapid decline as well.

Pre-financing mining development is just as important a concept to us as pre-developing underground reserves so production can be expanded.

In short, top management must be experienced, highly organized, realistic and somewhat opportunistic to successfully finance new mines.

12.2

A
CASE STUDY OF
THE STRATHCONA SOUND PROJECT
(A NON RECOURSE PROJECT FINANCING)

H.W. Schreiber
Behre Dolbear & Co.

G.R. Castle
Chemical Bank
New York, New York

INTRODUCTION

The small mining company faces the dilemma of how to finance the development of its properties. Many of these small companies don't have the financial resources to pay for development costs from their own funds and frequently they do not even have the financial strength to permit them to borrow the money needed for development. A project financing can sometimes be arranged in such a way as to solve this dilemma.

A typical project financing will not provide the answer to the "little guy's" problem because most project financings require that the sponsors of the project provide completion undertakings and sometimes other undertakings. These undertakings are often a prerequisite to obtaining financing, and if the sponsor is not financially strong enough to provide the comfort that the lenders require, then the sponsor will have difficulty obtaining financing no matter what undertakings are offered. However, with a little imagination and the right set of circumstances, a project financing can sometimes provide the solution to the small mining company's problem.

The Strathcona Sound Project is unusually suited as a case study because the sponsor, Mineral Resources International Limited,(MRI) was able to arrange a project financing for the development of the project even though the lenders probably would not have been willing to loan a similar amount of money direct to MRI based on the full faith and credit of MRI's balance sheet. The way that MRI was able to arrange this financing makes for an interesting story.

THE CASE

In early 1974, MRI owned 100% of a small to medium sized zinc - lead - silver massive sulfide deposit (Property) located 475 miles north of the Arctic Circle less than two miles from deep water on Strathcona Sound which is a shipping zone in the Canadian Arctic. The Property was situated at a desolate location except for an Eskimo settlement (Arctic Bay) which was about 20 miles west of the Property. In other words, the project required that complete infrastructure would need to be installed in this far northern location. Independent reserve and feasibility studies (Study) had been completed on the Property which confirmed a total of 6,825,000 tons of reserves classified and summarized as shown below on Table I:

TABLE I

(In Thousands)

	Tonnage	Percent Zn.	Percent Pb.	Oz. Per Ton Ag.
Proven	5,638	14.29	1.46	1.80
Probable	1,187	13.40	1.17	1.66
	6,825	14.12	1.40	1.77

These reserves were based on 229 diamond drill holes totalling 58,909 feet and were established on a cut-off grade of 7% zinc and a minimum mining height of seven feet. The overall mining recovery rate was estimated at not less than 85% and not more than 90%. The feasibility study indicated that it would cost an estimated $45,044,000 to develop an underground mine, a concentrator, a townsite, a power plant, storage facility, loading dock, a tailings disposal area and an airstrip on the Property with an annual capacity of 525,000 tons of ore per year. The construction period was expected to take two years and projected capital costs were broken down as set forth below on Table II:

TABLE II

(000's Omitted)

Item	Capital Cost
Mining	$ 2,682
Concentrator	4,805
Concentrate Storage and Handling	3,166
Power Plant	2,947
Fuel Storage	1,105
Water Supply	800
Service Buildings	987
Mobile Equipment	446
Dock	2,305
Airport	4,326
Roads	1,001
Tailings Disposal	254
Townsite	4,750
Ocean Freight	1,840
Parts Inventory	216
Insurance	448
Owners Costs	291
Unallocated Contingency	1,000
Final Engineering	1,927
Project and Construction Management	1,642
Sub Total	$ 36,938
Preproduction Interest and Financing Costs	3,106
Working Capital	5,000
Total	$ 45,044

Table III summarizes the economic forecast (Company Projection) that was included in the Study and listed below are the more important assumptions on which this Company Projection was based.

1. Commercial production from the Property commences June 30, 1976 and the mine ore tonnage and grade schedule are as follows:

Year	Tonnage	Percent Zn.	Percent Pb.	Oz./Ton Ag.
1	525,000	15.75	.54	1.76
2	525,000	17.47	1.01	2.05
3	525,000	16.67	1.67	2.00
4	525,000	15.32	1.48	1.97
5-8	525,000	15.34	1.37	2.00
9-13	525,000	11.03	1.70	1.42

The schedule of concentrate production is set forth below:

Year	Zn. Concentrate S. Tons @60% Zn.	Pb. Concentrate S. Tons @60% Pb.
1	130,900	4,200
2	145,200	7,900
3	138,500	13,100
4	127,300	11,600
5-8	127,500	10,700
9-13	91,600	13,300

2. The Economic Forecast has been based on the following market price assumptions:

   a) Zinc 22.7¢/lb. and 25.5¢/lb. yielding about 90% of the total revenue;
   b) lead 16.5¢/lb. yielding about 9% of the total revenue;
   c) silver 200¢/oz. yielding about 1% of the total revenue.

3. Revenues reflect a net smelter return of:

   a) for zinc, 50% of the gross contained metal value in concentrates;
   b) for lead, 53-1/2% of the gross contained metal value in concentrates;
   c) for silver, 22-1/2% of the gross contained metal value in concentrates.

MRI was a small Canadian Company which did not have the financial capacity to borrow $45 million as indicated by the financial information summarized in Tables IV and V below:

TABLE IV

Mineral Resources International Limited
and Subsidiary
Actual Balance Sheet
December 31, 1973

ASSETS	(Actual Dollars)
Current Assets	$ 114,547
Mining Properties	3,314,389
Oil & Gas Leases, Plant and Equipment	1,177,913
	$4,606,849

TABLE III

Company Analysis - Strathcona Sound Project
Economic Forecast for the Property
(000's Omitted)

Year	Net Smelter Return	Cash Operating Costs	Royalty	Shipping, Handling, Insurance & Admin. Expenses	Capital Cost Allowance and Depletion	Interest Expenses	Canadian Federal Income Taxes	Net Income	Cash Flow (Before Interest) Available For Debt Service		Present Worth Cash Flow Available For Debt Service	
									Annual	Cumulative	Annual	Cumulative
1976	$10,093	$ 4,159	$--	$ 1,615	$ 4,169	$ 150	$ --	$ --	$ 4,319	$ 4,319	$ 3,926	$3,926
1977	20,184	7,618	--	3,229	6,647	2,690	--	--	9,337	13,656	7,716	11,642
1978	20,184	7,618	--	3,229	7,258	2,079	--	--	9,337	22,993	7,015	18,657
1979	20,184	7,618	--	3,229	7,924	1,413	--	--	9,337	32,330	6,377	25,034
1980	20,184	7,618	--	3,229	8,063	1,274	--	--	9,337	41,667	5,798	30,832
1981	19,829	7,618	--	3,177	6,999	1,182	392	461	8,642	50,309	4,878	35,710
1982	18,915	7,618	1,274	3,044	2,142	1,091	1,723	2,023	5,256	55,565	2,697	38,407
1983	18,915	7,618	1,957	3,044	1,978	999	1,527	1,792	4,769	60,334	2,225	40,632
1984	18,915	7,618	1,967	3,044	2,047	907	1,533	1,799	4,753	65,087	2,016	42,648
1985	18,915	7,618	2,038	3,044	442	816	2,280	2,677	3,935	69,022	1,517	44,165
Sub-Total:	$186,318	$72,721	$ 7,236	$29,884	$47,669	$12,601	$ 7,455	$ 8,752	$69,022	---	$44,165	--
Thereafter:	47,149	22,854	3,627	7,734	702	1,898	4,754	5,580	8,180	$77,202	2,640	$46,805
Total:	$233,467	$95,575	$10,863	$37,618	$48,371	$14,499	$12,209	$14,332	$77,202		$46,805	

LIABILITIES AND SHAREHOLDERS EQUITY		
Secured Bank Loan	$	250,000
Current Portion Long Term Debt		14,130
Other Current Liabilities		85,626
Current Liabilities	$	349,756
7% Convertible Debentures		330,000
Advances Payable		325,000
Deferred Payable		7,500
		1,012,256
Capital Stock		23,681,449
Deficit		(20,086,856)
		$4,606,849

TABLE V

Mineral Resources International Limited
and Subsidiary
Statement of Income and deficit

	(Actual Dollars) Years Ended December December 31,	
	1973	1972
Oil and Gas Sales	$85,269	$100,815
Expenses		
Production Expense	$ 18,560	$ 27,064
General Overhead	88,505	137,005
Interest	37,049	32,428
Abandoned Properties & Related Expenses	33,008	15,588
	$177,122	$212,085
Loss from Production Before Undernoted Items	$(91,853)	$(111,270)
Depletion & Depreciation	54,160	78,773
Loss on Sale of Property, Plant & Equipment	8,309	35,981
Loss for the Year Before Extraordinary Items	$(154,322)	$(226,024)
Extraordinary Items		
Write-down of Investment in Guatemala	$ 185,650	-
Write-down of Helium Leases	-	477,400
Net Loss for the Year	$(339,972)	$(703,424)

Furthermore, Mineral Resources' financial condition was such that its Guaranty of Completion would not be adequate to support this kind of Project. An additional complicating factor is the location of this venture - 475 miles north of the Artic Circle. The hostile environment of this location would be enough to cause any prospective lender to hesitate. On the face of it even a project financing would seem like a "long shot". However, there were several unique circumstances associated with the venture.

1. The Canadian Government took an interest in the Project because it would provide jobs for the Eskimos that lived in the area and it would provide development for the area.

2. The deposit was a very rich grade of ore, and

3. The development occurred at a time when it looked like the zinc/lead price had a lot of potential.

As a result of these unusual circumstances, a project financing became a reality. MRI capitalized on these advantages that the Project had going for it.

First MRI negotiated an arrangement with the Canadian Government for a total of $16,544,000 of subordinated grants and loans to provide infrastructure to the Project. This represented 37% of the total $45,044,000 cost of the Project. However, in return for these grants and loans the Canadian Government received an 18% equity in the Project.

The next step toward arranging a financing was to capitalize on the favorable market for these minerals - zinc, lead and silver. MRI started negotiating with prospective purchasers to cover the sale of the output of the mine. However, these negotiations extended far beyond the normal terms that you would expect to find in a normal commercial sales contract because MRI needed provisions in their arrangements with these Purchasers that would support the financing of all or some part of the balance (that is $28,500,000) of the cost of developing the Project.

The results of these negotiations are summarized in the outline attached as Exhibit A at the end of this paper, and are also described graphically on Chart I.

Chart Showing Structure Of Financing

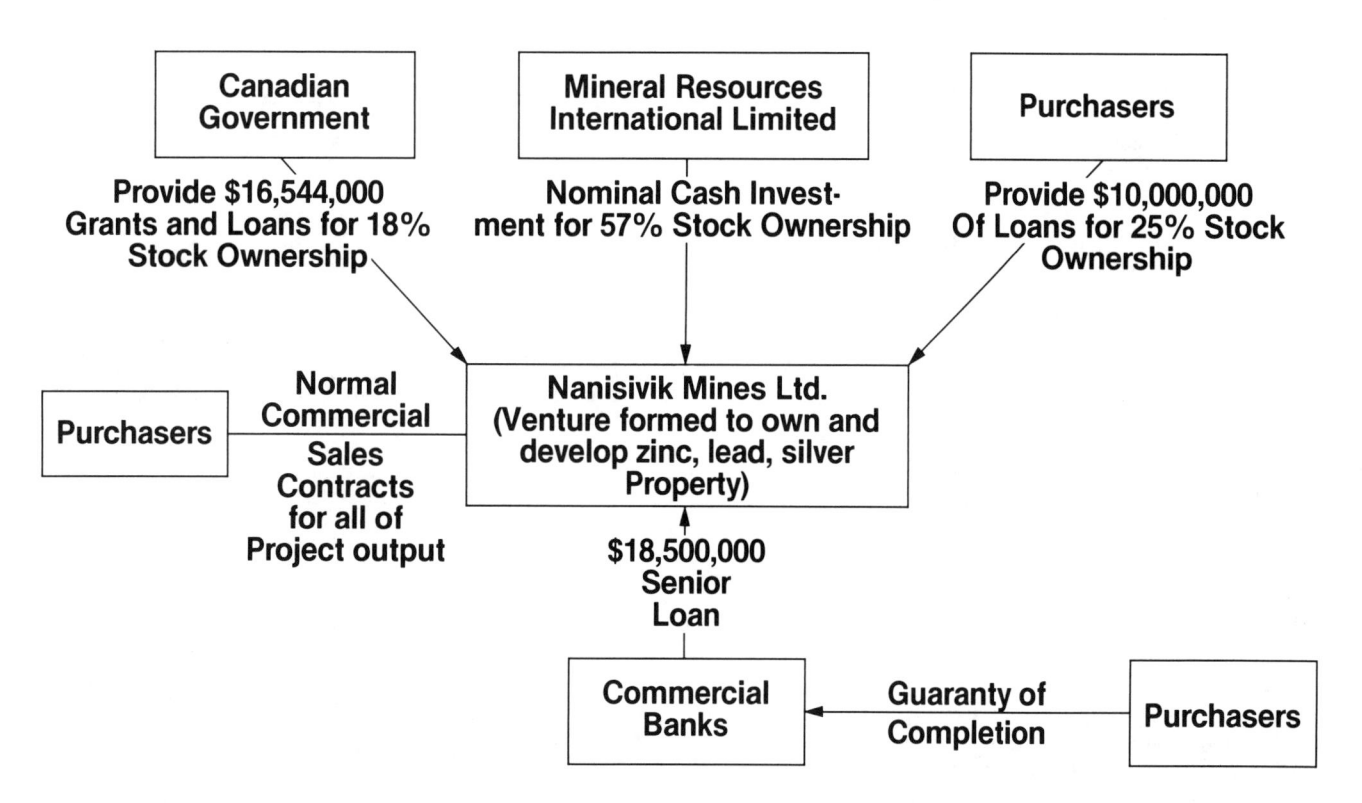

```
┌─────────────┐      ┌──────────────────┐      ┌─────────────┐
│  Canadian   │      │ Mineral Resources│      │ Purchasers  │
│ Government  │      │International Ltd.│      │             │
└─────────────┘      └──────────────────┘      └─────────────┘
       │                      │                       │
Provide $16,544,000    Nominal Cash Invest-    Provide $10,000,000
Grants and Loans for 18%  ment for 57% Stock   Of Loans for 25% Stock
  Stock Ownership          Ownership              Ownership
       │                      │                       │
       ▼                      ▼                       ▼
┌─────────────┐  Normal   ┌──────────────────────────────┐
│             │Commercial │     Nanisivik Mines Ltd.     │
│ Purchasers  │──Sales────│ (Venture formed to own and   │
│             │Contracts  │  develop zinc, lead, silver  │
│             │for all of │          Property)           │
└─────────────┘Project    └──────────────────────────────┘
               output                   ▲
                                   $18,500,000
                                    Senior
                                     Loan
                                        │
                              ┌──────────────┐  Guaranty of  ┌─────────────┐
                              │  Commercial  │◄──Completion──│ Purchasers  │
                              │    Banks     │               │             │
                              └──────────────┘               └─────────────┘
```

As you can see from Exhibit A and Chart I, MRI negotiated an arrangement with the Purchasers whereby the Purchasers made outright loans to the Project in the amount of $10,000,000 or 22% of the total estimated cost of the Project.

Finally, MRI approached banks for financing of the $18,500,000 balance of this development cost and Table VI shows a Revised Projection that the Banks prepared in the analysis of this loan request. This Revised Projection is based on proven reserves only and a zinc price of $22.7¢ lb.

These bank loans were senior to both the loans from the Purchaers and the grants and loans from the Government of Canada. Furthermore, the Purchasers were even willing to provide a guranty of completion (of the project) for the benefit of the Banks.

In return for this support to the Project the Purchasers received normal commercial sales contracts for all of the Project output plus a 25% equity interest in the Project. Thus MRI was able to retain a 57% interest in the Project and arrange a financing for 100% of the development cost.

The advantages to MRI are obvious. MRI is carried 100% but this cost MRI the loss of 43% of the equity in the Project. Nevertheless, this was still quite a feat.

ACTUAL PERFORMANCE AT THE STRATHCONA SOUND PROJECT

The original feasibility study for the Strathcona Sound Project estimated total capital expenditures for the Project at $40,044,000 (without working capital) but the definitive costs through August 31, 1976 came in at $53,777,000. This is a $13,733,000 or 34% overrun. Table VII shows a comparison of the actual costs with the budget. Some of the problems that caused this overrun include:

1. An accidental fire in a compressor room which contributed to a 5 to 6 month delay in getting the project on line.

2. The Government refused to permit the project to dump tailings into Strathcona Sound and this contributed to the delay.

3. This delay resulted in $4,362,000 of additional interest expense for the project over and above the original projections.

4. The Canadian Government wanted the town to include more and better facilities such as better recreation and other amenities and this accounted for the $2,430,000 overrun in the Townsite item and could also have been a factor in the construction camp operation increase of $2,238,000.

5. The mill capacity was increased 20% which accounts for the $2,315,000 variance in the concentrator item.

6. The high rate of inflation. Inflation was running at a rate of:

12% in 1975
11% in 1976
and
10.5% in 1977

The Purchasers of the output from the Project financed $12.5 million of this overrun and the Canadian Government covered the balance. However, as a consequence of these overrun financings, Mineral Resources had to give up an additional 3% of the project so that the final ownership ended up:

	Original	Final
Purchasers	25	28
Mineral Resources	57	54
Canadian Government	18	18
	100	100

During the first three years of operation, the Project processed more tonnage than was projected. The feasibility study contemplated 525,000 tons a year. The mill was overdesigned and processed 660,000 tons a year. Operating costs ran about as anticipated in the feasibility study. However, run of mine zinc grades were lower than originally projected because there was more dilution than expected. Also the lead content was higher. The slump in the price of zinc probably had the most serious impact on the Project. During this first three years no significant amount of additional reserves was found.

TABLE VI

Revised Projection - Strathcona Sound Project
(Based on Proven Reserves only at a Production Rate of 525,000 tons a Year
at 90% Confidence Level and Zinc Price - $22.7¢ lb)
(000's Omitted)

Year	Net Smelter Return	Cash Operating Costs	Shipping, Handling, Insurance & Admin. Expenses	Capital Cost Allowance and Depletion	Canadian Federal Income Taxes	Net Income	Cash Flow (Before Interest) Available For Debt Service Annual	Cumulative	Present Worth Cash Flow Available For Debt Service Annual	Cumulative
1976	$ 8,253	$ 4,159	$ 1,490	$ 2,604	$ -	$ -	$ 2,604	$ 2,604	$ 2,367	$ 2,367
1977	16,507	7,618	2,981	5,908	-	-	5,908	8,512	4,883	7,250
1978	16,507	7,618	2,981	5,908	-	-	5,908	14,420	4,439	11,689
1979	16,507	7,618	2,981	5,908	-	-	5,908	20,328	4,035	15,724
1980	16,507	7,618	2,981	5,908	-	-	5,908	26,236	3,568	19,392
1981	15,908	7,618	2,883	5,407	-	-	5,407	31,643	3,052	22,444
1982	15,365	7,618	2,794	4,953	-	-	4,953	36,596	2,542	24,986
1983	15,365	7,618	2,794	4,953	-	-	4,953	41,549	2,311	27,297
1984	15,365	7,618	2,794	4,953	-	-	4,953	46,502	2,101	29,398
1985	13,829	7,618	2,544	3,667	-	-	3,667	50,169	1,414	30,812
Sub-Total:	$150,113	$72,721	$27,223	$50,169	$ -	$ -	$50,169		$30,812	
Thereafter:	3,364	1,988	624	752	-	-	752	$ 50,921	264	$ 31,076
Total:	$153,477	$74,709	$27,847	$50,921	$ -	$ -	$50,921		$31,076	

Note: No royalty is earned at this level of earnings.

TABLE VII

THE STRATHCONA SOUND PROJECT

Comparison of Projected Costs with Actual Cost

Item	Projected Cost	Actual Cost	Variance	
Mining	$ 2,682	$ 2,748	$ 66	(2.5%)
Concentrator	4,805	7,120	2,315	(48.2%)
Concentrate Storage & Handling	3,166	3,753	587	(18.5%)
Power Plant	2,947	2,834	(113)	3.8%
Fuel Storage	1,105	1,430	325	(29.4%)
Water Supply	800	1,073	273	(34.1%)
Service Buildings	987	827	(160)	16.2%
Mobile Equipment	446	1,316	870	(195.1%)
Dock	2,305	2,705	400	(17.4%)
Airport	4,326	3,515	(811)	18.7%
Roads	1,001	1,687	686	(68.5%)
Tailings Disposal	254	474	220	(86.6%)
Townsite	4,750	7,180	2,430	(51.2%)
Ocean Freight	1,840	2,418	578	(31.4%)
Parts Inventory	216	500	284	(131.5%)
Insurance	448	494	46	(10.3%)
Owners Costs	791	308	(483)	61.1%
Unallocated Contingency	1,000	-	(1,000)	100.0%
Final Engineering	1,927	1,321	(606)	(31.5%)
Project & Construction Management	1,642	1,649	7	(.4%)
Preproduction Interest & Financing Costs	2,606	6,968	4,362	(167.4%)
Exploration	-	258	258	
Environmental Studies	-	515	515	
Communications	-	340	340	
Construction Camp Operation	-	2,238	2,238	
Training & Recruiting	-	106	106	
	$40,044	$53,777	$13,733	(34.3%)

Another factor that has had an adverse impact on the economics of the Project was the foreign exchange risk. Although the Senior Loan was in Canadian Dollars, the loans from the purchasers were in German Marks and Dutch Guilders and, as a result, the Project experienced some currency losses due to the decline in the value of the Canadian Dollar.

However, despite all of these problems, all loan repayments were made on time and there were no defaults under the documents. As a matter of fact, mandatory prepayment provisions in the loan documents which were payable out of excess cash flow kicked in and some of the loans prepaid.

EXHIBIT A

June 30, 1974

2 Year Standby Commitments
Followed
By
Term Loans
To
Nanisivik Mines Ltd.

Amounts:

Commercial Bank Senior Loan	Canadian Government Subordinated Grant And Loan	Subordinated Loans Made And Arranged By Purchasers	Total Financing
$18,500,000	$16,544,000	$10,000,000	$45,044,000

Purpose: Mineral Resources International Limited (MRI) owns an undeveloped zinc - lead - silver deposit (Property) which is located 475 miles north of the Arctic Circle on Strathcona Sound in the Canadian Arctic and MRI has formed Nanisivik Mines Ltd. (Nanisivik) for the purpose of owning and developing a mine, concentrator and other facilities (Project) on the Property. The development of the Property is expected to take two years and the total estimated Cost (Cost) of the development of the Project is $45,044,000.

MRI has negotiated certain Sales Contracts (Sales Contracts) with several Purchasers (Purchasers) that are interested in buying all of the output from the Project and in return for these Sales Contracts the Purchasers have agreed to either make direct loans or arrange for up to $10,000,000 of Loans (Shareholder Loans) to Nanisivik. Also as part of this negotiation with the Purchasers, MRI has given the Purchasers a 25% equity interest in Nanisivik.

Furthermore, the Canadian Government (Government) is anxious for the Project to be developed in order to provide jobs for the native Inuit Eskimos that live 20 miles from the Property in the settlement of Arctic Bay. As a result, the Government has agreed to make a total of $16,544,000 of Loans and outright grants (Government Financings) in order to cover a portion of the development Costs. The Government Financings will pay for part of the Infrastructure that will be needed at the Project. In order to obtain this Government Financing, MRI has agreed to give the Government an 18% equity interest in Nanisivik. The $18,500,000 balance of the Cost shall be financed by a Senior Loan from a group of commercial banks backed by a guaranty of completion from the Purchaers. The balance (57% of the equity has been retained by MRI.

Commitment
Period: Commencing June 30, 1974 and continuing through June 30, 1976.

Maturity: Senior Loan - Commencing December 31, 1977 the Senior Loan matured in five equal semi-annual principal installments each equal to $3,000,000 through December 31, 1979 and one final principal payment in the amount of $3,500,000 on June 30, 1980.

Government Financing and Shareholder Loans - Commencing December 31, 1980 the Government Financing and Shareholder Loans (Subordinated Financings) commenced to mature.

Borrowings:
: Borrowings under each of the three portions of the Total Financing were made pro rata in proportion to the amount of original commitment for each of the three portions of the Financings.

Prepayments:
: Voluntary - Permitted in whole or in part at Nanisivik's option in the inverse order of maturity without penalty applied against the different financings on a pro rata basis in proportion to the original commitment.

  Mandatory - Within 60 days after the end of each calendar year (commencing with the year ended December 31, 1977) Nanisivik was required to apply as a prepayment against all of the Financings described herein 50% of Excess Cash Flow earned in the previous year. Each such Prepayment was applied against the different Financings described herein in following proportions - 70% against the Senior Loan and 30% against the Subordinated Financing. Excess Cash Flow was defined as the sum of:

  1. Net Income, plus
  2. Capital Allowance and Depletion, plus
  3. All other non cash charges

  All determined after the deduction and deposit into an Escrow Account of an amount sufficient to meet the next two semi-annual regular princpal payments (Maturities) on the Senior Loan

Collateral:
: The Senior Loan was secured by a first Mortgage on the Project and the Property and an assignment of the Sales Contracts. Any Collateral for the Subordinated Financings was junior to the Collateral for the Senior Loan.

Guaranty of Completion:
: As part of the consideration for the Sales Contracts the Purchasers covenanted to:

  1. Cause the Project to be Completed prior to June 30, 1977.

     Completion was defined as:

     a) The operation of the Project at 100% of capacity for a continuous period of six months (Test Period) and during such Test Period:

        (i) the Mill was required to recover 95% of the design recovery rate;

        (ii) the grade of concentrate had to be equal to the level projected in the feasibility study;

        (iii) Cash Operating Costs could not exceed the projections contained in the feasibility study;

        (iv) the Project was required to have generated a Cash Flow before interest deductions of $4,500,000;

     b) A point in time when all costs of developing the Project have been paid.

     c) A point in time when there are no defaults under any of the Loan Documents.

  2. Pay for all overrun expenses in connection with the development of the Project and all advances to Company X to pay for such overruns shall be on a subordinated basis.

The obligations of each Purchaser under the Guaranty of Completion were several obligations pro rata in proportion to the percentage of production that that Purchaser took under the Sales Contracts.

Other Conditions of Lending:
Usual, plus prior to any Borrowing hereunder:

1. All governmental approvals and clearances had to have been obtained.

2. The Lenders needed to be satisfied that Nanisivik has good and marketable title to the Property free and clear of all liens.

3. All legal details were to be satisfactory to special counsel for the Lenders.

4. The Commercial banks needed to be satisfied with the terms of the Subordinated Financings including the terms of the Subordination provisions.

5. The Lenders needed to be satisfied with the terms of the Sales Contracts. These Contracts provided for the sale of all of the Project's output for at least the term of the Senior Loan. There was not any obligation for the Purchasers to make any payments under the Sales Contracts except for concentrate actually produced and shipped, and the price provided under the Contracts was the current market price at the time of sale. The contracts contained normal force majeure and default provisions.

Covenants of Nanisivik:

A. Nanisivik shall:

1. Maintain existence, pay taxes, royalties and liabilities when due, maintain the Project in good repair, and defend the Project against all claims.

2. Operate the Project in a good and workmanlike manner.

3. Maintain Insurance on the Project.

4. Furnish annual audited, quarterly interim unaudited statements and quarterly production reports.

5. Permit representatives of the Lenders to inspect the Project.

6. Give written notice of any claim or demand effecting title to the Project or any adverse governmental action against such title.

7. Complete the Project prior to June 30, 1977.

8. Pay all closing costs in connection with the Financings described herein.

9. Maintain working capital of at least $_____ at all times.

B. Nanisivik shall not:

1. Incur, create or permit to exist any indebtedness for borrowed money, except the Senior Loan, the Subordinated Financings and other loans which are subordinated to the Senior Loan.

2. Incur, create or permit to exist any mortgage, lien or any encumbrance (including CSA's, purchase money mortgages and other title retention agreements and including any production payments) on any of its assets, whether now owned or hereafter acquired, except for the usual exceptions and the Collateral provided for herein.

3. Guaranty, endorse or otherwise in any way become contingently liable

for any indebtedness of any other party (including take or pay or working capital maintenance agreements).

4. Incur or become liable for the payment of rent.

5. Acquire all or a substantial part of the assets or merge or consolidate with any other entity.

6. Sell all or a substantial part of its assets.

7. Make any loans or advances or investments in any other entity.

8. Make any distributions to the Stockholders of Nanisivik or retire for value any of Nanisivik's stock.

9. Amend the Sales Contracts.

10. Engage in any business other than the operation of the Project.

Events Of Default: In the event:

1. Of non payment of principal or interest when due.

2. Of misrepresentation.

3. Of voluntary or involuntary bankruptcy of Nanisivik or any of the Purchasers (prior to Completion only as to the Purchasers).

4. Of cross acceleration.

5. Nanisivik is no longer owned by MRI, the Purchasers and the Government.

6. Of final judgment against Nanisivik.

7. Of default under any Sales Contract.

8. Of default under any other provisions of the Loan Agreement.

9. Any Purchaser fails to perform its obligations under the Guaranty of Completion.

12.3

INTERNATIONAL MINING COMPANY CASE STUDY
INDUSTRIAL MINERA MEXICO, S.A.*

Tomek Ulatowski

Crocker National Bank
San Francisco, California

INTRODUCTION

In 1974, Industrial Minera Mexico, S.A. (IMMSA), agent for the borrowers, was the largest privately-held mining, smelting, and refining company in Mexico (accounting for 32% of Mexico's nonferrous metal and coal production in 1974), the fourth largest company in terms of sales, and the second most profitable company in Mexico, excluding government wholly-owned and operated industries. The company was engaged in the mining, smelting and refining of coal, copper, fluorite, gold, lead, silver, zinc and related by-products for domestic consumption and export.

A combination of favorable factors led IMMSA to propose an investment plan amounting to US$286 million for 1974 through 1978, of which US$150 million was to be financed. These factors included:

Substantial profits in 1973 and 1974

Positive near-term outlook for metal prices

Success of the reinvestment program in increasing productivity

Substantial additions to proved reserves during 1973-1974 and

Fiscal incentives provided by the Mexican government to stimulate investment in the mining industry.

The primary purpose of the investment program was to construct a new electrolytic zinc refinery and to expand and modernize many of the company's existing mining and refining facilities.

DISCLAIMER
Any opinions, findings, conclusions, or recommendations expressed in this case are those of the author and do not necessarily represent the views of Industrial Minera Mexico, S.A. or any financial institution.

THE BORROWER

Company History

Compania Minera Mexico, S.A. was incorporated in the city of Monterrey in 1918 as a wholly-owned subsidery of ASARCO Inc. (formerly American Smelting and Refining Company, New York) for the purpose of mining and processing non-ferrous metals and related products.

In compliance with the Government's Mexicanization program, ASARCO Inc. divested itself of 51% of its equity in Compania Minera Asarco, S.A. on June 30, 1965. Subsequently, ownership in the divested share was apportioned as follows:

15% ownership in Compania Minera Asarco, S.A. to a group of private investors, and
36% in trust for subsequent public sale.

The company's name was changed to Asarco Mexicana, S.A. in 1965.

Mexican ownership in Asarco Mexicana, S.A. was increased to 51% in 1967, following the purchase of the 36% share ownership held in trust, and then to 66% in 1974 when ASARCO Inc. divested itself of an additional 15% of its holdings in Asarco Mexicana, S.A. Following its 1974 divestiture of 15% of Asarco Mexicana, S.A., ASARCO Inc's ownership in Asarco Mexicana, S.A. was reduced to 34%. In 1974, the company changed its name for the second time to Industrial Minera Mexico, S.A. (IMMSA).

Fiscal Agreement

IMMSA entered into a fiscal agreement with the Mexican government in 1969. The agreement, which was renewed in 1974, provided for 100% of the production and export tax corresponding to the federal government with a minimum of 50% of the amount refunded to be reinvested in fixed assets. Further, the agreement permitted the company to

---
* Presented at the Minerals Resource Management Committee of SME - AIME Short Course on Project Finance in the Development of Mineral Resources, February, 1981.

depreciate 40% of any investment during the first year of acquisition and 8% in each subsequent year until 1978 when the agreement was to be terminated. A further provision of the agreement provided for 40% of income taxes paid at the rate of 42% of taxable income to be refundable, subject to a 42% tax in the following year. The net result of the fiscal agreement was to make the company's effective tax rate approximately 29.2% of its taxable income.

Management and Labor

IMMSA's management is well experienced in mining, smelting, and refining. A good balance had been struck between general management and technical expertise. Depth and continuity of management were reinforced further by in-house training, communications, and incentive policies.

IMMSA employed approximately 11,300 workers outside of the management category. This labor force operated under contracts that had been reviewed, modified and renewed in recent years. IMMSA's labor force is highly experienced, having an average tenure and subsequent experience level of close to 20 years.

Throughout the company's history, IMMSA's management had maintained good working relations with its unions and the Mexican government. By Mexican industrial standards, the company's labor compensation, working conditions, and living standards compared favorably with other heavy industries domiciled in Mexico.

Mining Operations in 1974

In 1974, IMMSA's mining operations consisted of eight active units. By industry standards these units were considered to be cost effective and efficient. Table 1 summarizes salient characteristics of the units, including: positive and inferred ore reserves, average metal assay, production capacity, and other relevant comments.

Smelting and Refining Operations in 1974

<u>Chihuahua Lead Smelter</u> The original lead smelter built by IMMSA in Chihuahua at the beginning of the century gradually had been expanded to a design capacity of 10,000 tons of lead per month in 1974.

In 1969, IMMSA management decided to modernize the Chihuahua operations in phases, rather than building a new smelter. The first stage in the modernization process was initiated in 1971 with the construction of a new sinter plant. The plant produced 95,740 mt of lead during 1974, an increase of 5,740 mt over the preceding year.

<u>Monterrey Lead Refinery</u> In 1974, lead production in the Monterrey refinery increased 12% over the 1973 production, having achieved a ten-year record production of 98,120 mt. Silver production increased 21% over 1973 production, achieving a production level of 646,000 kg during 1974.

TABLE 1

SUMMARY OF IMMSA's MINING OPERATIONS IN 1974

Unit	Positive Reserves**	Inferred Reserves**	% Zinc	% Lead	% Copper	grams of silver/Ton Ore	grams of gold Ton Ore	Current Production (MTPD*)	Milling Capacity (MTPD*)	Comments	Concentrate Type
Santa Barbara Unit	12.0	11.0	4.8%	3.0%	0.77%	150 gm	18 gm	2,400			lead-silver, copper-silver, zinc concentrates
Plomosas Mine	1.0		13.4%	7.3%						IMMSA's most profitable mine in 1974	zinc, lead, silver
Santa Eulalia Unit	1.0	1.2	9.9%	2.4%		114 gm		750		3 orebodies, one of which is nearly depleted	lead-zinc concentrates
San Martin Mine	5.5	4.2	5.9%	0.26%	1.1%	160 gm		700		One of IMMSA's most important mining units	copper-silver, zinc concentrates
Taxco Unit	6.6	1.3	6.6%	2.8%		145 gm			1,000		lead-silver, zinc concentrates
Ingusaran Mine	3.4	2.0			1.5%				2,800	o Began production in 1971 o Most modern and mechanized of all of IMMSA's operations	copper concentrates
Charcas Mine	3.4	1.6	7.4%	.7%	.31%	100 gm		900		----------	lead, silver, zinc, copper concentrates, with gold and silver
Parral Mine	.1		2.5%	1.1%	.8%	68 gm	0.3 gm	3,000 (through hydraulic mining)		o Company recovers flurospar from its tailings dams located in Parral, Chihuahua. o Ten million tons of tailings o Sales of approximately 6,000 metric tons of fluorspar per month at $72 per metric ton.	Zinc, copper, lead, silver, and gold

*Metric Tons Per Day
**Million Metric Tons

Source: Company data and IMMSA annual reports.

CASE STUDIES

<u>San Luis Potosi Copper Smelter</u> In 1974, the San Luis Potosi Smelter produced 34,741 mt of copper blister, compared with 33,638 mt the previous year. Sixty percent of the unit's feedstock was provided from IMMSA's mines, the remainder being purchased from outside suppliers. In addition to its copper blister facilities, the San Luis Potosi unit operated sulfuric acid and arsenic processing plants.

<u>Rosita Zinc Smelter</u> In 1974, IMMSA produced approximately 60,000 mtpy of prime western zinc from its horizontal retort located at Nueva Rosita, Coahuila. No plans to expand or modify the plant were included in the 1974-1978 investment program.

Coal and Coking Operations in 1974

The Rosita coking operations, located in Nueva Rosita, produced 360,000 tons of coke from 1 million tons of coal during 1974. Three-fourths of the coal utilized in the coking process was derived from IMMSA's mines located near Rosita and Sabinas, Coahuila, the remainder being purchased from sources outside of the company. Approximately 25% of the total coke produced in IMMSA's coking facilities was consumed by IMMSA's own units, the remainder being sold in Mexico. A doubling of the allowable ceiling price on coal by the Mexican government served as a further stimulus to expand proven reserves and to modernize the company's coal production facilities.

Exhibit 1 provides a location map of the major facilities maintained by the borrower.

Operating and Financial Statistics

Further insights into the borrowers' operating and financial performance are provided in Tables 2 through 5, and Exhibit 2. These tables and figures summarize, respectively, IMMSA's mine production (1974), smelting and refining production (1969-1974), percentage of metal processed by IMMSA's smelters and refineries deriving from IMMSA's mine production, and IMMSA audited financial statements for 1973 and 1974.

EXHIBIT 1
Location Map of IMMSA Facilities

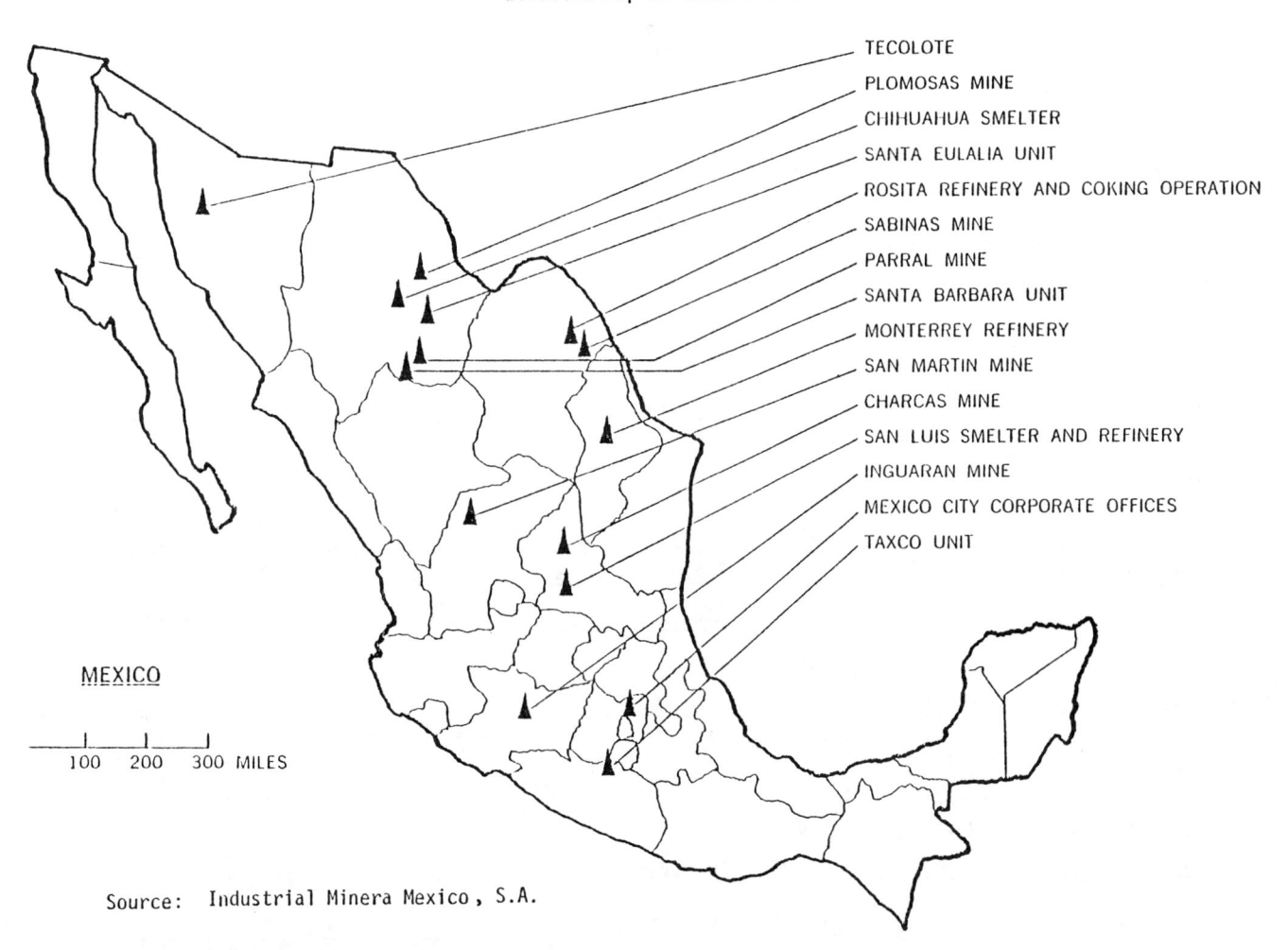

- TECOLOTE
- PLOMOSAS MINE
- CHIHUAHUA SMELTER
- SANTA EULALIA UNIT
- ROSITA REFINERY AND COKING OPERATION
- SABINAS MINE
- PARRAL MINE
- SANTA BARBARA UNIT
- MONTERREY REFINERY
- SAN MARTIN MINE
- CHARCAS MINE
- SAN LUIS SMELTER AND REFINERY
- INGUARAN MINE
- MEXICO CITY CORPORATE OFFICES
- TAXCO UNIT

Source: Industrial Minera Mexico, S.A.

Table 2

SUMMARY OF IMMSA's MINE PRODUCTION IN 1974

	Mined (Tons)	Milled (Tons)	Concentrate (Tons)	Gold (Kg)	Silver (Kg)	Lead (Tons)	Copper (Tons)	Zinc (Tons)
Charcas	313,654	312,875						
Lead			2,972	2.0	15,972	986	296	318
Copper			1,725	0.5	3,205	129	489	110
Zinc			28,833	2.9	3,734	80	209	16,769
Iron			5,816	10.1	857	13	14	344
			39,346	15.5	23,768	1,208	1,008	17,549
Parral	290,242	364,909						
Lead			8,695	108.1	18,377	4,953	778	580
Copper			1,174	8.2	815	111	368	22
Zinc			15,839	13.8	2,380	247	169	9,164
Special Lead			160	0.5	76	20	4	39
			25,868	130.6	21,648	5,331	1,319	9,805
Santa Barbara	741,974	742,195						
Lead			23,957	67.0	62,642	16,841	598	1,087
Copper			9,706	32.0	10,673	631	2,907	354
Zinc			60,669	17.0	7,557	336	765	34,104
			94,332	116.0	80,872	17,808	4,270	35,545
Plomosas	185,420	185,420						
Lead			16,455	-	3,900	10,967	-	1,288
Zinc			33,788	-	3,402	508	-	20,573
			50,243	-	7,302	10,967	-	21,961
San Martin	260,410	260,090						
Copper			7,803	-	25,691	424	1,975	613
Zinc			32,313	-	3,150	71	219	18,360
			40,116	-	28,841	495	2,194	18,973
San Eulalia	116,835	116,835						
Lead			3,936	1.2	9,525	2,386	135	123
Zinc			15,900	-	995	121	53	8,495
			19,836	1.2	10,520	2,507	188	8,618
Taxco	289,445	289,610						
Lead			12,608	21.6	31,732	6,678	158	608
Zinc			18,189	5.5	3,656	189	79	10,136
			30,797	27.1	35,388	6,867	237	10,739
Inguaran	845,120	845,980						
Copper			34,643	-	2,018	-	11,123	-
Fluorite		556,624	75,207					
	3,043,100	3,674,348	410,388	290.4	210,361	45,183	20,344	123,185

Source: Industrial Minera Mexico

Table 3

SUMMARY OF IMMSA's SMELTING AND REFINING PRODUCTION

Year	Gold (Kg)	Silver (Kg)	Lead (Tons)	Copper (Tons)	Zinc (Tons)	Zinc Fumes (Tons)
1969	2,207	519,624	74,880	23,477	58,848	18,399
1970	2,683	541,834	75,547	24,594	60,017	18,604
1971	1,799	476,925	66,609	23,007	58,960	15,488
1972	1,444	481,093	68,606	32,675	57,582	13,808
1973	1,412	535,752	87,599	33,638	45,708	5,514
1974	1,510	646,020	98,123	34,341	53,185	10,575

	1974 Production					
Unit	Gold (Kg)	Silver (Kg)	Lead (Tons)	Copper (Tons)	Zinc (Tons)	Zinc Fumes (Tons)
Monterrey Refinery	1,450	611,052	98,983			
San Luis Potosi-Copper Smelter	60	34,912		34,341		
Chihuahua-Lead Smelter		56	90,000			10,575
Rosita-Zinc Refinery					53,185	

Source: Industrial Minera Mexico, S.A.

Table 4

PERCENTAGE OF METAL PROCESSED BY IMMSA's SMELTERS AND REFINERIES EMANATING FROM IMMSA's MINE PRODUCTION

	Gold	Silver	Lead	Copper	Zinc
1970	29%	49%	69%	62%	100%
1971	20	40	51	64	100
1972	19	44	49	63	100
1973	16	35	38	67	100
1974	23	42	49	70	100
1975 (Est.)	24	35	45	60	100

Source: Industrial Minera Mexico, S.A.

TABLE 5

SUMMARY OF IMMSA's
DOMESTIC AND EXPORT SALES BY VOLUME

ZINC - Refined
(Metric Tons)

	1973	1974	1975 (est.)
Domestic	38,486	35,515	20,566
Export	8,139	12,165	23,300
Total	46,625	47,680	44,466

ZINC - Concentrate
(Metric Tons)

	1973	1974	1975 (est.)
Export	109,500	117,500	110,000

LEAD - Refined
(Metric Tons)

	1973	1974	1975 (est.)
Domestic	41,486	19,087	29,379
Export	37,248	55,032	54,800
Total	78,734	74,119	84,179

COPPER - Electrolytic
(Metric Tons)

	1973	1974	1975 (est.)
Domestic	17,531	22,725	21,850
Export	290	-0-	50
Total	17,821	22,725	21,900

COPPER - Blister
(Metric Tons)

	1973	1974	1975 (est.)
Export	8,400	3,100	13,000

SILVER
(Troy Ounces)

	1973	1974	1975 (est.)
Domestic*	6,880,000	8,180,000	3,000,000
Export	5,132,019	6,885,737	13,000,000
Total	12,012,019	15,065,739	16,000,000

* All domestic silver sales are made to Banco de Mexico and all export sales are registered with the same Central Banking Institution.

Source: Industrial Minera Mexico, S.A.

EXHIBIT 2 (a)

8. AUDITED FINANCIAL STATEMENTS

ARTHUR ANDERSEN & CO.

Mexico, D.F.

To the Stockholders of
Industrial Minera Mexico, S.A.:

We have examined the balance sheets of INDUSTRIAL MINERA MEXICO, S.A. (formerly ASARCO MEXICANA, S.A.) as of December 31, 1974 and 1973, and the related statements of income, retained earnings and changes in financial position for the years then ended. Our examination was made in accordance with generally accepted auditing standards and, accordingly, included such tests of the accounting records and such other auditing procedures as we considered necessary in the circumstances.

As explained in Note 2 to the financial statements, during the year 1974 the Company changed the method for pricing certain inventories (change with which we concur). As a result, the net income for that year was reduced by approximately $1,816,000. In addition, as explained in Notes 2, 3 and 6 to the financial statements, the Company has not recorded prepaid and deferred income taxes and prepaid employee profit participation, and the reserves provided from 1971 through 1974 for voluntary retirement compensation to unionized workers were not adequate to cover this liability.

In our opinion, except for the effect of the unrecorded prepaid and deferred income taxes and prepaid employee profit participation and the inadequacy of the reserves for voluntary retirement compensation to unionized workers as explained in the preceding paragraph, the accompanying balance sheets and statements of income, retained earnings and changes in financial position present fairly the financial position of Industrial Minera Mexico, S.A., as of December 31, 1974 and 1973, and the results of its operations and the changes in its financial position for the years then ended in conformity with generally accepted accounting principles which, except for the change in the method of pricing certain inventories in 1974 as explained above, were consistently applied during the periods.

On the basis of the examination described above, and in compliance with Article 57 of the Federal Commercial Revenues Tax Law, we report that during the years the Company has filed the tax returns required by such law.

ARTHUR ANDERSEN & CO.

Mexico, D.F.
March 7, 1975

Source: Industrial Minera Mexico Annual Reports

EXHIBIT 2 (b)

INDUSTRIAL MINERA MEXICO, S.A.
(Formerly Asarco Mexicana, S.A.)

STATEMENTS OF INCOME
FOR THE YEARS ENDED DECEMBER 31, 1974 AND 1973
(All Amounts in United States Dollars — 000's Omitted, Except Per Share Data)

	1974	1973
SALES OF PRODUCTS AND SERVICES	$293,065	$165,808
COSTS OF PRODUCTS AND SERVICES, EXCLUSIVE OF ITEMS DEDUCTED SEPARATELY BELOW (Note 4)	227,607	131,593
PROVISION FOR ADDITIONAL COST OF METAL TREATMENT (Note 2)	—	400
	227,607	131,993
	65,458	33,815
OTHER INCOME (net)	5,117	931
	70,575	34,746
DEDUCTIONS:		
Administrative Expenses	6,081	4,769
Depreciation, Depletion and Amortization (Note 5)	8,401	7,010
	14,482	11,779
INCOME BEFORE INCOME TAX AND EMPLOYEE PROFIT PARTICIPATION	56,093	22,967
PROVISIONS FOR:		
Income Tax (Note 4)	16,153	7,322
Employee Profit Participation	2,824	1,706
	18,977	9,028
NET INCOME FOR THE YEAR (Notes 1 and 4)	$ 37,116	$ 13,939
EARNINGS PER SHARE	$ 3.09	$ 1.16

The accompanying notes to financial statements are an integral part of these statements.

Source: Industrial Minera Mexico Annual Reports

EXHIBIT 2 (c)

INDUSTRIAL MINERA MEXICO, S.A.
(Formerly Asarco Mexicana, S.A.)

BALANCE SHEETS — DECEMBER 31, 1974 AND 1973
(All Amounts in United States Dollars — 000's Omitted)

ASSETS

	1974	1973
CURRENT ASSETS:		
Cash	$ 11,016	$ 2,617
Marketable securities, at cost, which approximates market value	27,192	2,192
Accounts and notes receivable, less reserve for doubtful accounts ($58 in 1974 and $198 in 1973)	39,343	34,774
Inventories of metals, manufactured products and by-products (Note 2)	50,692	57,674
Less — Reserve for additional cost of metal treatment	(1,306)	(1,993)
Materials and supplies, at average cost	16,535	12,913
Prepaid expenses, etc.	968	818
Total current assets (include $41,459 in 1974 and $21,864 in 1973)	144,440	108,995
MISCELLANEOUS ASSETS:		
Temporary investment in marketable securities, at cost, for use in the future acquisitions of equipment	8,752	—
Loans to workers and employees to purchase houses	1,715	1,489
Fund for sale of shares to employees (Note 9)	2,926	—
Deferred charges	235	667
Deposits, other long-term accounts receivable, etc.	431	189
	14,059	2,345
PROPERTY AND EQUIPMENT, at cost (Note 6):		
Mining concessions, land, buildings, machinery and equipment, etc.	106,025	92,885
Less — Accumulated depreciation, amortization and depletion	62,010	53,721
	44,015	39,164
INVESTMENT IN SUBSIDIARIES, at cost, (Note 7):		
Mexicana de Cobre, S.A.	—	11,831
Compania Minera La Loteria, S.A., less accumulated amortization of $3,330 in 1974 and $2,960 in 1973	2,220	2,590
Other investments in subsidiaries	2,272	1,222
	4,492	15,643
TOTAL ASSETS	$207,006	$166,147

The accompanying notes to financial statements are an integral part of these balance sheets.

Source: Industrial Minera Mexico Annual Reports

INDUSTRIAL MINERA MEXICO, S.A.
(Formerly Asarco Mexicana, S.A.)

BALANCE SHEETS — DECEMBER 31, 1974 AND 1973
(All Amounts in United States Dollars — 000's Omitted)

LIABILITIES

	1974	1973
CURRENT LIABILITIES:		
Notes and interest payable (Note 8)	$ 13,052	$ 26,075
Accounts payable and accrued liabilities	16,220	16,095
Current account with Mexicana de Cobre, S.A.	—	1,353
Current account with Compania Minera La Loteria, S.A.	10,651	7,422
Salaries, wages and employee profit participation payable	3,539	2,024
Taxes payable on production, exports and first-hand sales	13,818	6,289
Accrued income tax	4,170	2,601
Labor liabilities payable short-term (Note 3)	5,550	—
Total current liabilities (include $28,493 in 1974 and $33,496 in 1973)	67,000	61,859
RESERVE FOR LABOR LIABILITIES (Note 3)	4,748	3,101
LONG-TERM NOTES PAYABLE (Note 8)	33,162	18,616
STOCKHOLDERS' EQUITY (Note 1):		
Capital Stock — Represented by 12,000,000 shares authorized, subscribed and paid, with a par value of $4 each:		
Series A shares (ownership restricted to Mexican citizens only) 7,920,000 in 1974 and 6,120,000 in 1973	31,680	24,480
Series B shares (ownership not restricted) 4,080,000 in 1974 and 5,880,000 in 1973	16,320	23,520
	48,000	48,000
Retained earnings (see accompanying statements)	54,096	34,571
	102,096	82,571
TOTAL LIABILITIES AND STOCKHOLDERS' EQUITY	$207,006	$166,147

The accompanying notes to financial statements are an integral part of these balance sheets.

EXHIBIT 2 (d)

INDUSTRIAL MINERA MEXICO, S.A.
(Formerly Asarco Mexicana, S.A.)

STATEMENTS OF RETAINED EARNINGS
FOR THE YEARS ENDED DECEMBER 31, 1973 AND 1974 (Note 1)
(All Amounts in United States Dollars — 000's Omitted)

	Special Reinvestment Reserve	Legal Reserve	Reinvestment Reserve	Unappropriated Earnings	Total Retained Earnings
BALANCE, DECEMBER 31, 1972	$ 8,000	$ 2,985	$ 19,929	$ 9,558	$ 40,472
LESS — Appropriations approved by stockholders:					
Transfer to —					
Legal reserve	—	488	—	(488)	—
Reinvestment reserve	—	—	4,438	(4,438)	—
Dividends declared ($.48 per share on 8,000,000 shares outstanding)	—	—	—	(3,840)	(3,840)
Capitalization of reserves	(8,000)	—	(8,000)	—	(16,000)
ADD — Net income for the year (see accompanying statements)	—	—	—	13,939	13,939
BALANCE, DECEMBER 31, 1973	$ —	$ 3,473	$ 16,367	$ 14,731	$ 34,571
LESS — Appropriations approved by stockholders:					
Transfer to —					
Legal reserve	—	782	—	(782)	—
Reinvestment reserve	—	—	7,397	(7,397)	—
Dividends declared ($.48 per share on 12,000,000 shares outstanding)	—	—	—	(5,760)	(5,760)
Pro-rata distribution of shares of Mexicana de Cobre, S.A. (Note 7)	—	—	(11,342)	(489)	(11,831)
ADD — Net income for the year (see accompanying statements)	—	—	—	37,116	37,116
BALANCE, DECEMBER 31, 1974	$ —	$ 4,255	$ 12,422	$ 37,419	$ 54,096

The accompanying notes to financial statements are an integral part of these statements.

Source: Industrial Minera Mexico Annual Reports

EXHIBIT 2 (e)

INDUSTRIAL MINERA MEXICO, S. A.
(Formerly Asarco Mexicana, S. A.)

STATEMENTS OF CHANGES IN FINANCIAL POSITION
FOR THE YEARS ENDED DECEMBER 31, 1974 AND 1973
(All Amounts in United States Dollars — 000's Omitted)

	1974	1973
WORKING CAPITAL WAS PROVIDED FROM:		
Operations —		
Net income for the year	$37,116	$13,939
Expenses not requiring outlay of working capital in the current period —		
Depreciation, depletion and amortization	8,400	7,009
Amortization of investment in Compania Minera La Loteria, S. A.	370	370
Provision for labor liabilities	2,570	1,124
Total working capital provided from operations	48,456	22,442
Increase in long-term notes payable	14,546	682
Decrease in miscellaneous assets	—	1,191
Distribution of shares of Mexicana de Cobre, S.A. to stockholders (see below)	11,831	—
Total working capital provided	74,833	24,315
WORKING CAPITAL WAS APPLIED TO:		
Additions to property and equipment, less not book value of retirements	13,252	9,954
Dividends declared	5,760	3,840
Pro rata distribution of shares of Mexicana de Cobre, S. A.	11,831	—
Indemnities paid to workers and employees	922	702
Increase in investment in subsidiaries	1,050	5,500
Increase in miscellaneous assets	11,714	—
Total working capital applied	44,529	19,996
Net increase in working capital	$30,304	$ 4,319
CHANGES IN WORKING CAPITAL:		
Increase (decrease) in —		
Current assets —		
Cash	$ 8,399	$(6,845)
Marketable securities	25,000	1,976
Accounts and notes receivable	4,569	8,287
Inventories (Net)	(6,295)	7,018
Materials, supplies and prepaid expenses	3,772	2,388
	35,445	12,834
Current liabilities —		
Notes and interest payable	(13,023)	(3,498)
Accounts payable and accrued liabilities	9,067	11,073
Reserves for production, export, first-hand sales and income taxes	9,097	930
	5,141	8,505
Net increase in working capital	$30,304	$ 4,319

The accompanying notes to financial statements are an integral part of these statements.

Source: Industrial Minera Mexico Annual Reports

EXHIBIT 2 (f)

INDUSTRIAL MINERA MEXICO, S.A.
(Formerly Asarco Mexicana, S.A.)

NOTES TO FINANCIAL STATEMENTS
AS OF DECEMBER 31, 1974 AND 1973

(All Amounts in United States Dollars —
Converted at a Rate of 12.50 Mexican Pesos Per U.S. $1.00)

(1) Stockholders equity

At an extraordinary meeting held on April 26, 1974, the stockholders approved a change in the Company's name from Asarco Mexicana, S.A. to Industrial Minera Mexico, S.A. At the same stockholders' meeting, a modification in the Company's by-laws was approved, to the effect that at least 66% of the capital stock is to be represented by Series A shares.

Capital stock includes capitalized earnings of $35,169,040 on which no distributable profit or dividend taxes have been paid, since the payment of such taxes is deferred on capitalized earnings.

Retained earnings, except for $303,200, would be subject to a withholding tax of 15% to 20% upon distribution as dividends.

Net income for the year ended December 31, 1974, is subject to the legal requirement that 5% of each year's net income be appropriated to a segregated reserve until such reserve equals 20% of the outstanding capital stock. This legal reserve is not available for distribution to stockholders during the existence of the Company, except in the form of stock dividends.

(2) Valuation of inventories of metals, manufactured products and by-products

Inventories of primary gold, silver, copper, lead, zinc, coke and their by-products are valued at an amount equivalent to the purchase cost of such metals at each stage of processing using the last-in, first-out method (equivalent to the market price of the metal contents acquired, less deductions for the estimated cost of their pending processing). Up to December 31, 1973, secondary metals, manufactured products and other by-products were valued at their production cost, using the first-in, first-out method, except for bismuth and cadmium, for which sales commitments already existed and which were valued at amounts which approximated market. At December 31, 1974, with an authorization from the Treasury Department, the Company valued the inventories of bismuth, cadmium and concentrates at the mining units at their average production or extraction cost (since this practice is the most common in the mining industry). As a result of the change in the method of valuing such inventories, their book value was reduced by approximately $2,880,000 and the net income for the year 1974 by approximately $1,816,000.

In accordance with the authorization from the Treasury Department, only approximately $1,440,000 of the reduction in the inventory value mentioned in the above paragraph may be deducted from the 1974 taxable income; the remainder will be deducted from the 1975 taxable income. This partial deferring of the deduction results in a prepaid income tax and employee profit participation of approximately $468,000, which was charged against 1974 income.

(3) Labor liabilities

Unionized workers are entitled to voluntary retirement compensation of between 18 and 30 days' pay for each year of service, after working for the Company a required number of years, generally between 14 and 25. The reserves provided by the Company for voluntary retirement compensation of unionized workers who had already completed the required years of service were inadequate to cover this liability by approximately $2,208,000 at December 31, 1974 (as compared with $6,880,000 in 1973 and $5,928,000 in 1972) after considering the reduction in

Source: Industrial Minera Mexico Annual Reports

INDUSTRIAL MINERA MEXICO, S.A.
(Formerly Asarco Mexicana, S.A.)

NOTES TO FINANCIAL STATEMENTS (Continued)

(3) Labor liabilities (continued)

income tax and employee profit participation applicable upon payment of said compensation. It is not possible to determine to what extent the charges to income for the years ended December 31, 1974 and 1973 represent past service cost and represent the cost for the year.

No actuarial computation has been made of the contingent liability for voluntary retirement compensation accruing in favor of workers who have not yet completed the required years of service, nor does the Company have a reserve to cover this liability.

The Company is also liable under The Mexican Labor Law for indemnities to employees who are not entitled to voluntary retirement compensation, but who are separated from the Company under certain conditions. The Company follows the practice of charging these payments to expenses when made and has not provided a reserve to cover this liability.

(4) Tax subsidies

The Treasury Department has agreed to grant the Company subsidies on production, export and income taxes, until December 31, 1978, the proceeds of which can only be used to cover exploration expenses, construction of new mining units, smelters and refineries, and expansion of mining units, smelters and refineries presently in operation.

(5) Depreciation, depletion, and amortization of property and equipment

Depreciation, depletion, and amortization of property and equipment have been calculated at the following annual rates:

Assets	Rate
Acquired prior to December 31, 1967	10%
Acquired after December 31, 1967, as follows:	
Land	4%
Mining concessions	4%
Buildings and constructions	4%
Fixed equipment	10%
Mobile equipment	20%

(6) Deferred income tax

Up to December 31, 1968, the Company, for income tax purposes, computed depreciation, depletion and amortization on its property and equipment at the maximum rates allowed by the Mexican Income Tax Law and, therefore, the net book value of the fixed assets acquired prior to that date exceeds their net value for income tax purposes by approximately $2,304,000 (as compared to $3,112,000 in 1973 and $3,968,000 in 1972). This difference represents deductions for depreciation, depletion and amortization taken in the income tax returns of prior years in excess of the amounts charged to income for these items.

No provisions have been made for the income tax deferred as a result of this practice, which amounts to approximately $691,200 (as compared to $933,660 in 1973 and $1,190,400 in 1972).

(7) Investments in subsidiaries

In 1965, the Company acquired 51% of the capital stock of Compania Minera La Loteria, S.A. for $5,550,000. Starting in 1966, the original cost has been amortized at the rate of $370,000 per year (6⅔% annually). The underlying net book value of the Company's investment in Compania Minera La Loteria, S.A. at December 31, 1974 and 1973, amounted to $4,192,480 and $3,100,880 respectively, and its share of that Company's net income for the years then ended

INDUSTRIAL MINERA MEXICO, S.A.
(Formerly Asarco Mexicana, S.A.)

NOTES TO FINANCIAL STATEMENTS (Continued)

(7) Investments in subsidiaries (continued)

amounted, respectively, to $3,303 and $2,477,920. During the year 1974 the Company received dividends on this investment amounting to $2,176,400.

At the stockholders' meeting held on April 26, 1974, the stockholders approved a pro-rata distribution of shares of Mexicana de Cobre, S.A., which the Company then owned, with one share of Mexicana de Cobre, S.A. to be distributed for each 4.3125 shares of the Company's capital stock outstanding. As of December 31, 1974, 193,013 shares of Mexicana de Cobre, S.A., were still pending distribution to the Company's stockholders.

(8) Notes payable

Notes payable are represented by the following (all payable in U.S. dollars):

Payee	Payment Terms	1974 Portion Payable Within one year	1974 Portion Payable Over one year	1973 Portion Payable Within one year	1973 Portion Payable Over one year
ASARCO, INC.	Annual installments of $490,560 to April, 1979	$ 490,560	$ 1,962,240	$ —	$ —
Unsecured loans for construction or installations—					
Manufacturers Hanover Trust Company—					
"Sabinas Dos" mine and Santa Eulalia, Chih., mine unit	Semi-annual installments of $500,000 to June, 1976	1,000,000	500,000	1,000,000	1,500,000
United California Bank—					
Inguaran, Mich., mining unit	Semi-annual installments of $900,000 to March, 1976 ..	1,800,000	900,000	1,800,000	2,700,000
United California Bank and Manufacturers Hanover Trust Co.—					
San Luis Potosi, S.L.P., plant	Semi-annual installments of $416,000 to June, 1975	416,000	—	832,000	416,000
Manufacturers Hanover Trust Co. plus five other banks up to a total of $30,000,000—					
Taxco, Gro., mining unit, San Luis Potosi, S.L.P., plant, San Martin, Zac., mining unit and Chihuahua, Chih., plant	Payable in semi-annual installments of variable amounts from February, 1975, to August, 1981	1,200,000	28,800,000	—	1,200,000
Bank of America NT & SA—					
Chihuahua, Chih., plant	Semi-annual installments of $500,000 to December, 1976	1,000,000	1,000,000	1,000,000	2,000,000
Lines of credit to finance—					
Manufacturers Hanover Trust Co.—					
Exports sales up to $2,500,000	180 days	2,500,000	—	2,500,000	—
United California Bank—					
Exports sales up to $2,500,000	180 days	2,500,000	—	2,500,000	—
Other loans and short-term lines of credit		—	—	14,887,920	—
Compania Metalurgica Mexicana	Payable in 1974	—	—	666,400	—
Interest payable		2,145,680	—	888,640	—
		$13,052,240	$33,162,240	$26,074,960	$18,616,000

INDUSTRIAL MINERA MEXICO, S.A.
(Formerly Asarco Mexicana, S.A.)

NOTES TO FINANCIAL STATEMENTS (Continued)

(8) Notes payable (continued)

Certain of the above loan agreements provide that, unless approval is obtained from the banks, the Company's current assets shall not be less than 175% of its current liabilities, that total liabilities shall not exceed 150% of stockholders' equity, and that the Company may not reduce its capital stock nor pay dividends in an amount in excess of its annual net income.

(9) Benefits to employees

During the year 1974, the Company established a pension plan for non-unionized employees. A fund has been started to cover the cost of the pensions through the contribution, to a trust, of $1,295,600; future contributions will be determined based upon actuarial computations. Of the initial contribution, $833,200 represents the amortization, over a period of ten years, of past service cost. At December 31, 1974, the present value of vested benefits does not exceed the amount of the fund. The authorization for this plan from the Treasury Department is pending.

During the year 1974, the Company established a fund, by means of a trust, for $2,925,520, whose purpose is to acquire stock of the Company for sale to the Company's employees. At December 31, 1974, the fund held 450,000 shares of the capital stock of the Company for sale to the employees. The first sales of shares took place in March 1975.

(10) Contingent liability

In August, 1971, continuing its practice of helping the small-and medium-size miners, the Company signed an agreement with the Comision de Fomento Minero under which it guarantees the payment of a credit of up to $3,200,000 granted to the Comision by Nacional Financiera, S.A., destined to granting credits to small- and medium-size miners. At December 31, 1974, the Comision had granted credits of $897,448. In addition, the Company has guaranteed the payment by Sociedad Cooperativa de Rosita, S.C.L., of a loan made by Banco de Londres y Mexico, S.A., in the amount of $64,000, payable on February 3, 1975.

CASE STUDIES

CAPITAL INVESTMENT PROGRAM

In 1974, IMMSA proposed a US$286 million capital investment program for the years 1975 through 1978. Ten major projects were included in the investment program:

Project	Cost (US$ Millions)
Santa Barbara Mine (expansion)	49.7
San Martin Mine (expansion)	2.7
Taxco Mine (expansion)	10.0
Charcas Mine (expansion)	2.1
Tecolote Mine (new)	5.8
Coal Mines (expansion new)	18.3
Zinc Refinery (new, including a US$30.0 overrun cushion)	150.0
Chihuahua Smelter (modernization)	16.1
San Luis Smelter (expansion)	4.3
Replacement requirements	27.0
TOTAL	US$286.0

To support the proposed investment program, IMMSA prepared a detailed feasibility study for each project identified above.

Financing for the capital investment program was proposed as follows:

US$150.0 million syndicated term loan
US$136.0 million retained earnings
─────────────
US$286.0 million

Smelting and Refining Projects

New Zinc Refinery Based on the results of a feasibility study developed jointly by Lurgi and IMMSA, the company proposed to construct a major electrolytic zinc refinery in the city of San Luis Potosi, Mexico. Investment in the new refinery was expected to approximate US$120.0 million, including generous provisions for cost escalation plus an additional $30.0 million for cost overruns.

The refinery was designed to produce 113,900 mt of zinc and 1,398 tons of cadmium annually from zinc concentrates, zinc oxides, and cadmium fumes generated in the Chihuahua plant and other properties operated by the company. The proposed construction period for the new refinery was to be 30 months; construction was scheduled to commence in 1976 and to be completed during the first half of 1978.

IMMSA is a major producer of zinc concentrates. The proposed expansion of several of IMMSA's mines was planned to generate sufficient incremental production of zinc concentrates to supply both the existing horizontal retort at Rosita, as well as meeting the requirements of the proposed zinc refinery.

The proposed electrolytic zinc refinery was to be very similar to other electrolytic zinc refineries that were then being constructed in other parts of the world. An electrolytic-type refinery was selected as opposed to a "retort"-type plant because of the electrolytic refinery's higher recovery factors, greater market demand for high-grade zinc, ecological considerations, enhanced cost-effectiveness, reduced potential for damaging the environment and lower labor-intensity.

Major refining processes and components proposed for the new zinc refinery included: the Vielle-Montague/Base Process for the fluidized-bed roaster, the Jarosite Process for treating residues, the Bayer Double Contact Process for the proposed sulfuric acid plant, and the Mitsui Mining Company's Zinc Stripping Machine. All processes had been tried and proven in other commercial installations.

Any difficulties encountered in the new refinery were expected to be compensated by cash flows generated from direct sales of zinc concentrates in both domestic and international markets. Had cash flow problems eventuated, direct sales of zinc concentrates would have reduced the company's revenues by an amount equal to the value added in the refining process.

Chihuahua Smelter Upon its planned completion in 1977, the modernization program at the Chihuahua smelter was expected to increase the plant's design capacity from 8,000 to 10,000 tpm.

San Luis Smelter The San Luis Potosi copper smelter is one of the oldest facilities maintained by IMMSA. Most of the facility was completed in the mid-1920's. Only minor additions were made through 1974.

A 20% expansion in the San Luis Smelter at a cost of US$4.3 million was budgeted to allow IMMSA to be in a better position to handle programmed increases in mine production.

Mining

Santa Barbara In 1974, Santa Barbara was the most important mining unit in the company. Ore reserves at Santa Barbara were extensive: 12 million tons of positive reserves and 11 million tons of inferred reserves containing an average of 4.8% zinc, 0.77% copper, 3.0% lead, 150 gm of silver and 0.8 gm of gold per ton of ore. Production at Santa Barbara in 1974 was 2,400 tons of ore per day. The company's production plans envisioned doubling production at Santa Barbara by 1978 on an investment of $47.7 million.

Taxco In 1974, the Taxco unit was one of IMMSA's oldest operating units. The modernization and expansion program involved increasing Taxco's mill capacity of 1,000 tons of ore per day to 2,500 tons per day by the end of 1975. IMMSA's expansion program at Taxco was supported by ore reserves amounting to 6.6 million tons of proved reserves and 1.3 million tons of inferred reserves containing the average metal values of 6.6% zinc, 2.8% lead and 145 gm of silver per ton of ore.

Tecolote   IMMSA's 1974 through 1978 investment program included plans to develop the Tecolote mine, an orebody located in the State of Sonora containing 857,000 tons of proven reserves averaging 7.68% zinc, 2.02% copper, and 60 gm of silver per ton. Two years of construction and development work were scheduled in order to develop a mine and processing facilities capable of processing 350 tons of ore per day at a cost of $6.2 million. By year-end 1974, $400,000 had been spent on developing the Tecolote orebody.

San Martin   The investment program at San Martin envisioned increasing the unit's mid-1975 productive capacity of 700 tons of ore per day to 1,250 tons of ore per day by mid-1976. By 1974, the company had already expended US$1.9 million in routine expansion work at San Martin, leaving an additional US$2.7 million to be expended to bring the unit to its planned productive capacity.

Charcas   The proposed expansion program for Charcas was scheduled to increase the unit's productive capacity from 900 tons per day to 1,200 tons per day by year-end 1975. Of a proposed US$3.0 million budget, US$900,000 had been spent by the company by year-end 1975 in the Charcas expansion program.

Coal Mines   In 1975, IMMSA purchased approximately one-third of its total one million ton per annum coal requirements from sources outside of the company. Coke production in 1975 was 360,000 tons. The company's total positive and inferred coal reserves were approximately 28 million tons and 112 million tons, respectively. Coal and coke were in great demand in Mexico in 1975, with domestic demand far exceeding supplies. IMMSA's proposed US$18.3 million expenditure seemed quite justifiable in view of the strong market conditions and the company's excellent ore reserve position.

PRODUCT MARKET ANALYSIS

IMMSA derived the bulk of its revenues from the sale of metals produced in the company's mining and smelting operations. Combined sales of silver, lead, zinc and copper accounted for 85% of the company's total sales in 1974. Table 6 details IMMSA's revenues on a product-by-product basis for 1973 and 1974. Approximately 50% of IMMSA's metal production was sold in Mexico, the balance being exported, principally to the US.

Two characteristics of IMMSA's product mix and production life cycle offered the company a distinct advantage over many of its competitors:

The wide diversification in IMMSA's non-ferrous product portfolio minimized the company's sensitivity to price fluctuations in individual metals

IMMSA's production cycle was nearly vertically integrated, allowing the company to achieve significant economies of scale in its operations

Silver

Prior to 1974, the market price for silver had been quite volatile. In 1974, however, silver prices appeared to be relatively stable. With new silver production accounting for only 54% of silver consumption in 1974, existing stocks of metal were being depleted, suggesting that if new stocks of the metal were not discovered, continued upward pressure on prices could be expected.

Zinc

In 1974, the price of zinc was US$.35 per pound on the London Metal Exchange (LME) and US$.39 per pound for US producers. By historical standards,

Table 6

COMPOSITION OF IMMSA's MAJOR METAL PRODUCT SALES

	1973 Sales: US$	% of Sales	1974 Sales: US$	% of Sales
Silver	$29,603,019	17.9%	$65,944,353	22.5
Lead	31,086,393	18.8	57,168,533	19.5
Sub-products*	35,563,426	21.4	52,873,579	18.0
Copper	34,952,601	21.1	51,481,048	17.6
Zinc	21,064,088	12.7	47,045,654	16.1
Gold	4,517,930	2.7	7,958,842	2.7
Fluorite	2,006,374	1.2	4,746,272	1.6
Coke	5,836,630	3.5	3,669,715	1.3
Cadmium	1,417,137	0.7	2,176,978	0.7
	$166,047,901	100.0%	$293,064,974	100.00%

* includes sales of concentrates containing zinc, lead, and copper; all metals refer to sales of final metal products.

Source: Company Annual Reports

prices during the recessionary period of 1974 were quite firm and high. In 1974, world consumption of zinc declined by 7.9% to 4.3 million tons. Continued firming in world zinc prices was forecast as of 1974 for the following reasons:

An increase in world zinc consumption to 1973 levels was expected to result in a shortage of zinc as early as 1976

US zinc producers had curtailed production and announced plans to postpone expansion of new production facilities

Copper

Historically, world copper prices had been quite volatile. For example, LME copper prices rose from US$.50 per pound in January 1973 to US$1.50 per pound in April 1974. World refined copper production in 1974 (8.67 million tons) exceeded 1973 production levels by 2.2%, whereas refined copper consumption fell 5.2% in 1974 to 8.23 million tons.

Because most world copper was produced at a cost that was above US$.50 per pound, it was highly unlikely that world copper prices would fall below $.50 per pound for extended periods of time. Large inventories of copper in world markets were thought to exert a depressing effect on refined copper and concentrate. On the other hand, upward pressure on world copper prices was expected to result from world economic recovery as well as the cancellation of 1.5 million tons per year of new mine capacity since June 1974.

Lead

As was the case with zinc, demand for lead was closely tied to the automobile market. Any decline in lead demand by gasoline producers was thought to be compensated for by extensive consumption of lead in nuclear power plants and for insulation purposes in the construction industry. The slowdown in the automobile industry prior to 1974 resulted in a substantial decline in demand for lead.

FINANCIAL PROJECTIONS

Underlying Assumptions

Prices Reflecting the "bankers conservatism" doctrine, prices for the principal refined metal products that IMMSA expected to produce during the life of the term loan were held constant in the pro-forma financial statements. Prices for the four major metal products were as follows:

Silver   US$3.50/oz.
Zinc   .32/oz.
Lead   .16/lb.
Copper   .53/lb.

Sales Volume Projected sales volumes for the four major finished metal products produced by IMMSA for the period 1975-1983 were as follows:

Metal Units in Millions

	1975	1976	1977	1978	1979	Annually 1975-1983
Silver (oz.):	16.723	17.200	18.229	18.515	20.705	20.705
Lead (lbs.):	182.587	194.006	218.257	238.098	227.075	227.075
Copper (lbs.):	65.580	60.682	75.728	75.728	75.728	75.728
Zinc (lbs.):	95.673	101.197	145.615	227.737	341.054	364.627

Operating Costs The pro-forma financial statements assumed a linear relationship between metal product prices, cost of goods sold ("COGS"), and operating expenses. Ratios of COGS/gross sales and operating expenses/gross sales derived from the pro-forma financial projections were as follows:

	1975	1976	1977	1978	1979	Annually 1980-1983
COGS/Gross Sales	0.77	0.75	0.74	0.70	0.72	0.70
Operating Expenses/Gross Sales	2.4	2.1	1.8	1.8	1.6	1.9

Expenditures on fixed assets were assumed to occur during the term loan drawdown period and to reflect IMMSA's long-term capital investment plans. A schedule of fixed asset investments for the period 1975-1978, as presented in the pro-forma sources and uses of funds statement, was as follows:

(Millions of US$)

1975	1976	1977	1978
35.744	84.848	82.488	69.728

Depreciation The depreciation schedule utilized in the pro-forma financial statements had been prepared on a consistent basis according to IMMSA's established policy, as follows:

Book Depreciation:

Land, Mining Claims Buildings
 and Constructions   4% per annum
Stationary Equipment   10% per annum
Mobile & Transport Equipment 20% per annum

Depreciation for fiscal purposes throughout the life of the Fiscal Agreement (1978):

In year of investment   40%
Per annum thereafter   8%

Inventory The last-in-first-out inventory valuation method was utilized by IMMSA for its major metal products (coal, silver, copper, lead, zinc, and coke). Prices used in inventory valuation were those current at the level of processing achieved by the metal products at the time of sale. That is, product prices were equivalent to the market price of metal content in the product, less deductions for the estimated cost of completing final treatment and delivery to the market. Concentrate and by-product inventories at mines or plants were valued at average cost.

For the purpose of inventory valuation, the following purchase prices were used internally:

Gold: Monthly average "London Final" less 10c per troy ounce

Silver: Monthly arithmetic average of LME and Handy and Harman prices, less 1%

Lead: Weighted average of F.O.B. Monterrey refined lead of weekly sales made in international markets and in Mexico for the previous week

Zinc: Monthly average weighed price using US, and Mexico sales prices, less 1%

Copper: Weighed average using 60% of monthly average US producer price and 40% of monthly average LME wirebar settlement for the previous month, as recorded in *Metals Week*

Coke: Current price fixed by the Mexican government at 575 pesos per metric ton.

Long-Term Debt Projected drawdowns of long-term debt were scheduled in amount and timing to meet IMMSA's fixed-asset expansion and modernization program. An eight-year syndicated commercial bank term loan, amounting to US$150 million, was proposed to provide 52% of IMMSA's fixed asset modernization and expansion program. New debt funds were intended to supplement IMMSA's existing long-term debt ($46.208 million in 1974) and contributions of $136.0 million from the company's earnings. A debt-service schedule, including drawdown, principal, interest, and total payment for each year for the loan is presented in Exhibit 3d.

Pro-Forma Financial Statements

Exhibits 3-a through 3-f present, respectively, a pro-forma income statement, balance sheet, sources and uses of funds statement, debt service schedule, ratio analysis, and sensitivity coefficients for the period covered by the term loan. These projections incorporate the assumptions discussed in the first section of this paper.

THE MEXICAN ECONOMY

Exhibits 4 through 7 show, respectively, the international reserves, external funded debt of public sector, debt service requirement of public sector external funded debt, and balance of payments for the five-year period preceding the term loan.

The overall investment policy of the Mexican government was to encourage private investment, including foreign capital, while retaining certain sectors under state control. The Mexican government influenced the level and type of private investment through selective use of concessions for certain regulated industries such as mining, and import licenses for raw materials and capital goods. Other major incentives for investment provided by the Mexican government included temporary relief from income and other taxes and from export and import duties.

Most Mexican public utilities, communications, transportations and the oil industry are operated exclusively under state control. All other industries are open to private investment, providing that the state may participate.

Reflecting the effect of favorable government policies toward private investment in Mexico, in 1973 private investment totalled US$6 billion, or 58% of total investment, compared with US$4.1 billion or 62% in 1970.

SUMMARY OF THE LOAN AGREEMENT

Borrowers
Industrial Minera Mexico, S.A. (agent for borrowers)
Cia. Minera La Loteria, S.A.
Minera Picachas, S.A.
Minera Recancimiento,,S.A.
Carbonifera de Mexico, S.A.
Minera Septentrional, S.A.
Cia. Minera San Isidro y Anexas, S.A.
Aerominerales, S. de R.L.
Promotora Minera B.C., S.A.
Minera Selene, S.A.
Minera Normex. S.A.
Minera Surmex, S.A.
Cia. Minera y Beneficiadora de San Antonio y Anexas, S.A.

Lenders
Managers
Bank of America National Trust and Savings Association
Citibank, N.A.
Manufacturers Hanover Limited
Mellon Bank, N.A.
United California Bank

Co-Managers
Bank of California, N.A.
The Bank of New York
Republic National Bank of Dallas
Security Pacific National Bank
United California Bank

Agent
United California Bank

Amount
US$150 million

Loan Purpose
To provide partial funding (54%) for the purchase of equipment and materials and to defray construction costs and other expenditures in connection with IMMSA's 1974-1978 investment program.

Availability
Over a period of two years until August 31, 1978 in amounts of at least US$10 million and always in integral multiples of $50,000.

EXHIBIT 3 (a)

IMMSA
INCOME STATEMENT
($US Millions; dlrs. per pound/ounce)

PERIOD ENDING (ANNUALLY)	DEC. 1974	DEC. 1975	DEC. 1976	DEC. 1977	DEC. 1978	DEC. 1979	DEC. 1980	DEC. 1981	DEC. 1982	DEC. 1983
SILVER OZS	15.708	16.723	17.200	18.229	18.515	20.705	20.705	20.705	20.705	20.705
SILVER PRICE/OZ	4.198	3.500	3.500	3.500	3.500	3.500	3.500	3.500	3.500	3.500
TOTAL REVENUE A	65.945	58.530	60.200	63.801	64.802	72.467	72.467	72.467	72.467	72.467
LEAD LBS	230.587	182.587	194.006	218.257	238.098	227.075	227.075	227.075	227.075	227.075
LEAD PRICE/LB	0.247	0.160	0.160	0.160	0.160	0.160	0.160	0.160	0.160	0.160
TOTAL REVENUE B	56.983	29.214	31.041	34.921	38.096	36.332	36.332	36.332	36.332	36.332
COPPER LBS	56.996	65.580	60.682	75.728	75.728	75.728	75.728	75.728	75.728	75.728
COPPER PRICE/LB	0.903	0.530	0.530	0.530	0.530	0.530	0.530	0.530	0.530	0.530
TOTAL REVENUE C	51.479	34.757	32.161	40.136	40.136	40.136	40.136	40.136	40.136	40.136
ZINC REF LBS	110.387	95.673	101.197	145.615	227.737	341.054	364.627	364.627	364.627	364.627
ZINC PRICE/LB	0.426	0.320	0.320	0.320	0.320	0.320	0.320	0.320	0.320	0.320
TOTAL REVENUE D	47.043	30.615	32.383	46.597	72.876	109.137	116.681	116.681	116.681	116.681
OTHER PRODUCTS	71.616	83.867	114.943	126.033	98.730	84.479	84.479	84.479	84.479	84.479
GROSS SALES	293.064	236.984	270.728	311.488	314.640	342.552	349.696	349.696	349.696	349.696
NET SALES	293.064	236.984	270.728	311.488	314.640	342.552	349.696	349.696	349.696	349.696
COST OF GOODS SOLD	227.068	183.120	204.086	231.056	220.976	248.488	247.392	247.392	247.392	247.392
GROSS MARGIN	65.456	53.864	66.640	80.432	93.664	94.064	102.304	102.304	102.304	102.304
OPERATING EXPENSE	6.080	5.600	5.600	5.600	5.600	5.600	5.600	5.600	5.600	5.600
ADH EXPENSES	6.080	5.600	5.600	5.600	5.600	5.600	5.600	5.600	5.600	5.600
OPERATING INCOME	59.376	48.264	61.040	74.832	88.064	88.464	96.704	96.704	96.704	96.704
OTHER REVENUE	5.120	2.384	2.400	2.400	2.400	2.400	2.400	2.400	2.400	2.400
OTHER INCOME	0.000	2.384	2.400	2.400	2.400	2.400	2.400	2.400	2.400	2.400
NON RECURRING	5.120	0.000	0.000	0.000	0.000	0.000	0.000	0.000	0.000	0.000
TOTAL INCOME	64.496	50.648	63.440	77.232	90.464	90.864	99.104	99.104	99.104	99.104

IMMSA   INCOME STATEMENT (CONT.)

PERIOD ENDING (ANNUALLY)	DEC. 1974	DEC. 1975	DEC. 1976	DEC. 1977	DEC. 1978	DEC. 1979	DEC. 1980	DEC. 1981	DEC. 1982	DEC. 1983
DEPRECIATION	8.400	15.112	26.856	33.920	37.168	27.712	27.832	27.040	26.640	25.664
PLANT DEPRECIATION	8.400	15.112	26.856	33.920	37.168	27.712	27.832	27.040	26.640	25.664
TOTAL NON-CASH CHGS.	8.400	15.112	26.856	33.920	37.168	27.712	27.832	27.040	26.640	25.664
EARNING BEF INT/TAX	56.096	35.536	36.584	43.312	53.296	63.152	71.272	72.064	72.464	73.440
INTEREST ON ST DEBT	0.000	0.000	0.000	0.000	0.000	0.000	0.000	0.000	0.000	0.000
INTEREST ON LT DEBT	0.000	4.312	6.170	12.047	14.943	12.757	9.787	6.817	3.847	1.215
PROJECT LOAN	0.000	0.000	1.554	8.453	12.150	11.475	8.775	6.075	3.375	1.012
FCIA FIN	0.000	0.337	1.350	1.012	1.350	1.282	1.012	0.742	0.473	0.203
EXISTING LTD	0.000	3.975	3.266	2.581	1.443	0.000	0.000	0.000	0.000	0.000
INCOME BEFORE TAX	56.096	31.224	30.414	31.265	38.353	50.394	61.484	65.246	68.616	72.225
TAX EXPOSURE	16.152	7.432	4.912	5.712	7.152	22.512	26.824	28.264	29.632	31.552
PROFIT SHARING	2.824	0.320	0.400	0.640	4.136	5.104	5.376	5.640	5.760	5.760
INCOME AFTER TAX	37.120	23.472	25.102	24.913	27.065	22.778	29.284	31.342	33.224	34.913
COMMON DIVIDENDS	5.760	6.720	7.680	7.680	7.680	7.680	7.680	7.680	7.680	7.680
TO RETAINED EARNING	31.360	16.752	17.422	17.233	19.385	15.098	21.604	23.662	25.544	27.233
CASH FLOW	45.520	38.584	51.958	58.833	64.233	50.490	57.116	58.382	59.064	60.577
NET CASH FLOW	39.760	31.864	44.278	51.153	56.553	42.810	49.436	50.702	52.184	52.897

EXHIBIT 3 (b)

IMMSA
BALANCE SHEET

PERIOD ENDING (ANNUALLY)	DEC. 1974	DEC. 1975	DEC. 1976	DEC. 1977	DEC. 1978	DEC. 1979	DEC. 1980	DEC. 1981	DEC. 1982	DEC. 1983
CASH FOR OPERATIONS	46.960	21.256	20.216	20.000	20.000	20.000	20.000	20.000	20.000	20.000
CUMULATIVE FREE CASH	0.000	0.912	0.973	0.000	0.000	0.000	18.206	30.957	30.957	74.238
ACCOUNTS RECEIVABLE	39.344	43.200	48.720	56.000	56.640	61.600	61.600	61.600	61.600	61.600
INVENTORY	65.920	74.920	73.168	62.664	61.376	59.224	57.728	57.728	57.728	57.728
METAL INVENTORIES	49.384	55.184	53.168	41.864	39.376	36.024	33.328	33.328	33.328	33.328
MATERIALS & SUPPLS	16.536	19.736	20.000	20.800	22.000	23.200	24.400	24.400	24.400	24.400
OTHER CURRENT ASSETS	0.968	1.408	1.440	1.440	1.440	1.440	1.440	1.440	1.440	1.440
PREP EXP & OTHERS	0.968	1.408	1.440	1.440	1.440	1.440	1.440	1.440	1.440	1.440
TOTAL CURRENT ASSETS	153.192	141.696	144.517	140.104	139.456	142.264	158.974	171.725	185.797	215.006
INVESTS IN SUBS & OT	9.800	17.616	17.248	16.880	16.512	16.144	12.896	12.896	12.896	12.896
PRODUCTION FACILITY										
BUILDING/PLANT	106.024	141.768	226.616	321.167	391.991	401.690	412.856	412.856	418.632	424.440
ACCUM. DEPRECIATION	62.008	77.120	103.976	137.896	175.064	202.776	257.648	257.648	284.288	309.952
TOT. ACCUM. DEP	62.008	77.120	103.976	137.896	175.064	202.776	257.648	257.648	284.288	309.952
NEW PROD FACILITY	44.016	64.648	122.640	183.271	216.927	198.914	155.208	155.208	134.344	114.488
TOTAL ASSETS	207.008	223.960	284.405	340.255	372.895	357.322	339.829	339.829	333.037	342.390

IMMSA BALANCE SHEET (CONT.)

PERIOD ENDING (ANNUALLY)	DEC. 1974	DEC. 1975	DEC. 1976	DEC. 1977	DEC. 1978	DEC. 1979	DEC. 1980	DEC. 1981	DEC. 1982	DEC. 1983
ACCOUNTS PAYABLE	32.432	28.800	28.800	28.800	28.800	28.800	28.800	28.800	28.000	28.000
SHORT TERM DEBT	0.000	0.000	0.000	0.000	0.000	0.000	0.000	0.000	0.000	0.000
CURRENT LT DEBT	8.184	6.920	24.720	21.384	33.000	33.000	33.000	33.000	18.000	0.000
OTHER ST LIABTIES.	21.520	18.288	16.432	15.768	16.208	18.536	21.128	21.776	22.440	22.560
PROFIT PARTICIPATE	3.536	0.896	0.320	0.400	0.640	4.136	5.104	5.376	5.640	5.760
PROD & EXPORT TAX	13.816	16.000	15.200	14.400	14.400	14.400	14.400	14.400	14.400	14.400
ACCR INCOME TAX	4.168	1.392	0.912	0.968	1.168	0.000	1.624	2.400	2.400	2.400
TOTAL ST LIABTIES.	62.136	54.008	69.952	65.952	78.008	80.336	82.928	69.240	69.240	51.360
PROJECT DEBT	38.024	46.104	73.184	115.800	117.000	84.000	51.000	-0.000	-0.000	0.000
PROJECT LOAD	0.000	0.000	51.800	155.800	105.000	75.000	45.000	-0.000	-0.000	0.000
FCIA FIN	0.000	15.000	0.000	0.000	12.000	9.000	6.000	0.000	0.000	0.000
EXISTING LTD	38.024	31.104	21.384	0.000	0.000	0.000	0.000	0.000	0.000	0.000
OTHER LT LIABILITIES	4.752	5.000	5.000	5.000	5.000	5.000	5.000	5.000	5.000	5.000
RESERVE PER LABOR	4.752	5.000	5.000	5.000	5.000	5.000	5.000	5.000	5.000	5.000
TOTAL LT LIABTIES.	42.776	51.104	78.184	120.800	122.000	89.000	56.000	23.000	5.000	5.000
TOTAL LIABILITIES	104.912	105.112	148.136	186.752	200.008	169.336	138.928	106.576	74.240	56.360
EQUITY COMMON STOCK	48.000	48.000	48.000	48.000	48.000	48.000	48.000	48.000	48.000	48.000
RETAINED EARNINGS	54.096	70.848	88.269	105.503	124.887	139.986	161.590	185.253	210.797	238.030
CAPITAL SPECIF.	0.000	0.000	0.000	0.000	0.000	0.000	0.000	0.000	0.000	0.000
TOTAL EQUITY	102.096	118.848	136.269	153.503	172.887	187.986	209.590	233.253	258.797	286.030
TOTAL LIAB. & EQTY	207.008	223.960	284.405	340.255	372.895	357.322	348.518	339.829	333.037	342.390
WORKING CAPITAL	91.056	87.688	74.565	74.152	61.448	61.928	70.046	88.149	116.557	163.646

EXHIBIT 3 (c)

IMSA
SOURCES AND USES OF FUNDS

PERIOD ENDING (ANNUALLY)	DEC. 1974	DEC. 1975	DEC. 1976	DEC. 1977	DEC. 1978	DEC. 1979	DEC. 1980	DEC. 1981	DEC. 1982	DEC. 1983
CHANGES IN COMPONENTS OF WORKING CAPITAL:										
CASH FOR OPERATIONS	46.960	-25.704	-1.040	-0.216	0.000	0.000	0.000	0.000	0.000	0.000
CUMULTVE FREE CASH	0.000	0.912	0.062	-0.973	0.000	-0.000	18.206	12.750	14.073	29.209
ACCOUNTS RECEIVABLE	39.344	3.856	5.520	7.280	0.640	4.960	0.000	0.000	0.000	0.000
TOTAL INVENTORY	65.920	9.000	-1.752	-10.504	-1.288	-2.152	-1.496	0.000	0.000	0.000
OTHER CURRENT ASSETS	0.968	0.440	0.032	0.000	0.000	0.000	0.000	0.000	0.000	0.000
TOT CURRENT ASSETS	153.192	-11.496	2.822	-4.413	-0.648	2.808	16.710	12.750	14.073	29.209
ACCOUNTS PAYABLE	32.432	-3.632	0.000	0.000	0.000	0.000	0.000	0.000	0.000	0.000
CURRENT LT DEBT	8.184	-1.264	17.800	03.336	11.616	0.000	0.000	0.000	-15.000	-18.000
OTHER ST LIABIL	21.520	-3.232	-1.856	-0.664	0.440	2.328	2.592	0.648	0.664	0.120
TOT CURRENT LIABIL	62.136	-8.128	15.944	-4.000	12.056	2.328	2.592	0.648	-14.336	-17.880
NET CHANGE IN WC	91.056	-3.368	-13.122	-0.413	-12.704	0.480	14.118	12.102	28.408	47.089
SOURCES										
NET INCOME	59.856	23.472	25.102	24.913	27.065	22.778	29.284	31.342	33.224	34.913
TOT NON-CASH CHGS.	62.008	15.112	26.856	33.920	37.168	27.712	27.832	27.040	26.640	25.664
INC. LT LIABILITIES	42.776	8.328	27.080	42.616	1.200	0.000	0.000	0.000	0.000	0.000
DECR. OTHER ASSETS	0.000	0.000	0.368	0.368	0.368	0.368	3.248	0.000	0.000	0.000
INCR. EQUITY	48.000	0.000	0.000	0.000	0.000	0.000	0.000	0.000	0.000	0.000
DICR WORKING CAPITAL	0.000	3.368	13.122	0.413	12.704	0.000	0.000	0.000	0.000	0.000
TOTAL SOURCES:	212.640	50.280	92.528	102.231	78.505	50.859	60.364	58.382	59.864	60.577
USES										
COMMON DIVIDENDS	5.760	6.720	7.680	7.680	7.680	7.680	7.680	7.680	7.680	7.680
INCR. FIXED ASSETS	106.024	35.744	84.848	94.551	70.825	9.699	5.566	5.600	6.776	5.808
INC. OTHER ASSETS	9.800	7.816	0.000	0.000	0.000	0.000	0.000	0.000	0.000	0.000
DECR. LOT LIABIL	0.000	0.000	0.000	0.000	0.000	33.000	33.000	33.000	18.000	0.000
INC. WORKING CAPTL	91.056	0.000	0.000	0.000	0.000	0.480	14.118	12.102	28.408	47.089
TOTAL USES	212.640	50.280	92.528	102.231	78.505	50.859	60.365	58.382	59.864	60.577

EXHIBIT 3 (d)

IMMSA
DEBT SERVICE SCHEDULE

PERIOD ENDING (ANNUALLY)	DEC. 1974	DEC. 1975	DEC. 1976	DEC. 1977	DEC. 1978	DEC. 1979	DEC. 1980	DEC. 1981	DEC. 1982	DEC. 1983
PROJECT LOAN										
DRAWDOWN	0.000	0.000	51.800	64.000	19.200	0.000	0.000	0.000	0.000	0.000
PRINCIPAL	0.000	0.000	0.000	0.000	0.000	30.000	30.000	30.000	30.000	15.000
INTEREST	0.000	0.000	1.554	8.453	12.150	11.475	8.775	6.075	3.375	1.012
PAYMENT	0.000	0.000	1.554	8.453	12.150	41.475	38.775	36.075	33.375	16.012
FCIA FIN										
DRAWDOWN	0.000	15.000	0.000	0.000	15.000	0.000	0.000	0.000	0.000	0.000
PRINCIPAL	0.000	0.000	0.000	15.000	0.000	3.000	3.000	3.000	3.000	3.000
INTEREST	0.000	0.337	1.350	1.012	1.350	1.282	1.012	0.742	0.473	0.203
PAYMENT	0.000	0.337	1.350	16.012	1.350	4.282	4.012	3.742	3.472	3.202
EXISTING LTD										
DRAWDOWN	46.208	0.000	0.000	0.000	0.000	0.000	0.000	0.000	0.000	0.000
PRINCIPAL	0.000	8.184	6.920	9.720	21.384	0.000	0.000	0.000	0.000	0.000
INTEREST	0.000	3.975	3.266	2.581	1.443	0.000	0.000	0.000	0.000	0.000
PAYMENT	0.000	12.159	10.186	12.301	22.827	0.000	0.000	0.000	0.000	0.000
TOTALS:										
DRAWDOWNS	46.208	15.000	51.800	64.000	34.200	0.000	0.000	0.000	0.000	0.000
PRINCIPAL	0.000	8.184	6.920	24.720	21.384	33.000	33.000	33.000	33.000	18.000
INTEREST	0.000	4.312	6.170	12.047	14.943	12.757	9.787	6.817	3.847	1.215
PAYMENTS	0.000	12.496	13.090	36.767	36.327	45.757	42.787	39.817	36.847	19.215

EXHIBIT 3 (e)

IMMSA
RATIO ANALYSIS

PERIOD ENDING (ANNUALLY)	DEC. 1974	DEC. 1975	DEC. 1976	DEC. 1977	DEC. 1978	DEC. 1979	DEC. 1980	DEC. 1981	DEC. 1982	DEC. 1983
LIQUIDITY										
CURRENT RATIO	2.465	2.624	2.066	2.124	1.788	1.771	1.917	2.055	2.683	4.186
ACID TEST	1.389	1.210	0.999	1.152	0.982	1.016	1.204	1.347	1.829	3.034
RETURN ON SALES:										
EBIT/SALES	0.191	0.150	0.135	0.139	0.169	0.184	0.204	0.206	0.207	0.210
INC. AFTER TAX/SALES	0.127	0.099	0.093	0.080	0.086	0.066	0.084	0.090	0.095	0.100
GROSS MARGIN/SALES	0.223	0.227	0.246	0.258	0.298	0.275	0.293	0.293	0.293	0.293
SALES/AV. NET TOTAL ASSETS---		1.100	1.065	0.997	0.882	0.938	0.991	1.016	1.039	1.035
PROFITABILITY RATIOS:										
NET INC./AV. NET TOTAL ASSETS---		0.109	0.099	0.080	0.076	0.062	0.083	0.091	0.099	0.103
NET INC./AV. CR. TOTAL ASSETS---		0.082	0.073	0.058	0.053	0.041	0.051	0.053	0.055	0.055
NET INC./AV. TOTAL EQUITY--------		0.212	0.197	0.172	0.166	0.126	0.147	0.142	0.135	0.128
NET INC./AV. TOTAL LTD EQUITY----		0.149	0.131	0.102	0.095	0.080	0.108	0.120	0.128	0.126
SOLVENCY RATIOS:										
CASH FL/PRIN+CAP EXP	0.429	0.878	0.566	0.493	0.697	1.182	1.481	1.512	1.544	2.544
CASH FLOW/PRIN+PREF	0.000	4.715	7.508	2.380	3.004	1.530	1.731	1.769	1.814	3.365
EBIT/TOT INT EXP	0.000	8.241	5.929	3.595	3.567	4.950	7.282	10.570	18.834	60.444
CASH+INT-CAP EX/DEBT	0.000	0.572	-2.041	-0.644	0.230	12.170	1.434	1.497	1.572	2.914
DEBT – EQUITY RATIOS:										
TOT LIAB/TOT ASSETS	0.507	0.469	0.521	0.549	0.536	0.474	0.399	0.314	0.223	0.165
TOT LTD/TOT CAPITAL	0.207	0.228	0.2745	0.355	0.327	0.249	0.161	0.068	0.015	0.015
LTD/NET WORTH	0.419	0.430	0.574	0.787	0.706	0.473	0.267	0.099	0.019	0.017
TOT LIAB/NET WORTH	1.028	0.884	1.087	1.217	1.157	0.901	0.663	0.457	0.287	0.197

EXHIBIT 3 (f)

IMMSA
SENSITIVITY COEFFICIENTS

PERIOD ENDING (ANNUALLY)	DEC. 1974	DEC. 1975	DEC. 1976	DEC. 1977	DEC. 1978	DEC. 1979	DEC. 1980	DEC. 1981	DEC. 1982	DEC. 1983
SALES REVENUE	0.052	0.076	0.089	0.100	0.082	0.068	0.057	0.054	0.051	0.048
UNIT SALES	0.012	0.017	0.022	0.026	0.024	0.019	0.017	0.016	0.015	0.014
COST OF GOODS SOLD	0.041	0.059	0.067	0.074	0.058	0.049	0.040	0.038	0.036	0.034
OPERATING EXPENSES	0.001	0.002	0.002	0.002	0.001	0.001	0.001	0.001	0.001	0.001
NON-CASH CHARGES	0.001	0.005	0.009	0.011	0.010	0.005	0.005	0.004	0.004	0.004
FINANCIAL CHARGES	0.000	0.001	0.002	0.004	0.004	0.003	0.002	0.001	0.001	0.000
ASSUMPTIONS										
SALES GROWTH RATE	4.198	-0.191	0.142	0.151	0.010	0.089	0.021	0.000	0.000	0.000
SILVER PRICE/OZ	0.247	3.500	3.500	3.500	3.500	3.500	3.500	3.500	3.500	3.500
LEAD PRICE/LB	0.903	0.160	0.160	0.160	0.160	0.160	0.160	0.160	0.160	0.160
COPPER PRICE/LB	0.426	0.530	0.530	0.530	0.530	0.530	0.530	0.530	0.530	0.530
ZINC PRICE/lb		0.320	0.320	0.320	0.320	0.320	0.320	0.320	0.320	0.320
COGS/% OF SALES	0.777	0.773	0.54	0.742	0.702	0.725	0.707	0.707	0.707	0.707
OP EXP/% OF SALES	0.021	0.024	0.021	0.018	0.018	0.016	0.016	0.016	0.016	0.016
DEBT RATE:										
PROJECT LOAD	0.000	0.000	0.030	0.073	0.090	0.090	0.090	0.090	0.090	0.090
FCIA FIN	0.000	0.022	0.090	0.090	0.090	0.090	0.090	0.090	0.090	0.090
EXISTING LTD	0.000	0.090	0.090	0.090	0.090	0.090	0.090	0.090	0.090	0.090

DCF INTERNAL RATE OF RETURN

PERIOD OF DCF OUTFLOW
TOTAL NUMBER OF PERIODS   0.000

DIVIDENDS DISC. ON COMMON STOCK   0.000
CASH FLOW DISC. ON COMMON STOCK   0.000
NET INCOME DISC. ON COMMON STOCK 0.000

CASH FLOW DISC. ON LTD+EQ.   0.000
NET INCOME DISC. ON LTD+EQ. 0.000

CASE STUDIES

Exhibit 4
MEXICO'S INTERNATIONAL RESERVES
(US $ Millions; as of 31st December)

	1970	1971	1972	1973	1974
Gold	176	200	188	196	157
SDR's	48	96	139	154	158
Reserve Position in the IMF	135	106	106	118	120
Foreighn Exchange	385	550	731	888	960
	744	952	1,164	1,356	1,395

Source: International Financial Statistics

Exhibit 5
EXTERNALLY FUNDED DEBT OF PUBLIC SECTOR
(US $ Millions; as of 31st December)

	1970	1971	1972	1973	1974
Direct Debt of Federal Government	858	882	972	1,298	1,912
Debt Guaranteed by Federal Government	807	788	952	1,013	1,279
Debt of Other Agencies	1,595	1,884	2,331	3,421	4,749
	3,260	3,554	4,255	5,732	7,940

Source: Banco de Mexico, S.A.

Exhibit 6
DEBT SERVICE REQUIREMENTS OF
PUBLIC SECTOR EXTERNAL FUNDED DEBT
(US $ Millions; as of 31st December, 1974)

Due	Principal	Interest	Total Debt Service
1975	838.7	647.3	1,486.0
1976	680.4	595.0	1,275.4
1977	767.4	583.9	1,306.3
1978	790.0	480.9	1,270.9
1979	754.5	416.1	1,170.6
1980-84	3,013.9	1,168.2	4,182.1
1985-89	965.0	204.2	1,169.4
1990-94	102.0	21.4	123.4
1995-99	19.4	4.2	23.6
2000-2003	8.2	0.8	9.0
Total	7,939.5	4,077.2	12,016.7

Source: Banco de Mexico, S.A.

EXHIBIT 7

BALANCE OF PAYMENTS

($ Millions)

	1970	1971	1972	1973	1974
Trade Balance	-1,045.5	-890.5	-1,052.6	-1,749.5	-3,191.8
Tourism, Net	245.3	288.8	342.2	466.2	517.5
Passenger fares, net	-14.6	-7.0	-6.2	-9.2	-16.6
Border trade, net	293.9	354.4	407.7	512.7	554.9
Investment income, net	-586.7	-619.8	-713.3	-896.1	-1,271.2
Other	161.7	147.8	260.7	484.3	794.2
Balance on Current Account	-945.9	-726.4	-761.5	-1,191.6	-2,613.0
Direct foreign investment, net	200.7	196.1	189.8	287.3	272.6
Purchase of foreign enterprises	-	-	-10.0	-21.3	-
Securities transactions, net	7.2	52.0	6.2	-10.2	-59.6
Loans to public sector, net	263.1	286.4	359.7	1,046.6	2,039.3
Loans to private sector, net	61.1	164.2	186.3	299.6	338.6
Net changes in direct governmental external debt	-2.3	-28.9	37.8	69.9	-26.9
Mexican credits granted	-11.5	-0.7	-16.3	-18.9	-39.5
Balance on Long-Term Capital Account	503.9	669.1	753.5	1,635.5	2,524.5
Short-term capital and errors and omissions	498.7	217.7	233.5	-339.6	125.4
SDR's	45.4	39.6	39.2	-	-
Overall Balance of Payments	102.1	200.0	264.7	122.3	36.9

Source: Banco de Mexico, S.A.

Closings

Notice of loan and notice of disbursement

The total loan was to be made upon no less than ten working days prior written notice by IMMSA to agent specifying the amount requested to be dispersed (not in excess of the principal balance in the pledged accounts) and a date on or prior to the credit termination date upon which the disbursement was to occur. Notice of disbursement was made to the agent within 30 working days of any prior notice of disbursement.

Closings

Loan closings were to take place at the office of United California Bank International, 630 Fifth Avenue, New York, New York and the office of Banco de Mexico, S.A. on the date specified in the notice of loan. Each closing for a disbursement was to take place at the office of the agent. If IMMSA were to fail to borrow the total loan on the date specified in the notice of loan, whether voluntarily or by reason of the borrowers' failure to satisfy the conditions to closing required under the agreement, the borrowers were to pay all amounts as might have been necessary to compensate the banks for any loss incurred by them as a result of such failure, including, but not limited to, any loss or expense sustained or incurred in liquidating or employing fixed deposits or other funds from third parties acquired to make the total loan.

Interest

1-7/8% above the arithmetical average (rounded up to the nearest one-sixteenth of one percent) of the three- or six-month London interbank offered rate for relevant deposits, as quoted by the reference banks. Interest was to be payable quarterly from the date of first drawdown.

Principal Payment

The principal amount of the total loan was to be repaid in 10 semi-annual installments on the interest adjustment dates falling nearest February 28 and August 31 of each year, commencing August 31, 1978. The entire unpaid principal balance was to be due and payable on the interest adjustment date falling nearest February 28, 1983.

Final Maturity

Seven years from first drawdown, but not later than February 28, 1983.

Prepayments

Prepayment of any borrowing was to be permitted without penalty upon 30 days prior written notice on any interest adjustment date for such borrowing, provided, however, that any prepayment was to be in a minimum amount of US$5 million and partial prepayments of principal were to be applied in the inverse order of maturities of the respective borrowing.

Commitment Fee

Borrowers were to pay the agent for the account of the banks in the proportion of each bank's pro rata share, a commitment fee equal to 1/2% per annum of the unused portion of the total loan from and including the earlier of the date of execution of the loan until the loan closing date. The commitment fee was to be payable on the first day of March, June, September and December of each year on the loan closing date.

Management Fee

Borrowers were to pay the agent a management fee amounting to $1.125 million for the account of the total loan on the date of the execution by borrowers of this agreement. The management fee represented 3/4% of the amount of the total loan. The management fee was not to be affected by any reduction of the total loan or termination of obligation of any bank's party to the loan agreement.

Taxes

All payments of principal, interest, and fees were to be made free and clear of all taxes which might have been imposed thereon by the United Mexican States or any political subdivision thereof.

Reserve Requirements

The borrower was to reimburse any additional costs (considered unavoidable in the opinion of the lenders) as a result of the imposition of any reserve requirements with respect to the loan or resulting from the imposition of any interest equalization tax thereon.

Governing Law

The state of California, United States of America.

Conditions Precedent to Loan

Borrowers were to have executed and delivered to agent a note payable to the order of each bank with appropriate insertions regarding the amount of each bank's loan.

Agent was to receive a favorable written opinion, dated the loan closing date, of special Mexican counsel for borrowers, Messrs. Noriega y Escobedo, A.C. and of special Mexican counsel for agent and the banks, Messrs. Ramirez, de la Corte y Ritch with such changes and with any additional matters as the banks or special counsel for the banks might have approved or requested.

All representations and warranties made by the borrowers were to have been complied with as of and at the loan closing date.

Representations and Warranties

Typical contractual provisions regarding the following held: organization and powers, officers, corporate action, consents, contravening provisions, government appraisals, financial statements, title to properties, litigation, federal reserve regulations and taxes.

No defaults. No event had occurred from the making of the total loan or performance of any provision of this agreement which would constitute an event of default, or which, upon the lapse of time or giving of notice or both, would have become such an event of default hereunder.

Subsidiaries. Except as specified in a list attached to the loan agreement, IMMSA

had no other subsidiaries,

was the owner of all issued and outstanding shares of capital stock free and clear of any security interest, claims, liens, pledges, encumbrances or other restrictions whatsoever of each of its subsidiaries, and

had no subsidiaries with total assets of $800,000 or more other than as set forth in the list.

No event of default shall have transpired. Nor shall there have been an adverse change in the assets, properties, liabilities, business or condition, financial or otherwise of IMMSA and each of its subsidiaries from that shown in the financial statements.

Agent shall have received written acceptance of IMMSA's appointment as agent for the borrowers.

Borrowers shall have paid in full the commitment fee and management fee as specified in the loan agreement.

The form and legal sufficiency of the agreement, the notes and all other documents hereunder and all legal translations thereof shall have been judged satisfactory to special New York counsel to the banks, Messrs. Lord, Day & Lord.

Conditions to Disbursements

No event of default and no event which with the passage of time or the giving of notice or both would have constituted such an event of default shall have occurred or be continuing, and there shall have been no material adverse changes in the assets, properties, liabilities, business, or condition, financial or otherwise, of IMMSA and each of its subsidiaries from that shown in the financial statement.

In the case of the initial disbursement and all other disbursements made in the same calendar quarter as the initial disbursement, the aggregate amount of all such disbursements shall not have exceeded the sum of:

the total amount of projected asset expenditures for such quarter as shown in Exhibit 8 or such higher amount as was to be shown in the latest approved program progress report, and

an amount equal to 66-2/3% of the actual fixed expenditures for the program made through the quarter immediately preceding the quarter in which the initial disbursement closing date occurred.

Affirmative Covenants

Unless the banks waived compliance in writing, borrowers jointly and severally convenanted and agreed that until the full and final payment of the total loan and all indebtedness incurred thereunder, they would have:

Loan Repayment. Duly and punctually repaid the total loan and all notes evidencing the total loan, the interest thereon, and all amounts due hereunder when and at the same time they become due.

Use of Proceeds. Used all proceeds of the total loan and disbursement exclusively for purposes set forth in the statement of purpose.

Notice to Agent

Any event of default or any event which upon a lapse of time or giving of notice or both had become an event of default.

Any other matter which had resulted or might have resulted in a material adverse change in the financial condition or operations described in the financial statements.

Reports and Financial Statements. As soon as possible, but no later than 30 days from the close of each quarter of IMMSA's fiscal years, the following reports were to have been provided to the agent:

Actual and projected consolidated net income
   statement through 1983
Updated projected expenditures breakdown
Total program expenditures chart
Projected project status summary
Zinc refinery project status summary

If the projected expenditures program for any project for any quarter were to have been in excess of those shown in Exhibit 8, those banks whose aggregate pro rata shares were equal to 70% or more shall have had a period of 60 days in which to approve or disapprove the projected expenditures program or to have agreed with IMMSA on an approved modification thereto.

Negative Covenants

FINANCIAL TESTS. Permitted IMMSA's consolidated net current assets to have been less than 150% of its consolidated net current liabilities, or IMMSA's consolidated total liabilities to have exceeded 130% of consolidated tangible new worth, which was to have been equal to:

the par or nominal value of all classes of the issued and fully paid share capital of IMMSA

the capital surplus of IMMSA on a consolidated basis

the earned surplus of IMMSA on a consolidated basis, less asset revaluations after December 31, 1974

INDEBTEDNESS. Incurred indebtedness of any type in excess of US$1 million in total other than the indebtedness set forth in Exchibits 9a and 9b.

CAPITAL EXPENDITURES. Total capital expenditures by the borrower between December 31, 1974 and December 31, 1978 were not to have exceeded US$286 million for purposes of implementing the program and US$10 million for purposes other than the implementation of the program.

UTILIZATION OF THE TOTAL LOAN. Requested disbursements from the pledged accounts:

In the case of the quarter in which the initial disbursement was to be made so that the aggregate of all disbursements thereunder during such calendar quarter shall have exceeded the sum of:

projected fixed asset expenditures for the quarter shown in the expenditures breakdown or such higher amount as was to have been shown in the latest approved program report, plus

an amount equal to 66-2/3% of the actual fixed asset expenditures for the program made through the quarter immediately preceding the quarter in which the initial disbursement closing date was to have occurred.

In the case of all disbursements other than the initial disbursement, so that the aggregate of all disbursements thereunder during any calendar quarter was to have exceeded the total amount of projected fixed assset expenditures for the quarter as shown in the expenditures breakdown.

In the case of all disbursements, so that the aggregate amount of the disbursements in any two consecutive calendar quarters was to have been in excess of two-thirds of the total cumulative expenditures on the program during two such calendar quarters.

Events of Default

Any proceedings were to have begun or petition was to have been filed seeking reorganization of or the appointment of a receiver, trustee, liquidator or the like, of all or a substantial part of the assets of a borrower or guarantor.

Exhibit 8
BREAKDOWN OF FIXED ASSET EXPENDITURES BY QUARTERS
(US$ Millions)

	1975					1976					1977					1978					TOTAL
	1	2	3	4	Total	1	2	3	4	Total	1	2	3	4	Total	1	2	3	4	Total	75/78
Projects other than Zinc Refinery.....	$5.3	$7.6	$7.6	$10.9	$31.4	$13.0	$13.2	$9.2	$10.9	$46.3	$12.0	$10.9	$7.1	$7.5	$37.5	$5.6	$5.9	$6.2	$3.1	$20.8	$136.0
Zinc Refinery.....	1.4	.2	.3	2.4	4.3	5.5	11.0	11.1	11.0	38.6	13.6	14.8	16.3	12.4	57.1	21.8	18.3	9.5	.4	50.0	150.0
TOTAL........	$6.7	$7.8	$7.9	$13.3	$35.7	$18.5	$24.2	$20.3	$21.9	$84.9	$25.6	$25.7	$23.4	$19.9	$94.6	$27.4	$24.2	$15.7	$3.5	$70.8	$286.0

Source: Bank of America memoranda.

Any borrower or guarantor shall have failed to make payment of any principal of or interest on any obligation for borrowed money or for the deferred purchase price of any property or asset.

IMMSA was to have paid any dividend or to have made any distribution on any shares of capital stock to its stockholders (other than dividends payable in shares of stock of IMMSA) purchased, redeemed or otherwise acquired for consideration any shares of its capital stock if, after having given effect to such divident, distribution, purchase, redemption or other acquisition, the aggregate amount of expenditures by IMMSA for all such purposes

made in calendar year 1976 shall have exceeded US$7.68 million or

shall have resulted in an actual net cumulative cash flow which was less than the amount of projected net cumulative cash flow for the period beginning January 1, 1976 and ending on such December 1st as set forth in a schedule, or

shall have exceeded 50% of cumulative consolidated net income for IMMSA for the period January 1, 1976 and ending on such December 31st, or

shall have exceeded US$9 million in any one calendar year prior to January 1, 1980.

The updated projected total cost of the program was not to have exeeded US$286 million.

Any zinc refinery project status summary shall not have shown a total cost of the zinc refinery project in excess of US$150 million.

ACKNOWLEDGEMENT

The author wishes to extend his sincere appreciation to Industrial Minera Mexico, S.A. for making this case study possible.

Exhibit 9 (a)

LIST OF PROJECTED INDEBTEDNESS OF BORROWERS

United Calfornia Bank Insured by Foreign Credit Insurance Agency	For purchase of capital equipment (can be substitued).	Revolving 5 year term	U.S. $15,000,000
Lloyds Bank Ltd.	For purchase of capital equipment (can be substitued).	Pending	Pounds Sterling 3,000,000

* The Lender or Lenders. the terms and the amounts of the projected indebyedness may be changed witout consent of the Banks provided any indebtedness substitued therefore shall be for the same purpose and shall not, together with all other projected indebtedness set forth or substituted for projected indebtedness set forth above, exceed the aggregate amount of U.S. $21,000,000 or the equivalent in other currencies.

Exhibit 9 (b)

INDUSTRIAL MINERA MEXICO, S.A.
LIST OF EXISTING INDEBTEDNESS OF BORROWERS
(At April 30, 1976)

Borrower	Purpose	Amount	Repayment	Due Dates
Asarco Incorporated	Purchase of Industrial Minera Shares of Stock for sale to Employees	$ 2,452,800	U.S.$ 1,471,680	U.S.$ 490,560 per annum from 1975 — April 26, 1977, 1978, 1979
Asarco Incorporated	Purchase of La Loteria Shares of Stock	1,161,035	696,621	U.S.$ 232,207 per annum from 1975 — April 26, 1977, 1978, 1979
Empresarios Industriales de Mexico, S.A.	Purchase of 31.66% of La Loteria Stock	3,250,000	2,250,000	From 1976 U.S.$ 1,000,000 — February 7, 1977 1,250,000 — February 7, 1978
Bank of America	Construction of Chihuahua Sinter Plant	5,000,000	1,000,000	U.S.$ 500,000 per semester From 1971 — June 1, 1976 December 1, 1976
Manfacturers Hanover Trust Company	Sabinas #2 Coal Mine and Santa Eulalia Unit	5,000,000	500,000	U.S.$ 500,000 per semester From 1971 — June 30, 1976
Consortium Loan--Agent Manfacturers Hanover Trust Company	Construction at Taxco. San Martin, San Luis Potosi, and Chihuahua	30,000,000	27,600,006	From Feb. 1976 1,599,996 — August 21, 1976 3,999,993 — February 21, 1977 3,999,993 — August 21, 1977 3,750,000 — February 21, 1978 3,500,001 — August 21, 1978 3,250,002 — February 21, 1979 3,000,015 — August 21, 1979 1,350,000 — February 21, 1980 1,200,000 — August 21, 1980 1,050,000 — February 21, 1981 900,000 — August 21, 1981

T O T A L INDUSTRIAL MINERA MEXICO, S.A.   U.S.$ 33,518,830

CARBONIFERA DE MEXICO, S.A.

Borrower	Purpose	Amount	Repayment	Due Dates
Dowty Coporation	Supplier credit for purchase of Dowty Longwall Coal Cutting Equipment	$ 1,725,331	U.S.$ 1,552,798	U.S.$ 258,800 July 15, 1976 Balance in 20 quarterly payments beginning Oct. 15, 1976

Industrial Minera Mexico. S.A., in August 1971, signed an agreement with the Comision de Fomento Minero, a Governmental Agency, under which the Company guarantees payment to credits up to a maximum amount of MN$40,000,000, (U.S.$3,200), granted to the Comision by Nacional Financiera, S.A. Fomento Minero utilizes the credit in loans to small and medium size mining companies.

At April 30, 1976, Industrial Minera Mexico, S.A. acted as guarantor on loans totaling the peso equivalent of U.S.$ 1,831,000

12.4

FINANCING OF TECK'S INVESTMENT IN THE BULLMOOSE COAL PROJECT
N.R. MACMILLAN
SENIOR VICE PRESIDENT OF PROJECT FINANCE
BANK OF MONTREAL
TORONTO, ONTARIO. CANADA

INTRODUCTION

The Bullmoose Coal Project is part of a major development in northeastern British Columbia which comprises a new rail line, a new townsite, powerline, highway, the upgrading of the Canadian National Railway's main line, a new coal port at Prince Rupert and two new coal mines, the Bullmoose mine and the Quintette mine. The total investment by all parties is estimated to be in excess of $2.6 billion and made up approximately as follows :

Table No. 1

	$ Millions
Quintette Coal Limited	950
Teck-Bullmoose Coal Inc.	275
Canadian National Railway Company & British Columbia Railway Company	800
Port at Prince Rupert	280
Townsite, Roads & Power	350
	2,655

The new infrastructure is intended to serve not only the Bullmoose and Quintette coal projects but also eventually a number of other coal properties in northeastern British Columbia.

The Bullmoose project is structured on a joint venture basis as follows :

Table No. 2

Teck Corporation	51%
Lornex Mining Corporation Ltd.	39%
Nissho Iwai Coal Development (Canada) Ltd.	10%
	100%

The mine has a capacity of 2.3 million tonnes of clean coal per year. In the initial years 1.7 million tonnes will be metallurgical coal and 0.6 million tonnes thermal coal.

After the first three years of operations the annual production of metallurgical coal will increase to an average of approximately 2.1 million tonnes and the production of thermal coal will decrease to an average of approximately 0.2 million tonnes. At this level of 2.3 million tonnes annual production, the Bullmoose mine has sufficient reserves for at least 26 years of production.

History

Brameda Resources Ltd. commenced coal exploration in the Bullmoose-Sukunka area in 1969. Teck obtained an option on the Bullmoose coal licences in 1970 and started surface exploration and diamond drilling in 1971. Teck earned a 50% interest in the property, and thereafter expenses were jointly financed by Teck and Brameda. In 1981 Brameda merged with Teck making Teck the sole owner.

The initial phase of the exploration program, which was completed in 1975, included 7,200 meters of diamond drilling in 35 holes. During 1976 and 1977, detailed exploration and test work were carried out on the South Fork and West Fork areas where multiple seams close to surfaces indicated the potential for large tonnages of strippable coal. In those two years, work included road construction, detailed mapping and trenching to outline the coal seam traces at the surface, 6,300 meters of diamond drilling in 80 holes, and 300 meters of underground bulk sampling from seven adits. Washability tests and analysis work were carried out on the drill cores and adit samples, and pilot plant washability studies were undertaken.

In 1978 and 1979 Teck engineers prepared open pit mining plans and production schedules and Kaiser Engineering was retained to prepare a layout of the plant preparation facility.

A feasibility study prepared by Teck's engineering group in January 1982 was the base document used by the Lenders in assessing the technical and economic viability of the project.

Property Location

The property lies some 650 kilometers northeast of Vancouver, and 87 kilometers south by road from the village of Chetwynd. The Sukunka property (British Petroleum) adjoins to the north, the Mount Spieker property (Ranger Oil Option) adjoins to the east, and the Quintette property is located to the southeast.

The South Fork coal deposit contains sufficient reserves for over 16 years of production at the planned production schedule and will be mined first. It has a length of some 3 kilometers in a north-south direction and a width of 1.5 kilometers.

The West Fork has mineable coal reserves for a further 10 years. It has a length of some 5 kilometers (east-west) and a width of approximately 1 kilometer limited by increasing depth of cover to the north.

The preparation plant and associated facilities are located in the Bullmoose Creek Valley near the eastern boundary of the property.

A new highway from Chetwynd to the Tumbler Ridge townsite area was constructed. This route passes within 18 kilometers of the Bullmoose property, and the total distance from the site to the railhead near Tumbler Ridge is 35 kilometers. The Tumbler Ridge townsite is 38 kilometers from the Bullmoose property, three kilometers beyond the railhead.

Coal Reserves

The saleable coal produced is a medium volatile bituminous coal with low ash and sulphur levels and excellent coke strength. Some thermal coal will be produced from oxidized areas which contain coal of similar quality, but with low swelling properties.

The following table summarizes the measured the coal reserves for the South and West Fork Deposits :

Table No. 3
COAL RESERVES
(Millions of tonnes)

	Coal in Place	Clean Coal (8% Moisture)
Metallurgical		
South Fork	45.9	33.3
West Fork	19.4	14.3
Total :	65.3	47.6
Thermal		
South Fork	5.0	3.6
West Fork	11.6	7.8
Total :	16.6	11.4

Mine Development

Mine capacity is 2.3 million tonnes per year of saleable coal. Mining is carried out using a conventional truck/shovel operation.

The position of the coal seams close to the surface and parallel with the hill slope will result in fairly constant stripping ratios of approximately 6 bank cubic metres per tonne clean coal over the initial fifteen years. This compares favourably with other B.C. and Alberta open pit operations.

A detailed twenty year mining plan covering the entire South Fork and a portion of the West Fork coal reserves was prepared by Teck. It calls for an average run of mine coal production of 3 million tonnes per year.

All haulage in mining areas is handled by off-highway trucks. The majority of the waste material is loaded by conventional shovels. Waste material adjacent to coal seams and the coal itself is loaded using hydraulic shovels to maximize coal recovery and minimize dilution.

Coal Preparation System

The coal preparation facility located adjacent to the north end of the South Fork deposit of the mine in the Bullmoose Valley was designed by Phillips Barratt Kaiser Engineering Ltd. The plant has a capacity of 2.3 million mtpy (million tons per year) of saleable coal (8% total moisture).

Infrastructure.

Federal Government of Canada The Federal Government is involved in the development of northeastern B.C. coal through its commitment to transport coal from Prince George and to provide a new port facility at Prince Rupert. Federal government involvement is handled through the following agencies :

CANADIAN NATIONAL RAILWAY: The Canadian National Railway, which is owned 100% by the Federal Government, was responsible for the upgrading and maintenance of its railway between Prince George and Prince Rupert. Together with British Columbia Railway, it has provided the rolling stock necessary for the movement of northeastern B.C. coal. The estimated capital cost incurred in the provision of these facilities by the Canadian National Railway amounted to $250 million.

NATIONAL HARBOURS BOARD: Ridley Terminals Inc., a subsidiary of Ports Canada and Federal Commerce and Navigation Ltd., a private company, constructed a terminal on Ridley Island near Prince Rupert for the handling of northeastern B.C. coal. Port capacity is 12 million tonnes per annum with a peak ship loading rate of 8,000 tonnes per hour. The port capacity is considerably in excess of the initial combined annual capacity from Bullmoose and Quintette of 8.6 million tonnes. The port is able to handle ships with capacities up to 250,000 dwt (dead weight tons). The estimated cost of the port facility was $280 million.

Government of the Province of British Columbia The Provincial Government's commitment to the development of northeastern B.C. coal is handled through the following agencies :

THE NORTH EAST COAL DEVELOPMENT OFFICE: This office was created by the Provincial Government to co-ordinate all provincial infrastructure arrangements related to the development of northeastern B.C. coal. These infrastructure commitments included the upgrading of highways, construction of power lines (through the Provincial Government's power utility company, B.C. Hydro), and the establishment of a new town of Tumbler Ridge. The estimated aggregate cost of these facilities was $350 million.

BRITISH COLUMBIA RAILWAY: The British Columbia Railway is owned 100% by the Provincial Government. It was responsible for building a new 130 kilometer railway line from Anzac to Tumbler Ridge and the Quintette property, upgrading its track between Anzac and Prince George, and the necessary rolling stock. The cost of this program was approximated $550 million.

Marketing Arrangements

Agreement was reached in January 1981 for the purchase of metallurgical coal by a group of Japanese steel mills for a 15.5 year period commencing in October 1983. The contracts are renewable by mutual agreement for a certain number of years thereafter.

Published information on the sales contracts shows that they are structured on a 'take and pay' basis with the buyers having some limited rights to reduce volume offtake.

The initial contract price incorporated a premium to the price then being paid by the Japanese Mills for coal from the mines located in South East British Columbia. There is provision for the price to be escalated over time using relevant Canadian labour, materials and fuel indices.

Insurance Arrangements

Various insurance coverages have been arranged for the Bullmoose Coal Project. The following paragraphs briefly summarize the insurance arrangements.

Physical Damage-Builders All Risk Insurance This insurance covers the assets of the Project against all risks of physical damage. It includes business interruption insurance which covers, amongst other items, interest on the project debt. This insurance would be effective as a result of a delay caused by an insurable event such as damage caused by earthquake or fire.

Comprehensive - General Liability Insurance This insurance covers third party liability arising from construction of the Project.

Infrastructure Delay Insurance This insurance coverage is discussed under 'Financing Plan'.

CAPITAL COST

Capital Cost and Funding Requirements

The estimated capital cost figures used to develop the financing plan are summarized in the following table. These are based upon the 'best estimate' developed by Wright Engineers Ltd., one of the Lenders' Consultants in its report dated July 1982. This best estimate assumed completion of construction by December 31, 1983.

Table 4
ESTIMATED CAPITAL COST
(Millions of dollars, current)

Capital Cost 1982	85.7
1983	149.1
	234.8
Working Capital	18.5
	253.3
Teck's 51%	129.2
Interest during construction	17.6

TECK'S BASE CAPITAL REQUIREMENT   Cdn$146.8MM

It should be noted that Teck's original capital estimate set the base capital requirement at Cdn.$159 million which is 8.4 per cent higher than the above estimate which was developed by Wright Engineers Ltd. some months later.

FINANCING PLAN

The financing plan for Teck-Bullmoose's 51% share of the Project provides for a committed funding pool of $190.5 million consisting of:

Table No. 5

Commercial Bank Loan Facility	$158,000,000
Equity	$ 32,500,000
Total Funding Pool	$190,500,000

The funding pool incorporates a $43.7 million, or 29.8% contingency allowance to cover costs in excess of those provided for in the 'Base Capital Requirement'.

Pre Completion Risk

Lenders were willing to accept significant pre Completion risk for the following five reasons:
1. By mid 1982 there was limited capital investment in British Columbia, inflation was reducing and unemployment rising. All in all, it was a good economic background in which to be undertaking a $280 million project as part of $2.6 billion investment program.

2. Teck has an outstanding track record for bringing projects on line both within budget and a pre established time frame.

3. The Funding Pool incorporating a 30% overrun allowance was established using data developed by an independent technical consultant appointed by the lenders. The consultant undertook a detailed review of the capital budget and project construction schedule to determine the maximum financing required.

4. The lenders were particularly concerned with the possibility of Teck completing Bullmoose before the rail and port facilities were operational with the result that coal shipments would not commence in a timely manner. Teck was able to arrange insurance to cover this specific risk through the UK market. This was an imaginative approach to providing both Teck and the Lenders with significant financial protection against a risk which was totally outside Teck's control.

5. The construction period was short as reflected by the time from establishing the underwriting commitment in May 1982 to anticipated startup was 18 months.

The financial structure used was what is commonly known as the 'Pool of Funds'. The key aspect of this mechanism is that if the project costs (including Facility interest and initial working capital) had exceeded the Funding Pool, which in this case was $190.5 million, then neither the Lenders nor Teck were legally obligated to provide the additional funding required to achieve completion. A mechanism was provided for the two parties to consult and negotiate the provision of such additional funds in excess of $190.5 million if required.

In this case the Funding Pool was structured as follows:

Table No. 6

	Cdn$MM	%
Bank Facility Split into:		
Project Specific Tranche	98.0	51.4
Tranches guaranteed by Teck	60.0	31.5
Equity	32.5	17.1
	92.5	
	190.5	100.0

It should be noted that the project specific tranche represents only 51.4% of the committed available capital.

Marketing Risk Considerations

In establishing the initial amount of project specific debt at $98 million, the Lenders' analysis was focussed on the project economics. Not unsurprisingly two key variables were identified namely coal sales volume and price.

Turning first to the question of volume, the mine annual production capacity was established at 2.3 million tonnes, however, only 1.7 million tonnes/year of metallurgical coal are under contract to the Japanese steel mills. This is the sole underpinning on the volume side.

The contract price established in 1981 for Bullmoose coal incorporated a premium over similar quality coal in South East British Columbia and Australia in order to ensure a minimum acceptable return on and of the invested capital. It is understood that the Japanese steel mills were willing to pay that premium in order to increase diversification of their supply sources.

In over-supply situations the established mines in Australia and Canada with much if not all of their original capital repaid would be more competitive than the North East B.C. Mines on a FOB cost/tonne basis in selling surplus production tonnage. Consequently, the Lenders were only willing to accept volume risk on tonnages under contract in establishing the project specific debt capacity.

In order to reduce the vulnerability of the project to upward movements in interest rates, the debt was structured so that at least $50 million could be placed on a fixed rate basis.

Care was also taken to maximize the Canadian dollar component of the Bank debt so as to minimize currency exposure. A separate US dollar component had to be established to accomodate lenders that did not have access to a Canadian dollar funding base. However, a mechanism was incorporated to utilize forward exchange contracts to limit exposure to adverse Canadian, US dollar rate movements.

Taking the foregoing into account, the following structure was developed to facilitate the financing of Teck's share of Bullmoose project costs:

Table No. 7

	Cdn$MM
Tranche A - recourse solely to project assets and and cash flow	98
- guaranteed by Teck	32
Tranche B - guaranteed by Teck	28
	158

Draws of Tranche A were ordered so that the project specific amount was drawn down first. The reason for splitting the Tranche into two was because the $98 million project debt capacity was established based on the Japanese steel mill contracts of 1.7 million tonnes of met coal/year. The guarantee on the remaining $32 million will be released when acceptable long term sales contracts for the balance of the tonnage (ie. 600,000 tonnes/year) are put in place. Funds were advanced in the following order:

- Tranche A and Equity in an 80:20 ratio.
- To date there have been no advances on Tranche B and none are anticipated as capital costs have been contained well within the $162.5 million maximum of Tranche A + Equity.

Repayment Profile of Non-Recourse Debt.

The project debt incorporated a 12 year target term with provision for extension to 15 years if required. The potential for the operating cash flow available for repayment from year to year was recognized in establishing the repayment schedule. Repayments were geared to the lesser of "available cash flow" and a set repayment profile. To the extent that the set repayments could not be achieved, no cash distribution would be made to Teck until the loan balance is reduced in line with the profile as originally established. However, one years slippage in the target repayments is permitted before the debt will be in default. Through these various mechanisms provision was made for repayment of the project debt over a maximum 12½ year term after start up.

Syndication

In view of the political profile of the North East Coal Development in both Japan and Canada, structuring of the lending syndicate was an important mechanism for obtaining confirmation of the Japanese steel mills commitment to the Project. Japanese and Canadian banks were prominent in the syndicate representing 45% and 28% respectively of total commitments. The balance of the Facility is held by European and US Banks well regarded for their project finance capability.

CURRENT STATUS OF PROJECT AS AT JUNE 1, 1984

The Bullmoose Project was brought into production on schedule and under budget which is a major accomplishment for a project of this magnitude and complexity.

The first trainload of coal was shipped to the port on November 1, 1983 and the first shipload left Ridley Island on January 10, 1984. Since then the Project has produced and shipped approximately 140,000 tonnes per month of metallurgical coal corresponding to the contracted volume of 1.7 million tonnes/year. The price being paid by the Japanese steel mills is Cdn.$98.00 per tonne as of April 1, 1984. This is the price as computed under the contract terms and conditions.

As of June 1, 1984 formal Project Completion has yet to be certified but as construction is essentially completed, total project cost can be established with a high degree of certainty. Teck's total project cost is estimated at Cdn.$139 million or 2% below the base case capital requirement. The funding hereof is as follows :

Table No. 8

Equity	- Cdn.$ 27.8 million	20%
Debt	- Cdn.$111.2 million	80%
Total Funding	- Cdn.$139.0 million	100%

At present Cdn.$98 million of the debt component is non-recourse, while the remaining Cdn.$13.2 million is guaranteed by Teck with potential for redesignation to non-recourse when Teck obtains acceptable long-term sales contracts for the balance of the design tonnage of 2.3 million tonnes/year.

EDITOR'S NOTE

The following terms and conditions appeared in the October 16,1982 edition of the newsletter International Bondletter & Eurocurrency Financing Review published by Agefi.

"Teck-Bullmoose Coal Inc(Canada). A C$158m package over 12 years (December 31,1994) extendable for a further 18 months until June 30, 1996.

Tranche A, amounting to C$130m, comprises recourse and non-recourse portions. Teck Corporation is unconditionally guaranteeing the (recourse) amount until sales contracts are in place to provide backing. The (interest) spread is essentially the same for these segments but varies according to the finalisation of the project. Prior to completion it is either 3/4% above Canadian or US prime (bank lending) rates or 1 1/4% above Libor.* Post completion, the (interest) margin is: either 1/2% above Canadian or US prime rate or 7/8% above Libor. The project is estimated to be completed December 31,1983. Commitment fee (on the unutilized portion of the total Tranche A commitment):3/8% (p.a.).

Tranche B of C$28m is unconditionally guaranteed by Teck Corporation and carries a spread of 1/2% above the Canadian or US prime or 7/8% above Libor. Commitment fee is 1/8% until drawdown and 3/8% thereafter."

"The borrower is a wholly owned subsidiary of Teck Corporation, a Canadian resources company."

*Libor is the London Interbank Offered Rate, a common benchmark for Eurodollar borrowings.

CASE STUDIES

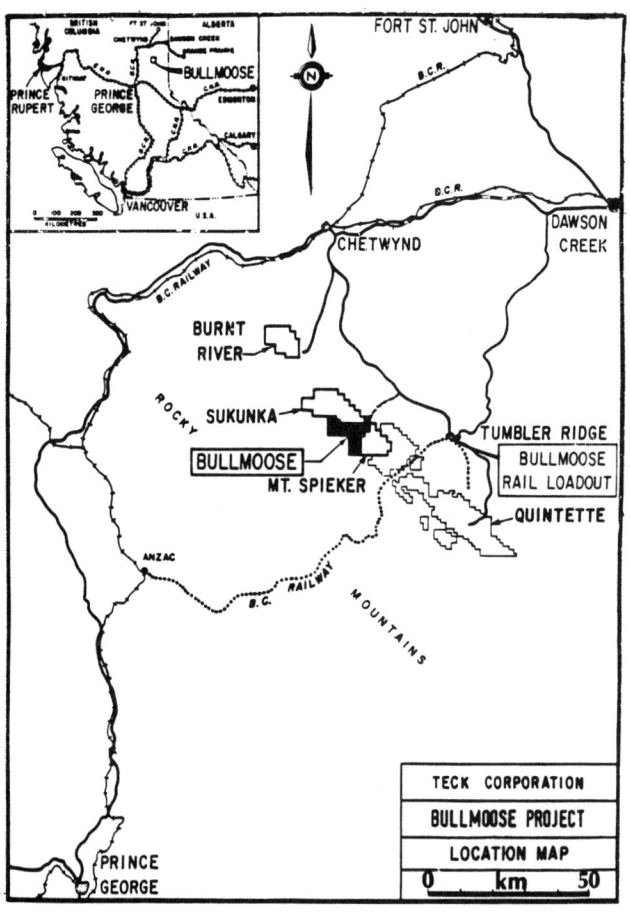

BULLMOOSE
FACTS AND STATISTICS

COAL SEAMS: Five seams totalling 13 metres in thickness - close to surface - gently sloping, relatively undisturbed.

MINE: Open pit with low strip ratio (waste to coal ratio - 4.5 b.c.m. waste per tonne of run-of-mine coal).
- South Fork pit, Year 1 to 16 plus
- West Fork pit, Year 17 to 28 plus

Equipment: 3 shovels, 2 excavators, 3 drills, 20 trucks, 4 loaders, 5 dozers, 2 graders, 5 scrapers.

Production: Waste: 12 million b.c.m./yr (30 million mtpy) Coal: (metallurgical) 2.3 million mtpy, feed to plant from mine.

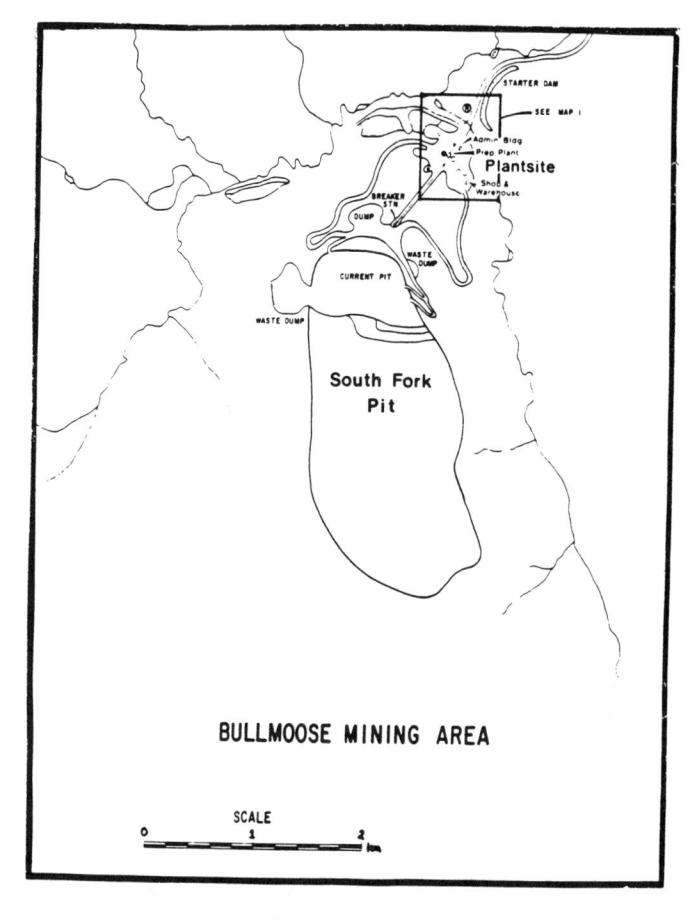

PLANT: Raw Coal Silo - 4,000 tonnes
Preparation Plant - Two, identical parallel circuits for flexibility and easy maintenance with by-pass conveyor. Heavy-media cyclone circuits and water-only cyclone circuits; vibrating sieve bends; dewatering centrifuges; flotation circuit; and disc filters.
Dryer - reduces moisture content of clean coal to less than 8%
Clean Coal Silo - 5,000 tonnes
Highway Truck Haul - 10 vehicles each pulling two trailers carry 40 tonne loads 35 km to rail loadout
Rail Loadout - 2 silos, each 11,000 tonne capacity - 100-car unit trains (9,000 tonnes) can be loaded in under 2 hours.

OPERATOR: Bullmoose Operating Corporation (Wholly owned by Teck)

MANPOWER: 425 employees

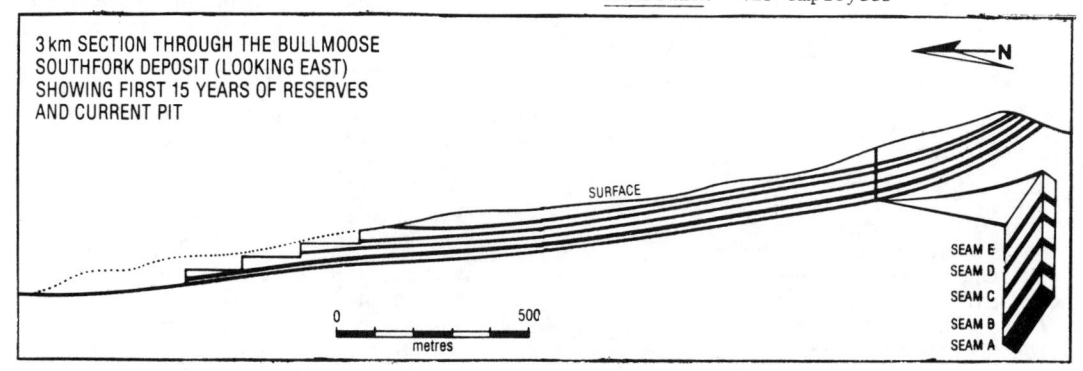

12.5

FINANCING THE OK TEDI MINE--CASE STUDY OF THE PROCESS FROM A GOVERNMENT PERSPECTIVE

Stuart McGill
State Solicitor's Office, Dept. of Justice

Government of Papua New Guinea
Papua, New Guinea

ABSTRACT

This case-study of the financing of the Ok Tedi project illustrates the nature of project financing and outlines the arrangements made in the case of this project; it also demonstrates that the limitation of financial exposure is an important goal for Governments involved in developing the resources sector. The account includes a chronological description of the process by which the sponsors prepared for the financing of the project and the lenders became involved in it. The type of security and loan documentation required for such financing is described, together with a broad analysis of their content and the manner in which the various risks involved in development are apportioned among the sponsors and lenders. The ability to apportion and limit financial liability is an important development for the owners of capital and resources.

INTRODUCTION

This paper provides a chronological analysis of the process by which the Ok Tedi gold-copper project was financed and comments upon the framework and content of the many agreements which link the entire financing deal together. A chronological format has been chosen in order to provide the reader with an overall perspective on the issues that arise and the sequence of events that occur in project financing.

The discovery of the Ok Tedi deposits by the Kennecott Copper Corporation in 1968 led to negotiations between it and the independent State of Papua New Guinea, hereafter called the State. These broke down in 1975. Negotiations then immediately began with an Australian company, Broken Hill Proprietary Ltd. (BHP), leading to the enactment of the Mining (Ok Tedi Agreement) Act 1976, (hereinafter referred to as the Principal Agreement). This agreement was conditional upon BHP forming a consortium to carry out a feasibility study during the following three years. A consortium of BHP, Amoco Minerals Company (a subsidiary of the Standard Oil Company of Indiana), and a West German group of companies was successfully formed and, on the basis of a feasibility study, proposals for development of the Ok Tedi deposits were presented to Government in November 1979. The sequence of events and decisions required by the Principal Agreement are summarized in Table 1.

Table 1

SEQUENCE OF EVENTS REQUIRED BY PRINCIPAL AGREEMENT

Time Period	Process
1976 to 1979	Feasibility study and presentation of proposals by Consortium.
December 1979 to February 1980	Consideration of proposals and their approval by the Government.
October 1980	Submission of financing and marketing strategy.
October 1980 to January 1981	Negotiation and approval of such strategies by Government.
December 1980	State elects to take a 20% equity.
February 1981	Nomination of OTML as project company.
August 1981	Initial grants of land
November 1981 to July 1984	Financing and construction of project

Three months after the submission of the proposals for development (and following detailed and arduous negotiations), the Government issued an Approval of Proposals for the development of Ok Tedi in three stages. The approved production plan and cost of each Stage are set out in Table 2.

Both the State and the consortium began to address the financing issue during the approval of proposals negotiations in February 1980. The consortium retained the services of the Bank of America to advise it on the likely availability of finance and the issues that would impact on it. The bank's work was the basis of the project's financing strategy which was to be submitted to the state for

---

The article is based on a paper presented 2 June 1982 to a seminar arranged by the facility of Law of the University of Papua New Guinea, Port Moresby and is published in Natural Resources Forum, Vol. 7, No. 2, United Nations, New York, 1983.

approval by 30 October 1980. At the same time, the State engaged the financial services of N.M. Rothschild and Sons Ltd. to assist it to evaluate the consortium's proposals for development, with particular emphasis on the project's financing. The sequence of the financing process is summarized in Table 3.

Table 2

PRODUCTION OF PLANT AND PROJECT COST

	Production Plan	Project Cost (US$ Million)
Stage 1 1984-86	15-20 tons gold produced per year. 22,500 tons of gold ore mined per day	702
Stage 2 1986-89	15-20 tons gold produced per year. 12,500 tons of copper and gold ore mined per day	410
Stage 3 1990-on	45,000 tons of copper and gold ore mined per day	139

Table 3

SEQUENCE OF OK TEDI FINANCING PROCESS

Time Period	Process
February 1980	Negotiation and settlement of State's limited loan support obligations for Stage I.
March 1980	Decision by project's Finance Committee to use export credit financing.
March 1980 to October 1980	Negotiation of shareholder agreement among sponsors.
October 1980	Submission of financing strategy by sponsors to State. Submission of Financial brochure to potential lenders.
November 1980	Verbal presentations to export credit agencies and invitations for a lead bank.
January 1981	Negotiation and approval of financing strategy (including limitation of State's loan support obligations for Stages II and III.) Lead bank selected.
February 1981 to March 1981	Sponsors agree on financing terms to be offered to lenders.
May 1981	General financial terms submitted to lenders.
May 1981 to January 1982	Negotiation of security and loan documentation
February 1982	Loans executed and drawdown procedures commence.

DEVELOPMENTS PRIOR TO FINANCING ARRANGEMENTS

Background on the Concept of Project Financing

Before discussing the developments outlined in Table 2, it is useful to define the concept of project financing.

"...the expression... cover(s) those transactions which involve the provision of non-equity funds for the development of mining and energy projects on the basis that, relying on the financier's assessment of the mining, technical, commercial and financial viability of the economic unit which constitutes the project, the financier is willing to look initially to the cash flow and, if necessary, assets of the project for his timely repayment" (Ladbury, 1981a).

Adamson (1979) cities three main reasons for choosing project financing. First, the project's developers desire to match the cash flows required for and generated by the project with loan drawdowns and repayments so that it is 'self-financing'. This allows the sponsors of a large project to protect their own credit standing without being unduly affected by the financial standing of the other sponsors or the project. Secondly, project financing can allow sponsors to achieve off-balance-sheet financing, that is, it does not show as a direct liability on the balance sheet. This protects corporate gearing and other commercial ratings. Thirdly, project financing allows sponsors to transfer some of the project risks, especially the political risk, onto the lenders.

With respect to risk exposure, it is worth noting that in a study of 29 project financings conducted by Chemical Bank, 82% were found to have experienced problems either during construction or operation. For instance, 71% had cost overruns, 59% experienced completion delays, and 35% had lower than anticipated cash flows (Castle, 1975, p. 16). It follows that all parties in a project financing are concerned about limiting the risks to which they are exposed.

When assessing risks, bankers usually approach the problem by identifying a number of specific risks which they then classify as either acceptable, (in their terms 'bankable'), or those which they require the project or project sponsors to assume. Such risks include: completion risk, political risk, reserve risk, production risk, and market risk. It is usual for the project sponsors to accept completion risk and for the bankers to accept at least some political risk, while the acceptance of the others varies according to the type of project.

Before proceeding, there is one more term with which the reader should be familiar. In any discussion of project financing, the terms 'non-recourse' and limited-recource' will be encountered. These refer to the extent to which the lenders can have recourse to the sponsors if the project meets financial difficulties. Although a sponsor's objective in project financing is to achieve non-recourse financing, such financing is rare as the project must be extremely attractive in all respects to achieve this. The typical case of project financing is that of a limited-recourse financing where the sponsors undertake to the lenders that they will complete the construction of the project and achieve a certain rate of production. At that point the completion guarantee usually terminates and the lender must rely on the project itself to satisfy the loan repayments. This type of financing should also be distinguished from what might be called 'full-recourse' financing, where the lenders can have resort to the sponsors during the entire life of the financing. This situation might be expected where the financing is on concessional terms such as at a less-than-market interest rate.

First Financing Issues

It was during the negotiations leading to the approval of proposals that the financing of the project became an issue between the State and the consortium. The first question was whether the consortim's proposals made optimal use of the gold ore. Behre Dolbear, a mining engineering firm appointed as the State's technical consultants, advised that the gold ore cut-off grade should be lowered to enable the processing of additional gold ore.

The issue made the State aware of the dual impact of the financing process on its bargaining ability. On the one hand, the State found that one of the formidable arguments against changing the proposals for development was the threat of losing the engineering endorsement of the construction firm Bechtel which had prepared the proposals for the consortium. The financing of the project could not proceed without their endorsement that the project was technically feasible. The delays would be costly and would defer revenue flows to the State if the State was to reject the proposals until the additional work was done. The issue was resolved by the amendment of the principal agreement to require the consortium to carry out a work program to assess the feasibility of the optimisation that the State required (Mining Agreement, 1980).

Another aspect of the financing process which came to the State's attention had a positive effect on its bargaining position. As much as the project needed an engineering endorsement for the purposes of financing, the consortium needed the State's approval for all aspects of the project's development before any financiers would lend. The strength of this potential sanction became fully apparent as the project slowly assembled all its various approvals from the government for submission to and approval by the financiers prior to the first drawdown of loans.

The second question was the issue of the extent of loan support or financial guarantees to be provided by the State to the project. On the assumption that the State would elect to exercise the option, contained in the principal agreement (mining Act 1976, clause 11), to take a 20% share in the project company, the State would be required to provide not only its share of equity capital but also its share of loan guarantees or such other loan support as the financing package would require. Following a study by Rothschilds of the debt-servicing capacity of Papua New Guinea and the effect of accepting loan support contingent liabilities, the State decided that it should attempt to avoid accepting such obligations. Its strategy was to bargain for such an exemption in return for other concessions which the consortium wanted. It should be noted that this financial risk issue is often overlooked when the ideological attractions of equity participation are being considered.

The deal that was struck with the consortium had three parts (Mining Agreement, 1980 Note 9, Clause 5). First, on the basis that the State was providing a loan to the project of up to A$50 million for the construction of an access road from the Fly River to the project, it was agreed that the State would not be required to provide any loan support for the first US$200 million (escalated to US$280 million) of borrowings by the project. Secondly, borrowings above that amount up to the level of initially committed finance (anticipated to be US$650 million) was to be provided proportionately by all shareholders, with a maximum level of State participation set at US$42 million (to be escalated). Thirdly, loan support for financial overruns above the initially committed finance was, unless the State elected to participate, to be provided by the corporate shareholders with the project company paying them an annual 5% guarantee fee. Such arrangements allowed the State to limit its risk exposure and those became the starting point for later discussions of loan security both among the Project's shareholders and with the lenders.

Developments in 1980

During the 12-month period prior to the nomination of Ok Tedi Mining Limited (OTML) as the project company in February 1981, the project's financing was the responsibility of a finance committee composed of representatives of the consortium and the State. After an initial survey the decision of the finance committee was to proceed with a plan to utilize export credit financing to the maximum extent.

As export credits are available at less-than-market interest rates, they represent an attractive form of financing providing that the goods which best suit the project are available in the countries offering such financing. However, it should be noted that more competitive prices in countries where export credits are not available can often offset the initial interest rate advantage.

Although in the Ok Tedi case competitive bidding has led to less use of export credit finance than was originally anticipated, the Finance Committee in 1980 had the entire procurement of goods for the project categorized into those countries from which they might be obtained. This work provided the basis for approaches to the export credit institutions of Australia, Austria, Canada, Japan, United Kingdom, France and the USA.

At the same time that the Finance Committee was meeting, a group of sponsor's legal and financial personnel were discussing the structure of a shareholder's agreement. Although such an agreement is quite normal, these negotiations assumed increasing importance due to the financial obligations involved and the complex nature of security undertakings. It was at this stage that the German group, being the smallest member of the group of sponsors, raised the issue of financial exposure.

The private sponsors reached an agreement that above a certain level of financial overruns for Stage I, a sponsor could elect not to proceed and to be correspondingly diluted. Other provisions of the shareholders agreement dealt with the creation of the project company, the obligation to provide finance in accordance with the financing plan, the guarantees by the corporate sponsor's parent companies, the occasions when shareholder approval of changes to the project was required, and the manner in which financial defaults by sponsors would be handled.

In the latter months of 1980, the formal submission of the consortium's financing and marketing strategies set the scene for the next negotiation between the State and the consortium. The marketing strategy outlined the broad marketing prospects for the project, noting that it was not necessary to enter into long-term contracts for the sale of gold bullion produced during Stage I and that it was too early to enter into sales contracts for the copper to be produced from Stages II and III. The development of the project by stages varied with the usual requirement that sales contracts be concluded prior to the arrangement of financing in order to provide security to the lenders. The only agreement which existed in relation to the marketing was a provision requested by the German group in the original consortium agreement whereby they were given an option (subject to normal transfer pricing safeguards) to purchase up to 50% of all products produced by the mine.

Initial objections by the State to the proposed marketing strategies led to the formulation of a more precise financing strategy and the resolution of several ancillary issues. The real gain for the State, however, was that its financial obligations were limited not only for Stage I but for Stages II and III as well. Accordingly, it was agreed to limit the State's liability to contribute equity with the provision that the state's share would be diluted without penalty (Ok Tedi Second Supplemental Agreement, Clause J). In addition, the Stage II loan support obligation of the State was limited, if it so elected, to the then estimated support requirement, with the corporate sponsors being required to contribute any overrun support in return for a 2.5% guarantee fee paid by Ok Tedi Mining Ltd. The outcome of these negotiations meant that the State's total financial obligations for financing the project were limited to those set out in Table 4.

Table 4

Financial Obligations of State
&
Private Sponsors*

Project Cost**	Support for Loan Finance		Equity	
	Private Sponsors	State	Private Sponsors	State
Stage I				
up to 400	280	–	96	24(a)
400 to 700	168	42	72	18
700 to 928	160	–	58/65	3/10(b)

over 928 — State has option to participate or not in loan support; if it does not participate, 5% P.A. guarantee fee is paid by OTML to the Private Sponsor guarantors.

up to $493/528(c)	296	74 (d)	97/132	25/32(e)

Stage II
over this figure — State has option to participate or not in loan support; if it does not participate 2½% P.A. guarantee fee is paid by OMTL to the Private Sponsor guarantors

Stage III — Private Sponsors and State contribute in proportion to shareholding.

* Reflecting State's limited obligations
** assuming 70:30 debt:equity

Source: Report to the State by N.M. Rothschild and Sons, Ltd.

a Some $19 million of this amount will be by way of equity credits.

b The State can decline to contribute equity above $45 million but is diluted accordingly.

c Figure is $493 million on basis of 75:25 debt:equity ratio allowed by the second supplemental agreement or $528 million if lenders insist on 70:30.

d This assumes that State has met full equity calls in Stage I and has not been diluted; if it has been diluted these figures will reflect the new shareholding percentages.

e State can decline to contribute equity above $26 million (either in form of new funds or capitalisation of retained earnings), but is diluted accordingly.

THE SECURITY DOCUMENTS

Delineation of the Lenders

The year 1981 represented the start of a new era in Ok Tedi financing in that the major negotiations between all the sponsors were completed. The focus of negotiations was now between the sponsors and their new project company OTML on the one hand and the various potential lenders to the project's financial brochure. This document consisted of three volumes. The first provided a financial summary of the project and included two computer financial models of the project under differing assumptions. The second consisted of copies of project documents such as the principal agreement, together with annual reports of and statistics on the corporate sponsors. And the third outlined the project's benefits to and impact on the Papua New Guinean economy.

Owing to the Finance Committee's decision to utilize export credits, contacts were made with both lenders that were in the export-credit group and with lenders that were in the commercial banking group. This approach proved extremely useful at a later stage of the negotiations, when the Export Finance and Insurance Corporation of Australia, being the largest export credit lender, co-ordinated the negotiation with all export-credit agencies. The main commercial lending contact was initiated when bids were invited from several large commercial banks for a so-called 'lead bank'.

This bank would be responsible for forming and coordinating a syndicate of approximately 15 banks which would each lend an amount to the project under the umbrella of one syndicate loan agreement between the project company and the lead bank. The Australian Citibank subsidiary, Citicorp, became the lead bank.

The first discussions with the export-credit lenders occurred when a group of sponsor representatives made a tour of such institutions in order to obtain an outline of the terms such lenders would offer (including the security they would require). From these discussions it became apparent to the sponsors that each of the major export credit agencies were in contact with each other upon the terms that they were offering and thus there was little opportunity for the project to bargain between the lenders.

It was about this time that the possibility of a third type of major lender to the project began to be discussed by the sponsors. The German sponsors proposed that a large borrowing should be made from the Kreditanstalt fur Wiederaufbau (KfW). The German sponsors had come to an arrangement with the German government whereby if the German sponsors provided loan support for the KfW loan, the German government, in order to assist its own corporations, in securing long-term supplies of raw materials, would provide a proportion of the loan support to the KfW loan in place of the German sponsors, on payment of a guarantee fee by such sponsors.

From the project's point of view the potential disadvantage of such a loan was that the project might be required to accept onerous terms and conditions which did not compete with those obtainable elsewhere. In addition, the other shareholders were concerned that the KfW arrangement would split the responsibility of the shareholders for loan support in a potentially uneven manner. Nevertheless, the American and Australian sponsors were prepared to accept the arrangement provided that the terms were acceptable, while the State actually saw an advantage for itself in the deal. It was aware that the Second Lome Convention of 1979 allowed the development bank of the European Economic Community, called the European Investment Bank (EIB), to make low-interest risk capital loans for the purpose of equity investment by certain developing countries in the Asia, Pacific and Caribbean regions. In the current circumstances, if the EIB was to make such a loan to the State then the EIB must also be able to make a substantial commercial loan to the project. Consequently, it was arranged that the EIB would be a part of the KfW loan syndicate, that the KfW would in fact lend to the project and that the EIB would make a concessional loan to the State. After taking into account the equity credits the State obtained under the principal agreement in recognition of bonds issued to the Kennecott Copper Corporation for exploration expenditures before Kennecott withdrew from the project, obtaining the EIB loan meant that the State had to contribute very little cash to OTML to purchase its equity share in Stage I of the project.

The Security Package Takes Shape

Deed of charge. During the first half of 1981, with the various proposed loans beginning to take shape as letters of commitment to lend were exchanged between OTML and the lenders, work also began to devise the form and content of the loan documentation, especially that of the loan security package. OTML engaged both an Australian and an American legal firm to advise it and the sponsors nominated some of their representatives from the Finance Committee to form a negotiating team. The team's first task was to circulate to lenders a paper entitled "General Financing Terms" which included the passages:

"...The borrower and its shareholders believe strongly that the financings must be done on a coordinated and consistent basis. The overall structure of the entire financing must permit both the various loans for Stage I as well as the financing required for Stages II and III. To the greatest extent possible terms of each loan agreement covering common issues must be identical so that the borrower is not subject to inconsistent or overlapping requirements....Also because of the need to achieve consistency, the borrower has assumed responsibility for drafting the Deed of Charge and shareholder undertakings referred to herein, realizing of course that each lender must be satisfied with the final document before proceeding to make its loan".

The first part of the loan security was to be a General Deed of Charge which would create a fixed and floating charge over the great majority of the property and other assets of OTML. A floating charge may be defined in these terms:

"A floating charge is shifting in its nature and is not initially attracted to any specific property. It hovers over and floats with the category of property which is intended to affect until some event occurs or some act is done which causes it to settle and fasten on specific property within reach..." (Ford, 1974).

The Deed of Charge was to be executed between OTML and a representative of the lenders. It was always assumed that the representative would be the lead bank for the main commercial syndicate. Hence the Australian Citibank subsidiary, Citi-corp, became the representative and the other lenders became parties to the general deed of charge by executing supplemental deeds. Needless to say, the general deed of charge is a complex document. The object of the deed is, in the event of OTML's default, to put all its assets into a common pool which is available for the equal and proportionate benefit of all lenders. In addition, there are provisions to ensure that a minority of lenders cannot unilaterally declare a default and 'collapse the house of cards', without good reason.

It should also be noted that the general deed of charge obliges OTML to enter into a proceeds agreement with Citicorp. Under this agreement, all of OTML's future sales proceeds are assigned, and the proceeds account there specified is established as a trust account as collateral security for the amounts secured by the deed of charge.

Completion and cash deficiency agreements. The second type of loan security document is called the Completion and Cash Deficiency Agreement. As its name indicates, such an agreement consists of two basic obligations, that of completion guarantee on the one hand and that of an arrangement to meet OTML cash deficiencies on the other.

The general requirements of completion guarantees have been described in these terms:

"In broad terms a completion guarantee is an undertaking by the project sponsor that the project will be completed and by a specified time..."

"....It is normal in the case of a completion guarantee to arrange for the appointment of some independent expert ....who can be called upon to adjudicate as to when the risk will pass from sponsor to the lender"

"There are four main aspects which must be considered for completion; firstly, that the anticipated maximum production levels can be met over a given period; secondly, that the project is capable of achieving a rather lower but constant rate of output over a considerably longer period so as to ensure the reliability of the project; thirdly, that the product is of the required quality; and finally, that the production expenses involved are not so high as to prevent the proper servicing of the project" (Davies, 1979).

The Ok Tedi completion test proceeds broadly along these lines. It is provided that completion will be achieved when four conditions are satisfied. First, the Project must satisfy a performance test by achieving a set rate of gold production from the Stage I plan in accordance with the time schedule in the Approved Proposals (subject to a reasonable contingency factor for delays). There must at that time then be representations by the sponsors as to their own financial health, together with a statement by OTML that is is not in default of any of its obligations. Lastly, OTML is required to satisfy a debt coverage test for each year the loans are outstanding. In broad terms such a coverage is computed by dividing cash flow in each year by the amount of required debt service (i.e. principal and interest) in that year. If there is any dispute as to whether the performance test or the debt coverage test have been satisfied, then these matters may be submitted to arbitration. The acceptance of such a test also implicitly means that the sponsors accept the vast majority of the reserve and marketing risks of the project.

The cash deficiency agreement was drawn up by the project's American lawyers, at the instigation of Standard Oil for whom it had certain advantages within the U.S.A. The agreement became necessary because neither the export-credit agencies nor the commercial banks were prepared to accept non-recourse financing. Thereafter, the issue centered around the extent to which each lender would have its recourse to the sponsors limited. The initial obligation in the agreement is that the sponsors will support OTML if there is a cash deficiency in the process of achieving completion. The underlying arrangement, however, is that if one of the sponsors or OTML meets financial difficulties, the sponsors each elect to proceed according to one of three options. They may, to the extent of their proportionate share of the total loan support obligations, either contribute in cash, or provide a guarantee, or substitute their own promissory notes.

Given the unconditional nature of these obligations, the critical question is the duration of each deficiency obligation. The answer, of course, determines the extent to which the recourse of the lenders is limited. It is here that there is a crucial distinction between the export-credit agency loans and the commercial bank loans. Due to the lower than commercial interest rates offered by the export-credit agencies, they were able to maintain that, subject to the acceptance of political risk in some cases and notwithstanding that completion had been achieved, the cash deficiency obligation remained in force for the entire life of the loan.

The commercial lenders, however, accepted that once completion had been achieved, provided that there was a commitment to Stage II of the project which assured them of ongoing revenue flows, the cash deficiency obligation dropped away and the lenders had no further recourse to the sponsors. The project was, however, required to pay a slightly higher interest rate after this point.

Due to this distinction between the commercial and export-credit loans, and because loan support obligations among the sponsors had already been partly divided up as a consequence of the German sponsor's Kfw arrangements, it was decided that only the two larger sponsors, namely BHP and Amoco, would be parties to the cash deficiency agreement with the commercial banks.

A final measure of protection is afforded to they lenders by the completion and cash deficiency guarantees in that these documents contractually subordinate the project's debts to the sponsors to the repayment of all monies due to the lenders. They further require that the sponsors not prove in any liquidation of the project company until the lenders, (being senior debtors), are repaid in full (ladbury, 1981, p. 157).

Political risk agreement. A central concern of the sponsors throughout the evolution of the security package was the extent to which they could avoid accepting the risk that political factors might interfere in the construction or preparation of the project. This risk was perceived by the private sponsors to be of particular importance due to the location of the mine only 18 km from the border between Indonesia and Papua New Guinea and due to the fact that the copper ore concentrates were to be transported along the Fly River, which itself marks part of the international boundary.

The concern of the private sponsors surfaced during their negotiations with the State over the financing strategy. Their initial suggestion was that the State should bear the project's political risk. This suggestion was no doubt based on an offer made by the State during the approval of proposals negotiations, that it would assume some contractual responsibility for risks which were under its control such as expropriation and currency convertibility. However, the consortium at that time did not accept the State's offer and the State was now not inclined to re-open the issue.

Hence the State refused to accept political risk although it did offer to provide the lenders with a so-called letter of awareness, in which it would simply assure the lenders that it was aware of the lenders' reliance on the undertakings of the State in the Principal Agreement (Castle, 1975, p22). This letter of awareness afforded some comfort in respect of political risks such as expropriation and foreign exchange availability, although it could not help with other political risks such as war and inability to transport goods on the Fly River due to disagreements with Indonesia.

The importance of the State refusing to accept overall political risks was borne out a few months later when a potential lender to the State who was under the false impression that the State had accepted overall political risk, refused (until its impression was corrected) to lend to the State. The Lender maintained that if the State had accepted such a liability, then its overall credit worthiness was severely affected.

In the end, even the offer of a letter of awareness was not in fact utilized and the discussions shifted to the extent to which the sponsors could transfer the political risk not onto the State but onto the lenders. It is worth pausing here for a moment to reflect on why in many cases lenders perceive that they are able to accept some or all of the political risk.

"Experience enables the banks to accept all of these political risks more readily than can a sponsor. Frequently, the banks have much more experience of doing business in the country and often the local partner has a relationship with international banks which has made it entirely credit worthy in the eyes of the banking community. In these circumstances, the banks are willing to accept risks because they perceive them as being bankable. This is one situation in which the commercial sponsor company is less well placed than a bank to analyse a project related risk.

Almost as important as experience is the perception of lender control. The assumption is that a host country will be less ready to expropriate foreign company assets or put up taxes to a penal level if the loss will be borne by the international banking community, not a foreign commercial company. The syndication of such financings tends to be done so that American, Canadian, British, German and French banks are included to increase the potential odium for the host country...

In concluding my remarks on political risks, I would point out that acceptance of these risks had been to the advantage of the banks as well as to the credit exposure limits. Banks would rather allocate the available credit to project loans offering an identifiable source of repayment, and with the involvement of a professional operator, than to balance of payments loans which, realistically, can only be repaid by refinancing." (Castle, G.R, 1975, p.4)

The sponsors' first attempt was to have the lenders bear the entire burden of a very widely defined political risk. It was partly over this issue that discussions broke down with the Export-Import Bank of the United States which indicated that it was not prepared to bear any political risk at all.

After further haggling with the lenders, political risk provisions were incorporated into the Completion and Cash Deficiency Agreements. Three types of political events were defined:

1. The inability to transport goods along the Fly River due to disagreements between the State and Indonesia.

2. The act of war, insurrection or similar disturbances.

3. The action of the State such as expropriation and refusal of currency convertibility.

The effect of a political event, provided it occurred after completion, was to either immediately terminate the loan support obligations if there was not potential for re-instating the project, or in any other case to suspend such obligations for up to two years, after which the support obligations were terminated.

Naturally, a sponsor could not benefit from the occurrence of a political event if such a sponsor had contributed to its occurrence by instigation, provocation, inefficiency, or failure to take reasonable remedial action. Because the State could not enter into an agreement where it benefited from its own acts such as expropriation and refusal to convert currency, and because it did not wish to compromise itself by being party to an agreement which contemplated such actions, the State decided to enter into its own political risk agreement with the lenders. This agreement, between the State and the export-credit agencies, only deals with transportation problems and war-like actions. The other sponsors also entered into political risk agreements with the export credit agencies and these contained references to all three political events. Meanwhile, Amoco and BHP, being the sponsors who provided the entire loan support to the commercial lenders, also had similar provisions in the Completion and Cash Deficiency Agreement between themselves and the commercial lenders.

In 1976, due to the importance placed by the sponsors upon friendly relations between PNG and Indonesia, BHP obtained an undertaking in the principal agreement that the State would use its best endeavors to negotiate a border agreement with Indonesia which recognized the project's right "to use...Fly River...without undue interference" (Mining Act, 1976, clause 45). Such an agreement was concluded in December 1979, containing general provisions on navigation and navigational aids, development of natural resources and protection of the environment. However, during the approval of proposals negotiations, the Consortium requested the State to make additional "arrangements" with Indonesia, containing more detailed provisions on the use of the river. It should be pointed out that the lenders also considered the successful conclusion of such border arrangements to be important aspect of their loan security.

Fortunately, after one round of discussions, such arrangements were concluded in November 1981, thus allowing the financing to be completed in early 1982. There arrangements, like all international documents, are couched in broad terms by the standards with which commercial lawyers are familiar.

The border arrangements confirm and regulate the right of free navigation of the Fly River without the normal incidents attendant upon entering another's territory, as well as allowing mooring and dredging, salvage operations, placement and servicing of navigational aids and emergency landing rights for aircraft. In return, the State accepted the responsibility for compensating for pollution by the project. The Principal Agreement was subsequently amended by a third supplemental agreement which required the project to reimburse the State for such compensation.

40.4 Agreement. The State had the responsibility for negotiating one other security related agreement. Due to the original termination provisions in the Principal Agreement, which provided that if the Project fell into default of that Agreement, it had 30 days to remedy the breach. If it was unable to do so, or at least was not taking reasonable steps to do whatever could be done by way of remedial action, then the Agreement could be terminated by the State with the entire Project then becoming the property of the State. During the Approval of Proposals negotiations, the Consortium maintained, on Bank of America's advice, that although for financing purposes the Principal Agreement permitted the assignment of the Project's assets to lenders by way of a deed of charge, such a security was worthless if the State could terminate all the Project's mining and other titles at 30 days notice. The State would therefore acquire all the Project's assets freed and discharged, by operation either of the common law or the terms of the original Agreement, from all The rights of the lenders.

To overcome this impediment to financing the Project, the State agreed in the First Supplemental Agreement, to give lenders the right to make their own attempt to restore the Project's cash flow if a default occurred. It was provided that lenders were to be kept informed of any potential defaults, and that if they then appointed a receiver and manager, the State could not terminate the agreement for a further period of 60 days. The fact that the lenders should only be concerned with restoration of cash flow, rather than seeking to recover their losses by realizing on the Project's (probably with less) assets, was implicitly recognized by the maintenance of the State's right to all Project assets at the expiration of the extended period.

In addition, sub-clause 40.4 was inserted into the Principal Agreement by the First Supplemental Agreement. It provided that the lenders and the State could directly make a separate agreement to further secure the rights of the lenders. The State foresaw that the lenders may wish to have a longer period to attempt to restore cash flow, (or sell the Project to some one who could) and it was envisaged that the 40.4 Agreement would provide for it. (It should be noted that in the State's now Standard Mining Development Agreement, the extension of time is incorporated directly into such agreements and thus clause 40.4 agreements will not be necessary in future). Thus, during the final stages of the ok Tedi financing negotiations, the lenders approached the State seeking such an extension of time, and after considerable debate, a period of two years with a possible extension of a further 12 months was permitted.

Export-Credit Support Agreement. With the security documentation virtually agreed between the sponsors and the lenders, the old issue of the apportionment of the loan support obligations among the sponsors arose once again. This led to the negotiation of an agreement known as the Export-Credit Support Agreement. Its object was to apportion among the lenders the extent of loan support each was required to offer under the Completion and Cash Deficiency Agreement with the export-credit agencies.

Such an agreement was necessary due to the manner in which the sponsors had agreed to offer loan support, and to the necessity for both sponsors and lenders to be aware at all times of the exact extent of their loan support obligations. This knowledge was especially important if at any time the sponsors had to elect to follow one of the three options specified in the Completion and Cash Deficiency Agreement. Unfortunately, calculating each sponsor's level of loan support was not simply a matter of computing that proportion of total debt outstanding which corresponded with each sponsor's shareholding. The first problem was that the State's support obligation was nil for the first US$280 million, and then was limited to US$42 million until a level of debt of US$650 million was reached. The second problem was the differing nature of the loans. For instance, the German shareholders were obligated to support the Kfw loan, which meant, as it had a longer term than the other loans, that at some time in the future they might be left bearing a disproportionate share of the loan support.

The solution devised was contained in four formulae incorporated in the Export Credit Support Agreement which calculated each sponsor's level of support at any one time. The results for various levels of debt are set out in the Table 5. The interrelationship of the various parties and the documentation/supports is given in the Appendix.

THE LOANS

Project Loans

Although the drafting of all security documents was done by the project's negotiating team, the actual loan agreements were, in conformity with general practice, drafted by the respective lending institutions. Much of the content of such documents was not, of course, of direct relevance to the State as a sovereign entity, and therefore it was prepared to conserve its relatively scarce resources and leave OTML and the private sponsors with the task of attempting to obtain the best deal for the project that they could negotiate with the lenders. Nevertheless, there were some matters, such as the extent to which the loan documents impinged on the project's ability to optimize production from the ore-body, which were of particular concern to the State.

The terms and conditions of project financing documentation are the most extensive and detailed of any corporate loan document. Such documentation is necessary because, with the lenders relying on cash flows from the project for repayment of the loans, they have an interest in ensuring that all information on a project is correct, that all technical proposals are feasible, that all financial estimates are realistic, that all titles and approvals are obtained, and that all other matters necessary for the success of the project are in hand.

Table 6 shows the final breakdown of the project's borrowing as it emerged in the latter part of 1981. With the total cost of Stage I of US$855 million and a ratio of project debt to sponsor's equity of 70:30, the amount of equity required of the sponsors was US$256 million. The remaining US$599 million was to be raised from the three banks and four export credit agencies shown in Table 6.

Table 5

DIVISION OF DEBT SUPPORT AMONG SPONSORS
(US$ million)

OTML debt - Stage I			
Export loans	$100	$300	$250
Commercial loans	100	200	100
KfW	50	100	50
Total	$250	$600	$400
Export debt support			
State	0%	14%	10%
Germans	0	13	11
Amoco and BHP	100	73	79

Table 6

LOANS TO THE PROJECT
(US$ Millions)

Citibank N.A. bank consortium	$150
Overseas Private Investment Corp. (OPIC)	50
Kreditanstalt fur Wiederaufbau (KfW)	100
Exports Finance and Investment Corp. of Australia (EFIC)	250
Exports Credits Guarantee Dept. of the United Kingdom (ECGD)	100
Export Development Corporation of Canada (EDC)	88
OKB of Austria (OKB)	50
Total of Credit Facilities	$ $788

The lenders have made available to the project potential borrowings amounting to US$788 million. This includes a commitment for US$189 million which may not be required. This leeway was necessary to allow for the uncertainty arising from the use of export-credit financing. Because it is not possible to know exactly what goods will be procured from which countries until the project is actually constructed, the additional commitments are necessary to allow for purchasing flexibility. If all export credits can be utilized, then less commercial finance will be required, and vice versa.

Loan Covenants: With reference to the contents of the loan documents, this paper is of course restricted by requirements of length and confidentiality. The following list is an extract of some of those covenants, which in due course, after the "wordsmiths" had hammered them into shape, appeared in the loan agreements:

"1. The Borrower will use its best effort to comply in all materials respects with the Principal Agreement.

2. The Borrower will use its best efforts to obtain and maintain all governmental approvals as described in (the) presentation...

3. The Borrower will not permit any Project assets to become subject to any liens other than the Deed of Charge, except as permitted by the Deed of Charge.

4. The Borrower will maintain its corporate existence and will not merge with or dispose of all or substantially all of its assets to any person, except for merges with or dispositions to any person who unconditionally assumes all of the Borrower's obligations under the Loan Agreement and the Deed of Charge and immediately after such merger or disposition there is no Event of Default under the Loan Agreement.

5. The Borrower will provide the Lender (i) within 120 days of the end of each of the Borrower's fiscal years, with audited annual financial statements, (ii) subject to appropriate confidentiality requirements (A) within 60 days of the end of each semi-annual period with unaudited semi-annual financial statements and a report on the progress of construction of the Project and (B) other information reasonably requested by the Lender and (iii) with an opportunity to visit the Project site at reasonable time.

6. The Borrower will pay all its taxes except when it is contesting a tax in good faith or where the failure to pay would not, in the good faith opinion of the Borrower, have a material adverse affect on its ability to perform the Loan Agreement.

7. The Borrower will maintain insurance on its assets against such risks and in such amounts as are, in the good faith opinion of the Borrower, customary for businesses of like nature. The policies will form part of the property secured by the deed of charge but so long as no event of default exists, proceeds under US$15 million will be payable directly to the Borrower and in any event and regardless of amount all proceeds will be applied to the extent necessary to the restoration of the Project.

8. The Borrower will maintain a debt-to-equity ratio of 70:30 (or if justified by historical and projected cash flow, 75:25) after completion of Stage I. Compliance will be tested only at semi-annual balance sheet dates...

9. The Borrower will not pay any dividends to any Share holder prior to the completion of Stage I or September 30, 1986, whichever is earlier."

The importance lenders attach to completion will again be noticed, for instance in paragraphs 8 and 9, where dividend restriction and a higher debt: equity ratio is imposed before completion in order to attempt to ensure additional financial stability during the initial stage of the Project. Other provisions peculiar to the Project also have this affect, sometimes even after completion. For example, the lenders required that OTML not only effectively be a single-purpose company, but also that it only have two types of subsidiaries - restricted and unrestricted. Restricted subsidiaries could only invest in project related activities while OTML's total investment in both types of subsidiaries could not at any time exceed 5% of the total assets of the Project.

Further covenants then sought to ensure that to the extent that OTML may be a financial success, the lenders should also benefit immediately without awaiting their normal repayment. For instance, OTML was required to commence a degree of early repayment of some of the loans if in any year it paid Additional Profits Tax to the State.

<u>Loan Stages</u>: The manner in which the development of the project is staged led to particular problems in financing it. Although the overall benefit of the staging mechanism is to allow an early cash-flow and maximum use of internally generated funds for its financing, there is also the problem that lenders, for several reasons, will be reluctant to relinquish sponsor loan support until most of the project is built and completed. At one level, these concerns relate to the fact that repayment only comes from cash flow. For instance, Citibank required that unless Stage II had been completed by a particular date, or at least been "committed to", then its 12 year repayment terms would be reduced to 5 years. It considered that in such circumstances, it must obtain maximum access to Stage I cash-flow, as Stage II cash-flow might not ever materialize. KfW also had similar requirements, although its main concern was to ensure that Stage II copper production occurred in order to obtain raw materials for West Germany. If it did not, then it was not interested in financing the project and wanted its money back.

This eagerness of some lenders to ensure that Stage II was completed led to a particular problem for State. This arose because the optimization of gold ore production which the State had required in earlier negotiations meant that Stage I may be extended from 1984-86 to 1987 or later. It also could entail alterations to the production plan for Stage II. Thus, the State had to ensure not only that there was not early cut-off date for Stage I but also that changes to the approved proposals for mine development were not prohibited.

Another issue of ongoing concern for the State is the lender's request for guarantees of foreign currency convertibility. This was resolved in the

Ok Tedi case by relying on what is effectively a foreign currency availability undertaking in Clause 26 of the Principal Agreement. However, the issue will not be easily resolved in the future, as the State's current policy, based on the strength of its stable financial and monetary history, is to offer only a guarantee of foreign currency sufficient to meet current payments, together with a non-discrimination undertaking.

Infrastructure Loan From The State

The State was requested by the consortium to provide considerable infrastructure, notwithstanding that the principal agreement provided that such expenditure was the responsibility of the project. The access road from the Fly River was the only item which the state did agree to fund, on the basis that the project repaid the loan at a commercial interest rate and that the loan was secured by the deed of charge. However, some concession was made as to when repayment could commence. The terms are set out in a schedule to the approved proposals and were subsequently embodied in an Infrastructure Agreement. The main benefit perceived by the State of such an arrangement was to allow construction of the road to commence earlier than planned so as to enhance communication with the people of the extremely under-developed area where the project is located. As mentioned earlier, it was also the reason advanced by the State for being exempted from its share of the first US$280 million worth of loan support for the project.

Under the infrastructure agreement, the State's obligation is to provide up to US$72 million (as escalated) for the purpose of constructing the road. In order to raise such monies, the State first approached concessionary finance agencies and was able to obtain one such loan for KfW. It then approached export-credit agencies, on the basis that about A$12 million worth of imported goods were required for the road. EFIC agreed to lend such an amount to the State.

CONCLUSION

The basic objective of project finance is to limit one's financial exposure. This leads to extremely complex security and loan documentation in respect of which the borrowers, particularly governments, should have access to the best expertise available.

In the case of the Ok Tedi project, the state's financial liabilities for equity contribution, loan support and infrastructure funds have been limited and obtained on the best available terms. In the case of equity, the concessional loan form the EIB, together with the issuing of bonds to the Kennecott Copper Corporation, largely meet this obligation. With respect to loan support, the state has a maximum loan support obligation for stage I of US$42 million, thereby avoiding the usual situation where it would be necessary to support another US$80 million in addition to overruns. Infrastructure funding is also limited, making maximum use of low-interest funds and ensuring a commercial return to the state. Such arrangements for the financing of a major project are as important to a country as the actual returns it obtains for a project.

Countries should resist the suggestion that they bear any risks associated with a project, particularly political risk. The Ok Tedi case demonstrates that the corporate sponsors of and the lenders to a major project are prepared to accommodate such limitations on the financial liability of a participating country. At the same time, this project is a good example of how sponsors and lenders apportion the risks of project financing among themselves, with the sponsors in this case realistically accommodating the requirements of the State. The sponsor also obtained a maximum of low-interest finance, retained maximum flexibility for itself and the project company while transferring the political risk onto the lenders.

Finally, project financing is believed to be viewed a logical innovation necessary to cope with the deployment of finance as more and more capital accumulates and concentrates. But it is also important for the holders of capital as well as the countries desiring resource development to introduce further refinements into such risk-sharing arrangements.

REFERENCES

Adamson, J.A. 1979a Proceedings of Energy Law Seminar. Intl. Bar Assoc (Committee on Energy and Natural Resources). Topic N, p. N1.3.
Castle G.R. 1975 Project financing guidelines for the commercial banker. Commercial Bank Lend. April, (note5) p. 16 and his comment on such a letter on p. 22.
Davies, R.G. 1979. Loans-A review of legal considerations from the standpoint of both lender and borrower. Proc. Energy Law Seminar, IBA (Committee on energy and Natural Resources), N 3.1-N 3.7.
Fewster, G.H. 1977 Financing mineral projects Aust. Min. Petrol. L.J. 1, 147-153.
Ford, H.A. J. 1974. Principles of Company Law. Butterworth, Austrailia, (2nd edn.) para 1211.
Jackson. R. 1982 Ok Tedi-Pot of Gold. World Publishing CO., Port Moresby.
Ladbury, R.A. 1981a. Current legal problems in project financing. Aust. Min. Petrol. L.A. 3:139, 157.
Mining Act 1976. See the schedule to the Mining (Ok Tedi Agreement) Act, clauses 11 and 45.
Mining Agreement 1980a. See the schedule to the Mining (Ok Tedi Supplemental Agreement) Act, clause 15, and Note 9 and clause 5. Supplemental Agreement 1981. See schedule to the Mining (Ok Tedi Second Supplemental Agreement) Act, clause 5.

APPENDIX

Figure 1 provides a complete picture by summarizing the relationship of all the corporate units involved in sponsoring and financing the project, together with most of the agreements among those organizations. When studying the tables the reader should start in the top left hand corner. The first two rectangles represent the principal

CASE STUDIES

agreement between the State and Damco, a subsidiary of BHP. Damco then assigned part of its rights to the other members of the consortium, MFD (a Standard Oil (Indiana) subsidiary) and Kupferexlorations-gesellschaft (KE, representing the German Group). Both of these companies then made further assignments until the State, BHP Minerals, Amoco PNG and Star Mountains Holdings took equity in OTML. The seven lenders to OTML are listed on the far right. These loans were supported by the two completion and Cash Deficiency Agreements and the two German support agreements. Other major agreements such as the shareholders' agreement, are shown in oval diagrammes. The connecting lines indicate the organizations that are parties to the various agreements.

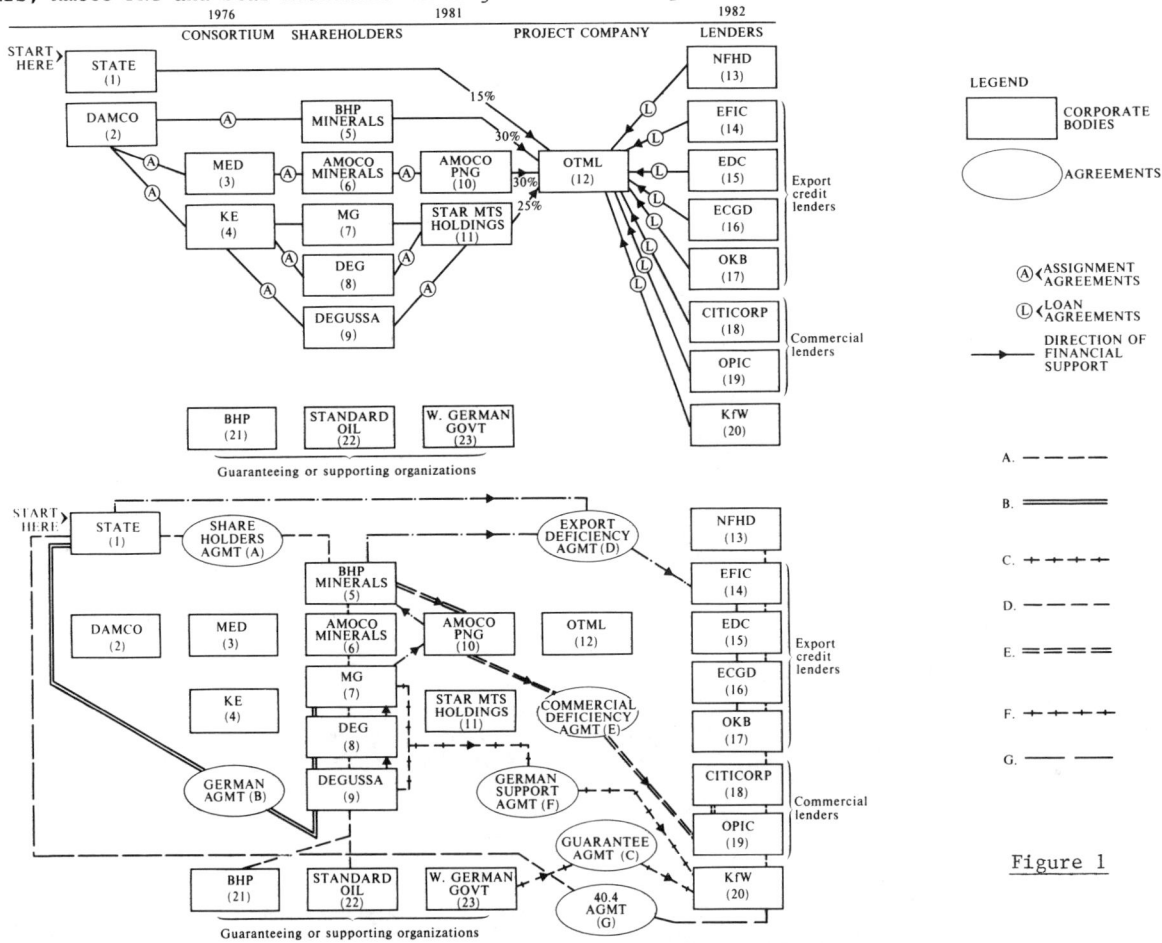

Figure 1

Key to fig 1:
1. Independent State of Papau New Guinea
2. Dampier Mining Company Limited (BHP subsidiary).
3. Mt. Fubilan Developemnt Company (formerly a Standard Oil group subsidiary).
4. Kupferexplorationsgesellschaft MBH.
5. BHP Minerals Holdings Pty. Limited.
6. Amoco Minerals Company
7. Metallgesellschaft AG.
8. Deutsche Gesellschaft fur Wirschaftliche Zusammenarbeit (Entwicklungsfesellschaft) MBH.
9. Degussa AG
10. Amoco Minerals (PNG) Company.
11. Star Mountains Holding Company Pty. Limited.
12. Ok Tedi Mining Limited.
13. North Fly Highway Development Company Pty. Ltd.
14. Export Finance and Insurance Corporation (Australia).
15. Export Development Corporation (Canada).
16. Export Credits Guarantee Department (U.K.)
17. Oesterreichische Kontrollbank Aktiengesellschaft (Austria).
18. Citicorp N.A.
19. Overseas Private Investment Corporation.
20. Kreditanstalt fur Weideraufbau.
21. Broken Hill Proprietary Company Limited.
22. Standard Oil Company (Indiana).
23. Federal Republic of Germany.

Description of Agreement
A. Main Shareholder Agreement between all sponsors together with the parent companies of BHP Minerals and Amoco Minerals.
B. German Group Shareholder Agreement (whereby State holds 5% of its OTML holding through Star Mountains Holding).
C. Guarantee Agreement (whereby W. German Government meets some of German Group's support obligations).
D. Completion and Cash Deficiency Agreement among all sponsors and all export credit agencies.
E. Completion and Cash Deficiency Agreement among commercial binders (Citicorp and OPIC) and Amoco PNG and BHP Minerals.
F. Loan support agreement among MG, Degussa and KfW.
G. 40.4 Agreement among the State and all lenders (on rights of lenders on termination of Principal Agreement).

12.6

INCO LIMITED'S SOROAKO NICKEL PROJECT:
A CASE STUDY IN FINANCING LARGE OVERSEAS MINING PROJECTS

Robert T. DeGavre

formerly Inco Ltd.
New York, New York

INTRODUCTION

The $645 million financing for Inco Ltd's Soroako nickel project in Indonesia not only represents a significant human achievement but also there are certain important lessons that can be derived from Inco's experience in financing the project.

The paper will not present a technical assessment of the project since Inco Ltd. submitted a series of papers on this subject at the International Laterite Symposium as part of the 108th Annual Meeting of the AIME in February 1979.

BRIEF HISTORY OF THE SOROAKO PROJECT

Table 1 summarizes the milestones achieved in the Soroako project and its financing.

Exploration Phase

A contract of Work was signed on July 27, 1968 by the Republic of Indonesia and P.T. International Nickel Indonesia (P.T. Inco) covering an extensive area of over 25,000 square miles on the two southeastern arms of the island of Sulawesi. The Contract provides that P.T. Inco has the sole responsibility for financing the project and determining the terms on which the financing should be obtained, that P.T. Inco has the right to receive and hold outside of Indonesia foreign exchange revenues and to freely dispose of such revenues, and that Indonesians have the right to purchase up to 20% of the equity of P.T. Inco over a period following start of production.

P.T. Inco began field investigations in the Contract Area in August 1968. Preliminary reconnaissance sampling with portable drills broadly outlined areas of favorable laterite terrain which were then systematically sampled on surveyed grids. By early 1971 sufficient exploration work had been completed to indicate that the laterite deposits near Soroako could sustain a major nickel-producing plant (figure 1). Pilot plant studies in 1970 and 1971 at Inco's research station in Port Colborne, Ontario, using two bulk samples of Soroako ore material, also confirmed that the ore could be upgraded and treated for the recovery of its metal values.

Feasibility Phase

Encouraged by the results of its exploration program and pilot-plant work, Inco Ltd. commissioned Toyo Engineering Company of Japan in August 1971 to prepare a capital cost estimate for a relatively small nickel project called "Stage I". Stage I was to be a single-line pyrometallurgical process plant with an annual capacity of approximately 15,000 metric tons of nickel contained in 75% nickel sulphide matte (see Figure 2). Electric power was to be supplied by a thermal power station.

It is important to recognize that Inco's strategy was to develop the Soroako ore deposits in several stages: first, construct a small project to gain experience and to reduce the initial capital outlay and then later expand in one or more steps to a much larger project to achieve maximum economies. The strategy reflected both Inco's financial position in 1971 as its sales and earning dropped sharply and the difficulties previously experienced by Inco in trying to launch in the late 1960's very large "greenfield" nickel projects in New Caledonia and Guatemala.

Toyo's Project Development Study was completed in January 1972. Based upon Toyo's cost estimate for physical facilities, Inco estimated the total capital cost of Stage I to be $115 million. For the next several months, Inco ran numerous in-house financial feasibility studies, using as inputs Toyo's cost estimate and the results of its exploration and pilot plant programs. In September 1972, Inco completed a project feasibility study which was based upon a capital cost estimate of $135 million. The feasibility study was immediately made available to potential partners and lenders.

Partner Negotiations

The Japanese nickel market had from the outset been central to Inco's planning for the Soroako project. That market had been growing dramatically at rates in excess of 20% per annum during the 1950's and 1960's. Japan had established, for rational security and other reasons, a domestic

Based on a paper presented at the SME/AIME Fall Meeting, Nassau, Bahamas, September 1978 and published in Mining Engineering, March 1979.

Table 1

SOROAKO NICKEL PROJECT
MILESTONE DATES

Background

July 1968	Contract of Work between Republic of Indonesia and P.T. Inco Indonesia signed.
August 1968	P.T. Inco begins Exploration in Contract Area.
May 1969	P.T. Inco begins exploration near Soroako in Contract Area.

Stage I Milestone Dates

August 1971	Toyo Engineering commissioned to undertake capital cost estimate for Project's physical facilities which was completed in January 1972. Total capital cost for Stage I estimated to be $115 million.
February 1972	Inco Ltd. enters into negotiations with Japanese participants and begins discussions with financial institutions.
August 1972	U.S. Eximbank issues preliminary financing commitment.
September 1972	Inco completes in-house feasibility study. Total capital cost estimated at $135 million
September 1972	Inco solicits proposals for engineering, procurement, and construction.
October 1972	Participation Agreement and Sales Contracts signed with Japanese Participants.
December 1972	Inco Ltd. Board approves capital authorization request to confirm capital estimate, order long delivery equipment, and begin construction work.
March 1973	Letter of Intent awarded to Dravo Corporation and contract signed in April 1973.
June 1973 - 1974	Seven Stage I Credit Agreements signed totalling U.S. $125 million equivalent.
October 1973	Dravo and P.T. Inco revise total capital cost estimate to 169 million.
October 1973	Construction on-site commences.
March 1977	Pre-operational testing and start-up commence.
	Formal inauguration ceremony for Project.
April 1978	P.T. Inco makes first delivery of product to Japanese buyers.

Stage II Milestone Dates

May 1974	Citicorp issues commitment to syndicate $170 million Eurodollar loan for Stage II.
July 1974	Inco Ltd. completes in-house feasibility study for Stage II based upon work by Dravo and Acres. Total capital cost estimate for Stages I and II $578 million.
July 1974	Inco Ltd. Board of Directors approves capital authorization request for Stage II.
September 1974	Bechtel engaged to prepare detailed plan for earliest completion of Stage I and for proposed expansion, including capital cost estimate.
November 1974	Bechtel and P.T. Inco revise total cost estimate to $650 million.
December 1974 - March 1975	Consents received from lenders to expand Project.
February 1975	Participation Agreement amended to limit Japanese equity interest to $11.25 million.
February 1975 - April 1975	Agreements and operating permits for Larona Hydro Project signed.
April 1975	P.T. Inco issues notice to Bechtel to proceed with Stage II.
May 1975	Letter of Intent awarded to Dravo for engineering of Stage II process Plant.
August 1975	Construction of Stage II on-site commences.
April 1975 - January 1977	Twelve additional credit agreements signed to finance Stage II and Stage I overrun.
November 1975	Bechtel and P.T. Inco revise total capital cost estimate to $820 million
March 1976	P.T. Inco authorized addition of third hydroelectric turbine to Larona River power plant. Capital cost estimate climbs to $833.7 million.
October 1977	Pre-operational testing of Stage II commences.
February 1978	Commercial power from first Larona unit.

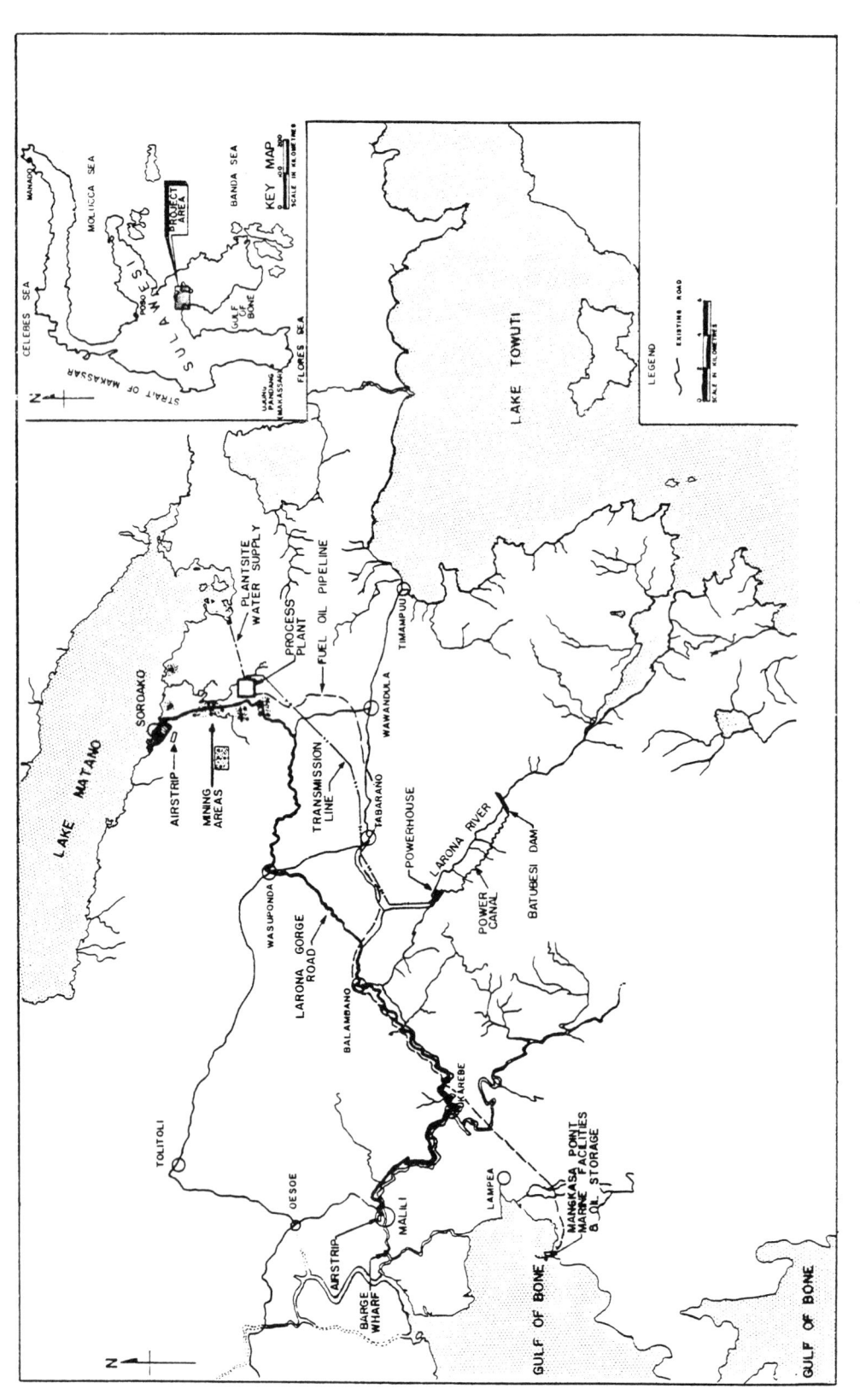

Figure 1

SIMPLIFIED FLOW DIAGRAM

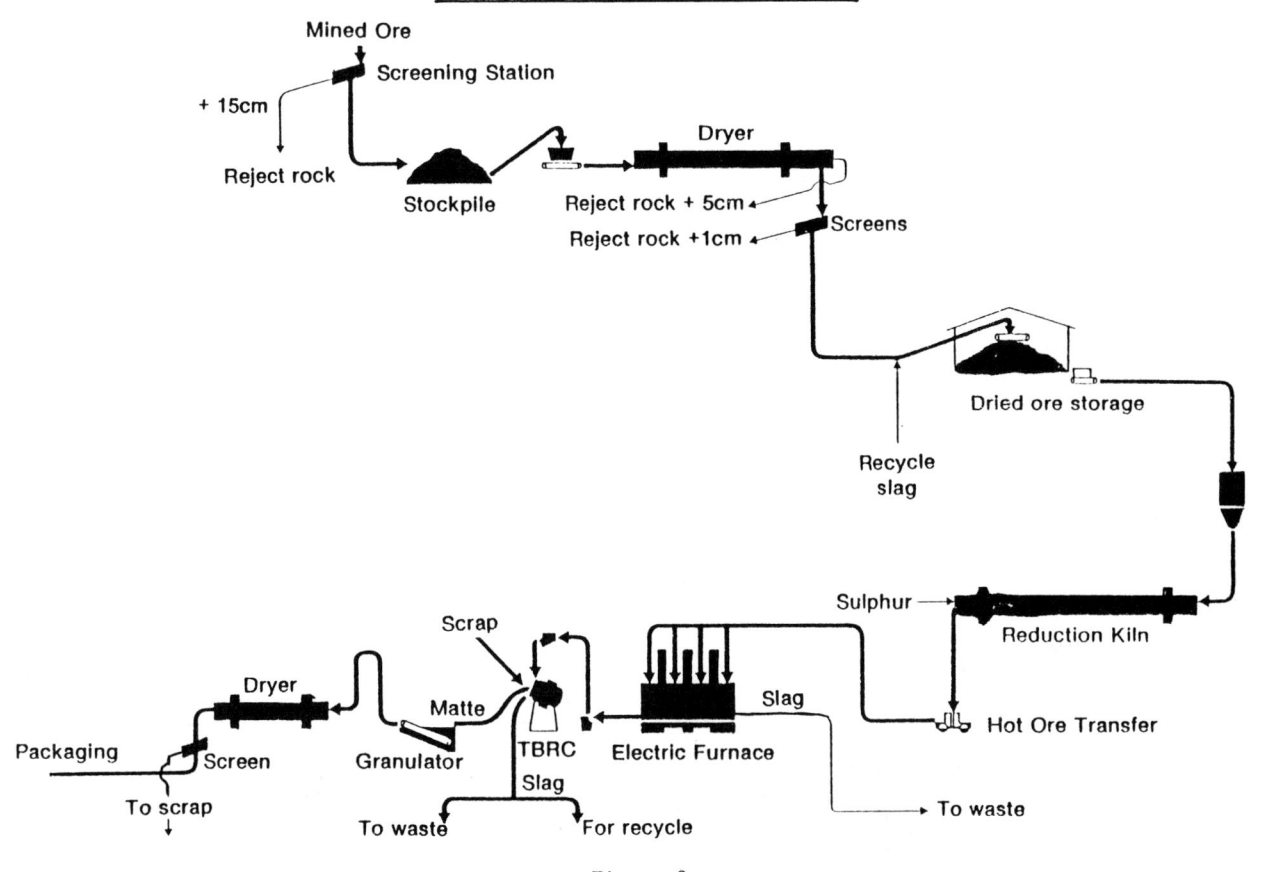

Figure 2

nickel refining industry based upon imported ore and semi-refined feed material. Inco had, however, been unable to supply Japan's feed material requirements in any substantial manner due to production limitations at its Canadian mines and to its inability to offer a long-term supply commitment to the Japanese as a result of the restrictions on the export of semi-refined materials contained in Ontario's and Manitoba's Mining Acts. Inco believed that the Soroako project should be well positioned to provide a long-term basis an economic source of supply for the Japanese nickel refining industry.

Discussions with potential Japanese partners in the project began, at the initiative of the Japanese, shortly after the award of the contract of work in July 1968. In May 1971 Inco submitted a formal written proposal to potential Japanese partners. These negotiations intensified early in 1972 following the completion of Toyo's study; and in October 1972 Inco with six Japanese companies - three nickel refining companies and three trading companies - entered into a participation agreement under which the Japanese participants purchased a 25% equity interest in P.T. Inco and two sales contracts under which the three refining companies agreed to purchase on a long-term basis the output form Stage I.

STAGE I FINANCING PLAN AND PROJECT IMPLEMENTATION

The financing plan for Stage I began to take shape following the completion of Toyo's Study in February 1972. At that time, the basic direction was toward a project engineered and constructed entirely by the Japanese and financed with Japanese credits. Toyo's study had been predicated upon this assumption. This direction was also reinforced by the policy of the Japanese government to reduce their mounting exchange reserves by exporting long-term capital to develop future raw material sources of supply for the Japanese economy. In 1971 Japan's overall external surplus was the largest in nominal terms ever recorded by a country through 1971.

Thus Inco's early financing discussions concentrated upon obtaining the necessary credits through the six Japanese participants. As these discussions progressed, it became clear that, if the Japanese participants were to provide the full debt financing, they would require a financial guarantee from Inco Ltd. This requirement was unacceptable to Inco for reasons, noted later, relating to non-financial objectives.

The impasse was broken on August 21, 1972 when the Export-Import Bank of the United States undertook in a preliminary commitment to provide (Eximbank) long-term credits to finance the U.S. exports required by the project and to accept as security the assignment of proceeds from the sale of the project's production. This preliminary commitment fundamentally altered both the financing and the procurement for the project, from a reliance upon Japanese sources to a multilateral financing and procurement program.

Eximbank's commitment reinforced Inco's own confidence in undertaking to arrange the multilateral credits which had been bolstered by the strong recovery in Inco's sales and earnings in 1972. Eximbank's commitment not only stemmed from the rapid, aggressive expansion in Eximbank's authorizations since 1969, but also reflected a growing positive sentiment towards Indonesia in the financial community. In 1971, a reluctant Eximbank agreed to reschedule its pre-July 1966 Loan to Indonesia as part of a general agreement among Indonesia's non-Communist creditors. It is notable that the P.T. Inco credit was one of the first Indonesian credits subsequent to the 1960's for which Eximbank did not require a guarantee from a third country party.

The Eximbank and Japanese commitments formed the cornerstone of the financing plan for Stage I. The Company then proceeded through early 1974 to broaden the financing plan to include additional creditors and to finalize the loan and related security documentation. In total, seven separate long-term credits for approximately $125 million were arranged (see Table 2). The first disbursement under these credits was made in November 1973. As will be discussed later, the project's estimated cost rose sharply in 1973 and 1974 causing major dislocations in the financing plan for Stage I.

The Project Expanded

In July 1974 Inco Ltd's Board of Directors authorized an expansion of the Soroako project by increasing the project's total capacity threefold. The expansion ("Stage II") involved the addition of two pyrometallurgical process lines and the construction of a 165 megawatt hydroelectric facility utilizing the Larona River to supply electric power to both Stages I and II. The expanded project (Stages I and II combined) will have a combined annual capacity of approximately 45,000 metric tons of nickel contained in a sulphide matte. In July 1974, engineering on Stage I was well-advanced, but construction work on-site was in a formative phase with less than 10% of the work completed.

The decision to expand before the completion of Stage I and to develop the hydroelectric potential of the Larona River was taken at a time when nickel prices and fuel oil prices were rising rapidly and when Inco's nickel deliveries were climbing to an all-time record and severely stretching the capacity of its Canadian operations. There was a mounting concern within the industrial world about the long-term scarcity of raw materials, a concern that was fueled by a global commodity boom of 1973/74, the trauma of the post-October 1973 energy crisis, and such writings as those of the Club of Rome on the limits to growth.

The Japanese shareholders declined to participate in the expansion. Confidence within Japan toward the future had been severely eroded by the events of 1973 and 1974. Japanese equity participation as a result has been progressively diluted from 25% to approximately 4%.

The strategy that was developed in late 1973 for financing the expanded project was (i) to duplicate the security arrangements for Stage I, with certain important modifications to provide for cost escalation, and to fold all the project's creditors, including Stage I senior lenders, into that single package, (ii) to obtain a large core loan from a commercial bank syndicate and to encourage the syndicate leader to assist us in implementing the financing plan for the expanded project, and (iii) following the commitment from the commercial bank syndicate, to obtain export credits to the maximum extent possible to provide the very long-term fixed rate monies required. This strategy reflected Inco's assessment that the Euro dollar financial markets were relatively liquid and the commercial banks aggressive, whereas the export-credit agencies, particularly Eximbank, were retrenching for a variety of reasons.

Discussions began in late 1973 with several commercial banks. In May 1974 P.T. Inco accepted a commitment from Citicorp International Bank Limited (CIBL), Citibank's merchant bank group, to syndicate a $170-million (later increased to $200 million) Eurodollar credit for the expanded project.

Shortly after the acceptance of the commitment, the tone of the Eurodollar financial markets fundamentally altered. Apprehension had begun to develop in the spring of 1974 about huge balance-of-payment deficits, inflation, erratic exchange rates, large foreign exchange losses reported by a few banks, the crisis of Britain's property-oriented fringe banks, and the general deterioration in the banks' capital to loan ratios. It was in this climate of unease that in late June 1974 the news about the failure of the Bankhaus I.D. Herstatt, a regional West German bank, hit the market. Worried bankers feared that a crisis of "unprecedented dimensions" might be touched off by the collapse. Spreads offered by bankers widened sharply and maturities shortened. Non-dollar based banks withdrew from making new medium-term U.S. dollar funding commitments. Faced with a sharply deteriorating market, CIBL and Inco agreed to split the commitment into two transcends, one for $140 million on the terms initially proposed and one from $60 million with shorter maturities. This commitment was subsequently successfully syndicated, to the credit of both the project and CIBL and signed in April 1975.

With CIBL's commitment in hand, we - together with CIBL in certain cases - then approached the export-credit agencies. In particular, a substantial and positive participation by Eximbank was essential for a variety of reasons. Eximbank's professionalism is well recognized within the

Table 2

SOURCES AND TERMS OF SENIOR DEBT

Lender/(guarantor)	Type of Facility	Date of Credit Agreement	Loan Principal (million)	Annual Interest Rate	Other Annual Charges	Principal Repayment Schedule
STAGE I CREDIT FACILITIES						
1. U.S. Eximbank	Buyer's Credit	Oct. 24, 73	US $13.	6.0%	0.5% p.a. c.f.	10 ESAI, Jan. 10, 82.
Bank of Montreal, Mercantile Bank	Buyer's Credit	Oct. 24, 73	US $18.	N.Y. prime + 0.5%	0.25% p.a. c.f. 0.5% p.a. g.f.	10 ESAI, Jan. 10, 77.
2. Export Development Corp.	Buyer's Credit	Nov. 20, 73	C $17.25	7.0%	0.5% p.a. c.f.	20 ESAI, Jan. 10, 77.
3. Wales International Lending Pty. (-Australia)	Buyer's Credit	Feb. 13, 74	A $10.	7.5% subject to review every 5 years	0.5% p.a. c.f.	20 ESAI, Mar. 31, 77.
4. Davy Ashmore Int.'l (-U.K.)	Supplier's Credit	Sept. 21, 73	UKL 0.36	6.0%	--	10 ESAI, commencing January 17, 77.
5. Elkem (Guaranti-Institute for Eksport Kredit	Supplier's Credit	July 23, 73	Nor. Kr. 6.07	Approx. 8.0% with first payment due Apr. 10, 76.	Insurance premium of approx. 1.0% p.a. due April 19, 1976.	21 ESAI, Apr. 10, 76.
6. Tokyo Nickel Co., Shimura Kako K.K., Sumitomo Metal Mining (M.I.T.I. Metal Mining Explore. Agency).	United, Production-Sharing Loan	June 12, 73	US $36.	8.0625%	0.55 pa.a. insurance premium payable for insurance obtained from M.I.T.I.	20 ESA, Jan. 10, 77.

Lender/(guarantor)	Type of Facility	Date of Credit Agreement	Loan Principal (million)	Annual Interest Rate	Other Annual Charges	Principal Repayment Schedule
7. Bank of Montreal and seven other Commercial Banks	Euro dlr. Loan	Nov. 29, 73	US $25.	LIBO + 1.5% on 365-day basis.	0.25% p.a. c.f.	13 ESAI, Oct. 10, 78.

STAGE II CREDIT FACILITIES

Lender/(guarantor)	Type of Facility	Date of Credit Agreement	Loan Principal (million)	Annual Interest Rate	Other Annual Charges	Principal Repayment Schedule
8. Citicorp Int'l Bank Ltd. and 6 other Commercial Banks.	Euro dlr. Loan (Tranche B)	Apr. 18, 75	US $145.	1. Up to Jan. 10, 77 LIBO + 1.0% 2. Jan.'77–Jan.10'85 LIBO + 1.25% 3. Jan. 10'80–Jan.10'85 LIBO + 1.5% 4. Jan. 10'85–Jan.10'85 LIBO + 1.75% on 360-day basis.	0.5% p.a. c.f.	ESAI, July 10, 79.
CIBL, as Agent and 7 other Commercial Banks.	Euro dlr. Loan (Tranche A)	Apr. 18, 75	US $55.	1. Up to Jan. 10'85 LIBO + 1.25% 2. Jan.10'80–Jan.10'85 LIBO + 1.5% on 360-day basis.	Effective 0.5% p.a. c.f.	12 ESAI, July 10, 70.
9. U.A. Eximbank	Buyer's Credit	Nov. 17, 75	US $35.10.	8.5%	0.5% p.a. c.f.	9 SAI, Jan. 10, 85.
Pefco U.S. Eximbank	Buyer's	Nov. 17, 76	US $35.	9.0%	0.625% p.a. c.f.	9 SAI, Jan. 10, 85.
10. Davy Ashmore Int.'l (-U.K.)	Supplier's Credit	Oct, 15, 73	UKL 2.5	7.5% with first payment due July 10, 77.	--	10 ESAI, July 10, 79.

CASE STUDIES

Lender/(guarantor)	Type of Facility	Date of Credit Agreement	Loan Principal (million)	Annual Interest Rate	Other Annual Charges	Principal Repayment Schedule
11. Eikem	Supplier's Credit	Aug. 1, 75	Nor. Kr. 27.	9.0% with first payment due July 10, 1979	Insurance premium of approx. 1.0% p.a. with first payment due July 10, 1979	18 ESAI, July 10, 79.
12. Hitachi (M.I.T.I)	Supplier's Credit	July 18, 75	Y 2,400	8.0%	--	20 ESAI, July 10, 79.
13. Export Development Corp. (Series A)	Buyer's Credit	Oct. 21, 76	C $20.	Canadian Prime + 1%	0.5% p.a. c.f.	10 ESAI, July 10, 79.
Export Development Corp. (Series B)	Buyer's Credit	Oct. 21, 76	C $20.	8.5%	0.5% p.a. c.f.	10 ESAI, July 10, 79.
14. Wales International Lending (EFIC)	Buyer's Credit	Mar. 19, 76	$ A12.	11.5% to July 31, 1978 after which either (i) 11.5%, or (ii) floating rate.	0.5% p.a. c.f.	12 ESAI, Sept. 79
EFIC Tranche A	Buyer's Credit	Mar. 19, 76	$ A8.	7.75%	0.5% p.a. c.f.	8 ESAI, Sept. 85.
15. CIBL, as agent, and 12 other Commercial Banks	Euro dlr. Loan	Oct. 15, 76	US$140.	LIBO + 2.25% pm 360-day basis	0.75% p.a. c.f.	16 ESAI, July 79.
16. EFIC	Buyer's Credit	Jan. 21, 77	$ A25.	9.5%	0.5% p.a. c.f.	20 ESA, Sept. 79

FINANCE FOR THE MINERALS INDUSTRY

Lender/(guarantor)	Type of Facility	Date of Credit Agreement	Loan Principal (million)	Annual Interest Rate	Other Annual Charges	Principal Repayment Schedule
17. BOECC	Supplier's Credit	Oct. 14, 76	UKL̄ 3,35	8.5%	1.519% one time charge covering various fees plus export premium.	10 ESAI, July 77.
18. Credit du Nord (COFACE-France)	Supplier's Credit	Oct. 27, 76	FF13.54	7.5%	0.3% c.f.	10 substantially ESAI Oct. 97.
19. National Industri (Guaranti-Instituttet for Eksport Kredit)	Supplier's Credit	Jan. 26, 76	Nor Kr. 0.82	8.25%	Insurance Premium of approx. 1.4% p.a. with first payment due July 1979.	18 substantially ESAI July 1979.

Veg

ESAI – Equal semi-annual instalments with repayment commencing on date specified.
c.f. – Commitment fee payable on undisbursed amount of the credit total.
g.f. – Guarantee fee payable on the disbursed loan amount drawn.
LIBO – London Inter-Bank offered rate, an international Eurodollar loan base.
MITI – Japan's Ministry of International Trade Industry.

fraternity of export credit agencies and this participation in a project based on certain security arrangements is often taken as a guide. The project's procurement was also heavily skewed towards the United States due to P.T. Inco's contracts with U.S. engineering and construction firms and to the competitive edge that the United States had traditionally enjoyed in heavy earth-moving equipment.

But during 1974, Eximbank began to impose new constraints and policies with respect to loan authorizations, emphasizing a greater role for private banking sector in the promotion of U.S. exports and conversely a lesser role for itself. As the U.S. trade position improved in 1973 and early 1974 and the cost of money skyrocketed, Congress became stringent in funding the agency. In July 1974, Eximbank was temporarily forced "out of business" as Congress let the agency's authorizing legislation lapse. In this atmosphere, P.T. Inco's application for export financing encountered rough-sledding; nevertheless, in January 1975, Eximbank agreed to provide loans and guarantees retailing $70 million. The financial terms of this commitment were substantially more onerous than Eximbank's earlier loan for Stage I.

Subsequently additional export or supplier credits were arranged in Canada, Australia, England, Norway, Japan, and France, totalling a further $120 million equivalent. By late 1975, the aggregate credit facilities available to the project had reached $500 million or an amount sufficient to finance a total project cost of up to about $700 million based upon a 2:1 debt-to-equity ratio.

The expanded project, like Stage I, underwent a series of substantial upward revisions in the estimated total capital cost. By early 1976, it had become necessary to seek additional financing. Again events external to the project adversely affected P.T. Inco's ability to raise new monies. In early 1975 Pertamina, the Indonesian state oil company, failed to make certain payments due foreign banks. As the details and the magnitude of Pertamina's financial difficulties were subsequently unravelled by government appointed auditors and aired publicly in the financial press, the credit worthiness of Indonesian borrowers was adversely affected through at least mid-1976.

In this market CIBL agreed in May 1976 to syndicate a further $140 million but with substantially higher margins. The credit was successfully syndicated by October 1976. The project's credit facilities now totalled $650 million equivalent, an amount sufficient to finance a project having a cost of up to slightly over $900 million.

AN APPRAISAL OF FINANCING PLAN

Most project financing plans are not static; they continue to evolve as the underlying project changes in response to a host of factors. One can make an appraisal of the Soroako financing plan in terms of the objectives Inco established for itself at the time.

Financial Objectives

Generally the principal financial objective of a project financing plan is readily assumed: that is, to maximize the shareholder's assured dividend return (i) by leveraging the returns through maximum prudent use of debt financing, and (ii) by minimizing and fixing the project's annual debt service costs.

The Soroako financing plan provides or a debt to equity ratio of 2:1. For example, when the project's cumulative net cash requirements reach $850 million, shareholders' paid-in capital will equal approximately $284 million and outstanding senior debt $566 million. This ratio of 2:1 was a relatively conservative capital structure for a mining project which was never seriously questioned. In 1978 however, the project's forecast net cash flows deteriorated, as its near-term production schedule was sharply lowered and as nickel prices declined both in real and nominal terms. The effect of using debt financing in a project with a low rate of return is to reverse lever or reduce the shareholders' dividend flow. If this situation were to continue over an extended period of time, the selection of the 2:1 ratio may in retrospect be questionable.

The second financial objective was to minimize and fix the project's annual debt service costs. This objective was achieved by obtaining loans which (i) bore fixed and low interest rates, (ii) carried long maturities, and (iii) were denominated in a currency matching the project's revenues.

The weighted average interest cost of the project's credits compared favorably with the incremental long-term borrowing costs of Inco Ltd. the parent. In addition, interest costs were insulated to a reasonable extent from changes in market interest rates by negating credits with fixed interest rates; A 1.0% increase or decrease in market interest rates would increase or decrease P.T. Inco's average interest costs by 0.62% (see Table 3).

Fixed Rates We were significantly more successful in securing fixed-rate, low-interest cost loans for the Stage I financing than for the expanded financing. By and large, this simply reflects the reduced proportions of government-related credits in the expanded financing plan and the more onerous terms for those official credits. Some 64% of the credits negotiated for Stage I bear fixed-interest rates, while only 32% of the additional credits negotiated for the expanded project bear fixed rates (see Table 4).

Maturity With respect to maturity terms, we were reasonably successful in negotiating a level annual principal repayment requirement over a ten-year period (see Figure 3). It can be noted from Figure 3 that because the availability of government-supported credits was reduced for the expansion, the repayment schedule for the additional credits for Stage II is shorter than for the Stage I credits.

Table 3

INTEREST COSTS
(% p.a.)

	1978 RATES*			1978 FLOATING RATE PLUS 1%		
	STAGE I	STAGE II	COMBINED	STAGE I	STAGE II	COMBINED
Buyer Credits						
Fixed Rate	6.81	9.45	8.86	6.81	9.45	8.86
Floating Rate	10.00	10.50	10.25	11.00	11.25	
Supplier Credits						
Fixed Rate	7.88	8.34	8.32	7.88	8.34	8.32
Euro Dollar Credits						
Floating Rate	9.94	10.10	10.09	10.94	11.10	11.09
Japanese P.S. Loan**						
Fixed Rate	8.6125	----	8.6125	8.6125	--	8.6125
Average Costs	8.48%	9.85%	9.58%	8.84%	10.52%	10.20%

* U.S. Prime   = 9.0%
  Canadian Prime = 9.25%
  6 month LIBO   = 8.44%

**Production-Sharing Loan

Table 4

CREDIT FACILITY MIX
(US$ Millions)

	STAGE 1		STAGE II		STAGES I & II COMBINED	
Commercial Bank Credits:						
Floating Rates	$ 25	(21%)	$340	(65%)	$365	(57%)
Government Related Credits:						
Fixed Rates	$ 78	(64%)	$168	(32%)	$246	(38%)
Floating Rates	$ 18	(15%)	$ 16	(3%)	$ 34	(5%)
TOTAL....	$121	(100%)	$524	(100%)	$645	(100%)

CASE STUDIES

Currency With respect to current exposure, Inco's preference was toward credits denominated in U.S. dollars since the project's revenues and financial statements are expressed in that currency. The currency risks inherent in no-U.S. dollar credits were reduced by obtaining a so-called "cocktail" of currencies, thereby spreading and hopefully offsetting the risks. By and large, this cocktail approach has been successful at least on a cumulative basis (see Table 5.) The project nevertheless does have a significant exposure to the Canadian dollar (which is somewhat mitigated by Canada's close historic economic ties to the United States) and more significantly to the Australian dollar.

Table 5

FOREIGN CURRENCY EXPOSURE
ON LONG TERM DEBT

Currency	Amount of Credits		1974	1975	1976	1977	1978	CUM.
Canadian Dollar	$Cdn.	57,250,000	-	.1	.1	2.1	.9	3.2
Australian Dollar	$A	55,000,000	.1	.3	3.8	(1.6)	(.3)	2.3
U.K. Pounds Sterling	£	6,168,915	-	.1	.1	(1.2)	.3	(.7)
Norwegian Krone	Nr. K.	44,370,676	-	.1	(.2)	(.1)	.4	.2
French Frank	FF	13,536,360	-	-	-	(.1)	(.1)	(.2)
Japanese Yen	Y	2,400,000,000**	-	-	-	(.9)	(1.5)	(2.4)
			.1	.6	3.8	(1.8)	(.3)	2.4

SOROAKO NICKEL PROJECT
PRINCIPAL REPAYMENT SCHEDULE

Millions U. S. $

Year	Stage I	Stage II	Total
1976	.01		
77	12.0		12.9
78	14.0		15.6
79	14.0	29.6	43.6
80	14.0	57.5	71.5
81	14.0	57.5	71.5
82	12.9	56.7	69.6
83	12.9	55.8	68.7
84	9.1	55.4	64.5
85	9.1	65.3	74.4
86	9.1	61.5	70.6
87		43.8	43.8
88		25.6	25.6
89		12.6	12.6

Figure 3

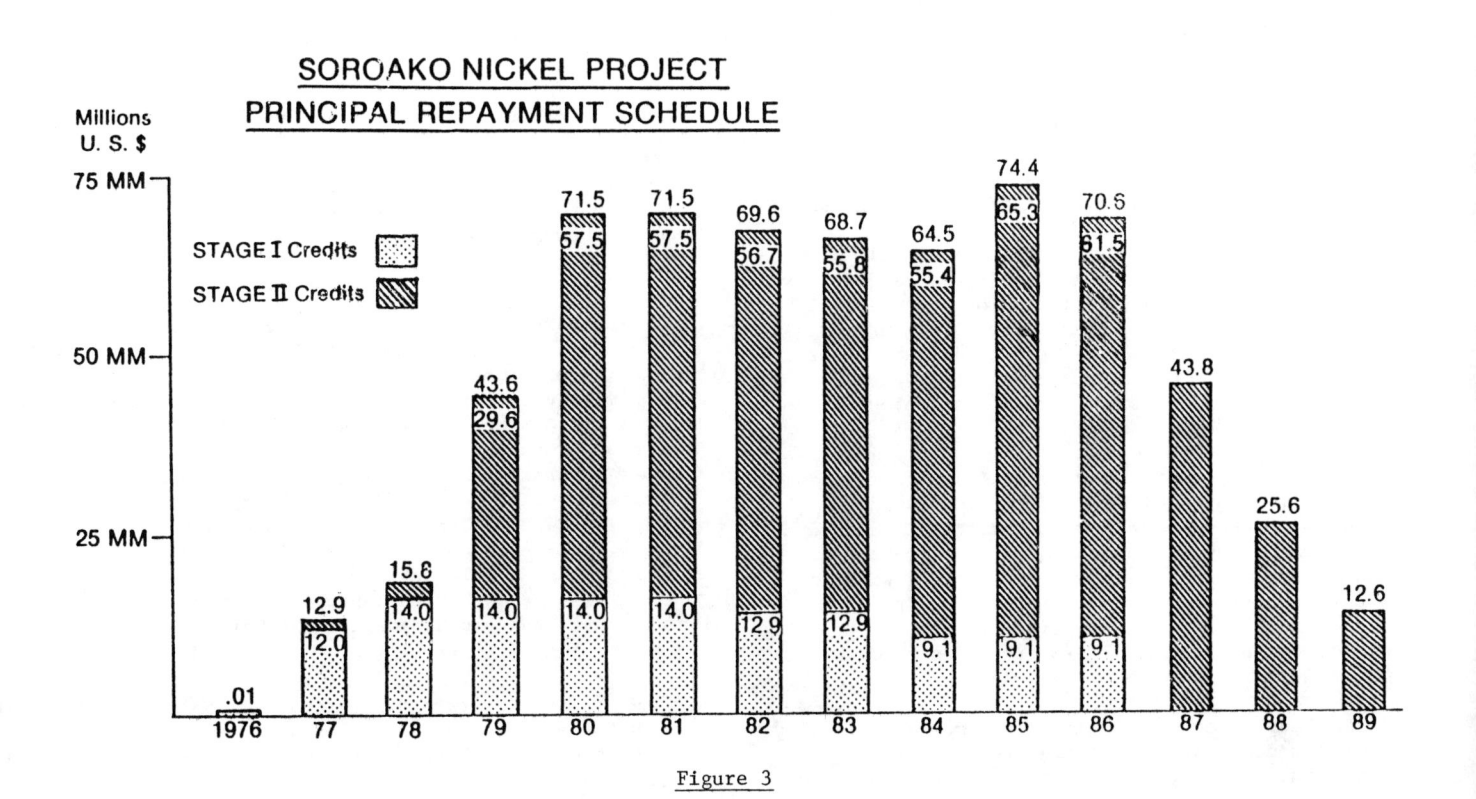

Non-Financial Objectives

It is the non-financial objectives of the sponsor that often make a project's financing plan and security documentation unique. there were three principal non-financial objectives which underlay the financing plan for the Soroako project: to share certain risks with the creditors, to strengthen the project's commercial ties with Japan, and to incorporate maximum uniformity, simplicity, and flexibility into the security arrangements.

Risk Sharing The first and principal non-financial objective was to transfer to the creditors the non-commerical (i.e. sovereign and certain environmental) risks associated with the debt financing for the project. It is this risk sharing characteristic for the financing plan which lets it properly be termed a "project financing plan".

Inco recognized at the outset that, given the size of the nickel resources in the contract area, Inco Limited's Soroako assets could one day represent a very significant portion of Inco Limited's consolidated global assets. Inco also, therefore, recognized that this concentration of assets could impinge severely upon the credit standing of parent and that the Company's financial exposure to this concentration must be minimized. In contrast, Inco felt that the Company could accept the commercial risks, including the technical risks, associated with the project, given our unparalleled experienced in the nickel industry. Inco sought to share the noncommerical risks both indirectly through multilateralizing the financing plan and more directly by establishing clear force majeure and other limitation on Inco Limited's obligations with respect to the project's debt.

Banks, of course, are not normally prepared to accept equity-type risks, in part because their potential returns are not of an equity nature. There is a fundamental precept among bankers that "risks should be allocated according to the distribution of reward". Certain financial institutions, however, are often ready to undertake sovereign risks. Large international commercial banks with diversified geographic and industry portfolios and with sophisticated internal controls on risk exposure can accept substantial risk exposure to a specific project. The analogy is to a responsible self-insurer.

This acceptance is often facilitated through the practice of syndicating large loan among numerous banks. In addition, government-related or supported financial institutions have been established in most industrial countries with the express purpose of undertaking or insuring non-commercial risks. The availability of non-recource financing of export sales through export-credit agencies is well known.

The project's security arrangements as finally negotiated transferred to the creditors the important non-commercial risks associated with the debt financing. No financial guarantee was required by Inco Ltd. The security arrangements consist of three principal documents: a completion guarantee agreement, an agreement of assignment and pledge, and the sales contracts.

COMPLETION: Under the completion guarantee agreement, Inco Limited agreed to provide (or see to the provisions of) funds to the extent required (i) for P.T. Inco to achieve project completion, (ii) for P.T. Inco to meet its financial obligations (including regularly-scheduled debt service) which fall due prior to project completion, and (iii) to cause the ratio of senior funded debt to paid-in capital not to exceed 2:1 through project completion. Inco Ltd's obligations under the agreement terminate at project completion, but could be terminated earlier if any senior lender accelerates its senior funded debt, or if an event of force majeure occurs which included certain natural disasters and political events. The completion guarantee agreement essentially ensures that, barring force majeure, funds will be available until project completion, a term that is defined in both a physical and an economic sense.

SALES PROCEEDS: The pledge agreement is among P.T. Inco, Morgan Guaranty Trust Company of New York as pledgee, and P.T. Inco's senior lenders. Under this agreement, P.T. Inco assigned to the pledge, for the benefit of the senior lenders, substantially all the proceeds of all its sales contract. The pledge has agreed to hold a portion of such proceeds in trust for the purpose of making debt service payments thus ensuring that the senior lenders will receive the sales proceeds effectively directly from the buyers in U.S. dollars, thereby minimizing the lenders' exchange risks and, to a lesser extent, their project risk.

The sales contracts are, by and large, normal commercial contracts under which the buyers take the nickel-containing sulphide matte at market prices if the material is available. Three Japanese buyers committed to take the production from Stage I and Inco Ltd. committed to take the production from Stage II.

SYNDICATE RISK: In addition to the security arrangements, the complexion of the senior lenders themselves provides security to the project. Approximately 80% of the credits made available for Stage I and 44% of the credits made available for the total project were provided or supported by various government institutions. The various countries represented include the United States, Australia, Canada, Japan, the United Kingdom, Norway, and France. The commercial bank syndicates also represent a multinational group of institutions, though with a heavy emphasis on Inco's traditional banks in Canada and the United States.

Japanese Link The second non-financial objective was that the financing plan should strengthen the project's commercial ties with the Japanese nickel market.

STAGE I: We were reasonable successful in securing Japanese financial participation in Stage I. The Japanese Participants contributed $11.25 million in equity and $36.0 million in long-term credits in the form of a "production-sharing loan". "Production-sharing loan", a type of credit which is repaid by direct deductions made by the Japanese buyers from the sales invoice, was important to the Stage I financing plan, since unlike an export credit it was not tied to procurement in any country.

STAGE II: Japanese commercial banks also participated in the $25 million Eurodollar loan syndicated by Bank of Montreal for Stage I. With respect to Stage II, however, we were disappointed in our efforts to secure any important Japanese participation except for a supplier credit relating to the purchase of hydroelectric equipment. The Japanese participants declined to participate any further in the project's equity requirements, and even the Japanese commercial banks were excluded from the syndicated Eurodollar credits due to administrative guidance then being exercised by the Ministry of Finance on Japanese bank's dollar loans.

Uniformity The third non-financial objective was to negotiate security arrangements that are uniform for all senior lenders and simple, and incorporate maximum flexibility to deal with the unexpected.

Uniformity and simplicity of documentation invaluably aids in the implementation, understand, and administration of any project financing plan, particularly when there are numerous creditors involved.

These views are often not shared by bankers or their counsel. For example, David O. Beim, then an Executive Vice President of Eximbank, gave a speech in March 1977 in which he set forth a model with respect to security arrangements for the new generation of projects. His model requires the participation in the financing plan by numerous unrelated parties with differing sets of priorities, a model which he concedes to be "more complex than before". While complex security arrangements have definite intellectual appeal, such arrangements can be so encumbered that it is difficult to get them off the ground, and, if they do indeed fly, there is a continuing threat of crash landing due to the sheer exhaustion or ill-will generated while trying to make the arrangements work as conditions evolve.

P.T. Inco's 19 credit agreements are essentially identical in their non-financial provisions and the security agreements are reasonably straightforward. This was achieved with substantial up-front investment in negotiating time and in the face of great reluctance expressed by various lenders and their counsel.

FLEXIBILITY: Flexibility was a more elusive goal since in any project financing plant, by definition, the creditors obtain a direct financial stake in the success of the project. Their natural tendency is to restrict the sponsor from changing the project, and thus potentially altering their security without first obtaining their consent. In particular, a major and unexpected problem arose in developing a financing plan that would automatically adjust to accommodate revisions to the project's capital-cost estimate.

The Stage I financing arrangements were notably inflexible in this regard. The senior lenders had required that senior debt could not exceed $100 million without their prior consent. Based on a debt to equity ratio of 2:1, such a ceiling would effectively permit the project's total cost to climb to $150 million, or $15 million more than initially estimated. Requirements in excess of this ceiling had to be financed either with equity or with quasi-equity in the form of deeply subordinate debt. In 1972 we were relatively confident that we had sufficient conservatism in the estimate and thus that the ceiling provided ample flexibility.

COST ESCALATION: Our confidence was rudely upset, when in October 1973 Dravo Corporation, the project's engineer/constructor at that time, completed its initial cost estimate for the physical facilities resulting in an increase in the project's total cost estimate to $169 million. Subsequent revisions to that estimate continued to drive the estimate higher, such that by March 1975 the total estimated cost of Stage I had increased to approximately $250 million or on a forward expenditure basis, by over 100%.

Much of this sharp increase was attributable to the unexpected inflationary surge which started in 1973 and continued through 1974, this surge coinciding with the procurement of the bulk of the equipment and materials for Stage I. Inflation began to moderate in 1975, braked in part by the collapse in the commodity boom and by the steep decline in industrial production in the OECD countries. Inflation was by no means the sole contributor to the overruns; in particular, field construction and project management costs rose enormously.

During 1974 we were unable to obtain the consent from the senior lenders to increase the debt ceiling. Thus, as the equity or quasi-equity commitments of the shareholders ballooned upwards, it was apparent that any financing plan for an expanded project would have to take into account the marked uncertainty surrounding initial capital cost estimates for mining projects in remote locations.

The major distinguishing feature of the financing plan for the expanded project was the agreement by the senior lenders to eliminate the absolute ceiling on senior debt. In return, Inco Ltd. agreed that, subject to force majeure, its obligations under the completion agreement would not terminate until the project had not only been physically completed but also demonstrated an ability to service its debt out of its operating cash flow. This novel arrangement permitted the financing plan to adjust automatically as the cost of the expanded project increased from $578 million to over $850 million.

SOROAKO NICKEL PROJECT

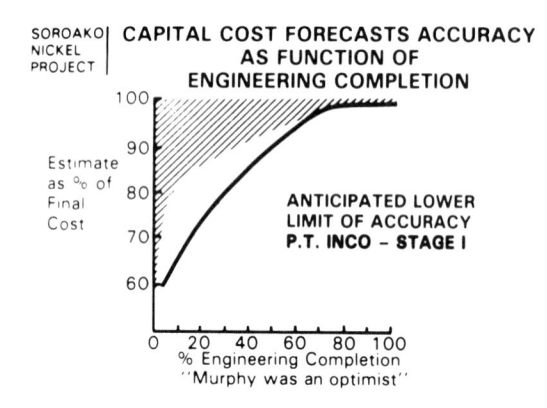

CAPITAL COST FORECASTS ACCURACY AS FUNCTION OF ENGINEERING COMPLETION

Estimate as % of Final Cost vs *% Engineering Completion*

ANTICIPATED LOWER LIMIT OF ACCURACY P.T. INCO - STAGE I

"Murphy was an optimist"

FACTORS IN INCO'S APPROACH

The financial community broadly understands the elements that go into a successful financing plan: a demonstrably viable project, a strong sponsor, balanced security arrangements, strong support from the host government at all levels, and so forth. There were, however, several other, perhaps unusual, factors which, in retrospect, bore heavily on the relative success of our efforts.

In-house The Indonesian project's financing plan was conceived and implemented in-house by Inco as the cooperative effort of the staff of the Treasurer's and Legal Departments. Outside financial advisers may however by necessary where there are several sponsors of roughly equal standing who need a unifying force or where the project has no element of sovereign risk in its financing plan and can therefore use the financial markets with which investment bankers are familiar.

Prototype The project was implemented in two stages. This schedule allowed us to launch a prototype financing plan on a relatively small scale and then, later, having experienced the weaknesses in that plan, to modify it an launch a much larger plan.

Timing The tenor of the times greatly assisted in underwriting the financing plan. The period 1972-1973, following on the heels of the 1960's, was a time frame in which it was relatively easy - both from a sponsor's and a creditor's perspective - to undertake and accept the risks associated with financing a large project. The years after 1973/1974 evidenced a sharp turn away from the trends of the 1960's, in terms of GNP growth, inflation, unemployment, and gross trade imbalances (see Figure 4).

Inco Ltd. were able to take a very long-term view of the demand or nickel in assessing the viability of the Soroako project, but I wonder if we would have the same luxury today. Bankers also in general have become more cautious - or more "skeptical" as one puts it - having witnessed since 1974 the tremendous downside risks experienced by mining projects which they have financed.

Unity of Interests The identity of interests among the sponsors and among creditors facilitated rapid decision-making concerning the financing plan, with respect to the sponsors, Inco Ltd. was dominant equity participant and, following the Japanese partners' decision not to participate in the expansion, became effectively the sole spokesman for the project. Today there is an increasing tendency to minimize the technical, economic, and financial risks associated with a large project by spreading those risks among several shareholders, often including from the outset the host government. Due to the varied arrangements, while they serve to spread risk, can severely and sometimes fatally complicate the efficient decision-making process necessary to implement the financing plan for a large project.

The Soroako project's creditors also evidenced a unity of interests which served to expedite documentation. While, unlike commercial banks, export-credit agencies have a dual focus-export promotion and adequate security to ensure repayment with interest - this dual focus did not complicate decision making. Commentators often praise the beneficial impact that the participation of an international or regional development bank, for example, the World Bank, can have upon a project's viability. I personally believe that the participation of such an institution is counterproductive to the prompt implementation of a project. Development banks have broad non-financial objectives in their lending programs; for example, to "accelerate institutional change to further development". Sometimes these goals are pursued to the direct detriment of their interest as a creditor.

Having enumerated certain factors underlying our relative success, it is only fair to review two factors which did not have that result:

1. Financial market timing was not one of our strengths. Our experience has been that a series of parallel negotiations - with shareholders, buyers, governments, engineering firms, and so forth-tend to converge and this confluence of forces thrusts the financial people in the market, for better or for worse. For example, 1973 was an excellent opportunity to approach Eximbank, while in contrast 1975 was the reverse. Where we had success in market timing, it was for the most part purely fortuitous.

2. During the early phase of the financing plan we did not have a lead lender, as we shifted away from an emphasis on Japanese credits to a multinational financing plan. This led to several bruising negotiations with certain of the project's creditors. In the expansion financing, however, we did provide for a lead lender of sufficient renown within the financial community - CIBL - to aid the senior lenders in reaching a consensus on the appropriate security arrangements.

CASE STUDIES

751

Figure 4

OECD countries: Growth, inflation, unemployment and external imbalance.

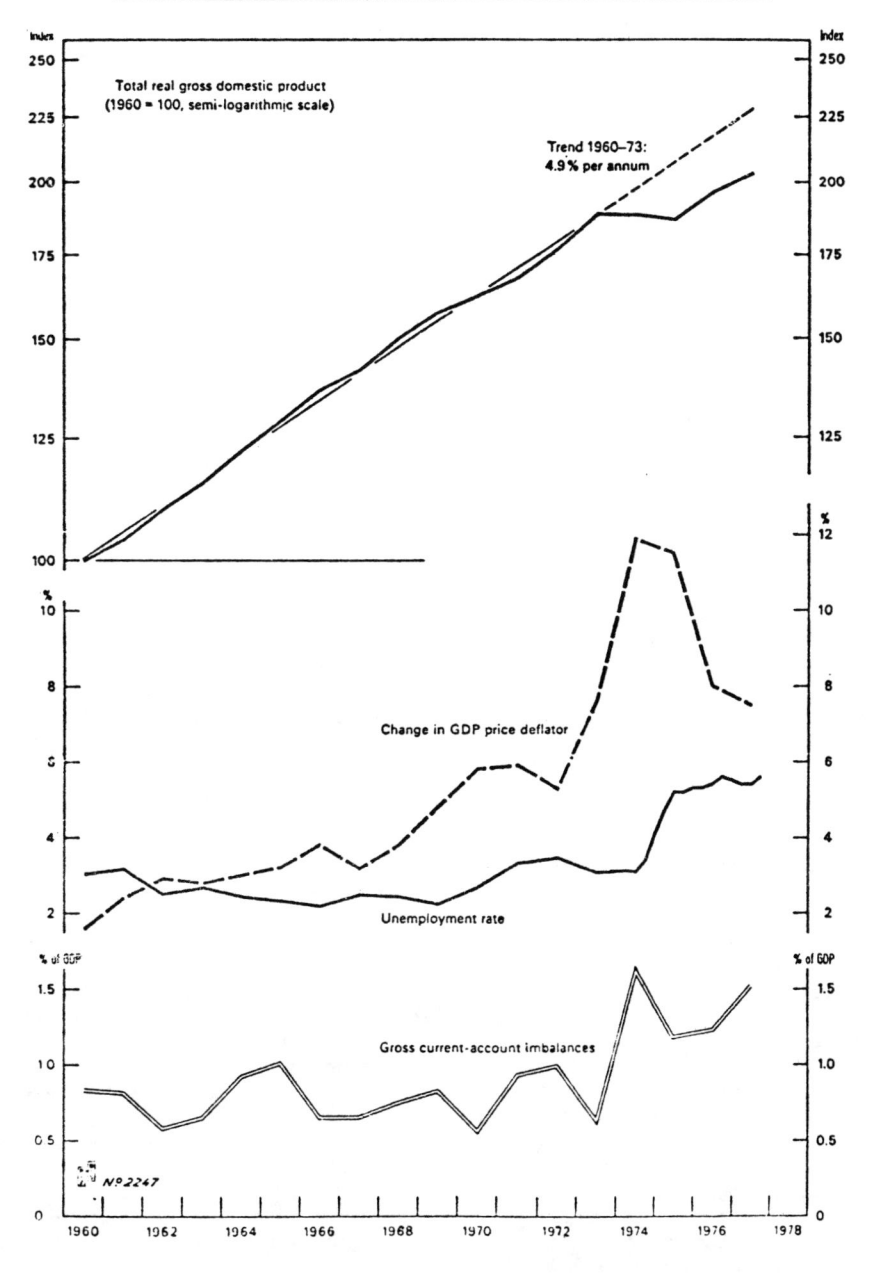

Total real gross domestic product
(1960 = 100, semi-logarithmic scale)

Trend 1960–73:
4.9% per annum

Change in GDP price deflator

Unemployment rate

Gross current-account imbalances

Extracted From 48th
Annual Report – BIS

CONTRIBUTIONS FROM THE ENGINEERING STAFFS

The sponsor's technical and engineering staffs play a vital role in the financing plan. Their professionalism - for example, in ascertaining ore reserves, in supervision the construction effort, or in training the operating team - is essential to underpin a banker's confidence in a project. The engineering staff, however plays two important direct roles in formulating and implementing the financing plan.

Capital Cost Estimates

Certain lessons, can be derived from the experience with the Soroako project. Firstly, the reliability of an estimate for a mining project in a remote area can only be reasonably assured if sufficient engineering has been completed. "Engineering" encompasses not only the preparation of construction drawings and purchase specifications, but also all the other detailed studies necessary for a thorough and balanced cost analysis of the project. Generally, if 2 to 3% of the project's total estimated engineering budget has been expended in engineering studies and preparing the estimate, the resulting "order of magnitude estimate" may have an accuracy of perhaps +/- 25% to 40%. If 10% of the engineering work has been completed, then the quality of the estimate is approaching a "preliminary budget" with an accuracy of +/- 15% to 20%. If 40% of the engineering dollars has been spent, then we have what is called a "definitive estimate" with an accuracy of +/- 10%.

Engineering costs represent very roughly 8% to 12% of the total capital cost of a project. Thus on a $850 million project, the owner must be prepared to spend some $35 million on engineering alone in order to obtain a "definitive estimate". What this inevitably implies is that on very large projects, sponsors will proceed with the project with a cost estimate having a reliability or quality less than a definitive estimate and thus effectively with an unreliable cost estimate. This general conclusion is supported by our experience both on the Soroako project and on Inco's nickel project in Guatemala.

Secondly, we must recognize that there is a difference between the financial and engineering staffs perception of the purpose of the estimate. Financial people generally see a capital cost estimate as being the total number of dollars it will take in the end to complete the project. In contrast, an engineer must use the capital cost estimate as the project budget, and thus as a measure of how much it should take to complete the project. The project budget as used by engineers is usually a tight number through which the owner's engineers discipline or control the project managers and the engineer/constructor. It is used first and foremost as a control document, not as a forecast of the total cash required. The result is that the numbers are tight and represent achievable, but superior targets. Therefore these budget control numbers in the early stages of evaluation are often lower in total than the actual cash requirements.

One should ask whether the estimate is being prepared to assess objectively the viability of the project, or is it being prepared with a specific audience in mind. The audience that an estimate can be directed toward can be quite wide: it can include the engineer/constructor who participates in preparing the estimate, owners' management, potential partners, potential lenders, and, importantly today for large projects requiring government support, the broad national public and political system.

In sum, it is the responsibility of the engineers to prepare for the financial staff a reasonably reliable cost estimate as a basis for developing a financing plan and to discuss candidly with your financial staff the probable magnitudes of errors in the estimate.

Procurement Plan

The second area in which the engineering staffs play a direct role is implementing the procurement plan in such a manner as to maximize the utilization of low-cost, fixed-rate, long maturity export credits. This may involve several steps:

1. The procurement staff must develop a procurement plan based on which the financial staff can seek to arrange better export credit facilities. We have found the initial procurement plans to be an imprecise, but necessary exercise.

2. The procurement staff will often be requested to make presentations to various export-credit agencies or to their ministries of industry in order to ensure that the project's procurement staff appreciate the potential and capability of the local suppliers.

3. In analyzing bids and negotiating purchase orders, the procurement staff must ensure that due recognition is given to the better export-credit facilities available to the project, and, if there are none to finance a specific order, to press the suppler to provide a supplier export credit.

In retrospect, our procurement staffs, in cooperation with Dravo and Bechtel, did a superb job in ensuring that the project's purchasing maximized in an economic manner the utilization of the export-credit facilities available to them.

CONCLUSION

The Stage I financing plan included a significant Japanese participation, but that plan suffered the unacceptable flaw of not providing flexibility for cost escalation. The financing plan for the expanded project nevertheless met our financial and non-financial objectives.

CASE STUDIES

SOROAKO NICKEL PROJECT

HISTORICAL UPDATE SINCE
SEPTEMBER 1978

by
Donald S. Small,
Senior Manager,
World Wide Treasury Operations
Inco Limited, New York

Project construction was essential completed during 1978, with the commissioning of the Larona hydroelectric system. Commercial production commenced in Soroako on January 1, 1979.

A series of mechanical and technical problems encountered during start-up substantially reduced the project's scheduled production in 1979, and production reached only 19 million pounds of nickel contained in matte. The most serious problem, which had not been apparent during pilot plant studies, resulted from the acidic nature of the higher grade ores that were planned to be processed in the early years of the project's life. These ores severely corroded the refractory linings of two of the three electric furnaces in the process plant, necessitating several major repairs.

Two changes were implemented in 1979 to address the corrosion problem. A second major mining area containing lower nickel grade and less acidic ores (originally scheduled to be mined in later years) was opened up to provide ores for blending with the higher nickel grade, more acidic ores. Modifications were also made to the furnaces; cooling devices were added to assist in protecting the refractory linings from corrosion. As a result of these changes, the project's original design capacity of about 100 million pounds per year of nickel contained in matte was reduced to 75-80 million pounds of nickel in matte. The project's current annual production capacity is about 65 million pounds of nickel in matte; this could be increased to the 75-80 million pound range for a cost of about $20 million.

Certain non-technical factors have also had a substantial negative impact on P.T. Inco's financial results, and consequently, Inco Limited's results. These include: (i) A sharp increase in operating costs in the late 1970's due to dramatic increases in the price of oil;

(ii) The rapid increase in interest rates after 1976 when the first drawdown was made under the CIBL-I Syndicated Credit Agreement; and

(iii) Weak world-wide nickel markets and, in particular, the Japanese nickel market which is P.T. Inco's principal market.

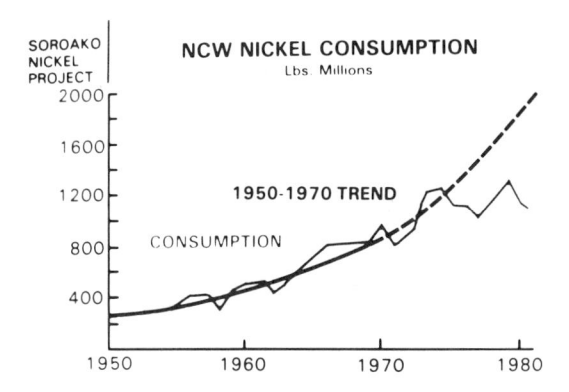

In order to provide P.T. Inco with the flexibility needed to improve its results, changes were made to its original financial structure. Together with Inco Limited, the Company completed three financial restructuring in 1979, 1981 and 1983, whereby the level of outstanding debt and interest expense was reduced, Inco Limited's common equity investment in P.T. Inco was increased, margins on Eurodollar-based credits were reduced, and certain security, selling and other arrangements were modified to recognize the fundamental changes that have occurred in the nickel industry and P.T. Inco since the early 1970's.

The changes have provided P.T. Inco with a greater financial flexibility and a more appropriate capital structure. It is encouraging to see that with a improving nickel market, particularly in Japan, P.T. Inco is currently achieving a positive operating cash flow at the 65 million pound annualized production rate, despite the continuing low nickel prices.

In accordance with P.T. Inco's Contract of Work, Indonesians have the right to purchase up to 20% of the equity of P.T. Inco at the rate of 2% per year. In 1980, an initial offer for sale of 2% of the equity of P.T. Inco was made to the Indonesian Government. This offer has been repeated each subsequent year, but to date, the Indonesian Government has not exercised its right to purchase an interest in P.T. Inco. In addition, in late 1982 P.T. Inco submitted a special proposal to the Indonesian Government for the purchase of a significant minority interest in P.T. Inco. Discussions of such proposal are continuing.

The following table sets forth P.T. Inco's production, deliveries, operating earnings and external debt service (excluding prepayments of debt in 1979, 1981 and 1983 of $65 million, $100 million and $60.5 million, respectively) for the years 1979-83.

P.T. INTERNATIONAL NICKEL INDONESIA

SUMMARY OF OPERATIONS & FINANCIAL RESULTS

Year	Production of nickel in matte	Deliveries of finished nickel to customers(1)
	(in millions of pounds)	
1979	19	13
1980	45	29
1981	44	34
1982	30	31
1983	40	54

Year	Operatings earnings (loss)(2)	External debt service
	(in millions of dollars)	
1979	$(27)	$73
1980	1	80
1981	(14)	84
1982	(38)	55
1983	(69)	48

(1) deliveries to customers of finished nickel refined from P.T. Inco's matte.
(2) reflects the consolidated results of P.T. Inco's operations.

SOROAKO NICKEL PROJECT — UTILIZATION OF VARIOUS CREDITS
U.S. $ Millions

	Amounts Disbursed Through 1978*	% of Fixed Rate	Average Interest Rate**	Average Final Maturity
Buyer Credits	$202.8	88%	9.30%	10 yrs.
Supplier Credits	30.2	100	8.20	8
Bank Eurodollar Credits	290.0	—0—	15.92	7.3/10.4***
Other	36.0	100	8.06	10
Orginal Plan	$559.0	42%	13.00%	
After Two Restructurings		62%	11.60%	

* Historical exchange rates
** Interest rates as of the end of March 1982
*** Before/after prepayment and rescheduling

12.7

DEVELOPMENT OF THE RANGER URANIUM FINANCING
FROM BANKS, CUSTOMERS, SHAREHOLDERS, AND THE STOCK MARKET

S. James Hodge
Consultant, Sydney, Australia

Norman Miskelly
Ord Minnett, Resources Research Partner
Sydney, Australia

C. Richard Tinsley
European Banking Company Limited, Assistant Director
London, England

ABSTRACT

Construction of the Ranger uranium project, located 230 kms east of Darwin in the Northern Territory, Australia, commenced in January 1979. Energy Resources of Australia Ltd (ERA) was incorporated in February 1980 to acquire all the rights in the project. The total cost to ERA of these rights was A$407 million, made up of A$125 million in cash to the Federal Government and the issue of 125 million A$1.00 shares at par, plus a payment of A$16 million in cash to each of Peko-Wallsend Ltd. (Peko) and EZ Industries Ltd. (EZ).

ERA's cash requirements to complete the project were estimated in October 1980 to be A$553 million. Project loans, secured by first ranking debenture stock, were arranged for a total of US$390 million, while second-ranking debenture loans totalling A$60 million were obtained from the vendors, Peko and EZ.

Overseas participants (power utilities who had agreed to purchase uranium yellowcake under contract) arranged to take up 25% of the equity capital for A$102.5 million; Peko and EZ were allotted 30.5% each as satisfaction of vendors' consideration; and 14% representing A$57.5 million was issued to Australian residents in an underwritten public flotation in November, 1980.

The loan and equity financing arrangements required the successful resolution of many complex and interlocking factors, including technical and economic feasibility, agreement with Aboriginal interests, compliance with Government policies, and securing of sales contracts.

INTRODUCTION

The financing design and corporate construction of Energy Resources of Australia Ltd (ERA) the entity responsible for the development and operation of the Ranger uranium resource, required the linkage of differing parties and their own interests from around the globe. For each segment of the ERA project there was a different "engineer" as final decision-maker, with his own strong ideas on where and when those links should be accomplished, how they should be designed, and how they should be financed - ideas that could change right up to the crucial moment when the links locked together.

HISTORY OF THE RANGER PROJECT PRIOR TO ERA

The Ranger uranium ore deposits were discovered in October 1969 by Peko's exploration arm, acting on behalf of Peko itself and EZ. By mid-1973 the project was almost ready to proceed. The two main orebodies had been sufficiently explored, the ore treatment methods developed and tested, the treatment plants defined, sufficient sales contracts signed with more in prospect, and finance assured. Only Government approval was needed, but that was not

Figure 1

LOCATION MAP

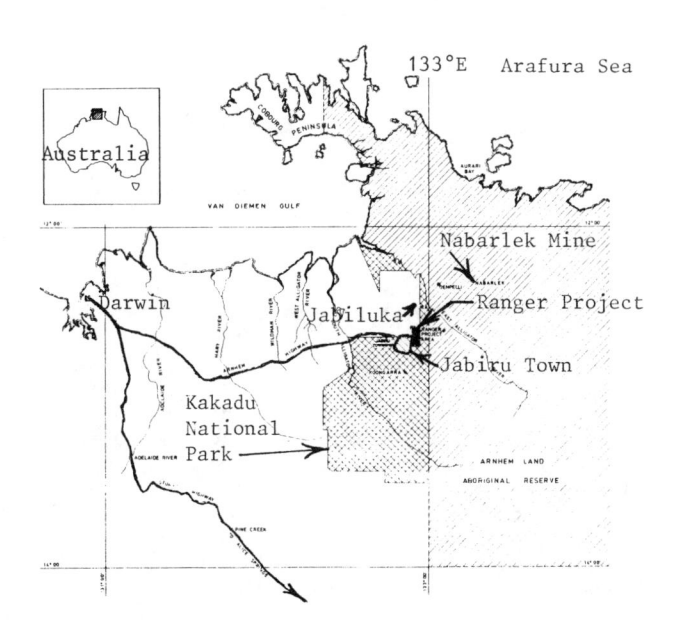

Based on a paper presented at the Australasian Institute of Mining and Metallurgy, Sydney Branch Project Development Symposium, November 1983.

Figure 2
OPEN-PIT GEOLOGICAL CROSS SECTION

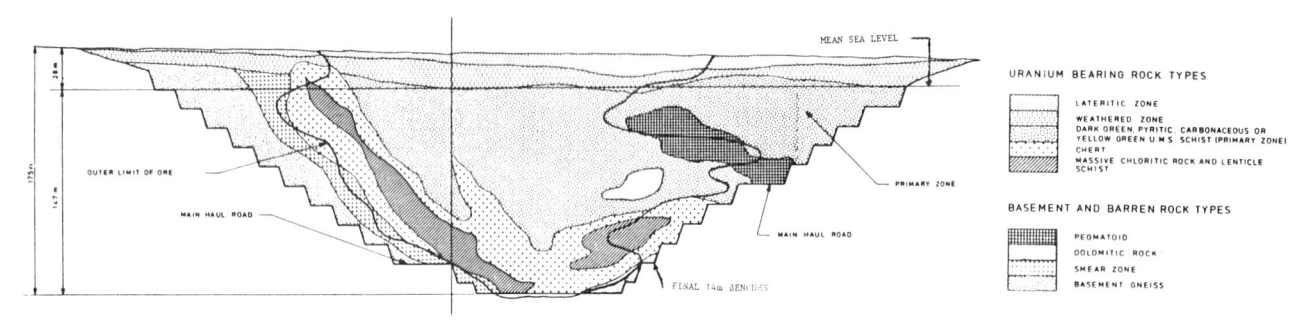

forthcoming.

Eighteen months later that hurdle seemed to have been overcome, though not in the way earlier envisaged. Following the Lodge Agreement, the Australian Federal government became a joint venturer in the development of Ranger. The Government made no payment to enter the project, but it had the obligation to subscribe 72.5% of the capital cost and pay 50% of the operating costs whilst being entitled to take and market 50% of the product. Peko and EZ each agreed to subscribe 13.75% of the capital cost, pay 25% of the operating costs, and take 25% of the product.

In the meantime, the Ranger Uranium Environmental Inquiry under Mr. Justice Fox had been instituted to, inter alia, enquire into and make recommendations as to whether the Ranger orebodies should be mined or not. It finally reported its conclusions in May 1977 and three months later the Federal Government policy was announced, permitting uranium mining and export.

By January 1979 the Lodge Agreement had been fully fleshed out and construction activities began on site. The Federal Government, Peko and EZ were still joint venturers, with the important difference being that Peko and EZ had some sales contracts, but the Government had none.

In August 1979, in an about face compared with previous policy, a new Federal Government invited offers for the purchase of its interest in the Ranger project. Peko made one of seventeen such offers. Essentially, Peko's offer consisted of the establishment of a new company, Energy Resources of Australia Ltd, to acquire both Peko's and the Government's shares. Peko's offer for the Government's share (A$125 million, plus the Government's expenditure on the project to the date of acquisition, plus interest on that expenditure) was accepted in December 1979. In February 1980 EZ joined this structure and the stage was set for ERA to become the sole developer for Ranger.

BRIEF PROJECT DESCRIPTION

Proven and probable ore reserves plus possible ore, before dilution and using a 0.1% cut-off grade were estimated at 44.65 million tonnes, containing 110,878 tonnes of uranium.

The initial open-pit accessed the No.1 ore zone. No.1 Orebody had been delineated by vertical diamond drill holes on 50-m spacing. Areas proposed for early production had been drilled on centres of 10-m separation, giving detailed information on the grade distribution. Its geological ore reserves calculated to a cut-off grade of 0.1% U_3O_8 were stated as being:

	Ore tonnes	Grade % U_3O_8	Tonnes Contained U_3O_8
Proven Ore Reserves	15,563,100	0.336	52,292
Probable Ore Reserves	307,000	0.191	586
Total Ore Reserves	15,870,100	0.333	52,878

The No.1 Orebody mining plan uses conventional open-cut methods, with heavy accent being placed on grade control and mining flexibility. In the initial years it was planned to mine at the rate of 4 million tonnes of material per annum, of which 1.15 million tonnes was ore, the balance being waste and material below ore grade.

The ore treatment plant is designed to process this ore, with 0.3% head grade, at 3,500 tonnes per day to produce 3,000 tonnes of U_3O_8 per annum. Its design is conventional, using proven technology and equipment. The production rate can readily be doubled without commensurate increase in capital or operating costs. (In fact, in the first six months of 1984 the plant (unaltered) achieved a production rate equivalent to 4,200 tonnes per annum.) Production of U_3O_8 was planned for October 1981. Engineering design work commenced in late 1978 and site construction work in January 1979.

The estimated capital cost of the project was A$345 million, plus interest during construction. In 1973 when the project first seemed ready to commence, the estimated cost was A$60 million. In 1973, it was common for plants estimated to cost A$60 million to in fact cost A$60 million. In 1979, it was equally common for plants estimated to cost A$345 million to actually cost a lot more than estimated, and for construction to take longer than planned.

Ranger was completed on time and within budget. (See Fixed Assets in Table 3).

THE ALTERNATIVES TO ERA

When, on 6 August 1979, the Federal Government called for offers to purchase its Ranger interest, Peko and EZ were each faced with these alternatives:

1. Assume a sale of the Government's share would be effected in which case

    a) they would be together in a 50/50 partnership with a party or parties as yet unknown, or

    b) they could separately or jointly attempt to acquire the Government's share,

    OR

2. Assume that no sale would take place, perhaps because no offer was sufficiently attractive politically to the Government.

For a number of reasons, alternative 2 appeared to be a realistic possibility. However, if Peko and EZ did nothing and in the event a sale did take place, they would find themselves irretrievably locked into the alternative 1(a) situation. As circumstances stood at the time, the pursuit of a strategy based upon allowing either 2 or 1(a) to occur could well have been viewed as the more prudent.

For example, Peko and EZ were each then entitled to 25% of the output for only 13.75% of the considerable capital cost, and subject to royalties of not more than 3%. The other share carried a 50% product entitlement, a contribution of 72.5% of the capital cost, and royalty charges of not less than 8%. Those shares of capital cost were important because, by Ranger's capital-intensive structure, depreciation and interest costs would greatly exceed production costs, while loan repayments would absorb much of the cash flow generated.

Add to that the cost of acquiring the Government's share, estimated at well over A$100 million in 1979 dollars, and the fact that no sales had been arranged for the Government's share of production, whereas Peko and EZ had arranged some sales of their respective shares and were confident of selling the balance, and it is easy to see why a strategy based on acquiring the Government's share could then seem to inject considerable, perhaps dangerous, risks into an otherwise sound Peko/EZ business proposition. However, within a month, the problems and attractions of the above alternatives had been roughly sorted out and Peko saw scope to optimise the acquisition strategy alternative 1(b).

PLANNING ERA AND ITS FINANCING

The reduction to an acceptable size of the considerable risks clearly involved in alternative 1(b) necessitated strong reasons for confidence that the participants could

1. secure the required mixture of equity and loan capital to both take over the Government's 72.5% capital expenditure commitment and pay to the Government for its share of the project a premium estimated to be over A$100 million;

2. write the firm uranium sales necessary to service that capital commitment; and

3. develop for Ranger a corporate structure giving real benefits to Australia and differentiating the bid from many others expected, while ensuring that overseas equity would be attracted in order to secure both capital contribution and sales contracts.

The first step involved more detailed modelling. This showed that economic viability depended upon

1. selling the full output of the plant from commencement of production at firm prices not much below the then current market price for long-term contracts, a price that had begun to fall;

2. raising equity capital in cash at least equal to the amount of the Government premium from other than Peko and EZ; and

3. loan financing on a non-recourse project basis 100% of construction costs and having the lenders also provide a rehabilitation guarantee of up to US$55 million.

PUTTING THE ERA PLAN TO WORK

Prior negotiations for the sale of Peko's 25% share of production and for borrowing its 13.75% of the project's capital cost had put its negotiating team and the project in touch with many potential buyers and lenders. That exposure gave confidence that the necessary sales and finance were available, but only if the strategy was executed purposefully and quickly. There was no room for tentatively exploring the possibilities with the Government and with a wide range of potential buyers, lenders, underwriters; etc..

Rather, the analysis pointed to one optimum overall plan, with essential dates and numbers that had to be achieved in all negotiations. Consequently, to initiate that plan, ERA negotiators contemporaneously

1. confirmed with J. Henry Schroder Wagg, the lead bank for Peko's share of financing, that the proposed non-recourse project finance package was bankable,

2. put to the Government a specific offer for purchase of its interest,

3. opened negotiations with certain potential buyers for specific annual purchases at stated price levels for contract durations exceeding the required loan period,

4. offered each of such buyers a specific equity participation in proportion to sales contracted, the total of such participations being within the Government guidelines,

5. invited the proposed Japanese customer/participants to subscribe a portion of the proposed project finance, and

6. arranged with the underwriting stockbrokers the provision of equity capital from the public.

ERA NEGOTIATIONS OVERALL

An essential early step towards achieving the plan was the positive identification of parties who had compatible interests and who would work confidently with ERA to execute a deal within the scope of the project. If the international, multi-faceted links were to be forged together, all negotiating parties must quickly establish one shared total "engineering" concept and together endeavour to achieve it. There was no place for extensively pursuing "ambit claims" or "feeling out" possibilities.

Because of constraints imposed by the ERA plan, there were naturally many tense moments in the negotiations - times when one side seemed to be offering "any colour you like as long as it's black." However, confidence progressively built up in each other's integrity in pursuing the overall objective and, as it did, both sides could see that a shade of charcoal could be accommodated with advantage. That confidence did not come without modification or accommodation of viewpoints at each stage of negotiations.

NEGOTIATING BANKABLE SALES CONTRACTS

Negotiations begun by Peko before EZ's entry into ERA in February 1980 were carried out with EZ's full knowledge, the door into ERA always being kept open. Upon their entry, Continental Illinois of Chicago also joined in the loan lead management.

The ERA team conducting all the negotiations was a very small one, working separately at many points around the world at one time. Each knew the others well, and knew also the essentials of the project's plan. For example, the essentials of the bankable sales contracts needed to finance this project were well understood by the team, as was an appreciation of the buyers' essential needs.

The buyers' negotiators appreciated the basic principles of project financing and the part that sales contracts must play in assuring adequate cash flow to meet both loan and interest commitments, but they needed facts and numbers to back up ERA's assertions. They wanted many of these on the run from ERA's sales negotiator during the negotiations, and they got them. Then they wanted to see and hear from the ERA finance negotiator, to get from him the overall picture and receive confirmation of the facts and numbers previously presented. They wanted assurance that the project was financially viable, that their purchase commitments would result in uranium deliveries in accordance with their contracts.

They also needed confidence that other matters were proceeding to ensure that those deliveries would be met, and in particular that

1. the negotiations with the Government on all matters would be satisfactorily concluded, including

    a) the terms of purchase of the Government interest

    b) the Government's requirements for mining and treatment and whether they would significantly alter the project as it had been described by them, or its operation,

    c) the internal and external political matters reflecting on ERA's ability to deliver in terms of their individual contracts,

    d) environmental matters,

    e) the Government Authority to mine uranium,

    f) royalties,

    g) the rehabilitation requirements and their effects.

2. Peko and EZ would proceed with the sale to ERA of their interests on the terms stated,

3. there was professional independent assurance that the ore resource at Ranger was as described and that the mining and treatment methods to be used would produce the planned results,

4. other sales contracts were being negotiated sufficient to meet the project's needs and that prices negotiated would not be less than theirs,

5. the debt and equity finance would be supplied,

6. they would be permitted to take up the shareholding proposed for each, and

7. they would, if politically practicable, have a preferred position for further purchases.

By September 1980, sales had been arranged in respect of almost 40,000 tonnes of U_3O_8 for delivery over a period of fifteen years. This represented an average for that period of some 88% of the initial designed capacity of the treatment plant, conservatively 3,000 tonnes per annum. The deliveries arranged, (see Marketing tables at end of paper), included

1. commitments by the overseas shareholders in West Germany and in Japan to purchase an aggregate of 30,647 tonnes over 15 years,

2. other sales aggregating 6,546 tonnes to power utilities in Japan, Korea and the USA, and

3. return of 1,011 tonnes to the Government stockpile, replacing U_3O_8 borrowed by Peko and EZ to fulfill earlier contracts.

Prices for concentrates sold to the overseas participant shareholders were fixed for the calendar year 1982, escalated for calendar years 1983 and 1984, and were to be determined thereafter until 1996 on a basis related to international market prices, but with a floor-price formula established by Government determination. Prices for other overseas sales were related to international market prices and were also subject to Government floor prices.

THE OVERALL PLAN CONTROLLED THE NEGOTIATIONS

To effectively negotiate with his opposite parties, each senior member of the small ERA negotiating team had to be sure of his up-to-the-minute knowledge of how the essentials of the project were coming together. Yet he had no time to give or listen to detailed ball-by-ball descriptions from other negotiators. So, except for the occasional assurance that all else was going well, he would tell others and be told only about any exceptions to the plan or its timetable likely to affect his own negotiations.

It helped that the negotiating team was very small, and that each member was personally negotiating or directing negotiations in several areas. That made the pressure on each more intensive, but greatly reduced the total workload required to ensure effective communication and decision making within the team. The result was that each member made almost all the decisions required in his areas, and was seen by the negotiators on the other side to be making them. The lenders, buyers and others were presented with a picture of a confident team that, negotiations concluded, could successfully manage ERA into the future and fulfill all contracts being negotiated.

The negotiating strategy used by ERA is not universal. Often the decision-making level of management has not been deeply enough involved in putting the project together, and neither it nor those negotiating for it or writing draft agreements have gone sufficiently into the real-life aspects of the project. As a result, they negotiate through several loose groups which individually do not have sufficient knowledge, purpose, or authority to make decisions and as a consequence their company or project misses many potential benefits.

SPECIAL ASPECTS

The timetable was one of the special aspects of Ranger and its financing that made it, some said, the most complex in Australian mining history. All negotiations were completed, agreements signed and the ERA prospectus issued within ten months of the Government's acceptance of Peko's offer to purchase its share. It is true that Peko and EZ had earlier opened negotiations with almost all the parties concerned and that under the Government-Peko-EZ joint venture, construction had begun on the project site some eleven months before and was proceeding on time and on budget. Still the tight timetable was a problem, but it was also an advantage as the project proceeded with a strong forward momentum seldom achieved elsewhere.

THE GOVERNMENT AGREEMENT

Some aspects were both complex and historically unique, including the Government Agreement and the Rehabilitation Guarantee.

The Federal Government had virtually accepted Peko's offer to purchase its share of the Ranger project, but the equivalent of a sales agreement still had to be put together. It will be recalled that Peko's offer was on behalf of a company (ERA) that did not then exist, had no money, and had few sales contracts. The Government was to receive for its share a premium of A$125 million, plus reimbursement of its expenditure (A$100 million plus) to the date of assignment.

Despite the Government's having acquired its share from Peko and EZ at no cost, it says a great deal for the flexibility of the Government and the public servants involved that they were willing to commit to selling their share in Ranger to a company such as ERA then was. Over A$225 million was involved and should ERA not be able to proceed with the deal the Government, through changing circumstances, might well lose its chance to sell, or might suffer a lower price. Imagine the political consequences. Of course, Peko provided very substantial evidence that willing sales, equity and loan backing for ERA was likely.

As a surety, ERA was required to make up-front deposits totalling A$35 million and would forfeit A$10 million of that should the balance of the purchase money not be paid on due dates in November and December 1980. In that event, the contracts with the Government would terminate and ERA as a company would fail. The amounts were paid on the due dates.

The Government was fully exposed until 12 September 1980, on which date the agreements for purchase of its share and to recompense it for past expenditure were signed and the deposits paid. Thereafter, until December 1980, they had only the protection of the potential A$10 million forfeiture.

Despite the flexibility demonstrated in the above, the negotiations with the Government were complex indeed. Most of the specific reasons were peculiar to Ranger but, as background to other mining project negotiations, it would be useful to read them in the ERA prospectus.

THE REHABILITATION GUARANTEE

The rehabilitation provisions of the Government Agreement and the financing of them have direct relevance to other future projects. The Federal Government requires that at all times there shall be a detailed plan of rehabilitation of the Ranger Project Area, with detailed costings, all approved by the appropriate Minister. ERA is required to provide security for rehabilitation in the amount of 10% above the estimated cost.

The initial security required was a guarantee from an Australian bank, for US$55 million before construction could continue and, of course, well before the project earned any income at all and years before ERA itself could accumulate the backing for such a guarantee.

As someone else had to provide the backing to this guarantee, ERA looked to the project bank lenders. The ERA negotiators said in effect:

"We may at any time, even during construction and before any income is derived, be required to cease production, sell whatever has a value, bury the rest on site, and restore the whole area. In that event, you will almost certainly lose all or most of the US$250 million you have lent us. Will you please in addition and, provided they do not exceed US$55 million, contract to pay the costs of the funeral?". They said, "Yes, we will".

There was a significant time interval and much thought and discussion between the asking of the question and the answer, but the response was never in serious doubt; in the total context the Rehabilitation Guarantee was bankable. Perhaps only just, but bankable.

Much of the negotiations with the Government, selection of lending banks, spreading the loans amongst a large number of them in many countries, selection of customers of standing able and sure to honour their contracts, contract provisions, associating both equity and more than one-third of the project loans with sales contracts, etc., were designed to make the project financing risks bankable. To illustrate, of all the banks offered participation in the loan, only one declined. Its participation was taken up within 48 hours by another bank that had not seen previously the loan documentation, all 30 centimetres thickness of it.

THE ESSENTIALS OF MAKING RANGER PROJECT FINANCABLE

Non-recourse project finance is not really a matter on which one negotiates solely with bankers. All the negotiation in the world with bankers will not make a poorly structured project bankable.

So what is such project finance about? It is about convincing the bank that it is likely to be repaid in terms of the loan. Likely, not certain. The bank will take risks if it can assess them as "bankable". The inherent nature of the project and how it will fit into an uncertain future, are the bank's first considerations.

It has seen the prices of virtually all mineral products go up and down, in many instances down to below the cash costs of production for much of the industry. It would like to feel assured that this project, with this mine, its customers and its contracts, is very unlikely to find itself for long with cash costs plus financing commitments exceeding its cash income. It would like to feel assured that in all but the worst case the project will even be able to pay a reasonable dividend. ERA set out to structure such a project.

In 1972 when Ranger was ready for launching, the spot price of uranium was about US$6 per pound. It subsequently reached a peak of US$46, and it later dropped to US$17 per pound.

From the outset, one of ERA's prime objectives was to insulate itself as much as practicable from the extreme vagaries of spot prices. This could be achieved only if the buyer by his contract were to feel so insulated in the long term from extremely high prices as to offset his contribution to insulating ERA from low prices. Complex pricing clauses were needed to achieve both objectives.

Mining and treatment plans were laid that tended to equalise over a number of years annual production costs in standard dollars. As interest costs on the declining project loans reduce, the sum of production costs plus interest, expressed in dollars of the day, could be expected to decline.

By providing this type of structure, and backing it with customers of the highest international standing, capable of weathering adverse contract prices over what could be a few years, ERA had the essential business ingredients for successful project finance.

Reasonably pessimistic estimates of its net operating cash flow (sales income less cash operating costs) provided reasonable cover, quarter by quarter, for principal loan repayments plus what then seemed pessimistic estimates of interest costs. (How perceptions change!).

Doubts were still aired over prices of uranium to be expected in about 1985-1986. Clouds hung over what the banks broadly classify as the political risk (national and international, including inflationary expectations, etc.).

On these two aspects, ERA provided as much objective documentary and other information as could be mustered. (One must always employ a devil's advocate and listen closely to him.) The banks also made their own investigations and were comforted to find that their results reasonably reflected ERA's analysis.

COMBINED EX-IM BANK/PROJECT FINANCING
LED TO COMPLICATIONS

Two complications in the project financing package needed special efforts to resolve. Of the total financing:

1. US$140 million was to be provided by Japanese customers, backed by a consortium of Japanese banks, including the Japanese Export-Import Bank.

2. US$250 million plus the Rehabilitation Guarantee, was to come from a consortium of banks organised by Schroders and Continental Illinois.

The ERA finance negotiator had prior experience of each type of borrowing but neither he nor any of the bankers in either group had previously encountered a dual Japanese/Western project loan sharing the one security without other recourse. Typically, as in ERA's case, the purpose of such a Japanese Ex-Im Bank loan is to support a foreign resource project from which Japan has contracted to purchase a substantial part of the output. As a vital condition of such loans, the Japanese purchasers deduct from each purchase invoice and pay to the Japanese lenders, an amount estimated to meet the financing costs, or at least the principal repayments, as they fall due. In this dual loan situation, the Japanese lenders would thus receive priority in payment over the Western banks, who had no such customer relationship. In the extreme, the latter might receive no payments until the Japanese were paid in full.

In such a circumstance, would you, as a Western banker, lend any part of the project finance, non-recourse except to such cash flow? You wouldn't; they wouldn't either.

There was always the possibility that a payment of interest or repayment of principal to the lenders could not be met in full when due from the project's then available cash. Such failure would probably arise in respect of the Western loans well before it occurred for the Japanese loans.

The Western banks' approach to the problem was, as normal, to write in specific deferment provisions, with equally specific pre-conditions. Such pre-conditions would include tests to give reasonable assurance at the time that any commitments so deferred could be met later. Payments so deferred would be postponed to beyond the original final repayment date, so extending the loan.

The Japanese tradition and approach seemed to be that should such circumstances occur, business commonsense and good faith would prevail. (Actually, behind all the rules, this is the Western approach also.) But the Japanese, as ERA read it, could not

1. agree to provisions whereby a payment due in respect of an Ex-Im Bank loan could be excused or the loan extended, and/or

2. agree to writing in advance precise rules which they felt were unlikely to meet the interests of the parties in the actual circumstances of the time.

Without common provisions in the two loan agreements, guaranteeing as far as practicable a common approach if trouble arose, neither group would have acceptable security, nor would ERA. A solution involving either group lending the full amount was not practicable.

The combined customer-lender-equity-holder position was basic to the Japanese consortium, but for a 10% shareholding they could not take over the whole project financing. To the Schroder/Continental consortium, it was also basic that the Japanese continue to hold this combined position in the project; basic both to the security of a large part of the cash flow from sales and to containment of their exposure, already including the whole of the Rehabilitation Guarantee. So neither would subscribe the total loan required.

These problems were raised early in the financial negotiations and were the last debt financing difficulties to be solved, not the least due to both banking groups, for good reason, having diametrically opposed positions. The solution must remain confidential. It was peculiar to the project, but, as projects that involve joint funding through the Japanese Ex-Im Bank and international commercial banks become more commonplace, so will solutions.

So, in August 1980, ERA had final, or nearly final, drafts of all the agreements with, amongst others,

1. the Federal Government,

2. the customers,

3. each of the lead lenders,

4. each of Peko and EZ, for its interests in the project,

5. each of Peko and EZ for a loan of A$30 million, subordinated to the project lenders' loans, but only from quarter to quarter and to the maintenance of certain ratios in the project lenders' favor, and

6. each of the overseas customer-shareholders.

RAISING THE EQUITY FUNDS

It was now time to flesh out understandings with the equity underwriters, those patient people who, with worried frowns had closely inspected the faces of the ERA negotiators at each meeting.

The equity underwriters, a consortium comprising Australian stockbrokers Ord Minett, Potter Partners, J.B. Were & Son, and Meares and Philips, had been first briefed in late 1979 after Peko made its offer to acquire the Government's interest in the project. For the next nine months or so the underwriting broking group were both interested spectators and direct participants in financing ERA. Spectators because there was little they could do to directly influence the negotiations with overseas customers for uranium sales or with bankers for debt finance, but direct participants since their presence and capabilities were essential ingredients for satisfactory completion of the whole project.

Well before the final agreements with all partners had been signed in September 1980, ERA with its advisers had begun preparation of the draft prospectus. With excellent co-operation from the Corporate Affairs Commission the final prospectus was ready almost contemporaneously with the final agreements.

The A$57.5 million raised was a record for a new Australian public float while the prospectus was without doubt the most comprehensive issued in Australia to that date. Arguably it had more input from the legal profession than any previously issued prospectus. The total equity capital of A$405.9 million was raised in A$1.00 par value shares (see table 1) as follows:

TABLE 1

	Shares Million	%
Peko	125.0	30.8
EZ	125.0	30.8
Overseas Participants	98.4	24.2
Public	57.5	14.2
	405.9	100.0

In addition, 4.1 million shares were held in reserve and later allotted to an overseas utility which, like the other overseas participants, entered into a long-term contract to purchase uranium concentrates. To give effect to the Government's desire to see the highest possible level of Australian ownership in the Ranger project, public shareholdings were restricted to those who had an address in Australia and who complied with the definition of "Persons within Australia."

To ensure as wide a spread of public shareholdings as possible, lead brokers to the issue were appointed in all States and in the Northern Territory. Evidence of the wide distribution is given by the statistical picture as at September 1, 1981, some nine months after public listing, as shown in Table 2.

THE EQUITY UNDERWRITING RISKS

The size of the public float, A$57.5 million, was nearly double that of any previous new equity raising in Australia. At a time of high interest rates the fact that the first dividend payment was several years away was a stock market negative.

The underwriters had also to assess the impact on public sentiment of the sometimes adverse media publicity associated with the Fox Ranger Uranium Enquiry, and with attendant antinuclear and conservationist opposition. Uranium is a controversial and sensitive stock market commodity in Australia. Furthermore, spot uranium prices had been in a downtrend for some time and such circumstance makes it more difficult to raise funds from the public who prefer to invest when commodity prices are rising.

There was uncertainty over the likely trend of exchange and interest rates, both

TABLE 2

Shareholding	Number of Shareholders	%	Number of Shares	%
1 - 1,000	31,979	86.1	13,358,955	23.3
1,001 - 5,000	4,436	12.0	10,541,903	18.3
5,000 -10,000	391	1.1	3,003,900	5.2
10,001 and over	311	0.8	30,595,242	53.2
	37,117	100.0	57,500,000	100.0

*equal to 14.02% of the issued capital, but excludes Peko and EZ vendor shares.

being major elements in the overall cost of operation. Apart from the normal financing risks, there was an additional political risk. A Federal election had been called for October 18, 1980 while subscription lists for the public share issue opened on October 15, and were to remain open until November 11, 1980. Public opinion polls at the time indicated a Labor Party victory was possible. Labor policy, as expressed in the prospectus, was opposed to the mining, processing and export of uranium and called for a moratorium on mining and treatment and repudiation of the commitments of non-Labor Governments. Some trade union interests were also strongly opposed to uranium mining.

Another factor which needed careful assessment was the fact that the Australian stock market, as measured by the All Ordinaries Index, had advanced by 120% from its 1978 low and some quarters believed a downward correction was possible, if not likely.

On the other hand, the reputation and technical standing of both Peko and EZ was a major advantage, as was the fact that these two participants had agreed to take up their vendor shares on a deferred dividend basis.

In the event, the public equity flotation was heavily over-subscribed and subscription lists closed early, thus enabling the last connecting link on the Ranger financing to be firmly bolted into place. One the stock market the A$1.00 shares on listing opened around A$2.40 and traded mostly between A$2.40 and A$2.00 during the following eight weeks. Subsequently most of the trading range has been between A$1.40 and A$1.80

CONCLUSION

Ranger is arguably the best uranium mine in the world. It could stand on its own feet in any market and prosper, given its head. It is notable because

1. All the customers who were offered uranium contracted to buy.

2. All but two of some thirty major world banks joined in the loans. One accepted for a smaller amount, one refused. The shortfall was placed within two days.

3. All overseas customers who were offered shares accepted.

4. The local flotation, the largest to that time, was heavily over-subscribed.

5. The project came in on time and to budget.

6. The plant has continued to produce and deliver to budget and as contracted, and costs, except for interest and exchange rates, have been within budget.

7. In addition to the project loans and the Peko/EZ loans, and without giving security, ERA has subsequently been able to borrow other moneys.

The fact that the Ranger uranium mine was immediately profitable from the commencement of operations is perhaps the exception rather than the rule in Australian mining history. In the initial nine months of operation to 30 June 1982, U_3O_8 production was 2,333 tonnes, while sales revenue generated was A$146 million. Pre-tax profit was A$45.5 million after interest of A$48.0 million and depreciation and amortisation charges of $16.6 million (See Table 5).

Energy Resources of Australia Ltd has ranked consistently within the top 15 stock exchange listed resource companies as measured by market capitalisation--evidence that the project was engineered, constructed and financed with professionalism.

TABLE 3

COMPARISON WITH INFORMATION MEMORANDUM
(DATED 29 SEPTEMBER, 1980) AND
BALANCE SHEET AT 30 JUNE, 1982

	A$ Million Projected	Actual
Assets		
Fixed Assets	402.8	377.6
Less: Depreciation	15.8	11.2
Net Fixed Assets	387.0	366.3
Debtors	17.3	59.2
Finished Product Stocks	11.0	13.9
Inventories/Spares at Cost	11.0	8.6
Ore and Work in Progress	-	5.1
Cash and Authorised Investments	26.7	26.1
Current Assets	66.0	112.8
Ranger Rights and Company Formation Expenses	408.7	407.0
Less: Amortization	9.3	5.4
	399.4	401.6
Deferred Expenditure	-	6.0
Total Assets	852.4	886.7
Liabilities and Shareholders Equity		
Issued Share Capital	410.0	410.0
Retained Earnings	14.7	24.5
Shareholders' Funds	424.7	434.5
Bank Overdraft	-	3.1
Creditors	4.5	24.7
Provision for Dividends	16.7	13.4
Current Liabilities	21.3	41.3
Loans	398.6	403.3
Provision for Deferred Income Tax less Future Tax Benefit	7.8	7.7
Total Liabilities and Shareholders' Equity	852.4	886.7

TABLE 4

COMPARISON WITH INFORMATION MEMORANDUM DATED 29 SEPTEMBER, 1980
CASH FLOW

Year Ended 30 June, 1982

	A$ Million Projected	Actual
Sources		
Gross Revenue	134.5	146.0
Less: Operating Costs	32.6	26.1
Royalties	8.0	8.0
Cash Flow from Operations	93.8	111.9
Return on Authorised Investments	0.2	1.7
Eurodollar Loan	25.5	76.6
Japanese Loan	20.9	23.2
Vendor Loans	6.5	6.6
	146.9	219.9
Application		
Capital Expenditure	35.9	45.9
Financing Fees	0.8	0.1
Interest on Loans	51.2	48.0
Eurodollar Loan Repayments	6.8	6.7
Japanese Loan Repayments	3.8	3.8
Vendor Loans Repayments	2.3	2.2
Rehabilitation Trust Fund Payments	2.7	2.9
Deposits	-	-
Increase in Working Capital		
Inventories/Spares	-	2.4
Finished Product Stocks	11.0	13.9
Debtors	17.3	57.3
Creditors	(4.5)	23.7
Increase (Decrease) in Cash and Authorised Investments	19.7	12.9
	146.9	219.9
Cash and Authorised Investments		
Beginning of Year	7.0	10.0
End of Year	26.7	22.9

TABLE 5

COMPARISON WITH INFORMATION MEMORANDUM DATED 29 SEPTEMBER, 1980
PROFIT AND LOSS

Year Ended 30 June, 1982

	A$ Million Projected	Actual
Gross Revenue	134.5	146.0
Less: Operating Costs	(32.6)	(26.1)
Royalties	(8.0)	(8.0)
Operating Profit	93.8	111.9
Interest and Financing Fees	27.0	48.1
Rehabilitation Provisions	2.7	2.9
	64.1	60.9
Plus: Net of Tax Investment Return	0.2	1.7
Less: Amortization of Ranger Rights	(9.3)	(5.8)
Book Depreciation	(15.8)	(11.2)
Profit Before Tax	39.3	45.6
Less: Provision for Future Tax	7.8	7.7
Profit After Tax	31.4	37.9

APPENDIX 1

HISTORY OF THE RANGER URANIUM PROJECT

1969	October	Peko and EZ discovered Ranger uranium deposits.
1971	Last quarter	Ranger judged viable, negotiations for right to mine and for sales contracts begin.
1972	Mid-year	Some sales contracts secured and approved.
1973	April	Large sales contracts and main finance secured but not approved.
1974	October	Lodge Agreement. Joint Venture established for Federal Government, Peko and EZ to mine and export uranium.
1975	July	Ranger Uranium Inquiry (Mr. Justice Fox) instituted by Federal Government.
	October	Memorandum of Understanding amplified Lodge Agreement.
1976	October	First Fox Inquiry report.
1977	May	Second Fox Inquiry report.
	August	Government announced Ranger Joint Venture to proceed.
1979	January	Joint Venture agreement with Government signed, construction commenced.
	August	Government announced its intention to sell its interest.
	December	Peko's offer to buy Government's interest on behalf of ERA accepted.
1980	February	EZ decided to assign its Ranger interest to ERA.
	September quarter	All Contracts signed.
	Last quarter	ERA floated to public.
1981	November	Project passed bank completion tests.

APPENDIX 2

PLANNED PROJECT COSTS AND FINANCING

Costs	A$ Million	Financing	A$ Million
Ranger Property Rights	$407	Equity Capital (Peko and EZ $250, Customer Shareholders $98.5, Australian Public $57.5)	$406
Capital Costs	336	Project Finance	350
Interest	67	Vendor Loans	73
Current Assets, Intangibles	43	Current Liabilities	24
	$853		$853
		Rehabilitation Guarantee	US$55

APPENDIX 3

EXTRACTS ADAPTED FROM PROSPECTUS
(1 October 1980)

FINANCING OF CONSTRUCTION AND ACQUISITION

Financing Arrangements

A summary of ERA's estimated cash requirements until Completion (1) of the Project is as follows:

	A$ Million
Purchase Price payable to the Commonwealth	125
Cash element of consideration for Pekos and EZs* (excluding Pekos/EZs Project Costs)	32
Development and Construction Costs (including Pekos/EZs Project Costs and AAEC**Project Costs) 336	
Working Capital 9	345
Financing, establishment, and Underwriters' Selling Expenses	6
Interest	44
Total	552

The cash requirements are intended to be satisfied as follow:

Peko/EZ Vendor Loans	60
Project Loans***(assumed rate at the time A$1 = US$1.1573)	337
Overseas Participants ("B" and "C" shares)	98
Public ("A" shares)	57
Total	552

* Pekos and EZs in this context means wholly owned subsidiaries of Peko-Wallsend Ltd and EZ Industries Ltd.
** AAEC = Australian Atomic Energy Commission (Government).
*** US$250 million from bank consortium, US$140 million from Japan, JAURD.

(1) Completion: in substance, is the stage at which Commissioning (see below) has taken place and when, in addition, certain financial tests required by the Consortium banks have been met by ERA; the security furnished to the Commonwealth for rehabilitation is certified by ERA to be adequate; and the Project Loans or Vendor Loans are not in danger of recall by reason of breach or other event designated in the Loan Agreements. (Over a six-week period commencing mid-September 1981, the mine and plant successfully passed the banks' Completion Test. Source: 1982 ERA Annual Report.)

The Project Loans are available to meet the following costs up to a maximum amount of the Loans (US$390 million) as follows:

a) AAEC Project Costs and AAEC Interest Charges,
b) Pekos/EZs Project Costs,
c) costs (other than royalties, compensation and like payments) incurred prior to Commissioning or as retention moneys, in relation to constructing, installing, equipping, completing, testing and commissioning the Project Facilities and in relation to mining, storing and processing Ranger uranium ore and Other Mineral Products,
d) costs of administration of those activities after 12 September 1980,
e) capitalising interest on the Project Loans.

In the case of the bank consortium's loan, lead by Schroder and Continental Illinois, not more than US$220 million may be drawn for items (a) through (d), and not more than 15% of advances for item (e). No such restrictions apply to the Japanese loan for US$140 million advanced by Japan Australia Uranium Resources Development Company (JAURD).

The Parent Companies' Undertakings require Peko and EZ to fund by way of Subordinated Debt cost overruns on construction and amounts, if any, by which conversion of the Project Loans to Australian currency realise less than A$337 million, if such amounts are required for construction, except in the circumstances mentioned in the Undertakings section below.

The Project Loans

ERA (Canberra) Ltd has negotiated the two Project Loans totalling up to US$390 million (presently approximately A$337 million) which are each repayable over a period of approximately eight years after Commissioning.

One Project Loan, for an amount of up to US$250 million was provided by a consortium of Australian and overseas banks managed by Schroder and Continental Illinois ("the Consortium"). It bears interest at margins above the London Inter-Bank Offer rate for three-month Eurodollar deposits. Both Project Loans are secured by first-ranking debenture stock issued by ERA (Canberra) Ltd, and repayment of principal and interest has been guaranteed by ERA.

The obligations of the lenders to provide funds under each Project Loan are not absolute but are subject to typical conditions of project financing, including force majeure. The Project Loan agreements incorporate certain tests which have to be satisfied before dividends can be declared. Subject to ERA's obligations in respect of principal and interest repayments, and to the observance of certain financial ratios, the loan agreements with the Vendors and with their holding companies do not restrict the payment of dividends.

The Vendor Loans

Peko-Wallsend (Canberra) Ltd and EZs have lent to ERA (Canberra) Ltd, Vendor Loans of A$30 million each (total A$60 million). The Vendor Loans are secured by second-ranking debenture stock issued by ERA (Canberra) Ltd and repayment of principal and interest has been guaranteed by ERA. Subject to priority of the Project Loans in the quarterly repayment of principal and payment of interest, the Vendor Loans are repayable over the same period as the Project Loans and bear interest at the rate of 12.75% per annum.

Security Documentation

This debenture stock and guarantees are secured by fixed and floating charges over the whole of the undertaking and uncalled capital of each of ERA and ERA (Canberra) Ltd (the borrowing vehicle).

Undertakings Provided by Peko and EZ

By the Parent Companies' Undertaking, Peko and EZ gave undertakings to the Project Lenders to the effect:

a) that ERA (Canberra) Ltd and ERA will perform their respective obligations (other than obligations to pay money and obligations to observe certain financial ratios) in the Project Loan Agreements. ERA (Canberra) Ltd is a wholly owned subsidiary of ERA formed for the usual reasons as the borrower under the Project Loans and the Vendor Loans;

b) that ERA will perform its obligations (other than obligations to pay money) under the various agreements with the government for the mine, stockpile repayment, and as may be required under other documents and loans connected with the Project; and

c) that to the extent that funds available to ERA for the purposes of the Project fall short of the funds required to enable ERA to cause Commissioning (as defined) of the Project (except as mentioned below) the shortfall will be provided by them to ERA in equal shares by way of Subordinated Debt (being Cost Overrun Loans),

to the intent that, subject to force majeure (as defined) Commissioning shall take place as planned. For Commissioning to occur, a construction threshold negotiated with the Government had been satisfied; the treatment plant, tailings dam and infrastructure has been certified to be practically completed; over specified periods, the mine and the treatment plant have met certain criteria of production; and certain rehabilitation work has been performed.

The exception to the undertaking in (c) above relates to the case where the shortfall is caused by a Consortium bank being relieved from participation in advances by reason of it becoming unlawful for such bank to make further advances, or to maintain advances that it has already made. If such an event should occur the Project Loan agreement with the Consortium contains provisions requiring Schroder and Continental Illinois to use all reasonable efforts to procure another bank or financial institution to take the place of the withdrawing bank, and ERA (Canberra) Ltd is also given rights to nominate a bank (at the same or a different interest rate).

Apart from the exceptional cases of force majeure and Consortium bank illegality, the parent companies of Peko and EZ are required to fund any deficiency in funds required to achieve Commissioning, including those resulting from cost overruns or exchange losses on conversion of the Project Loans to Australian currency.

Further, Peko and EZ undertake to the Project Lenders that their shares in ERA will not be sold, transferred, mortgaged or charged (except by floating charge) prior to Commissioning.

The Vendor Loans are to be repaid over a period of approximately eight years after Commissioning.

SHAREHOLDINGS

The two subsidiaries of Peko and EZ, here denoted as Pekos and EZs, received for their interests and commitments:

1. The allotment to each of them of 124,999,997 fully paid A$1 "A" Class shares, and

2. The payment to each of them in cash of A$35,939,495.70 (out of which Pekos and EZs have each provided Vendor Loans of A$30 million to ERA (Canberra) Ltd.)

In addition, Pekos and EZs were reimbursed for their proportions of Preliminary and Issue Expenses, the precise amount of which to be certified by ERA's auditors.

Special terms attach to 37,500,000 of the "A" Class shares allotted to each of Pekos and EZs, namely that such shares do not rank for dividend until ERA resolves, for the first time, to pay out of profits of a financial year dividends of not less than 12.5 cents per share on the whole of the issued capital of the Company, including such 75 million shares.

The "A" Class shares allotted to Pekos and EZs shall not be sold, assigned or transferred (otherwise than between Vendors or to Peko-Wallsend Ltd. or EZ Industries Ltd. or pursuant to a floating charge) until the later of 12 months after the shares offered to the public are granted official quotation or 12 months from the date of allotment of the shares in question.

The Overseas Participants agreed to take up a total of 98,400,000 "B" and "C" Class shares of A$1 each at par for cash, amounting to 24.2% of the Company's issued capital, and lodged applications for those shares supported by confirmed irrevocable letters of credit or guarantees as required by the Shareholders Agreement.

The Overseas Participants are divided into two groups, namely:

German Participants - "B" Class Shares

Australian subsidiaries of the following Germany organisations agreed to take up 57,400,000 "B" Class ordinary shares as follows:

Urangesellschaft mbh	16,400,000
Rheinische Braunkohlenwerke AG	25,625,000
Saarberg-Interplan Uran GmbH	15,375,000
	57,400,000

In early 1981 (after the float), the remaining 1% of ERA (4.1 million "B" Class shares) was taken up by OKG, a major Swedish utility.

Japanese Participant - "C" Class Shares

The Japanese participant, Japan Australia Uranium Resources Development Company Limited ("JAURD"), agreed to take up 41 million "C" Class ordinary shares.

The shares in JAURD are owned in the following proportions:

The Kansai Electric Power Company, Inc	50%
Kyushu Electric Power Co, Inc	25%
Shikoku Electric Power Co, Inc	15%
C. Itoh & Co., Ltd.	10%
	100%

Special Provisions Relating to "B" and "C" Class Shares

Limitations, restrictions or special rights attaching to or affecting the "B" and "B" Class shares are to be found in ERA's Articles of Association and in the shareholders agreement. A summary of the material provisions is set out below:

1. The "B" shares and "C" shares rank pari passu with the "A" shares, except in respect of the 75 million shares issued to Pekos and EZs which are subject to limitation on dividends, and as outlined above.

2. Any or all of the "B" and "C" shares may, at the request of their respective holders, be converted into "A" shares after 30 June 1984. Automatic conversion will also occur if the number of shares of the particular class becomes less than 2.5% of the issued ordinary capital.

3. "B" and "C" shares will not be listed for quotation on Australian Associated Stock Exchanges but, in the event of their conversion to "A" shares, then ERA will seek listing of the shares so converted.

4. Subject to limited exceptions:
   a) until after 30 June 1984 the "B" and "C" shareholders may not sell or assign their shares; and
   b) after 30 June 1984, if the shares are not converted to "A" shares, the consent of Pekos and EZs will be required to the assignment of any such shares to any person, provided that Pekos and EZs shall not unreasonably withhold such consent.

If the "B" and "C" Class shares are converted to "A" Class, each of Pekos and EZs have certain rights of first refusal.

5. The parties to the Shareholders Agreement have agreed to support a dividend policy of ERA with the aim that, so far as is prudent having regard to the contractual commitments and viability of ERA, it will distribute at least 35% of its after-tax profits for the year ending 30 June 1982, and at least 75% of its after-tax profits for each subsequent financial year.

6. Various other provisions cover voting, committees, and dilution (see Prospectus).

Through this shareholding structure, ERA achieved its objective of 75% Australian ownership (an Australian government guideline for uranium mines) while at the same time creating a special relationship with some of its principal customers through their taking up equity in ERA.

MARKETING OF URANIUM CONCENTRATES

Contained U_3O_8 for delivery shown in 000's of pounds avoirdupois
(tonnes in brackets)

Importing Country	1982(a)	1983	1984	1985	1986	1987 to 1992	1993 to 1996	Totals over 15 yr. period (1981/82 to 1996)	Percentage of Totals
Japan	2956 (1341)	1800 (816)	2100 (952)	2100 (952)	2300 (1043)	2000pa (907)	2000pa (907)	31,256	37.1%
West Germany	2100 (952)	2240 (1016)	2800 (1270)	2800 (1270)	2800 (1270)	2800pa (1270)	2800pa (1270)	40740	
Korea	-	500 (227)	500 (227)	500 (227)	500 (227)	500pa (227)	-	5000	
USA	500 (227)	500 (227)	500 (227)	500 (227)	500 (227)	500pa(b) (227)		5000	
Australia (Government stockpile replacement)	1000 (454)	1000 (454)	228 (103)	-	-	-	-	2228	
TOTAL	6556 (2974)	6040 (2740)	6128 (2779)	5900 (2676)	6100 (2766)	5800pa (2630)	4800pa (2177)	84224(e) (38,204)	
Total expressed as a percentage of estimated annual production of 6,612,000 lbs (3,000 tonnes).	99%(d)	91%	93%	89%	92%	88%	73%	85%	
Sales under negotiation	50 (23)	50 (23)	50 (23)	300 (136)	300 (136)	300pa (136)	300pa(c) (136)	3150 (1429)	
Total assuming negotiations are successful expressed as a percentage of initial annual production rate	100%(d)	92%	93%	94%	97%	92%	77%	88%	

(a) includes 1981 (b) to 1990 only plus 500,000 lb, subject to agreement on delivery schedule.
(c) to 1994 only (d) counting 1981 and 1982 as one year for this purpose.
(e) includes 500,000 lb as shown under note (b).

Although there is a high degree of uncertainty involved in the forecast of energy demand twenty years into the future, the Directors believe that in the longer term there will be an increase in demand for uranium as a fuel for power plant generation. The most recent joint report on uranium production and demand of the OECD Nuclear Energy Agency and the International Atomic Energy Agency concludes that, before the end of the century, there will be a need for additional sources of production beyond that from known resources.

MARKETING PERCENTAGES

Sales already negotiated by ERA or by Pekos and EZs (the benefit of which was acquired by ERA) and stockpile replacements account for some 85% of ERA's initial planned production of 3000 tonnes per annum and are summaried in the following table:

Importer	15 yr.total 000 lb.U_3O_8	Percentage of Contracts	Approx. Percentage of Capacity p.a.	Percentage Equity
Japan	31,256	37.1%	30%	10%
West Germany	40,740	48.4	42	14%
Korea	5,000	5.9	7.5	0
USA	5,000	5.9	7.5	0
Australian Government*	2,228	2.6	-	0
Totals	84,224	100.0%	87%	24%

* Government stockpile replacement.

12.8

FINANCING A GOVERNMENT-OWNED INDUSTRIAL MINERAL COMPANY

DOUGLAS A. KARVONEN

Potash Corporation of Saskatchewan
Senior Vice President, Corporate Development

INTRODUCTION

Although the economy in the Province of Saskatchewan has historically been agriculturally oriented, a major source of wealth has been realized through natural resources such as petroleum and natural gas, coal and uranium and since the 1960's - potash. Potash or potassium chloride, in the form in which it is presently produced in Saskatchewan, is used primarily as fertilizer, a source of potassium, one of three essential elements required for healthy plant growth, with about 5% of production sold to the chemical industry for use in such products as soap and liquid detergent, glass and ceramics, textiles and some pharmaceuticals. Saskatchewan is estimated to have approximately 40% of known world potash reserves and its ten active mines account for some 25% of annual world production.

POTASH CORPORATION OF SASKATCHEWAN

In an effort to guarantee a fair and lasting economic benefit to the Province of Saskatchewan, the Potash Corporation of Saskatchewan (PCS) was created on February 4, 1975 through an Order-in-Council under the Crown Corporation's Act and later on April 1, 1976, the Potash Corporation Act provided for its continuation. PCS was given the mandate to acquire ownership of a significant portion of the potash mining capacity in Saskatchewan and to establish an organization capable of operating the facilities, marketing the product, and conducting all of the other activities necessary for a successful corporation competing in the world potash industry.

The corporation is the largest potash producing and selling organization in the Western World - or third largest in the world. (See also Figure 1). PCS through its wholly-owned subsidiary, PCS Mining, has acquired a whole or partial interest in five geographically distinct producing properties or divisions, representing today more than 40% of Saskatchewan's potash producing capacity. The acquisition of each division was followed by a program designed to exploit and maximize the particular opportunities at each minesite for increased production.

To that end, PCS Mining has taken an active lead in expanding to meet future demand and has implemented expansion programs at all properties except for its Esterhazy Division where it has a long-term processing agreement with International Minerals & Chemical Corporation (Canada) Limited.

Through its wholly-owned subsidiaries, PCS Mining, PCS Sales and PCS Transport, this provincial crown corporation has the ability to operate as a producer and marketer of Saskatchewan's potash. In certain instances PCS, the parent company, assists in the financing of various projects undertaken by the subsidiaries and in other cases the subsidiary involved arranges its own financing.

POTASH RESERVES

As a background to the specific issue of financing, it is helpful to describe briefly the nature of the deposit which supports the industry - our collateral, if you like. The potash beds underlying Saskatchewan are sedimentary rock deposits formed by the gradual evaporation of an inland sea during the Middle Devonian period roughly 400 million years ago. Now known as the Prairie Evaporite, this formation stretches from North Dakota and Northeast Montana through Southern Saskatchewan into North-Central Alberta. These beds lie at mining depths ranging from 950 - 1000m (3,000 ft) below the surface, where conventional mining methods can be used, and to 1,650m (5,000 ft), where solution mining must be

---

Based on a paper presented at the SME/AIME Annual Meeting, Atlanta, March 1983.

practised. Because of the depth of potash in North Dakota for instance, technical and economic problems have prevented mining even using the solution technique currently employed at one mine in Saskatchewan.

In addition to variations in depth, the economics of mining are governed by the quality of the orebody. The Prairie Evaporite itself contains three potash-bearing members: the lowest or Esterhazy member, the middle or Belle Plains member, and the upper or Patience Lake member. (See Figure 2). As far as conventional mining is concerned, the distinction between these members is important since the processing of ore is facilitiated by the larger crystal size and lower clay content such as that found in the Esterhazy member compared to that of the Patience Lake member closer to Saskatoon. Since Saskatchewan has almost unlimited reserves of high-grade ore, the province's present advantage of lower production costs will likely continue in the future while those producers with lower grades of ore are rapidly using up their reserves and incurring increasingly higher production costs. The 40% of the world's known reserves of potash located in Saskatchewan is sufficient to last at least 2,000 years at the present rate of consumption.

POTASH MARKETS

Saskatchewan is currently the largest exporter of potash in the world, with approximately 95% of production sold outside Canada. The largest customer is still the United States, primarily in the northern midwest states where transportation costs allow us to compete favourably with New Mexico's Carlsbad mines and offshore imports from Europe and Israel.

Offshore, some of our major market areas are among the Northwest Pacific Rim nations including Japan, China, South Korea and Taiwan. (See Figure 3). In Latin America, Brazil is the largest consumer. Although several new markets have opened up offshore, China, India and Brazil are considered to have the greatest growth potential.

Demand. Overall world demand for potash is expected to grow at an average annual compounded rate of approximately 3% during the 1980's compared with 4.5% during the 1970's. Despite the overall growth in demand, the potash market is cyclical in nature and is affected by many factors such as worldwide economic conditions as well as the predominant influence of the agricultural environment, thus making the cycles very difficult to predict.

In the long-term, future potash demand must be viewed in the context of world food and population balance. (Table 1). Less than 30% of the world's population lives in countries whose food supplies are considered adequate, and these same countries are the world's main consumers of fertilizers. Before developing nations can make the best use of fertilizers, they must be assisted in educating their agricultural sector in its use, as well as in developing adequate transportation systems, distribution networks and financing arrangements consistent with the needs of their economies. So while we must assist in developing the demand in these countries for our product, we must also be prepared to meet this demand by expanding the supply.

PCS FINANCING REQUIREMENTS

For the financing requirements in the Corporation, there have been two primary periods of activity during its brief history where capital requirements have been the greatest. The first period was that of acquisition which took place between 1976 and 1978 and the second was the expansion period which started in a relatively small way upon the acquisition of the first operation in 1976 and has increased in scope up to the present time.

The original investment decision was made in each case, largely based upon discounting the cash-flow stream that could be expected from the potash operation. An independent consulting firm was commissioned to evaluate the properties and analyze the expected cash flow from the potash mines that were being considered for acquisition.

The acquisition process was funded primarily from the Provincial Heritage Fund which was established in April of 1978. Substantial amounts of cash had been accumulated by the province into this fund through a taxation process which was levied on the oil industry. A total of approximately C$530 million was paid out for the 4.8 million tonnes KCl in annual capacity in the five operating properties in which PCS Mining acquired 100% ownership or partial operating interest. Of this amount, C$418 million was funded through equity investment which was received from the Provincial Heritage Fund. The balance, approximately C$112 million, was largely financed through a bond issue incurred by the Province of Saskatchewan and transferred to the Corporation under the same terms for principal and interest as stated on the bonds together with a small amount of term financing from a Canadian chartered bank. There were also C$20 million in promissory notes which were due in 1981 payable to the previous owners of one of the mines.

PCS then made either an equity investment in PCS Mining to permit the acquisition of the properties or lent the money at certain interest rates to PCS Mining in order to complete the acquisition. At the end of June 1978, the Corporation had a debt/equity ratio of 25/75.

Acquisitions. The mandate given to PCS was to obtain about 50% of the productive capacity of the industry in the province. Approximately 38% of that was obtained through acquisitions. The

CASE STUDIES

Figure 1

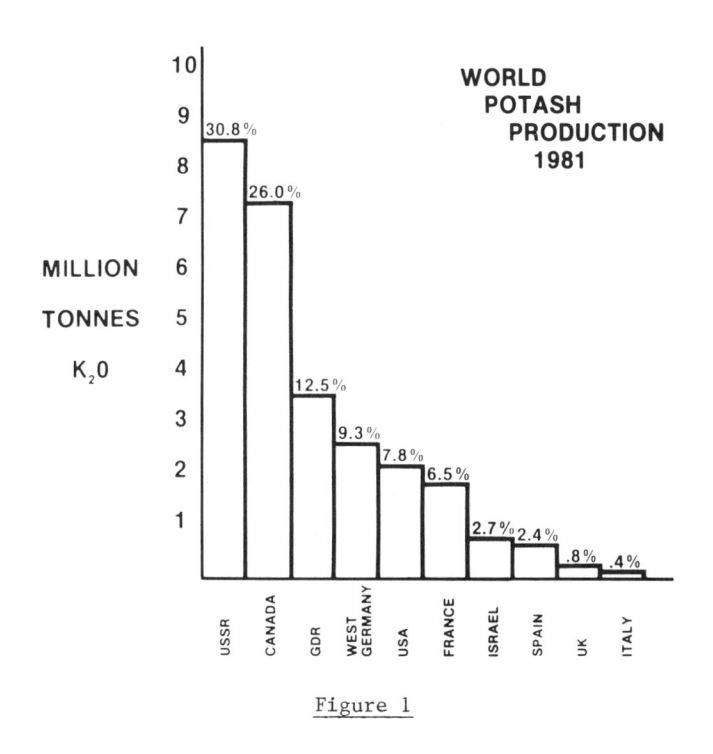

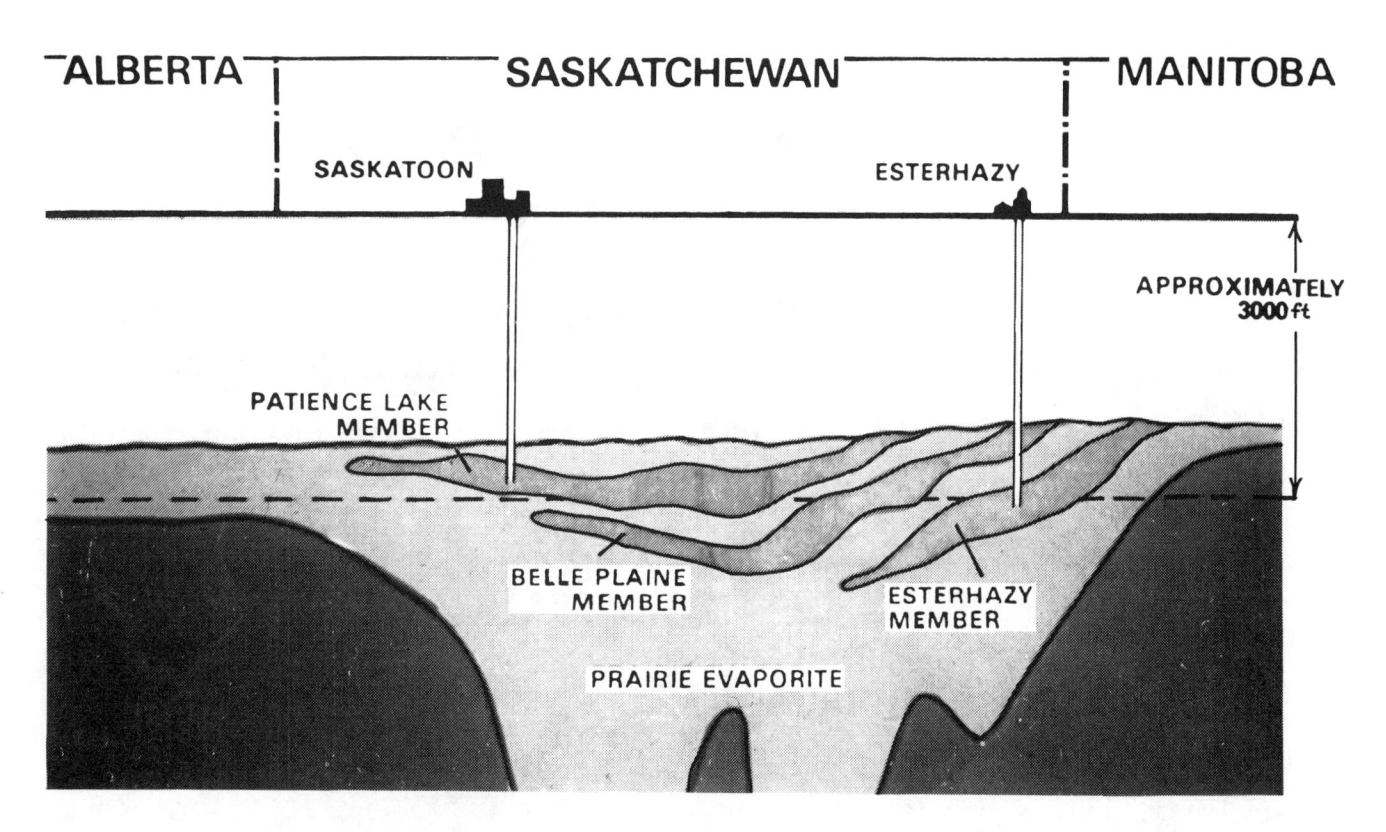

Figure 2

```
                    WORLD
                    POTASH
                    DEMAND
                    1981
```

MILLION TONNES K$_2$O

East Europe: 36.8%
West Europe: 23.7%
North America: 22.5%
Asia: 10.2%
Latin America: 4.8%
Africa: 1.2%
Oceania: 0.8%

Figure 3

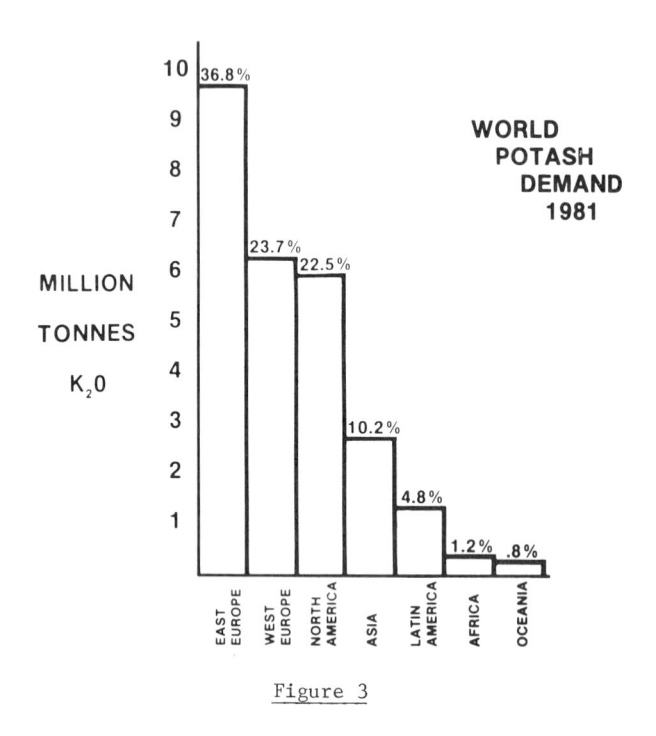

WORLD POPULATION PROJECTION
(billions of persons)

	1975	1980	1985	1990
World Total	4.09	4.47	4.89	5.34
Developed	1.13	1.17	1.22	1.25
Less Developed	2.96	3.30	3.67	4.09

Source: Chase Econometrics

Table 1

balance was to be achieved through expansion of the acquired operations and through the construction of a new mine. Immediately following the acquisition of the first three producing properties (Cory, Rocanville and Lanigan), PCS Mining began an expansion program which increased production capability at each operation between 15% and 25%. A similar program followed at its Allan Division, which is a 60% owned in a joint venture with Kidd Creek Mines (formerly Texasgulf Potash) following acquisition in 1978.

Expansions. These first Phase I expansions consisted in the main of streamlining and debottlenecking to increase capacity. This program added 0.8 million mtpy KC1 to PCS Mining capacity or nearly the equivalent in total of one average sized mine. During this period which extended primarily between 1978 and 1980 the Corporation was fortunate in that it was experiencing very strong market conditions and was able to fund this program of about C$65 million from internally generated cash flow. Gross income in 1980 was C$392 million with net income at C$167 million.

With the strength in world markets and a projected shortfall in supply the corporation prepared corporate plans which were to include several major expansions over a 10-year period. PCS Mining has many options for increasing capacity including adding new refineries and mining machines at most sites, to fully utilize existing shaft capacity, or Phase II expansions, as well as building totally new mine and refinery complexes complete with one or two new shafts. The first of these major Phase II expansions was implemented at the Rocanville Division between 1978 and 1981 and increased capacity from 1.3 to 1.9 million mtpy KC1 at a cost of approximately C$80 million. Once again the favourable market conditions allowed this program to be funded from internally generated funds although two long-term loans from the Heritage Fund totalling C$33.4 million were obtained during this period.

In 1980 the corporation began a second Phase II expansion at the Lanigan Division. This program, originally to cost C$475 million, was to be completed in 1983 and was designed to add approximately 2.1 million mtpy to the 1.1 million mtpy of capacity presently installed. From 1981 the potash industry experiences a downturn as a result of the worldwide recession which has affected the economies of many offshore markets and increased feed grain surpluses in the United States where 70% of Canadian Potash is sold. Because of the weaker market conditions the corporation slowed its expansion plans. The Lanigan Phase II expansion will now be completed in 1985 and other expansion plans originally slated for the mid 1980's, including that for a totally new mine and refinery have been deferred.

Debt Financing. Given the size of the Lanigan project and the less favourable market for potash, the corporation has proceeded to arrange additional debt financing. This program began in 1982 and has largely been accomplished utilizing the borrowing capabilities of the Province of Saskatchewan which has access to the world capital markets. The 1982 program was assumed in U.S. dollars, which results in foreign exchange exposure. However, the corporation has a natural hedge to some extent since most of its revenue stream is in U.S. dollars, while its payables and financial statements are expressed in Canadian dollars. As the cost of capital in the U.S. and Eurodollar markets has been less expensive than the Canadian market this has assisted in reducing the borrowing cost. These long-term borrowings by the Province were lent to the corporation by Order-in-Council and carry the same terms as stated on the bonds. In turn, PCS made certain loans to PCS Mining.

On a smaller scale the corporation through its subsidiaries has constructed warehouse facilities in the United States at a cost of about $5 million each to support the transportation and distribution network. The corporation has been able to make use of tax-exempt financing in the U.S. on two potash centers that have been built in Iowa and Indiana. This tax-exempt financing means basically that the revenue stream to the investor is non-taxable and this helps to subsidize the cost of funds and provides approximately a 2 - 3% interest-rate advantage over the taxable bond market.

Also on a small scale the corporation has in a few instances used leasing as a method of financing. This has been used to finance special pieces of equipment such as mining machines.

Short-term facilities. With respect to short-term arrangements the corporation has established credit facilities for bridge financing. The fact that the Bank Act has allowed many foreign banks to become chartered banks in Canada has increased competition in the banking industry and provides opportunities to compare rates and increases the availability of funds. The corporation has lines of credit with major Canadian Banks. A short term line of credit is also available from the Saskatchewan Department of Finance through access to funds raised from Government of Saskatchewan commercial paper. PCS has also used banker's acceptances as a short-term financing instrument.

FINANCIAL ANALYSIS

As stated earlier the original investment decisions to acquire the mines were based on discounted cash-flow analysis. Decisions to spend additional funds on expansion or other major projects are also made on the same basis. The internal rate of return is calculated as a part of a feasibility study which includes the preparation of capital cost estimates, project construction schedules, operating cost estimates, production and sales estimates and price forecasts. A cutoff rate of return has been established for purposes of making a decision to proceed with a project. This cutoff rate of return is generally higher for entirely new projects than it is for expansions at existing sites where risk is lower. A sensitivity analysis to major variables is also prepared and the payout period of the project determined. A decision to proceed is made after reviewing the analyses.

Implementation. After approval is received from the PCS Board of Directors, requests for capital and any borrowing programs proceeded to the Crown Investments Corporation, another Saskatchewan Government-owned corporation which was established primarily to act as a holding company for all of Saskatchewan's crown corporations to coordinate their capital requirements. If the capital request can be met from within the entire financing program for crown corporations, then ultimately an Order-in-Council will authorise the transfer of funds from a borrowing by the Department of Finance. If the request is for a short-term credit facility then an Order-in-Council will authorize the use of that facility and establish its limits.

SUMMARY

The Corporation has access to world capital markets through the Provincial Department of Finance. Where advantageous, it will look to a number of additional normal corporate financing techniques such as leasing and tax-exempt bonds for its long-term requirements. Common corporate methods of bank lines of credit and banker's acceptances are used for short-term financing. The corporation is consistently investigating additional sources of funds and where these sources can be shown to reduce costs they become incorporated into the program. The Act under which PCS is incorporated provides the flexibility to use whatever methods are reasonable and the Government of Saskatchewan has always been ready to make adjustments to policy if such actions can be seen to make good commercial sense.

CASE STUDIES

APPENDIX

Potash Corporation of Saskatchewan

Consolidated Statement of Financial Position
as at December 31
Assets

	(Thousands)	
	1983	**1982**
Current Assets		
Short-term investments, at cost	$ 38,924	$ 554
Accounts receivable	49,648	45,909
Due from the Province of Saskatchewan	3,939	23,377
Product inventory	18,781	44,561
Raw ore inventory	3,223	3,223
Materials and supplies inventory	19,752	20,478
Deposits and prepayments	10,382	19,739
	144,649	157,841
Property, plant and equipment (Note 4)	989,667	907,062
Deferred charges and investment (Note 5)	27,234	24,259
	$1,161,550	$1,089,162

Liabilities

Current Liabilities

	1983	1982
Bank indebtedness	$ 29,736	$ 34,724
Province of Saskatchewan advances	—	30,000
Accounts payable and accrued charges	63,140	44,971
Current portion of long-term debt	30,000	25,000
Current obligations under capital leases	1,021	—
Dividend payable	62,000	50,000
	185,897	184,695
Long-term debt (Note 6)	336,001	221,787
Obligations under capital leases (Note 7)	36,979	—
	558,877	406,482

Equity

Saskatchewan Heritage Fund	418,554	418,554
Retained earnings	184,119	264,126
	602,673	682,680
	$1,161,550	$1,089,162

Potash Corporation of Saskatchewan

Consolidated Statement of Income and Retained Earnings
For the Years Ended December 31

	(Thousands)	
	1983	**1982**
Income		
Sales	$215,737	$188,117
Interest and other	1,965	4,010
	217,702	192,127
Costs and expenses		
Operating costs other than those shown below	147,721	113,729
Depreciation and amortization	34,076	22,256
Selling, distribution and administrative	37,611	49,461
Interest expense (Note 8)	16,301	6,074
	235,709	191,520
Net income (loss)	(18,007)	607
Retained earnings, beginning of year	264,126	313,519
	246,119	314,126
Dividend (Note 9)	62,000	50,000
Retained earnings, end of year	$184,119	$264,126

Consolidated Statement of Changes in Financial Position
For the Years Ended December 31

	(Thousands)	
	1983	**1982**
Sources of cash		
Net income (loss)	$(18,007)	$ 607
Add expenses not requiring cash outlays		
— depreciation and amortization	28,334	24,100
	10,327	24,707
Decrease (increase) in		
— accounts receivable	15,699	25,359
— inventories	26,506	(23,435)
— other current assets	9,357	(6,803)
Increase (decrease) in		
— accounts payable	18,169	(11,976)
— other current liabilities	1,021	—
Cash generated internally	81,079	7,852
Increase in long-term debt	181,193	158,762
Total sources of cash	262,272	166,614
Application of cash		
Capital expenditures (net)	108,730	140,454
Repayment of long-term debt	25,000	25,000
Dividends paid	50,000	50,000
Additions to deferred charges	5,184	4,766
	188,914	220,220
Net decrease (increase) in short term borrowing	$ 73,358	$(53,606)

Potash Corporation of Saskatchewan

Notes to Consolidated Financial Statements
December 31, 1983

1. Significant Accounting Policies

The Corporation's accounting policies are in accordance with generally accepted accounting principles. The following policies are considered to be significant:

Principles of consolidation

The consolidated financial statements include the accounts of the Corporation and the wholly owned subsidiaries:

— Potash Corporation of Saskatchewan Sales Limited (PCS Sales)
  — PCS Sales (Iowa) Inc.
  — PCS Sales (Indiana) Inc.

— Potash Corporation of Saskatchewan Mining Limited (PCS Mining)

— Potash Corporation of Saskatchewan Transport Limited (PCS Transport)

PCS Transport has 64% interest in Canpotex Bulk Terminals Limited which is accounted for by the equity method.

Inventories

Inventories of product and raw ore are valued at the lower of average cost or net realizable value. Inventory of materials and supplies is valued at the lower of average cost or replacement cost.

Property, plant and equipment

Property, plant and equipment are carried at cost. Costs of additions, betterments and renewals are capitalized. Maintenance and repair expenditures which do not improve or extend productive life are expensed as incurred. Interest associated with major capital projects is capitalized during the construction period at the Corporation's average cost of long-term debt, and is included in the total cost of the project.

Depreciation and amortization

Depreciation and amortization are provided on a basis and at rates calculated to amortize the cost of the property, plant and equipment over their estimated useful lives, as set out below:

	Depreciation or Amortization Basis	Approximate Rate
Head Office		
Furniture and equipment	Straight-line	10%
Automotive equipment	Straight-line	20%
Leasehold improvements	Straight-line	Term of lease plus renewals
Mines		
All mine assets	Units of production	
Offsite Storage Facilities		
Buildings	Straight-line	4%
Equipment	Straight-line	8%
Roads and sites	Straight-line	4%

Research and development costs

Research costs are charged to expense as incurred.
Development costs are capitalized until a project is determined not to be technically feasible and no future benefit to the Corporation exists, at which time the costs are expensed. Development costs which are capitalized are included in Deferred Charges and will be amortized over the estimated useful life of the project to which they relate.

Foreign exchange translation

Foreign operating transactions are translated to Canadian dollars at the average exchange rate for each month. Foreign current assets and liabilities are translated on the basis of year-end exchange rates.
Non-current assets are translated at the exchange rate prevailing at the time of the transaction.
Non-current liabilities are translated

on the basis of year-end exchange rates. Unrealized exchange gains and losses relating to translation are deferred and amortized over the remaining life of the liability.

The Corporation has a foreign exchange hedging policy whereby it has entered into forward contracts based on a percentage of its forecasted United States dollar receipts.

Leases

Leases entered into are classified as either capital or operating leases. Leases that transfer substantially all of the benefits and risks of ownership of property to the Corporation are accounted for as capital leases. At the time a capital lease is entered into, an asset is recorded together with its related long term obligation to reflect the purchase and financing. Equipment acquired under capital leases is being depreciated on the same basis as described above. Gains or losses resulting from sale-leaseback transactions are deducted from the cost of equipment under capital lease and amortized on the same basis as the asset. Rental payments under operating leases are expensed as incurred.

2. Status of Potash Corporation of Saskatchewan

The Corporation was established by Order-In-Council on February 4, 1975 and continued under the Potash Corporation of Saskatchewan Act, proclaimed on April 1, 1976.

The Corporation's accounts are included in the annual consolidated financial statements of Crown Investments Corporation of Saskatchewan.

3. Income Taxes

By virtue of the fact that the Corporation is wholly owned by the Province of Saskatchewan, neither it not its wholly owned subsidiaries are subject to Federal or Provincial income taxes in Canada. Certain income and assets of its subsidiaries are subject to Federal and State taxes in the United States.

4. Property, Plant and Equipment

	(thousands)	
	1983	**1982**
Land	$ 19,465	$ 19,415
Buildings and improvements	128,181	120,044
Machinery and equipment	582,882	579,387
Plant under construction	313,080	247,191
Deferred exploration and development costs	69,907	68,265
Equipment under capital lease	29,517	—
	$1,143,032	$1,034,302
Less: Accumulated depreciation and amortization	153,365	127,240
	$ 989,667	$ 907,062

5. Deferred Charges and Investment

	(thousands)	
	1983	**1982**
Deferred charges, net of amortization		
Bond issue costs	$ 7,533	$ 2,954
Systems development and other	712	2,107
Preproduction costs	2,826	3,380
Bredenbury development	15,198	14,853
	26,269	23,294
Investment, at equity	965	965
	$27,234	$24,259

6. Long-Term Debt

	(thousands)	
	1983	**1982**
Province of Saskatchewan		
— Bearer bonds issued and sold by the Province and the proceeds transferred to the Corporation. Matures on January 28, 1984 with interest at 8 3/8% per annum, payable semi-annually.	$ 25,000	$ 50,000
— Bearer bonds issued and sold by the Province and the proceeds transferred to the Corporation. Matures on March 15, 1989 with interest at 16% per annum, payable annually. ($75,000 U.S.)	93,330	92,205

— Bearer bonds issued and sold by the Province and the proceeds transferred to the Corporation. Matures on August 15, 1992 with interest at 15% per annum, payable annually. ($50,000 U.S.)   62,220   61,470

— Bearer bonds issued and sold by the Province and the proceeds transferred to the Corporation. Matures March 15, 1990 with interest at 10.75% per annum, payable semi-annually. ($50,000 U.S.)   62,220   —

— Bearer bonds issued and sold by the Province and the proceeds transferred to the Corporation. Matures December 21, 1988 with interest at 10.75% per annum, payable semi-annually.   80,000   —

Saskatchewan Heritage Fund
— Loan capital matures in four equal annual installments on June 30, 1984 to 1987 inclusive with interest at 9.3% per annum, payable semi-annually.   20,000   20,000

— Loan capital matures $5,000 on December 31, 1988 and $8,400 on December 31, 1989 with interest at 13% per annum, payable semi-annually.   13,400   13,400

Webster County, Iowa
— Non-negotiable promissory note matures December 1, 1996 with interest at 13% per annum, payable semi-annually. ($3,900 U.S.)   4,853   4,795

City of Portage, Indiana
— Non-negotiable promissory note matures in eight equal annual installments on March 1, 1985 to 1992 inclusive with interest at 12.25% per annum, payable quarterly, secured by assets of PCS Sales (Indiana), Inc. ($4,000 U.S.)   4,978   4,917

	$366,001	$246,787
Less amounts due within one year	30,000	25,000
	$336,001	$221,787

Principal payments due on long-term debt for the next five years are as follows:

(thousands)
1984 - $30,000
1985 - $ 5,622
1986 - $ 5,622
1987 - $ 5,622
1988 - $85,662

7. Lease Commitments

The Corporation has long-term lease agreements, the longest of which expires in 1998. Future minimum lease payments under these capital and operating leases will be approximately as follows:

	(thousands)	
	Operating	**Capital**
Years ending December 31		
1984	$14,095	$ 5,337
1985	10,928	5,337
1986	8,180	5,337
1987	6,974	5,337
1988	6,509	5,337
Subsequent years	9,066	53,363
		80,048
Less — amount representing interest		42,048
Present value of minimum capital lease payments		38,000
Current portion		1,021
Long-term portion		$36,979

8. Interest Expense

	(thousands)	
	1983	**1982**
Interest on		
— Long-term debt	$36,914	$24,265
— Short-term debt	8,678	1,752
— Obligations under capital leases	189	—
	45,781	26,017
Less capitalized interest	29,480	19,943
	$16,301	$ 6,074

9. Dividend

During the year, a decision was made that the Corporation transfer a portion of retained earnings to the

Crown Investments Corporation of Saskatchewan. This transfer is recorded as payable at December 31, 1983 and is referred to as a dividend.

10. Research and Development Costs

Research and development costs charged to expense in 1983 amounted to $2,918,000 (1982 — $5,279,000).

11. Contractual Obligations and Commitments

The Corporation is expanding its Lanigan division at an estimated cost of $475,000,000 of which approximately $338,000,000 has been expended to date. The Corporation has previously announced its intention to develop a new mine near Bredenbury, subject to environmental approvals and on-going economic evaluations. No significant development is anticipated on the Bredenbury project in 1984.

PCS Sales is a participant in Canpotex Limited, an offshore marketing company. Should any operating losses be incurred by Canpotex Limited, they will be proportionately shared by its participants. PCS Sales is also a guarantor of the performance by Canpotex Limited for certain railcar leases.

PCS Sales and PCS Mining have guaranteed a mortgage loan not to exceed $4,300,000 for warehousing facilities in Danville, Illinois.

12. Directors and Officers

During the year ended December 31, 1983, twelve (1982 — twelve) persons served on the Corporation's Board of Directors and twenty-two (1982 — twenty-three) as senior officers. The aggregate 1983 remuneration to directors amounted to $52,000 (1982 — $47,000) and the aggregate 1983 remuneration to senior officers was $1,704,000 (1982 — $1,601,000).

13. Related Party Transactions

Included in these consolidated financial statements are expense amounts resulting from routine operating transactions conducted at prevailing market prices with various Saskatchewan Crown controlled departments, agencies and corporations with which the Corporation is related. Account balances resulting from these transactions are included in the Consolidated Statement of Financial Position and are settled on normal trade terms.

Other amounts due to or from related parties and the terms of settlement are described separately in the consolidated financial statements and the notes thereto.

14. Comparative Figures

Certain of the prior year's comparative figures have been reclassified to conform to the current year's presentation.

12.9

MINERA REAL DE ANGELES - A CASE STUDY

by

Mr. Donald J. Worth

Canadian Imperial Bank of Commerce
Vice-President, National Accounts Division
Toronto, Canada

INTRODUCTION

Minera Real de Angeles makes a very interesting case study in more ways than one. To begin with, it represents the first time a bulk silver deposit has been put into production in Mexico. Second, the feasibility and financing stage spanned a period when the silver price ranged from $4 to $40 per ounce. Third, there was a major devaluation of the peso and a financial crisis in Mexico at completion of the project. Fourth, the composition of the partnership (a major Canadian mining company, a major Mexican mining company and the host government's own mining entity) is worthy of note.

This paper gives considerable background on the project and how it came to fruition. The political risk aspects and how the major provider of funds perceived Mexican risk at the time of decision to participate and afterwards are highlighted.

LOCATION AND HISTORY

The Real de Angeles deposit is located at an elevation of 2300 meters in Central Mexico in the State of Zacatecas (Figure 1). The closest village is Noria de Angeles, three kilometers to the north. The closest cities are Aguascalientes, Zacatecas and San Luis Potosi. The first-mentioned is one hour to the southwest by automobile over good roads.

Exploitation of the vein outcrops probably began towards the end of the sixteenth century by the Spanish. Evidence of past mining activity is shown by numerous shafts, manways, dumps and ruins. Also, there were remnants of two rudimentary concentrating mills now obliterated by pit development and in the nearby town of Noria de Angeles the ruins of a small smelter still exist.

Compania Minera Gamma, an affiliate of a major international mining company, investigated Real de Angeles from 1969 to 1971. Its evaluation program included 20 drill holes and comprehensive geological engineering and metallurgical studies. The results were not encouraging and the option on the property was dropped. In late August 1973 geologists representing Placer Mexicana which was later renamed Explomin were advised of the availability of the property and liked the prospects sufficiently to make an option payment soon due and to agree to outright purchase by year-end.

A series of 34 holes were drilled by Explomin and a preliminary study indicated that the prospect had mine-making potential if the silver price was right.

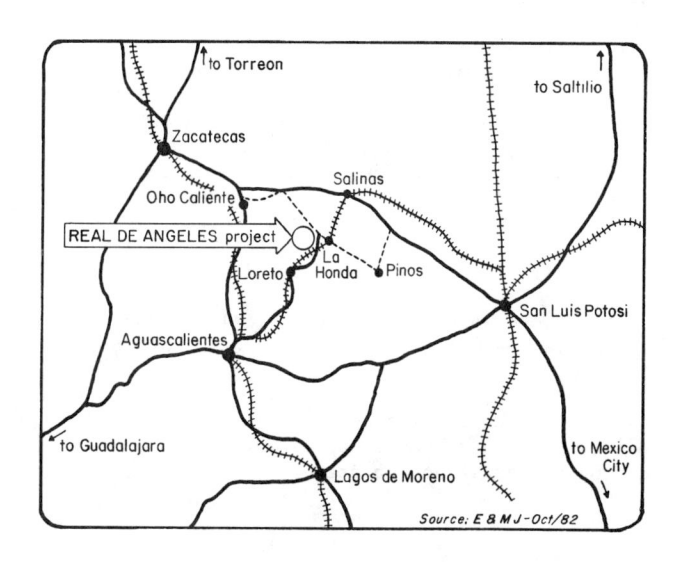

FIGURE 1. LOCATION MAP - REAL DE ANGELES PROJECT

DESCRIPTION OF THE PROJECT

Real de Angeles is best described as a large, low grade open pit silver mine because at today's prices approximately 70% of the realizable revenue is derived from that metal. One must also keep in mind that the operation produces two concentrates - a lead concentrate with high silver content and a zinc concentrate with cadmium and silver content.

At the commencement of production in mid-1982 mineable reserves were stated as 54 million t (metric tons) averaging 76 g/t (grams per metric ton) silver, 1.1% lead, 0.97% zinc and minor amounts of cadmium. The pit has an overall stripping ratio of 2.3 to 1 and a final slope of 40 degrees. Cut-off grade for silver is in the order of 30 g/t.

The pit is standard shovel-truck operation deploying 9 yard shovels and 75 ton trucks.

The mill is designed to handle 10,000 t/day (metric tons per day) or 3.5 million t/yr (metric ton per year). Annual output of the mill is approximately 50,000 t of lead concentrate grading 60% lead and containing over 4,000 g/t silver and 50,000 t of zinc concentrate grading approximately 50% zinc and 0.8% cadmium. The zinc concentrate is exported to European smelters.

The ore is not an easy one to treat. On the upper benches, it has been oxidized and this has resulted in lower silver recovery. Also, the ore is quite hard (work index 20) and it requires fine grinding to liberate the silver-bearing minerals. At present, silver recovery is in the neighbourhood of 80%.

Process water is obtained from a series of drilled wells a few kilometers from the mill. Power is supplied by Comision Federal de Electricidad via a 110 kilowatt line from a substation approximately 16 km east of the mine.

A large number of the mine employees were drawn from nearby villages. Skilled tradesmen and supervisory staff, however, were recruited in other parts of Mexico and a townsite was built at Loreto to provide housing and recreational facilities. There are approximately 480 salaried and hourly employees on the payroll.

First ore was put through the mill in June 1982 and by year-end the plant was operating at design tonnage.

The project experienced a slight overrun due mainly to higher than expected interest charges during the construction stage.

EARLY FINANCING ATTEMPTS

Placer first registered to do business in Mexico as a mining company in 1969 under the name of Placer Mexicana de C.V. As per Mexican law governing foreign investment in mining, the common shares created were divided into Series A and Series B shares. The Series A shares, comprising 51% of the ownership were transferred to Financiera Bancomer S.A., a wholly-owned subsidiary of Banco de Comercio, Mexico's second largest bank at that time. In March 1973, the Government of Mexico through Comision de Fomento Minero was invited to participate in Placer Mexicana and the name of the company was changed to Explomin S.A. de C.V.

After some negotiation, the ownership of Explomin became:

Banco de Comercio (via Financiera Bancomer S.A.)	33%
Comision de Fomento Minero	33%
Placer Development Limited	34%

In early 1975 Placer prepared a preliminary valuation study on the Real de Angeles property and by mid-1975 it commenced initial discussions with at least one Canadian bank on how to finance it to production.

At the time the commercial banks active in mine financing looked upon Mexico as an acceptable political risk. The economic risk of the project, however, posed a problem. Silver was selling at $4.50/oz but it was difficult for a bank to assume a base case of much more than $3.75/oz having in mind the historical behaviour of the price (figure 2). At that price the project could not support the level of borrowings sought by the Explomin's shareholders. It was estimated at the time that a 7,500 t/day operation could be put into production for approximately $80 million of which $60 million would be borrowed capital.

After some deliberation, the partners decided to put the project "on hold" in anticipation of better silver prices.

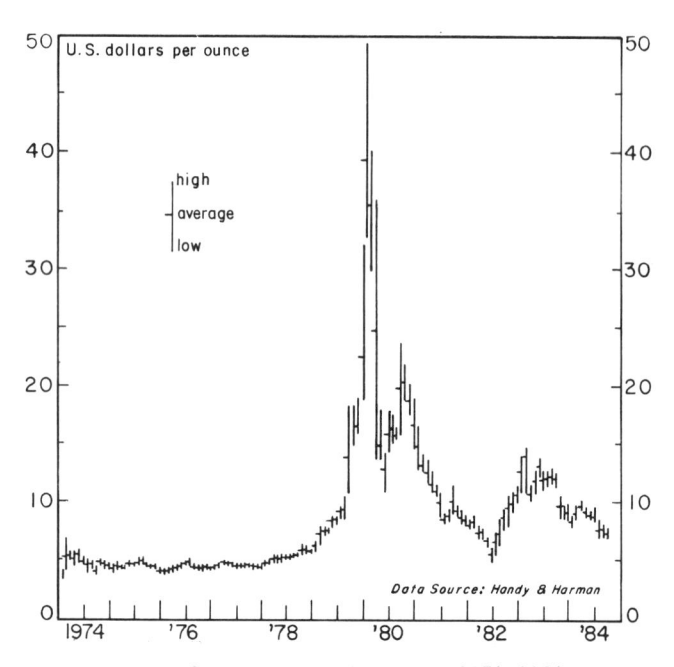

FIGURE 2. PRICE OF SILVER - 1974-1984

REVIVAL OF THE PROJECT AND NEGOTIATIONS WITH THE BANKS

In mid-1979 Banco de Comercio sold its interest in Explomin to Frisco S.A. de C.V., a strong and active Mexican mining company in which that bank had recently acquired a substantially larger holding. As a result, the ownership of Explomin became:

Frisco S.A. de C.V.	33%
Comision de Fomento Minero	33%
Placer Development Limited	34%

The entrance of Frisco brought to the fold several experienced mining people including Ing. Jorge Ordonez who was seconded to Explomin to become its general manager.

During the Autumn of 1979 many large international banks expressed an interest in financing the project. A feasibility study prepared by Placer for Explomin was handed to interested banks in October. The new study was on the basis of a 10,000 t/day or 3.5 million t/yr operation. This, coupled with the rapid inflation underway, had increased the capital cost from the $80 million figure mentioned previously to $150 million. Also, it was assumed that the average interest rate throughout the repayment period would be 13% rather than the 12% figure used in the 1975 study.

On the bright side the silver price had been climbing from approximately $6/oz at the start of 1979 to approximately $15/oz in October. Hence, it was not unreasonable for the base case to be set at $7.50/oz, double that assumed by at least one bank when considering the financing in 1975.

With the comfortable coverage ratios afforded by the higher price assumption, it was no great wonder that several banks came back to the company with financing proposals.

EMERGENCE OF INTERNATIONAL FINANCE CORP. IN THE MRA FINANCING

Before much further discussion ensued between the commercial banks and Explomin it became evident that International Finance Corporation (IFC), an affiliate of the World Bank would have a major role to play in the financing of the project. One obvious advantage was that IFC was immune to withholding tax whereas the commercial banks would have to pay it and perhaps pass on the expense to the borrower.

As enunciated in its annual report, IFC was established in 1956 to supplement the activities of the World Bank (the International Bank for Reconstruction and Development) by providing the type of financing and investment expertise particularly suited to attracting and lending confidence to private sector investors in developing countries.

IFC management felt from the outset that the Real de Angeles project was worthy of support from a social point of view. To begin with, the mine was located in a part of the State of Zacatecas where the land is not particularly suitable for agriculture. Secondly, the mining of silver has been an important factor in the Mexican economy for centuries.

The next step for IFC was to assure itself of the viability of the project. Accordingly, its in-house team commenced an engineering and economic evaluation of the project for presentation to the IFC board.

By mid-March IFC board approval had been obtained and an IFC term sheet was presented to Minera Real de Angeles S.A. de C.V. (MRA), the new name for Explomin. Documentation of the loan agreement commenced in earnest and by mid-June the participating banks had agreed to the draft agreement. In late August the loan agreement was signed. At the time of signing, silver was selling for $16/oz and the project was into the construction stage with the help of a bridging loan provided by one of the participating banks.

LOAN STRUCTURE

The tombstone for the MRA's financing as it appeared in the business press is given in Figure 3.

As one can observe there is a fixed rate portion and a floating rate portion to the loan. In the former, the funds were provided directly by IFC. In the latter, the funds were provided by the commercial banks through the purchase of participation certificates.

The participants had no recourse to IFC either before or after completion of the project. Accordingly, it was up to each bank to make its own assessments of the technical aspects, the silver price assumptions and the country risk.

The first repayment instalment was scheduled for December 15, 1982.

COUNTRY RISK ASPECTS

The term country risk describes the conditions in a foreign country that could cause losses to external private lenders. Among those conditions are the following:

- Confiscation or nationalization of borrowers' assets by the host country.

- Unfavourable and unpredictable changes in the tax laws of the host country.

- Disruption of essential services due to civil unrest.

- Unavailability of foreign exchange due to host government policies.

For some time, all the major international banks have had a country risk evaluation system for categorizing country risk which includes input from field and head office line and staff officers. Arising out of this exercise is a country limit proposal which is presented annually to the credit policy committee of the particular bank. The

> *This announcement appears as a matter of record only.*
>
> # U.S. $110,000,000
>
> Project Financing Term Loan
>
> ## Minera Real de Angeles, S.A. de C.V.
>
> Mexico
>
> ---
>
> ### U.S. $30,000,000
>
> Fixed Rate Funds
>
> Provided by
>
> International Finance Corporation
>
> ---
>
> ### U.S. $80,000,000
>
> Floating Rate IFC Participation Certificates
>
> Managed by
>
> Canadian Imperial Bank of Commerce   Bancomer, S.A.
>
> With Participation by
>
> Canadian Imperial Bank of Commerce   Bancomer, S.A.
> Citibank, N.A.   The First National Bank of Chicago
>
> Bank of British Columbia
>
> September, 1980

FIGURE 3. MRA FINANCING

approved amount establishes a maximum that can be loaned in a specific country. To go beyond that amount requires an upward revision of the limit, which would only be considered after a review of the country's rating or categorization.

During the late seventies, it was the general consensus among the major international banks that Mexico presented an acceptable country risk. The country had experienced a regular and orderly progression of presidents for decades. The rules under which business operated in the country were clear. The external debt although quite high was considered manageable given the economic prospects for the country and the fact that Mexico was a substantial net exporter of oil.

Hence, it was felt at the time of concluding the loan agreement that as long as Real de Angeles itself was a good credit risk, there was no reason not to go ahead with the deal.

The presence of IFC in the loan structure was probably not considered a very important factor by the participating banks in their assessments of the country risk. On the other hand, Placer stated that it took IFC's involvement into consideration in its assessment.

IMPACT OF THE MEXICAN FINANCIAL CRISIS

Since 1975 when financing of the Real de Angeles project was first considered, the Mexican peso has gone through a series of devaluations (Figure 4) which lowered its value in U.S. dollars from 8 cents to its present 2/3 of a cent.

As long as dollars were available to make scheduled interest and loan repayments, project finance bankers were for the most part not overly concerned with devaluation. After all, the mine was paying for wages, power and domestic supplies in pesos but receiving dollars for its output. This happy state of affairs changed when the Mexican liquidity crisis came to a head in late 1982.

On August 23 the Mexican government asked for and received a postponement of scheduled repayments on debt falling due to lending banks in the ensuing 90 days. The country's total foreign debt at that time was estimated at $80 billion of which an estimated $20 billion was to the private sector. In November 1982, the commercial banks agreed to a further 120 days' postponement. At that time, the Government of Mexico stated that it would soon present a more permanent debt rescheduling plan to bankers.

On December 10, the Mexican government sent a telex to all commercial bank lenders proposing a debt rescheduling plan that would allow the postponement of all public sector amortization payments falling due prior to December 31, 1984. In addition, the Mexican government asked the commercial banks for new financing of $5 billion. The total amount to be rescheduled was an estimated $20 billion and involved more than 1,400 commercial banks worldwide.

By December 15 more than 1,000 banks had responded favourably to Mexico's request for the rescheduling and for the $5 billion in new credits. This opened the way for the final approval of an International Monetary Fund (IMF) loan of $4 billion. In order to obtain the IMF loan the Mexican government pledged to hold the line on foreign borrowings, lift most retail price controls, boost taxes and interest rates, restrain wage increases, freeze hiring and cut the budget deficit. The austerity program has not been popular among Mexicans.

In January 1983 the governor of the central bank of Mexico, Banco de Mexico, announced a plan whereby private Mexican companies would deposit pesos in the central bank in the amount of the unpaid interest at a government-controlled exchange rate. The dollar liability would then become an obligation of the Mexican government. Under the new exchange system, the country had three exchange rates. First, there was the floating rate determined by the free market which has been since January 1983 close to 150 pesos to the dollar. Second, there was the controlled rate, kept lower than the floating rate and designated for importing raw materials and making interest payments. This rate opened at 95 pesos to the dollar on December 20, 1982 and has been depreciating by 13 centavos a day since then. Third, there was a special rate which was given to repay creditors who agreed to a stretched out repayment schedule. This rate was set at 70 pesos to the dollar.

An important detail in the rescheduling scheme was that it did not include debt owed to the World Bank or its affiliates. This was good news for the private banks participating in the MRA financing as it meant that they would be repaid as scheduled - provided, of course, that the mine continued to earn as predicted. All of a sudden it became quite comforting from a country risk point of view to be involved in a Mexican loan through IFC.

The devaluation of the peso caused a few uneasy moments. Although MRA had deposited sufficient pesos at the former exchange rate in December 1982, it found itself short of the dollars required to make payments due to the further devaluation.

By mid 1983, however, MRA had brought its schedule of repayments back into a current status and has maintained this situation to date. In fact, it has accumulated a cash surplus for which dividends to shareholders were paid prior to the end of 1984. As at the end of 1984, approximately half of the project debt had been repaid.

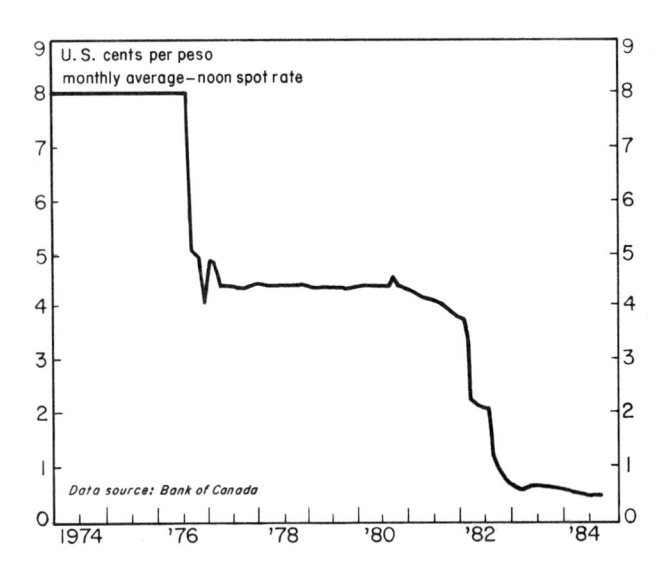

FIGURE 4. VALUE OF THE MEXICAN PESO IN TERMS OF THE U.S. DOLLAR - 1974-1984

SUMMARY

The perfect time for a lender to document a case study such as MRA would be on receipt of the final loan instalment. To wait for this to happen would have meant giving up the opportunity of preparing this article for insertion in this publication.

Notwithstanding a false start, two devaluations of the peso, a major financial crisis in Mexico, some start-up problems and tremendous swings in the silver price, the universe of Minera Real de Angeles continues to unfold as it should.

ACKNOWLEDGEMENT

The author is grateful to IFC as well as the MRA management and shareholders for their encouragement to prepare this paper.

REFERENCE

Friedman, Irving S. - The World Debt Dilemma: Managing Country Risk - 1983

12.10

MERGERS AND ACQUISITIONS IN THE MINING INDUSTRY (BRASCAN/NORANDA)
WITH PARTICULAR EMPHASIS ON THE HEDGING OF FINANCIAL RISK

J. TREVOR EYTON, Q.C.

President and Chief Executive Officer
Brascan Limited
Toronto, Ontario

This paper addresses some of the financial risk hedging mechanisms available to a corporation contemplating entering the mining business at a significant level of investment. The approach which I have adopted is that of a case study relating to the risk hedging aspects of Brascan Limited's acquisition of 46% of Noranda Inc., formerly Noranda Mines Limited, followed by some remarks on options open to mining companies for reducing financial risks.

To appreciate the rationale for Brascan committing over (Can.) $800 million to the natural resource sector, some knowledge of Brascan's parent company, Edper Investments Ltd., is required at the outset. Edper Investments, prior to acquiring a 48% interest in Brascan, controlled Trizec Corporation, one of the world's largest property companies. Edper also was involved in the financial services sector through Continental Bank of Canada and Hees International Corporation. For some considerable time prior to Edper's investment in Brascan, Edper felt a need to provide a better balance to its portfolio of property and financial services. Brascan was rich in cash, relatively unleveraged, and with important existing holdings in natural resources and consumer products. The acquisition of a controlling interest in Brascan provided an opportunity to balance Edper's portfolio of property and financial services with world class acquisitions in natural resources, supplemented by consumer products and financial services.

The end result of Brascan's activities over the last few years is illustrated by the following analysis of the book value of Brascan's assets at the end of December 1983.

	Can.$ Millions	%
Natural Resources	817	48
Consumer Products	654	38
Financial Services	157	9
Other Operations	89	5
	$1,717	100%

Source: Brascan 1983 Annual Report.

The large emphasis on natural resources came about from the fact that for Canadians operating mainly in Canada, natural resources were attractive. We were impressed by Canada's position in the world natural resource industry as illustrated by some of the following statistics:

Commodity	Canadian Estimated Production In 1983 As % of World Total
Newsprint	30
Potash	21
Uranium	18
Molybdenum	17
Zinc	17
Pulp	16
Nickel	16
Saw Logs	11
Silver	8
Copper	7
Lead	6
Gas	5
Oil	3

Source: U.S. Bureau of Mines and Brascan estimates.

In addition, the quality of Canada's reserves looked good to us. The polymetallic nature of many of Canada's orebodies offered a hedge against risk relating to any one metal. Put another way, the by-product metal credits offered a net operating cost advantage with respect to many of the metals mined. Our energy costs also appeared to provide Canada's natural resource industries with an advantage.

We considered Canadian mining expertise as amongst the world's finest, as were the exploration equipment and technology developed in Canada.

In addition, Canada offered a stable political climate and one we knew. Further, we felt the Canadian Government's future role in the industry would, on balance, be benign, partly because the Canadian content of the mining industry was very high in contrast to the petroleum industry. Granted, we could find cheaper sources of copper; however, to do so would involve venturing into regions where country risks were higher.

We were also attracted to the people in the industry who, for the most part, are entrepreneurial individuals deeply committed to the free enterprise ideal.

As investors, we felt we had an opportunity to buy a wide variety of good natural resource assets substantially below replacement costs.

We were not impressed by the trends in labour costs in the industry, but felt sanity would return sooner or later and this generally has occurred already. We also recognized currency movements could undermine Canada's competitiveness, as could indirect subsidies from international monetary organizations.

Having reviewed the various companies in the Canadian natural resource industry, it became obvious only Noranda supplied the broad base of mining, minerals, forest products and petroleum we wanted. Noranda was the world's largest mine producer of zinc, operated one of the world's biggest copper refiners, and through its investments in Fraser Inc., James Maclaren Industries, Northwood Pulp and MacMillan Bloedel was one of the world's largest forest products companies. Noranda was also a significant Canadian producer of natural gas with an excellent exploration record; so much so that the Elmworth gas field, discovered by Noranda's affiliate, Canadian Hunter, has the ability to satisfy about 20% of Canada's gas need.

The scope of Norada's metals operations, in both Canadian and world terms, is illustrated in the following table:

NORANDA GROUP MINE PRODUCTION AS A PERCENTAGE OF ESTIMATED TOTAL 1983 OUTPUT

Metal	Canada	World
Zinc	40.0%	10.1%
Silver	39.1	6.3
Lead	35.4	4.5
Molybdenum	29.8	5.0
Copper	15.7	1.8
Gold	9.6	0.8
Mercury	-	6.6

Source: Brascan estimates.

Brascan's initial purchase of Norada shares was a block of 7.9 million comprising about 10% of Noranda at market prices in October, 1979. This was followed shortly afterwards by open market purchases until a level of 14% was reached after taking into account dilution through the issue of treasury issues. This dilution was brought about by increases in Noranda's capitalization as treasury shares were issued to acquire MacMillan Bloedel and James Maclaren Industries, both forest products, and a further treasury issue to Zinor Holdings. Zinor was formed by Noranda as a defence mechanism. Its assets included Noranda shares amounting to about 24% of the company while its ownership lay in companies controlled by Noranda.

After the formation of Zinor, Brascan continued to accumulate Noranda shares by open market purchases; however, in the summer of 1981 Brascan joined with the Caisse de depot et placement du Quebec, the Province of Quebec pension fund, in forming Brascade Resources Inc. Prior to the formation of Brascade, Caisse had been a significant investor in Noranda. On the formation of Brascade, both Brascan and Caisse exchanged their Noranda shares, representing a combined 28% interest in Noranda, and $600 million in cash, for common shares of Brascade held 70% by Brascan and 30% by Caisse.

Brascade then announced its intention to acquire another 20 million Noranda common shares and 1.8 million convertible preferreds from the public. Before this offer was made an accord was reached with Noranda's board under which Brascade provided $500 million cash to Noranda in return for 12.5 million treasury common shares. Concurrently, Brascade made a modified offer to the public for 10 million Noranda common shares and 1.8 million convertible preferred shares in return for Brascade preferred shares of $311 million and cash $287 million. The coupon on the Brascade preferred shares was set at a level that could be serviced by means of a minimum level of Noranda dividends. Following these transactions, Brascade owned about 39% of Noranda on a fully diluted basis.

To summarize the risk hedging process in the above initiatives:

1. Brascade, through Noranda, had invested in a wide portfolio of natural resources.

2. Most of Brascan's purchases of Noranda common shares had been made at market prices or near market prices for an average cost of about $28 per Noranda common share.

3. As far as Brascan was concerned, $500 million in cash had been transferred from one controlled entity, Brascade, to another, Noranda, thereby strengthening Noranda's balance sheet in anticipation of the impending recession.

4. A significant and substantial partner had been obtained in the formation of Brascade.

5. Most of the additional cash Brascan required in the formation of Brascade was raised in straight preferred shares, thereby maintaining the underlying asset value of Brascan's common shares. Where Brascade's convertible preferred shares had been used, the conversion price reflected underlying asset values.

6. The remaining shares to bring Brascade to the 50% level could be acquired over time and for value. Brascan has since raised its Noranda investment to 46% through purchases at market prices close to book values, and thereby has significantly reduced its average cost per share. These additional purchases were mainly effected through the issue of additional

Brascade convertible preferred shares.

As a result of the accord, Brascade attained representation on the Noranda board proportionate to its interest. As part of the accord, Brascade agreed to limit its investment in Noranda to 50% plus one voting share.

This ownership limitation with respect to overall policy applied to investments in public companies. I should like to comment at this point on some aspects of Brascan's policy with respect to the degree of ownership of its public affiliates.

Brascan prefers to limit its shareholding in its principal operating companies to approximately 50%, notwithstanding traditional investment logic which favours acquiring 100% of high quality, high potential companies. This self-imposed limit results from Brascan's strong belief in the financial, management and risk hedging benefits to be derived from operating companies continuing as widely held public companies.

The public company status of Brascan's operating companies reinforces the autonomy of their managements who derive the major portion of their rewards in the same way as all shareholders, that is through the acquisition of important equity interests in their companies rather than by means of traditional remuneration programs.

Brascan expects its principal operating companies to be fundamentally selfsufficient, with quality managements having the requisite level of expertise. Given this, Brascan confines its participation in these companies by contributing ideas through the CEO or at the board level, seeking consensus on all major matters.

The limitation of Brascan's ownership of companies acquired to the 50% level also provides a risk hedging mechanism to Brascan as the presence of public shareholders reduces the overall risk to Brascan.

Our Noranda investment was achieved at the onset of the most severe recession in the resources industry since the early 1930's. Noranda recorded losses in 1982 and 1983, the first losses in its sixty year history. A recession was considered likely when we made the investment; however, the depth and duration of the recession that followed was more severe than we anticipated. The severity of the recession imposed a discipline on many of Noranda's operating units which has resulted in higher productivity and lower unit costs. In 1984 Noranda returned to profitability lean and trim with greater potential for earnings.

During 1983, Brascade acquired Brascan's interest in Westmin Resources Limited, so that all of Brascan's natural resource operations are now owned by Brascade.

Turning now to some of the various techniques available to mining companies themselves to hedge risk associated with their own capital investment decisions.

Project Finance.

This form of non-recourse financing reduces the risk experienced by common shareholders in that a loan to develop a property is obtained with the property itself as the sole collateral. With project finance the lender is repaid solely from the cash flow of the asset financed. Noranda are contemplating purusing this approach with respect to their Hemlo, Ontario gold properties while Westmin has used this financing method with respect to the H-W mine on Vancouver Island. Our experience has been that an attractive property, combined with hard bargaining with bankers, can result in a very high proportion of development costs being financed by project loans.

Debt Instruments With Coupons Linked to Commodity Prices.

The linking of interest rates on debt instruments to an index of prices of commodities produced can be a useful approach. The formula for calculating interest could be a form of call on the metal by the lender when the metal price increases. The attraction of that kind of instrument to a non-taxable financial institution, such as a Canadian pension fund, can be great because the institution benefits on a pre-tax basis. If an institution owns the common shares of the same mining company, its participation in increasing metal prices is on an after tax basis. If metal prices collapse the lending institution will enjoy a minimum return on its loan. The advantage to the mine is that it pays only a minimum level of interest in bad times. Of course, to the mine the cost of interest paid reduces taxable income, whereas interest paid is received by the institution tax free. All of this means that you have to give away some of the upside benefit of high metal prices; however, you are protected on the downside. This type of structure has conceptual appeal.

I believe there has been a tendency in the mining industry in the past to look at an ore deposit mainly from the mining and metallurgy aspects, whereas, in fact, financing aspects are just as important. The relationship between prime rate and mine profitability is currently inverse (i.e., high prime rate/low metal prices). A situation where interest costs drop in bad times and go up in good has a certain appeal.

An alternative approach is to link the redemption of the principal of a redeemable debt instrument to a certain physical quantity of a commodity. This gives investors a direct metal exposure, and when issued at a time when metal prices are depressed should result in a lower coupon. This form of financing was recently contemplated by Inco Limited.

Gold Financing

Under one approach a developing gold mine borrows say five years gold production repayable over say ten years from a financial institution which holds the gold in any event for reserve requirements. The interest rates on such borrowings are typically low. The gold is sold by the mine to provide development money or working capital. This is a form of financing with a known financing cost; however, the mine can gain or lose depending on future gold prices. Another form of gold financing was adopted by Dome Mines Ltd. recently when they issued common shares with warrants attached to purchase gold at a certain price within a specified time period. Echo Bay Mines Ltd. have also utilized this approach, using preferred shares and gold warrants. Lac Minerals Ltd. have used debentures with gold purchase warrants.

Long Term Take Or Pay Contracts With Customers

For most base metals long term contracts with customers with minimum prices sufficient to cover costs are now difficult to come by; however, it may be possible to negotiate long term smelter terms at attractive rates. The availability of a long term take or pay contract from a financially strong customer may increase the proportion of debt supplied by a lender in a project financing situation and/or reduce the coupon. Noranda recently sold its gas liquids production forward for seven years on such a basis while Westmin recently sold steam coal on a similar basis for the next 40 years. The disadvantage of this approach is that in return for price protection on the downside, you give away some of the upside benefit.

Forward Selling of Metals

A metal marketing division should be able to contribute materially to alleviating financial risk through forward selling of metal. I know I sleep better knowing that a substantial part of next year's production has been sold forward at attractive prices against known costs of production.

Utilizing Partners' Funds

The acquiring of a partner in some form, particularly with respect to large projects, has been an effective method for Brascan and Edper to reduce risk. It is also possible to finance a new mine or expand existing ones by selling publicly traded common shares when the market is favourable; knowing in a later period of poor metal prices you can purchase the shares back at prices lower than your issue price. This is a way of making money on the liabilities side of the balance sheet.

During a period when it is possible to obtain underlying asset values, the sale of a significant interest in a property can reduce exposure to a particular metal or situation. Brascan did this in 1981 when it sold 50% of its Brazilian tin operations to British Petroleum Ltd. for U.S. $50 million.

With respect to exploration, Noranda endeavours to reduce financial risk by having partners undertake the exploration drilling in return for a significant portion of a Noranda property; however, if an economic orebody is found then Noranda can bring the property to production at its own expense in return for a portion of the property that was originally given up during the exploration phase. This is a form of having your cake and eating it.

Similarly, Noranda and Westmin increasingly endeavour to participate in exploration plays where core drilling or other methods of exploration have already indicated the presence of encouraging levels of mineralization rather than engage in wildcatting exploration.

Incidentally, the existence of flow through shares in Canada, whereby the tax writeoffs of exploration accrue to the investor directly, has made it easier to find junior mining company partners to undertake the more risky exploration work.

The Quality of Orebodies

As a result of recent experience it is likely that Noranda and Westmin will only develop orebodies in the lower quartile or so of the world cost league after by product credits. This means orebodies such as Hemlo and the H-W, or good Brunswick or Geco type polymetallic orebodies with significant precious metal or base metal by product values. Because of the favourable country and economic risks associated with such deposits in the U.S. and Canada, very attractive financing can usually be arranged. Noranda and Westmin have found the availability of finance is not a problem. The problem relates to finding viable orebodies in areas of low country risks.

The Reduction of Unit Costs

While it is true that the high technology revolution has reduced the long term growth propsects for many base metals, so too has the high technology explosion provided the mining industry with the means to reduce unit costs and increase productivity and cash flow. This includes improved process computerization, microbiological leaching, etc. While the application of high technology to a high cost mine is unlikely to make it a low cost mine, improved technology can have an important impact on a mine's cash flow.

In the above comments, I have touched on only some of the risk hedging mechanisms available to the mining industry. The creative minds within the mining industry can blend these approaches into the appropriate combination to suit any individual situation. I believe acceptable financing packages can be constructed under most market conditions, provided that a company is prepared to be imaginative and flexible in providing investors with what they require at any particular time. Some of these financing packages may appear to be a return towards barter, but at least under the barter system you knew precisely what your obligations were and with imagination, you can be in a position to meet those obligations in the bad times while prospering in the good!

12.11

EVALUATION OF THE 16-to-1 MINE

AS A CANDIDATE FOR PROJECT FINANCING - A CASE STUDY

Hans W. Schreiber

David W. Neuhaus

Behre Dolbear & Company, Inc., New York, NY

INTRODUCTION

The Sunshine Mining Company's 16-to-1 silver deposit and mine project are located 384 kilometers (240 miles) southeast of Reno and 360 kilometers (225 miles northwest of Las Vegas, near Silver Peak in Esmeralda County, Nevada, at an elevation of about 2,300 meters (7,000 feet).

The deposit was being developed, in 1980, with an anticipated mining and milling rate of 150,000 tons per year (500 tons per day). Mining is by underground, blast-hole open stoping with rubber-tire haul-age out of the mine to a straight cya-nide plant about 5.5 kilometers (3.5 miles) from the mine. The product is a gold-silver dore which may be further refined at Sunshine's new refinery in Kellogg, Idaho.

The cash required for investment was obtained through the sale of $25 million worth of $6\frac{1}{2}\%$ silver-indexed bonds. Some $15.0 million were directed toward development of the 16-to-1 Mine. These bonds, which were underwritten by Drexel Burnham Lambert of Chicago in April of 1980, are redeemable on or after April 15, 1985 or payable at maturity on April 15, 1995 at the greater of either $1,000 or the market price of 50 ounces of silver.

Behre Dolbear acknowledges the kindness of the Sunshine Mining Company (Sunshine) for allowing the presentation of the subject material for this Case Study.

APPLICABILITY AS CASE STUDY

Reasons for Choosing 16-to-1

The reasons for choosing the 16-to-1 Mine as a case study follow:

1. The project is simple and compact.

2. The mine is small -- thus relevant to other contemplated precious metal operaions in the western states.

3. The areas of risk are readily identifiable.

4. A change to project financing was under consideration.

5. A substantial decline in the price of silver was experienced during development.

6. Behre Dolbear undertook one of the first feasibility studies of The 16-to-1 and thus has some familiarity with the project.

Historical Summary

The following history serves as a setting:

1. Late 1920s -- minor "production"(?) via shallow pitting.

2. Mid-1930s -- neighboring Nivlok Mine successfully put into production.

3. Late 1950s (?) -- Callahan Mining Company acquired rights and drilled -- results inconclusive; Callahan turned property back to Arthur Baker, who had originally interested Callahan in 16-to-1 on the basis of its similarity with Nivlok Mine.

The Case Study is based on a presentation of the Minerals Resource Management Committee of SME, Short Course on Project Financing in the Development of Mineral Resources SME-AIME Annual Meetings, Chicago, Illinois and Dallas, Texas, February 1981 and 1982.

4. During 1960s(?) -- Baker reached an agreement with the Mid-Continent Uranium Company (Mid-Continent), who drove an adit at the 2,300-meter (7,000-foot) elevation (upper portal).

5. January 1973 -- Mid-Continent, seeking a partner to commit further exploration/development funds, concluded an agreement with Sunshine.

6. Early 1973 through 1978 -- intermittent work by Sunshine consisting of mapping, drilling and cross-cutting -- dominantly to assess reserves.

7. August through December 1979 -- pulling together of Sunshine work by Behre Dolbear and preliminary feasibility determination (included in presentation -- dated January 1980).

8. April 1980 -- silver-backed bond financing secured through the Sunshine Mine (Drexel Burnham Lambert).

9. June 1980 -- commitment to production approved by Sunshine Directors.

10. September 1980 -- commencement of mine development to production.

11. April 1981 -- 16-to-1 structure intersected by lower portal adit; bulk sample submitted for testing.

12. May 1981 -- Mill construction underway.

13. March 1982 -- Expected commencement of production.

Significance of Example

The 16-to-1 Mine project presents some interesting aspects:

1. Upon completion of the preliminary feasibility study the project did not qualify for project financing despite the apparent attractiveness.

2. Further mine development was required to resolve uncertainties (reserve, processing method) which, in turn, would reduce risks to levels generally palatable for project financing.

3. Further mine development (driving of an adit and drifting along the mineralized zone below the 7,000-foot level) was financed by funds raised through the sale of silver-backed bonds. The silver backing the bonds was that of the developed reserves in the Sunshine Mine.

4. While funds from silver-backed bonds were used to finance the remaining installation (mine equipment, mill, infrastructure, etc.), consideration was being given to refinancing along project financing lines. The silver price current at that time worked against the project. However, the following serves as a strong counter-balance:

   a. High productivity, low operating cost, and highly competitive overall production costs -- (very low downside risk -- very high upside potential, subject to silver price performance).

   b. A simple, straight-forward, non-labor intensive mining and processing flow design.

   c. No development and construction risk (both were essentially completed -- there were no unknowns regarding overruns, unexpected water or ground problems, etc.).

   d. Completion tests were easy to set and implementation begun with the onset of production.

   e. There is universality in the market for silver and gold.

SUMMARY OF TECHNICAL DESCRIPTION

A schematic diagram of the mine is given as Figure 1.

Geology and Ore Reserves

The 16-to-1 ore deposit consists of a wide, steeply-dipping vein system situated in volcanic and sedimentary terrain. Ore grade mineralization occurs in a rather limited vertical zone with a leached cap near the surface and sub-economic mineralization and splitting of veins at depth. Diluted ore reserves are estimated by Sunshine to be 1,096,100 tons, grading 8.38 ounces Ag per ton and 0.03 ounces Au per ton. Percent confidence limits of these projections were estimated using modern geostatistical techniques and the following conclusions were made:

1. Probability of actual mine grade being less than 15% under projected grade = 4%.

2. Probability of actual mined tonnage being less than 15% under projected tonnage = 1%.

Reliability of projections, therefore, seems to be quite good while remaining consistent with the probable accuracy of other planning and cost projections.

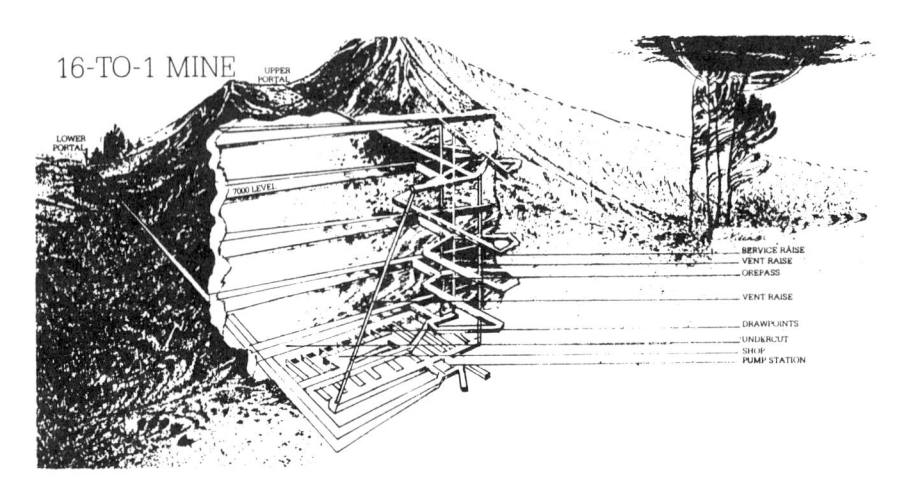

FIGURE 1

ISOMETRIC OF 16-TO-1 MINE DEVELOPMENT

Development and Mining

Due to the nature and characteristics of the vein and surrounding wall rocks and the necessity for a high efficiency type of mining system, large diameter blasthole stoping was chosen as the mining method. Briefly, this method involves drilling and blasting vertical slices of ore between sublevels into the void created by previous blasts, with the broken material being extracted from drawpoints at the bottom of the ore body (see Figure 2). Rubber-tired, diesel-powered equipment will be used for all phases of development and mining. Ore will be hauled to the surface via a 10% ramp by underground haulage trucks. Ore will then be transferred to the mill, located about 5.5 kilometers (3½ miles) away towards Silver Peak, by surface haulage trucks.

Preproduction underground mine development is divided into two parts. One portion, to be done by an outside contractor, will consist of driving a lower level decline ramp to the bottom of the ore body where drawpoint and undercut levels will be established (see Figure 3). Excavation of draw cones, shop and pump station will complete this portion of the work. The second part, to be done concurrently by Sunshine crews, will conist of slashing the present upper level adit, driving an access drift around the ore body and connecting the levels with a series of 15% decline ramps. The establishment of adequate surface facilities such as water supply shops, wastewater treatment facilities, and compressors is also very important and will preceed any underground development work.

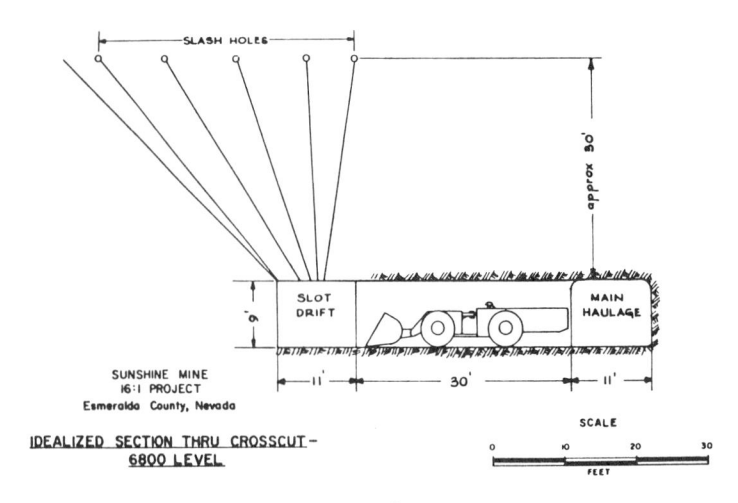

FIGURE 2

CASE STUDIES

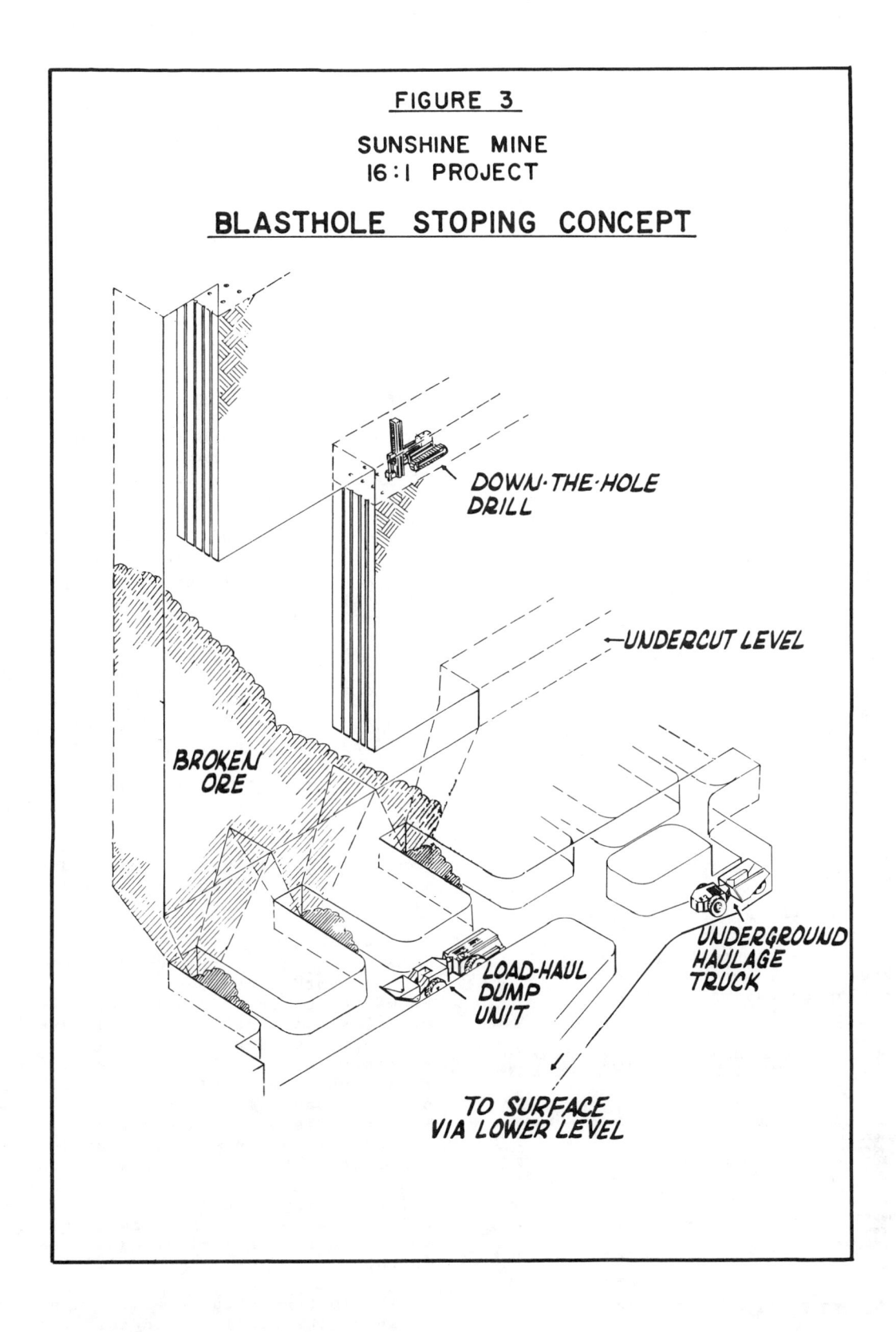

FIGURE 3

SUNSHINE MINE
16:1 PROJECT

BLASTHOLE STOPING CONCEPT

Preproduction development is estimated to take approximately 20 months, with the lower mine development being the critical path.

Milling

Ore will be processed at Sunshine's mill facilities by a straight cyanide process with counter-current decantation and zinc dust precipitation. By this method, a gold-silver doré will be produced at the plant. Other construction work at the millsite will include erection of main office, warehouse, shops, water system, dry, and tailings dam.

Other Development and Construction Work

Construction work, in addition to mine and mill development, will include the installation of a power supply line from Silver Peak to the mill and mine. Housing or trailer sites will also be made available to company employees in Silver Peak through a separate firm specializing in townsite development. Therefore, Sunshine will not be directly responsible for managing or upkeep on several company-owned housing units.

Scheduling

Preproduction development is estimated to take approximately 20 months, with the lower mine development being the critical path.

ESTIMATED FINANCIAL PERFORMANCE

Basis for Project Finance

A summary of the updating by Sunshine of the financial performance estimates by Behre Dolbear is interesting in that it illustrates a not unusual pattern (note the steadily decreasing DCF-ROR).

Item	Behre Dolbear (Dec. 1979)	Sunshine (May 1980)	Sunshine (March 1981)
Silver price, $/oz	15	13	13
Gold price, $/oz	300	500	500
Reserves, tons, '000	862.5	1,096	1,096
Reserves, Ag grade	10.0	8.38	8.38
Reserves, Au grade	0.03	0.03	0.03
Recovery, % Ag	89.4	90.8	89.9
Recovery, % Au	87.0	89.7	89.3
Capital cost, $'000	11,010	17,250	21,037
Operating cost (excluding depreciation), $/ton	23.93	26.80	29.94
DCF-ROR, %, project basis	42.5	38.2	30.2
Break-even Ag price, $/oz*	7.26	7.20	9.27

*Price at which DCF-ROR is reduced to 15 or 20%.

While Behre Dolbear does not know Sunshines assumptions as of February 1982, a guess is ventured that all of the technical and cost parameters remain about the same except, obviously, the DCF-ROR. The DCF-ROR would be significantly reduced, due to the precipitous decrease in silver prices. The DCF-ROR would likely be between 20 and 25% at the average 1981 silver price of $10.50 per ounce.

QUALIFYING FOR PROJECT FINANCING

At the time of completion of the preliminary feasibility study, it became apparent that there were three major areas of risk.

Reserve Risk

The continuity and grade of roughly one-half of the total reserves, located below the 2,230 to 2,270m (6,800 to 6.925 ft) elevation zone, were not definitively established (refer to Figure 4) The adit and drift at 2,300m (7,000 ft) had firmly established the reserves above a 2,270m (6,925 ft) elevation. Reserve tonnages were especially critical because of the contemplated plated mining rate, mining life was only about six years.

Mining Risk

The mining method presumed strong wall rock, allowing for vertical sections ranging from 33 to 82m (100 to 250 ft) to be taken in single slices -- viz blasthole stoping -- without undue dilution, viz not in excess of 10%. The mining method was critical to operating costs, since it was non-labor intensive and required far less working capital than more conventional stoping.

Processing Risk

While both flotation and cyanide were known to achieve adequate recovery, there were insufficient data on the nature of the ore body with depth (especially as regarded possible increase in lead and zinc sulfides) and there had not been bulk testing of any apparently representative sample. Thus a decision of adequate reliability between flotation and cyanidation, or a combination of the two, was not possible.

The driving of the decline, followed by drifting as part of mine development, resolved most of the risks enumerated above, in that:

1. The 16-to-1 structure was encountered at about the predicted location. Drifting along the structure yielded:

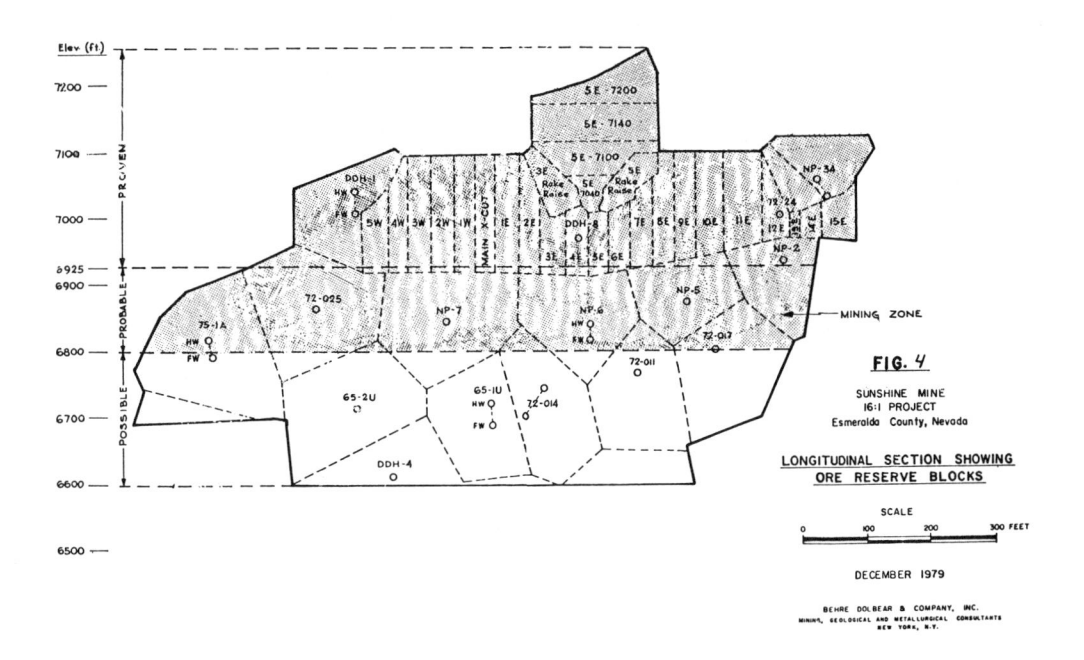

FIG. 4
SUNSHINE MINE
16:1 PROJECT
Esmeralda County, Nevada

LONGITUDINAL SECTION SHOWING ORE RESERVE BLOCKS

SCALE
0 100 200 300 FEET

DECEMBER 1979

BEHRE DOLBEAR & COMPANY, INC.
MINING, GEOLOGICAL AND METALLURGICAL CONSULTANTS
NEW YORK, N.Y.

a. Chip and channel samples which compared well with the grades indicated by diamond drilling.

b. A bulk sample (of 2 tons) for process testing.

2. No unexpected large quantities of water were encountered.

3. Both hanging and footwalls were found competent and not unduly fractured, reinforcing the presumption that dilution would not be a problem with blasthole stoping; that is, ground conditions continued at depth as they appeared at the 2,135m (7,000 ft) elevation adit and drift.

Bulk sample process testing confirmed the expected recoveries, and further indicated a slightly higher recovery using cyanidation rather than flotation. Silver ore from the adjacent Nivlok Mine had been treated by cyanidation. Finally, product marketability was far better as dore than as precious metal-rich sulfide concentrates and, thus, justified the somewhat greater capital cost of a cyanide plant over a flotation plant.

PROJECT FINANCE LOAN SIZE

Under the conditions and projections set forth in the cash flow of Behre Dolbear's original feasibility study, a commercial bank might consider making available up to $7.7 million, assuming that the lower portal decline and development drifting has been completed to render the reserve, mining method, and operating risks acceptable. The postulated amount of $7.7 million, made available under project finance terms and conditions, is derived as follows:

1. Term of loan: 6 years; 3-year takedown during construction, and a 3-year pay back -- about one-half the life of the proven ore reserves -- beginning in the first full year of productivity.

2. Ratio of total net cash flow to loan principal: $41,000,000 to $7,700,000 or about 5; minimum is usually 2.

3. Ratio of loan capital to effective equity capital: $7.7 million to $3.3 million or about 2.3; maximum is usually 4.

4. Ratio of operating cash flow to principal repayment over term of loan: about $21.9 million to $7.7 million or about 3.2; minimum is usually 2.

5. Ratio of gross operating profit to interest over operating period the loan is outstanding (note -- interest due over construction period is usually capitalized and is paid from equity capital): about $36.0 million to about $3.1 million (20% on $7.7 million repaid in equal installments over 3 years) or almost 12; 5 is often acceptoften able.

Base Case Assumptions

In the cash flow projections, Behre Dolbear has assumed that 70% of the cost of the buildings, equipment, land, and working capital would be debt financed and that the applicable interest rate would be the prime (as of early 1981) rate plus 2% or a total of 21%. The cash flows combine constant dollar revenues and costs with current dollar interest and loan repayments which, while technically incorrect, understates the benefits that would accrue to the project as a result of the debt financing.

For the actual financing assumptions used in the base case model, the ratios described above are as follows:

1. Total net project cash flow to loan principal: $48.01 million to $5.7 million or 8.32.

2. Loan capital to effective equity capital: $5.77 million to $5.48 million or 1.05.

3. Operating cash flow to principal repayment over term of loan: $30.8 million to $5.77 million or 5.3.

4. Gross operating profits to interest over operating period the loan is outstanding: $52.4 million to $4.5 million or 11.6.

Under conditions at the time of this writing (December 1984), postulating the amount Sunshine may be able to borrow under project finance arrangements is more difficult. Some $21 million has been expended and a depressed silver price will greatly reduce operating profits and cash flow from the projections of financial performance made at $13 per ounce silver and $500 per ounce gold. Conversely, the lender has somewhat less risk regarding completion and, at current precious metal prices, may reasonably expect an increase in prices to generate real profit and cash flow increases.

PROCESSING

Because of the alternatives in processing available to Sunshine at the time of feasibility determination and because installed processing capacity represent the single greatest item of capital cost, metallurgical aspects are presented in relatively greater detail than reserve and mining aspects.

Tables 1 and 2 show the projected metallurgical results from the flotation flow sheet and cyanidation flow sheet. Based on these projected metallurgical results, net smelter returns (NSR) were compared for each case. To determine the effect of metal prices, net smelter returns were calculated in each case for two price levels:

1. Silver $15/oz; Gold $300/oz.

2. Silver $20/oz; Gold $400/oz.

TABLE 1
AVERAGE PROJECTED PLANT METALLURGICAL RESULTS, FLOTATION FLOW SHEET

Case A: Flotation Flow Sheet, Lower Grade Concentrate

Product	% Weight	Tons/Yr	Assays Ag oz/ton	Au oz/ton	% Pb	% Zn	Distribution % Ag	% Au	%Pb	% Zn
plant feed	100.00	150,000	10.0	0.03	0.3	0.5	100.0	100.0	100.0	100.0
flotation conc	4.21	6,315	212.4	0.62	6.4	10.7	89.5	87.6	90.0	90.0
plant tailings	95.79	143,685	1.1	0.004	0.03	0.05	10.5	12.4	10.0	10.0

Case B: Flotation Flow Sheet, Higher Grade Concentrate

Product	% Weight	Tons/Yr	Assays Ag oz/ton	Au oz/ton	% Pb	% Zn	Distribution % Ag	% Au	%Pb	% Zn
plant feed	100.00	150,000	10.0	0.03	0.3	0.5	100.0	100.0	100.0	100.0
flotation conc	1.96	2,937	431.5	1.39	12.6	20.9	84.5	75.3	82.0	82.0
plant tailings	98.04	147,063	1.6	0.003	0.05	0.09	15.5	24.7	18.0	18.0

NOTE: Projections are based on Dawson Metallurgical Laboratories batch test results for stockpile ore sample.

TABLE 2
AVERAGE PROJECTED PLANT METALLURGICAL RESULTS, CYANIDATION FLOW SHEET

Product	% Weight	Tons/Yr	Assays, Oz/Ton Ag	Au	Distribution, % Ag	Au
plant feed	100.0	150,000	10.00	0.03	100.0	100.0
dore bullion (90% Ag+Au)	0.03	52*	**	**	90.8	89.7
plant tailings	99.97	149,953	0.92	0.003	9.2	10.3

*Includes added Zn dust, flux, etc.
**Equivalent to 1,362,000 oz silver/year and 4,036.5 oz gold/year.

NOTES: Projections are based on Dawson Metallurgical Laboratories batch test results for stockpile ore sample.

Overall recoveries of gold and silver into dore bullion provide for losses of solubles, etc. from treatment of the pregnant solution and the zinc precipitate.

Milling Equipment Selection

In order to compare capital estimates, the following qualifications apply to equipment selection:

1. State-of-the-art technology was applied throughout.

2. Consideration was given to the relatively short projected mine life (5.75 years) by selecting equipment which may have a high resale value and/or readily moved to another location. Mobile and portable equipment was used whenever possible to minimize engineering and installation costs. This resulted in the selection of equipment with higher prices than are to be expected in the mills of the flow sheet design.

3. Flotation equipment was selected for singleline operations to reduce capital and operating costs.

4. Four-stage leaching was chosen for cyanidation leaching to minimuze pulp short-circuiting and maximize Ag/Au extraction, yet avoid excessive capital and operating costs.

Capital Costs

The fixed capital costs were determined to be $5,500,000 for a flotation plant and $7,150,000 for a cyanidation plant. The additional capital for development and installation or a tailing dam would be about $500,000 for either type of process plant. Working capital was estimated to be $700,000 for either plant to cover necessary parts and spares and three months' operating cost.

Operating Costs

The estimated direct operating costs were determined to be $12.56 per ton of ore processed for a flotation plant and $15.77 per ton of ore processed for a cyanidation plant.

Comparative DCF-RORs

A rough comparison of the DCF-ROR values for the project as a whole was made as a function of processing alternatives (Table 3). The following additional rough parameters applied to the DCF-RORs:

1. Depletion rate used for silver only.

2. Capital cost drawdown of 40%, 50% and 10% over years 0, 1 and 2, respectively.

3. 75% of NSR during year 2 and 100% of NSR for each of years 3 through 7 thereafter.

4. Lower silver and gold recoveries in the case of the higher grade flotatation concentrate.

The choice between the processing alternatives is not clear cut. The relative DCF-RORs could easily change with small, subtle alterations in the nature and processing characteristics of the ore, in turn generating significant changes in recoveries or operating costs. But there was an indication that the favored processing alternative at the "state-of-knowledge" as of January 1980 was flotation producing a lower grade concentrate. This indication was supported by two factors which are not considered by the rough comparative financial profiles, namely:

1. Far easier and shorter permitting procedures for flotation than for cyanidation.

2. Non-recovery of zinc and lead values by cyanidation -- a disadvantage which could have been significant if mineralization below 2,230m (6,800 ft) level was economic because of increased lead and zinc sulfide grades.

The indicated choice of processing by flotation to produce a lower grade concentrate did not take into consideration the following:

1. Smelter tolling of concentrates.

2. Greater marketing flexibility and higher overall income with the production of dore bullion rather than with the production of concentrate.

TABLE 3

COMPARATIVE DISCOUNTED RATES OF RETURN FOR PROCESSING ALTERNATIVES
($ Million)

A. Parameters:

Processing Alernative	Cash Flow Parameter	Ag @ $ 15/oz Au @ $300/oz (Case A-1)	Ag @ $ 20/oz Au @ $400/oz (Case A-2)
flotation, lower grade concentrate	NSR capital cost operating cost/year	$19.8 11.9 5.0	$26.3 11.9 5.0
		(Case B-1)	(Case B-2)
flotation, higher grade concentrate	NSR capital cost operating cost/year	$19.4 11.9 5.0	$25.9 11.9 5.0
		(Case C-1)	(Case C-2)
cyanidation	NSR capital cost operating cost/year depreciation/year	$21.4 13.6 5.5 8.0	$28.6 13.6 5.5 8.0

B. Comparative DCF-ROR:

flotation	lower grade concentrate	59.6%	79.8%
flotation	higher grade concentrate	58.3%	78.8%
cyanidation	dore bullion	56.2%	76.1%

Ore Variability With Depth

The bulk sample which was tested by Hazen and by Dawson Laboratories may not have adequately represented the overall deposit, hence the selection of flotation as the most economically attractive process may prove to be incorrect upon closer study.

Ore above 2,100m (6,800 ft) level is likely to contain more oxidized mineralization and less sulfide values than ore below that level. The ore might therefore be less amenable to flotation and more amenable to cyanidation in terms of overall recovery, albeit with high consumption of cyanide. The fine particle size of some of the acanthite and native silver favor cyanidation. Should the ore exhibit fine particle size mineralization at depth, then cyanidation would probably give overall better recoveries of silver and gold. This holds true particularly if the silver minerals are not closely locked with other metallic sulfides.

At lower levels of the deposit, more of the silver materials (acanthite and native silver) may be more intimately associated with sulfide minerals and, thus, would be less amenable to cyanidation. With increasing abundance of sulfide minerals, the silver may occur, at least in part if not entirely, in solid solution within the sulfide minerals, particularly galena and, to a lesser extent, in pyrite. Recovery by flotation would then be clearly preferred and superior to cyanidation, which would not bring either galena or pyrite into solution with resulting lower recoveries of silver.

The variability and nature of occurrence of silver and gold with increasing depth, especially as regards occurrences associated with sulfides, must be determined before a final decision is made regarding the method of processing. Such a determination may likely require several bulk samples from the 2,100m (6,800 ft) level. The need to have determined the near-optimum processing method before operation is accentuated not by the possible capital expenditures required to make capital process changes, but by the limited reserves available for amortizing mistakes.

It is not unlikely that a combination of flotation of the ore followed by cyanidation of the flotation tailings may prove to be the most applicable method of processing. This would hold true if ore at depth were found to contain increasing quantities of galena and sphalerite in which the silver was present as very fine sized yet discrete particles. Consideration may also have to be given to the production of individual concentrates of lead and zinc, rather than to the production of a bulk lead and zinc concentrate.

ALTERNATIVE TO PLANT CONSTRUCTION

The Argentum Mill is located 72 km (45 miles by road) from the proposed 16-to-1 Mine adit and the primary crusher location. All but about 18 km (11 miles) is over excellent paved highway. The unpaved section at the Argentum Mill end is suitable for large truck or truck-trailer carriers. The 16-to-1 canyon road requires improvement estimated to cost about $60,000 to render the road passable to vehicles with a pay load of up to 30 tons.

Adaptation of the Argentum Mill to process 16-to-1 Mine ore would require much less capital investment than a new plant and facilities built at the 16-to-1 Mine. Operating costs may be higher, and concentrate grades as well as recoveries of silver and gold somewhat lower at Argentum since the mill was not designed and built specifically for treatment of 16-to-1 ore and is not a new installation. No state and federal permits would be required to process 16-to-1 ore through the Argentum Mill and production could possibly commence up to one year sooner, depending upon the 16-to-1 Mine development schedule and subject to any other commitments for the use of the Argentum Mill.

Therefore, while roughly estimated economics concerned with altering Mineral Management's Argentum Mill do not appear as attractive as those of a self-constructed concentrator, nonetheless the Argentum Mill presents an opportunity to decrease capital exposure and risk as compared with constructing a new concentrator. This is all the more so by the known need of Minerals Management for mill feed and ownership of bullion.

ON-SITE INSTALLATIONS AND INFRASTRUCTURE

The capital costs as of January 1980 for on-site installations and infrastructure are summarized as follows:

maintenance/repair facilities	$ 200,000
administration building	100,000
inventory, mill	230,000
inventory, mine	250,000
power and water	590,000
tailings disposal	500,000
townsite/living accommodations	780,000
Total	$2,650,000

PERMITTING AND SCHEDULING

Scheduling of the 16-to-1 Mine project will be closely tied to, and dependent upon, permit acquisition time. The two permitting procedures that appear most time consuming are:

1. Prevention of Significant Deterioration Permit, Federal EPA -- 18 months (after collection of baseline data for 12 months).

2. Air Quality Permit to Operate, State EP -- 12 months.

Both of these permits apply only to mill operations and, more specifically, to crusher dust production/emission.

ESTIMATE OF CASH FLOW GENERATION

This section is based upon information contained in the January 9, 1980 report to the Sunshine Mining Company, entitled "The 16-to-1 Mine, Esmeralda County, Nevada -- The Technical and Economic Feasibility of Production" by Behre Dolbear & Company, Inc. At the time of this analysis, Sunshine was still contemplating use of a flotation system rather than cyanidation. As a result, Behre Dolbear's cash flow analysis was performed using the assumptions and parameters associated with the flotation system. As previously analyzed in this paper, the effect of this difference on the overall project economics is relatively minor.

Net Smelter Return Schedule

Net smelter returns for a single concentrate containing payable values in silver, gold, lead and zinc were calculated according to prices at the Bunker Hill smelter at Kellogg, Idaho. The schedule presumes the sale of all payable metals to the smelter and is summarized below:

Metal	Payment Terms
Ag	pay 97% of the silver content at the Handy & Harmon carry-forward silver price less $0.05/oz refining charge
Au	pay 95% of gold content (but mining deduction of $0.03 oz/ton) at Engelhard price for unfabricated gold less $3.50/oz refining charge
Pb	pay 95% but minimum 3 units deduction at Metals Week average price less $0.045/lb refining charge
Zn	pay 25% of zinc content at Metals Week average price for prime western metal less $0.15/lb refining charge
penalties	arsenic, antimony and bismuth -- concentration of above metals is assumed below the limits attracting a penalty
smelting charge	$89.36 per dry short ton
shipping and insurance	Nevada to Idaho, $30.00/wet short ton; 10% moisture is assumed, yielding a cost of $33/dry short ton

Using this smelter return schedule at a production rate of 150,000 tons per year and using the basic technical and financial parameters set forth in following sections, the estimated net smelter return during a typical year is $19.98 million. The contributions to this return, by metal, are a follows:

Metal	Contribution	% of Total
silver	$18,621,000	93.2
gold	1,059,000	5.3
lead	220,000	1.1
zinc	80,000	0.4
	$19,980,000	100.0

Mid-Continent Agreement

With respect to the Mid-Continent/Sunshine ownership agreement, Sunshine's outlay of about $500,000 over the period January 1, 1973 through November 30, 1979 is taken into account, but outlays after November 30, 1979 to January 9, 1980 were not taken into account. An average interest rate of 10% on the cumulative balance of Mid-Continent's share of Sunshine's capital outlays has been applied. Mid-Continent's net proceeds interest has essentially been treated as a 33-1/3% royalty on operating cash flow, thus crediting Mid-Continent with its share of depreciation and depletion.

Schedule of Capital Outlay and Production Startup

The cash flow is based on a preproduction capital outlay schedule (Table 4). Production startup is predicated on the assumption that mine development will precede mill construction and will be completed before the mill is ready to operate. Ore will be stockpiled in preparation for mill operation.

TABLE 4

SCHEDULE OF CAPITAL OUTLAY
($'000)

	1980	1981	1982	Total
development (expenses)	$ 1,893	$ 1,400	$ 367	$ 3,660
buildings and equipment (capitalized)	2,430	2,750	2,080	7,260
land and working capital (recaptured)	388	299	298	985
	$ 4,711	$ 4,449	$ 2,745	$ 11,905

Percentage Distribution (excluding recaptured capital):

development	17.3	12.8	3.4	33.5
buildings and equipment	22.3	25.2	19.0	66.5
Total	39.6%	38.0%	22.4%	100.0%

Note: For further details, see Exhibit 2

Mill operation is scheduled to commence by July 1982. The mill is estimated to attain capacity throughput and design metal recoveries in concentrates by year end 1982. Over the second half of 1982, concentrate production is estimated at one-half of the production compared to production at full mill capacity over a six-month period. For 1982, then, concentrate production is estimated at one-fourth the production of each of the succeeding years.

Cash Flow Particulars and DCF-ROR

The base case (most likely) cash flow is presented in Table 5 on both an unleveraged and leveraged basis. Both financing situations are included because while it is common for a mining company to examine a property's economics on an unleveraged basis, financial institutions generally prefer to analyze a property under certain assumptions regarding its debt and equity situation. The critical parameters and assumptions used in the cash flow projections are summarized as follows:

price of silver	- $15/oz
run-of-mine grade	- 11 oz Ag/ton
mill recovery	- 89.4% for silver
mining reserves	- 862,500 tons
production rate	- 150,000 tons/year
capital cost	- $10,920,000 excluding (a) working capital of $895,000 and (b) land purchase totaling $90,000
operating cost	- $23.93/ton of ore processed excluding depreciation
taxes	- federal rate of 48%, depletion rate of 15%, investment tax credit of 10%
leverage	- 70% of equipment, buildings, land and working capital ($5,770,000), repayable in 3 equal annual installments, at an interest rate of 21% (prime +2%)

Of the total $11,905,000 preproduction capital investment, $3,660,000 are expensed, $7,260,000 are capitalized, and $985,000 are recaptured as working capital and land purchase outlays upon cessation of production. Expensed investment include:

mine site preparation and road improvement	$ 100,000
trailer hook-up	120,000
mine development	2,000,000
mill site preparation, power and water	590,000
permitting	100,000
metallurgical testing	50,000
engineering design	100,000
G & A salaries and supplies	120,000
parts inventory	480,000
Total	$3,660,000

Capitalized investments include:

trailers	$ 570,000
mine equipment and installation	1,100,000
concentrator	4,910,000
tailings dam	500,000
G & A installation	180,000
Total	$7,260,000

The Nevada state net proceeds and property tax appears on line 17 of the cash flow. The net proceeds are equal to the operating income (before depreciation and depletion) less a fixed figure for depreciation. The property tax is based on the appraised value of buildings, fixed equipment and mobile equipment. The appraised value has been taken as the purchase price for purposes of this study. Depreciation rates for buildings, fixed equipment and mobile equipment for state tax calculations are, respectively, 40, 20 and 10 years. The net proceeds from operations and the appraised value of the buildings and equipment are totaled, and are assessed at 35% of the total. The tax rate as of January 1980 was at $2.451 per $100 of the total assessed value.

Without consideration of any salvage value but with the recapture of cash outlays for working capital and land purchase, the varous financial indices (calculated on the net cash flow to Sunshine) for both the leveraged and unleveraged base case are summarized as follows:

Financial Index	Unleveraged Basis	Leveraged Basis
DCF-ROR, %	42.4	57.7
total net project cash flow, $'000s	32,066	31,043
net present value, $'000s:		
10%	15,217	15,580
15%	10,439	11,197
20%	6,990	8,027
payback period/ year of production	1.96	1.61

Sensitivity Analysis

Detailed sensitivity analyses were performed on both the leveraged and the unleveraged base case cashflows. In the case of the unleveraged

CASE STUDIES

TABLE 5

BASE CASE CASH FLOW

Unleveraged Basis:

	1980	1981	1982	1983	1984	1985	1986	1987	1988	TOTALS
NET SMELTER RETURN	0	0	4,995,000	19,980,000	19,980,000	19,980,000	19,980,000	19,980,000	9,990,000	114,885,000
OPERATING COSTS	0	0	1,794,750	3,589,500	3,589,500	3,589,500	3,589,500	3,589,500	1,794,750	21,537,000
OPERATING REVENUES	0	0	3,200,250	16,390,500	16,390,500	16,390,500	16,390,500	16,390,500	8,195,250	93,348,000
DEVELOPMENT	1,893,000	1,400,000	367,000	0	0	0	0	0	0	3,660,000
INTEREST	0	0	0	0	0	0	0	0	0	0
DEPRECIATION	0	0	1,037,000	1,037,000	1,037,000	1,237,000	1,237,000	1,237,000	1,037,000	7,859,000
MP & PROP. TAX	0	9,000	68,000	201,000	200,000	199,000	198,000	196,000	125,000	1,196,000
	-1,893,000	-1,409,000	1,728,250	15,152,500	15,153,500	14,954,500	14,955,500	14,957,500	7,033,250	80,633,000
LOSSES AVAILABLE	0	-1,893,000	-3,302,000	-1,573,750	0	0	0	0	0	-6,768,750
LOSS DEDUCTION	0	0	1,728,250	1,573,750	0	0	0	0	0	3,302,000
INC BEFORE DEPLETION	-1,893,000	-1,409,000	0	13,578,750	15,153,500	14,954,500	14,955,500	14,957,500	7,033,250	77,331,000
DEPLETION ALLOWANCE	0	0	0	2,997,000	2,997,000	2,997,000	2,997,000	2,997,000	1,498,500	16,483,500
TAXABLE INCOME	-1,893,000	-1,409,000	0	10,581,750	12,156,500	11,957,500	11,958,500	11,960,500	5,534,750	60,847,500
FEDERAL TAX	0	0	0	5,079,240	5,835,120	5,739,600	5,740,080	5,741,040	2,656,680	30,791,760
OPERATING NET INCOME	-1,893,000	-1,409,000	0	5,502,510	6,321,380	6,217,900	6,218,420	6,219,460	2,878,070	30,055,740
NON-CASH CHARGES	0	0	2,765,250	5,607,750	4,034,000	4,234,000	4,234,000	4,234,000	2,535,500	27,644,500
OPERATING CASHFLOW	-1,893,000	-1,409,000	2,765,250	11,110,260	10,355,380	10,451,900	10,452,420	10,453,460	5,413,570	57,700,240
CAPITAL ITEMS										
BUILDINGS & EQUIP	2,430,000	2,750,000	2,080,000	0	0	600,000	0	0	0	7,860,000
LAND & W.C.	388,000	299,000	298,000	0	0	0	0	0	-985,000	0
INVEST. TAX CRED.	0	0	0	-726,000	0	0	0	0	0	-726,000
	2,818,000	3,049,000	2,378,000	-726,000	0	600,000	0	0	-985,000	7,134,000
NET PROJ. CASH FLOW	-4,711,000	-4,458,000	387,250	11,836,260	10,355,380	9,851,900	10,452,420	10,453,460	6,398,570	50,566,240
LOAN CAPITAL										
ADDITIONS	0	0	0	0	0	0	0	0	0	0
PRINCIPAL PAYMENT	0	0	0	0	0	0	0	0	0	0
PRINCIPAL OUTSTAND	0	0	0	0	0	0	0	0	0	0
DEBT CASH FLOW	0	0	0	0	0	0	0	0	0	0
EQUITY CASH FLOW	-4,711,000	-4,458,000	387,250	11,836,260	10,355,380	9,851,900	10,452,420	10,453,460	6,398,570	50,566,240
ROYALTY CALCULATIONS										
EQUITY CAPITAL REQ	4,711,000	4,458,000	0	0	0	0	0	0	0	9,169,000
AVAILABLE CASH FLO	0	0	387,250	11,836,260	10,355,380	9,851,900	10,452,420	10,453,460	6,398,570	59,735,240
ADDITIONS	1,818,667	1,486,000	0	0	0	0	0	0	0	3,304,667
UNRECOVERED CAPITA	1,818,667	3,486,533	3,447,937	0	0	0	0	0	0	8,753,137
INTEREST	0	181,867	348,653	344,794	0	0	0	0	0	875,314
% CF SUBJECT TO MP	0%	0%	0%	68%	100%	100%	100%	100%	100%	
NET TO MID CONTINENT	0	0	0	2,678,495	3,448,342	3,280,683	3,480,656	3,481,002	2,130,724	18,499,901
NET TO SUNSHINE	-4,711,000	-4,458,000	387,250	9,157,765	6,907,038	6,571,217	6,971,764	6,972,458	4,267,846	32,066,339

Leveraged Basis:

	1980	1981	1982	1983	1984	1985	1986	1987	1988	TOTALS
NET SMELTER RETURN	0	0	4,995,000	19,980,000	19,980,000	19,980,000	19,980,000	19,980,000	9,990,000	114,885,000
OPERATING COSTS	0	0	1,794,750	3,589,500	3,589,500	3,589,500	3,589,500	3,589,500	1,794,750	21,537,000
OPERATING REVENUES	0	0	3,200,250	16,390,500	16,390,500	16,390,500	16,390,500	16,390,500	8,195,250	93,348,000
DEVELOPMENT	1,893,000	1,400,000	367,000	0	0	0	0	0	0	3,660,000
INTEREST	0	414,246	862,449	1,212,015	1,212,015	808,010	404,005	0	0	4,912,740
DEPRECIATION	0	0	1,037,000	1,037,000	1,037,000	1,237,000	1,237,000	1,237,000	1,037,000	7,859,000
MP & PROP. TAX	0	9,000	68,000	201,000	200,000	199,000	198,000	196,000	125,000	1,196,000
	-1,893,000	-1,823,246	865,801	13,940,485	13,941,485	14,146,490	14,551,495	14,957,500	7,033,250	75,720,260
LOSSES AVAILABLE	0	-1,893,000	-3,716,246	-2,850,445	0	0	0	0	0	-8,459,691
LOSS DEDUCTION	0	0	865,801	2,850,445	0	0	0	0	0	3,716,246
INC BEFORE DEPLETION	-1,893,000	-1,823,246	0	11,090,040	13,941,485	14,146,490	14,551,495	14,957,500	7,033,250	72,004,014
DEPLETION ALLOWANCE	0	0	0	2,997,000	2,997,000	2,997,000	2,997,000	2,997,000	1,498,500	16,483,500
TAXABLE INCOME	-1,893,000	-1,823,246	0	8,093,040	10,944,485	11,149,490	11,554,495	11,960,500	5,534,750	55,520,514
FEDERAL TAX	0	0	0	3,884,659	5,253,353	5,351,755	5,546,158	5,741,040	2,656,680	28,433,645
OPERATING NET INCOME	-1,893,000	-1,823,246	0	4,208,381	5,691,132	5,797,735	6,008,337	6,219,460	2,878,070	27,086,869
NON-CASH CHARGES	0	0	1,902,801	6,884,445	4,034,000	4,234,000	4,234,000	4,234,000	2,535,500	28,058,746
OPERATING CASHFLOW	-1,893,000	-1,823,246	1,902,801	11,092,826	9,725,132	10,031,735	10,242,337	10,453,460	5,413,570	55,145,615
CAPITAL ITEMS										
BUILDINGS & EQUIP	2,430,000	2,750,000	2,080,000	0	0	600,000	0	0	0	7,860,000
LAND & W.C.	388,000	299,000	298,000	0	0	0	0	0	-985,000	0
INVEST. TAX CRED.	0	0	0	-726,000	0	0	0	0	0	-726,000
	2,818,000	3,049,000	2,378,000	-726,000	0	600,000	0	0	-985,000	7,134,000
NET PROJ. CASH FLOW	-4,711,000	-4,872,246	-475,199	11,818,826	9,725,132	9,431,735	10,242,337	10,453,460	6,398,570	48,011,615
LOAN CAPITAL										
ADDITIONS	1,972,600	2,134,300	1,664,600	0	0	0	0	0	0	5,771,500
PRINCIPAL PAYMENT	0	0	0	1,923,833	1,923,833	1,923,833	0	0	0	5,771,500
PRINCIPAL OUTSTAND	1,972,600	4,106,900	5,771,500	5,771,500	3,847,667	1,923,833	0	0	0	23,394,000
DEBT CASH FLOW	-1,972,600	-2,134,300	-1,664,600	1,923,833	1,923,833	1,923,833	0	0	0	0
EQUITY CASH FLOW	-2,738,400	-2,737,946	1,189,401	9,894,992	7,801,299	7,507,901	10,242,337	10,453,460	6,398,570	48,011,615
ROYALTY CALCULATIONS										
EQUITY CAPITAL REQ	2,738,400	2,737,946	0	0	0	0	0	0	0	5,476,346
AVAILABLE CASH FLO	0	0	1,189,401	9,894,992	7,801,299	7,507,901	10,242,337	10,453,460	6,398,570	53,487,961
ADDITIONS	1,161,133	912,649	0	0	0	0	0	0	0	2,073,782
UNRECOVERED CAPITA	1,161,133	2,189,895	1,219,484	0	0	0	0	0	0	4,570,513
INTEREST	0	116,113	218,990	121,948	0	0	0	0	0	457,051
% CF SUBJECT TO MP	0%	0%	0%	86%	100%	100%	100%	100%	100%	
NET TO MID CONTINENT	0	0	0	2,848,336	2,597,833	2,500,131	3,410,698	3,481,002	2,130,724	16,968,724
NET TO SUNSHINE	-2,738,400	-2,737,946	1,189,401	7,046,657	5,203,466	5,007,770	6,831,639	6,972,458	4,267,846	31,042,892

cash flow, a more in-depth analysis was included to demonstrate the sensitivity of the cash flow to in-situ ore reserves and to show the change in certain parameters required to yield a DCF-ROR of 15%.

Unleveraged Cash Flow Sensitivity Analysis

The following parameters were examined for their individual effect on the DCF-ROR of the unleveraged base case cash flow:

1. price of silver;

2. in-situ grade of silver;

3. in-situ ore reserves (tons);

4. mill recovery of silver;

5. pre-production capital cost, excluding ing working capital;

6. overall operating cost.

Table 6 sets forth the resulting DCF-RORs and the number of years required for the return of capital (payback) for the various amounts used for each of the parameters and the amount for each parameter which would yield a DCF-ROR is 15%. The relative margins of safety between the parameter amounts of the base case cash flow and the parameter amounts yielding a minimum 15% DCF-ROR range from 2 to slightly over 3, excluding ore reserve tonnage. For the ore reserve tonnage, the safety margin drops to 1.3.

The DCF-ROR is most sensitive to the recovery of silver, price of silver and grade of silver ore. The ore reserve tonnage presents the greatest downside risk to non-achievement of the estimated, most likely financial performance, as set forth in the base case cash flow.

Table 7 confirms the indications of DCF-ROR greatest sensitivity to (1) mill recovery, (2) silver grade and (3) silver price, as shown by Table 6. The total all-equity cash flow -- net to Sunshine Mining -- over the life of the reserves divided by the total tons of ore processed yields the cash flow generated by each ton of ore processed. For each parameter, the difference between (i) the cash flow generated by change in the parameter and (ii) the base case parameter divided by (iii) the percentage change in that parameter yields the rate of change in the cash flow generated by each ton of ore processed. Referring to Table 7, the rate of increase (decrease) in cash flow per ton of ore processed for any given parameter is:

Column F minus Colum C; the result divided by the percentage difference between Column D and Column A, where Colum A is the original (base case) amount/figure.

The rates of increase/decrease in per ton cash flow generation are essentially straight line functions for all the parameters except for ore-reserve tonnage. Each percentage increase or decrease in the price, grade or recovery of silver yields

TABLE 6

DCF-ROR SENSITIVITY TO SELECTED PARAMETERS

Parameter	Base Case Cash Flow			Parameter 20% More Favorable Than Base Case			Parameter 20% Less Favorable Than Base Case			Figure/ Amount Which Yields DCF-ROR of 15%
	Figure/ Amount	ROR,%	Payback, Years	Figure/ Amount	ROR,%	Payback, Years	Figure/ Amount	ROR,%	Payback, Years	
1) Silver price, $/oz	15.00	42.48	1.46	18.00	50.37	1.28	12.00	33.47	1.80	7.26
2) Silver grade, oz/ton	11.5	42.48	1.46	13.8	50.59	1.28	9.2	33.18	1.81	5.7
3) Ore reserves, short tons '000	750	42.48	1.46	900	43.84	1.47	600	39.99	1.45	563
4) Mill recovery, %	89.4	42.48	1.46	98.34	46.64	1.36	80.46	38.06	1.60	44.5
5) Capital cost, $'000	10,920	42.48	1.46	8,736	49.87	1.30	13,104	36.88	1.64	30,125
6) Operating cost, $/ton ore processed	23.93	42.48	1.46	19.14	44.05	1.42	28.72	40.90	1.50	86.44

CASE STUDIES

TABLE 7

SENSITIVITY OF PARAMETERS AS A FUNCTION OF NET CASH FLOW GENERATION

Parameter	Base Case Cash Flow — Figure or Amount	Total Cash Flow, $'000	Cash Flow Per Ton Ore Processed	Parameter 20% More Favorable Than Base Case — Figure or Amount	Total Cash Flow, $'000	Cash Flow Per Ton Ore Processed	Cash Flow Increase Per Ton Ore Per 1% Parameter Amount Increase	Parameter 20% Less Favorable Than Base Case — Figure or Amount	Total Cash Flow, $'000	Cash Flow Per Ton Ore Processed	Cash Flow Decrease Per Ton Ore Per 1% Parameter Amount Decrease
1) Silver price, $/oz	15.00	32,423	37.59	18.00	41,192	47.76	0.51	12.00	23,505	27.25	0.52
2) Silver grade, oz/ton	11.5	32,423	37.59	13.8	41,445	48.05	0.52	9.2	23,241	26.95	0.53
3) Ore reserves, short tons '000	750	32,423	37.59	900	34,399	33.24	(0.22)	600	24,717	35.82	0.09
4) Mill recovery, %	89.4	32,423	37.59	98.34	36,934	42.82	0.52	80.46	27,886	32.33	0.53
5) Capital cost, $'000	10,920	32,423	37.59	8,736	33,431	38.76	0.06	13,104	31,389	36.39	0.06
6) Operating Cost, $/ton ore processed	23.93	32,423	37.59	19.14	33,909	39.31	0.09	28.72	30,938	35.87	0.09
Column Identifier	A	B	C	D	E	F	G	H	I	J	K

an increase or decrease of about $0.52 in the cash flow generated by each ton of ore processed.

The rate of increase/decrease in per-ton cash flow generation for ore reserve tonnage is parabolic function. The rate of increase is maximized for the given capital and operating costs somewhere between a reserve of 600,000 and 900,000 tons, although not necessarily at 750,000 tons. At a reserve tonnage somewhere below 560,000, the rate of decrease in per-ton cash flow generation will exceed that of the silver price, silver grade and mill recovery.

The multiplier effect of pairs of parameters on the DCF-ROR where the two parameters are both increasing or decreasing in tandem, while of interest, is relatively nominal at the estimated DCF-ROR level calculated for the 16-to-1 Mine project. The multiplier effect is quantified by Table 8.

TABLE 8

MULTIPLIER EFFECT ON DCF-ROR FOR SELECTED PAIRS OF PARAMETERS

		Silver Grade		
		Minus 20%	Base Case	Plus 20%
	Minus 10%	29.0	38.1	46.5
Silver	Base Case	33.2	42.5	50.6
Recovery	Plus 10%	36.5	46.6	55.5

		Silver Price		
		Minus 20%	Base Case	Plus 20%
Overall	Minus 20%	31.3	40.9	48.8
Operating	Base Case	33.5	42.5	50.4
Cost	Plus 20%	35.4	44.1	51.9

		Ore Reserves		
		Minus 20%	Base Case	Plus 20%
Pre-production	Minus 20%	34.7	36.9	38.6
Capital	Base Case	40.0	42.5	43.8
Cost	Plus 20%	47.0	49.9	50.5

Leverage Cash Flow Sensitivity Analysis

For the leveraged base case, a somewhat less detailed sensitivity analysis was performed examining the effect of 10% increases and decreases in the following parameters:

1. net smelter return
2. overall operating costs
3. capital costs
4. interest rate on borrowed capital
5. percentage of buildings, equipment, land and working capital financed by debt

The sensitivity analysis results for the leveraged base case are presented in Table 9. The analysis indicates that the net smelter return (a function of silver prices, metallurgical recovery and run-of-mine ore grade) exerts a greater influence on the project economics than the interest rate on debt capital and percentage of capital items which are debt financed.

These results are confirmed by the analysis presented in Table 10 where the sensitivity of the various parameters as a function of cash flow generation is presented: details of procedures used in this analysis are included in the previous subsection.

TABLE 9

SENSITIVITY OF DCF-ROR AND PAYBACK PERIOD TO CHANGES IN SELECTED PARAMETERS

Parameters Varied	Base Case Cash Flow			Parameter 10% More Favorable Than Base Case			Parameter 10% Less Favorable Than Base Case		
	Figure/Amount	DCF-ROR	Payback, Years	Figure/Amount	DCF-ROR	Payback, Years	Figure/Amount	DCF-ROR	Payback, Years
net smelter return per ton	133.20	57.7%	1.61	146.52	63.6	1.50	119.88	51.5	1.73
interest rate on leverage	21%	57.7%	1.61	18.9%	58.3	1.59	23.1%	57.1	1.62
percent of capital items financed	70%	57.7%	1.61	77.0%	60.2	1.56	63.0%	55.5	1.65
operating costs per ton	23.93	57.7%	1.61	21.54	58.9	1.45	26.32	56.5	1.63
capital costs	11.9 million	57.7%	1.61	13.1 million	60.2	1.56	10.7 million	55.4	1.66
Unleveraged Base Case Cash Flow		42.4%	2.0						

TABLE 10

SENSITIVITY OR PARAMETERS AS A FUNCTION OF NET CASH FLOW GENERATION

Parameter	Leveraged Base Case Cash Flow			Parameter 10% More Favorable than Base Case				Parameter 10% Less Favorable Than Base Case			
	Figure or Amount	Total Cash Flow, $'000	Cash Cash Per Ton Ore Processed	Figure or Amount	Total Cash Flow, $'000	Cash Flow/ Ton Ore Processed	Cash Flow Increase/ Ton Ore/1% Parameter Increase	Figure or Amount	Total Cash Flow, $'000	Cash Flow/ Ton Ore Processed	Cash Flow Decrease Ton Ore/1% Parameter Decrease
net smelter return per ton	133.20	31,043	35.99	146.52	35,539	41.20	0.52	119.88	26,547	30.78	0.52
interest rate on leverage	21%	31,043	35.99	18.9	31,219	36.20	0.02	23.1	30,867	35.79	0.02
percent of buildings/equipment, etc. financed	70%	31,043	35.99	77%	30,941	35.87	0.01	63%	31,145	36.11	0.01
operating costs per ton	23.93	31,043	35.99	21.54	31,784	36.85	0.09	26.32	30,302	35.13	0.07
capital costs	11.9 million	31,043	35.99	10.7 million	31,474	36.49	0.05	13.1 million	30,611	35.49	0.05
Unleveraged Base Case Cash Plan		32,066	37.18								
Column Identifier	A	B	C	D	E	F	G	H	I	J	K

EXHIBIT 1

SUMMARY AND SPECIAL CONSIDERATIONS FROM PROSPECTUS FOR SUNSHINE MINING COMPANY'S SILVER INDEXED BONDS

PROSPECTUS SUMMARY

This summary is qualified by the detailed information appearing elsewhere herein.

THE COMPANY

Sunshine Mining Company is the sole operator and principal owner of the Sunshine Mine, the largest mine in the United States which produces primarily silver. Since about 1880, the Sunshine Mine has produced more than 300 million troy ounces of silver. In 1979, 3,513,300 troy ounces of silver were produced at the Sunshine Mine. Independent mining engineers have estimated the Company's interest in the proven and probable silver reserves of the Sunshine Mine at 884,000 tons with a grade of 28.8 troy ounces per ton (approximately 24,348,000 estimated recoverable troy ounces of silver) as of January 1, 1980. See "Business — Reserves." Portions of the proceeds of this offering will be used to increase the Company's exploration, development and ownership of mining properties, construct a pilot facility for refining silver and increase mining and milling capacity.

SUMMARY OF OFFERING

Issue — $25,000,000 of 8½% Silver Indexed Bonds Due April 15, 1995 (the "Bonds").

Payment of Interest — April 15 and October 15.

Indexed Principal Amount — For each $1,000 Face Amount Bond, the greater of $1,000 or the Market Price of 50 Ounces of Silver (determined as provided in the Indenture).

Redemption — Callable at the Indexed Principal Amount on or after April 15, 1985, if such amount is $2,000 or more for a period of 30 consecutive calendar days.

Sinking Fund — In each year commencing April 15, 1982, the Company will call for redemption 7% of the Adjusted Original Issue. Holders whose Bonds have been called shall have the right to elect not to have their Bonds redeemed. The Face Amount of Bonds called but not redeemed shall be added to the amount called for redemption in the subsequent years; *provided, however*, that the Company will not be required to call for redemption in any year more than 14% of the Adjusted Original Issue, and the Face Amount of previously called Bonds which would otherwise have required more than 14% of the Adjusted Original Issue to be retired in any year will not be carried forward in subsequent years of the sinking fund. With respect to any sinking fund payment, the Company may deliver to the Trustee previously acquired Bonds only in satisfaction of that portion of such sinking fund payment, if any, which exceeds 7% of the Adjusted Original Issue.

Payments in Silver — If the Indexed Principal Amount is greater than $1,000, the Company, at its option, may deliver 50 Ounces of Silver per $1,000 Face Amount of Bonds to holders electing to accept such delivery in satisfaction of the Indexed Principal Amount.

Security Interest and Subordination — Secured by a security agreement entitling the Trustee, on behalf of the holders, to receive, upon default, 3.627% of the Annual Mining Production of the Sunshine Mine, limited in all events, however, to (i) a number of Ounces of Silver per year, determined at the date of acceleration, sufficient to supply, in equal annual installments from such date through April 15, 1995, the total number of Ounces of Silver specified in all Bonds outstanding at the date of acceleration, and, in any event, (ii) not more than 50 Ounces of Silver per $1,000 Face Amount of outstanding Bonds. See "Description of Bonds — Security for the Bonds." Subordinated to all Senior Indebtedness to the extent the Collateral is insufficient to satisfy fully the Company's obligations.

Restrictions — No restrictions upon the creation of Senior Indebtedness. The Company may not issue additional Silver Backed or Silver Related Securities except to the extent that, after giving effect to any such issue, the Company's Qualified Reserves would be equal to at least 500% of the total number of Ounces of Silver required by all of the Company's Silver Backed and Silver Related Securities. The Company must maintain Qualified Reserves equal to at least 400% of the aggregate amount of Silver required by all outstanding Silver Backed and Silver Related Securities. See "Description of Bonds — Certain Covenants." As of January 1, 1980, the Company had 25,418,000 Ounces of Qualified Reserves available for the issuance of Silver Backed or Silver Related Securities and the Company's Qualified Reserves were equal to approximately 2,034% of the aggregate amount of Silver represented by this issue.

Listing — The Company has applied to list the Bonds on the New York Stock Exchange.

SELECTED INFORMATION

Summary of Operating Results:

	Year Ended December 31.				
	1975	1976	1977	1978	1979
	(Dollars in Thousands)				
Mining revenue	$ 13,339	$ 3,360	$10,504	$ 16,450	$23,645
Income (loss) from continuing operations(1)	$ 2,230	$(1,793)	$ (559)	$ 1,019	$ 5,483
Income from discontinued operations(1)(2)	$ 1,989	$ 2,739	$ 2,118	$ 2,171	$ 2,815
Net income(2)	$ 4,219	$ 946	$ 1,559	$ 3,190	$ 8,298
Ratio of earnings to fixed charges:					
Historical	4.46	1.50	1.87	3.52	4.73
Pro forma					2.13
Ore milled (tons)(3)	126,300	25,800	87,000	113,500	97,100
Ounces of silver produced(3)	2,892,200	620,000	2,145,000	2,821,800	2,009,600

Balance Sheet Data:

	December 31, 1979	
	Actual	As Adjusted(4)
	(Dollars in Thousands)	
Bonds	$ —	$ 25,000
Subordinated Debt(5)	16,637	5,241
Stockholders' Equity	32,327	54,349

Price Range of Silver:(6)

	Year Ended December 31,			
	1976	1977	1978	1979
High	$5.14	$4.98	$6.32	$34.45
Low	3.83	4.31	4.81	5.92

(1) Net of provision (credit) for income taxes.

(2) Gain on disposal of discontinued oil and gas operations, which occurred during 1979, was $3,363,099 (net of income taxes) and is not included.

(3) For a discussion of the fluctuations in ore milled and ounces of silver produced, see "Management's Discussion and Analysis of Statements of Income."

(4) As adjusted to give effect to the sale of the Bonds offered hereby (excluding the Underwriters' over-allotment option), the sale of 500,000 shares of treasury stock in January 1980 at a purchase price of $21.25 per share, the reincorporation in the State of Delaware and the related issuance of five shares of common stock for each two shares of common stock held by each stockholder and the effect of conversion through April 3, 1980, of $11,396,377 of the 6½% convertible debentures into common stock of the Company.

(5) 6½% Convertible Subordinated Debentures Due April 15, 1989. The Company has called these debentures for redemption on April 11, 1980.

(6) Represents the Spot Settlement Price as quoted by the Commodity Exchange, Inc. For further information concerning the price range of silver, see "Price Range of Silver."

SPECIAL CONSIDERATIONS RELATING TO THE BONDS

Since the Bonds constitute a new type of investment security, potential investors should consider the following special considerations relating to the Bonds and consult their own investment adviser.

Certain Definitions

The following are certain definitions which are used throughout this Prospectus in describing the Bonds. For further information concerning the Bonds, see "Description of Bonds." All statements and definitions relating to the Bonds are qualified in their entirety by express reference to that certain Indenture, to be dated as of April 15, 1980 (the "Indenture"), between the Company and Continental Illinois National Bank and Trust Company of Chicago, Trustee (the "Trustee"), and that certain Mortgage and Security Agreement to be dated as of April 15, 1980, between the Company and the Trustee (the "Security Agreement"), copies of which have been filed as exhibits to the Registration Statement of which this Prospectus is a part.

"Adjusted Original Issue," with respect to any Sinking Fund payment, shall mean an aggregate Face Amount of Bonds equal to the difference between (i) the aggregate Face Amount of Bonds originally issued pursuant to the Indenture and (ii) the aggregate Face Amount of Bonds which, prior to the December 1 immediately preceding such Sinking Fund payment, have been acquired by the Company and retired by delivery thereof (other than in satisfaction of a Sinking Fund payment) to the Trustee pursuant to Section 309 of the Indenture.

"Annual Mining Production" shall mean Mining Production, exclusive of any related mining and production expenses, at the Sunshine Mine during each 12-month period (the "Annual Production Period"), the first such Annual Production Period beginning on the first day of the month immediately following the month in which the date of acceleration occurs pursuant to the provisions of the Indenture.

"Indexed Principal Amount," as of the date of determination thereof, shall mean the dollar amount payable at Stated Maturity or any Redemption Date, whether at the option of the Company or pursuant to the Sinking Fund, and, for each $1,000 Face Amount Bond, shall be equal to the greater of (i) $1,000, or (ii) the average Market Price for the fifteen Trading Days immediately preceding such date of determination of 50 Ounces of Silver.

"Market Price" shall mean, as of any Trading Day, the Spot Settlement Price published for such Trading Day by: (i) the Commodity Exchange, Inc.; (ii) in the event that the Commodity Exchange, Inc. does not publish such a quote on such day, the Chicago Board of Trade; (iii) in the event that the Chicago Board of Trade does not publish such a quote on such day, Handy & Harman; (iv) in the event that Handy & Harman does not publish such a quote on such day, the London Metal Exchange; and (v) in the event the London Metal Exchange does not publish such a quote on such day, such other Person who customarily publishes quotes for the purchase or sale (or both) of Silver and whose quotations for the price of Silver are generally accepted as representative of the price of Silver by Persons in the silver industry and who is so designated by an Officers' Certificate delivered to the Trustee, such designation to remain effective until the Company shall have designated a successor Person.

"Mining Production" shall mean the minerals, mineral rights, ore deposits and silver reserves located at, and the Silver, silver/copper residue and iron pyrite concentrates and any and all other concentrates or other minerals or ores, or any combination of the foregoing, produced through operation of or extracted from, the Sunshine Mine, and all rents, issues, profits, proceeds and products thereof and therefrom.

"Ounce" shall mean one troy ounce (31.10348 grams or approximately 110% of an ounce).

"Silver" shall mean an alloy or substance with a minimum fineness of 999 parts per 1,000 pure silver.

"Spot Month," as of any date, shall mean, with respect to a Silver futures contract, the current or nearest delivery month.

"Spot Settlement Price," as of any date, shall mean (i) with respect to a Silver futures contract, the settlement price per Ounce of Silver with respect to the Spot Month, as established daily at the close of trading on such date by the exchange on which such contract is traded, and (ii) with respect to a Person (other than an exchange on which futures contracts are traded) who customarily publishes quotes for the purchase or sale (or both) of Silver, the last price per Ounce published on or for such date by such Person for currently deliverable Silver.

"Trading Day" shall mean each day that the Person whose quotation is used as the Market Price is open during its normal business hours and each day for which such Person publishes price quotations for the purchase or sale (or both) of Silver.

Security Interest

The Bonds are secured by a mortgage and security agreement (the "Security Agreement") encumbering the Company's interest in the Sunshine Mine and entitling the Trustee, on behalf of the holders, to receive, upon default, 3.627% of the Annual Mining Production of the Sunshine Mine (equivalent to approximately 6½% of the Company's share of the Sunshine Mine's Annual Production), limited in all events, however, to (i) a number of Ounces of Silver per year, determined at the date of acceleration, sufficient to supply, in equal annual installments from such date through April 15, 1995, the total number of Ounces of Silver specified in all Bonds outstanding at the date of acceleration, and, in any event, (ii) not more than 50 Ounces of Silver per $1,000 Face Amount of outstanding Bonds. The result is that, although the Bonds may be accelerated and become immediately due and payable, under the Security Agreement the Trustee is only entitled to receive a fractional part in each year of the Ounces of Silver to which the Bonds are indexed. The Trustee cannot receive more Ounces of Silver in each year after default than would be sufficient to ratably provide, in annual installments through 1995, the Ounces to which the outstanding Bonds are indexed. Consequently, the Trustee may not receive the total number of Ounces to which the Trustee can be entitled in any one year, or ever, if the Mine's production is not great enough so that the Trustee's maximum interest of 3.627% of Annual Mining Production results in the necessary number of Ounces. However, the percentage interest in mining production represented by the Security Agreement will continue beyond 1995 if sufficient Silver to provide the Trustee 50 Ounces of Silver per $1,000 Face Amount of outstanding Bonds has not been produced by such date. It should be noted that the security interest relates to unmined Silver in the ground and that there can be no assurance that the Sunshine Mine will produce the full number of Ounces of Silver contemplated, and, even if such Silver were produced, the market value of such Silver may not be equal to the Face Amount of the outstanding Bonds.

If the Sunshine Mine is not being operated, neither the Trustee nor any other person having an interest in the Security Agreement has any right to mine and remove any of the minerals, including the silver reserves, located at the Sunshine Mine. Accordingly, the mortgage and security interest provided by the Security Agreement may not provide sufficient collateral to redeem the outstanding Bonds at acceleration of maturity unless the Trustee is able to sell the security interest in its entirety to a third party at a price sufficient to satisfy and discharge the outstanding Bonds. Further, there can be no assurance that, in the event of an acceleration of maturity resulting from a default by the Company, the Sunshine Mine will continue to be operated or that no material delays in production will occur. Any cessation of production at the Sunshine Mine will materially diminish and impair the value of the security interest to the holders of Bonds. Additionally, because the Trustee is only entitled to receive 50 Ounces of Silver per Bond outstanding at the acceleration date, in-

CASE STUDIES

creases in the cost of production at the Sunshine Mine and decreases in the Market Price of Silver will materially diminish and impair the value of such security interest. For additional information concerning the Security Agreement, including the opinion of counsel relating thereto, see "Description of Bonds — Security for the Bonds."

Delivery of Cash or Silver

The Company will satisfy its obligations with respect to the sinking fund, or at redemption or maturity, out of its production of Silver by delivering to the Trustee the required number of Ounces of Silver or the proceeds from the sale of its production of such number of Ounces of Silver if the Indexed Principal Amount exceeds $1,000. To the extent that this is precluded by events not within the control of the Company, the Company will satisfy its obligations from its other resources. In the event the Company determines to satisfy its obligations by the delivery of cash to the Trustee, the Company will instruct the persons or entities to whom or which it sells its production to pay the proceeds of such sale directly to the Trustee. The Trustee will hold such funds in trust for the benefit of holders of the Bonds in accordance with the provisions of the Indenture.

With respect to any delivery of Silver by the Company, the Company may deliver Silver appropriately smelted, refined or minted in any size (whole ounces), shape, form or configuration (including, without limitation, medallions or other commemorative-type emblems), and such Silver may be inscribed with such brands or other marks of identification as the Company may determine. See "Description of Bonds — Delivery of Silver."

Silver is traded on exchanges in accordance with specialized requirements and only Silver meeting these specialized requirements is eligible for delivery pursuant to a Silver futures contract traded on a particular exchange. There may not be a market for quantities smaller than those traded and, even if such a market exists, holders who have received less than traded quantities may experience a substantial discount in the sales of such smaller quantities.

There can be no assurance that a holder who elects to receive Silver will receive Silver in a form and quantity meeting the requirements of a particular exchange or will have an available market in which to sell such Silver. Even if the holder were able to sell the Silver received, the net proceeds received may not approximate the Market Price. Proceeds received from any sale of Silver may also be reduced by applicable storage, insurance, transportation and assaying charges to date of sale and any brokerage, sales commissions or dealer spreads resulting from the sale.

Holders who elect to receive Silver may incur a sales tax with respect to the receipt of Silver and any subsequent disposition thereof. See "Description of Bonds — Certain Tax Consequences."

Although holders of Bonds who have elected to receive payment in Silver rather than cash will receive a silver alloy with a minimum fineness of 999 parts per 1,000 pure silver, the presence of residual trace elements of other materials such as antimony, which is present in the Company's Silver, could render it inappropriate for certain commercial uses.

Silver Prices

The Indexed Principal Amount of the Bonds is determined by reference to the Market Price of Silver. Market Price is defined as the Spot Settlement Price quoted by the Commodity Exchange, Inc. ("COMEX"), or, if such entity is not publishing quotations on such date, certain other entities which customarily publish such quotations.

On January 21, 1979, COMEX suspended the trading of Silver futures contracts except for the liquidation of existing contracts. Since that time, COMEX has taken additional actions affecting Silver futures contracts trading based on its determination that an emergency condition exists in this market. It is impossible to predict at this time the effects of these COMEX decisions on the availability of future Silver price quotations for use in determining Indexed Principal Amount, since COMEX markets continue to be significantly restricted.

Other sources of price quotations for Silver, which may be utilized in determining the Market Price, are the Chicago Board of Trade (which has also recently imposed certain trading restrictions), Handy & Harman (a major silver fabricator) and the London Metal Exchange. The existence of these sources for future quotations, however, cannot be assured. If these sources are unavailable, the Company is authorized under the Indenture to specify, in connection with the determination of Market Price, any entity customarily publishing quotes for the purchase or sale of Silver and which quotations for the price of Silver are generally accepted in the silver industry.

In addition, because of the nature of the Bonds and the calculation of Indexed Principal Amount, the price of the Bonds as traded may be related to the price of Silver. As a result, the trading price of the Bonds may be subject to wide fluctuations and may be affected by psychological and other factors relating to the Silver markets. Silver and instruments indexed to Silver have been in the past, and could be in the future, subject to government actions and regulations intended to limit the price or restrict the ownership of precious metals such as Silver.

The Company expects that market prices of the Bonds may reflect the market price of Silver. If this occurs, a sharp decline in the Market Price of Silver would probably result in the Bonds trading similarly to straight debt obligations. In view of the 8½% rate of interest to be paid on the Face Amount of the Bonds, they may, therefore, trade at a substantial discount.

Recently, the trading prices of Silver have been characterized by unprecedented volatility. Following a very sharp increase in Silver prices, particularly during the last quarter of 1979 and January 1980, the price of Silver declined sharply from its high-point. On March 26, 1980, the price of Silver declined $4.40 to $15.80 per Ounce, and on March 27, 1980, declined $5.00 to a low Spot Settlement Price on the COMEX of $10.80 per Ounce, a decline of approximately 50% during a two-day period. It has been publicly reported that such sharp fluctuations may have been the result of accumulations and subsequent financial difficulties of a few major speculators in Silver. There can be no assurance that current prices reflect current levels of supply of and demand for Silver. Additionally, such volatility may continue in the future. See "Price Range of Silver."

It is not the Company's policy to take positions in the Silver futures market except as a "hedge" in connection with its silver mining activities. However, it is possible that the Company's operations, including mining and selling its silver, could affect the Market Price of Silver and, therefore, the Indexed Principal Amount. Underwriters and dealers participating in this offering may have recently effected transactions in the Silver cash or futures markets for themselves or customers, and may hereafter do so. Such activities of the Company, underwriters or dealers may be reflected in the current Market Price of Silver and could affect the Market Price of Silver in the future, including the Market Price used to calculate the Indexed Principal Amount.

Employee Relations

On March 15, 1980, employees at the Sunshine Mine commenced a strike against the Company. Such strike is continuing and accordingly the Company's mining production is substantially impaired and its revenues and income are being adversely affected. See "Business — Additional Information — Employee Relations."

Accounting Treatment

The Bonds will be stated at their principal amount. Interest will be accrued and paid at the stated rate.

The Company has dedicated the amount of silver production necessary to retire the Bonds. See "Delivery of Cash or Silver." If the Market Price of Silver is above the Face Amount of Bonds, the Bonds will be fully satisfied by such silver production or the proceeds therefrom. As the concentrates designated for retirement of the Bonds are shipped to the smelter, revenue will be recognized similar in timing to the Company's customary method of revenue recognition. If the Market Price of Silver exceeds $20 per Ounce, the amount of revenue recognized will be $20 per Ounce. In the event and to the extent that such Silver is sold to meet a cash requirement for the sinking fund, the person to whom the Silver is sold will be instructed to pay the proceeds directly to the Trustee and no gain over the price specified in the Bonds will be recognized.

In the event the Market Price of Silver is less than $20 per Ounce, all the Company's production will be sold in the normal course and the sinking fund will be satisfied by cash.

In the event that the Company is unable to produce the amount of Silver required in any given year due to the occurrence of an unforeseen event, such as a prolonged strike or earthquake, such occurrence and the effect thereof on the Company will be accounted for as an event occurring during that particular year. Consequently, in those circumstances, the Company may be required to provide the sinking fund deposit from its general resources other than production, and the cost of such deposit over the Face Amount of Bonds to be redeemed with that deposit will be charged to income in that period.

If at any reporting date the estimated cost to mine and process a sufficient amount of silver ore into a deliverable form in satisfaction of the then outstanding Face Amount of Bonds exceeds the Face Amount of such Bonds outstanding, the liability for the Bonds will be increased by such amount and operations will be charged for the amount of such increase.

The Company has discussed the foregoing accounting treatment with its auditors, Coopers & Lybrand, and they have concurred in the accounting described.

Other Consequences

The purchase of Bonds may result in certain tax and other legal consequences. See "Description of Bonds — Certain Tax Consequences" and "Description of Bonds — Usury Laws."

EXHIBIT 2

Summary of The 16-to-1 Mine Complex Estimated Capital and Pre-Tax Operating Costs

		Annual Operating Cost	
Item	Capital Cost ($'000)	Total ($'000)	$/Ton of Ore Processed
1. mine site preparation and road improvement	$ 100	$ --	$ --
2. townsite land, trailer hook-ups (30 units)	780	164.3	1.10
3. mine development (by contract)	2,000	--	--
4. mine equipment, installation and operation	1,100	763.5	5.09**
5. processing: a. site preparation, power and water .... $ 590 b. permitting .... 100 c. metallurgical testing. 50 d. engineering design ... 100 e. concentrator .... 4,910 f. tailings dam 500	6,250	1,884.0	12.56
6. transport, mine mouth to mill by contract	--	45.0	.30
7. G&A installations, salaries, supplies	300	731.4	4.88
8. parts inventory: a. mine .... 250 b. mill 230	480	--	--
Sub-Totals	$ 11,010	$ 3,588.2	$ 23.93
9. working capital: a. mine, 3 months .... 200 b. mill, 3 months .... 471 c. G&A, townsite 224	895	--	--
10. depreciation:* ($11,010,000 less land cost of $90,000 = $10,920,000 over 862,500 tons)	--	1,889.0	12.66
TOTALS	$ 11,905	$ 5,487.2	$ 36.59

*Theoretical -- disregards fact that some items are expensed.
**For buildup, see "Summary of Mine Operating Cost Estimate".

Summary of Mine Operation Cost Estimate

Item	Quantity (feet)	Unit Cost ($/foot)	Total Cost	Cost/Ton Processed
Stope Development (over life of mine):				
haul drift (6,800 ft)	600	$150	$ 90,000	$0.104
crosscut (6,800 ft)	420	150	63,000	0.073
slot drift (6,800 ft)	800	150	120,000	0.139
drill drifts (7,000 & 7,100 ft)	2,700	150	405,000	0.470
raises (all levels)	600	200	120,000	0.139
Rounded Totals and Costs	5,120 ft	$156/ft	$798,000	$0.930

Blast Hole and Vertical Crater Retreat Stoping:

average stoping cost	$1.32	
haulage, including operator & loaded salary of $28,000	0.61	$1.93
stope development and stoping/hauling		2.86
Contingency @ 20%		0.57
Total (with Contingency)		$3.43

Mine Support and Supervision (per year):

Item	Quantity	Unit Cost	Total Cost	Cost/Ton Processed
mine superintendent	1	$ 36,000	$ 36,000	$0.24
master mechanic	1	36,000	36,000	0.24
mechanics	3	24,000	72,000	0.48
mine engineer	1	24,000	24,000	0.16
clerk	1	16,000	16,000	0.11
fringe benefits @ 35%	–	9,200	64,400	0.43
Totals	7	$ 35,485	$ 248,400	$1.66

GRAND TOTAL, MINE OPERATING COST $5.09

12.12

THE IMPACT OF A RECESSIONARY ENVIRONMENT ON PRIVATE COMPANY FINANCING

William J. Potter and Roger N. Pyle

Prudential-Bache Securities Inc.

New York, New York

Introduction

The primary ingredient for survival of mining companies during a recessionary period is to be innovative and resourceful in the structuring and financing of operations. Over the past decade, many mining companies have had to retrench and restructure their operations and to redeploy their assets just to survive. During the same period, traditional sources of financing for mining companies -- commercial banks, subsidized export credit loans and long term straight debt financing in the public markets -- became less available and more expensive. Meanwhile, new markets and new financing instruments provided new alternatives for mining companies. Those companies which exploited these alternatives greatly enhanced their capacity for survival and even for profitability.

This article considers the strategies adopted by some companies to adapt to changes in this recessionary environment; the inadequacy of traditional sources of financing; certain of the new financing instruments, structures and markets which have become available; and how several mining companies have utilized these new instruments, structures and financial markets to improve their competitive positions.

THE IMPACT OF RECESSION ON MINING COMPANIES AND TRADITIONAL FINANCING SOURCES

In its 1983 year-end article "The New Economic Order" Mining Journal characterized many of the large private sector mining companies with operations in the industrialized countries as being "gripped between the jaws of high working costs and stagnating metal commodity prices." The article went on to characterize the Canadian mining industry in particular as being "in the difficult position of having to seek to sell the bulk of its output in a world market characterized by over-capacity, where production costs were determined by internal political and economic factors while prices were set by external competitive forces of supply and demand." Miners of a wide range of commodities in the industrialized countries have come to realize that their domestic mineral resources are in many cases no longer competitive with global sources of supply. To compete and survive they would have to rely on their technical and financial expertise in mineral exploration and development and in mining and processing.

Strategies For Survival

Global competition has compelled management in the extractive industries to improve their worldwide cost position. This is the only means whereby a mining company can differentiate its product and attain a substantial competitive advantage. During the past decade, for example, ASARCO Inc. and Inco Ltd. chopped their payrolls, postponed expansion plans and otherwise retrenched. ASARCO reduced its capital expenditures and cash requirements for operations in 1984 to a level almost equal to its internally generated cash. The company's finances were also strengthened by increasing total lines of credit from outside sources, by selling common stock, and by reducing annual pension funding through a change in actuarial assumptions. Other companies returned to their traditional organizational structure where they had a demonstrated competitive strength. Consolidated Gold Fields, for instance, resumed its operation as a mining finance house. Its capital investment programs are now devoted solely to resource development. In mid-1984, at a time of firming zinc prices, Union Miniere S.A. used the opportunity to integrate its smelting capacity and to direct investment to certain subsidiaries at the expense of less viable operations.

Smaller corporations realized that it was not enough just to be the low-cost producer. For example, Drummond Coal Company in Alabama, which dominates its regional market due to its low-costs, has also developed a world-wide trading capability and has placed incremental tonnage from its own mines as well as brokered coal produced by others, thereby minimizing capital risk. In Canada, Campbell Resources has exploited difficult times by acquiring assets cheaply. In the process it improved its capacity to withstand cyclical economic swings by diversifying from solely a copper and gold producer to a company based upon asbestos, coal, copper, gold and natural gas.

Other companies entered into joint ventures with other companies or investors in order to continue or expand mining operations. In a period of twelve months, Wharf Resources, another Canadian company, capitalized on the recession by moving from having virtually no revenue-producing properties to controlling almost 45000 oz. per annum of gold production. This was accomplished by joint venturing with a series of financial institutions and by purchasing significant stakes in Rayrock Resources, Limited and Lacana Mining Corporation. Financial institutions were attracted by the opportunity to co-venture with an operating management team capable of providing a substantial yield on its capital plus the opportunity for a capital gain.

This generic partnership of miner and investor was expanded through successful and inexpensive financial placements which included commodity options for Lac Minerals, Ltd., Dome Mines, Limited, Homestake and a number of other precious metal producers; the success of a number of gold funds in Canada, the U.S., Australia and Europe; and the development of a number of successful exploration funds. The C.M.P. Mineral Partnership is an example of a coventure of miners and investors to enable a group of public Canadian mining companies to continue mining operations at a time when economic conditions in the industry precluded individual companies from continuing such operations. These companies developed a financing vehicle to take advantage of certain Canadian tax regulations available to mining ventures which allow limited partners, primarily individuals holding "flow-through shares," to obtain income tax deductions related to the cost of mineral exploration. In July 1984, C.M.P. raised U.S. $60 million for exploration through the issuance of 600,000 limited partnership units having the benefit of such deductions.

Inadequacy of Traditional Financing Sources

The impact of the recession upon mining sector lenders and financial intermediaries was equally severe. Lending problems and strained margins occurred at a time when financial deregulation had created a new and more competitive environment Insurance companies, in an effort to match revenue streams altered due to inflation, became eager players in funding capital projects. Investment bankers, in the face of more intense competition, mustered greater amounts of capital to support underwritings and committed substantial new resources to enhancing marketing capability and new product development. The rules of the game for the financial sector were changed at a time when many institutions were least able to deal with change. Bank failures in 1982-83 reached new post-depression heights in the United States and government nationalization became commonplace in lesser developed countries such as Mexico.

As a result of the foregoing and other factors, the traditional sources of finance for mineral development on a project finance basis became less available and more expensive. Commercial bankers began to impose severe credit restraints while they worked out their own problems. The change by OECD countries in 1983 to eliminate the subsidy element in export credits rendered that source of finance in many cases more expensive than market rates set by commercial banks. Finally, for companies preferring to finance on their own corporate credit, costs of straight long-term debt reached new highs and investors increasingly declined to purchase long-term obligations of lesser credits.

Increased Receptivity of Investors to New Instruments

Many investors had also suffered from the recessionary environment and the increased volatility of the credit markets. For example, traditional portfolio investors accustomed to preserving principal by investing in fixed income instruments found that substantial increases in rate levels for long term instruments could dramatically reduce the value of their portfolios. Greater market volatility made investors more receptive to financial instruments which provided them with flexibility or a hedge against sudden market changes.

NEW ALTERNATIVES FOR FINANCING FLEXIBILITY

In the face of adverse industry and capital market conditions, creative bankers and financial intermediaries sought to create financing instruments and strategies designed to satisfy both the need of the mining companies for capital and the requirement of investors for security, adequate yield and flexibility. Other developments also provided new alternatives. The following is a summary of certain major recent developments in the capital markets which have given mining companies new options and greater flexibility:

Globalization of Capital Markets

The market for capital has become global. The U.S. public securities market has expanded in volume and depth, and many new types of companies have been able to enter that market. The Euro-markets have greatly expanded in volume and many new types of instruments have been successfully introduced. The Japanese capital market is just opening for non-Japanese borrowers and equity risk capital, while Japanese authorities are in the process of lifting many restrictions which have protected their domestic institutions. This capital market promises to provide substantial competition for Europe and North America in the future. Large sources of capital have also emerged in centers such as Singapore and Hong Kong as well as in the Middle East. As a result, the mining financial executive now has access to many alternate capital sources and currencies, and can fund projects in many markets on varying terms.

More Merger and Acquisition Opportunities

The facilities of investment banks, merchant banks and commercial banks for merger, acquisition and divestiture services have dramatically expanded. The mining executive therefore has opportunities to acquire new assets or businesses to strengthen existing operations as well as to diversify into contracyclical businesses, or to divest entities or assets which are inconsistent with future strategic plans.

Opportunities to Restructure the Corporate Balance Sheet

The mining finance executive now has many new opportunities to restructure the corporate balance sheet and to insulate the company's financial position from interest rate swings, currency shifts or changes in raw material costs. For example, through currency and interest rate swaps a company may shift interest rate exposure from a fixed rate to a floating rate or visa versa, or shift exposure from one currency to another. This market has mushroomed as nearly $80 billion in interest rate swaps have been effected in the past five years. Through the use of currency, interest rate or raw material options, it is now possible to reduce forward commodity or other economic risks. The recent establishment and growth of option trading exchanges in various parts of the world has greatly expanded the availability of such alternatives. Finally, through transactions such as debt-for-equity swaps or economic defeasance of debt (the economic equivalent of retirement of debt), the liability side of the balance sheet may now be restructured.

New Opportunities for Asset-Based Finance

There are new opportunities in asset-based finance which enable a company to borrow, often on an off-balance sheet basis, on the credit of a particular asset or group of assets rather than the economic condition of the company as a whole (e.g. the Ok Tedi financing already profiled elsewhere in this book). Accordingly, the company can raise funds on improved credit terms by limiting the lender's exposure to an asset package which represents a better credit than the company as a whole. At first, asset-based finance was applied primarily by banks and other lending institutions to finance heavy equipment. However, creative bankers have found ways to finance on the basis of other assets (even intangible assets such as receivables) and to do asset-based financings in the public (as opposed to the private institutional) markets.

Increased Use of Hybrid Securities

The use of so-called "hybrid" securities has increased. These securities enable a company to raise funds more cheaply by offering the investor the opportunity at specified times to convert into another security, thereby providing a "hedge" for the investor in volatile credit markets in return for which the issuer receives a benefit. Historically, companies obtained lower debt rates by offering bonds convertible into stock. Recently, an increased number of floating-to-fixed rate debt and other similar debt securities have been offered which enable the issuer to obtain lower rates without offering equity. For example, by giving the investor in a floating rate debt security the option to convert into a fixed rate security at a time several years after issuance, the issuer provides the investor with a valuable hedge. In return for such option, the investor is prepared to provide the issuer an attractive floating rate. As another example, a number of mining companies in negative tax-paying positions have realized savings by issuing convertible exchangeable preferred stock which is exchangeable into subordinated debt at the option of the issuer when it returns to a taxable position.

Increased Use Of Credit Supports

The use of financial credit support in connection with borrowings in the public and private markets has increased, and should increase significantly in the future. In cases where it is cost effective (that is, the rate savings outweigh the premium or cost of credit support), the use of a credit support can also broaden the market for an issuer's securities and, in some cases, improve investor perception of the issuer. Although initially bank standby letters of credit were the most frequently utilized credit support, as investor perception of bank credit and bank ratings declined, surety bonds or other credit supports issued by insurance companies or other financial institutions have been used more frequently.

Increase of Short-Term Financing in Public U. S. Securities Markets

An increasing volume of short-term financing has been done in the public markets. Accordingly, companies not only have access to an alternative to the commercial banks for short-term financing, but in many instances have obtained better rates in the public markets. A member of new instruments have been developed to access public short-term credit markets, including floating rate notes, commercial paper and public market note facilities.

Opportunities to Reduce Project Risk

Various financial institutions are now more prepared to absorb projects risk than previously. Banks, for competitive reasons, have begun to assume a greater share of risk in project-related financings. Many insurance companies have increasingly been prepared to insure non-financial risks (i.e. environmental, legislative) in connection with projects. The result is that mining companies can now undertake large projects and limit their exposure to major risks in a project to a much greater extent than before.

Yield Curve Finance

Investors have sought to obtain lower rates for long-term borrowing by accessing the short end of the yield curve. For example, traditional issuers of, say, thirty year bonds have issued bonds with a thirty year maturity, but which may be put back to the issuer at par in the fifth year. Because of the put feature, the bonds bear interest at a 5 year interest rate despite their thirty year maturity. Other issuers have obtained even lower rates by issuing long-term bonds which may be put to the issuer at very short intervals (i.e. daily, weekly, monthly), with provision for remarketing bonds which are put. In either case, lower short-term rates are obtained for what is, in effect, long-term financing.

New Balance Between Issuer and Investor Interests

New alternative instruments have become available at least in part because creative financial intermediaries were able to balance the needs of issuers and investors in ways more complex than just simply rate and term. An example of such new balance is the retractable debenture which was popular during 1984. The following is a summary of the complementary advantages to issuers and investor of a retractable debenture with a ten year maturity and a put back to the issuer at par in the fourth year:

Issuer Advantages

(1) The bonds bear interest at a rate which is below the rate at which the issuer could currently issue four year fixed rate debt.
(2) If the put option is not exercised at the end of the fourth year, the issuer has issued fixed rate debt below the current cost of ten year debt.
(3) The issuer will save costs of reissuance if the put is not exercised.

Investor Advantages

(1) If rates decline over the first four years investors retain an above-market rate for the remaining six years.
(2) If rates increase in the first four years, investors can exercise the put and reinvest at higher rates.
(3) Investors do not incur transaction costs with the put.

Debt Financing Alternatives

The variations and complexity of the traditional note or bond alternatives for the mining financial executive in a specific by the end of 1984 is illustrated in TABLE 1. This chart was prepared for a major multinational resource company heavily dependent upon base metal revenues and subject to severe revenue swings. The chart highlights the costs associated with a bond issue of $50 million issued publicly in the United States or Europe, or to a group of select institutional lenders (a private placement). The seven alternatives (contrasted with what would probably have been two alternatives available in a similar context five years earlier) include a traditional debt placement for a U.S. (BBB) rated corporation; an interest rate swap (which is a contract to exchange interest rate exposure between an institution with fixed rate debt wanting floating rate debt, such as a bank, and an institution wanting the opposite (such as a mining company); "hybrids" of the notes and bonds; or the surety bond-supported instruments already discussed.

CASE STUDIES

The use and diversity of application of some of the innovative instruments and structures developed for mining companies may be illustrated by two case examples.

Floating Rate Notes--Broken Hill Proprietary

One major development in the financial markets designed to deal with the uncertainties and pressures outlined above for both the miner and the investor/lender has been the floating rate note market. Faced with a need to reschedule billions of dollars worth of lesser developed nation debts in the late 1970's and early 1980's commercial banks needed greater liquidity within their loan portfolios and they were prepared to pay for this liquidity. Mining company treasurers faced with volatile interest rates and debt-laden balance sheets found it attractive to enter into borrowing relationships on a floating rate note basis, provided that they could spread the maturities of the principal repayments or "hedge" their financial risk through some other mechanism.

In 1984, using project financings structured after those used to finance mineral deposits in Brazil and Australia, a floating rate note (FRN) non-recourse project finance package was organized for Broken Hill Proprietary (BHP) to refinance the acquisition of the Utah/General Electric mining assets. BHP achieved a lower cost by utilizing the FRN while limiting its corporate risk exposure to the assets to be financed.

BHP initially financed the acquisition of the Utah/GE assets through credit facilities to two joint ventures formed to acquire such assets. The facilities, which totalled U.S. $785 million, were provided on a production loans basis with corporate deficiency guarantees and other credit supports. The two joint ventures were the Coca Joint Venture, which owns and operates five coal mines in central Queensland, Australia and a nearby port, and the Gregory Joint Venture, which owns and operates Gregory Coal Mine, located in central Queensland. Long-term sales agreements were in place for sales to South Korea, Great Britain, France, Italy, Spain, Holland and Taiwan. The credit facilities were secured by the assets of the respective joint ventures and bore interest at a margin over the London Interbank rate.

In part because the financing under the foregoing arrangement was adversely impacting feasibility studies, BHP sought to refinance the joint ventures at better rates available in the public FRN market. For this purpose, Queensland Coal Finance Limited was incorporated in January 1984 for the sole purpose of issuing the FRNs and onlending the proceeds to the joint ventures. Queensland Coal Finance issued 12-year FRNs of (U.S.) $46 million and (U.S.) $450 million in the Euromarkets in April 1984. The FRNs were supported by a letter of credit issued by Bank of America in the former case and an unconditional guarantee from Bank of Tokyo in the latter. The FRNs were issued at substantially lower rates from those changed under the original production loan financing due to the standing of the supporting banks and the fact that the FRNs are short-term in nature and therefore not subject to Australian withholding tax.

The FRN market has matured in 1984 to the point where in a seven day period over $2.5 billion

Prudential-Bache Securities
Investment Banking Division

TABLE 1

Summary of Financing Alternatives based on October 31, 1984 quotes

	Financing Alternative	Public Market (Assuming $50MM principal)				Private Marketing (assuming $25MM principal for straight debt and $50MM for domestic surety backed bond and retractable debenture)			
		Base Note	Gross Spread	Credit Support Fee	All-in Rate	Base Rate	Placement Fee	Credit Support Fee	All-in Rate
I.	Straight Debt BBB equivalent credit (Refer to Exhibit II)	11.38% (175 basis points over 30 year	7/8% (converts to 12 basis points)	N.A.	13.50%	13.78% (200 basis point over	1% (converts to 19 basis the 10 year points)	N.A.	13.97%
II.	Domestic surety backed bond, assuming Aetna backing 30 year maturity for public, 15 for private	12.38% (75 basis points over 30 year Treasury)	3/4% (converts to 10 basis points)	1/2% (converts to 6 basis points)	12.54%	12.78% (100 basis point over 10 year Treasury (a)	1.25% (converts to 23 basis points)	3/4% (converts to 14 basis points)	13.15%
III.	International surety backed bond, assuming Aetna backing (Refer to Exhibit III 7 year maturity	11.83% (15 basis points over 7 year Treasury)	1 7/8% (converts to 41 basis points)	1/2% (converts to 11 basis points)	12.35%	N.A.	N.A.	N.A.	N.A.
IV.	Retractable Debenture with 4 year put, 10 year with final maturity (Refer to Exhibit IV) 10 year maturity	11.84% (40 basis points over 4 year Treasury)	7/8% (converts to 28 basis points)	N.A.	12.12%	12.94% (150 basis points over 4 year Treasury)	1.25% (converts to 41 basis points)	N.A.	13.35%
V.	Single Exchange Debenture (SED) (Refer to Exhibit V) 5 year maturity	10.20% floating (present floating rate is 85 basis points over 91-day T-bills; converts to 3 year final obligation at 75 over comparable Treasury)	7/8% (converts to 23 basis points)	N.A.	10.43% floating	N.A.	N.A.	N.A.	N.A.
VI.	3 Year Extendable Debenture, 10 year maturity	11.85% (55 basis points over 3 year Treasury compared with 70 basis points over for a 102 basis points over comparable 10 year Treasury	5/8% (converts to 25 basis points)	N.A.	12.10%	N.A.	N.A.	N.A.	N.A.
VII.	Interest Rate Swap 7 year maturity (Assumes floating Libor debt at a 75 basis point spread)	12.23% (55 basis points over 7 year Treasury)	75 basis point	N.A.	12.98%	N.A.	N.A.	N.A.	N.A.

(a) On a private basis, both the straight debt and the domestic surety bonds are over the 10 year Treasury with a 5 year sinking fund.
N.A. (not applicable)

of new debt was issued in Europe. This is an interesting figure when contrasted with 1983 yearly international bond levels of (U.S.) $75 billion, (U.S.) $50 billion less than 1981. Its use has become commonplace within a period of thirty-six months for projects as diverse as the Ashton diamond project in Australia and the ALCAN and Alcoa aluminum projects in Brazil and Australia.

Bullion Financing--Lac Minerals and Gold Company of America

The degree of financial flexibility which has developed during this latest economic cycle as well as the development of new instruments balancing issuer and investor needs can be further illustrated in the mining context by the numerous world-wide bullion financing deals. These deals address a problem inherent in project financing for commodity companies, namely the interrelationship between the financial condition of the borrower and the frequently volatile price of the commodity. Bullion financing deals seek to stabilize this relationship by tying repayment of the borrowings to the shifting price of the commodity. At the same time, investors obtain a hedge against erosion of the real value of their instruments.

In March of 1984 Lac Minerals Ltd. obtained funds for the exploration and development of their Hemlo, Ontario gold mine operations by selling 50,000 units each consisting of one $1,000 8% debenture, due April 15, 1989, and four gold purchase warrants, in the Euromarkets. Each warrant was exercisable upon completion of distribution and entitled the holder to purchase 0.5 Troy ounces of gold from the company, at a purchase price of US $230 (equal to a price of U.S. $460 per Troy ounce). The issue allowed the company to borrow at approximately 33% to 50% less than if it had sold debentures conventionally or borrowed from commercial banks, in exchange for agreeing to deliver gold at a price approximately 25% higher than when the units were issued. Investors thereby obtain the opportunity to benefit from an upward movement in the price of gold, but the company (through a flexible redemption and purchase clause) retained the right, after the exchange, to repurchase the debentures at par - thereby tempering the possible dilution on earnings if a rapid movement in gold occurred.

A somewhat similar bullion security was recently issued by Gold Company of America, a California limited partnership. That company recently offered depository units with distributions in the form of: gold certificates issued by Bank of Nova Scotia; bullion; or the cash equivalent of gold bullion. The distribution in gold were deemed to be tax free to U.S. investors and distributions were tied to the gold price. The partners in this venture effectively sold their production forward in order to provide front-end financing for their project. Moreover, by giving the investor a portion of the upside in these financings, the partnership effectively covered a portion of its own downside exposure.

CONCLUSION

The global restructuring of the mining and metals industry brought on by severe and turbulent economic conditions has been painful, particularly for those producers in the midst of capital expansion or having ties to the U.S. or Canadian dollar. The industry's balance sheet deteriorated markedly in the 80-83' period in which historically high debt levels and debt service exerted a significant drag on cash flow. As a result, the number of new base metal mines being developed has declined significantly. Only precious metals properties with short payback periods are being developed in volume. As we write in the fall of 1984, high real interest rates combined with a large U.S. budget deficit and stable or declining energy prices suggest that the world's mining industry could well encounter further economic difficulties or even a recession in the mid-80's. If such difficulties occur, then the fundamental changes miners have endured during the past decade will continue.

The lesson of this past recession upon the private mining company should not be lost. A strategy involving a global commodity can take many different forms and should afford flexibility as fundamental variables change. An appropriate strategy should include utilization of the global capital markets with the understanding that the same forces affecting the organizational restructuring of basic industries such as mining also affect the financial sector. A period of flux like the past decade presents opportunities for the innovative and disappointment for the inflexible.

REFERENCE

(1) The Mining Journal Ltd.; "Mining, the New Economic Order", December 30, 1983 Pages 465-67.

12.13

RISK CAPITAL: FINANCING JUNIOR MINING COMPANIES

MURRAY PEZIM

CHAIRMAN, INTERNATIONAL CORONA RESOURCES LTD.
Vancouver, British Columbia

In the good old days of the folklore of the mining industry, grizzled prospectors obtained a few loose dollars from friends and neighbours then wandered off to eventually return with the great discovery which became the core of the new corporate empires.

Alternatively, merchant banking barons obtained exclusive rights to just about anything of future value over hugh nation-size territories.

Some of this probably still goes on, likely with about the same ineffectiveness and waste of effort and money as in the past.

CURRENT CONCERNS IN MINING

In today's difficult climate in the mining industry, for that matter in all industry, **effectiveness** is the vital consideration. Cost effectiveness -- more bang for the buck -- is especially important in such an area of high risk as mineral exploration.

And there are other factors to consider.

Since minerals are depleting resources, the mining inddustry is constantly confronted with the reality that those resources extracted today must be immediately replaced if a resource enterprise is to continue. And if growth is to be achieved, even more resources must be found to support future expansion.

The current and immediately foreseeable climate internationally for the mining industry creates a special need and urgency for aggressive and successful exploration, certainly in Canada, if not also in the United States. Depressed world markets and prices have turned yesterday's economic reserves into today's waste, hopefully to be converted into economic ore again in some distant time. One way of replenishing reserves and maintaining a viable mining industry is to find and develop mines which are economic under short-term depressed price and market conditions. This probably means higher grades. A new copper discovery with the right combination of characteristics to be an attractive investment even at today's pices for copper would still merit development into production. Exploration is the only way to find such mines for the difficult short-term conditions.

Another important consideration is that normally many years stretch from the first thought of conducting exploration to production -- in those rare instances where an economic discovery is made. If discoveries are not made today, then there will not be any projects to develop when business conditions are more attrative.

This industry and the vast number of skilled professionals who are part of mining, are faced with two key problems that underlie the need for a junior exploration segment.

One consideration is how to maintain exploration during those years, indeed eras, of recession and depression. We are in such a time now. Throughout the world, major companies have sharply cut back, if not eliminated completely, exploration expenditures other than what is vitally necessary and easily and inexpensively done at the known and producing deposits.

And related to that consideration, even more important today than during happier times, is how to finance such necessary ongoing exploration if major companies and governments back away because of financial stringency.

CANADIAN SUCCESS

In Canada, such problems are being tackled with a success which is capturing worldwide attention. The ability of mineral explorers to obtain willing, eager, high-risk capital is achieving even more outstanding results than in the past. These results are reflected in the widespread continuing exploration ativity in Canada which even spreads into the United States, with some outstanding discovery successes.

The most notable exploration success from junior mining activity is the discovery of the Hemlo gold camp in northwestern Ontario, a source of gold which promises to rank with the South

CASE STUDIES

African Rand. In the short period since the 1981 discovery by International Corona Resources, the three major mine projects in the Hemlo area have reported 75 million tons of reserves containing about 19 million ounces of gold. Much more is expected to be found in the area.

Hemlo has an additional significance beyond gold mining -- in leading to a new and most promising trend in how junior exploration companies are financed.

These trends in Canadian junior company financing as well as some of the key considerations which make for an effective system of junior mining financing and exploration-development activity will be examined in this discussion.

A key point to keep in mind is that junior mining is a distinct and separate part of the Canadian mining industry developed to its present important status since the silver discoveries of Cobalt early in the century.

The Prospectors and Developers Association which brings together the many disparate interests in junior mining organizations last year marked its fiftieth anniversary. The fact that such an organization with some 3,500 members is an active part of the Canadian economic system illustrates the importance of junior mining. The membership illustrates the many necessary elements which go into successful junior mining organizations. It includes independent prospectors, financiers, bankers, government officials from resources departments, stockbrokers, investment bankers, geologists, geochemists, geophysicists, drillers, stakers, linecutters, assayers and others.

Canada has some excellent reasons for having such a strong junior mining sector. The mining industry is a most important part of Canadian economic life. Canada is the largest exporter of minerals in the world. Minerals account for a larger share of gross national product than agriculture and the minerals share is triple the contribution of the large automotive industry. Mining in Canada is sophisticated and technologically advanced. Average weekly wages exceed $555 Canadian and are among the highest of any Canadian industry.

The potential future contribution of junior mining to the Canadian economic was discussed by a former senior official in the important government of Ontario. G.A. Jewett, in a paper published in "Policy Options Politque," a magazine of considerable influence in government, noted:

"In this decade, the (junior mining) sector could invest as much as one billion dollars, in each of several provinces, with the probability that other Hemlos, Pine Points and Chibougamaus will be discovered. Most certainly widespread employment in our Northern regions will be created, if we let this investment happen."

Mr. Jewett recalled that the junior sector, equity financed, searching for gold during the depths of the depression fifty years ago also carried out the preliminary work which led later to major base metal, uranium and other developments.

Gold may be the objective today, as it was in the thirties, but the search for gold invariably turns up other minerals.

VANCOUVER A KEY BASE

Junior Canadian mining is certainly well on the way to meeting these predictions. The Vancouver Stock Exchange is a useful indicator. That stock exchange takes pride in being a marketplace where risk capital is mobilized. It is probably the most efficient mechanism yet created for the raising of high risk money for resources exploration and development.

Vancouver has taken over that role from the well known Toronto Stock Exchange which has its origins in the financing of junior mining ventures starting with Cobalt silver, expanding with gold in the thirties and extending into the 1950s and 1960s booms with uranium and base metals. As the then central marketplace for those who would venture into the Canadian Shield and the Cordillera, Toronto based juniors grew into such resources giants as Noranda, Falconbridge and Denison.

In 1983 some $200 million of high risk equity capital was raised in Vancouver, employing the mechanism created and watched over by the stock exchange authorities and the provincial government authorities. It is not unusual, during those brief intervals when public willingness to speculate is especially eager, for Vancouver-based junior companies to draw in risk capital at rates exceeding $20 million monthly. This is an outstanding result for such a fickle source of financing.

The Pezim group is proud to have accounted for the largest single share of the risk capital obtained in 1983 in Vancouver.

BROADER ENCOURAGEMENT

Vancouver's success has not escaped the envy of Toronto and Montreal and the governments of the province of Ontario and Quebec. Where formerly these important mineral provinces placed effectively prohibitive obstacles in the way of junior resource financing, today they are taking actions which indicate encouragement.

And even the federal government in Ottawa, through special tax legislation is providing practical, financial encouragement.

The prediction of "as much as one billion" of junior mining risk capital spending in each of several provinces -- probably with Ontario, Quebec and British Columbia leading -- certainly looks

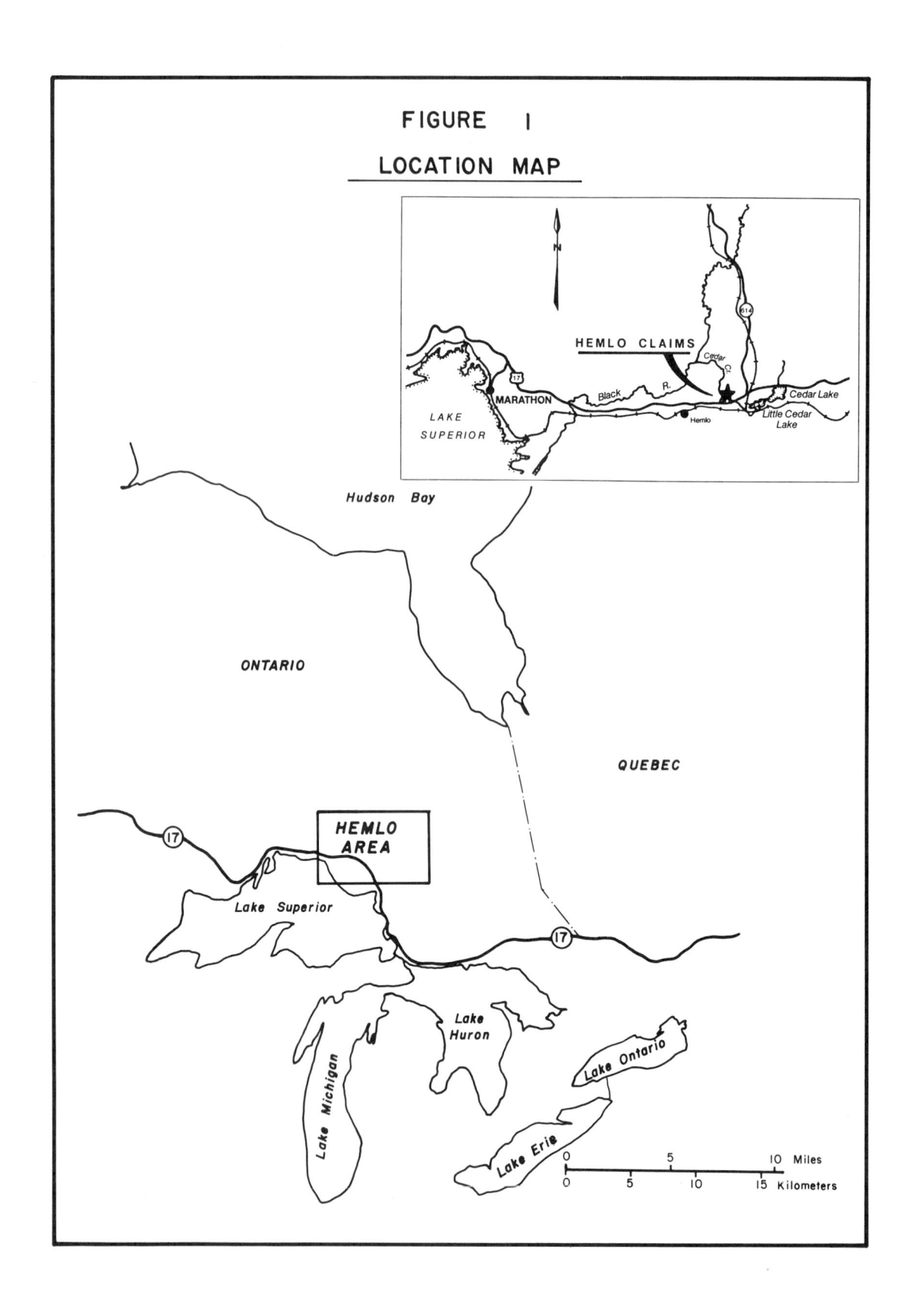

FIGURE 1

LOCATION MAP

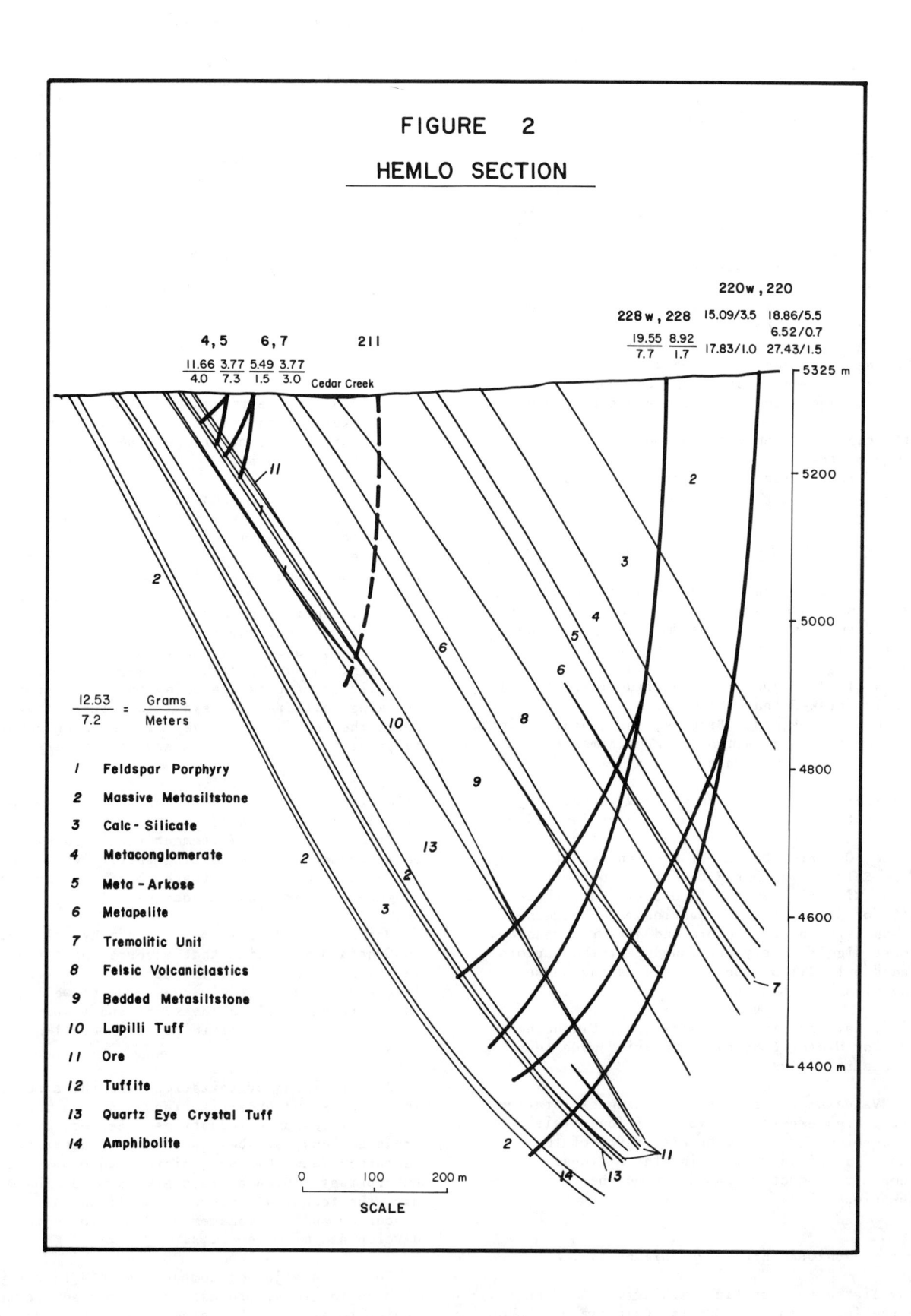

FIGURE 2

HEMLO SECTION

$\dfrac{12.53}{7.2} = \dfrac{\text{Grams}}{\text{Meters}}$

1. Feldspar Porphyry
2. Massive Metasiltstone
3. Calc-Silicate
4. Metaconglomerate
5. Meta-Arkose
6. Metapelite
7. Tremolitic Unit
8. Felsic Volcaniclastics
9. Bedded Metasiltstone
10. Lapilli Tuff
11. Ore
12. Tuffite
13. Quartz Eye Crystal Tuff
14. Amphibolite

realizeable.

Risk capital as mobilized and employed by the junior mining industry has several significant characteristics which help explain why, in Canada, both the established mining majors and the authorities are providing increasing encouragement.

Such risk capital is entrepreneurial and voluntary. There is no stronger force in society than the entrepreneurial drive. The American society is beginning to again recognize this force, judging by the management literature of the past several years. It is now generally accepted, as a result of research, that small enterprises create more jobs than are created by large industry and government. The uniqueness of the enterpreneur was recognized many years ago by the great Austrian economist Joseph Schumpeter. Schumpeter described the entrepreneur as having:

"the dream and the will to found a private kingdom; the will to conquer - for the sake, not of the fruits of success, but of success itself."

RISK CAPITAL IS EFFECTIVE

Studies carried out in relation to the mining industry in Canada have clearly established that the Canadian Economy obtains "more bang for the buck" from junior mining than from the majors. Quite simply, the prospectors, grubstakers and all those involved in junior mining find more mineral per dollar risked than the majors. They find more mines than the majors. Broadly, the junior mining end of the industry is two to three times as effective in exploration as the majors.

Consider data from a study a few years ago carried out by a couple of Canadian professors on behalf of the government of the largest mining province, Ontario. Dr. A. J. Freyman, in his study on "The Role of Smaller Enterprises in the Canadian Mineral Industry in Ontario" attributed 80% (27 of 33) of the discoveries made between 1951 and 1974 to prospectors and junior companies. And most significantly, he found that the juniors expended only 29% of the funds in making these discoveries.

B. A. Kalymon, in his analysis of "Financing of the Junior Mining Company in Ontario" made this cost/benefit analysis:

"Value added from junior mining exploration activities exceeds the total of funds raised by a factor of 22.0 ... The efficiency of the funds raised in stimulating the national economic product of Canada cannot be questioned."

JUNIORS' EFFECTIVE ALTERNATIVE

Now listen to a pratical mining man looking into the future for the British Columbia and Yukon Chamber of Mines. John Brock predicted:

"There will be few effective alternatives for exploration financing for the first half of this decade other than use of the publicly financed junior mining company."

And, he added, putting the Ontario studies on a national and long-term scale:

"Junior companies consistently show their ability to find more ore than the major resource companies, and, more importantly, for less dollars spent. The statistic for the past thirty years, is roughly 40% of the expenditures for 60% of the discoveries."

A couple of other figures on risk are worth noting.

In 1970, before the major inflation, the majors of the mining industry calculated that an expenditure of $30 million was necessary to find one mineable deposit.

The risk, in the Canadian experience, is that for 10,000 mineral examinations, possibly 1,000 might reach the stage of drill testing, of which only one might be developed into production to return something on the investment made.

The financial stringencies under which major companies and governments are now forced to live with are obvious. How can they justify today employing in such a high risk way as exploration what little profits are being made from production or being collected as taxes? In effect, they would force the shareholder-investor and the general taxpayer to become involuntary risk takers.

Junior mining mobilizes the resources of voluntary risk takers -- those of us in society willing to venture a few dollars to join vicariously in the excitement generated by the entrepreneurial free-booting spirits who eagerly tramp through insect-infested bush and explore dangerous deserts and mountains.

Major company managements and many politicians in Canada now realize that encouraging (or at least, not preventing) voluntary risk taking is far preferable for their own personal well being than forcing restless investors and taxpayers to become involuntary participants in drill-hole bets.

Junior mining organizations and the associated individuals and expertise represent the initiative sector of resource industries. Senior, or major organizations, are best suited to converting the resources found by the juniors into commercial enterprises. Once a discovery is made, the majors have the technical experts, the financial resources and the management depth to define, develop and mine deposits.

Sometimes a junior company is able to carry through the total process and thus graduate from junior to senior. But generally, a marriage of convenience between junior and senior is the preferred path.

HEMLO DISCOVERY FINANCING

The gold discovery at Hemlo illustrates what happens. International Corona Resources was the company through which voluntary risk takers were brought together to finance exploration on a property staked by a two-man prospecting partnership. Donald D. McKinnon and John Larche were attracted to the area because of gold indications known from exploration carried out sporadically since the turn of the century.

Promoters (in the best sense of that word) provided the initial money and arranged public financing through stockbrokers and the stock exchange, under the broader supervision of the British Columbia equivalent of a securities commission.

Extensive exploration was carried out. This included continuing drilling starting in January 1981 until the key "discovery" holes began to establish a major gold deposit in a new geological environment.

The breakthrough drill hole was Number 76. That illustrates the persistence that a junior company can demonstrate. Further financing became only slightly easier. Another 50 holes had to be drilled until substantial public attention could be attracted. And it was only at that stage, late in 1981, that a major mining company entered the picture. An arrangement was worked out with Teck Corporation where Teck assumed responsibility for continuing development. Under this arrangement, effectively International Corona has a 45% continuing ownership without being required to provide any additional financing to bring the Corona property into production 1985.

In total, International Corona Resources provided about $2 million of exploration money for the Hemlo program. Of this, some $600,000 was raised and provided through a public offering of shares and warrants before the discovery hole. Exercise of warrants and a further private financing provided the remainder of the funds to detail the gold discovery until the Teck organization undertook to finance and carry out more advanced work. As was to be expected, the outstanding success at Hemlo facilitated further financing for other projects and programs. International Corona subsequently raised a further $12 million through sale of shares and convertible debentures to Royex Gold Mining Corporation.

Corona's discovery, the entry of Teck and other major mining companies, facilitated exploration financing for the many other companies which poured into this mining boom area on the north shore of Lake Superior.

The Pezim organization which financed Corona, as the leader, raised from the public more than $25 million for other Hemlo area exploration programs alone. And other junior mining groups also raised money for Hemlo properties.

These and other voluntarily-provided venture funds for exploration in Hemlo and many other areas are provided by thousands of people who become shareholders in the junior companies. Their backgrounds and personalities range from inveterate lottery ticket buyers to the most prestigious of British merchant bankers, Wall Street barons and Hollywood stars.

The Vancouver risk-capital mobilizing mechanism has built an international reputation. The first, the startup, money, is relatively local. Amounts raised can vary upwards from a modest private **grubstake** of as little as $50,000. More typical are amounts from $500,000 to $1 million at the stage where public financing is being sought. Under such public financings, the authorities require that the amount of money raised must be sufficient to carry out the recommended exploration program.

Broadly, as the public financing mechanism and exploration activity attracts wider interest, British Columbia residents provide about 30% of this junior exploration company money. Eastern Canada and the United States probably account for another 25% each. European investors looking for interesting speculations account for the remaining 15%. Interestingly, investors in the United States, United Kingdom and West Germany are showing growing interest in the efforts of Canadian junior companies, probably because of the fair treatment and respect their funds receive -- and the frequently outstanding results.

SOME KEY CONDITIONS

Risk capital is the most difficult to obtain, for any enterprise, let alone mining exploration. An effective system depends on many important factors which can be summed up in one word -- fairness.

In Canada, with a few exceptions, the mining industry is treated reasonably fairly in certain areas vital for effective growth of the mining industry.

Access is a key operative word. Those areas where the geological environment is favourable for discovery of mines are readily accessible -- in the legal sense rather than in the physical sense. In Canada some 80% of all land areas are held by the Crown -- by government. Prospectors, junior companies and others can, at a modest cost, quickly lay an initial claim to minerals on these lands. They do not have the frequently large expense of lengthy negotiations with landowners.

Access to explore and stake claims in Canada carries with it the right to extract minerals should they be found. This tenure, or legal right to hold, develop and mine, has proven vital to encouraging Canadians to finance exploration projections.

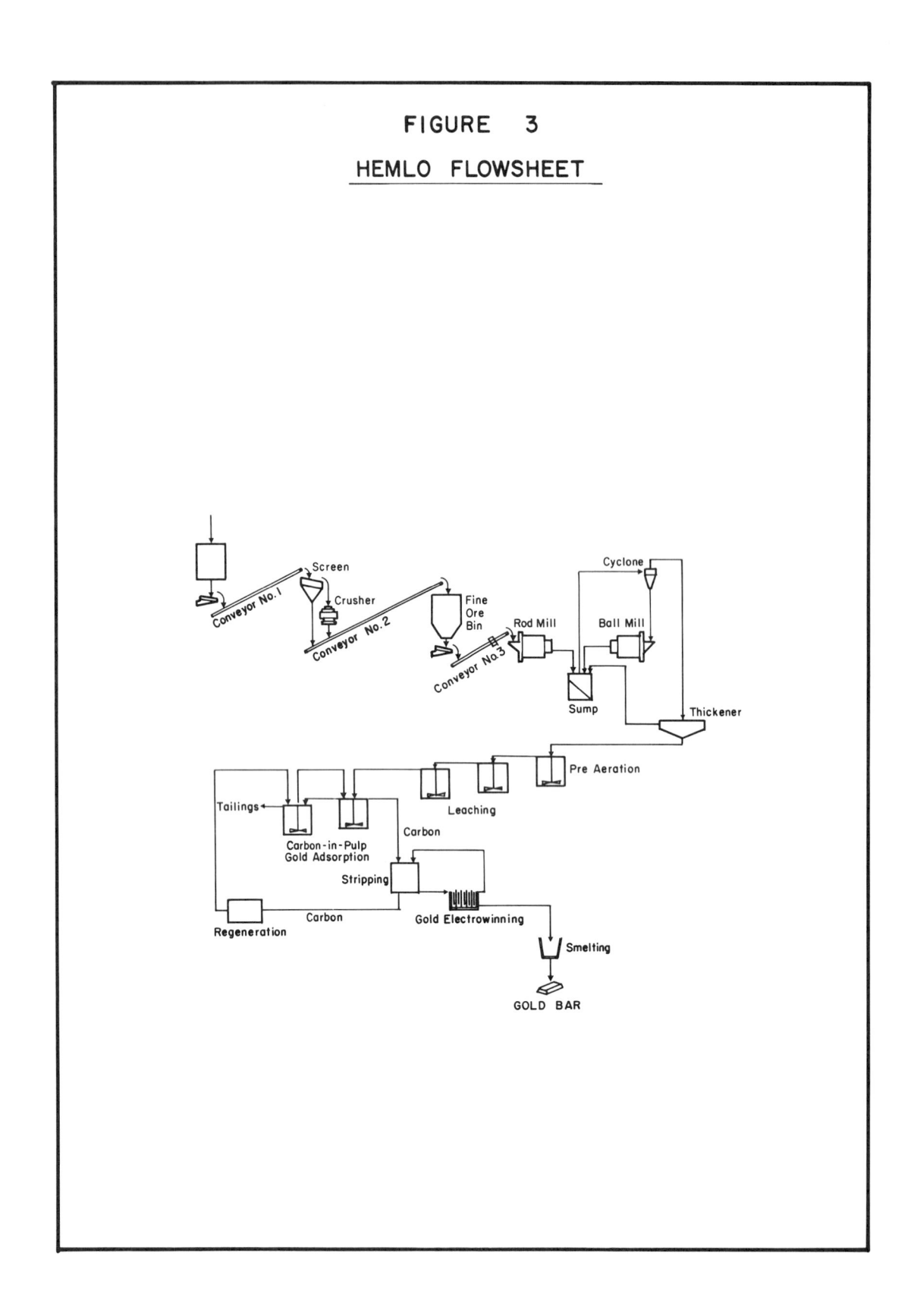

FIGURE 3

HEMLO FLOWSHEET

Canada is the only country in the world where access to explore and the right to extract go together. The normal practice in other countries is a two-stage process where the discoverer of an economic mineral deposit must negotiate a second approval before the right to mine is granted.

Further, this accessibility and right to extract package is available to foreign and domestic capital alike, with the single exception of uranium where the ultimate level of foreign ownership of a mine is limited to 33% and sales outside Canada are controlled. Compare this favourable situation with such an exciting mineral nation as Australia where all mining projects must be Australianized.

Contrast the generally favourable Canadian conditions for mining exploration with the United States where only two states, Nevada and Utah, have over 50% of their land area as state or federal government lands open for exploration. It is significant that exploration activity in the United States is concentrated in the ten western states where 20% or more of the land area is state or federal. Three states border on the mining-vibrant province of Ontario and have similar favourable geological environments. Yet very little exploration is carried out in Minnesota, Wisconsin and Michigan. Coincidently, there is no guarantee of the right to lease land where a discovery has been made.

ACCESS TO FINANCING

Access is also the appropriate word to describe how mining exploration fits into the Canadian financial community. Junior mining companies have access to the general public to varying degrees ranging from impossible to excellent, depending on the attitude of the particular provincial government. Vancouver today is a centre of mining finance and exploration companies because the government of the province of British Columbia set out to create such a financial centre as a spur to building industry -- mining and other.

In Canada the provinces have jurisdiction over the securities field. How provincial securities commissions regulate the attitudes and practices of stock exchanges determines the ability to raise exploration money from the public. While there are problems, especially in the time taken for approval of financing, the system certainly works in British Columbia.

Such regulation goes one step further than the much more complex and expensive United States securities industry regulation where the principle of full disclosure seems to be the sole criteria. That extra step in Canada is an effective effort to make certain that exploration funds raised are employed wisely for that purpose.

In Vancouver, some 90% of the money raised goes into the ground on the specific project or projects proposed at the time the public was asked to buy shares. Special professional consultants are used by the authorities to help make certain the information presented to the public is accurate and that the project isn't simply a worthless piece of "moose pasture" with no mineral possibilities.

New companies (mining or industrial) are quickly listed for trading on a special section of the stock exchange so that shareholders can have an open and public market available.

In the United States, only some 3,000 of 20,000 junior companies of all kinds trade on an exchange. The British Columbia system works so effectively that an increasing number of U.S. companies are appearing on the Vancouver Stock Exchange list. One consideration these small U.S. companies have is the high expense involved in the U.S. regulatory system.

Australia is a nation of **punters** always ready to gamble. The Australian method for financing junior enterprises appears to have an even more open system in some respects. Mining companies are regulated under companies legislation, not securities laws which regulate stock exchanges, brokers, and trading. For a mining company, disclosure is the only requirement with stock exchanges vetting a prospectus.

QUALITY PROJECTS

The Canadian emphasis on quality in an exploration project probably is the single most important underlying influence in earning public financial participation. Quite simply, the risk taker is enjoying better odds than ever in better deals.

This is happening because major mining organizations are joining in a more active partnership with junior organizations. Major mining companies now want the general public to take the initial risks. To win that public participation, they're prepared to leave a larger share of ownership.

Entrepreneurial organizations experienced in raising risk capital have been making arrangements with majors to provide money for continuing exploration on properties of merit in the major's inventory. The large mining company contributes the property while the junior provides the early voluntary risk money. The exploration department of the major is paid to manage the project.

Should success be achieved and advanced work warranted, the major then also takes over the financing responsibility -- a relatively easy task if the initial high risk has disappeared. Frequently, the major can back in to end up with controlling ownership of both the orebody and the junior company. For example, Pezamerica Resources, in the mining area, is far more than a holding company; it has its own exploration arm involved in two high-potential, highly unusual exploration projects; audacious and low-risk at one and the same time.

Operation Pezamerica consists of an intensive search for gold over a vast 50,000 acres, 40 miles east of Hemlo. Pezamerica will act as operator and retain a 25% interest in the project while Galveston Petroleum, Tri-Basin Petroleum and Zep Energy each pick up one-third of the development costs for a 25% interest. The area has never been systematically explored but a positive test for nickel was obtained in sulfide mineralization covered by the claims as well as a major showing of molybdenite. The claims were acquired because volcanogenic sediments host the Hemlo deposit and predominate on the Pezamerica claims as well. Initial exploration will consist of a geophysical investigation and diamond drilling.

The Norex Joint Venture is an investigation of 33,000 acres of claims containing a very promising gold/iron formation in the Northwest Territories. Noranda Exploration Co. Ltd. turned over a prime exploration prospect to Pezamerica and retained the right for a 50% back-in interest. The reason was quite simply. Noranda lost $82.9 million in 1982, its first loss ever, and was forced to cut subsidiary Norex's budget. Thus, Noranda President Adam Zimmerman, who admires "Pezim's tremendous capacity for raising capital," broke with precedent. Traditionally, junior firms find the play and senior firms like Noranda spend the money to make them feasible.

Pezamerica has underwritten the Norex Joint Venture for $500,000 to investigate known gold mineralization within an arsenical iron formation, according to Noranda geologist Robert Cluffs, appears "very similar if not identical to the gold mineralization at Echo Bay's Lupin Mine." The Lupin Mine, 100 miles away, contains 3.9 million tons of 0.393 ounce of gold per ton to a depth of 1,280 feet. In addition to the Lupin, this type of iron formation hosts the Homestake Mine, South Dakota, and the rich Cullaton Lake orebody in the Northwest Territories.

More typical in Canadian mining is the case where a junior company makes an important discovery on its own property. In the past major companies could acquire such discoveries on an 80%-20% or 70%-30% ownership split with the 80% for the large company. And all too frequently, after some work, the property would be placed on the shelf for years.

Now the terms are much more favourable for the risk-taking shareholders of the junior company. Mine ownership proportions such as the 55%-45% in the International Corona instance are more typical. The public knows a better deal can be obtained if it provides more funds to carry the discovery evaluation process into a more advanced stage. And so junior companies find financing facilitated.

Where junior companies are providing funds for a project developed by a major company, amounts of money raised can become quite substantial. Depending on financial market conditions and how a financing is structured, the entrepreneurs directing the junior companies might aim for several million dollars. One important influencing factor is the availability of Canadian tax shelter legislation for exploration programs. Under such provisions, junior companies have reached for $10 million and more where projects have gone well beyond the grass roots stage.

Overall, there is a circle of mutually-beneficial interests which result in happiness for all parties -- the government and society obtains new economic assets, the major mining companies can still achieve growth while taking fewer risks, and those willing individuals who support junior exploration companies receive better rewards more commensurate with the high risks taken.

INTERNATIONAL CORONA RESOURCES LTD.
(Incorporated under the Company Act of British Columbia)

Consolidated Balance Sheet
As at January 31, 1984

	1984	1983
Assets		
Current		
Cash and short term deposits	$11,449,934	$ 312,102
Accounts receivable	412,479	73,718
	11,862,413	385,820
Debenture receivable (Note 2)	2,536,500	1,000,000
Investments (Note 3)	5,977,516	1,293,334
Fixed (Note 4)	175,209	285,118
Resource properties (Note 5)	6,068,394	3,916,433
Deferred charges (Note 6)	2,079,000	—
	$28,699,032	$6,880,705
Liabilities		
Current		
Accounts payable and accrued liabilities	$ 295,828	$ 79,699
Debentures payable (Note 7)	7,700,000	—
	7,995,828	79,699
Shareholders' Equity		
Capital stock (Notes 8 and 11)		
Authorized		
20,000,000 (1983 - 9,250,000) common shares without par value		
Issued		
11,009,949 (1983 - 3,955,832) common shares	21,848,447	7,570,224
Contributed surplus	7,500	7,500
Deficit	(1,152,743)	(776,718)
	20,703,204	6,801,006
	$28,699,032	$6,880,705

Approved by the Directors

[signature]
Director

[signature]
Director

See accompanying notes to consolidated financial statements.

INTERNATIONAL CORONA RESOURCES LTD.
Consolidated Statement of Loss and Deficit
For the year ended January 31, 1984

	1984	1983
Revenue		
Interest	$ 337,785	$ 79,916
Gain on sale of resource property interest	250,000	—
Gain on sale of investments	16,441	2,602
Gain on sale of fixed assets	13,425	—
	617,651	82,518
Administrative expenses		
Administration	30,500	36,000
Audit and accounting	27,650	26,642
Directors' fees	9,000	3,500
Filing and listing fees	5,506	5,064
General exploration	31,744	8,693
Interest	100,869	—
Legal	262,895	160,036
Office	70,834	21,863
Printing and shareholder information	118,535	116,677
Transfer agent fees	43,471	35,710
Public relations and travel	208,636	138,579
Wages and benefits	84,036	30,729
	993,676	583,493
Net loss for the year	376,025	500,975
Deficit at beginning of year	776,718	275,743
Deficit at end of year	$1,152,743	$776,718
Loss per share	$ 0.04	$ 0.07

See accompanying notes to consolidated financial statements.

INTERNATIONAL CORONA RESOURCES LTD.
Consolidated Statement of Resource Properties
For the year ended January 31, 1984

	Balance January 31 1983	Expenditures	Recoveries	Balance January 31 1984
Molson Lake Area	$3,301,673	$1,949,828	$ —	$5,251,501
Rous Lake Area	515,000	61,893	—	576,893
Interlake Joint Venture	—	225,000	—	225,000
Operation Wawa	99,760	607,587	692,347	15,000
	$3,916,433	$2,844,308	$692,347	$6,068,394

See accompanying notes to consolidated financial statements.

INTERNATIONAL CORONA RESOURCES LTD.
Consolidated Statement of Changes in Financial Position
For the year ended January 31, 1984

	1984	1983
Source of funds		
Issue of capital stock	$14,278,223	$5,150,724
Proceeds from sale of fixed assets	120,000	21,739
Proceeds from sale of investments	200,237	48,551
Debentures payable	7,700,000	—
	22,298,460	5,221,014
Application of funds		
Net loss for the year	376,025	500,975
Items not requiring an outlay of funds		
Depreciation	(8,858)	—
Gain on sale of investments	16,441	2,602
Gain on sale of fixed assets	13,425	—
Write-off of incorporation costs	—	(500)
	397,033	503,077
Resource property expenditures, net	2,151,961	2,038,896
Purchase of fixed assets	5,524	111,500
Purchase of investments	4,867,978	1,339,283
Debenture receivable	1,536,500	1,000,000
Deferred charges	2,079,000	—
	11,037,996	4,992,756
Increase in working capital	11,260,464	228,258
Working capital at beginning of year	306,121	77,863
Working capital at end of year	$11,566,585	$ 306,121

See accompanying notes to consolidated financial statements.

INTERNATIONAL CORONA RESOURCES LTD.
Notes to Consolidated Financial Statements
January 31, 1984

1. **Summary of significant accounting policies**

   a. Basis of consolidation

   These consolidated financial statements include the accounts of the Company and its wholly-owned U.S. subsidiary, Corona American Resources Inc. All material inter-company transactions have been eliminated.

   b. Resource properties

   Acquisition costs of resource properties together with direct exploration and development expenditures thereon are deferred in the accounts. When production commences, depletion will be provided on the unit of production method over estimated proven reserves. When an undeveloped property is abandoned or proved non-productive all associated costs are written off.

   c. Fixed assets and depreciation

   Fixed assets are recorded at cost and depreciation is provided on the declining balance basis at the following annual rates:

Buildings	5%
Furniture and fixtures	20%

   d. Deferred charges

   Deferred debenture issue expenses and unamortized interest expense are amortized on a straight-line basis to maturity of the debt.

2. **Debenture receivable**

   The debenture receivable of $2,536,500, which includes accrued interest of $286,500, is due from Windmill Enterprises Ltd., matures on December 17, 1984, is secured by a first charge on property near Tod Inlet, Vancouver Island and bears interest at the prime interest rate of a Canadian chartered bank plus 2%. The Company has the right to convert the debenture plus accrued interest into shares of Windmill Enterprises Ltd. at the rate of $0.50 per share. (Note 10.d)

3. **Investments**

   Investments are recorded at cost and consist of:

	1984	1983
a. Common shares of public companies		
900,000 Pezamerica Resources Corporation	**$3,155,080**	$ —
824,700 Tri-Basin Resources Ltd.	**2,256,533**	988,029
60,000 British Columbia Resources Investment Corporation	**230,894**	—
50,000 P.M. Industries Inc.	**220,509**	121,509
11,250 Interlake Development Corp.	**112,500**	—
16,000 Teck Corporation	**—**	183,796
	5,975,516	1,293,334
b. Other	**2,000**	—
	$5,977,516	$1,293,334

At January 31, 1984 the quoted market value of the shares in public companies was $6,221,413.

Other investments consist of 200 preferred shares of a private company acquired for $2,000.

4. **Fixed assets**

	1984			1983
	Cost	Accumulated Depreciation	Net	Net
Land	$ 55,136	$ —	$ 55,136	$ 73,514
Buildings	128,918	11,331	117,587	211,604
Furniture and fixtures	2,762	276	2,486	—
	$186,816	$11,607	$175,209	$285,118

5. **Resource properties**

   a. The amounts shown for resource properties represent costs to date and do not necessarily reflect present or future values. The recovery of such costs is dependent upon the future commercial success of the properties or the proceeds from disposition.

   b. The Molson Lake Area claims (seventeen Hemlo discovery claims), located in the Thunder Bay Mining Division of the Province of Ontario, were acquired for $40,000 cash and 400,000 shares. The Company is required to issue an additional 200,000 shares and pay a 3% net smelter return royalty to the optionor when the claims are in commercial production.

   The Company granted to Teck Corporation the right to acquire a 55% interest in the property. Pursuant to the agreement with Teck Corporation, expenditures of $1,930,203 (1983 - $69,797) on the Molson Lake Area by Teck Corporation have been converted into 196,272 (1983 - 3,728) shares of the Company.

   Teck Corporation has elected to place the property into commercial production and the Company has agreed to assign a 55% interest in the claims to Teck Corporation. Teck Corporation has the obligation to arrange for the provision of all money or credit necessary to bring about commercial production and has commenced shaft sinking and mill construction on the property.

   c. The Rous Lake Area claims (sixty additional Hemlo claims), located in the Thunder Bay Mining Division of the Province of Ontario, were acquired for 200,000 common shares. Galveston Petroleums Ltd. has agreed to spend $133,500 for a geological survey and issue 50,000 common shares to the Company for a 25% interest in the claims. At January 31, 1984, Galveston Petroleums Ltd. had expended $34,000 for the survey. (Notes 10.a and 11.c)

d. On March 4, 1983, International Corona Resources Ltd., Teck Corporation and Noranda Exploration Company Limited entered into an option agreement to acquire thirty-one unpatented mining claims located in the Molson Lake Area, Thunder Bay Mining Division, Ontario, from Interlake Development Corp. International Corona Resources Ltd. has a 22.5% participating interest in that agreement.

The optionees collectively are required to:
  i) Purchase 50,000 common shares of Interlake Development Corp.;
  ii) Incur expenditures of $1,000,000 by March 31, 1984;
  iii) Incur additional expenditures of $2,000,000 by March 31, 1985;
  iv) Incur additional expenditures of $3,000,000 by March 31, 1986;
  v) Provide notice of production commitment by March 31, 1987;
  vi) Commence commercial production by October 1, 1988.

When production commences, a 40% net profits royalty is payable to the optionor. At January 31, 1984 the optionees had incurred expenditures of $914,965.

The Company advanced 225,000 and purchased 11,250 common shares of Interlake Development Corp. pursuant to items i and ii above. (Notes 10.a and 11.a).

e. The Company has obtained the exclusive right to explore certain properties in the District of Algoma, Sault Ste. Marie Mining District, Ontario (Operation Wawa), from Algoma Central Railway for $15,000 cash. If the Company has spent its committed amount of $500,000 by December 31, 1985, it may exercise its option to acquire leases on the properties and Algoma Central Railway will participate in further development as a 25% joint venturer.

The Compay entered into an agreement on July 15, 1983, amended November 30, 1983, with Galveston Petroleums Ltd., Berle Resources Ltd. and Tri-Basin Resources Ltd., whereby the participants will pay 33⅓% of the costs up to the time of election to bring the property into commercial production to acquire a 25% interest in the property. Thereafter each company, inclusive of International Corona Resources Ltd., will share costs equally to earn a 25% working interest. (Note 10.a)

On December 21, 1983 the Company entered into an assignment agreement with Teck Corporation wherein the Company sold 50% of its interest in the above property for $250,000.

6. **Deferred charges**

    Unamortized 1984 debenture expenses (note 7):

	1984	1983
Interest expense	$1,754,000	$ —
Issue expenses	325,000	—
	$2,079,000	$ —

7. **Debentures payable**

   On March 15, 1983, the Company issued a $1,200,000 convertible debenture (1983 debenture) due March 15, 1984 bearing interest at 9% per annum payable semi-annually. The principal and interest, at the election of the holders, are convertible into one common share for each $5.00 owing. (Note 11.d)

   On January 31, 1984, the Company issued a $6,500,000 convertible debenture (1984 debenture) due January 31, 1987 bearing interest at 9% per annum. After December 31, 1986 the principal, at the election of the holder, is convertible into one common share of the Company for each $9.25 owing. The interest for the term of the debenture has been paid in advance by the issue of 200,000 shares at a deemed price of $8.77 per share.

   The debentures constitute a floating charge on the undertakings of the Company.

8. **Capital stock**

   a. During the year the Company subdivided its authorized and issued capital stock on a two-for-one basis. Thereafter, the Company increased its authorized capital to 20,000,000 common shares.

   The current year's share transactions are summarized, as if the two-for-one share subdivision had occurred at the beginning of the year, as follows:

	Number of shares	Amount
Outstanding at beginning of year (before share subdivision 3,955,832 shares)	7,911,664	$ 7,570,224
Issuance of shares		
For cash	900,000	6,653,750
Series "A" warrants	862,464	1,780,984
Series "B" warrants	862,549	1,940,719
Exercise of directors' options	77,000	186,050
Exploration expenditures	196,272	1,962,720
Debenture interest	200,000	1,754,000
Outstanding at end of year	11,009,949	$21,848,447

   b. On March 4, 1983, pursuant to a private placement for 200,000 shares the Company issued a warrant to purchase up to 200,000 shares at $8.50 per share to March 4, 1985.

   c. On January 31, 1984, pursuant to a private placement for 700,000 shares the Company issued warrants to purchase up to 850,000 shares at $8.75 per share to January 31, 1985 and $9.00 per share thereafter to January 31, 1986.

   d. At January 31, 1984 the following stock options to officers, directors and employees were outstanding:

Shares	Exercise price	Expiry Date
848,000	$7.50	August 17, 1988
50,000	6.25	August 17, 1988
20,000	8.32	August 31, 1988
25,000	6.25	October 5, 1988
25,000	7.57	October 5, 1988
23,000	6.25	November 18, 1988
991,000		

9. **Directors' and senior officers' remuneration**

    During the year the Company paid $9,000 (1983 - $3,500) as directors' fees and $83,000 (1983 - $30,000) as remuneration to directors and senior officers.

10. **Related party transactions**

    a. The Company has agreements with the following companies which are related by way of common directors.
       - Berle Resources Ltd.   (Note 5.e)
       - Galveston Petroleums Ltd.   (Note 5.c and 5.e)
       - Interlake Development Corp.   (Note 5.d)
       - Tri-Basin Resources Ltd.   (Note 5.e)

       At January 31, 1984, certain of these companies owed the Company $355,659 (1983 - $5,210).

    b. During the year administration fees of $30,500 (1983 - $36,000) and office expense of $2,000 (1983 - Nil) were paid to a company controlled by certain directors.

    c. The private company in which the Company holds preferred shares is related by way of common directors (Note 3.b).

    d. The debenture receivable is due from a company related by way of common directors.

11. **Subsequent events**

    a. Subsequent to the year end the Company agreed to purchase an additional 2.5% participating interest in the Interlake joint venture agreement by paying $25,000 cash and agreeing to purchase 1,250 shares of Interlake Development Corp. for $12,500.

    b. Subsequent to the year end 70,000 shares of the Company were issued pursuant to directors' and employees' options for $525,000 cash.

    c. The Company and Galveston Petroleums Ltd. entered into an amending agreement on March 9, 1984 whereby Galveston agreed to issue 50,000 shares of capital stock and a promissory note for $99,500 due April 1, 1985, bearing interest at the prime interest rate of a Canadian chartered bank plus 1%, to earn an undivided 25% interest in the Rous Lake Area claims.

    d. On March 15, 1984 the holders of the 1983 debenture converted principal and accrued interest of $1,254,000 into 250,800 common shares of the Company.

AUTHOR INDEX

Affleck, E.L., 501
Allen, P., 454
Allen, P.A., 667
Arne, K.G., 527
Arnold, R.A., 365
Berry, C., 472
Biaz III, C.F., 614
Bispham, T.P., 498
Boettcher, J.H., 53, 592
Born, C.A., 433
Boulay, R.A., 129
Bradbury, M., 624
Brock, J.S., 584
Brooks, F.H., 590
Bush, W.R., 363
Buttazzoni, F.A., 401
Caldon, J.R., 150
Caraghiaur, G., 65
Cash, L.H., 639
Castle, G.R., 103, 461, 559, 611, 671
Castleman, B.B., 377
Chender, M., 579
Cockburn, D.C.N., 47
Cole, N.H., 140
Coopers & Lybrand, 274
Coplin, W.D., 445
Cork, E.K., 42
Cramer, M., 75
Cruft, E.F., 574
Davidoff, R.L., 231
De Gavre, R.T., 736
del Castillo, N., 263
Drechsler, H.D., 295, 530
Emerson, M.E., 169, 229, 483
Eppler, W.D., 53
Ernst & Whinney, 171
Eyton, J.T., 786
Fletcher, P.H., 300
Forbes, V.R., 283
Gibbs, N.J., 493
Gillham, R., 611
Glanville, R., 114
Gluschke, W.O., 313, 315
Gorval, B.J., 600
Gustafson, M.A., 94
Haldane, L.P., 466
Hammes, J.K., 5, 305
Handelsman, S.D., 293, 657, 665
Higgins, D.M., 103

Hodge, S.J., 755
Karvonen, D.A., 769
Kemeny, R.L., 600
Klimley, B.J., 103
Kovisars, L., 33, 369
Langdon, W.L., 35
Lizaur, P.L., 326
Lohden, W.P., 134
Macmillan, N.R., 718
Mansbach, B.T., 457
Maxworthy, P.J., 390
McCabe, J.J., 278
McCarthy, D.J., 254
McGill, S., 724
McNulty, J.E., 427
Mills, R.D., 160
Miskelly, N., 755
Munita, J.C., 401
Neuhaus, D.W., 790
O'Leary, M.K., 445
Parsons, R.B., 241
Pezim, M., 816
Poole, J.L., 440
Potter, W.J., 810
Pralle, G.E., 353
Pyle, R.N., 810
Ravenscroft, R.G., 645
Rides, E.A., 630
Robinson, E.L., 10
Robison III, J.C., 126, 417
Rowe, R.M.M., 657
Sandri, Jr., H.J., 383
Schenck, G.H.K., 77
Schreiber, H.W., 488, 671, 790
Sotelino, F.B., 94
Sroka, J., 493
Stevenson, W.G., 501
Strauss, S.D., 17
Stuermer, J.R., 454
Szabo, P.J., 436
Thomason, G.P., 507
Tinsley, C.R., 3, 419, 481, 512, 519, 549, 618, 755
Tuke, A., 23
Ulatowski, T., 521, 683
Walde, T.W., 342
Wightman, K., 568
Worth, D.J., 781
Yablonsky, V., 611
Zorn, S., 331

SUBJECT INDEX

A

Accelerated Cost Recovery System (ACRS), 265, 281
Acceptances, 623
Accounting, 7
   and tax analysis, 77
   financial, 51
   for mining companies, 169
   implications, leases, 157
   policy disclosures, 210
Accounts receivable, financing, 622
Acquisitions
   analysis, 383
   in mining industry, 786
   opportunities, 811
ACRS. *See* Accelerated Cost Recovery System.
Agreements
   development, 313, 315
   foreign investment, 316
   management, 315
   mineral
      Chile, 342
      Colombia, 343
      exploration, 313, 315
      Guyana, 344
      legal status, 349
      South America, 342
Amalgamated Larder, 582
Amortization, 191
Annual reports, 530
Apportionment, 280
Appraisals, post-installation, theory, practice of, 160
ARCO, 466
Argentina, project deferral in, 354
Argyle diamond project, Australia, 572
ASARCO, 530
Asset-based finance, 812
Assets, 532
   depletion, 384
   management, 44
   replacement, 383
   valuation, 385
      model, 386
Australia
   Argyle diamond project, 572
   Blair Athol coal project, 572
   Energy Resources of Australia (ERA), 755
   Hamersley Iron Ore mines, 26
   nationalization, 354
   Ranger uranium project, 755
   Tarong coal project, 570
   taxes in, 283, 300, 301

B

Balance sheet, 36, 531
   see also Off-balance-sheet.
Banks
   approval processes, 512
   commercial
      guidelines for, project financing, 559
      international, 526
      see also International Finance Corp. (IFC).
   evaluation process, 513
   presentation to, 515
BEP. *See* Cash Flow Breakeven Price.
BHP. *See* Broken Hill Proprietary.
Blair Athol coal project, Australia, 572
Bonds, 538
   see also Industrial Revenue Bond (IRB).
Bougainville, 27
Brascan Ltd., 786
British Columbia, Bullmoose coal project, 718
Broken Hill Proprietary (BHP), 813
Broker, role of, 586
Bullmoose coal project, British Columbia, 718
Business
   combinations, 178
   cycles, 43, 75
   strategy, 363, 377
      assessment, 377
Buy decision, 150
   vs. lease, 156
Byproduct credits, 372

C

Caisse de depot et placement du Quebec, 787
California, Gold Co. of America, 814
Canada
   Export Development Corp. (EDC), 636, 637
   Hemlo project, 669
   La Mine Doyen, 668
   Mine de Bousquet, 667
   mining taxation, 241
   nationalization, 354
   Quebec Sturgeon River Mines, 583
Capital
   and operating costs, trade-off between, 119
   cost of, 78
   environmental pressure on, 21
   gains, 265
   markets, globalization of, 811
   requirements
      of mineral industry, 17
      per annum, 4
   risk, 816
   risk-adjusted cost of, 33
   see also Capital Asset(s) Pricing Model (CAPM).
   structure, 10
   alternatives, solutions, 34
   introduction to, 35
   mineral corporations, mineral projects, 33
   venture, 519, 595
      companies, 69
      structuring, pricing, 593
Capital Asset(s) Pricing Model (CAPM), 88, 143
Capital-importing countries, 276
Capital-market problems, 4
Capitalization Rate (CR), 80
Capitalization, cost, 186
CAPM. *See* Capital Asset(s) Pricing Model.
Carried interests, 588
Cash
   budget analysis, 77
   flow, 78, 81
      model, discounted, 386
      timing, 81
   option payments, 588
   profile, 381
   return on investable, 134
   see also Management, cash.
Cash Flow Breakeven Price (BEP), 60
Chile
   mineral agreements in, 342
   nationalization, 353
Coal
   Blair Athol project, Australia, 572
   mine acquisition, financial evaluation of, 53
   reserve estimates, 417
   risks in, 427
Colombia, mineral agreements in, 343
Colorado
   Equity Gold Inc., 581
   Mt. Gunnison mine, 466
Commercial paper, 537, 611
   markets, 520
Competition, impact of structural changes on, 373
Competitive analysis, 369
Completion risks. *See* Risks, completion.
Concession, mining exploration, 316
Construction, 418
   delays, 462
   planning, 469
Consultants
   independent engineering, 498
   role of, 489
Consulting firms, independent, role of in financing, 488
Contracts
   management, 317
   negotiations, international, 326
Contractual mechanisms, 313
Control processes, 365
Conzinc Riotinto of Australia (CRA), 27, 519, 568
Copper mining, long-range planning, 401
Corporations, major mining, 69

Costs
  analysis, 369
  capital, operating
    trade-off between, 119
  capitalization, 186
  cash, 370
  control, operating, 417
  depletion, 279
  estimation, engineering, 370
  indirect, 369
  labor, 371
  materials, supplies, 371
  operating, 21
  overruns, 560
    risks, 466
  production, 403
  risks, 417, 434
  structure, competitive, 75
  variations in capital, operating, debt, 129
Country
  credits, 3
  risk analysis, 417, 454, 783
Coverage, 13
CR. *See* Capitalization Rate.
CRA. *See* Conzinc Riotinto of Australia.
Credits
  byproduct, 372
  consumer, 336
  country, 3
  export, 335, 524, 640.2, 731
    insurance, 637
    programs, 520, 630, 636
    see also Export Credit Agencies (ECAS).
  see also Letters of Credit (L/C's).
  supports, 812
  trade, 536, 622
Currency
  foreign, translation, 184
  inconvertibility, 457
  swap, 441
Cut-off grade determination, impact of, 124

D

Data sources, 372
DCR. *See* Debt Coverage Ratio.
Debentures
  retractable, advantages of, 813
  sinking fund, 538
Debt
  burden, 33
  currency mix, 400
  documentation, group, 52
  financing, 528
    alternatives, 813
  level, 397
  long-term, 537
  rescheduling, 454
Debt Coverage Ratio (DCR), 60, 97
Debt:equity ratio, 555
Deferment period, 81
Depletion, 191, 257, 265
  asset, 384
  cost, 279
  percentage, 279
Depreciation, 191, 246
  U.S., 257
Development
  agreements, 313, 315
  alternatives in mining industry, 408
  expenditures, 279

expenses, U.S. tax treatment of, 254
Discounted models
  cash flow, 386
  income, 385
Discounting, 78
Diversified companies, financial management of, 390

E

Earn-in formulas, 34
Earning level, 588
ECAs. *See* Export Credit Agencies.
Economic model, 229
Economics
  managerial, 77
  project, tailoring financing decision to, 94
  see also Rent, economic.
Economies of scale, 380
EDC. *See* Export Development Corp.
Edper Investments Ltd., 786
Elliot Lake uranium mines, 24
Employment Effect Ratio, 327
Energy Resources of Australia Ltd. (ERA), 755
Engineering
  integrity, 490
  multiple program, 637
Environment, recessionary, impact of on financing, 810
Environmental risk, 553
Equipment, purchased, leased, 76
Equity Gold Inc., Colorado, 581
Equity, 614
  information requirements for, 501
  investments, unlevered, 33
  underwriting, 762
ERA. *See* Energy Resources of Australia Ltd.
Eurobond market, 522
Europe, insurance, 481
Evaluations
  geologic, 78
  project, 76
  triumvirate of, financial, technical, economic, 75
Exchange rate risks, 417, 440
Eximbank. *See* U.S. Eximbank.
Expansion, *unlevered no, levered*, 95
Expenditures
  development, 279
  exploration, 278
Exploration
  agreements, mineral, 313, 315
  and economic rent, 297
  concession, 316
  expenditures, 278
  expenses, U.S. tax treatment of, 254
  grass roots, 67
  on-site, 68
  technology, 3
Export Credit Agencies (ECAs), 630
  application to mining industry, 633
  form of insurance offered by, 631
Export Development Corp. (EDC), Canada, 636, 637
Export finance in free market, forfaiting, 624
Export-credit programs, 520
Exports. *See* Credits, export.
Expropriation, 457

F

Feasibility
  report, 493
  studies, 79, 418, 484, 499, 533
    reliability, 461
Finance
  acquisition, 576
  asset-based, 812
  compensation trade, 7
  consumer credits, 336
  export, forfaiting in free market, 624
  from mining companies, 587
  insurance companies, 339
  mining, 297, 530
    alternatives for developing country, 331
    defined, 5
  off-balance-sheet, 575
  petrodollars, 338
  production payments, 336
  risk analysis for, 417
  see also Credits, export.
  sources of, 519
    risk vs. reward, 519
  what it means for mining industry, 5
  yield curve, 812
Financial
  analysis
    methodologies of, 76
    results of, 120
  evaluation for minerals industry, 75
  futures, 443
  objectives, 7
    mining company, 42
  planning, 363
  reports, 370
  see also Management, financial.
  sources, European, 575
  statements
    analysis, 176
    elements, concepts, 171
  strategy, 363
  structure of mining industry, changes in, 10
Financier, qualities of, 574
Financing
  accounts receivable, 622
  analysis, 77
  bank, non-bank institutional, 519
  debt, 528
    alternatives, 813
  flexibility, new alternatives for, 811
  gold, 667, 788
  government-owned industrial mineral company, 769
  history of for multinational mining company, 23
  impact of
    mineral development, operating agreements on, 313
    recessionary environment on private company, 810
  industrial minerals industry, 614
  international mine, case studies illustrating, 665
  junior mining companies, 579, 816
  Mexico, 683
  mineral prospects, 574
  multiparty, 665
  of Ok Tedi mine, Papua New Guinea, 724
  of small, independent mining enterprise, 65

INDEX

public, source of funding for Canadian mineral industry, 600
see also Project, financing.
sources of, 527
    inadequacy of traditional, 811
    matching risk, reward, 527
structures, *non-recourse*, 33
trade, 622
transactions, 198
World Bank Group, 639
Forecasting, 411
    market price, 373
Foreign exchange
    risk, 554
    see also Net Foreign Exchange Effect.
Foreign Investment in Real Property Tax Act (FIRPTA), 266
Forfaiting, export finance in free market, 624
Forward sales, gold, 667
Functional units, 409
Funding
    Canadian mineral industry, 600
    considerations, 49
    promotional exploration company, 584
        government sources, 589
    sources of for mineral projects, 521
Funds, sourcing on global basis, 665

G

Geologic evaluation, 78
Go/no go decision, 95
Gold
    financing, 667, 788
    forward sales, 667
    loan, 788
        as financing mechanism, 657
        as lease, 658
        covenants, 658
    purchase agreement, 658
    warrant, 669
Gold Co. of America, California, 814
Government Agreement, 759
Governments, relations between in developing countries, 313
Grenoble Energy, 582
Group debt documentation, 52
Guyana, mineral agreements in, 344

H

Hamersley Iron Ore mines, Australia, 26
Head office, role of, 390
Hedging, 623
Hemlo project, Canada, 669, 816
Holding company, 267
Hurdle rates, impact of inflation on, 140

I

IFC. *See* International Finance Corp.
IMMSA. *See* Industrial Minera Mexico, S.A.
Inco Ltd., 736
Income
    after-tax, 81
    model, discounted, 385
    net, 39

periodic, 268
statements, 533
Indonesia, Soroako nickel project, 736
Industrial minerals
    industry
        financing, 614
        structure, 618
    markets, 619
Industrial Minera Mexico, S.A. (IMMSA), 683
Industrial minerals
    project finance for, 618
Industrial Revenue Bond (IRB), 136
Inflation, 78, 146
    global, 76
    impact of on hurdle rates, 140
Information memorandum, preparing for loan request, 498
Insurance, 720
    Europe, 481
    export-credit, 637
    Overseas Private Investment Corp. (OPIC), 457
    private, 475
    risk
        conventional, nonconventional, 418, 472
        political, 474
Interest rate
    fluctuations, 129
    risks, 417, 440
    swaps, 443
Internal Rate of Return (IRR), 80
Internal Revenue Code of 1954, 278
International Corona Resources, 821
International Finance Corp. (IFC), 641, 783
    and commercial banks, 643
    and World Bank, 481, 639
Inventories
    loans, 622
    measurement of, 195
    valuation of, 195, 264
Investments
    analysis, 77, 140
        considerations of, 78
        for minerals industries, 77
    foreign
        agreement, 316
        effect of host government attitude upon, 353
        in Peru, 353
        see also Foreign Investment in Real Property Tax Act.
        tax structure of in U.S., 263
    long-term, 178
    unlevered, equity, 33
IRB. *See* Industrial Revenue Bond.
Ireland, nationalization, 355
IRR. *See* Internal Rate of Return.

J

Japan, U.S. Eximbank, 761
Joint ventures, 47
    international, 316

K

Kerr Addison, 582

L

L/C's. *See* Letters of Credit.
La Mine Doyen, Canada, 668
Lac Minerals Ltd., 667, 814
LDCs. *See* Lesser Developed Countries.
Leases, 150, 525, 539
    accounting implications, 157
    gold loan as, 658
    tax considerations, 151
    vs. buy decision, 156
Leasing, 137, 520
    benefits, 652
    discounted cash flow, 152
    leveraged, 158, 645
        cross border, 158
        sources of, 655
Legal advisors, 313
Lending agencies, international, 523
Lesser Developed Countries (LDCs), 3
Letters of Credit (L/C's), 611, 622
Leverage, 13
    theory of financial, 37
Leveraged
    see also Leasing, leveraged.
    studies, 134
Liabilities, 532
    management, 45
Loans
    applications, information needed for project, 493
    approval process, 507
    covenants, 492, 545, 567, 679
    inventory, 622
    request, project, 498
    see also Gold, loan.
    short-term, 536
    structure, Minera Real de Angeles, Mexico, 781
    syndication, 481
Lornex, 26
Low cost position, 382

M

Management
    agreement, 315
    cash, 391
    centralized, 390
    contracts, 317
    decentralized, 390
    financial, 6
        diversified companies, 390
        risk, 391
    issue, 367
    liability, 45
    portfolio, 367
    project, 468
    surprise, 367
    treasury, 390
Market
    risks, 434, 552, 615
    to book ratios, 378
Market-Maker, 585
Marketing, 720
Mergers
    in mining industry, 786
    opportunities, 811
Mexico
    financing, 683
    impact of Mexican financial crisis, 784
    Minera Real de Angeles, 781
        loan structure, 783

nationalization, 353
Mine de Bousquet, Canada, 667
Minera Real de Angeles, Mexico, 781
  loan structure, 783
Mineral Resources International Ltd. (MRI), 671
Minerals
  agreements. *see also* Agreements, mineral.
  industry
    financial evaluation for, 75
    financial requirements of, 3
    investment analysis for, 77
    taxation, 229
    trends, 45
Mines, high-grade/low tonnage, optimum production rate, 114
Mining industry, changes in financial structure of, 10
Models
  asset valuation, 386
  discounted
    cash flow, 386
    income, 385
  economic, 229
  Prince, 450
MRI. *See* Mineral Resources International Ltd.
Mt. Gunnison mine, Colorado, 466
Multiparty structures, 47

N

Nameplate capacity, 372
National Profitability (NP), 326
National Wealth Effect (NWE), 327
Nationalism, 22
Nationalization
  Australia, 354
  Canada, 354
  Chile, 353
  creeping, 289
  Ireland, 355
  Mexico, 353, 683
  Philippines, 354
  Zambia, 354
  Zimbabwe, 354
Net Foreign Exchange Effect, 327
Net income, 39
Net Present Value (NPV), 80
  analysis, 75, 143
Nevada, Sunshine Mining Co., 790
Non-recourse financing structures, 33
Noranda Inc., 786, 824
Norddeutsche Affinerie, 27
NP. *See* National Profitability.
NPV. *See* Net Present Value.
NWE. *See* National Wealth Effect.

O

Off-balance-sheet
  finance, 575
  treatment, 33
Ok Tedi mine, Papua New Guinea, 724
OML. *See* Optimum Mine Life.
Operating
  risk. *see also* Risks, operating.
  units, role of, 390
OPIC. *See* Overseas Private Investment Corp.

Optimum Mine Life (OML), 114
Ore
  possible, 387
  probable, 387
  production, 371
  proven, 386
Overseas Private Investment Corp. (OPIC), 417
  financing, 460
  insurance programs, 457

P

Palabora mine, South Africa, 25
Paley report, 17
Paper, commercial, 537, 611
  markets, 520
Papua New Guinea
  Ok Tedi mine, 724
  tax, 300
Partnerships, 47, 575
  investment, 68
  limited, 579
Payback Period (PbP), 80
PCS. *See* Potash Corp. of Saskatchewan.
Pefco. *See* Private Export Funding Corp.
Peru, foreign investment in, 353
Petrodollars, 338
Petroleum
  basin assessment, 380
  life cycle, 381
Pezamerica Resources, 823
Philippines, nationalization, 354
Planning
  analysis techniques, 368
  construction, 469
  cycle, 404
  guidelines, 408
  long-range, 366
    copper mining, 401
  systems, 403
  process, evolution of in U.S., 365
  see also Strategic planning.
  system, 365
Political risk, 457, 553, 560, 624, 730
  analysis, 445
  insurance, 474
  minerals producing countries, 447
  Prince Model, 450
Portfolio
  analysis, 363
  management, 367
Potash Corp. of Saskatchewan (PCS), 769
  financial analysis, 773
Present Value Ratio (PVR), 80
Present worth coverage ratio, 112
Private Export Funding Corp. (Pefco), 637
Private placements, 521, 579
Production
  capacity, effective, 372
  metal, 371
  ore, 371
  rate, optimum
    definition of, 114
    for high-grade/low tonnage mines, 114
Profitability, 12, 417
Profits
  and economic rent, 293
  economist's concept of, 306
  from mineral resources, 293
Project
  companies, 48

deferral in Argentina, 354
economics, tailoring financing decision to, 94
evaluation, 76, 129
finance, 11, 578
  a borrower's perception, 568
  advantages, 549
  benefits, 569
  for industrial minerals industry, 618
  loan covenants, 567
  nonrecourse, 561
  structure, 566
financing, 3, 75, 334, 481, 671, 790
  definition of, 559
  development, banking needs for, 512
  engineering study information for, 483
  guidelines for commercial banker, 559
  information requirements for, 481
  large overseas mining projects, 736
  of Ok Tedi mine, Papua New Guinea, 724
  risk sharing by, 420
  role of independent consulting firm in, 488
inefficiency, 560
loan
  applications, information needed for, 493
  request, 498
management, 468
ranking, 141
risk, 812
selection, impact of inflation on hurdle rates for, 140
Projects
  financed, 33
  financing development of small, 590
  mineral, sources of funding for, 521
  mining, sensitivity of to cost variations, 129
  post-installation appraisals of, 160
  resource
    development plan, 104
    measuring economic viability of, 103
    tests used by banks for loan determination, 111
Promoters, 584
  role of, 609
Property deals, 581
Prospectus, 604
  contents of, 503
Public offerings, 521
  costs of, 580
PVR. *See* Present Value Ratio.

Q

Quebec Sturgeon River Mines, Canada, 583

R

Ranger uranium project, Australia, 755
Rapid Mining Co., Inc., 53
Rate of return, internal, 75
Ratios
  debt:equity, 555
  Employment Effect, 327
  market to book, 378
  present worth coverage, 112

INDEX

reserve coverage, 112
Recapture, 278
Receivables
see also Accounts receivable.
untaxed, 269
Regulation risks, 435
Rehabilitation Guarantee, 760
Rent, economic, 293, 300
and exploration, 297
determining, 306
division of, 298
in international mineral development finance, 305
source of, 295
Reserve coverage ratio, 112
Reserve risk. See Risks, reserve.
Resource allocation studies, 379
Restructuring, 379
Return
internal rate of, 75
targeted, 592
Return on Equity (ROE), 377
Return on Investment (ROI), 80
Revenue
estimation, 78
recognition of, 204
Rio Tinto-Zinc Corp., 23
Risks
allocation, 425
analysis
country, 417, 454, 783
for mine finance, 417
political, 445
capital, 816
categories, 420
classification, 419
completion, 551, 560, 720
cost, 417, 434
overrun, 466
definition, 419
discreet, 75
engineering, 551
environmental, 553
force majeure, 554
foreign exchange, 554
funding, 555
infrastructure, 553
insurance, 472
political, 474
interest, exchange rate, 440
investor, 528
legal, 555
market, 434, 552, 615
mining, 794
operating, 433, 436, 549
climate, 438
equipment availability, 438
labor, 437
maintenance, 438
management, 437
participant, 551
processing, 794
project, 812
regulation, 435
reserve, 427, 552, 560, 794
see also Exchange rate risks.
sharing, 748
analysis of, 419
by project financing, 420
syndication, 554
taxation, regulation, 435
ROE. See Return on Equity.
ROI. See Return on Investment.
Royalties, 280, 293, 299
taxation, 320

S

Sales
contracts, negotiating bankable, 758
forward, gold, 667
Saskatchewan, Potash Corp. of Saskatchewan (PCS), 769, 773
SEC. See U.S. Securities and Exchange Commission.
Securities
analysis, financial, 77
borrowings, 400
hybrid, 812
Sensitivities, 129, 491
capital expenditure, 128
cost over-run, 127
environmental, 127
foreign exchange, 128
inflation, 128
interest, 127
operating, 127
price, 127
production, 127
reserve, 126
tax, 127
Sensitivity analysis, 76, 112, 126, 129, 232, 800
Share offerings, types of, 605
Shareholder value, 378
Silver, Minera Real de Angeles, Mexico, 781
Sinking fund debentures, 538
Soroako nickel project, Indonesia, 736
South Africa, Palabora mine, 25
South America
mineral agreements in, 342
taxes in, 345
Stock
common, 539
exchange
Toronto, 601
Vancouver, 501, 600, 817
issues, public, 69
market, 481
penny, 580
performance, 14
preferred, 540
Strategic planning, 363, 366
alternatives, 367
Strathcona Sound project, 671
Structural changes, impact of on competition, 373
Sunshine Mining Co., Nevada, 790

T

Targeted return, 592
Tarong coal project, Australia, 570
Tax Equity & Fiscal Responsibility Act of 1982 (TEFRA), 265
Tax Reform Act of 1984, 267, 272, 278
Tax-Loss Funds, 528
Taxable Income From the Property (TIFP), 279
Taxation
Australian, 283
Canadian mining, 241
developing countries, 320
double, 270
government share of income, 284
minerals industry, 229
risks, 435
royalties, 320
U.S., of mining companies, 254
Taxes
and accounting analysis, 77
Australia, 301
benefits, 278
burden, comparative, Canada, 242
considerations, joint structures, multiparty structures, 48
credits
foreign, 279
investment, 265
income, 207, 299
Canada, 244
increasing cash flow through, 78
laws, *capital exporting* countries, 229
Papua New Guinea, 300
preference, 278
property, 231
regimes, comparison of international mining, 274
resource rent, proposals in Australia, 300
severance, 231
South America, 345
U.S.
branch, 268
comparison, 276
corporate, 263
joint venture, 270
local, state, federal, 231
multiple corporations, 278
partnership, 268
tax structure of foreign mining investments in, 263
treaties, 271
treatment of exploration, development expenses, 254
unitary, 266
witholding
on foreign distributions, 269
rates, 271
Teck Corp., 582, 718, 821
TEFRA. See Tax Equity & Fiscal Responsibility Act.
Test completion, 564
TIFP. See Taxable Income From the Property.
Trade
credits, 536, 622
financing, 622
Transactions, financing, 198
Trends, minerals industry, 45
Trusts, 48

U

U.S. Eximbank, 636, 740
Japan, 761
U.S. Real Property Interests (USRPI), 266
U.S. Securities and Exchange Commission (SEC), 506
U.S. Uniform Partnership Acts, 269
Underwriters, role of, 608
United Nations, 313
Units
functional, 409
operating, role of, 390
Uranium mines, Elliot lake, 24
USRPI. See U.S. Real Property Interests.

V

Value creation, 379
Ventures. *See* Joint ventures.

W

West Elk Coal Co., 466
Westmin Resources Ltd., 788
Work commitments, 588
World Bank, 639
    and International Finance Corp. (IFC), 481, 639
    cofinancing, 640.2
    economic priority, 640
    financing, 639
    market, marketing, 640.1
    types of projects, 640

Z

Zambia, nationalization, 354
Zimbabwe, nationalization, 354